Climate Zones after C. Troll/KH. Paffen

I = Polar and subpolar Zones

I_1 = High - polar ice - cap climates

I_2 = Polar climates

I_3 = Subarctic tundra climates

I_4 = Highly oceanic subpolar climates

II = Cold - temperate Boreal Zones

II_1 = Oceanic boreal climates

II_2 = Continental boreal climates

II_3 = Highly continental boreal climates

III = Cool - temperate Zones

III_1 = Highly oceanic climates

III_2 = Oceanic climates

III_3 = Sub - oceanic climates

III_4 = Sub - continental climates

III_5 = Continental climates

III_6 = Highly continental climates

III_7 = Humid - and - warm - summer climates
a = with humid winters

III_8 = Permanently humid, warm - summer climates

III_9 = Humid steppe climates with cold winters
a = with mild winters

III_{10} = Dry steppe climates with cold winters
a = with mild winters

III_{11} = Humid - summer steppe climates with cold winters

III_{12} = Semi - desert and desert climates with cold winters
a = with mild winters

IV = Warm - temperate Subtropical Zones

IV_1 = Dry - summer Mediterranean climates with humid winters

IV_2 = Dry - summer steppe climates with humid winters

IV_3 = Steppe climates with short summer humidity

IV_4 = Dry - winter climates with long summer humidity

IV_5 = Semi - desert and desert climates

IV_6 = Permanently humid grassland climates

IV_7 = Permanently humid climates with hot summers

V = Tropical Zone

V_1 = Tropical rainy climates

V_2 = Tropical humid - summer climates
a = with humid winters

V_3 = Wet and dry tropical climates

V_4 = Tropical dry climates
a = with humid winters

V_5 = Tropical semi desert and desert climates

= **Climatic altitudes of mountains**

Selected climatic data

Tasks for vegetation science 5

Series Editor
HELMUT LIETH
University of Osnabrück, F.R.G.

K L I M A P L O T   P R O G R A M M E

Enclosure to: Manfred J. Müller - Selected climatic data
              for a global set of standard stations for
              vegetation science (Series Tasks for
              Vegetation Science, Volume 5, published by
              Dr. W. Junk Publishers, The Hague, The
              Netherlands, © 1982, ISBN 90 6193 945 3)

See page XXIV - section 2.3

```
C*************************************************************
C**                    K L I M A P L O T                   **
C** FORTRAN-PROGRAMM ZUM AUTOMATISCHEN ZEICHNEN VON KLIMA-  **
C** DIAGRAMMEN.                                             **
C** GESCHRIEBEN VON BERTRAM OSTENDORF, HORST LEHKER         **
C** OSNABRUECK                                              **
C*************************************************************
C****** SKALIERUNGSWUNSCH DES BENUTZERS ******
      REAL FUNCTION NIESKA(NIEDER,OPT)
      REAL NIEDER
      INTEGER OPT
C--- DIE NIEDERSCHLAGSWERTE WERDEN IN INCH EINGEGEBEN
C    UND IN MM UMGERECHNET
      NIESKA=NIEDER
      IF(OPT.EQ.0) RETURN
      NIESKA= NIEDER/0.03937
      RETURN
      END
      REAL FUNCTION TEMSKA(TEMP,OPT)
      INTEGER OPT
      REAL TEMP
C--- DIE TEMPERATURWERTE WERDEN VON GRAD FAHRENHEIT
C    NACH GRAD CELSIUS UMGERECHNET
      TEMSKA=TEMP
      IF (OPT.EQ.0) RETURN
      TEMSKA= 10*(TEMP-32)/18.
      RETURN
      END
C
C****** SCHRAFFUREN ********
      SUBROUTINE SCHRAF(K,L,KNICK,OPT2,HELPL)
      REAL HELP,K(12),L(12),DIFFK,DIFFL,FELDK(5),FELDL(5),X1,Y1,DX,Y
      REAL KNICK(8),L1,L2,HELPL
      INTEGER I,J,MARK,ANZAHL,INT,M,OPT2
      MARK=0
C--- PUNKTIERUNG DES ERSTEN U./ODER LETZTEN HALBSCHRITTES
      HELPK=(K(1)+K(12))/2.
      IF (HELPL.LT.HELPK .OR. L(1).LT.K(1))GOTO 310
      IF (HELPL.LT.HELPK .OR. L(12).LT.K(12))GOTO 320
      GOTO 209
310   FELDK(1)=HELPK+    (K(1)-HELPK)/5.
      FELDL(1)=HELPL+    (L(1)-HELPL)/5.
      FELDK(2)=HELPK+3.*(K(1)-HELPK)/5.
      FELDL(2)=HELPL+3.*(L(1)-HELPL)/5.
      FELDK(3)=K(1)
      FELDL(3)=L(1)
      X=0.1
      J=3
301   DO 303 I=1,J
      FELDK(I)=FELDK(I)*5.
      INT=FELDK(I)
      FELDK(I)=INT/5.
302   IF (FELDL(I).LE.FELDK(I))GOTO 333
      GOTO 303
333   CALL SYMBOL(X,FELDK(I),.05,1,0.0,-1)
      FELDK(I)=FELDK(I)-.2
      GOTO 302
303   X=X+.2
      IF (MARK.EQ.2 .OR.(HELPL.GT.HELPK.AND.L(12).GT.K(12))) GOTO 209
320   FELDK(1)=K(12)+2.*(HELPK-K(12))/5.
```

```
            FELDL(1)=L(12)+2.*(HELPL-L(12))/5.
            FELDK(2)=K(12)+4.*(HELPK-K(12))/5.
            FELDL(2)=L(12)+4.*(HELPL-L(12))/5.
            X=11.7
            J=2
            MARK=2
            GOTO 301
C--- HAUPTPUNKTIERUNG
C
209         MARK=0
            DO 210 I=1,12
            IF (ABS(K(I)-L(I)).LT.1E-4) GOTO 300
            IF (L(I).LT.K(I)) GOTO 205
100         CALL PLOT(I-.5,AMIN1(L(I),13.99),3)
            CALL PLOT(I-.5,AMAX1(K(I),0.0),2)
            HELP=L(I)
            IF(OPT2.GE.3 .AND. L(I).GE.5.) HELP=5.
            HELP=HELP*5.
            INT=HELP
            HELP=INT/5.0
230         IF(HELP.LT.AMAX1(K(I),0.0)) GOTO 300
            CALL PLOT(I-.4,HELP,3)
            CALL PLOT(I-.6,HELP,2)
            HELP=HELP-.2
            GOTO 230
300         IF(MARK.EQ.0)GOTO 210
205         IF(I.GT.1)GOTO 206
            MARK=1
            GOTO 210
206         DIFFK=(K(I)-K(I-1))/5.
            L2 = L(I)
            IF(OPT2.GE.3 .AND. L(I).GT.5.) L2=(L(I)-4.)*5.
            L1 = L(I-1)
            IF(OPT2.GE.3 .AND. L(I-1).GT.5.) L1=(L(I-1)-4.)*5.
            DIFFL=(L2-L1)/5.
            DO 1 J=1,5
            FELDK(J)=K(I-1)+ J*DIFFK
            FELDL(J)=L1+ J*DIFFL
            IF(FELDL(J).LT.FELDK(J)) GOTO 17
            GOTO 1
17          FELDK(J)=FELDK(J)*5.0
            INT=FELDK(J)
            FELDK(J)=INT/5.0
10          IF (FELDL(J).LE.FELDK(J) ) GOTO 11
            GOTO 1
11          CALL SYMBOL(I+J*.2-1.5,FELDK(J),.05,1,0.0,-1)
            FELDK(J)=FELDK(J)-.2
            GOTO 10
1           CONTINUE
            MARK=0
            IF (I.EQ.12) GOTO 210
            IF (L(I+1) .LT. K(I+1)) GOTO 210
            MARK=1
210         CONTINUE
            IF(OPT2.LT.3)RETURN
C-----------------------------------------------------------
C--- WAAGERECHTE SCHRAFFUR BEI MASSTABSVEAENDERUNG
            Y=0.1
777         IF(KNICK(1).LT.0.0) GOTO 40
            IF(KNICK(1).GT.11.5) GOTO 30
```

```
        GOTO 77
C--- SCHRAFFUR VOR DEM ERSTEN WERT
40      IF   (KNICK(2).LT.0.5) GOTO 50
        GOTO 60
50      CALL PLOT (0.,5.,3)
        CALL PLOT(KNICK(2),5.0,2)
        ANZAHL=(AMIN1(HELPL,13.99)-5.)*10.
        IF (ANZAHL.LT.1)GOTO 2000
        DX=0.1*KNICK(2)/(HELPL-5.)
        DO 51 M=1,ANZAHL
51      CALL PLOTL (0.0,5.+M*Y,KNICK(2)-M*DX,M)
        GOTO 2000
C
C--- SCHRAFFUR ZWISCHEN L(12) UND ENDE DES DIAGRAMMS
30      CALL PLOT(KNICK(1),5.0,3)
        CALL PLOT(12.,5.0,2)
        ANZAHL=(AMIN1(HELPL,13.99)-5.)*10.
        IF(ANZAHL.LT.1) RETURN
        DX=0.1*(12.-KNICK(1))/(HELPL-5.)
        DO 31 M=1,ANZAHL
31      CALL PLOTL(KNICK(1)+M*DX,5.+M*Y, 12.,M)
        RETURN
C--- SCHRAFFUR DES ERSTEN OD. LETZTEN HALBEN SCHRITTES
60      CALL PLOT(0.,5.,3)
        CALL PLOT(KNICK(2),5.,2)
        A=HELPL
        B=L(1)
        X=0.0
        MARK=0
        GOTO 600
61      A=L(12)
        B=HELPL
        X=11.5
        MARK=2
C
600     IF(B.GE.A-0.1) GOTO 650
C--- NEG. STEIGUNG
        ANZAHL=(AMIN1(B,13.99)-5.)*10.
        IF (ANZAHL.LT.1) GOTO 602
        DO 601 M=1,ANZAHL
601     CALL PLOTL (X,5.+M*Y,X+0.5,M)
602     J=ANZAHL+1
        ANZAHL=(AMIN1(A,13.99)-5.)*10.
        IF(ANZAHL.LT.J) GOTO 690
        INT=B*10
        Y1=INT/10.+0.1-B
        X1=Y1*0.5/(A-B)
        DX=0.1*X1/Y1
        X1=X+0.5-X1 +DX*J
        DO 606 M=J,ANZAHL
606     CALL PLOTL (X,5.+M*Y,X1-M*DX,M)
        GOTO 690
C--- POS. STEIGUNG
650      ANZAHL=(AMIN1(A,B,13.99)-5.)*10.
        IF(ANZAHL.LT.1) GOTO 652
        DO 651 M=1,ANZAHL
651     CALL PLOTL (X,5.+M*Y,X+0.5,M)
652     IF(B.LE.A+0.1)GOTO 690
        J=ANZAHL+1
        ANZAHL=(AMIN1(B,13.99)-5.)*10
```

```
        IF (ANZAHL.LT.J)GOTO 690
        INT=A*10.
        Y1=INT/10.+0.1-A
        X1=Y1*0.5/(B-A)
        DX=0.1*0.5/(B-A)
        X1=X+X1-J*DX
        DO 653 M=J,ANZAHL
653     CALL PLOTL (X1+M*DX,5.+M*Y,X+0.5,M)
C--- ENDE
C--- REGION DER HAUPTSCHRAFFUR
690     IF (MARK.EQ.2) RETURN
        I=1
        GOTO 11111
77      I=KNICK(1)+1.5
C--- SCHRAFFUR DER SPITZE NACH DEM LINKSBEGRENZENDEN KNICK
        ANZAHL=(AMIN1 (L(I),13,99)-5.)*10.
        DX=0.1*(I-0.5-KNICK(1))/(L(I)-5.)
        CALL PLOT (KNICK(1),5.,3)
        CALL PLOT (AMIN1(KNICK(2),12.),5.,2)
        DO 7 M=1,ANZAHL
7       CALL PLOTL (KNICK(1)+M*DX,5.+M*Y,I-0.5,M)
C--- MACHE ABFRAGE
11111   I=I+1
        IF(I.EQ.13 .AND. KNICK(2).LE.12.) GOTO 103
        IF(I.EQ.13 .AND. KNICK(2).GT.12.) GOTO 61
        IF(I.EQ.13) RETURN
        IF(L(I).LE.5.)GOTO 103
        IF(L(I).GE.L(I-1)-0.1)GOTO 101
C--- SCHRAFFIERE UNTER STEIGENDEM ODER KONSTANTEM GRAPHEN
        IF(L(I).LT.L(I-1)-0.1)GOTO 102
C--- SCHRAFFIERE UNTER NEG. STEIGENDEM GRAPHEN
C
C--- POSITIVE ODER GARKEINE STEIGUNG DES GRAPHEN
101     ANZAHL=(AMIN1(L(I-1),L(I),13,99)-5.)*10.
        IF (ANZAHL.LT.1)GOTO 500
        DO2 M=1,ANZAHL
2       CALL PLOTL(I-1.5,5.+M*Y,I-0.5,M)
C--- SPRUNG BEI KEINER ODER ZU GERINGER STEIGUNG
500     IF (L(I).LE.L(I-1)+0.1)GOTO 11111
        J=ANZAHL+1
        INT=L(I-1)*10.
        Y1=INT/10.+0.1-L(I-1)
        X1=Y1/(L(I)-L(I-1))
        DX=0.1/(L(I)-L(I-1))
        X1=I-1.5+X1-J*DX
        ANZAHL=(AMIN1 (L(I),13,99)-5.)*10.
        DO3 M=J,ANZAHL
3       CALL PLOTL (X1+M*DX,5.+M*Y,I-0.5,M)
        GOTO 11111
C--- SCHRAFFIERE UNTER NEG. STEIGENDEM GRAPHEN
102     ANZAHL=(AMIN1 (L(I),13,99)-5.)*10.
        IF (ANZAHL.LT.1) GOTO 1000
        DO4 M=1,ANZAHL
4       CALL PLOTL (I-1.5,5.+M*Y,I-0.5,M)
1000    J=ANZAHL+1
        ANZAHL=(AMIN1(13,99,L(I-1))-5.)*10.
        IF(ANZAHL.LE.J) GOTO 11111
        INT=L(I)*10.
        Y1=INT/10.+0.1-L(I)
        X1=Y1/(L(I-1)-L(I))
```

```
        DX=0.1/(L(I-1)-L(I))
        X1=(I-0.5-X1)+DX*J
        DO 5 M=J,ANZAHL
5       CALL PLOTL(I-1.5,5.+M*Y,X1-M*DX,M)
        GOTO 11111
C--- SCHRAFFIERE BIS ZUM RECHTSBEGRENZENDEN KNICK
103     ANZAHL=(AMIN1(L(I-1),13.99)-5.)*10.
        IF (ANZAHL.LT.1) GOTO 2000
        X1=(KNICK(2)-I+1.5)
        DX=0.1*X1/(L(I-1)-5.)
        DO 6 M=1,ANZAHL
6       CALL PLOTL (I-1.5,5.+M*Y,KNICK(2)-M*DX,M)
C--- ENDE DER SCHRAFFUR ZWISCHEN ZWEI KNICKS
C--- UEBERPRUEFUNG OB WEITERE SCHRAFFUR
2000    IF(KNICK(3).GT.-10.) GOTO 70
        IF(KNICK(5).GT.-10.) GOTO 71
        IF(KNICK(7).GT.-10.) GOTO 72
        RETURN
70      KNICK(1)=KNICK(3)
        KNICK(2)=KNICK(4)
        KNICK(3)=-11.
        GOTO 777
71      KNICK(1)=KNICK(5)
        KNICK(2)=KNICK(6)
        KNICK(5)=-11.
        GOTO 777
72      KNICK(1)=KNICK(7)
        KNICK(2)=KNICK(8)
        KNICK(7)=-11.
        GOTO 777
        END
C
C****** BESCHRIFTUNG DES DIAGRAMMS ******
C--- RECHTSBUENDIGKEIT DER BESCHRIFTUNG
        REAL FUNCTION ORT(X)
        IF(ABS(X).LT.10.)GOTO 1
        GOTO 2
1       ORT=-1.8
        GOTO 3
2       ORT=-2.2
3       RETURN
        END
        SUBROUTINE SCHRIF(F,Y)
        IF(F.GE.1000)RETURN
        IF(F.LT.0.0) GOTO 1
        GOTO 2
1       CALL NUMBER(ORT(F),Y,0.4,F,0.0,1)
        RETURN
2       CALL SYMBOL(ORT(F),Y,0.4,1H+,0.0,1)
        CALL NUMBER(999.,Y,0.4,F,0.0,1)
        END
C
        SUBROUTINE BESCHR(A,B,C,D,E,F,G,H,I,J,MAXI,K,OPT2)
        INTEGER A(20),B,C,LEER,MAXI,OPT2
        REAL D,E,F,G,H,I,J,K,Y,HOEHE
        DATA LEER /4H    /
        HOEHE=AMAXO(MAXI,10)
        IF (OPT2.NE.1) HOEHE=AMAXO(MAXI,5)
        Y=HOEHE+2.
        CALL SYMBOL(0.2,Y,.5,A,0,0,80)
```

```
      Y=HOEHE+1.
      CALL SYMBOL(0.2,Y,.5,1HC,0.0,1)
      CALL SYMBOL(999.,Y,.5,B,0.0,2)
      IF(C.NE.LEER) GOTO 1000
 2000 CALL SYMBOL(999.,Y,.5,2HJ ,0.0,2)
      CALL NUMBER(999.,Y,.5,D,0.0,1)
      CALL SYMBOL(999.,Y+0.32,.18,1HC,0.0,1)
      CALL SYMBOL(999.,Y,.5,2HC ,0.0,2)
      CALL NUMBER(999.,Y,.5,E,0.0,-1)
      CALL SYMBOL(999.,Y,.5,2HMM,0.0,2)
      Y=.2
      CALL SCHRIF(F,Y)
      Y=-.4
      CALL SCHRIF(G,Y)
      Y=AMAX1( K-0.18,1.)
      IF (OPT2.EQ.1) Y=AMAX1( K/2,-.18,1.)
      CALL SCHRIF(J,Y)
      Y=HOEHE+0.5
      CALL SCHRIF(H,Y)
      Y=HOEHE+1.5
      CALL SCHRIF(I,Y)
      Y=HOEHE+2.5
      CALL SYMBOL(-.7,Y+.26,0.14,1HC,0.0,1)
      CALL SYMBOL(999.,Y,0.4,1HC,0.0,1)
      CALL SYMBOL(12.2,-1.,0.4,2HMM,0.,2)
      GOTO 999
 1000 CALL SYMBOL(999.,Y,.5,1H-,0.0,1)
      CALL SYMBOL(999.,Y,.5,C,0.0,2)
      GOTO2000
      GOTO2000
      RETURN
  999 END
C
C****** SCHRAFFUR FUER TEMPERATURMINIMA < O ********
      SUBROUTINE TMIN (TO12)
      INTEGER TO12 (12)
      DO 1 I=1,12
      IF (TO12(I) .EO.0) GOTO 1
      IF (TO12(I) .EO.1) GOTO 20
      IF (TO12(I) .EQ.2) GOTO 10
      GOTO 1
C--- ZUSAETZLICHE SENKRECHTE SCHRAFFIERUNG
 10   CALL PLOT( I-1. ,0.0,-3)
      DO 11 J=1,4
      CALL PLOT (J*0.2,0.0,3)
 11   CALL PLOT (J*0.2,-0.5,2)
      CALL PLOT(-(I-1.),0.0,-3)
C--- WAAGERECHTE SCHRAFFIERUNG
 20   CALL PLOT( I-1. ,0.0,-3)
      CALL PLOT(0.,-0.5,2)
      CALL PLOT(1.,-0.5,2)
      CALL PLOT(1.,0.0,2)
      CALL PLOT(1.,-0.5/3.,3)
      CALL PLOT(0.,-0.5/3.,2)
      CALL PLOT(0.,-1./3.,3)
      CALL PLOT(1.,-1./3.,2)
      CALL PLOT(-(I-1.) ,0.0,-3)
 1    CONTINUE
      RETURN
      END
```

```
C
C****** BESCHRIFTUNG DES ACHSENKREUZES ********
      SUBROUTINF ACHSB(OPT2,MAXI,MINI,K,JOT,MIN)
      INTEGER OPT2,MAXI,MINI,M,I,J
      REAL K,JOT,Y,MIN
      IF (OPT2.EQ.1) GOTO 101
      GOTO 102
101   M=2
      IF(MIN.LT.-10.)M=1
      DO 10 I=M,8
      J=(I-3)*2
      CALL PLOT ( 0.2,1.*J,3)
      CALL PLOT (-0.2,1.*J,2)
      IF (JOT.GT.1000) GOTO 7
      IF (J.LT.K-0.53  .OR. J.GT.K+0.32 ) GOTO 7
      GOTO 10
7     IF (J.NE.0) CALL NUMBER (-1.,1.*J,0.25,J*5,0.0,-1)
10    CONTINUE
      DO 11 I=1,11
      CALL PLOT (12.2,I-1.,3)
      CALL PLOT (11.8,I-1.,2)
11    CALL NUMBER (12.5,I-1.,0.25,(I-1)*10,0.0,-1)
      GOTO 999
102   Y=MINI+1.
      M=ABS(Y)+6
      DO 20 I=1,M
      CALL PLOT ( 0.2,Y,3)
      CALL PLOT (-0.2,Y,2)
      IF (JOT.GT.1000)GOTO 8
      IF (Y.LT. K-0.53 .OR. Y.GT.K+0.32) GOTO 8
      GOTO 20
8     IF (ABS(Y).GT.1.5) CALL NUMBER (-1.,Y,0.25,Y*10,0.0,-1)
20    Y=Y+1
C
      IF (OPT2.GE.3) GOTO 103
      M=MAXO(MAXI,5)+1
      DO 21 I=1,M
      CALL PLOT (12.2,I-1.,3)
      CALL PLOT (11.8,I-1.,2)
21    CALL NUMBER (12.5,I-1.,0.25,(I-1)*20,0.0,-1)
      GOTO 999
C
103   M=MAXO(MAXI,5)+1
      DO 31 I=1,M
      CALL PLOT (12.2,I-1.,3)
31    CALL PLOT (11.8,I-1.,2)
      DO 32 I=1,6
32    CALL NUMBER (12.5,I-1.,0.25,(I-1)*20.,0.0,-1)
      M=MAXI-4
      DO 33 I=2,M
33    CALL NUMBER (12.5,I+4.,0.25,I*100.,0.0,-1)
999   RETURN
      END
C
C****** ZEICHNEN DES ACHSENKREUZES ******
      SUBROUTINE ACHSE1(OPT2,MAXI,MINI)
      INTEGER OPT2,MAXI,MINI,HOEHE
      HOEHE=MAXO(MAXI,10)
      IF (OPT2.NE.1) HOEHE=MAXO(MAXI,5)
      CALL PLOT(0.,HOEHE+3.,3)
```

```
            CALL PLOT(0.,MINI*1.,2)
            CALL PLOT(0.,0.,3)
            DO 20 I=1,11
            CALL PLOT(I*1.,0.0,3)
            IF (I.EQ.6) CALL PLOT (I*1.,-0.35,2)
20          CALL PLOT(I*1.,-0.2,2)
            CALL PLOT(12.,0.,3)
            CALL PLOT(0.,0.,2)
            CALL PLOT(12.,HOEHE+3.,3)
            CALL PLOT(12.,MINI*1.,2)
            Y=10.
            IF (OPT2.EQ.1)GOTO 10
            Y=5.
10          CALL PLOT(0.,Y,3)
            DO 1 I=1,30
            CALL PLOT((I-1)*0.4+0.2,Y,2)
1           CALL PLOT((I-1)*0.4+0.4,Y,3)
            RETURN
            END
C
C****** ZEICHNEN EINER KURVE DER MONATSMESSDATEN  ******
            SUBROUTINE KURVE(FELD,HELP)
            DIMENSION FELD(12)
            REAL HELP,FELD(12)
            INTEGER I
            HELP=(FELD(12)+FELD(1))/2
            CALL PLOT (0.,HELP,3)
            DO 3 I=1,12
3            CALL PLOT(I-0.5,FELD(I),2)
            CALL PLOT(12.,HELP,2)
            RETURN
            END
C****** ZEICHNEN EINER KURVE BEI MASSTABSVERAENDERUNG ******
            SUBROUTINE KNIKUR(L,KNICK,HELP)
            REAL HELP,L(12),KNICK(8),X,ZAELER,HELP1,HELP2
            INTEGER I,KZAEHL
            KZAEHL=1
            HELP1 =L(1)
            HELP2 =L(12)
            IF (L(1) .GT.5.) HELP1 =5.+ (L(1) -5.)*5.
            IF (L(12).GT.5.) HELP2 =5.+ (L(12)-5.)*5.
            HELP=(HELP1+HELP2)/2.
            IF (HELP.GT.5.) HELP= 5.+(HELP-5)/5.
            CALL PLOT (0.,HELP,3)
            ZAELER=2.
C--- BEHANDLUNG DES ERSTEN HALBSCHRITTES
            IF (L(1).GT.5. .AND. HELP.LE.5.) GOTO 50
            IF (L(1).LE.5. .AND. HELP.GT.5.) GOTO 55
            IF (L(1).GT.5. .AND. HELP.GT.5.) GOTO 56
            GOTO 51
50          KNICK(1)=(5.-HELP)/((L(1)-5.)*5.+5.-HELP)*0.5
            KZAEHL=2
            CALL PLOT (KNICK(1),5.,2)
            GOTO 51
55          KNICK(1)=-1.
            KNICK(2)=(HELP-5.)*5./((HELP-5.)*5.+5.-L(1))*0.5
            CALL PLOT (KNICK(2),5.,2)
            KZAEHL=3
            GOTO 51
56          KNICK(1) =-1.
```

```fortran
      KZAEHL=2
      GOTO 51
C--- VERBINDEN DER DATENPUNKTE
51    DO 1 I=1,12
      IF(I.GT.1) GOTO 52
      IF(L(I).LE.14.) GOTO 53
      GOTO 11
53    CALL PLOT (0.5,L(I),2)
      GOTO 1
52    IF (L(I).GT.5.0 .AND. L(I-1).LE.5.0) GOTO 100
      IF (L(I).LE.5.0 .AND. L(I-1).GT.5.0 .AND. L(I-1).LE.14.0)GOTO 200
      IF (L(I).LE.5.0 .AND. L(I-1).GT.14.) GOTO 12
      GOTO 3
100   KNICK(KZAEHL)=(5.-L(I-1)) / ((L(I)-5)*5, +5.-L(I-1)) +I-1.5
      CALL PLOT (KNICK(KZAEHL),5.,2)
      KZAEHL=KZAEHL+1
      GOTO 3
200   KNICK(KZAEHL)=((L(I-1)-5.)*5.) / ((L(I-1)-5.)*5, +5.-L(I)) +I-1.5
      CALL PLOT (KNICK(KZAEHL),5.,2)
      KZAEHL=KZAEHL+1
      IF(L(I).LE.5. .AND. L(I-1).GT.14) GOTO 2
3     IF(L(I).LE.14. .AND. L(I-1).LE.14.) GOTO 2
C--- BEHANDLUNG DER FAELLE FUER L>14
      IF(L(I).GT.14. .AND. L(I-1).LE.14.) GOTO 10
      IF(L(I).GT.14. .AND. L(I-1).GT.14.) GOTO 11
      IF(L(I).LE.14. .AND. L(I-1).GT.14.) GOTO 12
C--- POS. STEIGUNG
10    X=(L(I)-14.)/(L(I)-L(I-1))
      CALL PLOT (I-0.5-X,14.,2)
C--- PUNKT BESCHRIFTEN
11    CALL PLOT(I-0.5,14.,3)
      CALL PLOT(ZAELER*0.3,14.5,2)
      CALL NUMBER((ZAELER-1.)*0.3,14.6,0.3,(L(I)-4.)*100.,0.0,-1)
      CALL PLOT(I-0.5,14.,3)
      ZAELER = ZAELER + 6.
      GOTO1
C--- NEG. STEIGUNG
12    X=(L(I-1)-14.)/(L(I-1)-L(I))
      CALL PLOT(I-1.5+X,14.,3)
      IF(L(I).LE.5. .AND. L(I-1).GT.14.0) GOTO 200
2     CALL PLOT (I-0.5,L(I),2)
1     CONTINUE
C--- BEHANDLUNG DES LETZTEN HALBSCHRITTES
      I=12
      IF (HELP.GT.5. .AND. L(I).LE.5.) GOTO 60
      IF (HELP.LE.5. .AND. L(I).GT.5.) GOTO 65
      IF (HELP.GT.5. .AND. L(I).GT.5.) GOTO 66
      IF (HELP.LE.5. .AND. L(I).LE.5.) GOTO 1000
60    KNICK(KZAEHL)=(5.-L(12))/((HELP-5.)*5.+5.-L(12))*0.5+11.5
      CALL PLOT (KNICK(KZAEHL),5.,2)
      KZAEHL=KZAEHL+1
      KNICK(KZAEHL)=13.
      GOTO 1000
65    KNICK(KZAEHL)= ((L(12)-5.)*5.)/((L(12)-5.)*5.+5.-HELP)*0.5+11.5
      CALL PLOT (KNICK(KZAEHL),5.,2)
      GOTO 1000
66    KNICK(KZAEHL)=13.
1000  IF(HELP.LE.14.) CALL PLOT (12.,HELP,2)
      RETURN
      END
```

```
C
C****** SKALIERUNG DER FEUCHTE-WERTE
C       BERECHNUNG DER KOORDINATEN ******
        REAL FUNCTION LSKALI(L,OPT2)
        REAL L
        INTEGER OPT2
        IF (OPT2.EQ.1)GOTO 1
        IF (OPT2.EQ.2)GOTO 2
        IF (L.LT.100.0)GOTO 2
        LSKALI=5.+((L-100.)/100.)+1E-10
        RETURN
1       LSKALI=L/10.
        RETURN
2       LSKALI=L/20.
        RETURN
        END
C
C****** SKALIERUNG DER TEMPERATURWERTE
C       BERECHNUNG DER KOORDINATEN ******
        REAL FUNCTION KSKALI(K,OPT2)
        REAL K
        INTEGER OPT2
        IF (OPT2.EQ.1)GOTO 1
        KSKALI=K/10.
        RETURN
1       KSKALI=K/5.
        RETURN
        END
C
C
C*********** RAHMENPROGRAMM **********
        INTEGER A(20),B,C,I1,OPT,OPT2,MINI,MAXI,TO12(12)
        REAL D,E,F,G,H,I,J,TEMSKA,NIESKA,KSKALI,LSKALI,K(12),L(12)
        REAL MAX,MIN,LMAX,KMAX,KNICK(8),HELP
111     LMAX=0.
        KMAX=0.
        MIN=0.
        DATA KNICK /8*-11./
C       (SETZEN DES FELDES KNICK AUF -11.0)
        READ(5,1,END=99999) A
        READ(5,2) OPT,OPT2
        READ(5,3)B,C,D,E,F,G,H,I,J
        D=TEMSKA(D,OPT)
        E=NIESKA(E,OPT)
        F=TEMSKA(F,OPT)
        G=TEMSKA(G,OPT)
        H=TEMSKA(H,OPT)
        I=TEMSKA(I,OPT)
        J=TEMSKA(J,OPT)
        READ(5,4)K(1),K(2),K(3),K(4),K(5),K(6)
        READ(5,4)K(7),K(8),K(9),K(10),K(11),K(12)
        DO 10 I1=1,12
        K(I1)=TEMSKA(K(I1),OPT)
        KMAX=AMAX1(KMAX,K(I1))
10      MIN=AMIN1(MIN,K(I1))
        READ(5,4)L(1),L(2),L(3),L(4),L(5),L(6)
        READ(5,4)L(7),L(8),L(9),L(10),L(11),L(12)
        DO 20 I1=1,12
        L(I1)=NIESKA(L(I1),OPT)
20      LMAX=AMAX1(LMAX,L(I1))
```

```
          READ (5,5) TO12
C--- BERECHNUNG VON OPT2, MINI, MAXI
          IF (OPT2.GT.0) GOTO 770
          IF (LMAX.GT.200.0)GOTO 773
          IF (LMAX.GT.100.0)GOTO 772
          IF (MIN.LT.-20.0)GOTO 772
          OPT2=1
          GOTO 770
773       OPT2=3
          GOTO 770
772       OPT2=2
770       XX= LSKALI(LMAX,OPT2)
          XXXX= KSKALI(KMAX,OPT2)
          MAX=AMAX1(XX,XXXX)
          MIN=KSKALI(MIN,OPT2)
          MAXI=MIN1(MAX,14.)+1
          MINI=MIN1(MIN,-2.)-1
          DO 21 I1=1,12
          L(I1)=LSKALI(L(I1),OPT2)
21        K(I1)=KSKALI(K(I1),OPT2)
C         CALL FACTOR(0.75)
          CALL PLOT (9.,9.,-3)
          CALL ACHSE1(OPT2,MAXI,MINI)
          CALL ACHSB(OPT2,MAXI,MINI,K(1),J,MIN)
          CALL BESCHR(A,B,C,D,E,F,G,H,I,J,MAXI,K(1),OPT2)
          CALL KURVE(K,HELP)
          IF (OPT2.GE.3)GOTO66
          GOTO 67
66        CALL KNIKUR (L,KNICK,HELP)
          GOTO 68
67        CALL KURVE(L,HELP)
68        CALL SCHRAF(K,L,KNICK,OPT2,HELP)
          CALL TMIN(TO12)
          CALL PLOT (-9.,-9.,-3)
          GOTO 111
99999 STOP
C******* FORMATE*************
1         FORMAT(20A4)
2         FORMAT(2I1)
3         FORMAT(2A4,X,X,7F10.2)
4         FORMAT(6F10.2)
5         FORMAT(12I1)
C===========================
          END
C*** ZEICHNET WAAGERECHTE LINIEN (ALLE 10MM NIEDERSCHLAG GESTRICHELT)
          SUBROUTINE PLOTL (A,B,C,M)
          REAL=M/10.
          INT=M/10
          IF ((REAL-INT).LT.1E-2) GOTO 40
          GOTO 41
C--- GESTRICHELT
40        CALL PLOT (A,B,3)
          DO 1 I=1,100,2
          IF ((A+(0.06*I)).GT.C) RETURN
          CALL PLOT (A+0.06*I,B,2)
1         CALL PLOT (A+0.06*(I+1),B,3)
          RETURN
C--- NORMALE VERBINDUNG
41        CALL PLOT (A,B,3)
          CALL PLOT(C,B,2)
          END
```

MANFRED J. MÜLLER

Selected climatic data
for a global set of standard stations
for vegetation science

DR W. JUNK PUBLISHERS   THE HAGUE / BOSTON / LONDON 1982

Distributors:

for the United States and Canada

Kluwer Boston, Inc.
190 Old Derby Street
Hingham, MA 02043
USA

for all other countries

Kluwer Academic Publishers Group
Distribution Center
P.O. Box 322
3300 AH Dordrecht
The Netherlands

Library of Congress Cataloging in Publication Data   CIP

Müller, Manfred J.
    Selected climatic data for a global set of
standard stations for vegetation science.

    (Tasks for vegetation sciences ; 5)
    Includes index.
    1. Climatology--Observations. 2. Vegetation
and climate--Observations. I. Title. II. Series.
QC982.M84 1982   551.6   82-8958
                                    AACR2

ISBN-13: 978-94-009-8042-6   e-ISBN-13: 978-94-009-8040-2
DOI: 10.1007/978-94-009-8040-2

SERIES EDITOR'S PREFACE

The present volume, Nr. 5 in the T:VS series is an example of a handbook volume for working in vegetation science. Anyone working and teaching in this field knows the difficulties in obtaining basic environmental data needed for research and interpretation. There are regional publications and there are other data sets available. In both cases the distribution is limited. The present volume by Dr. Müller intends to provide a selection of climatic parameters as they are commonly needed for the work of the vegetation scientist. The same set of monthly mean values are provided for about 1000 stations distributed as evenly as possible over the global land surfaces.

The tabular presentation of data from the individual stations is put into a geographical context through various means. Climatological classifications according to KÖPPEN/GEIGER and TROLL/PAFFEN are included as well as a revised system of climate diagrams from WALTER and LIETH. In the present form the global standard climate data set should prove to be useful for vegetation science, biometeorology, agriculture, and forestry as well as general geography.

I have to thank Dr. Müller for providing this valuable work to the T:VS series. I am sure that this selection of climatic data for the special needs of vegetation scientists will help many colleagues in different parts of the world.

Osnabrück, November 1981   H. Lieth

PREFACE OF THE AUTHOR

Whoever is concerned with climate geography, vegetation science or geoecological questions will have repeatedly felt the absence of an extensive but easily accessible collection of climate data from selected stations on the earth. In a series of handbooks of various scientific disciplines climate tables from the most important stations can indeed be found but unfortunately they mostly only contain the average values of the air temperature and precipitation.

There is no need to give particular reasons why these average values are not sufficient for climate geographical or vegetational observations of an area. Therefore, we attempted to record as many additional values of the individual climate elements as possible. In spite of great progress in the publication of climate data in the past few years it was not all that easy to fill the tables designed for this handbook. Only a few stations register all the values of the climate elements given here — that is the reason for the gaps. From the beginning it was clear to us that an entirety in this respect was by no means to be achieved. The aim of our work was rather to compile information from all the climate regions of the earth by means of station data and thereby to pay particular attention to the regional aspect. For this reason the climate zones according to KÖPPEN/GEIGER and TROLL/PAFFEN were also displayed cartographically and the location of the stations was recorded in the maps.

Without the active help of E. Lutz and Th. Boullion (cartography), E. Krames, M.A., H. Willger, R. Strasser, M. Rosen and J. Wolfers this work would not have been accomplished. Therefore, I would especially like to thank them here. In addition I would also like to thank Dr. K. Albrecht, Dachau, Dr. A. Bronger, Kiel, Dr. A.G. Benzing, Villingen-Schwenningen, Prof. Dr. M. Domrös, Mainz, Dr. J. v. Lengerke, Heidelberg, Prof. Dr. C. Rathjens, Saarbrücken, Prof. Dr. H. Reinhard, Greifswald, Prof. Dipl.-Met. H. Schirmer, Offenbach, Dipl.-Met. H. Skade, Enkirch and Prof. Dr. H. Walter, Stuttgart, for their advice, stimulus and criticism.

Special thanks are owing to Prof. Dr. KH. Paffen, Kiel, who put his revised map of the seasonal climates of 1969 and 1980 at our disposal and helped with the correction of the maps we used.

I thank Prof. Dr. H. Lieth, Osnabrück, for accepting this project into the series. I also thank him and Mr. Ostendorf for providing the climate diagrams and the chapter on producing these diagrams on computer facilities.

Trier, November 1981   Manfred J. Müller

V

CONTENTS

INTRODUCTION

The climate data had be compiled from various sources in order to achieve a meaningful geographical distribution. In the process we could not avoid to use in part different measurement periods. In most of the cases with the average temperatures and the average amount of precipitation we have a long dependable series of measurements at our disposal. Our standard series lasts from 1931 to 1960, as accepted at the International Meteorological Conference in Washington in 1957. However, a large number of the climate stations which we included were initiated in the 1940's or 1950's only. Therefore many of the measurement series are shorter and the values compiled in this manual cannot be compared with one another without reservations. For reasons of space it was not possible to state the exact years of the measurement period in years under the heading "p" (period). If the figure 30 appears there it can normally be assumed that it concerns the measurement period from 1931 to 1960. Values of shorter measurement periods mostly come from the 1960's and 1970's or have undergone interruptions during the Second World War. If no details were to be found with regard to the measurement period the column was left empty.

SELECTION OF THE CLIMATE STATIONS

The number of the stations had to be limited for various reasons. The handbook was not to become too extensive but was to retain the character of a handy reference book. Stations which only measure the temperature and precipitation were not considered as they do not provide enough information. The stations in the Polar regions are an exception. Europe has purposely been over-represented with regard to the number of stations. To prevent this from effecting the vegetation-geographical comparison of the regions of the earth the suitable stations have been marked. An essential criterion in the selection of the stations was the traceability in the current atlases. Small locations which are not marked in the atlases were only included in the survey if no comparable station of a larger location could be called upon. However, as the exact location can be ascertained from the climate maps this disadvantage does not seem to be of particular importance. Moreover, a corresponding reference is to be found in the index for small locations which are close to larger towns and for suburbs. In addition, attention was paid to achieving as uniform a distribution as possible across the earth. This could not always be realized because for example, very few stations which can reveal a longer series of measurements are situated in the large dry regions or the Central Asian heights. Very short series, as, for example, in the case of the stations of the Polar regions, were only brought in in exceptional cases, even if a comparison of these data with those from longer measurement periods is not possible. Furthermore it seemed important to us for every country to be represented by stations. Unfortunately this demand could not completely be fulfilled. In addition we took care that all types of climate are represented by characteristic stations and that topographical-orographical dependent peculiarities are not lost.

In the writing of place names there are problems, especially for languages with non-Latin characters. To make the work easier for the user we have taken over the method of writing of the Rand McNally Atlas.

THE STRUCTURE OF THE TABLES

The location of the stations can indeed be determined from the corresponding maps, however, it seemed meaningful to additionally state the exact situation in the grid of the earth. With calculations of the continentality of a place it is necessary, for example, to take into consideration the geographical latitude if one uses the demarcation formulae by ZENKER, GORCZYNSKI or IWANOW. For each station the types of climate are presented according to the climate classifications of KÖPPEN and TROLL. KÖPPEN has marked smaller deviations from the type criteria by an apostrophe added to the climate formulae, e.g. Aw'. For reasons of simplicity this was omitted. Normally the climate types can be found again in the appertaining climate zone. In a few exceptional cases stations with a certain type of climate are situated in a zone of another type of climate. Then it is mostly a question of very small regions which have fallen a victim to generalization. The following stations are quoted as examples:

According to KÖPPEN Popayán (30) in Columbia has a Csa-climate but is situated in a climate zone with a Cw-climate.

Adelaide (15) in Australia has a Cfa-climate according to KÖPPEN but lies in a Cs-climate zone. Simla (162) in India has the climate formula V,1 according to TROLL but lies in a V, 3-climate zone. Hyderabad (172) in India has a V, 3-climate according to TROLL and lies in a V, 4-climate zone. Bangalore (190) in India has a V, 2-climate according to TROLL although it lies in the climate zone V, 3. Almost all the named stations have a more humid climate than the region in which they lie. The explanation for the deviations is therefore relatively simple. The increased rainfall is due to the orographic conditions. This is most clear in the case of Simla. It is situated at a heigt of 2.202 m in the Siwalik mountains at the foot of the Himalayas, against which the west winds (SW monsoon) are forced to rise and bring increasing rainfall.

Mean temperature

We could not avoid the fact that on the one hand numbers appear in these lines which are rounded to whole values but on the other hand an exactness of one decimal place is given. In the sources we found very varyingly prepared figures. Admittedly, we could have rounded all the values up or down to whole numbers, particularly as decimal values from completely different series of measurement feign an exactness which is not really given, but on the other hand there is a decisive reason against such a procedure: A large number of data were gained by converting from Fahrenheit to Celsius. If we had rounded these values the same value in Fahrenheit would not have been gained again when converting back. Unfortunately the calculation of the daily mean temperature is not carried out by all the national meteorological offices according to the same principle. The so-called "real daily mean temperature" is gained from hourly measurements. However, that only occurs in a few cases because the work expenditure of reading off the hourly measurements is too great for this method to be able to be carried out in the large number of stations which are in part managed by volunteers or as a subsidiary occupation. In many countries, as in Germany, measurements are taken at 7 a.m., 2 p.m. and 9 p.m. The value of 9 p.m. is counted double in the calculation.

(1) Daily mean temperature = $t_{7a.m.} + t_{2p.m.} + 2\ t_{9p.m.}/4$

Comparisons have shown that daily mean temperatures gained in this way come very close to the "real daily mean temperature". In other countries (e.g. USA, France, Morocco) the daily mean temperature is calculated from the daily extreme values:

(2) t max. + t min./2

Certain deviations result from this.

The average daily mean temperatures form the basis for the monthly mean temperature and the average monthly mean temperatures are that for the annual mean temperature.

Average maximum of the temperature
Average minimum of the temperature
Absolute maximum of the temperature
Absolute minimum of the temperature

What is stated in the previous paragraph is equally valid for the average and the absolute extreme values. All the average extreme values are average daily extreme values. Unfortunately a discrepency was revealed in the sources with regard to these data. In volume 10, Africa, of the 15 volume collective work published by H.E. LANDSBERG the average daily extreme values of the temperature are quoted, whilst in the other volumes the average monthly values are to be found. Unfortunately no reference to this is made in volume 10. The main source for the average daily extreme values is the 6 volume publication of the Meteorological Office in London. However, here all the temperature values are given in $^\circ$ F which had to be converted in $^\circ$ C.

Mean relative humidity

With most of the climate stations the mean values of the relative humidity are available. However, it is never stated in the sources how the mean value has been gained. Some stations measure the relative humidity only once a day, in the morning, others measure it in the morning and at noon or even in the morning, at noon and in the evening. A certain inaccuracy results from this situation. It would have been best to take over the various values each time giving the corresponding time of measurement. This was not possible due to standardized tables. Therefore, we decided to form the mean values ourselves if several measurements were available. If only the morning value was available as for example, in various stations in Australia, this has been noted. No note has been made about whether the mean value was formed from 2 or 3 average monthly values. It seemed of minor importance to us for the very reason that no reference was found in the sources available about how the average values were calculated in this case.

To enable a better evaluation of the data we shall briefly go into the normal daily procedure of the relative humidity of the atmosphere. It generally proceeds as follows: the maximum is present in the morning when the air temperature is still low. In the course of the day the relative air humidity sinks with the rise of the air temperature so that in the afternoon the minimum is reached. Towards evening the relative humidity naturally increases again with the cooling of the air. This normal daily course is frequently varied by influences, like air mass change or by the land-sea wind-phenomena.

Mean precipitation

The values of mean amount of precipitation together with those of the mean air temperature are those values which can most easily be compared with one another. They are nearly always based on long series of measurements from which the mean values have been formed. As precipitation and temperature can be measured with simple instruments the network of stations in most regions of the earth is unusually dense (see F. L. WERNSTEDT 1972). As with the temperature values conversions had to be carried out with a large number of stations of Anglo-Saxon countries or previous colonies of England. However, in this case we omitted a decimal place in spite of the resulting inaccuracy when converting back into inches.

Maximum precipitation
Minimum precipitation

What has been said about the mean precipitation is not always equally valid for the monthly extreme values of the precipitation. Here omissions could not be prevented. With some stations in arid regions the abbreviation tr (= traces) sometimes occurs in columns 7, 8 and 9 (average, maximum and minimum precipitation). This stands for amounts of precipitation which are so small that they cannot be measured.

Maximum precipitation in 24 hours

Now and again in this column higher values are to be found for 24 hours than in column 8 (maximum precipitation) for a whole month. In such cases stem the values from the different measurement periods.

Days with precipitation

Due to the fact that the measurement units employed in meteorology and climatology are still not uniform the comparability of these data is unfortunately not always guaranteed. In some countries, e.g. in the USA and Great Britain, measurements are still made in inches. Those days count as days with precipitation which yield more than 0.01, 0.04 or 0.1 inches, that is 0.25, 1.0 or 2.5 mm. In most countries the days with more than 0.1 mm precipitation are counted. The consequence of these different measurements is a higher number of rainy days in the countries which measure according to the metric system.

Sunshine duration

Relatively few stations measure the sunshine duration all the year round and over long periods of time. Often the sunshine hours are only estimated, as, for example, in Australia. However, for vegetation science and for certain questions of agricultural geography the number of sunshine hours is not insignificant. Therefore they were included in this handbook although not all data are available as yet.

Amount of radiation

In meteorology the term "global radiation" is mostly used. It was given in Ly/day (Ly = cal/cm^2). This physical unit is named after the American astrophysicist S. P. LANGLEY (1834 to 1906).

After the reform of the measurement units one is supposed to use $Watt/m^2$. However, in the sources all figures are given in Ly/day. A conversion would have meant an extremely great expenditure of time and was therefore not carried out. If required the reader can carry out the conversion himself for each desired station. The conversion factor is 0.485 for 1 Ly/day = 0.485 W/m^2.

Potential evaporation

A decidedly difficult problem in climatology is the assessment of the evaporation. There are two different methods, measurement and calculation.

Measurements are carried out with the help of differing equipment. The simplest ones measure the actual evaporation of open water surfaces. They are water vessels, flat evaporation pans, damp absorbent papers or even evaporation balances lowered into the surface of the earth. They all have the disadvantage that they do not consider the conditions in nature where soil differences and vegetation play an important role.

To compensate these disadvantages one has gone over to using lysimetres. These are vessels filled with soil and lowered into the surface of the earth. They are taken out daily and weighed so that one can draw up a total water budget. The advantages are obvious for the soil conditions are taken into consideration and the vegetation can also be partly included. However, the work expenditure is not insignificant. The results obtained in this way are fairly dependable but it is problematic to generalize them and to relate them to larger areas. The large number of factors which influence the evapotranspiration can lead to the results turning out quite differently in a small area. Alongside the manifold measuring methods there is the possibility of calculating potential evapotranspiration with the help of empirical formulae. Nowadays difficulties arise not so much with complicated calculating procedures, for the electronic calculators help here, but rather with the large number of different meteorological data which have to be taken into consideration in these formulae. By no means all climate stations supply these data, so that there are limits for a handbook in which as uniform a distribution of stations as possible across the earth is strived for. The semi-empirical evapotranspiration formula recognized by most authors nowadays is that of PENMAN, H. L. (1948). It is applicable worldwide and due to its large number of parameters it attains quite a close approximate value to the natural conditions.

(3) $$E_{pw} = \frac{\triangle \cdot Ho + \gamma \cdot Ea}{\triangle + \gamma}$$

$\triangle$ = gradient of the vapour pressure curve in mm Hg per degree

Ho = radiation budget, it is composed of the radiation in the solar climate, the albedo for water, the real sunshine duration, the vapour pressure in mm Hg and $\delta \cdot T4$ = Boltzmann equation

γ = psychrometre constants

Ea = ventilation and humidity; they are composed of the wind velocity at a height of 2 m in km/h and the saturation deficit of the air at a height of 2 m in mm Hg

The explanation of the individual parameter from which the formula is constructed shows the difficulties which occur for this handbook. In the sources there are only a few stations in which all the necessary data are present. So if we had wanted to fill column 14 with evaporation values according to PENMAN this column of the tables would have inevitably remained empty in the case of a large number of stations.

The evapotranspiration formula of THORNTHWAITE, C.W. (1948) is much simpler as it is based on remarkably fewer measured climate data.

(4) $PE = 1.6 \, L \frac{10 \, t^a}{I}$

PE = Potential Evaporation

L = correction factor for different latitudes and month lengths

t = mean temperature of the selected period of time

I = an index calculated from the twelve monthly mean temperatures

a = a complicated geometric function of I

MATHER, J.R. (1963/65) and his colleagues have calculated the Potential Evapotranspiration for a remarkable large number of climate stations of the earth according to the formula of THORNTHWAITE and published it in the series "Publications in Climatology". The fact shall not be concealed that THORNTHWAITE's evapotranspiration formula is disputed. JÄTZOLD, R. (1962) asserts the following objections: "The summer months appear too dry and the winter months too wet. The PE values used as a guide-line for determining the dryness are too high in the summer as a large amount of water evaporates from soils saturated with water and densely covered with hygrophile vegetation ... On the other hand in the winter the PE values are too low as THORNTHWAITE, contrary to the lysimetre measurement, fixes the potential evapotranspiration at nought for monthly mean temperatures under 0° C and correspondingly low for 0 to 5° C (approx. 1—15 mm mthly.) whereby, similar to the case of GAUSSEN, too many months would appear arid in regions with cold winters" (JÄTZOLD, R. 1962, p. 93).

In spite of these weakness we decided on THORNTHWAITE's evapotranspiration formula for the following reasons:

1. No other formula would have given us values for each station because the necessary measurement data are lacking.

2. This handbook is intended to be a reference book which shall contain as many data as possible. Therefore concessions were made to exactitude in favour of completeness. For complicated geobotanical or vegetation-sociological statements, especially in the dry regions, one will have to refer back in every case to more wide-scoped data material and investigations on the spot.

Mean wind velocity
Predominant direction of the wind

Wind velocity and direction of the wind are recorded from numerous climate stations. However, the interpretation of the data from many stations is problematic because the local influences of the relief and of the ruggedness of the ground can be greater than those of the atmospheric circulation. If a climate station is situated in a valley one can assume with fair certainty that the predominant direction of the wind is also the direction of the valley. In the case of the wind velocity the location of anemometre is of decisive importance. In spite of these reservations the data about air movement given in the handbook are by all means suitable for carrying out a comparison of stations which lie in different wind belts. A few examples are given to make this easier to understand.

The station Alger/Algeria (7) reflects very clearly the interchange between the influence of the west wind zone in the winter and the north-east trade-wind in the summer, which makes itself noticeable both in the increase of the temperature and in the slackening of the rainfall.

In the station Dakar/Senegal (62) it can be read that the predominant wind direction changes from N to W with the beginning of the rainy period in June, i.e. when the station comes under the influence of the ITC. The beginning of the SW monsoon can be observed very well in the example of Nakhon Ratchasima/Thailand (278). In February the NE monsoon still reigns. The average air temperature is around 26.6° C, the average precipitation amounts to 37 mm. March already has west winds as the predominant wind direction. In April they veer to SW and bring increased precipitation (85 mm). At the same time the maximum is reached. In the course of the rainy period the temperatures fall due to increased cloudiness, the rainfall remains high and reaches its maximum in September. In the course of October the NE monsoon begins again and the rainfall decreases.

In the case of stations in climate zones with frequently occuring circulating winds as, for example, in the Tropics or in the Polar regions the sign "var" for varying wind directions appears now and again in column 16.

The stations at the Poles give the wind direction partly in angle degrees because the points of the compass cannot be exactly determined here. At the South Pole itself every direction is north.

THE CLIMATIC MAPS

Of the numerous climate classifications which have been drawn up, by climatologists and climate geographers, two of the most well known have been selected.

The classification by KÖPPEN with the map of KÖPPEN/GEIGER was taken over above all due to its clarity and its wide distribution — especially in Anglo Saxon countries. In spite of the known workness it is still of great didactical value. The short "climate formulae" can be easily remembered and make the general view across the distribution of the climates of the earth easier for the beginner. The classification of TROLL/PAFFEN impresses because of its exactness. The advantage of this classification is its consideration of the vegetation, the daily and seasonal differences as well as the maritimity or continentality with the help of the annual amplitude of the temperature.

By means of the large number of different types of climate — there are 38 altogether — this climate classification can naturally not be as easily remembered as that of KÖPPEN. However, one can ascertain that it is clear and relatively easy to follow. The judgement made by J. BLÜTHGEN (1980 p. 683) expresses the quality of the climate classification best of all: "This represents a climax of classification achievements, close to the limit beyond which a further subdivision would be bound to lead to unclarity and more difficult comparability".

Contrary to the works we used for reference the climate zones were not represented on a world map but were divided into continents on the scale 1:36 mill. The original map by TROLL has the scale 1:45 mill. Some difficulties arose with the demarcation of the climate zones as a result of the enlargement and the selection of a different projection. One effect of the greater generalization of the world maps was that on the maps with the larger scale some stations which lie on the boundary of two climate zones were sometimes caused to slip into the wrong zone. In order not to cause any confusion the boundaries were correspondingly shifted in such cases. Especially with KÖPPEN's map some considerable alterations had to be made to the course of the boundaries as, for example, in the Taiga of South Jakutia. On account of the station values in Jakutsk (49) and Aldan (45) the Df climate extends remarkably further to the east than is marked by KÖPPEN. In East

Siberia a region with a Df climate had to be marked in the valley of the upper Kolyma on the Western edge of the Jukagiren plateau because both of the stations situated here, Zyr'anka (54) and Sejmtčan (46), have a Df climate and not a Dw climate according to KÖPPEN's definition. The reason for this inexactitude on the maps is surely to be found in the lack of climate data which KÖPPEN and also TROLL had at their disposal when they were drawing up their maps. It is only in very few cases that stations lie in climate zones to which they really do not belong. Then it is usually a case of regions which are too small to be marked on the map. These stations are listed on page VIII.

We regard the entry and numeration of all the stations mentioned in the handbook as an essential advantage compared with the original maps. This makes the topographical co-ordination easier for the user. Besides this, with the help of the maps, he can very quickly select stations of individual climate zones for global comparisons.

Unforunately no completely uniform principle could be stuck to in the numeration of the stations. On the whole they were numbered according to states from north to south and from west to east.

THE CLIMATE CLASSIFICATIONS

Although the demarcation criteria for climate types of the climate classifications by W. KÖPPEN and C. TROLL can be looked up in almost every handbook they are given again here, slightly shortened. This will make the work easier for the user.

The climate classification by W. KÖPPEN (1923, 1931)

KÖPPEN devides the climates of the earth into 5 climate zones which he marks with capital letters.

First letter

A-climates = Tropical rain climate without cool season. All monthly mean temperatures $> 18^0$ C.

B-climates = Dry climates. Boundaries determined by formula using mean annual temperature (T) and mean annual precipitation (R).

    a) Precipitation concentrated in summer (high-sun) season
       $R = 2 (T + 14)$

    b) Precipitation evenly distributed throughout year
       $R = 2 (T + 7)$

    c) Precipitation concentrated in winter (low-sun) season
       $R = 2 T$

The dry climates are divided into Steppe-climate and Desert-climate. KÖPPEN calculates the boundary between the two by omitting the factor 2 in the formulae.

C-climates = Warm temperate climates. Mean temperature of coldest month between 18^0 C and -3^0 C

D-climate = Snow climates. Warmest month mean $> 10^0$ C coldest month $< -3^0$ C. Only on the northern hemisphere.

E-climates = Ice climates. Warmest month mean $< 10^0$ C. The E-climates are divided in ET (tundra climate) and EF (climate of permanent frost). The warmest month mean is under 0^0 C.

The further subdivision of the climate zones leads to eleven main types of climate as listed in the legend of the KÖPPEN map. They are subdivided according to the rainfall distribution and the amount of rainfall. In the case of the climates A to D this is expressed by the second letter. The third letter, or the second in the case of E-climates, differentiates according to the summer warmth or the winter cold. KÖPPEN has drawn up the following demarcation criteria individually.

Second letter

f   =   Sufficient moisture in all months.

m   =   Rainforest despite a dry season (i.e. monsoon cycle)

s   =   Dry season in summer of the respective hemisphere

w   =   Dry season in winter of the respective hemisphere

Third letter

a   =   Hot summers. Warmest month $> 22^o$ C

b   =   Warm summers. Warmest month $< 22^o$ C. At least 4 months have means $> 10^o$ C.

c   =   Cool summers. Fewer than 4 months (1−3) with means $> 10^o$ C.

d   =   Severely cold winters. Fewer than 4 months (1−3) with means $> 10^o$ C, but coldest month $< -38^o$ C.

B-Climates only

h   =   Hot. Mean annual temperature $> 18^o$ C

k   =   Cold. Mean annual temperature $< 18^o$ C.

Climate Classification of C. TROLL and KH. PAFFEN

The climatic levels of mountains should be interpreted as altitudinal variations of the climatic zone concerned.

I. Polar and Subpolar Zones

1. High-polar ice-cap climates: polar ice-deserts.
2. Polar climates with little solar heat (warmest month below $+6^o$ C); polar frost-debris belt.
3. Subarctis tundra climates with cool summers (warmest month 6° to 10° C) and great winter cold (coldest month below -8° C): tundra.
4. Highly oceanic sub-polar climates with moderately cold winters, poor in snow (coldest month -8° to $+2^o$ C) and cool summers (warmest month $+5^o$ to $+12^o$ C); annual fluctuation $< 13^o$ C, often $< 10^o$ C): sub-polar tussock grassland and moors.

II. Cold-temperate Boreal Zone

1. Oceanic boreal climates (annual fluctuation 13° to 19° C) with moderately cold winter, with, however, relatively prolific in snow (coldest month $+2^o$ to -3° C   winter precipitation maximum), moderately warm summers (warmest month $+10^o$ to $+15^o$ C) and a vegetation period of 120 to 180 days: oceanic humid coniferous woods.
2. Continental boreal climates (annual fluctuation 20° to 40° C) with long, very cold winters, prolific in snow, but short relatively warm summers (warmest month $+10^o$ to $+20^o$ C) and a vegetation period of 100 to 150 days: continental coniferous woods.
3. Highly continental boreal climates (annual fluctuation $> 40^o$ C) with permanently frozen soils, very long, extremely cold and dry winters·(coldest month below -25° C) short, but sufficient warming up in summertime (warmest month $+10^o$ to $+20^o$ C) and deep thawing soils: highly continental dry coniferous woods.

III. Cool-temperate Zones
Woodland Climates

1. Highly oceanic climates (annual fluctuation < 10° C) with very mild winters (coldest month +2° to +10° C) high winter precipitation maximum and cool to moderately warm summers (warmest month below +15° C): evergreen broad-leaved and mixed woods.

2. Oceanic climates (annual fluctuation < 16° C) with mild winters (coldest month above +2° C), autumn and winter maxima of precipitation and moderately warm summers (warmest month below +20° C): oceanic deciduous broad-leaved and mixed woods.

3. Sub-oceanic climates (annual fluctuation 16° to 25° C) with mild to moderately cold winters (coldest month +2° to −3° C), autumn to summer maxima of precipitation, moderately warm to warm and long summers and a period of vegetation of more than 200 days: sub-oceanic deciduous broad-leaved and mixed woods.

4. Sub-continental climates (annual fluctuation 20° to 30° C) with cold winters (coldest month −3° to −13° C) and distinct winter break in vegetative process, with moderately warm summers (warmest month generally below +20° C), summer maximum of precipitation and vegetation period of 160 to 210 days: sub-continental deciduous broad-leaved and mixed woods.

5. Continental climates with cold, slightly dry winters (annual fluctuation 30° to 40°; coldest month −10° to −20° C) and moderately warm and moderately humid summers (warmest month 15° to 20° C) and a vegetation period of 150 to 180 days: continental deciduous broad-leaved and mixed wood as well as wooded steppe.

6. Highly continental climates with cold and dry winters (annual fluctuation generally > 40° C; coldest month −10° to −30° C) and short, warm and humid summers (warmest month above 20° C): highly continental deciduous broad-leaved and mixed woods as well as wooded steppe.

7. Humid-and-warm-summer climates (annual fluctuation 25° to 35° C) with moderately cold, but dry winters (coldest month 0° to −8° C; warmest month 20° to 26° C): deciduous broad-leaved and mixed wood and wooded steppe favoured by warmth, but withstanding cold and aridity in winter.

7a. Dry-and-warm-summer climates with a mild to moderately cold, but slightly humid winter half year (coldest month +2° to −6° C; warmest month 20° to 26° C): thermophile dry wood and wooded steppe which withstands moderate to hard winters.

8. Permanently humid, warm summer climates (annual fluctuation 20° to 30° C) with mild to moderately cold winters (coldest month +2° to −6° C; warmest month 20° to 26° C): humid deciduous broad-leaved and mixed wood which favours warmth.

Steppe Climates

9. Humid steppe climates with cold winters and 6 or more humid months, vegetation period in spring and early summer (coldest month below 0° C): high grass-steppe with perennial herbs.

9a. Humid steppe climates with mild winters (coldest month above 0° C).

10. Steppe climates with cold winters, arid summers and less than 6 months of humidity (coldest month below 0° C): short grass-, or dwarf shrub-, or thorn-steppe.

10a. Dry steppe climates with cold winters and arid summers (coldest month 0° to +6° C): steppe with short grass, dwarf shrubs and thorns.

11. Humid-summer steppe climates with cold and dry winters (coldest month below 0° C): Central and East-Asian grass- and dwarf shrub- steppe.

12. Semi-desert and desert climates with cold winters (coldest month below 0° C): semi-deserts and deserts with cold winters.

12a. Semi-desert and desert climates with mild winters (coldest month 0° to +6° C): semi-deserts and deserts with mild winters.

IV. Warm-temperate Sub-tropical Zones

All plains and hill country climates with mild winters, coldest month +2º to +13º C, from +6º to +13º C in the southern hemisphere.

1. Dry-summer Mediterranean climates with humid winters (mostly more than 5 humid months): sub-tropical hard-leaved and coniferous wood.
2. Dry-summer steppe climates with humid winters (mostly less than 5 humid months): sub-tropical grass- and shrub-steppe.
3. Steppe climates with short summer humidity and dry winters (less than 5 humid months): sub-tropical thorn- and succulents-steppe.
4. Dry-winter climates with long summer humidity (generally 6 to 9 humid months): sub-tropical steppe with short grass, hard-leaved monsoon wood and wooded-steppe.
5. Semi-desert and desert climates without hard winters, but frequent transient or night frosts (generally less than 2 humid months): sub-tropical semi-deserts and deserts.
6. Permanently humid grassland-climates of the southern hemisphere (10 to 12 humid months): sub-tropical high-grassland.
7. Permanently humid climates with hot summers and a maximum of precipitation in summer: sub-tropical humid forests (laurel and coniferous forests).

V. Tropical Zone

1. Tropical rainy climates with or without short interruptions of the rainy season (12 to 9 1/2 humid months): evergreen tropical rain forest and half deciduous transition wood.
2. Tropical humid-summer climates with 9 1/2 to 7 humid and 2 1/2 to 5 arid months: rain-green humid forest and humid grass-savannah.
2a. Tropical humid-winter climates with 9 1/2 to 7 humid and 2 1/2 to 5 arid months: half deciduous transition wood.
3. Wet and dry tropical climates with 7 to 4 1/2 humid and 5 to 7 1/2 arid months: rain-green dry wood and dry savannah.
4. Tropical dry climates with 4 1/2 to 2 humid and 7 1/2 to 10 arid months: tropical thorn-succulent wood and savannah.
4a. Tropical dry climates with humid months in winter.
5. Tropical semi-desert and desert climates with less than 2 humid and more than 10 arid months: tropical semi-deserts and deserts.

IV/V. Littoral Climates with Seasonal Mists

Seasonally atmosperically humid coastal climates in regions of tropical — sub-tropical desert climates and alternately humid climates caused by coastal mist

a) in summer,
b) in winter: types of coastal and mountainous coastal vegetation abundant in epiphytes, mist-green to evergreen, more humid than in the corresponding regional climate.

THE COMPUTER DRAWN CLIMATE DIAGRAMS

Bertram Ostendorf[1] u. Helmut Lieth
FB5 University of Osnabrück, 4500 Osnabrück, FRG

Abstract

A set of typical diagrams in this book were computer drawn using a plotting routine elaborated by OSTENDORF, LIETH, and LEHKER (1981). This computer program called climaplot is described here. The inputs needed are explained together with format and hardware requirements. The listing of the program is provided. If mounted at a computer center the data provided for each station in this book can be used to produce diagrams as shown in this book. If auxiliary subroutines are added, diagrams can be drawn from computer tapes as available for a world wide set of climatic stations provided by the US weather service in Ashville, N.C. The full address of this source is given in the literature cited.

1. The content of the diagrams

Climate diagrams facilitate the evaluation of several phytogeographically important climatic elements simultaneously. They make it possible to compare meteorological stations from distant geographical locations. One of the best known diagrammatic presentations of climate is the climate diagram, as designed by WALTER and LIETH in the Climate Diagram World Atlas (1961—1967). The computer drawn diagrams described here follow basically the conventions adopted in that atlas. Changes made are either due to the mechanization of the drawing process or to improvements in the numerical information implemented through various graphical elements of the diagrams.

The main conventions of the computer drawn diagrams are explained in fig. 1A, the diagram for the meteorological station of the city of Trier.

The alphanumeric entries into the diagram are:

a — Station name
b — Elevation above sea level
c — Years of observation, 2 entries when different, and then first figure for temperature readings, and second for precipitation.
d — Mean annual temperature in °C
e — Mean annual sum of precipitation in mm
f — Mean daily minimum of the coldest month
g — Absolute temperature minimum recorded
h — Mean maximum of the warmest month
i — Absolute highest temperature recorded

Other entries are:

k — Dashed line indicating 50 °C temperature, equal to 100 mm precipitation
l — Solid line connecting mean monthly precipitation sums for the 12 months entered at the center position for each month
m — Solid line connecting mean monthly temperatures plotted at the centre position for each month
n — Squared blocks below the baseline of a month indicate mean temperature minima of that month below zero
p — Distance lines between temperature as low point and precipitations as high point in the center position for each month divided by tick marks for each 4 mm of precipitation.

---

1   Supported by a stipend to H. Lieth from the Deutsche Forschungsgemeinschaft.

The monthly means for temperature and precipitation are scaled such that 10 °C correspond to 20 mm of precipitation. At this ratio the precipitation level in humid climates is above the temperature values. If the precipitation is lower than 200 mm, the distance line between temperature and precipitations is divided in 4 mm, equal to 2 °C intervals (p). If the precipitation exceeds 100 mm at any given month, or if the monthly mean temperature values fall below — 20 °C, the scale below 100 mm is reduced by the factor 2 and marked equally as in the diagrams with values lower than 100 mm (see fig. 1 B and C). Diagrams in which the precipitation values exceed 200 mm in any month have the precipitation above 100 mm reduced by the factor 10 compared to diagrams of type 1 A and by the factor 5 compared to diagrams of types 1 B and 1 C. The 100 mm line is horizontally dashed (k) and above 100 mm each 10 mm is entered as an horizontally solid line (q in fig. 1 D) and full hundred mm precipitation are marked as dotted line (r in fig. 1 D) — example Calcutta.

In arid and semi-arid climates the precipitation is often so low that the monthly precipitation values on our scales lie below the respective temperature levels. In such cases the distance between precipitation value and temperature value is marked in dots 2 °C (2 mm precipitation equivalent) apart (sees in fig. 1 B and D). In order to allow better reading, the 2 °C distances are marked as dotted horizontal lines. Each 5 dots cover one month, the values entered at the center of each month can be read above and below the 3rd dot.

Of special ecological significance are temperatures below zero centigrades as well as freeze free periods. Squared (n) or horizontally lined (o) blocks below the base line, stand for mean monthly minimum temperatures below zero (h) and absolute monthly minimum below zero (o). Freeze free months are indicated by the absence of blocks. Missing minimum values cause also the blocks to be missing. When this is the case, the mean minimum value of the coldest month (f in fig. 1 A) is also missing.

2.   Description of the program

As computer language Fortran IV was chosen, since a Fortran compiler is usually available in every large computer center and has also fast running times (approx. 2-5 sec per diagram).
Conversion routines are built into the program to facilitate the use of non metric measures.

2.1   Description of the diagram formates

Three drawing formates are provided which are selected automatically or by the user.

Format 1:   Example shown in fig. 1 A TRIER
Base line 12 cm
100 mm precipitation 10 cm
(for stations with monthly precipitation values below
100 mm, example Trier).

Format 2:   Examples shown in fig. 1 B and C
Base line 12 cm
100 mm precipitation 5 cm
(for stations with monthly precipitation values above
100 mm but below 200 mm, example Mexico City, or temperatures below — 20 °C, example Verchojansk).

Format 3:   Example shown in fig. 1 D
Base line 12 cm
100 mm precipitation 5 cm
every additional 100 mm—step 1 cm
(for stations with monthly precipitation values above
200 mm or temperatures below — 20 °C, example Calcutta).

The connecting lines between two entry points for either temperature or precipitation are rectilinear. If two monthly precipitation values belong in format 3 to different scaling ranges, the line is bent between the two points at the intercept between the 100 mm line and the connecting line calculated as if the higher value were positioned in standard scaling.

Monthly precipitation values above 1000 mm are entered with the numeric value just above the 1000 mm level.

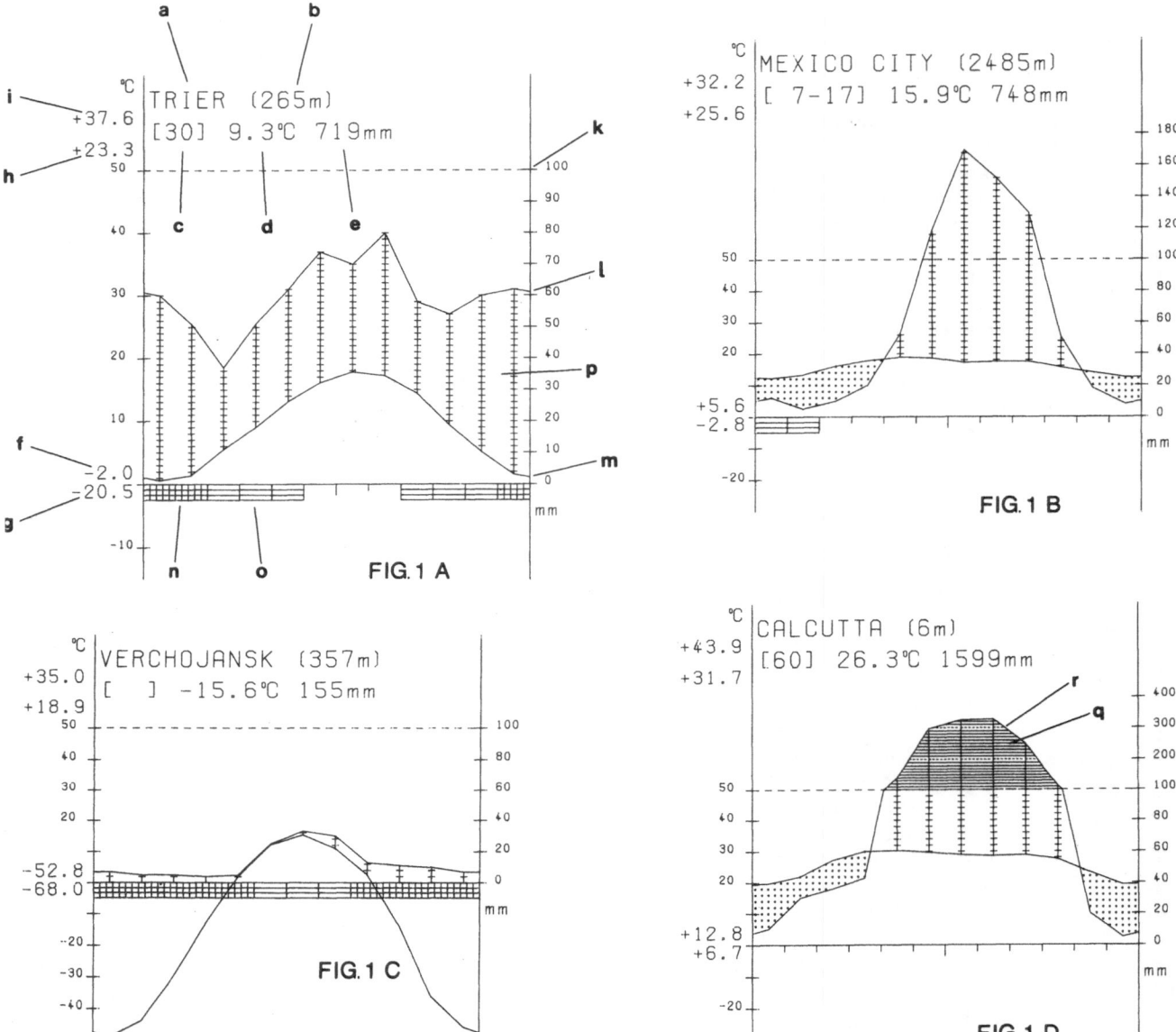

Fig. 1: Samples of climate diagrams, plotted with the KLIMAPLOT routine.

1 A,	TRIER:	Format 1 —	standard
1 B,	MEXICO CITY:	Format 2 —	because precipitation values exceed 100 mm but not 200 mm
1 C,	VERCHOJANSK:	Format 2 —	because temperature drops below — 20 °C
1 D,	CALCUTTA:	Format 3 —	because precipitation values exceed 200 mm

The meaning of the diagram elements pointed to by letters is given in the text on page XIX

2.2 Data input

For each diagram one needs 8 data cards - data set. Any number of data sets can be put behind one another.

Fig. 2: Images of the 8 punched cards for the diagram TRIER
(fig. 1 A).

	1	2	3	4	5	6	7	8
	12345678901234567890123456789012345678901234567890123456789012345678901234567890							
1:	TRIER (265m)							
2:	00							
3:	30	9.3	719.	−2.	−20.5	23.3	37.6	10000.
4:	0.6	1.4	5.5	9.0	13.1	16.1		
5:	17.8	17.2	14.4	9.4	5.1	1.6		
6:	60.	51.	37.	51.	62.	74.		
7:	70.	80.	58.	54.	60.	62.		
8:	221110001112							

1. Card

column 1 — 80: Text for labelling the diagram: Please note that the frame of the diagram provides for 22 characters regular size only. Longer words will be plotted up to 80 characters but will all appear in one line.

2. Card

column 1: Input of Ø or 1 for conversion of input datas.
Ø: The data are put in mm and oC. No conversion follows.
1: Data are put in in inch and degree Fahrenheit. Conversion follows in °C and mm.

column 2: Format-Parameter
Input of figures Ø, 1, 2, 3.
With this value the user can determine the diagram format. The formats are described in chapter 5.
Ø: The program finds out the most convenient formats 1—3. The respective criteria are described in chapter 5.
1: Format 1 is selected
2: Format 2
3: Format 3

3. Card

column 1— 2: Number of observation years for temperature
5— 6: Number of observation years for precipitation, in case they deviate from the observation period for temperature.
11—20: Mean yearly temperature
21—30: Mean yearly precipitation sum

31—40:	Mean daily minimum of the coldest month; if no test data is available, insert the number 10000.0
41—50:	Absolute minimum; if no test data available, insert the number 10000.0
51—60:	Mean daily maximum of warmest month; if no test data available, insert 10000.0
61—70:	Absolute maximum; if no test data available insert 10000.0.
71—80:	Mean daily temperature fluctuation; if no test data available, insert 10000.0

4. Card

mean monthly temperatures

columns	northern hemisphere	southern hemisphere
1—10	January	July
11—20	February	August
•	•	•
•	•	•
•	•	•
51—60	June	December

5. Card

mean monthly temperatures

columns	northern hemisphere	southern hemisphere
1—10	July	January
11—20	August	February
21—30	•	•
•	•	•
•	•	•
51—60	December	June

6. Card

mean monthly precipitation sums

columns	northern hemisphere	southern hemisphere
1—10	January	July
11—20	February	August
•	•	•
•	•	•
•	•	•
51—60	June	December

7. Card

mean monthly precipitation sums

columns	northern hemisphere	southern hemisphere
1—10	July	January
11—20	August	February
•	•	•
•	•	•
•	•	•
51—60	December	June

8. Card

Input of co led values for temperature minima of individual months

possible values:		**effect:**
Ø: = absolu te temperature minimum		no
1: = absolu te temperature minimum		horizontal
2: = mean temperature minimum		

For the northern hemisphere the values are listed under column
1—12, from January till December, and for the southern hemisphere
under column 1—12 from July till June.
If no values are available, a Ø is listed for every month.

2.3 *Listing of the program and diagram samples prepared from data presented in this book.*

The following pages contain the listing of the KLIMAPLOT program. A card deck, or computer tape to specifications of the user may be obtained from the authors at the University of Osnabrück at cost[2]. A similar program was developed by the authors in BASIC.
The listing of the program is followed by a set of diagrams from data presented in this book by Dr. Müller.

REFERENCES

OSTENDORF, B., H. LIETH and H. LEHKER, 1981: KLIMAPLOT, a computer routine to plot climate diagrams in H. Lieth, E. Landolt and R.K. Peet ed. Contributions to the knowledge of Flora and Vegetation in the Carolinas. Vol. 3, Veröffentl. des Geobotan. Inst. der ETH, Stiftung Rübel, Heft 70, Zürich

SPANGLER, W.M.L., JENNE, R.L., 1980: World monthly surface station climatology. TD 9645 MASTER NCAR Boulder Colorado, available through National Climatic Center, Federal Building, Ashville N.C. 28801, USA

WALTER, H. and LIETH, H., 1961—1967: Climate Diagram World Atlas. Fischer Verlag Jena

2   The program is free of any copyright.

REFERENCES

BEE, Ooi Jin and SIEN, Chia Lin (1974)
   The Climate of West Malaysia and Singapore. — Singapore
BERLAGE, H.P. (—)
   Observations Made at Secondary Stations in the Netherlands Indies. — Meteorological and Geophysical Service Department of Public Works and Communications. (= Royal Magnetic and Meteorological Observatory, Vol. **XIX B**) Batavia
BLÜTHGEN, J. and WEISCHET, W. (1980)
   Allgemeine Klimageographie. — (Lehrbuch der Allgemeinen Geographie, **2**). 3rd Ed., Berlin
BOEREMA, J. (1937)
   Observations Made at Secondary Stations in the Netherlands Indies. — (= Royal Magnetical and Meteorological Observatory, Vol. **XIX A**) Batavia
BORISOV, A. (1965)
   Climates of the UdSSR. — Edinburgh
BRAAK, C. (1972)
   Klimakunde von Hinterindien und Insulinde. — Nendeln
CHANDLER, T. J. and GREGORY, S. (1976)
   The climate of the British Isles. — London
Computer Section General Meteorological Directorate Civil Aviation and Tourism Authority: (—)
   Monthly Climatological Data. — Kabul
DALE, W. L. (1964)
   Sunshine in Malaya. — Journal of Tropical Geography, **19**, 20—26
Department of Irrigation, Hydrology and Meteorology, Ministry of Food, Agriculture and Irrigation (1977)
   Climatological Records of Nepal. — 3 Vols., Kathmandu
Department of Meteorology (Ed.) (1978)
   Climatological Table of Observatories in Sri Lanka. — Colombo
EIMERN, van J. (1964)
   Zum Begriff und zur Messung der potentiellen Evapotranspiration. — Meteorol. Rundschau, **17**, 33—42
FELKEL, H. (1978)
   Klimadaten von Österreich. — Wien
FLOHN, H. (1960)
   Zur Didaktik der allgemeinen Zirkulation der Atmosphäre. — Geogr. Rundschau, **12**, 129—142; 189—195
FLOHN, H. (1971)
   Arbeiten zur allgemeinen Klimatologie. — Darmstadt
GORCZYNSKI, W. (1920)
   Sur le calcul du degré de continentalisme et son application dans la climatologie. — Geogr. Annaler, **2**, 324—331
HARE, F. K. and MORLEY, K. T. (1974)
   Climate Canada. — Toronto
HERMAN, N. M. (1965)
   Le climat de l'Afghanistan. — (= Monogr. de la Météorologie Nationale, **52**), Paris
IWANOW, N. N. (1959)
   Belts of Continentality on the Globe. — Izwest. Wsesoj. Geogr. Obschtsch, **91**, 410—423
HÖLLER, E. (1975)
   Manual Climatologico America Latina. — 2nd Ed., Hamburg
HÖLLER, E. and STRANZ, D. (1975)
   Climatological Handbook East Asia. — Hamburg

HÖLLER, E. (1979)
  Climatological Handbook Australia, New Zealand, South Pacific. – Hamburg
HÖLLER, E. and STRANZ, D. (1979)
  Climatological Handbook Near and Middle East Asia. – 2nd Ed., Hamburg
HÖLLER, E. and STRANZ, D. (1981)
  Climatological Handbook Africa. – 2nd Ed., Hamburg
India Meteorological Department (Ed.) (–)
  Climatological Tables of Observatories in India (1931–1960). – New Delhi
JÄTZOLD, R. (1962)
  Die Dauer der ariden und humiden Zeiten des Jahres als Kriterium für Klimaklassifikationen. – In: Herm.
  v. Wissmann-Festschrift, 89–108
Klimatologische Forschung (1973)
  Festschrift Hermann Flohn zum 60. Geburtstag. – (= Bonner Meteorologische Abhandlungen, **17**), Bonn
KÖPPEN, W. (1931)
  Die Klimate der Erde, Grundriß der Klimakunde. – 2nd Ed., Berlin and Leipzig
KÖPPEN, W. and GEIGER, R. (1928)
  Klimakarte der Erde. – Gotha
KÖPPEN, W. and GEIGER, R. (1930/39)
  Handbuch der Klimatologie. – 5 Vols., Berlin
KUTSCH, H.G. (1978)
  Le Pouvoir d'Evaporation du Climat Marocain. – Ministère de l'Agriculture et de la Réforme Agraire Direction de la Recherche Agronomique (D. R. A.), Rabat
LANDSBERG, H.E. (Editor in Chief) (1969/81)
  World Survey of Climatology. – 15 Vols., Amsterdam, London, New York
  Volume 2   (1969) General Climatology. – Flohn, H. (Ed.)
  Volume 4   (1969) Climate of the Free Atmosphere. – Rex, D.F. (Ed.)
  Volume 5   (1970) Climates of Northern and Western Europe. – Wallén, C.C. (Ed.)
  Volume 6   (1977) Climates of Central and Southern Europe. – Wallén, C.C. (Ed.)
  Volume 7   (1977) Climates of the Soviet Union. – Lydolph, P.E. (Ed.)
  Volume 8   (1969) Climates of Northern and Eastern Asia. – Arakawa, H. (Ed.)
  Volume 9   (1981) Climates of Southern and Western Asia. – Takahashi, K. and Arakawa, H. (Eds.)
  Volume 10 (1972) Climates of Africa. – Griffiths, J.F. (Ed.)
  Volume 11 (1974) Climates of North America. – Bryson, R.A. and Hare, F.K. (Eds.)
  Volume 12 (1976) Climates of Central and South America. – Schwerdtfeger, W. (Ed.)
  Volume 13 (1971) Climates of Australia and New Zealand. – Gentilli, J. (Ed.)
  Volume 14 (1970) Climates of the Polar Regions. – Orvig, S. (Ed.)
LEBEDEV, A.N. (Ed.)
  The climate of Africa. – Jerusalem
LENGERKE, H.J. v. (1977)
  The Nilgiris. Weather and Climate of a Mountain Area in South India. – (= Beiträge zur Südasienforschung, **32**) Wiesbaden
MATHER, J.R.
–   (1962) Average Climatic Water Balance Data of the Continents. Part I, Africa. – Publications in Climatology, Vol. **XV**, No. **2**)
–   (1963) Average Climatic Water Balance Data of the Continents. Part II, Asia (excluding U.S.S.R.). – (= Publications in Climatology, Vol. **XVI**, No. **1**)
–   (1963) Average Climatic Water Balance Data of the Continents. Part III, U.S.S.R. – (=Publications in Climatology, Vol. **XVI**, No. **2**)

— (1963) Average Climatic Water Balance Data of the Continents. Part IV, Australia, New Zealand and Oceania. — (= Publications in Climatology, Vol. **XVI**, No. **3**)

— (1964) Average Climatic Water Balance Data of the Continents. Part V, Europe. — (= Publications in Climatology, Vol. **XVII**, No. **1**)

— (1964) Average Climatic Water Balance Data of the Continents. Part VI, North America (excluding United States). — (= Publications in Climatology, Vol. **XVII**, No. **2**)

— (1964) Average Climatic Water Balance Data of the Continents. Part VII, United States. — (= Publications in Climatology, Vol. **XVII**, No. **3**)

— (1965) Average Climatic Water Balance Data of the Continents. Part VIII, South America. — (= Publications in Climatology, Vol. **XVIII**, No. **2**)

Meteorological Department, Ministry of Communications (1977)

Climatological Data of Thailand, 25 Year Period (1951—1975). — Bangkok

Meteorological Office (1967)

Investigations Division Climatological Report No. **113**, China Part I and II. —

Meteorological Office Air Ministry (1961)

Investigations Revision Climatological Report No. **105**, Burma. —

Meteorological Office Air Ministry (1962)

Investigations Division Climatological Report No. **110**, Indian Ocean Islands, Parts I and II. —

Meteorological Office

— (1975) Tables of temperature, relative humidity and precipitation for the world. — Part I, North America, Greenland and the North Pacific Ocean, 2nd Ed., London

— (1977) Tables of temperature, relative humidity and precipitation for the world. — Part II, Central and South America, the West Indies and Bermuda, 6th Ed., London

— (1972) Tables of temperature, relative humidity, precipitation and sunshine for the world. — Part III, Europe and the Azores, London

— (1967) Tables of temperature, relative humidity and precipitation for the world. — Part IV, Africa, the Atlantic Ocean south of 35° N and the Indian Ocean, London

— (—) Tables of temperature, relative humidity and precipitation for the world. — Part V, Asia, London

— (1975) Tables of temperature, relative humidity and precipitation for the world. — Part VI, Austral-asia and the South Pacific Ocean, including the corresponding sectors of Antarctica. 2nd Ed., London

Meteorological Service Singapore (1973):

Climatological Summaries Singapore. —

MÜLLER, M.J., BALTES, K. and WERLE, D. (1980):

Handbuch ausgewählter Klimastationen der Erde. — (= Forschungsstelle Bodenerosion der Universität Trier, Mertesdorf Ruwertal, **5**) 2nd Ed., Trier

NIEUWOLT, S. (1969):

Klimageographie der Malaiischen Halbinsel. — (= Mainzer Geographische Studien, **2**) Mainz

PENMAN, H. L. (1948):

Natural Evaporation from Open Water, Bare Soils and Grass. — Proc. Roy. Soc. (A), **193**

PENMAN, H. L. (1956):

Evapotranspiration — an Introductory Survey. — Neth. Journ. of Agr. Sciences, **4**, 9—29

PERARD, J. u. YACONO, D. (1976):

Les Îles et presqu'îles de l'Asie du Sud-Est. — Etat de la documentation météorologique. — (= Cahier No. **5** du Centre de Recherches de Climatologie) Dijon

PERKHIDMATAN KAJICUACA MALAYSIA (—)

Records of Temperature and Relative Humidity. —

Royal Observatory (1952):

Hongkong Meteorological Records and Climatological Notes 60 Years 1884—1939, 1947—1950. — (Edited by Peacock, J. E.) Hongkong

RUDLOFF, W. (1981):
  World-Climates with Tables of Climatic Data and Practical Suggestions. — Stuttgart
TEICH, S. (1975):
  Neue Spitzenwerte des Niederschlags in Cherrapunji. — Meteorologische Rundschau, **28**, 94—95
TROLL, C. u. PAFFEN, KH. (1980):
  Jahreszeitenklimate der Erde. — Berlin
THOMPSON, B.W. (1970):
  The climate of Africa. — London
THORNTHWAITE, C.W. u. MATHER, J.R. (1957):
  Instructions and Tables for Computing Potential Evapotranspiration and the Water Balance. — (= Publications in Climatology, Vol. **X**, No. **3**)
THORNTHWAITE, C.W., MATHER, J.R. u. CARTER, D.B. (1958):
  Three Water Balance Maps of Southwest Asia. — (= Publications in Climatology, Vol. **XI**, 1—57)
United States Department of Commerce (1968):
  Weather Atlas of the United States. — Repr. 1975, Detroit
United States Navy Department, Hydrographic Department (Ed.) (1943):
  Weather Summary for H.O. Pub. No. 276 Naval Air Pilot New Guinea Area. — Washington
United States Navy Department, Hydrographical Office (Ed.) (1943):
  Weather Summary for H.O. Pub. No. 272 Naval Air Pilot Southwest Pacific, Fiji and Samoa Area. — Washington
United States Navy Department, Hydrographic Office (Ed.) (1943):
  Weather Summary for H.O. Pub. No. 275 Naval Air Pilot Southwest Pacific Solomon Islands Area. — Washington
WALTER, H. (1979):
  Vegetation und Klimazonen. — 4th Ed., Stuttgart
WALTER, H., HARNICKEL, E. and MÜLLER-DOMBOIS, D. (1975):
  Klimadiagramm-Karten der einzelnen Kontinente und die ökologische Klimagliederung der Erde. — Stuttgart
WALTER, H. u. LIETH, H. (1967):
  Klimadiagramm-Weltatlas. — Jena
WERNSTEDT, F.L. (1972):
  World Climatic Data. — Lemont, Pennsylvania
World Meteorological Organization (1971):
  Manual on Codes. — Vol. **1**, International Codes, Genf
World Meteorological Organization (1972):
  Manual on Codes. — Vol. **2**, Regional Codes and National Coding Practices, Genf

EUROPE

ICELAND

1 Akureyri*
2 Hallormstadur
3 Reykyavik*

NORWAY

4 Jan Mayen*
5 Vardø*
6 Tromsø
7 Mo í Rana*
8 Trondheim*
9 Dombås
10 Lillehammer*
11 Bergen
12 Oslo
13 Dalen
14 Kristiansand*

SWEDEN

15 Haparanda*
16 Stensele
17 Östersund*
18 Härnösand
19 Karlstad*
20 Stockholm
21 Jönköping*
22 Göteborg (Gothenburg)
23 Visby*
24 Kalmar
25 Malmö*

FINLAND

26 Sodankylä*
27 Oulu
28 Kajaani
29 Vaasa (Vasa)
30 Punkaharju*
31 Tampere*
32 Helsingfors (Ilmala)*
33 Mariehamn (Maariahamina)

DENMARK

34 Hoyvik (Farøer-Islands)*
35 Tylstrup (near Ålburg)*
36 Studsgård (near Herning)
37 København (Copenhagen)*
38 Sandvig (Bornholm)*

UNITED KINGDOM

39 Lerwik (Shetland Islands)*
40 Stornoway (Hebrides)*
41 Rattray Head (Scotland)*
42 Achnashellach (Scotland)*
43 Dalwhinnie (Scotland)
44 Eskdalemuir (Scotland)*
45 Tynemouth (England)
46 Belfast (Northern Ireland)*
47 Douglas (Isle of Man)*
48 Kingston-upon-Hull (England)
49 Cromer (England)
50 Birmingham (England)*
51 Aberstwyth (Wales)
52 Kew (near London) (England)*
53 Dover (England)
54 Plymouth (England)*
55 St. Mary's (Scilly Islands)*

IRELAND

56 Malin Head (Ceann Malainn)
57 Belmullet (Béal an Mhuirthid)*
58 Dublin (Baile Atha Cliath)
59 Kilkenny (Cill Choinningh)*
60 Valentia (Dairbhre)*

NETHERLANDS

61 Eelde
62 Den Helder*
63 De Bilt (near Utrecht)*

BELGIUM

64 Bruxelles (Brussels)*
65 Botrange-Robertville
   (near Malmedy)

LUXEMBOURG

66 Luxembourg-City

FEDERAL REPUBLIC OF GERMANY

67 Helgoland
68 Hamburg*
69 Lüchow
70 Berlin-Dahlem*
71 Hannover-Langenhagen*
72 Essen
73 Kassel
74 Bad Ems
75 Hof*
76 Frankfurt/Main (Airport)*
77 Trier*
78 Nürnberg
79 Neustadt/Weinstraße*
80 Regensburg*
81 Stuttgart*
82 München (Munich)*
83 Friedrichshafen
84 Zugspitze*

GERMAN DEMOCRATIC REPUBLIC

85 Greifswald*
86 Brocken (Harz)
87 Dresden*
88 Erfurt

POLAND

89 Gdynia*
90 Suwalki
91 Poznán*
92 Warszawa (Warsaw)*
93 Wroclaw (Breslau)
94 Zamość
95 Kraków*

CZECHOSLOVAKIA

96 Praha*
97 Prešov*
98 Ceské Budějovice

FRANCE

99 Cherbourg*

100 Reims
101 Paris (Le Bourget)*
102 Strasbourg
103 Brest*
104 Rennes
105 Tours
106 Dijon*
107 Nantes
108 Limoges*
109 Clermont-Ferrand*
110 Lyon
111 Grenoble*
112 Bordeaux*
113 Toulouse*
114 Marseille*
115 Pic-du-Midi
116 Perpignan
117 Bastia (Corsica)*

MONACO

118 Monaco*

ANDORRA

119 Les Escaldes

SWITZERLAND

120 Zürich*
121 Säntis
122 Genève
123 Lugano*

AUSTRIA

124 Wien (Vienna)*
125 Salzburg*
126 Innsbruck
127 Sonnblick
128 Graz

HUNGARY

129 Debrecen
130 Budapest*
131 Pécs*

ROMANIA

132 Iaşi*
133 Cluj*
134 Timişoara
135 Sibiu
136 Bucareşti (Bucharest)*

PORTUGAL

137 Braganca*
138 Porto
139 Coimbra
140 Campo Major (near Badajoz)*
141 Lisboa (Lisbon)*
142 Ponta Degada, Sao Miguel
    (Azores)
143 Praia da Rocha

SPAIN

144 Santander*
145 La Coruña
146 Valladolid*
147 Zaragoza
148 Barcelona*
149 Madrid*
150 Palma (Balearic Islands)*
151 Valencia*
152 Ciudad Real
153 Murcia
154 Sevilla
155 Granada*
156 Almeria
157 Las Palmas (Canary Islands)*

UNITED KINGDOM

158 Gibraltar Town (Gibraltar)*

ITALY

159 Bolzano*
160 Trieste
161 Milano (Milan)*
162 Venezia (Venice)*
163 Genova (Genoa)
164 Firenze (Florence)*
165 Ancona
166 Roma (Rome)*

167 Foggia*
168 Napoli (Naples)
169 Potenza
170 Taranto
171 Cagliari (Sardinia)*
172 Messina (Sicily)
173 Palermo (Sicily)*
174 Caltanissetta (Sicily)*

MALTA

175 Valetta*

YUGOSLAVIA

176 Zagreb
177 Beograd (Belgrade)*
178 Sarajevo*
179 Split*
180 Mostar
181 Skopje*

BULGARIA

182 Pleven
183 Varna*
184 Sofija (Sofia)*
185 Plovdiv

ALBANIA

186 Tiranë*

GREECE

187 Alexandroupolis
188 Thessaloniki (Salonika)*
189 Límnos*
190 Kérkira
191 Tríkkala*
192 Athínai (Athens)*
193 Tripolis
194 Náxos*
195 Iráklion (Kriti)*

CYPRUS

196 Levkosia (Nicosia)*

EUROPE

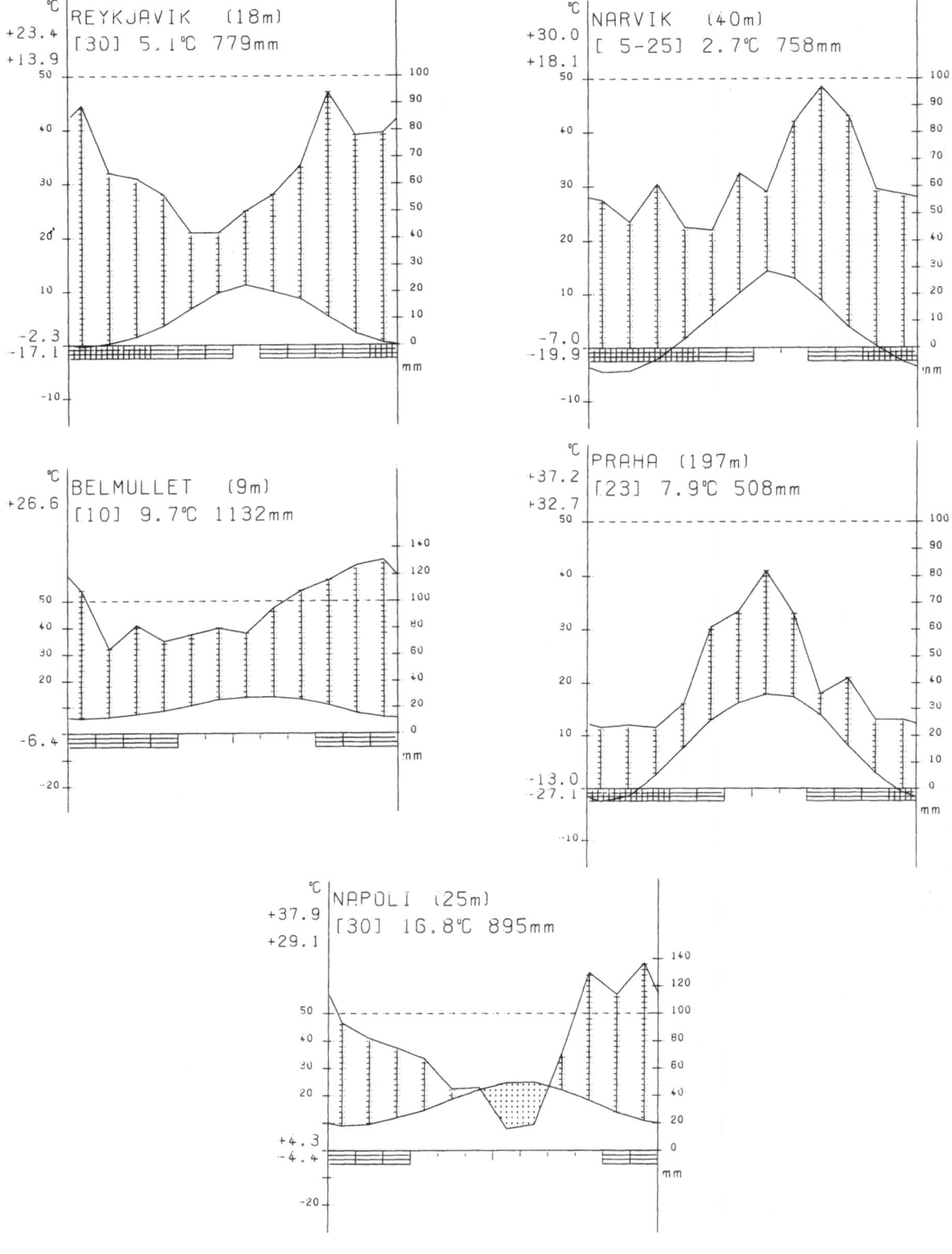

1 Station / Country Akureyri / Iceland

Location 65°41'N / 18°15'W Height above sealevel 7 m Climate symbol: Köppen ET Troll I,3

		J	F	M	A	M	J	J	A	S	O	N	D	year	P	
1 Mean daily temperature	in °C	-1,6	-1,5	-0,4	1,2	6,3	9,4	11,0	10,2	7,7	3,9	0,7	-1,1	3,8	30	1
2 Mean daily maximum temperature	in °C	1,0	0,9	2,6	3,2	9,7	12,9	14,1	13,3	10,7	6,6	3,3	1,4	6,6	30	2
3 Mean daily minimum temperature	in °C	-4,2	-3,9	-3,4	-1,2	2,8	5,9	7,8	7,1	4,7	1,2	-1,9	-3,5	0,9	30	3
4 Absolute maximum temperature	in °C	13,6	12,6	16,0	16,0	21,8	28,6	26,3	25,0	22,0	17,8	17,6	15,1	28,6	30	4
5 Absolute minimum temperature	in °C	-20,3	-20,0	-22,9	-16,6	-9,0	-3,0	0,5	-2,2	-8,4	-11,2	-18,3	-20,2	-22,9	30	5
6 Mean relative humidity	in %	86	86	85	81	74	72	77	80	78	82	86	87	81	10	6
7 Mean precipitation	in mm	45	40	41	31	18	24	32	34	44	49	46	53	457	30	7
8 Maximum precipitation	in mm															8
9 Minimum precipitation	in mm															9
10 Maximum precipitation in 24 h	in mm	27	21	26	28	24	42	24	52	92	30	27	33	92	30	10
11 Mean number of days with precipitation	> 0,1 mm	14	13	12	11	8	9	11	11	12	14	14	16	145	30	11
12 Mean duration of sunshine	in h	7	33	79	107	163	168	154	116	78	55	13	0	973	30	12
13 Mean quantity of radiation	in ly / day															13
14 Mean potential evaporation	in mm	0	0	0	9	55	90	100	84	57	25	0	0	420		14
15 Mean windspeed	in m / sec															15
16 Mean predominent direction of the wind																16

2 Station / Country Hallormsstaður / Iceland

Location 65°05'N / 14°39'W Height above sealevel 40 m Climate symbol: Köppen Cfc Troll I,4

		J	F	M	A	M	J	J	A	S	O	N	D	year	P	
1 Mean daily temperature	in °C	-1,5	-1,3	0,1	1,9	5,6	8,9	10,9	10,1	7,6	4,3	1,2	-1,0	3,9	29	1
2 Mean daily maximum temperature	in °C	1,1	1,6	3,3	5,2	9,5	13,0	14,9	13,7	11,1	7,2	3,8	1,5	7,2	29	2
3 Mean daily minimum temperature	in °C	-4,0	-4,1	-3,1	-1,4	1,7	4,7	6,8	6,4	4,1	1,4	-1,4	-3,4	0,8	29	3
4 Absolute maximum temperature	in °C	13,0	13,2	16,5	17,5	24,4	26,0	30,0	27,0	24,4	18,5	17,0	12,4	30,0	20	4
5 Absolute minimum temperature	in °C	-19,5	-20,6	-20,9	-17,4	-9,9	-3,8	-1,0	-2,4	-7,9	-11,9	-17,3	-18,5	-20,9	29	5
6 Mean relative humidity	in %	75	75	73	71	67	68	73	77	74	75	76	79	74	9	6
7 Mean precipitation	in mm	104	62	41	39	25	28	39	40	51	65	75	87	656	29	7
8 Maximum precipitation	in mm															8
9 Minimum precipitation	in mm															9
10 Maximum precipitation in 24 h	in mm	68	51	35	40	31	36	37	34	30	52	59	48	68	29	10
11 Mean number of days with precipitation	> 0,1 mm	13	11	11	11	8	9	12	12	12	12	13	13	137	29	11
12 Mean duration of sunshine	in h	1	33	89	126	175	175	169	112	93	53	6	0	1032	14	12
13 Mean quantity of radiation	in ly / day															13
14 Mean potential evaporation	in mm	0	0	0	21	56	82	103	83	60	33	11	0	448	29	14
15 Mean windspeed	in m / sec															15
16 Mean predominent direction of the wind																16

3 Station / Country Reykjavik / Iceland

Location 64°08'N / 21°56'W Height above sealevel 18 m Climate symbol: Köppen Cfc Troll I,4

		J	F	M	A	M	J	J	A	S	O	N	D	year	P	
1 Mean daily temperature	in °C	-0,3	0,3	1,5	3,6	6,8	9,8	11,4	10,0	8,6	5,3	2,2	0,5	5,1	30	1
2 Mean daily maximum temperature	in °C	1,8	2,5	4,0	6,2	9,8	12,4	13,9	13,6	11,0	7,4	4,1	2,4	7,4	30	2
3 Mean daily minimum temperature	in °C	-2,3	-2,0	-1,0	0,9	3,7	7,1	8,8	8,1	6,1	3,1	0,2	-1,5	2,6	30	3
4 Absolute maximum temperature	in °C	10,0	10,1	14,2	15,2	20,6	20,7	23,4	21,4	20,1	15,7	11,5	11,4	23,4	30	4
5 Absolute minimum temperature	in °C	-17,1	-13,6	-14,3	-12,7	-7,2	-0,2	1,4	-0,4	-3,5	-10,2	-11,6	-16,8	-17,1	30	5
6 Mean relative humidity	in %	80	77	75	77	71	75	77	76	78	81	80	81	77	10	6
7 Mean precipitation	in mm	89	64	62	56	42	42	50	56	67	94	78	79	779	30	7
8 Maximum precipitation	in mm															8
9 Minimum precipitation	in mm															9
10 Maximum precipitation in 24 h	in mm	42	40	24	23	19	22	31	35	49	37	29	55	55	30	10
11 Mean number of days with precipitation	> 0,1 mm	20	17	18	18	16	15	15	16	19	21	18	20	213	30	11
12 Mean duration of sunshine	in h	24	56	111	138	182	182	180	166	106	72	34	9	1258	30	12
13 Mean quantity of radiation	in ly / day															13
14 Mean potential evaporation	in mm	0	0	0	29	35	47	100	88	58	32	10	0	399		14
15 Mean windspeed	in m / sec															15
16 Mean predominent direction of the wind																16

4 Station / Country Jan Mayen / Norway

Location 70°59'N / 8°20'W Height above sealevel 23 m Climate symbol: Köppen Dfc Troll I,2

		J	F	M	A	M	J	J	A	S	O	N	D	year	P	
1 Mean daily temperature	in °C	-3,9	-5,0	-4,6	-3,3	-0,2	2,7	5,4	5,8	4,0	1,2	-0,9	-2,8	-0,2	4	1
2 Mean daily maximum temperature	in °C	-1,4	-2,2	-1,8	-0,7	1,9	4,8	7,3	7,7	5,9	3,7	1,7	-0,2	2,2	14	2
3 Mean daily minimum temperature	in °C	-6,3	-7,8	-7,4	-5,8	-2,3	0,6	3,3	3,9	2,1	-1,3	-3,5	-5,4	-2,5	30	3
4 Absolute maximum temperature	in °C	6,6	6,8	4,8	7,0	9,4	18,1	16,2	15,7	13,4	11,6	10,0	6,8	18,1	30	4
5 Absolute minimum temperature	in °C	-26,6	-27,9	-24,8	-21,4	-11,6	-4,7	-1,6	-2,3	-5,0	-13,4	-15,2	-22,5	-27,9	30	5
6 Mean relative humidity	in %	84	83	84	82	85	88	91	89	86	84	85	84	85	30	6
7 Mean precipitation	in mm	66	47	48	44	25	27	36	54	74	75	67	65	628	30	7
8 Maximum precipitation	in mm															8
9 Minimum precipitation	in mm															9
10 Maximum precipitation in 24 h	in mm	60	30	36	40	51	19	87	35	50	47	32	32	87	30	10
11 Mean number of days with precipitation	> 0,1 mm	22	20	21	19	15	14	16	18	19	21	21	22	228	30	11
12 Mean duration of sunshine	in h															12
13 Mean quantity of radiation	in ly / day															13
14 Mean potential evaporation	in mm	0	0	0	0	0	50	73	81	59	0	0	0	263	11	14
15 Mean windspeed	in m / sec															15
16 Mean predominent direction of the wind																16

5 Station / Country Vardø / Norway

Location 70°22'N/31°06'E Height above sealevel 10 m Climate symbol: Köppen ET Troll I,3

		J	F	M	A	M	J	J	A	S	O	N	D	year	P	
1 Mean daily temperature	in °C	-4,8	-5,3	-4,0	-1,0	2,7	6,2	9,3	9,8	8,8	2,4	-0,7	-3,0	1,6	21	1
2 Mean daily maximum temperature	in °C	-2,1	-2,7	-1,5	1,2	4,8	8,7	11,5	12,0	8,6	4,3	1,4	-0,8	3,8	21	2
3 Mean daily minimum temperature	in °C	-7,0	-7,8	-6,4	-3,1	0,5	4,0	7,1	7,5	4,9	0,5	-2,7	-5,2	-0,6	30	3
4 Absolute maximum temperature	in °C	5,5	5,9	8,4	10,0	17,9	23,8	23,5	24,0	19,8	12,8	8,4	6,9	24,0	30	4
5 Absolute minimum temperature	in °C	-20,8	-21,9	-20,6	-12,1	-8,6	-2,5	-1,3	1,8	-4,8	-8,5	-13,4	-18,2	-21,9	30	5
6 Mean relative humidity	in %	86	87	86	83	80	83	86	85	84	84	85	86	85	28	6
7 Mean precipitation	in mm	44	46	47	36	36	37	41	52	63	56	43	43	544	30	7
8 Maximum precipitation	in mm															8
9 Minimum precipitation	in mm															9
10 Maximum precipitation in 24 h	in mm	20	18	15	14	17	29	33	29	34	26	29	18	34	25	10
11 Mean number of days with precipitation	> 0,1 mm	20	18	20	17	16	14	13	16	19	20	18	19	210	30	11
12 Mean duration of sunshine	in h															12
13 Mean quantity of radiation	in ly / day															13
14 Mean potential evaporation	in mm	0	0	0	0	39	73	96	88	60	24	0	0	380	41	14
15 Mean windspeed	in m/sec															15
16 Mean predominent direction of the wind																16

6 Station / Country Tromsø / Norway

Location 69°36'N/18°57'E Height above sealevel 115 m Climate symbol: Köppen Dfc Troll II,1

		J	F	M	A	M	J	J	A	S	O	N	D	year	P	
1 Mean daily temperature	in °C	-3,5	-4,0	-2,7	0,3	4,1	8,8	12,4	11,0	7,2	3,0	-0,1	-1,9	2,9		1
2 Mean daily maximum temperature	in °C	-1,9	-2,9	-0,4	3,0	6,9	12,2	16,0	14,1	9,9	5,0	1,7	-0,5	5,3	24	2
3 Mean daily minimum temperature	in °C	-5,7	-6,3	-5,2	-2,3	1,4	5,8	9,1	8,3	4,9	1,1	-2,0	-4,0	0,4	30	3
4 Absolute maximum temperature	in °C	7,4	8,2	8,6	13,9	22,6	27,7	28,5	26,6	22,4	15,4	11,3	8,8	28,5		4
5 Absolute minimum temperature	in °C	-15,8	-15,0	-15,5	-11,2	-6,2	-2,0	1,7	1,1	-3,2	-8,0	-12,5	-14,9	-15,8		5
6 Mean relative humidity	in %	79	79	76	72	71	73	77	80	80	80	81	80	77	30	6
7 Mean precipitation	in mm	96	79	91	65	61	59	58	80	109	115	88	95	994		7
8 Maximum precipitation	in mm															8
9 Minimum precipitation	in mm															9
10 Maximum precipitation in 24 h	in mm	38	30	28	22	29	22	30	29	45	34	36	42	45		10
11 Mean number of days with precipitation	> 0,1 mm	18	17	19	18	18	18	15	19	21	21	18	19	221		11
12 Mean duration of sunshine	in h	2	35	101	176	172	202	226	176	97	52	7	0	1246	10	12
13 Mean quantity of radiation	in ly / day	0	21	103	216	294	401	397	237	120	41	5	0	153		13
14 Mean potential evaporation	in mm	0	0	0	0	44	80	105	91	56	22	0	0	398	41	14
15 Mean windspeed	in m/sec	3,9	3,8	3,5	3,0	2,9	2,5	2,1	2,1	2,6	3,1	3,1	3,7	3,0	10	15
16 Mean predominent direction of the wind		SSW	SSW	SSW	SW	SW	NE	NE	NE	SW	SSW	SSW	SSW		10	16

7 Station / Country Mo i Rana / Norway

Location 66°21'N/14°08'E Height above sealevel 20 m Climate symbol: Köppen Dfc Troll II,2

		J	F	M	A	M	J	J	A	S	O	N	D	year	P	
1 Mean daily temperature	in °C	-7,6	-7,2	-3,1	1,5	6,3	10,9	14,2	13,2	8,7	3,4	-1,4	-4,5	2,9	10	1
2 Mean daily maximum temperature	in °C	-3,8	-3,9	1,3	5,1	10,5	15,2	18,3	17,2	12,1	6,1	1,3	-1,6	6,5	10	2
3 Mean daily minimum temperature	in °C	-11,3	-10,4	-7,5	-2,2	2,1	6,6	10,1	9,1	5,2	0,6	-4,0	-7,3	-0,7	25	3
4 Absolute maximum temperature	in °C	7,5	9,7	15,0	24,5	20,3	30,6	30,1	23,3	16,2	11,0	8,4		30,6	25	4
5 Absolute minimum temperature	in °C	-29,1	-30,4	-29,0	-18,5	-9,2	-1,7	-0,1	-2,0	-5,2	-17,6	-22,1	-29,5	-30,4	25	5
6 Mean relative humidity	in %	84	81	79	73	70	72	74	79	83	83	83	84	79	16	6
7 Mean precipitation	in mm	124	117	133	99	65	84	71	93	131	164	116	140	1336	25	7
8 Maximum precipitation	in mm															8
9 Minimum precipitation	in mm															9
10 Maximum precipitation in 24 h	in mm	78	109	87	58	36	36	32	53	58	52	51	59	109	25	10
11 Mean number of days with precipitation	> 0,1 mm	15	16	15	14	11	14	14	16	17	19	15	17	183	25	11
12 Mean duration of sunshine	in h															12
13 Mean quantity of radiation	in ly / day															13
14 Mean potential evaporation	in mm	0	0	0	14	56	98	115	98	60	22	0	0	463	25	14
15 Mean windspeed	in m/sec															15
16 Mean predominent direction of the wind																16

8 Station / Country Trondheim / Norway

Location 63°25'N/10°27'E Height above sealevel 133 m Climate symbol: Köppen Dfc Troll II,2

		J	F	M	A	M	J	J	A	S	O	N	D	year	P	
1 Mean daily temperature	in °C	-3,8	-3,1	-0,5	3,6	8,3	11,8	15,0	13,9	9,9	5,3	1,3	-1,3	5,0	24	1
2 Mean daily maximum temperature	in °C	-1,3	-0,3	3,0	7,1	12,2	15,5	18,8	17,6	13,1	8,0	3,6	1,1	8,2	24	2
3 Mean daily minimum temperature	in °C	-6,3	-5,8	-3,9	0,1	4,3	8,0	11,1	10,2	6,8	2,8	-1,0	-3,6	1,9	30	3
4 Absolute maximum temperature	in °C	10,4	10,8	12,5	17,5	25,0	29,5	32,9	29,6	24,0	20,5	14,0	12,2	32,9	30	4
5 Absolute minimum temperature	in °C	-24,5	-25,0	-22,0	-14,4	-4,3	-0,3	3,8	1,4	-3,0	-11,1	-16,6	-23,8	-25,0	30	5
6 Mean relative humidity	in %	78	77	75	75	73	75	77	79	82	82	80	79	78	30	6
7 Mean precipitation	in mm	71	71	70	63	47	64	69	77	93	99	67	79	870	30	7
8 Maximum precipitation	in mm															8
9 Minimum precipitation	in mm															9
10 Maximum precipitation in 24 h	in mm	34	41	65	40	29	31	38	61	50	81	43	39	81	30	10
11 Mean number of days with precipitation	> 0,1 mm	16	15	16	17	15	18	18	17	20	19	15	17	203	30	11
12 Mean duration of sunshine	in h	16	52	124	162	204	180	181	174	119	72	29	8	1321		12
13 Mean quantity of radiation	in ly / day	17	54	156	263	413	412	436	313	184	90	26	9	198		13
14 Mean potential evaporation	in mm	0	0	0	28	66	91	111	94	60	30	0	0	480	41	14
15 Mean windspeed	in m/sec															15
16 Mean predominent direction of the wind																16

9 Station / Country Dombås / Norway

Location 62°04'N / 9°07'E Height above sealevel 643 m Climate symbol: Köppen Dfc Troll II,2

		J	F	M	A	M	J	J	A	S	O	N	D	year	P	
1 Mean daily temperature	in °C	-9,4	-8,2	-3,9	1,0	6,4	9,5	13,3	12,1	7,7	2,1	-3,2	-6,5	1,7	24	1
2 Mean daily maximum temperature	in °C	-6,3	-4,5	0,8	5,3	11,5	15,3	18,1	16,9	11,8	5,4	-0,4	-3,5	5,9	24	2
3 Mean daily minimum temperature	in °C	-12,5	-11,9	-8,6	-3,4	1,3	5,4	8,5	7,3	3,5	-1,2	-6,0	-9,4	-2,3	30	3
4 Absolute maximum temperature	in °C	9,4	9,3	12,3	17,5	25,0	28,9	30,0	28,4	23,4	16,4	11,2	8,7	30,0	30	4
5 Absolute minimum temperature	in °C	-26,3	-29,6	-26,5	-20,5	-12,3	-4,0	-0,3	-1,8	-8,8	-19,5	-21,9	-26,6	-29,8	30	5
6 Mean relative humidity	in %	83	79	72	68	63	65	70	72	77	80	85	85	75	30	6
7 Mean precipitation	in mm	30	29	18	16	20	44	69	54	36	30	30	33	409	30	7
8 Maximum precipitation	in mm															8
9 Minimum precipitation	in mm															9
10 Maximum precipitation in 24 h	in mm	33	23	26	20	23	26	41	63	72	21	22	32	72	25	10
11 Mean number of days with precipitation	> 0,1 mm	13	11	9	9	7	11	15	14	11	11	13	13	137	30	11
12 Mean duration of sunshine	in h															12
13 Mean quantity of radiation	in ly / day															13
14 Mean potential evaporation	in mm	0	0	0	14	61	91	111	91	53	16	0	0	437	30	14
15 Mean windspeed	in m / sec															15
16 Mean predominent direction of the wind																16

10 Station / Country Lillehammer / Norway

Location 61°06'N / 10°26'E Height above sealevel 226 m Climate symbol: Köppen Dfc Troll II,2

		J	F	M	A	M	J	J	A	S	O	N	D	year	P	
1 Mean daily temperature	in °C	-9,1	-8,7	-3,2	2,8	9,0	13,7	15,8	12,3	9,6	3,8	-1,4	-5,5	3,3	8	1
2 Mean daily maximum temperature	in °C	-6,0	-6,0	1,4	7,3	14,8	19,6	21,0	19,5	14,1	7,1	1,2	-2,7	7,6	8	2
3 Mean daily minimum temperature	in °C	-12,1	-11,4	-7,7	-1,8	3,1	7,8	10,5	9,1	5,0	0,4	-4,0	-8,2	-0,8	30	3
4 Absolute maximum temperature	in °C	10,6	10,6	13,0	18,2	27,8	31,5	31,8	29,0	23,6	17,8	12,5	9,3	31,8	30	4
5 Absolute minimum temperature	in °C	-29,9	-30,3	-32,9	-16,2	-7,1	-2,1	1,7	-1,0	-4,4	-16,2	-21,0	-23,8	-32,9	30	5
6 Mean relative humidity	in %	87	82	75	70	63	65	71	76	80	83	88	90	78	30	6
7 Mean precipitation	in mm	44	31	25	39	42	77	96	87	70	61	61	58	691	30	7
8 Maximum precipitation	in mm															8
9 Minimum precipitation	in mm															9
10 Maximum precipitation in 24 h	in mm	20	16	17	24	49	61	48	43	51	29	33	28	61	24	10
11 Mean number of days with precipitation	> 0,1 mm	15	13	10	11	10	15	18	17	15	15	16	16	171	30	11
12 Mean duration of sunshine	in h															12
13 Mean quantity of radiation	in ly / day															13
14 Mean potential evaporation	in mm	0	0	0	21	70	101	119	94	58	22	0	0	485	41	14
15 Mean windspeed	in m / sec															15
16 Mean predominent direction of the wind																16

11 Station / Country Bergen / Norway

Location 60°12'N / 5°19'E Height above sealevel 45 m Climate symbol: Köppen Cfb Troll III,2

		J	F	M	A	M	J	J	A	S	O	N	D	year	P	
1 Mean daily temperature	in °C	1,5	1,3	3,1	5,8	10,2	12,6	15,0	14,7	12,0	8,3	5,5	3,3	7,8		1
2 Mean daily maximum temperature	in °C	3,1	3,3	6,1	9,2	14,2	15,9	18,8	18,6	15,2	11,1	7,5	5,0	10,7	24	2
3 Mean daily minimum temperature	in °C	-0,5	-0,9	0,4	3,0	6,7	9,8	12,2	12,0	9,5	6,1	3,4	1,3	5,4	30	3
4 Absolute maximum temperature	in °C	13,3	11,2	19,8	22,1	27,1	31,8	30,5	29,7	26,0	19,5	15,4	16,4	31,8		4
5 Absolute minimum temperature	in °C	-13,5	-10,9	-10,0	-5,8	-0,1	0,8	5,2	5,4	1,3	-3,1	-5,6	-8,4	-13,5		5
6 Mean relative humidity	in %	79	77	73	74	71	77	80	80	81	81	79	80	78	30	6
7 Mean precipitation	in mm	179	139	109	140	83	126	141	167	228	236	207	203	1958		7
8 Maximum precipitation	in mm															8
9 Minimum precipitation	in mm															9
10 Maximum precipitation in 24 h	in mm	60	54	66	55	44	67	79	58	61	91	76	99	99		10
11 Mean number of days with precipitation	> 0,1 mm	21	18	16	19	15	18	21	20	22	24	22	23	239		11
12 Mean duration of sunshine	in h	22	54	127	142	191	175	167	132	102	69	29	13	1223	10	12
13 Mean quantity of radiation	in ly / day	29	67	168	251	426	400	371	297	172	97	38	16	194		13
14 Mean potential evaporation	in mm	9	9	18	39	70	95	107	92	67	40	20	11	577	41	14
15 Mean windspeed	in m / sec	3,2	3,5	3,1	3,6	3,1	2,9	2,4	2,8	3,2	3,3	3,3	3,6	3,2	10	15
16 Mean predominent direction of the wind		S	S	S	S	NNW	NNW	NNW	NNW	S	S	S	S		10	16

12 Station / Country Oslo / Norway

Location 59°56'N / 10°44'E Height above sealevel 96 m Climate symbol: Köppen Dfb Troll III,4

		J	F	M	A	M	J	J	A	S	O	N	D	year	P	
1 Mean daily temperature	in °C	-4,7	-4,0	-0,5	4,8	10,7	14,7	17,3	15,9	11,3	5,9	1,1	-2,0	5,9		1
2 Mean daily maximum temperature	in °C	-2,4	-1,1	3,8	9,8	15,9	19,9	22,3	20,9	15,7	9,1	3,2	0,1	9,8	24	2
3 Mean daily minimum temperature	in °C	-7,3	-7,1	-4,1	0,9	5,9	10,1	13,0	11,9	7,8	3,1	-0,8	-4,1	2,4	30	3
4 Absolute maximum temperature	in °C	9,9	13,8	15,5	21,5	28,4	33,7	32,8	30,9	25,5	20,1	11,7	10,8	33,7		4
5 Absolute minimum temperature	in °C	-26,0	-21,9	-21,3	-14,9	-3,4	1,8	3,7	3,9	-2,0	-7,6	-15,7	-20,2	-26,0		5
6 Mean relative humidity	in %	84	79	77	66	60	62	67	70	76	80	86	86	74	30	6
7 Mean precipitation	in mm	49	35	26	44	44	71	84	96	83	76	69	63	740		7
8 Maximum precipitation	in mm															8
9 Minimum precipitation	in mm															9
10 Maximum precipitation in 24 h	in mm	33	21	16	29	43	43	46	52	38	50	32	24	52		10
11 Mean number of days with precipitation	> 0,1 mm	15	13	9	11	10	13	15	14	14	14	16	17	161		11
12 Mean duration of sunshine	in h	45	83	152	182	233	244	219	183	138	87	41	25	1632	10	12
13 Mean quantity of radiation	in ly / day	26	74	189	292	434	422	396	300	182	113	32	16	206		13
14 Mean potential evaporation	in mm	0	0	0	30	74	106	123	101	64	29	0	0	527	41	14
15 Mean windspeed	in m / sec	2,1	1,9	2,0	2,5	2,5	2,7	2,5	2,3	2,2	2,1	2,4	1,8	2,2	10	15
16 Mean predominent direction of the wind		NNE	NNE	NNE	S	S	S	S	S	S	NNE	NNE	NNE		10	16

13 Station / Country Dalen / Norway

Location 59°27'N/8°00'E   Height above sealevel 77 m   Climate symbol: Köppen Dfb   Troll II,2

		J	F	M	A	M	J	J	A	S	O	N	D	year	P	
1 Mean daily temperature	in °C	-4,6	-5,1	-0,9	4,6	10,2	14,6	17,6	15,1	11,4	6,0	0,9	-2,0	5,6	4	1
2 Mean daily maximum temperature	in °C	-1,6	-2,5	2,6	8,9	15,7	20,3	21,3	19,5	15,3	9,1	2,8	-0,2	9,3	4	2
3 Mean daily minimum temperature	in °C	-7,5	-7,6	-4,3	0,3	4,7	8,8	11,8	10,9	7,4	2,9	-1,1	-4,2	1,8	30	3
4 Absolute maximum temperature	in °C	12,6	12,8	13,7	19,6	27,4	31,5	32,4	29,4	22,8	19,0	13,6	12,2	32,4	30	4
5 Absolute minimum temperature	in °C	-24,6	-22,8	-20,5	-13,6	-2,8	1,3	4,6	3,3	-1,2	-8,1	-11,2	-17,8	-24,6	30	5
6 Mean relative humidity	in %	85	80	73	69	64	66	70	74	79	83	89	88	77	30	6
7 Mean precipitation	in mm	60	45	30	40	50	71	100	113	92	93	87	78	859	30	7
8 Maximum precipitation	in mm															8
9 Minimum precipitation	in mm															9
10 Maximum precipitation in 24 h	in mm	32	18	17	23	35	35	69	63	47	44	38	43	69	24	10
11 Mean number of days with precipitation	> 0,1 mm	13	11	8	9	8	10	13	12	13	13	14	15	139	30	11
12 Mean duration of sunshine	in h															12
13 Mean quantity of radiation	in ly/day															13
14 Mean potential evaporation	in mm	0	0	0	30	72	103	119	98	61	27	0	0	510	41	14
15 Mean windspeed	in m/sec															15
16 Mean predominent direction of the wind																16

14 Station / Country Kristiansand / Norway

Location 58°10'N/7°59'E   Height above sealevel 23 m   Climate symbol: Köppen Cfb   Troll III,3

		J	F	M	A	M	J	J	A	S	O	N	D	year	P	
1 Mean daily temperature	in °C	-1,9	-1,5	1,3	5,8	10,5	14,1	16,5	15,6	12,3	8,0	3,9	1,2	7,2	21	1
2 Mean daily maximum temperature	in °C	0,9	1,8	4,9	10,0	15,4	18,9	21,1	19,7	16,2	11,4	6,4	3,9	10,9	21	2
3 Mean daily minimum temperature	in °C	-4,7	-4,7	-2,4	1,5	5,8	9,2	11,8	11,4	8,4	4,6	1,4	-1,5	3,4	26	3
4 Absolute maximum temperature	in °C	11,5	12,1	18,3	21,5	27,7	30,2	32,1	31,4	24,2	21,0	13,8	13,1	32,1	26	4
5 Absolute minimum temperature	in °C	-25,5	-20,8	-21,6	-15,0	-3,7	1,3	4,2	2,7	-1,8	-5,0	-10,6	-17,4	-25,5	26	5
6 Mean relative humidity	in %	82	79	75	72	66	68	71	76	79	81	83	85	76	21	6
7 Mean precipitation	in mm	128	100	62	73	58	75	104	143	156	153	176	173	1401	26	7
8 Maximum precipitation	in mm															8
9 Minimum precipitation	in mm															9
10 Maximum precipitation in 24 h	in mm	63	76	37	41	42	45	73	89	74	72	76	56	89	24	10
11 Mean number of days with precipitation	> 0,1 mm	19	16	12	13	10	12	14	15	18	18	20	21	188	26	11
12 Mean duration of sunshine	in h															12
13 Mean quantity of radiation	in ly/day															13
14 Mean potential evaporation	in mm	0	0	9	38	76	102	119	101	70	41	18	17	592	26	14
15 Mean windspeed	in m/sec															15
16 Mean predominent direction of the wind																16

15 Station / Country Haparanda / Sweden

Location 65°50'N/24°09'E   Height above sealevel 7 m   Climate symbol: Köppen Dfc   Troll II,2

		J	F	M	A	M	J	J	A	S	O	N	D	year	P	
1 Mean daily temperature	in °C	-10,9	-11,1	-7,8	-1,1	5,8	12,3	16,3	14,0	8,5	2,8	-2,9	-7,0	1,6	30	1
2 Mean daily maximum temperature	in °C	-6,7	-7,0	-2,9	3,0	10,3	16,6	20,8	18,4	12,1	4,8	-0,1	-3,8	5,5	30	2
3 Mean daily minimum temperature	in °C	-15,1	-15,2	-12,7	-5,2	1,3	7,9	11,7	9,5	4,8	0,9	-5,7	-10,4	-2,4	30	3
4 Absolute maximum temperature	in °C	6,4	7,0	10,3	15,5	25,9	30,0	31,5	28,6	24,0	17,0	10,5	6,5	31,5	30	4
5 Absolute minimum temperature	in °C	-40,8	-36,4	-33,0	-28,0	-7,1	-1,0	3,2	-1,8	-8,0	-23,0	-26,5	-37,3	-40,8	30	5
6 Mean relative humidity	in %	88	88	81	76	64	65	68	75	80	86	90	89	79	30	6
7 Mean precipitation	in mm	40	37	24	34	29	41	54	71	65	53	58	46	552	30	7
8 Maximum precipitation	in mm															8
9 Minimum precipitation	in mm															9
10 Maximum precipitation in 24 h	in mm	21	23	20	24	23	25	47	47	56	32	22	22	56	30	10
11 Mean number of days with precipitation	> 0,1 mm	15	14	11	11	9	11	11	12	14	14	17	16	153	30	11
12 Mean duration of sunshine	in h	24	50	151	209	269	299	308	214	154	98	34	8	1818		12
13 Mean quantity of radiation	in ly/day															13
14 Mean potential evaporation	in mm	0	0	0	0	49	98	123	98	56	13	0	0	437	30	14
15 Mean windspeed	in m/sec															15
16 Mean predominent direction of the wind																16

16 Station / Country Stensele / Sweden

Location 65°04'N/17°10'E   Height above sealevel 330 m   Climate symbol: Köppen Dfc   Troll II,2

		J	F	M	A	M	J	J	A	S	O	N	D	year	P	
1 Mean daily temperature	in °C	-12,5	-11,4	-7,0	-0,7	5,4	10,7	14,0	12,0	7,1	1,0	-4,5	-8,8	0,5	30	1
2 Mean daily maximum temperature	in °C	-8,1	-6,3	-0,9	4,2	10,8	15,9	19,3	17,0	11,2	4,2	-1,5	-4,8	5,1	30	2
3 Mean daily minimum temperature	in °C	-16,9	-16,4	-13,1	-5,8	0,0	6,6	8,6	7,0	3,0	-2,3	-7,4	-12,4	-4,2	30	3
4 Absolute maximum temperature	in °C	7,2	7,2	11,1	17,2	25,0	29,0	31,0	26,7	25,2	15,1	9,2	7,0	31,0	30	4
5 Absolute minimum temperature	in °C	-43,4	-37,9	-35,1	-27,3	-12,5	-3,9	-0,8	-3,0	-9,1	-18,8	-31,3	-40,3	-43,4	30	5
6 Mean relative humidity	in %	86	84	80	74	63	64	68	75	80	85	89	89	78	30	6
7 Mean precipitation	in mm	30	23	21	25	33	57	80	67	47	37	38	36	494	30	7
8 Maximum precipitation	in mm															8
9 Minimum precipitation	in mm															9
10 Maximum precipitation in 24 h	in mm	16	12	16	19	23	46	49	39	44	35	17	21	49	30	10
11 Mean number of days with precipitation	> 0,1 mm	15	14	11	11	9	13	15	15	13	13	15	16	160	30	11
12 Mean duration of sunshine	in h															12
13 Mean quantity of radiation	in ly/day															13
14 Mean potential evaporation	in mm	0	0	0	0	55	97	118	91	52	3	0	0	416	30	14
15 Mean windspeed	in m/sec															15
16 Mean predominent direction of the wind																16

17 Station/Country: Östersund/Sweden

Location 63°10'N/14°40'E   Height above sealevel 328 m   Climate symbol: Köppen Dfc   Troll II,2

		J	F	M	A	M	J	J	A	S	O	N	D	year	P	
1	Mean daily temperature in °C	-8,4	-7,1	-4,1	1,2	6,7	11,3	14,7	13,4	8,9	3,8	-0,8	-4,5	2,9		1
2	Mean daily maximum temperature in °C	-5,2	-3,9	0,8	5,9	12,6	16,9	20,0	18,1	12,5	5,9	0,8	-2,2	6,8	30	2
3	Mean daily minimum temperature in °C	-11,9	-10,0	-8,3	-2,5	2,3	7,1	10,3	9,1	5,4	0,7	-3,5	-7,6	-0,8	30	3
4	Absolute maximum temperature in °C	9,8	9,2	16,0	18,8	26,5	33,5	32,5	31,3	25,2	16,4	12,2	8,5	33,5		4
5	Absolute minimum temperature in °C	-33,8	-31,4	-30,0	-18,0	-6,8	-1,5	2,5	0,2	-5,2	-15,2	-20,4	-31,0	-33,8		5
6	Mean relative humidity in %	85	84	78	74	64	67	70	74	80	83	88	88	78	30	6
7	Mean precipitation in mm	34	23	23	29	31	69	77	74	51	43	42	36	532		7
8	Maximum precipitation in mm															8
9	Minimum precipitation in mm															9
10	Maximum precipitation in 24 h in mm	19	13	17	32	22	39	72	87	54	19	23	32	72		10
11	Mean number of days with precipitation >0,1 mm	15	13	12	12	11	14	15	14	14	14	15	14	163		11
12	Mean duration of sunshine in h +	26	61	121	180	252	240	249	204	126	74	29	10	1572		12
13	Mean quantity of radiation in ly/day															13
14	Mean potential evaporation in mm	0	0	0	13	60	95	115	93	57	22	0	0	455	30	14
15	Mean windspeed in m/sec	2,0	2,1	2,0	2,1	2,3	2,7	2,4	2,1	2,4	2,3	2,1	2,0	2,2		15
16	Mean predominant direction of the wind	SE	SE	NW	NW	NW	NW	NW	NW	NW	SE	SE	SE			16

+ Gisselås (63°42'N/15°22'E, 320 m)

18 Station/Country: Härnösand/Sweden

Location 62°28'N/17°57'E   Height above sealevel 8 m   Climate symbol: Köppen Dfb   Troll II,2

		J	F	M	A	M	J	J	A	S	O	N	D	year	P	
1	Mean daily temperature in °C	-6,4	-5,8	-2,9	2,1	7,7	12,8	16,4	15,1	10,4	5,0	0,7	-2,8	4,4		1
2	Mean daily maximum temperature in °C	-2,5	-1,6	1,9	7,1	13,1	18,5	22,2	21,7	15,4	8,8	3,3	0,3	9,0	30	2
3	Mean daily minimum temperature in °C	-9,8	-9,9	-7,3	-2,1	2,8	7,8	11,5	10,4	6,4	1,8	-2,1	-5,7	0,3	30	3
4	Absolute maximum temperature in °C	10,2	10,4	17,0	20,0	27,4	31,6	31,6	30,0	26,0	19,6	11,5	10,3	31,6		4
5	Absolute minimum temperature in °C	-30,0	-30,6	-31,0	-17,5	-5,6	-2,7	4,8	0,2	-5,8	-11,2	-19,0	-27,0	-31,0		5
6	Mean relative humidity in %	84	81	77	72	63	65	69	73	77	80	85	85	76	30	6
7	Mean precipitation in mm	32	37	32	46	34	52	58	78	68	63	87	80	697		7
8	Maximum precipitation in mm															8
9	Minimum precipitation in mm															9
10	Maximum precipitation in 24 h in mm	47	30	23	78	28	49	74	108	59	39	51	43	108		10
11	Mean number of days with precipitation >0,1 mm	15	12	10	10	8	10	11	12	11	12	15	16	142		11
12	Mean duration of sunshine in h	54	70	144	195	259	251	267	217	145	90	52	27	1771		12
13	Mean quantity of radiation in ly/day															13
14	Mean potential evaporation in mm	0	0	0	15	57	97	119	100	63	29	0	0	480	30	14
15	Mean windspeed in m/sec	2,0	1,9	2,0	2,4	2,8	3,6	2,9	2,6	2,6	2,1	2,0	2,1	2,4		15
16	Mean predominant direction of the wind	N	N	N	S	S	S	S	S	S	S	S	S			16

19 Station/Country: Karlstad/Sweden

Location 59°22'N/13°28'E   Height above sealevel 47 m   Climate symbol: Köppen Dfb   Troll III,4

		J	F	M	A	M	J	J	A	S	O	N	D	year	P	
1	Mean daily temperature in °C	-4,4	-4,2	-1,1	4,7	10,4	14,9	17,6	16,3	12,0	6,7	2,1	-1,1	6,2	30	1
2	Mean daily maximum temperature in °C	-1,2	-0,5	3,4	9,3	15,7	19,7	22,2	20,8	15,9	9,9	4,5	1,5	10,1	30	2
3	Mean daily minimum temperature in °C	-7,5	-7,8	-5,5	0,0	5,0	10,0	12,9	11,8	8,0	3,4	-0,4	-3,8	2,2	30	3
4	Absolute maximum temperature in °C	9,0	11,2	17,0	22,6	27,4	32,4	34,0	31,8	24,0	19,8	11,2	11,2	34,0	30	4
5	Absolute minimum temperature in °C	-29,4	-28,0	-27,0	-18,4	-4,8	0,8	4,4	2,5	-4,2	-10,2	-19,9	-27,0	-29,4	30	5
6	Mean relative humidity in %	86	83	78	69	62	65	69	73	77	80	87	88	76	30	6
7	Mean precipitation in mm	47	31	25	38	38	49	64	81	70	61	69	51	624	30	7
8	Maximum precipitation in mm															8
9	Minimum precipitation in mm															9
10	Maximum precipitation in 24 h in mm	28	27	24	31	68	36	67	64	43	36	38	33	68	30	10
11	Mean number of days with precipitation >0,1 mm	14	11	8	11	10	12	12	13	13	13	15	15	147	30	11
12	Mean duration of sunshine in h	47	70	141	189	272	299	290	237	162	84	44	25	1860	30	12
13	Mean quantity of radiation in ly/day															13
14	Mean potential evaporation in mm	0	0	0	28	73	107	126	103	66	33	7	0	543	30	14
15	Mean windspeed in m/sec															15
16	Mean predominant direction of the wind															16

20 Station/Country: Stockholm/Sweden

Location 59°21'N/18°04'E   Height above sealevel 44 m   Climate symbol: Köppen Cfb   Troll III,4

		J	F	M	A	M	J	J	A	S	O	N	D	year	P	
1	Mean daily temperature in °C	-2,9	-3,1	-0,7	4,4	10,1	14,9	17,8	16,6	12,2	7,1	2,8	0,1	6,6		1
2	Mean daily maximum temperature in °C	-0,9	-0,9	2,5	8,3	14,4	19,2	21,9	20,2	15,3	9,4	4,5	1,8	9,6	30	2
3	Mean daily minimum temperature in °C	-5,1	-5,4	-3,6	1,0	6,0	10,8	14,1	13,3	9,4	4,8	1,0	-1,8	3,7	30	3
4	Absolute maximum temperature in °C	9,6	11,8	15,2	20,2	28,0	32,2	34,6	31,0	25,7	17,4	12,4	12,2	34,6		4
5	Absolute minimum temperature in °C	-28,2	-24,9	-22,0	-11,3	-3,3	1,0	8,0	4,8	0,1	-6,5	-11,0	-16,3	-28,2		5
6	Mean relative humidity in %	84	80	75	68	60	62	67	73	78	82	87	78	75	30	6
7	Mean precipitation in mm	43	30	26	31	34	45	61	76	60	48	53	48	555		7
8	Maximum precipitation in mm															8
9	Minimum precipitation in mm															9
10	Maximum precipitation in 24 h in mm	26	23	17	20	36	30	42	47	44	38	39	33	47		10
11	Mean number of days with precipitation >0,1 mm	16	13	10	11	11	13	13	14	14	15	16	17	163		11
12	Mean duration of sunshine in h	41	76	151	208	292	318	295	248	174	103	41	26	1973		12
13	Mean quantity of radiation in ly/day	27	81	179	208	374	442	408	328	210	102	41	19	202		13
14	Mean potential evaporation in mm	0	0	0	26	69	102	123	103	66	35	8	0	532	30	14
15	Mean windspeed in m/sec	3,9	3,6	3,6	3,8	4,0	4,3	3,7	3,8	3,9	4,1	3,9	3,9	3,8		15
16	Mean predominant direction of the wind	WSW	W	WSW	SW	NNE	WSW	SW	SW	WSW	SW	SW	SW			16

21 Station / Country Jönköping / Sweden

Location 57°46'N / 14°11'E Height above sealevel 92 m Climate symbol: Köppen **Cfb** Troll III,3

		J	F	M	A	M	J	J	A	S	O	N	D	year	P	
1	Mean daily temperature	in °C	-2,9	-3,1	-1,1	4,6	9,2	13,7	17,4	15,3	11,5	6,7	2,7	-0,2	6,2	30	1
2	Mean daily maximum temperature	in °C	0,0	0,4	3,7	9,4	15,0	19,4	21,4	20,4	16,0	10,4	5,2	2,3	10,3	30	2
3	Mean daily minimum temperature	in °C	-5,8	-6,5	-4,9	-0,3	3,3	8,0	10,7	10,1	6,9	2,9	0,1	-2,7	1,8	30	3
4	Absolute maximum temperature	in °C	8,8	12,9	18,5	22,4	28,7	33,2	33,4	31,8	26,8	20,2	13,4	12,6	33,4	30	4
5	Absolute minimum temperature	in °C	-33,0	-33,3	-29,3	-18,1	-7,2	-2,0	0,2	0,0	-5,8	-9,8	-15,0	-24,2	-33,3	30	5
6	Mean relative humidity	in %	86	84	79	72	68	68	73	76	80	81	86	88	78	30	6
7	Mean precipitation	in mm	35	25	25	31	41	54	71	69	57	50	43	37	538	30	7
8	Maximum precipitation	in mm															8
9	Minimum precipitation	in mm															9
10	Maximum precipitation in 24 h	in mm	26	15	19	25	43	83	65	64	36	25	38	19	83	30	10
11	Mean number of days with precipitation	> 0,1 mm	15	12	10	11	11	12	14	14	14	14	15	16	158	30	11
12	Mean duration of sunshine	in h +	29	54	116	152	227	233	218	192	125	76	29	19	1470	30	12
13	Mean quantity of radiation	in ly / day															13
14	Mean potential evaporation	in mm	0	0	4	29	70	102	119	100	67	38	12	0	541	30	14
15	Mean windspeed	in m / sec															15
16	Mean predominent direction of the wind																16

+ Flahult (57°42'N / 14°08'E, 224 m)

---

22 Station / Country Göteborg(Gothenburg) / Sweden

Location 57°42'N / 11°58'E Height above sealevel 31 m Climate symbol: Köppen **Cfb** Troll III,3

		J	F	M	A	M	J	J	A	S	O	N	D	year	P	
1	Mean daily temperature	in °C	-1,1	-1,2	1,0	5,6	11,0	14,5	17,0	16,3	12,9	8,8	4,2	1,6	7,6		1
2	Mean daily maximum temperature	in °C	1,0	0,9	4,1	9,2	15,5	18,9	21,1	20,1	16,0	10,8	6,2	3,5	10,6	30	2
3	Mean daily minimum temperature	in °C	-3,1	-3,8	-1,7	2,5	7,2	11,5	14,0	14,0	10,1	6,1	2,5	-0,1	4,9	30	3
4	Absolute maximum temperature	in °C	8,0	9,0	17,0	22,0	28,3	32,0	32,0	30,0	25,0	20,0	13,0	10,9	32,0		4
5	Absolute minimum temperature	in °C	-26,0	-20,0	-19,2	-11,0	-3,0	3,0	8,0	5,2	-0,5	-6,0	-7,8	-15,8	-26,0		5
6	Mean relative humidity	in %	83	81	75	68	62	66	70	72	76	78	83	85	75	30	6
7	Mean precipitation	in mm	51	34	29	39	34	54	86	84	75	65	62	57	670		7
8	Maximum precipitation	in mm															8
9	Minimum precipitation	in mm															9
10	Maximum precipitation in 24 h	in mm	28	15	29	22	28	33	66	41	45	48	30	23	66		10
11	Mean number of days with precipitation	> 0,1 mm	15	12	10	12	10	12	14	14	16	15	16	17	163		11
12	Mean duration of sunshine	in h	48	76	151	201	274	286	285	245	178	108	47	29	1928		12
13	Mean quantity of radiation	in ly / day															13
14	Mean potential evaporation	in mm	0	0	11	37	78	105	122	104	70	40	16	4	587	30	14
15	Mean windspeed	in m / sec	4,1	4,0	3,7	4,2	3,8	4,3	3,9	3,8	3,9	4,3	4,0	4,1	4,0		15
16	Mean predominent direction of the wind		ESE	E	ESE	WSW	SW	W	W	W	W	WSW	SE	SE			16

---

23 Station / Country Visby / Sweden

Location 57°39'N / 18°18'E Height above sealevel 28 m Climate symbol: Köppen **Cfb** Troll III,3

		J	F	M	A	M	J	J	A	S	O	N	D	year	P	
1	Mean daily temperature	in °C	-0,9	-1,5	0,2	4,6	9,4	14,2	17,3	16,7	13,1	8,3	4,3	1,7	7,3	30	1
2	Mean daily maximum temperature	in °C	1,0	0,5	2,8	7,9	13,5	18,2	20,9	20,0	15,8	10,5	6,1	3,3	10,0	30	2
3	Mean daily minimum temperature	in °C	-2,7	-3,5	-2,4	1,3	5,3	10,2	13,7	13,4	10,3	6,1	2,0	0,0	4,5	30	3
4	Absolute maximum temperature	in °C	10,0	9,5	14,8	22,2	27,5	29,5	32,0	30,2	27,5	20,0	11,8	10,8	32,0	30	4
5	Absolute minimum temperature	in °C	-25,0	-20,9	-21,0	-12,3	-3,1	0,0	8,0	3,6	0,8	-2,9	-6,4	-13,3	-25,0	30	5
6	Mean relative humidity	in %	85	83	81	75	71	73	75	78	81	82	86	86	80	30	6
7	Mean precipitation	in mm	53	44	29	31	30	32	51	56	51	51	48	53	529	30	7
8	Maximum precipitation	in mm															8
9	Minimum precipitation	in mm															9
10	Maximum precipitation in 24 h	in mm	21	29	18	22	29	37	45	48	43	35	30	22	48	30	10
11	Mean number of days with precipitation	> 0,1 mm	18	15	12	10	9	9	11	12	12	14	16	18	156	30	11
12	Mean duration of sunshine	in h	28	56	128	208	292	317	302	255	186	108	30	22	1932	30	12
13	Mean quantity of radiation	in ly / day															13
14	Mean potential evaporation	in mm	0	0	5	28	65	98	115	101	71	44	20	8	573	30	14
15	Mean windspeed	in m / sec															15
16	Mean predominent direction of the wind																16

---

24 Station / Country Kalmar / Sweden

Location 56°39'N / 16°23'E Height above sealevel 12 m Climate symbol: Köppen **Cfb** Troll III,3

		J	F	M	A	M	J	J	A	S	O	N	D	year	P	
1	Mean daily temperature	in °C	-1,4	-1,6	0,2	4,7	9,5	14,5	17,4	16,7	13,2	8,6	4,3	1,4	7,3	30	1
2	Mean daily maximum temperature	in °C	0,8	0,9	3,1	7,6	12,8	17,9	20,7	19,9	16,3	11,1	6,2	3,3	10,1	30	2
3	Mean daily minimum temperature	in °C	-3,5	-4,1	-2,7	1,7	6,1	11,0	14,0	13,5	10,1	6,1	2,4	-0,5	4,5	30	3
4	Absolute maximum temperature	in °C	9,8	13,5	16,5	21,0	27,7	31,0	31,5	30,0	26,0	21,0	12,6	12,5	31,5	30	4
5	Absolute minimum temperature	in °C	-31,0	-23,0	-22,5	-12,7	-3,0	2,9	7,4	5,5	0,0	-4,6	-9,7	-14,0	-31,0	30	5
6	Mean relative humidity	in %	88	86	84	79	76	75	77	79	81	82	87	88	82	30	6
7	Mean precipitation	in mm	37	27	25	28	36	36	56	55	47	43	43	38	471	30	7
8	Maximum precipitation	in mm															8
9	Minimum precipitation	in mm															9
10	Maximum precipitation in 24 h	in mm	22	16	22	21	29	40	43	57	51	37	27	19	57	30	10
11	Mean number of days with precipitation	> 0,1 mm	14	12	10	10	10	10	12	12	11	12	13	14	140	30	11
12	Mean duration of sunshine	in h +	38	60	128	180	261	285	281	246	183	103	42	28	1833	30	12
13	Mean quantity of radiation	in ly / day															13
14	Mean potential evaporation	in mm	0	0	4	28	65	98	121	105	72	42	16	4	555	30	14
15	Mean windspeed	in m / sec															15
16	Mean predominent direction of the wind																16

+ Ölands s. udde(56°12'N / 16°24'E, 4 m)

Station/Country Malmö/Sweden

Location 55°26'N/13°03'E Height above sealevel 8 m Climate symbol: Köppen Cfb Troll III,3

		J	F	M	A	M	J	J	A	S	O	N	D	year	P	
1	Mean daily temperature	in °C	-0,7	-0,9	1,4	6,3	11,1	14,9	17,3	16,8	13,5	8,7	4,8	1,9	7,9	30	1
2	Mean daily maximum temperature	in °C	1,6	1,7	4,7	10,3	16,1	19,7	21,7	21,0	17,4	11,9	7,0	3,9	11,4	30	2
3	Mean daily minimum temperature	in °C	-3,0	-3,4	-1,6	2,2	6,1	10,1	12,9	12,5	9,8	5,5	2,5	-0,1	4,4	30	3
4	Absolute maximum temperature	in °C	9,0	12,2	17,5	23,0	29,8	34,0	31,8	31,1	27,1	19,6	13,2	11,5	34,0	30	4
5	Absolute minimum temperature	in °C	-28,0	-23,1	-23,3	-8,2	-4,5	-0,1	2,5	3,0	-2,2	-8,5	-15,0	-19,0	-28,0	30	5
6	Mean relative humidity	in %	86	85	81	74	68	68	73	76	79	82	86	88	79	30	6
7	Mean precipitation	in mm	49	36	30	33	38	46	65	62	53	53	44	44	553	30	7
8	Maximum precipitation	in mm															8
9	Minimum precipitation	in mm															9
10	Maximum precipitation in 24 h	in mm	18	18	27	32	34	40	79	46	36	31	19	19	79	30	10
11	Mean number of days with precipitation	> 0,1 mm	17	13	10	11	10	11	14	13	13	14	15	16	157	30	11
12	Mean duration of sunshine	in h +	39	66	125	177	289	263	253	215	172	84	35	20	1718	30	12
13	Mean quantity of radiation	in ly/day															13
14	Mean potential evaporation	in mm	0	0	0	15	68	102	121	95	57	22	0	0	480	30	14
15	Mean windspeed	in m/sec															15
16	Mean predominent direction of the wind																16

+ Svalöv (55°55'N/13°07'E, 72 m)

Station/Country Sodankylä/Finland

Location 67°22'N/26°36'E Height above sealevel 178 m Climate symbol: Köppen Dfc Troll II,2

		J	F	M	A	M	J	J	A	S	O	N	D	year	P	
1	Mean daily temperature	in °C	-13,5	-13,0	-9,0	-2,1	4,9	11,3	14,7	12,0	6,2	-0,5	-5,8	-9,8	-0,4		1
2	Mean daily maximum temperature	in °C	-9,1	-8,9	-3,4	2,6	9,3	16,3	20,2	17,4	10,3	2,4	-2,8	-6,1	4,0	30	2
3	Mean daily minimum temperature	in °C	-19,0	-18,9	-15,8	-8,0	-0,3	5,9	8,8	6,5	2,0	-4,1	-10,0	-14,7	-5,6	30	3
4	Absolute maximum temperature	in °C	6,0	5,6	9,0	16,4	25,4	30,7	31,5	28,6	24,0	14,5	7,9	5,5	31,5		4
5	Absolute minimum temperature	in °C	-45,6	-41,6	-38,9	-31,8	-13,7	-5,0	-1,1	-5,1	-10,6	-25,9	-40,3	-43,1	-45,6		5
6	Mean relative humidity	in %	87	85	78	71	61	60	65	74	81	87	91	89	77	30	6
7	Mean precipitation	in mm	27	26	20	31	31	56	74	71	57	43	39	31	508		7
8	Maximum precipitation	in mm															8
9	Minimum precipitation	in mm															9
10	Maximum precipitation in 24 h	in mm	13	12	14	21	32	33	40	48	36	26	17	17	48		10
11	Mean number of days with precipitation	> 0,1 mm	19	17	14	15	13	16	15	17	17	17	19	19	198		11
12	Mean duration of sunshine	in h	9	49	140	191	222	277	278	184	102	60	19	<1	1532		12
13	Mean quantity of radiation	in ly/day	3	43	153	312	372	445	452	284	136	51	10	<1	189		13
14	Mean potential evaporation	in mm	0	0	0	0	47	97	116	90	48	0	0	0	398		14
15	Mean windspeed	in m/sec	2,8	2,9	3,3	3,4	3,2	3,2	2,9	2,6	3,1	2,9	2,4	2,8	3,0		15
16	Mean predominent direction of the wind		S	S	S	S	N	S	NNE	S	S	S	S	S			16

Station/Country Oulu/Finland

Location 65°01'N/25°29'E Height above sealevel 17 m Climate symbol: Köppen Dfc Troll II,2

		J	F	M	A	M	J	J	A	S	O	N	D	year	P	
1	Mean daily temperature	in °C	-9,8	-10,0	-6,8	0,2	6,7	12,8	16,5	14,6	9,0	2,6	-2,3	-6,2	2,3	30	1
2	Mean daily maximum temperature	in °C	-6,3	-6,4	-2,4	4,2	11,3	17,3	21,1	18,9	12,5	5,2	0,1	-3,3	6,0	30	2
3	Mean daily minimum temperature	in °C	-13,3	-13,6	-11,2	-3,8	2,1	8,3	11,8	10,2	5,5	0,0	-4,6	-9,1	-1,5	30	3
4	Absolute maximum temperature	in °C	5,5	5,4	10,0	20,1	28,0	32,0	33,3	29,2	26,0	14,3	10,2	6,3	33,3	30	4
5	Absolute minimum temperature	in °C	-39,5	-33,9	-34,0	-26,6	-8,5	-0,7	3,0	-0,2	-6,0	-18,9	-27,0	-33,0	-39,5	30	5
6	Mean relative humidity	in %	87	87	82	78	67	68	70	76	81	86	90	88	80	24	6
7	Mean precipitation	in mm	33	28	23	34	32	49	70	65	57	48	41	36	514	30	7
8	Maximum precipitation	in mm															8
9	Minimum precipitation	in mm															9
10	Maximum precipitation in 24 h	in mm	17	18	16	22	27	23	85	37	42	28	15	15	85	30	10
11	Mean number of days with precipitation	> 0,1 mm	16	15	12	13	10	13	14	14	16	17	18	18	176	30	11
12	Mean duration of sunshine	in h	22	63	153	167	259	289	289	178	114	70	22	6	1632	5	12
13	Mean quantity of radiation	in ly/day															13
14	Mean potential evaporation	in mm	0	0	0	0	56	98	123	99	57	17	0	0	450		14
15	Mean windspeed	in m/sec															15
16	Mean predominent direction of the wind																16

Station/Country Kajaani/Finland

Location 64°17'N/27°41'E Height above sealevel 134 m Climate symbol: Köppen Dfc Troll II,2

		J	F	M	A	M	J	J	A	S	O	N	D	year	P	
1	Mean daily temperature	in °C	-10,6	-10,5	-6,7	0,3	6,9	12,9	16,1	14,0	8,3	2,1	-2,6	-7,0	1,9		1
2	Mean daily maximum temperature	in °C	-7,4	-7,2	-2,1	4,7	11,8	17,8	21,2	18,7	11,9	4,5	0,6	-4,4	5,8	30	2
3	Mean daily minimum temperature	in °C	-14,2	-14,4	-11,6	-4,2	1,5	7,7	11,0	9,5	4,9	-0,3	-5,0	-10,1	-2,1	30	3
4	Absolute maximum temperature	in °C	6,4	5,5	11,8	21,2	27,1	30,8	31,8	29,4	25,2	15,0	9,7	6,0	31,8		4
5	Absolute minimum temperature	in °C	-40,1	-39,0	-35,6	-25,8	-10,9	-2,8	0,2	-3,5	-7,6	-21,5	-25,8	-36,8	-40,1		5
6	Mean relative humidity	in %	88	85	79	72	63	64	68	75	82	87	90	89	79	29	6
7	Mean precipitation	in mm	34	27	24	35	38	67	72	72	63	53	43	36	563		7
8	Maximum precipitation	in mm															8
9	Minimum precipitation	in mm															9
10	Maximum precipitation in 24 h	in mm	14	19	17	26	41	49	34	43	58	25	30	22	58		10
11	Mean number of days with precipitation	> 0,1 mm	19	17	14	15	13	15	15	15	16	18	19	19	180		11
12	Mean duration of sunshine	in h	18	58	157	186	237	272	269	188	112	56	17	6	1576	10	12
13	Mean quantity of radiation	in ly/day															13
14	Mean potential evaporation	in mm	0	0	0	0	59	99	121	95	55	13	0	0	442		14
15	Mean windspeed	in m/sec															15
16	Mean predominent direction of the wind																16

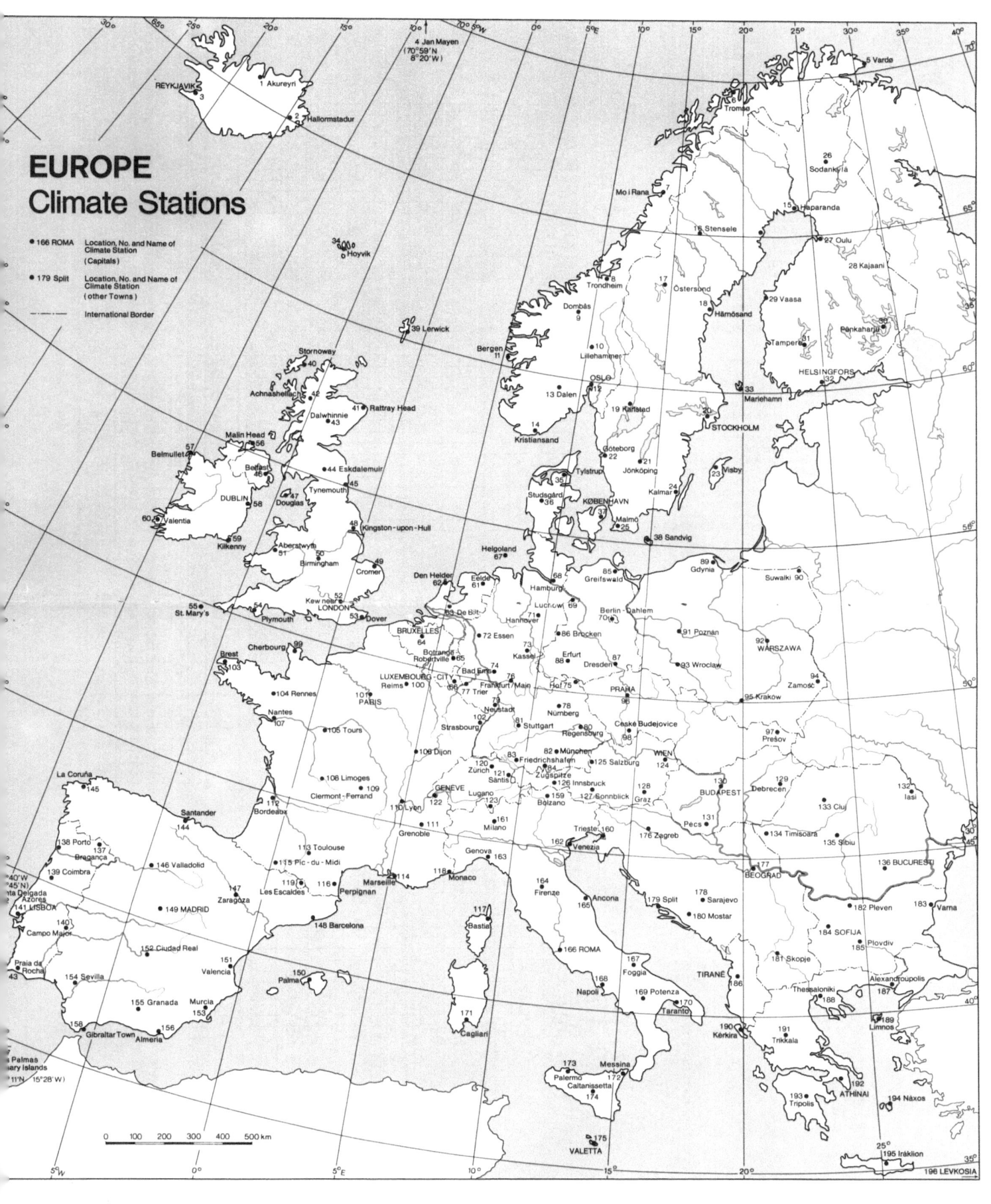

EUROPE
Climate Stations

● 166 ROMA Location, No. and Name of
　　　　　　 Climate Station
　　　　　　 (Capitals)

● 179 Split Location, No. and Name of
　　　　　　 Climate Station
　　　　　　 (other Towns)

———— International Border

29 Station/Country Vaasa(Vasa)/Finland

Location 63°03'N/21°46'E Height above sealevel 6 m Climate symbol: Köppen Dfc Troll II,2

		J	F	M	A	M	J	J	A	S	O	N	D	year	P	
1 Mean daily temperature	in °C	−7,3	−7,5	−4,7	1,3	7,5	12,8	16,2	14,6	9,6	3,8	−0,5	−3,7	3,5		1
2 Mean daily maximum temperature	in °C	−4,3	−4,2	−0,5	5,8	12,5	17,6	21,3	19,4	13,4	6,9	1,8	−1,2	7,4	30	2
3 Mean daily minimum temperature	in °C	−11,0	−11,2	−10,0	−3,2	2,2	7,5	10,8	9,5	5,3	0,7	−3,2	−8,7	−0,8	30	3
4 Absolute maximum temperature	in °C	7,5	7,0	11,4	19,7	29,0	31,8	30,6	29,3	27,5	15,4	10,5	7,9	31,8		4
5 Absolute minimum temperature	in °C	−32,1	−33,4	−34,1	−19,1	−7,1	−1,9	1,8	0,5	−5,0	−14,7	−19,2	−30,2	−34,1		5
6 Mean relative humidity	in %	89	87	82	76	66	68	71	77	82	86	90	90	80	29	6
7 Mean precipitation	in mm	35	21	20	31	30	45	63	65	67	65	53	39	532		7
8 Maximum precipitation	in mm															8
9 Minimum precipitation	in mm															9
10 Maximum precipitation in 24 h	in mm	19	34	14	21	26	33	50	40	51	38	27	24	51		10
11 Mean number of days with precipitation	> 0,1 mm	15	13	11	11	9	11	12	13	14	15	16	17	157		11
12 Mean duration of sunshine	in h	30	77	168	194	267	312	272	206	141	82	35	18	1802	10	12
13 Mean quantity of radiation	in ly/day				.											13
14 Mean potential evaporation	in mm	0	0	0	9	60	95	121	100	63	28	0	0	476		14
15 Mean windspeed	in m/sec	3,8	3,9	3,9	4,3	3,8	3,9	3,4	3,2	4,2	3,9	3,9	4,0	3,8		15
16 Mean predominent direction of the wind		SW	SW	SW	SW	N	W	N	N	SSW	SW	S	E			16

30 Station/Country Punkaharju/Finland

Location 61°48'N/29°17'E Height above sealevel 88 m Climate symbol: Köppen Dfb Troll II,2

		J	F	M	A	M	J	J	A	S	O	N	D	year	P	
1 Mean daily temperature	in °C	−9,7	−9,8	−5,8	1,7	8,4	14,1	17,4	15,7	10,2	3,9	−1,2	−6,0	3,3	30	1
2 Mean daily maximum temperature	in °C	−6,4	−6,1	−0,7	6,2	13,7	19,2	22,3	20,1	13,8	6,4	0,8	−3,4	7,2	30	2
3 Mean daily minimum temperature	in °C	−13,0	−13,4	−10,5	−2,9	3,0	9,0	12,4	11,2	6,6	1,4	−3,2	−8,8	−0,7	30	3
4 Absolute maximum temperature	in °C	5,5	7,5	11,2	20,2	28,3	32,0	33,0	30,5	25,0	18,0	10,3	7,6	33,0	30	4
5 Absolute minimum temperature	in °C	−37,7	−38,2	−32,5	−24,7	−10,2	−2,7	3,8	−2,9	−4,7	−9,7	−20,7	−39,5	−39,5	30	5
6 Mean relative humidity	in %	87	84	76	71	63	65	69	75	81	85	89	89	78	30	6
7 Mean precipitation	in mm	40	30	25	32	38	52	67	72	61	60	48	40	565	30	7
8 Maximum precipitation	in mm															8
9 Minimum precipitation	in mm															9
10 Maximum precipitation in 24 h	in mm	15	12	15	22	22	28	48	33	29	35	27	19	48	30	10
11 Mean number of days with precipitation	> 0,1 mm	20	17	13	13	13	15	16	16	19	20	20	20	202	30	11
12 Mean duration of sunshine	in h															12
13 Mean quantity of radiation	in ly/day															13
14 Mean potential evaporation	in mm	0	0	0	10	64	102	122	100	60	22	0	0	480	50	14
15 Mean windspeed	in m/sec															15
16 Mean predominent direction of the wind																16

31 Station/Country Tampere/Finland

Location 61°28'N/23°46'E Height above sealevel 84 m Climate symbol: Köppen Dfb Troll II,2

		J	F	M	A	M	J	J	A	S	O	N	D	year	P	
1 Mean daily temperature	in °C	−7,8	−7,8	−4,2	2,5	8,9	14,0	17,3	15,5	10,5	4,6	−0,2	−4,0	4,0	30	1
2 Mean daily maximum temperature	in °C	−4,6	−4,4	0,4	7,0	14,4	19,2	22,4	20,3	14,1	7,2	1,9	−1,5	8,0	30	2
3 Mean daily minimum temperature	in °C	−11,0	−11,2	−8,8	−2,1	3,3	8,8	12,2	10,6	6,6	1,9	−2,2	−6,4	−0,2	30	3
4 Absolute maximum temperature	in °C	7,4	9,0	12,2	21,5	28,3	31,0	32,8	31,7	26,7	17,1	10,4	9,1	32,8	30	4
5 Absolute minimum temperature	in °C	−36,2	−36,0	−30,5	−21,9	−7,0	−2,3	1,1	−0,6	−5,5	−14,1	−21,6	−31,9	−36,2	30	5
6 Mean relative humidity	in %	87	84	78	73	64	65	69	75	80	85	89	89	78	30	6
7 Mean precipitation	in mm	38	30	25	35	42	48	76	75	57	57	49	41	573	30	7
8 Maximum precipitation	in mm															8
9 Minimum precipitation	in mm															9
10 Maximum precipitation in 24 h	in mm	20	25	22	23	26	42	44	48	35	26	27	21	48	30	10
11 Mean number of days with precipitation	> 0,1 mm	17	14	11	12	11	12	14	14	15	16	18	18	172	30	11
12 Mean duration of sunshine	in h															12
13 Mean quantity of radiation	in ly/day				.											13
14 Mean potential evaporation	in mm	0	0	0	17	69	103	126	100	61	26	0	0	502		14
15 Mean windspeed	in m/sec															15
16 Mean predominent direction of the wind																16

32 Station/Country Helsingfors(Ilmala)/Finland

Location 60°12'N/24°55'E Height above sealevel 45 m Climate symbol: Köppen Dfb Troll III,4

		J	F	M	A	M	J	J	A	S	O	N	D	year	P	
1 Mean daily temperature	in °C	−6,1	−6,6	−3,4	2,0	8,8	14,0	17,2	16,0	11,1	5,4	1,0	−2,6	4,8		1
2 Mean daily maximum temperature	in °C	−3,4	−3,9	0,1	6,4	13,5	18,6	21,6	20,1	15,0	8,1	2,9	−0,5	8,2	30	2
3 Mean daily minimum temperature	in °C	−8,5	−9,3	−6,8	−0,9	4,3	9,2	12,5	11,7	7,5	2,7	−1,0	−4,8	1,4	30	3
4 Absolute maximum temperature	in °C	6,8	11,8	15,0	20,5	26,1	31,2	33,1	30,1	24,3	17,8	10,7	9,4	33,1		4
5 Absolute minimum temperature	in °C	−33,2	−30,2	−26,0	−13,4	−5,5	−0,3	5,4	3,5	−4,1	−10,0	−16,3	−27,8	−33,2		5
6 Mean relative humidity	in %	88	86	78	74	64	66	70	75	81	85	88	90	79	30	6
7 Mean precipitation	in mm	57	42	36	44	41	51	68	72	71	73	68	66	692		7
8 Maximum precipitation	in mm															8
9 Minimum precipitation	in mm															9
10 Maximum precipitation in 24 h	in mm	21	18	22	35	31	56	88	60	57	50	35	40	88		10
11 Mean number of days with precipitation	> 0,1 mm	20	18	14	13	12	13	14	15	15	18	19	20	191		11
12 Mean duration of sunshine	in h	31	63	138	184	270	294	295	251	152	76	30	18	1799		12
13 Mean quantity of radiation	in ly/day	19	70	197	291	401	478	450	304	187	84	23	12	210		13
14 Mean potential evaporation	in mm	0	0	0	18	65	99	125	104	65	31	3	0	510		14
15 Mean windspeed	in m/sec	4,3	4,1	3,9	3,8	3,8	3,7	3,3	3,4	3,8	3,9	4,2	4,0	3,8		15
16 Mean predominent direction of the wind		S	N	WNW	S	SSW	SSW	S	S	WNW	S	S	S			16

12

33 Station/Country Mariehamn(Maarianhamina)/Finland

Location 60°07'N/19°54'E   Height above sealevel 4 m   Climate symbol: Köppen Dfb   Troll III,4

		J	F	M	A	M	J	J	A	S	O	N	D	year	P	
1 Mean daily temperature	in °C	-3,3	-4,5	-2,5	2,7	7,9	12,6	16,3	15,6	11,3	6,4	2,7	-0,2	5,4	30	1
2 Mean daily maximum temperature	in °C	-0,8	-1,5	1,1	6,2	12,4	16,9	20,5	19,5	14,8	9,0	4,6	1,8	8,7	30	2
3 Mean daily minimum temperature	in °C	-5,8	-7,5	-6,0	-0,9	3,4	8,3	12,1	11,6	7,8	3,7	0,8	-2,2	2,1	30	3
4 Absolute maximum temperature	in °C	8,4	11,2	12,8	18,7	24,9	28,6	31,3	31,0	25,0	17,4	10,5	10,3	31,3	30	4
5 Absolute minimum temperature	in °C	-32,4	-32,0	-25,9	-16,6	-6,9	-3,2	0,1	-0,5	-5,4	-11,9	-18,1	-21,9	-32,4	30	5
6 Mean relative humidity	in %	88	85	83	79	71	72	75	79	83	84	88	89	81	29	6
7 Mean precipitation	in mm	48	31	26	33	32	37	45	66	64	57	62	57	558	30	7
8 Maximum precipitation	in mm															8
9 Minimum precipitation	in mm															9
10 Maximum precipitation in 24 h	in mm	32	24	17	21	24	25	56	62	40	29	20	22	62	30	10
11 Mean number of days with precipitation	> 0,1 mm	20	16	13	12	10	11	12	13	16	18	19	20	180	30	11
12 Mean duration of sunshine	in h															12
13 Mean quantity of radiation	in ly/day															13
14 Mean potential evaporation	in mm	0	0	0	20	59	92	118	104	66	36	11	0	506		14
15 Mean windspeed	in m/sec															15
16 Mean predominent direction of the wind																16

34 Station/Country Høyvik(Farøer–Islands)/Denmark

Location 62°02'N/6°45'W   Height above sealevel 20 m   Climate symbol: Köppen Cfc   Troll III,2

		J	F	M	A	M	J	J	A	S	O	N	D	year	P	
1 Mean daily temperature	in °C	3,9	3,5	4,4	5,3	7,4	9,3	11,2	11,4	10,0	7,6	6,0	4,8	7,1	27	1
2 Mean daily maximum temperature	in °C	6,2	5,8	6,6	7,6	9,6	11,6	13,3	13,6	12,1	9,8	8,1	7,0	9,3	30	2
3 Mean daily minimum temperature	in °C	1,5	1,2	2,2	3,0	5,1	7,0	9,0	9,2	7,8	5,4	3,9	2,5	4,9	30	3
4 Absolute maximum temperature	in °C	14,8	11,9	12,7	13,3	18,9	18,8	22,1	21,6	18,1	16,5	14,6	12,0	22,1	27	4
5 Absolute minimum temperature	in °C	-9,9	-10,4	-8,5	-6,7	-4,5	0,5	2,2	3,1	0,0	-4,3	-5,0	-10,0	-10,4	30	5
6 Mean relative humidity	in %	82	82	82	81	82	83	85	86	85	83	84	83	83	30	6
7 Mean precipitation	in mm	149	136	114	106	67	74	79	96	132	157	156	167	1433	30	7
8 Maximum precipitation	in mm															8
9 Minimum precipitation	in mm															9
10 Maximum precipitation in 24 h	in mm	73	51	49	51	43	47	55	40	55	51	53	56	73	30	10
11 Mean number of days with precipitation	> 0,1 mm	25	22	23	22	16	16	18	20	21	24	24	26	257	30	11
12 Mean duration of sunshine	in h	15	41	46	113	137	140	109	102	82	58	22	7	902	30	12
13 Mean quantity of radiation	in ly/day															13
14 Mean potential evaporation	in mm	13	16	28	41	64	69	90	79	64	44	27	1	537	30	14
15 Mean windspeed	in m/sec															15
16 Mean predominent direction of the wind																16

35 Station/Country Tylstrup near Ålburg/Denmark

Location 57°11'N/9°57'E   Height above sealevel 13 m   Climate symbol: Köppen Cfb   Troll III,3

		J	F	M	A	M	J	J	A	S	O	N	D	year	P	
1 Mean daily temperature	in °C	-0,5	-0,9	1,1	5,8	10,8	14,2	16,4	15,9	12,6	8,0	4,4	1,8	7,5		1
2 Mean daily maximum temperature	in °C	1,8	2,0	4,8	10,2	16,1	19,5	21,7	20,9	17,1	11,8	7,0	4,0	11,4	30	2
3 Mean daily minimum temperature	in °C	-3,3	-4,1	-2,3	1,4	5,0	8,7	11,2	11,0	8,3	4,5	1,6	-0,7	3,4	30	3
4 Absolute maximum temperature	in °C	10,6	13,4	17,5	23,1	31,1	31,9	34,0	33,6	27,8	19,8	14,5	12,7	34,0		4
5 Absolute minimum temperature	in °C	-27,2	-23,5	-21,2	-9,6	-5,6	1,0	2,1	1,0	-1,0	-7,0	-7,8	-18,2	-27,2		5
6 Mean relative humidity	in %	89	86	83	75	67	69	72	75	80	84	89	90	80	30	6
7 Mean precipitation	in mm	49	34	27	38	34	51	76	72	73	70	65	48	637		7
8 Maximum precipitation	in mm															8
9 Minimum precipitation	in mm															9
10 Maximum precipitation in 24 h	in mm	28	44	17	23	20	40	62	56	61	27	23	17	62		10
11 Mean number of days with precipitation	> 0,1 mm	15	11	10	11	10	11	13	14	15	15	17	16	158		11
12 Mean duration of sunshine	in h	40	71	134	190	266	273	264	229	165	102	46	28	1808		12
13 Mean quantity of radiation	in ly/day															13
14 Mean potential evaporation	in mm	2	1	16	38	78	106	118	99	67	40	17	6	588	41	14
15 Mean windspeed	in m/sec	4,4	4,4	4,4	4,6	4,2	4,4	4,1	3,9	4,1	4,1	4,1	4,2	4,2		15
16 Mean predominent direction of the wind																16

36 Station/Country Studsgård near Herning/Denmark

Location 56°05'N/8°55'E   Height above sealevel 54 m   Climate symbol: Köppen Cfb   Troll III,3

		J	F	M	A	M	J	J	A	S	O	N	D	year	P	
1 Mean daily temperature	in °C	-0,8	-0,9	3,6	6,0	10,7	14,0	16,1	15,6	12,9	8,3	4,4	1,7	7,7	30	1
2 Mean daily maximum temperature	in °C	1,9	2,2	5,2	10,6	16,4	19,3	21,1	20,7	17,3	11,9	7,0	·4,1	11,5	30	2
3 Mean daily minimum temperature	in °C	-3,4	-4,0	-2,1	1,4	5,0	8,6	11,1	10,9	8,4	4,6	1,7	-0,8	3,4	30	3
4 Absolute maximum temperature	in °C	10,0	12,9	18,0	23,5	31,0	33,9	35,3	33,0	29,5	20,2	14,2	12,3	35,3	30	4
5 Absolute minimum temperature	in °C	-25,4	-22,9	-19,3	-6,7	-4,3	-0,8	3,5	2,6	-2,3	-6,9	-8,1	-19,1	-25,4	30	5
6 Mean relative humidity	in %	93	90	87	81	72	72	76	79	84	89	93	94	84	29	6
7 Mean precipitation	in mm	68	45	39	41	39	59	91	94	86	90	72	66	782	30	7
8 Maximum precipitation	in mm															8
9 Minimum precipitation	in mm															9
10 Maximum precipitation in 24 h	in mm	28	21	22	27	23	57	85	51	42	39	27	23	85	30	10
11 Mean number of days with precipitation	> 0,1 mm	16	13	11	12	11	13	15	15	16	16	18	17	173	30	11
12 Mean duration of sunshine	in h +	41	75	126	185	226	229	221	191	167	92	41	25	1619	10	12
13 Mean quantity of radiation	in ly/day															13
14 Mean potential evaporation	in mm	0	0	17	38	76	102	115	101	70	39	18	6	582	30	14
15 Mean windspeed	in m/sec															15
16 Mean predominent direction of the wind																16

+ Lyngvig(56°03'N/8°06'E, 4 m)

37 Station / Country København (Copenhagen) / Denmark

Location 55°41'N/12°33'E — Height above sealevel 9 m — Climate symbol: Köppen Cfb — Troll III,3

		J	F	M	A	M	J	J	A	S	O	N	D	year	P	
1 Mean daily temperature	in °C	0,1	-0,1	1,9	6,6	11,8	15,6	17,8	17,3	13,9	9,3	5,4	2,5	8,5		1
2 Mean daily maximum temperature	in °C	2,0	2,1	5,0	10,4	16,1	19,4	21,8	21,2	17,5	12,1	7,3	4,2	11,6	30	2
3 Mean daily minimum temperature	in °C	-2,0	-2,5	-0,8	3,1	7,5	11,2	13,6	13,5	10,5	6,7	3,3	0,7	5,4	30	3
4 Absolute maximum temperature	in °C	9,9	14,0	18,5	22,1	27,7	32,7	30,7	30,5	26,7	19,9	14,4	12,3	32,7	30	4
5 Absolute minimum temperature	in °C	-24,2	-19,6	-17,8	-8,8	-1,5	3,0	7,5	5,6	0,9	-4,0	-6,8	-11,4	-24,2	30	5
6 Mean relative humidity	in %	87	85	82	74	65	65	68	70	80	81	86	88	78	30	6
7 Mean precipitation	in mm	49	39	32	38	43	47	71	66	62	59	48	49	603	30	7
8 Maximum precipitation	in mm															8
9 Minimum precipitation	in mm															9
10 Maximum precipitation in 24 h	in mm	24	24	25	38	38	37	77	41	50	47	34	30	77	30	10
11 Mean number of days with precipitation	> 0,1 mm	17	13	12	13	11	13	14	14	15	16	16	17	171	30	11
12 Mean duration of sunshine	in h	36	55	118	161	245	245	239	207	157	87	34	19	1603	30	12
13 Mean quantity of radiation	in ly/day															13
14 Mean potential evaporation	in mm	2	1	10	35	76	103	118	103	71	43	19	8	589	40	14
15 Mean windspeed	in m/sec	2,7	2,7	2,5	2,3	2,3	2,2	2,0	2,0	2,0	2,3	2,3	2,3	2,3		15
16 Mean predominent direction of the wind																16

38 Station / Country Sandvig (Bornholm) / Denmark

Location 55°17'N/14°47'E — Height above sealevel 11 m — Climate symbol: Köppen Cfb — Troll III,3

		J	F	M	A	M	J	J	A	S	O	N	D	year	P	
1 Mean daily temperature	in °C	0,5	-0,1	1,5	5,4	9,5	14,4	17,5	17,3	14,3	9,8	5,7	2,9	8,2	30	1
2 Mean daily maximum temperature	in °C	2,3	1,9	3,8	8,3	12,7	17,6	20,4	19,9	16,6	11,9	7,5	4,5	10,6	30	2
3 Mean daily minimum temperature	in °C	-1,3	-2,0	-0,9	2,5	6,2	11,2	14,5	14,7	11,9	7,7	3,9	1,9	5,8	30	3
4 Absolute maximum temperature	in °C	10,8	11,1	15,6	23,5	26,6	30,1	30,3	32,6	27,9	19,6	15,5	12,1	32,6	30	4
5 Absolute minimum temperature	in °C	-20,5	-14,6	-14,5	-6,6	-1,6	3,7	8,4	8,7	4,2	-2,1	-6,8	-9,1	-20,5	30	5
6 Mean relative humidity	in %	87	87	85	80	79	79	80	82	82	83	85	88	83	30	6
7 Mean precipitation	in mm	49	37	29	31	32	42	58	58	55	59	50	53	553	30	7
8 Maximum precipitation	in mm															8
9 Minimum precipitation	in mm															9
10 Maximum precipitation in 24 h	in mm	18	17	15	33	21	38	34	70	54	29	23	44	70	30	10
11 Mean number of days with precipitation	> 0,1 mm	15	13	11	10	9	9	10	11	12	14	15	15	144	30	11
12 Mean duration of sunshine	in h +	38	55	121	189	271	284	257	224	181	98	36	26	1780	11	12
13 Mean quantity of radiation	in ly/day															13
14 Mean potential evaporation	in mm	2	0	6	35	64	102	123	109	76	47	23	11	597	30	14
15 Mean windspeed	in m/sec															15
16 Mean predominent direction of the wind																16

+ Bornholm (55°06'N/11°55'E, 22 m)

39 Station / Country Lerwick (Shetland Islands) / United Kingdom

Location 60°09'N/1°10'W — Height above sealevel 82 m — Climate symbol: Köppen Cfb — Troll III,2

		J	F	M	A	M	J	J	A	S	O	N	D	year	P	
1 Mean daily temperature	in °C	3,1	3,0	3,9	5,5	7,8	10,0	12,1	12,1	10,7	8,2	6,0	4,5	7,2	30	1
2 Mean daily maximum temperature	in °C	5,0	5,0	6,0	8,1	10,5	12,8	14,4	14,4	12,9	10,1	7,8	6,2	9,4	30	2
3 Mean daily minimum temperature	in °C	1,2	0,9	1,8	2,8	5,1	7,4	9,7	9,7	8,4	6,2	4,1	2,7	5,0	30	3
4 Absolute maximum temperature	in °C	11,7	11,7	12,8	16,1	19,4	23,3	22,8	20,8	19,4	16,1	14,4	12,2	23,3	30	4
5 Absolute minimum temperature	in °C	-8,9	-7,2	-8,3	-5,4	-2,2	-0,6	3,9	2,8	-0,6	-3,3	-4,4	-7,2	-8,9	30	5
6 Mean relative humidity	in %	87	86	84	81	81	83	87	87	87	87	89	87	86	16	6
7 Mean precipitation	in mm	109	87	69	68	52	55	72	71	87	104	111	118	1003	30	7
8 Maximum precipitation	in mm															8
9 Minimum precipitation	in mm															9
10 Maximum precipitation in 24 h	in mm	27	36	40	23	40	29	86	44	47	26	66	37	86	30	10
11 Mean number of days with precipitation	> 0,25 mm	25	22	20	21	15	16	17	17	19	23	24	25	243	30	11
12 Mean duration of sunshine	in h	25	51	90	132	165	158	125	117	105	67	33	14	1082	30	12
13 Mean quantity of radiation	in ly/day															13
14 Mean potential evaporation	in mm	19	18	24	38	57	76	89	81	63	43	27	20	555	35	14
15 Mean windspeed	in m/sec															15
16 Mean predominent direction of the wind																16

40 Station / Country Stornoway (Hebrides) / United Kingdom

Location 58°13'N/6°20'W — Height above sealevel 3 m — Climate symbol: Köppen Cfb — Troll III,2

		J	F	M	A	M	J	J	A	S	O	N	D	year	P	
1 Mean daily temperature	in °C	4,4	4,4	5,7	7,0	9,4	11,8	13,3	13,3	11,8	9,4	7,0	5,5	8,6	30	1
2 Mean daily maximum temperature	in °C	6,3	6,6	8,2	9,8	12,4	14,8	15,9	16,0	14,4	11,7	9,1	7,3	11,0	30	2
3 Mean daily minimum temperature	in °C	2,4	2,2	3,1	4,2	6,3	8,6	10,6	10,6	9,1	7,0	4,8	3,6	6,0	30	3
4 Absolute maximum temperature	in °C	12,2	12,8	17,2	16,7	21,7	24,4	25,0	24,4	21,7	19,4	15,0	16,1	25,0	30	4
5 Absolute minimum temperature	in °C	-12,2	-10,6	-8,3	-3,3	-2,8	0,6	2,8	1,7	0,6	-4,4	-7,8	-10,6	-12,2	30	5
6 Mean relative humidity	in %	87	85	83	80	78	79	82	82	83	84	86	87	83	16	6
7 Mean precipitation	in mm	152	104	80	86	61	74	93	104	121	142	131	149	1297	30	7
8 Maximum precipitation	in mm															8
9 Minimum precipitation	in mm															9
10 Maximum precipitation in 24 h	in mm	52	30	28	44	25	46	40	70	49	35	26	37	70	30	10
11 Mean number of days with precipitation	> 0,25 mm	24	21	19	20	17	17	21	20	21	24	23	25	252	30	11
12 Mean duration of sunshine	in h	35	62	108	142	195	173	128	133	111	76	45	26	1234	30	12
13 Mean quantity of radiation	in ly/day															13
14 Mean potential evaporation	in mm	18	18	27	43	64	83	93	84	65	41	26	18	580	35	14
15 Mean windspeed	in m/sec															15
16 Mean predominent direction of the wind																16

41 Station/Country: Rattray Head (Scotland)/United Kingdom

Location 57°37'N/1°50'W — Height above sealevel 26 m — Climate symbol: Köppen Cfb — Troll III,2

		J	F	M	A	M	J	J	A	S	O	N	D	year	P	
1	Mean daily temperature	in °C	3,2	3,4	4,9	6,8	8,8	11,1	13,0	12,9	11,7	9,3	6,8	4,4	8,0	13	1
2	Mean daily maximum temperature	in °C	5,2	5,7	7,3	9,8	11,3	13,7	15,7	15,4	14,5	11,8	8,7	6,4	10,4	13	2
3	Mean daily minimum temperature	in °C	1,1	1,0	2,5	3,8	5,9	8,4	10,3	10,3	8,9	6,8	4,5	2,4	5,5	13	3
4	Absolute maximum temperature	in °C	12,8	15,0	17,8	18,9	20,8	28,3	24,4	25,8	23,3	20,0	14,4	13,3	28,3	13	4
5	Absolute minimum temperature	in °C	-12,2	-6,7	-9,4	-2,8	0,0	1,7	3,3	5,0	1,1	-2,2	-3,3	-6,1	-12,2	13	5
6	Mean relative humidity	in %	84	84	83	77	81	81	82	84	82	82	85	84	82	13	6
7	Mean precipitation	in mm	36	27	28	21	23	29	35	38	30	39	43	40	387	13	7
8	Maximum precipitation	in mm															8
9	Minimum precipitation	in mm															9
10	Maximum precipitation in 24 h	in mm	31	16	25	22	29	34	50	27	31	32	29	28	50	13	10
11	Mean number of days with precipitation	>0,25 mm	21	17	15	16	15	15	15	18	17	19	21	20	209	13	11
12	Mean duration of sunshine	in h	49	74	102	169	185	162	167	150	127	93	51	45	1374	12	12
13	Mean quantity of radiation	in ly/day															13
14	Mean potential evaporation	in mm	16	16	31	45	64	86	99	90	70	47	30	19	611	13	14
15	Mean windspeed	in m/sec															15
16	Mean predominent direction of the wind																16

42 Station/Country: Achnashellach (Scotland)/United Kingdom

Location 57°29'N/5°16'W — Height above sealevel 67 m — Climate symbol: Köppen Cfb — Troll III,2

		J	F	M	A	M	J	J	A	S	O	N	D	year	P	
1	Mean daily temperature	in °C	2,8	3,2	5,2	6,9	10,1	12,4	13,8	13,8	11,8	8,7	5,7	4,1	8,2	30	1
2	Mean daily maximum temperature	in °C	6,0	6,4	8,8	10,9	14,7	16,8	17,5	17,5	15,4	12,1	8,9	7,0	11,8	30	2
3	Mean daily minimum temperature	in °C	-0,4	-0,1	1,5	2,9	5,4	8,0	9,8	9,8	7,8	5,3	2,4	1,1	4,4	30	3
4	Absolute maximum temperature	in °C	14,4	15,6	21,1	22,8	27,2	30,0	31,1	27,8	26,1	23,9	19,4	18,3	31,1	30	4
5	Absolute minimum temperature	in °C	-12,8	-16,1	-12,2	-6,7	-3,9	-0,6	-0,6	0,8	-2,2	-4,4	-10,6	-13,9	-16,1	30	5
6	Mean relative humidity	in %															6
7	Mean precipitation	in mm	232	177	127	156	93	114	148	150	189	234	196	242	2058	30	7
8	Maximum precipitation	in mm															8
9	Minimum precipitation	in mm															9
10	Maximum precipitation in 24 h	in mm	59	60	58	58	51	64	45	69	73	56	75	66	75	30	10
11	Mean number of days with precipitation	>0,25 mm	23	21	19	20	18	19	22	21	22	24	23	24	256	30	11
12	Mean duration of sunshine	in h															12
13	Mean quantity of radiation	in ly/day															13
14	Mean potential evaporation	in mm	13	16	31	45	72	94	99	90	67	44	25	17	612	30	14
15	Mean windspeed	in m/sec															15
16	Mean predominent direction of the wind																16

43 Station/Country: Dalwhinnie (Scotland)/United Kingdom

Location 56°56'N/4°14'W — Height above sealevel 359 m — Climate symbol: Köppen Cfb — Troll III,2

		J	F	M	A	M	J	J	A	S	O	N	D	year	P	
1	Mean daily temperature	in °C	0,3	0,7	2,8	5,1	8,0	11,0	12,5	12,2	10,0	6,8	3,7	1,9	6,3	30	1
2	Mean daily maximum temperature	in °C	2,7	3,4	6,2	9,3	12,8	15,8	16,7	16,4	13,7	9,8	6,2	4,0	9,7	30	2
3	Mean daily minimum temperature	in °C	-2,1	-2,0	-0,7	0,8	3,1	6,1	8,2	8,0	6,2	3,8	1,2	-0,3	2,7	30	3
4	Absolute maximum temperature	in °C	13,3	11,7	17,2	19,4	25,6	28,8	30,0	27,2	23,3	20,6	15,6	13,3	30,0	30	4
5	Absolute minimum temperature	in °C	-20,6	-21,7	-17,8	-10,0	-7,2	-5,6	-1,7	-2,8	-6,7	-11,7	-17,2	-17,2	-21,7	30	5
6	Mean relative humidity	in %	88	88	85	78	73	73	78	80	83	85	87	88	82	26	6
7	Mean precipitation	in mm	133	97	70	78	69	66	90	85	103	143	125	158	1217	30	7
8	Maximum precipitation	in mm															8
9	Minimum precipitation	in mm															9
10	Maximum precipitation in 24 h	in mm	59	41	30	34	36	43	48	35	40	56	54	68	68	30	10
11	Mean number of days with precipitation	>0,25 mm	20	19	18	18	16	16	19	18	18	20	20	21	223	30	11
12	Mean duration of sunshine	in h	27	51	92	118	158	150	121	115	91	61	31	17	1032	30	12
13	Mean quantity of radiation	in ly/day															13
14	Mean potential evaporation	in mm	2	5	21	38	73	94	99	90	67	39	23	11	559	30	14
15	Mean windspeed	in m/sec															15
16	Mean predominent direction of the wind																16

44 Station/Country: Eskdalemuir (Scotland)/United Kingdom

Location 55°19'N/3°12'W — Height above sealevel 242 m — Climate symbol: Köppen Cfb — Troll III,2

		J	F	M	A	M	J	J	A	S	O	N	D	year	P	
1	Mean daily temperature	in °C	1,5	2,0	3,8	6,1	9,2	12,0	13,5	13,2	11,1	7,9	4,9	3,0	7,4	30	1
2	Mean daily maximum temperature	in °C	3,8	4,7	7,4	10,4	14,2	16,7	17,8	17,4	14,9	11,2	7,9	5,1	10,9	30	2
3	Mean daily minimum temperature	in °C	-0,9	-0,8	0,2	1,7	4,2	7,2	9,2	8,9	7,2	4,2	2,2	0,8	3,7	30	3
4	Absolute maximum temperature	in °C	11,7	13,3	19,1	20,2	25,0	28,3	29,2	27,3	23,3	18,3	15,0	12,3	29,2	30	4
5	Absolute minimum temperature	in °C	-17,0	-16,1	-14,1	-8,5	-6,0	-4,4	-1,6	-0,6	-4,8	-8,2	-10,9	-12,8	-17,0	30	5
6	Mean relative humidity	in %	88	85	80	73	70	73	77	80	82	85	89	90	81	16	6
7	Mean precipitation	in mm	175	112	97	97	87	108	131	120	136	149	153	162	1527	30	7
8	Maximum precipitation	in mm															8
9	Minimum precipitation	in mm															9
10	Maximum precipitation in 24 h	in mm	56	45	44	42	45	108	57	54	58	61	64	65	106	30	10
11	Mean number of days with precipitation	>0,25 mm	22	18	17	17	15	17	19	18	18	20	21	22	224	30	11
12	Mean duration of sunshine	in h	44	65	95	130	174	169	140	129	100	78	50	36	1210	30	12
13	Mean quantity of radiation	in ly/day															13
14	Mean potential evaporation	in mm	13	14	22	38	68	87	101	89	63	40	19	13	667		14
15	Mean windspeed	in m/sec															15
16	Mean predominent direction of the wind																16

45 Station / Country — Tynemouth (England)/United Kingdom

Location 55°01'N/1°25'W — Height above sealevel 33 m — Climate symbol: Köppen Cfb — Troll III,2

		J	F	M	A	M	J	J	A	S	O	N	D	year	P	
1	Mean daily temperature	in °C	4,1	4,4	5,7	7,7	9,7	12,9	15,1	15,0	13,3	10,4	7,4	5,9	9,3	30	1
2	Mean daily maximum temperature	in °C	5,9	6,4	8,1	10,4	12,2	15,7	17,9	17,6	15,9	12,7	9,2	7,0	11,8	30	2
3	Mean daily minimum temperature	in °C	2,2	2,4	3,3	5,0	7,2	10,1	12,3	12,3	10,6	8,1	5,5	3,7	6,9	30	3
4	Absolute maximum temperature	in °C	14,4	15,0	20,0	21,7	22,8	30,0	26,7	27,2	27,2	21,1	18,3	15,0	30,0	30	4
5	Absolute minimum temperature	in °C	-8,9	-7,8	-7,8	-2,2	3,0	3,9	7,8	5,6	0,6	-1,1	-3,3	-5,6	-8,9	30	5
6	Mean relative humidity	in %	82	79	78	71	76	75	75	77	76	80	81	82	78	16	6
7	Mean precipitation	in mm	62	45	39	38	48	47	64	74	55	61	64	53	650	30	7
8	Maximum precipitation	in mm															8
9	Minimum precipitation	in mm															9
10	Maximum precipitation in 24 h	in mm	30	63	29	23	35	36	57	51	37	49	37	26	63	30	10
11	Mean number of days with precipitation	>0,25 mm	17	15	13	13	13	13	15	14	14	18	17	17	177	30	11
12	Mean duration of sunshine	in h	42	65	98	152	179	178	169	146	124	91	51	35	1330	25	12
13	Mean quantity of radiation	in ly / day															13
14	Mean potential evaporation	in mm	14	18	26	40	64	86	102	92	67	43	24	15	589	35	14
15	Mean windspeed	in m / sec															15
16	Mean predominent direction of the wind																16

46 Station / Country — Belfast (Northern Ireland)/United Kingdom

Location 54°39'N/6°13'W — Height above sealevel 67 m — Climate symbol: Köppen Cfb — Troll III,2

		J	F	M	A	M	J	J	A	S	O	N	D	year	P	
1	Mean daily temperature	in °C	3,8	4,2	6,0	7,9	10,5	13,4	14,7	14,5	12,7	9,7	6,6	4,9	9,1	30	1
2	Mean daily maximum temperature	in °C	6,0	6,8	9,2	11,8	14,9	17,5	18,4	18,3	16,1	12,8	9,1	6,9	12,3	30	2
3	Mean daily minimum temperature	in °C	1,5	1,5	2,7	3,9	6,1	9,2	11,0	10,7	9,2	6,8	4,1	2,9	5,8	30	3
4	Absolute maximum temperature	in °C	13,3	13,9	19,4	20,6	26,1	28,3	29,4	27,8	25,6	21,1	16,1	14,4	29,4	30	4
5	Absolute minimum temperature	in °C	-12,8	-11,7	-12,2	-4,4	-3,3	-0,6	3,9	1,1	-2,2	-4,4	-6,1	-10,6	-12,8	30	5
6	Mean relative humidity	in %	90	86	81	76	73	76	79	81	84	86	89	91	83	16	6
7	Mean precipitation	in mm	80	52	50	48	52	68	94	77	80	83	72	90	845	30	7
8	Maximum precipitation	in mm															8
9	Minimum precipitation	in mm															9
10	Maximum precipitation in 24 h	in mm	28	22	26	21	25	36	45	44	43	38	28	35	45	30	10
11	Mean number of days with precipitation	>0,25 mm	20	17	16	16	15	16	19	17	18	19	19	21	213	30	11
12	Mean duration of sunshine	in h	45	65	104	149	196	179	136	135	107	81	53	35	1285	30	12
13	Mean quantity of radiation	in ly / day															13
14	Mean potential evaporation	in mm	16	19	28	45	71	93	101	92	67	42	24	17	615	35	14
15	Mean windspeed	in m / sec															15
16	Mean predominent direction of the wind																16

47 Station / Country — Douglas (Isle of Man)/United Kingdom

Location 54°10'N/4°28'W — Height above sealevel 87 m — Climate symbol: Köppen Cfb — Troll III,2

		J	F	M	A	M	J	J	A	S	O	N	D	year	P	
1	Mean daily temperature	in °C	4,8	4,7	5,9	7,8	10,4	13,1	14,5	14,7	13,1	10,5	7,8	6,2	9,5	30	1
2	Mean daily maximum temperature	in °C	6,8	6,9	8,3	10,7	13,6	16,3	17,4	17,5	15,6	12,7	9,9	8,2	12,0	30	2
3	Mean daily minimum temperature	in °C	2,7	2,4	3,4	4,9	7,2	9,9	11,5	11,8	10,6	8,2	5,7	4,1	6,9	30	3
4	Absolute maximum temperature	in °C	12,8	13,3	18,3	18,3	23,9	26,7	27,8	26,7	22,8	20,6	16,1	12,8	27,8	30	4
5	Absolute minimum temperature	in °C	-7,2	-4,4	-5,6	-1,7	0,0	4,4	4,4	5,6	1,7	-3,9	3,9	-3,3	-7,2	30	5
6	Mean relative humidity	in %															6
7	Mean precipitation	in mm	125	81	72	63	69	72	81	91	111	122	124	128	1139	30	7
8	Maximum precipitation	in mm															8
9	Minimum precipitation	in mm															9
10	Maximum precipitation in 24 h	in mm	49	36	41	30	53	60	50	84	53	57	39	46	84	30	10
11	Mean number of days with precipitation	>0,25 mm	20	16	15	14	14	14	15	15	16	18	19	21	197	30	11
12	Mean duration of sunshine	in h	57	71	120	177	223	221	188	181	138	102	61	45	1584	30	12
13	Mean quantity of radiation	in ly / day															13
14	Mean potential evaporation	in mm	18	19	27	41	67	87	97	88	67	43	27	20	601	35	14
15	Mean windspeed	in m / sec															15
16	Mean predominent direction of the wind																16

48 Station / Country — Kingston-upon-Hull (England)/United Kingdom

Location 53°45'N/0°16'W — Height above sealevel 2 m — Climate symbol: Köppen Cfb — Troll III,2

		J	F	M	A	M	J	J	A	S	O	N	D	year	P	
1	Mean daily temperature	in °C	3,7	4,3	6,0	8,7	11,4	14,7	16,8	16,5	14,4	10,7	7,1	4,9	9,9	30	1
2	Mean daily maximum temperature	in °C	6,0	6,9	9,2	12,6	15,6	19,0	21,0	20,7	18,2	14,0	9,5	7,1	13,3	30	2
3	Mean daily minimum temperature	in °C	1,4	1,6	2,8	4,8	7,2	10,3	12,6	12,3	10,6	7,3	4,6	2,7	6,5	30	3
4	Absolute maximum temperature	in °C	13,9	17,2	21,1	25,6	27,2	32,8	32,8	32,2	30,6	25,6	17,8	15,0	32,8	30	4
5	Absolute minimum temperature	in °C	-11,1	-7,8	-10,0	-2,8	-0,5	1,7	5,6	2,8	0,6	-3,3	-4,4	-7,2	-11,1	30	5
6	Mean relative humidity	in %															6
7	Mean precipitation	in mm	61	49	38	42	48	50	60	64	52	58	66	54	642	30	7
8	Maximum precipitation	in mm															8
9	Minimum precipitation	in mm															9
10	Maximum precipitation in 24 h	in mm	28	23	31	28	42	44	47	50	47	57	30	44	57	30	10
11	Mean number of days with precipitation	>0,25 mm	18	15	13	13	12	12	14	14	13	15	17	18	174	30	11
12	Mean duration of sunshine	in h	39	61	96	143	176	188	176	163	130	94	47	32	1345	30	12
13	Mean quantity of radiation	in ly / day															13
14	Mean potential evaporation	in mm	11	14	28	45	72	98	111	101	73	47	23	15	637	35	14
15	Mean windspeed	in m / sec															15
16	Mean predominent direction of the wind																16

49 Station / Country Cromer (England)/United Kingdom

Location 52°56'N/1°17'E   Height above sealevel 54 m   Climate symbol: Köppen Cfb   Troll III,2

		J	F	M	A	M	J	J	A	S	O	N	D	year	P	
1 Mean daily temperature	in °C	3,7	4,0	5,9	8,4	11,1	14,3	16,5	16,4	14,7	11,1	7,3	5,1	9,9	30	1
2 Mean daily maximum temperature	in °C	6,3	6,7	9,2	12,0	14,9	18,3	20,5	20,2	18,2	14,2	9,9	7,5	13,1	30	2
3 Mean daily minimum temperature	in °C	1,1	1,2	2,5	4,7	7,3	10,3	12,4	12,6	11,1	7,9	4,7	2,8	6,5	30	3
4 Absolute maximum temperature	in °C	13,9	18,3	23,3	26,7	28,3	32,2	34,4	33,3	32,2	26,1	18,9	14,4	33,3	30	4
5 Absolute minimum temperature	in °C	-7,8	-8,9	-6,7	-1,1	-1,1	3,9	6,7	7,2	6,1	0,0	-3,9	-6,1	-8,9	30	5
6 Mean relative humidity	in %															6
7 Mean precipitation	in mm	58	46	37	39	48	39	63	56	54	61	64	53	618	30	7
8 Maximum precipitation	in mm															8
9 Minimum precipitation	in mm															9
10 Maximum precipitation in 24 h	in mm	27	37	22	25	44	34	48	72	30	29	35	34	72	30	10
11 Mean number of days with precipitation	>0,25 mm	18	16	13	13	11	11	13	12	14	16	18	18	173	30	11
12 Mean duration of sunshine	in h	55	73	125	162	197	203	195	180	150	111	60	50	1561	30	12
13 Mean quantity of radiation	in ly / day															13
14 Mean potential evaporation	in mm	11	14	28	45	68	94	111	101	76	47	23	15	632	30	14
15 Mean windspeed	in m / sec															15
16 Mean predominent direction of the wind																16

50 Station / Country Birmingham (England)/United Kingdom

Location 52°29'N/1°56'W   Height above sealevel 136 m   Climate symbol: Köppen Cfb   Troll III,2

		J	F	M	A	M	J	J	A	S	O	N	D	year	P	
1 Mean daily temperature	in °C	4,0	3,8	5,9	8,5	11,5	14,6	16,3	16,1	13,8	10,2	6,8	4,7	9,7	30	1
2 Mean daily maximum temperature	in °C	5,3	6,0	9,1	12,2	15,6	18,8	20,2	20,0	17,2	13,0	8,8	6,4	12,7	30	2
3 Mean daily minimum temperature	in °C	1,6	1,5	2,7	4,7	7,3	10,4	12,3	12,1	10,3	7,3	4,7	2,9	6,5	30	3
4 Absolute maximum temperature	in °C	13,3	15,6	20,6	23,9	29,4	30,6	32,2	32,8	27,2	25,0	19,4	14,4	32,8	30	4
5 Absolute minimum temperature	in °C	-11,7	-8,9	-7,2	-1,7	-1,1	2,8	6,1	6,1	2,8	-2,2	-4,4	-6,1	-11,7	30	5
6 Mean relative humidity	in %	86	83	77	67	66	67	69	72	76	81	85	87	76	13	6
7 Mean precipitation	in mm	74	54	50	53	64	50	69	69	61	69	84	67	764	30	7
8 Maximum precipitation	in mm															8
9 Minimum precipitation	in mm															9
10 Maximum precipitation in 24 h	in mm	42	26	33	38	33	33	36	49	39	41	38	47	49	30	10
11 Mean number of days with precipitation	>0,25 mm	17	15	13	13	14	13	15	14	14	15	17	18	178	30	11
12 Mean duration of sunshine	in h	43	58	98	139	167	180	166	159	117	86	48	38	1299	30	12
13 Mean quantity of radiation	in ly / day															13
14 Mean potential evaporation	in mm	11	14	24	43	72	95	108	95	69	41	22	13	607	35	14
15 Mean windspeed	in m / sec															15
16 Mean predominent direction of the wind																16

51 Station / Country Aberystwyth (Wales)/United Kingdom

Location 52°25'N/4°03'W   Height above sealevel 138 m   Climate symbol: Köppen Cfb   Troll III,2

		J	F	M	A	M	J	J	A	S	O	N	D	year	P	
1 Mean daily temperature	in °C	4,5	4,3	6,2	8,1	10,9	13,5	14,9	15,2	13,5	10,7	7,6	5,8	9,6	30	1
2 Mean daily maximum temperature	in °C	6,9	6,8	9,3	11,3	14,5	16,8	17,8	18,1	16,4	13,4	10,0	8,1	12,5	30	2
3 Mean daily minimum temperature	in °C	2,1	1,8	3,1	4,8	7,3	10,2	12,0	12,2	10,6	7,9	5,2	3,5	6,7	30	3
4 Absolute maximum temperature	in °C	13,9	15,0	20,0	22,8	25,6	30,6	31,1	29,4	25,6	25,0	17,2	15,0	31,1	30	4
5 Absolute minimum temperature	in °C	-11,1	-8,9	-6,7	-2,8	-1,1	3,9	6,1	5,0	2,2	-2,2	-2,8	-5,6	-11,1	30	5
6 Mean relative humidity	in %															6
7 Mean precipitation	in mm	97	72	60	56	65	76	99	93	108	118	111	96	1051	30	7
8 Maximum precipitation	in mm															8
9 Minimum precipitation	in mm															9
10 Maximum precipitation in 24 h	in mm	37	33	41	30	34	41	52	37	41	38	38	36	52	30	10
11 Mean number of days with precipitation	>0,25 mm	21	17	16	16	16	16	19	18	19	20	20	22	220	30	11
12 Mean duration of sunshine	in h	55	75	125	158	197	195	156	164	128	98	57	47	1455	30	12
13 Mean quantity of radiation	in ly / day											.				13
14 Mean potential evaporation	in mm	17	18	31	48	70	92	104	98	73	49	29	19	648	35	14
15 Mean windspeed	in m / sec															15
16 Mean predominent direction of the wind																16

52 Station / Country Kew near London (England)/United Kingdom

Location 51°28'N/0°19'W   Height above sealevel 5 m   Climate symbol: Köppen Cfb.   Troll III,2

		J	F	M	A	M	J	J	A	S	O	N	D	year	P	
1 Mean daily temperature	in °C	4,3	5,1	6,7	9,4	12,5	16,0	17,7	17,3	14,9	11,1	7,7	5,4	10,7	30	1
2 Mean daily maximum temperature	in °C	6,3	6,9	10,1	13,3	16,7	20,3	21,8	21,4	18,5	14,2	10,1	7,3	13,9	30	2
3 Mean daily minimum temperature	in °C	2,2	2,2	3,3	5,5	8,2	11,6	13,5	13,2	11,3	7,9	5,3	3,5	7,3	30	3
4 Absolute maximum temperature	in °C	14,3	16,1	21,4	25,5	30,2	32,7	33,8	33,1	29,9	25,6	19,0	15,1	33,8	30	4
5 Absolute minimum temperature	in °C	-9,5	-9,4	-7,7	-2,1	-1,0	4,8	7,0	6,2	3,0	-3,6	-5,0	-7,0	-9,5	30	5
6 Mean relative humidity	in %	82	79	73	64	64	64	65	64	73	78	83	84	73	16	6
7 Mean precipitation	in mm	54	40	37	37	46	45	57	59	49	57	64	48	593	30	7
8 Maximum precipitation	in mm															8
9 Minimum precipitation	in mm															9
10 Maximum precipitation in 24 h	in mm	29	22	28	24	25	38	60	56	46	35	35	26	60	30	10
11 Mean number of days with precipitation	>0,25 mm	15	13	11	12	12	11	12	11	13	13	15	15	153	30	11
12 Mean duration of sunshine	in h	46	64	113	160	199	213	198	188	142	98	53	40	1514	30	12
13 Mean quantity of radiation	in ly / day															13
14 Mean potential evaporation	in mm	13	16	31	48	76	106	119	105	76	47	25	17	680	30	14
15 Mean windspeed	in m / sec															15
16 Mean predominent direction of the wind																16

53 Station / Country   Dover (England) / United Kingdom

Location   51°07'N/1°19'E   Height above sealevel   6 m   Climate symbol: Köppen   Cfb   Troll   III,2

		J	F	M	A	M	J	J	A	S	O	N	D	year	P	
1 Mean daily temperature	in °C	4,5	4,5	6,3	9,0	11,9	14,9	16,8	17,1	15,4	11,9	8,3	5,8	10,4	30	1
2 Mean daily maximum temperature	in °C	6,8	6,9	9,2	12,2	15,3	18,4	20,1	20,5	18,7	14,8	10,8	8,1	13,5	30	2
3 Mean daily minimum temperature	in °C	2,2	2,1	3,4	5,7	8,4	11,4	13,5	13,8	12,1	8,9	5,8	3,5	7,5	30	3
4 Absolute maximum temperature	in °C	13,3	15,8	20,0	24,4	27,8	28,9	30,6	31,1	26,7	23,9	17,8	15,0	31,1	30	4
5 Absolute minimum temperature	in °C	-9,4	-11,1	-7,8	-1,1	-0,5	4,4	4,4	5,0	0,0	-1,7	-4,4	-7,2	-11,1	30	5
6 Mean relative humidity	in %															6
7 Mean precipitation	in mm	79	58	48	43	46	41	62	60	62	88	94	72	751	30	7
8 Maximum precipitation	in mm															8
9 Minimum precipitation	in mm															9
10 Maximum precipitation in 24 h	in mm	38	36	28	25	36	27	42	46	46	86	52	44	86	30	10
11 Mean number of days with precipitation	>0,25 mm	17	14	12	12	11	10	11	12	12	14	15	17	157	30	11
12 Mean duration of sunshine	in h	60	77	136	189	217	236	218	203	162	119	82	50	1709	23	12
13 Mean quantity of radiation	in ly / day															13
14 Mean potential evaporation	in mm	16	16	31	48	76	102	115	105	83	58	30	19	698	30	14
15 Mean windspeed	in m / sec															15
16 Mean predominent direction of the wind																16

54 Station / Country   Plymouth (England) / United Kingdom

Location   50°21'N/4°07'W   Height above sealevel   27 m   Climate symbol: Köppen   Cfb   Troll   III,2

		J	F	M	A	M	J	J	A	S	O	N	D	year	P	
1 Mean daily temperature	in °C	6,2	5,8	7,4	9,2	11,8	14,5	16,0	16,2	14,7	11,8	9,0	7,2	10,8	30	1
2 Mean daily maximum temperature	in °C	8,2	8,1	10,1	12,3	15,1	17,7	19,0	19,3	17,7	14,8	11,3	9,2	13,6	30	2
3 Mean daily minimum temperature	in °C	4,1	3,5	4,6	6,1	8,4	11,3	13,0	13,0	11,7	9,2	6,8	5,1	8,0	30	3
4 Absolute maximum temperature	in °C	13,9	15,0	19,4	22,2	26,1	27,8	28,9	31,1	27,2	23,3	17,2	14,4	31,1	30	4
5 Absolute minimum temperature	in °C	-8,9	-8,3	-5,0	-1,7	-0,8	1,7	7,2	3,9	2,8	-1,7	-3,9	-5,0	-8,9	30	5
6 Mean relative humidity	in %	85	83	80	74	74	77	78	79	81	83	84	86	80	16	6
7 Mean precipitation	in mm	99	74	69	53	63	53	70	77	78	91	113	110	950	30	7
8 Maximum precipitation	in mm															8
9 Minimum precipitation	in mm															9
10 Maximum precipitation in 24 h	in mm	38	36	38	25	39	35	68	77	43	42	50	43	77	30	10
11 Mean number of days with precipitation	>0,25 mm	19	15	14	12	12	12	14	14	15	16	17	18	178	30	11
12 Mean duration of sunshine	in h	60	80	133	182	219	222	198	198	152	114	67	52	1677	30	12
13 Mean quantity of radiation	in ly / day															13
14 Mean potential evaporation	in mm	17	19	29	46	72	94	107	96	71	46	27	20	644	35	14
15 Mean windspeed	in m / sec															15
16 Mean predominent direction of the wind																16

55 Station / Country   St. Mary's (Scilly Islands) / United Kingdom

Location   49°56'N/6°18'W   Height above sealevel   50 m   Climate symbol: Köppen   Cfb   Troll   III,2

		J	F	M	A	M	J	J	A	S	O	N	D	year	P	
1 Mean daily temperature	in °C	7,8	7,4	8,5	9,8	11,8	14,4	16,1	16,5	15,2	12,7	10,3	8,8	11,6	30	1
2 Mean daily maximum temperature	in °C	9,2	9,0	10,5	12,1	14,3	17,0	18,7	19,2	17,5	14,8	11,8	10,0	13,7	30	2
3 Mean daily minimum temperature	in °C	6,3	5,7	6,5	7,4	9,2	11,7	13,4	12,9	10,8	8,7	7,2		9,4	30	3
4 Absolute maximum temperature	in °C	13,9	13,3	16,1	19,4	21,7	26,1	25,6	27,8	24,4	20,6	16,1	13,9	27,8	30	4
5 Absolute minimum temperature	in °C	-3,9	-3,3	-1,7	1,7	2,8	6,1	9,4	8,9	6,7	3,9	2,2	0,0	-3,9	30	5
6 Mean relative humidity	in %	83	82	84	81	82	82	83	82	84	84	83	83	83	16	6
7 Mean precipitation	in mm	91	71	69	46	56	49	61	64	67	80	96	94	844	30	7
8 Maximum precipitation	in mm															8
9 Minimum precipitation	in mm															9
10 Maximum precipitation in 24 h	in mm	33	61	30	50	27	48	34	43	33	51	41	43	61	30	10
11 Mean number of days with precipitation	>0,25 mm	22	17	16	13	14	14	16	15	16	17	19	21	200	30	11
12 Mean duration of sunshine	in h	62	81	130	192	235	228	207	208	155	121	76	57	1752	30	12
13 Mean quantity of radiation	in ly / day															13
14 Mean potential evaporation	in mm	22	23	37	48	68	90	103	94	73	52	32	25	667	30	14
15 Mean windspeed	in m / sec															15
16 Mean predominent direction of the wind																16

56 Station / Country   Malin Head (Ceann Malainn) / Ireland

Location   55°22'N/7°20'W   Height above sealevel   20 m   Climate symbol: Köppen   Cfb   Troll   III,2

		J	F	M	A	M	J	J	A	S	O	N	D	year	P	
1 Mean daily temperature	in °C	5,2	5,4	6,6	8,0	10,2	12,5	14,0	14,2	13,0	10,5	7,9	6,4	9,5	30	1
2 Mean daily maximum temperature	in °C															2
3 Mean daily minimum temperature	in °C															3
4 Absolute maximum temperature	in °C	13,9	13,8	17,3	19,8	23,3	27,2	26,7	27,1	24,4	21,1	18,9	16,8	27,2	30	4
5 Absolute minimum temperature	in °C	-6,1	-6,1	-4,4	-1,1	0,2	3,3	6,7	4,8	2,8	0,6	2,6	-3,9	-6,1	30	5
6 Mean relative humidity	in %	82	80	79	77	77	78	82	81	79	80	82	84	80	5	6
7 Mean precipitation	in mm	101	67	58	54	53	71	94	82	99	102	98	103	980	30	7
8 Maximum precipitation	in mm															8
9 Minimum precipitation	in mm															9
10 Maximum precipitation in 24 h	in mm	39	46	21	16	31	38	39	35	42	40	31	26	52	30	10
11 Mean number of days with precipitation	>1,0 mm	18	15	12	13	11	13	17	14	16	17	17	19	182	30	11
12 Mean duration of sunshine	in h	41	63	109	154	207	177	142	167	114	72	51	31	1328	30	12
13 Mean quantity of radiation	in ly / day															13
14 Mean potential evaporation	in mm	18	20	29	43	65	83	93	88	67	44	28	20	589	35	14
15 Mean windspeed	in m / sec															15
16 Mean predominent direction of the wind																16

57 Station / Country Belmullet (Béal an Mhuirthid) / Ireland

Location 54°14'N/10°00'W   Height above sealevel 9 m   Climate symbol: Köppen Cfb   Troll III2

		J	F	M	A	M	J	J	A	S	O	N	D	year	P	
1 Mean daily temperature	in °C	5.1	6.1	7.2	8.6	10.7	13.1	13.9	14.0	13.1	10.9	7.8	6.2	9.7	10	1
2 Mean daily maximum temperature	in °C															2
3 Mean daily minimum temperature	in °C															3
4 Absolute maximum temperature	in °C	12.8	12.8	19.4	19.1	23.3	26.6	22.2	21.9	23.2	21.7	14.8	12.9	26.6	10	4
5 Absolute minimum temperature	in °C	-6.4	-6.3	-5.5	-1.1	1.4	1.9	5.1	4.4	3.5	1.6	-1.0	-5.1	-6.4	10	5
6 Mean relative humidity	in %	84	81	80	78	76	79	81	82	82	82	83	83	81	10	6
7 Mean precipitation	in mm	108	64	82	70	75	80	76	95	108	116	127	131	1132	10	7
8 Maximum precipitation	in mm															8
9 Minimum precipitation	in mm															9
10 Maximum precipitation in 24 h	in mm	28	21	26	25	20	25	33	31	27	45	34	26	45	10	10
11 Mean number of days with precipitation	> 1,0 mm	18	13	16	15	14	12	14	17	16	18	20	22	195	10	11
12 Mean duration of sunshine	in h	58	71	105	155	216	181	142	158	118	91	58	40	1393	10	12
13 Mean quantity of radiation	in ly / day															13
14 Mean potential evaporation	in mm	10	21	34	48	72	86	95	86	67	50	27	21	624	10	14
15 Mean windspeed	in m / sec															15
16 Mean predominent direction of the wind																16

58 Station / Country Dublin (Baile Atha Cliath) / Ireland

Location 53°26'N/6°15'W   Height above sealevel 68 m   Climate symbol: Köppen Cfb   Troll III2

		J	F	M	A	M	J	J	A	S	O	N	D	year	P	
1 Mean daily temperature	in °C	4.5	4.8	6.5	8.4	10.5	13.5	15.0	14.8	13.1	10.5	7.2	5.8	9.8	21	1
2 Mean daily maximum temperature	in °C															2
3 Mean daily minimum temperature	in °C															3
4 Absolute maximum temperature	in °C	14.9	15.3	20.1	19.7	22.4	27.3	25.9	25.6	23.5	20.6	17.6	17.1	27.3	21	4
5 Absolute minimum temperature	in °C	-10.7	-10.9	-7.7	-2.2	-2.5	1.4	5.2	4.4	-0.2	-2.2	-3.6	-5.1	-10.9	21	5
6 Mean relative humidity	in %	84	81	78	73	73	73	75	78	79	81	83	84	79	13	6
7 Mean precipitation	in mm	71	52	51	43	62	55	66	80	77	68	67	77	769	21	7
8 Maximum precipitation	in mm															8
9 Minimum precipitation	in mm															9
10 Maximum precipitation in 24 h	in mm	31	33	37	28	28	45	43	54	52	50	33	40	54	21	10
11 Mean number of days with precipitation	> 1,0 mm	11	10	9	10	10	10	12	12	12	10	12	14	132	21	11
12 Mean duration of sunshine	in h	63	78	112	166	209	200	164	162	128	102	65	54	1503	20	12
13 Mean quantity of radiation	in ly / day															13
14 Mean potential evaporation	in mm	18	20	31	44	69	91	105	94	62	46	25	19	624	35	14
15 Mean windspeed	in m / sec															15
16 Mean predominent direction of the wind																16

59 Station / Country Kilkenny (Cill Choinningh) / Ireland

Location 52°40'N/7°16'W   Height above sealevel 67 m   Climate symbol: Köppen Cfb   Troll III2

		J	F	M	A	M	J	J	A	S	O	N	D	year	P	
1 Mean daily temperature	in °C	3.9	5.5	6.4	8.8	10.8	13.5	14.7	14.5	12.6	9.9	6.2	4.6	9.3	10	1
2 Mean daily maximum temperature	in °C	6.9	8.4	9.9	12.7	15.4	18.2	18.9	19.0	17.3	13.9	9.8	7.9	13.2	10	2
3 Mean daily minimum temperature	in °C	0.7	2.2	2.8	4.4	6.1	8.8	10.4	9.9	7.9	5.9	2.6	1.2	5.2	10	3
4 Absolute maximum temperature	in °C	12.8	14.5	18.5	19.6	24.8	26.3	26.6	25.6	26.4	23.7	17.2	13.6	26.6	10	4
5 Absolute minimum temperature	in °C	-11.6	-11.1	-6.7	-4.1	-3.7	0.5	2.3	1.2	-1.2	-3.5	-6.7	-10.8	-11.6	10	5
6 Mean relative humidity	in %	87	81	78	74	71	72	74	76	79	83	85	86	79	10	6
7 Mean precipitation	in mm	80	59	60	67	70	58	75	67	91	93	76	89	885	10	7
8 Maximum precipitation	in mm															8
9 Minimum precipitation	in mm															9
10 Maximum precipitation in 24 h	in mm	25	33	18	19	26	30	33	24	39	26	21	39	39	10	10
11 Mean number of days with precipitation	> 1,0 mm	13	12	12	12	13	11	13	12	12	12	12	14	146	10	11
12 Mean duration of sunshine	in h	60	70	104	144	194	176	144	156	120	91	72	58	1389	18	12
13 Mean quantity of radiation	in ly / day															13
14 Mean potential evaporation	in mm	16	19	29	45	71	93	104	91	67	41	23	16	615	35	14
15 Mean windspeed	in m / sec															15
16 Mean predominent direction of the wind																16

60 Station / Country Valentia (Dairbhre) / Ireland

Location 51°56'N/10°15'W   Height above sealevel 9 m   Climate symbol: Köppen Cfb   Troll III2

		J	F	M	A	M	J	J	A	S	O	N	D	year	P	
1 Mean daily temperature	in °C	7.0	6.8	8.3	9.4	11.5	13.8	15.0	15.4	14.0	11.6	9.1	7.8	10.8	30	1
2 Mean daily maximum temperature	in °C	9.4	9.3	11.1	12.5	14.8	16.8	17.7	18.2	16.8	14.2	11.5	10.1	13.5	30	2
3 Mean daily minimum temperature	in °C	4.5	4.3	5.4	6.3	8.1	10.8	12.3	12.6	11.2	8.9	6.6	5.5	8.0	30	3
4 Absolute maximum temperature	in °C	13.8	16.8	19.9	23.9	26.1	27.2	28.6	29.8	26.7	23.5	18.2	15.7	29.8	30	4
5 Absolute minimum temperature	in °C	-7.2	-5.4	-3.3	-1.8	0.2	1.7	6.2	4.4	1.7	-2.1	-2.3	-4.8	-7.2	30	5
6 Mean relative humidity	in %	82	80	77	74	74	79	81	81	81	82	82	83	80	13	6
7 Mean precipitation	in mm	165	107	103	75	86	81	107	95	122	146	151	168	1400	30	7
8 Maximum precipitation	in mm															8
9 Minimum precipitation	in mm															9
10 Maximum precipitation in 24 h	in mm	47	76	41	47	48	44	56	37	60	85	59	59	85	30	10
11 Mean number of days with precipitation	> 1,0 mm	20	15	14	13	13	13	15	15	16	17	18	21	190	30	11
12 Mean duration of sunshine	in h	49	70	109	155	202	177	145	151	115	88	59	41	1361	30	12
13 Mean quantity of radiation	in ly / day															13
14 Mean potential evaporation	in mm	23	23	36	46	68	86	96	88	67	48	29	23	633	35	14
15 Mean windspeed	in m / sec															15
16 Mean predominent direction of the wind																16

61 Station / Country Eelde/Netherlands

Location 53°08'N/6°35'E Height above sealevel 4,9 m Climate symbol: Köppen Cfb Troll III.3

		J	F	M	A	M	J	J	A	S	O	N	D	year	P	
1	Mean daily temperature	in °C	0,9	1,3	3,9	7,6	11,6	14,7	16,5	16,4	13,8	9,4	5,4	2,5	8,7		1
2	Mean daily maximum temperature	in °C	3,7	3,8	7,8	12,5	16,7	19,6	21,0	20,9	18,6	13,7	8,4	5,4	12,7	15	2
3	Mean daily minimum temperature	in °C	-1,0	-1,6	0,5	3,3	6,8	9,8	11,9	11,7	9,8	6,0	2,8	0,6	5,1	15	3
4	Absolute maximum temperature	in °C	12,2	16,0	20,2	24,4	29,7	33,3	33,2	31,5	31,8	24,5	15,5	13,4	33,3		4
5	Absolute minimum temperature	in °C	-18,7	-22,9	-15,6	-3,9	-3,3	1,5	3,7	3,6	0,8	-3,7	-7,6	-14,6	-22,9		5
6	Mean relative humidity	in %	88	85	81	72	70	71	74	76	79	83	89	91	80	15	6
7	Mean precipitation	in mm	67	49	41	48	53	56	91	87	73	74	73	65	776		7
8	Maximum precipitation	in mm															8
9	Minimum precipitation	in mm															9
10	Maximum precipitation in 24 h	in mm	60	41	25	32	26	38	35	76	54	55	42	38	76		10
11	Mean number of days with precipitation	> 0,1 mm	20	18	16	16	14	14	17	17	18	20	21	21	212		11
12	Mean duration of sunshine	in h	47	64	111	158	209	208	187	179	139	94	46	38	1479		12
13	Mean quantity of radiation	in ly/day															13
14	Mean potential evaporation	in mm	4	5	18	41	76	102	115	105	73	44	21	8	613	15	14
15	Mean windspeed	in m/sec	6,3	6,0	5,9	6,3	5,3	4,7	4,7	4,4	4,1	5,1	5,4	5,5	5,3		15
16	Mean predominant direction of the wind		SW	SW	E	SWW	NE	W	W	SW	SW	SW	SW	SW			16

62 Station / Country Den Helder/Netherlands

Location 52°58'N/4°45'E Height above sealevel 5,8 m Climate symbol: Köppen Cfb Troll III.2

		J	F	M	A	M	J	J	A	S	O	N	D	year	P	
1	Mean daily temperature	in °C	2,5	2,4	4,4	7,6	11,3	14,5	16,7	17,1	15,2	11,2	7,2	4,3	9,5		1
2	Mean daily maximum temperature	in °C	4,5	4,3	6,9	10,2	14,2	17,5	19,5	19,5	18,6	13,6	9,1	6,2	12,0	30	2
3	Mean daily minimum temperature	in °C	0,7	0,4	2,2	5,4	8,7	12,0	14,3	14,5	12,8	8,9	5,1	2,4	7,3	30	3
4	Absolute maximum temperature	in °C	12,4	14,9	20,0	24,3	28,5	30,8	33,9	32,8	32,6	24,8	17,6	13,0	33,9		4
5	Absolute minimum temperature	in °C	-14,4	-18,5	-14,5	-3,7	-2,1	2,0	6,8	6,4	3,0	-4,6	-10,6	-12,3	-18,5		5
6	Mean relative humidity	in %	86	84	82	79	79	76	77	77	77	80	85	87	81	30	6
7	Mean precipitation	in mm	65	46	39	40	38	37	64	71	75	90	83	65	714		7
8	Maximum precipitation	in mm															8
9	Minimum precipitation	in mm															9
10	Maximum precipitation in 24 h	in mm	84	27	36	52	30	57	49	80	55	67	52	37	84		10
11	Mean number of days with precipitation	> 0,1 mm	21	18	16	15	13	12	15	15	18	20	21	22	206		11
12	Mean duration of sunshine	in h	54	74	127	181	227	238	217	207	151	102	48	40	1665		12
13	Mean quantity of radiation	in ly/day															13
14	Mean potential evaporation	in mm	10	10	22	40	74	98	115	105	78	48	24	13	637	40	14
15	Mean windspeed	in m/sec	7,7	7,3	6,6	7,0	6,4	6,2	6,3	6,1	6,5	7,0	7,0	7,3	6,8		15
16	Mean predominant direction of the wind		S	SW	E	SW	NE	SW	SW	SW	SW	SSW	S	S			16

63 Station / Country De Bilt (near Utrecht)/Netherlands

Location 52°06'N/5°11'E Height above sealevel 2,9 m Climate symbol: Köppen Cfb Troll III.3

		J	F	M	A	M	J	J	A	S	O	N	D	year	P	
1	Mean daily temperature	in °C	1,7	2,0	5,0	8,5	12,4	15,5	17,0	16,8	14,3	10,0	5,9	3,9	9,3		1
2	Mean daily maximum temperature	in °C	4,3	5,3	9,5	13,4	17,8	20,9	22,1	21,9	19,2	14,1	8,9	5,4	13,6	30	2
3	Mean daily minimum temperature	in °C	-0,8	-0,8	1,2	4,3	7,6	10,8	12,7	12,5	10,2	6,6	3,3	0,6	5,7	30	3
4	Absolute maximum temperature	in °C	13,0	18,5	22,5	27,4	33,6	36,8	35,6	35,8	34,2	26,6	19,3	14,4	36,8		4
5	Absolute minimum temperature	in °C	-24,8	-21,6	-12,3	-6,0	-3,7	0,7	3,4	3,8	-0,8	-7,8	-14,4	-20,8	-24,8		5
6	Mean relative humidity	in %	86	83	76	70	67	67	72	74	77	81	87	88	77	30	6
7	Mean precipitation	in mm	68	52	45	49	52	58	77	88	71	72	70	63	765		7
8	Maximum precipitation	in mm															8
9	Minimum precipitation	in mm															9
10	Maximum precipitation in 24 h	in mm	46	54	30	33	48	51	66	74	44	56	39	42	74		10
11	Mean number of days with precipitation	> 0,1 mm	21	19	16	16	14	14	17	18	19	20	21	21	216		11
12	Mean duration of sunshine	in h	56	69	127	164	211	223	199	186	146	102	50	41	1572		12
13	Mean quantity of radiation	in ly/day	58	110	201	315	395	424	368	317	242	140	66	42	223		13
14	Mean potential evaporation	in mm	7	9	24	47	88	112	123	109	76	43	18	8	664	40	14
15	Mean windspeed	in m/sec	3,8	3,7	3,5	3,6	3,1	2,9	2,9	2,8	2,9	3,1	3,4	3,6	3,3		15
16	Mean predominant direction of the wind		SW	SW	NE	SW	NE	SW	SW	SW	SW	SW	SSW	SSW			16

64 Station / Country Bruxelles (Brussels) / Belgium

Location 50°48'N/4°21'E Height above sealevel 100 m Climate symbol: Köppen Cfb Troll III.2

		J	F	M	A	M	J	J	A	S	O	N	D	year	P	
1	Mean daily temperature	in °C	2,2	2,8	6,0	9,2	13,0	16,0	17,5	17,3	14,7	10,3	6,2	3,3	9,9		1
2	Mean daily maximum temperature	in °C	4,3	6,7	10,3	14,2	18,4	22,0	22,7	22,3	20,5	15,4	8,9	5,6	14,3	10	2
3	Mean daily minimum temperature	in °C	-1,2	0,3	2,2	5,1	7,9	10,9	12,1	12,2	10,6	7,3	3,1	0,2	5,9	10	3
4	Absolute maximum temperature	in °C	15,3	20,0	23,1	28,7	33,8	38,8	37,1	36,5	31,7	26,0	20,4	16,1	38,8		4
5	Absolute minimum temperature	in °C	-18,7	-16,7	-9,4	-3,6	-0,3	5,2	4,8	0,2	-6,4	-16,0	-18,7		-18,7		5
6	Mean relative humidity	in %	89	87	83	81	78	76	80	81	82	85	89	89	83	10	6
7	Mean precipitation	in mm	83	67	47	53	42	55	97	83	69	90	64	67	817		7
8	Maximum precipitation	in mm															8
9	Minimum precipitation	in mm															9
10	Maximum precipitation in 24 h	in mm	38	24	24	38	39	42	75	43	48	34	62	34	75		10
11	Mean number of days with precipitation	> 0,1 mm	23	17	15	16	15	14	17	18	14	19	19	19	206		11
12	Mean duration of sunshine	in h	54	77	119	158	219	196	198	198	160	110	60	36	1585		12
13	Mean quantity of radiation	in ly/day															13
14	Mean potential evaporation	in mm	9	11	27	45	84	101	115	102	73	45	18	10	840	30	14
15	Mean windspeed	in m/sec	4,3	4,1	4,0	3,8	3,6	3,4	3,4	3,4	3,4	3,6	3,9	4,2	3,8		15
16	Mean predominant direction of the wind		SW	SW	SW	SW	SW	SW	SW	SW	SW	SW	SW	SW			16

20

65 Station/Country Botrange-Robertville near Malmedy/Belgium

Location 50°30'N/6°06'E Height above sealevel 694 m Climate symbol: Köppen **Cfb** Troll III,3

		J	F	M	A	M	J	J	A	S	O	N	D	year	P	
1 Mean daily temperature	in °C	-2.1	-1.6	1.4	4.8	8.4	11.8	13.1	13.2	11.0	6.6	2.2	-0.8	5.7		1
2 Mean daily maximum temperature	in °C	-0.5	2.1	4.8	9.1	13.1	17.1	17.3	16.9	15.4	10.8	4.1	1.8	9.3	10	2
3 Mean daily minimum temperature	in °C	-4.4	-2.7	-1.7	1.6	4.7	8.1	9.3	9.2	7.5	4.3	0.0	-3.0	2.7	10	3
4 Absolute maximum temperature	in °C	15.9	16.4	19.8	26.2	29.2	33.0	31.4	33.3	33.0	22.7	16.4	14.8	33.3		4
5 Absolute minimum temperature	in °C	-23.6	-23.1	-14.4	-7.4	-6.4	-1.5	1.2	2.9	-0.5	-9.8	-10.0	-20.7	-23.6		5
6 Mean relative humidity	in %	90	91	88	87	83	81	86	88	88	90	93	91	88	10	6
7 Mean precipitation	in mm	145	141	85	93	114	116	161	150	133	128	96	148	1510		7
8 Maximum precipitation	in mm															8
9 Minimum precipitation	in mm															9
10 Maximum precipitation in 24 h	in mm	61	48	61	56	81	77	101	73	77	128	63	73	128		10
11 Mean number of days with precipitation	> 0,1 mm	21	16	16	18	18	16	19	20	15	19	17	18	213		11
12 Mean duration of sunshine	in h															12
13 Mean quantity of radiation	in ly/day															13
14 Mean potential evaporation	in mm	0	0	12	41	68	90	103	94	73	41	14	0	536	10	14
15 Mean windspeed	in m/sec	5.2	5.1	4.5	4.5	4.2	3.9	4.0	4.3	4.3	4.6	4.6	5.2	4.5		15
16 Mean predominant direction of the wind																16

66 Station/Country Luxembourg-City/Luxembourg

Location 49°37'N/6°03'E Height above sealevel 334 m Climate symbol: Köppen **Cfb** Troll III,3

		J	F	M	A	M	J	J	A	S	O	N	D	year	P	
1 Mean daily temperature	in °C	0.3	1.0	4.9	8.5	12.8	15.7	17.4	16.7	13.8	9.0	4.8	1.3	8.8		1
2 Mean daily maximum temperature	in °C	2.5	4.2	9.7	13.8	18.3	21.2	22.9	21.8	18.8	13.1	6.8	3.7	13.1	12	2
3 Mean daily minimum temperature	in °C	-1.4	-0.7	1.4	4.2	7.9	11.0	13.0	12.3	10.2	6.2	2.7	0.3	5.6	12	3
4 Absolute maximum temperature	in °C	11.7	17.2	22.5	29.4	30.2	36.8	36.8	33.8	33.0	23.5	17.0	14.0	36.8	15	4
5 Absolute minimum temperature	in °C	-16.0	-19.8	-10.5	-3.9	-1.6	3.7	5.3	4.3	1.0	-4.5	-7.2	-15.2	-19.8	15	5
6 Mean relative humidity	in %	89	85	76	72	73	75	75	77	80	86	90	93	81	12	6
7 Mean precipitation	in mm	73	56	43	54	60	64	68	74	63	55	64	68	740		7
8 Maximum precipitation	in mm															8
9 Minimum precipitation	in mm															9
10 Maximum precipitation in 24 h	in mm	28	23	23	28	67	33	50	51	44	28	42	42	67		10
11 Mean number of days with precipitation	> 0,1 mm	19	15	13	13	14	13	15	14	14	16	18	18	182		11
12 Mean duration of sunshine	in h	50	75	145	165	210	215	225	200	160	104	45	35	1630		12
13 Mean quantity of radiation	in ly/day															13
14 Mean potential evaporation	in mm	3	6	23	46	80	106	119	102	72	42	14	6	619	30	14
15 Mean windspeed	in m/sec															15
16 Mean predominant direction of the wind		SW	SW	E	E	E	S	SW	SW	S	E	S	SW		15	16

67 Station/Country Helgoland/Federal Republic of Germany

Location 54°11'N/7°54'E Height above sealevel 4 m Climate symbol: Köppen **Cfb** Troll III,3

		J	F	M	A	M	J	J	A	S	O	N	D	year	P	
1 Mean daily temperature	in °C	2.2	1.7	2.9	6.2	10.3	13.8	16.3	17.0	15.3	11.2	7.5	4.7	9.1	30	1
2 Mean daily maximum temperature	in °C	3.8	3.0	4.5	8.1	12.9	15.9	18.4	19.1	16.9	12.9	9.0	6.1	10.9	30	2
3 Mean daily minimum temperature	in °C	0.5	0.1	1.1	4.3	7.8	11.7	14.3	14.8	13.2	9.3	5.8	3.0	7.2	30	3
4 Absolute maximum temperature	in °C	9.7	11.5	17.1	24.0	25.8	30.2	31.8	28.7	28.9	21.3	15.5	11.0	31.8	55	4
5 Absolute minimum temperature	in °C	-15.3	-15.8	-10.8	-4.1	0.0	5.3	7.2	9.0	3.6	-1.7	-9.2	-10.0	-15.8	55	5
6 Mean relative humidity	in %	85	87	85	84	81	82	82	81	78	82	84	86	83	10	6
7 Mean precipitation	in mm	54	43	35	39	43	44	81	89	80	82	63	55	708	10	7
8 Maximum precipitation	in mm															8
9 Minimum precipitation	in mm															9
10 Maximum precipitation in 24 h	in mm															10
11 Mean number of days with precipitation	> 1,0 mm	11	10	10	9	8	8	9	13	11	13	13	14	129		11
12 Mean duration of sunshine	in h	50	67	129	198	244	246	229	196	165	101	47	36	1708	10	12
13 Mean quantity of radiation	in ly/day															13
14 Mean potential evaporation	in mm	7	8	15	35	69	94	109	101	77	50	25	13	601		14
15 Mean windspeed	in m/sec															15
16 Mean predominant direction of the wind		SW	SW	SW	E	NW	NW	W	W	SW	SW	SW	SW		36	16

68 Station/Country Hamburg/Federal Republic of Germany

Location 53°38'N/10°00'E Height above sealevel 14 m Climate symbol: Köppen **Cfb** Troll III,3

		J	F	M	A	M	J	J	A	S	O	N	D	year	P	
1 Mean daily temperature	in °C	0.0	0.3	3.3	7.5	12.0	15.3	17.0	16.8	13.5	9.1	4.9	1.8	8.4	30	1
2 Mean daily maximum temperature	in °C	2.3	3.0	7.3	12.5	17.4	20.5	22.2	22.0	18.7	12.9	7.3	3.9	12.5	30	2
3 Mean daily minimum temperature	in °C	-2.6	-2.5	-0.3	3.1	6.5	9.9	12.3	12.0	9.2	5.5	2.4	-0.5	4.6	30	3
4 Absolute maximum temperature	in °C	14.4	17.2	21.1	27.8	32.1	34.5	35.1	35.7	32.3	25.1	17.3	13.1	35.7	80	4
5 Absolute minimum temperature	in °C	-22.8	-29.1	-14.3	-7.1	-5.0	1.3	3.4	2.4	-1.2	-5.9	-13.5	-16.4	-29.1	80	5
6 Mean relative humidity	in %	87	85	78	73	69	70	74	76	78	83	88	89	79	30	6
7 Mean precipitation	in mm	57	47	38	52	55	64	82	84	61	59	57	58	714	30	7
8 Maximum precipitation	in mm	197	104	134	112	120	152	185	183	153	145	125	142	1069		8
9 Minimum precipitation	in mm	19	10	10	2	7	18	21	24	9	8	3	8	538		9
10 Maximum precipitation in 24 h	in mm	30	27	35	35	38	47	68	42	65	61	44	25	68	30	10
11 Mean number of days with precipitation	> 0,1 mm	18	16	13	14	14	14	17	16	15	17	18	18	190	30	11
12 Mean duration of sunshine	in h	51	84	131	188	230	222	220	183	171	100	44	28	1630	30	12
13 Mean quantity of radiation	in ly/day															13
14 Mean potential evaporation	in mm	1	4	18	44	82	106	119	103	73	40	16	6	612		14
15 Mean windspeed	in m/sec	5.0	4.3	4.6	4.1	4.0	3.9	4.0	3.8	3.8	4.0	4.3	5.0	4.2	30	15
16 Mean predominant direction of the wind		SW	SW	E	NW	W	W	W	SW	SW	SW	SE	SW		30	16

69 Station/Country Lüchow/Federal Republic of Germany

Location 52°58'N/11°10'E   Height above sealevel 21 m   Climate symbol: Köppen **Cfb**   Troll III,3

		J	F	M	A	M	J	J	A	S	O	N	D	year	P	
1 Mean daily temperature	in °C	-0.4	0.0	3.2	8.0	12.9	16.3	17.9	17.2	13.7	8.8	4.8	1.3	8.8	30	1
2 Mean daily maximum temperature	in °C	2.1	3.2	7.8	13.2	18.6	21.7	23.3	22.8	19.3	13.2	7.2	3.7	13.0	30	2
3 Mean daily minimum temperature	in °C	-3.2	-3.3	-0.8	2.8	6.5	9.8	12.1	11.8	8.4	4.9	1.8	-1.5	4.1	30	3
4 Absolute maximum temperature	in °C	11.8	16.8	24.0	30.1	29.7	33.1	36.5	35.9	29.9	26.5	21.1	14.0	36.5	17	4
5 Absolute minimum temperature	in °C	-22.7	-26.1	-19.0	-6.2	-3.5	1.9	4.4	2.1	-0.1	-5.4	-19.6	-19.0	-26.1	17	5
6 Mean relative humidity	in %	87	85	79	73	71	72	74	78	80	85	90	90	80	10	6
7 Mean precipitation	in mm	37	33	33	40	50	62	67	66	45	46	43	41	563	30	7
8 Maximum precipitation	in mm															8
9 Minimum precipitation	in mm															9
10 Maximum precipitation in 24 h	in mm															10
11 Mean number of days with precipitation	> 0,1 mm	15	13	11	13	12	12	13	14	12	14	14	14	157	30	11
12 Mean duration of sunshine	in h	53	72	139	195	240	237	226	193	182	107	52	36	1732	10	12
13 Mean quantity of radiation	in ly/day															13
14 Mean potential evaporation	in mm	0	2	16	43	84	107	119	103	70	38	14	3	599	40	14
15 Mean windspeed	in m/sec															15
16 Mean predominent direction of the wind																16

70 Station/Country Berlin–Dahlem/Federal Republic of Germany

Location 52°28'N/13°18'E   Height above sealevel 51 m   Climate symbol: Köppen **Cfb**   Troll III,3

		J	F	M	A	M	J	J	A	S	O	N	D	year	P	
1 Mean daily temperature	in °C	-0.6	-0.3	3.6	8.7	13.8	17.0	18.5	17.7	13.9	8.9	4.5	1.1	8.9	30	1
2 Mean daily maximum temperature	in °C	1.7	2.9	7.8	13.5	19.1	22.3	23.8	23.3	18.5	13.0	6.9	3.1	13.1	30	2
3 Mean daily minimum temperature	in °C	-3.5	-3.1	-0.3	3.8	9.1	11.1	13.3	12.8	9.3	5.3	1.9	-1.4	4.7	30	3
4 Absolute maximum temperature	in °C	13.0	16.7	25.1	30.9	33.2	35.0	37.8	36.6	34.2	26.5	19.5	15.4	37.8	62	4
5 Absolute minimum temperature	in °C	-21.0	-26.0	-16.5	-6.7	-2.9	1.4	5.7	4.7	-0.5	-9.8	-13.5	-20.2	-26.0	62	5
6 Mean relative humidity	in %	84	82	73	68	66	70	74	77	80	83	87	88	78	10	6
7 Mean precipitation	in mm	43	40	31	41	46	62	70	68	46	47	48	41	581	30	7
8 Maximum precipitation	in mm															8
9 Minimum precipitation	in mm															9
10 Maximum precipitation in 24 h	in mm	32	22	27	27	36	53	48	125	40	36	34	20	125	30	10
11 Mean number of days with precipitation	> 0,1 mm	17	15	12	13	12	12	14	14	12	14	16	15	166	30	11
12 Mean duration of sunshine	in h	56	78	151	193	239	244	242	212	194	123	50	36	1818	10	12
13 Mean quantity of radiation	in ly/day															13
14 Mean potential evaporation	in mm	0	0	16	45	88	110	125	105	72	39	13	2	615		14
15 Mean windspeed	in m/sec	3.5	3.5	3.8	3.3	3.1	3.0	2.9	2.8	2.8	2.9	3.3	3.3	3.2	20	15
16 Mean predominent direction of the wind		W,SW	W,SW	SE,W	NW,SE	NW,SE	NW,W	NW,W	W,NW	NW,SE	SE	SE	SE,W			16

71 Station/Country Hannover–Langenhagen/Federal Republic of Germany

Location 52°20'N/9°43'E   Height above sealevel 53 m   Climate symbol: Köppen **Cfb**   Troll III,3

		J	F	M	A	M	J	J	A	S	O	N	D	year	P	
1 Mean daily temperature	in °C	0.1	0.5	3.6	8.1	12.8	15.8	17.4	17.0	13.8	9.1	5.1	1.8	8.7	30	1
2 Mean daily maximum temperature	in °C	2.8	3.8	8.0	13.0	17.8	20.7	22.4	22.2	18.8	13.1	7.5	4.0	12.8	30	2
3 Mean daily minimum temperature	in °C	-2.5	-2.4	-0.1	3.2	6.9	10.0	12.2	11.9	9.2	5.4	2.3	-0.8	4.6	30	3
4 Absolute maximum temperature	in °C	14.3	18.8	23.8	29.7	34.7	34.7	36.4	38.0	35.6	27.2	20.6	15.4	38.0	90	4
5 Absolute minimum temperature	in °C	-25.0	-24.3	-17.3	-8.1	-2.9	0.5	3.3	3.3	-1.3	-7.8	-17.1	-20.9	-25.0	90	5
6 Mean relative humidity	in %	86	85	80	73	72	73	76	78	80	84	87	89	90	10	6
7 Mean precipitation	in mm	48	46	38	48	52	64	84	73	54	56	52	46	661	30	7
8 Maximum precipitation	in mm															8
9 Minimum precipitation	in mm															9
10 Maximum precipitation in 24 h	in mm	28	37	27	25	34	54	64	69	55	33	85	22	85	30	10
11 Mean number of days with precipitation	> 0,1 mm	17	15	12	15	13	13	15	15	13	14	16	15	173	30	11
12 Mean duration of sunshine	in h	48	69	120	184	227	214	206	188	165	105	52	34	1610	10	12
13 Mean quantity of radiation	in ly/day															13
14 Mean potential evaporation	in mm	2	6	20	45	85	107	121	104	72	41	17	7	627		14
15 Mean windspeed	in m/sec															15
16 Mean predominent direction of the wind		SW	SW	SW	W	W	W	W	W	SW	SW	SW	SW		45	16

72 Station/Country Essen/Federal Republic of Germany

Location 51°24'N/6°58'E   Height above sealevel 154 m   Climate symbol: Köppen **Cfb**   Troll III,3

		J	F	M	A	M	J	J	A	S	O	N	D	year	P	
1 Mean daily temperature	in °C	1.5	1.9	5.3	8.9	13.1	16.0	17.5	17.3	14.6	10.0	5.8	2.8	9.6	30	1
2 Mean daily maximum temperature	in °C	3.7	4.6	9.1	13.2	17.8	20.7	22.1	22.0	18.9	13.8	8.2	4.8	13.2	30	2
3 Mean daily minimum temperature	in °C	-1.1	-0.8	1.9	4.7	8.2	11.2	13.4	13.2	10.8	6.9	3.4	0.5	6.0	30	3
4 Absolute maximum temperature	in °C	13.8	19.4	23.2	28.9	34.3	32.8	35.1	35.0	33.0	27.2	19.8	16.7	35.1	65	4
5 Absolute minimum temperature	in °C	-24.0	-20.4	-12.5	-6.3	-2.8	1.0	4.4	3.9	0.3	-7.8	-11.0	-16.0	-24.0	65	5
6 Mean relative humidity	in %	85	83	76	70	69	72	76	79	78	82	84	80	78	10	6
7 Mean precipitation	in mm	73	63	47	61	63	75	86	90	66	67	72	66	829	30	7
8 Maximum precipitation	in mm															8
9 Minimum precipitation	in mm															9
10 Maximum precipitation in 24 h	in mm															10
11 Mean number of days with precipitation	> 0,1 mm	19	17	14	16	14	14	17	16	15	17	19	19	197	30	11
12 Mean duration of sunshine	in h	40	61	119	173	212	204	179	171	150	98	53	34	1494	10	12
13 Mean quantity of radiation	in ly/day															13
14 Mean potential evaporation	in mm	4	8	23	46	85	106	118	103	72	42	19	8	634	40	14
15 Mean windspeed	in m/sec															15
16 Mean predominent direction of the wind																16

22

73 Station / Country Kassel/Federal Republic of Germany

Location 51°19'N/9°29'E Height above sealevel 158 m Climate symbol: Köppen Cfb Troll III,3

		J	F	M	A	M	J	J	A	S	O	N	D	year	P	
1 Mean daily temperature	in °C	0,1	0,9	4,7	8,9	13,3	16,5	17,9	17,4	14,3	9,2	5,0	1,6	9,2	30	1
2 Mean daily maximum temperature	in °C	2,3	3,9	9,0	13,8	18,4	21,7	23,3	23,0	19,7	13,5	7,6	3,6	13,3	30	2
3 Mean daily minimum temperature	in °C	-2,4	-2,1	0,8	4,1	7,8	11,1	12,8	12,4	9,6	5,6	2,4	-0,7	5,1	30	3
4 Absolute maximum temperature	in °C	12,8	20,2	24,2	29,0	36,2	34,8	37,0	36,5	33,9	28,3	20,4	15,5	37,0	90	4
5 Absolute minimum temperature	in °C	-26,6	-23,5	-17,6	-6,3	-3,0	0,3	3,5	4,5	-1,3	-5,4	-14,6	-20,0	-26,6	90	5
6 Mean relative humidity	in %	84	81	75	70	69	70	73	77	78	82	84	86	77	10	6
7 Mean precipitation	in mm	47	42	33	47	60	64	70	66	52	53	49	46	629	30	7
8 Maximum precipitation	in mm															8
9 Minimum precipitation	in mm															9
10 Maximum precipitation in 24 h	in mm															10
11 Mean number of days with precipitation	> 0,1 mm	18	15	13	15	13	14	15	14	14	14	16	16	177	30	11
12 Mean duration of sunshine	in h	48	73	137	188	221	213	203	181	150	103	51	28	1596	10	12
13 Mean quantity of radiation	in ly/day															13
14 Mean potential evaporation	in mm	0	4	20	46	82	106	118	103	70	40	15	4	608		14
15 Mean windspeed	in m/sec															15
16 Mean predominent direction of the wind																16

74 Station / Country Bad Ems/Federal Republic of Germany

Location 50°20'N/7°43'E Height above sealevel 77 m Climate symbol: Köppen Cfb Troll III,3

		J	F	M	A	M	J	J	A	S	O	N	D	year	P	
1 Mean daily temperature	in °C	0,9	1,6	5,2	9,2	13,6	16,8	18,4	17,7	14,6	9,5	5,6	2,2	9,6	30	1
2 Mean daily maximum temperature	in °C	3,6	5,1	10,3	15,0	19,4	22,7	24,2	23,6	20,3	13,8	8,5	4,7	14,3	30	2
3 Mean daily minimum temperature	in °C	-2,1	-1,6	0,8	4,6	7,8	10,7	12,5	11,9	9,4	5,5	2,7	-0,4	5,1	30	3
4 Absolute maximum temperature	in °C	18,4	17,1	25,8	31,1	32,4	38,3	38,2	35,5	33,5	23,8	20,6	15,9	38,3	23	4
5 Absolute minimum temperature	in °C	-19,6	-24,0	-12,7	-5,7	-1,7	1,0	3,8	5,2	1,0	-3,5	-9,8	-16,7	-24,0	18	5
6 Mean relative humidity	in %	86	82	79	76	74	76	77	80	82	87	86	88	81	10	6
7 Mean precipitation	in mm	56	49	42	50	58	73	74	77	55	54	56	53	697	30	7
8 Maximum precipitation	in mm															8
9 Minimum precipitation	in mm															9
10 Maximum precipitation in 24 h	in mm															10
11 Mean number of days with precipitation	> 0,1 mm	16	15	13	15	14	13	15	15	13	14	15	15	173	30	11
12 Mean duration of sunshine	in h															12
13 Mean quantity of radiation	in ly/day															13
14 Mean potential evaporation	in mm	5	9	26	50	90	109	123	109	75	42	18	8	664		14
15 Mean windspeed	in m/sec															15
16 Mean predominent direction of the wind																16

75 Station / Country Hof/Federal Republic of Germany

Location 50°19'N/11°53'E Height above sealevel 567 m Climate symbol: Köppen Cfb Troll III,3

		J	F	M	A	M	J	J	A	S	O	N	D	year	P	
1 Mean daily temperature	in °C	-3,4	-2,5	1,3	5,7	10,5	13,9	15,6	14,9	11,7	6,7	1,9	-1,8	6,2	30	1
2 Mean daily maximum temperature	in °C	-0,9	0,6	5,9	10,9	15,9	19,3	21,0	20,6	17,2	11,1	4,5	0,5	10,6	30	2
3 Mean daily minimum temperature	in °C	-6,4	-5,9	-2,6	1,2	5,2	8,4	10,6	10,0	7,2	3,2	-0,4	4,3	2,2	30	3
4 Absolute maximum temperature	in °C	9,5	15,1	21,3	26,8	28,5	31,1	34,9	32,4	31,0	22,9	16,7	13,4	34,9	25	4
5 Absolute minimum temperature	in °C	-24,5	-27,0	-23,9	-11,6	-5,1	1,6	1,2	0,8	-4,0	-10,5	-14,7	-25,4	-27,0	25	5
6 Mean relative humidity	in %	87	86	80	75	74	75	77	78	81	84	89	90	81	10	6
7 Mean precipitation	in mm	52	47	41	48	61	74	85	69	52	52	46	50	677	10	7
8 Maximum precipitation	in mm															8
9 Minimum precipitation	in mm															9
10 Maximum precipitation in 24 h	in mm															10
11 Mean number of days with precipitation	> 0,1 mm	17	16	14	14	14	14	15	14	13	13	14	15	173	30	11
12 Mean duration of sunshine	in h	50	79	134	179	212	202	208	193	168	122	47	40	1634	10	12
13 Mean quantity of radiation	in ly/day															13
14 Mean potential evaporation	in mm	0	0	10	39	81	105	115	98	66	35	9	0	558	40	14
15 Mean windspeed	in m/sec															15
16 Mean predominent direction of the wind												~				16

76 Station / Country Frankfurt/Main (Airport) / Federal Republic of Germany

Location 50°07'N/8°39'E Height above sealevel 103 m Climate symbol: Köppen Cfb Troll III,3

		J	F	M	A	M	J	J	A	S	O	N	D	year	P	
1 Mean daily temperature	in °C	0,0	1,0	5,0	9,4	13,8	17,1	18,7	17,9	14,5	9,2	4,6	1,2	9,4	30	1
2 Mean daily maximum temperature	in °C	2,9	4,6	10,3	14,9	19,6	22,6	24,3	23,7	20,2	13,9	7,6	3,7	14,0	30	2
3 Mean daily minimum temperature	in °C	-3,0	-2,5	0,0	3,8	7,9	11,1	13,0	12,6	9,6	5,2	1,6	-1,6	4,8	30	3
4 Absolute maximum temperature	in °C	13,8	17,8	24,7	30,0	32,5	33,7	37,6	36,7	32,6	26,4	18,8	15,8	37,6	31	4
5 Absolute minimum temperature	in °C	-28,0	-20,4	-13,0	-6,8	-4,0	0,2	3,1	2,5	-1,9	-6,3	-11,5	-25,2	-28,0	31	5
6 Mean relative humidity	in %	84	82	74	67	66	69	70	74	78	83	85	88	77	10	6
7 Mean precipitation	in mm	57	44	36	43	54	72	68	77	56	50	53	53	663	30	7
8 Maximum precipitation	in mm	93	86	102	101	114	135	124	214	109	169	116	118	890	30	8
9 Minimum precipitation	in mm	12	1	3	0	4	13	1	16	2	4	14	11	359	30	9
10 Maximum precipitation in 24 h	in mm	34	20	36	27	55	55	48	46	36	28	41	41	55	29	10
11 Mean number of days with precipitation	> 0,1 mm	17	14	12	14	14	14	14	14	13	14	16	16	172	30	11
12 Mean duration of sunshine	in h	46	69	144	188	230	211	218	196	162	103	44	29	1640	10	12
13 Mean quantity of radiation	in ly/day															13
14 Mean potential evaporation	in mm	2	7	24	50	92	114	126	109	73	40	15	5	657		14
15 Mean windspeed	in m/sec															15
16 Mean predominent direction of the wind		SW	SW	SW	SW,NE	SW	SW	SW	SW	SW	SW	SW	SW		30	16

77 Station / Country Trier / Federal Republic of Germany

Location 49°45'N / 6°40'E — Height above sealevel 265 m — Climate symbol: Köppen Cfb — Troll III,3

		J	F	M	A	M	J	J	A	S	O	N	D	year	P	
1 Mean daily temperature	in °C	0,8	1,4	5,5	9,0	13,1	16,1	17,8	17,2	14,4	9,4	5,1	1,8	9,3	30	1
2 Mean daily maximum temperature	in °C	3,0	4,8	10,2	14,3	18,8	21,8	23,3	22,9	19,6	13,7	7,8	3,8	13,7	30	2
3 Mean daily minimum temperature	in °C	-2,0	-1,6	1,1	4,3	7,8	11,0	12,8	12,3	10,0	5,8	2,7	-0,7	5,3	30	3
4 Absolute maximum temperature	in °C	13,6	19,3	23,4	29,6	32,8	34,8	37,4	37,6	32,7	25,3	19,4	17,7	37,6	32	4
5 Absolute minimum temperature	in °C	-20,5	-19,3	-11,3	-4,0	-1,6	0,9	4,3	4,4	-0,5	-5,4	-8,5	-16,5	-20,5	32	5
6 Mean relative humidity	in %	86	82	74	68	69	74	73	77	78	83	86	90	78	10	6
7 Mean precipitation	in mm	60	51	37	51	62	74	70	80	58	54	60	62	719	30	7
8 Maximum precipitation	in mm															8
9 Minimum precipitation	in mm															9
10 Maximum precipitation in 24 h	in mm	22	32	28	22	58	35	38	33	50	31	34	36	58	20	10
11 Mean number of days with precipitation	> 0,1 mm	17	14	12	13	12	13	13	13	12	14	15	15	163	30	11
12 Mean duration of sunshine	in h	41	73	133	187	219	203	207	184	157	102	42	26	1574	10	12
13 Mean quantity of radiation	in ly/day															13
14 Mean potential evaporation	in mm	4	8	24	48	88	113	125	108	72	40	17	6	653		14
15 Mean windspeed	in m/sec	3,2	3,6	3,7	4,1	3,3	3,0	2,8	2,8	2,8	2,9	4,1	3,3	3,3	5	15
16 Mean predominent direction of the wind		SW	SW	SW	SW	SW	SW	SW	SW	SW	SW	SW	SW		20	16

78 Station / Country Nürnberg / Federal Republic of Germany

Location 49°30'N / 11°06'E — Height above sealevel 310 m — Climate symbol: Köppen Cfb — Troll III,3

		J	F	M	A	M	J	J	A	S	O	N	D	year	P	
1 Mean daily temperature	in °C	-1,4	-0,4	3,7	8,2	13,0	16,5	18,0	17,3	13,8	8,4	3,7	0,0	8,4	30	1
2 Mean daily maximum temperature	in °C	1,6	3,7	9,3	14,1	19,0	22,1	23,7	23,1	19,9	13,7	7,1	3,0	13,4	30	2
3 Mean daily minimum temperature	in °C	-4,8	-4,5	-1,2	2,6	6,6	10,3	12,2	11,5	8,2	3,9	0,5	-3,0	3,5	30	3
4 Absolute maximum temperature	in °C	14,3	18,7	23,7	30,1	31,2	34,3	37,7	34,7	31,0	27,3	20,4	15,1	37,7	29	4
5 Absolute minimum temperature	in °C	-27,6	-22,4	-18,3	-9,2	-4,4	0,0	3,1	0,6	-3,5	-6,2	-12,7	-23,0	-27,6	29	5
6 Mean relative humidity	in %	84	81	75	71	68	70	71	74	77	82	86	88	77	10	6
7 Mean precipitation	in mm	43	39	35	40	55	71	90	75	46	46	41	42	623	30	7
8 Maximum precipitation	in mm	109	80	89	118	119	152	148	135	117	138	108	140	916	30	8
9 Minimum precipitation	in mm	14	2	2	2	21	15	10	5	8	0	4	8	380	30	9
10 Maximum precipitation in 24 h	in mm	23	22	25	32	55	56	64	110	29	53	32	22	110	20	10
11 Mean number of days with precipitation	> 0,1 mm	17	15	13	14	14	16	16	14	13	13	15	16	176	30	11
12 Mean duration of sunshine	in h	55	81	153	189	231	221	229	214	175	125	54	39	1766	10	12
13 Mean quantity of radiation	in ly/day															13
14 Mean potential evaporation	in mm	0	2	19	46	88	111	123	107	71	38	13	2	620	40	14
15 Mean windspeed	in m/sec															15
16 Mean predominent direction of the wind		SE	SE	SW	NW	NW	NW	NW	SW	SW	SE	SE	SE		45	16

79 Station / Country Neustadt (Weinstraße) / Federal Republic of Germany

Location 49°22'N / 8°08'E — Height above sealevel 161 m — Climate symbol: Köppen Cfb — Troll III,3

		J	F	M	A	M	J	J	A	S	O	N	D	year	P	
1 Mean daily temperature	in °C	1,0	2,1	5,8	10,0	14,3	17,5	19,2	18,5	15,2	10,0	5,5	2,0	10,1	30	1
2 Mean daily maximum temperature	in °C	3,8	5,4	10,8	15,4	20,0	22,9	24,9	24,2	20,9	14,5	8,4	4,5	14,6	30	2
3 Mean daily minimum temperature	in °C	-2,0	-1,4	1,4	5,0	8,7	12,0	13,9	13,3	10,2	6,1	2,7	-0,6	5,8	30	3
4 Absolute maximum temperature	in °C	15,8	19,9	24,6	31,0	32,2	35,4	39,6	35,4	33,8	26,8	20,2	16,8	39,6	34	4
5 Absolute minimum temperature	in °C	-16,2	-21,7	-11,6	-5,0	-1,5	2,8	6,6	6,0	0,0	-5,2	-8,3	-15,0	-21,7	34	5
6 Mean relative humidity	in %	81	78	71	63	63	66	66	70	74	80	83	84	73	10	6
7 Mean precipitation	in mm	54	45	37	48	50	71	59	59	51	46	48	46	614	30	7
8 Maximum precipitation	in mm															8
9 Minimum precipitation	in mm															9
10 Maximum precipitation in 24 h	in mm	36	25	30	33	39	36	40	51	47	21	34	34	51	20	10
11 Mean number of days with precipitation	> 0,1 mm	16	13	12	14	13	14	14	12	12	14	15	14	163	30	11
12 Mean duration of sunshine	in h	48	70	151	189	227	203	231	215	183	117	8	32	1712	10	12
13 Mean quantity of radiation	in ly/day															13
14 Mean potential evaporation	in mm	3	6	22	49	90	114	127	109	74	40	15	5	654	40	14
15 Mean windspeed	in m/sec	1,9	2,6	2,7	3,0	2,6	2,8	2,5	2,4	2,3	2,5	2,7	2,8	2,6	4	15
16 Mean predominent direction of the wind																16

80 Station / Country Regensburg / Federal Republic of Germany

Location 49°01'N / 12°04'E — Height above sealevel 376 m — Climate symbol: Köppen Cfb — Troll III,3

		J	F	M	A	M	J	J	A	S	O	N	D	year	P	
1 Mean daily temperature	in °C	-2,7	-1,6	3,2	8,0	12,9	16,2	18,0	17,2	13,8	8,3	3,0	-0,9	8,0	30	1
2 Mean daily maximum temperature	in °C	-0,1	2,0	8,3	13,8	18,8	22,0	23,6	23,1	19,4	12,4	5,7	1,3	12,5	30	2
3 Mean daily minimum temperature	in °C	-5,4	-4,0	-1,0	2,9	6,8	10,6	12,4	11,8	8,7	4,4	0,2	-3,1	3,7	30	3
4 Absolute maximum temperature	in °C	14,9	16,0	22,7	29,2	31,8	36,0	36,8	35,2	32,6	29,0	19,6	17,2	36,8	87	4
5 Absolute minimum temperature	in °C	-28,8	-28,4	-17,3	-8,4	-4,0	0,8	3,6	2,6	-1,4	-6,7	-15,1	-24,1	-28,8	87	5
6 Mean relative humidity	in %	85	83	77	72	71	73	74	77	79	83	87	89	79	10	6
7 Mean precipitation	in mm	46	41	33	40	59	83	93	74	52	44	39	42	646	30	7
8 Maximum precipitation	in mm															8
9 Minimum precipitation	in mm															9
10 Maximum precipitation in 24 h	in mm															10
11 Mean number of days with precipitation	> 0,1 mm	17	15	12	14	14	15	15	15	13	12	15	16	173	30	11
12 Mean duration of sunshine	in h	54	71	150	181	225	212	231	209	171	115	41	32	1692	10	12
13 Mean quantity of radiation	in ly/day															13
14 Mean potential evaporation	in mm	0	0	17	44	87	110	122	105	69	35	9	0	598	40	14
15 Mean windspeed	in m/sec															15
16 Mean predominent direction of the wind		SE	SE	W	W	W	W	W	W	W	W	W	SE		43	16

Station / Country Stuttgart / Federal Republic of Germany

Location 48°42'N/9°12'E   **Height above sealevel** 401 m   **Climate symbol:** Köppen Cfb   **Troll** III,3

		in	J	F	M	A	M	J	J	A	S	O	N	D	year	P	
1	Mean daily temperature	°C	-0,8	0,4	4,5	8,5	12,7	15,8	17,6	17,0	13,9	8,8	3,9	0,3	8,5	30	1
2	Mean daily maximum temperature	°C	2,2	4,1	9,2	13,6	18,1	21,2	23,2	22,8	19,5	13,4	7,3	3,1	13,1	30	2
3	Mean daily minimum temperature	°C	-3,7	-2,9	0,1	3,8	7,5	10,7	12,4	12,0	9,3	4,7	1,1	-2,3	4,4	30	3
4	Absolute maximum temperature	°C	15,0	19,2	22,9	28,3	32,3	33,5	37,0	36,9	32,1	28,2	21,8	18,8	37,0	90	4
5	Absolute minimum temperature	°C	-25,7	-24,9	-18,2	-7,6	-5,1	2,1	3,6	4,0	-2,3	-7,4	-16,0	-25,8	-25,8	90	5
6	Mean relative humidity	%	83	81	73	68	69	74	73	76	78	83	86	87	78	10	6
7	Mean precipitation	mm	46	39	38	49	73	92	80	75	64	47	46	38	687	30	7
8	Maximum precipitation	mm															8
9	Minimum precipitation	mm															9
10	Maximum precipitation in 24 h	mm	51	36	24	40	58	69	44	54	40	32	37	26	69	29	10
11	Mean number of days with precipitation > 0,1 mm		16	13	12	14	14	15	15	14	14	12	15	13	167	30	11
12	Mean duration of sunshine	h	70	90	150	181	225	204	236	218	178	136	66	60	1814	10	12
13	Mean quantity of radiation	ly/day															13
14	Mean potential evaporation	mm	2	7	24	49	88	113	126	109	74	41	16	6	655	40	14
15	Mean windspeed	m/sec															15
16	Mean predominent direction of the wind		SW	SW	SW	SW	NW	NW	SW	SW	SW	SW	SW	SW		41	16

Station / Country München (Munich) / Federal Republic of Germany

Location 48°09'N/11°42'E   **Height above sealevel** 527 m   **Climate symbol:** Köppen Cfb   **Troll** III,3

		in	J	F	M	A	M	J	J	A	S	O	N	D	year	P	
1	Mean daily temperature	°C	-2,4	-1,2	3,0	7,6	12,2	15,4	17,2	16,6	13,3	7,6	2,9	-0,9	7,6	30	1
2	Mean daily maximum temperature	°C	1,1	2,9	8,3	13,2	17,6	20,8	22,9	22,4	19,2	13,1	6,5	2,3	12,5	30	2
3	Mean daily minimum temperature	°C	-5,7	-4,9	-1,3	2,9	6,8	10,3	12,1	11,8	8,9	3,9	0,1	-3,8	3,4	30	3
4	Absolute maximum temperature	°C	15,5	19,8	22,5	27,6	28,9	34,1	34,7	35,2	30,4	25,6	22,6	17,5	35,2	21	4
5	Absolute minimum temperature	°C	-29,6	-29,8	-18,0	-7,8	-2,8	0,5	3,4	3,9	-2,5	-6,2	-14,7	-23,2	-29,6	21	5
6	Mean relative humidity	%	83	83	77	72	73	73	73	75	78	82	86	86	79	10	6
7	Mean precipitation	mm	59	55	51	82	107	125	140	104	87	67	57	50	984	30	7
8	Maximum precipitation	mm															8
9	Minimum precipitation	mm															9
10	Maximum precipitation in 24 h	mm	52	40	37	38	155	51	81	47	73	40	53	25	155	29	10
11	Mean number of days with precipitation > 0,1 mm		16	15	13	14	15	16	16	15	13	12	14	14	173	30	11
12	Mean duration of sunshine	h	65	76	147	179	224	206	232	220	180	137	60	45	1771	10	12
13	Mean quantity of radiation	ly/day															13
14	Mean potential evaporation	mm	0	0	15	44	83	106	119	105	70	38	9	0	589	40	14
15	Mean windspeed	m/sec															15
16	Mean predominent direction of the wind		SW	SW	SW	SW	SW	SW	SW	SW	SW	SW	SW	SW		45	16

Station / Country Friedrichshafen / Federal Republic of Germany

Location 47°40'N/9°30'E   **Height above sealevel** 401 m   **Climate symbol:** Köppen Cfb   **Troll** III,3

		in	J	F	M	A	M	J	J	A	S	O	N	D	year	P	
1	Mean daily temperature	°C	-1,0	0,2	4,1	8,6	13,2	16,7	18,4	17,6	14,3	8,9	4,2	0,5	8,8	30	1
2	Mean daily maximum temperature	°C	1,7	3,7	9,1	13,8	18,5	21,8	23,6	22,9	19,5	13,2	7,2	2,9	13,2	30	2
3	Mean daily minimum temperature	°C	-4,6	-4,3	-1,2	2,3	6,1	9,1	11,0	10,7	8,4	4,3	0,2	-3,1	3,2	30	3
4	Absolute maximum temperature	°C	15,8	20,0	21,6	27,3	34,3	34,3	34,2	34,0	32,5	27,1	22,6	17,2	34,3	88	4
5	Absolute minimum temperature	°C	-20,8	-24,0	-15,4	-6,4	-2,5	1,2	5,0	3,7	-0,5	-4,4	-14,4	-18,5	-24,0	88	5
6	Mean relative humidity	%	84	83	77	73	70	73	74	76	79	85	86	87	79	10	6
7	Mean precipitation	mm	63	56	53	60	95	112	137	113	93	65	59	54	960	30	7
8	Maximum precipitation	mm															8
9	Minimum precipitation	mm															9
10	Maximum precipitation in 24 h	mm															10
11	Mean number of days with precipitation > 0,1 mm		16	15	13	14	15	16	16	15	14	14	14	15	177	30	11
12	Mean duration of sunshine	h	51	69	158	186	232	216	239	225	177	107	57	39	1756	10	12
13	Mean quantity of radiation	ly/day															13
14	Mean potential evaporation	mm	0	2	19	46	85	109	122	105	71	39	15	3	616	40	14
15	Mean windspeed	m/sec															15
16	Mean predominent direction of the wind		NE	NE	NE	NE	NE	SW	SW	NE	NE	NE	NE	NE		45	16

Station / Country Zugspitze / Federal Republic of Germany

Location 47°23'N/10°59'E   **Height above sealevel** 2960 m   **Climate symbol:** Köppen ET   **Troll** III,3

		in	J	F	M	A	M	J	J	A	S	O	N	D	year	P	
1	Mean daily temperature	°C	-11,6	-11,6	-9,5	-6,9	-2,5	0,5	2,5	2,4	0,6	-3,2	-7,0	-10,0	-4,7	30	1
2	Mean daily maximum temperature	°C	-9,2	-9,2	-7,0	-4,2	0,3	3,3	5,3	5,0	3,0	-0,9	-4,8	-7,7	-2,2	30	2
3	Mean daily minimum temperature	°C	-14,0	-13,9	-11,8	-9,2	-4,9	-1,8	0,4	0,3	1,5	-5,2	-9,1	-12,2	-6,9	30	3
4	Absolute maximum temperature	°C	1,6	5,8	4,7	8,4	14,7	16,0	17,9	17,2	10,8	5,5	3,8	1,8	17,9	71	4
5	Absolute minimum temperature	°C	-34,6	-35,6	-28,2	-23,8	-18,2	-12,5	-8,7	-9,9	-14,7	-18,2	-25,9	-31,1	-35,6	71	5
6	Mean relative humidity	%	74	73	75	80	85	89	87	86	81	72	70	72	79	10	6
7	Mean precipitation	mm	175	160	146	169	169	191	209	179	142	134	134	138	1946	30	7
8	Maximum precipitation	mm															8
9	Minimum precipitation	mm															9
10	Maximum precipitation in 24 h	mm	61	65	49	69	72	62	91	103	64	52	63	72	103	30	10
11	Mean number of days with precipitation > 0,1 mm		18	17	18	20	21	21	20	20	17	14	15	17	218	30	11
12	Mean duration of sunshine	h	116	120	164	163	170	136	167	171	178	176	136	118	1815	10	12
13	Mean quantity of radiation	ly/day															13
14	Mean potential evaporation	mm	0	0	0	0	0	0	77	67	0	0	0	0	144	40	14
15	Mean windspeed	m/sec															15
16	Mean predominent direction of the wind		NW	NW	NW	NW	NW	NW	NW	NW	NW	NW	NW	NW		25	16

85 Station/Country Greifswald/German Democratic Republic

Location 54°06'N/13°27'E — Height above sealevel 2 m — Climate symbol: Köppen Cfb — Troll III,3

		J	F	M	A	M	J	J	A	S	O	N	D	year	P	
1 Mean daily temperature	in °C	-1,0	-0,6	2,4	7,1	12,3	16,1	18,1	17,7	14,4	9,2	4,5	1,0	8,3	30	1
2 Mean daily maximum temperature	in °C	2,3	2,3	6,2	11,3	16,4	20,1	22,0	21,6	18,5	12,6	6,9	3,4	12,0	23	2
3 Mean daily minimum temperature	in °C	-2,3	-3,2	-0,6	3,2	7,5	10,9	13,2	13,0	10,0	6,0	2,2	-0,5	4,9	23	3
4 Absolute maximum temperature	in °C	11,8	15,2	21,4	25,9	30,3	33,7	33,8	32,5	30,3	22,8	15,8	13,5	33,8	30	4
5 Absolute minimum temperature	in °C	-22,8	-27,2	-18,5	-10,6	-3,2	2,5	5,0	5,2	1,2	-3,6	-8,1	-17,4	-27,2	30	5
6 Mean relative humidity	in %	87	85	82	77	74	74	77	79	80	83	89	90	81	20	6
7 Mean precipitation	in mm	40	33	30	39	45	55	69	55	59	51	36	41	553	30	7
8 Maximum precipitation	in mm															8
9 Minimum precipitation	in mm															9
10 Maximum precipitation in 24 h	in mm	18	22	25	27	34	39	68	52	43	31	23	20	68	30	10
11 Mean number of days with precipitation	> 0,1 mm	17	15	12	14	12	13	15	14	15	15	15	16	173	23	11
12 Mean duration of sunshine	in h	50	66	136	181	268	270	252	221	181	112	51	46	1834	30	12
13 Mean quantity of radiation	in ly/day															13
14 Mean potential evaporation	in mm	0	0	14	39	78	106	121	104	71	40	15	4	593	40	14
15 Mean windspeed	in m/sec	5,8	5,5	6,0	5,8	5,3	4,9	4,4	4,4	4,6	4,9	5,1	5,8	5,3	30	15
16 Mean predominent direction of the wind		SW	SW	W	W	NE	W	W	SW	SW	SW	SW	SW		30	16

86 Station/Country Brocken(Harz)/German Democratic Republic

Location 51°48'N/10°37'E — Height above sealevel 1142 m — Climate symbol: Köppen Dfb — Troll III,3

		J	F	M	A	M	J	J	A	S	O	N	D	year	P	
1 Mean daily temperature	in °C	-4,8	-4,7	-2,0	1,2	5,7	9,1	10,8	10,7	7,9	3,6	-0,3	-3,0	2,9	30	1
2 Mean daily maximum temperature	in °C	-1,9	-2,5	0,8	3,7	9,1	12,3	13,7	13,5	10,7	6,0	2,2	-0,5	5,6	18	2
3 Mean daily minimum temperature	in °C	-6,5	-7,1	-4,4	-1,7	2,9	6,2	8,1	8,0	5,4	2,1	-2,1	-4,7	0,5	18	3
4 Absolute maximum temperature	in °C	14,0	12,0	12,9	20,3	22,6	24,4	27,3	27,6	24,4	19,6	15,1	12,9	27,8	30	4
5 Absolute minimum temperature	in °C	-25,9	-28,4	-17,0	-11,3	-10,5	-2,8	-0,1	0,0	-4,5	-10,3	-15,2	-24,8	-28,4	30	5
6 Mean relative humidity	in %	90	89	83	85	81	82	86	87	87	90	90	92	87	18	6
7 Mean precipitation	in mm	158	126	94	105	96	115	143	117	105	122	115	126	1422	30	7
8 Maximum precipitation	in mm															8
9 Minimum precipitation	in mm															9
10 Maximum precipitation in 24 h	in mm	154	53	49	82	30	124	75	75	63	60	66	69	154	18	10
11 Mean number of days with precipitation	> 0,1 mm	24	21	19	19	18	17	20	20	19	22	21	23	243	18	11
12 Mean duration of sunshine	in h	56	63	112	122	170	181	160	156	128	83	45	48	1324	30	12
13 Mean quantity of radiation	in ly/day															13
14 Mean potential evaporation	in mm	0	0	0	10	63	84	97	85	59	29	0	0	427		14
15 Mean windspeed	in m/sec	14,2	14,2	11,3	11,3	8,8	8,8	9,0	9,0	10,4	12,0	12,9	13,3	11,3	30	15
16 Mean predominent direction of the wind		SW	W	W	W	SW	W	W	W	SW	SW	SW	SW		30	16

87 Station/Country Dresden/German Democratic Republic

Location 51°07'N/13°41'E — Height above sealevel 246 m — Climate symbol: Köppen Cfb — Troll III,3

		J	F	M	A	M	J	J	A	S	O	N	D	year	P	
1 Mean daily temperature	in °C	-1,2	-0,7	3,2	8,2	13,0	16,5	18,1	17,8	14,4	9,1	4,3	0,4	8,6	30	1
2 Mean daily maximum temperature	in °C	1,8	3,0	8,0	13,7	18,6	22,3	23,8	23,4	19,7	13,4	7,5	3,3	13,2	25	2
3 Mean daily minimum temperature	in °C	-3,6	-3,4	-0,2	3,8	7,6	11,3	13,4	12,9	9,8	5,3	2,1	-1,7	4,8	25	3
4 Absolute maximum temperature	in °C	14,1	16,6	20,8	28,7	31,7	34,0	36,6	36,8	33,0	25,6	19,1	13,2	36,8	30	4
5 Absolute minimum temperature	in °C	-23,8	-27,0	-14,7	-5,8	-3,3	5,3	6,8	5,3	1,0	-6,7	-9,4	-20,3	-27,0	30	5
6 Mean relative humidity	in %	81	79	75	70	69	69	72	72	79	82	82	83	75	25	6
7 Mean precipitation	in mm	38	36	37	46	63	68	109	72	48	52	42	37	648	30	7
8 Maximum precipitation	in mm															8
9 Minimum precipitation	in mm															9
10 Maximum precipitation in 24 h	in mm	26	47	30	41	70	48	114	62	37	33	37	31	114	30	10
11 Mean number of days with precipitation	> 0,1 mm	16	15	14	14	15	13	15	13	13	13	14	14	169	25	11
12 Mean duration of sunshine	in h	61	71	129	166	225	230	227	214	167	117	60	54	1721	30	12
13 Mean quantity of radiation	in ly/day	68	121	218	323	421	441	408	365	271	168	74	48	244	30	13
14 Mean potential evaporation	in mm	1	2	16	41	85	109	122	106	70	37	13	3	605		14
15 Mean windspeed	in m/sec	5,8	5,5	5,1	5,1	4,2	3,8	4,2	4,0	4,4	4,9	5,1	5,3	4,9	30	15
16 Mean predominent direction of the wind		W,SE	W	W	W	W	W	W	W	W	W,SE	W	W,SE		30	16

88 Station/Country Erfurt/German Democratic Republic

Location 50°59'N/10°58'E — Height above sealevel 315 m — Climate symbol: Köppen Cfb — Troll III,3

		J	F	M	A	M	J	J	A	S	O	N	D	year	P	
1 Mean daily temperature	in °C	-1,8	-0,8	2,8	7,5	12,1	15,5	17,3	16,5	13,1	8,0	3,8	-0,1	7,8	30	1
2 Mean daily maximum temperature	in °C	2,4	2,8	6,1	13,2	18,2	21,4	23,0	22,7	19,4	13,1	7,1	3,1	12,9	23	2
3 Mean daily minimum temperature	in °C	-3,0	-3,7	-0,7	3,3	7,5	10,9	12,7	12,3	9,4	5,1	1,7	-2,1	4,5	23	3
4 Absolute maximum temperature	in °C	13,5	17,6	21,5	30,2	31,5	33,1	36,5	37,0	33,1	28,6	20,1	17,4	37,0	30	4
5 Absolute minimum temperature	in °C	-24,4	-22,7	-19,4	-8,1	-3,8	2,0	5,2	5,0	-0,3	-8,0	-9,3	-23,8	-24,4	30	5
6 Mean relative humidity	in %	84	83	77	69	68	68	70	70	72	79	85	87	76	30	6
7 Mean precipitation	in mm	33	31	28	34	58	67	71	55	46	45	34	30	532	30	7
8 Maximum precipitation	in mm															8
9 Minimum precipitation	in mm															9
10 Maximum precipitation in 24 h	in mm	18	29	26	29	39	49	47	34	59	35	21	18	59	30	10
11 Mean number of days with precipitation	> 0,1 mm	17	15	13	14	14	13	14	13	13	14	15	15	170	30	11
12 Mean duration of sunshine	in h	50	70	123	157	200	215	211	198	158	105	52	42	1581	30	12
13 Mean quantity of radiation	in ly/day															13
14 Mean potential evaporation	in mm	0	0	18	44	85	107	119	104	70	39	13	2	601		14
15 Mean windspeed	in m/sec	3,6	3,2	2,8	3,0	2,4	2,6	2,6	2,4	2,6	2,8	2,8	3,4	2,8	30	15
16 Mean predominent direction of the wind		SW	SW	SW	SW	NE	W	SW	SW	SW	SW	SW			30	16

89 Station / Country Gdynia / Poland

Location 54°31'N / 18°33'E   Height above sealevel 15 m   Climate symbol: Köppen Cfb   Troll III,3

#	Parameter	unit	J	F	M	A	M	J	J	A	S	O	N	D	year	P
1	Mean daily temperature	in °C	-1,3	-0,9	1,5	6,1	11,0	15,3	17,9	17,6	14,0	9,1	4,3	1,1	7,9	30
2	Mean daily maximum temperature	in °C	1,4	1,4	4,2	8,7	15,1	18,8	21,3	21,0	17,7	12,5	6,6	3,4	11,0	14
3	Mean daily minimum temperature	in °C	-3,0	-3,8	-1,3	2,2	7,2	11,1	14,3	13,9	10,6	6,7	2,2	-0,7	4,9	14
4	Absolute maximum temperature	in °C	10,8	13,6	20,5	25,8	29,9	31,8	36,0	32,5	30,7	22,4	14,7	12,2	36,0	10
5	Absolute minimum temperature	in °C	-25,2	-28,8	-12,9	-4,3	-0,5	3,0	8,2	8,2	1,7	-6,5	-8,2	-14,8	-28,8	10
6	Mean relative humidity	in %	85	84	81	76	72	72	73	75	77	81	85	87	79	30
7	Mean precipitation	in mm	34	30	30	34	44	59	77	79	55	54	39	42	576	30
8	Maximum precipitation	in mm														
9	Minimum precipitation	in mm														
10	Maximum precipitation in 24 h	in mm														
11	Mean number of days with precipitation	> 0,1 mm	15	14	13	13	11	11	13	13	13	14	13	15	158	30
12	Mean duration of sunshine	in h	38	70	134	163	224	259	236	225	174	105	45	32	1705	10
13	Mean quantity of radiation	in ly / day														
14	Mean potential evaporation	in mm	0	0	5	29	77	99	123	108	72	44	19	4	580	6
15	Mean windspeed	in m / sec	4,2	3,8	4,0	3,9	3,6	3,0	3,2	3,0	3,3	3,4	3,4	3,9	3,6	10
16	Mean predominant direction of the wind		S	S	S	N	N	N	N	W	W	S	S	S		10

90 Station / Country Suwalki / Poland

Location 54°06'N / 22°55'E   Height above sealevel 170 m   Climate symbol: Köppen Dfb   Troll III,4

#	Parameter	unit	J	F	M	A	M	J	J	A	S	O	N	D	year	P
1	Mean daily temperature	in °C	-5,6	-4,8	-1,3	5,7	12,2	15,8	17,7	16,8	12,3	6,5	1,5	-2,4	6,2	30
2	Mean daily maximum temperature	in °C	-2,6	-2,4	2,6	9,5	17,3	21,6	22,0	21,5	17,2	11,2	3,6	0,1	10,1	9
3	Mean daily minimum temperature	in °C	-8,3	-8,7	-4,9	0,9	6,3	10,5	12,3	11,9	7,7	4,0	-0,7	-4,2	2,2	9
4	Absolute maximum temperature	in °C	7,0	6,5	16,2	25,4	30,8	31,2	35,3	32,9	29,8	22,1	14,5	10,3	35,3	10
5	Absolute minimum temperature	in °C	-32,0	-32,0	-21,9	-8,3	-4,2	1,4	6,2	3,9	-0,5	-14,2	-20,7	-22,9	-32,0	10
6	Mean relative humidity	in %	89	86	80	75	67	69	74	76	80	86	90	91	80	30
7	Mean precipitation	in mm	29	28	30	37	42	77	93	81	56	47	38	36	594	30
8	Maximum precipitation	in mm														
9	Minimum precipitation	in mm														
10	Maximum precipitation in 24 h	in mm	11	17	18	23	20	34	50	86	37	64	30	16	86	12
11	Mean number of days with precipitation	> 0,1 mm	17	14	13	13	11	13	16	14	12	13	14	16	166	30
12	Mean duration of sunshine	in h	28	55	147	162	227	254	234	217	158	86	31	22	1621	10
13	Mean quantity of radiation	in ly / day														
14	Mean potential evaporation	in mm	0	0	0	32	82	113	127	105	68	32	0	0	552	49
15	Mean windspeed	in m / sec	4,9	4,3	4,3	4,1	4,0	3,5	3,5	3,4	3,5	4,0	4,5	4,6	4,1	10
16	Mean predominant direction of the wind		SW	SW	SE	SE	NW	NW	W	W	W	SW	SE	SW		10

91 Station / Country Poznań / Poland

Location 52°25'N / 16°50'E   Height above sealevel 92 m   Climate symbol: Köppen Cfb   Troll III,3

#	Parameter	unit	J	F	M	A	M	J	J	A	S	O	N	D	year	P
1	Mean daily temperature	in °C	-2,2	-1,6	2,3	7,9	13,6	17,0	18,8	17,9	14,0	8,5	3,7	0,2	8,3	10
2	Mean daily maximum temperature	in °C	0,5	1,1	6,5	12,0	19,5	22,8	24,2	23,0	19,4	13,2	6,0	2,9	12,6	14
3	Mean daily minimum temperature	in °C	-4,5	-5,3	-1,3	2,5	8,1	11,1	13,8	12,7	9,0	4,9	0,6	-2,0	4,1	14
4	Absolute maximum temperature	in °C	11,2	13,7	18,8	28,5	31,8	34,4	38,2	38,2	24,8	24,8	12,5		38,2	10
5	Absolute minimum temperature	in °C	-22,7	-28,0	-14,8	-6,2	-3,0	0,6	3,8	5,2	-1,6	-6,0	-10,1	-14,7	-28,0	10
6	Mean relative humidity	in %	87	85	81	73	68	67	69	73	77	83	88	89	78	30
7	Mean precipitation	in mm	30	29	28	28	59	60	73	64	41	40	34	33	528	30
8	Maximum precipitation	in mm														
9	Minimum precipitation	in mm														
10	Maximum precipitation in 24 h	in mm	11	12	15	25	46	47	54	51	30	35	17	13	54	14
11	Mean number of days with precipitation	> 0,1 mm	16	15	12	13	12	12	12	15	12	14	15	16	165	30
12	Mean duration of sunshine	in h	51	64	142	168	255	241	221	218	184	112	55	40	1751	7
13	Mean quantity of radiation	in ly / day														
14	Mean potential evaporation	in mm	0	0	9	43	85	117	130	112	74	39	9	0	618	49
15	Mean windspeed	in m / sec	4,6	4,2	4,5	4,0	4,0	3,6	3,8	3,4	3,5	3,5	3,7	4,2	3,9	30
16	Mean predominant direction of the wind		SW	W	W	W	W	W	W	W	W	SW	SW	SW		30

92 Station / Country Warszawa (Warsaw) / Poland

Location 52°09'N / 20°59'E   Height above sealevel 107 m   Climate symbol: Köppen Cfb   Troll III,4

#	Parameter	unit	J	F	M	A	M	J	J	A	S	O	N	D	year	P
1	Mean daily temperature	in °C	-3,5	-2,5	1,4	8,0	14,0	17,5	19,2	18,2	13,9	8,1	3,0	-0,6	8,1	30
2	Mean daily maximum temperature	in °C	-0,4	0,3	5,6	11,8	19,6	22,6	24,1	22,9	19,0	12,9	5,6	1,9	12,2	14
3	Mean daily minimum temperature	in °C	-5,5	-6,0	-2,0	2,9	9,0	12,3	14,6	13,5	9,5	5,1	0,7	-2,5	4,3	14
4	Absolute maximum temperature	in °C	10,7	12,0	18,3	27,0	30,8	32,1	35,1	35,1	31,4	25,2	16,7	11,9	35,1	10
5	Absolute minimum temperature	in °C	-27,1	-26,1	-19,0	-6,9	-3,0	2,3	5,2	5,4	-0,3	-8,0	-10,0	-18,9	-27,1	10
6	Mean relative humidity	in %	86	85	77	73	68	69	74	74	77	82	86	88	78	10
7	Mean precipitation	in mm	23	26	24	36	44	62	79	65	41	35	37	30	502	30
8	Maximum precipitation	in mm														
9	Minimum precipitation	in mm														
10	Maximum precipitation in 24 h	in mm	8	13	12	25	22	49	40	27	22	50	29	17	50	30
11	Mean number of days with precipitation	> 0,1 mm	14	14	9	11	11	12	14	12	12	10	13	15	147	30
12	Mean duration of sunshine	in h	21	51	131	138	192	198	203	208	171	106	36	10	1465	10
13	Mean quantity of radiation	in ly / day														
14	Mean potential evaporation	in mm	0	0	5	43	88	121	130	112	72	36	6	0	613	49
15	Mean windspeed	in m / sec	5,2	4,7	4,9	4,2	4,0	3,4	3,3	3,1	3,5	3,6	4,3	4,7	4,1	10
16	Mean predominant direction of the wind		W	W	E	W	W	W	W	W	W	W	SE	W		10

93 Station / Country Wrocław (Breslau) / Poland

Location 51°08'N / 16°59'E   Height above sealevel 119 m   Climate symbol: Köppen **Cfb**   Troll III,3

		J	F	M	A	M	J	J	A	S	O	N	D	year	P	
1 Mean daily temperature	in °C	-2,0	-1,1	2,8	8,3	13,6	17,0	18,8	17,9	14,1	8,7	4,1	0,3	8,5	30	1
2 Mean daily maximum temperature	in °C	0,5	1,9	7,6	12,5	18,5	22,3	23,6	23,3	19,8	13,9	7,3	3,5	12,9	17	2
3 Mean daily minimum temperature	in °C	-5,9	-5,7	-1,1	2,6	7,0	10,9	13,1	12,2	8,8	4,8	0,7	-2,7	3,7	17	3
4 Absolute maximum temperature	in °C	13,0	16,0	22,1	27,6	30,0	32,8	36,6	35,6	31,6	25,7	18,4	15,4	36,6	10	4
5 Absolute minimum temperature	in °C	-23,3	-32,0	-19,0	-6,8	-2,4	1,4	4,3	4,6	-1,8	-6,2	-10,3	-17,6	-32,0	10	5
6 Mean relative humidity	in %	83	80	75	69	66	66	67	69	73	78	83	84	74	30	6
7 Mean precipitation	in mm	31	30	31	39	60	72	81	73	48	42	34	33	574	30	7
8 Maximum precipitation	in mm															8
9 Minimum precipitation	in mm															9
10 Maximum precipitation in 24 h	in mm															10
11 Mean number of days with precipitation	> 0,1 mm	16	14	14	14	14	13	15	13	12	13	14	17	169	30	11
12 Mean duration of sunshine	in h	48	67	124	146	200	204	202	199	156	112	48	38	1644	10	12
13 Mean quantity of radiation	in ly / day															13
14 Mean potential evaporation	in mm	0	0	15	45	90	114	130	110	74	40	12	2	632	50	14
15 Mean windspeed	in m / sec	4,0	3,8	3,7	3,2	3,1	2,9	3,1	2,7	3,0	2,8	3,3	3,8	3,2	10	15
16 Mean predominent direction of the wind		W	W	W	NW	NW	NW	NW	W	NW	SE	SE	SE		10	16

94 Station / Country Zamość / Poland

Location 50°44'N / 23°15'E   Height above sealevel 219 m   Climate symbol: Köppen **Dfb**   Troll III,4

		J	F	M	A	M	J	J	A	S	O	N	D	year	P	
1 Mean daily temperature	in °C	-4,4	-3,3	0,7	7,6	13,5	16,9	18,6	17,6	13,3	7,6	2,6	-1,3	7,4	30	1
2 Mean daily maximum temperature	in °C															2
3 Mean daily minimum temperature	in °C															3
4 Absolute maximum temperature	in °C	11,6	14,3	19,0	26,7	31,6	32,2	35,7	35,7	31,2	25,4	17,5	14,1	35,7	10	4
5 Absolute minimum temperature	in °C	-27,2	-27,6	-21,6	-5,6	-1,4	0,3	5,1	5,0	-1,1	-6,4	-13,9	-16,2	-27,6	10	5
6 Mean relative humidity	in %															6
7 Mean precipitation	in mm	29	32	31	39	63	76	93	94	55	58	42	32	644	30	7
8 Maximum precipitation	in mm															8
9 Minimum precipitation	in mm															9
10 Maximum precipitation in 24 h	in mm															10
11 Mean number of days with precipitation	> 1,0 mm	9	8	8	7	9	9	10	9	8	8	8	10	103		11
12 Mean duration of sunshine	in h	42	60	128	157	201	232	237	213	168	120	37	36	1631	10	12
13 Mean quantity of radiation	in ly / day															13
14 Mean potential evaporation	in mm	0	0	3	41	92	118	127	113	70	33	9	0	607	30	14
15 Mean windspeed	in m / sec	3,9	3,4	3,4	2,8	2,7	2,4	2,3	2,3	2,7	2,8	3,1	3,6	3,0		15
16 Mean predominent direction of the wind		SW	W	E	NW	NW	NW	W	W	W	SW	SE	SW		10	16

95 Station / Country Kraków / Poland

Location 50°05'N / 20°01'E   Height above sealevel 213 m   Climate symbol: Köppen **Cfb**   Troll III,3

		J	F	M	A	M	J	J	A	S	O	N	D	year	P	
1 Mean daily temperature	in °C	-2,9	-1,4	2,6	8,6	14,1	17,5	19,3	18,4	14,4	8,8	3,8	-0,2	8,6	10	1
2 Mean daily maximum temperature	in °C	0,2	1,2	7,0	12,5	19,6	22,4	24,3	23,0	19,1	13,6	6,4	2,9	12,7	14	2
3 Mean daily minimum temperature	in °C	-5,4	-5,3	-0,9	3,3	9,0	12,3	14,5	13,5	9,7	5,3	0,6	-2,4	4,5	14	3
4 Absolute maximum temperature	in °C	11,3	15,8	21,2	27,1	32,8	32,5	35,2	35,7	30,7	26,4	20,9	16,6	35,7	10	4
5 Absolute minimum temperature	in °C	-22,8	-26,6	-18,1	-7,1	-2,2	3,0	6,2	4,2	-1,5	-5,0	-13,0	-17,1	-26,6	10	5
6 Mean relative humidity	in %															6
7 Mean precipitation	in mm	34	34	35	42	57	86	95	83	56	46	42	34	645	30	7
8 Maximum precipitation	in mm															8
9 Minimum precipitation	in mm															9
10 Maximum precipitation in 24 h	in mm	17	16	21	21	38	63	51	65	43	34	25	19	65	14	10
11 Mean number of days with precipitation	> 0,1 mm	17	15	13	14	13	15	15	14	12	13	16	15	172	30	11
12 Mean duration of sunshine	in h	43	55	113	147	189	199	220	210	151	111	48	36	1522	10	12
13 Mean quantity of radiation	in ly / day															13
14 Mean potential evaporation	in mm	0	0	9	44	88	116	129	111	73	40	8	0	618	49	14
15 Mean windspeed	in m / sec	3,2	3,2	3,5	2,7	2,8	2,4	2,4	2,2	2,5	2,3	2,7	3,0	2,7	10	15
16 Mean predominent direction of the wind		W	W	W	W	W	W	W	W	W	W	W	W		10	16

96 Station / Country Praha / Czechoslovakia

Location 50°05'N / 14°25'E   Height above sealevel 197 m   Climate symbol: Köppen **Cfb**   Troll III,3

		J	F	M	A	M	J	J	A	S	O	N	D	year	P	
1 Mean daily temperature	in °C	-2,6	-1,6	2,7	7,8	12,9	16,2	17,9	17,4	13,9	8,2	3,1	-0,8	7,9	23	1
2 Mean daily maximum temperature	in °C	9,5	11,4	17,5	22,5	27,9	30,9	32,7	31,8	28,7	21,7	13,8	10,2	21,6	23	2
3 Mean daily minimum temperature	in °C	-13,0	-12,3	-8,0	-1,7	2,0	6,8	9,3	8,2	3,5	-1,6	-4,7	-9,8	-1,8	23	3
4 Absolute maximum temperature	in °C	13,3	13,2	19,6	27,6	32,4	37,2	35,5	35,0	32,1	27,0	19,5	13,6	37,2	25	4
5 Absolute minimum temperature	in °C	-21,8	-27,1	-21,2	-5,5	-1,2	4,0	8,5	6,9	0,9	-5,4	-7,2	-20,4	-27,1	25	5
6 Mean relative humidity	in %	86	83	77	70	69	70	72	71	75	81	87	89	78	23	6
7 Mean precipitation	in mm	23	24	23	32	61	67	82	66	38	42	26	26	508	23	7
8 Maximum precipitation	in mm															8
9 Minimum precipitation	in mm															9
10 Maximum precipitation in 24 h	in mm	16	17	25	40	56	61	87	58	46	29	25	44	87	50	10
11 Mean number of days with precipitation	> 0,1 mm	13	12	12	13	13	13	13	13	10	13	12	13	150	50	11
12 Mean duration of sunshine	in h	55	86	153	189	242	264	265	245	191	117	53	42	1902	25	12
13 Mean quantity of radiation	in ly / day															13
14 Mean potential evaporation	in mm	0	0	19	46	92	116	130	111	73	39	13	0	639		14
15 Mean windspeed	in m / sec															15
16 Mean predominent direction of the wind																16

28

Station / Country Prešov/Czechoslovakia

Location 49°00'N/21°15'E   Height above sealevel 270 m   Climate symbol: Köppen Dfb   Troll III,4

		in	J	F	M	A	M	J	J	A	S	O	N	D	year	P	
1	Mean daily temperature	in °C	-3,9	-2,1	3,2	8,8	14,3	17,2	19,1	18,2	14,3	8,7	3,1	-1,2	8,3	50	1
2	Mean daily maximum temperature	in °C															2
3	Mean daily minimum temperature	in °C															3
4	Absolute maximum temperature	in °C															4
5	Absolute minimum temperature	in °C															5
6	Mean relative humidity	in %	84	81	73	67	66	67	68	70	72	78	82	84	74	25	6
7	Mean precipitation	in mm	31	27	31	46	65	80	91	77	59	49	42	33	631	50	7
8	Maximum precipitation	in mm															8
9	Minimum precipitation	in mm															9
10	Maximum precipitation in 24 h	in mm	23	20	25	57	44	48	58	65	66	68	35	18	68	18	10
11	Mean number of days with precipitation	> 0,1 mm	10	9	9	11	11	13	13	12	9	10	10	11	128	50	11
12	Mean duration of sunshine	in h	61	79	158	190	247	256	274	247	203	136	58	47	1956	25	12
13	Mean quantity of radiation	in ly / day															13
14	Mean potential evaporation	in mm	0	0	14	47	91	113	126	112	72	38	11	0	624		14
15	Mean windspeed	in m / sec	3,8	4,1	4,2	3,9	3,5	3,2	3,0	2,9	2,9	3,2	3,8	3,8	3,5	10	15
16	Mean predomint direction of the wind		N	N	N	N	N	N	N	N	S	S	S	N		8	16

98 Station / Country České Budějovice/Czechoslovakia

Location 48°59'N/14°28'E   Height above sealevel 383 m   Climate symbol: Köppen Cfb   Troll III,3

		in	J	F	M	A	M	J	J	A	S	O	N	D	year	P	
1	Mean daily temperature	in °C	-2,1	-1,1	3,1	7,5	12,8	15,8	17,4	16,8	13,0	7,8	2,9	-0,7	7,8	50	1
2	Mean daily maximum temperature	in °C	1,7	2,4	7,9	13,6	18,8	22,0	23,5	23,4	20,1	13,6	6,4	3,0	13,0	13	2
3	Mean daily minimum temperature	in °C	-5,3	-5,7	-1,3	3,0	6,7	10,1	12,1	11,0	7,9	2,8	0,8	-2,5	3,3	13	3
4	Absolute maximum temperature	in °C	15,1	18,2	21,0	27,6	31,2	36,8	36,7	36,5	33,5	30,1	21,6	13,5	36,8	25	4
5	Absolute minimum temperature	in °C	-31,3	-39,7	-27,3	-18,2	-4,8	-0,2	3,0	2,9	-3,4	-12,2	-13,2	-29,9	-39,7	25	5
6	Mean relative humidity	in %	83	80	76	73	73	73	64	74	78	80	84	85	78	25	6
7	Mean precipitation	in mm	25	28	29	46	67	85	102	73	54	46	33	32	620	50	7
8	Maximum precipitation	in mm															8
9	Minimum precipitation	in mm															9
10	Maximum precipitation in 24 h	in mm	23	33	30	34	37	67	57	128	36	43	39	30	128	50	10
11	Mean number of days with precipitation	> 0,1 mm	10	9	10	11	12	13	13	13	10	10	10	11	132	50	11
12	Mean duration of sunshine	in h	46	82	136	164	207	226	238	219	174	108	55	36	1691	25	12
13	Mean quantity of radiation	in ly / day															13
14	Mean potential evaporation	in mm	0	0	15	41	83	109	122	104	69	36	11	0	541	50	14
15	Mean windspeed	in m / sec															15
16	Mean predomint direction of the wind																16

99 Station / Country Cherbourg/France

Location 49°39'N/1°38'W   Height above sealevel 8 m   Climate symbol: Köppen Cfb   Troll III,2

		in	J	F	M	A	M	J	J	A	S	O	N	D	year	P	
1	Mean daily temperature	in °C	6,4	6,3	7,8	9,6	12,2	14,9	16,6	16,8	15,6	12,9	9,7	7,5	11,4	30	1
2	Mean daily maximum temperature	in °C	8,4	8,4	10,3	12,2	15,0	17,8	19,4	19,5	18,5	15,3	11,9	9,5	13,9	30	2
3	Mean daily minimum temperature	in °C	4,4	4,1	5,2	7,0	9,3	12,0	13,8	14,1	13,1	10,4	7,5	5,4	8,9	30	3
4	Absolute maximum temperature	in °C	14,4	18,4	22,5	24,1	30,0	31,4	31,8	32,9	30,3	25,8	18,8	16,8	32,9	21	4
5	Absolute minimum temperature	in °C	-8,1	-10,0	-4,1	0,3	3,5	6,0	7,9	9,0	6,2	3,2	-0,8	-5,7	-10,0	21	5
6	Mean relative humidity	in %	81	79	79	78	79	80	80	82	80	79	80	82	80	17	6
7	Mean precipitation	in mm	109	75	62	49	41	39	55	71	79	99	133	119	931	30	7
8	Maximum precipitation	in mm															8
9	Minimum precipitation	in mm															9
10	Maximum precipitation in 24 h	in mm	38	28	45	28	21	38	36	52	62	44	43	35	62	15	10
11	Mean number of days with precipitation	> 0,1 mm	19	15	13	12	11	10	12	12	15	16	17	19	171	30	11
12	Mean duration of sunshine	in h	56	80	143	183	204	214	233	218	151	74	45	47	1608	5	12
13	Mean quantity of radiation	in ly / day															13
14	Mean potential evaporation	in mm	19	20	32	45	73	92	108	100	76	52	30	21	668	40	14
15	Mean windspeed	in m / sec															15
16	Mean predomint direction of the wind																16

100 Station / Country Reims/France

Location 49°18'N/4°02'E   Height above sealevel 94 m   Climate symbol: Köppen Cfb   Troll III,3

		in	J	F	M	A	M	J	J	A	S	O	N	D	year	P	
1	Mean daily temperature	in °C	2,0	3,0	6,5	9,6	13,3	16,4	18,3	18,1	15,4	10,6	6,2	3,1	10,2	30	1
2	Mean daily maximum temperature	in °C	5,0	6,5	11,6	15,0	19,1	22,2	23,9	23,6	20,6	15,0	9,3	5,0	14,8	30	2
3	Mean daily minimum temperature	in °C	-1,0	-0,6	1,4	4,2	7,4	10,6	12,6	12,5	10,1	6,1	3,0	0,3	5,6	30	3
4	Absolute maximum temperature	in °C	15,1	21,6	24,0	29,4	32,4	38,3	37,7	36,2	33,0	27,0	20,0	16,7	38,3	21	4
5	Absolute minimum temperature	in °C	-19,2	-19,9	-10,4	-7,7	-2,1	-0,4	3,1	3,8	-2,2	-6,6	-8,5	-16,0	-19,9	21	5
6	Mean relative humidity	in %	87	83	77	73	73	74	73	76	79	82	87	89	79	17	6
7	Mean precipitation	in mm	49	44	36	41	52	51	57	63	58	47	48	52	598	30	7
8	Maximum precipitation	in mm															8
9	Minimum precipitation	in mm	19	18	34	27	32	32	35	29	35	35	28	47	47	15	9
10	Maximum precipitation in 24 h	in mm	17	15	12	13	13	13	12	13	12	13	16	16	164	30	10
11	Mean number of days with precipitation	> 0,1 mm	65	67	155	171	213	219	221	201	165	122	59	51	1729	30	11
12	Mean duration of sunshine	in h															12
13	Mean quantity of radiation	in ly / day	5	10	26	48	86	107	124	109	75	43	18	9	680	40	13
14	Mean potential evaporation	in mm															14
15	Mean windspeed	in m / sec															15
16	Mean predomint direction of the wind																16

101 Station / Country Paris (Le Bourget)/France

Location 48°58'N/2°27'E — Height above sealevel 52 m — Climate symbol: Köppen Cfb — Troll III,2

		J	F	M	A	M	J	J	A	S	O	N	D	year	P	
1 Mean daily temperature	in °C	3,1	3,8	7,2	10,3	14,0	17,1	19,0	18,5	15,9	11,1	6,8	4,1	10,9		1
2 Mean daily maximum temperature	in °C	6,0	7,4	12,2	15,8	19,7	22,9	24,6	24,0	21,1	15,8	10,0	6,8	15,5	30	2
3 Mean daily minimum temperature	in °C	0,9	1,3	3,6	6,3	9,5	12,7	14,5	14,3	11,9	7,9	4,5	2,0	7,5	30	3
4 Absolute maximum temperature	in °C	15,6	20,8	24,7	31,9	33,1	36,2	39,8	36,6	34,8	27,2	20,3	16,2	39,8	15	4
5 Absolute minimum temperature	in °C	-17,0	-16,8	-7,8	-3,7	-1,8	1,7	4,9	5,1	0,1	-4,6	-8,3	-13,2	-17,0	15	5
6 Mean relative humidity	in %	84	80	74	68	69	71	70	74	78	81	85	86	77	15	6
7 Mean precipitation	in mm	54	43	32	38	52	50	55	62	51	49	50	49	585		7
8 Maximum precipitation	in mm															8
9 Minimum precipitation	in mm															9
10 Maximum precipitation in 24 h	in mm	18	17	16	26	29	35	56	37	31	31	27	25	56	15	10
11 Mean number of days with precipitation	> 0,1 mm	17	14	12	13	12	12	13	13	13	14	15	16	164		11
12 Mean duration of sunshine	in h	64	83	152	185	223	233	231	204	166	122	63	53	1779	15	12
13 Mean quantity of radiation	in ly/day	74	128	244	353	439	478	454	383	289	174	84	58	263		13
14 Mean potential evaporation	in mm	9	13	29	51	87	109	124	110	78	44	19	11	684	40	14
15 Mean windspeed	in m/sec	4,4	4,3	4,2	4,5	3,9	3,4	3,7	3,3	3,4	3,3	3,7	4,1	3,9	10	15
16 Mean predoment direction of the wind		W	W	E	NNE	NNE	W	W	W	W	W	S	S		10	16

No.:2,3,8,= Montsouris 48°49'N/2°20'E,75 m / No.:12,13=Pasc–St.Maur 48°49'N/2°30'E,50 m

102 Station / Country Strasbourg/France

Location 48°33'N/7°38'E — Height above sealevel 149 m — Climate symbol: Köppen Cfb — Troll III,3

		J	F	M	A	M	J	J	A	S	O	N	D	year	P	
1 Mean daily temperature	in °C	0,4	1,5	5,6	9,8	14,0	17,2	19,0	18,3	15,1	9,5	4,9	1,3	9,7		1
2 Mean daily maximum temperature	in °C	3,4	5,4	11,1	15,5	19,7	23,0	24,8	24,2	20,8	14,4	8,1	3,9	14,5	30	2
3 Mean daily minimum temperature	in °C	-2,3	-1,7	1,2	4,5	8,2	11,5	13,4	13,0	10,3	6,7	2,3	-1,0	5,4	30	3
4 Absolute maximum temperature	in °C	15,1	20,8	23,5	29,7	31,5	37,0	37,4	35,3	33,4	25,8	21,0	16,5	37,4		4
5 Absolute minimum temperature	in °C	-22,0	-22,2	-15,7	-5,6	-2,4	2,8	5,8	5,1	0,4	-7,6	-10,0	-21,0	-22,2		5
6 Mean relative humidity	in %	84	81	75	71	72	72	71	75	79	83	86	87	78	17	6
7 Mean precipitation	in mm	39	33	30	39	60	77	77	80	58	42	41	31	607		7
8 Maximum precipitation	in mm															8
9 Minimum precipitation	in mm															9
10 Maximum precipitation in 24 h	in mm	25	36	26	28	33	54	44	46	22	24	30	23	54	15	10
11 Mean number of days with precipitation	> 0,1 mm	15	13	12	13	13	14	14	13	12	12	13	14	158		11
12 Mean duration of sunshine	in h	49	68	148	188	211	206	225	216	188	121	49	36	1685	15	12
13 Mean quantity of radiation	in ly/day															13
14 Mean potential evaporation	in mm	2	6	26	51	88	113	126	109	75	41	15	5	657	40	14
15 Mean windspeed	in m/sec	2,8	2,8	2,6	2,8	2,4	1,8	1,9	1,8	1,9	1,7	1,9	2,3	2,2	10	15
16 Mean predoment direction of the wind		S	S	NNE	NNE	NNE	S	W	S	S	S	S	S		10	16

103 Station / Country Brest/France

Location 48°27'N/4°25'W — Height above sealevel 98 m — Climate symbol: Köppen Cfb — Troll III,2

		J	F	M	A	M	J	J	A	S	O	N	D	year	P	
1 Mean daily temperature	in °C	6,1	5,8	7,6	9,2	11,6	14,4	15,6	16,0	14,7	12,0	9,0	7,0	10,8		1
2 Mean daily maximum temperature	in °C	8,8	8,8	11,6	13,0	15,4	18,1	19,4	19,7	18,2	15,2	11,5	9,4	14,1	30	2
3 Mean daily minimum temperature	in °C	3,7	3,3	4,7	5,7	8,0	10,8	12,0	12,4	11,3	8,7	6,3	4,3	7,6	30	3
4 Absolute maximum temperature	in °C	14,5	19,3	20,1	26,5	28,6	28,8	35,2	33,7	29,0	25,3	18,8	16,2	35,2	15	4
5 Absolute minimum temperature	in °C	-14,0	-13,4	-4,4	-0,7	0,0	4,3	6,3	6,6	4,0	-0,8	-4,0	-5,0	-14,0	15	5
6 Mean relative humidity	in %	86	85	83	82	83	84	85	84	85	85	87	88	85	15	6
7 Mean precipitation	in mm	133	96	83	69	68	56	62	80	87	104	138	150	1126		7
8 Maximum precipitation	in mm															8
9 Minimum precipitation	in mm															9
10 Maximum precipitation in 24 h	in mm	42	34	47	55	38	39	26	48	37	33	48	57	57	15	10
11 Mean number of days with precipitation	> 0,1 mm	22	16	15	15	14	13	14	15	16	19	20	22	201		11
12 Mean duration of sunshine	in h	66	85	142	189	220	209	210	207	156	120	69	56	1729	15	12
13 Mean quantity of radiation	in ly/day															13
14 Mean potential evaporation	in mm	20	22	34	48	75	93	108	101	78	52	30	23	684	40	14
15 Mean windspeed	in m/sec	5,5	5,4	5,7	5,4	5,2	4,5	4,8	4,5	4,5	4,2	4,7	5,2	5,0	10	15
16 Mean predoment direction of the wind		SW	W	NE	NE	NE	NE	W	SW	SW	SW	W	W		10	16

104 Station / Country Rennes/France

Location 48°04'N/1°43'W — Height above sealevel 35 m — Climate symbol: Köppen Cfb — Troll III,2

		J	F	M	A	M	J	J	A	S	O	N	D	year	P	
1 Mean daily temperature	in °C	4,8	5,4	8,1	10,2	13,2	16,3	18,1	18,1	16,0	12,0	8,0	5,5	11,3	30	1
2 Mean daily maximum temperature	in °C	7,8	8,8	12,6	15,0	18,4	21,8	23,4	23,3	20,8	16,2	11,4	8,3	15,6	30	2
3 Mean daily minimum temperature	in °C	1,8	1,9	3,6	5,4	8,0	11,0	12,7	12,8	11,2	7,7	4,6	2,6	6,9	30	3
4 Absolute maximum temperature	in °C	15,0	19,5	22,4	25,4	30,8	35,9	38,4	37,8	33,0	25,7	19,9	17,8	38,4	21	4
5 Absolute minimum temperature	in °C	-12,1	-11,2	-6,0	-2,5	-1,0	3,1	5,4	4,0	2,0	-4,6	-7,5	-11,2	-12,1	21	5
6 Mean relative humidity	in %	87	82	80	77	78	78	77	78	82	84	87	89	82	17	6
7 Mean precipitation	in mm	68	51	48	43	47	45	51	57	58	64	66	71	669	30	7
8 Maximum precipitation	in mm															8
9 Minimum precipitation	in mm															9
10 Maximum precipitation in 24 h	in mm	20	35	25	26	24	27	38	27	45	31	30	25	45	15	10
11 Mean number of days with precipitation	> 0,1 mm	18	14	14	12	13	11	12	12	13	15	16	18	168	30	11
12 Mean duration of sunshine	in h	68	90	160	190	219	229	231	212	163	120	69	54	1805	30	12
13 Mean quantity of radiation	in ly/day															13
14 Mean potential evaporation	in mm	14	17	32	49	80	100	115	105	77	48	23	16	674	40	14
15 Mean windspeed	in m/sec															15
16 Mean predoment direction of the wind																16

105 Station/Country Tours/France

Location 47°25'N/0°46'E Height above sealevel 98 m Climate symbol: Köppen Cfb Troll III.2

		J	F	M	A	M	J	J	A	S	O	N	D	year	P	
1 Mean daily temperature	in °C	3,7	4,7	8,0	10,7	14,1	17,4	19,3	19,0	16,5	11,9	7,4	4,5	11,4	30	1
2 Mean daily maximum temperature	in °C	6,8	8,0	12,8	15,7	19,2	22,8	24,8	24,4	21,4	16,1	10,5	7,2	15,8	30	2
3 Mean daily minimum temperature	in °C	0,8	1,3	3,4	5,7	8,9	12,0	13,7	13,6	11,5	7,6	4,2	1,8	7,0	30	3
4 Absolute maximum temperature	in °C	15,3	20,2	26,4	29,3	32,1	37,3	41,4	36,7	34,8	28,0	22,8	18,9	41,4	21	4
5 Absolute minimum temperature	in °C	-13,5	-13,2	-6,0	-2,0	0,4	2,8	6,2	6,7	2,0	-4,4	-9,0	-14,0	-14,0	21	5
6 Mean relative humidity	in %	88	83	77	72	73	73	71	74	78	81	87	89	79	17	6
7 Mean precipitation	in mm	64	55	49	48	63	49	49	61	59	60	64	68	689	30	7
8 Maximum precipitation	in mm															8
9 Minimum precipitation	in mm															9
10 Maximum precipitation in 24 h	in mm	24	25	41	28	23	27	48	57	74	26	53	42	74	15	10
11 Mean number of days with precipitation	> 0,1 mm	16	13	12	12	13	11	11	12	13	13	15	16	157	30	11
12 Mean duration of sunshine	in h	67	86	160	194	214	229	237	218	175	128	84	54	1826	30	12
13 Mean quantity of radiation	in ly/day															13
14 Mean potential evaporation	in mm	10	14	30	50	84	107	125	110	78	47	20	12	687	40	14
15 Mean windspeed	in m/sec	4,2	4,4	4,1	4,2	3,8	3,2	3,8	3,2	3,0	2,9	3,8	4,0	3,7	10	15
16 Mean predominent direction of the wind		W	W	E	NE	W	W	W	W	W	W	E	W		10	16

106 Station/Country Dijon/France

Location 47°16'N/5°06'E Height above sealevel 220 m Climate symbol: Köppen Cfb Troll III.3

		J	F	M	A	M	J	J	A	S	O	N	D	year	P	
1 Mean daily temperature	in °C	1,3	2,8	6,9	10,4	14,3	17,7	19,6	19,0	15,9	10,5	5,7	2,1	10,5		1
2 Mean daily maximum temperature	in °C	4,4	6,5	12,3	15,8	19,8	23,0	25,1	24,7	21,3	15,3	9,1	4,8	15,2	30	2
3 Mean daily minimum temperature	in °C	-1,6	-0,9	2,0	5,1	8,7	12,1	13,0	13,5	10,9	6,4	2,7	-0,3	6,0	30	3
4 Absolute maximum temperature	in °C	16,5	19,7	23,5	29,0	33,0	36,0	37,2	36,0	33,0	26,0	21,8	16,0	37,2	15	4
5 Absolute minimum temperature	in °C	-16,6	-19,6	-15,3	-5,3	-3,3	0,8	5,3	5,4	0,0	-3,8	-9,4	-20,4	-20,4	15	5
6 Mean relative humidity	in %	86	82	74	67	68	70	67	70	75	80	85	88	76	15	6
7 Mean precipitation	in mm	64	42	42	46	64	81	58	77	72	60	76	57	739	30	7
8 Maximum precipitation	in mm															8
9 Minimum precipitation	in mm															9
10 Maximum precipitation in 24 h	in mm	26	21	33	27	30	110	62	83	57	53	52	38	110	15	10
11 Mean number of days with precipitation	> 0,1 mm	16	13	10	11	12	12	11	11	11	12	14	14	147	30	11
12 Mean duration of sunshine	in h	68	94	189	210	248	251	271	244	211	148	70	48	2049	15	12
13 Mean quantity of radiation	in ly/day															13
14 Mean potential evaporation	in mm	4	10	28	50	87	111	126	112	80	45	18	8	679	40	14
15 Mean windspeed	in m/sec	4,2	4,4	4,2	4,2	3,8	3,2	3,8	3,2	3,0	2,9	3,8	4,0	3,7	10	15
16 Mean predominent direction of the wind		W	W	E	NE	W	W	W	W	W	W	E	W		10	16

107 Station/Country Nantes/France

Location 47°10'N/1°37'W Height above sealevel 26 m Climate symbol: Köppen Cfb Troll III.2

		J	F	M	A	M	J	J	A	S	O	N	D	year	P	
1 Mean daily temperature	in °C	5,0	5,3	8,4	10,8	13,9	17,2	18,8	18,8	16,4	12,2	8,2	5,5	11,7		1
2 Mean daily maximum temperature	in °C	13,3	15,9	19,9	22,8	26,4	29,5	32,0	29,7	27,5	23,4	17,2	13,9	22,8	15	2
3 Mean daily minimum temperature	in °C	-5,2	-5,0	-1,9	1,0	3,3	7,2	9,5	9,3	6,5	0,9	-2,6	-3,4	1,8	15	3
4 Absolute maximum temperature	in °C	15,4	21,4	22,8	27,5	32,7	36,8	40,3	38,0	33,8	27,4	21,1	18,6	40,3	15	4
5 Absolute minimum temperature	in °C	-11,4	-15,6	-5,7	-1,5	0,1	4,2	5,8	5,6	2,8	-2,5	-5,3	-10,8	-15,6	15	5
6 Mean relative humidity	in %	89	85	81	77	78	78	77	79	82	85	88	90	82	15	6
7 Mean precipitation	in mm	71	58	52	45	54	45	47	60	73	76	81	79	741	30	7
8 Maximum precipitation	in mm															8
9 Minimum precipitation	in mm															9
10 Maximum precipitation in 24 h	in mm	23	27	33	24	51	35	24	32	43	43	62	30	62	15	10
11 Mean number of days with precipitation	> 0,1 mm	18	14	14	12	14	12	13	12	14	15	16	18	172	30	11
12 Mean duration of sunshine	in h	76	98	163	201	235	247	250	233	176	135	77	63	1952	30	12
13 Mean quantity of radiation	in ly/day															13
14 Mean potential evaporation	in mm	15	18	32	49	79	102	117	106	78	48	23	20	687	40	14
15 Mean windspeed	in m/sec	4,2	4,2	4,1	4,1	3,7	3,2	3,5	3,3	3,1	3,1	3,4	3,8	3,6	10	15
16 Mean predominent direction of the wind		NE	NE	NE	NE	W	W	W	W	W	W	NE	W		10	16

108 Station/Country Limoges/France

Location 45°49'N/1°17'E Height above sealevel 282 m Climate symbol: Köppen Cfb Troll III.2

		J	F	M	A	M	J	J	A	S	O	N	D	year	P	
1 Mean daily temperature	in °C	3,1	3,9	7,4	9,9	13,3	16,8	18,4	17,3	15,3	10,7	6,7	3,8	10,6		1
2 Mean daily maximum temperature	in °C	14,3	16,4	20,4	24,0	26,8	31,0	32,7	31,8	27,8	24,0	18,9	14,4	23,5	15	2
3 Mean daily minimum temperature	in °C	-9,9	-9,8	-5,8	-2,9	-0,3	3,8	5,4	5,4	2,1	-2,8	-5,9	-7,4	-2,3	15	3
4 Absolute maximum temperature	in °C	17,4	22,3	24,8	29,7	32,2	36,7	38,8	38,2	32,3	27,0	25,3	19,6	38,8	15	4
5 Absolute minimum temperature	in °C	-18,2	-21,7	-9,4	-5,3	-3,9	0,2	3,2	2,7	-0,8	-8,2	-11,2	-17,0	-21,7	15	5
6 Mean relative humidity	in %	83	79	74	71	72	73	72	75	78	80	82	85	77	15	6
7 Mean precipitation	in mm	89	76	66	65	80	67	71	74	84	80	88	94	934		7
8 Maximum precipitation	in mm															8
9 Minimum precipitation	in mm															9
10 Maximum precipitation in 24 h	in mm	30	33	31	28	43	48	38	46	62	50	48	30	62	15	10
11 Mean number of days with precipitation	> 0,1 mm	17	14	13	13	14	12	12	12	12	14	15	17	165		11
12 Mean duration of sunshine	in h	75	96	151	188	207	225	240	221	185	154	84	59	1885	15	12
13 Mean quantity of radiation	in ly/day	100	187	288	362	451	515	490	411	309	206	108	72	288		13
14 Mean potential evaporation	in mm	9	14	28	47	81	103	120	109	79	47	22	13	673	40	14
15 Mean windspeed	in m/sec	2,9	3,2	3,0	2,9	2,5	2,1	2,3	2,1	1,9	1,9	2,8	2,9	2,5	10	15
16 Mean predominent direction of the wind		SSW	W	S	W	W	W	W	W	W	W	S	S		10	16

109 Station/Country **Clermont–Ferrand/France**

Location **45°48'N/3°09'E** Height above sealevel **329 m** Climate symbol: Köppen **Cfb** Troll **III,3**

		J	F	M	A	M	J	J	A	S	O	N	D	year	P	
1 Mean daily temperature	in °C	2,7	3,5	7,3	10,1	13,7	17,2	19,2	18,8	16,1	11,0	6,7	3,5	10,9		1
2 Mean daily maximum temperature	in °C	15,0	17,3	21,0	24,9	27,7	31,8	34,1	33,7	30,4	25,0	19,0	16,0	24,6	21	2
3 Mean daily minimum temperature	in °C	-10,2	-10,8	-6,9	-3,1	0,4	5,2	6,8	6,4	2,7	-2,9	-5,8	-9,1	-2,3	21	3
4 Absolute maximum temperature	in °C	19,5	25,9	26,3	31,3	33,0	37,4	39,3	39,6	34,6	29,1	23,4	19,4	39,6		4
5 Absolute minimum temperature	in °C	-22,0	-19,8	-21,3	-6,4	-4,2	1,8	3,8	2,8	-2,8	-6,2	-11,8	-25,8	-25,8		5
6 Mean relative humidity	in %	79	73	69	67	69	66	68	72	72	75	79	79	72	17	6
7 Mean precipitation	in mm	25	25	29	43	67	72	51	68	61	49	40	33	563		7
8 Maximum precipitation	in mm															8
9 Minimum precipitation	in mm															9
10 Maximum precipitation in 24 h	in mm	17	22	23	28	48	45	45	55	41	40	32	43	55	15	10
11 Mean number of days with precipitation	> 0,1 mm	12	11	9	12	12	12	9	10	10	11	12	12	132		11
12 Mean duration of sunshine	in h	82	108	167	190	215	221	259	235	197	158	86	72	1990	15	12
13 Mean quantity of radiation	in ly/day															13
14 Mean potential evaporation	in mm	7	13	29	47	80	103	119	109	76	47	20	12	662	40	14
15 Mean windspeed	in m/sec	3,7	3,9	3,8	3,6	3,0	2,5	2,6	2,6	2,5	2,7	3,3	3,5	3,1	10	15
16 Mean predominant direction of the wind		S	S	S	N	N	N	N	N	S	S	S	S		10	16

---

110 Station/Country **Lyon/France**

Location **45°43'N/4°57'E** Height above sealevel **200 m** Climate symbol: Köppen **Cfb** Troll **III,3**

			J	F	M	A	M	J	J	A	S	O	N	D	year	P	
1 Mean daily temperature	in °C	+	2,1	3,3	7,7	10,9	14,9	18,5	20,7	20,1	16,9	11,4	6,7	3,1	11,4		1
2 Mean daily maximum temperature	in °C		5,4	7,4	12,8	16,3	20,3	24,1	26,8	26,0	22,8	16,2	6,0	6,1	16,1	30	2
3 Mean daily minimum temperature	in °C		-0,9	-0,3	2,9	5,7	9,3	12,8	14,8	14,4	11,9	7,3	3,6	0,3	6,8	30	3
4 Absolute maximum temperature	in °C		17,7	21,9	23,0	30,1	34,2	36,8	39,5	39,7	35,6	27,9	22,9	18,9	39,7		4
5 Absolute minimum temperature	in °C		-20,7	-21,4	-10,0	-4,4	-3,8	2,3	6,4	4,6	0,7	-4,5	-8,0	-24,6	-24,6		5
6 Mean relative humidity	in %		85	80	74	70	70	69	65	70	75	81	85	85	76	17	6
7 Mean precipitation	in mm		52	46	53	56	69	85	58	89	93	77	80	57	813		7
8 Maximum precipitation	in mm																8
9 Minimum precipitation	in mm																9
10 Maximum precipitation in 24 h	in mm		37	44	43	56	47	71	40	58	89	47	81	81	89	15	10
11 Mean number of days with precipitation	> 0,1 mm		15	12	11	11	13	11	10	11	11	12	14	14	145	15	11
12 Mean duration of sunshine	in h		64	94	181	213	250	282	293	257	205	144	61	48	2072	15	12
13 Mean quantity of radiation	in ly/day		86	159	278	390	458	513	545	438	316	193	95	61	294		13
14 Mean potential evaporation	in mm		6	12	30	54	92	114	135	122	82	48	20	9	724	40	14
15 Mean windspeed	in m/sec		3,1	3,5	3,4	4,0	3,5	2,9	2,8	2,6	2,4	2,4	3,0	2,8	3,0	15	15
16 Mean predominant direction of the wind			N	N,S	S	N	N	N	N	N	S	S	S	S		10	16

+ Saint–Genis–Laval (45°42'N/4°47'E ,299 m)

---

111 Station/Country **Grenoble/France**

Location **45°10'N/5°44'E** Height above sealevel **223 m** Climate symbol: Köppen **Cfb** Troll **III,3**

		J	F	M	A	M	J	J	A	S	O	N	D	year	P	
1 Mean daily temperature	in °C	1,5	3,2	7,7	10,7	14,5	17,8	20,1	19,5	16,7	11,5	6,5	2,3	11,0	30	1
2 Mean daily maximum temperature	in °C	5,5	7,9	13,3	16,3	20,3	23,6	26,4	25,8	22,2	16,2	10,3	5,8	16,1	30	2
3 Mean daily minimum temperature	in °C	-2,5	-1,5	2,0	5,0	8,6	12,0	13,8	13,4	11,2	6,7	2,7	-1,2	5,8	30	3
4 Absolute maximum temperature	in °C	17,2	22,2	23,6	27,6	34,8	36,3	39,4	38,2	33,0	25,3	19,4	38,4		15	4
5 Absolute minimum temperature	in °C	-18,8	-20,0	-10,3	-3,6	-0,4	2,6	6,7	5,6	1,6	-4,2	-7,2	-11,0	-20,0	15	5
6 Mean relative humidity	in %	82	77	73	72	74	73	72	75	79	82	83	84	77	15	6
7 Mean precipitation	in mm	80	75	60	65	80	90	70	95	100	95	95	80	985	30	7
8 Maximum precipitation	in mm															8
9 Minimum precipitation	in mm															9
10 Maximum precipitation in 24 h	in mm	43	37	32	33	55	67	135	92	85	44	56	68	135	10	10
11 Mean number of days with precipitation	> 0,1 mm	14	11	11	12	14	11	10	11	11	12	13	14	144	30	11
12 Mean duration of sunshine	in h	80	115	179	194	236	258	291	259	199	136	74	64	2085	14	12
13 Mean quantity of radiation	in ly/day															13
14 Mean potential evaporation	in mm	3	8	29	51	86	112	127	114	79	48	19	8	684	40	14
15 Mean windspeed	in m/sec															15
16 Mean predominant direction of the wind																16

---

112 Station/Country **Bordeaux/France**

Location **44°50'N/0°42'W** Height above sealevel **47 m** Climate symbol: Köppen **Cfb** Troll **III,2**

		J	F	M	A	M	J	J	A	S	O	N	D	year	P	
1 Mean daily temperature	in °C	5,2	5,9	9,3	11,7	14,7	18,0	19,8	19,5	17,1	12,7	8,4	5,7	12,3		1
2 Mean daily maximum temperature	in °C	9,2	10,5	14,9	17,3	20,3	23,7	25,4	25,7	23,1	18,1	12,7	9,3	17,5	30	2
3 Mean daily minimum temperature	in °C	1,7	2,0	4,2	6,1	9,1	12,2	13,7	13,6	12,0	8,2	4,2	2,5	7,5	30	3
4 Absolute maximum temperature	in °C	18,7	22,4	25,5	30,9	33,8	38,4	38,8	37,1	36,2	30,4	23,9	21,2	38,8		4
5 Absolute minimum temperature	in °C	-14,4	-15,2	-8,1	-4,8	-1,8	2,5	5,1	4,7	-1,7	-5,3	-6,4	-13,4	-15,2		5
6 Mean relative humidity	in %	87	82	78	76	76	77	76	77	82	84	88	89	81	17	6
7 Mean precipitation	in mm	90	75	63	48	61	65	56	70	84	83	96	109	900		7
8 Maximum precipitation	in mm															8
9 Minimum precipitation	in mm															9
10 Maximum precipitation in 24 h	in mm	39	34	32	33	40	49	42	54	44	41	47	38	54	15	10
11 Mean number of days with precipitation	> 0,1 mm	16	13	13	13	14	11	11	12	13	14	15	17	162		11
12 Mean duration of sunshine	in h	81	103	174	210	229	253	262	243	195	158	84	60	2052	15	12
13 Mean quantity of radiation	in ly/day															13
14 Mean potential evaporation	in mm	13	18	34	51	83	106	124	115	85	53	25	16	723	40	14
15 Mean windspeed	in m/sec	3,3	3,5	3,8	3,6	3,2	3,1	3,1	2,8	2,5	2,5	2,9	3,3	3,1	10	15
16 Mean predominant direction of the wind		W	W	W	W	W	W	W	W	W	W	WSW			10	16

32

113 Station / Country Toulouse / France

Location 43°37'N/1°22'E — Height above sealevel 151 m — Climate symbol: Köppen Cfb — Troll III,3

	in	J	F	M	A	M	J	J	A	S	O	N	D	year	P	
1 Mean daily temperature	in °C	4,5	5,4	9,0	11,4	14,8	18,6	20,8	20,7	18,0	13,0	8,3	5,3	12,5		1
2 Mean daily maximum temperature	in °C	8,5	10,1	14,4	16,9	20,3	24,4	26,8	26,8	23,7	18,3	12,7	8,7	17,8	30	2
3 Mean daily minimum temperature	in °C	0,8	1,3	4,1	6,1	9,4	13,0	14,7	14,9	12,9	8,4	4,8	2,1	7,7	30	3
4 Absolute maximum temperature	in °C	21,2	21,2	24,0	30,0	32,8	39,8	39,0	40,2	33,0	31,5	22,9	20,3	40,2	21	4
5 Absolute minimum temperature	in °C	-17,0	-19,2	-5,3	-3,0	-0,8	6,2	8,2	7,3	3,8	-3,0	-6,3	-10,5	-19,2	21	5
6 Mean relative humidity	in %	87	80	77	78	76	74	73	73	78	82	86	87	79	17	6
7 Mean precipitation	in mm	49	46	53	50	75	61	44	54	64	45	51	67	659		7
8 Maximum precipitation	in mm															8
9 Minimum precipitation	in mm															9
10 Maximum precipitation in 24 h	in mm	24	57	38	32	43	62	78	47	40	39	36	40	78	15	10
11 Mean number of days with precipitation	> 0,1 mm	14	12	11	12	13	10	9	9	10	11	12	15	138		11
12 Mean duration of sunshine	in h	80	117	183	199	223	234	262	254	206	168	92	62	2080	15	12
13 Mean quantity of radiation	in ly / day															13
14 Mean potential evaporation	in mm	11	16	32	49	81	110	129	120	86	52	24	14	724	40	14
15 Mean windspeed	in m / sec	3,5	3,8	4,5	4,8	3,8	3,5	3,5	3,3	3,1	3,1	3,1	3,7	3,8	10	15
16 Mean predominent direction of the wind		W	W	SE	W	W	W	W	W	SE	W	W	W		10	16

114 Station / Country Marseille / France

Location 43°27'N/5°13'E — Height above sealevel 3 m — Climate symbol: Köppen Csa — Troll IV,1

	in	J	F	M	A	M	J	J	A	S	O	N	D	year	P	
1 Mean daily temperature	in °C	5,5	6,6	10,0	13,0	16,8	20,8	23,3	22,8	19,8	15,0	10,2	6,9	14,2		1
2 Mean daily maximum temperature	in °C	10,0	11,5	15,0	17,9	21,8	26,1	28,9	28,3	25,1	19,8	14,7	10,9	19,2	30	2
3 Mean daily minimum temperature	in °C	1,5	2,1	5,1	7,8	11,1	14,7	17,1	17,0	14,7	10,4	6,0	3,0	9,2	30	3
4 Absolute maximum temperature	in °C	20,0	21,9	24,0	28,5	33,0	38,0	39,0	36,8	34,3	30,2	22,8	20,2	39,0		4
5 Absolute minimum temperature	in °C	-10,7	-16,8	-10,0	-2,4	0,0	5,4	7,8	8,6	1,0	-2,2	-5,4	-12,8	-16,8		5
6 Mean relative humidity	in %	75	71	69	67	66	61	57	62	68	73	76	76	68		6
7 Mean precipitation	in mm	43	32	43	42	46	24	11	34	60	76	69	66	546		7
8 Maximum precipitation	in mm															8
9 Minimum precipitation	in mm															9
10 Maximum precipitation in 24 h	in mm	48	81	80	47	45	55	26	46	68	86	44	46	86	15	10
11 Mean number of days with precipitation	> 0,1 mm	8	6	7	6	4	2	4	6	6	8	10		76		11
12 Mean duration of sunshine	in h	134	157	208	251	281	323	368	324	253	191	151	123	2784	15	12
13 Mean quantity of radiation	in ly / day															13
14 Mean potential evaporation	in mm	15	18	33	52	86	114	137	122	88	57	29	17	788	40	14
15 Mean windspeed	in m / sec	4,0	4,3	4,6	5,2	4,6	4,8	5,0	4,4	4,0	3,9	4,1	4,2	4,4	10	15
16 Mean predominent direction of the wind		NNW	NW	NW	NNW	W	NW	W	NW	W	NW	NW	E		10	16

115 Station / Country Pic-du-Midi / France

Location 42°58'N/0°09'E — Height above sealevel 2860 m — Climate symbol: Köppen ET — Troll III,3

	in	J	F	M	A	M	J	J	A	S	O	N	D	year	P	
1 Mean daily temperature	in °C	-7,3	-7,7	-5,3	-3,8	-0,5	3,8	7,1	6,8	4,2	-0,4	-4,1	-6,9	-1,2	30	1
2 Mean daily maximum temperature	in °C	-4,6	-4,8	-2,3	-0,5	2,7	7,0	10,5	10,1	7,2	2,3	-1,3	-4,4	1,8	30	2
3 Mean daily minimum temperature	in °C	-10,0	-10,6	-8,3	-7,1	-3,9	0,6	3,7	3,5	1,2	-3,0	-6,8	-9,4	-4,2	30	3
4 Absolute maximum temperature	in °C	10,1	10,0	7,9	10,6	12,8	18,0	19,3	18,0	16,4	16,8	9,1	7,3	19,3	15	4
5 Absolute minimum temperature	in °C	-26,8	-32,9	-21,3	-20,8	-14,3	-11,4	-7,0	-6,5	-9,8	-18,2	-18,4	-23,4	-32,9	15	5
6 Mean relative humidity	in %	68	69	71	74	79	79	69	73	74	74	68	72	73	15	6
7 Mean precipitation	in mm	124	74	74	80	61	71	52	90	82	82	98	125	1013	15	7
8 Maximum precipitation	in mm															8
9 Minimum precipitation	in mm															9
10 Maximum precipitation in 24 h	in mm	65	35	43	41	25	31	38	73	43	73	50	57	73	15	10
11 Mean number of days with precipitation	> 0,1 mm	18	14	14	16	16	13	11	11	14	15	16	17	175	30	11
12 Mean duration of sunshine	in h	135	141	181	178	188	207	282	257	202	167	145	120	2204	14	12
13 Mean quantity of radiation	in ly / day															13
14 Mean potential evaporation	in mm	0	0	0	0	0	50	77	68	44	0	0	0	239	15	14
15 Mean windspeed	in m / sec															15
16 Mean predominent direction of the wind																16

118 Station / Country Perpignan / France

Location 42°44'N/2°52'E — Height above sealevel 43 m — Climate symbol: Köppen Csa — Troll IV,1

	in	J	F	M	A	M	J	J	A	S	O	N	D	year	P	
1 Mean daily temperature	in °C	7,5	8,4	11,3	13,9	17,1	21,1	23,8	23,3	20,5	15,9	11,5	8,6	12,5		1
2 Mean daily maximum temperature	in °C	11,8	12,9	15,9	18,7	21,8	25,9	28,8	28,4	25,2	20,3	15,8	12,5	19,8	30	2
3 Mean daily minimum temperature	in °C	3,7	4,5	7,1	9,5	12,7	16,3	18,8	18,7	16,5	12,1	7,9	5,2	11,1	30	3
4 Absolute maximum temperature	in °C	22,6	26,4	26,2	32,4	33,1	36,8	38,0	38,7	36,8	29,1	25,4	25,0	38,7		4
5 Absolute minimum temperature	in °C	-6,9	-11,0	-2,7	1,7	5,5	9,1	12,0	11,6	7,4	1,2	-2,1	-5,0	-11,0		5
6 Mean relative humidity	in %	69	66	67	65	66	63	58	63	67	69	70	70	66	17	6
7 Mean precipitation	in mm	39	52	66	39	51	38	24	31	82	74	58	87	639		7
8 Maximum precipitation	in mm															8
9 Minimum precipitation	in mm															9
10 Maximum precipitation in 24 h	in mm	59	109	42	39	59	91	45	37	186	97	55	114	186	15	10
11 Mean number of days with precipitation	> 0,1 mm	7	6	8	7	9	7	5	6	7	8	8	9	85		11
12 Mean duration of sunshine	in h	161	172	209	245	255	279	322	282	238	189	158	138	2644	15	12
13 Mean quantity of radiation	in ly / day															13
14 Mean potential evaporation	in mm	15	20	34	54	84	118	139	127	92	57	33	19	790	40	14
15 Mean windspeed	in m / sec	5,4	5,8	4,9	6,3	5,0	4,8	5,0	4,0	3,9	4,3	4,7	5,4	4,9	10	15
16 Mean predominent direction of the wind		NW	NW	NW	NW	NW	NW	NW	NW	NW	NW	NW	NW		10	16

117 Station/Country Bastia(Corsica)/France

Location 42°33'N/9°29'E Height above sealevel 10 m Climate symbol: Köppen **Csa** Troll IV,1

		J	F	M	A	M	J	J	A	S	O	N	D	year	P
1 Mean daily temperature	in °C	7,9	8,8	10,3	12,7	16,2	20,2	23,0	23,0	20,4	16,0	11,9	9,1	14,9	
2 Mean daily maximum temperature	in °C														
3 Mean daily minimum temperature	in °C														
4 Absolute maximum temperature	in °C	23,6	22,0	23,8	24,2	29,7	32,8	35,8	36,0	34,0	27,6	23,8	24,0	36,0	15
5 Absolute minimum temperature	in °C	-4,6	-5,0	-3,8	0,5	1,3	8,2	10,2	11,8	7,8	3,0	1,2	-1,8	-5,0	15
6 Mean relative humidity	in %														
7 Mean precipitation	in mm	75	65	60	65	50	20	10	25	65	110	95	95	735	
8 Maximum precipitation	in mm														
9 Minimum precipitation	in mm														
10 Maximum precipitation in 24 h	in mm	75	120	44	47	32	44	21	201	156	135	66	57	201	15
11 Mean number of days with precipitation	> 0,1 mm	10	8	10	9	9	4	1	2	6	10	11	11	91	
12 Mean duration of sunshine	in h	137	132	188	223	258	306	364	313	249	197	128	110	2603	15
13 Mean quantity of radiation	in ly/day														
14 Mean potential evaporation	in mm	17	20	34	50	79	115	139	130	97	60	34	21	796	
15 Mean windspeed	in m/sec	2,5	2,7	2,4	2,2	1,9	2,1	2,2	2,1	2,0	2,1	2,2	2,4	2,2	10
16 Mean predominent direction of the wind		SW	W	SW	SE	SE	SE	SE	SE,E	SE	SE	SW	SW		10

118 Station/Country Monaco/Monaco

Location 43°43'N/7°25'E Height above sealevel 55 m Climate symbol: Köppen **Csa** Troll IV,1

		J	F	M	A	M	J	J	A	S	O	N	D	year	P
1 Mean daily temperature	in °C	10,3	10,5	11,9	14,0	17,1	20,8	23,5	23,8	21,8	17,9	14,2	11,8	16,4	30
2 Mean daily maximum temperature	in °C	12,3	12,5	13,8	16,0	19,1	22,8	25,5	25,8	23,6	19,9	16,1	13,6	18,4	30
3 Mean daily minimum temperature	in °C	8,3	8,4	9,9	12,0	15,1	18,8	21,5	21,7	19,8	15,9	12,2	9,8	14,4	30
4 Absolute maximum temperature	in °C	21,0	19,0	20,0	25,2	29,0	34,0	33,0	34,0	30,6	26,8	22,3	19,9	34,0	30
5 Absolute minimum temperature	in °C	0,0	-1,3	1,0	4,0	7,8	12,2	13,8	14,0	11,0	7,0	5,0	0,8	-1,3	30
6 Mean relative humidity	in %	67	70	74	75	77	77	75	74	74	72	72	72	73	13
7 Mean precipitation	in mm	61	58	71	65	64	33	21	22	66	113	123	99	796	30
8 Maximum precipitation	in mm														
9 Minimum precipitation	in mm														
10 Maximum precipitation in 24 h	in mm	63	65	46	94	67	52	91	43	69	97	111	101	111	18
11 Mean number of days with precipitation	> 0,1 mm	5	5	7	5	5	4	1	2	4	7	7	8	58	18
12 Mean duration of sunshine	in h	143	149	167	190	220	248	294	271	212	176	139	122	2331	23
13 Mean quantity of radiation	in ly/day														
14 Mean potential evaporation	in mm	22	23	35	50	81	115	139	132	97	65	37	29	825	40
15 Mean windspeed	in m/sec														
16 Mean predominent direction of the wind															

119 Station/Country Las Escaldes/Andorra

Location 42°30'N/1°31'E Height above sealevel 1080 m Climate symbol: Köppen **Cfb** Troll III,3

		J	F	M	A	M	J	J	A	S	O	N	D	year	P
1 Mean daily temperature	in °C	2,3	3,1	7,0	9,0	11,4	16,4	19,3	18,0	15,7	10,8	6,1	2,5	10,1	9
2 Mean daily maximum temperature	in °C	5,9	7,3	12,3	14,2	16,9	23,0	26,1	24,2	21,8	15,8	10,3	5,8	15,3	9
3 Mean daily minimum temperature	in °C	-1,3	-1,1	1,7	3,8	5,9	9,7	12,4	11,8	9,6	5,7	1,8	-0,8	4,9	9
4 Absolute maximum temperature	in °C	15,0	17,0	20,0	25,0	29,0	36,0	35,0	33,0	31,0	27,0	20,0	13,0	36,0	9
5 Absolute minimum temperature	in °C	-13,0	-18,0	-9,0	-4,0	0,0	2,0	5,0	4,0	2,0	-5,0	-5,0	-11,0	-18,0	9
6 Mean relative humidity	in %														
7 Mean precipitation	in mm	34	37	46	63	105	69	65	98	81	73	68	69	808	9
8 Maximum precipitation	in mm														
9 Minimum precipitation	in mm														
10 Maximum precipitation in 24 h	in mm	25	19	37	35	41	97	31	61	56	77	75	55	97	9
11 Mean number of days with precipitation	> 0,1 mm	4	6	6	10	15	9	8	10	9	8	6	7	98	9
12 Mean duration of sunshine	in h														
13 Mean quantity of radiation	in ly/day														
14 Mean potential evaporation	in mm	7	10	31	44	64	100	120	104	78	43	22	7	661	9
15 Mean windspeed	in m/sec														
16 Mean predominent direction of the wind															

120 Station/Country Zürich/Switzerland

Location 47°23'N/8°34'E Height above sealevel 569 m Climate symbol: Köppen **Cfb** Troll III,3

		J	F	M	A	M	J	J	A	S	O	N	D	year	P
1 Mean daily temperature	in °C	-1,1	0,3	4,5	8,8	12,7	15,9	17,8	17,0	14,0	8,8	3,1	0,1	8,5	30
2 Mean daily maximum temperature	in °C	2,4	5,0	10,4	14,9	19,4	22,7	24,5	23,9	20,4	13,7	7,2	3,0	14,0	30
3 Mean daily minimum temperature	in °C	-3,1	-2,3	1,0	4,3	8,2	11,7	13,5	13,2	10,5	6,0	1,9	-1,5	5,3	30
4 Absolute maximum temperature	in °C	16,0	18,2	22,3	29,2	31,8	34,9	36,4	34,7	31,4	25,7	22,8	16,1	36,4	64
5 Absolute minimum temperature	in °C	-18,8	-24,8	-11,5	-6,7	-2,0	3,3	5,8	4,1	-0,4	-6,1	-11,8	-19,3	-24,8	64
6 Mean relative humidity	in %	81	77	71	66	66	66	67	69	74	78	82	83	73	30
7 Mean precipitation	in mm	74	70	66	80	107	136	143	131	108	80	76	65	1136	30
8 Maximum precipitation	in mm														
9 Minimum precipitation	in mm														
10 Maximum precipitation in 24 h	in mm	54	48	45	47	90	81	60	65	103	38	47	50	103	80
11 Mean number of days with precipitation	> 1,0 mm	12	10	9	11	13	13	13	13	10	10	10	10	134	30
12 Mean duration of sunshine	in h	48	79	149	173	207	220	238	219	186	108	51	37	1693	30
13 Mean quantity of radiation	in ly/day														
14 Mean potential evaporation	in mm	0	2	21	48	83	106	117	104	72	40	15	2	608	40
15 Mean windspeed	in m/sec	2,8	3,0	3,1	3,2	2,9	2,9	2,7	2,8	2,4	2,3	2,8	3,2	2,8	10
16 Mean predominent direction of the wind		W,SW	W	W	W	W	W	W	W	W	W	W	W		10

121 Station/Country Säntis/Switzerland

Location 47°15'N/9°20'E **Height above sealevel** 2500 m **Climate symbol: Köppen** ET **Troll** III,3

		J	F	M	A	M	J	J	A	S	O	N	D	year	P	
1 Mean daily temperature	in °C	-9.0	-9.0	-6.6	-4.1	0.4	3.6	5.6	5.5	3.5	-0.8	-4.5	-7.6	-1.9	30	1
2 Mean daily maximum temperature	in °C	-6.9	-6.7	-4.3	-1.7	2.9	6.3	8.4	8.2	6.0	1.5	-2.8	-5.6	0.5	30	2
3 Mean daily minimum temperature	in °C	-11.3	-11.3	-8.8	-6.4	-2.2	1.1	3.0	3.2	1.3	-2.7	-8.5	-9.8	-4.2	30	3
4 Absolute maximum temperature	in °C	4.3	5.7	7.2	11.6	17.8	17.0	20.5	18.7	16.0	12.5	10.4	6.1	20.5	60	4
5 Absolute minimum temperature	in °C	-32.0	-30.4	-23.7	-19.9	-15.3	-9.4	-5.6	-5.7	-13.0	-16.8	-21.4	-30.1	-32.0	60	5
6 Mean relative humidity	in %	75	75	74	79	77	79	79	78	76	73	74	73	76	30	6
7 Mean precipitation	in mm	202	180	164	166	197	249	302	278	208	183	190	168	2487	30	7
8 Maximum precipitation	in mm															8
9 Minimum precipitation	in mm															9
10 Maximum precipitation in 24 h	in mm	125	88	128	129	172	183	110	112	96	91	168	115	183	66	10
11 Mean number of days with precipitation	> 1.0 mm	14	13	13	15	15	17	17	16	13	12	12	13	170	30	11
12 Mean duration of sunshine	in h	112	123	166	160	184	174	196	186	170	183	129	117	1880	30	12
13 Mean quantity of radiation	in ly/day															13
14 Mean potential evaporation	in mm	0	0	0	0	0	71	89	80	52	0	0	0	292	40	14
15 Mean windspeed	in m/sec	8.0	7.5	6.5	5.9	5.4	5.7	6.3	6.8	6.2	6.6	7.0	7.7	6.6	30	15
16 Mean predominant direction of the wind		WSW	WSW	WSW	WSW	WSW	WSW	WSW	WSW	WSW	WSW	WSW	WSW		30	16

122 Station/Country Genève/Switzerland

Location 48°12'N/6°09'E **Height above sealevel** 405 m **Climate symbol: Köppen** Cfb **Troll** III,3

		J	F	M	A	M	J	J	A	S	O	N	D	year	P	
1 Mean daily temperature	in °C	1.1	2.2	6.1	10.0	14.1	17.8	19.9	19.1	15.8	10.3	5.7	2.1	10.3	30	1
2 Mean daily maximum temperature	in °C	3.6	5.5	10.4	14.8	19.1	22.9	25.2	24.4	20.5	14.3	8.3	4.4	14.5	30	2
3 Mean daily minimum temperature	in °C	-1.5	-0.9	2.1	5.4	9.2	12.7	14.5	14.3	11.6	6.9	3.0	-0.3	6.4	30	3
4 Absolute maximum temperature	in °C	17.3	20.6	22.9	27.5	32.0	35.7	38.3	38.5	34.8	26.0	23.2	20.8	38.5	60	4
5 Absolute minimum temperature	in °C	-14.9	-18.3	-11.4	-4.5	-1.7	3.8	5.3	4.8	1.2	-3.7	-8.5	-16.0	-18.3	60	5
6 Mean relative humidity	in %	83	79	73	68	69	68	67	71	76	80	82	84	75	30	6
7 Mean precipitation	in mm	63	56	55	51	67	89	64	94	99	72	83	59	852	30	7
8 Maximum precipitation	in mm	147	130	189	244	159	170	203	250	196	212	186	194	1271	40	8
9 Minimum precipitation	in mm	2	2	1	0	17	25	2	14	15	10	4	9	457	40	9
10 Maximum precipitation in 24 h	in mm	46	42	49	45	45	58	63	71	80	61	76	54	80	60	10
11 Mean number of days with precipitation	> 1.0 mm	9	8	8	8	9	10	7	9	9	9	9	9	104	30	11
12 Mean duration of sunshine	in h	54	98	169	206	243	269	297	266	198	131	61	44	2036	30	12
13 Mean quantity of radiation	in ly/day															13
14 Mean potential evaporation	in mm	3	6	23	46	84	109	123	110	75	41	21	5	646	40	14
15 Mean windspeed	in m/sec	3.0	3.6	3.5	3.2	2.9	2.8	2.7	2.3	2.2	2.7	3.5		2.9	10	15
16 Mean predominant direction of the wind		SW	SW	NE	NE	NE	NE	SW	SW	NE	NE	NE	SW		30	16

123 Station/Country Lugano/Switzerland

Location 48°00'N/8°57'E **Height above sealevel** 276 m **Climate symbol: Köppen** Cfb **Troll** III,3

		J	F	M	A	M	J	J	A	S	O	N	D	year	P	
1 Mean daily temperature	in °C	1.9	3.6	7.5	11.5	15.4	19.3	21.4	20.5	17.4	12.1	6.9	3.1	11.7	30	1
2 Mean daily maximum temperature	in °C	6.0	8.7	13.1	17.1	20.8	24.9	27.4	26.6	23.1	16.4	10.8	6.7	16.8	30	2
3 Mean daily minimum temperature	in °C	-1.7	-0.7	2.5	6.5	10.2	13.7	15.8	15.4	12.8	8.1	3.4	-0.3	7.1	30	3
4 Absolute maximum temperature	in °C	24.6	24.8	27.0	31.4	32.6	34.4	38.0	36.4	36.0	28.2	22.8	24.6	38.0	60	4
5 Absolute minimum temperature	in °C	-12.5	-14.0	-6.8	-2.7	0.5	3.3	8.0	6.5	2.0	-2.4	-6.0	-9.6	-14.0	60	5
6 Mean relative humidity	in %	66	63	63	62	65	62	61	65	68	71	71	69	66	30	6
7 Mean precipitation	in mm	62	67	98	148	214	198	185	196	159	173	147	95	1742	30	7
8 Maximum precipitation	in mm															8
9 Minimum precipitation	in mm															9
10 Maximum precipitation in 24 h	in mm	70	80	107	92	125	134	129	263	139	155	100	86	263	60	10
11 Mean number of days with precipitation	> 1.0 mm	8	8	7	11	13	12	10	10	9	9	8	7	107	30	11
12 Mean duration of sunshine	in h	117	143	171	186	191	234	268	243	189	147	110	102	2101	30	12
13 Mean quantity of radiation	in ly/day															13
14 Mean potential evaporation	in mm	4	7	26	51	88	116	133	118	79	45	18	7	692	40	14
15 Mean windspeed	in m/sec	2.0	2.0	2.1	2.1	2.0	1.9	1.9	1.8	1.8	1.8	1.9	1.9	1.9	30	15
16 Mean predominant direction of the wind																16

124 Station/Country Wien (Vienna)/Austria

Location 48°15'N/16°22'E **Height above sealevel** 203 m **Climate symbol: Köppen** Cfb **Troll** III,3

		J	F	M	A	M	J	J	A	S	O	N	D	year	P	
1 Mean daily temperature	in °C	-1.4	0.4	4.7	10.3	14.8	18.1	19.9	19.3	15.8	9.8	4.8	1.0	9.8	30	1
2 Mean daily maximum temperature	in °C	0.9	3.2	8.4	14.5	19.2	22.6	24.6	23.8	20.1	13.5	7.0	2.8	13.4	30	2
3 Mean daily minimum temperature	in °C	-3.6	-2.5	0.9	5.7	10.0	13.5	15.3	14.7	11.4	6.5	2.8	-1.0	6.1	30	3
4 Absolute maximum temperature	in °C	13.2	18.5	24.0	27.3	32.6	36.1	38.3	34.2	31.6	27.8	19.6	16.5	38.3	30	4
5 Absolute minimum temperature	in °C	-21.9	-22.6	-11.2	-3.2	-0.3	4.1	8.8	8.0	-0.1	-3.1	-8.8	-15.3	-22.6	30	5
6 Mean relative humidity	in %	79	76	71	66	68	67	68	70	74	79	81	82	74	50	6
7 Mean precipitation	in mm	39	44	44	45	70	67	84	72	42	56	52	45	660	30	7
8 Maximum precipitation	in mm	93	110	208	148	180	155	203	242	181	212	136	120	988	50	8
9 Minimum precipitation	in mm	5	5	2	2	11	8	14	24	8	1	1	6	404	50	9
10 Maximum precipitation in 24 h	in mm	21	44	41	38	93	66	67	76	65	54	45	50	93	30	10
11 Mean number of days with precipitation	> 0.1 mm	15	14	13	13	13	14	13	13	10	13	14	15	160	30	11
12 Mean duration of sunshine	in h	57	84	138	184	235	249	266	250	199	129	55	45	1891	30	12
13 Mean quantity of radiation	in ly/day															13
14 Mean potential evaporation	in mm	0	2	21	49	89	112	127	111	75	40	15	2	643	50	14
15 Mean windspeed	in m/sec	3.2	3.2	3.2	3.1	3.1	3.1	3.2	2.8	2.7	2.8	3.1	2.9	3.0	30	15
16 Mean predominant direction of the wind		W	W	W	W	W	W	W	W	W	W	W	W		30	16

125 Station / Country Salzburg/Austria

Location 47°48′N/13°00′E — Height above sealevel 435 m — Climate symbol: Köppen Cfb — Troll III,3

		J	F	M	A	M	J	J	A	S	O	N	D	year	P	
1 Mean daily temperature	in °C	-2,5	-1,1	3,7	8,3	13,2	16,0	17,8	17,1	14,0	8,4	3,3	-0,9	8,1	30	1
2 Mean daily maximum temperature	in °C	1,6	3,7	9,3	13,8	18,5	21,9	23,5	23,1	20,0	14,1	7,6	2,7	13,3	26	2
3 Mean daily minimum temperature	in °C	-5,6	-4,9	-0,9	3,5	7,5	11,1	13,0	12,6	9,5	4,6	0,3	-3,6	3,9	26	3
4 Absolute maximum temperature	in °C	15,2	19,4	24,2	30,0	31,2	35,0	36,2	36,3	32,6	26,0	21,5	16,3	36,3	30	4
5 Absolute minimum temperature	in °C	-30,4	-30,8	-18,1	-9,2	-3,4	0,2	5,2	2,0	-3,0	-8,3	-16,7	-27,7	-30,8	30	5
6 Mean relative humidity	in %	80	78	73	71	72	72	72	74	77	80	82	84	76	30	6
7 Mean precipitation	in mm	73	70	70	89	127	167	191	163	111	82	70	65	1278	30	7
8 Maximum precipitation	in mm															8
9 Minimum precipitation	in mm															9
10 Maximum precipitation in 24 h	in mm	77	55	49	64	104	103	135	120	62	50	56	47	135	30	10
11 Mean number of days with precipitation	> 0,1 mm	16	14	13	16	17	18	18	17	14	13	14	14	184	29	11
12 Mean duration of sunshine	in h	62	99	151	154	173	154	180	186	172	146	91	71	1659	30	12
13 Mean quantity of radiation	in ly/day															13
14 Mean potential evaporation	in mm	0	0	20	47	88	109	122	106	74	39	12	0	617	50	14
15 Mean windspeed	in m/sec	1,1	1,8	1,5	1,7	1,5	1,3	1,3	1,3	1,5	1,4	1,2	1,1	1,3	30	15
16 Mean predominent direction of the wind		S	N	N	N	N	N	N	N	N	N	N	S		30	16

126 Station / Country Innsbruck/Austria

Location 47°16′N/11°24′E — Height above sealevel 582 m — Climate symbol: Köppen Cfb — Troll III,3

		J	F	M	A	M	J	J	A	S	O	N	D	year	P	
1 Mean daily temperature	in °C	-2,8	-0,5	4,8	9,3	13,8	16,7	18,1	17,4	14,6	9,0	3,4	-1,1	8,6	30	1
2 Mean daily maximum temperature	in °C	1,1	4,2	10,7	15,5	20,2	23,5	24,8	24,0	20,8	14,7	7,5	2,3	14,1	30	2
3 Mean daily minimum temperature	in °C	-6,5	-4,5	-0,1	3,9	7,9	11,1	12,8	12,4	9,6	4,5	0,1	-4,2	3,9	30	3
4 Absolute maximum temperature	in °C	18,5	17,5	24,8	28,6	32,6	35,9	36,9	34,5	30,7	24,6	17,9	16,9	36,9	30	4
5 Absolute minimum temperature	in °C	-26,8	-26,9	-16,9	-4,7	-2,0	0,8	4,2	3,3	-0,8	-4,2	-15,2	-24,8	-26,9	30	5
6 Mean relative humidity	in %	77	72	65	63	62	66	69	70	72	73	77	79	70	23	6
7 Mean precipitation	in mm	57	52	43	55	77	114	140	113	84	71	57	48	911	30	7
8 Maximum precipitation	in mm															8
9 Minimum precipitation	in mm															9
10 Maximum precipitation in 24 h	in mm	58	34	39	39	61	54	65	78	52	49	39	44	78	30	10
11 Mean number of days with precipitation	> 0,1 mm	13	13	11	14	15	19	19	17	14	12	12	13	172	30	11
12 Mean duration of sunshine	in h	73	105	158	166	189	190	210	199	177	145	86	68	1766	30	12
13 Mean quantity of radiation	in ly/day															13
14 Mean potential evaporation	in mm	0	0	23	50	88	110	120	105	73	41	12	0	622	50	14
15 Mean windspeed	in m/sec	1,2	1,3	1,4	1,6	1,5	1,4	1,2	1,2	1,2	1,2	1,2	1,2	1,3	30	15
16 Mean predominent direction of the wind		N	W	E	E	E	E	E	E	E	E	W	W		30	16

127 Station / Country Sonnblick/Austria

Location 47°03′N/12°57′E — Height above sealevel 3107 m — Climate symbol: Köppen ET — Troll III,3

		J	F	M	A	M	J	J	A	S	O	N	D	year	P	
1 Mean daily temperature	in °C	-13,3	-13,1	-11,2	-8,1	-3,8	-0,7	1,6	1,5	-0,5	-4,4	-8,4	-11,4	-6,0	30	1
2 Mean daily maximum temperature	in °C	-10,8	-10,6	-8,8	-5,6	-1,6	1,5	3,9	3,7	1,6	-2,3	-6,2	-9,2	-3,7	30	2
3 Mean daily minimum temperature	in °C	-15,7	-15,6	-13,5	-10,5	-6,0	-2,8	-0,7	-0,7	-2,5	-6,5	-10,5	-13,6	-8,2	30	3
4 Absolute maximum temperature	in °C	1,0	3,4	1,7		9,4	12,0	12,8	12,0	9,9	8,6	4,8	1,4	12,8	30	4
5 Absolute minimum temperature	in °C	-31,3	-36,8	-30,2	-23,2	-19,8	-10,8	-10,5	-10,0	-15,5	-18,6	-23,2	-30,4	-36,8	30	5
6 Mean relative humidity	in %	77	76	80	84	89	91	90	89	85	80	80	78	83	23	6
7 Mean precipitation	in mm	115	108	112	153	136	142	154	134	104	118	108	111	1495	30	7
8 Maximum precipitation	in mm															8
9 Minimum precipitation	in mm															9
10 Maximum precipitation in 24 h	in mm	76	65	63	50	108	51	34	49	37	57	49	62	108	30	10
11 Mean number of days with precipitation	> 0,1 mm	17	15	16	19	20	20	20	19	15	15	15	15	205	30	11
12 Mean duration of sunshine	in h	109	118	147	139	148	150	187	185	161	147	112	108	1671	30	12
13 Mean quantity of radiation	in ly/day															13
14 Mean potential evaporation	in mm	0	0	0	0	0	0	20	18	0	0	0	0	38	30	14
15 Mean windspeed	in m/sec	7,3	7,1	6,7	5,8	5,2	5,1	5,1	5,2	5,5	6,2	6,8	6,8	6,0	30	15
16 Mean predominent direction of the wind		N	N	SW	SW	SW	N	SW	SW	SW	SW	SW	SW		30	16

128 Station / Country Graz/Austria

Location 46°59′N/15°27′E — Height above sealevel 342 m — Climate symbol: Köppen Dfb — Troll III,3

		J	F	M	A	M	J	J	A	S	O	N	D	year	P	
1 Mean daily temperature	in °C	-3,8	-1,5	3,4	9,0	13,7	17,1	19,0	18,0	14,3	8,6	3,3	-1,3	8,3	30	1
2 Mean daily maximum temperature	in °C	0,8	3,8	9,0	14,8	19,3	22,9	24,8	24,1	20,3	14,2	7,1	2,4	13,6	29	2
3 Mean daily minimum temperature	in °C	-5,0	-4,0	0,1	4,8	9,2	12,5	14,2	13,8	10,5	5,6	1,4	-2,4	5,0	29	3
4 Absolute maximum temperature	in °C	13,6	18,5	23,7	27,5	32,5	36,0	37,1	35,8	32,9	27,0	20,0	15,0	37,1	30	4
5 Absolute minimum temperature	in °C	-23,0	-20,0	-14,8	-3,8	-1,4	3,2	4,9	4,1	-0,3	-6,4	-9,0	-19,0	-23,0	30	5
6 Mean relative humidity	in %	78	74	68	65	67	67	67	70	73	78	81	82	73	22	6
7 Mean precipitation	in mm	31	35	34	52	93	126	114	91	80	79	57	48	840	30	7
8 Maximum precipitation	in mm															8
9 Minimum precipitation	in mm															9
10 Maximum precipitation in 24 h	in mm	29	41	21	40	55	90	105	49	70	56	44	28	105	30	10
11 Mean number of days with precipitation	> 0,1 mm	9	9	9	11	15	15	15	13	11	12	12	11	142	29	11
12 Mean duration of sunshine	in h	75	107	151	178	210	234	259	240	188	133	70	61	1906	30	12
13 Mean quantity of radiation	in ly/day															13
14 Mean potential evaporation	in mm	0	0	20	49	90	114	126	107	72	40	12	0	630	50	14
15 Mean windspeed	in m/sec	1,2	1,4	1,6	1,7	1,8	1,8	1,6	1,4	1,4	1,2	1,2	1,1	1,4	30	15
16 Mean predominent direction of the wind		S	S	N	S	S	NW	NW	NW	NW	S	S	N		30	16

129 Station/Country Debrecen/Hungary

Location 47°33'N/21°37'E — Height above sealevel 123 m — Climate symbol: Köppen Cfb — Troll III,3

		J	F	M	A	M	J	J	A	S	O	N	D	year	P	
1 Mean daily temperature	in °C	-2.7	-0.8	4.5	11.0	16.5	19.8	21.8	20.8	16.4	10.2	4.9	0.5	10.3	30	1
2 Mean daily maximum temperature	in °C	0.3	3.1	9.9	16.4	21.8	25.1	27.2	26.9	23.0	16.2	8.8	3.4	15.2	28	2
3 Mean daily minimum temperature	in °C	-6.0	-3.9	0.1	5.0	9.8	13.1	14.9	14.0	10.2	5.1	2.0	-2.2	5.2	28	3
4 Absolute maximum temperature	in °C	13.8	17.9	25.8	33.6	32.7	37.0	38.5	39.0	36.0	29.5	21.4	16.0	39.0	50	4
5 Absolute minimum temperature	in °C	-30.2	-26.0	-17.8	-7.1	-3.0	-0.4	5.2	2.7	-2.9	-14.9	-19.0	-28.0	-30.2	50	5
6 Mean relative humidity	in %	85	82	74	67	67	69	67	70	73	78	85	87	75	30	6
7 Mean precipitation	in mm	35	36	30	36	61	80	59	64	41	49	53	40	584	30	7
8 Maximum precipitation	in mm															8
9 Minimum precipitation	in mm															9
10 Maximum precipitation in 24 h	in mm	23	29	23	36	65	45	62	57	34	49	31	30	65	44	10
11 Mean number of days with precipitation	> 0,1 mm	8	7	7	8	9	10	8	7	7	8	9	9	97	50	11
12 Mean duration of sunshine	in h	61	82	145	191	258	271	297	268	201	144	67	45	2030	50	12
13 Mean quantity of radiation	in ly/day															13
14 Mean potential evaporation	in mm	0	0	19	50	98	122	138	114	75	40	11	7	674	30	14
15 Mean windspeed	in m/sec	3.3	3.2	3.5	3.5	3.2	2.8	2.7	2.5	2.5	2.8	2.5	3.6	3.0	5	15
16 Mean predominent direction of the wind		SW	SW	SW	SW	NE	SW	SW	SW	NE	SW	SW	NE		50	16

130 Station/Country Budapest/Hungary

Location 47°31'N/19°02'E — Height above sealevel 120 m — Climate symbol: Köppen Cfa — Troll III,3

		J	F	M	A	M	J	J	A	S	O	N	D	year	P	
1 Mean daily temperature	in °C	-1.1	1.0	5.8	11.8	16.8	20.2	22.2	21.4	17.4	11.3	5.8	1.5	11.2	30	1
2 Mean daily maximum temperature	in °C	1.2	4.1	10.1	16.8	21.9	25.5	27.7	27.3	23.2	15.9	8.1	3.6	15.5	26	2
3 Mean daily minimum temperature	in °C	-4.0	-2.3	1.5	6.6	11.3	14.7	16.4	15.7	11.9	6.9	3.1	-1.0	6.7	26	3
4 Absolute maximum temperature	in °C	15.1	18.0	25.4	30.2	32.4	39.5	38.4	39.0	35.2	30.8	22.6	15.7	39.5	50	4
5 Absolute minimum temperature	in °C	-21.7	-23.4	-13.8	-4.2	0.0	3.0	8.9	7.0	1.2	-9.5	-11.9	-19.1	-23.4	50	5
6 Mean relative humidity	in %	81	76	67	60	62	62	60	62	65	74	81	83	69	30	6
7 Mean precipitation	in mm	42	44	39	45	72	76	54	51	34	56	69	48	630	30	7
8 Maximum precipitation	in mm															8
9 Minimum precipitation	in mm															9
10 Maximum precipitation in 24 h	in mm	34	40	37	44	94	55	64	62	62	46	40	37	94	50	10
11 Mean number of days with precipitation	> 0,1 mm	8	7	7	7	9	8	7	6	8	8	9	9	91	50	11
12 Mean duration of sunshine	in h	59	83	136	186	252	289	297	270	195	134	67	40	1988	50	12
13 Mean quantity of radiation	in ly/day															13
14 Mean potential evaporation	in mm	0	2	23	51	100	124	140	122	77	43	13	2	697	30	14
15 Mean windspeed	in m/sec	2.1	2.5	2.7	2.6	2.5	2.5	2.5	2.5	2.0	1.8	1.7	2.2	2.3	5	15
16 Mean predominent direction of the wind		NW	NW	NW	NW	NW	NW	NW	NW	NW	NW	NW	NW		50	16

131 Station/Country Pécs/Hungary

Location 48°05'N/18°15'E — Height above sealevel 141 m — Climate symbol: Köppen Cfa — Troll III,3

		J	F	M	A	M	J	J	A	S	O	N	D	year	P	
1 Mean daily temperature	in °C	-0.7	1.3	6.1	11.9	16.9	20.4	22.6	21.9	17.9	11.8	6.2	1.8	11.5	30	1
2 Mean daily maximum temperature	in °C	2.4	4.4	10.1	16.6	21.7	25.1	27.5	27.6	23.5	16.5	9.0	4.5	15.7	20	2
3 Mean daily minimum temperature	in °C	-3.7	-3.1	0.4	5.5	10.3	13.6	15.8	15.0	11.6	6.8	2.8	-1.4	6.1	20	3
4 Absolute maximum temperature	in °C	17.1	18.7	24.2	29.9	33.8	38.9	41.3	39.6	34.4	29.4	22.4	16.3	41.3	29	4
5 Absolute minimum temperature	in °C	-27.0	-23.4	-14.1	-7.6	-2.6	3.1	3.1	3.4	0.9	-5.8	-9.8	-16.8	-27.0	29	5
6 Mean relative humidity	in %	83	78	70	64	66	64	62	63	67	75	81	83	71	30	6
7 Mean precipitation	in mm	41	46	41	58	66	69	64	55	47	64	71	45	667	30	7
8 Maximum precipitation	in mm															8
9 Minimum precipitation	in mm															9
10 Maximum precipitation in 24 h	in mm	40	38	54	39	45	72	41	58	59	69	59	37	72	18	10
11 Mean number of days with precipitation	> 0,1 mm	8	7	8	9	10	8	8	7	7	8	9	9	98	50	11
12 Mean duration of sunshine	in h	67	93	136	179	249	266	299	278	193	141	73	51	2025	50	12
13 Mean quantity of radiation	in ly/day															13
14 Mean potential evaporation	in mm	0	1	23	52	95	124	141	128	80	44	15	3	706	30	14
15 Mean windspeed	in m/sec	3.2	3.5	4.0	4.1	3.2	3.0	2.9	2.8	2.7	3.0	3.1	3.4	3.3	5	15
16 Mean predominent direction of the wind		NW	NW	NW	NW	W	N	N	N	NE	NE	NE	NE		40	16

132 Station/Country Iaşi/Romania

Location 47°10'N/27°36'E — Height above sealevel 101 m — Climate symbol: Köppen Dfb — Troll III,4

		J	F	M	A	M	J	J	A	S	O	N	D	year	P	
1 Mean daily temperature	in °C	-4.1	-2.3	2.5	10.0	16.0	19.5	21.6	20.7	16.2	10.0	4.0	-1.0	9.4		1
2 Mean daily maximum temperature	in °C	-0.4	1.7	7.6	16.1	22.5	25.8	28.4	27.6	22.9	15.7	7.8	2.2	14.8	30	2
3 Mean daily minimum temperature	in °C	-7.8	-6.1	-2.0	4.2	9.7	13.4	15.4	14.5	10.4	5.2	0.7	-4.1	4.5	30	3
4 Absolute maximum temperature	in °C	15.4	18.2	25.5	31.5	36.4	38.9	39.0	39.6	38.0	33.9	21.1	18.2	39.6		4
5 Absolute minimum temperature	in °C	-29.0	-30.0	-22.7	-7.0	-2.1	3.5	8.5	6.0	-1.0	-6.5	-18.0	-29.5	-30.0		5
6 Mean relative humidity	in %	83	80	72	62	61	62	60	63	66	73	81	85	71		6
7 Mean precipitation	in mm	32	30	21	37	48	73	62	64	36	35	39	29	506		7
8 Maximum precipitation	in mm															8
9 Minimum precipitation	in mm															9
10 Maximum precipitation in 24 h	in mm	35	24	28	38	49	83	111	51	33	55	47	20	111		10
11 Mean number of days with precipitation	> 0,1 mm	10	11	9	9	11	12	10	9	7	8	10	10	116		11
12 Mean duration of sunshine	in h	72	75	140	175	229	266	299	278	205	147	62	60	2008	13	12
13 Mean quantity of radiation	in ly/day															13
14 Mean potential evaporation	in mm	0	0	12	49	96	121	136	123	77	17	9	0	640	60	14
15 Mean windspeed	in m/sec															15
16 Mean predominent direction of the wind		NW	NW	NW	NW	NW	NW	NW	NW	NW	NW	SE	NW			16

133 Station/Country Cluj/Romania

Location 46°47'N/23°42'E Height above sealevel 313 m Climate symbol: Köppen **Dfb** Troll III,4

		J	F	M	A	M	J	J	A	S	O	N	D	year	P	
1 Mean daily temperature	in °C	-3,4	-2,7	3,2	8,9	14,0	18,1	19,9	19,3	14,8	8,7	3,5	-0,3	8,8		1
2 Mean daily maximum temperature	in °C	-0,1	2,3	9,5	15,5	21,4	24,1	26,7	25,6	21,8	15,9	8,1	2,5	14,4	24	2
3 Mean daily minimum temperature	in °C	-7,8	-6,4	-1,5	3,8	8,7	12,0	13,5	12,6	8,4	3,8	0,3	-3,7	3,6	24	3
4 Absolute maximum temperature	in °C	12,2	16,6	24,6	25,1	31,6	31,7	36,2	38,0	31,6	32,6	18,6	16,3	38,0		4
5 Absolute minimum temperature	in °C	-29,6	-27,7	-21,0	-5,8	-2,9	0,4	5,6	4,7	-2,7	-7,2	-12,7	-22,4	-29,8		5
6 Mean relative humidity	in %	86	84	70	66	67	68	63	66	68	75	83	88	74		6
7 Mean precipitation	in mm	27	36	20	49	72	73	64	71	34	30	31	31	538		7
8 Maximum precipitation	in mm															8
9 Minimum precipitation	in mm															9
10 Maximum precipitation in 24 h	in mm	21	22	23	27	26	65	40	51	46	25	26	18	64		10
11 Mean number of days with precipitation	> 0,1 mm	13	13	12	14	16	14	11	11	9	9	11	15	148		11
12 Mean duration of sunshine	in h	64	88	170	176	222	233	286	271	209	163	80	54	2016	13	12
13 Mean quantity of radiation	in ly/day															13
14 Mean potential evaporation	in mm	0	0	13	46	92	114	126	110	71	39	8	.0	619	35	14
15 Mean windspeed	in m/sec															15
16 Mean predominent direction of the wind		W	W	NW	NW	NW	NW	NW	NW	NW	SE	SE	E			16

134 Station/Country Timişoara (Temeswar)/Romania

Location 45°47'N/21°13'E Height above sealevel 90 m Climate symbol: Köppen **Cfb** Troll III,3

		J	F	M	A	M	J	J	A	S	O	N	D	year	P	
1 Mean daily temperature	in °C	-1,6	0,4	5,5	11,4	16,4	19,7	21,7	21,0	17,2	11,3	6,0	1,5	10,9		1
2 Mean daily maximum temperature	in °C															2
3 Mean daily minimum temperature	in °C															3
4 Absolute maximum temperature	in °C	16,7	18,6	27,0	32,0	34,5	38,4	39,6	40,0	39,7	33,8	23,0	18,5	40,0		4
5 Absolute minimum temperature	in °C	-27,0	-29,2	-20,0	-5,2	-5,0	3,6	7,3	5,8	-1,0	-5,1	-10,6	-24,5	-29,2		5
6 Mean relative humidity	in %	85	81	75	65	66	65	62	63	67	75	83	86	73		6
7 Mean precipitation	in mm	45	43	40	43	71	76	56	50	40	53	59	49	625		7
8 Maximum precipitation	in mm															8
9 Minimum precipitation	in mm															9
10 Maximum precipitation in 24 h	in mm	33	30	25	29	55	83	73	38	54	39	40	25	83		10
11 Mean number of days with precipitation	> 0,1 mm	11	11	10	11	13	12	9	9	7	10	13	12	128		11
12 Mean duration of sunshine	in h															12
13 Mean quantity of radiation	in ly/day															13
14 Mean potential evaporation	in mm	0	24	22	51	98	128	141	123	77	44	13	4	725	45	14
15 Mean windspeed	in m/sec	2,0	2,2	2,4	2,6	2,3	2,2	2,2	2,2	2,6	2,6	1,8	2,0	2,0	63	15
16 Mean predominent direction of the wind		E	N	N	N	N	N	N	E	E	E	E	E		20	16

135 Station/Country Sibiu/Romania

Location 45°47'N/24°09'E Height above sealevel 407 m Climate symbol: Köppen **Dfb** Troll III,4

		J	F	M	A	M	J	J	A	S	O	N	D	year	P	
1 Mean daily temperature	in °C	-4,0	-1,6	3,7	9,6	14,6	17,8	19,9	19,3	15,4	9,6	4,2	-0,9	9,0	30	1
2 Mean daily maximum temperature	in °C	0,2	2,8	8,7	15,1	20,4	23,5	26,1	25,6	21,5	15,5	8,1	2,7	14,2	30	2
3 Mean daily minimum temperature	in °C	-8,2	-6,0	-1,3	4,0	8,8	12,0	13,6	12,9	9,1	4,1	0,2	-4,4	3,7	30	3
4 Absolute maximum temperature	in °C	15,0	19,0	30,6	30,1	31,4	33,8	36,4	37,4	38,2	32,5	23,6	19,3	37,4	30	4
5 Absolute minimum temperature	in °C	-30,4	-29,4	-22,5	-6,7	-1,4	2,6	5,6	4,5	-3,4	-7,5	-20,0	-29,3	-30,4	30	5
6 Mean relative humidity	in %	85	81	78	69	71	73	72	72	74	78	82	85	76	30	6
7 Mean precipitation	in mm	30	30	30	59	76	118	82	69	48	47	33	30	647	30	7
8 Maximum precipitation	in mm															8
9 Minimum precipitation	in mm															9
10 Maximum precipitation in 24 h	in mm	27	30	24	67	42	71	81	61	33	37	43	19	71	30	10
11 Mean number of days with precipitation	> 0,1 mm	11	11	10	12	16	15	11	11	9	10	10	11	137	30	11
12 Mean duration of sunshine	in h	69	89	151	167	200	218	263	255	212	162	84	67	1935	13	12
13 Mean quantity of radiation	in ly/day															13
14 Mean potential evaporation	in mm	0	0	17	47	95	115	129	113	74	42	9	0	641	45	14
15 Mean windspeed	in m/sec															15
16 Mean predominent direction of the wind																16

136 Station/Country Bucureşti (Bucharest)/Romania

Location 44°25'N/26°06'E Height above sealevel 82 m Climate symbol: Köppen **Cfa** Troll III,3

		J	F	M	A	M	J	J	A	S	O	N	D	year	P	
1 Mean daily temperature	in °C	-2,7	-0,8	4,8	11,7	17,0	20,9	23,3	22,7	18,3	12,0	5,5	0,4	11,1		1
2 Mean daily maximum temperature	in °C	0,9	3,5	9,8	17,9	23,4	27,3	29,8	29,7	25,3	18,1	9,6	3,8	16,6	29	2
3 Mean daily minimum temperature	in °C	-7,0	-5,1	-0,9	4,8	10,3	13,7	15,8	15,0	11,1	6,0	1,5	-3,1	5,2	29	3
4 Absolute maximum temperature	in °C	16,6	20,8	28,8	34,4	36,6	37,2	39,3	41,1	39,6	35,5	23,5	19,7	41,1		4
5 Absolute minimum temperature	in °C	-30,0	-23,6	-13,3	-4,0	0,8	7,1	8,6	7,1	0,0	-4,0	-11,3	-19,9	-30,0		5
6 Mean relative humidity	in %	86	82	71	63	62	61	58	57	61	73	84	87	70		6
7 Mean precipitation	in mm	43	36	35	47	69	87	55	49	30	44	43	41	549		7
8 Maximum precipitation	in mm															8
9 Minimum precipitation	in mm															9
10 Maximum precipitation in 24 h	in mm	50	34	36	45	62	88	87	135	57	42	35	34	135		10
11 Mean number of days with precipitation	> 0,1 mm	10	10	10	10	13	12	9	8	8	10	11	11	115		11
12 Mean duration of sunshine	in h	65	85	150	191	242	272	326	301	233	161	73	60	2159	26	12
13 Mean quantity of radiation	in ly/day	124	198	292	417	519	563	558	496	378	263	117	82	334	6	13
14 Mean potential evaporation	in mm	0	0	16	51	98	126	144	128	83	46	11	0	703	60	14
15 Mean windspeed	in m/sec	2,2	2,3	2,4	2,4	2,0	1,5	1,4	1,5	1,5	1,8	2,0	2,0	2,0	60	15
16 Mean predominent direction of the wind		W	W	E	E	E	E	E	E	E	E	E	W		20	16

137 Station/Country Bragança/Portugal

Location 41°49'N/8°46'W   Height above sealevel 720 m   Climate symbol: Köppen Csb   Troll IV,1

		J	F	M	A	M	J	J	A	S	O	N	D	year	P	
1	Mean daily temperature	in °C	3,8	5,8	7,9	10,2	12,9	17,3	20,2	20,3	16,9	12,0	7,8	4,4	11,8	30	1
2	Mean daily maximum temperature	in °C	7,6	10,5	12,7	15,8	18,5	23,9	27,8	28,1	23,5	17,5	11,9	8,0	17,1	30	2
3	Mean daily minimum temperature	in °C	0,0	0,8	3,1	4,6	7,2	10,7	12,7	12,5	10,2	6,5	3,2	0,7	6,0	30	3
4	Absolute maximum temperature	in °C	18,0	21,4	25,0	28,5	31,2	36,2	39,5	38,2	35,6	28,6	22,2	16,8	39,5	30	4
5	Absolute minimum temperature	in °C	-12,0	-11,0	-7,0	-4,3	-2,0	0,8	3,0	0,5	-1,0	-4,0	-6,4	-10,0	-12,0	30	5
6	Mean relative humidity	in %	85	81	77	71	71	67	62	62	69	77	82	86	74	30	6
7	Mean precipitation	in mm	149	104	133	73	69	42	15	16	39	79	110	144	973	30	7
8	Maximum precipitation	in mm															8
9	Minimum precipitation	in mm															9
10	Maximum precipitation in 24 h	in mm	209	116	97	41	50	66	72	27	52	85	66	124	209	30	10
11	Mean number of days with precipitation	> 0,1 mm	15	12	15	10	11	7	3	4	7	10	14	17	125	30	11
12	Mean duration of sunshine	in h	121	156	184	241	268	318	384	347	254	197	142	112	2724	30	12
13	Mean quantity of radiation	in ly/day															13
14	Mean potential evaporation	in mm	11	16	30	46	69	100	119	113	81	46	21	10	662	11	14
15	Mean windspeed	in m/sec															15
16	Mean predominent direction of the wind																16

138 Station/Country Porto/Portugal

Location 41°08'N/8°36'W   Height above sealevel 95 m   Climate symbol: Köppen Csb   Troll IV,1

		J	F	M	A	M	J	J	A	S	O	N	D	year	P	
1	Mean daily temperature	in °C	9,0	9,6	11,9	13,6	15,2	18,0	19,6	19,6	18,6	15,6	12,2	9,6	14,4		1
2	Mean daily maximum temperature	in °C	13,2	14,2	16,3	16,6	19,6	22,6	24,7	25,0	23,7	20,8	16,7	13,7	19,1	30	2
3	Mean daily minimum temperature	in °C	4,7	5,0	7,5	8,8	10,8	13,4	14,6	14,8	13,6	10,8	7,8	5,4	9,8	30	3
4	Absolute maximum temperature	in °C	21,7	29,0	28,5	31,9	33,3	36,6	40,1	39,4	37,2	34,4	25,7	21,9	40,1		4
5	Absolute minimum temperature	in °C	-4,1	-3,8	-1,9	0,6	3,6	6,8	8,8	8,8	5,6	1,5	-1,3	-3,7	-4,1		5
6	Mean relative humidity	in %	78	75	73	69	71	70	67	68	72	73	77	78	73	30	6
7	Mean precipitation	in mm	159	112	147	86	87	41	20	26	51	105	148	168	1153		7
8	Maximum precipitation	in mm															8
9	Minimum precipitation	in mm															9
10	Maximum precipitation in 24 h	in mm	95	56	44	41	85	69	43	36	72	61	101	62	101		10
11	Mean number of days with precipitation	> 0,1 mm	18	15	17	13	13	7	5	7	11	15	18	18	157		11
12	Mean duration of sunshine	in h	142	163	183	249	275	298	330	311	238	194	151	134	2668		12
13	Mean quantity of radiation	in ly/day	158	233	340	512	589	651	665	583	439	299	186	142	400		13
14	Mean potential evaporation	in mm	24	27	42	54	72	92	107	105	79	58	34	22	716	11	14
15	Mean windspeed	in m/sec	5,8	5,6	5,7	5,6	5,3	4,9	4,7	4,6	4,4	4,6	5,4	5,7	5,2		15
16	Mean predominent direction of the wind		E	E	E	NW	NW	NW	NW	NW	NW	E	E	E			16

139 Station/Country Coimbra/Portugal

Location 40°12'N/8°25'W   Height above sealevel 141 m   Climate symbol: Köppen Csa   Troll IV,1

		J	F	M	A	M	J	J	A	S	O	N	D	year	P	
1	Mean daily temperature	in °C	9,7	10,8	13,2	15,1	16,8	19,6	21,9	22,2	20,7	17,4	13,3	10,3	15,9	30	1
2	Mean daily maximum temperature	in °C	14,0	15,7	18,2	20,9	22,5	26,1	28,9	29,3	27,2	23,0	17,8	14,4	21,5	30	2
3	Mean daily minimum temperature	in °C	5,4	5,8	8,2	9,3	11,0	13,6	14,9	15,0	14,1	11,1	8,7	6,1	10,3	30	3
4	Absolute maximum temperature	in °C	22,9	26,1	30,5	35,6	36,9	40,8	45,6	43,7	39,6	36,7	28,4	21,6	45,8	30	4
5	Absolute minimum temperature	in °C	-3,6	-4,0	-2,2	0,8	4,2	6,8	9,0	8,9	5,5	2,2	-0,9	-2,8	-4,0	30	5
6	Mean relative humidity	in %	75	68	67	60	62	61	57	57	61	65	72	76	65	30	6
7	Mean precipitation	in mm	132	95	131	76	76	38	13	16	48	87	105	142	901	30	7
8	Maximum precipitation	in mm															8
9	Minimum precipitation	in mm															9
10	Maximum precipitation in 24 h	in mm	76	57	61	64	55	48	36	36	123	58	55	83	123	30	10
11	Mean number of days with precipitation	> 0,1 mm	15	13	16	13	13	8	4	5	9	13	14	16	138	30	11
12	Mean duration of sunshine	in h	143	164	184	234	252	275	324	308	236	198	151	138	2605	30	12
13	Mean quantity of radiation	in ly/day															13
14	Mean potential evaporation	in mm	23	27	41	52	77	101	116	111	88	60	33	22	751	30	14
15	Mean windspeed	in m/sec															15
16	Mean predominent direction of the wind																16

140 Station/Country Campo Major near Badajoz/Portugal

Location 39°01'N/7°04'W   Height above sealevel 280 m   Climate symbol: Köppen Csa   Troll IV,1

		J	F	M	A	M	J	J	A	S	O	N	D	year	P	
1	Mean daily temperature	in °C	8,7	10,1	12,6	15,1	17,6	22,1	25,1	25,0	22,4	17,7	12,6	9,2	16,5		1
2	Mean daily maximum temperature	in °C															2
3	Mean daily minimum temperature	in °C															3
4	Absolute maximum temperature	in °C	22,5	26,0	29,0	34,3	38,0	41,1	45,6	44,1	40,1	35,8	26,9	22,2	45,6		4
5	Absolute minimum temperature	in °C	-4,2	-5,0	-1,3	1,4	3,1	6,5	8,2	9,0	6,3	3,1	-1,0	-3,0	-5,0		5
6	Mean relative humidity	in %															6
7	Mean precipitation	in mm	64	55	81	44	38	20	2	4	29	53	62	68	520		7
8	Maximum precipitation	in mm															8
9	Minimum precipitation	in mm															9
10	Maximum precipitation in 24 h	in mm	41	41	51	64	37	40	6	26	92	50	93	58	93		10
11	Mean number of days with precipitation	> 0,1 mm	11	9	12	8	8	3	<1	<1	3	8	9	11	84		11
12	Mean duration of sunshine	in h	161	183	218	261	281	348	384	362	277	223	171	181	2930		12
13	Mean quantity of radiation	in ly/day															13
14	Mean potential evaporation	in mm	16	22	35	52	82	120	149	145	104	62	29	17	831	30	14
15	Mean windspeed	in m/sec	2,3	2,2	2,3	2,5	2,4	2,5	2,7	2,8	2,2	2,0	2,1	2,0	2,3		15
16	Mean predominent direction of the wind		var	var	W	W	W	W	W	W	W	W	W	W			16

39

141 Station/Country Lisboa (Lisbon)/Portugal

Location 38°43'N/9°09'W   Height above sealevel 77 m   Climate symbol: Köppen Csa   Troll IV,1

		J	F	M	A	M	J	J	A	S	O	N	D	year	P	
1 Mean daily temperature	in °C	10,8	11,8	13,6	15,6	17,2	20,1	22,2	22,5	21,2	18,2	14,4	11,5	16,6		1
2 Mean daily maximum temperature	in °C															2
3 Mean daily minimum temperature	in °C															3
4 Absolute maximum temperature	in °C	20,6	25,4	27,4	31,0	34,3	37,7	39,9	40,3	35,5	35,3	26,7	20,9	40,3		4
5 Absolute minimum temperature	in °C	-0,5	-1,2	2,8	4,4	6,4	9,8	12,1	13,3	10,3	6,7	3,6	0,0	-1,2		5
6 Mean relative humidity	in %	78	72	71	63	63	60	55	57	62	67	75	78	67	30	6
7 Mean precipitation	in mm	111	76	109	54	44	16	3	4	33	62	93	103	708		7
8 Maximum precipitation	in mm															8
9 Minimum precipitation	in mm															9
10 Maximum precipitation in 24 h	in mm	59	59	70	35	50	36	18	25	50	64	88	78	88		10
11 Mean number of days with precipitation	> 0,1 mm	15	12	14	10	10	5	2	2	6	9	13	15	113		11
12 Mean duration of sunshine	in h	161	182	206	265	301	330	378	357	279	231	174	159	3023		12
13 Mean quantity of radiation	in ly/day	195	276	380	527	604	674	705	623	478	339	225	177	434		13
14 Mean potential evaporation	in mm	26	30	43	54	77	89	117	115	92	69	38	27	773	30	14
15 Mean windspeed	in m/sec	4,1	4,1	4,2	4,3	4,2	4,5	4,8	4,4	4,0	3,6	3,8	3,9	4,1		15
16 Mean predominent direction of the wind		N	N	N	N	N	N	N	N	N	N	N	N			16

142 Station/Country Ponta Delgada,São Miguel(Azores)/Portugal

Location 37°45'N/25°40'W   Height above sealevel 35 m   Climate symbol: Köppen Csa   Troll IV,1

		J	F	M	A	M	J	J	A	S	O	N	D	year	P	
1 Mean daily temperature	in °C	14,4	14,2	14,5	15,1	16,5	18,8	20,9	22,0	21,1	19,1	16,8	15,3	17,4	30	1
2 Mean daily maximum temperature	in °C	17,2	17,1	17,4	18,3	19,9	22,4	24,7	25,9	24,8	22,4	19,8	18,2	20,7	30	2
3 Mean daily minimum temperature	in °C	11,5	11,3	11,5	11,9	13,1	15,2	17,0	18,0	17,3	15,7	13,6	12,4	14,1	30	3
4 Absolute maximum temperature	in °C	22,0	20,9	21,5	23,1	25,2	28,0	29,5	30,7	30,5	28,0	25,2	24,4	30,7	30	4
5 Absolute minimum temperature	in °C	2,8	4,4	4,5	6,0	7,4	9,5	11,0	11,5	11,0	9,0	7,8	6,0	2,8	30	5
6 Mean relative humidity	in %	80	78	77	75	76	77	74	74	74	76	79	80	76	30	6
7 Mean precipitation	in mm	120	100	105	67	62	42	27	29	81	103	120	102	958	30	7
8 Maximum precipitation	in mm															8
9 Minimum precipitation	in mm															9
10 Maximum precipitation in 24 h	in mm	61	65	59	62	63	35	83	42	75	219	147	97	219	30	10
11 Mean number of days with precipitation	> 0,1 mm	20	18	19	14	14	12	10	10	14	18	19	20	188	30	11
12 Mean duration of sunshine	in h	80	91	121	137	158	162	192	204	171	136	93	81	1626	30	12
13 Mean quantity of radiation	in ly/day															13
14 Mean potential evaporation	in mm	40	38	47	55	73	92	110	112	92	71	50	43	823	30	14
15 Mean windspeed	in m/sec															15
16 Mean predominent direction of the wind																16

143 Station/Country Praia da Rocha(Algarve)/Portugal

Location 37°07'N/8°32'W   Height above sealevel 19 m   Climate symbol: Köppen Csa   Troll IV,1

		J	F	M	A	M	J	J	A	S	O	N	D	year	P	
1 Mean daily temperature	in °C	11,8	12,1	13,6	15,4	16,8	20,4	22,8	23,0	31,3	18,3	15,1	12,5	16,9		1
2 Mean daily maximum temperature	in °C															2
3 Mean daily minimum temperature	in °C															3
4 Absolute maximum temperature	in °C	20,8	22,8	25,5	30,3	33,9	35,1	40,6	37,7	34,5	32,2	27,5	21,4	40,6		4
5 Absolute minimum temperature	in °C	0,0	-1,9	3,5	5,0	6,1	9,7	12,2	13,6	10,5	7,2	4,2	0,0	-1,9		5
6 Mean relative humidity	in %															6
7 Mean precipitation	in mm	59	38	69	31	24	7	2	<1	18	44	58	66	414		7
8 Maximum precipitation	in mm															8
9 Minimum precipitation	in mm															9
10 Maximum precipitation in 24 h	in mm	58	27	48	53	43	30	34	9	57	38	84	58	84		10
11 Mean number of days with precipitation	> 0,1 mm	10	8	11	7	5	2	<1	<1	3	7	9	10	74		11
12 Mean duration of sunshine	in h	175	191	219	276	322	352	387	363	281	235	181	180	3162		12
13 Mean quantity of radiation	in ly/day															13
14 Mean potential evaporation	in mm	23	28	40	56	70	103	131	126	173	67	41	27	884		14
15 Mean windspeed	in m/sec	4,6	4,5	4,6	4,5	4,1	4,1	3,9	3,8	3,8	3,8	4,2	4,3	4,2		15
16 Mean predominent direction of the wind		NE	NE	NW	NW	NW	NW	NW	NW	NW	NW	N	N			16

144 Station/Country Santander/Spain

Location 43°48'N/3°49'W   Height above sealevel 68 m   Climate symbol: Köppen Cfb   Troll III,2

		J	F	M	A	M	J	J	A	S	O	N	D	year	P	
1 Mean daily temperature	in °C	9,3	9,2	11,5	12,3	14,2	16,9	18,8	19,3	18,2	15,3	12,2	9,9	13,9	27	1
2 Mean daily maximum temperature	in °C	11,9	12,1	14,4	15,3	16,9	19,8	21,5	22,2	21,2	18,3	14,8	12,5	16,7	27	2
3 Mean daily minimum temperature	in °C	6,7	6,7	8,1	9,5	11,3	14,2	16,1	16,4	14,9	12,4	9,5	7,8	11,1	27	3
4 Absolute maximum temperature	in °C	21,0	22,8	30,0	33,4	31,0	34,0	34,6	40,2	34,0	29,6	23,8	21,4	40,2		4
5 Absolute minimum temperature	in °C	-2,6	-3,8	0,4	2,0	3,6	7,5	11,0	12,0	7,7	4,4	1,8	-0,2	-3,8		5
6 Mean relative humidity	in %	75	76	74	77	80	81	80	81	80	78	76	76	78	21	6
7 Mean precipitation	in mm	119	89	74	82	88	66	59	84	114	134	134	155	1198		7
8 Maximum precipitation	in mm															8
9 Minimum precipitation	in mm															9
10 Maximum precipitation in 24 h	in mm	61	62	73	65	73	68	86	78	138	51	58	55	138		10
11 Mean number of days with precipitation	> 0,1 mm	16	14	13	14	15	13	12	14	14	14	16	18	173		11
12 Mean duration of sunshine	in h	85	99	140	165	187	199	212	197	158	134	97	76	1747		12
13 Mean quantity of radiation	in ly/day															13
14 Mean potential evaporation	in mm	21	29	39	48	76	93	110	106	85	62	36	28	733	30	14
15 Mean windspeed	in m/sec	4,5	2,5	4,2	3,8	3,4	3,4	3,1	2,8	2,8	3,4	4,2	4,5	3,8		15
16 Mean predominent direction of the wind		S	S,NW	S,NW	NW	NW	NW	NW	NW	NW	NW	S	S			16

145 Station / Country La Coruña / Spain

Location 43°22'N/8°25'W		Height above sealevel 67 m				Climate symbol: Köppen Csb				Troll III,2					
		J	F	M	A	M	J	J	A	S	O	N	D	year	P
1 Mean daily temperature	in °C	9,9	9,8	11,5	12,4	14,0	16,5	18,2	18,9	17,8	15,3	12,4	10,2	13,9	27
2 Mean daily maximum temperature	in °C	12,8	13,0	15,0	16,0	17,5	20,1	21,9	22,8	21,8	18,7	15,4	13,2	17,3	27
3 Mean daily minimum temperature	in °C	7,1	6,8	8,1	9,1	10,8	12,9	14,8	14,9	14,1	11,8	9,4	7,5	10,8	27
4 Absolute maximum temperature	in °C	20,4	27,4	26,6	29,6	28,5	30,5	33,5	33,6	30,5	31,0	25,0	19,8	33,6	27
5 Absolute minimum temperature	in °C	-2,0	-3,0	1,0	2,0	3,4	7,2	9,9	9,4	6,7	4,5	1,0	-1,0	-3,0	27
6 Mean relative humidity	in %	78	77	79	75	75	77	78	78	80	79	79	81	78	21
7 Mean precipitation	in mm	118	78	95	71	58	47	30	44	76	89	128	139	971	
8 Maximum precipitation	in mm														
9 Minimum precipitation	in mm														
10 Maximum precipitation in 24 h	in mm	50	39	44	38	41	60	33	75	83	60	73	91	91	
11 Mean number of days with precipitation	> 0,1 mm	18	14	16	13	13	10	8	9	12	14	17	19	162	
12 Mean duration of sunshine	in h	95	117	148	192	218	223	264	248	191	151	111	84	2042	
13 Mean quantity of radiation	in ly/day														
14 Mean potential evaporation	in mm	28	30	40	50	73	87	102	96	77	59	36	31	709	30
15 Mean windspeed	in m/sec	6,4	6,4	5,9	5,9	5,9	5,3	4,8	4,8	4,2	4,8	5,3	5,6	5,3	
16 Mean predominent direction of the wind		SW	SW	SW	NE	NE	NE	NE	N	NE	NE	SW	SW		

146 Station / Country Valladolid / Spain

Location 41°39'N/4°43'W		Height above sealevel 715 m				Climate symbol: Köppen BSk				Troll IV,2					
		J	F	M	A	M	J	J	A	S	O	N	D	year	P
1 Mean daily temperature	in °C	3,3	5,1	8,6	11,0	14,1	18,5	21,3	20,4	17,8	12,9	7,7	4,4	12,1	
2 Mean daily maximum temperature	in °C	7,6	10,2	14,3	17,1	20,1	25,6	29,4	28,6	24,9	18,5	12,3	7,9	18,0	27
3 Mean daily minimum temperature	in °C	0,2	0,0	2,9	4,8	7,7	11,5	13,5	13,4	11,3	8,0	2,9	1,0	6,3	27
4 Absolute maximum temperature	in °C	17,9	24,8	26,0	31,0	33,2	35,8	39,0	38,3	36,1	29,8	24,8	16,9	39,0	
5 Absolute minimum temperature	in °C	-10,6	-11,6	-5,2	-2,7	-1,2	4,3	6,2	7,0	3,6	-2,0	-4,6	-7,2	-11,6	
6 Mean relative humidity	in %	84	78	69	64	62	57	51	55	63	71	80	85	68	21
7 Mean precipitation	in mm	30	26	42	30	35	33	13	13	28	34	40	40	364	
8 Maximum precipitation	in mm														
9 Minimum precipitation	in mm														
10 Maximum precipitation in 24 h	in mm	24	18	30	26	19	32	15	35	14	35	44	44	44	
11 Mean number of days with precipitation	> 0,1 mm	8	8	11	8	10	7	4	4	6	9	8	10	91	
12 Mean duration of sunshine	in h	105	142	108	231	263	318	345	341	243	197	131	88	2584	
13 Mean quantity of radiation	in ly/day														
14 Mean potential evaporation	in mm	8	14	29	45	78	104	128	124	82	48	21	10	691	30
15 Mean windspeed	in m/sec	3,1	3,4	3,4	3,4	3,4	3,1	3,1	3,1	3,4	2,5	2,8	3,1	3,1	
16 Mean predominent direction of the wind		N	S	S	N	N	N	N	N	N	N	N	N		

147 Station / Country Zaragoza / Spain

Location 41°39'N/0°53'W		Height above sealevel 237 m				Climate symbol: Köppen BSk				Troll IV,2					
		J	F	M	A	M	J	J	A	S	O	N	D	year	P
1 Mean daily temperature	in °C	6,1	7,9	11,2	13,8	17,0	21,3	24,1	23,7	20,9	15,4	10,2	6,8	14,9	27
2 Mean daily maximum temperature	in °C	9,9	12,5	16,6	19,3	22,6	27,4	30,8	30,0	26,5	20,3	14,3	10,1	20,0	27
3 Mean daily minimum temperature	in °C	2,2	3,3	5,8	8,2	11,4	15,2	17,5	17,4	15,2	10,5	6,1	3,4	9,7	27
4 Absolute maximum temperature	in °C	21,0	22,8	28,5	31,6	35,8	40,0	40,0	42,0	36,2	29,2	23,8	19,2	42,0	27
5 Absolute minimum temperature	in °C	-10,5	-8,6	-6,8	0,0	4,3	6,0	11,0	11,0	7,8	1,0	-3,2	-6,2	-10,5	27
6 Mean relative humidity	in %	74	78	69	64	62	57	51	55	63	71	80	85	67	21
7 Mean precipitation	in mm	17	17	27	34	49	36	15	19	30	34	28	31	337	27
8 Maximum precipitation	in mm														
9 Minimum precipitation	in mm														
10 Maximum precipitation in 24 h	in mm	36	27	55	34	47	41	65	37	39	49	104	55	104	27
11 Mean number of days with precipitation	> 0,1 mm	6	5	7	8	8	7	3	3	6	6	5	7	71	27
12 Mean duration of sunshine	in h	149	182	215	245	289	305	361	330	257	209	158	124	2824	8
13 Mean quantity of radiation	in ly/day														
14 Mean potential evaporation	in mm	11	17	33	49	88	118	143	137	93	55	25	14	783	30
15 Mean windspeed	in m/sec														
16 Mean predominent direction of the wind															

148 Station / Country Barcelona / Spain

Location 41°24'N/2°09'E		Height above sealevel 95 m				Climate symbol: Köppen Csa				Troll IV,1					
		J	F	M	A	M	J	J	A	S	O	N	D	year	P
1 Mean daily temperature	in °C	9,4	9,9	12,3	14,6	17,7	21,6	24,4	24,2	21,7	17,5	13,5	10,2	16,4	
2 Mean daily maximum temperature	in °C	12,7	13,7	15,6	18,2	21,4	25,4	27,9	27,7	25,0	20,5	16,4	13,1	19,8	27
3 Mean daily minimum temperature	in °C	6,3	7,1	8,9	11,0	14,0	18,1	20,7	20,8	18,7	14,5	10,5	7,5	13,2	27
4 Absolute maximum temperature	in °C	20,8	21,1	24,3	27,8	32,2	34,5	35,4	36,1	32,6	27,5	24,5	20,0	36,1	
5 Absolute minimum temperature	in °C	-2,4	-6,7	0,8	3,9	4,8	11,0	14,3	13,2	10,4	5,0	2,8	-2,5	-6,7	
6 Mean relative humidity	in %	68	65	68	66	66	64	65	69	73	71	70	67	68	21
7 Mean precipitation	in mm	33	42	46	47	52	43	29	48	77	80	49	47	593	
8 Maximum precipitation	in mm														
9 Minimum precipitation	in mm														
10 Maximum precipitation in 24 h	in mm	63	143	53	52	76	56	65	82	107	106	70	92	143	
11 Mean number of days with precipitation	> 0,1 mm	5	5	6	8	8	6	4	6	9	9	5	8	79	
12 Mean duration of sunshine	in h	150	164	175	213	252	280	313	274	202	175	150	132	2480	
13 Mean quantity of radiation	in ly/day														
14 Mean potential evaporation	in mm	20	23	37	52	81	113	141	138	99	65	34	24	827	30
15 Mean windspeed	in m/sec	2,5	2,5	2,2	2,2	2,2	2,2	2,2	2,0	2,0	2,0	2,2	2,5	2,2	
16 Mean predominent direction of the wind		W	W	S	SW	S	S	S	S	S	SW	W.N	NW		

149 Station/Country Madrid/Spain

Location 40°25'N/3°41'W   Height above sealevel 667 m   Climate symbol: Köppen Csb   Troll IV,1

		J	F	M	A	M	J	J	A	S	O	N	D	year	P	
1	Mean daily temperature	in °C	4,9	6,5	10,0	13,0	15,7	20,6	24,2	23,8	19,8	14,0	8,9	5,8	13,9		1
2	Mean daily maximum temperature	in °C	8,5	11,0	14,8	18,4	21,2	26,9	30,8	29,5	25,0	18,5	12,8	8,8	18,9	27	2
3	Mean daily minimum temperature	in °C	1,5	2,2	5,2	7,4	10,2	14,6	17,4	17,1	14,1	9,5	5,3	2,2	8,9	27	3
4	Absolute maximum temperature	in °C	18,0	22,0	25,8	30,1	33,4	38,1	39,1	38,9	35,1	29,8	22,1	17,2	39,1		4
5	Absolute minimum temperature	in °C	-10,1	-9,1	-3,5	-0,8	0,6	6,4	8,5	9,2	5,0	-0,4	-3,0	-6,5	-10,1		5
6	Mean relative humidity	in %	79	73	68	64	61	54	46	49	59	70	75	78	65	21	6
7	Mean precipitation	in mm	38	34	45	44	44	27	12	14	32	53	47	48	438		7
8	Maximum precipitation	in mm															8
9	Minimum precipitation	in mm															9
10	Maximum precipitation in 24 h	in mm	33	40	56	36	41	48	30	39	53	54	65	30	65		10
11	Mean number of days with precipitation	> 0,1 mm	7	8	10	9	9	6	2	3	6	8	9	9	84		11
12	Mean duration of sunshine	in h	153	173	187	235	279	317	382	352	256	206	157	136	2824		12
13	Mean quantity of radiation	in ly/day	128	191	281	361	464	551	559	438	311	176	104	91	303		13
14	Mean potential evaporation	in mm	10	15	30	48	79	113	144	135	87	52	22	11	746	30	14
15	Mean windspeed	in m/sec	2,6	2,6	2,8	3,1	2,7	2,7	2,9	2,9	2,5	2,4	2,3	2,5	2,7		15
16	Mean predominent direction of the wind		NE	NE	SW	NE	SW	NE	NE	NE	NE	NE	NE	NE			16

150 Station/Country Palma (Balearic Islands)/Spain

Location 39°33'N/2°39'E   Height above sealevel 28 m   Climate symbol: Köppen Csa   Troll IV,1

		J	F	M	A	M	J	J	A	S	O	N	D	year	P	
1	Mean daily temperature	in °C	10,1	10,5	12,2	14,5	17,4	21,4	24,1	24,5	22,6	18,4	14,3	11,8	16,8		1
2	Mean daily maximum temperature	in °C	14,1	14,8	16,8	18,9	21,9	26,0	28,9	28,8	26,9	22,5	18,1	15,1	21,1	27	2
3	Mean daily minimum temperature	in °C	6,3	6,4	7,9	10,4	12,8	16,9	19,6	20,2	18,1	13,9	7,6	7,2	12,5	27	3
4	Absolute maximum temperature	in °C	24,0	22,8	25,0	26,0	31,0	37,8	37,8	35,0	34,6	31,4	25,5	23,6	37,8		4
5	Absolute minimum temperature	in °C	-3,0	-4,0	-1,0	0,5	4,6	8,4	12,0	11,0	9,1	1,6	0,8	-1,5	-4,0		5
6	Mean relative humidity	in %	78	76	75	72	72	68	66	70	74	77	78	77	74	20	6
7	Mean precipitation	in mm	39	34	36	28	27	20	4	23	56	77	56	51	451		7
8	Maximum precipitation	in mm															8
9	Minimum precipitation	in mm															9
10	Maximum precipitation in 24 h	in mm	55	32	39	26	83	50	20	44	133	97	66	36	133		10
11	Mean number of days with precipitation	> 0,1 mm	8	6	6	5	5	3	1	3	6	9	8	9	71		11
12	Mean duration of sunshine	in h	158	171	200	229	298	311	355	331	239	196	166	141	2795		12
13	Mean quantity of radiation	in ly/day															13
14	Mean potential evaporation	in mm	20	22	36	50	82	116	150	145	106	70	37	25	859	30	14
15	Mean windspeed	in m/sec	2,5	2,6	3,1	3,1	2,6	2,6	2,5	2,5	2,5	2,5	2,5	2,6	2,6		15
16	Mean predominent direction of the wind		N	NW	SW	SW	SW	SW	SW	SW	N	N	NW	NW			16

151 Station/Country Valencia/Spain

Location 39°29'N/0°23'W   Height above sealevel 13 m   Climate symbol: Köppen Csa   Troll IV,1

		J	F	M	A	M	J	J	A	S	O	N	D	year	P	
1	Mean daily temperature	in °C	10,3	11,2	13,2	15,0	17,9	21,5	24,2	24,6	22,6	18,3	14,3	11,4	17,0	26	1
2	Mean daily maximum temperature	in °C	15,0	16,2	18,1	19,9	22,7	26,2	28,5	29,3	27,2	23,1	19,1	16,0	21,8	26	2
3	Mean daily minimum temperature	in °C	5,6	6,2	8,2	10,1	13,0	16,8	19,9	19,8	18,0	13,4	9,5	6,7	12,3	26	3
4	Absolute maximum temperature	in °C	24,4	25,5	30,8	35,2	34,1	34,8	41,7	38,8	37,2	33,5	29,8	24,9	41,7	26	4
5	Absolute minimum temperature	in °C	-6,5	-7,2	-0,4	3,2	5,0	10,2	11,8	12,5	9,2	4,1	1,7	-2,2	-7,2	26	5
6	Mean relative humidity	in %	67	65	66	64	66	67	69	71	72	70	68	67	68	20	6
7	Mean precipitation	in mm	32	32	30	31	31	55	9	26	56	75	38	37	422	26	7
8	Maximum precipitation	in mm															8
9	Minimum precipitation	in mm															9
10	Maximum precipitation in 24 h	in mm	88	91	45	38	45	121	31	53	125	151	281	85	281	26	10
11	Mean number of days with precipitation	> 0,1 mm	5	5	6	7	7	5	2	3	6	7	6	5	64	26	11
12	Mean duration of sunshine	in h	154	161	184	218	258	270	310	283	224	184	149	143	2538	11	12
13	Mean quantity of radiation	in ly/day															13
14	Mean potential evaporation	in mm	22	24	38	56	82	117	144	139	102	67	35	25	851	30	14
15	Mean windspeed	in m/sec															15
16	Mean predominent direction of the wind																16

152 Station/Country Ciudad Real/Spain

Location 38°59'N/3°56'W   Height above sealevel 628 m   Climate symbol: Köppen Csa   Troll IV,1

		J	F	M	A	M	J	J	A	S	O	N	D	year	P	
1	Mean daily temperature	in °C	5,5	7,4	10,4	13,0	15,7	21,3	25,8	24,7	21,0	14,6	9,5	6,3	14,6	25	1
2	Mean daily maximum temperature	in °C	10,2	13,4	16,6	19,6	22,7	29,6	34,5	33,4	28,8	21,4	14,7	10,6	21,3	25	2
3	Mean daily minimum temperature	in °C	0,7	1,3	4,2	6,3	8,9	13,0	16,8	16,0	13,1	8,3	4,3	1,8	7,9	25	3
4	Absolute maximum temperature	in °C	20,2	25,0	30,0	34,0	37,4	41,2	44,2	42,0	40,0	32,4	24,0	20,2	44,2	25	4
5	Absolute minimum temperature	in °C	-8,4	-10,0	-5,2	-1,0	0,5	5,0	6,2	8,2	5,6	-1,2	-4,8	-7,0	-10,0	25	5
6	Mean relative humidity	in %	78	75	73	68	66	63	62	64	67	71	76	77	70	19	6
7	Mean precipitation	in mm	37	40	46	51	50	22	2	8	25	43	39	46	409	23	7
8	Maximum precipitation	in mm															8
9	Minimum precipitation	in mm															9
10	Maximum precipitation in 24 h	in mm	35	37	38	34	68	40	10	57	29	55	29	32	68	23	10
11	Mean number of days with precipitation	> 0,1 mm	6	7	8	8	7	3	1	1	3	6	6	7	63	23	11
12	Mean duration of sunshine	in h	138	161	177	232	295	343	398	357	260	194	152	117	2681	2	12
13	Mean quantity of radiation	in ly/day															13
14	Mean potential evaporation	in mm	10	14	29	46	81	115	149	144	93	54	24	12	771	30	14
15	Mean windspeed	in m/sec															15
16	Mean predominent direction of the wind																16

153 Station/Country Murcia/Spain

Location 37°59'N/1°08'W — Height above sealevel 44 m — Climate symbol: Köppen BSh — Troll IV.2

		J	F	M	A	M	J	J	A	S	O	N	D	year	P	
1 Mean daily temperature	in °C	10.0	11.2	14.0	16.2	19.4	23.8	26.4	26.4	23.7	18.9	14.3	11.4	18.0	24	1
2 Mean daily maximum temperature	in °C	15.7	17.1	20.3	22.4	25.7	30.4	33.4	33.4	30.4	25.2	20.4	16.9	24.3	24	2
3 Mean daily minimum temperature	in °C	4.3	5.2	7.7	10.0	13.0	17.1	19.3	19.3	17.0	12.5	8.1	5.9	11.8	24	3
4 Absolute maximum temperature	in °C	25.5	29.5	31.4	37.5	37.0	40.0	44.9	43.0	39.0	37.5	30.5	25.1	44.9	24	4
5 Absolute minimum temperature	in °C	-5.0	-10.0	-3.0	2.0	4.0	11.0	9.8	11.8	8.0	-4.2	-1.4	-5.0	-10.0	24	5
6 Mean relative humidity	in %															6
7 Mean precipitation	in mm	24	18	23	43	39	10	1	8	30	45	26	37	304	24	7
8 Maximum precipitation	in mm															8
9 Minimum precipitation	in mm															9
10 Maximum precipitation in 24 h	in mm	63	52	38	59	54	30	11	33	67	110	64	105	110	24	10
11 Mean number of days with precipitation	> 0,1 mm	4	4	5	6	6	2	1	1	4	5	4	5	47	24	11
12 Mean duration of sunshine	in h	183	182	205	242	300	327	367	332	242	203	174	184	2941	7	12
13 Mean quantity of radiation	in ly/day															13
14 Mean potential evaporation	in mm	20	24	40	57	91	126	160	156	108	69	35	22	908	30	14
15 Mean windspeed	in m/sec															15
16 Mean predominent direction of the wind																16

154 Station/Country Sevilla/Spain

Location 37°24'N/6°00'W — Height above sealevel 30 m — Climate symbol: Köppen Csa — Troll IV.1

		J	F	M	A	M	J	J	A	S	O	N	D	year	P	
1 Mean daily temperature	in °C	10.5	12.3	14.8	17.2	19.9	24.8	27.9	27.8	24.8	19.8	15.0	11.4	18.8		1
2 Mean daily maximum temperature	in °C	15.1	17.4	20.3	23.6	26.6	32.0	36.4	36.0	32.1	25.8	19.9	15.7	25.1	27	2
3 Mean daily minimum temperature	in °C	5.5	6.5	9.0	11.1	13.4	17.2	19.6	19.8	17.7	13.7	9.9	6.8	12.5	27	3
4 Absolute maximum temperature	in °C	25.0	29.0	30.8	36.0	39.5	43.5	45.8	47.0	40.8	39.0	32.9	24.0	47.0		4
5 Absolute minimum temperature	in °C	-2.8	-3.2	0.0	2.6	10.2	12.5	12.1	9.0	5.2	0.2	-2.8	-3.2			5
6 Mean relative humidity	in %	81	77	78	71	67	61	55	58	63	74	74	81	70	21	6
7 Mean precipitation	in mm	64	62	57	59	39	9	1	4	20	66	70	84	535		7
8 Maximum precipitation	in mm															8
9 Minimum precipitation	in mm															9
10 Maximum precipitation in 24 h	in mm	80	73	60	100	37	51	7	37	49	69	78	69	100		10
11 Mean number of days with precipitation	> 0,1 mm	8	8	9	7	5	1	0	0	2	5	6	8	57		11
12 Mean duration of sunshine	in h	182	190	189	235	292	332	380	328	242	207	166	155	2878		12
13 Mean quantity of radiation	in ly/day	174	273	360	476	532	567	558	505	413	294	196	154	375		13
14 Mean potential evaporation	in mm	17	22	39	57	98	140	180	176	121	71	33	19	973	30	14
15 Mean windspeed	in m/sec	1.4	1.7	2.0	1.7	1.7	1.7	1.7	1.4	1.4	1.4	1.4	1.4	1.7		15
16 Mean predominent direction of the wind		NE	SW	SW	SW	SW	SW	SW	SW	SW	SW	NE	SW			16

155 Station/Country Granada/Spain

Location 37°09'N/3°35'W — Height above sealevel 689 m — Climate symbol: Köppen Csa — Troll IV.1

		J	F	M	A	M	J	J	A	S	O	N	D	year	P	
1 Mean daily temperature	in °C	7.1	8.5	10.9	13.5	16.3	22.1	25.7	25.3	21.7	16.4	11.7	7.9	15.6	23	1
2 Mean daily maximum temperature	in °C	12.0	14.1	16.5	19.5	22.7	29.7	34.2	33.3	28.9	22.5	17.1	12.4	21.9	23	2
3 Mean daily minimum temperature	in °C	2.1	2.8	5.2	7.4	9.9	14.5	17.2	17.2	14.5	10.2	6.3	3.3	9.2	23	3
4 Absolute maximum temperature	in °C	23.8	29.5	28.0	31.5	34.4	38.2	42.8	40.2	37.8	33.4	27.2	21.2	42.8	23	4
5 Absolute minimum temperature	in °C	-7.4	-10.4	-5.8	-0.2	1.0	5.8	9.4	9.4	7.1	0.6	-5.0	-5.2	-10.4	23	5
6 Mean relative humidity	in %	78	70	71	68	64	54	48	53	59	69	73	79	66	15	6
7 Mean precipitation	in mm	54	49	62	53	44	8	3	8	25	49	48	70	473	23	7
8 Maximum precipitation	in mm															8
9 Minimum precipitation	in mm															9
10 Maximum precipitation in 24 h	in mm	47	50	35	37	42	12	13	29	42	55	54	39	55	23	10
11 Mean number of days with precipitation	> 0,1 mm	8	8	10	10	8	2	1	1	4	7	8	10	77	23	11
12 Mean duration of sunshine	in h	149	161	175	201	232	251	331	324	243	189	152	124	2532	11	12
13 Mean quantity of radiation	in ly/day															13
14 Mean potential evaporation	in mm	14	18	30	47	79	114	154	150	100	60	27	16	805	30	14
15 Mean windspeed	in m/sec															15
16 Mean predominent direction of the wind																16

156 Station/Country Almeria/Spain

Location 36°50'N/2°28'W — Height above sealevel 7 m — Climate symbol: Köppen BSh — Troll IV.2

		J	F	M	A	M	J	J	A	S	O	N	D	year	P	
1 Mean daily temperature	in °C	11.7	11.8	14.1	16.1	18.4	22.0	24.7	25.3	23.4	19.4	15.6	12.8	18.0		1
2 Mean daily maximum temperature	in °C	15.8	16.1	17.8	19.8	22.0	25.7	28.5	29.1	27.1	22.9	19.2	16.6	21.7	27	2
3 Mean daily minimum temperature	in °C	8.0	8.5	10.5	12.5	14.9	18.3	20.9	21.6	19.9	15.8	12.0	9.2	14.3	27	3
4 Absolute maximum temperature	in °C	22.6	25.7	26.6	29.7	34.8	35.9	37.7	37.4	36.0	31.5	26.7	25.3	37.7		4
5 Absolute minimum temperature	in °C	1.9	0.2	2.8	5.3	8.2	12.7	14.6	15.5	10.1	7.6	4.5	2.5	0.2		5
6 Mean relative humidity	in %	74	74	73	73	72	72	72	74	74	74	75	74	73	21	6
7 Mean precipitation	in mm	31	21	20	28	17	4	<1	5	16	26	27	38	232		7
8 Maximum precipitation	in mm															8
9 Minimum precipitation	in mm															9
10 Maximum precipitation in 24 h	in mm	52	24	60	54	34	25	2	42	98	58	74	59	98		10
11 Mean number of days with precipitation	> 0,1 mm	7	4	5	5	4	2	1	1	3	5	6		48		11
12 Mean duration of sunshine	in h	189	190	227	280	309	331	362	336	284	227	185	173	3053		12
13 Mean quantity of radiation	in ly/day	215	298	402	502	559	588	596	539	439	340	242	185	409		13
14 Mean potential evaporation	in mm	28	28	41	56	88	117	151	157	115	81	41	29	932	30	14
15 Mean windspeed	in m/sec	3.1	3.4	3.6	3.6	3.6	3.4	2.8	3.1	2.8	2.8	2.8	3.1	3.1		15
16 Mean predominent direction of the wind		W	W	W	W	W	W	E.W	E	S	WSW	WSWE	N			16

43

157 Station / Country Las Palmas (Canary Islands)/Spain

Location 28°11'N/15°28'W **Height above sealevel** 6 m **Climate symbol: Köppen** BSh **Troll** IV,2

		J	F	M	A	M	J	J	A	S	O	N	D	year	P	
1 Mean daily temperature	in °C	17.8	18.1	18.3	18.9	19.7	21.2	22.2	23.6	23.3	22.8	21.1	18.9	20.5	45	1
2 Mean daily maximum temperature	in °C	21.4	21.2	21.6	20.8	26.7	24.2	23.5	24.3	25.2	24.7	22.6	20.8	23.1	39	2
3 Mean daily minimum temperature	in °C	16.3	16.6	16.2	17.6	18.1	18.8	20.6	21.0	20.4	21.2	18.5	16.4	18.5	39	3
4 Absolute maximum temperature	in °C	30.0	28.9	30.0	32.8	31.1	31.7	35.0	37.8	35.6	35.0	31.1	29.4	37.8	45	4
5 Absolute minimum temperature	in °C	7.8	8.3	8.3	10.0	12.2	14.4	15.6	16.7	15.0	13.3	11.1	8.3	7.8	45	5
6 Mean relative humidity	in %	72	73	73	73	72	74	77	76	75	75	74	73	74	19	6
7 Mean precipitation	in mm	36	23	23	13	5	<2	<2	<2	5	28	53	41	233	48	7
8 Maximum precipitation	in mm															8
9 Minimum precipitation	in mm															9
10 Maximum precipitation in 24 h	in mm	117	38	61	51	64	8	28	15	18	69	239	145	239	48	10
11 Mean number of days with precipitation	>0.1 mm	8	5	5	3	1	<1	<1	<1	1	5	7	8	46	48	11
12 Mean duration of sunshine	in h															12
13 Mean quantity of radiation	in ly/day															13
14 Mean potential evaporation	in mm	52	48	60	66	84	96	107	118	107	95	72	59	964	35	14
15 Mean windspeed	in m/sec															15
16 Mean predominent direction of the wind																16

158 Station / Country Gibraltar Town (Gibraltar)/United Kingdom

Location 36°06'N/5°21'W **Height above sealevel** 27 m **Climate symbol: Köppen** Csa **Troll** IV,1

		J	F	M	A	M	J	J	A	S	O	N	D	year	P	
1 Mean daily temperature	in °C	12.5	12.4	14.4	16.2	18.7	21.7	23.8	24.4	22.8	19.2	15.1	13.0	17.8	5	1
2 Mean daily maximum temperature	in °C	16.3	16.3	18.2	20.9	23.4	26.6	28.2	28.7	26.7	23.1	18.4	16.7	21.9	5	2
3 Mean daily minimum temperature	in °C	8.7	8.5	10.6	11.4	13.9	16.7	19.4	20.1	18.5	15.2	11.7	9.3	13.7	5	3
4 Absolute maximum temperature	in °C	22.2	21.7	23.9	27.8	30.6	33.3	36.1	35.6	34.4	28.3	25.6	21.1	36.1	5	4
5 Absolute minimum temperature	in °C	1.7	0.6	3.9	7.8	10.0	11.1	14.4	16.1	12.2	7.8	5.6	1.7	0.6	5	5
6 Mean relative humidity	in %	74	72	75	71	69	67	70	70	72	74	76	75	72	5	6
7 Mean precipitation	in mm	109	92	125	99	30	5	<1	<1	18	85	149	151	863	5	7
8 Maximum precipitation	in mm															8
9 Minimum precipitation	in mm															9
10 Maximum precipitation in 24 h	in mm	125	54	50	75	36	13	<1	1	19	56	53	69	125	5	10
11 Mean number of days with precipitation	>1.0 mm	7	8	11	7	4	2	<1	<1	2	5	13	11	70	5	11
12 Mean duration of sunshine	in h +	180	187	217	249	302	326	353	330	261	224	178	177	2984	12	12
13 Mean quantity of radiation	in ly/day															13
14 Mean potential evaporation	in mm	30	32	44	58	81	106	128	128	100	70	45	32	854	88	14
15 Mean windspeed	in m/sec															15
16 Mean predominent direction of the wind																16

+ Windmill Hill (36°06'N/5°21'W, 122 m)

159 Station / Country Bolzano/Italy

Location 46°30'N/11°21'E **Height above sealevel** 271 m **Climate symbol: Köppen** Dfa **Troll** III,3

		J	F	M	A	M	J	J	A	S	O	N	D	year	P	
1 Mean daily temperature	in °C	0.2	3.7	8.8	13.1	16.9	20.1	22.0	21.1	18.4	12.8	6.0	1.7	12.1	16	1
2 Mean daily maximum temperature	in °C	4.8	8.7	14.5	18.8	23.0	26.0	28.2	27.3	24.5	18.8	10.8	5.6	17.5	16	2
3 Mean daily minimum temperature	in °C	-4.5	-1.4	3.0	7.4	10.8	14.2	15.8	14.9	12.3	6.6	1.4	-2.7	6.5	16	3
4 Absolute maximum temperature	in °C	16.4	20.3	27.2	30.4	34.0	35.0	37.2	35.9	33.0	27.1	21.5	17.0	37.2	16	4
5 Absolute minimum temperature	in °C	-15.4	-10.2	-5.6	-1.0	1.4	2.8	8.8	7.5	3.9	-2.2	-7.8	-13.7	-15.4	16	5
6 Mean relative humidity	in %	64	71	66	66	67	68	68	71	72	74	74	79	70	10	6
7 Mean precipitation	in mm	18	29	30	60	69	93	100	113	71	75	80	43	781	16	7
8 Maximum precipitation	in mm															8
9 Minimum precipitation	in mm															9
10 Maximum precipitation in 24 h	in mm	37	29	25	49	37	45	58	62	61	68	55	34	68	10	10
11 Mean number of days with precipitation	>0.1 mm	3	4	5	7	8	11	10	10	8	8	5	8	84	16	11
12 Mean duration of sunshine	in h	111	125	158	171	193	210	234	206	178	154	81	88	1913	7	12
13 Mean quantity of radiation	in ly/day	100	146	277	368	441	449	456	401	320	218	114	78	280		13
14 Mean potential evaporation	in mm	1	7	31	59	97	126	147	127	83	46	16	2	742	17	14
15 Mean windspeed	in m/sec															15
16 Mean predominent direction of the wind																16

160 Station / Country Trieste/Italy

Location 45°39'N/13°46'E **Height above sealevel** 11 m **Climate symbol: Köppen** Cfa **Troll** III,3

		J	F	M	A	M	J	J	A	S	O	N	D	year	P	
1 Mean daily temperature	in °C	4.7	5.6	8.9	13.5	17.9	21.5	23.7	23.1	20.1	15.6	10.6	7.1	14.4	18	1
2 Mean daily maximum temperature	in °C	6.7	7.9	11.7	16.7	21.4	25.3	27.5	26.9	23.5	18.3	12.8	9.0	17.3	18	2
3 Mean daily minimum temperature	in °C	2.8	3.2	6.1	10.3	14.3	17.7	19.7	19.3	16.7	12.9	8.4	5.1	11.4	16	3
4 Absolute maximum temperature	in °C	15.5	16.5	23.5	28.6	32.0	34.1	34.8	32.7	30.3	26.5	21.4	16.5	34.8	16	4
5 Absolute minimum temperature	in °C	-9.1	-13.6	-5.7	2.6	3.9	8.9	11.1	12.7	9.1	3.5	-3.0	-6.1	-13.6	16	5
6 Mean relative humidity	in %	69	71	65	65	68	66	64	64	67	68	73	73	68	10	6
7 Mean precipitation	in mm	68	62	62	84	77	93	92	91	104	94	107	89	1032	18	7
8 Maximum precipitation	in mm	304	212	209	228	282	274	206	268	279	352	352	277	1611	111	8
9 Minimum precipitation	in mm	0	0	0	7	7	7	8	10	1	6	1	0	806	111	9
10 Maximum precipitation in 24 h	in mm	44	55	58	46	52	55	59	59	55	60	59	56	60	10	10
11 Mean number of days with precipitation	>0.1 mm	7	7	7	8	8	9	7	7	7	8	8	9	93	16	11
12 Mean duration of sunshine	in h	108	109	145	185	237	260	301	272	204	178	81	85	2165	7	12
13 Mean quantity of radiation	in ly/day															13
14 Mean potential evaporation	in mm	8	11	29	52	95	126	149	135	89	54	25	13	786	30	14
15 Mean windspeed	in m/sec															15
16 Mean predominent direction of the wind																16

44

181 Station / Country Milano (Milan) / Italy

Location 45°28'N/9°11'E   Height above sealevel 147 m   Climate symbol: Köppen Cfa   Troll III,3

		J	F	M	A	M	J	J	A	S	O	N	D	year	P	
1	Mean daily temperature	in °C	2.0	4.0	8.6	12.7	17.9	21.7	24.1	23.5	19.1	13.1	7.1	3.8	13.1	30	1
2	Mean daily maximum temperature	in °C	4.5	7.8	13.1	18.2	23.1	26.5	29.1	27.8	23.8	17.2	10.4	5.9	17.3	16	2
3	Mean daily minimum temperature	in °C	-0.2	1.9	5.9	9.7	13.9	17.2	19.6	18.9	16.1	11.0	6.1	1.7	10.1	16	3
4	Absolute maximum temperature	in °C	16.4	22.0	25.0	28.9	35.0	35.7	38.3	36.8	35.1	26.8	19.7	17.4	38.3	30	4
5	Absolute minimum temperature	in °C	-10.2	-14.1	-3.3	-0.5	3.8	6.3	11.3	10.3	6.3	-2.1	-6.2	-7.0	-14.1	30	5
6	Mean relative humidity	in %	88	81	77	73	72	71	74	77	84	88	91		79	30	6
7	Mean precipitation	in mm	55	62	93	94	83	79	63	67	75	113	93	86	963	30	7
8	Maximum precipitation	in mm	208	261	264	282	339	270	233	338	343	375	337	324	1671	187	8
9	Minimum precipitation	in mm	0	0	0	1	15	1	3	3	0	0	6	0	584	187	9
10	Maximum precipitation in 24 h	in mm	37	41	53	63	50	56	37	73	52	58	47	65	73	10	10
11	Mean number of days with precipitation	> 0,1 mm	8	6	9	10	8	8	6	5	8	9	9	9	92	30	11
12	Mean duration of sunshine	in h	62	95	158	186	223	240	282	254	180	121	51	47	1899	30	12
13	Mean quantity of radiation	in ly/day	75	133	223	320	384	426	433	370	279	166	79	60	246	30	13
14	Mean potential evaporation	in mm	2	7	26	55	97	132	153	134	89	47	17	5	764	163	14
15	Mean windspeed	in m/sec															15
16	Mean predominent direction of the wind		WSW	WSW	NE-E	E.SW	E.SW	E.SW	ENE	NNE	E.SW	E.SW	E.SW	SW		30	16

162 Station / Country Venezia (Venice) / Italy

Location 45°27'N/12°19'E   Height above sealevel 1 m   Climate symbol: Köppen Cfa   Troll III,3

		J	F	M	A	M	J	J	A	S	O	N	D	year	P	
1	Mean daily temperature	in °C	3.0	4.7	8.4	13.0	17.3	20.9	23.1	22.6	19.9	15.0	9.3	5.1	13.5	16	1
2	Mean daily maximum temperature	in °C	5.5	7.6	11.8	16.5	21.0	24.7	27.2	26.8	23.8	18.5	11.8	7.5	16.9	16	2
3	Mean daily minimum temperature	in °C	0.5	1.7	5.0	9.5	13.5	17.1	18.9	18.4	15.9	11.4	6.7	2.6	10.1	16	3
4	Absolute maximum temperature	in °C	14.0	17.8	22.4	27.4	32.9	33.0	34.4	34.0	30.5	26.8	20.7	15.2	34.4	16	4
5	Absolute minimum temperature	in °C	-7.9	-9.4	-4.7	1.9	4.9	8.3	11.6	12.6	9.3	3.3	-2.1	-4.3	-9.4	16	5
6	Mean relative humidity	in %	81	78	77	77	77	74	73	74	76	78	82	84	78	10	6
7	Mean precipitation	in mm	37	48	61	78	65	69	52	69	59	77	94	61	770	16	7
8	Maximum precipitation	in mm															8
9	Minimum precipitation	in mm															9
10	Maximum precipitation in 24 h	in mm	34	53	54	48	39	37	33	99	43	50	81	39	99	10	10
11	Mean number of days with precipitation	> 0,1 mm	8	6	7	9	8	8	7	7	5	7	9	8	87	16	11
12	Mean duration of sunshine	in h	94	102	156	180	241	255	304	262	199	158	72	81	2104	7	12
13	Mean quantity of radiation	in ly/day	97	144	260	377	503	518	538	452	342	228	117	70	304		13
14	Mean potential evaporation	in mm	4	9	26	54	97	129	155	136	92	53	21	8	784	85	14
15	Mean windspeed	in m/sec															15
16	Mean predominent direction of the wind																16

163 Station / Country Genova (Genoa) / Italy

Location 44°25'N/8°55'E   Height above sealevel 54 m   Climate symbol: Köppen Cfa   Troll III,3

		J	F	M	A	M	J	J	A	S	O	N	D	year	P	
1	Mean daily temperature	in °C	7.9	8.3	11.0	13.7	17.5	21.3	23.7	24.1	21.2	16.8	12.0	9.2	15.6	30	1
2	Mean daily maximum temperature	in °C	10.5	11.2	13.9	17.1	20.8	23.9	26.7	26.5	24.2	20.3	14.8	12.1	18.5	16	2
3	Mean daily minimum temperature	in °C	5.1	5.9	8.1	11.1	14.6	17.9	20.8	20.4	18.2	14.5	9.9	7.9	12.8	16	3
4	Absolute maximum temperature	in °C	18.7	19.9	23.7	26.7	30.2	32.8	36.9	35.0	34.0	29.2	23.7	20.4	36.9	30	4
5	Absolute minimum temperature	in °C	-5.2	-8.0	-1.0	3.1	7.3	11.0	14.3	11.9	11.0	3.3	-1.0	-2.8	-8.0	30	5
6	Mean relative humidity	in %	64	65	68	69	70	70	67	69	69	67	68	66	68	30	6
7	Mean precipitation	in mm	99	108	144	92	77	72	49	53	110	181	174	136	1297	30	7
8	Maximum precipitation	in mm	435	340	452	286	247	250	216	344	440	776	733	648	2451	120	8
9	Minimum precipitation	in mm	1	0	0	6	4	0	0	0	4	5	3	1	572	120	9
10	Maximum precipitation in 24 h	in mm	73	127	67	116	46	56	121	41	123	175	99	151	175	10	10
11	Mean number of days with precipitation	> 0,1 mm	7	7	10	9	7	6	3	4	6	9	9	9	86	30	11
12	Mean duration of sunshine	in h	127	126	167	189	236	252	298	270	264	171	105	112	2217	30	12
13	Mean quantity of radiation	in ly/day	109	156	245	329	406	445	460	397	296	199	113	92	271	30	13
14	Mean potential evaporation	in mm	15	18	33	53	88	151	146	136	96	61	30	18	817	50	14
15	Mean windspeed	in m/sec	6.9	6.5	5.5	4.5	3.9	3.1	3.0	3.2	4.1	5.5	6.4	6.7	4.9		15
16	Mean predominent direction of the wind		N	N	N	N.S	N.S	N.S	N.S	N.S	N.S	N	N	N		30	16

164 Station / Country Firenze (Florence) / Italy

Location 43°46'N/11°15'E   Height above sealevel 76 m   Climate symbol: Köppen Cfa   Troll IV,1

		J	F	M	A	M	J	J	A	S	O	N	D	year	P	
1	Mean daily temperature	in °C	5.1	6.3	9.6	13.1	17.7	21.8	24.2	24.1	20.2	15.1	9.8	5.7	14.4	30	1
2	Mean daily maximum temperature	in °C	9.0	11.0	14.4	18.8	23.2	27.0	30.0	29.5	25.7	20.3	14.0	10.6	19.5	16	2
3	Mean daily minimum temperature	in °C	1.9	3.0	5.1	8.3	11.7	15.2	17.9	17.3	14.7	10.9	7.0	3.7	9.7	16	3
4	Absolute maximum temperature	in °C	18.8	19.5	25.2	29.8	32.6	35.0	38.9	40.2	34.3	29.8	23.2	18.5	40.2	16	4
5	Absolute minimum temperature	in °C	-9.8	-10.8	-5.3	-1.5	1.6	7.3	10.5	10.7	4.7	1.9	-5.3	-5.4	-10.8	16	5
6	Mean relative humidity	in %	77	75	74	73	71	69	65	67	71	75	81	82	73	30	6
7	Mean precipitation	in mm	61	58	72	67	62	70	26	36	68	97	101	78	796	30	7
8	Maximum precipitation	in mm	156	203	191	128	163	141	85	99	272	349	284	243	1487	30	8
9	Minimum precipitation	in mm	6	6	5	10	7	0	0	3	4	6	0	18	553	30	9
10	Maximum precipitation in 24 h	in mm	38	45	29	25	91	36	61	68	47	53	132	52	132	10	10
11	Mean number of days with precipitation	> 0,1 mm	7	7	8	9	7	7	3	3	5	9	9	9	83	30	11
12	Mean duration of sunshine	in h	134	134	153	205	300	305	348	322	219	177	97	96	2488	4	12
13	Mean quantity of radiation	in ly/day	93	139	251	379	484	551	565	476	344	213	107	72	306		13
14	Mean potential evaporation	in mm	8	13	30	52	90	126	151	137	93	54	25	12	791	26	14
15	Mean windspeed	in m/sec															15
16	Mean predominent direction of the wind		N	N	N	N.W	N.W	W	W	W	N.W	N	N	N		30	16

45

165 Station/Country Ancona/Italy

Location 43°37'N/13°31'E   Height above sealevel 17 m   Climate symbol: Köppen Cfa   Troll IV,1

		J	F	M	A	M	J	J	A	S	O	N	D	year	P	
1 Mean daily temperature	in °C	5,3	6,2	9,8	13,3	17,3	22,0	24,4	24,3	20,5	16,0	11,6	7,9	14,9	30	1
2 Mean daily maximum temperature	in °C	7,5	9,5	12,6	16,3	20,1	24,1	26,8	26,6	23,4	19,0	14,2	10,0	17,5	10	2
3 Mean daily minimum temperature	in °C	2,5	4,3	7,2	10,5	14,1	18,0	20,4	20,4	17,7	14,1	9,8	5,4	12,0	10	3
4 Absolute maximum temperature	in °C	18,0	20,6	21,0	25,2	31,0	35,0	36,0	39,0	34,2	27,0	26,0	19,0	39,0	30	4
5 Absolute minimum temperature	in °C	-7,6	-7,5	-2,5	0,5	4,2	9,0	8,3	9,2	6,5	4,3	0,8	-3,8	-7,6	30	5
6 Mean relative humidity	in %	78	74	74	72	73	69	65	66	71	76	78	80	73	30	6
7 Mean precipitation	in mm	65	43	40	58	54	49	37	38	89	95	64	77	709	30	7
8 Maximum precipitation	in mm															8
9 Minimum precipitation	in mm															9
10 Maximum precipitation in 24 h	in mm	43	24	31	42	32	56	62	34	80	58	46	63	80	10	10
11 Mean number of days with precipitation	> 0,1 mm	8	6	8	8	8	6	5	4	8	9	10	89	30	11
12 Mean duration of sunshine	in h	74	104	146	201	267	270	322	295	213	161	75	65	2193	30	12
13 Mean quantity of radiation	in ly/day	100	171	288	392	479	509	527	466	354	232	118	85	308	30	13
14 Mean potential evaporation	in mm	9	13	28	53	91	133	165	145	100	60	26	12	835	40	14
15 Mean windspeed	in m/sec															15
16 Mean predominent direction of the wind		W,NW	W	W,N	W,ESE	W,ESE	W,ESE	W,N	W,E	W,N	W,N	W	W		30	16

166 Station/Country Roma(Rome)/Italy

Location 41°54'N/12°29'E   Height above sealevel 46 m   Climate symbol: Köppen Csa   Troll IV,1

		J	F	M	A	M	J	J	A	S	O	N	D	year	P	
1 Mean daily temperature	in °C	6,9	7,7	10,8	13,9	18,1	22,1	24,7	24,5	21,1	16,4	11,7	8,5	15,6	30	1
2 Mean daily maximum temperature	in °C	11,1	12,6	15,2	18,8	23,4	27,6	30,4	29,8	26,3	21,5	16,1	12,6	20,5	16	2
3 Mean daily minimum temperature	in °C	4,5	5,4	7,2	9,8	13,3	17,2	19,4	19,6	16,9	12,8	9,3	6,4	11,8	16	3
4 Absolute maximum temperature	in °C	18,1	20,7	25,3	29,8	32,8	34,9	40,1	39,2	34,2	28,2	24,6	19,3	40,1	30	4
5 Absolute minimum temperature	in °C	-5,0	-5,4	-1,2	0,3	2,1	9,2	11,9	13,2	8,0	2,1	-2,4	-5,0	-5,4	30	5
6 Mean relative humidity	in %	77	73	71	70	67	62	58	59	66	72	77	79	69	30	6
7 Mean precipitation	in mm	76	88	77	72	63	48	14	22	70	128	116	106	874	30	7
8 Maximum precipitation	in mm	179	189	157	144	130	86	55	66	138	425	254	246	1056	30	8
9 Minimum precipitation	in mm	12	1	0	10	5	0	0	0	0	19	21	12	369	30	9
10 Maximum precipitation in 24 h	in mm	43	25	44	31	25	35	33	22	97	91	72	38	97	10	10
11 Mean number of days with precipitation	> 0,1 mm	6	6	8	8	6	4	2	2	5	6	8	8	70	30	11
12 Mean duration of sunshine	in h	133	132	205	210	267	282	336	307	243	198	123	102	2537	30	12
13 Mean quantity of radiation	in ly/day	143	198	286	387	477	517	532	465	355	252	149	114	323	30	13
14 Mean potential evaporation	in mm	12	17	32	53	91	124	152	140	99	60	30	15	825	43	14
15 Mean windspeed	in m/sec															15
16 Mean predominent direction of the wind		NE	NE	NE	SW	NE	SW	NE	NE,W	NE,W	NE,W	NE,S	NE		30	16

167 Station/Country Foggia/Italy

Location 41°28'N/15°33'E   Height above sealevel 74 m   Climate symbol: Köppen Csa   Troll IV,1

		J	F	M	A	M	J	J	A	S	O	N	D	year	P	
1 Mean daily temperature	in °C	6,8	7,9	10,4	13,8	18,3	22,6	25,7	25,7	22,3	12,1	12,3	6,6	15,8	16	1
2 Mean daily maximum temperature	in °C	10,4	12,2	15,1	19,3	24,4	29,0	32,2	31,9	27,8	21,5	16,0	12,7	21,0	16	2
3 Mean daily minimum temperature	in °C	3,1	3,6	5,6	8,3	12,1	16,2	19,2	19,4	16,7	12,6	8,8	5,5	10,9	16	3
4 Absolute maximum temperature	in °C	20,1	22,5	26,4	29,7	35,1	39,0	40,6	41,8	37,4	29,8	27,0	21,9	41,8	16	4
5 Absolute minimum temperature	in °C	-6,9	-8,9	-4,1	-1,2	3,5	5,0	10,9	11,0	7,0	3,8	-3,0	-4,0	-8,9	16	5
6 Mean relative humidity	in %	80	79	72	68	66	53	53	54	64	76	78	80	69	10	6
7 Mean precipitation	in mm	55	35	31	39	38	29	15	23	27	64	60	49	465	16	7
8 Maximum precipitation	in mm															8
9 Minimum precipitation	in mm															9
10 Maximum precipitation in 24 h	in mm	52	19	23	32	29	45	31	71	23	61	20	41	71	10	10
11 Mean number of days with precipitation	> 0,1 mm	8	6	6	6	6	4	2	3	4	7	8	9	69	16	11
12 Mean duration of sunshine	in h	121	128	176	217	262	279	326	304	238	199	135	113	2498	7	12
13 Mean quantity of radiation	in ly/day	142	235	261	325	472	500	489	453	360	209	133	102	306		13
14 Mean potential evaporation	in mm	12	13	29	50	90	130	164	150	102	62	22	16	840	26	14
15 Mean windspeed	in m/sec															15
16 Mean predominent direction of the wind																16

+ Amendola 41°32'N/15°43'E 56 m

168 Station/Country Napoli(Naples)/Italy

Location 40°51'N/14°15'E   Height above sealevel 25 m   Climate symbol: Köppen Csa   Troll IV,1

		J	F	M	A	M	J	J	A	S	O	N	D	year	P	
1 Mean daily temperature	in °C	9,0	9,8	12,0	14,6	18,7	22,2	24,8	25,0	22,1	18,3	13,9	10,9	16,8	30	1
2 Mean daily maximum temperature	in °C	11,7	12,6	14,8	18,2	22,2	26,3	29,1	29,0	26,0	21,7	17,0	13,6	20,2	16	2
3 Mean daily minimum temperature	in °C	4,3	4,8	6,4	8,9	12,3	16,0	18,1	18,1	15,9	12,4	9,0	6,4	11,1	16	3
4 Absolute maximum temperature	in °C	18,3	21,5	24,7	27,5	32,2	35,2	37,9	36,6	33,8	29,6	24,9	20,8	37,9	30	4
5 Absolute minimum temperature	in °C	-4,4	-4,0	-3,8	0,9	2,9	7,1	11,0	12,7	8,0	3,1	-1,8	-4,4	-4,4	16	5
6 Mean relative humidity	in %	73	73	71	72	71	67	65	66	69	71	75	76	62	30	6
7 Mean precipitation	in mm	93	82	75	67	45	46	16	19	71	130	114	137	895	30	7
8 Maximum precipitation	in mm	190	197	201	99	198	99	73	115	202	211	276	287	1188	30	8
9 Minimum precipitation	in mm	9	7	3	4	11	0	0	0	4	10	26	43	520	30	9
10 Maximum precipitation in 24 h	in mm	50	53	37	30	35	31	31	44	59	79	89	82	89	30	10
11 Mean number of days with precipitation	> 0,1 mm	10	11	10	9	7	5	2	3	7	10	11	13	98	30	11
12 Mean duration of sunshine	in h	118	128	161	198	254	279	322	307	243	198	123	93	2422	30	12
13 Mean quantity of radiation	in ly/day	118	166	226	315	397	441	447	392	302	219	131	97	263	30	13
14 Mean potential evaporation	in mm	17	18	32	50	88	118	145	136	98	63	34	22	821	60	14
15 Mean windspeed	in m/sec															15
16 Mean predominent direction of the wind		N	N,S	N,S	N,S	N,S	N,S	N,S	N,S	N,S	N,S	N,S	N		30	16

46

169 Station/Country Potenza/Italy

Location 40°38'N/15°48'E Height above sealevel 826 m Climate symbol: Köppen Csb Troll IV,1

		J	F	M	A	M	J	J	A	S	O	N	D	year	P	
1 Mean daily temperature	in °C	3.2	3.8	6.3	9.5	14.0	17.9	20.7	20.8	17.4	12.8	8.0	5.1	11.8	30	1
2 Mean daily maximum temperature	in °C	6.4	8.1	10.6	14.5	18.5	23.8	27.3	26.9	23.1	17.7	12.8	8.5	16.5	16	2
3 Mean daily minimum temperature	in °C	1.2	1.5	3.2	6.2	9.8	13.5	15.9	15.9	13.3	9.9	6.2	2.8	8.3	16	3
4 Absolute maximum temperature	in °C	21.0	21.2	25.5	25.3	31.5	34.5	36.2	38.0	33.0	28.6	24.8	19.0	38.0	16	4
5 Absolute minimum temperature	in °C	-10.0	-8.8	-8.0	-3.9	-0.4	4.5	5.4	9.2	3.5	1.8	-4.8	-9.2	-10.0	16	5
6 Mean relative humidity	in %	80	71	69	64	66	60	57	64	64	72	75	79	68	10	6
7 Mean precipitation	in mm	89	85	67	82	72	58	29	34	64	91	115	106	892	30	7
8 Maximum precipitation	in mm															8
9 Minimum precipitation	in mm															9
10 Maximum precipitation in 24 h	in mm	55	32	27	39	30	43	21	38	43	69	57	29	69	10	10
11 Mean number of days with precipitation	> 0,1 mm	9	8	9	9	9	6	3	3	7	9	10	10	92	30	11
12 Mean duration of sunshine	in h															12
13 Mean quantity of radiation	in ly/day															13
14 Mean potential evaporation	in mm	7	10	23	41	76	101	124	115	80	53	25	13	668	26	14
15 Mean windspeed	in m/sec															15
16 Mean predominent direction of the wind																16

170 Station/Country Taranto/Italy

Location 40°28'N/17°13'E Height above sealevel 16 m Climate symbol: Köppen Csa Troll IV,1

		J	F	M	A	M	J	J	A	S	O	N	D	year	P	
1 Mean daily temperature	in °C	8.8	9.2	11.7	14.4	18.6	22.7	25.4	25.7	22.8	18.3	13.8	10.7	16.8	30	1
2 Mean daily maximum temperature	in °C	11.8	12.8	14.8	18.0	22.3	26.6	29.8	29.9	26.3	21.8	17.5	14.1	20.5	16	2
3 Mean daily minimum temperature	in °C	6.4	6.9	8.4	11.1	14.9	18.9	21.6	21.9	18.9	15.5	11.8	8.8	13.8	16	3
4 Absolute maximum temperature	in °C	18.0	19.6	24.8	27.1	30.5	34.2	39.0	38.4	32.5	28.5	23.1	20.6	39.0	16	4
5 Absolute minimum temperature	in °C	-3.8	-3.0	-2.5	0.9	6.0	11.8	15.5	15.5	12.2	7.9	3.4	-2.2	-3.8	16	5
6 Mean relative humidity	in %	77	73	70	73	65	58	56	57	64	73	76	77	68	10	6
7 Mean precipitation	in mm	55	44	47	33	31	17	15	20	26	52	70	59	469	16	7
8 Maximum precipitation	in mm															8
9 Minimum precipitation	in mm															9
10 Maximum precipitation in 24 h	in mm	35	50	63	33	26	38	15	37	53	42	82	37	82	10	10
11 Mean number of days with precipitation	> 0,1 mm	8	6	6	5	5	3	2	2	3	6	7	6	58	16	11
12 Mean duration of sunshine	in h	128	131	156	221	284	316	341	327	248	197	140	110	2595	4	12
13 Mean quantity of radiation	in ly/day	137	223	361	486	604	663	674	591	441	292	166	148	399		13
14 Mean potential evaporation	in mm	17	17	32	50	84	128	163	150	108	70	42	23	884	26	14
15 Mean windspeed	in m/sec															15
16 Mean predominent direction of the wind																16

171 Station/Country Cagliari (Sardinia)/Italy

Location 39°13'N/9°06'E Height above sealevel 75 m Climate symbol: Köppen Csa Troll IV,1

		J	F	M	A	M	J	J	A	S	O	N	D	year	P	
1 Mean daily temperature	in °C	9.4	9.9	11.8	14.2	17.8	21.7	24.5	24.7	22.1	18.3	14.0	11.0	16.6	30	1
2 Mean daily maximum temperature	in °C	14.2	14.8	16.5	19.1	23.4	27.1	30.3	29.9	27.2	23.3	18.8	15.6	21.7	16	2
3 Mean daily minimum temperature	in °C	6.9	7.4	8.8	10.6	13.7	17.7	20.7	20.8	18.7	15.1	11.3	8.7	13.3	16	3
4 Absolute maximum temperature	in °C	21.2	21.2	24.5	28.1	35.0	37.4	40.2	37.9	35.2	28.8	26.2	24.2	40.2	16	4
5 Absolute minimum temperature	in °C	-2.0	-1.2	0.7	4.2	6.9	10.1	14.2	15.4	11.7	7.7	4.2	-1.0	-2.0	16	5
6 Mean relative humidity	in %	80	78	77	76	75	70	70	72	74	77	77	80	76	10	6
7 Mean precipitation	in mm	50	50	45	31	26	13	1	10	32	54	72	67	451	16	7
8 Maximum precipitation	in mm															8
9 Minimum precipitation	in mm															9
10 Maximum precipitation in 24 h	in mm	34	52	27	28	23	48	6	44	47	51	53	75	75	10	10
11 Mean number of days with precipitation	> 0,1 mm	8	7	7	5	4	1		3	6	9	9		60	16	11
12 Mean duration of sunshine	in h	128	122	187	208	269	274	327	304	229	193	122	105	2468	7	12
13 Mean quantity of radiation	in ly/day	156	212	302	384	467	484	518	458	356	253	161	122	321	30	13
14 Mean potential evaporation	in mm	19	21	35	52	82	118	148	138	103	68	39	24	845	26	14
15 Mean windspeed	in m/sec															15
16 Mean predominent direction of the wind		NNW	NW	NW	NW	NW.S	NW.S	NW.S	NW.S	NW.S	NW.S	NNW	NNW		30	16

172 Station/Country Messina (Sicily)/Italy

Location 38°12'N/15°33'E Height above sealevel 54 m Climate symbol: Köppen Csa Troll IV,1

		J	F	M	A	M	J	J	A	S	O	N	D	year	P	
1 Mean daily temperature	in °C	11.5	11.4	12.4	14.9	18.4	22.6	25.4	25.9	23.4	19.7	15.9	13.0	17.9	30	1
2 Mean daily maximum temperature	in °C	13.5	14.3	15.6	18.2	22.3	26.1	29.4	29.6	26.6	22.8	19.0	15.4	21.1	11	2
3 Mean daily minimum temperature	in °C	9.3	9.5	10.6	12.6	16.0	19.6	22.6	23.2	20.7	17.5	14.4	11.2	15.6	11	3
4 Absolute maximum temperature	in °C	21.4	21.8	24.0	26.0	30.2	32.8	37.6	36.8	34.0	28.4	25.2	22.2	37.6	11	4
5 Absolute minimum temperature	in °C	0.4	2.8	1.0	7.0	10.4	14.0	16.2	18.8	15.4	11.0	7.8	0.6	0.4	11	5
6 Mean relative humidity	in %	71	71	69	68	67	62	63	64	68	73	71	70	68	10	6
7 Mean precipitation	in mm	98	92	81	67	41	28	10	24	58	120	122	117	854	30	7
8 Maximum precipitation	in mm															8
9 Minimum precipitation	in mm															9
10 Maximum precipitation in 24 h	in mm	59	56	46	40	136	24	46	38	106	66	67	73	136	10	10
11 Mean number of days with precipitation	> 0,1 mm	14	9	10	8	4	3	2	2	6	9	10	14	91	11	11
12 Mean duration of sunshine	in h	115	137	171	210	257	285	329	310	237	189	129	102	2471	30	12
13 Mean quantity of radiation	in ly/day	132	198	261	358	428	486	465	415	316	225	148	114	295	30	13
14 Mean potential evaporation	in mm	23	24	36	53	85	122	159	156	115	77	43	28	921	41	14
15 Mean windspeed	in m/sec															15
16 Mean predominent direction of the wind		W	W	W.E	W.E	W.E	W.E	W.E	W.E	W.E	W.E	W	W		30	16

173 Station / Country Palermo (Sicily) / Italy

Location 38°07'N / 13°21'E — Height above sealevel 71 m — Climate symbol: Köppen Csa — Troll IV,1

		J	F	M	A	M	J	J	A	S	O	N	D	year	P
1 Mean daily temperature	in °C	10,2	10,8	12,8	15,1	18,3	22,2	24,8	25,1	23,1	19,1	15,3	11,9	17,4	30
2 Mean daily maximum temperature	in °C	15,8	16,4	17,4	20,0	23,6	27,0	29,7	30,0	28,2	25,0	21,4	17,7	22,7	16
3 Mean daily minimum temperature	in °C	8,0	8,1	8,9	11,1	14,2	17,8	20,5	21,1	18,9	15,8	12,4	9,7	13,9	16
4 Absolute maximum temperature	in °C	30,4	26,8	29,6	34,0	36,0	39,5	41,2	41,8	41,1	35,2	31,9	25,7	41,8	16
5 Absolute minimum temperature	in °C	0,3	0,2	1,0	3,5	7,8	11,0	13,6	16,0	11,0	9,3	4,5	2,0	0,3	16
6 Mean relative humidity	in %	72	68	66	65	65	61	58	58	60	67	69	70	60	10
7 Mean precipitation	in mm	71	43	50	49	19	9	2	18	41	77	71	62	512	16
8 Maximum precipitation	in mm	179	165	154	145	91	119	26	41	108	189	186	185	797	30
9 Minimum precipitation	in mm	6	6	15	0	1	0	0	0	1	3	13	19	306	30
10 Maximum precipitation in 24 h	in mm	29	19	60	55	21	23	22	32	42	70	28	63	70	10
11 Mean number of days with precipitation	> 0,1 mm	12	8	8	8	3	2	0	2	4	8	8	10	71	16
12 Mean duration of sunshine	in h	140	146	189	225	288	303	350	322	252	202	156	121	2694	30
13 Mean quantity of radiation	in ly / day	189	232	309	408	466	474	498	496	367	260	188	135	334	30
14 Mean potential evaporation	in mm	20	23	36	53	82	117	148	140	108	73	42	27	867	50
15 Mean windspeed	in m / sec														
16 Mean predominant direction of the wind															

174 Station / Country Caltanissetta (Sicily) / Italy

Location 37°29'N / 14°04'E — Height above sealevel 570 m — Climate symbol: Köppen Csa — Troll IV,1

		J	F	M	A	M	J	J	A	S	O	N	D	year	P
1 Mean daily temperature	in °C	7,3	8,2	9,7	12,6	17,1	21,9	24,8	24,8	21,6	16,9	12,3	9,0	15,5	16
2 Mean daily maximum temperature	in °C	10,5	11,9	14,0	17,5	22,5	27,8	31,0	30,8	27,0	21,1	15,9	12,1	20,2	16
3 Mean daily minimum temperature	in °C	4,0	4,4	5,3	7,7	11,8	15,9	18,8	18,8	16,1	12,6	8,7	5,8	10,8	16
4 Absolute maximum temperature	in °C	20,3	23,8	24,4	28,2	32,0	37,7	40,0	40,2	35,8	31,3	25,1	20,8	40,2	16
5 Absolute minimum temperature	in °C	-4,5	-4,1	-4,2	-0,2	3,5	9,0	12,8	11,0	9,0	5,5	1,0	-1,8	-4,5	16
6 Mean relative humidity	in %	74	75	70	65	58	43	42	44	54	63	70	72	60	
7 Mean precipitation	in mm	78	47	52	47	38	18	15	8	42	60	73	74	550	16
8 Maximum precipitation	in mm														
9 Minimum precipitation	in mm														
10 Maximum precipitation in 24 h	in mm	40	51	33	65	59	38	43	12	83	53	45	37	83	10
11 Mean number of days with precipitation	> 0,1 mm	10	7	8	6	5	2	1	2	4	7	8	10	70	
12 Mean duration of sunshine	in h														
13 Mean quantity of radiation	in ly / day														
14 Mean potential evaporation	in mm	12	14	27	44	79	117	148	136	98	62	31	17	785	26
15 Mean windspeed	in m / sec														
16 Mean predominant direction of the wind															

175 Station / Country Valetta / Malta

Location 35°54'N / 14°31'E — Height above sealevel 70 m — Climate symbol: Köppen Csa — Troll IV,2

		J	F	M	A	M	J	J	A	S	O	N	D	year	P
1 Mean daily temperature	in °C	12,3	12,5	13,7	15,7	18,8	22,7	25,5	26,1	24,4	21,4	17,7	14,1	18,7	17
2 Mean daily maximum temperature	in °C	14,4	14,7	16,1	18,3	21,6	25,9	28,9	29,3	27,1	23,8	19,7	16,1	21,3	17
3 Mean daily minimum temperature	in °C	10,2	10,3	11,2	13,1	15,9	19,4	22,1	22,9	21,6	18,9	15,6	12,0	16,1	17
4 Absolute maximum temperature	in °C	22,8	24,4	26,1	27,8	34,4	39,4	39,4	40,0	36,7	32,8	26,1	22,2	40,0	17
5 Absolute minimum temperature	in °C	5,0	5,0	5,0	7,2	11,7	13,9	17,8	16,7	16,1	16,1	9,4	5,8	5,0	17
6 Mean relative humidity	in %	72	71	72	71	69	66	65	69	68	71	73	73	70	17
7 Mean precipitation	in mm	90	60	39	15	12	2	0	8	29	63	91	110	519	17
8 Maximum precipitation	in mm														
9 Minimum precipitation	in mm														
10 Maximum precipitation in 24 h	in mm	60	124	93	37	26	9	1	67	69	132	80	92	132	17
11 Mean number of days with precipitation	> 0,1 mm	12	8	5	2	2	0	0	1	3	6	9	13	61	17
12 Mean duration of sunshine	in h	170	177	225	264	305	344	386	354	283	232	172	154	3066	17
13 Mean quantity of radiation	in ly / day														
14 Mean potential evaporation	in mm	26	28	37	56	83	124	156	153	117	87	54	33	954	10
15 Mean windspeed	in m / sec														
16 Mean predominant direction of the wind															

176 Station / Country Zagreb / Yugoslavia

Location 45°49'N / 15°58'E — Height above sealevel 163 m — Climate symbol: Köppen Cfa — Troll III,3

		J	F	M	A	M	J	J	A	S	O	N	D	year	P
1 Mean daily temperature	in °C	0,2	2,2	6,8	12,0	16,4	19,9	22,0	21,3	17,7	11,8	6,6	2,4	11,8	
2 Mean daily maximum temperature	in °C	2,6	5,5	10,7	16,8	21,2	25,0	27,3	26,6	22,2	15,4	8,9	4,5	15,6	29
3 Mean daily minimum temperature	in °C	-2,4	-1,2	2,9	7,6	11,7	15,1	16,9	16,4	13,2	8,1	4,0	0,1	7,7	29
4 Absolute maximum temperature	in °C	17,9	19,5	23,5	27,4	30,3	37,0	37,1	34,7	32,1	28,4	23,0	18,3	37,1	16
5 Absolute minimum temperature	in °C	-24,6	-30,5	-17,6	-5,0	-2,2	0,2	5,2	4,4	-1,1	-4,3	-15,6	-26,3	-30,5	16
6 Mean relative humidity	in %	80	75	67	64	67	67	66	67	71	78	83	83	72	
7 Mean precipitation	in mm	56	54	47	59	86	95	79	74	70	88	89	67	864	
8 Maximum precipitation	in mm														
9 Minimum precipitation	in mm														
10 Maximum precipitation in 24 h	in mm	27	25	33	42	53	54	84	117	46	83	63	30	117	16
11 Mean number of days with precipitation	> 0,1 mm	13	10	12	13	17	14	11	12	11	14	14	16	157	16
12 Mean duration of sunshine	in h	60	91	143	182	212	241	282	269	200	131	57	47	1913	29
13 Mean quantity of radiation	in ly / day														
14 Mean potential evaporation	in mm	1	5	24	54	92	122	141	122	81	46	22	2	712	16
15 Mean windspeed	in m / sec														
16 Mean predominant direction of the wind		E	NE	NE	NE	NE	NE	NE	NE	NE	NE	NE	NE		16

177 Station / Country Beograd (Belgrade) / Yugoslavia

Location 44°48'N / 20°27'E Height above sealevel 132 m Climate symbol: Köppen Cfa Troll III,3

		J	F	M	A	M	J	J	A	S	O	N	D	year	P	
1 Mean daily temperature	in °C	-0,2	1,6	6,2	12,2	17,1	20,5	22,5	22,0	18,3	12,5	6,8	2,5	11,8		1
2 Mean daily maximum temperature	in °C	2,9	5,4	11,0	17,5	22,5	26,0	28,3	28,1	24,3	17,6	10,5	5,4	16,6	30	2
3 Mean daily minimum temperature	in °C	-3,2	-1,9	2,1	7,3	12,0	15,1	16,9	16,5	13,2	8,2	3,9	-0,2	7,5	30	3
4 Absolute maximum temperature	in °C	19,8	20,1	26,3	30,9	33,2	36,7	39,4	39,2	35,4	34,7	29,3	20,3	39,4	18	4
5 Absolute minimum temperature	in °C	-19,5	-25,5	-14,4	-6,1	-1,4	4,8	9,3	8,3	1,7	-1,0	-5,0	-19,3	-25,5	18	5
6 Mean relative humidity	in %	81	77	68	62	65	65	62	62	64	72	80	82	70		6
7 Mean precipitation	in mm	48	46	46	54	75	96	60	55	50	55	61	55	701		7
8 Maximum precipitation	in mm															8
9 Minimum precipitation	in mm															9
10 Maximum precipitation in 24 h	in mm	23	27	41	31	69	78	32	88	54	48	30	29	88	18	10
11 Mean number of days with precipitation	> 0,1 mm	13	11	12	13	15	13	9	10	10	12	12	15	146	18	11
12 Mean duration of sunshine	in h	71	95	148	191	224	259	295	280	229	107	84	69	2112	12	12
13 Mean quantity of radiation	in ly/day	129	193	280	394	489	535	564	470	369	271	135	82	326	6	13
14 Mean potential evaporation	in mm	0	2	20	54	95	124	143	123	84	48	22	2	717	18	14
15 Mean windspeed	in m/sec															15
16 Mean predominent direction of the wind		SE	SE	SE	SE	SE	W	W	W	SE	SE	SE	SE		18	16

178 Station / Country Sarajevo / Yugoslavia

Location 43°52'N / 18°26'E Height above sealevel 537 m Climate symbol: Köppen Cfb Troll III,3

		J	F	M	A	M	J	J	A	S	O	N	D	year	P	
1 Mean daily temperature	in °C	-1,4	0,7	4,9	9,8	14,3	17,4	19,5	19,7	16,0	10,2	5,4	1,7	9,8		1
2 Mean daily maximum temperature	in °C	3,0	5,3	9,9	15,1	19,5	23,5	25,9	26,8	22,5	15,9	9,5	6,0	15,2	12	2
3 Mean daily minimum temperature	in °C	-4,0	-3,4	0,1	4,5	8,3	11,9	13,3	13,1	10,1	5,8	2,6	-0,6	5,1	12	3
4 Absolute maximum temperature	in °C	16,5	18,0	23,6	30,0	30,8	35,3	36,9	38,1	33,7	32,2	24,7	16,9	38,1	16	4
5 Absolute minimum temperature	in °C	-20,9	-23,4	-15,9	-6,4	-2,3	2,8	5,5	4,6	-2,2	-3,1	-7,9	-22,4	-23,4	16	5
6 Mean relative humidity	in %	80	75	68	68	68	68	66	63	69	79	80	81	72		6
7 Mean precipitation	in mm	71	69	50	69	84	86	68	62	71	84	98	87	889		7
8 Maximum precipitation	in mm															8
9 Minimum precipitation	in mm															9
10 Maximum precipitation in 24 h	in mm	36	50	38	35	54	61	27	42	87	49	47	33	87		10
11 Mean number of days with precipitation	> 0,1 mm	12	10	13	13	16	13	9	10	11	13	13	15	148		11
12 Mean duration of sunshine	in h	63	89	125	136	168	207	240	243	180	110	69	55	1686	16	12
13 Mean quantity of radiation	in ly/day															13
14 Mean potential evaporation	in mm	0	0	20	48	84	108	127	109	75	47	24	0	642	16	14
15 Mean windspeed	in m/sec	1,0	1,6	1,6	1,8	1,5	1,2	1,3	1,6	1,5	1,5	1,2	0,9	1,4	16	15
16 Mean predominent direction of the wind		E	E	E	E	E	W	E,W	E	E	E	E	E		16	16

179 Station / Country Split / Yugoslavia

Location 43°31'N / 16°26'E Height above sealevel 128 m Climate symbol: Köppen Csa Troll IV,1

		J	F	M	A	M	J	J	A	S	O	N	D	year	P	
1 Mean daily temperature	in °C	7,8	8,1	10,3	14,0	18,6	22,9	25,6	25,4	21,6	16,8	12,3	10,1	16,1		1
2 Mean daily maximum temperature	in °C	9,9	10,8	13,6	17,8	22,8	26,9	30,2	30,1	25,9	20,1	15,1	11,8	19,6	18	2
3 Mean daily minimum temperature	in °C	4,9	5,0	7,4	11,1	15,5	18,9	21,8	21,8	18,7	14,0	10,1	7,2	13,0	18	3
4 Absolute maximum temperature	in °C	17,7	19,5	21,2	26,8	33,2	36,3	37,8	37,1	34,9	30,8	22,0	17,9	37,8	25	4
5 Absolute minimum temperature	in °C	-8,3	-8,1	-3,3	1,9	4,8	10,7	12,2	11,7	7,4	3,0	-4,5	-6,0	-8,3	25	5
6 Mean relative humidity	in %	62	63	63	60	60	57	52	50	58	63	66	67	60		6
7 Mean precipitation	in mm	76	74	53	62	60	53	40	32	55	71	110	130	816		7
8 Maximum precipitation	in mm															8
9 Minimum precipitation	in mm															9
10 Maximum precipitation in 24 h	in mm	39	44	53	55	54	89	44	51	103	76	86	78	103	16	10
11 Mean number of days with precipitation	> 0,1 mm	13	9	11	11	14	9	5	5	8	11	13	13	122	16	11
12 Mean duration of sunshine	in h	148	158	195	206	255	323	354	345	246	185	128	115	2658		12
13 Mean quantity of radiation	in ly/day															13
14 Mean potential evaporation	in mm	13	14	28	50	89	132	169	146	98	61	33	14	847	16	14
15 Mean windspeed	in m/sec	3,5	3,6	3,2	3,3	3,3	2,8	2,6	2,7	2,7	3,0	3,7	3,1	3,1		15
16 Mean predominent direction of the wind		NE	NE	NE	NE	SE	NE,SW	NE	NE	NE	NE	NE	NE		16	16

180 Station / Country Mostar / Yugoslavia

Location 43°20'N / 17°49'E Height above sealevel 99 m Climate symbol: Köppen Csa Troll IV,1

		J	F	M	A	M	J	J	A	S	O	N	D	year	P	
1 Mean daily temperature	in °C	5,4	6,5	9,7	13,8	18,3	22,4	25,4	25,5	21,4	15,8	10,7	7,8	15,2	12	1
2 Mean daily maximum temperature	in °C	8,7	10,5	14,3	19,1	24,0	28,4	31,7	32,1	27,2	20,6	14,2	11,1	20,2	12	2
3 Mean daily minimum temperature	in °C	2,1	2,5	5,1	8,5	12,6	16,3	19,0	18,8	15,5	10,9	7,1	4,5	10,2	12	3
4 Absolute maximum temperature	in °C	18,8	22,2	25,8	29,6	34,8	40,0	43,0	40,2	38,2	30,3	22,9	19,2	43,0	12	4
5 Absolute minimum temperature	in °C	-10,2	-6,9	6,0	1,2	3,3	8,4	11,0	9,6	7,1	4,0	-4,0	-4,3	-10,2	12	5
6 Mean relative humidity	in %	66	64	60	59	59	56	49	48	64	70	71		60	12	6
7 Mean precipitation	in mm	136	131	116	107	104	71	38	52	102	171	201	226	1455	22	7
8 Maximum precipitation	in mm															8
9 Minimum precipitation	in mm															9
10 Maximum precipitation in 24 h	in mm	79	66	69	76	100	69	50	70	75	151	134	96	151	22	10
11 Mean number of days with precipitation	> 0,1 mm	13	12	11	11	11	9	6	5	8	12	14	16	128	12	11
12 Mean duration of sunshine	in h	109	119	177	192	228	272	340	315	232	160	106	90	2338	10	12
13 Mean quantity of radiation	in ly/day															13
14 Mean potential evaporation	in mm	9	11	28	56	91	132	168	150	96	56	30	11	838	16	14
15 Mean windspeed	in m/sec															15
16 Mean predominent direction of the wind																16

181 Station/Country Skopje/Yugoslavia

Location 42°00'N/21°06'E Height above sealevel 245 m Climate symbol: Köppen Csa Troll III,3

		J	F	M	A	M	J	J	A	S	O	N	D	year	P	
1	Mean daily temperature	in °C	1.1	2.9	6.5	12.1	17.0	21.8	23.8	23.7	18.6	11.9	7.2	2.9	12.4		1
2	Mean daily maximum temperature	in °C	4.7	8.3	11.9	19.3	23.3	28.0	30.8	31.1	26.0	18.5	11.7	7.4	18.4	12	2
3	Mean daily minimum temperature	in °C	-2.9	-2.5	0.6	5.3	10.1	13.4	15.2	14.3	11.1	5.9	2.9	-1.1	6.0	12	3
4	Absolute maximum temperature	in °C	19.0	18.1	25.0	30.2	33.5	37.0	41.2	40.5	38.5	34.2	24.8	19.4	41.2	16	4
5	Absolute minimum temperature	in °C	-23.0	-23.9	-18.4	-4.5	-0.6	3.3	5.5	5.7	-3.2	-4.4	-9.7	-21.8	-23.9	16	5
6	Mean relative humidity	in %	85	79	73	66	69	63	58	57	67	80	85	87	72		6
7	Mean precipitation	in mm	48	41	38	34	52	49	35	37	42	58	71	43	548		7
8	Maximum precipitation	in mm															8
9	Minimum precipitation	in mm															9
10	Maximum precipitation in 24 h	in mm	32	23	60	25	29	67	22	32	40	41	37	59	67	16	10
11	Mean number of days with precipitation	> 0,1 mm	7	6	8	8	11	8	4	5	4	8	6	11	86	16	11
12	Mean duration of sunshine	in h	66	116	135	197	220	265	318	306	220	153	72	60	2128	11	12
13	Mean quantity of radiation	in ly/day	146	220	297	413	521	552	574	514	404	289	150	100	348	6	13
14	Mean potential evaporation	in mm	1	2	22	53	92	128	155	133	88	49	21	2	746	16	14
15	Mean windspeed	in m/sec															15
16	Mean predominent direction of the wind		W	W	SE,W	SE	SE	W	W	W	W	W	W	W			16

182 Station/Country Pleven/Bulgaria

Location 43°36'N/24°35'E Height above sealevel 110 m Climate symbol: Köppen Csa Troll III,3

		J	F	M	A	M	J	J	A	S	O	N	D	year	P	
1	Mean daily temperature	in °C	-2.0	0.4	5.4	12.4	17.4	21.0	23.5	23.4	18.8	12.4	5.9	0.8	11.6		-1
2	Mean daily maximum temperature	in °C	1.4	4.3	11.1	17.7	23.0	26.9	29.4	28.6	24.6	19.1	9.9	3.3	16.6	9	2
3	Mean daily minimum temperature	in °C	-5.3	-3.5	1.3	6.2	11.5	14.8	17.1	16.0	12.2	8.4	2.4	-2.5	6.7	9	3
4	Absolute maximum temperature	in °C	19.4	21.6	34.2	32.0	37.0	38.6	40.6	41.8	40.8	37.3	23.7	22.4	41.8		4
5	Absolute minimum temperature	in °C	-25.5	-28.3	-15.9	-3.8	0.6	2.8	8.3	8.9	2.5	-3.4	-11.1	-23.1	-28.3		5
6	Mean relative humidity	in %	86	82	73	66	67	65	63	63	67	73	84	87	73	9	6
7	Mean precipitation	in mm	37	30	30	50	65	83	63	33	33	48	43	41	556		7
8	Maximum precipitation	in mm															8
9	Minimum precipitation	in mm															9
10	Maximum precipitation in 24 h	in mm	24	20	22	38	48	58	60	74	60	39	33	35	74		10
11	Mean number of days with precipitation	> 0,1 mm	11	9	9	10	12	10	8	5	5	8	10	10	107		11
12	Mean duration of sunshine	in h	75	103	144	209	247	287	332	320	240	154	72	63	2246	20	12
13	Mean quantity of radiation	in ly/day															13
14	Mean potential evaporation	in mm	0	1	20	54	96	125	144	128	82	45	13	2	710	30	14
15	Mean windspeed	in m/sec	1.5	2.0	2.3	2.2	1.9	1.6	1.5	1.6	1.4	1.5	1.6	1.4	1.7		15
16	Mean predominent direction of the wind		W	W	W	W	W	W	W	W	W	W	W	W			16

183 Station/Country Varna/Bulgaria

Location 43°12'N/27°55'E Height above sealevel 3 m Climate symbol: Köppen Cfa Troll III,3

		J	F	M	A	M	J	J	A	S	O	N	D	year	P	
1	Mean daily temperature	in °C	1.2	2.4	5.0	10.0	15.5	20.2	22.9	22.6	18.9	14.0	8.6	4.1	12.1		1
2	Mean daily maximum temperature	in °C	5.8	6.2	10.8	15.7	21.6	26.1	29.8	29.3	25.5	20.5	13.0	7.1	17.6	9	2
3	Mean daily minimum temperature	in °C	-1.2	-1.1	2.3	6.8	11.9	15.9	18.5	17.9	14.4	10.9	5.9	1.0	8.6	9	3
4	Absolute maximum temperature	in °C	20.4	21.4	27.5	29.5	34.7	35.4	38.7	39.4	35.4	32.4	24.2	21.0	39.4		4
5	Absolute minimum temperature	in °C	-23.5	-15.8	-9.7	-2.3	2.4	7.2	10.1	9.8	0.0	-1.6	-8.2	-12.8	-23.5		5
6	Mean relative humidity	in %	85	81	78	76	76	73	68	70	73	78	80	84	77	9	6
7	Mean precipitation	in mm	36	31	26	35	40	56	39	38	25	43	49	56	474		7
8	Maximum precipitation	in mm															8
9	Minimum precipitation	in mm															9
10	Maximum precipitation in 24 h	in mm	59	38	39	31	37	64	93	258	80	66	36	38	258		10
11	Mean number of days with precipitation	> 0,1 mm	10	9	8	9	10	9	6	4	4	7	10	10	96		11
12	Mean duration of sunshine	in h	72	96	134	188	255	275	336	311	247	176	91	72	2253	7	12
13	Mean quantity of radiation	in ly/day															13
14	Mean potential evaporation	in mm	3	7	22	46	84	118	141	129	90	56	23	32	751	30	14
15	Mean windspeed	in m/sec	3.6	3.6	3.5	3.0	2.8	2.9	2.7	2.8	3.1	3.2	3.5	3.4	3.2		15
16	Mean predominent direction of the wind		NW	NW	NW	E	E	NW	NW	NW	NW	NW	NW	NW			16

184 Station/Country Sofija(Sofia)/Bulgaria

Location 42°42'N/23°20'E Height above sealevel 550 m Climate symbol: Köppen Cfa Troll III,3

		J	F	M	A	M	J	J	A	S	O	N	D	year	P	
1	Mean daily temperature	in °C	-1.7	0.6	4.6	10.6	15.5	19.0	21.3	20.7	17.0	11.1	5.5	0.7	10.4		1
2	Mean daily maximum temperature	in °C	1.8	4.1	10.4	15.8	20.7	24.4	27.1	26.0	22.1	17.4	9.1	3.5	15.2	9	2
3	Mean daily minimum temperature	in °C	-4.0	-2.9	0.8	5.3	10.2	13.5	15.6	14.9	11.3	7.9	2.5	-2.1	6.1	9	3
4	Absolute maximum temperature	in °C	16.9	20.7	30.8	28.5	32.1	34.0	36.7	37.3	37.5	33.2	24.2	20.0	37.5		4
5	Absolute minimum temperature	in °C	-27.5	-24.5	-14.9	-4.8	-1.5	2.5	6.9	6.1	-1.5	-3.3	-10.7	-20.3	-27.5		5
6	Mean relative humidity	in %	84	78	72	66	68	67	62	61	68	75	83	85	72		6
7	Mean precipitation	in mm	42	31	37	55	71	90	60	43	42	55	52	44	622		7
8	Maximum precipitation	in mm															8
9	Minimum precipitation	in mm															9
10	Maximum precipitation in 24 h	in mm	37	23	32	38	55	72	60	50	59	37	30	28	72		10
11	Mean number of days with precipitation	> 0,1 mm	14	13	14	15	17	15	11	9	8	11	13	14	154		11
12	Mean duration of sunshine	in h	55	91	138	187	221	261	314	304	233	155	75	49	2083		12
13	Mean quantity of radiation	in ly/day															13
14	Mean potential evaporation	in mm	0	1	20	49	89	111	128	116	79	44	13	1	651	30	14
15	Mean windspeed	in m/sec	2.0	2.4	2.4	2.0	2.1	1.9	1.9	1.8	1.6	1.7	1.8	1.6	2.0		15
16	Mean predominent direction of the wind		W	W	W	~W	W	W	W	W	E,W	E	E	W			16

50

185 Station/Country Plovdiv/Bulgaria

Location 42°09'N/24°45'E Height above sealevel 160 m Climate symbol: Köppen Cfa Troll III,3

	J	F	M	A	M	J	J	A	S	O	N	D	year	P	
1 Mean daily temperature in °C	-0,4	2,1	6,0	12,2	17,2	21,0	23,4	22,8	18,3	12,6	7,1	2,3	12,0		1
2 Mean daily maximum temperature in °C	4,8	7,0	12,4	18,4	23,4	27,6	30,8	29,9	25,7	20,5	12,0	5,9	18,2	9	2
3 Mean daily minimum temperature in °C	-3,3	-2,2	0,9	5,2	10,2	13,7	16,2	15,1	11,2	7,5	2,7	-1,6	6,3	9	3
4 Absolute maximum temperature in °C	19,4	23,6	28,4	30,7	35,3	38,5	39,3	41,3	36,5	32,7	23,9	22,1	41,3		4
5 Absolute minimum temperature in °C	-31,5	-29,1	-17,5	-4,0	-0,3	6,0	8,2	5,6	0,2	-5,8	-9,1	-18,0	-31,5		5
6 Mean relative humidity in %	84	78	74	68	67	63	59	61	67	75	81	83	72	9	6
7 Mean precipitation in mm	41	34	40	43	55	67	47	31	33	45	53	52	541		7
8 Maximum precipitation in mm															8
9 Minimum precipitation in mm															9
10 Maximum precipitation in 24 h in mm	25	37	31	32	39	80	50	53	42	53	29	58	80		10
11 Mean number of days with precipitation > 0,1 mm	7	5	6	7	9	8	6	4	4	5	7	6	74		11
12 Mean duration of sunshine in h	81	105	149	203	234	271	328	321	241	167	87	75	2262		12
13 Mean quantity of radiation in ly/day															13
14 Mean potential evaporation in mm	1	6	25	53	96	124	146	131	84	49	17	5	737	30	14
15 Mean windspeed in m/sec															15
16 Mean predominent direction of the wind	W	W	W	W	W	W	W	W	W	W	W	W			16

186 Station/Country Tiranë/Albania

Location 41°18'N/19°48'E Height above sealevel 114 m Climate symbol: Köppen Csa Troll IV,1

	J	F	M	A	M	J	J	A	S	O	N	D	year	P	
1 Mean daily temperature in °C	7,3	8,3	10,6	14,4	18,4	22,4	25,0	24,9	21,8	17,4	12,9	9,2	16,0		1
2 Mean daily maximum temperature in °C	11,6	12,2	15,2	18,4	23,1	27,6	30,6	31,4	27,4	22,8	17,4	13,5	20,9	10	2
3 Mean daily minimum temperature in °C	2,0	2,0	4,9	8,1	11,8	15,5	17,2	16,9	14,2	10,2	8,2	4,6	9,6	10	3
4 Absolute maximum temperature in °C	21,4	23,7	27,2	31,2	35,3	38,0	42,0	39,8	39,7	35,5	28,4	21,9	42,0		4
5 Absolute minimum temperature in °C	-10,0	-9,0	-10,5	-1,0	3,1	5,6	8,0	8,5	3,0	0,1	-5,8	-8,0	-10,5		5
6 Mean relative humidity in %	70	70	68	67	69	62	58	58	64	70	74	70	67		6
7 Mean precipitation in mm	132	120	100	87	99	60	28	39	73	157	152	142	1189		7
8 Maximum precipitation in mm															8
9 Minimum precipitation in mm															9
10 Maximum precipitation in 24 h in mm	79	103	67	62	91	101	43	46	38	91	153	66	153	10	10
11 Mean number of days with precipitation > 0,1 mm	12	10	11	11	10	6	4	4	6	11	13	12	110		11
12 Mean duration of sunshine in h	131	122	161	202	250	298	354	335	269	208	103	93	2526	10	12
13 Mean quantity of radiation in ly/day															13
14 Mean potential evaporation in mm	16	18	32	53	90	125	152	138	98	67	34	18	841		14
15 Mean windspeed in m/sec	1,5	1,7	1,6	1,5	1,5	1,6	1,7	1,5	1,4	1,4	1,4	1,4	1,5		15
16 Mean predominent direction of the wind	SE	SE	NW	NW	NW	NW	NW	NW	NW	NW	SE	SE			16

187 Station/Country Alexandroúpolis/Greece

Location 40°51'N/25°53'E Height above sealevel 7 m Climate symbol: Köppen Csa Troll IV,1

	J	F	M	A	M	J	J	A	S	O	N	D	year	P	
1 Mean daily temperature in °C	5,5	6,8	8,6	12,8	17,3	21,5	24,8	24,8	20,9	16,3	11,3	7,9	14,9	18	1
2 Mean daily maximum temperature in °C	9,1	10,6	12,9	17,9	22,6	27,3	30,9	30,9	26,7	21,5	15,4	11,6	19,8	19	2
3 Mean daily minimum temperature in °C	1,9	2,9	4,2	7,7	11,9	15,7	18,7	18,6	15,0	11,1	7,2	4,1	9,9	18	3
4 Absolute maximum temperature in °C	21,0	21,6	24,8	31,6	32,2	35,8	39,0	39,0	34,0	32,8	24,2	21,2	39,0	19	4
5 Absolute minimum temperature in °C	-13,2	-11,8	-8,6	-1,2	2,6	7,2	8,4	12,0	5,4	-0,2	-9,3	-13,5	-13,5	18	5
6 Mean relative humidity in %	77	75	71	68	68	61	54	53	58	68	74	78	67	19	6
7 Mean precipitation in mm	81	45	46	30	32	36	18	15	25	47	86	97	558	19	7
8 Maximum precipitation in mm															8
9 Minimum precipitation in mm															9
10 Maximum precipitation in 24 h in mm	65	50	31	48	43	47	39	63	51	95	89	87	95	19	10
11 Mean number of days with precipitation > 0,1 mm	10	8	8	7	8	7	3	2	3	6	8	11	81	19	11
12 Mean duration of sunshine in h															12
13 Mean quantity of radiation in ly/day															13
14 Mean potential evaporation in mm	10	12	25	50	86	121	152	143	97	58	30	17	800	19	14
15 Mean windspeed in m/sec															15
16 Mean predominent direction of the wind															16

188 Station/Country Thessaloniki(Salonika)/Greece

Location 40°39'N/23°07'E Height above sealevel 2 m Climate symbol: Köppen Csa Troll IV,1

	J	F	M	A	M	J	J	A	S	O	N	D	year	P	
1 Mean daily temperature in °C	5,5	7,1	9,6	14,5	19,6	24,7	27,3	26,8	22,5	17,1	12,0	7,5	16,1		1
2 Mean daily maximum temperature in °C	9,4	11,7	14,4	19,7	24,8	29,4	32,4	32,1	27,7	21,7	15,9	11,4	20,9	27	2
3 Mean daily minimum temperature in °C	1,9	2,5	5,2	9,5	14,2	18,3	20,9	20,7	17,2	12,8	8,6	4,1	11,3	27	3
4 Absolute maximum temperature in °C	19,5	24,2	30,1	30,0	37,8	37,8	41,8	40,0	37,4	32,7	24,2	21,2	41,8	27	4
5 Absolute minimum temperature in °C	-10,3	-8,9	-4,7	-1,0	5,2	9,7	14,4	10,3	8,1	3,8	-2,8	-7,4	-10,3	27	5
6 Mean relative humidity in %	78	71	69	67	66	56	51	52	60	69	76	78	66		6
7 Mean precipitation in mm	44	34	35	36	40	33	20	14	28	55	58	54	449		7
8 Maximum precipitation in mm															8
9 Minimum precipitation in mm															9
10 Maximum precipitation in 24 h in mm	28	53	45	57	61	56	48	59	51	45	44	64	64	27	10
11 Mean number of days with precipitation > 0,1 mm	6	6	7	7	6	6	4	3	4	6	7	6	70		11
12 Mean duration of sunshine in h	117	149	169	227	277	309	367	345	253	182	118	110	2624	27	12
13 Mean quantity of radiation in ly/day															13
14 Mean potential evaporation in mm	8	11	29	56	99	140	166	155	105	62	27	13	871	36	14
15 Mean windspeed in m/sec															15
16 Mean predominent direction of the wind															16

189 Station/Country Límnos/Greece

Location 39°53'N/25°04'E — Height above sealevel 2 m — Climate symbol: Köppen Csa — Troll IV,1

		J	F	M	A	M	J	J	A	S	O	N	D	year	P	
1 Mean daily temperature	in °C	8,3	8,8	9,9	14,1	18,5	22,8	25,4	25,7	22,1	18,0	13,3	10,7	16,9	19	1
2 Mean daily maximum temperature	in °C	11,0	12,0	13,1	18,0	22,8	27,3	30,2	30,3	26,2	21,5	16,3	13,4	20,2	19	2
3 Mean daily minimum temperature	in °C	5,6	5,5	6,7	10,1	14,2	18,3	20,6	21,1	18,0	14,5	10,6	8,1	12,8	19	3
4 Absolute maximum temperature	in °C	18,8	21,4	23,6	29,5	31,4	35,0	37,0	38,0	34,0	31,8	24,8	20,6	38,0	19	4
5 Absolute minimum temperature	in °C	-5,6	-5,0	-2,4	0,6	5,3	11,0	11,8	13,2	8,4	5,6	-1,0	-4,2	-5,6	19	5
6 Mean relative humidity	in %	77	73	72	70	68	62	58	58	63	72	76	77	69	20	6
7 Mean precipitation	in mm	96	51	62	28	21	14	11	12	20	50	72	88	525	20	7
8 Maximum precipitation	in mm															8
9 Minimum precipitation	in mm															9
10 Maximum precipitation in 24 h	in mm	142	70	73	40	69	51	85	41	36	50	50	59	142	20	10
11 Mean number of days with precipitation	> 0,1 mm	13	8	8	6	5	2	1	1	2	6	9	11	72	20	11
12 Mean duration of sunshine	in h															12
13 Mean quantity of radiation	in ly/day															13
14 Mean potential evaporation	in mm	15	15	25	50	89	131	160	149	103	66	38	44	885	20	14
15 Mean windspeed	in m/sec															15
16 Mean predominent direction of the wind																16

190 Station/Country Kérkira/Greece

Location 39°37'N/19°55'E — Height above sealevel 25 m — Climate symbol: Köppen Csa — Troll IV,1

		J	F	M	A	M	J	J	A	S	O	N	D	year	P	
1 Mean daily temperature	in °C	10,0	10,4	11,8	14,5	18,2	22,5	25,3	25,4	22,5	18,8	14,9	11,8	17,2	20	1
2 Mean daily maximum temperature	in °C	14,0	14,6	16,0	19,1	23,3	28,2	31,4	31,5	27,8	23,4	19,1	15,9	22,0	20	2
3 Mean daily minimum temperature	in °C	5,9	6,1	7,5	9,9	13,1	16,8	19,2	19,2	17,2	14,2	10,7	7,7	12,3	20	3
4 Absolute maximum temperature	in °C	20,0	21,4	23,8	28,7	32,1	36,6	40,2	40,7	37,8	33,2	25,0	21,0	40,7	20	4
5 Absolute minimum temperature	in °C	-4,4	-3,3	-2,8	-0,1	3,6	10,2	12,0	13,0	6,8	3,8	-2,5	-2,5	-4,4	20	5
6 Mean relative humidity	in %	74	73	73	74	71	65	62	62	71	71	74	76	70	20	6
7 Mean precipitation	in mm	196	132	100	70	41	14	4	20	95	184	237	259	1352	20	7
8 Maximum precipitation	in mm															8
9 Minimum precipitation	in mm															9
10 Maximum precipitation in 24 h	in mm	79	61	83	47	49	31	26	77	183	191	158	132	191	20	10
11 Mean number of days with precipitation	> 0,1 mm	15	14	12	10	6	3	1	2	6	11	15	17	112	20	11
12 Mean duration of sunshine	in h															12
13 Mean quantity of radiation	in ly/day															13
14 Mean potential evaporation	in mm	18	21	35	57	97	135	169	157	113	73	40	26	941	21	14
15 Mean windspeed	in m/sec															15
16 Mean predominent direction of the wind																16

191 Station/Country Trikkala/Greece

Location 39°33'N/21°46'E — Height above sealevel 149 m — Climate symbol: Köppen Csa — Troll IV,1

		J	F	M	A	M	J	J	A	S	O	N	D	year	P	
1 Mean daily temperature	in °C	4,6	7,2	10,2	14,7	18,7	23,5	27,0	26,5	22,6	18,4	11,9	7,1	16,0	6	1
2 Mean daily maximum temperature	in °C	8,7	12,5	16,2	21,0	25,3	30,8	34,7	34,1	29,8	24,7	16,5	10,8	22,1	6	2
3 Mean daily minimum temperature	in °C	0,4	1,9	4,2	8,3	12,1	16,2	19,2	18,8	15,3	12,1	7,2	3,5	9,9	6	3
4 Absolute maximum temperature	in °C	19,9	21,7	25,9	31,3	34,5	40,8	43,5	41,5	40,1	36,5	25,5	20,7	43,5	6	4
5 Absolute minimum temperature	in °C	-11,8	-7,6	-2,1	-1,1	4,0	9,5	10,1	11,8	8,1	2,0	-1,5	-7,9	-11,8	6	5
6 Mean relative humidity	in %	78	70	65	62	59	51	44	45	55	65	75	81	63	6	6
7 Mean precipitation	in mm	84	69	59	80	61	51	19	12	27	80	10	125	677	6	7
8 Maximum precipitation	in mm															8
9 Minimum precipitation	in mm															9
10 Maximum precipitation in 24 h	in mm	70	39	41	68	40	45	40	42	59	39	76	100	100	6	10
11 Mean number of days with precipitation	> 0,1 mm	13	11	10	9	9	7	4	3	5	9	11	17	108	6	11
12 Mean duration of sunshine	in h															12
13 Mean quantity of radiation	in ly/day															13
14 Mean potential evaporation	in mm	10	12	29	57	99	142	180	165	106	58	24	11	893	36	14
15 Mean windspeed	in m/sec															15
16 Mean predominent direction of the wind																16

192 Station/Country Athínai (Athens)/Greece

Location 37°58'N/23°43'E — Height above sealevel 107 m — Climate symbol: Köppen Csa — Troll IV,1

		J	F	M	A	M	J	J	A	S	O	N	D	year	P	
1 Mean daily temperature	in °C	9,3	9,9	11,3	15,3	20,0	24,6	27,6	27,4	23,5	19,0	14,7	11,0	17,8		1
2 Mean daily maximum temperature	in °C	12,9	13,9	15,5	20,2	25,0	29,9	33,2	33,1	29,0	23,8	18,6	14,6	22,5	30	2
3 Mean daily minimum temperature	in °C	6,4	6,7	7,8	11,3	15,9	20,0	22,8	22,8	19,3	15,4	11,7	8,2	14,0	30	3
4 Absolute maximum temperature	in °C	20,9	22,5	27,8	32,2	36,2	41,9	42,3	42,6	38,4	36,5	27,7	22,2	42,6	30	4
5 Absolute minimum temperature	in °C	-4,4	-5,7	-0,7	-0,3	6,2	13,6	16,0	15,5	11,6	7,2	-1,1	-3,7	-5,7	30	5
6 Mean relative humidity	in %	74	70	67	63	59	53	47	47	56	67	73	75	63		6
7 Mean precipitation	in mm	62	36	38	23	23	14	6	7	15	51	56	71	402		7
8 Maximum precipitation	in mm															8
9 Minimum precipitation	in mm															9
10 Maximum precipitation in 24 h	in mm	47	61	42	30	50	49	24	39	143	67	57	48	143	30	10
11 Mean number of days with precipitation	> 0,1 mm	12	11	10	8	7	5	2	3	4	8	12	12	93		11
12 Mean duration of sunshine	in h	149	156	190	215	232	292	364	340	272	210	129	108	2655		12
13 Mean quantity of radiation	in ly/day	156	226	308	441	529	560	590	520	408	287	183	140	362	6	13
14 Mean potential evaporation	in mm	16	17	30	52	91	136	174	160	109	70	36	22	913	36	14
15 Mean windspeed	in m/sec	1,9	2,2	2,7	1,8	1,8	1,8	2,2	2,2	1,9	1,8	2,3	2,1	2,0		15
16 Mean predominent direction of the wind		NE	NE,SW	NE	NE	S	SW	NE	NE	NE	NE	NE	NE			16

193 Station/Country Trípolis/Greece

Location 37°31'N/22°21'E   Height above sealevel 661 m   Climate symbol: Köppen Csa   Troll IV,1

		J	F	M	A	M	J	J	A	S	O	N	D	year	P	
1 Mean daily temperature	in °C	5,3	6,1	7,7	11,5	15,4	20,1	23,1	22,9	19,3	15,5	10,5	7,1	13,7	19	1
2 Mean daily maximum temperature	in °C	9,1	10,6	12,8	17,5	22,3	27,5	30,5	30,4	26,0	21,6	15,1	11,0	19,5	19	2
3 Mean daily minimum temperature	in °C	1,5	1,6	2,5	5,5	8,5	12,7	15,6	15,4	12,6	9,3	5,8	3,1	7,8	19	3
4 Absolute maximum temperature	in °C	18,0	21,0	29,6	29,0	33,2	37,6	39,8	40,2	36,4	33,4	26,0	19,0	40,2	19	4
5 Absolute minimum temperature	in °C	-12,4	-16,4	-5,6	-1,0	1,0	5,5	9,0	9,0	4,0	0,6	-5,0	-10,2	-16,4	19	5
6 Mean relative humidity	in %	78	74	69	61	57	48	41	42	53	65	75	78	62	18	6
7 Mean precipitation	in mm	127	104	94	62	51	36	20	13	37	82	133	178	932	18	7
8 Maximum precipitation	in mm															8
9 Minimum precipitation	in mm															9
10 Maximum precipitation in 24 h	in mm	66	60	53	35	94	45	34	30	51	73	61	122	122	18	10
11 Mean number of days with precipitation	> 0,1 mm	14	10	11	9	7	5	2	2	4	7	12	15	98	18	11
12 Mean duration of sunshine	in h															12
13 Mean quantity of radiation	in ly/day															13
14 Mean potential evaporation	in mm	9	13	26	47	81	115	146	137	96	57	29	16	772	36	14
15 Mean windspeed	in m/sec															15
16 Mean predominent direction of the wind																16

194 Station/Country Náxos/Greece

Location 37°06'N/25°25'E   Height above sealevel 3 m   Climate symbol: Köppen Csa   Troll IV,1

		J	F	M	A	M	J	J	A	S	O	N	D	year	P	
1 Mean daily temperature	in °C	12,2	12,2	13,4	16,4	19,5	22,7	24,8	25,0	22,9	20,9	17,3	14,3	18,5	10	1
2 Mean daily maximum temperature	in °C	14,5	14,8	16,2	19,5	22,7	25,6	27,3	27,6	25,5	24,0	19,9	16,6	21,2	10	2
3 Mean daily minimum temperature	in °C	9,9	9,5	10,6	13,3	16,3	19,8	22,3	22,4	20,3	17,8	14,7	11,9	15,7	10	3
4 Absolute maximum temperature	in °C	21,2	24,6	27,2	30,0	32,8	34,5	38,0	34,7	34,2	32,2	28,0	22,0	38,0	10	4
5 Absolute minimum temperature	in °C	2,0	0,0	3,3	7,5	10,9	15,0	17,3	16,5	12,0	11,3	8,2	1,2	0,0	10	5
6 Mean relative humidity	in %	72	72	71	70	69	69	69	71	70	72	72	72	71	10	6
7 Mean precipitation	in mm	91	73	69	19	12	11	2	1	11	45	48	93	475	10	7
8 Maximum precipitation	in mm															8
9 Minimum precipitation	in mm															9
10 Maximum precipitation in 24 h	in mm	47	55	64	33	15	31	20	3	27	56	56	62	64	10	10
11 Mean number of days with precipitation	> 0,1 mm	12	9	8	4	3	1	<1	<1	1	4	6	12	60	10	11
12 Mean duration of sunshine	in h															12
13 Mean quantity of radiation	in ly/day															13
14 Mean potential evaporation	in mm	27	27	40	58	89	125	150	143	105	76	48	33	921	36	14
15 Mean windspeed	in m/sec															15
16 Mean predominent direction of the wind																16

195 Station/Country Iráklion(Kríti)/Greece

Location 35°21'N/25°08'E   Height above sealevel 29 m   Climate symbol: Köppen Csa   Troll IV,1

		J	F	M	A	M	J	J	A	S	O	N	D	year	P	
1 Mean daily temperature	in °C	12,3	12,6	13,5	16,1	19,0	23,0	25,4	25,6	23,2	20,4	17,3	14,2	18,6	20	1
2 Mean daily maximum temperature	in °C	15,8	16,2	17,2	20,3	23,4	27,1	29,3	29,3	27,0	24,3	20,8	17,7	22,4	20	2
3 Mean daily minimum temperature	in °C	8,8	8,9	9,7	11,9	14,6	18,9	21,5	21,9	19,4	16,5	13,8	10,7	14,7	20	3
4 Absolute maximum temperature	in °C	23,5	29,2	34,8	36,0	39,0	41,2	40,8	40,7	38,5	36,2	31,9	24,2	41,2	20	4
5 Absolute minimum temperature	in °C	2,2	-0,5	2,6	5,6	7,0	8,7	15,0	17,0	11,7	9,4	5,5	2,4	-0,5	20	5
6 Mean relative humidity	in %	71	68	65	61	62	58	57	60	62	66	70	71	64	20	6
7 Mean precipitation	in mm	95	46	43	26	13	3	1	1	11	64	71	79	453	20	7
8 Maximum precipitation	in mm															8
9 Minimum precipitation	in mm															9
10 Maximum precipitation in 24 h	in mm	76	35	33	34	29	15	4	4	57	123	49	47	123	20	10
11 Mean number of days with precipitation	> 0,1 mm	14	9	10	6	4	1	1	1	2	6	11	14	77	20	11
12 Mean duration of sunshine	in h															12
13 Mean quantity of radiation	in ly/day															13
14 Mean potential evaporation	in mm	24	23	38	57	88	131	162	153	110	76	45	29	936	36	14
15 Mean windspeed	in m/sec															15
16 Mean predominent direction of the wind	+	S	S	S	+	+	NE	NE	NE	NE	NE	S	S			16

+ var/NE

196 Station/Country Levkosia(Nicosia)/Cyprus

Location 35°09'N/33°17'E   Height above sealevel 218 m   Climate symbol: Köppen Csa   Troll IV,1

		J	F	M	A	M	J	J	A	S	O	N	D	year	P	
1 Mean daily temperature	in °C	10,0	10,3	12,5	16,7	22,0	25,6	28,4	28,4	25,6	20,8	16,0	12,0	18,9	40	1
2 Mean daily maximum temperature	in °C	14,4	15,0	18,3	23,3	28,3	32,8	36,1	36,1	32,8	27,2	22,2	16,7	25,0	40	2
3 Mean daily minimum temperature	in °C	5,6	5,6	6,7	10,0	15,6	18,3	20,6	20,6	18,3	14,4	10,6	7,2	12,8	40	3
4 Absolute maximum temperature	in °C	21,1	24,4	31,1	34,4	42,8	40,6	46,7	42,2	41,1	40,6	35,0	24,4	46,7	39	4
5 Absolute minimum temperature	in °C	-3,9	-5,0	-2,6	0,0	4,4	9,4	11,1	13,9	9,4	4,4	-3,3	-1,7	-5,0	39	5
6 Mean relative humidity	in %	75	72	64	54	45	43	39	43	48	57	64	74	57	12	6
7 Mean precipitation	in mm	97	66	33	20	15	5	0	0	5	33	56	109	439	30	7
8 Maximum precipitation	in mm	188	94	137	47	76	82	28	24	140	138	91	279	503	30	8
9 Minimum precipitation	in mm															9
10 Maximum precipitation in 24 h	in mm	79	43	38	33	28	33	18	25	20	61	101	122	122	30	10
11 Mean number of days with precipitation	> 0,1 mm	10	8	6	3	3	1	0	0	1	5	6	8	47	42	11
12 Mean duration of sunshine	in h	171	218	236	279	341	378	397	378	324	279	216	171	3388	30	12
13 Mean quantity of radiation	in ly/day															13
14 Mean potential evaporation	in mm	16	15	28	59	113	152	185	174	130	81	44	23	1020	40	14
15 Mean windspeed	in m/sec	3,1	3,1	3,1	3,1	3,1	3,1	3,1	3,1	3,1	2,6	2,6	2,6	3,1	30	15
16 Mean predominent direction of the wind		SE	E	E	NW	NW	NW	NW	NW	NW	NW	E	E	NW	30	16

53

U.S.S.R. AND ASIA

U.S.S.R.

1 Ostrov Rudolfa
2 Ostrov Domashniy
3 Mys Čel' Uskin
4 Ostrov Kotel'nyi
5 Ostrov Vrangel'a
6 Malyje Karmakuly
7 Dikson
8 Bulun
9 Čokurdach
10 Mys Šmidta
11 Narjan-Mar
12 Velen
13 Anadyr'
14 Apuka
15 Murmansk
16 Louchi
17 Kem'
18 Petrozavodsk
19 Archangel'sk
20 Vologda
21 Syktyvkar
22 Kirov
23 Perm'
24 Serov
25 Sverdlovsk
26 Salechard
27 Tobol'sk
28 Surgut
29 Omsk
30 Kolpaševo
31 Podkamennaja /Tunguska
32 Jenisejsk
33 Minusinsk
34 Irkutsk
35 Nikolajevsk'-na-Amure
36 Cholmsk
37 Petropavlovsk-Kamčatskij
38 Ust'-Kamčatsk
39 Dudinka
40 Turuchansk
41 Tura
42 Chatanga
43 Jerbogačon
44 Olen'ok
45 Aldan
46 Sejmčan
47 Kirensk
48 Vil'ujsk
49 Jakutsk
50 Bomnak
51 Verchojansk

52 Ojm'akon
53 Ochotsk
54 Zyr'anka
55 Kaliningrad (Königsberg)
56 Tbilisi
57 Tallinn (Reval)
58 Rīga
59 Kaunas
60 Užgorod
61 Minsk
62 Kišin'ov (Kishinev)
63 Leningrad
64 Kijev
65 Moskva (Moscow)
66 Char'kov
67 Penza
68 Kazan'
69 Kujbyšev
70 Ufa
71 Kustanaj
72 Novosibirsk
73 Krasnojarsk
74 Odessa
75 Simferopol
76 Zaporožje
77 Rostov-na-Donu
78 P'atigorsk
79 Turgay
80 Karaganda
81 Volgograd
82 Kuška
83 Termez
84 Taškent
85 Chorog
86 Alma-Ata
87 Semipalatinsk
88 Gurjev
89 Balchaš
90 Fort-Ševčenko
91 Kzyl-Orda
92 Krasnovodsk
93 Ašchabad
94 Jerevan
95 Dušanbe
96 Fergana
97 Ulan-Ude
98 Krasnodar
99 Soči
100 Kutaisi
101 Blagoveščensk
102 Chabarovsk
103 Rudnaja Pristan
104 Vladivostok

TURKEY

105 Zonguldak
106 Samsun
107 Trabzon
108 Kars
109 Bursa
110 Ankara
111 Erzurum
112 Sivas
113 Van
114 Izmir
115 Konya
116 Urfa
117 Adana
118 Antalya

SYRIA

119 Halab (Aleppo)
120 Dayr-Az-Zawr
121 Dimashq (Damās)

LEBANON

122 Bayrūt (Beirut)

ISRAEL

123 Hefa (Haifa)
124 Yerushalayim (Jerusalem)
125 Elat

JORDAN

126 'Ammān

IRAQ

127 Ar-Rutbah
128 Baghdād

SAUDI ARABIA

129 Hā'il
130 Ar-Riyād
131 Juddah (Jidda)

P.D.R. YEMEN

132 Kamārān Island
133 Riyan
134 Khormaksar
135 Barīm Island

OMAN

136 Salālah
137 Al-Masīrah
138 Masqat

UNITED ARAB EMIRATES

139 Ash-Shāriqah (Sharjah)

BAHRAIN

140 Al-Manāmah

KUWAIT

141 Al-Kuwayt

IRAN

142 Mashhad
143 Tehrān
144 Kermānshāh
145 Esfahān
146 Seistan
147 Ābādān
148 Kermān
149 Būshehr
150 Bandar 'Abbas
151 Chāh Bahār

AFGHANISTAN

152 Herāt
153 Kābul
154 South Salang
155 Quandahār

PAKISTAN

156 Peshāwar
157 Lahore
158 Multān

159 Quetta
160 Hyderabad
161 Karāchi

INDIA

162 Simla
163 Darjeeling
164 Kānpur
165 Patna
166 Allāhābād
167 Jabalpur
168 New Delhi
169 Ahmadābād
170 Verāval
171 Pune (Poona)
172 Hyderabad
173 Sholāpur
174 Bellary
175 Leh
176 Bikaner
177 Jodhpur
178 Dwārka
179 Jamshedpur
180 Indore
181 Calcutta
182 Raipur
183 Surat
184 Nāgpur
185 Puri
186 Jagdalpur
187 Bombay
188 Vishākhapatnam
189 Kakinada
190 Bangalore
191 Madras
192 Amīndīvi (Amīndīvi Islands)
193 Nagappattinam
194 Trivandrum
195 Tezpur
196 Dibrugarh
197 Dhubri
198 Cherrapunji
199 Marmagoa (near Goa)
200 Mangalore
201 Port Blair (Andaman Islands)
202 Cochin
203 Car Nicobar (Nicobar Islands)
204 Pāmban Island
322 Srinagar (Kashmir)
323 Kodaikanal

SRI LANKA

205 Trincomalee

206 Nuwara-Eliya
207 Colombo

NEPAL

208 Kātmāndu
324 Pokhara Airport

BHUTAN

209 Yatung

BANGLADESH

210 Narayangany (near Dacca)
211 Chittagong

BURMA

212 Bhamo
213 Lashio
214 Mandalay
215 Sittwe (Akyab)
216 Rangoon
217 Diamond Island
218 Tavoy
219 Mergui

MONGOLIA

220 Ulaanbaatar (Ulan Bator)

CHINA

221 Kucha
222 Baotou
223 Jiuquan
224 Kashgar
225 Taiyuan
226 Xining
227 Lanzhou
228 Haerbin
229 Changchun
230 Wulumuqi (Urumchi)
231 Shenyang
232 Beijing (Peking)
233 Tianjin (Tientsin)
234 Lüda
235 Jinan
236 Qingdao (Tsingtao)
237 Xuzhou
238 Changdu

239 Chengdu
240 Lasa
241 Chongqing
242 Guangzhou (Canton)
243 Xi'an
244 Nanjing (Nanking)
245 Shanghai
246 Wuhan
247 Nanchang
248 Changsha
249 Wenzhou
250 Guiyang
251 Guilin (Kweilin)
252 Kunming
253 Xiamen (Amoy)
254 Nanning
255 Yulin (Hainando Island)

UNITED KINGDOM

256 Hongkong

TAIWAN

257 T'aipei
258 T'ainan

NORTH KOREA

259 Wŏnsan
260 P'yongyang

SOUTH KOREA

261 Soul (Seoul)
262 Pusan

JAPAN

263 Wakkanai
264 Sapporo
265 Aomori

266 Sendai
267 Niigata
268 Kanazawa
269 Tokyo
270 Nagoya
271 Osaka
272 Hiroshima
273 Kochī
274 Nagasaki
275 Kagoshima
276 Naha (Okinawa)

THAILAND

277 Chiang Mai
278 Nakhon Ratchasima
279 Bangkok
280 Surat Thani (Ban Don)
281 Phuket

LAOS

282 Louang Prabang
283 Vientiane

CAMBODIA

284 Stung Treng
285 Bătdâmbâng
286 Phnum Pénh

VIETNAM

287 Ha-noi
288 Guang Tri
289 Qui-nhon
290 Ho Chi Minh City (Sai-gon)

MALAYSIA

291 Sandakan (Borneo)

292 Pinang
293 Kuala Trengganu
294 Cameron Highlands
295 Malacca
296 Kuching (Borneo)
297 Pulau Labuan

SINGAPORE

298 Singapore

INDONESIA

299 Tarakan (Borneo)
300 Takingeun (Sumatra)
301 Medan (Sumatra)
302 Manado (Celebes)
303 Pontianak (Borneo)
304 Padang (Sumatra)
305 Balikapan (Borneo)
306 Tandjungpandan (Belitung)
307 Ambon (Moluccas)
308 Makasar (Celebes)
309 Djakarta (Java)
310 Surabaja (Java)
311 Surakarta (Java)
312 Tombora (Sumbawa)
313 Kupang (Timor)

PORTUGAL

314 Dili (Timor)

PHILIPPINES

315 Aparri
316 Baguio
317 Manila
318 Legazpi
319 Iloilo
320 Davao
321 Zamboanga

U.S.S.R. AND ASIA

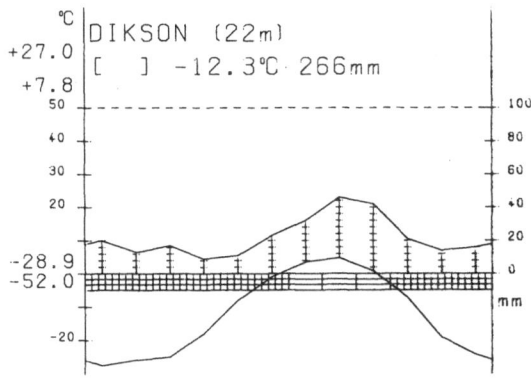

DIKSON (22m)
[ ] -12.3℃ 266mm

℃
+27.0
+7.8
50
40
30
20
-28.9
-52.0
-20

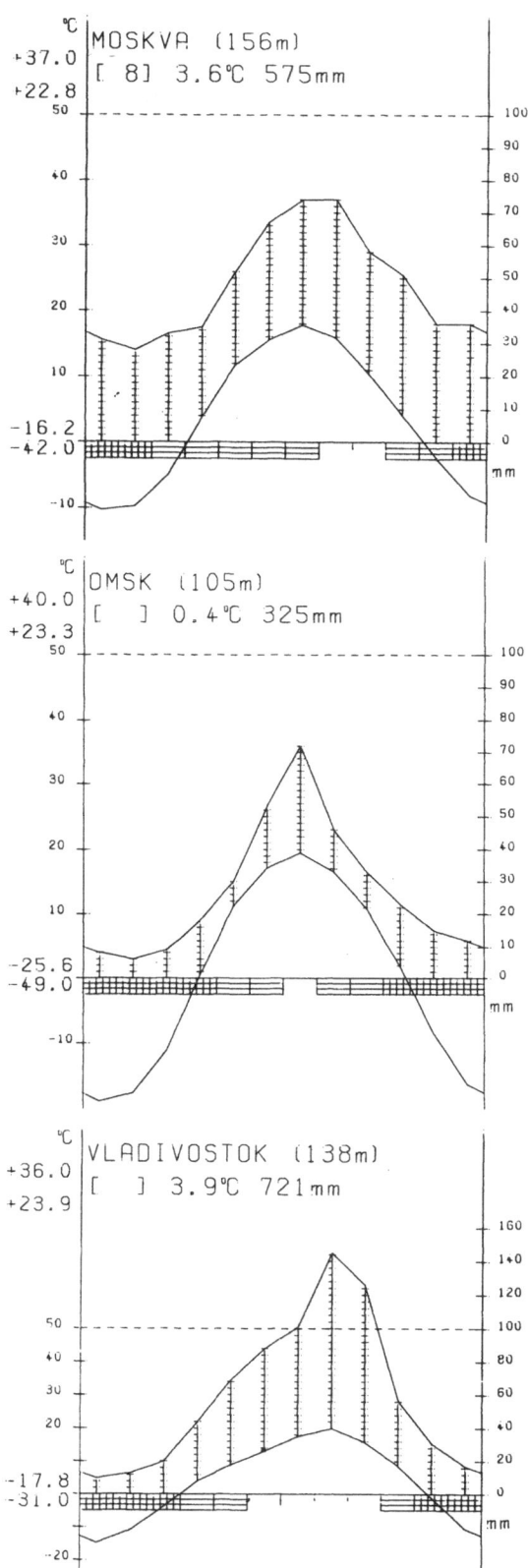

MOSKVA (156m)
[8] 3.6℃ 575mm

℃
+37.0
+22.8
50
40
30
20
10
-16.2
-42.0
-10

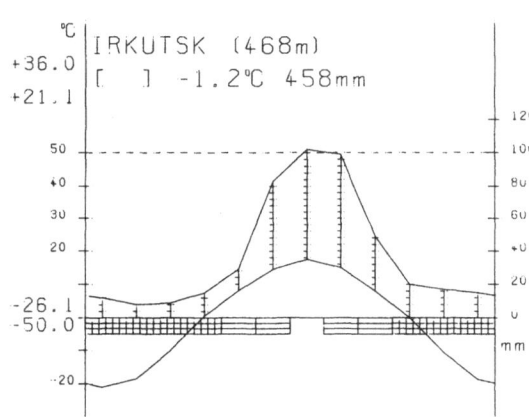

IRKUTSK (468m)
[ ] -1.2℃ 458mm

℃
+36.0
+21.1
50
40
30
20
-26.1
-50.0
-20

OMSK (105m)
[ ] 0.4℃ 325mm

℃
+40.0
+23.3
50
40
30
20
10
-25.6
-49.0
-10

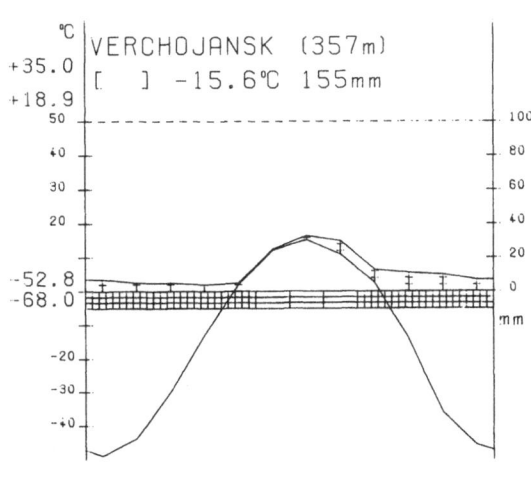

VERCHOJANSK (357m)
[ ] -15.6℃ 155mm

℃
+35.0
+18.9
50
40
30
20
-52.8
-68.0
-20
-30
-40

VLADIVOSTOK (138m)
[ ] 3.9℃ 721mm

℃
+36.0
+23.9
50
40
30
20
-17.8
-31.0
-20

U.S.S.R. AND ASIA

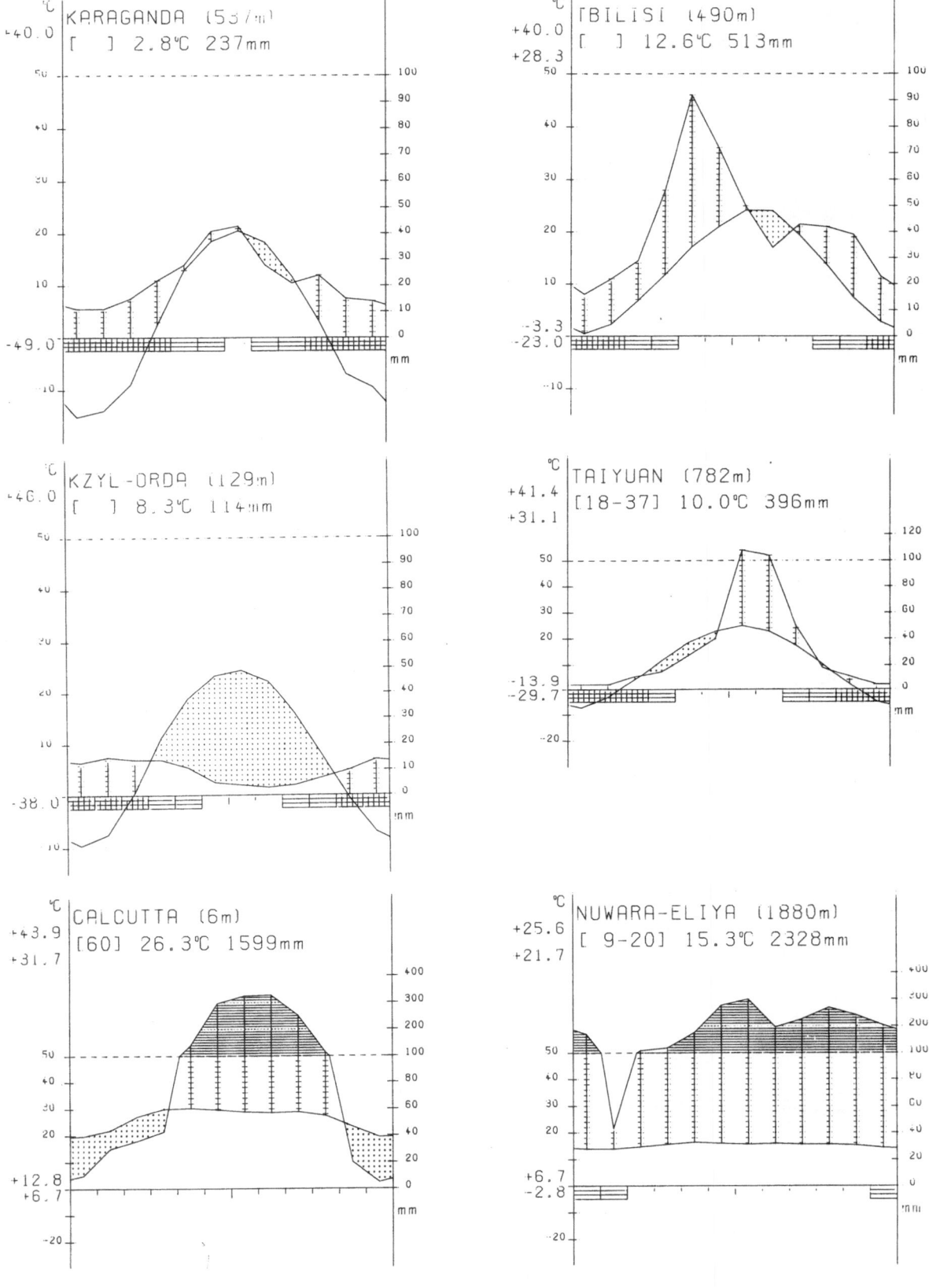

KARAGANDA (537m)
[　] 2.8℃ 237mm

TBILISI (490m)
[　] 12.6℃ 513mm

KZYL-ORDA (129m)
[　] 8.3℃ 114mm

TAIYUAN (782m)
[18-37] 10.0℃ 396mm

CALCUTTA (6m)
[60] 26.3℃ 1599mm

NUWARA-ELIYA (1880m)
[9-20] 15.3℃ 2328mm

1 Station / Country Ostrov Rudolfa / S.S.S.R.

Location 81°48'N / 57°57'E Height above sealevel 48 m Climate symbol: Köppen ET Troll I,2

		J	F	M	A	M	J	J	A	S	O	N	D	year	P	
1 Mean daily temperature	in °C	-20,0	-20,6	-23,3	-18,3	-8,9	-1,7	0,6	0,0	-5,0	-10,6	-16,1	-18,9	-11,9		1
2 Mean daily maximum temperature	in °C															2
3 Mean daily minimum temperature	in °C															3
4 Absolute maximum temperature	in °C	0,5	-0,6	0,0	-0,5	0,0	7,7	11,7	7,2	5,0	1,6	0,5	-0,5	11,7		4
5 Absolute minimum temperature	in °C	-37,2	-42,8	-37,2	-35,6	-20,6	-8,3	-4,4	-6,1	-22,2	-32,8	-36,7	-35,6	-42,8		5
6 Mean relative humidity	in %	76	82	90	88	90	93	97	95	93	92	91	82	89		6
7 Mean precipitation	in mm	5	3	3	3	5	10	20	15	18	8	5	3	98		7
8 Maximum precipitation	in mm															8
9 Minimum precipitation	in mm															9
10 Maximum precipitation in 24 h	in mm	9	7	8	9	12	12	13	18	18	15	13	9	143		10
11 Mean number of days with precipitation	> 0,1 mm															11
12 Mean duration of sunshine	in h															12
13 Mean quantity of radiation	in ly / day															13
14 Mean potential evaporation	in mm	0	0	0	0	0	0	49	33	0	0	0	0	82		14
15 Mean windspeed	in m / sec	8,9	7,6	6,3	6,7	5,8	4,9	3,6	4,5	5,8	6,7	8,0	8,0	6,4		15
16 Mean predominent direction of the wind																16

2 Station / Country Ostrov Domashniy / S.S.S.R.

Location 79°30'N / 91°08'E Height above sealevel 3 m Climate symbol: Köppen ET Troll I,2

		J	F	M	A	M	J	J	A	S	O	N	D	year	P	
1 Mean daily temperature	in °C	-25,6	-25,0	-27,8	-22,2	-10,0	-1,7	1,1	0,0	-3,3	-10,6	-18,9	-25,0	-14,0	24	1
2 Mean daily maximum temperature	in °C															2
3 Mean daily minimum temperature	in °C															3
4 Absolute maximum temperature	in °C	1,7	-1,7	-1,6	-5,5	0,6	4,4	6,1	5,6	5,0	0,6	-0,6	0,0	6,1	24	4
5 Absolute minimum temperature	in °C	-46,7	-47,2	-45,6	-40,0	-26,7	-10,0	-4,4	-8,3	-17,8	-31,1	-35,6	-41,7	-47,2	24	5
6 Mean relative humidity	in %															6
7 Mean precipitation	in mm	5	5	3	3	5	8	28	13	15	5	3	5	98	24	7
8 Maximum precipitation	in mm															8
9 Minimum precipitation	in mm															9
10 Maximum precipitation in 24 h	in mm															10
11 Mean number of days with precipitation	> 0,1 mm	13	8	7	7	6	8	11	14	12	10	9	9	114	24	11
12 Mean duration of sunshine	in h															12
13 Mean quantity of radiation	in ly / day															13
14 Mean potential evaporation	in mm	0	0	0	0	0	0	44	0	0	0	0	0	44		14
15 Mean windspeed	in m / sec	6,3	6,3	4,9	4,9	5,8	5,8	4,9	5,4	7,2	6,3	6,3	5,4	5,8		15
16 Mean predominent direction of the wind																16

3 Station / Country Mys Čel'uskin / S.S.S.R.

Location 77°43'N / 104°17'E Height above sealevel 6 m Climate symbol: Köppen ET Troll I,2

		J	F	M	A	M	J	J	A	S	O	N	D	year	P	
1 Mean daily temperature	in °C	-31,1	-29,7	-28,5	-22,6	-11,2	-2,8	0,8	0,6	-2,9	-10,3	-22,6	-27,4	-15,6		1
2 Mean daily maximum temperature	in °C															2
3 Mean daily minimum temperature	in °C															3
4 Absolute maximum temperature	in °C	0	0	-3	1	8	12	24	20	13	4	0	-1	24		4
5 Absolute minimum temperature	in °C	-49	-46	-42	-42	-26	-13	-6	-7	-16	-30	-39	-44	-49		5
6 Mean relative humidity	in %	84	86	85	85	88	91	93	94	91	87	86	84	88		6
7 Mean precipitation	in mm	12	14	18	21	24	25	27	28	24	21	16	14	294		7
8 Maximum precipitation	in mm							61	58	76	55					8
9 Minimum precipitation	in mm							2	2	7	3					9
10 Maximum precipitation in 24 h	in mm	3	1	4	4	4	24	24	33	29	10	3	2	33		10
11 Mean number of days with precipitation	> 0,1 mm	13	11	9	8	8	8	11	16	17	17	15	14	147		11
12 Mean duration of sunshine	in h	0	2	95	223	197	172	187	91	28	7	0	0	1002		12
13 Mean quantity of radiation	in ly / day	0	3	81	279	496	501	388	202	75	12	0	0	170		13
14 Mean potential evaporation	in mm	0	0	0	0	0	62	30	0	0	0	0	0	92		14
15 Mean windspeed	in m / sec	7,2	6,4	6,6	5,7	6,1	6,1	6,0	6,2	6,7	6,6	6,9	7,6	6,5		15
16 Mean predominent direction of the wind		SW	SW	E	E,SW	E	E	E	W	SW	SW	SW	SW			16

4 Station / Country Ostrov Kotel'nyi / S.S.S.R.

Location 76°00'N / 137°54'E Height above sealevel 11 m Climate symbol: Köppen ET Troll I,2

		J	F	M	A	M	J	J	A	S	O	N	D	year	P	
1 Mean daily temperature	in °C	-29,5	-29,9	-27,0	-20,6	-9,1	-0,2	2,5	2,0	-1,5	-10,5	-21,9	-26,4	-14,3		1
2 Mean daily maximum temperature	in °C															2
3 Mean daily minimum temperature	in °C															3
4 Absolute maximum temperature	in °C	-9	-11	-10	-2	8	21	22	19	11	8	-4	-3	22		4
5 Absolute minimum temperature	in °C	-47	-49	-47	-42	-30	-12	-4	-6	-20	-29	-40	-42	-49		5
6 Mean relative humidity	in %	79	79	78	79	80	75	69	79	83	85	81	80	79		6
7 Mean precipitation	in mm	8	8	6	9	10	11	20	19	11	11	10	8	131		7
8 Maximum precipitation	in mm						25	67	57	36						8
9 Minimum precipitation	in mm						1	6	2							9
10 Maximum precipitation in 24 h	in mm	5	3	3	3	7	12	17	26	9	5	2	2	26		10
11 Mean number of days with precipitation	> 0,1 mm	8	8	8	8	8	9	12	14	15	15	10	9	124		11
12 Mean duration of sunshine	in h	0	8	125	237	190	172	146	98	41	10	0	0	1027		12
13 Mean quantity of radiation	in ly / day	0	6	96	294	502	468	344	214	90	19	0	0	169		13
14 Mean potential evaporation	in mm	0	0	0	0	0	85	59	0	0	0	0	0	144		14
15 Mean windspeed	in m / sec	5,7	5,7	5,7	5,6	6,2	6,2	6,2	6,7	6,8	6,4	5,8	5,4	6,0		15
16 Mean predominent direction of the wind		SW	SW	SW	SE	SE	E	N,W	W	SE	SE	SE	SE			16

5 Station/Country Ostrov Vrangel'a / S.S.S.R.

Location 70°58'N / 178°32'W   Height above sealevel 3 m   Climate symbol: Köppen ET   Troll I,2

		J	F	M	A	M	J	J	A	S	O	N	D	year	P	
1	Mean daily temperature	in °C	-23,9	-25,6	-23,3	-17,2	-8,3	0,6	2,8	2,2	-1,7	-8,3	-17,2	-21,1	-11,7	10	1
2	Mean daily maximum temperature	in °C	-19,4	-21,7	-18,9	-13,3	-5,6	2,8	5,6	4,4	0,0	-6,1	-13,9	-17,2	-8,3	9	2
3	Mean daily minimum temperature	in °C	-27,8	-29,4	-27,8	-21,1	-12,2	-2,2	0,0	-0,6	-3,9	-11,1	-20,6	-24,4	-15,0	9	3
4	Absolute maximum temperature	in °C	-5,6	-1,7	-1,1	0,6	9,4	17,2	18,3	16,7	8,3	3,3	1,7	0,0	18,3	10	4
5	Absolute minimum temperature	in °C	-41,7	-41,8	-45,6	-36,7	-25,0	-10,0	-5,0	-4,4	-14,4	-23,3	-33,3	-36,7	-45,6	10	5
6	Mean relative humidity	in %	86	84	84	84	85	89	75	92	88	86	84	87	85	9	6
7	Mean precipitation	in mm	5	5	5	5	5	10	16	23	13	10	3	5	104	10	7
8	Maximum precipitation	in mm															8
9	Minimum precipitation	in mm															9
10	Maximum precipitation in 24 h	in mm	10	8	8	8	5	18	15	13	10	10	5	5	18	8	10
11	Mean number of days with precipitation	> 0,1 mm	8	7	7	8	8	8	9	12	10	10	7	7	99	10	11
12	Mean duration of sunshine	in h															12
13	Mean quantity of radiation	in ly / day															13
14	Mean potential evaporation	in mm	0	0	0	0	0	39	75	63	0	0	0	0	177		14
15	Mean windspeed	in m/sec	5,4	4,9	4,9	4,0	4,0	3,1	4,0	3,6	5,8	6,7	6,7	5,4	4,9	10	15
16	Mean predominent direction of the wind																16

6 Station/Country Malyje Karmakuly / S.S.S.R.

Location 72°23'N / 52°44'E   Height above sealevel 16 m   Climate symbol: Köppen ET   Troll I,3

		J	F	M	A	M	J	J	A	S	O	N	D	year	P	
1	Mean daily temperature	in °C	-15,0	-14,5	-15,4	-10,8	-4,5	1,4	6,4	6,3	2,7	-2,7	-9,0	-13,0	-5,7		1
2	Mean daily maximum temperature	in °C															2
3	Mean daily minimum temperature	in °C															3
4	Absolute maximum temperature	in °C	1	1	1	6	13	20	24	20	17	10	3	2	24		4
5	Absolute minimum temperature	in °C	-41	-40	-44	-32	-24	-17	-10	-1	-13	-18	-34	-36	-44		5
6	Mean relative humidity	in %	80	78	78	80	80	82	82	84	83	82	82	82	81		6
7	Mean precipitation	in mm	26	18	19	18	20	24	30	36	41	35	28	24	317		7
8	Maximum precipitation	in mm							59	82	95	88					8
9	Minimum precipitation	in mm							2	2	7	8					9
10	Maximum precipitation in 24 h	in mm							28	27	27	28					10
11	Mean number of days with precipitation	> 0,1 mm	20	14	18	15	14	13	13	17	18	17	15	18	194		11
12	Mean duration of sunshine	in h	0	21	95	155	175	178	194	146	76	44	2	0	1086		12
13	Mean quantity of radiation	in ly / day	3	25	102	240	338	378	378	245	117	47	21	16	159		13
14	Mean potential evaporation	in mm	0	0	0	0	0	50	92	85	53	0	0	0	280		14
15	Mean windspeed	in m/sec	10,3	10,4	9,2	8,1	7,2	6,9	6,3	6,9	7,1	7,9	9,0	10,4	8,3		15
16	Mean predominent direction of the wind		SE	SE	SE	SE	N	N	N	N	SE	SE	SE	SE			16

7 Station/Country Dikson / S.S.S.R.

Location 73°30'N / 80°14'E   Height above sealevel 22 m   Climate symbol: Köppen ET   Troll I,3

		J	F	M	A	M	J	J	A	S	O	N	D	year	P	
1	Mean daily temperature	in °C	-27,5	-25,9	-25,0	-18,1	-8,2	-1,0	3,6	4,8	0,8	-7,2	-19,2	-24,3	-12,3		1
2	Mean daily maximum temperature	in °C	-20,6	-19,4	-20,6	-14,4	-5,6	1,7	7,8	7,8	3,9	-4,4	-12,8	-18,3	-7,7	12	2
3	Mean daily minimum temperature	in °C	-28,9	-28,3	-27,8	-13,3	-11,1	-1,7	2,2	2,8	0,0	-8,9	-14,4	-27,2	-12,8	16	3
4	Absolute maximum temperature	in °C	0	0	0	7	11	22	24	27	14	8	2	0	27		4
5	Absolute minimum temperature	in °C	-51	-52	-50	-40	-33	-15	-3	-3	-19	-36	-40	-48	-52		5
6	Mean relative humidity	in %	86	86	84	86	87	90	90	89	89	87	88	86	87		6
7	Mean precipitation	in mm	20	13	17	9	11	23	32	46	42	21	14	18	266		7
8	Maximum precipitation	in mm							67	84	82	83					8
9	Minimum precipitation	in mm							1	4	8	7					9
10	Maximum precipitation in 24 h	in mm	3	4	4	6	9	13	25	31	24	11	26	7	31		10
11	Mean number of days with precipitation	> 0,1 mm	13	12	11	11	11	12	13	17	19	18	15	13	165		11
12	Mean duration of sunshine	in h	0	16	129	211	152	149	196	122	59	22	0	0	1058		12
13	Mean quantity of radiation	in ly / day	0	14	124	297	459	417	378	229	96	31	0	0	170		13
14	Mean potential evaporation	in mm	0	0	0	0	0	85	81	45	0	0	0	0	211	22	14
15	Mean windspeed	in m/sec	8,4	8,1	7,4	7,3	7,1	6,8	6,6	6,7	7,3	8,1	8,1	8,0	7,5		15
16	Mean predominent direction of the wind		S	S	S	S	NE	NE	NE	NE	S	S	S	S			16

8 Station/Country Bulun / S.S.S.R.

Location 70°45'N / 127°47'E   Height above sealevel 37 m   Climate symbol: Köppen Dwd   Troll I,3

		J	F	M	A	M	J	J	A	S	O	N	D	year	P	
1	Mean daily temperature	in °C	-41,4	-37,2	-27,8	-17,2	-5,6	6,4	10,8	7,2	2,0	-11,1	-26,4	-34,2	-14,5	11	1
2	Mean daily maximum temperature	in °C	-38,9	-34,4	-23,9	-13,3	-2,8	9,4	13,9	11,7	3,3	-10,0	-28,3	-35,0	-12,2	18	2
3	Mean daily minimum temperature	in °C	-42,8	-40,0	-31,7	-21,1	-8,3	3,3	7,8	5,6	-1,7	-14,4	-32,8	-39,4	-17,8	18	3
4	Absolute maximum temperature	in °C	-3,9	-11,1	-5,0	0,6	10,6	23,3	30,6	27,2	20,0	7,2	-5,6	-16,1	30,6	4	4
5	Absolute minimum temperature	in °C	-47,8	-49,4	-43,9	-34,4	-25,6	-7,8	0,0	-2,8	-7,8	-30,6	-42,8	-45,6	-49,4	4	5
6	Mean relative humidity	in %	78	82	76	79	81	81	82	83	78	82	83	76	80	4	6
7	Mean precipitation	in mm	3	3	3	1	3	13	43	20	20	5	5	3	122	4	7
8	Maximum precipitation	in mm															8
9	Minimum precipitation	in mm															9
10	Maximum precipitation in 24 h	in mm	1	1	3	1	3	8	23	15	18	3	3	1	23	4	10
11	Mean number of days with precipitation	> 0,1 mm	9	9	7	5	4	10	11	12	14	12	9	6	108	4	11
12	Mean duration of sunshine	in h															12
13	Mean quantity of radiation	in ly / day															13
14	Mean potential evaporation	in mm	0	0	0	0	0	86	115	90	29	0	0	0	320	11	14
15	Mean windspeed	in m/sec															15
16	Mean predominent direction of the wind																16

9 Station/Country Čokurdach / S.S.S.R.

Location 70°37'N/147°53'E   Height above sealevel 20 m   Climate symbol: Köppen ET   Troll 1,3

		J	F	M	A	M	J	J	A	S	O	N	D	year	P	
1	Mean daily temperature	in °C	-36,2	-34,1	-29,2	-18,3	-6,3	6,1	10,2	7,0	1,0	-12,3	-24,9	-32,8	-14,2		1
2	Mean daily maximum temperature	in °C															2
3	Mean daily minimum temperature	in °C															3
4	Absolute maximum temperature	in °C	-8	-3	-2	1	18	29	32	29	20	7	-4	-4	32		4
5	Absolute minimum temperature	in °C	-52	-50	-46	-40	-30	-8	0	-4	-14	-34	-44	-48	-52		5
6	Mean relative humidity	in %	86	86	86	85	84	79	75	81	84	87	85	85	84		6
7	Mean precipitation	in mm	7	8	6	8	10	13	22	24	15	12	11	9	145		7
8	Maximum precipitation	in mm						40	60	65	48						8
9	Minimum precipitation	in mm						1	6	5	2						9
10	Maximum precipitation in 24 h	in mm															10
11	Mean number of days with precipitation	> 0,1 mm	7	10	8	7	8	10	11	13	12	16	11	9	122		11
12	Mean duration of sunshine	in h	0	48	179	284	269	283	278	134	94	43	4	0	1616		12
13	Mean quantity of radiation	in ly/day															13
14	Mean potential evaporation	in mm	0	0	0	0	0	53	82	60	6	0	0	0	202		14
15	Mean windspeed	in m/sec	3,7	3,9	4,4	4,9	5,5	6,0	6,0	5,6	4,2	4,1	3,9	3,8	4,7		15
16	Mean predominent direction of the wind		W	W	W	NE	E	NE	NE	NE	W	SW	SW	SW			16

10 Station/Country Mys Šmidta / S.S.S.R.

Location 68°55'N/179°29'W   Height above sealevel 7 m   Climate symbol: Köppen ET   Troll 1,3

		J	F	M	A	M	J	J	A	S	O	N	D	year	P	
1	Mean daily temperature	in °C	-26,4	-27,3	-25,7	-17,6	-7,8	1,3	3,6	2,8	-0,3	-7,7	-16,5	-23,4	-12,1		1
2	Mean daily maximum temperature	in °C															2
3	Mean daily minimum temperature	in °C															3
4	Absolute maximum temperature	in °C	5	5	3	5	11	23	26	26	17	12	6	4	26		4
5	Absolute minimum temperature	in °C	-47	-48	-45	-40	-30	-12	-5	-7	-14	-30	-39	-43	-47		5
6	Mean relative humidity	in %	85	86	86	88	90	89	90	92	90	86	87	85	88		6
7	Mean precipitation	in mm	23	21	17	17	19	25	33	42	30	25	20	24	296		7
8	Maximum precipitation	in mm						36	78	125	161						8
9	Minimum precipitation	in mm						2	2	14	3						9
10	Maximum precipitation in 24 h	in mm	5	6	6	7	6	16	26	39	21	14	19	6	39		10
11	Mean number of days with precipitation	> 0,1 mm	9	9	11	9	9	9	12	16	15	15	13	11	138		11
12	Mean duration of sunshine	in h	1	45	146	200	169	225	197	103	74	45	4	0	1209		12
13	Mean quantity of radiation	in ly/day	0	31	161	315	471	488	381	239	129	53	6	0	188		13
14	Mean potential evaporation	in mm	0	0	0	0	0	54	85	69	0	0	0	0	208		14
15	Mean windspeed	in m/sec	6,7	6,1	6,5	5,1	4,8	4,8	4,8	5,0	5,5	7,0	7,3	6,5	5,8		15
16	Mean predominent direction of the wind		W	W	W,SW	SE	SE	SE	SE	NW	NW	NW	W	W			16

11 Station/Country Narjan-Mar/S.S.S.R.

Location 67°39'N/53°01'E   Height above sealevel 7 m   Climate symbol: Köppen Dfc   Troll 1,3

		J	F	M	A	M	J	J	A	S	O	N	D	year	P	
1	Mean daily temperature	in °C	-17,3	-16,7	-14,2	-7,2	-0,6	7,1	12,0	10,2	5,5	-1,7	-9,2	-14,5	-3,9		1
2	Mean daily maximum temperature	in °C	-15,7	-13,6	-9,9	-2,8	3,5	11,4	17,8	14,5	8,6	0,3	-6,5	-11,4	-0,3	8	2
3	Mean daily minimum temperature	in °C	-25,4	-22,1	-19,7	-11,6	-4,1	2,7	8,7	6,9	2,7	-5,0	-13,8	-19,6	-8,4	8	3
4	Absolute maximum temperature	in °C	5	5	5	15	26	33	33	33	23	14	6	7	33		4
5	Absolute minimum temperature	in °C	-51	-47	-47	-37	-24	-7	-2	-3	-11	-26	-45	-48	-51		5
6	Mean relative humidity	in %	85	84	83	82	78	74	76	82	76	88	88	86	83		6
7	Mean precipitation	in mm	20	14	17	21	24	38	41	59	52	39	41	22	378		7
8	Maximum precipitation	in mm															8
9	Minimum precipitation	in mm															9
10	Maximum precipitation in 24 h	in mm	13	5	8	10	16	15	27	19	30	15	15	7	30		10
11	Mean number of days with precipitation	> 0,1 mm	19	16	17	15	16	14	12	17	19	21	20	19	205		11
12	Mean duration of sunshine	in h	3	43	146	185	189	230	269	160	74	39	10	0	1348		12
13	Mean quantity of radiation	in ly/day															13
14	Mean potential evaporation	in mm	0	0	0	0	0	36	58	61	42	0	0	0	196		14
15	Mean windspeed	in m/sec	5,4	5,0	5,2	5,0	5,2	5,5	4,8	4,4	4,8	4,8	4,8	5,1	5,0		15
16	Mean predominent direction of the wind		S	S	SW	SW	NE	NE	NE	NE	S	SW	SW	SW			16

12 Station/Country Velen/S.S.S.R.

Location 66°10'N/169°50'E   Height above sealevel 7 m   Climate symbol: Köppen Dwd   Troll 1,3

		J	F	M	A	M	J	J	A	S	O	N	D	year	P	
1	Mean daily temperature	in °C	-21,7	-21,5	-20,9	-13,7	-4,7	1,6	5,4	5,0	2,7	-1,9	-10,1	-18,1	-8,2		1
2	Mean daily maximum temperature	in °C															2
3	Mean daily minimum temperature	in °C															3
4	Absolute maximum temperature	in °C	6	4	3	5	7	18	21	19	14	9	6	2	21		4
5	Absolute minimum temperature	in °C	-45	-44	-42	-38	-28	-8	-2	-3	-7	-25	-38	-40	-45		5
6	Mean relative humidity	in %	84	83	83	86	90	91	91	92	90	87	86	85	87		6
7	Mean precipitation	in mm	26	27	24	23	25	27	36	57	45	36	35	33	394		7
8	Maximum precipitation	in mm						46	62	150	128						8
9	Minimum precipitation	in mm						0	4	18	11						9
10	Maximum precipitation in 24 h	in mm	17	16	13	12	13	16	17	49	36	53	24	26	53		10
11	Mean number of days with precipitation	> 0,1 mm	8	9	9	9	10	7	12	15	16	16	13	11	135		11
12	Mean duration of sunshine	in h	9	53	137	146	112	210	173	102	58	30	7	1	1038		12
13	Mean quantity of radiation	in ly/day	6	30	153	261	369	423	341	236	111	47	9	3	166		13
14	Mean potential evaporation	in mm	0	0	0	0	0	20	70	60	32	0	0	0	182		14
15	Mean windspeed	in m/sec	5,8	5,2	5,0	5,3	4,6	4,8	6,6	5,9	6,6	7,3	7,8	6,0	5,9		15
16	Mean predominent direction of the wind		N	N	N	N	N	S	N	N	N	N	N	N			16

13 Station/Country Anadyr'/S.S.S.R.

Location 64°47'N/177°34'E   **Height above sealevel** 62 m   **Climate symbol:** Köppen Dfc   **Troll** I,3

		J	F	M	A	M	J	J	A	S	O	N	D	year	P	
1 Mean daily temperature	in °C	-22,7	-21,6	-20,2	-13,3	-3,5	4,8	10,5	9,6	3,9	-5,0	-14,2	-20,9	-7,7		1
2 Mean daily maximum temperature	in °C	-22,8	-20,6	-18,3	-12,2	-1,1	5,6	12,2	11,1	5,6	-3,9	-14,4	-21,1	-6,7	23	2
3 Mean daily minimum temperature	in °C	-27,8	-26,7	-25,0	-20,0	-7,7	1,1	7,8	6,7	0,6	-7,7	-17,8	-25,0	-11,7	17	3
4 Absolute maximum temperature	in °C	6	3	2	7	10	26	28	24	17	12	5	4	28		4
5 Absolute minimum temperature	in °C	-51	-51	-41	-36	-24	-8	0	-4	-11	-29	-40	-44	-51		5
6 Mean relative humidity	in %	82	81	81	85	84	81	81	81	81	84	84	83	82		6
7 Mean precipitation	in mm	21	14	16	13	9	13	39	46	26	27	17	19	260		7
8 Maximum precipitation	in mm	124	54	61	60	34	82	84	101	73	101	87	101	436		8
9 Minimum precipitation	in mm	0	0	0	0	0	0	7	5	5	3	0	0	16		9
10 Maximum precipitation in 24 h	in mm	13	5	5	5	10	26	46	40	29	18	8	16	46		10
11 Mean number of days with precipitation	> 0,1 mm	10	10	8	8	8	8	12	12	11	10	11	10	118		11
12 Mean duration of sunshine	in h	32	84	185	207	232	277	278	188	149	110	57	21	1820		12
13 Mean quantity of radiation	in ly/day															13
14 Mean potential evaporation	in mm	0	0	0	0	0	64	107	94	47	0	0	0	312		14
15 Mean windspeed	in m/sec	7,3	8,2	6,8	6,6	5,4	5,7	6,0	5,9	5,9	7,1	8,1	7,2	6,7		15
16 Mean predominent direction of the wind		NW	NW	NW	NW	NW	SE	SE	SE	NW	NW	NW	NW			16

14 Station/Country Apuka/S.S.S.R.

Location 60°26'N/169°40'E   **Height above sealevel** 10 m   **Climate symbol:** Köppen Dfc   **Troll** I,3

		J	F	M	A	M	J	J	A	S	O	N	D	year	P	
1 Mean daily temperature	in °C	-11,5	-13,0	-12,0	-5,1	0,7	6,1	10,0	10,6	6,6	-1,2	-7,9	-10,5	-2,3		1
2 Mean daily maximum temperature	in °C															2
3 Mean daily minimum temperature	in °C															3
4 Absolute maximum temperature	in °C	3	5	6	10	15	20	28	23	20	14	11	2	28		4
5 Absolute minimum temperature	in °C	-42	-42	-35	-29	-18	-4	-1	-3	-10	-22	-32	-38	-42		5
6 Mean relative humidity	in %	78	76	74	81	84	87	89	88	83	78	76	76	81		6
7 Mean precipitation	in mm	36	20	22	27	18	27	60	62	53	47	35	24	431		7
8 Maximum precipitation	in mm						99	55	142	125	192	123				8
9 Minimum precipitation	in mm					0	1	9	10	2	2					9
10 Maximum precipitation in 24 h	in mm						7	9	15	15	13	14				10
11 Mean number of days with precipitation	> 0,1 mm	10	9	8	11	9	10	15	14	12	12	11	12	133		11
12 Mean duration of sunshine	in h	32	84	129	160	182	136	125	142	157	107	53	45	1362		12
13 Mean quantity of radiation	in ly/day	31	73	223	324	428	408	378	307	213	109	42	19	213		13
14 Mean potential evaporation	in mm	0	0	0	0	12	61	95	90	54	0	0	0	318		14
15 Mean windspeed	in m/sec	5,9	6,6	6,5	4,5	3,0	3,1	3,2	3,3	3,5	4,4	6,1	6,4	4,7		15
16 Mean predominent direction of the wind		NE	NE	NE	NE	NE	SE	SE	W	NE	NE	NE	NE			16

15 Station/Country Murmansk/S.S.R.

Location 68°58'N/33°03'E   **Height above sealevel** 46 m   **Climate symbol:** Köppen Dfc   **Troll** II,2

		J	F	M	A	M	J	J	A	S	O	N	D	year	P	
1 Mean daily temperature	in °C	-9,9	-9,9	-7,0	-1,2	3,5	8,9	12,8	10,9	6,4	0,3	-5,1	-8,6	0,1		1
2 Mean daily maximum temperature	in °C	-8,0	-9,1	-4,7	1,7	6,9	13,1	15,7	14,5	9,4	2,8	-1,1	-5,8	2,9	8	2
3 Mean daily minimum temperature	in °C	-14,9	-15,7	-12,2	-4,8	0,3	4,9	8,1	7,8	4,2	-1,5	-6,4	-12,2	-3,5	8	3
4 Absolute maximum temperature	in °C	7	6	9	17	27	31	33	30	24	14	9	7	33		4
5 Absolute minimum temperature	in °C	-37	-38	-36	-27	-12	-4	0	-1	-10	-21	-32	-37	-38		5
6 Mean relative humidity	in %	83	82	78	73	69	68	73	77	81	83	84	84	78		6
7 Mean precipitation	in mm	19	16	18	19	25	40	54	60	44	30	28	33	376		7
8 Maximum precipitation	in mm					67	134	136	139	108	90					8
9 Minimum precipitation	in mm					7	4	10	16	18	8					9
10 Maximum precipitation in 24 h	in mm	10	13	9	8	20	24	31	21	16	23	13	14	31	8	10
11 Mean number of days with precipitation	> 0,1 mm	19	16	17	15	15	15	14	17	17	18	19	18	200		11
12 Mean duration of sunshine	in h	1	32	121	203	197	248	236	146	73	43	3	0	1297		12
13 Mean quantity of radiation	in ly/day															13
14 Mean potential evaporation	in mm	0	0	0	0	47	94	116	94	58	5	0	0	414		14
15 Mean windspeed	in m/sec	5,7	5,3	4,7	4,2	4,4	4,1	3,5	3,4	4,0	4,2	4,7	5,0	4,4		15
16 Mean predominent direction of the wind		S	S	S	S	N	N	N	S	S	S	S	S			16

16 Station/Country Louchi/S.S.S.R.

Location 66°05'N/32°59'E   **Height above sealevel** 94 m   **Climate symbol:** Köppen Dfc   **Troll** II,2

		J	F	M	A	M	J	J	A	S	O	N	D	year	P	
1 Mean daily temperature	in °C	-12,0	-12,2	-8,6	-1,9	4,4	10,8	14,4	12,3	7,0	0,6	-5,0	-8,9	0,6		1
2 Mean daily maximum temperature	in °C															2
3 Mean daily minimum temperature	in °C															3
4 Absolute maximum temperature	in °C	7	6	12	21	29	31	31	29	24	16	9	8	31		4
5 Absolute minimum temperature	in °C	-44	-46	-42	-36	-14	-7	-3	-6	-11	-24	-36	-42	-46		5
6 Mean relative humidity	in %	86	85	78	72	66	66	70	78	83	86	88	87	79		6
7 Mean precipitation	in mm	27	23	20	25	34	50	64	68	51	43	35	27	467		7
8 Maximum precipitation	in mm	61	45	47	60	66	118	170	166	120	120	65	64	650		8
9 Minimum precipitation	in mm	10	7	2	2	8	10	15	10	17	11	13	12	324		9
10 Maximum precipitation in 24 h	in mm	13	8	10	15	29	29	47	60	41	44	16	13	60		10
11 Mean number of days with precipitation	> 0,1 mm	19	16	14	13	13	14	13	16	16	17	19	19	189		11
12 Mean duration of sunshine	in h	6	47	142	197	232	269	286	208	114	56	9	0	1566		12
13 Mean quantity of radiation	in ly/day	25	45	149	234	378	420	397	267	147	71	27	0	180		13
14 Mean potential evaporation	in mm	0	0	0	0	44	90	115	98	54	3	0	0	403		14
15 Mean windspeed	in m/sec	3,0	3,0	3,2	3,1	3,3	3,4	3,0	2,8	3,1	3,3	3,4	3,2	3,2		15
16 Mean predominent direction of the wind		SW	SW	SW	SW	SW	SW	NE	SW	SW	SW	SW	SW			16

64

17 Station / Country Kem'/S.S.S.R.

Location 65°00'N / 34°48'E Height above sealevel 10 m Climate symbol: Köppen Dfc Troll II,2

			J	F	M	A	M	J	J	A	S	O	N	D	year	P	
1	Mean daily temperature	in °C	-10.7	-10.9	-7.4	-0.8	4.8	10.5	14.3	12.3	7.5	1.5	-4.0	-8.1	0.8		1
2	Mean daily maximum temperature	in °C	-9.7	-9.1	-4.1	2.7	8.5	14.3	17.3	16.4	11.5	4.9	-0.2	-5.7	3.9	8	2
3	Mean daily minimum temperature	in °C	-17.4	-16.5	-12.9	-4.4	1.1	6.6	10.1	9.9	5.6	-0.1	-4.7	-12.1	-2.9	8	3
4	Absolute maximum temperature	in °C	6	5	10	20	26	30	32	30	25	16	10	7	32		4
5	Absolute minimum temperature	in °C	-43	-40	-32	-26	-12	-4	1	-2	-7	-19	-31	-34	-43		5
6	Mean relative humidity	in %	87	87	81	77	74	75	78	81	83	85	88	87	82		6
7	Mean precipitation	in mm	24	19	18	21	28	53	56	60	53	39	30	24	425		7
8	Maximum precipitation	in mm	58	54	63	65	100	145	127	146	132	148	77	63	714		8
9	Minimum precipitation	in mm	8	8	5	5	3	3	9	20	17	10	6	10	345		9
10	Maximum precipitation in 24 h	in mm	9	11	14	23	32	60	53	51	62	58	14	14	62		10
11	Mean number of days with precipitation	> 0,1 mm	18	16	16	13	13	14	14	15	18	18	19	19	193		11
12	Mean duration of sunshine	in h	14	52	140	196	242	296	299	225	124	64	21	6	1679		12
13	Mean quantity of radiation	in ly / day															13
14	Mean potential evaporation	in mm	0	0	0	0	51	92	116	95	56	13	0	0	423		14
15	Mean windspeed	in m / sec	5.3	5.0	5.2	4.7	4.8	4.8	4.4	4.3	5.0	5.5	5.7	5.3	5.0		15
16	Mean predominent direction of the wind		W	S.W	W	SW	N	N	NE	SE	W	W	SW,W	SW,W			16

18 Station / Country Petrozavodsk / S.S.S.R.

Location 61°49'N / 34°16'E Height above sealevel 40 m Climate symbol: Köppen Dfc Troll II,2

			J	F	M	A	M	J	J	A	S	O	N	D	year	P	
1	Mean daily temperature	in °C	-9.7	-9.9	-5.8	-1.3	7.7	13.3	16.6	14.3	9.1	3.2	-2.2	-7.0	2.6		1
2	Mean daily maximum temperature	in °C	-10.3	-8.7	-2.3	5.3	12.7	17.9	19.4	18.6	12.8	6.3	0.0	-5.2	5.5	8	2
3	Mean daily minimum temperature	in °C	-17.6	-15.5	-10.0	-3.6	3.3	8.2	10.7	10.4	6.2	1.1	-4.3	-11.2	-1.9	8	3
4	Absolute maximum temperature	in °C	5	6	12	24	30	33	35	33	27	18	11	8	35		4
5	Absolute minimum temperature	in °C	-40	-38	-33	-25	-10	-2	2	1	-7	-16	-27	-37	-40		5
6	Mean relative humidity	in %	86	84	76	70	65	68	73	78	83	85	88	88	79		6
7	Mean precipitation	in mm	35	26	17	27	51	54	77	73	74	58	39	30	561		7
8	Maximum precipitation	in mm															8
9	Minimum precipitation	in mm															9
10	Maximum precipitation in 24 h	in mm	17	12	21	21	64	53	73	71	46	27	22	15	73		10
11	Mean number of days with precipitation	> 0,1 mm	20	18	17	15	14	13	15	16	17	20	20	23	208		11
12	Mean duration of sunshine	in h	19	52	162	200	249	300	294	219	128	58	26	12	1719		12
13	Mean quantity of radiation	in ly / day	19	48	180	261	372	426	415	291	147	62	21	6	187		13
14	Mean potential evaporation	in mm	0	0	0	0	64	102	132	105	60	22	0	0	485		14
15	Mean windspeed	in m / sec	4.4	4.1	4.4	3.9	3.8	3.8	3.5	3.1	3.7	4.2	4.3	4.2	3.9		15
16	Mean predominent direction of the wind		SW	SW	SW	SW	E	SW	SW	SW	SW	SW	SW	SW			16

19 Station / Country Archangel'sk / S.S.S.R.

Location 64°30'N / 40°30'E Height above sealevel 4 m Climate symbol: Köppen Dfc Troll II,2

			J	F	M	A	M	J	J	A	S	O	N	D	year	P	
1	Mean daily temperature	in °C	-12.8	-12.1	-8.0	-0.8	5.5	12.1	15.6	13.4	8.0	1.3	-5.1	-10.2	-0.6		1
2	Mean daily maximum temperature	in °C	-12.3	-10.3	-3.8	5.0	11.6	17.1	20.0	18.5	11.9	4.3	-1.5	-7.7	4.4	8	2
3	Mean daily minimum temperature	in °C	-20.4	-17.7	-13.4	-4.3	1.9	6.4	10.1	9.7	5.0	-0.6	-6.5	-14.6	-3.7	8	3
4	Absolute maximum temperature	in °C	5	4	10	23	30	32	34	33	28	17	10	4	34		4
5	Absolute minimum temperature	in °C	-45	-41	-37	-27	-14	-4	1	0	-7	-20	-36	-43	-45		5
6	Mean relative humidity	in %	88	86	82	76	71	69	71	78	85	88	89	89	81		6
7	Mean precipitation	in mm	33	28	28	28	39	59	63	67	58	55	44	39	539		7
8	Maximum precipitation	in mm	64	66	85	66	92	140	154	137	132	124	91	103	775		8
9	Minimum precipitation	in mm	7	8	7	6	4	5	2	9	13	14	6	8	279		9
10	Maximum precipitation in 24 h	in mm	21	16	16	15	43	45	47	55	49	25	18	22	55		10
11	Mean number of days with precipitation	> 0,1 mm	22	19	19	15	14	14	12	14	19	21	21	23	213		11
12	Mean duration of sunshine	in h	7	40	121	187	224	283	302	230	102	59	19	2	1576		12
13	Mean quantity of radiation	in ly / day	6	36	164	288	388	429	424	298	129	50	12	3	186		13
14	Mean potential evaporation	in mm	0	0	0	0	52	98	122	98	55	10	0	0	435		14
15	Mean windspeed	in m / sec	4.6	4.5	4.4	4.1	4.5	4.3	3.8	4.1	4.8	5.0	5.1	4.5	4.4		15
16	Mean predominent direction of the wind		NW	NW	NE	NW	SW	SE	SE	NW	NE	NE	NE	NW			16

20 Station / Country Vologda / S.S.S.R.

Location 59°17'N / 39°52'E Height above sealevel 118 m Climate symbol: Köppen Dfc Troll II,2

			J	F	M	A	M	J	J	A	S	O	N	D	year	P	
1	Mean daily temperature	in °C	-11.7	-11.0	-6.2	2.4	9.8	14.5	17.1	14.6	9.0	2.7	-3.5	-9.2	2.4		1
2	Mean daily maximum temperature	in °C	-11.1	-8.5	-1.6	7.9	16.2	19.8	21.5	20.7	14.4	6.6	-0.3	-6.5	6.6	8	2
3	Mean daily minimum temperature	in °C	-19.5	-17.6	-11.1	-1.6	5.1	8.1	10.8	9.8	5.7	0.7	-5.2	-12.8	-2.3	8	3
4	Absolute maximum temperature	in °C	5	3	13	28	31	32	35	35	29	23	11	6	35		4
5	Absolute minimum temperature	in °C	-48	-43	-36	-24	-11	-4	1	-2	-6	-25	-32	-40	-48		5
6	Mean relative humidity	in %	86	83	78	73	67	70	75	79	84	87	88	88	80		6
7	Mean precipitation	in mm	2	2	4	10	10	48	118	104	38	9	6	4	374		7
8	Maximum precipitation	in mm	95	44	82	79	138	150	181	175	146	113	75	77	852		8
9	Minimum precipitation	in mm	8	3	5	2	2	18	11	11	12	7	5	3	382		9
10	Maximum precipitation in 24 h	in mm	9	10	17	17	28	26	37	24	28	18	14	10	37		10
11	Mean number of days with precipitation	> 0,1 mm	19	14	16	13	14	13	14	14	16	18	19	21	193		11
12	Mean duration of sunshine	in h	27	59	134	187	264	271	292	241	124	55	26	13	1693		12
13	Mean quantity of radiation	in ly / day	28	73	223	285	381	414	431	319	165	71	33	19	204		13
14	Mean potential evaporation	in mm	0	0	0	17	78	110	129	101	56	16	0	0	507		14
15	Mean windspeed	in m / sec	5.3	5.2	5.2	4.6	4.7	4.3	3.6	3.5	4.1	4.8	5.2	5.3	4.7		15
16	Mean predominent direction of the wind		SW	SE	SW	SW	NW	NW	NE	SW	SW	SW	SW	SW			16

21 Station / Country Syktyvkar / S.S.S.R.

Location 61°40'N / 50°51'E — Height above sealevel 96 m — Climate symbol: Köppen Dfc — Troll II,2

		J	F	M	A	M	J	J	A	S	O	N	D	year	P	
1	Mean daily temperature in °C	-15.2	-13.8	-7.8	1.1	7.7	13.8	16.6	13.8	7.8	0.3	-7.2	-13.6	0.3		1
2	Mean daily maximum temperature in °C	-14.9	-10.5	-3.1	6.9	14.1	18.8	22.2	19.5	12.6	3.1	-4.2	-10.3	4.5	8	2
3	Mean daily minimum temperature in °C	-23.0	-19.0	-12.6	-3.5	3.4	7.6	11.8	9.4	4.7	-2.1	-9.6	-16.9	-4.2	8	3
4	Absolute maximum temperature in °C	3	3	13	26	30	35	35	35	29	20	10	3	35		4
5	Absolute minimum temperature in °C	-51	-45	-39	-27	-15	-5	-1	-2	-9	-30	-44	-45	-51		5
6	Mean relative humidity in %	85	82	76	69	64	64	69	77	84	86	86	86	78		6
7	Mean precipitation in mm	24	19	23	30	48	55	68	57	60	49	30	29	492		7
8	Maximum precipitation in mm	55	49	70	80	118	132	166	156	124	111	96	50	791		8
9	Minimum precipitation in mm	6	4	0	1	6	8	13	13	3	14	8	7	340		9
10	Maximum precipitation in 24 h in mm	21	15	15	16	23	28	37	44	31	21	13	11	44		10
11	Mean number of days with precipitation > 0,1 mm	20	16	17	13	14	14	14	13	18	21	19	21	200		11
12	Mean duration of sunshine in h	18	52	100	168	212	272	292	220	95	39	17	11	1496		12
13	Mean quantity of radiation in ly/day															13
14	Mean potential evaporation in mm	0	0	0	7	67	106	126	99	51	0	0	0	456		14
15	Mean windspeed in m/sec	4.5	4.1	4.3	3.8	4.3	3.9	3.3	3.2	3.8	4.3	4.7	4.5	4.1		15
16	Mean predominent direction of the wind	SW	S	SW	SW	SW	SW	N	SW	SW	SW	SW	SW			16

22 Station / Country Kirov / S.S.S.R.

Location 67°39'N / 53°01'E — Height above sealevel 7 m — Climate symbol: Köppen Dfc — Troll II,2

		J	F	M	A	M	J	J	A	S	O	N	D	year	P	
1	Mean daily temperature in °C	-14.4	-13.0	-7.1	1.9	9.9	15.1	18.0	15.1	8.8	1.5	-6.1	-12.3	1.4		1
2	Mean daily maximum temperature in °C	-12.7	-9.7	-2.8	7.6	16.8	20.3	22.8	20.7	14.2	4.7	-1.5	-8.9	6.0	8	2
3	Mean daily minimum temperature in °C	-20.2	-17.3	-10.2	-1.0	6.2	9.8	13.2	10.8	6.2	-0.8	-6.1	-14.8	-2.0	8	3
4	Absolute maximum temperature in °C	3	4	11	27	31	37	35	36	29	21	11	4	37		4
5	Absolute minimum temperature in °C	-41	-41	-34	-20	-11	-2	3	0	-8	-23	-40	-45	-45		5
6	Mean relative humidity in %	88	84	79	72	62	62	69	74	81	86	88	88	78		6
7	Mean precipitation in mm	33	24	26	28	45	58	72	69	54	56	38	35	538		7
8	Maximum precipitation in mm															8
9	Minimum precipitation in mm															9
10	Maximum precipitation in 24 h in mm	7	11	16	17	19	28	137	49	27	22	18	13	137		10
11	Mean number of days with precipitation > 0,1 mm	21	17	16	12	14	15	14	16	17	19	21	21	203		11
12	Mean duration of sunshine in h	31	70	130	138	258	287	286	252	126	55	31	23	1747		12
13	Mean quantity of radiation in ly/day															13
14	Mean potential evaporation in mm	0	0	0	14	76	112	133	103	53	7	0	0	498		14
15	Mean windspeed in m/sec	5.2	5.3	5.4	4.9	5.1	4.7	4.0	3.9	4.6	5.2	5.4	5.1	4.9		15
16	Mean predominent direction of the wind	SE	SE	SW.W	W	NW	W	NW	NW	NW	W	W	SW			16

23 Station / Country Perm / S.S.S.R.

Location 57°57'N / 56°13'E — Height above sealevel 170 m — Climate symbol: Köppen Dfb — Troll II,2

		J	F	M	A	M	J	J	A	S	O	N	D	year	P	
1	Mean daily temperature in °C	-15.4	-13.4	-7.2	2.2	10.0	15.6	18.0	15.3	9.2	1.6	-6.7	-13.2	1.3		1
2	Mean daily maximum temperature in °C	-12.9	-9.3	-2.3	8.5	16.5	20.0	23.7	20.0	14.4	4.4	-2.1	-9.1	6.0	8	2
3	Mean daily minimum temperature in °C	-20.3	-17.7	-11.2	-1.3	5.3	8.8	12.9	10.1	5.7	-1.3	-7.5	-16.0	-2.7	8	3
4	Absolute maximum temperature in °C	4	6	14	27	35	36	37	37	30	22	12	3	37		4
5	Absolute minimum temperature in °C	-45	-41	-35	-24	-13	-3	2	-1	-8	-21	-36	-44	-45		5
6	Mean relative humidity in %	82	78	75	68	60	62	68	72	78	83	83	83	74		6
7	Mean precipitation in mm	38	27	31	35	47	64	68	62	59	55	43	41	570		7
8	Maximum precipitation in mm	77	73	108	99	103	137	179	183	120	113	151	128	912		8
9	Minimum precipitation in mm	15	4	0	1	16	15	19	18	17	13	20	11	451		9
10	Maximum precipitation in 24 h in mm	15	15	28	22	48	72	55	48	46	26	24	25	72		10
11	Mean number of days with precipitation > 0,1 mm	21	17	15	11	13	15	15	16	18	19	21	21	200		11
12	Mean duration of sunshine in h	24	75	146	210	268	303	309	236	130	48	33	17	1799		12
13	Mean quantity of radiation in ly/day															13
14	Mean potential evaporation in mm	0	0	0	18	77	114	134	105	56	7	0	0	511		14
15	Mean windspeed in m/sec	3.4	3.3	3.4	3.1	3.6	3.5	2.7	2.8	3.1	3.6	3.5	3.3	3.3		15
16	Mean predominent direction of the wind	S	SE.SW	SW	SW	SE	W	SE	N	S.SW	SW	SW	SW			16

24 Station / Country Serov / S.S.S.R.

Location 59°36'N / 60°32'E — Height above sealevel 132 m — Climate symbol: Köppen Dfc — Troll II,2

		J	F	M	A	M	J	J	A	S	O	N	D	year	P	
1	Mean daily temperature in °C	-16.9	-15.1	-7.9	2.0	8.0	14.5	16.7	14.4	8.1	0.0	-8.2	-15.3	0.0		1
2	Mean daily maximum temperature in °C															2
3	Mean daily minimum temperature in °C															3
4	Absolute maximum temperature in °C	2	6	18	26	34	36	35	34	30	22	12	4	36		4
5	Absolute minimum temperature in °C	-45	-45	-40	-27	-12	-4	-1	-3	-10	-25	-45	-51	-51		5
6	Mean relative humidity in %	80	76	70	62	59	61	69	74	77	78	77	80	72		6
7	Mean precipitation in mm	18	15	19	20	41	64	84	65	46	35	30	24	446		7
8	Maximum precipitation in mm															8
9	Minimum precipitation in mm															9
10	Maximum precipitation in 24 h in mm															10
11	Mean number of days with precipitation > 0,1 mm															11
12	Mean duration of sunshine in h															12
13	Mean quantity of radiation in ly/day															13
14	Mean potential evaporation in mm	0	0	0	17	72	110	127	101	57	0	0	0	485		14
15	Mean windspeed in m/sec	2.7	3.0	3.4	3.5	3.9	3.6	3.1	2.8	3.1	3.4	3.3	2.7	3.2		15
16	Mean predominent direction of the wind	W	W	W	W	W	W	W	W	W	W	W	W			16

25 Station / Country Sverdlovsk / S.S.S.R.

Location 56°44'N/61°04'E   Height above sealevel 282 m   Climate symbol: Köppen Dfc   Troll II,2

		J	F	M	A	M	J	J	A	S	O	N	D	year	P	
1 Mean daily temperature	in °C	-15,6	-13,8	-7,4	2,1	9,9	15,2	17,3	14,8	9,0	1,2	-7,1	-13,6	1,0		1
2 Mean daily maximum temperature	in °C	-14,4	-10,0	-3,9	5,6	13,9	18,3	21,1	18,3	12,2	2,8	-6,7	-12,2	3,9	29	2
3 Mean daily minimum temperature	in °C	-20,6	-17,2	-12,2	-3,3	3,9	9,4	12,2	10,0	5,0	-2,2	-11,7	-18,3	-3,9	21	3
4 Absolute maximum temperature	in °C	4	8	17	27	33	35	38	37	30	25	14	6	38		4
5 Absolute minimum temperature	in °C	-42	-42	-39	-21	-14	-2	2	-1	-9	-22	-39	-43	-43		5
6 Mean relative humidity	in %	80	77	71	65	58	61	68	72	75	77	79	82	72		6
7 Mean precipitation	in mm	15	17	17	20	40	59	80	82	49	29	25	27	462		7
8 Maximum precipitation	in mm	56	39	62	73	115	160	221	212	119	106	56	95	743		8
9 Minimum precipitation	in mm	1	1	0	0	2	6	21	10	5	5	5	2	314		9
10 Maximum precipitation in 24 h	in mm	9	12	15	14	15	15	14	14	14	13	9	8	15		10
11 Mean number of days with precipitation	> 0,1 mm	14	11	10	9	13	14	15	15	14	15	15	16	161		11
12 Mean duration of sunshine	in h	52	90	144	202	245	269	267	220	133	73	46	40	1781		12
13 Mean quantity of radiation	in ly / day	47	98	260	336	434	474	456	350	201	102	51	31	237		13
14 Mean potential evaporation	in mm	0	0	0	14	80	118	132	105	57	8	0	0	514		14
15 Mean windspeed	in m / sec	4,5	4,4	4,8	4,7	4,3	4,0	3,8	3,9	4,4	5,1	5,1	4,4	4,4		15
16 Mean predominent direction of the wind		W	W	W	W	W	W	W	W	W	W	W	W			16

26 Station / Country Salechard / S.S.S.R.

Location 66°32'N/66°32'E   Height above sealevel 35 m   Climate symbol: Köppen Dfc   Troll II,2

		J	F	M	A	M	J	J	A	S	O	N	D	year	P	
1 Mean daily temperature	in °C	-24,4	-21,9	-17,9	-10,2	-2,1	7,1	13,8	11,2	5,2	-1,1	-15,8	-21,5	-6,7		1
2 Mean daily maximum temperature	in °C	-25,0	-20,6	-15,6	-7,7	0,0	8,3	16,1	13,3	6,7	-3,3	-16,7	-22,2	-5,6	27	2
3 Mean daily minimum temperature	in °C	-29,4	-27,2	-22,8	-15,6	-5,6	3,3	9,4	7,2	2,8	-6,7	-20,6	-26,7	-11,1	18	3
4 Absolute maximum temperature	in °C	2	2	5	11	24	30	31	30	22	15	5	2	31		4
5 Absolute minimum temperature	in °C	-50	-54	-47	-33	-26	-11	0	-2	-10	-28	-40	-52	-54		5
6 Mean relative humidity	in %	84	84	81	80	77	71	70	77	83	85	85	84	80		6
7 Mean precipitation	in mm	24	20	24	32	39	51	57	57	46	31	29	29	464		7
8 Maximum precipitation	in mm	45	46	39	71	108	114	135	165	130	71	71	53	696		8
9 Minimum precipitation	in mm	4	3	5	5	10	2	9	12	11	14	1	7	265		9
10 Maximum precipitation in 24 h	in mm	6	6	18	12	32	44	48	42	25	22	11	12	48		10
11 Mean number of days with precipitation	> 0,1 mm	19	17	16	11	13	13	11	14	12	16	18	20	180		11
12 Mean duration of sunshine	in h	5	46	145	191	223	245	299	190	95	53	20	0	1512		12
13 Mean quantity of radiation	in ly / day	6	40	174	321	415	450	496	291	114	56	12	9	199		13
14 Mean potential evaporation	in mm	0	0	0	0	0	80	122	96	48	0	0	0	346		14
15 Mean windspeed	in m / sec	3,9	4,0	4,4	5,1	5,7	6,0	5,0	5,2	4,5	5,5	4,2	4,2	4,8		15
16 Mean predominent direction of the wind		S,SW	NE	NE	NE	NE	NE	NE	NE	NE	SW	SW	SW			16

27 Station / Country Tobol'sk / S.S.S.R.

Location 58°12'N/69°14'E   Height above sealevel 64 m   Climate symbol: Köppen Dfc   Troll II,2

		J	F	M	A	M	J	J	A	S	O	N	D	year	P	
1 Mean daily temperature	in °C	-18,3	-15,9	-9,1	1,0	9,1	16,0	18,2	15,4	9,5	0,7	-9,3	-16,4	0,1		1
2 Mean daily maximum temperature	in °C	-18,3	-13,3	-5,6	3,9	12,2	18,9	21,7	18,9	12,8	2,8	-8,9	-15,0	2,8	22	2
3 Mean daily minimum temperature	in °C	-24,4	-20,0	-13,9	-3,3	3,9	9,4	12,8	10,6	5,0	-3,9	-13,3	-20,6	-5,0	12	3
4 Absolute maximum temperature	in °C	3	2	8	24	31	32	34	33	27	22	7	3	34		4
5 Absolute minimum temperature	in °C	-42	-38	-40	-22	-14	-1	4	0	-6	-22	-40	-45	-47		5
6 Mean relative humidity	in %	82	79	75	72	64	68	74	78	81	81	83	83	77		6
7 Mean precipitation	in mm	10	16	21	26	42	60	73	63	54	39	36	26	475		7
8 Maximum precipitation	in mm	44	40	50	90	112	200	219	198	126	88	65	48	706		8
9 Minimum precipitation	in mm	1	0	2	0	15	6	10	7	18	6	7	6	288		9
10 Maximum precipitation in 24 h	in mm	9	10	11	40	50	65	53	48	38	24	14	17	65		10
11 Mean number of days with precipitation	> 0,1 mm	15	12	12	10	13	12	15	15	15	16	17	16	168		11
12 Mean duration of sunshine	in h	38	88	150	184	239	298	267	217	138	65	38	26	1748		12
13 Mean quantity of radiation	in ly / day															13
14 Mean potential evaporation	in mm	0	0	0	10	72	118	134	103	57	5	0	0	499	50	14
15 Mean windspeed	in m / sec	3,8	4,1	4,4	4,1	4,5	4,1	3,6	3,4	3,2	3,9	4,0	3,6	3,9		15
16 Mean predominent direction of the wind		SE	SE	SE	SE	NW	NW	N	NW	W	W	W	SE			16

28 Station / Country Surgut / S.S.S.R.

Location 61°15'N/73°30'E   Height above sealevel 40 m   Climate symbol: Köppen Dfc   Troll II,2

		J	F	M	A	M	J	J	A	S	O	N	D	year	P	
1 Mean daily temperature	in °C	-22,2	-19,3	-12,8	-4,4	3,6	12,6	16,6	13,9	7,4	-1,7	-13,3	-20,2	-3,3		1
2 Mean daily maximum temperature	in °C	-22,8	-17,2	-9,4	-1,7	6,1	14,4	19,4	17,8	10,6	-0,6	-13,3	-20,0	-1,7	25	2
3 Mean daily minimum temperature	in °C	-28,3	-23,9	-18,9	-10,6	-0,6	7,8	12,2	10,6	4,4	-5,0	-18,3	-26,1	-8,3	19	3
4 Absolute maximum temperature	in °C	3	6	10	22	32	34	34	30	27	21	8	2	34		4
5 Absolute minimum temperature	in °C	-52	-55	-49	-37	-22	-7	-1	-4	-10	-30	-47	-55	-55		5
6 Mean relative humidity	in %	79	78	74	69	67	66	70	78	81	83	82	81	76		6
7 Mean precipitation	in mm	24	19	23	30	48	55	68	57	60	49	30	29	492		7
8 Maximum precipitation	in mm	62	45	64	73	121	139	189	186	128	92	68	62	665		8
9 Minimum precipitation	in mm	6	2	4	2	5	8	21	0	8	16	8	10	288		9
10 Maximum precipitation in 24 h	in mm	11	6	12	19	37	24	68	87	34	21	15	12	87		10
11 Mean number of days with precipitation	> 0,1 mm	23	16	16	14	16	15	15	17	16	20	22	23	213		11
12 Mean duration of sunshine	in h	30	80	151	193	217	253	275	204	120	59	38	14	1632		12
13 Mean quantity of radiation	in ly / day															13
14 Mean potential evaporation	in mm	0	0	0	0	40	106	132	101	54	0	0	0	433	50	14
15 Mean windspeed	in m / sec	4,7	5,2	5,4	5,2	5,4	5,0	4,3	4,5	5,2	5,8	5,8	5,2	5,2		15
16 Mean predominent direction of the wind		SW	W	W	W	NW	N	N	N	SW,W	SW	W	SW,W			16

## 29

Station/Country Omsk/S.S.S.R.

Location 54°26'N/73°24'E   **Height above sealevel** 105 m   **Climate symbol:** Köppen **Dfb**   **Troll** III,5

		J	F	M	A	M	J	J	A	S	O	N	D	year	P	
1	Mean daily temperature	in °C	-18,9	-17,6	-11,1	0,5	11,2	17,1	19,5	18,5	10,8	1,8	-8,5	-16,3	0,4		1
2	Mean daily maximum temperature	in °C	-18,3	-14,4	-7,2	3,9	15,0	20,6	23,3	21,1	15,6	4,4	-7,7	-15,0	4,4	22	2
3	Mean daily minimum temperature	in °C	-25,6	-22,8	-17,8	-6,1	4,4	10,6	13,3	11,1	5,0	-2,8	-13,3	-21,7	-5,6	19	3
4	Absolute maximum temperature	in °C	4	5	14	29	35	40	40	37	32	24	15	3	40		4
5	Absolute minimum temperature	in °C	-47	-49	-43	-27	-13	-2	2	-2	-8	-28	-41	-43	-49		5
6	Mean relative humidity	in %	81	80	80	71	54	59	68	71	70	75	82	81	73		6
7	Mean precipitation	in mm	8	6	9	18	30	53	72	46	33	23	15	12	325		7
8	Maximum precipitation	in mm	35	23	28	58	90	134	199	121	108	56	52	29	524		8
9	Minimum precipitation	in mm	4	0	0	0	3	2	20	5	5	1	4	3	211		9
10	Maximum precipitation in 24 h	in mm	12	13	6	24	21	61	29	40	25	29	11	10	40		10
11	Mean number of days with precipitation	> 0,1 mm	11	8	9	7	10	11	13	12	9	12	14	13	129		11
12	Mean duration of sunshine	in h	82	122	192	249	290	318	299	252	191	97	71	60	2923		12
13	Mean quantity of radiation	in ly/day	65	109	273	375	456	502	456	372	237	121	63	43	248		13
14	Mean potential evaporation	in mm	0	0	0	0	80	121	137	109	60	6	0	0	513		14
15	Mean windspeed	in m/sec	4,7	5,0	5,2	5,0	5,0	4,5	3,7	3,6	4,0	4,7	4,8	4,9	4,6		15
16	Mean predominent direction of the wind		SW	SW	SW	SW	SW	N	NW	NW	SW	SW	SW	SW			16

## 30

Station/Country Kolpaševo / S.S.S.R.

Location 58°18'N/82°54'E   **Height above sealevel** 76 m   **Climate symbol:** Köppen **Dfc**   **Troll** II,2

		J	F	M	A	M	J	J	A	S	O	N	D	year	P	
1	Mean daily temperature	in °C	-20,8	-17,8	-10,5	-1,3	6,8	15,5	18,3	14,9	9,0	-0,2	-11,7	-16,6	-1,4		1
2	Mean daily maximum temperature	in °C															2
3	Mean daily minimum temperature	in °C															3
4	Absolute maximum temperature	in °C	3	7	11	24	33	36	36	32	29	23	10	5	36		4
5	Absolute minimum temperature	in °C	-51	-51	-42	-31	-19	-4	1	-2	-9	-28	-38	-54	-54		5
6	Mean relative humidity	in %	80	78	73	66	62	66	72	79	79	79	82	81	75		6
7	Mean precipitation	in mm	18	12	15	21	51	58	77	79	52	37	31	24	475		7
8	Maximum precipitation	in mm	36	38	41	52	110	113	183	158	109	74	63	55	624		8
9	Minimum precipitation	in mm	5	3	5	2	14	10	17	14	7	14	15	11	311		9
10	Maximum precipitation in 24 h	in mm															10
11	Mean number of days with precipitation	> 0,1 mm															11
12	Mean duration of sunshine	in h	48	83	153	197	238	273	276	205	136	57	42	23	1731		12
13	Mean quantity of radiation	in ly/day	40	84	248	354	381	468	450	363	201	74	42	25	228		13
14	Mean potential evaporation	in mm	0	0	0	0	52	118	132	101	57	0	0	0	460		14
15	Mean windspeed	in m/sec	3,5	3,8	4,3	4,0	4,6	3,8	2,9	2,9	3,5	4,6	4,3	4,0	3,8		15
16	Mean predominent direction of the wind		S	S	S	NW	NW	NW	S	NW	S,SW	SW	SW	S,SW			16

## 31

Station/Country Podkamennaja Tunguska / S.S.S.R.

Location 61°36'N/90°00'E   **Height above sealevel** 60 m   **Climate symbol:** Köppen **Dfc**   **Troll** II,2

		J	F	M	A	M	J	J	A	S	O	N	D	year	P	
1	Mean daily temperature	in °C	-24,2	-21,1	-13,9	-2,4	4,9	13,6	17,4	13,8	7,5	-1,8	-16,0	-23,6	-3,8		1
2	Mean daily maximum temperature	in °C															2
3	Mean daily minimum temperature	in °C															3
4	Absolute maximum temperature	in °C	2	4	10	21	33	36	37	34	26	22	7	4	37		4
5	Absolute minimum temperature	in °C	-58	-54	-51	-36	-25	-6	-1	-3	-13	-33	-33	-50	-58		5
6	Mean relative humidity	in %	78	76	72	62	63	67	72	79	80	82	82	80	74		6
7	Mean precipitation	in mm	28	17	18	27	48	66	70	73	56	52	40	30	527		7
8	Maximum precipitation	in mm															8
9	Minimum precipitation	in mm															9
10	Maximum precipitation in 24 h	in mm															10
11	Mean number of days with precipitation	> 0,1 mm															11
12	Mean duration of sunshine	in h	26	76	144	194	230	266	303	211	108	42	35	11	1646		12
13	Mean quantity of radiation	in ly/day															13
14	Mean potential evaporation	in mm	0	0	0	0	44	110	132	94	57	0	0	0	437		14
15	Mean windspeed	in m/sec	2,8	2,9	3,4	3,3	3,3	3,0	2,4	2,5	2,9	3,8	3,3	3,0	3,0		15
16	Mean predominent direction of the wind		SW	E	SW	SW	NW	NW	E	E	SW	SW	SW	E			16

Station/Country Jenisejsk/S.S.S.R.

Location 58°27'N/92°10'E   **Height above sealevel** 78 m   **Climate symbol:** Köppen **Dfc**   **Troll** II,2

		J	F	M	A	M	J	J	A	S	O	N	D	year	P	
1	Mean daily temperature	in °C	-22,0	-19,3	-10,9	-1,8	6,5	14,4	17,8	14,7	7,9	-0,9	-12,1	-20,9	-2,2		1
2	Mean daily maximum temperature	in °C															2
3	Mean daily minimum temperature	in °C															3
4	Absolute maximum temperature	in °C	4	7	13	23	33	36	37	34	29	22	9	6	37		4
5	Absolute minimum temperature	in °C	-59	-53	-47	-35	-17	-4	1	-3	-13	-34	-39	-53	-59		5
6	Mean relative humidity	in %	80	77	71	63	61	67	74	79	80	79	81	80	74		6
7	Mean precipitation	in mm	24	17	17	19	44	57	60	60	53	42	41	36	470		7
8	Maximum precipitation	in mm	54	44	36	52	102	152	115	174	95	90	76	63	633		8
9	Minimum precipitation	in mm	5	1	2	5	9	10	18	15	18	11	14	12	308		9
10	Maximum precipitation in 24 h	in mm	11	12	9	19	18	40	41	50	37	12	20	9	50		10
11	Mean number of days with precipitation	> 0,1 mm	21	14	14	12	15	15	15	18	19	20	22	21	206		11
12	Mean duration of sunshine	in h	27	87	145	208	224	262	278	194	118	61	30	16	1650		12
13	Mean quantity of radiation	in ly/day	37	92	233	345	391	474	471	329	198	96	42	25	228		13
14	Mean potential evaporation	in mm	0	0	0	0	60	110	136	109	57	0	0	0	472		14
15	Mean windspeed	in m/sec	2,2	1,8	2,7	2,8	3,1	2,3	1,7	1,9	2,2	2,8	2,8	2,3	2,4		15
16	Mean predominent direction of the wind		SE	SE	SW	SW	W	SW	SW	SW	SW	SW	SW	SE			16

33 Station/Country Minusinsk / S.S.S.R.

Location 53°42'N / 91°42'E Height above sealevel 251 m Climate symbol: Köppen Dfb Troll II,2

		J	F	M	A	M	J	J	A	S	O	N	D	year	P	
1 Mean daily temperature	in °C	-20,3	-19,9	-10,9	2,2	10,3	17,1	19,7	16,6	9,5	1,0	-9,0	-17,8	-0,1		1
2 Mean daily maximum temperature	in °C															2
3 Mean daily minimum temperature	in °C															3
4 Absolute maximum temperature	in °C	6	8	21	27	34	37	39	38	32	26	15	7	39		4
5 Absolute minimum temperature	in °C	-52	-50	-47	-32	-18	-4	2	-3	-12	-31	-48	-53	-53		5
6 Mean relative humidity	in %	77	78	74	63	56	62	67	70	75	73	76	77	71		6
7 Mean precipitation	in mm	9	9	10	15	32	54	66	43	33	16	17	13	316		7
8 Maximum precipitation	in mm	18	23	22	34	72	99	137	118	85	59	43	24	493		8
9 Minimum precipitation	in mm	0	0	0	1	7	6	18	10	7	1	1	0	155		9
10 Maximum precipitation in 24 h	in mm	8	7	15	12	25	52	60	49	29	15	12	28	60		10
11 Mean number of days with precipitation	> 0,1 mm					7	12	13	14	13						11
12 Mean duration of sunshine	in h	57	92	148	164	199	232	251	225	162	100	51	37	1718		12
13 Mean quantity of radiation	in ly/day															13
14 Mean potential evaporation	in mm	0	0	0	15	74	122	141	113	57	6	0	0	528		14
15 Mean windspeed	in m/sec	2,0	1,5	2,2	3,2	3,3	2,3	1,9	2,0	2,2	2,8	3,3	2,2	2,4		15
16 Mean predominent direction of the wind		SW	SW	SW	SW	SW	SW	SW	SW	SW	SW	SW	SW			16

34 Station/Country Irkutsk / S.S.S.R.

Location 52°16'N / 104°19'E Height above sealevel 468 m Climate symbol: Köppen Dwc Troll II,2

		J	F	M	A	M	J	J	A	S	O	N	D	year	P	
1 Mean daily temperature	in °C	-20,9	-18,5	-10,0	0,6	8,1	14,5	17,5	15,0	8,0	0,1	-10,7	-18,7	-1,2		1
2 Mean daily maximum temperature	in °C	-16,1	-12,2	-3,9	5,6	13,3	20,0	21,1	20,0	13,9	5,0	-6,7	-15,6	3,9	10	2
3 Mean daily minimum temperature	in °C	-26,1	-25,0	-16,7	-6,7	0,6	6,7	10,0	8,9	1,7	-6,1	-16,7	-24,4	-7,7	10	3
4 Absolute maximum temperature	in °C	2	6	16	29	33	35	36	34	29	26	14	4	36		4
5 Absolute minimum temperature	in °C	-50	-45	-37	-32	-14	-4	0	-3	-12	-30	-40	-46	-50		5
6 Mean relative humidity	in %	80	74	68	59	56	66	74	78	78	75	80	85	73		6
7 Mean precipitation	in mm	12	8	9	15	29	83	102	99	49	20	17	15	458		7
8 Maximum precipitation	in mm	29	23	34	46	109	303	226	173	140	73	40	71	797		8
9 Minimum precipitation	in mm	2	1	1	2	5	11	13	20	10	2	5	2	251		9
10 Maximum precipitation in 24 h	in mm	9	7	11	14	43	49	68	76	34	21	13	30	76		10
11 Mean number of days with precipitation	> 0,1 mm	12	9	7	8	11	12	14	15	12	10	12	14	136		11
12 Mean duration of sunshine	in h	86	132	205	221	239	247	244	222	178	141	83	50	2046		12
13 Mean quantity of radiation	in ly/day	62	112	264	342	450	468	474	406	255	158	78	43	259		13
14 Mean potential evaporation	in mm	0	0	0	5	65	109	129	104	52	0	0	0	464		14
15 Mean windspeed	in m/sec	2,1	2,3	2,7	3,3	3,4	2,7	2,2	2,2	2,5	2,6	2,3	1,8	2,5		15
16 Mean predominent direction of the wind		SE	SE	SE	NW	NW	NW	SE	SE	NW	SE	NW	NW			16

35 Station/Country Nikolajevsk'-na-Amure / S.S.S.R.

Location 53°09'N / 140°42'E Height above sealevel 46 m Climate symbol: Köppen Dfb Troll II,2

		J	F	M	A	M	J	J	A	S	O	N	D	year	P	
1 Mean daily temperature	in °C	-25,8	-20,3	-11,1	0,7	8,8	14,8	19,6	18,0	12,3	3,4	-10,0	-22,4	-1,1		1
2 Mean daily maximum temperature	in °C	-21,7	-16,7	-8,3	1,1	6,7	13,9	18,9	18,9	15,0	5,0	-7,2	-17,2	0,6	24	2
3 Mean daily minimum temperature	in °C	-28,9	-26,7	-18,9	-8,3	-0,6	6,7	11,7	11,7	6,1	-2,2	-13,9	-23,9	-7,2	26	3
4 Absolute maximum temperature	in °C	0	2	10	20	29	32	34	35	30	22	11	5	35		4
5 Absolute minimum temperature	in °C	-47	-46	-38	-29	-12	-4	1	1	-6	-25	-34	-44	-47		5
6 Mean relative humidity	in %	77	77	74	75	78	78	80	82	82	76	76	79	78		6
7 Mean precipitation	in mm	19	20	18	32	43	41	55	72	73	55	43	31	503		7
8 Maximum precipitation	in mm	94	82	76	85	100	105	208	170	200	212	152	112	703		8
9 Minimum precipitation	in mm	0	0	0	5	1	0	1	6	2	10	3	0	271		9
10 Maximum precipitation in 24 h	in mm	38	24	24	29	24	55	60	74	54	82	40	60	82		10
11 Mean number of days with precipitation	> 0,1 mm	12	10	12	12	11	10	11	13	15	14	12	12	144		11
12 Mean duration of sunshine	in h	122	166	217	182	206	216	214	224	166	164	121	106	2106		12
13 Mean quantity of radiation	in ly/day															13
14 Mean potential evaporation	in mm	0	0	0	0	35	91	126	111	71	15	0	0	449		14
15 Mean windspeed	in m/sec	4,0	3,7	3,8	3,7	3,8	3,8	3,6	3,3	3,3	3,5	4,1	4,2	3,7		15
16 Mean predominent direction of the wind		W	W	W	E	E	E	E	E	E	W	W	W			16

36 Station/Country Cholmsk / S.S.S.R.

Location 47°03'N / 142°03'E Height above sealevel 29 m Climate symbol: Köppen Dfb Troll II,2

		J	F	M	A	M	J	J	A	S	O	N	D	year	P	
1 Mean daily temperature	in °C	-9,5	-8,4	-4,2	2,2	6,5	10,9	15,6	17,8	14,1	7,7	-0,2	-6,0	3,1		1
2 Mean daily maximum temperature	in °C															2
3 Mean daily minimum temperature	in °C															3
4 Absolute maximum temperature	in °C	10	6	15	20	24	28	30	30	28	22	16	10	30		4
5 Absolute minimum temperature	in °C	-29	-22	-22	-14	-6	-1	3	5	-1	-9	-15	-22	-29		5
6 Mean relative humidity	in %	72	70	72	74	75	80	84	82	79	72	69	71	75		6
7 Mean precipitation	in mm	40	26	32	50	59	63	90	92	98	88	76	63	777		7
8 Maximum precipitation	in mm	108	66	83	143	156	121	207	248	177	199	158	116	1092		8
9 Minimum precipitation	in mm	11	5	2	10	5	3		8	22	36	21	8	22	524	9
10 Maximum precipitation in 24 h	in mm	21	27	33	50	51	60	105	137	59	73	58	36	137		10
11 Mean number of days with precipitation	> 0,1 mm	22	17	16	13	14	15	15	14	15	15	19	24	199		11
12 Mean duration of sunshine	in h	54	96	155	168	188	165	148	165	178	156	86	45	1604		12
13 Mean quantity of radiation	in ly/day															13
14 Mean potential evaporation	in mm	0	0	0	17	51	83	112	117	79	42	0	0	500		14
15 Mean windspeed	in m/sec	6,8	6,3	6,1	6,0	5,2	3,9	3,5	3,8	5,3	6,1	7,1	7,8	5,7		15
16 Mean predominent direction of the wind		NE	SE	SE	SE	SW	SW	SW	SE	SE	SE	NW	NW			16

37 Station / Country Petropavlovsk-Kamčatskij / S.S.R.

Location 52°58'N / 158°45'E | Height above sealevel 32 m | Climate symbol: Köppen Dfc | Troll II,2

		J	F	M	A	M	J	J	A	S	O	N	D	year	P	
1	Mean daily temperature	in °C	-8.3	-8.7	-5.3	-0.6	3.9	8.7	12.8	13.5	10.4	4.9	-1.6	-6.1	2.0		1
2	Mean daily maximum temperature	in °C	-5.0	-5.0	-2.8	1.7	6.1	10.6	13.3	16.7	13.3	7.8	0.6	-2.8	4.4	7	2
3	Mean daily minimum temperature	in °C	-11.7	-12.2	-8.9	-3.9	0.0	4.4	8.3	9.4	6.1	1.1	-5.0	-9.4	-1.7	24	3
4	Absolute maximum temperature	in °C	6	6	10	15	22	27	31	27	24	19	13	6	31		4
5	Absolute minimum temperature	in °C	-34	-28	-24	-16	-10	-1	2	3	-1	-8	-16	-26	-34		5
6	Mean relative humidity	in %	64	65	67	74	76	79	81	82	78	66	68	67	72		6
7	Mean precipitation	in mm	111	88	174	107	76	58	73	106	102	143	182	115	1335		7
8	Maximum precipitation	in mm	432	229	657	186	156	138	182	178	239	378	733	554	2170		8
9	Minimum precipitation	in mm	25	7	4	14	3	5	20	26	18	27	23	6	875		9
10	Maximum precipitation in 24 h	in mm	26	25	39	29	26	20	26	34	41	57	59	36	59		10
11	Mean number of days with precipitation	> 0,1 mm	12	11	13	11	11	10	13	14	12	11	11	12	141		11
12	Mean duration of sunshine	in h	83	103	154	182	191	177	179	175	158	149	103	80	1734		12
13	Mean quantity of radiation	in ly/day	71	118	273	363	397	414	353	329	258	149	72	56	238		13
14	Mean potential evaporation	in mm	0	0	0	3	81	117	136	110	60	6	0	0	513		14
15	Mean windspeed	in m/sec	2.3	2.8	3.6	2.9	2.5	2.5	2.2	2.1	2.1	3.1	3.0	3.5	2.8		15
16	Mean predominent direction of the wind		N	N	N	NW	S	S	S	N,S	N	N	N	N			16

38 Station / Country Ust'-Kamčatsk / S.S.R.

Location 56°14'N / 162°28'E | Height above sealevel 6 m | Climate symbol: Köppen Dfc | Troll II,2

		J	F	M	A	M	J	J	A	S	O	N	D	year	P	
1	Mean daily temperature	in °C	-12.6	-12.8	-9.9	-3.8	1.6	6.8	11.2	12.2	9.1	2.6	-4.4	-10.6	-0.9		1
2	Mean daily maximum temperature	in °C	-8.9	-8.3	-2.8	0.6	4.4	10.6	13.9	15.0	12.2	5.6	-1.7	-6.7	2.8	4	2
3	Mean daily minimum temperature	in °C	-16.7	-17.2	-13.3	-7.7	-1.7	4.4	8.9	10.0	5.0	-1.7	-8.9	-15.6	-4.4	4	3
4	Absolute maximum temperature	in °C	5	5	7	12	20	29	30	26	23	16	12	7	30		4
5	Absolute minimum temperature	in °C	-42	-40	-37	-29	-16	-2	3	1	-3	-13	-25	-38	-42		5
6	Mean relative humidity	in %	81	82	80	82	86	87	88	86	84	77	78	81	83		6
7	Mean precipitation	in mm	82	74	51	39	34	28	52	58	49	59	66	78	670		7
8	Maximum precipitation	in mm	196	221	130	98	79	107	125	108	178	160	152		892		8
9	Minimum precipitation	in mm	11	3	9	6	3	2	10	10	4	10	3	9	353		9
10	Maximum precipitation in 24 h	in mm	14	14	10	9	10	9	15	19	13	16	15	12	19		10
11	Mean number of days with precipitation	> 0,1 mm	19	17	15	13	12	10	13	13	13	11	13	17	166		11
12	Mean duration of sunshine	in h	46	72	130	144	158	147	130	144	142	124	79	47	1363		12
13	Mean quantity of radiation	in ly/day															13
14	Mean potential evaporation	in mm	0	0	0	0	35	78	104	104	64	23	0	0	408		14
15	Mean windspeed	in m/sec	5.9	5.6	5.2	4.4	3.9	4.5	4.1	3.8	4.0	4.2	4.9	5.4	4.7		15
16	Mean predominent direction of the wind		NW	N	N	N	S	S	S	S	NW	NW	NW				16

39 Station / Country Dudinka / S.S.R.

Location 69°24'N / 86°10'E | Height above sealevel 20 m | Climate symbol: Köppen Dfc | Troll II,3

		J	F	M	A	M	J	J	A	S	O	N	D	year	P	
1	Mean daily temperature	in °C	-29.5	-25.7	-22.5	-16.0	-6.4	3.8	12.0	10.4	3.2	-8.4	-21.8	-26.9	-10.7		1
2	Mean daily maximum temperature	in °C	-30.6	-21.1	-22.8	-14.4	-3.3	6.7	15.0	15.6	5.6	-7.2	-22.8	-28.9	-8.9	5	2
3	Mean daily minimum temperature	in °C	-35.0	-25.6	-28.9	-23.3	-10.0	2.8	8.3	8.3	1.1	-11.7	-27.2	-33.3	-15.0	5	3
4	Absolute maximum temperature	in °C	-2	-1	4	9	16	28	30	30	24	11	3	0	30		4
5	Absolute minimum temperature	in °C	-57	-55	-52	-42	-36	-15	-1	-2	-20	-40	-48	-54	-57		5
6	Mean relative humidity	in %	78	79	79	80	81	79	71	78	80	86	82	79	79		6
7	Mean precipitation	in mm	12	11	9	10	12	29	32	49	33	28	18	13	267		7
8	Maximum precipitation	in mm						86	129	120	114						8
9	Minimum precipitation	in mm						6	1	10	22						9
10	Maximum precipitation in 24 h	in mm															10
11	Mean number of days with precipitation	> 0,1 mm	18	16	14	14	12	15	14	15	19	19	17	16	189		11
12	Mean duration of sunshine	in h	0	35	131	204	238	243	320	219	85	38	5	0	1518		12
13	Mean quantity of radiation	in ly/day															13
14	Mean potential evaporation	in mm	0	0	0	0	0	58	112	99	40	0	0	0	309	22	14
15	Mean windspeed	in m/sec	6.4	7.2	6.6	7.5	7.4	6.3	6.2	6.2	6.7	7.1	6.3	6.4	6.7		15
16	Mean predominent direction of the wind		SE	SE	ESE	E	N	NE	NE	NE	NE	SE	SE	SE			16

40 Station / Country Turuchansk / S.S.R.

Location 65°47'N / 87°57'E | Height above sealevel 45 m | Climate symbol: Köppen Dfc | Troll II,3

		J	F	M	A	M	J	J	A	S	O	N	D	year	P	
1	Mean daily temperature	in °C	-28.4	-23.8	-17.2	-9.6	-0.8	8.8	15.4	12.6	5.1	-6.0	-19.9	-27.4	-7.6		1
2	Mean daily maximum temperature	in °C	-28.3	-21.1	-13.9	-6.7	1.7	10.6	18.3	15.6	7.2	-5.6	-20.6	-27.2	-5.6	22	2
3	Mean daily minimum temperature	in °C	-33.3	-27.2	-22.8	-16.7	-5.0	5.0	11.1	8.3	2.2	-8.9	-25.0	-32.8	-12.2	13	3
4	Absolute maximum temperature	in °C	-2	1	10	14	28	32	34	31	24	14	6	1	34		4
5	Absolute minimum temperature	in °C	-60	-61	-53	-41	-29	-8	0	-6	-17	-43	-57	-60	-61		5
6	Mean relative humidity	in %	80	79	76	69	67	65	70	77	80	83	82	80	76		6
7	Mean precipitation	in mm	23	16	20	24	32	55	67	74	67	55	32	31	496		7
8	Maximum precipitation	in mm	55	41	48	78	90	116	183	163	146	108	72	64	692		8
9	Minimum precipitation	in mm	0	0	1	3	2	14	6	9	14	23	0	12	293		9
10	Maximum precipitation in 24 h	in mm	15	6	10	16	23	37	41	41	33	15	11	8	41		10
11	Mean number of days with precipitation	> 0,1 mm	18	16	17	13	14	14	14	16	19	22	20	18	202		11
12	Mean duration of sunshine	in h	4	59	137	222	252	253	322	203	107	42	17	0	1618		12
13	Mean quantity of radiation	in ly/day	6	45	164	345	431	450	459	313	102	68	15	3	200		13
14	Mean potential evaporation	in mm	0	0	0	0	0	83	130	100	43	0	0	0	356		14
15	Mean windspeed	in m/sec	4.0	3.9	4.1	4.0	3.8	3.9	3.4	3.6	4.0	4.8	3.9	4.0	4.0		15
16	Mean predominent direction of the wind		S	E	SE	NW	NW	NW	NW	NW	S	S	S	S			16

41 Station / Country Tura / S.S.S.R.

Location 64°10'N / 100°04'E — Height above sealevel 130 m — Climate symbol: Köppen Dfc — Troll II,3

		J	F	M	A	M	J	J	A	S	O	N	D	year	P	
1 Mean daily temperature	in °C	-36.8	-29.0	-19.8	-8.2	3.8	12.2	15.8	12.2	5.1	-7.0	-24.5	-34.4	-9.2		1
2 Mean daily maximum temperature	in °C															2
3 Mean daily minimum temperature	in °C															3
4 Absolute maximum temperature	in °C	-2	3	11	16	29	34	35	32	25	17	4	-1	35		4
5 Absolute minimum temperature	in °C	-66	-67	-56	-41	-30	-8	-2	-7	-21	-47	-59	-65	-67		5
6 Mean relative humidity	in %	77	78	72	64	62	64	70	75	77	80	79	78	73		6
7 Mean precipitation	in mm	10	10	9	12	22	48	62	48	34	24	20	18	317		7
8 Maximum precipitation	in mm	21	20	14	26	42	117	132	160	83	45	44	30	424		8
9 Minimum precipitation	in mm	5	2	3	4	4	11	15	13	8	15	9	4	270		9
10 Maximum precipitation in 24 h	in mm															10
11 Mean number of days with precipitation	> 0,1 mm	13	12	11	10	10	13	14	14	13	16	15	15	156		11
12 Mean duration of sunshine	in h	3	65	168	229	232	249	253	177	98	56	22	0	1552		12
13 Mean quantity of radiation	in ly / day	9	53	180	339	415	447	403	298	141	74	21	3	199		13
14 Mean potential evaporation	in mm	0	0	0	0	40	106	132	98	36	0	0	0	413		14
15 Mean windspeed	in m / sec	1.5	1.4	1.8	2.5	2.7	2.4	2.0	2.0	2.0	2.3	1.8	1.7	2.0		15
16 Mean predominent direction of the wind		E	NW	NW	NW	NW	NW	NW	W	W	W	E.NW	E.NW			16

42 Station / Country Chatanga / S.S.S.R.

Location 71°50'N / 102°28'E — Height above sealevel 24 m — Climate symbol: Köppen Dfc — Troll II,3

		J	F	M	A	M	J	J	A	S	O	N	D	year	P	
1 Mean daily temperature	in °C	-34.9	-31.4	-29.0	-19.2	-7.2	3.9	11.8	9.1	1.7	-11.5	-28.0	-31.2	-13.8		1
2 Mean daily maximum temperature	in °C															2
3 Mean daily minimum temperature	in °C															3
4 Absolute maximum temperature	in °C	-4	0	4	9	15	27	34	29	20	12	2	-2	34		4
5 Absolute minimum temperature	in °C	-61	-61	-57	-44	-37	-14	-1	-2	-23	-44	-51	-57	-61		5
6 Mean relative humidity	in %	79	79	78	79	80	75	69	79	83	85	81	80	79		6
7 Mean precipitation	in mm	17	13	12	12	16	27	38	48	39	31	24	19	296		7
8 Maximum precipitation	in mm						78	134	122	56						8
9 Minimum precipitation	in mm						5	7	8	5						9
10 Maximum precipitation in 24 h	in mm															10
11 Mean number of days with precipitation	> 0,1 mm	12	11	9	11	11	13	14	18	14	18	12	12	155		11
12 Mean duration of sunshine	in h	0	24	154	257	247	265	320	196	99	46	1	0	1609		12
13 Mean quantity of radiation	in ly / day															13
14 Mean potential evaporation	in mm	0	0	0	0	45	111	79	19	0	0	0	0	254		14
15 Mean windspeed	in m / sec	4.2	3.8	4.2	5.1	4.7	5.0	4.8	4.6	4.4	4.1	4.2	4.2	4.4		15
16 Mean predominent direction of the wind		SW	SW	NE	NE	NE	NE	NE	NE	SW	SW	SW	SW			16

43 Station / Country Jerbogačon / S.S.S.R.

Location 61°16'N / 108°01E — Height above sealevel 287 m — Climate symbol: Köppen Dfc — Troll II,3

		J	F	M	A	M	J	J	A	S	O	N	D	year	P	
1 Mean daily temperature	in °C	-31.2	-25.6	-15.8	-5.4	4.8	14.0	17.0	13.0	5.6	-5.0	-21.6	-29.9	-6.7		1
2 Mean daily maximum temperature	in °C															2
3 Mean daily minimum temperature	in °C															3
4 Absolute maximum temperature	in °C	0	4	11	22	32	34	37	35	27	18	4	2	37		4
5 Absolute minimum temperature	in °C	-61	-58	-51	-42	-23	-8	-6	-11	-20	-44	-52	-64	-64		5
6 Mean relative humidity	in %	79	78	66	59	56	60	68	74	76	79	80	80	71		6
7 Mean precipitation	in mm	14	10	8	16	24	42	60	49	30	32	21	17	323		7
8 Maximum precipitation	in mm	21	23	17	27	58	127	113	118	88	63	38	28	446		8
9 Minimum precipitation	in mm	6	2	2	5	6	22	16	2	3	14	7	6	246		9
10 Maximum precipitation in 24 h	in mm	6	6	6	8	18	71	45	36	16	16	7	8	71		10
11 Mean number of days with precipitation	> 0,1 mm	19	15	11	10	12	12	12	12	13	19	19	20	174		11
12 Mean duration of sunshine	in h	40	108	198	228	228	289	298	222	127	73	50	16	1878		12
13 Mean quantity of radiation	in ly / day	25	63	214	312	372	420	437	298	174	99	33	9	205		13
14 Mean potential evaporation	in mm	0	0	0	0	48	110	127	98	41	0	0	0	424		14
15 Mean windspeed	in m / sec	1.4	1.5	2.1	2.8	3.0	2.6	2.2	2.1	2.1	2.4	2.0	1.7	2.2		15
16 Mean predominent direction of the wind		SW	SW	SW	SW	SW	SW	NE	NE	SW	SW	SW	SW			16

44 Station / Country Olen'ok / S.S.S.R.

Location 68°30'N / 112°36'E — Height above sealevel 130 m — Climate symbol: Köppen Dfd — Troll II,3

		J	F	M	A	M	J	J	A	S	O	N	D	year	P	
1 Mean daily temperature	in °C	-40.9	-35.8	-26.7	-13.0	-0.9	10.8	14.1	9.7	2.5	-11.2	-31.0	-37.4	-13.3		1
2 Mean daily maximum temperature	in °C															2
3 Mean daily minimum temperature	in °C															3
4 Absolute maximum temperature	in °C	-2	-1	5	12	27	34	36	33	25	13	1	-2	36		4
5 Absolute minimum temperature	in °C	-65	-64	-57	-44	-29	-15	-4	-12	-24	-44	-57	-62	-65		5
6 Mean relative humidity	in %	73	74	70	64	61	60	64	73	77	80	77	75	71		6
7 Mean precipitation	in mm	10	7	7	10	20	38	54	55	26	22	13	13	275		7
8 Maximum precipitation	in mm	22	12	18	32	49	88	110	114	63	43	23	27	362		8
9 Minimum precipitation	in mm	3	1	2	1	4	8	2	10	5	6	3	3	145		9
10 Maximum precipitation in 24 h	in mm	6	4	8	8	20	34	60	42	19	14	4	5	60		10
11 Mean number of days with precipitation	> 0,1 mm	12	10	11	10	9	13	14	16	13	18	13	14	153		11
12 Mean duration of sunshine	in h	0	48	187	268	276	302	323	194	97	50	10	0	1755		12
13 Mean quantity of radiation	in ly / day	0	31	164	333	465	483	508	285	126	62	9	0	206		13
14 Mean potential evaporation	in mm	0	0	0	0	0	98	115	79	22	0	0	0	314		14
15 Mean windspeed	in m / sec	0.8	0.8	1.4	2.4	3.0	3.2	2.6	2.3	2.0	1.8	1.0	1.0	1.9		15
16 Mean predominent direction of the wind		N	N	N	N	N	N	N.NE	N.NE	N	N	N	N			16

45 Station/Country Aldan/S.S.S.R.

Location 58°37'N/125°22'E — Height above sealevel 680 m — Climate symbol: Köppen Dfc — Troll II,3

		J	F	M	A	M	J	J	A	S	O	N	D	year	P	
1 Mean daily temperature	in °C	-27.6	-23.9	-16.1	-6.1	3.5	13.0	17.1	13.3	5.6	-6.4	-20.7	-26.8	-6.3		1
2 Mean daily maximum temperature	in °C															2
3 Mean daily minimum temperature	in °C															3
4 Absolute maximum temperature	in °C	-2	2	6	13	30	33	34	33	26	16	7	-2	34		4
5 Absolute minimum temperature	in °C	-51	-50	-40	-33	-16	-4	0	-2	-14	-30	-44	-48	-51		5
6 Mean relative humidity	in %	78	76	69	63	63	60	67	73	75	76	78	79	71		6
7 Mean precipitation	in mm	19	15	15	27	63	73	88	83	70	49	28	21	551		7
8 Maximum precipitation	in mm	41	38	31	84	104	189	191	217	139	88	38	50	779		8
9 Minimum precipitation	in mm	5	2	2	8	15	16	23	28	18	7	7	6	357		9
10 Maximum precipitation in 24 h	in mm	9	6	12	26	35	75	56	64	37	20	10	11	75		10
11 Mean number of days with precipitation	> 0,1 mm	23	18	15	14	14	13	14	15	15	20	20	21	202		11
12 Mean duration of sunshine	in h	72	106	218	225	232	283	271	199	138	96	54	48	1942		12
13 Mean quantity of radiation	in ly/day	34	81	220	390	459	468	409	313	214	124	48	25	232		13
14 Mean potential evaporation	in mm															14
15 Mean windspeed	in m/sec	2.9	2.6	3.3	3.5	3.2	2.8	2.6	2.5	2.5	3.2	3.2	2.7	2.9		15
16 Mean predominent direction of the wind		SW	SW	SW	SW	SW	SW	SW	SW	SW.W	SW	SW	SW			16

46 Station/Country Sejmčan / S.S.S.R.

Location 62°5'N/152°25'E — Height above sealevel 211 m — Climate symbol: Köppen Dfd — Troll II,3

		J	F	M	A	M	J	J	A	S	O	N	D	year	P	
1 Mean daily temperature	in °C	-39.5	-34.5	-27.1	-11.9	2.1	12.7	15.5	11.8	3.9	-11.8	-27.5	-36.3	-11.9		1
2 Mean daily maximum temperature	in °C															2
3 Mean daily minimum temperature	in °C															3
4 Absolute maximum temperature	in °C	-14	-3	3	8	27	35	37	35	25	14	6	-5	37		4
5 Absolute minimum temperature	in °C	-62	-61	-60	-45	-26	-6	-4	-9	-22	-40	-55	-62	-62		5
6 Mean relative humidity	in %	73	75	69	66	59	59	67	71	73	80	80	78	71		6
7 Mean precipitation	in mm	28	21	11	8	11	34	41	39	25	20	31	27	296		7
8 Maximum precipitation	in mm	61	41	25	34	27	74	80	95	83	40	55	50	409		8
9 Minimum precipitation	in mm	10	5	0	0	2	0	13	16	2	2	12	5	66		9
10 Maximum precipitation in 24 h	in mm	9	10	8	11	14	24	37	25	30	19	18	10	37		10
11 Mean number of days with precipitation	> 0,1 mm	18	16	11	8	7	10	11	10	9	13	18	18	149		11
12 Mean duration of sunshine	in h	20	74	214	267	301	292	277	229	162	106	45	15	2002		12
13 Mean quantity of radiation	in ly/day															13
14 Mean potential evaporation	in mm	0	0	0	0	20	102	123	94	32	0	0	0	371		14
15 Mean windspeed	in m/sec	1.1	1.2	1.5	2.3	2.5	2.4	2.3	2.0	1.9	1.5	1.3	1.1	1.8		15
16 Mean predominent direction of the wind																16

47 Station/Country Kirensk/S.S.S.R.

Location 57°46'N/108°07'E — Height above sealevel 256 m — Climate symbol: Köppen Dfc — Troll II,3

		J	F	M	A	M	J	J	A	S	O	N	D	year	P	
1 Mean daily temperature	in °C	-26.9	-22.4	-11.5	-2.1	6.8	15.0	18.8	15.2	7.2	-2.3	-14.8	-25.1	-3.8		1
2 Mean daily maximum temperature	in °C	-25.6	-17.8	-6.1	3.3	11.7	20.6	23.3	20.6	11.7	1.1	-12.2	-23.3	0.6	19	2
3 Mean daily minimum temperature	in °C	-33.3	-29.4	-21.7	-9.4	0.0	7.2	10.6	8.9	1.7	-6.7	-20.0	-30.6	-10.0	18	3
4 Absolute maximum temperature	in °C	3	7	13	24	32	36	37	36	28	21	8	4	37		4
5 Absolute minimum temperature	in °C	-58	-53	-47	-35	-15	-4	0	-5	-11	-34	-48	-57	-58		5
6 Mean relative humidity	in %	78	77	71	65	60	68	74	78	80	79	80	79	74		6
7 Mean precipitation	in mm	19	11	7	14	27	66	67	56	36	28	25	25	381		7
8 Maximum precipitation	in mm	40	26	25	30	58	115	142	165	89	53	42	49	538		8
9 Minimum precipitation	in mm	4	0	2	3	3	10	10	2	7	8	11	5	243		9
10 Maximum precipitation in 24 h	in mm	12	8	6	8	23	66	58	61	29	13	10	8	66		10
11 Mean number of days with precipitation	> 0,1 mm	18	14	11	11	12	14	12	14	14	16	19	17	174		11
12 Mean duration of sunshine	in h	42	88	184	200	211	258	276	184	127	77	46	22	1715		12
13 Mean quantity of radiation	in ly/day	43	95	245	321	375	420	450	316	204	109	54	28	222		13
14 Mean potential evaporation	in mm	0	0	0	0	60	118	140	109	49	0	0	0	474		14
15 Mean windspeed	in m/sec	2.1	2.0	2.3	2.7	2.9	2.4	2.0	2.0	2.2	2.6	2.6	2.0	2.3		15
16 Mean predominent direction of the wind		SW	SW	SW	SW	SW	SW	NE	N	SW	SW	SW	SW			16

48 Station/Country Vil'ujsk/S.S.S.R.

Location 63°46'N/121°37'E — Height above sealevel 107 m — Climate symbol: Köppen Dfc — Troll II,3

		J	F	M	A	M	J	J	A	S	O	N	D	year	P	
1 Mean daily temperature	in °C	-38.2	-31.2	-19.7	-7.6	4.2	14.3	18.0	13.9	5.4	-7.7	-25.9	-35.8	-9.2		1
2 Mean daily maximum temperature	in °C															2
3 Mean daily minimum temperature	in °C															3
4 Absolute maximum temperature	in °C	-5	0	8	19	32	36	37	35	28	18	3	-3	37		4
5 Absolute minimum temperature	in °C	-61	-60	-50	-40	-23	-4	0	-6	-15	-38	-53	-58	-61		5
6 Mean relative humidity	in %	76	76	71	62	56	57	63	69	73	79	80	77	70		6
7 Mean precipitation	in mm	9	7	6	11	18	23	43	36	24	18	13	10	226		7
8 Maximum precipitation	in mm	24	28	17	36	67	82	106	135	70	36	28	31	353		8
9 Minimum precipitation	in mm	2	0	1	0	0	1	1	2	0	4	2	4	127		9
10 Maximum precipitation in 24 h	in mm	6	6	6	8	27	60	49	71	20	11	6	11	71		10
11 Mean number of days with precipitation	> 0,1 mm	13	11	9	9	8	9	10	9	11	15	15	13	132		11
12 Mean duration of sunshine	in h	36	90	224	266	281	321	299	280	142	74	63	3	2079		12
13 Mean quantity of radiation	in ly/day															13
14 Mean potential evaporation	in mm	0	0	0	0	44	114	136	101	41	0	0	0	436		14
15 Mean windspeed	in m/sec	1.8	1.7	2.1	2.6	2.8	2.5	2.3	2.1	2.2	2.4	2.1	1.8	2.2		15
16 Mean predominent direction of the wind		SW	SW	SW	SW	N	N	N	N	SW	SW	SW	SW			16

72

49 Station / Country: Jakutsk / S.S.S.R.

Location: 62°05'N / 129°45'E   Height above sealevel: 100 m   Climate symbol: Köppen Dfd   Troll II,3

			J	F	M	A	M	J	J	A	S	O	N	D	year	P	
1	Mean daily temperature	in °C	-43,2	-35,8	-22,0	-7,4	5,6	15,4	18,8	14,8	6,2	-7,8	-27,7	-39,6	-10,2		1
2	Mean daily maximum temperature	in °C															2
3	Mean daily minimum temperature	in °C															3
4	Absolute maximum temperature	in °C	-8	-7	6	21	33	35	38	35	27	19	2	-2	38		4
5	Absolute minimum temperature	in °C	-63	-64	-55	-41	-21	-4	-1	-9	-12	-41	-55	-60	-64		5
6	Mean relative humidity	in %	74	74	70	61	54	54	60	67	70	78	78	75	68		6
7	Mean precipitation	in mm	7	6	5	7	16	31	43	38	22	16	13	9	213		7
8	Maximum precipitation	in mm	21	21	13	24	44	117	131	118	71	41	31	18	333		8
9	Minimum precipitation	in mm	0	1	0	0	1	2	4	4	3	0	1	0	104		9
10	Maximum precipitation in 24 h	in mm	7	10	7	13	17	47	49	37	22	11	6	8	49		10
11	Mean number of days with precipitation	> 0,1 mm	13	11	7	6	8	9	10	10	9	13	14	15	125		11
12	Mean duration of sunshine	in h	40	126	247	286	293	344	338	268	182	93	61	16	2294		12
13	Mean quantity of radiation	in ly / day	22	70	239	384	453	504	462	366	207	105	36	12	238		13
14	Mean potential evaporation	in mm	0	0	0	0	48	118	140	109	41	0	0	0	456		14
15	Mean windspeed	in m / sec	1,5	1,5	1,9	2,7	3,5	3,1	2,9	2,8	2,6	2,6	1,9	1,5	2,4		15
16	Mean predominent direction of the wind		N	N	N	NW	NW	NW	NW	NW	NW	NW	N	N			16

50 Station / Country: Bomnak / S.S.S.R.

Location: 54°43'N / 128°56'E   Height above sealevel: 357 m   Climate symbol: Köppen Dwc   Troll II,3

			J	F	M	A	M	J	J	A	S	O	N	D	year	P	
1	Mean daily temperature	in °C	-32,4	-24,4	-13,6	-1,5	7,9	15,0	18,2	15,4	8,6	-2,7	-20,0	-30,2	-4,9		1
2	Mean daily maximum temperature	in °C															2
3	Mean daily minimum temperature	in °C															3
4	Absolute maximum temperature	in °C	-9	2	8	23	31	34	35	32	29	21	6	-1	35		4
5	Absolute minimum temperature	in °C	-51	-49	-40	-28	-11	-2	0	-1	-10	-35	-46	-52	-52		5
6	Mean relative humidity	in %	74	66	58	53	55	67	75	81	76	67	67	76	68		6
7	Mean precipitation	in mm	4	3	9	26	45	82	113	130	90	33	23	8	546		7
8	Maximum precipitation	in mm	17	15	32	80	134	182	406	304	175	102	59	31	896		8
9	Minimum precipitation	in mm	0	0	0	2	9	13	16	26	6	2	5	0	226		9
10	Maximum precipitation in 24 h	in mm	5	5	21	29	45	56	93	101	57	29	17	7	101		10
11	Mean number of days with precipitation	> 0,1 mm	6	4	5	7	10	12	13	14	12	8	10	8	109		11
12	Mean duration of sunshine	in h	111	154	209	204	205	219	228	185	152	141	101	82	1991		12
13	Mean quantity of radiation	in ly / day															13
14	Mean potential evaporation	in mm	0	0	0	0	61	106	133	109	51	0	0	0	460		14
15	Mean windspeed	in m / sec	0,4	0,9	1,7	2,6	2,6	2,3	2,1	1,9	2,0	1,4	1,1	0,6	1,6		15
16	Mean predominent direction of the wind		NE	NE	NE	E	E	E	E	E	E	E	E	NE			16

51 Station / Country: Verchojansk / S.S.S.R.

Location: 67°33'N / 133°23'E   Height above sealevel: 137 m   Climate symbol: Köppen Dwd   Troll II,3

			J	F	M	A	M	J	J	A	S	O	N	D	year	P	
1	Mean daily temperature	in °C	-48,9	-43,7	-29,9	-13,0	2,0	12,2	15,3	11,0	2,6	-14,1	-36,1	-45,6	-15,6		1
2	Mean daily maximum temperature	in °C	-47,8	-40,6	-25,0	-7,2	5,6	15,6	18,9	14,4	6,1	-11,1	-35,0	-46,7	-12,8	24	2
3	Mean daily minimum temperature	in °C	-52,8	-48,9	-39,4	-23,3	-5,0	8,9	8,3	4,4	-2,8	-19,4	-40,0	-48,9	-21,7	40	3
4	Absolute maximum temperature	in °C	-12	-11	0	5	14	31	34	35	33	25	14	1	35		4
5	Absolute minimum temperature	in °C	-66	-68	-60	-54	-29	-7	-3	-10	-22	-45	-57	-64	-68		5
6	Mean relative humidity	in %	75	75	71	65	58	58	62	70	74	80	80	78	70		6
7	Mean precipitation	in mm	7	5	5	4	5	25	33	30	13	11	10	7	155		7
8	Maximum precipitation	in mm	16	10	10	19	23	87	74	65	44	40	25	17	237		8
9	Minimum precipitation	in mm	0	0	0	0	0	3	0	5	0	2	0	0	51		9
10	Maximum precipitation in 24 h	in mm	5	9	3	16	12	28	33	46	28	20	10	3	46		10
11	Mean number of days with precipitation	> 0,1 mm	9	8	7	5	6	10	9	9	8	10	12	11	104		11
12	Mean duration of sunshine	in h	1	79	215	298	300	309	300	232	126	73	20	0	1953		12
13	Mean quantity of radiation	in ly / day	3	20	174	351	453	480	440	319	156	81	9	0	207		13
14	Mean potential evaporation	in mm	0	0	0	0	32	113	129	90	25	0	0	0	389		14
15	Mean windspeed	in m / sec	0,7	0,8	1,0	1,8	2,9	3,2	2,6	2,2	1,8	1,4	0,8	0,7	1,7		15
16	Mean predominent direction of the wind		SW	SW	SW	NE	NE	NE	NE	SW	SW	SW	SW	SW			16

52 Station / Country: Ojm'akon / S.S.S.R.

Location: 63°16'N / 143°09'E   Height above sealevel: 740 m   Climate symbol: Köppen Dwd   Troll II,3

			J	F	M	A	M	J	J	A	S	O	N	D	year	P	
1	Mean daily temperature	in °C	-50,1	-44,3	-32,0	-14,8	1,7	11,4	14,5	10,4	2,4	-14,8	-26,1	-47,1	-16,5		1
2	Mean daily maximum temperature	in °C															2
3	Mean daily minimum temperature	in °C															3
4	Absolute maximum temperature	in °C	-16	-11	4	11	28	31	33	31	24	11	3	-8	33		4
5	Absolute minimum temperature	in °C	-70	-71	-64	-57	-30	-10	-5	-11	-25	-48	-62	-68	-71		5
6	Mean relative humidity	in %	74	74	72	67	59	60	67	72	74	78	77	76	71		6
7	Mean precipitation	in mm	7	6	5	4	10	37	40	37	20	12	11	9	193		7
8	Maximum precipitation	in mm	24	19	14	19	33	67	90	78	65	29	32	38	316		8
9	Minimum precipitation	in mm	2	0	0	0	5	17	5	3	2	1	1	0	29		9
10	Maximum precipitation in 24 h	in mm	4	5	4	16	14	25	29	29	16	16	11	15	29		10
11	Mean number of days with precipitation	> 0,1 mm	12	10	8	5	8	12	12	12	10	11	15	12	127		11
12	Mean duration of sunshine	in h	29	115	224	268	261	268	273	214	140	75	42	2	1911		12
13	Mean quantity of radiation	in ly / day	19	70	251	399	496	459	450	403	201	112	33	9	242		13
14	Mean potential evaporation	in mm	0	0	0	0	24	98	119	86	22	0	0	0	350		14
15	Mean windspeed	in m / sec	0,3	0,3	0,8	1,5	2,4	2,4	1,9	1,5	1,6	1,5	0,6	0,3	1,3		15
16	Mean predominent direction of the wind		SE	SE	SE	SE	E	E	E	NW	NW	NW	NW	SE			16

53 Station/Country Ochotsk/S.S.S.R.

Location 59°22'N/143°12'E   Height above sealevel 6 m   Climate symbol: Köppen Dwc   Troll II,2

		J	F	M	A	M	J	J	A	S	O	N	D	year	P	
1 Mean daily temperature	in °C	-24,5	-20,5	-14,2	-5,7	0,7	5,6	11,9	12,9	8,3	-2,2	-14,7	-21,2	-5,3		1
2 Mean daily maximum temperature	in °C	-21,1	-18,3	-9,4	-1,7	3,3	7,8	13,9	15,0	11,7	0,6	-12,2	-18,3	-2,2	19	2
3 Mean daily minimum temperature	in °C	-27,2	-26,1	-20,9	-12,2	-2,8	1,1	8,9	9,4	3,9	-6,1	-17,8	-23,9	-9,4	19	3
4 Absolute maximum temperature	in °C	3	0	4	9	26	31	31	32	24	18	5	1	32		4
5 Absolute minimum temperature	in °C	-40	-45	-37	-34	-16	-2	2	1	-9	-21	-37	-40	-45		5
6 Mean relative humidity	in %	65	63	67	74	83	89	89	87	81	66	64	63	74		6
7 Mean precipitation	in mm	11	6	14	17	38	44	65	55	54	39	25	10	378		7
8 Maximum precipitation	in mm	61	25	60	50	100	117	284	148	270	100	48	96	671		8
9 Minimum precipitation	in mm	0	0	0	0	2	4	11	5	2	0	0	0	165		9
10 Maximum precipitation in 24 h	in mm	15	16	16	18	70	60	76	52	80	34	24	36	80		10
11 Mean number of days with precipitation	> 0,1 mm	4	3	4	7	7	8	10	9	11	6	4	4	77	15	11
12 Mean duration of sunshine	in h	102	156	228	229	214	148	170	201	173	187	118	84	2010		12
13 Mean quantity of radiation	in ly/day	40	90	260	351	440	378	332	301	213	140	51	25	218		13
14 Mean potential evaporation	in mm	0	0	0	0	27	73	109	103	63	0	0	0	375		14
15 Mean windspeed	in m/sec	4,4	3,8	3,7	3,6	3,4	3,5	3,4	3,4	3,5	4,5	4,9	4,7	3,9		15
16 Mean predominent direction of the wind		N	N	N	N	SE	SE	SE	SE	N	N	N	N			16

54 Station/Country Zyr'anka/S.S.S.R.

Location 65°44'N/150°54'E   Height above sealevel 43 m   Climate symbol: Köppen Dfd   Troll II,3

		J	F	M	A	M	J	J	A	S	O	N	D	year	P	
1 Mean daily temperature	in °C	-38,8	-34,7	-26,2	-12,5	1,3	12,8	15,3	11,2	3,8	-11,7	-28,1	-36,2	-12,0		1
2 Mean daily maximum temperature	in °C															2
3 Mean daily minimum temperature	in °C															3
4 Absolute maximum temperature	in °C	-10	-4	5	13	29	34	36	33	28	15	5	-8	36		4
5 Absolute minimum temperature	in °C	-60	-61	-55	-51	-28	-6	1	-4	-13	-37	-50	-59	-61		5
6 Mean relative humidity	in %	77	77	72	69	61	62	70	74	76	83	81	78	73		6
7 Mean precipitation	in mm	14	13	8	8	8	28	52	47	21	19	19	17	254		7
8 Maximum precipitation	in mm	36	35	21	24	27	80	167	88	65	48	49	48	371		8
9 Minimum precipitation	in mm	2	5	0	1	0	4	12	13	4	4	3	2	194		9
10 Maximum precipitation in 24 h	in mm	7	7	6	10	13	33	38	39	21	18	9	10	39		10
11 Mean number of days with precipitation	> 0,1 mm	18	16	11	8	5	9	12	12	10	15	17	17	150		11
12 Mean duration of sunshine	in h	9	69	214	296	349	344	294	241	166	83	29	2	2096		12
13 Mean quantity of radiation	in ly/day															13
14 Mean potential evaporation	in mm	0	0	0	0	20	106	127	90	35	0	0	0	378		14
15 Mean windspeed	in m/sec	1,9	1,8	2,2	3,0	3,4	3,4	3,4	3,2	2,8	2,6	2,1	1,9	2,6		15
16 Mean predominent direction of the wind		NW	NW	NW	N,NW	N	N	N	N,NW	N,NW	NW	NW	NW			16

55 Station/Country Kaliningrad (Königsberg)/S.S.S.R.

Location 54°42'N/20°37'E   Height above sealevel 27 m   Climate symbol: Köppen Cfb   Troll III,3

		J	F	M	A	M	J	J	A	S	O	N	D	year	P	
1 Mean daily temperature	in °C	-2,7	-2,1	0,8	6,2	11,8	15,0	17,3	16,4	12,9	7,7	2,6	-1,0	7,1		1
2 Mean daily maximum temperature	in °C	-1,7	-0,9	3,2	11,1	16,4	21,0	21,9	21,5	18,2	12,5	5,5	-0,1	10,7	8	2
3 Mean daily minimum temperature	in °C	-7,4	-6,9	-3,4	2,2	6,7	10,8	12,8	12,3	9,6	5,6	0,9	-5,2	3,1	8	3
4 Absolute maximum temperature	in °C	8	9	23	27	27	34	34	33	28	26	19	10	34		4
5 Absolute minimum temperature	in °C	-33	-31	-22	-10	-4	1	4	6	-1	-7	-14	-22	-33		5
6 Mean relative humidity	in %	86	84	79	76	72	73	77	80	83	84	88	87	81		6
7 Mean precipitation	in mm	56	39	27	43	39	55	90	84	70	52	59	697			7
8 Maximum precipitation	in mm	116	109	84	91	108	127	201	250	179	233	183	149	931		8
9 Minimum precipitation	in mm	11	4	7	4	8	10	12	14	12	3	8	12	481		9
10 Maximum precipitation in 24 h	in mm	21	24	20	35	48	67	83	110	51	57	41	25	110		10
11 Mean number of days with precipitation	> 0,1 mm	18	16	15	14	13	13	15	16	16	17	18	19	190		11
12 Mean duration of sunshine	in h	42	59	130	184	256	276	252	230	180	107	39	31	1786		12
13 Mean quantity of radiation	in ly/day															13
14 Mean potential evaporation	in mm	0	0	6	39	82	108	122	105	70	38	11	0	579	50	14
15 Mean windspeed	in m/sec	5,1	5,0	4,8	4,2	4,2	4,0	3,6	3,6	3,5	4,0	4,6	5,0	4,3		15
16 Mean predominent direction of the wind		SW	SW	SE	SE	W	W	W	SW,W	SW	SW	SW	SW			16

56 Station/Country Tbilisi/S.S.S.R.

Location 41°14'N/44°57'E   Height above sealevel 490 m   Climate symbol: Köppen Cfa   Troll III,3

		J	F	M	A	M	J	J	A	S	O	N	D	year	P	
1 Mean daily temperature	in °C	0,5	2,3	6,8	11,8	17,1	21,0	24,2	24,1	19,4	13,8	7,4	2,8	12,6		1
2 Mean daily maximum temperature	in °C	3,9	7,2	10,6	16,1	21,1	26,1	28,3	28,3	23,3	17,8	10,6	6,1	16,7	10	2
3 Mean daily minimum temperature	in °C	-3,3	-0,6	1,7	6,7	11,7	15,8	18,3	18,3	13,9	8,9	2,8	-0,6	7,8	10	3
4 Absolute maximum temperature	in °C	18	22	29	32	35	38	40	40	38	33	27	22	40		4
5 Absolute minimum temperature	in °C	-23	-14	-13	-4	1	7	9	9	1	-5	-7	-19	-23		5
6 Mean relative humidity	in %	74	69	65	61	63	60	57	57	64	72	77	75	66		6
7 Mean precipitation	in mm	18	22	29	56	92	72	50	34	43	42	39	23	513		7
8 Maximum precipitation	in mm	68	87	88	130	198	220	175	203	179	139	126	83	767		8
9 Minimum precipitation	in mm	0	0	1	5	5	3	1	0	1	4	1	0	241		9
10 Maximum precipitation in 24 h	in mm	29	44	30	99	65	68	80	65	130	67	76	46	130		10
11 Mean number of days with precipitation	> 0,1 mm	6	7	8	12	16	11	10	8	9	8	9	7	112		11
12 Mean duration of sunshine	in h	104	110	149	170	211	253	272	284	206	170	110	93	2112		12
13 Mean quantity of radiation	in ly/day	136	174	273	366	499	510	533	502	345	264	141	112	321		13
14 Mean potential evaporation	in mm	0	4	21	46	89	125	155	147	94	52	18	5	756	35	14
15 Mean windspeed	in m/sec	2,8	3,4	3,6	3,7	3,1	3,4	3,5	3,2	3,1	2,7	2,5	2,5	3,1		15
16 Mean predominent direction of the wind		NW	NW	SE	SE	NE	N	N	N	NW	SE,NW	SE	NW			16

57 Station / Country Tallinn (Reval) / S.S.S.R.

Location 59°25'N / 24°48'E Height above sealevel 44 m Climate symbol: Köppen Dfb Troll III,4

		J	F	M	A	M	J	J	A	S	O	N	D	year	P	
1 Mean daily temperature	in °C	-5,0	-5,8	-3,0	2,6	8,4	13,1	16,4	15,0	11,0	5,6	0,8	-3,0	4,7		1
2 Mean daily maximum temperature	in °C	-4,1	-4,0	0,0	7,5	13,6	18,7	20,1	19,5	15,3	9,7	3,1	-1,4	8,2	8	2
3 Mean daily minimum temperature	in °C	-10,4	-10,5	-7,3	-0,1	4,6	9,9	11,9	11,2	8,5	4,4	-1,4	-6,6	1,2	8	3
4 Absolute maximum temperature	in °C	7	8	15	24	30	31	33	33	28	21	12	10	33		4
5 Absolute minimum temperature	in °C	-32	-32	-24	-17	-5	1	5	4	-2	-11	-17	-27	-32		5
6 Mean relative humidity	in %	85	83	80	78	74	77	79	80	83	84	87	87	81		6
7 Mean precipitation	in mm	33	26	24	32	41	49	71	68	75	65	45	39	568		7
8 Maximum precipitation	in mm	86	107	81	75	91	135	139	173	177	155	153	104	813		8
9 Minimum precipitation	in mm	5	5	2	2	8	12	19	6	1	3	10	7	363		9
10 Maximum precipitation in 24 h	in mm	28	25	18	24	44	82	41	57	47	40	24	24	82		10
11 Mean number of days with precipitation	> 0,1 mm	17	15	12	12	12	12	13	15	15	17	17	17	174		11
12 Mean duration of sunshine	in h	23	51	148	192	262	287	281	234	156	75	28	16	1753		12
13 Mean quantity of radiation	in ly / day															13
14 Mean potential evaporation	in mm	0	0	0	20	66	99	122	104	67	32	5	0	515	70	14
15 Mean windspeed	in m / sec	6,3	5,4	5,3	5,4	5,1	5,0	4,8	4,7	5,0	5,9	6,3	6,4	5,5		15
16 Mean predominent direction of the wind		S	SW	SW	S,SW	W	W	W	SW,W	SW	SW	SW	SE			16

58 Station / Country Rīga / S.S.S.R.

Location 56°58'N / 24°04'E Height above sealevel 3 m Climate symbol: Köppen Dfb Troll III,4

		J	F	M	A	M	J	J	A	S	O	N	D	year	P	
1 Mean daily temperature	in °C	-5,0	-4,8	-2,0	4,6	10,7	14,3	17,1	15,7	11,7	6,2	1,5	-2,6	5,6		1
2 Mean daily maximum temperature	in °C	-3,8	-2,9	1,6	10,2	16,1	20,8	21,7	21,0	17,1	11,1	3,9	-1,5	9,8	8	2
3 Mean daily minimum temperature	in °C	-9,9	-9,7	-6,6	1,0	5,6	9,4	11,3	11,1	8,3	4,2	-0,9	-6,8	1,4	8	3
4 Absolute maximum temperature	in °C	7	11	17	25	30	32	34	33	28	22	14	12	34		4
5 Absolute minimum temperature	in °C	-32	-35	-30	-13	-7	-2	3	2	-5	-11	-22	-26	-35		5
6 Mean relative humidity	in %	85	83	77	74	71	73	76	80	83	83	86	87	80		6
7 Mean precipitation	in mm	32	38	24	33	42	60	78	71	60	53	47	38	566		7
8 Maximum precipitation	in mm	79	86	91	127	107	173	249	167	144	131	142	100	891		8
9 Minimum precipitation	in mm	12	4	2	11	0	13	14	17	0	12	13	11	436		9
10 Maximum precipitation in 24 h	in mm	26	21	24	25	34	44	61	50	41	36	33	20	61		10
11 Mean number of days with precipitation	> 0,1 mm	19	15	12	13	12	13	14	15	16	16	17	18	180		11
12 Mean duration of sunshine	in h	36	61	140	197	288	282	276	235	166	91	35	25	1812		12
13 Mean quantity of radiation	in ly / day	28	62	180	270	425	441	443	332	195	99	33	19	211		13
14 Mean potential evaporation	in mm	0	0	0	33	78	107	127	105	69	34	8	0	561	30	14
15 Mean windspeed	in m / sec	4,2	3,9	3,8	3,8	3,6	3,5	3,1	3,0	3,0	3,6	3,9	4,0	3,6		15
16 Mean predominent direction of the wind		S	S	S,SE	S	N	N	N	SW	S	S	S	S			16

59 Station / Country Kaunas / S.S.S.R.

Location 54°53'N / 23°53'E Height above sealevel 75 m Climate symbol: Köppen Dfb Troll III,4

		J	F	M	A	M	J	J	A	S	O	N	D	year	P	
1 Mean daily temperature	in °C	-4,7	-4,0	-0,2	6,3	12,7	15,6	17,8	16,4	12,6	7,0	1,2	-2,8	6,5		1
2 Mean daily maximum temperature	in °C	-4,0	-2,8	1,8	11,0	17,5	21,6	22,8	22,0	18,0	11,5	4,2	-2,0	10,1	8	2
3 Mean daily minimum temperature	in °C	-9,9	-9,5	-5,5	2,1	7,0	10,4	11,8	11,3	8,5	4,5	-0,2	-6,8	2,0	8	3
4 Absolute maximum temperature	in °C	5	7	20	25	28	31	34	35	29	24	17	6	35		4
5 Absolute minimum temperature	in °C	-34	-33	-26	-12	-3	0	5	4	-3	-10	-18	-27	-34		5
6 Mean relative humidity	in %	87	84	79	75	69	70	75	78	81	85	88	88	80		6
7 Mean precipitation	in mm	33	35	29	40	49	69	98	92	57	48	39	36	625		7
8 Maximum precipitation	in mm	52	66	66	91	160	158	218	242	109	120	109	83	915		8
9 Minimum precipitation	in mm	8	6	4	5	11	8	18	20	8	8	8	3	408		9
10 Maximum precipitation in 24 h	in mm	22	24	20	29	75	48	77	71	45	34	22	25	77		10
11 Mean number of days with precipitation	> 0,1 mm	16	15	14	13	13	13	15	15	13	14	16	16	173		11
12 Mean duration of sunshine	in h	41	57	137	185	258	276	272	232	167	98	37	30	1790		12
13 Mean quantity of radiation	in ly / day	40	92	220	243	400	420	403	319	204	105	36	25	209		13
14 Mean potential evaporation	in mm	0	0	0	34	90	110	127	106	68	35	11	0	581	12	14
15 Mean windspeed	in m / sec	4,8	4,2	3,8	3,6	2,2	3,0	2,8	2,8	3,0	3,6	4,2	4,6	3,6		15
16 Mean predominent direction of the wind		SW	SE	SE	SE	N	SW,NW	SW,W	SW	SW	S,SW	SE	S			16

60 Station / Country Užgorod / S.S.S.R.

Location 48°38'N / 22°16'E Height above sealevel 118 m Climate symbol: Köppen Dfb Troll III,4

		J	F	M	A	M	J	J	A	S	O	N	D	year	P	
1 Mean daily temperature	in °C	-2,9	-1,4	4,3	10,0	15,4	17,9	19,9	19,0	15,0	10,1	4,3	-0,2	9,3		1
2 Mean daily maximum temperature	in °C	-1,6	2,5	7,2	16,5	21,4	24,5	25,9	25,1	21,7	16,2	9,6	1,3	14,2	8	2
3 Mean daily minimum temperature	in °C	-8,3	-4,8	-0,2	5,7	9,9	13,2	14,4	13,8	10,7	5,2	3,0	-4,0		8	3
4 Absolute maximum temperature	in °C	13	17	27	32	33	37	39	40	34	31	29	17	40		4
5 Absolute minimum temperature	in °C	-28	-28	-24	-12	-2	3	6	4	-1	-18	-22	-25	-28		5
6 Mean relative humidity	in %	81	79	69	63	64	67	66	68	71	75	81	84	72		6
7 Mean precipitation	in mm	55	50	31	33	60	98	69	91	49	44	49	76	705		7
8 Maximum precipitation	in mm	116	95	110	100	136	233	187	224	157	169	133	126	1068		8
9 Minimum precipitation	in mm	5	6	5	9	3	30	16	12	9	0	3	6	416		9
10 Maximum precipitation in 24 h	in mm	32	24	27	42	47	75	50	67	44	52	43	30	75		10
11 Mean number of days with precipitation	> 0,1 mm	15	13	11	11	12	14	13	12	11	11	13	15	151		11
12 Mean duration of sunshine	in h	52	67	141	215	258	249	285	261	222	166	61	35	2040		12
13 Mean quantity of radiation	in ly / day															13
14 Mean potential evaporation	in mm	0	0	16	51	99	118	134	115	76	42	11	0	662	29	14
15 Mean windspeed	in m / sec	2,4	2,5	2,9	2,8	2,8	2,5	2,2	2,2	1,9	2,0	2,2	2,2	2,4		15
16 Mean predominent direction of the wind		SE	SE	SE	SE	NE	NW	NW	NW	SE	SE	SE	SE			16

61 Station/Country Minsk/S.S.S.R.

Location 53°52'N/27°32'E   Height above sealevel 234 m   Climate symbol: Köppen Dfb   Troll III,4

		J	F	M	A	M	J	J	A	S	O	N	D	year	P	
1 Mean daily temperature	in °C	-6,8	-6,2	-2,1	5,1	12,5	15,8	17,6	15,9	11,4	5,7	-0,2	-4,7	5,3		1
2 Mean daily maximum temperature	in °C	-7,2	-4,0	0,8	10,5	18,0	21,5	22,7	21,8	17,3	10,5	3,3	-3,1	9,3	8	2
3 Mean daily minimum temperature	in °C	-13,3	-10,9	-6,9	1,9	7,6	10,8	12,3	11,7	7,9	3,8	-1,2	-7,8	1,3	8	3
4 Absolute maximum temperature	in °C	6	8	19	26	31	33	35	34	29	25	16	10	35		4
5 Absolute minimum temperature	in °C	-39	-35	-39	-18	-6	0	5	2	-5	-20	-27	-31	-39		5
6 Mean relative humidity	in %	88	85	80	74	67	69	73	76	80	85	90	89	80		6
7 Mean precipitation	in mm	34	30	26	42	59	72	83	81	62	47	40	30	606		7
8 Maximum precipitation	in mm	101	70	87	109	120	221	206	158	129	108	140	94	896		8
9 Minimum precipitation	in mm	6	15	6	9	12	18	13	19	5	3	5	7	369		9
10 Maximum precipitation in 24 h	in mm	21	24	18	33	62	61	57	61	40	30	43	26	62		10
11 Mean number of days with precipitation	> 0,1 mm	19	17	15	15	14	15	16	15	14	16	19	20	195		11
12 Mean duration of sunshine	in h	40	60	146	193	261	286	272	238	170	92	32	25	1815		12
13 Mean quantity of radiation	in ly/day	50	81	223	270	428	447	437	338	207	115	45	28	222		13
14 Mean potential evaporation	in mm	0	0	0	25	88	114	125	105	65	28	0	0	550		14
15 Mean windspeed	in m/sec	4,9	5,0	4,7	4,4	4,1	3,9	3,6	3,5	3,7	4,3	4,8	5,0	4,3		15
16 Mean predominent direction of the wind		SW	SE	W	SE	NW	NW	NW	W.NW	W	W	SE	SE			16

62 Station/Country Kišin'ov (Kishinev)/S.S.S.R.

Location 47°01'N/28°52'E   Height above sealevel 95 m   Climate symbol: Köppen Dfb   Troll III,4

		J	F	M	A	M	J	J	A	S	O	N	D	year	P	
1 Mean daily temperature	in °C	-3,8	-2,5	2,7	9,2	15,9	19,0	21,2	20,3	15,8	10,3	3,8	-1,2	9,2		1
2 Mean daily maximum temperature	in °C	-1,3	1,2	6,0	14,2	22,9	25,7	27,4	27,2	23,0	17,0	9,9	1,7	14,7	8	2
3 Mean daily minimum temperature	in °C	-7,6	-4,8	-1,5	5,5	11,3	14,2	15,9	15,3	11,1	6,6	2,7	-3,7	5,4	8	3
4 Absolute maximum temperature	in °C	15	18	25	32	36	37	39	39	37	33	28	16	39		4
5 Absolute minimum temperature	in °C	-30	-32	-23	-9	-2	4	8	7	-1	-16	-22	-22	-32		5
6 Mean relative humidity	in %	84	81	79	67	62	66	63	64	68	76	84	86	73		6
7 Mean precipitation	in mm	26	35	17	32	39	79	36	49	43	25	56	34	471		7
8 Maximum precipitation	in mm	154	83	107	104	155	221	307	153	123	151	146	97	853		8
9 Minimum precipitation	in mm	0	2	0	1	3	9	2	4	0	0	0	0	279		9
10 Maximum precipitation in 24 h	in mm	28	28	24	34	68	91	91	84	58	40	77	41	91		10
11 Mean number of days with precipitation	> 0,1 mm	12	12	12	9	11	12	10	8	7	8	11	12	123		11
12 Mean duration of sunshine	in h	70	79	146	201	258	297	329	307	232	168	74	54	2215		12
13 Mean quantity of radiation	in ly/day	81	115	242	312	446	489	549	428	315	195	72	62	276		13
14 Mean potential evaporation	in mm	0	0	9	43	90	121	140	122	76	41	9	0	651		14
15 Mean windspeed	in m/sec	4,2	4,2	4,2	4,3	3,7	3,2	3,0	2,7	3,1	2,8	3,9	3,8	3,6		15
16 Mean predominent direction of the wind		NW	NW	NW	NW	NW	NW	NW	NW	NW	NW	NW	NW			16

63 Station/Country Leningrad/S.S.S.R.

Location 59°58'N/30°18'E   Height above sealevel 4 m   Climate symbol: Köppen Dfb   Troll III,4

		J	F	M	A	M	J	J	A	S	O	N	D	year	P	
1 Mean daily temperature	in °C	-7,5	-7,9	-4,1	2,9	9,6	14,5	17,7	15,7	10,7	4,7	-0,6	-5,3	4,2		1
2 Mean daily maximum temperature	in °C	-7,1	-5,4	-0,2	8,0	15,1	19,8	21,2	20,3	15,4	8,8	1,8	-3,1	7,9	8	2
3 Mean daily minimum temperature	in °C	-13,4	-11,9	-7,9	0,4	5,8	10,5	12,7	12,5	8,5	3,7	-2,1	-7,9	0,9	8	3
4 Absolute maximum temperature	in °C	6	8	13	24	31	32	33	32	28	21	12	9	33		4
5 Absolute minimum temperature	in °C	-36	-35	-28	-17	-6	0	6	3	-3	-13	-17	-33	-36		5
6 Mean relative humidity	in %	86	84	79	73	66	68	71	76	81	74	87	88	79		6
7 Mean precipitation	in mm	36	32	25	34	41	54	69	77	58	52	45	38	559		7
8 Maximum precipitation	in mm	92	67	72	72	115	146	154	203	178	110	106	93	825		8
9 Minimum precipitation	in mm	5	8	1	8	3	11	5	1	11	11	9	11	417		9
10 Maximum precipitation in 24 h	in mm	23	13	19	24	56	42	56	76	34	28	28	17	76		10
11 Mean number of days with precipitation	> 0,1 mm	21	18	14	13	13	14	14	16	16	17	19	21	196		11
12 Mean duration of sunshine	in h	17	38	111	166	253	263	277	212	130	66	21	9	1563		12
13 Mean quantity of radiation	in ly/day	12	42	146	231	360	399	394	270	162	62	15	6	175		13
14 Mean potential evaporation	in mm	0	0	0	21	72	109	127	105	64	27	0	0	525		14
15 Mean windspeed	in m/sec	3,4	3,1	3,0	2,9	2,8	2,9	2,6	2,4	2,7	3,2	3,3	3,4	3,0		15
16 Mean predominent direction of the wind		S	S	W	W	W	W	W	W	W	W	S	SW			16

64 Station/Country Kijev/S.S.S.R.

Location 50°24'N/30°27'E   Height above sealevel 179 m   Climate symbol: Köppen Dfb   Troll III,4

		J	F	M	A	M	J	J	A	S	O	N	D	year	P	
1 Mean daily temperature	in °C	-5,9	-5,3	-0,5	7,1	14,7	17,4	19,3	18,2	13,6	7,7	1,1	-3,7	7,0		1
2 Mean daily maximum temperature	in °C	-4,4	-2,1	2,5	13,5	20,7	23,7	24,9	24,2	19,8	13,2	5,5	-1,3	11,7	8	2
3 Mean daily minimum temperature	in °C	-10,2	-8,4	-3,8	4,9	10,7	13,6	15,0	14,3	9,9	5,5	-0,2	-5,8	3,8	8	3
4 Absolute maximum temperature	in °C	9	12	21	29	33	34	39	39	33	27	23	13	39		4
5 Absolute minimum temperature	in °C	-33	-34	-26	-11	-3	2	5	3	-4	-19	-23	-31	-34		5
6 Mean relative humidity	in %	87	84	79	69	63	65	67	70	73	80	86	88	76		6
7 Mean precipitation	in mm	43	39	35	46	58	66	70	72	47	47	53	41	615		7
8 Maximum precipitation	in mm	88	113	88	142	144	239	195	223	149	141	127	103	925		8
9 Minimum precipitation	in mm	3	5	2	1	4	7	5	5	2	1	2	5	405		9
10 Maximum precipitation in 24 h	in mm	19	30	28	37	50	85	74	48	57	36	41	32	74		10
11 Mean number of days with precipitation	> 0,1 mm	17	15	14	12	13	13	13	12	10	12	15	18	163		11
12 Mean duration of sunshine	in h	42	64	112	162	257	273	287	252	189	123	51	31	1843		12
13 Mean quantity of radiation	in ly/day	68	109	220	278	428	488	490	397	261	155	60	50	248		13
14 Mean potential evaporation	in mm	0	0	0	38	96	118	133	115	70	32	2	0	604		14
15 Mean windspeed	in m/sec	4,3	4,5	4,3	4,3	3,8	3,7	3,5	3,5	3,7	3,9	4,3	4,2	4,0		15
16 Mean predominent direction of the wind		W	NW	NW	NW	N	NW	NW	NW	NW	W	SE	SE,W			16

65 Station / Country Moskva (Moscow) / S.S.R.

Location 55°45'N / 37°34'E   Height above sealevel 156 m   Climate symbol: Köppen Dfb   Troll III,4

		J	F	M	A	M	J	J	A	S	O	N	D	year	P	
1 Mean daily temperature	in °C	-10,3	-9,7	-5,0	3,7	11,7	15,4	17,8	15,8	10,4	4,1	-2,3	-8,0	3,6		1
2 Mean daily maximum temperature	in °C	-9,3	-5,7	0,1	10,2	18,7	21,0	22,8	22,0	16,3	9,0	1,5	-4,5	8,5	8	2
3 Mean daily minimum temperature	in °C	-16,2	-13,6	-7,8	1,3	7,9	10,6	12,9	11,9	7,3	2,9	-3,3	-9,4	0,4	8	3
4 Absolute maximum temperature	in °C	4	6	15	28	32	35	37	37	32	24	13	8	37		4
5 Absolute minimum temperature	in °C	-42	-40	-32	-19	-7	-2	4	1	-5	-20	-33	-39	-42		5
6 Mean relative humidity	in %	85	82	77	71	64	66	69	74	79	82	85	86	77		6
7 Mean precipitation	in mm	31	28	33	35	52	67	74	74	58	51	36	36	575		7
8 Maximum precipitation	in mm	67	75	98	100	103	174	169	164	131	143	114	82	819		8
9 Minimum precipitation	in mm	8	7	6	3	2	5	25	1	7	6	11	7	354		9
10 Maximum precipitation in 24 h	in mm	20	21	21	25	33	51	79	63	52	42	22	26	79		10
11 Mean number of days with precipitation	> 0,1 mm	17	15	14	13	12	15	16	16	17	16	17	19	187		11
12 Mean duration of sunshine	in h	30	58	113	161	242	256	258	218	136	73	32	20	1597		12
13 Mean quantity of radiation	in ly / day	43	81	226	284	403	468	474	386	201	93	48	31	225		13
14 Mean potential evaporation	in mm	0	0	0	28	88	114	132	109	64	22	0	0	556		14
15 Mean windspeed	in m / sec	5,0	4,9	5,2	4,7	4,5	3,9	3,5	3,5	4,3	4,7	4,9	4,7	4,5		15
16 Mean predominent direction of the wind		W	SE	W	SE	N	NW	NW	NW	W	W	SW	S			16

66 Station / Country Char'kov / S.S.R.

Location 49°56'N / 36°17'E   Height above sealevel 152 m   Climate symbol: Köppen Dfb   Troll III,4

		J	F	M	A	M	J	J	A	S	O	N	D	year	P	
1 Mean daily temperature	in °C	-7,4	-7,0	-1,6	7,1	15,0	18,1	20,3	18,9	13,5	7,2	0,4	-5,2	6,6		1
2 Mean daily maximum temperature	in °C	-4,9	-3,2	2,2	13,7	22,1	24,6	26,7	26,2	21,0	12,9	5,9	-1,1	12,1	8	2
3 Mean daily minimum temperature	in °C	-11,8	-10,0	-4,4	3,9	9,2	12,9	15,1	14,3	9,1	4,1	-0,6	-6,2	3,1	8	3
4 Absolute maximum temperature	in °C	10	10	19	29	33	38	39	38	37	30	21	12	39		4
5 Absolute minimum temperature	in °C	-36	-36	-32	-13	-5	-1	5	1	-4	-18	-28	-31	-36		5
6 Mean relative humidity	in %	87	84	82	69	58	62	65	65	68	75	84	89	74		6
7 Mean precipitation	in mm	36	33	32	33	50	60	75	48	34	42	39	37	519		7
8 Maximum precipitation	in mm	114	75	100	105	117	133	148	137	97	111	149	122	744		8
9 Minimum precipitation	in mm	5	4	3	0	6	4	7	1	0	0	2	3	331		9
10 Maximum precipitation in 24 h	in mm	35	26	37	28	44	68	83	66	43	48	35	34	83		10
11 Mean number of days with precipitation	> 0,1 mm	16	15	13	10	10	10	10	9	8	11	12	15	139		11
12 Mean duration of sunshine	in h	40	67	116	182	254	279	290	260	195	122	54	28	1887		12
13 Mean quantity of radiation	in ly / day															13
14 Mean potential evaporation	in mm	0	0	0	38	100	122	140	120	70	30	2	0	622		14
15 Mean windspeed	in m / sec	5,6	5,9	5,9	5,4	5,0	4,3	4,1	3,8	3,9	4,6	5,4	5,8	5,0		15
16 Mean predominent direction of the wind		SE	E	E	SE,E	E	NW	NW	NW(	NW	E	E,SE	SE			16

67 Station / Country Penza / S.S.R.

Location 53°11'N / 45°01'E   Height above sealevel 235 m   Climate symbol: Köppen Dfb   Troll III,4

		J	F	M	A	M	J	J	A	S	O	N	D	year	P	
1 Mean daily temperature	in °C	-12,0	-11,5	-5,6	4,3	13,5	17,5	18,8	17,9	11,7	4,5	-3,3	-9,6	3,9		1
2 Mean daily maximum temperature	in °C															2
3 Mean daily minimum temperature	in °C															3
4 Absolute maximum temperature	in °C	4	5	14	30	33	38	38	37	35	25	15	7	38		4
5 Absolute minimum temperature	in °C	-43	-39	-26	-17	-8	-1	5	2	-4	-18	-36	-35	-43		5
6 Mean relative humidity	in %	86	83	80	70	60	61	66	68	70	79	84	87	74		6
7 Mean precipitation	in mm	40	34	36	32	52	55	64	56	50	48	46	46	559		7
8 Maximum precipitation	in mm	103	132	98	71	138	121	188	157	124	129	104	129	781		8
9 Minimum precipitation	in mm	7	5	2	2	4	2	11	7	7	2	5	10	344		9
10 Maximum precipitation in 24 h	in mm	25	28	30	43	71	69	100	49	50	41	38	29	100		10
11 Mean number of days with precipitation	> 0,1 mm	17	14	14	10	11	12	13	12	12	13	16	18	162		11
12 Mean duration of sunshine	in h	45	79	123	184	214	274	288	238	152	95	49	36	1807		12
13 Mean quantity of radiation	in ly / day															13
14 Mean potential evaporation	in mm	0	0	0	24	92	122	132	116	64	22	0	0	572		14
15 Mean windspeed	in m / sec	4,9	4,8	5,0	4,6	4,5	4,0	3,7	3,7	4,5	5,0	4,7	4,9	4,5		15
16 Mean predominent direction of the wind		S	S	S	S	NW	NW	NW	NW	NW	NW	S	S			16

68 Station / Country Kazan' / S.S.R.

Location 55°47'N / 49°11'E   Height above sealevel 64 m   Climate symbol: Köppen Dfb   Troll III,4

		J	F	M	A	M	J	J	A	S	O	N	D	year	P	
1 Mean daily temperature	in °C	-13,0	-12,1	-6,2	3,9	13,0	17,6	20,0	17,6	11,3	3,9	-3,0	-10,5	3,5		1
2 Mean daily maximum temperature	in °C	-11,1	-10,0	-2,7	8,5	18,7	21,9	24,2	22,9	16,7	6,7	-0,6	-7,2	7,3	8	2
3 Mean daily minimum temperature	in °C	-16,6	-17,2	-10,2	0,0	8,0	11,1	14,1	12,4	7,8	0,9	-5,0	-13,4	-0,8	8	3
4 Absolute maximum temperature	in °C	4	8	12	31	34	38	39	38	36	23	16	6	39		4
5 Absolute minimum temperature	in °C	-44	-37	-31	-19	-6	-1	5	2	-5	-23	-37	-39	-44		5
6 Mean relative humidity	in %	84	81	80	74	62	62	67	70	75	81	84	85	75		6
7 Mean precipitation	in mm	16	17	18	25	46	50	66	63	45	45	21	23	435		7
8 Maximum precipitation	in mm															8
9 Minimum precipitation	in mm															9
10 Maximum precipitation in 24 h	in mm	19	16	23	26	38	75	121	74	101	45	24	33	121		10
11 Mean number of days with precipitation	> 0,1 mm	17	14	13	10	11	12	12	12	13	15	16	18	163		11
12 Mean duration of sunshine	in h	61	118	151	216	285	307	271	295	155	89	35	38	2021		12
13 Mean quantity of radiation	in ly / day															13
14 Mean potential evaporation	in mm	0	0	0	24	92	122	140	113	60	19	0	0	570		14
15 Mean windspeed	in m / sec	4,8	4,6	4,7	4,1	4,4	3,7	3,4	3,5	4,0	4,5	4,5	4,8	4,2		15
16 Mean predominent direction of the wind		S	S	S	S	W	W	NW	N	S	S	S	S			16

69 Station / Country Kujbyšev / S.S.R.

Location 53°15'N/50°27'E Height above sealevel 44 m Climate symbol: Köppen **Dfb** Troll III,5

		J	F	M	A	M	J	J	A	S	O	N	D	year	P	
1 Mean daily temperature	in °C	-13,8	-13,0	-6,8	4,6	14,0	18,7	20,7	19,0	12,4	4,2	-4,1	-10,7	3,8		1
2 Mean daily maximum temperature	in °C	-10,3	-8,8	-1,7	11,0	20,7	23,1	25,6	24,6	18,6	8,9	0,6	-6,4	8,8	8	2
3 Mean daily minimum temperature	in °C	-17,4	-16,5	-8,6	1,8	9,5	12,3	15,2	13,7	8,3	1,3	-4,1	-12,3	0,3	8	3
4 Absolute maximum temperature	in °C	4	4	13	31	34	38	39	38	37	26	15	6	39		4
5 Absolute minimum temperature	in °C	-43	-37	-29	-25	-8	0	5	2	-7	-21	-37	-38	-43		5
6 Mean relative humidity	in %	85	82	82	69	54	56	62	62	68	76	84	86	72		6
7 Mean precipitation	in mm	33	24	30	32	43	40	50	44	41	46	33	33	449		7
8 Maximum precipitation	in mm	85	84	92	67	100	168	159	96	82	108	90	80	649		8
9 Minimum precipitation	in mm	6	1	3	1	6	3	1	2	7	7	1	4	293		9
10 Maximum precipitation in 24 h	in mm	25	32	30	41	30	59	55	54	72	27	34	26	72		10
11 Mean number of days with precipitation	> 0,1 mm	17	13	13	9	10	10	10	9	10	12	13	17	143		11
12 Mean duration of sunshine	in h	59	92	140	209	285	329	317	278	188	109	69	38	2113		12
13 Mean quantity of radiation	in ly/day	65	112	236	338	471	486	484	400	246	136	69	47	257		13
14 Mean potential evaporation	in mm	0	0	0	24	92	126	144	120	64	17	0	0	587		14
15 Mean windspeed	in m/sec	4,7	5,0	5,3	4,6	4,5	3,8	3,5	3,5	3,8	4,5	4,5	4,8	4,4		15
16 Mean predominent direction of the wind		E	E	SW	SW	W	W	N,W	E	W	W	SW	SW			16

70 Station / Country Ufa / S.S.R.

Location 54°45'N/56°00'E Height above sealevel 197 m Climate symbol: Köppen **Dfb** Troll III,5

		J	F	M	A	M	J	J	A	S	O	N	D	year	P	
1 Mean daily temperature	in °C	-14,6	-13,7	-7,4	3,2	12,5	17,7	19,0	17,0	10,9	2,7	-5,6	-11,9	2,5		1
2 Mean daily maximum temperature	in °C	-11,5	-8,5	-1,6	10,1	19,7	22,6	25,3	22,8	17,6	7,2	-0,2	-7,7	8,0	8	2
3 Mean daily minimum temperature	in °C	-20,4	-20,0	-11,9	-1,5	6,9	10,4	13,7	10,3	5,7	-0,5	-7,0	-16,1	-2,5	8	3
4 Absolute maximum temperature	in °C	4	9	14	31	36	38	39	36	32	25	14	6	39		4
5 Absolute minimum temperature	in °C	-43	-39	-34	-30	-10	-1	4	0	-5	-22	-38	-44	-44		5
6 Mean relative humidity	in %	82	80	78	71	59	62	71	72	75	81	83	85	75		6
7 Mean precipitation	in mm	22	16	19	27	33	48	57	44	44	47	32	30	419		7
8 Maximum precipitation	in mm															8
9 Minimum precipitation	in mm															9
10 Maximum precipitation in 24 h	in mm	17	13	15	16	21	39	32	28	28	17	17	14	39	8	10
11 Mean number of days with precipitation	> 0,1 mm	20	15	15	11	12	13	14	14	15	17	20	22	188		11
12 Mean duration of sunshine	in h	60	108	130	226	247	282	285	295	170	95	59	54	1991	4	12
13 Mean quantity of radiation	in ly/day	47	87	195	327	397	462	422	326	204	105	48	31	221		13
14 Mean potential evaporation	in mm	0	0	0	15	92	125	141	109	60	16	0	0	558		14
15 Mean windspeed	in m/sec	3,9	3,8	3,8	3,0	3,4	3,0	2,3	2,4	2,9	3,6	3,6	4,4	3,3		15
16 Mean predominent direction of the wind		S	S	SW	SW	SW	SW	N	SW	SW	SW	SW	S			16

71 Station / Country Kustanaj / S.S.R.

Location 53°13'N/63°37'E Height above sealevel 171 m Climate symbol: Köppen **Dfb** Troll III,5

		J	F	M	A	M	J	J	A	S	O	N	D	year	P	
1 Mean daily temperature	in °C	-17,8	-17,0	-10,7	1,8	12,9	18,4	20,4	18,1	11,9	3,0	-6,4	-14,9	1,6		1
2 Mean daily maximum temperature	in °C															2
3 Mean daily minimum temperature	in °C															3
4 Absolute maximum temperature	in °C	6	2	13	31	37	40	42	39	34	28	18	3	42		4
5 Absolute minimum temperature	in °C	-51	-48	-37	-28	-9	-4	3	0	-9	-22	-38	-44	-51		5
6 Mean relative humidity	in %	81	81	82	72	56	57	61	62	66	74	81	81	71		6
7 Mean precipitation	in mm	10	9	9	18	26	35	46	34	25	28	15	13	268		7
8 Maximum precipitation	in mm				76	73	111	143	104	117	69					8
9 Minimum precipitation	in mm				0	2	6	4	4	0	1					9
10 Maximum precipitation in 24 h	in mm				24	42	54	69	55	36	28					10
11 Mean number of days with precipitation	> 0,1 mm				7	10	10	11	10	9	10					11
12 Mean duration of sunshine	in h	76	107	162	220	282	307	293	261	186	110	75	58	2137		12
13 Mean quantity of radiation	in ly/day	78	120	285	342	496	498	512	388	237	130	66	53	267		13
14 Mean potential evaporation	in mm	0	0	0	4	89	122	140	116	63	11	0	0	545		14
15 Mean windspeed	in m/sec	4,8	4,9	4,8	4,6	5,4	4,2	3,5	3,5	4,2	4,7	4,6	4,4	4,5		15
16 Mean predominent direction of the wind		S	S	SW	SW	N,NW	N	N	SW	SW	SW	SW	SW			16

72 Station / Country Novosibirsk / S.S.R.

Location 55°02'N/82°54'E Height above sealevel 162 m Climate symbol: Köppen **Dfb** Troll III,5

		J	F	M	A	M	J	J	A	S	O	N	D	year	P	
1 Mean daily temperature	in °C	-19,0	-17,2	-10,7	-0,2	10,0	16,3	18,7	16,0	9,9	1,5	-9,7	-16,9	-0,2		1
2 Mean daily maximum temperature	in °C															2
3 Mean daily minimum temperature	in °C															3
4 Absolute maximum temperature	in °C	6	6	10	28	36	38	38	35	33	27	11	7	38		4
5 Absolute minimum temperature	in °C	-50	-47	-41	-33	-17	-2	2	-2	-9	-29	-46	-48	-50		5
6 Mean relative humidity	in %	80	78	78	70	59	66	72	76	76	77	82	82	75		6
7 Mean precipitation	in mm	16	12	13	22	34	60	74	60	45	35	30	24	425		7
8 Maximum precipitation	in mm	52	60	34	44	79	100	140	128	66	81	90	60	512		8
9 Minimum precipitation	in mm	14	3	4	7	15	18	11	8	13	12	24	19	339		9
10 Maximum precipitation in 24 h	in mm	14	14	14	24	32	95	55	45	44	22	31	15	95		10
11 Mean number of days with precipitation	> 0,1 mm	17	14	14	12	13	14	16	16	16	16	20	19	197		11
12 Mean duration of sunshine	in h	67	107	166	213	264	302	304	245	170	100	58	45	2041		12
13 Mean quantity of radiation	in ly/day	59	118	260	351	434	471	431	369	252	124	69	53	249		13
14 Mean potential evaporation	in mm	0	0	0	0	59	112	133	104	56	0	0	0	464		14
15 Mean windspeed	in m/sec	4,1	3,2	4,6	4,2	4,2	3,4	2,9	3,1	3,4	4,2	4,4	4,3	3,8		15
16 Mean predominent direction of the wind		SW	SW	SW	SW	SW	N	SW	SW	SW	SW	SW	SW			16

78

73 Station/Country Krasnojarsk/S.S.S.R.

Location 56°00′N/92°53′E Height above sealevel 156 m Climate symbol: Köppen Dfc Troll III,5

		J	F	M	A	M	J	J	A	S	O	N	D	year	P	
1 Mean daily temperature	in °C	−17,4	−16,0	−8,0	−1,6	9,5	16,7	19,9	16,6	9,6	1,6	−8,3	−15,9	0,8		1
2 Mean daily maximum temperature	in °C	−16,1	−15,0	−7,7	1,1	9,4	16,1	19,4	16,1	10,0	1,1	−10,0	−15,6	0,8	20	2
3 Mean daily minimum temperature	in °C	−23,3	−20,0	−13,3	−5,0	3,3	10,0	12,8	11,7	5,0	−3,3	−12,8	−20,6	−4,4	10	3
4 Absolute maximum temperature	in °C	6	8	20	28	34	38	41	37	32	26	13	10	41		4
5 Absolute minimum temperature	in °C	−49	−43	−40	−28	−17	−4	4	−1	−9	−32	−43	−45	−49		5
6 Mean relative humidity	in %	72	72	66	58	54	62	71	76	75	68	72	72	68		6
7 Mean precipitation	in mm	12	9	10	22	38	58	83	65	47	34	25	17	419		7
8 Maximum precipitation	in mm	28	27	25	68	84	108	171	157	88	78	91	59	585		8
9 Minimum precipitation	in mm	2	0	0	1	4	15	17	14	15	3	4	3	280		9
10 Maximum precipitation in 24 h	in mm	13	9	12	21	45	45	67	47	32	22	32	8	67		10
11 Mean number of days with precipitation	> 0,1 mm	11	8	8	8	12	14	13	15	13	11	13	12	138		11
12 Mean duration of sunshine	in h	54	84	163	191	221	280	267	212	160	97	47	30	1806		12
13 Mean quantity of radiation	in ly/day				·											13
14 Mean potential evaporation	in mm	0	0	0	8	70	117	138	109	58	5	0	0	505	30	14
15 Mean windspeed	in m/sec	2,5	1,8	2,4	2,7	2,7	2,1	1,7	1,7	2,2	2,5	2,8	2,3	2,3		15
16 Mean predominent direction of the wind		SW	SW	SW	SW	SW	SW	SW	SW	SW	SW	SW	SW			16

74 Station/Country Odessa/S.S.S.R.

Location 46°29′N/30°38′E Height above sealevel 64 m Climate symbol: Köppen Cfa Troll III,9

		J	F	M	A	M	J	J	A	S	O	N	D	year	P	
1 Mean daily temperature	in °C	−2,8	−2,3	2,0	8,0	15,0	19,2	22,1	21,4	16,7	11,5	4,9	0,0	9,6		1
2 Mean daily maximum temperature	in °C	0,0	1,5	4,9	12,4	18,9	23,4	26,1	25,7	21,4	15,6	9,9	3,7	13,6	8	2
3 Mean daily minimum temperature	in °C	−5,7	−3,8	−0,5	5,9	11,9	15,6	17,7	17,5	13,6	8,6	4,2	−1,6	6,9	8	3
4 Absolute maximum temperature	in °C	15	19	25	31	34	37	38	38	34	32	27	16	38		4
5 Absolute minimum temperature	in °C	−28	−30	−18	−8	−2	4	6	6	−3	−16	−16	−22	−30		5
6 Mean relative humidity	in %	86	83	80	75	72	69	63	66	70	77	84	86	76		6
7 Mean precipitation	in mm	28	26	20	27	34	45	34	37	29	35	43	31	389		7
8 Maximum precipitation	in mm	90	102	58	58	103	113	125	139	139	194	150	98	599		8
9 Minimum precipitation	in mm	1	1	0	0	1	2	1	0	0	1	0	0	192		9
10 Maximum precipitation in 24 h	in mm	26	17	25	23	57	66	40	46	49	54	32	29	66		10
11 Mean number of days with precipitation	> 0,1 mm	11	10	10	9	9	9	7	6	6	8	10	11	106		11
12 Mean duration of sunshine	in h	70	80	143	208	277	305	349	322	250	175	69	60	2308		12
13 Mean quantity of radiation	in ly/day	87	129	270	351	502	519	567	471	348	217	81	62	300		13
14 Mean potential evaporation	in mm	0	0	7	36	85	118	139	126	80	46	13	0	650		14
15 Mean windspeed	in m/sec	6,2	6,1	6,2	5,2	4,7	4,4	4,2	4,2	4,6	5,6	6,3	6,5	5,4		15
16 Mean predominent direction of the wind		N	NW	N,NW	S	S	N	N,NW	NW	N	N,E,NW	E	NE			16

75 Station/Country Simferopol'/S.S.S.R.

Location 45°01′N/33°59′E Height above sealevel 205 m Climate symbol: Köppen Cfb Troll III,9

		J	F	M	A	M	J	J	A	S	O	N	D	year	P	
1 Mean daily temperature	in °C	−0,7	−0,6	3,6	8,5	14,3	18,0	20,6	20,0	15,3	11,1	5,3	1,7	9,6		1
2 Mean daily maximum temperature	in °C	2,9	4,8	8,6	15,6	21,6	25,3	27,7	27,7	23,3	17,3	12,3	6,9	16,1	8	2
3 Mean daily minimum temperature	in °C	−4,5	−3,1	−0,5	4,9	10,1	13,7	15,7	15,3	11,5	6,6	4,2	0,1	6,2	8	3
4 Absolute maximum temperature	in °C	20	23	29	32	36	37	40	40	38	34	23	22	40		4
5 Absolute minimum temperature	in °C	−26	−29	−20	−10	−3	3	6	6	−3	−10	−18	−22	−29		5
6 Mean relative humidity	in %	85	83	77	67	67	67	63	63	66	75	83	85	74		6
7 Mean precipitation	in mm	44	40	35	36	49	75	58	34	31	41	44	41	528		7
8 Maximum precipitation	in mm	102	103	84	91	111	221	312	101	114	114	139	162	768		8
9 Minimum precipitation	in mm	2	0	0	0	2	8	4	2	2	1	2	3	318		9
10 Maximum precipitation in 24 h	in mm	29	31	20	43	39	101	122	40	58	42	52	43	122		10
11 Mean number of days with precipitation	> 0,1 mm	13	13	11	9	9	10	8	6	7	9	11	13	119		11
12 Mean duration of sunshine	in h	83	90	154	214	282	315	357	334	255	190	117	78	2469		12
13 Mean quantity of radiation	in ly/day															13
14 Mean potential evaporation	in mm	0	0	14	42	87	115	140	123	76	45	14	4	660		14
15 Mean windspeed	in m/sec	2,9	3,3	3,7	3,5	2,8	2,5	2,2	2,2	2,3	2,9	3,3	3,2	2,8		15
16 Mean predominent direction of the wind		NE	SW	NE	NE,SE	SE	E,SE	SE	SE	SE	SE	NE,SE	SE			16

76 Station/Country Zaporožje/S.S.S.R.

Location 47°48′N/35°15′E Height above sealevel 86 m Climate symbol: Köppen Dfa Troll III,9

		J	F	M	A	M	J	J	A	S	O	N	D	year	P	
1 Mean daily temperature	in °C	−4,9	−4,2	1,0	9,0	16,4	20,1	22,8	21,6	16,0	9,3	2,8	−2,3	9,0		1
2 Mean daily maximum temperature	in °C															2
3 Mean daily minimum temperature	in °C															3
4 Absolute maximum temperature	in °C	14	16	24	31	35	38	39	41	37	33	24	15	41		4
5 Absolute minimum temperature	in °C	−32	−34	−25	−9	−2	3	8	6	−3	−18	−22	−28	−34		5
6 Mean relative humidity	in %	86	84	80	66	60	61	58	58	64	75	84	87	72		6
7 Mean precipitation	in mm	31	27	26	35	39	57	50	45	30	30	36	37	443		7
8 Maximum precipitation	in mm	94	90	84	81	116	166	142	150	90	79	91	102	660		8
9 Minimum precipitation	in mm	3	1	3	0	0	1	4	0	1	2	4	3	282		9
10 Maximum precipitation in 24 h	in mm	20	24	27	51	45	104	84	61	62	47	38	30	104		10
11 Mean number of days with precipitation	> 0,1 mm	13	12	10	9	10	10	8	7	7	9	11	13	119		11
12 Mean duration of sunshine	in h															12
13 Mean quantity of radiation	in ly/day	90	140	242	381	515	522	543	477	354	236	111	65	306		13
14 Mean potential evaporation	in mm	0	0	3	41	98	128	149	129	79	36	7	0	670		14
15 Mean windspeed	in m/sec	4,2	4,4	4,6	4,2	4,0	3,3	3,2	3,1	3,0	3,4	4,4	4,3	3,8		15
16 Mean predominent direction of the wind		NE	NE	NE	NE,E	NE,E	N,NE	N	N	N	NE	E	E			16

79

77 Station / Country Rostov-na-Donu / S.S.S.R.

Location 47°15'N/39°49'E   Height above sealevel 77 m   Climate symbol: Köppen **Dfa**   Troll III,9

		J	F	M	A	M	J	J	A	S	O	N	D	year	P	
1 Mean daily temperature	in °C	-6,3	-5,5	0,2	8,8	15,9	19,8	22,7	21,8	15,8	9,3	2,0	-3,5	8,4		1
2 Mean daily maximum temperature	in °C	-1,9	-0,9	5,0	15,8	23,2	26,7	29,3	28,8	23,3	15,2	8,0	1,8	14,5	8	2
3 Mean daily minimum temperature	in °C	-8,2	-7,3	-1,8	5,7	12,3	15,5	17,5	16,9	11,8	5,4	1,8	-2,9	5,5	8	3
4 Absolute maximum temperature	in °C	15	19	28	32	34	38	40	40	36	33	25	15	40		4
5 Absolute minimum temperature	in %	-33	-30	-28	-6	-2	0	8	5	-4	-10	-23	-28	-33		5
6 Mean relative humidity	in %	87	85	80	67	59	61	58	58	62	75	82	87	72		6
7 Mean precipitation	in mm	38	41	32	39	38	58	49	37	32	44	40	37	483		7
8 Maximum precipitation	in mm	109	125	94	111	157	176	186	112	123	105	138	134	758		8
9 Minimum precipitation	in mm	4	1	1	4	1	4	1	1	0	0	2	1	30		9
10 Maximum precipitation in 24 h	in mm	26	25	28	39	78	100	78	57	49	49	35	31	100		10
11 Mean number of days with precipitation	> 0,1 mm	13	13	12	10	9	10	8	7	6	9	10	13	120		11
12 Mean duration of sunshine	in h	47	68	132	189	270	297	330	304	245	152	79	36	314		12
13 Mean quantity of radiation	in ly / day															13
14 Mean potential evaporation	in mm	0	0	2	41	101	132	156	137	78	37	6	0	690		14
15 Mean windspeed	in m / sec	6,5	7,0	6,8	6,4	5,8	4,8	4,3	4,2	4,4	5,4	7,0	7,0	5,8		15
16 Mean predominant direction of the wind		E	E	E	E	E	E	W	NE,E	E	E	E	E			16

78 Station / Country P'atigorsk/S.S.S.R.

Location 44°03'N/43°02'E   Height above sealevel 573 m   Climate symbol: Köppen **Dfb**   Troll III,9

		J	F	M	A	M	J	J	A	S	O	N	D	year	P	
1 Mean daily temperature	in °C	-4,3	-3,2	-1,8	8,1	14,7	18,7	21,8	20,8	15,5	9,6	2,7	-2,0	8,8		1
2 Mean daily maximum temperature	in °C															2
3 Mean daily minimum temperature	in °C															3
4 Absolute maximum temperature	in °C	18	18	34	32	33	40	41	40	40	36	27	20	41		4
5 Absolute minimum temperature	in °C	-31	-28	-24	-13	-3	2	7	5	-3	-13	-28	-33	-33		5
6 Mean relative humidity	in %	85	84	82	70	69	68	65	65	73	79	86	85	76		6
7 Mean precipitation	in mm	13	15	25	37	65	77	70	50	51	39	25	15	482		7
8 Maximum precipitation	in mm															8
9 Minimum precipitation	in mm															9
10 Maximum precipitation in 24 h	in mm															10
11 Mean number of days with precipitation	> 0,1 mm	14	14	13	11	13	13	11	10	9	12	14	14	148		11
12 Mean duration of sunshine	in h	77	74	106	136	185	230	255	244	171	133	83	62	1756		12
13 Mean quantity of radiation	in ly / day															13
14 Mean potential evaporation	in mm	0	0	6	36	86	118	140	118	75	41	7	0	625		14
15 Mean windspeed	in m / sec	3,2	3,8	4,0	4,1	3,3	2,8	2,9	2,9	3,0	3,3	3,5	3,5	3,3		15
16 Mean predominant direction of the wind		E	E	E	E	E	NW	E	E	E	E	E				16

79 Station / Country Turgaj/S.S.S.R.

Location 49°38'N/63°30'E   Height above sealevel 123 m   Climate symbol: Köppen **BSk**   Troll III,9

		J	F	M	A	M	J	J	A	S	O	N	D	year	P	
1 Mean daily temperature	in °C	-17,2	-16,0	-8,8	5,1	15,9	21,8	24,2	21,8	14,9	5,6	-4,4	-13,0	4,2		1
2 Mean daily maximum temperature	in °C															2
3 Mean daily minimum temperature	in °C															3
4 Absolute maximum temperature	in °C	7	5	24	31	39	42	43	42	36	30	19	7	43		4
5 Absolute minimum temperature	in °C	-44	-41	-35	-24	-6	0	6	3	-6	-20	-33	-40	-44		5
6 Mean relative humidity	in %	81	82	82	65	50	43	46	48	48	64	78	84	64		6
7 Mean precipitation	in mm	12	10	9	17	16	18	27	14	9	19	12	14	177		7
8 Maximum precipitation	in mm	76	62	41	63	52	65	116	69	36	72	49	61	318		8
9 Minimum precipitation	in mm	1	0	0	0	0	0	0	0	0	0	0	0	78		9
10 Maximum precipitation in 24 h	in mm	23	15	22	31	23	35	93	59	25	19	27	9	93		10
11 Mean number of days with precipitation	> 0,1 mm	10	8	7	6	6	5	6	5	4	6	7	10	79		11
12 Mean duration of sunshine	in h	93	129	166	234	314	326	336	312	242	157	104	78	2491		12
13 Mean quantity of radiation	in ly / day															13
14 Mean potential evaporation	in mm	0	0	0	18	98	141	162	131	72	17	0	0	639		14
15 Mean windspeed	in m / sec	5,2	5,8	5,8	5,6	5,2	4,7	4,5	4,5	4,8	4,8	4,7	4,8	5,0		15
16 Mean predominant direction of the wind		N,NE	N	NE	NE	NE	N,NE	N	N	W	W	SW	SW			16

80 Station / Country Karaganda/S.S.S.R.

Location 49°48'N/73°08'E   Height above sealevel 537 m   Climate symbol: Köppen **BSk**   Troll III,9

		J	F	M	A	M	J	J	A	S	O	N	D	year	P	
1 Mean daily temperature	in °C	-15,2	-14,0	-8,9	2,4	13,0	18,5	20,6	18,3	11,8	3,2	-6,9	-9,4	2,8		1
2 Mean daily minimum temperature	in °C															2
3 Mean daily minimum temperature	in °C															3
4 Absolute maximum temperature	in °C	7	7	22	28	34	38	39	40	35	28	18	6	40		4
5 Absolute minimum temperature	in °C	-49	-46	-35	-25	-10	-2	3	-1	-6	-25	-38	-43	-49		5
6 Mean relative humidity	in %	80	79	81	70	55	55	55	55	57	69	81	81	68		6
7 Mean precipitation	in mm	11	11	15	22	28	41	43	28	21	24	15	14	273		7
8 Maximum precipitation	in mm															8
9 Minimum precipitation	in mm															9
10 Maximum precipitation in 24 h	in mm															10
11 Mean number of days with precipitation	> 0,1 mm	12	10	12	7	10	10	11	8	6	11	13	15	125		11
12 Mean duration of sunshine	in h	99	127	159	217	281	302	303	291	226	145	88	66	2302		12
13 Mean quantity of radiation	in ly / day															13
14 Mean potential evaporation	in mm	0	0	0	14	88	126	144	116	60	14	0	0	562		14
15 Mean windspeed	in m / sec	4,8	4,5	5,2	4,8	4,8	4,0	3,9	3,8	3,5	4,7	5,0	5,0	4,5		15
16 Mean predominant direction of the wind		SW	SW	SW	SW	SW	NE	NE	NE	SW	SW	SW	SW			16

81 Station / Country Volgograd / S.S.S.R.

Location 48°42'N / 44°31'E Height above sealevel 42 m Climate symbol: Köppen BSk Troll III,10

		J	F	M	A	M	J	J	A	S	O	N	D	year	P	
1 Mean daily temperature	in °C	-9,6	-8,9	-2,6	8,2	17,0	21,4	24,2	22,7	15,9	8,2	0,2	-6,3	7,5		1
2 Mean daily maximum temperature	in °C															2
3 Mean daily minimum temperature	in °C															3
4 Absolute maximum temperature	in °C	11	10	23	31	35	40	42	43	36	32	22	12	43		4
5 Absolute minimum temperature	in °C	-35	-31	-26	-14	-4	4	9	6	-2	-14	-25	-31	-35		5
6 Mean relative humidity	in %	85	85	84	65	56	49	47	51	57	71	82	86	68		6
7 Mean precipitation	in mm	23	20	18	19	27	40	33	23	27	23	34	31	318		7
8 Maximum precipitation	in mm	60	63	67	63	79	137	103	116	74	66	76	117	571		8
9 Minimum precipitation	in mm	1	1	4	0	0	0	0	0	1	0		3	156		9
10 Maximum precipitation in 24 h	in mm	19	15	14	9	7	10	9	8	8	10	13	18	140		10
11 Mean number of days with precipitation	> 0,1 mm															11
12 Mean duration of sunshine	in h	54	82	135	209	279	313	332	311	242	139	83	46	2225		12
13 Mean quantity of radiation	in ly / day	81	118	270	354	505	507	524	456	309	189	96	53	289		13
14 Mean potential evaporation	in mm	0	0	0	38	103	141	162	138	79	30	0	0	690		14
15 Mean windspeed	in m / sec	6,1	7,1	5,3	5,2	4,8	4,9	4,2	4,6	4,3	4,6	5,0	5,8	5,1		15
16 Mean predominent direction of the wind		NE	NE	NE	NE	E	NE	N,NW	N	N,SE	W	NE	NE			16

82 Station / Country Kuška / S.S.S.R.

Location 35°17'N / 62°21'E Height above sealevel 630 m Climate symbol: Köppen BSk Troll III,10 a

		J	F	M	A	M	J	J	A	S	O	N	D	year	P	
1 Mean daily temperature	in °C	2,3	4,8	8,7	14,3	20,5	25,3	27,6	25,3	19,3	13,4	7,7	3,9	14,4		1
2 Mean daily maximum temperature	in °C															2
3 Mean daily minimum temperature	in °C															3
4 Absolute maximum temperature	in °C	27	31	32	37	44	45	46	43	42	37	35	32	46		4
5 Absolute minimum temperature	in °C	-33	-26	-26	-6	1	4	10	7	-4	-9	-19	-24	-33		5
6 Mean relative humidity	in %	75	74	74	63	47	33	30	29	35	46	57	70	53		6
7 Mean precipitation	in mm	40	42	60	30	12	2	0	0	1	0	25	25	246		7
8 Maximum precipitation	in mm															8
9 Minimum precipitation	in mm															9
10 Maximum precipitation in 24 h	in mm															10
11 Mean number of days with precipitation	> 0,1 mm	9	9	10	7	3	<1	0	0	0	2	4	7	51		11
12 Mean duration of sunshine	in h	128	116	145	228	320	377	386	368	319	266	190	141	2986		12
13 Mean quantity of radiation	in ly / day	242	280	428	510	651	717	713	682	540	434	306	248	479		13
14 Mean potential evaporation	in mm	3	4	24	54	115	162	185	154	86	46	19	8	860		14
15 Mean windspeed	in m / sec	1,9	1,9	2,0	1,8	2,2	3,1	3,5	3,3	2,4	1,8	1,6	1,6	2,3		15
16 Mean predominent direction of the wind		SW	SW	NE	NE	N,NE	NE	N	N	N	NE,SW	NE,NW	SW			16

83 Station / Country Termez / S.S.S.R.

Location 37°17'N / 67°19'E Height above sealevel 302 m Climate symbol: Köppen BSk Troll III,10 a

		J	F	M	A	M	J	J	A	S	O	N	D	year	P	
1 Mean daily temperature	in °C	2,8	5,7	11,5	18,5	24,5	29,3	31,4	29,6	23,3	16,9	10,1	4,8	17,4		1
2 Mean daily maximum temperature	in °C															2
3 Mean daily minimum temperature	in °C															3
4 Absolute maximum temperature	in °C	24	30	32	41	46	50	50	50	41	40	36	27	50		4
5 Absolute minimum temperature	in °C	-24	-20	-14	-3	4	7	11	10	2	-9	-17	-25	-25		5
6 Mean relative humidity	in %	79	74	66	60	47	40	34	33	39	50	62	75	55		6
7 Mean precipitation	in mm	21	23	30	19	10	1	0	0	0	3	10	17	133		7
8 Maximum precipitation	in mm	48	56	59	56	71	10	2	0	1	12	74	84	257		8
9 Minimum precipitation	in mm	0	0	1	0	0	0	0	0	0	0	0	0	62		9
10 Maximum precipitation in 24 h	in mm	17	30	29	24	28	10	2	0	0	25	16	22	30		10
11 Mean number of days with precipitation	> 0,1 mm	8	7	8	5	3	<1	0	0	0	1	3	7	42		11
12 Mean duration of sunshine	in h	140	148	176	235	328	373	388	362	312	361	184	138	3043		12
13 Mean quantity of radiation	in ly / day	174	213	329	429	564	600	608	570	393	344	207	152	382		13
14 Mean potential evaporation	in mm	2	5	22	89	148	203	206	193	111	49	22	5	1035		14
15 Mean windspeed	in m / sec	2,1	3,0	3,8	3,4	2,9	2,7	2,6	2,5	2,0	2,0	2,1	2,0	2,6		15
16 Mean predominent direction of the wind		SW	NE	NE	NE,SW	SW	SW	SW	SW	SW	SW	NE	NE			16

84 Station / Country Taškent / S.S.S.R.

Location 41°16'N / 69°16'E Height above sealevel 479 m Climate symbol: Köppen Csa Troll III,10

		J	F	M	A	M	J	J	A	S	O	N	D	year	P	
1 Mean daily temperature	in °C	-1,1	1,5	7,8	14,7	20,2	25,3	27,4	25,5	19,7	12,7	6,7	1,8	13,5		1
2 Mean daily maximum temperature	in °C	2,8	6,7	11,7	18,3	25,6	30,6	33,3	31,7	26,7	18,3	11,7	6,7	18,9	19	2
3 Mean daily minimum temperature	in °C	-6,1	-2,8	2,8	8,3	13,3	16,7	17,8	15,6	11,1	5,0	1,7	-1,7	6,7	19	3
4 Absolute maximum temperature	in °C	22	26	33	35	42	44	44	43	40	38	31	24	44		4
5 Absolute minimum temperature	in °C	-28	-26	-20	-6	0	4	8	7	0	-11	-22	-30	-30		5
6 Mean relative humidity	in %	74	69	67	60	55	44	40	44	46	56	67	75	58		6
7 Mean precipitation	in mm	49	51	81	58	32	12	4	3	3	23	44	57	417		7
8 Maximum precipitation	in mm	107	110	165	154	89	67	36	22	34	117	148	164	643		8
9 Minimum precipitation	in mm	8	0	6	0	0	0	0	0	0	0	1	6	141		9
10 Maximum precipitation in 24 h	in mm	30	32	40	38	50	33	25	10	19	30	47	37	50		10
11 Mean number of days with precipitation	> 0,1 mm	11	10	11	10	7	4	1	1	1	5	8	10	79		11
12 Mean duration of sunshine	in h	116	125	165	229	312	359	390	371	304	233	156	110	2820		12
13 Mean quantity of radiation	in ly / day	149	182	276	393	558	579	601	533	405	291	165	115	354		13
14 Mean potential evaporation	in mm	0	2	21	61	115	160	180	150	86	42	14	2	833		14
15 Mean windspeed	in m / sec	1,3	1,6	1,7	1,6	1,5	1,4	1,3	1,3	1,4	1,2	1,2	1,2	1,4		15
16 Mean predominent direction of the wind		NE	NE	NE	NE	NE	NE	NE	NE	NE	NE	NE	NE			16

85 Station / Country Chorog / S.S.S.R.

Location 37°30'N / 71°30'E   Height above sealevel 2080 m   Climate symbol: Köppen **Csa**   Troll III,10

		J	F	M	A	M	J	J	A	S	O	N	D	year	P	
1	Mean daily temperature	in °C	-7,9	-5,8	0,8	9,2	14,9	19,0	22,8	22,6	18,3	10,9	3,4	-3,8	8,7		1
2	Mean daily maximum temperature	in °C															2
3	Mean daily minimum temperature	in °C															3
4	Absolute maximum temperature	in °C	10	14	21	29	34	37	38	38	36	30	23	14	38		4
5	Absolute minimum temperature	in °C	-32	-32	-23	-10	-1	2	5	5	0	-14	-18	-27	-32		5
6	Mean relative humidity	in %	70	69	63	51	44	39	35	30	28	28	51	64	48		6
7	Mean precipitation	in mm	31	32	39	42	24	9	3	0	1	11	19	24	235		7
8	Maximum precipitation	in mm	88	174	123	198	101	145	25	11	20	67	84	91	431		8
9	Minimum precipitation	in mm	2	0	2	0	2	0	0	0	0	0	0	0	97		9
10	Maximum precipitation in 24 h	in mm	31	32	29	70	28	16	17	11	12	23	20	28	70		10
11	Mean number of days with precipitation	> 0,1 mm	8	9	9	8	7	3	1	<1	<1	3	5	7	61		11
12	Mean duration of sunshine	in h	93	101	144	168	218	266	299	282	244	188	119	95	2217		12
13	Mean quantity of radiation	in ly / day															13
14	Mean potential evaporation	in mm	0	0	0	41	105	150	186	150	80	24	0	0	736		14
15	Mean windspeed	in m / sec	1,6	1,8	2,6	2,7	2,3	2,6	2,8	2,8	2,4	2,0	1,9	1,8	2,3		15
16	Mean predominent direction of the wind		NE	NE	NE	NE	W	W	W	W	W	W	E	NE			16

86 Station / Country Alma-Ata / S.S.S.R.

Location 43°14'N / 76°56'E   Height above sealevel 848 m   Climate symbol: Köppen **Dfa**   Troll III,10

		J	F	M	A	M	J	J	A	S	O	N	D	year	P	
1	Mean daily temperature	in °C	-8,8	-7,4	-0,1	9,9	15,7	19,8	22,2	21,0	15,3	7,4	-0,8	-6,2	7,3		1
2	Mean daily maximum temperature	in °C	-5,0	-3,3	3,9	13,3	20,0	24,4	27,2	26,7	21,7	12,8	3,9	-1,7	12,2	27	2
3	Mean daily minimum temperature	in °C	-13,9	-12,8	-5,6	3,3	10,0	13,9	15,6	13,9	8,3	1,7	-5,0	-9,4	1,7	19	3
4	Absolute maximum temperature	in °C	11	15	27	33	38	40	42	41	37	33	24	13	42		4
5	Absolute minimum temperature	in °C	-43	-48	-31	-14	-9	-1	5	•0	-5	-18	-40	-39	-48		5
6	Mean relative humidity	in %	74	74	73	59	55	51	55	55	44	45	70	74	60		6
7	Mean precipitation	in mm	26	32	64	89	99	59	35	23	25	46	48	35	581		7
8	Maximum precipitation	in mm	81	68	134	190	209	152	102	75	70	139	113	87	923		8
9	Minimum precipitation	in mm	4	1	12	17	13	3	0	0	0	0	4	4	296		9
10	Maximum precipitation in 24 h	in mm	26	33	35	46	62	54	55	33	44	38	46	33	62		10
11	Mean number of days with precipitation	> 0,1 mm	8	8	11	12	11	10	9	6	5	7	9	9	105		11
12	Mean duration of sunshine	in h	115	119	140	195	242	282	316	296	249	198	129	111	2392		12
13	Mean quantity of radiation	in ly / day	140	168	245	336	471	492	521	453	360	260	138	109	308		13
14	Mean potential evaporation	in mm	0	0	1	47	95	129	148	129	75	31	1	0	654		14
15	Mean windspeed	in m / sec	1,3	1,6	2,0	2,2	2,6	2,7	2,7	2,2	2,0	1,9	1,6	1,4	2,0		15
16	Mean predominent direction of the wind		SE	SE	SE	SE	SE	SE	SE	SE	SE	SE	SE	SE			16

87 Station / Country Semipalatinsk / S.S.S.R.

Location 50°24'N / 80°13'E   Height above sealevel 202 m   Climate symbol: Köppen **BSk**   Troll III,10

		J	F	M	A	M	J	J	A	S	O	N	D	year	P	
1	Mean daily temperature	in °C	-16,2	-15,6	-8,9	3,8	13,9	20,0	22,1	19,9	13,2	4,8	-6,1	-13,5	3,1		1
2	Mean daily maximum temperature	in °C	-13,3	-11,7	-6,1	7,2	20,0	25,0	27,2	25,0	18,9	7,8	-3,9	-9,4	7,2	10	2
3	Mean daily minimum temperature	in °C	-21,7	-20,6	-15,6	-3,3	7,2	12,2	13,9	12,2	6,7	-1,1	-12,2	-16,1	-3,3	10	3
4	Absolute maximum temperature	in °C	6	7	24	31	37	41	42	42	37	28	17	8	42		4
5	Absolute minimum temperature	in °C	-47	-45	-41	-24	-10	-1	5	1	-8	-18	-49	-46	-49		5
6	Mean relative humidity	in %	77	77	78	64	51	51	55	57	59	66	76	77	66		6
7	Mean precipitation	in mm	14	15	17	19	22	30	32	23	21	22	27	22	264		7
8	Maximum precipitation	in mm	56	36	63	62	82	91	120	69	54	71	83	58	418		8
9	Minimum precipitation	in mm	3	0	1	0	3	3	1	0	0	3	3	6	142		9
10	Maximum precipitation in 24 h	in mm	16	13	14	22	26	40	38	26	27	21	30	15	40		10
11	Mean number of days with precipitation	> 0,1 mm	12	9	10	6	8	8	8	8	7	9	13	13	111		11
12	Mean duration of sunshine	in h	113	137	197	238	302	314	321	305	251	153	103	89	2523		12
13	Mean quantity of radiation	in ly / day	115	162	326	315	453	519	502	409	345	177	87	84	290		13
14	Mean potential evaporation	in mm	0	0	0	15	93	133	151	123	68	15	0	0	598		14
15	Mean windspeed	in m / sec	4,1	3,8	4,2	4,0	3,7	3,4	3,1	2,9	3,2	3,6	4,4	4,0	3,7		15
16	Mean predominent direction of the wind		E	E	E	E,W	W	W	W	W	E	E	E	E			16

88 Station / Country Gurjev / S.S.S.R.

Location 47°01'N / 51°51'E   Height above sealevel 23 m   Climate symbol: Köppen **BWk**   Troll III,12

		J	F	M	A	M	J	J	A	S	O	N	D	year	P	
1	Mean daily temperature	in °C	-10,4	-9,4	-2,5	8,2	17,7	22,6	25,4	23,2	16,2	8,2	0,2	-6,2	7,8		1
2	Mean daily maximum temperature	in °C	-4,3	-2,9	4,9	16,9	26,2	29,7	32,8	30,9	24,0	14,3	6,4	-1,4	14,8	8	2
3	Mean daily minimum temperature	in °C	-11,9	-11,7	-4,0	4,1	12,3	16,0	18,5	17,0	10,6	3,2	-2,3	-8,5	3,6	8	3
4	Absolute maximum temperature	in °C	7	15	23	32	37	42	45	45	37	29	19	12	45		4
5	Absolute minimum temperature	in °C	-38	-38	-32	-12	-4	4	8	4	-6	-13	-30	-38	-38		5
6	Mean relative humidity	in %	82	81	78	65	54	52	53	52	58	70	76	84	67		6
7	Mean precipitation	in mm	14	12	12	12	16	19	20	11	8	16	10	14	164		7
8	Maximum precipitation	in mm	49	42	36	56	56	134	114	99	68	69	55	64	321		8
9	Minimum precipitation	in mm	0	0	0	0	0	0	0	0	0	0	0	1	74		9
10	Maximum precipitation in 24 h	in mm	20	11	17	30	27	36	87	45	47	24	20	18	87		10
11	Mean number of days with precipitation	> 0,1 mm	10	8	8	6	5	5	4	4	4	7	9	12	83		11
12	Mean duration of sunshine	in h	74	107	158	241	325	340	343	338	266	192	127	68	2579		12
13	Mean quantity of radiation	in ly / day															13
14	Mean potential evaporation	in mm	0	0	0	35	104	148	189	137	75	28	0	0	696		14
15	Mean windspeed	in m / sec	6,3	6,4	6,4	6,0	6,1	5,4	4,8	4,6	4,5	4,9	5,5	5,6	5,5		15
16	Mean predominent direction of the wind		SE	E	SE	E,SE	SE	W	W	W	NW	SE	SE	SE			16

89 Station / Country Balchaš / S.S.R.

Location 46°54'N/75°00'E   Height above sealevel 423 m   Climate symbol: Köppen BWk   Troll III,12

		J	F	M	A	M	J	J	A	S	O	N	D	year	P	
1 Mean daily temperature	in °C	-15,6	-13,0	-5,7	6,2	16,1	21,3	23,9	21,7	15,0	6,5	-5,4	-11,6	5,1		1
2 Mean daily maximum temperature	in °C															2
3 Mean daily minimum temperature	in °C															3
4 Absolute maximum temperature	in °C	4	8	24	28	34	41	41	32	37	28	17	7	41		4
5 Absolute minimum temperature	in °C	-46	-40	-30	-14	-7	1	8	3	-5	-20	-33	-41	-46		5
6 Mean relative humidity	in %	78	78	76	58	47	44	44	44	48	60	74	80	61		6
7 Mean precipitation	in mm	10	8	10	11	9	19	11	9	4	8	9	12	115		7
8 Maximum precipitation	in mm															8
9 Minimum precipitation	in mm															9
10 Maximum precipitation in 24 h	in mm															10
11 Mean number of days with precipitation	> 0,1 mm	8	6	5	5	4	4	5	3	2	4	7	9	62		11
12 Mean duration of sunshine	in h															12
13 Mean quantity of radiation	in ly/day															13
14 Mean potential evaporation	in mm	0	0	0	29	98	137	158	132	72	24	0	0	650		14
15 Mean windspeed	in m/sec	5,7	5,6	5,2	5,0	5,2	4,9	4,7	4,8	4,6	4,5	4,8	4,9	5,0		15
16 Mean predominent direction of the wind		NE	NE	NE	NE	NE	NE	NE	NE	NE	NE	NE	NE			16

90 Station / Country Fort-Ševčenko / S.S.R.

Location 44°33'N/50°17'E   Height above sealevel -23 m   Climate symbol: Köppen BWk   Troll III,12

		J	F	M	A	M	J	J	A	S	O	N	D	year	P	
1 Mean daily temperature	in °C	-3,6	-2,7	2,2	10,0	17,8	22,7	25,8	24,5	19,4	12,0	4,9	-0,2	11,0		1
2 Mean daily maximum temperature	in °C															2
3 Mean daily minimum temperature	in °C															3
4 Absolute maximum temperature	in °C	14	18	24	32	38	40	43	43	38	30	22	18	43		4
5 Absolute minimum temperature	in °C	-22	-26	-19	-8	2	6	13	10	3	-3	-17	-21	-26		5
6 Mean relative humidity	in %	77	77	73	64	62	63	63	60	58	63	68	73	67		6
7 Mean precipitation	in mm	7	8	8	14	11	17	15	7	14	12	9	8	130		7
8 Maximum precipitation	in mm	60	38	33	67	61	87	71	85	68	58	40	63	290		8
9 Minimum precipitation	in mm	0	0	0	0	0	0	0	0	0	0	0	0	57		9
10 Maximum precipitation in 24 h	in mm															10
11 Mean number of days with precipitation	> 0,1 mm	9	6	7	6	5	4	4	4	5	6	7	10	73		11
12 Mean duration of sunshine	in h	69	110	161	222	318	343	341	263	198	123	78		2563		12
13 Mean quantity of radiation	in ly/day	90	137	248	375	512	501	502	468	393	229	111	87	304		13
14 Mean potential evaporation	in mm	0	0	6	41	100	143	169	145	94	42	12	0	752		14
15 Mean windspeed	in m/sec	8,7	9,5	8,9	7,9	7,4	6,7	6,4	7,2	8,3	8,9	9,8	9,7	8,3		15
16 Mean predominent direction of the wind		SE	SE	SE	SE	N.SE	N	N	N	SE	SE	SE	SE			16

91 Station / Country Kzyl-Orda / S.S.R.

Location 44°46'N/65°32'E   Height above sealevel 129 m   Climate symbol: Köppen BWk   Troll III,12

		J	F	M	A	M	J	J	A	S	O	N	D	year	P	
1 Mean daily temperature	in °C	-9,6	-7,5	0,5	11,2	18,8	23,6	24,6	22,5	15,8	7,6	-0,6	-7,0	8,3		1
2 Mean daily maximum temperature	in °C															2
3 Mean daily minimum temperature	in °C															3
4 Absolute maximum temperature	in °C	15	24	30	36	41	44	48	44	42	34	26	17	48		4
5 Absolute minimum temperature	in °C	-38	-33	-30	-11	-2	3	8	5	-6	-12	-28	-33	-38		5
6 Mean relative humidity	in %	80	78	70	49	39	38	38	39	44	54	69	78	56		6
7 Mean precipitation	in mm	13	15	14	14	11	5	4	3	4	7	10	14	114		7
8 Maximum precipitation	in mm	36	55	43	44	63	34	45	28	30	31	47	45	187		8
9 Minimum precipitation	in mm	1	0	0	0	0	0	0	0	0	0	0	1	46		9
10 Maximum precipitation in 24 h	in mm	16	16	24	25	41	27	35	22	23	20	30	20	41		10
11 Mean number of days with precipitation	> 0,1 mm	8	7	7	5	4	2	1	1	2	3	5	7	52		11
12 Mean duration of sunshine	in h	113	143	204	253	348	374	424	375	328	240	150	112	3062		12
13 Mean quantity of radiation	in ly/day	152	202	338	432	601	615	620	527	420	295	156	87	370		13
14 Mean potential evaporation	in mm	0	0	0	51	111	151	161	134	75	20	0	0	712		14
15 Mean windspeed	in m/sec	5,5	5,7	5,7	6,1	5,5	4,8	4,1	4,1	4,1	4,2	4,5	5,0	4,9		15
16 Mean predominent direction of the wind		NE	NE	NE	NE	NE	NE	NE	NE	NE	NE	NE	NE			16

92 Station / Country Krasnovodsk / S.S.R.

Location 40°02'N/52°59'E   Height above sealevel -10 m   Climate symbol: Köppen BWk   Troll III,12a

		J	F	M	A	M	J	J	A	S	O	N	D	year	P	
1 Mean daily temperature	in °C	2,1	3,0	6,0	12,6	19,3	24,1	27,6	27,5	22,3	15,3	8,2	4,1	14,3		1
2 Mean daily maximum temperature	in °C	3,3	5,6	10,6	16,1	23,9	28,9	32,2	31,7	26,7	19,4	12,2	7,8	18,3	25	2
3 Mean daily minimum temperature	in °C	0,0	1,7	4,4	9,4	16,1	21,1	24,4	23,9	18,9	12,8	6,7	3,3	11,7	15	3
4 Absolute maximum temperature	in °C	18	22	30	34	38	43	44	43	39	32	27	21	44		4
5 Absolute minimum temperature	in °C	-17	-13	-10	-4	8	10	12	14	7	0	-13	-16	-17		5
6 Mean relative humidity	in %	77	75	71	54	56	50	48	48	60	65	74		61		6
7 Mean precipitation	in mm	11	13	15	12	7	2	2	3	3	6	9	11	92		7
8 Maximum precipitation	in mm															8
9 Minimum precipitation	in mm															9
10 Maximum precipitation in 24 h	in mm	17	20	21	29	28	77	46	17	14	26	18	15	46		10
11 Mean number of days with precipitation	> 0,1 mm	7	6	8	6	3	2	1	1	1	3	5	8	51		11
12 Mean duration of sunshine	in h															12
13 Mean quantity of radiation	in ly/day															13
14 Mean potential evaporation	in mm	2	4	20	51	114	164	197	179	110	62	22	6	931		14
15 Mean windspeed	in m/sec	3,4	3,4	3,5	4,0	4,0	4,5	4,6	4,5	3,9	3,2	3,0	3,0	2,7		15
16 Mean predominent direction of the wind		E	E	NW	NW	NW	NW	NW	NW	N	N	E	E			16

93 Station/Country Ašchabad / S.S.R.

Location 37°58'N/58°20'E Height above sealevel 219 m Climate symbol: Köppen BSk Troll III,12a

		J	F	M	A	M	J	J	A	S	O	N	D	year	P	
1 Mean daily temperature	in °C	2,1	4,7	8,8	16,3	23,3	28,6	31,2	29,3	23,5	15,9	7,7	2,8	16,2		1
2 Mean daily maximum temperature	in °C	3,3	8,3	12,8	21,1	22,8	33,3	36,1	35,0	30,0	22,2	13,9	8,3	20,6	19	2
3 Mean daily minimum temperature	in °C	-3,9	-0,6	3,9	9,4	15,6	19,4	21,7	19,4	14,4	7,8	3,3	0,0	9,4	18	3
4 Absolute maximum temperature	in °C	29	33	38	39	46	46	47	46	44	40	33	31	47		4
5 Absolute minimum temperature	in °C	-23	-26	-21	-4	5	8	12	10	3	-5	-15	-22	-26		5
6 Mean relative humidity	in %	78	71	70	57	45	34	32	32	37	48	65	77	54		6
7 Mean precipitation	in mm	22	21	44	38	28	6	2	1	3	11	15	19	210		7
8 Maximum precipitation	in mm	50	61	106	128	71	50	35	28	17	77	71	55	377		8
9 Minimum precipitation	in mm	0	2	10	0	3	0	0	0	0	0	0	0	136		9
10 Maximum precipitation in 24 h	in mm	23	22	55	47	29	49	25	21	13	24	23	24	55		10
11 Mean number of days with precipitation	> 0,1 mm	10	9	10	9	7	2	1	1	1	4	6	9	69		11
12 Mean duration of sunshine	in h	111	122	150	195	289	327	358	353	293	244	159	111	2712		12
13 Mean quantity of radiation	in ly/day	152	190	254	342	490	510	558	536	429	316	174	115	339		13
14 Mean potential evaporation	in mm	0	5	24	61	132	186	206	173	105	38	15	5	950		14
15 Mean windspeed	in m/sec	1,7	2,2	2,2	2,3	2,4	2,6	2,4	2,3	2,0	1,9	1,7	1,7	2,1		15
16 Mean predominent direction of the wind		E	E	NW	NW	NW	NW	NW	NW	E	E	E	E			16

94 Station/Country Jerevan / S.S.R.

Location 40°08'N/44°28'E Height above sealevel 907 m Climate symbol: Köppen Dfa Troll III,7a

		J	F	M	A	M	J	J	A	S	O	N	D	year	P	
1 Mean daily temperature	in °C	-4,0	-1,3	5,4	11,8	17,0	21,1	25,1	24,9	20,1	13,6	6,2	-0,9	11,6		1
2 Mean daily maximum temperature	in °C	1,2	3,3	12,2	17,8	24,4	31,8	32,7	27,8	20,5	13,2	5,9		18,3	8	2
3 Mean daily minimum temperature	in °C	-7,5	-4,9	1,2	6,7	11,3	14,4	18,5	17,9	13,1	7,7	2,5	-1,8	6,5	8	3
4 Absolute maximum temperature	in °C	15	20	24	30	34	39	41	40	39	34	26	16	41		4
5 Absolute minimum temperature	in °C	-31	-23	-17	-7	-1	5	7	7	-2	-7	-14	-24	-31		5
6 Mean relative humidity	in %	78	70	64	55	56	50	45	44	49	60	72	78	60		6
7 Mean precipitation	in mm	23	24	29	42	50	26	13	9	12	26	28	22	304		7
8 Maximum precipitation	in mm	74	84	103	104	181	76	47	68	86	133	79	62	465		8
9 Minimum precipitation	in mm	2	0	3	5	13	0	0	0	0	0	0	0	128		9
10 Maximum precipitation in 24 h	in mm	21	21	34	29	42	31	29	26	39	35	36	28	42		10
11 Mean number of days with precipitation	> 0,1 mm	9	9	8	11	13	8	5	3	3	7	7	8	91		11
12 Mean duration of sunshine	in h	89	118	169	212	283	334	359	352	300	246	144	90	2696		12
13 Mean quantity of radiation	in ly/day	146	190	322	414	574	609	620	577	435	341	174	118	377		13
14 Mean potential evaporation	in mm	0	0	11	50	91	131	163	151	98	52	12	0	759	17	14
15 Mean windspeed	in m/sec	0,8	1,3	1,7	1,9	1,8	2,3	2,9	2,4	1,6	1,0	0,8	0,8	1,6		15
16 Mean predominent direction of the wind		SW	SW	SE	SE	S	NE	NE	NE	NE	SW	E	SW			16

95 Station/Country Dušanbe / S.S.R.

Location 38°35'N/68°47'E Height above sealevel 824 m Climate symbol: Köppen Csa Troll III,7a

		J	F	M	A	M	J	J	A	S	O	N	D	year	P	
1 Mean daily temperature	in °C	1,4	3,2	8,7	14,9	19,6	24,2	26,2	26,8	21,6	14,9	9,5	4,7	14,8		1
2 Mean daily maximum temperature	in °C															2
3 Mean daily minimum temperature	in °C															3
4 Absolute maximum temperature	in °C	21	23	31	36	38	43	43	44	41	36	32	24	44		4
5 Absolute minimum temperature	in °C	-28	-22	-17	-8	2	8	12	9	3	-6	-15	-23	-28		5
6 Mean relative humidity	in %	75	71	67	62	57	43	33	32	35	46	59	70	54		6
7 Mean precipitation	in mm	79	74	108	111	73	19	1	3	1	19	45	71	604		7
8 Maximum precipitation	in mm	141	165	248	229	203	84	23	36	12	144	129	144	969		8
9 Minimum precipitation	in mm	5	29	30	43	5							19	389		9
10 Maximum precipitation in 24 h	in mm	36	45	58	83	68	49	14	30	7	43	40	33	83		10
11 Mean number of days with precipitation	> 0,1 mm	13	14	13	12	10	3	1	<1	<1	3	8	12	90		11
12 Mean duration of sunshine	in h	112	120	146	213	287	332	363	342	295	229	158	116	2713		12
13 Mean quantity of radiation	in ly/day	143	207	307	369	542	603	605	555	420	329	189	130	367		13
14 Mean potential evaporation	in mm	0	3	22	60	100	145	189	163	100	52	20	5	859		14
15 Mean windspeed	in m/sec	1,4	1,8	2,2	2,0	1,7	1,6	1,4	1,5	1,6	1,6	1,6	1,4	1,6		15
16 Mean predominent direction of the wind		E	E	E	E	E	E	E,SW	W	SW	SW	E	E			16

96 Station/Country Fergana / S.S.R.

Location 40°23'N/71°45'E Height above sealevel 578 m Climate symbol: Köppen BSk Troll III,7a

		J	F	M	A	M	J	J	A	S	O	N	D	year	P	
1 Mean daily temperature	in °C	-2,7	0,2	7,5	15,2	20,4	24,5	26,4	24,8	19,0	11,8	5,4	0,6	12,8		1
2 Mean daily maximum temperature	in °C															2
3 Mean daily minimum temperature	in °C															3
4 Absolute maximum temperature	in °C	16	21	30	35	41	43	42	41	38	33	26	19	43		4
5 Absolute minimum temperature	in °C	-28	-24	-24	-5	2	8	10	8	2	-10	-23	-28	-28		5
6 Mean relative humidity	in %	84	83	71	63	52	49	49	49	52	61	73	80	64		6
7 Mean precipitation	in mm	20	16	27	18	19	10	5	3	2	12	19	18	169		7
8 Maximum precipitation	in mm	73	74	116	66	49	40	48	52	26	51	155	80	342		8
9 Minimum precipitation	in mm	1	0	0	1	0	0	0	0	0	0	0	0	68		9
10 Maximum precipitation in 24 h	in mm	24	32	46	33	51	22	18	10	24	37	67	19	67		10
11 Mean number of days with precipitation	> 0,1 mm	6	5	7	6	7	4	2	1	1	3	5	5	52		11
12 Mean duration of sunshine	in h	112	127	158	218	285	331	362	352	298	234	152	99	2728		12
13 Mean quantity of radiation	in ly/day	152	193	288	414	524	570	583	558	423	307	185	102	358		13
14 Mean potential evaporation	in mm	0	0	20	67	118	164	184	153	92	38	9	0	843		14
15 Mean windspeed	in m/sec	1,2	1,5	2,0	2,0	2,2	1,8	1,8	1,6	1,4	1,3	1,3	1,2	1,6		15
16 Mean predominent direction of the wind		SE	SE	SE	W	W	SE	NW	SE	SE	SE	SE				16

97 Station / Country Ulan-Ude / S.S.S.R.

Location 51°48'N/107°26'E Height above sealevel 510 m Climate symbol: Köppen Dwc Troll III,11

		J	F	M	A	M	J	J	A	S	O	N	D	year	P
1 Mean daily temperature	in °C	-25.4	-20.9	-10.6	1.2	8.8	16.2	19.4	16.5	8.8	-0.1	-12.7	-21.9	-1.7	1
2 Mean daily maximum temperature	in °C														2
3 Mean daily minimum temperature	in °C														3
4 Absolute maximum temperature	in °C	0	1	15	28	32	39	39	40	30	23	10	5	40	4
5 Absolute minimum temperature	in °C	-51	-45	-40	-28	-15	-4	1	-4	-11	-28	-38	-49	-51	5
6 Mean relative humidity	in %	75	72	64	54	49	57	65	69	69	68	75	78	66	6
7 Mean precipitation	in mm	6	2	3	6	13	32	69	62	27	8	9	9	246	7
8 Maximum precipitation	in mm	13	8	15	19	47	113	185	153	80	30	43	34	413	8
9 Minimum precipitation	in mm	0	0	0	0	1	9	2	3	1	0	0		153	9
10 Maximum precipitation in 24 h	in mm	7	3	12	12	28	42	92	55	41	12	15	9	92	10
11 Mean number of days with precipitation	> 0,1 mm	8	4	3	4	6	9	12	11	8	6	9	10	90	11
12 Mean duration of sunshine	in h														12
13 Mean quantity of radiation	in ly/day	99	176	304	369	450	447	434	369	264	192	120	62	274	13
14 Mean potential evaporation	in mm	0	0	0	5	63	118	138	109	52	0	0	0	485	14
15 Mean windspeed	in m/sec	1.9	1.9	2.7	3.7	3.8	3.4	2.8	2.6	2.5	2.5	2.6	2.1	2.7	15
16 Mean predominent direction of the wind		W	W	E	NW	NW	NW	NW	NW	NW	E	E	SW		16

98 Station / Country Krasnodar / S.S.S.R.

Location 45°02'N/39°09'E Height above sealevel 33 m Climate symbol: Köppen Cfa Troll IV,7

		J	F	M	A	M	J	J	A	S	O	N	D	year	P
1 Mean daily temperature	in °C	-2.1	-1.1	4.0	10.7	16.5	20.2	22.9	22.5	17.2	11.5	5.0	0.2	10.6	1
2 Mean daily maximum temperature	in °C														2
3 Mean daily minimum temperature	in °C														3
4 Absolute maximum temperature	in °C	21	22	28	36	37	37	39	41	38	34	26	22	41	4
5 Absolute minimum temperature	in °C	-36	-34	-22	-10	-3	3	8	4	-3	-9	-24	-29	-36	5
6 Mean relative humidity	in %	85	83	78	69	69	68	66	66	70	77	84	85	75	6
7 Mean precipitation	in mm	52	52	50	50	58	64	58	45	38	51	58	64	640	7
8 Maximum precipitation	in mm	153	131	143	182	177	158	164	155	200	181	201	175	1020	8
9 Minimum precipitation	in mm	7	6	1	9	4	10	0	2	2	5	4	3	458	9
10 Maximum precipitation in 24 h	in mm	43	34	55	60	99	68	67	62	58	49	51	47	99	10
11 Mean number of days with precipitation	> 0,1 mm	14	13	13	11	11	11	9	8	10	12	14		133	11
12 Mean duration of sunshine	in h	64	80	123	174	239	289	322	294	235	170	95	61	2146	12
13 Mean quantity of radiation	in ly/day	118	148	276	336	512	507	530	498	357	248	138	90	313	13
14 Mean potential evaporation	in mm	0	0	16	46	95	124	153	136	83	47	13	2	715	14
15 Mean windspeed	in m/sec	3.9	4.4	5.3	4.8	3.9	3.5	3.1	3.0	3.2	3.4	3.9	4.2	3.9	15
16 Mean predominent direction of the wind		E	E	E	E	SW	SW	SW	NE	NE	NE	E	E		16

99 Station / Country Soči / S.S.S.R.

Location 43°35'N/39°43'E Height above sealevel 31 m Climate symbol: Köppen Cfa Troll IV,7

		J	F	M	A	M	J	J	A	S	O	N	D	year	P
1 Mean daily temperature	in °C	5.7	5.7	8.4	11.5	16.0	19.9	22.7	23.0	19.7	16.1	11.3	7.9	14.0	1
2 Mean daily maximum temperature	in °C	10.3	10.4	12.5	16.1	20.5	24.1	26.4	26.6	24.5	20.1	16.9	13.2	18.5	8 2
3 Mean daily minimum temperature	in °C	3.3	3.7	5.3	8.7	12.8	16.1	19.1	19.1	16.4	11.7	9.5	6.3	11.0	8 3
4 Absolute maximum temperature	in °C	21	24	30	31	34	35	35	38	36	34	29	23	38	4
5 Absolute minimum temperature	in °C	-14	-14	-11	-2	4	9	11	10	3	-5	-5	-9	-14	5
6 Mean relative humidity	in %	70	71	70	74	78	78	78	76	74	72	70	69	73	6
7 Mean precipitation	in mm	145	126	99	92	71	78	94	84	127	127	143	170	1356	7
8 Maximum precipitation	in mm	354	310	274	273	205	255	269	581	380	350	407	425	2762	8
9 Minimum precipitation	in mm	2	10	15	13	3	8	2	5	3	0	14	8	1100	9
10 Maximum precipitation in 24 h	in mm	145	126	99	92	71	78	94	84	127	127	143	170	170	10
11 Mean number of days with precipitation	> 0,1 mm	15	14	15	14	13	11	10	8	9	11	13	15	148	11
12 Mean duration of sunshine	in h	84	98	128	159	223	283	313	305	252	194	121	94	2253	12
13 Mean quantity of radiation	in ly/day	102	134	189	300	437	495	527	468	315	217	114	84	282	13
14 Mean potential evaporation	in mm	12	12	24	44	80	116	140	130	90	60	31	18	758	14
15 Mean windspeed	in m/sec	2.5	2.8	2.5	2.3	2.0	2.0	1.9	2.0	2.2	2.3	2.4	2.8	2.3	15
16 Mean predominent direction of the wind		NE	NE	SE	SE	SE	NE	NE	NE	NE	NE	NE	NE		16

100 Station / Country Kutaisi / S.S.S.R.

Location 42°16'N/42°38'E Height above sealevel 116 m Climate symbol: Köppen Cfa Troll IV,7

		J	F	M	A	M	J	J	A	S	O	N	D	year	P
1 Mean daily temperature	in °C	4.7	5.5	9.1	12.9	17.7	20.7	22.9	23.4	20.7	16.9	11.5	7.4	14.4	1
2 Mean daily maximum temperature	in °C														2
3 Mean daily minimum temperature	in °C														3
4 Absolute maximum temperature	in °C	22	26	32	34	37	40	40	40	40	35	29	25	40	4
5 Absolute minimum temperature	in °C	-17	-14	-10	-3	2	7	10	10	3	-3	-11	-14	-17	5
6 Mean relative humidity	in %	67	66	64	66	68	71	71	74	70	68	68	68	69	6
7 Mean precipitation	in mm	106	129	100	112	85	105	106	86	118	108	141	139	1333	7
8 Maximum precipitation	in mm	310	324	267	245	168	271	483	205	326	305	397	321	1971	8
9 Minimum precipitation	in mm	20	23	20	10	8	33	16	2	2	25	11		578	9
10 Maximum precipitation in 24 h	in mm	12	14	14	13	12	12	14	12	11	10	12	15	151	10
11 Mean number of days with precipitation	> 0,1 mm														11
12 Mean duration of sunshine	in h														12
13 Mean quantity of radiation	in ly/day														13
14 Mean potential evaporation	in mm	9	14	29	50	94	121	147	138	97	65	29	16	809	31 14
15 Mean windspeed	in m/sec	5.6	5.6	5.9	5.7	4.6	3.7	3.0	3.4	3.6	4.8	7.2	6.7	5.0	15
16 Mean predominent direction of the wind		E	E	E	E	W	W	W	W	E	E	E	E		16

101 Station / Country Blagoveščensk / S.S.S.R.

Location 50°16'N / 127°30'E   Height above sealevel 132 m   Climate symbol: Köppen **Dwb**   Troll III,6

		J	F	M	A	M	J	J	A	S	O	N	D	year	P	
1	Mean daily temperature	in °C	-24,2	-18,5	-9,5	2,4	19,8	17,5	21,5	19,2	12,2	1,8	-11,3	-21,7	0,0		1
2	Mean daily maximum temperature	in °C	-18,3	-12,2	-3,3	7,8	17,2	23,3	27,2	26,1	18,3	8,3	-6,7	-17,2	6,1	7	2
3	Mean daily minimum temperature	in °C	-28,3	-24,4	-15,0	-2,8	4,4	11,7	16,1	14,4	6,1	-3,3	-16,7	-27,2	-5,6	7	3
4	Absolute maximum temperature	in °C	-1	5	14	28	33	38	41	37	34	26	11	4	41		4
5	Absolute minimum temperature	in °C	-42	-45	-36	-16	-8	0	8	4	-6	-25	-33	-41	-45		5
6	Mean relative humidity	in %	72	70	66	58	58	70	76	77	74	65	70	73	69		6
7	Mean precipitation	in mm	5	3	9	21	45	100	120	106	80	26	13	6	534		7
8	Maximum precipitation	in mm	19	15	38	66	143	235	284	325	183	79	39	20	785		8
9	Minimum precipitation	in mm	0	0	0	1	0	29	2	7	13	0	0	0	260		9
10	Maximum precipitation in 24 h	in mm	6	11	29	32	47	67	113	122	48	28	17	8	122		10
11	Mean number of days with precipitation	> 0,1 mm	6	4	4	7	10	15	14	14	13	6	7	7	107		11
12	Mean duration of sunshine	in h	151	185	215	199	215	228	235	220	173	170	148	127	2266		12
13	Mean quantity of radiation	in ly / day	102	176	322	354	415	426	391	360	255	195	120	93	267		13
14	Mean potential evaporation	in mm	0	0	0	15	72	121	148	119	67	0	0	0	542		14
15	Mean windspeed	in m / sec	3,0	3,2	4,7	5,8	5,0	4,0	3,4	3,4	3,6	4,9	3,6	2,9	4,0		15
16	Mean predominent direction of the wind		NW	NW	NW	NW	N,NW	S	S	NW	NW	NW	NW	NW			16

102 Station / Country Chabarovsk / S.S.S.R.

Location 48°31'N / 135°07'E   Height above sealevel 86 m   Climate symbol: Köppen **Dwb**   Troll III,6

		J	F	M	A	M	J	J	A	S	O	N	D	year	P	
1	Mean daily temperature	in °C	-22,7	-17,6	-8,8	-2,8	11,2	17,1	21,0	19,9	13,9	-4,8	-8,0	-18,6	0,0		1
2	Mean daily maximum temperature	in °C															2
3	Mean daily minimum temperature	in °C															3
4	Absolute maximum temperature	in °C	0	6	12	25	31	35	40	36	29	25	15	7	40		4
5	Absolute minimum temperature	in °C	-43	-41	-30	-17	-4	2	5	7	-4	-15	-29	-38	-43		5
6	Mean relative humidity	in %	76	72	68	62	63	72	78	79	77	66	69	72	71		6
7	Mean precipitation	in mm	10	7	12	32	53	74	111	118	82	37	20	13	569		7
8	Maximum precipitation	in mm	37	32	41	107	103	152	301	308	304	93	49	40	834		8
9	Minimum precipitation	in mm	1	0	0	6	15	12	9	30	18	8	1	1	334		9
10	Maximum precipitation in 24 h	in mm	11	7	16	32	47	67	83	85	99	42	21	17	99		10
11	Mean number of days with precipitation	> 0,1 mm	8	6	7	9	12	13	12	14	13	3	8	7	112		11
12	Mean duration of sunshine	in h	165	191	222	195	220	239	245	217	196	189	158	153	2390		12
13	Mean quantity of radiation	in ly / day	127	182	329	366	465	465	409	388	277	202	132	109	287		13
14	Mean potential evaporation	in mm	0	0	0	16	73	113	137	122	72	20	0	0	553		14
15	Mean windspeed	in m / sec	3,1	3,3	4,1	4,6	4,3	3,7	3,5	3,4	4,5	5,4	5,5	4,5	4,1		15
16	Mean predominent direction of the wind		SW	SW	SW	SW	SW	SW	SW	SW	SW	SW	SW	SW			16

103 Station / Country Rudnaja Pristan / S.S.S.R.

Location 44°22'N / 135°51'E   Height above sealevel 7 m   Climate symbol: Köppen **Dwb**   Troll III,6

		J	F	M	A	M	J	J	A	S	O	N	D	year	P	
1	Mean daily temperature	in °C	-12,9	-9,3	-3,6	2,9	6,9	10,9	15,8	18,4	14,1	6,9	-2,4	-10,1	3,1		1
2	Mean daily maximum temperature	in °C															2
3	Mean daily minimum temperature	in °C															3
4	Absolute maximum temperature	in °C	8	12	17	24	30	33	38	36	31	26	18	10	38		4
5	Absolute minimum temperature	in °C	-34	-30	-28	-13	-4	0	3	3	-3	-12	-20	-30	-34		5
6	Mean relative humidity	in %	46	48	58	70	77	86	89	87	77	64	52	47	67		6
7	Mean precipitation	in mm	12	12	27	51	71	85	117	110	125	70	42	20	742		7
8	Maximum precipitation	in mm															8
9	Minimum precipitation	in mm															9
10	Maximum precipitation in 24 h	in mm															10
11	Mean number of days with precipitation	> 0,1 mm															11
12	Mean duration of sunshine	in h	194	196	223	191	186	132	123	141	183	216	184	186	2155		12
13	Mean quantity of radiation	in ly / day															13
14	Mean potential evaporation	in mm	0	0	0	20	53	81	113	119	78	40	0	0	505		14
15	Mean windspeed	in m / sec	6,8	5,3	3,9	3,0	2,7	2,5	2,1	2,3	3,0	3,4	4,2	4,7	3,7		15
16	Mean predominent direction of the wind		W	W	W	W	E	E	E	E	W	W	W	W			16

104 Station / Country Vladivostok / S.S.S.R.

Location 43°07'N / 135°54'E   Height above sealevel 138 m   Climate symbol: Köppen **Dwb**   Troll III,6 (III,8)

		J	F	M	A	M	J	J	A	S	O	N	D	year	P	
1	Mean daily temperature	in °C	-14,7	-10,9	-3,9	4,1	8,9	13,0	17,5	20,0	15,8	8,7	-1,1	-10,5	3,9		1
2	Mean daily maximum temperature	in °C	-10,6	-5,6	0,6	7,8	12,8	17,2	21,7	23,9	20,0	12,8	2,2	-6,7	7,8	14	2
3	Mean daily minimum temperature	in °C	-17,8	-14,4	-7,2	1,1	6,1	11,1	15,6	17,8	12,8	5,0	-4,4	-13,3	1,1	26	3
4	Absolute maximum temperature	in °C	5	10	15	21	30	32	36	32	29	23	18	9	36		4
5	Absolute minimum temperature	in °C	-31	-29	-22	-9	-1	4	8	10	2	-8	-20	-28	-31		5
6	Mean relative humidity	in %	64	64	67	69	77	88	92	88	78	68	62	63	73		6
7	Mean precipitation	in mm	10	13	20	44	69	88	101	145	126	57	31	17	721		7
8	Maximum precipitation	in mm	53	57	78	122	191	271	234	423	361	159	211	83	1076		8
9	Minimum precipitation	in mm	0	0	0	6	11	16	10	27	5	0	0	0	371		9
10	Maximum precipitation in 24 h	in mm	48	36	35	69	71	138	108	153	178	78	127	38	178		10
11	Mean number of days with precipitation	> 0,1 mm	5	4	7	9	13	16	17	14	11	8	6	5	115		11
12	Mean duration of sunshine	in h	192	194	206	186	178	136	125	163	204	205	169	173	2131		12
13	Mean quantity of radiation	in ly / day	167	213	329	354	409	348	322	338	312	254	171	146	280		13
14	Mean potential evaporation	in mm	0	0	0	25	59	88	119	125	86	45	0	0	547		14
15	Mean windspeed	in m / sec	8,1	7,5	6,9	7,0	6,6	6,7	6,4	6,3	6,5	7,3	7,8	8,0	7,1		15
16	Mean predominent direction of the wind		N	N	N	SE	SE	SE	SE	SE	SE	N	N	N			16

105 Station/Country Zonguldak/Turkey

Location 41°27'N/31°48'E Height above sealevel 42 m Climate symbol: Köppen Csb Troll IV.1

		J	F	M	A	M	J	J	A	S	O	N	D	year	P	
1 Mean daily temperature	in °C	5.6	6.4	6.4	10.0	14.7	18.3	20.9	20.6	18.1	15.6	12.0	7.8	13.0	8	1
2 Mean daily maximum temperature	in °C	8.9	9.4	9.4	13.3	18.3	22.2	25.0	24.4	22.2	19.4	15.6	11.1	16.7	8	2
3 Mean daily minimum temperature	in °C	2.2	3.3	3.3	6.7	11.1	14.4	16.7	16.7	13.9	11.7	8.3	4.4	9.4	8	3
4 Absolute maximum temperature	in °C	20.6	21.1	31.7	28.9	36.7	40.6	37.2	40.0	30.6	29.4	27.8	24.4	40.6	8	4
5 Absolute minimum temperature	in °C	-7.7	-2.8	-4.4	0.6	2.8	7.8	12.2	10.0	7.8	3.3	0.0	-7.2	-7.7	8	5
6 Mean relative humidity	in %	78	78	77	80	79	81	78	78	79	80	80	78	79	8	6
7 Mean precipitation	in mm	145	97	104	84	51	53	46	64	84	142	152	157	1179	9	7
8 Maximum precipitation	in mm															8
9 Minimum precipitation	in mm															9
10 Maximum precipitation in 24 h	in mm	36	46	33	124	28	94	46	69	71	81	76	41	124	9	10
11 Mean number of days with precipitation	>1,0 mm	13	12	11	8	7	5	5	6	7	10	12	14	110	13	11
12 Mean duration of sunshine	in h															12
13 Mean quantity of radiation	in ly/day															13
14 Mean potential evaporation	in mm	13	15	21	40	79	110	130	118	82	59	34	18	719	10	14
15 Mean windspeed	in m/sec															15
16 Mean predominent direction of the wind																16

106 Station/Country Samsun/Turkey

Location 41°17'N/36°19'E Height above sealevel 40 m Climate symbol: Köppen Cfa Troll IV.1

		J	F	M	A	M	J	J	A	S	O	N	D	year	P	
1 Mean daily temperature	in °C	6.7	7.0	8.3	11.1	15.6	19.4	22.2	22.5	20.0	17.0	13.1	9.4	14.4	24	1
2 Mean daily maximum temperature	in °C	10.0	10.6	12.2	15.0	19.4	23.3	26.1	23.9	20.6	16.7	12.8		18.3	24	2
3 Mean daily minimum temperature	in °C	3.3	3.3	4.4	7.2	11.7	15.6	18.7	18.3	16.1	13.3	9.4	6.1	10.6	24	3
4 Absolute maximum temperature	in °C	22.2	25.0	32.2	34.4	37.2	35.0	39.4	38.9	34.4	35.0	32.2	24.4	39.4	22	4
5 Absolute minimum temperature	in °C	-6.7	-6.7	-6.7	-2.2	2.2	7.8	10.6	9.4	6.7	3.3	-2.8	-5.0	-6.7	22	5
6 Mean relative humidity	in %	69	72	75	77	79	75	73	72	74	75	72	68	73	19	6
7 Mean precipitation	in mm	74	66	69	58	46	38	38	33	61	81	89	86	739	27	7
8 Maximum precipitation	in mm															8
9 Minimum precipitation	in mm															9
10 Maximum precipitation in 24 h	in mm	66	53	43	46	79	41	91	64	76	84	64	99	99	21	10
11 Mean number of days with precipitation	>1,0 mm	10	10	11	9	8	6	4	4	6	7	8	9	92	18	11
12 Mean duration of sunshine	in h	87	96	112	138	211	276	313	288	195	158	123	93	2090		12
13 Mean quantity of radiation	in ly/day															13
14 Mean potential evaporation	in mm	14	14	22	42	77	112	139	143	96	69	38	20	786	14	14
15 Mean windspeed	in m/sec															15
16 Mean predominent direction of the wind																16

107 Station/Country Trabzon/Turkey

Location 41°00'N/39°43'E Height above sealevel 108 m Climate symbol: Köppen Cfa Troll IV.1

		J	F	M	A	M	J	J	A	S	O	N	D	year	P	
1 Mean daily temperature	in °C	7.2	7.0	7.8	11.1	15.6	19.7	22.5	23.0	20.3	17.5	13.3	9.1	14.5	14	1
2 Mean daily maximum temperature	in °C	10.0	10.0	11.1	14.4	18.9	22.8	25.6	26.1	23.3	20.6	16.1	12.2	17.8	14	2
3 Mean daily minimum temperature	in °C	4.4	3.9	4.4	7.8	12.8	16.7	19.4	20.0	17.2	14.4	10.6	6.1	11.7	14	3
4 Absolute maximum temperature	in °C	21.7	23.3	31.1	31.7	38.3	33.3	32.8	38.3	32.2	32.2	32.8	26.1	38.3	17	4
5 Absolute minimum temperature	in °C	-6.1	-7.2	-5.6	-0.6	4.4	10.0	15.0	13.3	7.2	4.4	0.0	-3.3	-7.2	17	5
6 Mean relative humidity	in %	71	73	75	77	80	79	77	76	76	74	74	70	75	14	6
7 Mean precipitation	in mm	71	69	58	56	43	48	46	41	69	81	101	76	759	19	7
8 Maximum precipitation	in mm	463	360	300	190	250	200	270	370	400	480	450	470	2877	30	8
9 Minimum precipitation	in mm															9
10 Maximum precipitation in 24 h	in mm	79	41	23	41	66	33	61	46	51	97	69	43	97	16	10
11 Mean number of days with precipitation	>1,0 mm	10	9	8	8	8	7	5	8	8	10	10	10	102	16	11
12 Mean duration of sunshine	in h	96	101	115	141	180	231	239	226	153	177	111	105	1875	30	12
13 Mean quantity of radiation	in ly/day															13
14 Mean potential evaporation	in mm	16	14	22	42	77	111	139	137	94	69	38	21	780	11	14
15 Mean windspeed	in m/sec															15
16 Mean predominent direction of the wind																16

108 Station/Country Kars/Turkey

Location 40°36'N/43°05'E Height above sealevel 1.750 m Climate symbol: Köppen Dfb Troll III.4

		J	F	M	A	M	J	J	A	S	O	N	D	year	P	
1 Mean daily temperature	in °C	-12.2	-10.0	-5.0	3.9	10.3	13.6	17.2	17.8	13.1	7.5	0.9	-7.2	4.2	18	1
2 Mean daily maximum temperature	in °C	-6.1	-3.9	1.1	10.0	17.2	21.1	25.0	26.1	21.7	15.0	6.7	-1.7	11.1	18	2
3 Mean daily minimum temperature	in °C	-18.3	-16.1	-11.1	-2.2	3.3	6.1	9.4	9.4	4.4	0.0	-5.0	-12.8	-2.8	18	3
4 Absolute maximum temperature	in °C	5.0	6.7	18.9	23.9	26.7	29.4	34.4	34.4	32.2	25.0	21.1	11.1	34.4	18	4
5 Absolute minimum temperature	in °C	-35.6	-37.2	-33.9	-22.8	-7.2	-1.1	0.6	0.6	-4.4	-17.2	-24.4	-35.0	-37.2	18	5
6 Mean relative humidity	in %	65	68	71	70	69	67	63	60	61	69	72	71	67	18	6
7 Mean precipitation	in mm	28	28	28	43	86	74	53	53	30	41	30	25	519	18	7
8 Maximum precipitation	in mm															8
9 Minimum precipitation	in mm															9
10 Maximum precipitation in 24 h	in mm	13	33	15	28	46	25	41	71	28	23	18	18	71	18	10
11 Mean number of days with precipitation	>1,0 mm	7	7	8	9	15	12	8	7	5	7	6	7	98	18	11
12 Mean duration of sunshine	in h															12
13 Mean quantity of radiation	in ly/day															13
14 Mean potential evaporation	in mm	0	0	0	23	65	92	114	102	74	39	0	0	509	27	14
15 Mean windspeed	in m/sec															15
16 Mean predominent direction of the wind																16

109 Station/Country Bursa/Turkey

Location 40°11'N/29°05'E — Height above sealevel 161 m — Climate symbol: Köppen Csa — Troll IV,1

		J	F	M	A	M	J	J	A	S	O	N	D	year	P
1 Mean daily temperature	in °C	5,3	6,1	8,3	13,1	17,2	21,4	23,9	23,9	20,3	16,4	12,2	7,5	14,8	18
2 Mean daily maximum temperature	in °C	8,9	10,0	13,3	18,9	22,8	27,8	30,6	30,6	26,7	22,2	16,7	11,1	20,0	18
3 Mean daily minimum temperature	in °C	1,7	2,2	3,3	7,2	11,7	15,0	17,2	17,2	13,9	10,6	7,8	3,9	9,4	18
4 Absolute maximum temperature	in °C	21,7	22,2	32,8	33,3	37,2	40,6	41,1	42,8	37,8	35,6	30,6	23,9	42,8	18
5 Absolute minimum temperature	in °C	-12,8	-19,4	-7,7	-1,7	1,7	6,1	8,3	7,8	5,0	0,0	-8,3	-17,8	-19,4	18
6 Mean relative humidity	in %	76	74	69	68	68	62	58	59	65	69	75	74	68	18
7 Mean precipitation	in mm	89	89	74	61	61	33	51	23	48	61	79	97	766	18
8 Maximum precipitation	in mm														
9 Minimum precipitation	in mm														
10 Maximum precipitation in 24 h	in mm	58	53	38	38	48	43	208	56	104	38	43	48	208	18
11 Mean number of days with precipitation	> 1,0 mm	11	11	9	8	7	4	3	2	3	7	9	11	85	18
12 Mean duration of sunshine	in h	87	107	149	186	254	312	357	341	252	186	126	96	2453	
13 Mean quantity of radiation	in ly/day														
14 Mean potential evaporation	in mm	8	9	23	49	89	122	196	136	91	55	26	12	816	14
15 Mean windspeed	in m/sec														
16 Mean predominent direction of the wind															

110 Station/Country Ankara/Turkey

Location 39°57'N/32°53'E — Height above sealevel 861 m — Climate symbol: Köppen Csa — Troll III,10

		J	F	M	A	M	J	J	A	S	O	N	D	year	P
1 Mean daily temperature	in °C	-0,3	1,1	5,0	10,9	16,1	18,6	22,5	22,8	18,3	13,6	8,3	2,2	11,6	26
2 Mean daily maximum temperature	in °C	3,9	5,6	10,6	17,2	22,8	26,7	30,0	30,6	25,6	20,6	13,9	6,1	17,8	26
3 Mean daily minimum temperature	in °C	-4,4	-3,3	-0,6	4,4	9,4	11,7	15,0	15,0	11,1	6,7	2,8	-1,7	5,6	26
4 Absolute maximum temperature	in °C	15,0	17,8	26,7	31,7	34,4	36,7	37,8	37,8	35,6	31,7	25,6	17,2	37,8	26
5 Absolute minimum temperature	in °C	-25,0	-24,4	-16,7	-6,7	-0,6	1,7	6,7	4,4	-1,7	-2,8	-17,8	-25,0	-25,0	26
6 Mean relative humidity	in %	78	76	67	56	53	49	43	40	47	55	67	79	59	16
7 Mean precipitation	in mm	33	30	33	33	48	25	13	10	18	23	30	48	344	24
8 Maximum precipitation	in mm	85	83	90	96	110	77	103	51	96	77	67	121	501	30
9 Minimum precipitation	in mm														
10 Maximum precipitation in 24 h	in mm	20	20	28	28	38	41	48	25	23	20	28	69	69	22
11 Mean number of days with precipitation	> 1,0 mm	8	8	7	7	7	5	2	1	3	5	6	9	68	22
12 Mean duration of sunshine	in h	96	118	174	210	282	336	384	369	291	226	159	96	2741	30
13 Mean quantity of radiation	in ly/day														
14 Mean potential evaporation	in mm	0	0	12	43	77	124	128	119	87	49	17	2	658	17
15 Mean windspeed	in m/sec														
16 Mean predominent direction of the wind															

111 Station/Country Erzurum/Turkey

Location 39°54'N/41°16'E — Height above sealevel 1951 m — Climate symbol: Köppen Dfb — Troll III,4

		J	F	M	A	M	J	J	A	S	O	N	D	year	P
1 Mean daily temperature	in °C	-8,9	-6,7	-2,2	5,0	10,9	14,4	18,6	19,2	15,0	11,4	2,8	-4,7	6,2	16
2 Mean daily maximum temperature	in °C	-4,4	-2,2	1,7	10,0	16,7	21,1	25,6	26,7	22,2	15,0	7,2	-0,6	11,7	16
3 Mean daily minimum temperature	in °C	-13,3	-11,1	-7,7	0,0	5,0	7,8	11,7	11,7	7,8	2,8	-1,7	-8,9	0,6	16
4 Absolute maximum temperature	in °C	5,6	9,4	17,8	21,7	25,6	30,0	33,9	33,9	31,7	26,1	18,9	12,2	33,9	16
5 Absolute minimum temperature	in °C	-30,0	-27,2	-25,0	-18,3	-5,6	1,1	2,8	1,1	-3,9	-12,2	-23,3	-25,6	-30,0	16
6 Mean relative humidity	in %	76	74	75	63	58	52	44	44	48	55	67	70	61	5
7 Mean precipitation	in mm	36	41	51	64	79	53	33	23	28	58	48	28	540	16
8 Maximum precipitation	in mm														
9 Minimum precipitation	in mm														
10 Maximum precipitation in 24 h	in mm	41	23	25	41	23	38	43	25	38	36	28	36	43	16
11 Mean number of days with precipitation	> 1,0 mm	8	8	8	10	11	9	5	4	4	8	8	8	89	16
12 Mean duration of sunshine	in h	105	136	158	198	257	318	363	347	285	236	156	115	2674	
13 Mean quantity of radiation	in ly/day														
14 Mean potential evaporation	in mm	0	0	0	27	71	85	124	118	85	40	7	0	557	14
15 Mean windspeed	in m/sec														
16 Mean predominent direction of the wind															

112 Station/Country Sivas/Turkey

Location 39°44'N/36°59'E — Height above sealevel 1185 m — Climate symbol: Köppen Dsb — Troll III,7a

		J	F	M	A	M	J	J	A	S	O	N	D	year	P
1 Mean daily temperature	in °C	-5,0	-3,1	1,1	8,1	12,5	15,6	18,6	18,9	15,3	10,9	5,6	-1,1	8,1	18
2 Mean daily maximum temperature	in °C	-0,6	1,1	6,1	14,4	19,4	23,3	27,2	27,8	23,9	18,3	10,6	3,3	14,4	18
3 Mean daily minimum temperature	in °C	-9,4	-7,2	-3,9	1,7	5,6	7,8	10,0	10,0	6,7	3,3	0,6	-5,6	1,7	18
4 Absolute maximum temperature	in °C	10,6	13,3	24,4	25,6	31,1	34,4	36,7	36,1	33,9	28,9	23,9	15,6	36,7	18
5 Absolute minimum temperature	in °C	-31,1	-34,4	-23,9	-8,3	-5,6	-0,6	2,8	3,3	-2,8	-20,6	-30,0	-34,4	-34,4	18
6 Mean relative humidity	in %	77	77	72	64	61	57	53	52	54	62	72	76	65	18
7 Mean precipitation	in mm	46	41	41	51	58	30	8	5	20	38	43	41	422	18
8 Maximum precipitation	in mm														
9 Minimum precipitation	in mm														
10 Maximum precipitation in 24 h	in mm	25	20	20	38	38	33	18	10	38	53	41	28	53	18
11 Mean number of days with precipitation	> 1,0 mm	10	8	8	9	9	5	1	1	3	6	7	9	76	18
12 Mean duration of sunshine	in h														
13 Mean quantity of radiation	in ly/day														
14 Mean potential evaporation	in mm	0	0	6	35	79	100	109	116	89	55	16	0	605	14
15 Mean windspeed	in m/sec														
16 Mean predominent direction of the wind															

113 Station/Country Van/Turkey

Location 38°28'N/43°21'E   Height above sealevel 1732 m   Climate symbol: Köppen **Dsb**   Troll **III,7a**

		J	F	M	A	M	J	J	A	S	O	N	D	year	P	
1 Mean daily temperature	in °C	-3,3	-3,3	-0,3	6,1	12,2	17,0	21,1	21,1	17,0	10,9	6,9	-1,1	8,7	9	1
2 Mean daily maximum temperature	in °C	1,1	1,7	4,4	11,1	18,3	23,9	28,3	28,3	24,4	16,7	11,1	3,3	14,4	9	2
3 Mean daily minimum temperature	in °C	-7,7	-8,3	-5,0	1,1	6,1	10,0	13,9	13,9	9,4	5,0	0,6	-5,6	2,8	10	3
4 Absolute maximum temperature	in °C	8,3	13,3	17,2	20,6	26,7	33,3	33,9	35,0	32,8	27,8	16,7	12,8	35,0	9	4
5 Absolute minimum temperature	in °C	-23,9	-22,8	-20,0	-10,0	-3,3	2,8	7,2	5,0	3,9	-13,9	-16,1	-20,6	-23,9	10	5
6 Mean relative humidity	in %	73	73	73	69	60	51	45	41	43	61	69	70	61	10	6
7 Mean precipitation	in mm	56	41	51	58	36	15	5	3	8	51	38	. 33	395	10	7
8 Maximum precipitation	in mm															8
9 Minimum precipitation	in mm															9
10 Maximum precipitation in 24 h	in mm	33	28	23	23	23	18	18	8	18	36	38	20	38	10	10
11 Mean number of days with precipitation	> 1,0 mm	9	8	9	16	8	2	1	1	1	7	7	7	70	10	11
12 Mean duration of sunshine	in h	130	147	161	222	282	339	360	313	234	177	150	133	2648		12
13 Mean quantity of radiation	in ly/day															13
14 Mean potential evaporation	in mm	0	0	0	32	76	106	144	134	87	55	18	0	652	19	14
15 Mean windspeed	in m/sec															15
16 Mean predominent direction of the wind																16

114 Station/Country Izmir/Turkey

Location 38°27'N/27°15'E   Height above sealevel 28 m   Climate symbol: Köppen **Csa**   Troll **IV,1**

		J	F	M	A	M	J	J	A	S	O	N	D	year	P	
1 Mean daily temperature	in °C	8,3	9,2	11,7	15,3	19,7	23,9	27,0	27,0	23,1	18,6	14,4	10,0	17,4	39	1
2 Mean daily maximum temperature	in °C	12,8	13,9	17,2	21,1	26,1	30,6	33,3	33,3	29,4	24,4	19,4	14,4	22,8	39	2
3 Mean daily minimum temperature	in °C	3,9	4,4	6,1	9,4	13,3	17,2	20,6	20,6	16,7	12,8	9,4	5,6	11,7	39	3
4 Absolute maximum temperature	in °C	22,2	22,2	28,9	32,8	41,1	40,6	42,2	41,7	39,4	37,6	31,7	26,1	42,2	39	4
5 Absolute minimum temperature	in °C	-11,1	-11,1	-7,2	-1,1	2,8	10,0	11,1	11,7	5,6	-0,6	-7,2	-6,7	-11,1	39	5
6 Mean relative humidity	in %	69	63	62	59	55	48	42	47	53	60	68	71	58	10	6
7 Mean precipitation	in mm	112	84	76	43	33	15	5	5	20	53	84	122	652	58	7
8 Maximum precipitation	in mm	397	296	193	184	118	42	29	20	58	130	201	358	1117	30	8
9 Minimum precipitation	in mm															9
10 Maximum precipitation in 24 h	in mm	84	76	76	81	43	46	28	43	84	231	81	117	231	28	10
11 Mean number of days with precipitation	> 1,0 mm	10	8	7	5	4	2	<1	<1	2	4	8	10	60	29	11
12 Mean duration of sunshine	in h	124	157	195	252	307	354	388	378	312	239	168	133	3007	30	12
13 Mean quantity of radiation	in ly/day															13
14 Mean potential evaporation	in mm	13	16	51	54	97	148	179	163	104	67	35	18	945	14	14
15 Mean windspeed	in m/sec	2,6	2,1	2,6	2,1	1,5	1,5	2,1	1,0	1,0	1,5	2,1	2,6	2,1	30	15
16 Mean predominent direction of the wind		SE	SE	SE	SE	SE	SE	NW		NE	SE	SE	SE		30	16

115 Station/Country Konya/Turkey

Location 37°51'N/32°30'E   Height above sealevel 1025 m   Climate symbol: Köppen **Csa**   Troll **III,7a**

		J	F	M	A	M	J	J	A	S	O	N	D	year	P	
1 Mean daily temperature	in °C	-0,6	1,4	4,7	10,0	15,3	19,2	22,5	22,0	17,8	13,1	7,2	1,7	11,2	18	1
2 Mean daily maximum temperature	in °C	3,9	6,7	11,1	16,7	22,2	26,7	30,0	29,4	25,6	20,6	13,3	6,1	17,8	18	2
3 Mean daily minimum temperature	in °C	-5,0	-3,9	-1,7	3,3	8,3	11,7	15,0	14,4	10,0	5,6	1,1	-2,8	4,4	18	3
4 Absolute maximum temperature	in °C	16,1	19,4	25,6	30,6	33,9	35,0	37,8	36,7	35,0	31,7	25,6	20,0	37,8	18	4
5 Absolute minimum temperature	in °C	-28,3	-22,8	-16,1	-6,7	-0,6	1,7	6,7	5,0	0,0	-8,3	-18,9	-26,1	-28,3	18	5
6 Mean relative humidity	in %	76	72	62	55	53	48	41	39	47	58	70	78	58	18	6
7 Mean precipitation	in mm	46	33	28	33	36	23	8	3	13	33	36	41	333	18	7
8 Maximum precipitation	in mm															8
9 Minimum precipitation	in mm															9
10 Maximum precipitation in 24 h	in mm	28	74	33	28	33	48	18	15	30	56	43	28	74	18	10
11 Mean number of days with precipitation	> 1,0 mm	7	5	5	5	6	3	1	1	2	4	4	7	50	18	11
12 Mean duration of sunshine	in h	105	144	167	222	295	342	388	375	294	239	177	112	2860		12
13 Mean quantity of radiation	in ly/day															13
14 Mean potential evaporation	in mm	0	3	18	49	87	91	144	134	88	46	20	3	683	14	14
15 Mean windspeed	in m/sec															15
16 Mean predominent direction of the wind																16

116 Station/Country Urfa/Turkey

Location 37°07'N/38°46'E   Height above sealevel 540 m   Climate symbol: Köppen **Csa**   Troll **IV,1**

		J	F	M	A	M	J	J	A	S	O	N	D	year	P	
1 Mean daily temperature	in °C	5,0	7,0	10,3	15,3	21,7	27,2	31,6	30,6	26,4	20,3	13,9	7,5	18,1	13	1
2 Mean daily maximum temperature	in °C	8,9	11,7	15,6	21,7	28,9	35,0	38,9	38,3	33,9	26,7	19,4	11,7	24,4	13	2
3 Mean daily minimum temperature	in °C	1,1	2,2	5,0	8,9	14,4	19,4	23,3	22,8	18,9	13,9	8,3	3,3	11,7	13	3
4 Absolute maximum temperature	in °C	17,2	20,6	28,9	33,3	39,4	42,2	46,1	42,8	41,7	35,6	30,6	20,6	46,1	13	4
5 Absolute minimum temperature	in °C	-10,6	-12,2	-5,6	-1,7	3,9	8,3	8,9	17,2	10,0	1,7	-2,2	-5,6	-12,2	13	5
6 Mean relative humidity	in %	71	67	60	53	41	29	26	30	34	45	59	69	49	13	6
7 Mean precipitation	in mm	119	64	56	41	13	2	2	2	3	23	46	76	447	15	7
8 Maximum precipitation	in mm															8
9 Minimum precipitation	in mm															9
10 Maximum precipitation in 24 h	in mm	53	56	48	61	43	33	30	2	8	30	46	69	69	15	10
11 Mean number of days with precipitation	> 1,0 mm	12	8	8	5	3	<1	<1	0	<1	3	6	7	55	15	11
12 Mean duration of sunshine	in h	155	171	205	258	335	390	409	384	327	285	189	143	3251		12
13 Mean quantity of radiation	in ly/day															13
14 Mean potential evaporation	in mm	3	7	18	51	97	181	216	196	155	78	27	71	1100	9	14
15 Mean windspeed	in m/sec															15
16 Mean predominent direction of the wind																16

117 Station / Country   Adana / Turkey

Location   36°59'N / 35°18'E   Height above sealevel 25 m   Climate symbol: Köppen Csa   Troll IV,1

		J	F	M	A	M	J	J	A	S	O	N	D	year	P	
1	Mean daily temperature	in °C	8.9	10.0	13.1	17.0	21.7	25.3	27.8	28.3	25.9	21.7	16.7	11.1	19.0	21	1
2	Mean daily maximum temperature	in °C	13.9	15.0	18.9	23.3	28.3	31.7	33.9	34.4	32.8	28.9	22.8	16.1	25.0	21	2
3	Mean daily minimum temperature	in °C	3.9	5.0	7.2	10.6	15.0	18.9	21.7	22.2	18.9	14.4	10.6	6.1	12.8	21	3
4	Absolute maximum temperature	in °C	22.2	24.4	30.6	33.9	41.1	42.8	41.1	42.8	42.8	41.7	34.4	26.7	42.8	23	4
5	Absolute minimum temperature	in °C	-7.2	-6.7	-5.0	0.0	7.2	9.4	13.9	15.0	11.1	3.3	0.0	-4.4	-7.6	24	5
6	Mean relative humidity	in %	60	59	56	59	59	61	60	60	55	49	50	60	57	5	6
7	Mean precipitation	in mm	109	102	64	41	51	18	5	5	18	48	61	97	619	31	7
8	Maximum precipitation	in mm	265	378	175	106	196	66	23	32	109	130	265	259	786	30	8
9	Minimum precipitation	in mm															9
10	Maximum precipitation in 24 h	in mm	107	69	81	66	66	43	36	30	64	58	94	122	122	35	10
11	Mean number of days with precipitation	> 1,0 mm	8	7	6	5	4	2	<1	<1	1	4	5	7	51	34	11
12	Mean duration of sunshine	in h	155	168	208	243	310	348	372	363	306	260	192	149	3074	30	12
13	Mean quantity of radiation	in ly / day															13
14	Mean potential evaporation	in mm	6	17	35	66	104	147	180	177	128	83	42	18	1003	14	14
15	Mean windspeed	in m / sec	2.6	2.6	2.6	2.1	2.1	1.5	1.5	1	1.5	2.1	2.6	2.6	2.1	30	15
16	Mean predominant direction of the wind		NE	N	N	N	NE			NE	NE	N	NE	NE		30	16

118 Station / Country   Antalya / Turkey

Location   36°53'N / 30°42'E   Height above sealevel 40 m   Climate symbol: Köppen Csa   Troll IV,1

		J	F	M	A	M	J	J	A	S	O	N	D	year	P	
1	Mean daily temperature	in °C	10.6	11.1	12.8	16.1	20.6	24.7	28.3	27.8	25.0	20.9	16.4	12.2	18.9	18	1
2	Mean daily maximum temperature	in °C	15.0	15.6	17.8	21.1	25.6	30.0	33.9	33.3	30.6	26.7	21.7	16.7	23.9	18	2
3	Mean daily minimum temperature	in °C	6.1	6.7	7.8	11.1	15.6	19.4	22.8	22.2	19.4	15.0	11.1	7.8	13.9	19	3
4	Absolute maximum temperature	in °C	20.6	22.2	27.2	32.8	38.9	41.7	43.3	42.2	42.8	38.9	36.1	22.8	43.3	18	4
5	Absolute minimum temperature	in °C	-4.4	-4.4	-1.1	3.3	6.1	11.7	15.0	13.9	10.6	7.8	0.0	-7.7	-7.7	19	5
6	Mean relative humidity	in %	70	69	65	67	68	64	60	61	59	64	68	70	65	18	6
7	Mean precipitation	in mm	259	175	79	38	33	13	3	3	15	53	119	267	1057	21	7
8	Maximum precipitation	in mm	611	493	287	182	194	79	32	50	99	201	408	635	1645	30	8
9	Minimum precipitation	in mm															9
10	Maximum precipitation in 24 h	in mm	224	96	94	94	119	64	18	8	89	150	182	292	292	19	10
11	Mean number of days with precipitation	> 1,0 mm	11	9	8	4	3	1	<1	<1	1	4	5	11	57	19	11
12	Mean duration of sunshine	in h	149	185	223	255	326	366	397	375	321	267	198	158	3220	30	12
13	Mean quantity of radiation	in ly / day															13
14	Mean potential evaporation	in mm	16	19	31	58	102	146	190	174	127	77	40	21	1001	10	14
15	Mean windspeed	in m / sec	3.1	2.6	2.1	2.1	2.1	2.6	2.6	2.1	2.1	1.5	2.1	2.6	2.1	30	15
16	Mean predominant direction of the wind		N	N	N	NW	NW	NW	NW	NW	NE	NE	N	N		30	16

119 Station / Country   Halab (Aleppo) / Syria

Location   36°14'N / 37°08'E   Height above sealevel 390 m   Climate symbol: Köppen Csa   Troll IV,1

		J	F	M	A	M	J	J	A	S	O	N	D	year	P	
1	Mean daily temperature	in °C	5.6	8.1	10.9	16.4	21.4	25.9	26.7	26.7	24.7	19.7	13.3	7.8	17.3	8	1
2	Mean daily maximum temperature	in °C	10.0	13.3	17.8	23.9	29.4	34.4	36.1	36.1	33.3	27.2	19.4	12.2	24.4	8	2
3	Mean daily minimum temperature	in °C	1.1	2.8	3.9	8.9	13.3	17.2	20.6	20.6	16.1	12.2	7.2	3.3	10.6	8	3
4	Absolute maximum temperature	in °C	17.2	20.6	30.6	33.9	40.6	47.2	46.1	43.3	41.1	37.2	30.0	18.3	47.2	8	4
5	Absolute minimum temperature	in °C	-12.8	-10.0	-7.2	-2.2	0.0	8.9	15.6	15.0	6.7	5.0	-2.8	-7.7	-12.8	8	5
6	Mean relative humidity	in %	71	71	71	70	67	64	62	61	61	64	64	70	66	30	6
7	Mean precipitation	in mm	89	64	38	28	8	3	0	2	2	25	56	84	399	10	7
8	Maximum precipitation	in mm	155	110	154	80	68	10	1	21	20	150	230	163	547	30	8
9	Minimum precipitation	in mm															9
10	Maximum precipitation in 24 h	in mm	38	58	33	28	18	13	0	2	5	18	58	41	58	10	10
11	Mean number of days with precipitation	> 1,0 mm	11	10	7	4	2	<1	0	<1	<1	4	8	10	59	10	11
12	Mean duration of sunshine	in h	149	151	195	225	307	363	369	350	278	254	198	146	2983	30	12
13	Mean quantity of radiation	in ly / day															13
14	Mean potential evaporation	in mm	8	12	24	56	105	154	187	176	131	74	34	13	974	13	14
15	Mean windspeed	in m / sec															15
16	Mean predominant direction of the wind																16

120 Station / Country   Dayr az-Zwar / Syria

Location   35°21'N / 40°09'E   Height above sealevel 213 m   Climate symbol: Köppen BWh   Troll IV,5

		J	F	M	A	M	J	J	A	S	O	N	D	year	P	
1	Mean daily temperature	in °C	6.7	8.9	13.3	18.9	24.7	29.1	33.1	32.2	28.1	21.7	15.0	8.7	20.0	5	1
2	Mean daily maximum temperature	in °C	11.7	14.4	21.1	26.7	33.3	37.2	40.6	40.0	36.1	30.0	22.2	14.4	27.2	5	2
3	Mean daily minimum temperature	in °C	1.7	3.3	5.6	11.1	16.1	21.1	25.6	24.4	20.0	13.3	7.8	2.8	12.8	5	3
4	Absolute maximum temperature	in °C	22.2	22.2	32.8	39.4	40.6	43.9	45.6	45.0	43.9	36.1	32.2	20.0	45.6	5	4
5	Absolute minimum temperature	in °C	-8.9	-7.7	-4.4	2.8	7.8	7.2	19.4	20.0	8.9	6.1	-3.9	-8.3	-8.9	5	5
6	Mean relative humidity	in %	80	73	65	61	45	36	29	38	39	48	51	60	52	2	6
7	Mean precipitation	in mm	41	20	8	20	3	2	0	0	0	5	38	23	160	8	7
8	Maximum precipitation	in mm															8
9	Minimum precipitation	in mm															9
10	Maximum precipitation in 24 h	in mm	97	23	36	30	8	3	0	0	0	10	48	28	97	8	10
11	Mean number of days with precipitation	> 1,0 mm	6	5	3	4	<1	<1	0	0	0	2	5	4	30	8	11
12	Mean duration of sunshine	in h															12
13	Mean quantity of radiation	in ly / day															13
14	Mean potential evaporation	in mm	5	9	26	65	142	190	216	202	153	88	33	9	1138	27	14
15	Mean windspeed	in m / sec															15
16	Mean predominant direction of the wind																16

121 Station / Country Dimashq (Damascus) / Syria

Location 33°30'N/36°20'E Height above sealevel 720 m Climate symbol: Köppen BSh Troll IV,1

		J	F	M	A	M	J	J	A	S	O	N	D	year	P	
1 Mean daily temperature	in °C	7,0	8,9	12,0	16,7	20,9	24,4	26,7	27,5	24,2	19,7	13,9	8,9	17,6	13	1
2 Mean daily maximum temperature	in °C	11,7	13,9	18,3	23,9	28,9	32,8	35,6	37,2	32,8	27,2	19,4	13,3	24,4	13	2
3 Mean daily minimum temperature	in °C	2,2	3,9	5,6	9,4	12,8	16,1	17,2	17,8	15,6	12,2	8,3	4,4	10,6	13	3
4 Absolute maximum temperature	in °C	20,6	30,0	28,3	35,0	38,3	38,9	42,2	45,0	38,9	33,9	30,0	20,6	45,0	9	4
5 Absolute minimum temperature	in °C	-6,1	-5,0	-2,2	0,6	6,7	8,9	12,8	12,8	10,0	5,6	-2,2	-5,0	-6,1	8	5
6 Mean relative humidity	in %	69	66	52	41	35	34	31	34	35	43	60	70	48	5	6
7 Mean precipitation	in mm	43	43	8	13	3	2	2	0	18	10	41	41	224	7	7
8 Maximum precipitation	in mm +	85	85	123	91	50	0	0	0	10	33	50	120	275	30	8
9 Minimum precipitation	in mm															9
10 Maximum precipitation in 24 h	in mm	23	20	8	13	18	3	2	0	13	28	36	43	43	5	10
11 Mean number of days with precipitation	> 1,0 mm	7	6	2	3	1	<1	0	0	2	2	5	5	33	7	11
12 Mean duration of sunshine	in h	161	182	211	264	313	372	403	360	309	257	195	152	3179	30	12
13 Mean quantity of radiation	in ly/day															13
14 Mean potential evaporation	in mm	13	17	29	65	104	138	159	151	111	78	36	19	920	12	14
15 Mean windspeed	in m/sec															15
16 Mean predominant direction of the wind																16

+ No. 7 = Dayr az-Zawr (35°21'N/40°09'E)

122 Station / Country Bayrūt / Lebanon

Location 33°54'N/35°28'E Height above sealevel 34 m Climate symbol: Köppen Csa Troll IV,1

		J	F	M	A	M	J	J	A	S	O	N	D	year	P	
1 Mean daily temperature	in °C	13,6	13,9	15,6	18,3	21,7	24,4	26,2	27,5	26,4	23,9	19,4	15,6	20,5	62	1
2 Mean daily maximum temperature	in °C	16,7	17,2	18,9	22,2	25,6	28,3	30,6	31,7	30,0	27,2	22,8	18,3	23,9	62	2
3 Mean daily minimum temperature	in °C	10,6	10,6	12,2	14,4	17,8	20,6	22,8	23,3	22,8	20,6	16,1	12,8	17,8	62	3
4 Absolute maximum temperature	in °C	25,0	30,6	36,1	37,2	41,7	40,0	36,7	37,2	37,2	38,3	32,8	28,9	41,7	62	4
5 Absolute minimum temperature	in °C	-0,6	-1,1	2,2	6,1	10,0	13,3	17,8	16,7	15,6	11,1	5,6	-1,1	-1,1	62	5
6 Mean relative humidity	in %	71	71	71	70	67	64	62	61	61	64	64	70	66	30	6
7 Mean precipitation	in mm	191	157	94	56	18	3	2	2	5	51	132	185	893	71	7
8 Maximum precipitation	in mm															8
9 Minimum precipitation	in mm															9
10 Maximum precipitation in 24 h	in mm	101	89	91	94	41	61	10	8	53	140	119	99	140	58	10
11 Mean number of days with precipitation	> 1,0 mm	15	12	9	5	2	<1	<1	<1	1	4	8	12	71	20	11
12 Mean duration of sunshine	in h	149	151	195	225	307	363	369	350	276	254	198	146	2983	30	12
13 Mean quantity of radiation	in ly/day															13
14 Mean potential evaporation	in mm	30	27	43	64	104	132	161	164	107	65	43		1075	15	14
15 Mean windspeed	in m/sec	3,1	3,1	2,6	3,1	3,1	3,1	3,1	2,6	2,6	2,1	2,1	2,6	2,6	30	15
16 Mean predominant direction of the wind		SW	SW	SW	SW	SW	SW	SW	SW	SW	SW	SW	SW		30	16

123 Station / Country Hefa / Israel

Location 32°48'N/34°59'E Height above sealevel 10 m Climate symbol: Köppen Csa Troll IV,1

		J	F	M	A	M	J	J	A	S	O	N	D	year	P	
1 Mean daily temperature	in °C	13,9	14,7	16,7	19,7	23,3	25,6	27,5	28,3	27,2	24,4	20,0	15,9	21,4	16	1
2 Mean daily maximum temperature	in °C	18,3	19,4	21,7	25,0	28,3	29,4	31,1	32,2	31,1	29,4	25,6	20,0	26,1	21	2
3 Mean daily minimum temperature	in °C	9,4	10,0	11,7	14,4	18,3	21,7	23,9	24,4	23,3	20,0	15,6	11,7	17,2	16	3
4 Absolute maximum temperature	in °C	26,1	30,6	40,0	42,8	44,4	42,8	35,6	37,2	41,7	41,1	36,1	29,4	44,4	40	4
5 Absolute minimum temperature	in °C	-1,7	-2,8	0,6	4,4	10,0	13,3	17,2	18,3	16,1	8,3	6,7	0,6	-2,8	36	5
6 Mean relative humidity	in %	61	61	59	59	61	67	69	70	67	66	59	61	63	10	6
7 Mean precipitation	in mm	175	109	41	25	5	2	2	2	3	25	94	185	668	30	7
8 Maximum precipitation	in mm															8
9 Minimum precipitation	in mm															9
10 Maximum precipitation in 24 h	in mm	91	130	69	41	38	8	8	3	28	56	137	272	272	59	10
11 Mean number of days with precipitation	> 1,0 mm	13	11	7	4	1	<1	<1	<1	<1	2	7	11	58	59	11
12 Mean duration of sunshine	in h	174	175	291	300	363	372	384	378	324	307	249	208	3525		12
13 Mean quantity of radiation	in ly/day															13
14 Mean potential evaporation	in mm	24	27	45	69	117	151	174	174	148	113	66	32	1138	14	14
15 Mean windspeed	in m/sec															15
16 Mean predominant direction of the wind																16

124 Station / Country Yerushalayim (Jerusalem) / Israel

Location 31°47'N/35°13'E Height above sealevel 757 m Climate symbol: Köppen Csa Troll IV,1

		J	F	M	A	M	J	J	A	S	O	N	D	year	P	
1 Mean daily temperature	in °C	8,9	9,4	13,1	16,4	20,6	22,5	23,9	24,2	23,1	21,1	16,4	11,1	17,6	19	1
2 Mean daily maximum temperature	in °C	12,8	13,3	18,3	22,8	27,2	29,4	30,6	30,6	29,4	27,2	21,1	15,0	22,8	19	2
3 Mean daily minimum temperature	in °C	5,0	5,6	7,8	10,6	13,9	15,6	17,8	17,8	15,0	11,7	7,2	11,7	11,7	19	3
4 Absolute maximum temperature	in °C	25,0	26,7	30,6	38,9	39,4	41,7	37,8	39,4	39,4	36,1	31,1	26,1	41,7	18	4
5 Absolute minimum temperature	in °C	-3,3	-2,8	-1,1	2,2	5,6	8,3	10,0	11,1	10,0	8,3	3,9	-2,8	-3,3	18	5
6 Mean relative humidity	in %	72	66	59	49	40	40	44	47	49	48	58	67	53	9	6
7 Mean precipitation	in mm	132	132	64	28	3	2	0	0	2	13	71	86	533	20	7
8 Maximum precipitation	in mm															8
9 Minimum precipitation	in mm															9
10 Maximum precipitation in 24 h	in mm	99	86	36	38	13	3	0	0	10	23	56	76	99	16	10
11 Mean number of days with precipitation	> 1,0 mm	9	11	3	3	1	<1	0	0	<1	1	4	7	41	16	11
12 Mean duration of sunshine	in h	180	182	231	290	346	412	410	400	339	288	211	185	3474		12
13 Mean quantity of radiation	in ly/day															13
14 Mean potential evaporation	in mm	16	18	29	56	105	118	133	130	102	82	52	24	865	10	14
15 Mean windspeed	in m/sec															15
16 Mean predominant direction of the wind																16

125 Station / Country Elat/Israel

Location 29°33′N/34°57′E — Height above sealevel 2 m — Climate symbol: Köppen BWh — Troll IV,5

#			J	F	M	A	M	J	J	A	S	O	N	D	year	P
1	Mean daily temperature	in °C	15,6	17,0	20,0	24,2	26,1	31,1	32,8	33,1	31,1	27,2	22,0	17,5	24,8	7
2	Mean daily maximum temperature	in °C	21,1	22,8	26,1	30,6	35,6	38,3	39,4	40,0	37,2	33,3	27,8	23,3	31,1	7
3	Mean daily minimum temperature	in °C	10,0	11,1	13,9	17,8	16,7	23,9	26,1	26,1	25,0	21,1	16,1	11,7	18,3	7
4	Absolute maximum temperature	in °C	27,2	30,6	33,9	40,6	44,4	44,4	46,7	45,6	43,3	39,4	36,7	31,1	46,7	4
5	Absolute minimum temperature	in °C	2,8	2,8	8,3	10,6	15,6	20,6	22,2	23,3	21,1	15,6	7,8	5,0	2,8	4
6	Mean relative humidity	in %	50	51	47	38	35	29	25	32	40	45	47	50	41	7
7	Mean precipitation	in mm	2	8	8	5	2	0	0	0	0	2	2	8	37	10
8	Maximum precipitation	in mm														
9	Minimum precipitation	in mm														
10	Maximum precipitation in 24 h	in mm	3	23	18	33	10	0	0	0	0	2	18	15	33	4
11	Mean number of days with precipitation	> 1,0 mm	1	1	2	1	<1	0	0	0	0	<1	1	1	7	10
12	Mean duration of sunshine	in h	229	235	251	273	319	324	347	347	291	282	246	217	3361	30
13	Mean quantity of radiation	in ly/day														
14	Mean potential evaporation	in mm	20	23	52	111	179	194	210	202	172	133	69	30	1395	7
15	Mean windspeed	in m/sec														
16	Mean predominent direction of the wind															

126 Station / Country Ammān/Jordan

Location 31°57′N/35°57′E — Height above sealevel 777 m — Climate symbol: Köppen BSk — Troll IV,1

#			J	F	M	A	M	J	J	A	S	O	N	D	year	P
1	Mean daily temperature	in °C	8,1	10,0	13,1	11,4	21,7	23,3	25,0	25,3	23,9	20,6	15,6	10,3	17,4	25
2	Mean daily maximum temperature	in °C	12,2	13,3	15,6	22,8	28,3	30,6	31,7	32,2	31,1	27,2	21,1	15,0	23,3	25
3	Mean daily minimum temperature	in °C	3,9	4,4	6,1	9,4	13,9	16,1	18,3	18,3	16,7	13,9	10,0	5,6	11,1	25
4	Absolute maximum temperature	in °C	24,4	29,4	32,2	39,4	40,6	42,0	40,0	42,8	39,4	37,2	32,8	25,0	42,8	25
5	Absolute minimum temperature	in °C	-6,1	-5,0	-3,3	1,1	5,0	7,8	13,3	12,8	11,1	6,7	1,7	-3,9	-6,1	25
6	Mean relative humidity	in %	68	65	51	44	34	34	36	38	42	42	53	65	48	30
7	Mean precipitation	in mm	69	74	30	15	5	0	0	0	2	5	33	46	279	25
8	Maximum precipitation	in mm	142	142	169	55	31	tr	tr	tr	15	55	137	180	477	30
9	Minimum precipitation	in mm														
10	Maximum precipitation in 24 h	in mm	61	79	36	41	23	0	0	0	15	41	79	53	79	25
11	Mean number of days with precipitation	> 1,0 mm	8	8	4	3	<1	0	0	0	<1	1	4	5	35	25
12	Mean duration of sunshine	in h	195	210	298	306	353	411	428	400	360	307	240	192	3700	30
13	Mean quantity of radiation	in ly/day														
14	Mean potential evaporation	in mm	12	13	31	57	106	134	146	144	109	80	40	17	889	15
15	Mean windspeed	in m/sec														
16	Mean predominent direction of the wind															

127 Station / Country Ar-Rutbah/Iraq

Location 33°02′N/40°17′E — Height above sealevel 946 m — Climate symbol: Köppen BWh — Troll IV,5

#			J	F	M	A	M	J	J	A	S	O	N	D	year	P
1	Mean daily temperature	in °C	7,0	8,9	11,4	17,8	23,3	27,2	29,7	30,0	26,7	21,4	14,4	8,6	18,3	22
2	Mean daily maximum temperature	in °C	12,8	15,6	20,0	25,6	31,7	36,1	38,3	38,9	36,1	30,0	21,7	14,4	26,7	22
3	Mean daily minimum temperature	in °C	1,1	2,2	2,8	10,0	15,0	18,3	21,1	21,1	17,2	12,8	7,2	2,8	11,1	22
4	Absolute maximum temperature	in °C	25,0	31,7	35,0	38,3	42,2	44,4	46,1	45,6	45,0	37,8	35,0	24,4	46,1	22
5	Absolute minimum temperature	in °C	-14,4	-8,3	-6,1	0,0	5,6	12,2	14,4	15,0	8,9	0,6	-5,0	-8,9	-14,4	22
6	Mean relative humidity	in %	65	59	49	42	35	28	28	26	29	34	52	64	43	19
7	Mean precipitation	in mm	15	15	15	18	8	2	2	2	2	5	10	23	117	22
8	Maximum precipitation	in mm														
9	Minimum precipitation	in mm														
10	Maximum precipitation in 24 h	in mm	20	33	18	43	23	2	2	2	5	30	25	46	46	22
11	Mean number of days with precipitation	> 1,0 mm	3	3	3	3	1	<1	0	0	<1	1	2	3	19	22
12	Mean duration of sunshine	in h														
13	Mean quantity of radiation	in ly/day														
14	Mean potential evaporation	in mm	7	12	29	64	124	168	196	186	141	81	34	11	1053	19
15	Mean windspeed	in m/sec														
16	Mean predominent direction of the wind															

128 Station / Country Baghdād/Iraq

Location 33°20′N/44°24′E — Height above sealevel 34 m — Climate symbol: Köppen BWh — Troll IV,5

#			J	F	M	A	M	J	J	A	S	O	N	D	year	P
1	Mean daily temperature	in °C	9,7	11,7	15,3	21,7	27,8	31,7	33,9	33,9	30,6	24,7	17,8	11,7	22,5	15
2	Mean daily maximum temperature	in °C	15,6	17,8	21,7	29,4	36,1	40,6	43,3	43,3	40,0	33,3	25,0	17,8	30,6	15
3	Mean daily minimum temperature	in °C	3,9	5,6	8,9	13,9	19,4	22,8	24,4	24,4	21,1	16,1	10,6	5,6	15,0	15
4	Absolute maximum temperature	in °C	25,0	30,0	32,2	40,0	44,4	48,3	49,4	48,9	46,7	41,7	34,4	26,1	49,4	15
5	Absolute minimum temperature	in °C	-7,7	-5,0	-2,8	2,8	10,6	14,4	16,7	17,8	10,6	3,9	-1,7	-6,7	-7,7	15
6	Mean relative humidity	in %	68	60	55	49	33	24	22	23	27	36	55	68	43	15
7	Mean precipitation	in mm	23	25	28	13	3	2	2	2	2	3	20	25	148	15
8	Maximum precipitation	in mm														
9	Minimum precipitation	in mm														
10	Maximum precipitation in 24 h	in mm	36	33	56	20	15	2	2	2	2	15	36	33	56	15
11	Mean number of days with precipitation	> 1,0 mm	4	3	4	3	1	0	0	0	0	1	3	5	24	15
12	Mean duration of sunshine	in h	192	204	245	258	301	348	347	353	315	273	213	195	3244	30
13	Mean quantity of radiation	in ly/day														
14	Mean potential evaporation	in mm	5	12	33	78	168	209	222	209	173	117	33	12	1271	11
15	Mean windspeed	in m/sec														
16	Mean predominent direction of the wind															

129 Station/Country Hā'il/Saudi Arabia

Location 27°30'N/42°02'E — Height above sealevel 971 m — Climate symbol: Köppen BWh — Troll IV,5

		J	F	M	A	M	J	J	A	S	O	N	D	year	P	
1	Mean daily temperature	in °C	10,3	11,4	15,9	19,4	25,3	29,7	30,6	30,3	28,0	24,4	18,1	11,4	21,2	2	1
2	Mean daily maximum temperature	in °C	16,7	18,9	23,9	27,8	33,3	38,3	38,3	38,9	36,7	32,8	23,9	17,2	28,9	2	2
3	Mean daily minimum temperature	in °C	3,9	3,9	7,8	11,1	17,2	21,1	22,8	21,7	19,4	16,1	12,2	5,6	13,3	2	3
4	Absolute maximum temperature	in °C	26,7	30,0	32,2	36,1	42,8	42,8	44,4	42,2	40,6	37,2	33,9	23,3	44,4	2	4
5	Absolute minimum temperature	in °C	-2,2	-5,6	0,0	1,7	7,2	11,7	15,6	10,0	15,6	12,2	4,4	0,6	-5,6	2	5
6	Mean relative humidity	in %	47	37	33	27	29	17	13	14	14	20	43	52	29	2	6
7	Mean precipitation	in mm	10	15	13	5	10	0	0	0	0	2	33	13	101	2	7
8	Maximum precipitation	in mm															8
9	Minimum precipitation	in mm															9
10	Maximum precipitation in 24 h	in mm	64	41	20	8	20	0	0	0	0	2	13	13	64	2	10
11	Mean number of days with precipitation	> 1,0 mm	3	1	2	1	2	0	0	0	0	0	5	3	17	2	11
12	Mean duration of sunshine	in h															12
13	Mean quantity of radiation	in ly/day															13
14	Mean potential evaporation	in mm	11	14	39	68	142	186	195	185	152	108	49	15	1164	2	14
15	Mean windspeed	in m/sec															15
16	Mean predominent direction of the wind																16

130 Station/Country Ar-Riyād/Saudi Arabia

Location 24°39'N/46°42'E — Height above sealevel 591 m — Climate symbol: Köppen BWh — Troll IV,5

		J	F	M	A	M	J	J	A	S	O	N	D	year	P	
1	Mean daily temperature	in °C	14,4	15,9	20,6	24,7	30,0	33,3	33,6	32,8	30,6	25,3	20,9	15,3	24,8	3	1
2	Mean daily maximum temperature	in °C	21,1	22,8	27,8	31,7	37,8	41,7	41,7	41,7	38,9	34,4	28,9	21,1	32,2	3	2
3	Mean daily minimum temperature	in °C	7,8	8,9	13,3	17,8	22,2	25,0	25,6	23,9	22,2	16,1	12,8	9,4	17,2	3	3
4	Absolute maximum temperature	in °C	30,0	32,8	38,3	40,0	43,3	45,0	45,0	44,4	43,9	38,3	34,4	30,6	45,0	3	4
5	Absolute minimum temperature	in °C	-7,2	-1,7	0,6	2,2	15,0	19,4	19,4	16,7	17,2	10,0	1,7	0,0	-7,2	3	5
6	Mean relative humidity	in %	57	50	51	49	41	39	26	27	33	36	47	64	43	3	6
7	Mean precipitation	in mm	3	20	23	25	10	2	0	2	0	0	2	2	89	3	7
8	Maximum precipitation	in mm															8
9	Minimum precipitation	in mm															9
10	Maximum precipitation in 24 h	in mm	5	58	61	51	18	2	0	2	0	0	2	2	61	3	10
11	Mean number of days with precipitation	> 1,0 mm	1	1	3	4	1	0	0	<1	0	0	0	0	11	3	11
12	Mean duration of sunshine	in h															12
13	Mean quantity of radiation	in ly/day															13
14	Mean potential evaporation	in mm	17	22	60	111	185	204	210	197	169	118	55	21	1367		14
15	Mean windspeed	in m/sec															15
16	Mean predominent direction of the wind																16

131 Station/Country Juddah (Jidda)/Saudi Arabia

Location 21°28'N/39°10'E — Height above sealevel 6 m — Climate symbol: Köppen BWh — Troll V,5

		J	F	M	A	M	J	J	A	S	O	N	D	year	P	
1	Mean daily temperature	in °C	23,9	23,6	24,4	27,0	29,2	30,0	31,7	32,0	30,3	28,9	27,2	24,7	27,7	5	1
2	Mean daily maximum temperature	in °C	28,9	28,9	29,4	32,8	35,0	36,1	37,2	37,2	35,6	35,0	32,8	30,0	33,3	5	2
3	Mean daily minimum temperature	in °C	18,9	18,3	19,4	21,1	23,3	23,9	26,1	26,7	25,0	22,8	21,7	19,4	22,2	5	3
4	Absolute maximum temperature	in °C	33,3	35,0	38,3	40,0	42,2	47,2	42,2	42,2	40,6	40,6	40,6	33,9	47,2	5	4
5	Absolute minimum temperature	in °C	9,4	11,1	12,8	12,2	12,8	19,4	21,1	22,8	21,1	20,0	17,2	10,0	9,4	5	5
6	Mean relative humidity	in %	58	52	52	54	53	56	53	55	63	61	57	55	56	5	6
7	Mean precipitation	in mm	5	2	2	2	0	2	2	2	2	25	30	76	5		7
8	Maximum precipitation	in mm	14	tr	4	16	20	tr	tr	tr	tr	tr	83	19	96		8
9	Minimum precipitation	in mm															9
10	Maximum precipitation in 24 h	in mm	13	3	3	3	2	0	2	2	2	2	56	140	140	5	10
11	Mean number of days with precipitation	> 1,0 mm	<1	<1	<1	<1	0	0	0	0	0	0	2	1	7	5	11
12	Mean duration of sunshine	in h															12
13	Mean quantity of radiation	in ly/day															13
14	Mean potential evaporation	in mm	67	59	106	144	170	177	191	188	165	152	126	92	1637	10	14
15	Mean windspeed	in m/sec	4,6	4,1	4,1	4,1	4,1	4,1	3,6	3,6	3,6	3,1	3,1	3,6	3,6		15
16	Mean predominent direction of the wind																16

132 Station/Country Kamarān Island/P.D.R. of Yemen

Location 15°20'N/42°37'E — Height above sealevel 6 m — Climate symbol: Köppen BWh — Troll V,5

		J	F	M	A	M	J	J	A	S	O	N	D	year	P	
1	Mean daily temperature	in °C	25,6	25,9	27,5	28,9	31,4	32,5	33,1	32,8	32,5	30,9	28,1	26,1	29,6	26	1
2	Mean daily maximum temperature	in °C	27,8	28,3	30,0	31,7	35,0	36,1	36,7	36,1	36,1	33,9	30,6	28,3	32,8	26	2
3	Mean daily minimum temperature	in °C	23,3	23,3	25,0	26,1	27,8	28,9	29,4	29,4	28,9	27,8	25,6	23,9	26,7	26	3
4	Absolute maximum temperature	in °C	31,1	31,7	34,4	36,7	38,9	40,0	40,6	39,4	40,0	38,9	34,4	32,2	40,6	26	4
5	Absolute minimum temperature	in °C	18,9	19,4	21,1	22,8	23,3	23,9	22,2	22,2	23,3	22,8	20,0	20,0	18,9	26	5
6	Mean relative humidity	in %	74	71	70	68	63	61	58	61	65	62	69	73	66	13	6
7	Mean precipitation	in mm	5	5	3	3	3	2	13	18	3	3	10	23	91	21	7
8	Maximum precipitation	in mm															8
9	Minimum precipitation	in mm															9
10	Maximum precipitation in 24 h	in mm	3	18	8	23	10	8	28	76	3	25	43	48	76	23	10
11	Mean number of days with precipitation	> 1,0 mm	<1	<1	<1	<1	<1	<1	2	1	<1	<1	<1	2	14	13	11
12	Mean duration of sunshine	in h															12
13	Mean quantity of radiation	in ly/day															13
14	Mean potential evaporation	in mm	112	119	145	164	189	189	199	192	178	171	142	127	1927	8	14
15	Mean windspeed	in m/sec															15
16	Mean predominent direction of the wind																16

133 Station/Country Riyan/P.D.R. of Yemen

Location 14°39'N/49°23'E Height above sealevel 25 m Climate symbol: Köppen BWh Troll V,5

		J	F	M	A	M	J	J	A	S	O	N	D	year	P	
1 Mean daily temperature	in °C	23,6	24,2	25,3	27,2	28,9	30,6	29,2	28,6	28,9	27,6	25,3	24,2	27,0	13	1
2 Mean daily maximum temperature	in °C	27,8	28,3	29,4	31,1	32,8	34,4	33,3	32,8	32,2	31,1	30,0	28,3	31,1	13	2
3 Mean daily minimum temperature	in °C	19,4	20,0	21,1	23,3	25,0	26,7	25,0	24,4	25,6	22,2	20,8	20,0	22,8	13	3
4 Absolute maximum temperature	in °C	30,6	33,3	36,1	33,9	40,0	43,9	36,7	37,2	31,7	38,9	34,4	32,8	43,9	13	4
5 Absolute minimum temperature	in °C	13,9	15,0	14,4	18,3	20,6	22,2	20,0	19,4	21,7	17,2	16,1	15,6	13,9	13	5
6 Mean relative humidity	in %	71	69	75	76	79	77	75	76	81	78	70	75	6		6
7 Mean precipitation	in mm	8	1	15	5	2	3	3	3	2	2	18	8	67	13	7
8 Maximum precipitation	in mm															8
9 Minimum precipitation	in mm															9
10 Maximum precipitation in 24 h	in mm	23	8	91	18	5	20	25	18	2	3	104	18	104	13	10
11 Mean number of days with precipitation	> 1,0 mm	1	<1	<1	<1	<1	<1	<1	<1	<1	<1	1	1	12	13	11
12 Mean duration of sunshine	in h															12
13 Mean quantity of radiation	in ly/day															13
14 Mean potential evaporation	in mm	81	88	120	146	171	180	177	166	158	136	108	95	1626		14
15 Mean windspeed	in m/sec															15
16 Mean predominent direction of the wind																16

134 Station/Country Khormaksar/P.D.R. Yemen

Location 12°50'N/45°01'E Height above sealevel 7 m Climate symbol: Köppen BWh Troll V,5

		J	F	M	A	M	J	J	A	S	O	N	D	year	P	
1 Mean daily temperature	in °C	25,0	25,6	27,2	28,3	30,6	32,8	32,2	31,7	32,0	28,6	26,4	25,6	28,8	6	1
2 Mean daily maximum temperature	in °C	27,8	28,3	30,0	31,7	33,9	36,7	36,1	35,6	35,6	32,8	30,0	28,3	32,2	6	2
3 Mean daily minimum temperature	in °C	22,2	22,8	24,4	25,0	27,2	28,9	28,3	27,8	28,3	24,4	22,8	25,6	25,6	6	3
4 Absolute maximum temperature	in °C	30,6	30,0	35,0	37,2	39,4	41,1	40,6	38,3	38,3	37,8	32,8	30,6	41,1	6	4
5 Absolute minimum temperature	in °C	16,1	17,2	19,4	20,0	23,9	26,1	22,8	23,3	25,0	18,9	18,3	16,7	16,1	6	5
6 Mean relative humidity	in %	71	72	74	75	75	64	63	64	67	68	69	69	69	6	6
7 Mean precipitation	in mm	5	2	5	2	2	2	5	3	2	2	2	5	37	6	7
8 Maximum precipitation	in mm															8
9 Minimum precipitation	in mm															9
10 Maximum precipitation in 24 h	in mm	10	5	25	2	2	2	8	8	2	8	2	8	25	6	10
11 Mean number of days with precipitation	> 0,1 mm	1	<1	<1	<1	<1	<1	1	<1	<1	<1	<1	2	13	6	11
12 Mean duration of sunshine	in h	233	196	217	240	279	300	295	279	270	295	285	248	3137		12
13 Mean quantity of radiation	in ly/day															13
14 Mean potential evaporation	in mm	103	101	145	156	182	189	191	185	177	155	124	109	1817	6	14
15 Mean windspeed	in m/sec															15
16 Mean predominent direction of the wind																16

135 Station/Country Barīm Island/P.D.R. of Yemen

Location 12°39'N/43°24'E Height above sealevel 27 m Climate symbol: Köppen BWh Troll V,5

		J	F	M	A	M	J	J	A	S	O	N	D	year	P	
1 Mean daily temperature	in °C	26,4	26,7	27,8	29,2	31,7	33,1	33,6	33,3	32,5	30,3	28,3	27,0	30,8	8	1
2 Mean daily maximum temperature	in °C	28,9	29,4	30,6	32,2	35,0	36,7	37,2	36,7	36,1	33,3	31,1	29,4	32,8	8	2
3 Mean daily minimum temperature	in °C	23,9	23,9	25,0	26,1	28,3	29,4	30,0	30,0	28,9	27,2	25,6	24,4	27,2	8	3
4 Absolute maximum temperature	in °C	31,7	32,2	32,8	35,0	39,4	40,6	41,1	41,1	40,6	37,8	33,3	31,7	41,1	9	4
5 Absolute minimum temperature	in °C	21,7	21,7	21,1	23,3	26,1	25,6	26,1	23,3	23,3	23,3	22,8	21,1	21,1	9	5
6 Mean relative humidity	in %	71	72	73	74	72	71	69	68	74	72	71	73	72	4	6
7 Mean precipitation	in mm	5	3	5	3	2	2	3	5	13	3	3	3	50	9	7
8 Maximum precipitation	in mm															8
9 Minimum precipitation	in mm															9
10 Maximum precipitation in 24 h	in mm	18	13	33	10	<1	28	8	10	28	8	33	43	43	9	10
11 Mean number of days with precipitation	> 1,0 mm	<1	<1	1	<1	<1	<1	<1	1	<1	<1	<1	<1	12	9	11
12 Mean duration of sunshine	in h															12
13 Mean quantity of radiation	in ly/day															13
14 Mean potential evaporation	in mm	114	108	143	158	183	185	192	188	176	164	140	131	1882	19	14
15 Mean windspeed	in m/sec															15
16 Mean predominent direction of the wind																16

136 Station/Country Salālah/Oman

Location 17°03'N/54°06'E Height above sealevel 17 m Climate symbol: Köppen BWh Troll V,5

		J	F	M	A	M	J	J	A	S	O	N	D	year	P	
1 Mean daily temperature	in °C	22,5	23,3	25,3	27,0	28,6	28,9	25,9	25,3	26,1	25,6	25,0	23,9	25,6	12	1
2 Mean daily maximum temperature	in °C	27,2	27,8	30,0	31,1	32,2	31,7	27,8	27,2	28,9	30,6	30,0	28,3	29,4	12	2
3 Mean daily minimum temperature	in °C	17,8	18,9	20,6	22,2	25,0	26,1	23,9	23,3	23,3	20,8	20,0	19,4	21,7	12	3
4 Absolute maximum temperature	in °C	32,2	34,4	38,3	40,0	42,2	47,2	32,2	30,6	31,7	37,2	37,8	34,4	47,2	12	4
5 Absolute minimum temperature	in °C	11,7	11,7	15,6	17,2	20,6	22,8	21,1	21,1	18,9	16,1	15,6	13,9	11,7	12	5
6 Mean relative humidity	in %	59	59	63	67	76	81	90	90	80	66	60	58	71	6	6
7 Mean precipitation	in mm	2	2	2	2	2	5	28	25	3	13	2	8	94	12	7
8 Maximum precipitation	in mm															8
9 Minimum precipitation	in mm															9
10 Maximum precipitation in 24 h	in mm	5	5	10	5	2	8	8	8	5	43	15	28	43	12	10
11 Mean number of days with precipitation	> 1,0 mm	1	<1	<1	<1	<1	1	11	11	1	<1	<1	1	32	12	11
12 Mean duration of sunshine	in h															12
13 Mean quantity of radiation	in ly/day															13
14 Mean potential evaporation	in mm	76	80	118	144	161	173	148	129	134	122	109	91	1485	18	14
15 Mean windspeed	in m/sec															15
16 Mean predominent direction of the wind																16

Station / Country Al-Masïrah / Oman

Location 20°41'N / 58°54'E Height above sealevel 16 m Climate symbol: Köppen BWh Troll V,5

		J	F	M	A	M	J	J	A	S	O	N	D	year	P	
1 Mean daily temperature	in °C	22,0	22,5	25,3	28,3	30,6	30,0	27,2	26,4	26,1	26,7	25,0	23,3	26,1	11	1
2 Mean daily maximum temperature	in °C	25,6	26,1	29,4	33,3	35,6	34,4	31,1	30,0	30,0	31,1	28,9	26,7	30,0	11	2
3 Mean daily minimum temperature	in °C	18,3	18,9	21,1	23,3	25,6	25,6	23,3	22,8	22,2	22,2	21,1	20,0	22,2	12	3
4 Absolute maximum temperature	in °C	31,1	32,8	37,2	40,0	42,8	43,3	37,2	37,8	36,1	37,2	35,0	31,7	43,3	11	4
5 Absolute minimum temperature	in °C	12,2	11,7	15,0	16,7	21,7	22,2	20,6	20,0	18,9	16,7	16,7	14,4	11,7	12	5
6 Mean relative humidity	in %	66	67	66	63	65	70	76	77	77	69	69	67	69	6	6
7 Mean precipitation	in mm	2	2	5	0	2	2	2	2	2	2	2	10	33	12	7
8 Maximum precipitation	in mm															8
9 Minimum precipitation	in mm															9
10 Maximum precipitation in 24 h	in mm	10	2	41	0	5	10	3	2	2	15	8	53	53	12	10
11 Mean number of days with precipitation	> 1,0 mm	<1	<1	<1	0	<1	<1	<1	<1	0	<1	<1	<1	10	12	11
12 Mean duration of sunshine	in h															12
13 Mean quantity of radiation	in ly / day															13
14 Mean potential evaporation	in mm	64	69	117	159	187	180	162	150	136	137	107	83	1551	16	14
15 Mean windspeed	in m / sec															15
16 Mean predominent direction of the wind																16

Station / Country Masqat / Oman

Location 23°37'N / 58°35'E Height above sealevel 5 m Climate symbol: Köppen BWh Troll V,5

		J	F	M	A	M	J	J	A	S	O	N	D	year	P	
1 Mean daily temperature	in °C	22,0	22,2	25,3	28,9	33,3	34,4	33,3	31,1	31,1	30,3	26,4	23,0	28,4	23	1
2 Mean daily maximum temperature	in °C	25,0	25,0	28,3	32,2	36,7	37,8	36,1	33,3	33,9	30,0	26,1		31,7	24	2
3 Mean daily minimum temperature	in °C	18,9	19,4	22,2	25,6	30,0	31,1	30,6	28,9	28,3	26,7	22,8	20,0	25,6	23	3
4 Absolute maximum temperature	in °C	30,6	32,2	41,7	40,6	44,4	46,7	45,0	42,2	41,7	40,6	35,6	33,3	46,7	24	4
5 Absolute minimum temperature	in °C	10,6	11,7	16,7	18,9	23,9	25,6	25,0	23,9	22,8	20,6	16,7	15,6	10,6	23	5
6 Mean relative humidity	in %	72	73	71	66	59	72	77	81	76	72	71	71	72	7	6
7 Mean precipitation	in mm	28	18	10	10	2	3	2	2	0	3	10	18	106	38	7
8 Maximum precipitation	in mm	143	99	65	98	4	64	8	15	0	44	77	116	285		8
9 Minimum precipitation	in mm															9
10 Maximum precipitation in 24 h	in mm	79	56	43	51	5	61	5	10	0	38	53	56	79	38	10
11 Mean number of days with precipitation	> 0,25 mm	2	1	1	1	<1	<1	<1	<1	0	<1	1	2	14	38	11
12 Mean duration of sunshine	in h	295	277	279	333	360	357	285	322	321	326	288	267	3710		12
13 Mean quantity of radiation	in ly / day															13
14 Mean potential evaporation	in mm	51	51	105	161	202	203	205	186	169	158	119	67	1677	4	14
15 Mean windspeed	in m / sec															15
16 Mean predominent direction of the wind																16

Station / Country Ash-Shāriqah (Sharjah) / United Arab Emirates

Location 25°20'N / 55°24'E Height above sealevel 5 m Climate symbol: Köppen BWh Troll V,4

		J	F	M	A	M	J	J	A	S	O	N	D	year	P	
1 Mean daily temperature	in °C	17,8	18,9	21,1	24,2	28,1	30,6	32,8	33,6	31,1	27,5	24,2	20,0	25,8	11	1
2 Mean daily maximum temperature	in °C	23,3	23,9	26,7	30,0	33,9	36,1	37,8	39,4	37,2	33,3	30,6	25,6	31,7	11	2
3 Mean daily minimum temperature	in °C	12,2	13,9	15,6	18,3	22,2	25,0	27,8	27,8	25,0	21,7	17,8	14,4	20,0	11	3
4 Absolute maximum temperature	in °C	29,4	32,8	40,0	39,4	42,8	44,4	47,8	45,0	40,0	36,1	31,1		47,8	11	4
5 Absolute minimum temperature	in °C	2,8	7,8	7,8	11,7	16,1	19,4	22,8	22,8	20,6	17,8	12,2	8,3	2,8	11	5
6 Mean relative humidity	in %	71	72	68	65	62	65	64	65	69	70	69	72	68	10	6
7 Mean precipitation	in mm	23	23	10	5	0	0	0	0	0	0	10	30	101	12	7
8 Maximum precipitation	in mm	59	106	48	11	6	0	1	<1	<1	2	113	63	229		8
9 Minimum precipitation	in mm															9
10 Maximum precipitation in 24 h	in mm	33	64	23	30	0	0	0	0	0	0	109	38	109	10	10
11 Mean number of days with precipitation	> 0,1 mm	2	2	1	<1	0	0	0	0	0	0	<1	2	9	12	11
12 Mean duration of sunshine	in h															12
13 Mean quantity of radiation	in ly / day															13
14 Mean potential evaporation	in mm	31	37	58	100	171	185	205	200	170	144	86	47	1432		14
15 Mean windspeed	in m / sec	3,1	3,1	3,1	3,1	3,1	3,1	3,1	3,1	2,6	2,6	2,6	2,6	3,1		15
16 Mean predominent direction of the wind																16

Station / Country Al-Manāmah / Bahrain

Location 26°12'N / 50°30'E Height above sealevel 5 m Climate symbol: Köppen BWh Troll IV,5

		J	F	M	A	M	J	J	A	S	O	N	D	year	P	
1 Mean daily temperature	in °C	17,0	18,1	20,6	25,0	29,4	31,7	33,3	33,6	31,4	28,1	23,7	18,6	26,5	16	1
2 Mean daily maximum temperature	in °C	20,0	21,1	23,9	28,9	33,3	35,6	37,2	37,8	35,6	32,2	27,8	21,7	29,4	16	2
3 Mean daily minimum temperature	in °C	13,9	15,0	17,2	21,1	25,6	27,8	29,4	29,4	27,2	23,9	20,6	15,6	22,2	16	3
4 Absolute maximum temperature	in °C	29,4	34,4	35,0	40,6	42,2	43,9	44,4	45,0	44,4	39,4	36,1	31,1	45,0	16	4
5 Absolute minimum temperature	in °C	5,0	7,2	10,6	13,3	18,9	21,1	23,9	23,9	21,7	18,9	14,4	8,9	5,0	16	5
6 Mean relative humidity	in %	78	77	75	71	67	67	68	70	70	73	75	81	73	9	6
7 Mean precipitation	in mm	8	18	13	8	2	0	0	0	0	0	18	18	85	16	7
8 Maximum precipitation	in mm	32	69	44	65	7	0	0	0	0	<1	147	63	169		8
9 Minimum precipitation	in mm															9
10 Maximum precipitation in 24 h	in mm	30	33	38	5	0	0	0	0	0		71	43	71	16	10
11 Mean number of days with precipitation	> 0,25 mm	1	2	1	1	<1	0	0	0	0	0	1	2	8	16	11
12 Mean duration of sunshine	in h	183	193	250	264	329	396	375	372	360	319	231	198	3470		12
13 Mean quantity of radiation	in ly / day															13
14 Mean potential evaporation	in mm	24	29	52	111	182	197	208	200	174	146	87	37	1447	4	14
15 Mean windspeed	in m / sec	5,1	5,7	6,2	5,2	5,7	6,4	5,4	4,9	4,6	4,1	4,4	4,9	5,4		15
16 Mean predominent direction of the wind																16

141 Station/Country Al-Kuwayt/Kuwait

Location 29°21'N/48°00'E Height above sealevel 5 m Climate symbol: Köppen BWh Troll IV.5

		J	F	M	A	M	J	J	A	S	O	N	D	year	P	
1 Mean daily temperature	in °C	12,8	14,4	18,6	24,2	29,7	32,2	34,7	35,0	32,5	27,8	20,6	15,0	24,6	14	1
2 Mean daily maximum temperature	in °C	16,1	18,3	22,2	28,3	34,4	36,7	39,4	40,0	37,8	32,8	25,0	18,3	29,4	14	2
3 Mean daily minimum temperature	in °C	9,4	10,6	15,0	20,0	25,0	27,8	30,0	30,0	27,2	22,8	16,7	11,7	20,8	15	3
4 Absolute maximum temperature	in °C	27,8	25,6	32,2	39,4	42,8	48,3	47,8	46,1	47,2	40,6	37,8	26,1	48,3	16	4
5 Absolute minimum temperature	in °C	0,3	2,2	4,4	12,2	15,6	22,2	25,6	20,0	19,4	13,9	6,1	2,2	0,3	16	5
6 Mean relative humidity	in %	69	65	67	61	61	56	43	48	52	62	63	71	60	3	6
7 Mean precipitation	in mm	23	23	28	5	2	0	0	0	0	3	15	28	127	10	7
8 Maximum precipitation	in mm															8
9 Minimum precipitation	in mm															9
10 Maximum precipitation in 24 h	in mm	25	43	38	13	8	0	0	0	0	25	56	30	56	10	10
11 Mean number of days with precipitation	> 0,25 mm	2	2	2	<1	<1	0	0	0	0	<1	1	3	13	10	11
12 Mean duration of sunshine	in h	248	252	279	240	310	300	310	341	300	310	240	217	3347		12
13 Mean quantity of radiation	in ly/day															13
14 Mean potential evaporation	in mm	11	15	41	105	187	198	214	206	179	139	54	16	1365	3	14
15 Mean windspeed	in m/sec															15
16 Mean predominent direction of the wind		SE	W	W	S	SW	NW	NW	W	N	NW	NW	SE			16

142 Station/Country Mashhad/Iran

Location 36°17'N/59°36'E Height above sealevel 946 m Climate symbol: Köppen BSk Troll III,10a

		J	F	M	A	M	J	J	A	S	O	N	D	year	P	
1 Mean daily temperature	in °C	0,6	3,1	7,8	13,9	19,7	23,6	25,3	23,6	19,4	13,9	8,6	3,9	13,6	25	1
2 Mean daily maximum temperature	in °C	6,7	8,3	13,9	20,6	26,7	31,7	33,3	32,2	28,3	22,2	16,1	9,4	20,6	25	2
3 Mean daily minimum temperature	in °C	-5,6	-2,2	1,7	7,2	12,8	15,6	17,2	15,0	10,6	5,6	1,1	-1,7	6,1	26	3
4 Absolute maximum temperature	in °C	22,2	26,7	28,9	32,8	38,3	40,0	40,6	40,0	36,7	34,4	32,2	23,9	40,6	25	4
5 Absolute minimum temperature	in °C	-23,9	-18,9	-17,8	-7,7	1,7	5,6	8,3	5,0	-1,7	-5,6	-19,4	-23,9		25	5
6 Mean relative humidity	in %	71	69	65	55	50	38	35	34	39	51	61	68	53	9	6
7 Mean precipitation	in mm	20	25	56	46	30	8	3	2	2	10	15	18	235	26	7
8 Maximum precipitation	in mm	87	109	133	114	101	52	32	21	22	47	60	60	459		8
9 Minimum precipitation	in mm															9
10 Maximum precipitation in 24 h	in mm	38	36	61	46	43	25	33	5	3	36	36	25	61	28	10
11 Mean number of days with precipitation	> 2,5 mm	3	3	5	5	3	1	<1	<1	<1	1	1	2	27	26	11
12 Mean duration of sunshine	in h	171	146	174	180	279	342	363	360	300	226	195	167	2903		12
13 Mean quantity of radiation	in ly/day															13
14 Mean potential evaporation	in mm	1	5	23	57	100	137	155	130	87	50	21	6	772	37	14
15 Mean windspeed	in m/sec	0,5	0,5	0,5	0,5	0,5	0,5	0,5	0,5	0,5	0,5	0,5	0,5	0,5		15
16 Mean predominent direction of the wind																16

143 Station/Country Tehrān/Iran

Location 35°41'N/51°25'E Height above sealevel 1220 m Climate symbol: Köppen BSk Troll IV.2

		J	F	M	A	M	J	J	A	S	O	N	D	year	P	
1 Mean daily temperature	in °C	2,2	5,0	9,4	15,6	21,1	26,4	29,7	28,9	25,0	18,0	11,7	5,6	16,6	22	1
2 Mean daily maximum temperature	in °C	7,2	10,0	15,0	21,7	27,8	33,9	37,2	36,1	32,2	24,4	17,2	10,6	22,8	22	2
3 Mean daily minimum temperature	in °C	-2,8	0,0	3,9	9,4	14,4	18,9	22,2	21,7	17,8	11,7	6,1	0,6	10,6	22	3
4 Absolute maximum temperature	in °C	18,3	19,4	29,4	32,8	37,2	42,8	42,8	38,3	32,2	32,2	20,0		42,8	25	4
5 Absolute minimum temperature	in °C	-20,6	-15,6	-8,9	-2,2	3,9	10,6	15,0	13,9	8,3	3,3	-7,2	-12,2	-20,6	24	5
6 Mean relative humidity	in %	76	66	50	47	51	50	46	47	49	54	65	76	56	4	6
7 Mean precipitation	in mm	46	38	46	36	13	3	3	3	3	8	20	30	246	33	7
8 Maximum precipitation	in mm															8
9 Minimum precipitation	in mm															9
10 Maximum precipitation in 24 h	in mm	64	38	25	25	25	13	5	10	8	18	28	23	64	25	10
11 Mean number of days with precipitation	> 2,5 mm	4	4	5	4	2	1	<1	<1	<1	1	3	4	30	24	11
12 Mean duration of sunshine	in h	183	188	233	222	267	348	347	341	303	236	207	195	3070		12
13 Mean quantity of radiation	in ly/day															13
14 Mean potential evaporation	in mm	1	5	22	55	110	166	198	179	127	65	28	6	960	27	14
15 Mean windspeed	in m/sec															15
16 Mean predominent direction of the wind		W	W	W	W	W	W	W	E	W	W	W				16

144 Station/Country Kermānshāh/Iran

Location 34°21'N/47°06'E Height above sealevel 1306 m Climate symbol: Köppen Csa Troll IV.1

		J	F	M	A	M	J	J	A	S	O	N	D	year	P	
1 Mean daily temperature	in °C	1,1	2,5	6,7	11,7	18,4	20,3	25,3	24,4	19,7	14,7	9,4	3,6	13,0	15	1
2 Mean daily maximum temperature	in °C	7,2	8,9	13,9	20,0	26,1	32,2	37,2	36,1	32,2	26,1	17,8	10,0	22,2	15	2
3 Mean daily minimum temperature	in °C	-5,0	-3,9	-0,6	3,3	6,7	8,3	13,3	12,8	7,2	3,3	1,1	-2,8	3,9	15	3
4 Absolute maximum temperature	in °C	15,6	23,9	25,0	29,4	34,4	39,4	41,7	40,6	39,4	33,3	28,9	18,9	41,7	15	4
5 Absolute minimum temperature	in °C	-25,0	-25,0	-11,1	-10,0	-3,3	-0,8	3,3	5,6	-0,6	-8,3	-12,2	-17,2	-25,0	15	5
6 Mean relative humidity	in %	79	77	67	57	55	36	33	28	27	33	57	78	45	15	6
7 Mean precipitation	in mm	66	58	71	56	41	2	2	2	2	10	51	61	422	15	7
8 Maximum precipitation	in mm															8
9 Minimum precipitation	in mm															9
10 Maximum precipitation in 24 h	in mm	46	46	41	43	43	2	15	3	5	25	56	38	56	15	10
11 Mean number of days with precipitation	> 1,0 mm	9	7	9	8	5	0	0	0	<1	1	6	7	53	15	11
12 Mean duration of sunshine	in h	140	153	205	204	242	351	360	344	291	217	180	158	2843		12
13 Mean quantity of radiation	in ly/day															13
14 Mean potential evaporation	in mm	0	3	18	41	82	126	163	150	100	66	23	4	776	17	14
15 Mean windspeed	in m/sec															15
16 Mean predominent direction of the wind																16

145 Station / Country Esfahān / Iran

Location 32°34'N/51°44'E Height above sealevel 1773 m Climate symbol: Köppen BWk Troll IV,2

		J	F	M	A	M	J	J	A	S	O	N	D	year	P	
1 Mean daily temperature	in °C	2,0	5,0	9,4	15,0	20,3	25,0	28,1	26,7	22,8	16,4	10,0	4,7	15,6	22	1
2 Mean daily maximum temperature	in °C	8,3	11,7	16,1	22,2	28,3	33,3	36,7	35,6	32,2	25,0	17,2	11,1	23,3	22	2
3 Mean daily minimum temperature	in °C	-4,4	-1,7	2,8	7,8	12,2	16,7	19,4	17,8	13,3	7,8	2,8	-1,7	7,8	22	3
4 Absolute maximum temperature	in °C	18,3	23,3	27,8	31,1	35,6	43,3	41,7	42,2	37,8	33,3	25,0	22,8	43,3	26	4
5 Absolute minimum temperature	in °C	-19,4	-13,9	-11,1	-3,3	2,8	8,9	8,9	11,7	5,6	-1,1	-8,9	-12,8	-19,4	27	5
6 Mean relative humidity	in %	64	54	45	40	39	30	28	29	32	38	50	59	42	5	6
7 Mean precipitation	in mm	15	10	25	15	5	2	2	2	2	3	15	20	116	21	7
8 Maximum precipitation	in mm															8
9 Minimum precipitation	in mm															9
10 Maximum precipitation in 24 h	in mm	25	25	43	36	23	13	33	8	20	8	23	38	43	28	10
11 Mean number of days with precipitation	> 2,5 mm	3	3	2	2	1	<1	<1	<1	<1	<1	2	1	19	20	11
12 Mean duration of sunshine	in h	217	206	270	249	298	363	353	344	297	254	243	220	3314		12
13 Mean quantity of radiation	in ly/day															13
14 Mean potential evaporation	in mm	5	8	20	45	80	122	165	148	109	74	36	14	826	13	14
15 Mean windspeed	in m/sec															15
16 Mean predominent direction of the wind																16

146 Station / Country Seistan / Iran

Location 31°00'N/61°30'E Height above sealevel 610 m Climate symbol: Köppen BWh Troll IV,5

		J	F	M	A	M	J	J	A	S	O	N	D	year	P	
1 Mean daily temperature	in °C	7,5	10,9	15,6	22,0	27,0	30,9	33,1	31,4	27,0	20,9	14,2	8,9	20,8	30	1
2 Mean daily maximum temperature	in °C	13,9	17,8	22,8	29,4	34,4	38,3	40,0	38,3	34,4	29,4	22,2	16,1	28,3	30	2
3 Mean daily minimum temperature	in °C	1,1	3,9	8,3	14,4	19,4	23,3	24,4	24,4	19,4	12,2	6,1	1,7	13,3	30	3
4 Absolute maximum temperature	in °C	25,6	31,7	37,8	40,0	45,6	48,3	47,8	45,0	42,2	39,4	33,3	27,2	48,3	30	4
5 Absolute minimum temperature	in °C	-8,3	-6,7	-3,9	-1,1	5,6	3,9	17,2	16,7	9,4	-0,6	-5,0	-11,1	-11,1	30	5
6 Mean relative humidity	in %	60	58	52	48	44	34	31	31	32	37	45	55	44	10	6
7 Mean precipitation	in mm	10	10	15	3	2	2	2	0	0	15	13	8	80	30	7
8 Maximum precipitation	in mm															8
9 Minimum precipitation	in mm															9
10 Maximum precipitation in 24 h	in mm	15	28	46	13	8	3	2	0	0	10	5	58	58	30	10
11 Mean number of days with precipitation	> 2,5 mm	2	1	2	<1	<1	<1	0	0	0	<1	<1	1	11	30	11
12 Mean duration of sunshine	in h															12
13 Mean quantity of radiation	in ly/day															13
14 Mean potential evaporation	in mm	5	16	40	97	163	197	212	195	142	79	26	8	1180	30	14
15 Mean windspeed	in m/sec															15
16 Mean predominent direction of the wind																16

147 Station / Country Ābādān / Iran

Location 30°21'N/48°16'E Height above sealevel 2 m Climate symbol: Köppen BWh Troll IV,2

		J	F	M	A	M	J	J	A	S	O	N	D	year	P	
1 Mean daily temperature	in °C	12,0	14,7	18,6	24,4	30,9	33,9	36,1	36,1	32,2	27,2	20,6	14,4	25,1	10	1
2 Mean daily maximum temperature	in °C	17,2	20,0	24,4	31,1	38,3	42,2	44,4	45,0	41,7	36,1	27,2	20,0	32,2	10	2
3 Mean daily minimum temperature	in °C	6,7	9,4	12,8	17,8	23,3	25,6	27,8	27,2	22,8	18,3	13,9	8,9	17,8	10	3
4 Absolute maximum temperature	in °C	25,0	28,3	33,9	42,8	46,7	47,8	50,0	50,6	47,8	43,3	36,1	28,9	50,6	12	4
5 Absolute minimum temperature	in °C	-3,3	-2,8	2,2	7,2	15,6	19,4	22,8	21,7	15,6	11,7	0,6	-4,4	-4,4	12	5
6 Mean relative humidity	in %	77	75	59	45	33	25	25	29	33	39	60	75	48	10	6
7 Mean precipitation	in mm	38	43	15	20	3	0	0	0	0	3	25	46	193	10	7
8 Maximum precipitation	in mm															8
9 Minimum precipitation	in mm															9
10 Maximum precipitation in 24 h	in mm	28	36	20	30	18	0	0	0	0	13	56	58	58	10	10
11 Mean number of days with precipitation	> 1,0 mm	6	5	3	3	1	0	0	0	0	<1	3	4	26	10	11
12 Mean duration of sunshine	in h	202	192	257	246	245	306	307	285	255	254	209	217	2975		12
13 Mean quantity of radiation	in ly/day															13
14 Mean potential evaporation	in mm	10	11	34	124	198	213	222	210	183	145	46	11	1407		14
15 Mean windspeed	in m/sec															15
16 Mean predominent direction of the wind																16

148 Station / Country Kermān / Iran

Location 30°21'N/57°05'E Height above sealevel 1859 m Climate symbol: Köppen BWk Troll IV,2

		J	F	M	A	M	J	J	A	S	O	N	D	year	P	
1 Mean daily temperature	in °C	5,9	7,5	10,9	16,7	23,1	28,3	28,3	25,9	22,0	17,5	11,7	6,7	17,0	7	1
2 Mean daily maximum temperature	in °C	14,4	15,6	18,3	25,0	32,8	38,3	38,3	36,7	33,3	29,4	22,8	14,4	26,7	7	2
3 Mean daily minimum temperature	in °C	-2,8	-0,6	3,3	8,3	13,3	18,3	18,3	15,0	10,8	5,6	0,6	-1,1	7,2	7	3
4 Absolute maximum temperature	in °C	23,9	27,8	28,9	32,2	40,0	42,8	44,4	42,2	39,4	35,0	31,7	25,6	44,4	7	4
5 Absolute minimum temperature	in °C	-13,9	-10,0	-15,0	-2,2	0,6	9,4	8,3	7,2	1,7	-1,7	-7,2	-11,7	-15,0	7	5
6 Mean relative humidity	in %	63	58	58	49	35	30	34	33	41	47	51	67	47	8	6
7 Mean precipitation	in mm	13	23	23	18	3	5	0	0	2	2	13	36	138	5	7
8 Maximum precipitation	in mm															8
9 Minimum precipitation	in mm															9
10 Maximum precipitation in 24 h	in mm	8	13	13	20	8	10	0	0	3	3	20	15	20	5	10
11 Mean number of days with precipitation	> 2,5 mm	2	3	3	2	<1	<1	0	0	<1	<1	1	6	21	5	11
12 Mean duration of sunshine	in h	220	184	218	225	288	300	294	335	306	282	246	223	3121		12
13 Mean quantity of radiation	in ly/day															13
14 Mean potential evaporation	in mm	5	8	23	58	125	164	185	144	128	76	32	11	959	12	14
15 Mean windspeed	in m/sec															15
16 Mean predominent direction of the wind																16

Station / Country Büshehr / Iran

Location 28°59'N / 50°49'E   Height above sealevel 4 m   Climate symbol: Köppen BSh   Troll IV,2

		J	F	M	A	M	J	J	A	S	O	N	D	year	P	
1	Mean daily temperature	in °C	14,2	15,0	18,9	23,3	28,1	30,3	31,9	32,5	30,3	26,7	21,4	15,9	24,0	53	1
2	Mean daily maximum temperature	in °C	17,8	18,3	22,8	27,2	31 7	33,3	35,0	36,1	34,4	31,1	25,6	20,0	27,8	53	2
3	Mean daily minimum temperature	in °C	10,6	11,7	15,0	19,4	24 4	27,2	28,9	28,9	26,1	22,2	17,2	12,8	20,6	53	3
4	Absolute maximum temperature	in °C	26,7	29,4	40,6	39,4	41,7	44,4	44,4	46,1	41,7	38,3	33,9	30,6	46,1	53	4
5	Absolute minimum temperature	in °C	0,0	2,8	5,6	8,3	14,4	19,4	23,3	20,6	17,2	12,8	5,6	2,8	0,0	53	5
6	Mean relative humidity	in %	80	79	74	62	74	73	74	72	70	68	73	80	74	8	6
7	Mean precipitation	in mm	74	48	46	10	2	0	0	2	0	3	41	81	307	53	7
8	Maximum precipitation	in mm	328	169	119	54	14	0	0	5	0	52	297	337	676		8
9	Minimum precipitation	in mm															9
10	Maximum precipitation in 24 h	in mm	140	84	66	38	15	0	0	5	0	43	74	135	140	53	10
11	Mean number of days with precipitation	>0,25 mm	5	3	2	1	<1	0	0	<1	0	<1	3	4	21	53	11
12	Mean duration of sunshine	in h	223	190	220	237	288	330	316	313	297	279	252	220	3165		12
13	Mean quantity of radiation	in ly / day															13
14	Mean potential evaporation	in mm	17	20	49	98	173	190	205	198	169	133	62	27	1341	4	14
15	Mean windspeed	in m / sec	3,1	3,1	3,7	3,1	3,1	4,2	3,1	2,6	2,6	2,1	2,6	2,6	3,1		15
16	Mean predominent direction of the wind																16

Station / Country Bandar 'Abbas / Iran

Location 27°11'N / 57°17'E   Height above sealevel 9 m   Climate symbol: Köppen BWh   Troll V,5

		J	F	M	A	M	J	J	A	S	O	N	D	year	P	
1	Mean daily temperature	in °C	18,3	19,4	22 5	26,4	30,3	32,8	33,6	33,3	32,0	29,4	24,7	19,7	26,9	7	1
2	Mean daily maximum temperature	in °C	23,3	23,9	27,2	31,7	36,1	38,3	37,8	37,8	36,7	35,0	30,6	25,0	31,7	7	2
3	Mean daily minimum temperature	in °C	13,3	15,0	17,8	21,1	24,4	27,2	29,4	28,9	27,2	23,9	18,9	14,4	21,7	7	3
4	Absolute maximum temperature	in °C	28,3	30,6	33,9	39,4	41,7	48,3	45,6	45,0	42,2	39,4	37,2	32,8	48,3	7	4
5	Absolute minimum temperature	in °C	3,3	8,9	10,6	11,7	20,6	23,3	26,1	25,6	22,8	15,6	13,3	7,2	3,3	7	5
6	Mean relative humidity	in %	68	70	65	63	62	63	68	68	73	71	65	67	67	7	6
7	Mean precipitation	in mm	61	30	8	5	3	0	0	0	2	2	2	41	154	8	7
8	Maximum precipitation	in mm															8
9	Minimum precipitation	in mm															9
10	Maximum precipitation in 24 h	in mm	43	46	25	38	25	0	0	0	8	2	13	94	94	7	10
11	Mean number of days with precipitation	> 0,25 mm	2	2	<1	<1	<1	0	0	0	<1	<1	<1	2	12	8	11
12	Mean duration of sunshine	in h	251	209	248	240	322	312	288	282	270	291	264	145	3222		12
13	Mean quantity of radiation	in ly / day															13
14	Mean potential evaporation	in mm	33	34	80	138	191	204	212	200	177	157	92	41	1559	7	14
15	Mean windspeed	in m / sec															15
16	Mean predominent direction of the wind																16

Station / Country Chāh Bahār / Iran

Location 25°17'N / 60°37'E   Height above sealevel 8 m   Climate symbol: Köppen BWh   Troll V,5

		J	F	M	A	M	J	J	A	S	O	N	D	year	P	
1	Mean daily temperature	in °C	19,2	20,6	23,3	26,4	29,2	30,9	30,3	28,6	28,1	27,0	24,2	21,1	25,7	12	1
2	Mean daily maximum temperature	in °C	23,3	24,4	27,2	30,6	32,8	.33,9	32,8	31,1	31,1	28,9	25,6	22,2	29,4	12	2
3	Mean daily minimum temperature	in °C	15,0	16,7	19,4	22,2	25,6	27,8	27,8	26,1	25,0	22,8	19,4	16,7	22,2	14	3
4	Absolute maximum temperature	in °C	29,4	30,0	34,4	36,7	39,4	40,6	44,4	38,9	40,0	37,2	36,1	30,6	44,4	12	4
5	Absolute minimum temperature	in °C	5,0	7,2	10,0	16,1	17,2	23,9	22,8	22,2	20,6	15,0	13,9	7,8	5,0	14	5
6	Mean relative humidity	in %	67	70	70	73	77	78	77	77	77	76	66	63	73	10	6
7	Mean precipitation	in mm	43	20	5	3	0	2	3	0	0	0	3	25	104	17	7
8	Maximum precipitation	in mm															8
9	Minimum precipitation	in mm															9
10	Maximum precipitation in 24 h	in mm	76	30	48	10	0	8	13	0	0	0	10	69	76	10	10
11	Mean number of days with precipitation	> 2,5 mm	2	1	<1	<1	0	<1	<1	0	0	0	<1	1	9	17	11
12	Mean duration of sunshine	in h															12
13	Mean quantity of radiation	in ly / day															13
14	Mean potential evaporation	in mm	41	50	84	139	181	189	192	170	150	137	91	61	1485	8	14
15	Mean windspeed	in m / sec															15
16	Mean predominent direction of the wind																16

Station / Country Herāt / Afghanistan

Location 34°18'N / 62°13'E   Height above sealevel 964 m   Climate symbol: Köppen BSk   Troll IV,2

		J	F	M	A	M	J	J	A	S	O	N	D	year	P	
1	Mean daily temperature	in °C	1,7	4,9	10,3	15,4	21,7	27,2	29,5	27,5	22,3	15,8	8,8	4,2	15,8	18	1
2	Mean daily maximum temperature	in °C	8,4	12,3	17,4	23,2	29,7	34,9	36,5	35,0	31,1	25,1	18,1	12,1	23,6	18	2
3	Mean daily minimum temperature	in °C	-3,8	-0,5	3,6	8,9	13,3	18,3	21,3	19,0	12,9	6,5	0,8	-1,8	8,2	18	3
4	Absolute maximum temperature	in °C	28,8	28,8	31,0	36,2	39,7	43,2	44,4	42,7	39,3	34,4	30,0	24,8	44,4	18	4
5	Absolute minimum temperature	in °C	-26,7	-20,5	-13,3	-2,3	3,0	9,7	14,7	8,4	1,3	-5,6	-12,8	-22,2	-26,7	18	5
6	Mean relative humidity	in %	71	67	61	57	44	35	33	32	34	42	52	65	49	15	6
7	Mean precipitation	in mm	48	47	54	48	8	0	0	0	0	2	12	33	250	18	7
8	Maximum precipitation	in mm	118	107	152	109	91	tr	0	0	tr	14	39	126	405	18	8
9	Minimum precipitation	in mm	6	3	7	8	0	0	0	0	0	0	0	0	131	18	9
10	Maximum precipitation in 24 h	in mm	32	30	49	42	31	tr	0	0	tr	5	21	43	49	18	10
11	Mean number of days with precipitation	> 0,1 mm	7	7	7	8	2	0	0	0	0	1	2	5	39	18	11
12	Mean duration of sunshine	in h	143	174	201	245	345	381	385	348	.334	288	254	158	3254	6	12
13	Mean quantity of radiation	in ly / day															13
14	Mean potential evaporation	in mm	5	9	32	65	115	157	193	173	101	58	22	6	936		14
15	Mean windspeed	in m / sec	2,6	2,9	3,6	3,9	3,5	4,4	5,8	5,4	3,7	2,7	2,2	2,2	3,5	13	15
16	Mean predominent direction of the wind		S	S,W	S	W	N	N	N	N	N	W	S,W	S		13	16

153 Station/Country Kābul/Afghanistan

Location 34°33'N/69°13'E   Height above sealevel 1791 m   Climate symbol: Köppen Csa   Troll III,10

		J	F	M	A	M	J	J	A	S	O	N	D	year	P	
1	Mean daily temperature	in °C	-2,7	-0,3	8,3	12,2	17,2	22,8	25,0	24,1	19,5	12,8	5,5	0,2	11,9	18	1
2	Mean daily maximum temperature	in °C	4,5	6,1	12,5	18,8	24,2	29,8	32,1	31,9	28,3	22,5	15,2	8,3	19,5	18	2
3	Mean daily minimum temperature	in °C	-13,8	-5,1	0,5	5,8	8,7	12,3	14,9	14,1	9,1	3,7	-1,3	-5,0	4,1	18	3
4	Absolute maximum temperature	in °C	18,8	20,0	26,7	27,8	33,5	36,0	37,7	36,7	33,5	31,8	24,4	17,5	37,7	18	4
5	Absolute minimum temperature	in °C	-25,5	-23,0	-21,8	-2,1	0,5	3,1	7,5	6,0	0,3	-3,0	-9,4	-18,8	-25,5	18	5
6	Mean relative humidity	in %	68	68	64	62	49	35	35	37	37	40	49	63	51	15	6
7	Mean precipitation	in mm	33	61	60	89	22	1	4	<1	2	2	16	19	309	18	7
8	Maximum precipitation	in mm	76	107	101	177	105	5	17	7	7	22	40	61	474	18	8
9	Minimum precipitation	in mm	tr	3	10	29	5	0	0	0	0	0	0	5	177	18	9
10	Maximum precipitation in 24 h	in mm	37	39	89	45	24	5	13	7	12	8	32	29	69	18	10
11	Mean number of days with precipitation	> 1,0 mm	6	6	10	12	8	0	1	0	0	1	3	5	52	18	11
12	Mean duration of sunshine	in h	174	180	200	223	307	357	356	343	307	285	240	206	3186		12
13	Mean quantity of radiation	in ly/day															13
14	Mean potential evaporation	in mm	0	0	28	58	90	127	150	133	83	44	15	0	730	18	14
15	Mean windspeed	in m/sec	1,7	1,8	2,2	2,2	2,8	3,6	3,5	2,8	2,2	1,8	1,7	1,5	2,3		15
16	Mean predominent direction of the wind		NW	N,NW	var	N,NW	N	N	N	N	N	N,NW	N	N,S,NW		13	16

154 Station/Country South Salang/Afghanistan

Location 35°18'N/69°04'E   Height above sealevel 3172 m   Climate symbol: Köppen BSh   Troll IV,5

		J	F	M	A	M	J	J	A	S	O	N	D	year	P	
1	Mean daily temperature	in °C	-8,8	-7,0	-2,8	1,7	5,4	10,5	12,8	12,3	8,9	4,1	-1,1	-6,0	2,5	16	1
2	Mean daily maximum temperature	in °C	-5,0	-2,6	1,6	5,3	8,6	14,2	16,4	16,3	13,7	8,3	2,4	-4,0	6,3	16	2
3	Mean daily minimum temperature	in °C	-11,0	-9,8	-6,0	-1,2	2,4	7,5	9,5	9,1	5,7	1,4	-3,7	-8,2	-0,4	16	3
4	Absolute maximum temperature	in °C	5,8	13,2	12,0	14,7	20,6	22,4	23,7	23,0	20,9	18,5	12,2	8,1	23,7	16	4
5	Absolute minimum temperature	in °C	-25,3	-23,6	-22,8	-11,8	-5,6	0,3	3,7	1,4	-7,7	-9,7	-14,8	-18,0	-25,3	16	5
6	Mean relative humidity	in %	53	53	58	61	55	42	41	36	37	43	43	50	48	15	6
7	Mean precipitation	in mm	120	171	228	242	81	5	4	1	4	19	64	108	1045	16	7
8	Maximum precipitation	in mm	270	358	418	412	335	23	22	5	26	82	174	215	1437	16	8
9	Minimum precipitation	in mm	21	22	76	60	2	0	0	0	0	0	5	42	699	16	9
10	Maximum precipitation in 24 h	in mm	55	86	90	88	73	19	10	4	25	61	58	50	90	16	10
11	Mean number of days with precipitation	> 1,0 mm	12	11	13	13	11	2	1	<1	<1	3	4	10	81	11	11
12	Mean duration of sunshine	in h	141	118	172	190	274	344	351	340	287	240	193	144	2814	14	12
13	Mean quantity of radiation	in ly/day															13
14	Mean potential evaporation	in mm															14
15	Mean windspeed	in m/sec	3,7	4,0	3,9	4,0	4,9	5,3	5,9	4,7	4,2	3,6	3,9	3,6	4,3	13	15
16	Mean predominent direction of the wind		W	W	W	W	W	W	W	W	W	W	W	W			16

155 Station/Country Qandahār/Afghanistan

Location 31°30'N/65°51'E   Height above sealevel 1010 m   Climate symbol: Köppen BWh   Troll IV,2

		J	F	M	A	M	J	J	A	S	O	N	D	year	P	
1	Mean daily temperature	in °C	4,8	7,7	14,3	19,9	25,3	29,9	31,8	29,1	23,4	17,2	10,7	6,7	18,4	14	1
2	Mean daily maximum temperature	in °C	12,1	14,7	22,4	27,9	34,4	39,1	40,1	38,2	34,4	28,0	21,1	15,3	27,3	14	2
3	Mean daily minimum temperature	in °C	-0,6	2,1	7,5	12,1	15,3	19,1	22,2	19,3	12,9	8,2	2,8	0,5	10,1	13	3
4	Absolute maximum temperature	in °C	25,0	26,2	32,0	36,8	43,0	45,0	46,5	44,0	40,1	34,8	31,5	25,2	46,5	14	4
5	Absolute minimum temperature	in °C	-12,1	-9,1	-6,0	2,0	5,6	8,5	13,5	9,0	5,2	-2,2	-5,6	-11,4	-12,1	14	5
6	Mean relative humidity	in %	57	59	48	42	28	23	25	24	23	29	40	52	37	14	6
7	Mean precipitation	in mm	52	34	26	14	2	0	3	1	0	<1	6	25	164	14	7
8	Maximum precipitation	in mm	163	86	112	42	13	0	38	13	tr	2	30	94	282	14	8
9	Minimum precipitation	in mm	tr	tr	<1	0	0	0	0	0	0	0	0	tr	57	14	9
10	Maximum precipitation in 24 h	in mm	51	30	34	13	7	0	20	13	tr	2	23	37	51	14	10
11	Mean number of days with precipitation	> 1,0 mm	7	6	5	4	<1	0	<1	<1	0	<1	<1	3	28	14	11
12	Mean duration of sunshine	in h	190	191	240	249	351	376	352	349	329	317	270	222	3456		12
13	Mean quantity of radiation	in ly/day															13
14	Mean potential evaporation	in mm	8	15	42	89	156	189	206	177	110	63	18	8	1077	2	14
15	Mean windspeed	in m/sec	4,0	3,9	3,7	3,1	3,2	3,0	2,7	2,5	2,4	2,2	2,0	2,8	2,9	13	15
16	Mean predominent direction of the wind		NE	NE	W	W	W	W	S,W	W	W	W	N,W	NE		13	16

156 Station/Country Peshāwar/Pakistan

Location 34°01'N/71°34'E   Height above sealevel 354 m   Climate symbol: Köppen BSh   Troll IV,2

		J	F	M	A	M	J	J	A	S	O	N	D	year	P	
1	Mean daily temperature	in °C	10,9	12,5	17,5	22,5	28,9	33,1	32,8	31,4	28,6	22,8	16,4	11,7	22,4	33	1
2	Mean daily maximum temperature	in °C	17,2	18,9	23,9	29,4	36,7	41,1	39,4	37,2	35,6	31,1	25,0	19,4	29,4	33	2
3	Mean daily minimum temperature	in °C	4,4	6,1	11,1	15,6	21,1	25,0	26,1	25,6	21,7	14,4	7,8	3,9	15,0	33	3
4	Absolute maximum temperature	in °C	24,4	30,0	33,9	42,2	47,8	48,9	50,0	47,8	43,3	38,3	32,8	28,3	50,0	30	4
5	Absolute minimum temperature	in °C	-3,3	-0,6	2,2	6,8	11,1	18,3	20,6	20,0	14,4	11,1	0,6	-2,2	-3,3	30	5
6	Mean relative humidity	in %	59	59	56	49	35	34	50	58	52	46	52	58	51	26	6
7	Mean precipitation	in mm	36	38	61	46	20	8	33	51	20	5	8	18	344	60	7
8	Maximum precipitation	in mm															8
9	Minimum precipitation	in mm															9
10	Maximum precipitation in 24 h	in mm	48	41	152	36	25	56	38	30	23	13	20	38	152	39	10
11	Mean number of days with precipitation	> 2,5 mm	3	3	5	4	2	1	2	3	2	1	1	2	29	60	11
12	Mean duration of sunshine	in h	174	176	217	234	288	306	276	276	258	273	243	195	2916		12
13	Mean quantity of radiation	in ly/day															13
14	Mean potential evaporation	in mm	11	13	43	92	187	212	216	198	158	84	31	10	1255	33	14
15	Mean windspeed	in m/sec	0,6	0,8	0,9	1,0	1,1	1,1	1,0	0,9	0,8	0,5	0,5	0,5	0,8		15
16	Mean predominent direction of the wind		S	S	S	N	NW	N	N	N	N	S	S	S			16

157 Station/Country Lahore/Pakistan

Location 31°35′N/74°20′E   Height above sealevel 214 m   Climate symbol: Köppen **BSh**   Troll IV.3

		J	F	M	A	M	J	J	A	S	O	N	D	year	P	
1	Mean daily temperature	in °C	12,5	14,4	20,0	26,1	31,1	33,8	32,2	30,9	29,4	25,0	35,0	13,8	23,3	23	1
2	Mean daily maximum temperature	in °C	20,8	22,2	28,3	35,0	40,0	41,1	37,8	36,1	36,1	35,0	38,3	22,2	31,7	23	2
3	Mean daily minimum temperature	in °C	4,4	6,7	11,7	17,2	22,2	26,7	25,8	22,8	22,8	15,0	8,3	4,4	16,1	23	3
4	Absolute maximum temperature	in °C	25,8	32,2	37,8	48,1	47,2	47,2	45,0	42,2	41,7	40,0	34,4	27,8	47,2	10	4
5	Absolute minimum temperature	in °C	−2,2	0,0	5,6	10,0	16,1	18,9	20,6	19,4	17,2	9,4	2,8	0,6	−2,2	10	5
6	Mean relative humidity	in %	65	58	44	34	28	35	59	65	54	43	51	65	50	8	6
7	Mean precipitation	in mm	28	23	23	13	18	43	140	135	61	8	3	10	505	23	7
8	Maximum precipitation	in mm	95	111	84	76	29	153	284	292	526	80	33	71	851		8
9	Minimum precipitation	in mm															9
10	Maximum precipitation in 24 h	in mm	74	53	30	41	18	89	89	114	89	23	10	20	114	10	10
11	Mean number of days with precipitation	> 2,5 mm	2	2	2	2	1	3	6	8	3	<1	<1	1	30	23	11
12	Mean duration of sunshine	in h	195	212	257	288	304	279	223	254	270	301	270	226	3079		12
13	Mean quantity of radiation	in ly/day															13
14	Mean potential evaporation	in mm	11	17	59	132	201	212	210	190	163	114	38	14	1361	60	14
15	Mean windspeed	in m/sec	0,8	0,9	1,2	1,2	1,2	1,3	1,3	1,0	0,8	0,8	0,5	0,5	0,9		15
16	Mean predominent direction of the wind		NW	NW	NW	NW	N	SE	SE	SE	SE	NW	NW	NW			16

158 Station/Country Multān/Pakistan

Location 30°12′N/71°31′E   Height above sealevel 126 m   Climate symbol: Köppen **BWh**   Troll IV.5

		J	F	M	A	M	J	J	A	S	O	N	D	year	P	
1	Mean daily temperature	in °C	13,8	15,9	22,2	28,3	33,8	35,8	35,8	33,3	31,7	27,0	20,8	15,0	26,0	33	1
2	Mean daily maximum temperature	in °C	21,1	23,3	30,0	36,7	41,7	42,2	40,0	38,3	38,3	35,6	29,4	22,8	33,3	33	2
3	Mean daily minimum temperature	in °C	6,1	8,3	14,1	20,0	25,6	28,9	28,9	28,3	25,0	18,3	11,7	7,2	18,9	33	3
4	Absolute maximum temperature	in °C	28,3	32,8	42,2	45,6	49,4	49,4	48,3	44,4	43,9	41,1	40,0	34,4	49,4	33	4
5	Absolute minimum temperature	in °C	−1,7	0,0	5,0	10,6	16,1	15,0	22,8	22,2	16,7	10,0	1,1	−1,1	−1,7	33	5
6	Mean relative humidity	in %	56	54	44	36	34	41	56	60	52	43	45	54	48	10	6
7	Mean precipitation	in mm	10	10	10	5	8	10	51	43	10	2	3	8	170	33	7
8	Maximum precipitation	in mm															8
9	Minimum precipitation	in mm															9
10	Maximum precipitation in 24 h	in mm	30	18	20	28	36	41	155	71	109	28	10	20	155	30	10
11	Mean number of days with precipitation	> 2,5 mm	1	1	1	1	1	1	2	2	1	<1	<1	1	14	33	11
12	Mean duration of sunshine	in h															12
13	Mean quantity of radiation	in ly/day															13
14	Mean potential evaporation	in mm	8	15	68	162	211	215	218	203	178	136	46	13	1473	60	14
15	Mean windspeed	in m/sec	0,5	0,7	0,8	0,9	0,9	1,2	1,1	1,0	0,9	0,5	0,5	0,5	0,8		15
16	Mean predominent direction of the wind		N	N	N	N	N	SW	SW	SW	SE	N	N	N			16

159 Station/Country Quetta/Pakistan

Location 30°10′N/67°01′E   Height above sealevel 1730 m   Climate symbol: Köppen **BSk**   Troll IV.2

		J	F	M	A	M	J	J	A	S	O	N	D	year	P	
1	Mean daily temperature	in °C	3,8	5,8	10,6	15,3	20,0	24,4	26,7	25,3	20,3	14,4	8,9	5,0	15,0	30	1
2	Mean daily maximum temperature	in °C	10,0	12,2	17,8	23,3	28,9	33,9	35,0	33,9	30,8	25,0	18,3	12,8	23,3	30	2
3	Mean daily minimum temperature	in °C	−2,8	−1,1	3,3	7,2	11,1	15,0	18,3	16,7	10,0	3,9	−0,6	−2,8	6,7	30	3
4	Absolute maximum temperature	in °C	23,9	26,7	28,3	32,8	36,7	39,4	40,0	39,4	36,7	32,8	27,2	23,3	40,0	30	4
5	Absolute minimum temperature	in °C	−14,4	−13,3	−7,2	−3,9	1,7	6,7	8,3	7,2	0,6	−5,6	−11,1	−19,4	−19,4	30	5
6	Mean relative humidity	in %	65	60	51	43	37	36	43	42	35	35	43	53	45	11	6
7	Mean precipitation	in mm	48	51	43	25	10	5	13	6	2	3	6	25	241	60	7
8	Maximum precipitation	in mm															8
9	Minimum precipitation	in mm															9
10	Maximum precipitation in 24 h	in mm	48	41	152	36	25	56	36	30	13	23	20	38	152	30	10
11	Mean number of days with precipitation	> 2,5 mm	5	5	5	3	1	<1	1	1	<1	<1	1	3	28	60	11
12	Mean duration of sunshine	in h	223	232	254	261	328	354	341	341	345	347	287	234	3525		12
13	Mean quantity of radiation	in ly/day															13
14	Mean potential evaporation	in mm	5	10	32	63	101	137	161	141	91	49	22	10	822		14
15	Mean windspeed	in m/sec	1,1	1,3	1,4	1,4	1,2	1,1	1,2	1,0	0,8	0,8	0,8	0,9	1,1		15
16	Mean predominent direction of the wind		SE	SE	SE	SE	SE	SE	SE	S	S	S	SE	SE			16

160 Station/Country Hyderabad/Pakistan

Location 25°23′N/68°25′E   Height above sealevel 29 m   Climate symbol: Köppen **BWh**   Troll IV.5

		J	F	M	A	M	J	J	A	S	O	N	D	year	P	
1	Mean daily temperature	in °C	17,5	20,0	25,9	30,6	33,6	34,2	32,2	30,9	30,3	28,9	23,3	18,9	27,2	20	1
2	Mean daily maximum temperature	in °C	24,4	27,2	33,9	38,9	41,7	40,6	37,2	35,6	36,1	36,7	31,7	26,1	33,9	20	2
3	Mean daily minimum temperature	in °C	10,6	12,8	17,8	22,2	25,6	27,8	27,2	26,1	24,4	21,1	15,0	11,7	20,0	20	3
4	Absolute maximum temperature	in °C	31,1	39,4	43,3	45,0	48,3	50,0	43,9	41,7	43,9	45,0	40,0	38,1	50,0	10	4
5	Absolute minimum temperature	in °C	1,1	5,6	10,0	13,9	21,7	22,8	23,9	23,9	21,7	14,4	5,6	5,6	1,1	10	5
6	Mean relative humidity	in %	46	44	38	38	45	54	62	65	59	45	41	44	48	10	6
7	Mean precipitation	in mm	5	5	5	2	5	10	76	51	15	2	2	3	181	60	7
8	Maximum precipitation	in mm															8
9	Minimum precipitation	in mm															9
10	Maximum precipitation in 24 h	in mm	13	20	43	18	46	38	191	130	157	23	15	20	191	30	10
11	Mean number of days with precipitation	> 2,5 mm	<1	<1	<1	<1	<1	<1	3	3	<1	<1	<1	<1	16	60	11
12	Mean duration of sunshine	in h															12
13	Mean quantity of radiation	in ly/day															13
14	Mean potential evaporation	in mm	20	25	113	173	199	195	124	131	150	153	70	24	1377	60	14
15	Mean windspeed	in m/sec	2,3	2,5	2,7	3,0	3,4	6,6	6,1	5,1	3,5	2,5	2,2	2,1	3,5		15
16	Mean predominent direction of the wind		ESE	E	SE	SW	W	W	W	W	WNW	WNW	E	E			16

100

161 Station/Country Karāchi/Pakistan

Location 24°48'N/66°59'E — Height above sealevel 4 m — Climate symbol: Köppen BWh — Troll V,5

		J	F	M	A	M	J	J	A	S	O	N	D	year	P	
1 Mean daily temperature	in °C	18,9	20,3	24,4	27,5	30,0	30,9	30,0	28,6	28,1	27,5	24,2	20,3	25,9	43	1
2 Mean daily maximum temperature	in °C	25,0	26,1	29,4	32,2	33,9	33,9	32,8	31,1	31,1	32,8	30,6	26,7	30,6	43	2
3 Mean daily minimum temperature	in °C	12,8	14,4	19,4	22,8	26,1	27,8	27,2	26,1	25,0	22,2	17,8	13,9	21,1	43	3
4 Absolute maximum temperature	in °C	31,7	33,9	41,1	43,9	47,8	45,6	43,3	37,2	41,1	42,2	37,8	32,8	47,8	43	4
5 Absolute minimum temperature	in °C	4,4	6,1	8,3	13,9	18,3	20,0	22,8	22,8	20,6	13,9	8,9	3,9	3,9	43	5
6 Mean relative humidity	in %	54	61	68	75	78	78	81	82	80	70	59	55	70	29	6
7 Mean precipitation	in mm	13	10	8	3	18	84	41	13	2	3	5		203	59	7
8 Maximum precipitation	in mm	69	51	56	131	60	183	392	428	252	69	41	66	676		8
9 Minimum precipitation	in mm															9
10 Maximum precipitation in 24 h	in mm	41	28	43	104	30	183	201	137	206	13	23	46	206	59	10
11 Mean number of days with precipitation	> 2,5 mm	1	1	1	<1	<1	1	2	2	1	<1	<1	1	14	59	11
12 Mean duration of sunshine	in h	279	244	295	306	319	213	118	130	225	301	279	273	2982		12
13 Mean quantity of radiation	in ly/day															13
14 Mean potential evaporation	in mm	36	51	91	136	178	184	181	165	143	135	75	44	1419	4	14
15 Mean windspeed	in m/sec	3,3	3,6	4,3	5,1	6,1	6,7	6,7	6,2	5,2	3,5	2,8	3,1	4,7		15
16 Mean predominent direction of the wind		NE	SW	W	W	SW	SW	SW	SW	W	SW	SW	SW			16

162 Station/Country Simla/India

Location 31°06'N/77°10'E — Height above sealevel 2202 m — Climate symbol: Köppen Cwb — Troll V,1

		J	F	M	A	M	J	J	A	S	O	N	D	year	P	
1 Mean daily temperature	in °C	5,3	5,9	10,3	14,7	18,3	19,4	18,1	17,2	16,7	13,9	10,6	7,5	13,2	30	1
2 Mean daily maximum temperature	in °C	8,3	8,9	13,9	18,3	22,2	22,8	20,6	19,4	19,4	17,2	13,9	10,6	16,1	30	2
3 Mean daily minimum temperature	in °C	2,2	2,8	6,7	11,1	14,4	16,1	15,6	15,0	13,9	10,6	7,2	4,4	10,0	30	3
4 Absolute maximum temperature	in °C	17,2	19,4	23,9	28,3	30,0	30,6	27,8	25,6	24,4	23,9	19,4	20,0	30,6	30	4
5 Absolute minimum temperature	in °C	-9,4	-7,7	-5,6	0,0	4,4	7,8	10,0	11,1	5,0	3,9	0,0	-6,1	-9,4	30	5
6 Mean relative humidity	in %	54	55	41	36	38	60	88	91	76	50	40	47	56	10	6
7 Mean precipitation	in mm	61	69	61	53	66	175	424	434	160	33	13	28	1577	75	7
8 Maximum precipitation	in mm	255	230	231	197	277	472	758	1080	454	263	125	207	2787	80	8
9 Minimum precipitation	in mm	0	0	0	<1	3	24	133	120	12	0	0	0	993	80	9
10 Maximum precipitation in 24 h	in mm	43	58	58	38	48	89	188	132	104	84	33	76	188	30	10
11 Mean number of days with precipitation	> 2,5 mm	4	5	5	4	5	10	20	19	9	2	1	2	86	75	11
12 Mean duration of sunshine	in h															12
13 Mean quantity of radiation	in ly/day															13
14 Mean potential evaporation	in mm	14	14	34	64	99	107	99	90	75	57	35	19	707	60	14
15 Mean windspeed	in m/sec	3,9	4,3	4,7	4,5	4,4	3,7	2,8	2,4	2,7	3,1	3,1	3,5	3,6	30	15
16 Mean predominent direction of the wind		SE,S	S	S	S	S,SW	S,SW	S	NE,S	NE,S	NE,S	NE,S	NE,S		28	16

163 Station/Country Darjeeling/India

Location 27°03'N/88°16'E — Height above sealevel 2127 m — Climate symbol: Köppen Cwb — Troll V,1

		J	F	M	A	M	J	J	A	S	O	N	D	year	P	
1 Mean daily temperature	in °C	5,0	5,6	9,7	13,1	14,7	15,9	16,7	16,1	15,3	13,1	8,9	6,1	11,7	25	1
2 Mean daily maximum temperature	in °C	8,3	8,9	13,9	16,7	17,8	18,3	18,9	18,3	18,1	16,1	12,2	9,4	15,0	25	2
3 Mean daily minimum temperature	in °C	1,7	2,2	5,6	9,4	11,7	13,3	14,4	13,9	12,8	10,0	5,6	2,8	8,3	25	3
4 Absolute maximum temperature	in °C	16,1	16,7	23,3	23,9	25,0	23,3	25,0	25,0	25,0	23,3	19,4	16,7	25,0	30	4
5 Absolute minimum temperature	in °C	-2,8	-2,2	-0,6	1,1	5,6	8,3	8,9	10,6	10,0	4,4	2,2	-1,1	-2,8	30	5
6 Mean relative humidity	in %	83	82	73	78	88	93	95	95	93	87	79	78	85	11	6
7 Mean precipitation	in mm	13	28	43	104	216	589	798	638	447	130	23	8	3037	59	7
8 Maximum precipitation	in mm	117	97	156	286	498	1402	1480	1022	1190	913	373	64	4024	80	8
9 Minimum precipitation	in mm	0	0	0	11	75	217	206	338	143	0	0	0	2271	80	9
10 Maximum precipitation in 24 h	in mm	28	43	71	135	94	246	196	236	213	335	119	30	335	30	10
11 Mean number of days with precipitation	> 2,5 mm	1	3	4	7	14	21	26	24	17	5	1	1	124	59	11
12 Mean duration of sunshine	in h															12
13 Mean quantity of radiation	in ly/day															13
14 Mean potential evaporation	in mm	16	18	40	60	74	85	91	87	72	57	35	20	655	60	14
15 Mean windspeed	in m/sec	0,8	1,0	1,1	1,4	1,1	1,0	0,8	0,9	0,7	0,6	0,5	0,5	0,9	30	15
16 Mean predominent direction of the wind		E,W	E,W	E,W	E,W	E,W	E,W	E	E	E	E,W	E,W	E,W		28	16

164 Station/Country Kānpur/India

Location 26°28'N/80°21'E — Height above sealevel 127 m — Climate symbol: Köppen Cwa — Troll V,3

		J	F	M	A	M	J	J	A	S	O	N	D	year	P	
1 Mean daily temperature	in °C	15,8	18,1	24,2	30,0	33,8	33,8	30,8	29,4	28,2	26,4	20,9	16,1	25,6	32	1
2 Mean daily maximum temperature	in °C	23,3	25,6	32,2	38,3	41,1	38,9	34,4	32,8	33,3	33,3	28,9	23,9	32,2	32	2
3 Mean daily minimum temperature	in °C	8,3	10,6	16,1	21,7	26,1	28,3	26,7	26,1	25,0	19,4	12,8	8,3	18,9	32	3
4 Absolute maximum temperature	in °C	30,0	35,6	42,2	45,0	46,7	47,2	43,9	40,0	40,0	39,4	35,0	30,6	47,2	20	4
5 Absolute minimum temperature	in °C	1,1	2,8	8,9	12,8	18,3	20,6	22,2	21,7	20,6	12,8	6,7	2,8	1,1	19	5
6 Mean relative humidity	in %	61	52	35	29	33	51	74	78	73	53	53	63	54	9	6
7 Mean precipitation	in mm	15	13	8	5	10	79	254	254	142	30	5	8	823	77	7
8 Maximum precipitation	in mm	75	98	77	94	47	379	704	664	528	335	92	80	1587	70	8
9 Minimum precipitation	in mm	0	0	0	0	0	11	56	122	17	0	0	0	421	70	9
10 Maximum precipitation in 24 h	in mm	36	58	28	25	25	135	168	236	234	109	38	38	236	30	10
11 Mean number of days with precipitation	> 2,5 mm	1	1	1	1	1	4	12	12	7	1	<1	1	43	77	11
12 Mean duration of sunshine	in h															12
13 Mean quantity of radiation	in ly/day															13
14 Mean potential evaporation	in mm	17	30	97	167	208	207	191	174	158	125	48	20	1440	60	14
15 Mean windspeed	in m/sec	1,8	2,4	3,0	3,4	3,8	3,7	3,2	2,7	2,5	1,7	1,4	1,4	2,8		15
16 Mean predominent direction of the wind		W	W	W	W	W	E,W	E	E	W	W,N	W	W		28	16

165 Station / Country   Patna / India

Location 25°37'N/85°10'E — Height above sealevel 53 m — Climate symbol: Köppen Cwa — Troll V.2

		J	F	M	A	M	J	J	A	S	O	N	D	year	P	
1	Mean daily temperature	in °C	16.7	19.4	25.3	30.3	32.2	31.4	29.7	29.2	29.2	27.2	22.2	17.8	25.9	30	1
2	Mean daily maximum temperature	in °C	22.8	25.6	32.2	37.2	38.3	36.1	32.8	31.7	32.2	31.7	28.3	23.9	31.1	30	2
3	Mean daily minimum temperature	in °C	10.6	13.3	18.3	23.3	26.1	26.7	26.7	26.7	26.1	22.8	16.1	11.7	20.6	30	3
4	Absolute maximum temperature	in °C	28.9	32.2	40.6	42.8	45.6	46.1	40.0	37.8	37.8	36.1	33.3	28.3	46.1	30	4
5	Absolute minimum temperature	in °C	3.3	5.6	11.1	16.1	17.2	21.7	21.1	21.7	21.7	15.0	8.3	6.7	3.3	30	5
6	Mean relative humidity	in %	63	57	37	34	49	67	79	83	79	66	60	65	62	10	6
7	Mean precipitation	in mm	15	18	10	8	36	180	295	333	218	71	8	2	1194	60	7
8	Maximum precipitation	in mm	105	91	72	50	244	733	565	769	650	371	71	45	1959	80	8
9	Minimum precipitation	in mm	0	0	0	0	0	13	95	105	33	0	0	0	642	80	9
10	Maximum precipitation in 24 h	in mm	33	38	28	23	56	188	152	165	366	117	64	74	366	30	10
11	Mean number of days with precipitation	> 2,5 mm	1	2	1	1	2	8	14	14	9	3	1	1	57	60	11
12	Mean duration of sunshine	in h	279	266	276	294	322	231	180	202	210	267	285	270	3082		12
13	Mean quantity of radiation	in ly/day															13
14	Mean potential evaporation	in mm	24	36	116	172	197	194	185	174	160	138	64	26	1486	60	14
15	Mean windspeed	in m/sec	1,4	1,7	2,1	2,4	2,7	2,5	2,4	2,1	1,9	1,3	1,0	1,1	1,9	30	15
16	Mean predominent direction of the wind		SW,W	SW,W	SW,W	NE,W	NE	NE	E	E	E	SW	SW	SW,W		28	16

166 Station / Country   Allāhābād / India

Location 25°17'N/81°44'E — Height above sealevel 98 m — Climate symbol: Köppen Cwa — Troll V.3

		J	F	M	A	M	J	J	A	S	O	N	D	year	P	
1	Mean daily temperature	in °C	16.1	18.3	24.7	30.6	34.2	33.9	30.0	28.9	28.9	25.6	20.3	16.4	25.7	30	1
2	Mean daily maximum temperature	in °C	23.9	26.1	33.3	39.4	41.7	39.4	33.3	31.7	32.8	32.2	28.3	24.4	32.2	30	2
3	Mean daily minimum temperature	in °C	8.3	10.6	16.1	21.7	26.7	28.3	26.7	26.1	25.0	19.4	12.2	8.3	18.9	30	3
4	Absolute maximum temperature	in °C	31.1	33.9	41.1	45.0	46.7	47.2	42.8	38.9	37.8	36.7	34.4	30.6	47.2	9	4
5	Absolute minimum temperature	in °C	2.2	5.0	10.0	15.0	19.4	22.2	22.8	22.8	20.6	13.3	7.2	3.3	2.2	9	5
6	Mean relative humidity	in %	62	52	32	25	29	51	77	82	76	58	55	73	56	9	6
7	Mean precipitation	in mm	23	15	15	5	15	127	320	254	213	58	8	8	1061	60	7
8	Maximum precipitation	in mm	126	104	81	44	57	527	782	915	470	577	118	68	1936	80	8
9	Minimum precipitation	in mm	0	0	0	0	0	<1	26	53	<1	0	0	0	516	80	9
10	Maximum precipitation in 24 h	in mm	36	36	23	25	30	94	140	124	267	74	36	20	267	20	10
11	Mean number of days with precipitation	> 2,5 mm	1	2	1	<1	1	4	15	15	9	3	<1	<1	54	60	11
12	Mean duration of sunshine	in h	251	258	273	294	304	213	155	150	207	270	276	270	2921		12
13	Mean quantity of radiation	in ly/day															13
14	Mean potential evaporation	in mm	22	35	105	175	207	205	190	175	159	122	46	19	1460	30	14
15	Mean windspeed	in m/sec	1,3	1,5	1,9	2,1	2,4	3,6	2,3	2,1	1,8	1,3	0,9	1,0	1,8	30	15
16	Mean predominent direction of the wind		W	W	W,NW	W,NW	W,NW	W,NW	E	E,W	W	W	W	W		28	16

167 Station / Country   Jabalpur / India

Location 23°10'N/79°59'E — Height above sealevel 393 m — Climate symbol: Köppen Cwa — Troll V.3

		J	F	M	A	M	J	J	A	S	O	N	D	year	P	
1	Mean daily temperature	in °C	17.0	19.4	24.7	29.7	33.6	31.4	27.2	26.4	26.7	24.2	19.7	16.4	24.7	33	1
2	Mean daily maximum temperature	in °C	25.0	27.8	33.3	38.3	41.1	36.7	30.6	29.4	30.6	30.6	27.8	25.0	31.1	33	2
3	Mean daily minimum temperature	in °C	8.9	11.1	16.1	21.1	26.1	26.1	23.9	23.3	22.8	17.8	11.7	7.8	18.3	33	3
4	Absolute maximum temperature	in °C	32.2	35.0	40.6	45.0	45.6	45.6	37.8	33.9	36.7	33.9	32.8	32.8	45.6	9	4
5	Absolute minimum temperature	in °C	1.1	1.7	8.3	13.9	18.9	21.1	21.1	20.6	17.8	10.6	5.0	2.2	1.1	9	5
6	Mean relative humidity	in %	59	49	32	22	24	56	83	84	77	61	56	58	55	9	6
7	Mean precipitation	in mm	18	18	10	8	13	196	467	419	201	41	10	8	1409	33	7
8	Maximum precipitation	in mm	211	141	143	59	161	608	1030	1094	810	266	155	125	2407	80	8
9	Minimum precipitation	in mm	0	0	0	0	0	16	154	150	9	0	0	0	774	80	9
10	Maximum precipitation in 24 h	in mm	43	43	33	51	43	124	229	140	147	66	23	28	229	9	10
11	Mean number of days with precipitation	> 1,0 mm	2	3	1	2	3	12	24	21	13	4	1	1	87	10	11
12	Mean duration of sunshine	in h															12
13	Mean quantity of radiation	in ly/day															13
14	Mean potential evaporation	in mm	31	42	106	167	204	191	163	142	139	101	43	24	1353	60	14
15	Mean windspeed	in m/sec	0,9	1,1	1,3	1,5	2,0	2,4	2,3	2,1	1,6	1,0	0,8	0,7	1,5	30	15
16	Mean predominent direction of the wind		SE,NE	SE,NE	SE	SE,NW	SW,W	SW,W	SW,W	SW,W	SW,W	SE,NE	SE,NE	SE,NE		28	16

168 Station / Country   New Delhi / India

Location 28°35'N/77°12'E — Height above sealevel 218 m — Climate symbol: Köppen BSh — Troll V.4

		J	F	M	A	M	J	J	A	S	O	N	D	year	P	
1	Mean daily temperature	in °C	13.9	16.7	22.5	28.1	33.3	33.6	31.4	30.0	28.9	26.1	20.0	15.3	25.0	10	1
2	Mean daily maximum temperature	in °C	21.1	23.9	30.6	36.1	40.6	38.9	35.6	33.9	33.9	33.9	28.9	22.8	31.7	10	2
3	Mean daily minimum temperature	in °C	6.7	9.4	14.4	20.0	26.1	28.3	27.2	26.1	23.9	18.3	11.1	7.8	18.3	10	3
4	Absolute maximum temperature	in °C	28.9	31.7	39.4	45.6	46.1	46.1	45.0	40.0	40.6	39.4	33.9	28.3	46.1	10	4
5	Absolute minimum temperature	in °C	-0.6	0.0	7.2	11.7	18.3	18.9	21.7	22.2	17.8	10.6	5.0	1.1	-0.6	10	5
6	Mean relative humidity	in %	57	51	36	27	28	45	67	72	62	44	41	56	49	10	6
7	Mean precipitation	in mm	23	18	13	8	13	74	180	173	117	10	3	10	642	75	7
8	Maximum precipitation	in mm	173	153	78	105	83	415	464	583	492	238	31	134	1533	80	8
9	Minimum precipitation	in mm	0	0	0	0	0	0	4	0	0	0	0	0	263	80	9
10	Maximum precipitation in 24 h	in mm	38	104	15	13	15	236	130	178	165	56	3	33	236	10	10
11	Mean number of days with precipitation	> 2,5 mm	2	2	1	1	2	4	8	8	4	1	<1	1	35	75	11
12	Mean duration of sunshine	in h	239	249	257	282	301	225	202	194	228	282	279	257	2995		12
13	Mean quantity of radiation	in ly/day	341	430	518	584	627	569	443	432	474	454	386	323	485	12	13
14	Mean potential evaporation	in mm	16	24	91	169	207	209	201	183	134	49	23		1469		14
15	Mean windspeed	in m/sec	2,3	2,8	3,0	3,0	3,6	4,0	2,9	2,5	2,7	1,7	1,8	2,1	2,7	23	15
16	Mean predominent direction of the wind		W	W	W	W	W	W	E,SE	W,SE	W	W,N	W,N	W		28	16

102

169 Station / Country **Ahmadābād / India**

Location **23°04'N / 72°38'E** — Height above sealevel **55 m** — Climate symbol: Köppen **BSh** — Troll **V,4**

		J	F	M	A	M	J	J	A	S	O	N	D	year	P	
1	Mean daily temperature	in °C	22,0	23,1	27,8	31,7	33,9	32,8	30,0	28,6	29,2	29,2	26,1	22,5	28,1	23	1
2	Mean daily maximum temperature	in °C	29,4	31,1	36,1	40,0	41,7	38,3	33,9	32,2	33,9	36,1	33,9	30,0	35,0	23	2
3	Mean daily minimum temperature	in °C	14,4	15,0	19,4	23,3	26,1	27,2	28,1	25,0	24,4	22,2	18,3	15,0	21,7	23	3
4	Absolute maximum temperature	in °C	35,0	37,8	41,1	43,3	45,0	43,3	40,0	37,8	40,0	39,4	38,3	35,6	45,0	10	4
5	Absolute minimum temperature	in °C	5,6	10,0	11,7	16,7	21,1	22,2	22,2	22,8	21,7	16,1	14,4	9,4	5,6	10	5
6	Mean relative humidity	in %	36	35	31	33	41	57	72	74	67	44	37	37	47	8	6
7	Mean precipitation	in mm	2	3	3	2	10	109	284	206	94	15	3	2	733	23	7
8	Maximum precipitation	in mm	57	26	17	26	107	298	953	589	637	181	53	14	1998	65	8
9	Minimum precipitation	in mm	0	0	0	0	0	0	4	1	0	0	0	0	120	65	9
10	Maximum precipitation in 24 h	in mm	8	3	13	3	30	76	290	140	89	30	10	2	290	10	10
11	Mean number of days with precipitation	> 2,5 mm	<1	<1	<1	<1	<1	5	12	11	5	<1	<1	<1	41	23	11
12	Mean duration of sunshine	in h	298	286	288	300	329	264	143	133	201	296	291	298	3125		12
13	Mean quantity of radiation	in ly / day	410	483	583	623	650	585	413	405	480	469	420	387	492	7	13
14	Mean potential evaporation	in mm	59	87	151	181	205	200	188	171	159	154	113	63	1711	23	14
15	Mean windspeed	in m / sec	1,5	1,4	2,0	2,2	2,8	3,0	3,0	2,3	2,0	1,3	1,1	0,7	1,9	30	15
16	Mean predominent direction of the wind		NE,NW	NE,NW	NW	NW,W	SW,W	SW,W	SW,W	SW,W	SW,W	NE	NE	NE		28	16

170 Station / Country **Verāval / India**

Location **20°55'N / 70°22'E** — Height above sealevel **8 m** — Climate symbol: Köppen **BSh** — Troll **V,4**

		J	F	M	A	M	J	J	A	S	O	N	D	year	P	
1	Mean daily temperature	in °C	21,7	22,0	24,4	26,1	28,1	28,9	27,8	27,0	27,0	27,2	25,6	23,3	25,8	50	1
2	Mean daily maximum temperature	in °C	27,8	27,8	30,0	30,0	30,0	30,6	28,9	28,3	28,9	31,7	31,7	29,4	29,4	50	2
3	Mean daily minimum temperature	in °C	15,6	16,1	18,9	22,2	26,1	27,2	26,7	25,6	25,0	22,8	20,0	17,2	21,7	50	3
4	Absolute maximum temperature	in °C	34,4	35,6	40,6	40,6	40,0	33,9	33,9	31,7	36,7	38,9	37,2	35,0	40,6	30	4
5	Absolute minimum temperature	in °C	4,4	4,4	9,4	13,9	18,9	23,3	21,1	22,8	20,6	16,1	13,9	7,8	4,4	30	5
6	Mean relative humidity	in %	54	64	68	77	83	83	87	87	84	70	60	56	73	10	6
7	Mean precipitation	in mm	2	3	3	2	8	119	198	91	64	15	5	3	513	50	7
8	Maximum precipitation	in mm	14	19	22	138	181	506	719	697	458	283	142	28	1327	70	8
9	Minimum precipitation	in mm	0	0	0	0	<1	10	5	0	0	0	0	0	69	70	9
10	Maximum precipitation in 24 h	in mm	13	18	18	2	124	198	259	269	277	173	91	25	277	47	10
11	Mean number of days with precipitation	> 2,5 mm	<1	<1	<1	<1	<1	4	9	6	4	1	<1	<1	31	30	11
12	Mean duration of sunshine	in h															12
13	Mean quantity of radiation	in ly / day															13
14	Mean potential evaporation	in mm	85	61	99	133	168	171	169	152	140	141	116	76	1491	50	14
15	Mean windspeed	in m / sec	3,4	3,7	4,1	4,6	4,8	6,2	7,9	6,7	4,2	3,4	2,9	3,1	4,6	30	15
16	Mean predominent direction of the wind		NE,W	N,W	N,W	NW,W	W	W	W	W	W	N,W	NE,S	NE,W		28	16

171 Station / Country **Pune (Poona) / India**

Location **18°32'N / 73°51'E** — Height above sealevel **559 m** — Climate symbol: Köppen **BSh** — Troll **V,3**

		J	F	M	A	M	J	J	A	S	O	N	D	year	P	
1	Mean daily temperature	in °C	21,1	22,8	26,4	29,2	29,7	27,2	25,3	24,7	25,0	25,3	22,8	20,6	25,0	24	1
2	Mean daily maximum temperature	in °C	30,6	32,8	36,1	38,3	37,2	31,7	28,3	27,8	29,4	31,7	30,6	29,4	32,2	24	2
3	Mean daily minimum temperature	in °C	11,7	12,8	16,7	20,0	22,2	22,8	22,2	21,7	20,6	18,9	15,0	11,7	17,8	24	3
4	Absolute maximum temperature	in °C	35,0	38,9	42,8	43,3	43,3	41,7	35,6	35,0	35,6	37,8	36,1	35,0	43,3	24	4
5	Absolute minimum temperature	in °C	1,7	3,9	7,2	10,6	13,9	17,2	18,9	17,2	16,1	11,1	7,2	4,4	1,7	24	5
6	Mean relative humidity	in %	47	44	38	78	47	70	80	79	78	64	56	52	61	24	6
7	Mean precipitation	in mm	2	2	2	15	28	114	168	89	135	89	28	3	675	60	7
8	Maximum precipitation	in mm	35	26	38	70	182	384	509	277	310	441	209	71	1242	80	8
9	Minimum precipitation	in mm	0	0	0	0	0	1	31	19	10	4	0	0	289	80	9
10	Maximum precipitation in 24 h	in mm	20	15	23	51	84	97	114	64	132	147	97	33	147	24	10
11	Mean number of days with precipitation	> 2,5 mm	1	<1	<1	1	2	7	13	8	8	5	2	<1	50	60	11
12	Mean duration of sunshine	in h	301	288	304	300	295	186	93	115	165	254	273	288	2862		12
13	Mean quantity of radiation	in ly / day	445	524	578	611	620	514	388	392	448	487	435	409	487	12	13
14	Mean potential evaporation	in mm	82	79	143	166	183	158	130	114	111	118	78	59	1401	60	14
15	Mean windspeed	in m / sec	0,8	1,0	1,2	1,5	2,0	2,5	2,5	2,1	1,6	1,0	0,8	0,7	1,5	30	15
16	Mean predominent direction of the wind		W	W	W	W	W	W	W	W	W	W	E	E		28	16

172 Station / Country **Hyderābād / India**

Location **17°26'N / 78°27'E** — Height above sealevel **542 m** — Climate symbol: Köppen **BSh** — Troll **V,4**

		J	F	M	A	M	J	J	A	S	O	N	D	year	P	
1	Mean daily temperature	in °C	22,2	24,7	28,8	31,4	33,3	29,7	26,7	26,7	26,4	26,9	23,1	21,7	26,8	30	1
2	Mean daily maximum temperature	in °C	28,9	31,7	36,1	38,3	40,0	35,0	30,6	30,6	30,6	31,1	28,9	28,3	32,2	30	2
3	Mean daily minimum temperature	in °C	15,6	17,8	21,1	24,4	26,7	24,4	22,8	22,8	22,2	20,6	17,2	15,0	21,1	30	3
4	Absolute maximum temperature	in °C	35,0	37,2	41,1	43,3	44,4	43,9	37,2	36,1	36,1	36,1	33,3	33,3	44,4	30	4
5	Absolute minimum temperature	in °C	8,3	11,1	15,6	16,1	19,4	17,8	19,4	19,4	17,8	13,9	7,8	7,8	7,8	30	5
6	Mean relative humidity	in %	57	49	41	44	44	63	73	74	76	63	58	57	58	8	6
7	Mean precipitation	in mm	8	10	13	30	28	112	152	135	165	84	28	8	752	45	7
8	Maximum precipitation	in mm	132	96	114	141	116	324	365	334	499	355	229	69	1431	65	8
9	Minimum precipitation	in mm	0	0	0	0	0	17	31	25	32	0	0	0	455	65	9
10	Maximum precipitation in 24 h	in mm	94	38	104	61	58	122	109	71	101	107	97	43	122	30	10
11	Mean number of days with precipitation	> 2,5 mm	<1	1	1	2	2	7	11	10	9	4	2	<1	51	45	11
12	Mean duration of sunshine	in h	313	288	291	288	282	210	117	140	174	228	258	295	2882		12
13	Mean quantity of radiation	in ly / day	409	505	521	533	579	461	334	352	339	392	347	394	431	3	13
14	Mean potential evaporation	in mm	66	102	157	177	198	171	157	145	130	126	80	56	1565	50	14
15	Mean windspeed	in m / sec	2,2	2,5	2,7	3,0	3,4	6,6	6,1	5,1	3,5	2,5	2,2	2,0	3,5	30	15
16	Mean predominent direction of the wind		SE,E	SE,E	SE,E	SE	W	W	W	W	W	NE	NE	E		28	16

173 Station / Country Sholāpur / India

Location 17°40'N / 75°54'E — Height above sealevel 479 m — Climate symbol: Köppen BSh — Troll V.3

		J	F	M	A	M	J	J	A	S	O	N	D	year	P	
1 Mean daily temperature	in °C	23,1	25,8	28,9	32,0	32,8	28,9	27,0	26,7	26,7	26,7	23,9	22,2	27,0	30	1
2 Mean daily maximum temperature	in °C	30,8	33,3	36,7	39,4	40,6	35,0	31,7	31,7	31,7	32,8	30,8	29,4	33,9	30	2
3 Mean daily minimum temperature	in °C	15,8	17,8	21,1	24,4	25,0	23,3	22,2	21,7	21,7	20,6	17,2	15,0	20,6	30	3
4 Absolute maximum temperature	in °C	35,0	37,8	41,1	43,3	45,6	45,6	37,8	36,1	36,1	37,2	36,1	33,9	45,6	30	4
5 Absolute minimum temperature	in °C	7,2	10,0	13,9	17,2	18,9	17,2	18,3	18,9	18,3	15,6	10,6	7,8	7,2	30	5
6 Mean relative humidity	in %	37	30	25	28	33	61	69	67	65	49	41	39	45	5	6
7 Mean precipitation	in mm	3	3	5	8	23	112	107	109	193	76	33	10	682	60	7
8 Maximum precipitation	in mm	59	47	86	100	141	342	319	519	447	261	210	105	1236	80	8
9 Minimum precipitation	in mm	0	0	0	0	0	12	3	8	2	0	0	0	325	80	9
10 Maximum precipitation in 24 h	in mm	36	33	33	13	43	119	89	191	99	119	79	28	191	30	10
11 Mean number of days with precipitation	> 2,5 mm	<1	<1	1	1	2	7	8	7	9	4	2	1	44	60	11
12 Mean duration of sunshine	in h															12
13 Mean quantity of radiation	in ly / day															13
14 Mean potential evaporation	in mm	78	104	162	182	197	173	157	151	136	127	86	65	1618		14
15 Mean windspeed	in m / sec	2,2	2,4	2,3	2,7	3,3	3,5	3,6	3,2	2,6	2,4	2,5	2,3	2,7	30	15
16 Mean predominent direction of the wind		SE	SE	NE	NE	NW	W	W	W	W	NE	NE	SE		28	16

174 Station / Country Bellary / India

Location 15°09'N / 76°51'E — Height above sealevel 449 m — Climate symbol: Köppen BSh — Troll V.3

		J	F	M	A	M	J	J	A	S	O	N	D	year	P	
1 Mean daily temperature	in °C	23,9	26,7	30,0	32,2	32,2	29,4	28,1	27,8	27,5	26,7	25,0	23,3	27,7	30	1
2 Mean daily maximum temperature	in °C	30,6	33,9	37,2	38,9	38,9	34,4	32,2	32,2	32,2	31,7	30,6	29,4	33,3	30	2
3 Mean daily minimum temperature	in °C	17,2	19,4	22,8	25,6	25,6	24,4	23,9	23,3	22,8	21,7	19,4	17,2	21,7	30	3
4 Absolute maximum temperature	in °C	35,0	38,9	41,7	42,8	43,3	42,2	38,3	36,7	37,8	36,7	35,0	35,6	43,3	30	4
5 Absolute minimum temperature	in °C	11,7	13,9	16,7	19,4	20,0	20,0	19,4	19,4	16,1	16,1	12,8	10,6	10,6	30	5
6 Mean relative humidity	in %	47	38	31	36	39	55	60	62	62	59	54	51	50	5	6
7 Mean precipitation	in mm	3	5	5	13	48	43	41	61	124	107	51	3	504	60	7
8 Maximum precipitation	in mm	46	90	50	86	211	137	224	285	336	404	223	61	949	80	8
9 Minimum precipitation	in mm	0	0	0	0	0	0	2	0	2	<1	0	0	208	80	9
10 Maximum precipitation in 24 h	in mm	13	61	28	53	163	86	71	99	104	107	94	33	163	30	10
11 Mean number of days with precipitation	> 2,5 mm	<1	<1	1	2	3	3	3	4	7	6	3	<1	35	60	11
12 Mean duration of sunshine	in h	310	294	310	297	288	198	149	171	195	242	252	288	2994		12
13 Mean quantity of radiation	in ly / day															13
14 Mean potential evaporation	in mm	85	125	175	176	193	175	171	162	152	145	121	88	1768	60	14
15 Mean windspeed	in m / sec	1,4	1,4	1,6	1,9	2,8	3,8	4,1	3,8	3,1	1,5	1,2	1,2	2,3	30	15
16 Mean predominent direction of the wind		SE	SE	SE	NW,SE	NW	W	W	W	NW	NW	SE	SE		28	16

175 Station / Country Leh / India

Location 34°09'N / 77°34'E — Height above sealevel 3506 m — Climate symbol: Köppen BWk — Troll III.11

		J	F	M	A	M	J	J	A	S	O	N	D	year	P	
1 Mean daily temperature	in °C	-7,2	-5,9	0,6	6,1	8,3	13,3	17,5	17,0	13,3	7,2	0,9	-4,2	5,6	23	1
2 Mean daily maximum temperature	in °C	-1,1	0,6	7,2	13,3	16,1	20,0	25,0	23,9	21,1	15,0	8,3	2,2	12,8	23	2
3 Mean daily minimum temperature	in °C	-13,3	-12,2	-6,1	-1,1	0,6	6,7	10,0	10,0	5,6	-0,6	-6,7	-10,6	-1,1	23	3
4 Absolute maximum temperature	in °C	5,6	12,8	21,7	21,7	25,0	28,9	31,1	28,3	28,3	23,9	14,4	10,6	31,1	15	4
5 Absolute minimum temperature	in °C	-26,1	-30,6	-17,2	-7,7	-3,9	-5,0	0,6	3,3	-0,6	-7,7	-12,8	-30,6	-30,6	15	5
6 Mean relative humidity	in %	57	54	53	49	39	42	54	58	50	46	57	56	51	23	6
7 Mean precipitation	in mm	10	8	8	5	5	5	13	15	8	2	3	5	87	60	7
8 Maximum precipitation	in mm	41	42	39	32	69	28	75	112	69	88	16	35	231	80	8
9 Minimum precipitation	in mm	0	0	0	0	0	0	0	0	0	0	0	0	25	80	9
10 Maximum precipitation in 24 h	in mm	13	10	15	36	13	8	18	51	13	25	5	10	51	30	10
11 Mean number of days with precipitation	> 2,5 mm	1	1	1	1	1	1	1	2	1	<1	<1	1	13	60	11
12 Mean duration of sunshine	in h															12
13 Mean quantity of radiation	in ly / day															13
14 Mean potential evaporation	in mm	0	0	0	37	65	91	110	104	73	34	0	0	514	60	14
15 Mean windspeed	in m / sec	0,9	1,1	1,5	1,9	1,9	1,8	1,4	1,3	1,3	1,4	1,4	1,1	1,4	30	15
16 Mean predominent direction of the wind		NE,SW	NE,SW	S,SW	S,SW	S,SW	S,SW	SW	SW	SW	SW	SW	SW		24	16

176 Station / Country Birkaner / India

Location 28°00'N / 73°18'E — Height above sealevel 224 m — Climate symbol: Köppen BWh — Troll IV.5

		J	F	M	A	M	J	J	A	S	O	N	D	year	P	
1 Mean daily temperature	in °C	14,7	18,3	24,2	29,7	34,4	35,3	33,3	31,1	30,9	28,1	21,7	16,4	26,5	30	1
2 Mean daily maximum temperature	in °C	22,2	25,6	32,2	37,2	41,7	41,7	38,9	36,1	36,7	35,6	30,0	24,4	33,3	30	2
3 Mean daily minimum temperature	in °C	7,2	11,1	16,1	22,2	27,2	28,9	27,8	26,1	25,0	20,6	13,3	8,3	19,4	30	3
4 Absolute maximum temperature	in °C	31,1	36,7	42,8	47,2	49,4	47,8	46,1	42,2	43,9	42,2	37,2	31,7	49,4	30	4
5 Absolute minimum temperature	in °C	-1,1	-0,6	6,7	10,6	16,7	18,9	20,6	21,7	19,4	8,3	0,6	0,0	-1,1	30	5
6 Mean relative humidity	in %	41	37	27	22	26	39	53	59	51	30	31	42	38	10	6
7 Mean precipitation	in mm	8	8	5	5	15	30	84	91	33	3	2	3	287	60	7
8 Maximum precipitation	in mm	61	81	69	35	120	148	311	337	223	111	44	38	771	80	8
9 Minimum precipitation	in mm	0	0	0	0	0	0	0	0	0	0	0	0	29	80	9
10 Maximum precipitation in 24 h	in mm	18	41	13	30	46	58	135	118	132	97	13	23	135	30	10
11 Mean number of days with precipitation	> 2,5 mm	1	1	1	1	1	2	5	5	2	<1	<1	1	22	60	11
12 Mean duration of sunshine	in h															12
13 Mean quantity of radiation	in ly / day															13
14 Mean potential evaporation	in mm	17	29	108	181	211	212	205	193	172	147	147	17	1639	45	14
15 Mean windspeed	in m / sec	1,3	1,4	1,8	2,0	2,8	3,7	3,6	3,0	2,6	1,5	1,0	1,0	2,1	23	15
16 Mean predominent direction of the wind		SE,NE	SE,NW	SE,NW	SW,NW	SW	SW	SW	SW	SW	SW	SE,NW	SE,NE		28	16

177 Station/Country Jodhpur/India

Location 26°18'N/73°01'E   Height above sealevel 238 m   Climate symbol: Köppen BWh   Troll IV.5

		J	F	M	A	M	J	J	A	S	O	N	D	year	P	
1	Mean daily temperature	in °C	16,7	18,9	24,2	29,2	33,3	33,9	31,4	29,4	29,2	26,7	22,0	18,1	26,1	23	1
2	Mean daily maximum temperature	in °C	24,4	26,7	32,2	37,2	40,6	40,0	36,1	33,9	34,4	35,0	31,1	26,1	33,3	23	2
3	Mean daily minimum temperature	in °C	8,9	11,1	16,1	21,1	26,1	27,8	26,7	25,0	23,9	18,3	12,8	10,0	18,9	23	3
4	Absolute maximum temperature	in °C	32,8	37,8	41,1	45,0	48,9	46,1	44,4	40,6	42,2	41,1	36,7	32,2	48,9	10	4
5	Absolute minimum temperature	in °C	-1,1	2,8	9,4	17,2	18,9	22,2	22,2	22,8	20,6	13,9	8,9	3,9	-1,1	10	5
6	Mean relative humidity	in %	36	33	25	23	30	45	60	67	57	32	29	36	39	8	6
7	Mean precipitation	in mm	3	5	3	3	10	36	101	122	61	8	3	3	358	50	7
8	Maximum precipitation	in mm	57	49	48	30	99	183	391	544	305	183	41	23	1177	70	8
9	Minimum precipitation	in mm	0	0	0	0	0	0	0	0	0	0	0	0	24	70	9
10	Maximum precipitation in 24 h	in mm	25	23	8	13	18	61	79	165	38	38	18	10	165	10	10
11	Mean number of days with precipitation	> 2,5 mm	<1	1	<1	1	1	2	5	6	3	1	<1	<1	24	50	11
12	Mean duration of sunshine	in h	279	270	279	300	322	285	208	198	249	310	282	288	3270		12
13	Mean quantity of radiation	in ly/day	399	469	557	627	651	619	513	485	520	497	421	375	511	9	13
14	Mean potential evaporation	in mm	13	18	108	175	207	209	200	177	161	145	73	32	1518	10	14
15	Mean windspeed	in m/sec	2,5	2,4	2,7	2,8	4,2	5,1	4,6	3,6	3,0	1,8	1,6	2,0	3,0	30	15
16	Mean predominent direction of the wind		NE	NE	NE,W	SW,W	SW	SW	SW	SW	SW	NE,SW	NE	NE		28	16

178 Station/Country Dwārka/India

Location 22°22'N/69°05'E   Height above sealevel 11 m   Climate symbol: Köppen BWh   Troll V,4

		J	F	M	A	M	J	J	A	S	O	N	D	year	P	
1	Mean daily temperature in °C	20,3	21,7	24,7	26,9	29,2	29,7	28,9	27,5	27,5	27,2	25,6	22,0	25,9	40	1
2	Mean daily maximum temperature in °C	25,6	26,1	27,8	29,4	31,1	31,7	30,6	29,4	29,4	30,6	30,6	27,2	28,9	40	2
3	Mean daily minimum temperature in °C	15,0	17,2	21,7	24,4	27,2	27,8	27,2	26,1	25,0	23,9	20,6	16,7	22,8	40	3
4	Absolute maximum temperature in °C	33,3	35,6	38,3	41,1	42,2	36,7	35,6	31,7	39,4	39,4	37,2	33,9	42,2	40	4
5	Absolute minimum temperature in °C	6,1	8,3	7,8	17,2	20,0	22,8	22,8	21,7	22,2	17,8	12,2	8,3	6,1	40	5
6	Mean relative humidity in %	54	65	71	79	80	79	81	83	81	74	60	53	72	10	6
7	Mean precipitation in mm	3	5	3	2	2	51	175	66	38	8	3	3	359	40	7
8	Maximum precipitation in mm	25	73	59	24	8	307	813	357	390	158	379	42	1080	60	8
9	Minimum precipitation in mm	0	0	0	0	0	0	1	<1	0	0	0	0	26	60	9
10	Maximum precipitation in 24 h in mm	20	64	48	25	8	193	274	302	135	140	25	28	302	40	10
11	Mean number of days with precipitation > 2,5 mm	<1	1	<1	<1	<1	2	6	5	2	<1	<1	<1	23	20	11
12	Mean duration of sunshine in h															12
13	Mean quantity of radiation in ly/day															13
14	Mean potential evaporation in mm	53	55	106	144	177	177	178	157	144	141	112	63	1507	40	14
15	Mean windspeed in m/sec	3,7	4,1	4,5	4,6	5,5	6,5	7,5	6,1	4,2	3,3	3,4	3,4	4,7	30	15
16	Mean predominent direction of the wind	NE,NW	N,NW	N,NW	W	W	SW,W	SW,W	SW,W	SW,W	N,NW	NE,NW	NE,NW		28	16

179 Station/Country Jamshedpur/India

Location 22°49'N/86°11'E   Height above sealevel 129 m   Climate symbol: Köppen Aw   Troll V,2

		J	F	M	A	M	J	J	A	S	O	N	D	year	P	
1	Mean daily temperature in °C	19,2	21,4	26,7	31,1	33,1	31,7	28,6	28,6	28,6	26,7	22,2	18,6	26,4	20	1
2	Mean daily maximum temperature in °C	26,7	28,9	35,0	38,9	40,0	36,7	31,7	31,7	32,2	31,7	29,4	26,1	32,2	20	2
3	Mean daily minimum temperature in °C	11,7	13,9	18,3	23,3	26,1	26,7	25,6	25,6	25,0	21,7	15,0	11,1	20,6	20	3
4	Absolute maximum temperature in °C	32,8	35,0	41,7	44,4	46,1	47,2	40,6	35,6	36,1	36,1	33,3	31,1	47,2	20	4
5	Absolute minimum temperature in °C	3,9	5,0	10,0	16,1	19,4	21,1	21,7	21,7	21,7	13,3	6,1	5,0	3,9	20	5
6	Mean relative humidity in %	81	74	54	53	62	72	85	86	85	80	78	81	74	20	6
7	Mean precipitation in mm	18	30	18	23	69	191	417	366	301	71	20	8	1424	20	7
8	Maximum precipitation in mm	87	112	108	128	203	578	636	630	404	225	110	23	1863	30	8
9	Minimum precipitation in mm	0	0	0	0	0	72	127	148	64	<1	0	0	748	30	9
10	Maximum precipitation in 24 h in mm	38	46	51	41	61	124	165	160	91	74	66	43	165	20	10
11	Mean number of days with precipitation > 2,5 mm	2	3	2	2	5	10	18	16	11	4	1	<1	74	20	11
12	Mean duration of sunshine in h															12
13	Mean quantity of radiation in ly/day															13
14	Mean potential evaporation in mm	36	54	135	178	200	192	179	170	155	138	65	34	1536	10	14
15	Mean windspeed in m/sec	0,9	1,1	1,4	1,7	2,2	2,3	2,2	2,0	1,6	1,1	0,9	0,8	1,5	30	15
16	Mean predominent direction of the wind	W	W	W	W	W,E	W,E	W,E	W,E	W,E	W,E	W	W		28	16

180 Station/Country Indore/India

Location 22°43'N/75°54'E   Height above sealevel 556 m   Climate symbol: Köppen Aw   Troll V,4

		J	F	M	A	M	J	J	A	S	O	N	D	year	P	
1	Mean daily temperature in °C	18,3	20,0	25,6	29,2	32,0	30,0	26,1	25,0	25,3	24,2	20,9	18,3	24,6	30	1
2	Mean daily maximum temperature in °C	26,7	28,3	33,3	37,2	39,4	35,6	29,4	28,3	29,4	31,7	28,9	26,7	31,1	30	2
3	Mean daily minimum temperature in °C	10,0	11,7	17,8	21,1	24,4	24,4	22,8	21,7	21,1	16,7	12,8	10,0	17,8	30	3
4	Absolute maximum temperature in °C	32,2	36,7	40,6	42,8	45,6	43,9	38,3	33,3	36,1	37,2	35,0	31,7	45,6	30	4
5	Absolute minimum temperature in °C	-1,1	-2,8	7,2	12,8	17,8	19,4	19,4	18,9	17,2	10,0	5,6	1,1	-2,8	30	5
6	Mean relative humidity in %	41	32	23	20	30	57	78	79	74	46	43	45	48	10	6
7	Mean precipitation in mm	5	3	2	3	13	147	282	206	165	30	15	8	879	60	7
8	Maximum precipitation in mm	105	40	35	57	108	409	690	707	767	174	217	92	1743	80	8
9	Minimum precipitation in mm	0	0	0	0	0	0	56	16	8	0	0	0	400	60	9
10	Maximum precipitation in 24 h in mm	81	15	8	18	36	117	292	211	137	79	64	46	292	30	10
11	Mean number of days with precipitation > 2,5 mm	1	1	<1	<1	1	7	13	11	8	2	1	1	48	60	11
12	Mean duration of sunshine in h															12
13	Mean quantity of radiation in ly/day															13
14	Mean potential evaporation in mm	40	45	105	167	195	177	148	124	114	102	57	39	1313	60	14
15	Mean windspeed in m/sec	2,7	3,0	3,6	4,3	6,6	7,5	7,3	6,0	5,1	2,6	2,1	2,0	4,4	30	15
16	Mean predominent direction of the wind	NE	NE,W	NE,W	W	W	W	W	W	W	NE	NE	NE		28	16

181 Station / Country **Calcutta / India**

Location 22°32'N / 88°20'E   Height above sealevel 6 m   Climate symbol: Köppen **Aw**   Troll **V.2**

		J	F	M	A	M	J	J	A	S	O	N	D	year	P	
1 Mean daily temperature	in °C	19.7	22.0	27.2	30.0	30.3	29.7	28.9	28.6	28.9	27.5	23.3	19.4	26.3	60	1
2 Mean daily maximum temperature	in °C	26.7	28.9	33.9	36.1	35.8	33.3	31.7	31.7	32.2	31.7	28.9	26.1	31.7	60	2
3 Mean daily minimum temperature	in °C	12.8	15.0	20.6	23.9	25.0	26.1	26.1	25.6	25.6	23.3	17.8	12.8	21.1	60	3
4 Absolute maximum temperature	in °C	31.7	36.7	40.0	41.7	42.2	43.9	36.7	35.6	36.1	35.6	33.3	30.8	43.9	60	4
5 Absolute minimum temperature	in °C	6.7	7.8	10.0	16.1	18.3	21.1	22.8	23.3	22.2	17.2	10.6	7.2	6.7	60	5
6 Mean relative humidity	in %	69	64	63	66	70	79	83	85	84	79	68	73	20		6
7 Mean precipitation	in mm	10	30	36	43	140	297	325	328	251	114	20	5	1599	60	7
8 Maximum precipitation	in mm	66	202	159	155	435	791	644	673	1157	474	239	65	2501	80	8
9 Minimum precipitation	in mm	0	0	0	0	11	41	115	122	56	0	0	0	909	80	9
10 Maximum precipitation in 24 h	in mm	43	81	69	107	155	302	183	254	368	173	84	53	368	60	10
11 Mean number of days with precipitation	> 2,5 mm	1	2	2	3	7	13	18	18	13	6	1	<1	85	60	11
12 Mean duration of sunshine	in h	245	241	254	258	254	132	118	121	135	195	249	254	2456		12
13 Mean quantity of radiation	in ly/day	357	417	480	528	542	413	394	392	367	381	388	350	417	12	13
14 Mean potential evaporation	in mm	41	58	141	171	185	179	180	168	153	135	72	37	1520		14
15 Mean windspeed	in m/sec	0.8	1.0	1.4	2.0	2.4	1.9	1.7	1.5	1.2	0.9	0.7	0.7	1.4	30	15
16 Mean predominent direction of the wind		N,NW	NW	SW	SW,S	SW,S	SW,S	SW,S	SW,S	SW,S	NE	N	N		28	16

182 Station / Country **Raipur / India**

Location 21°14'N / 81°39'E   Height above sealevel 296 m   Climate symbol: Köppen **Aw**   Troll **V.3**

		J	F	M	A	M	J	J	A	S	O	N	D	year	P	
1 Mean daily temperature	in °C	20.6	23.1	27.5	32.1	34.7	31.4	27.2	27.0	27.5	26.4	22.5	20.0	26.7	30	1
2 Mean daily maximum temperature	in °C	27.8	30.0	35.0	39.4	41.7	36.7	30.6	30.0	31.1	31.7	28.9	27.2	32.2	30	2
3 Mean daily minimum temperature	in °C	13.3	16.1	20.0	24.4	27.8	26.1	23.9	23.9	23.9	21.1	16.1	12.8	20.8	30	3
4 Absolute maximum temperature	in °C	33.3	37.2	41.7	46.1	47.2	47.2	38.9	34.4	35.6	37.2	35.6	32.2	47.2	30	4
5 Absolute minimum temperature	in °C	7.2	6.1	12.8	12.7	17.8	18.3	20.6	20.0	19.4	13.9	8.9	5.6	5.6	30	5
6 Mean relative humidity	in %	53	47	33	29	30	60	84	84	80	68	57	56	57	10	6
7 Mean precipitation	in mm	10	23	18	15	23	231	381	363	198	56	13	5	1334	60	7
8 Maximum precipitation	in mm	134	119	108	131	147	639	989	795	556	246	138	69	2181	80	8
9 Minimum precipitation	in mm	0	0	0	0	0	19	144	101	8	0	0	0	689	80	9
10 Maximum precipitation in 24 h	in mm	48	58	36	23	66	198	203	236	124	114	71	25	236	30	10
11 Mean number of days with precipitation	> 2,5 mm	1	2	2	2	2	10	16	15	10	3	1	<1	65	60	11
12 Mean duration of sunshine	in h															12
13 Mean quantity of radiation	in ly/day															13
14 Mean potential evaporation	in mm	48	75	151	182	207	190	162	154	147	122	63	20	1521	60	14
15 Mean windspeed	in m/sec	1.4	1.7	1.9	2.3	3.0	3.4	3.3	2.9	2.0	1.7	1.1	1.2	2.2	30	15
16 Mean predominent direction of the wind		NE	N	NE,W	SW,W	SW,NW	SW	SW,W	SW,W	SW,W	NE	NE	NE		27	16

183 Station / Country **Surat / India**

Location 21°12'N / 72°50'E   Height above sealevel 12 m   Climate symbol: Köppen **Aw**   Troll **V.4**

		J	F	M	A	M	J	J	A	S	O	N	D	year	P	
1 Mean daily temperature	in °C	22.5	23.6	27.5	30.3	31.1	30.3	28.1	27.8	28.1	28.3	25.9	23.1	27.2	60	1
2 Mean daily maximum temperature	in °C	30.6	33.3	36.7	39.4	40.6	35.0	31.7	31.7	31.7	32.8	30.6	29.4	33.9	30	2
3 Mean daily minimum temperature	in °C	15.6	17.8	21.1	24.4	25.0	23.3	22.2	21.7	21.7	20.6	17.2	15.0	20.6	30	3
4 Absolute maximum temperature	in °C	37.8	39.4	43.3	45.0	45.0	45.6	38.9	37.2	38.9	41.1	38.9	36.7	45.6	60	4
5 Absolute minimum temperature	in °C	4.4	5.6	10.6	15.0	19.4	21.7	20.6	21.1	20.6	14.4	10.6	6.7	4.4	60	5
6 Mean relative humidity	in %	47	48	44	49	60	71	81	79	74	55	47	47	59	10	6
7 Mean precipitation	in mm	3	3	2	2	5	218	442	191	150	38	8	3	1065	60	7
8 Maximum precipitation	in mm	92	43	10	149	128	782	1064	492	480	429	190	42	2284	80	8
9 Minimum precipitation	in mm	0	0	0	0	0	2	22	19	0	0	0	0	383	80	9
10 Maximum precipitation in 24 h	in mm	43	38	5	33	48	259	457	229	244	256	79	43	457	60	10
11 Mean number of days with precipitation	> 2,5 mm	<1	<1	<1	<1	<1	8	16	13	7	2	<1	<1	53	43	11
12 Mean duration of sunshine	in h	301	288	307	315	319	222	124	133	186	291	291	295	3072		12
13 Mean quantity of radiation	in ly/day															13
14 Mean potential evaporation	in mm	64	78	145	172	191	184	171	163	147	149	108	71	1643	60	14
15 Mean windspeed	in m/sec	1.7	1.7	1.8	2.1	3.0	3.4	3.3	3.0	2.0	1.8	1.7	1.7	2.2	30	15
16 Mean predominent direction of the wind		NE,NW	NE,NW	NE,NW	SW	SW	SW	SW	SW	SW,NE	SW,NE	NE,E	NE,NW		28	16

184 Station / Country **Nagpur / India**

Location 21°09'N / 79°09'E   Height above sealevel 312 m   Climate symbol: Köppen **Aw**   Troll **V.3**

		J	F	M	A	M	J	J	A	S	O	N	D	year	P	
1 Mean daily temperature	in °C	20.9	23.6	28.3	32.5	35.3	31.4	27.5	27.2	27.5	26.7	22.5	19.7	27.3	28	1
2 Mean daily maximum temperature	in °C	28.3	29.4	31.7	33.9	36.1	36.7	35.6	34.4	33.9	31.7	29.4	27.8	32.2	43	2
3 Mean daily minimum temperature	in °C	21.7	22.8	24.4	26.1	26.7	26.1	26.1	25.0	25.0	24.4	23.3	22.2	24.4	43	3
4 Absolute maximum temperature	in °C	35.0	38.9	45.0	46.1	47.8	47.2	40.6	37.8	38.9	38.3	35.6	33.3	47.8	28	4
5 Absolute minimum temperature	in °C	5.0	6.7	8.3	16.1	19.4	21.1	21.1	20.6	18.9	13.9	3.9	5.6	3.9	28	5
6 Mean relative humidity	in %	48	40	32	28	27	56	78	78	73	54	49	48	51	11	6
7 Mean precipitation	in mm	10	18	15	15	20	224	371	290	203	56	20	13	1255	60	7
8 Maximum precipitation	in mm	105	157	104	129	165	501	597	595	627	267	162	128	1931	80	8
9 Minimum precipitation	in mm	0	0	0	0	<1	90	47	52	0	0	0	0	365	80	9
10 Maximum precipitation in 24 h	in mm	41	51	41	58	56	145	124	185	175	107	41	41	185	30	10
11 Mean number of days with precipitation	> 2,5 mm	1	1	1	1	2	10	17	13	11	3	1	1	62	60	11
12 Mean duration of sunshine	in h	264	280	270	287	287	162	87	96	156	251	287	264	2611		12
13 Mean quantity of radiation	in ly/day	420	500	543	583	598	505	379	367	433	488	442	390	471	9	13
14 Mean potential evaporation	in mm	53	79	153	184	207	192	166	157	147	83	66	49	1538	60	14
15 Mean windspeed	in m/sec															15
16 Mean predominent direction of the wind		N,E	N,NE	N,NE	NW	NW	W	W	W	W	N,NE	N,NE	N,E		28	16

EURASIA
Climate Zones after W. Köppen/R. Geiger

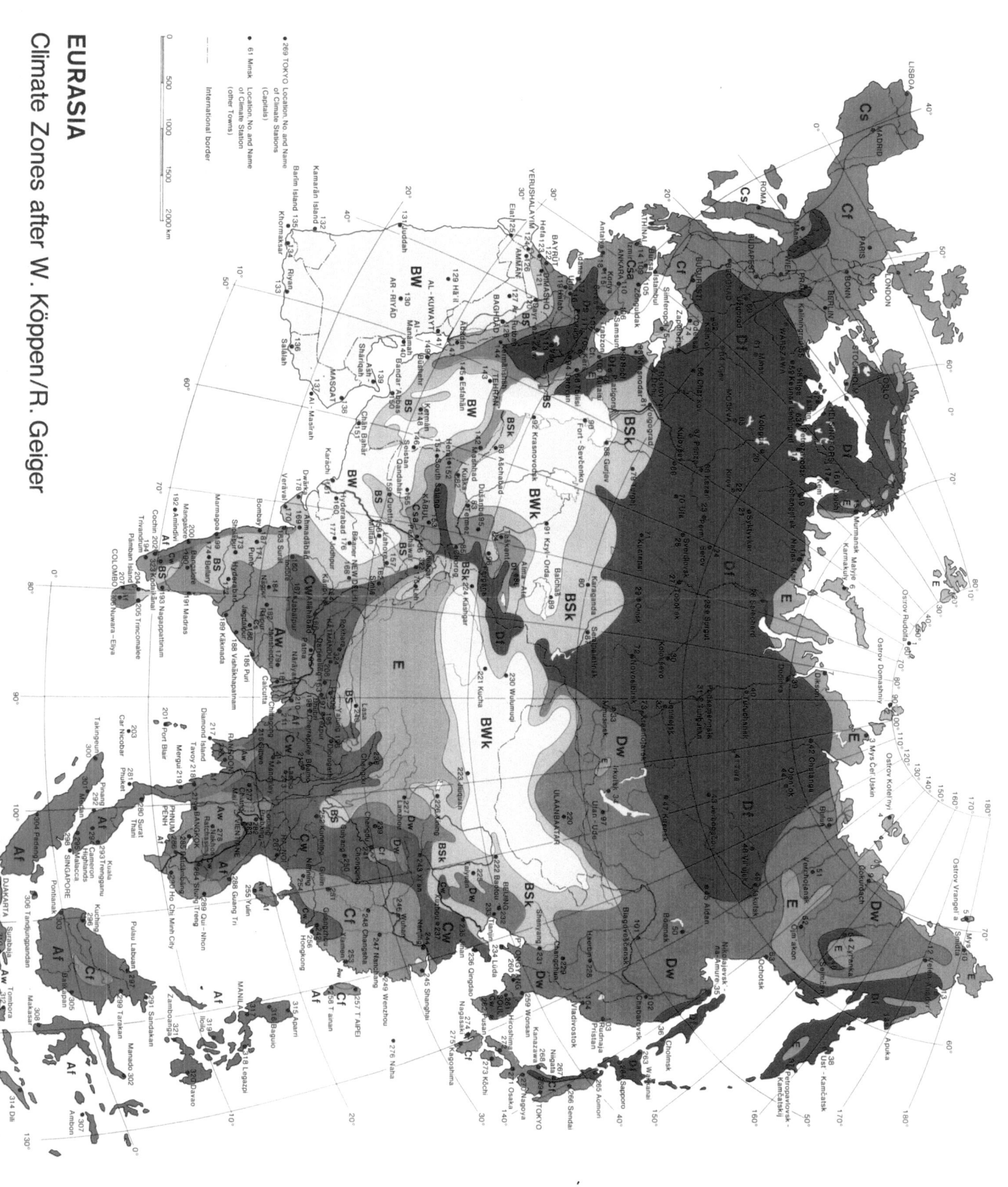

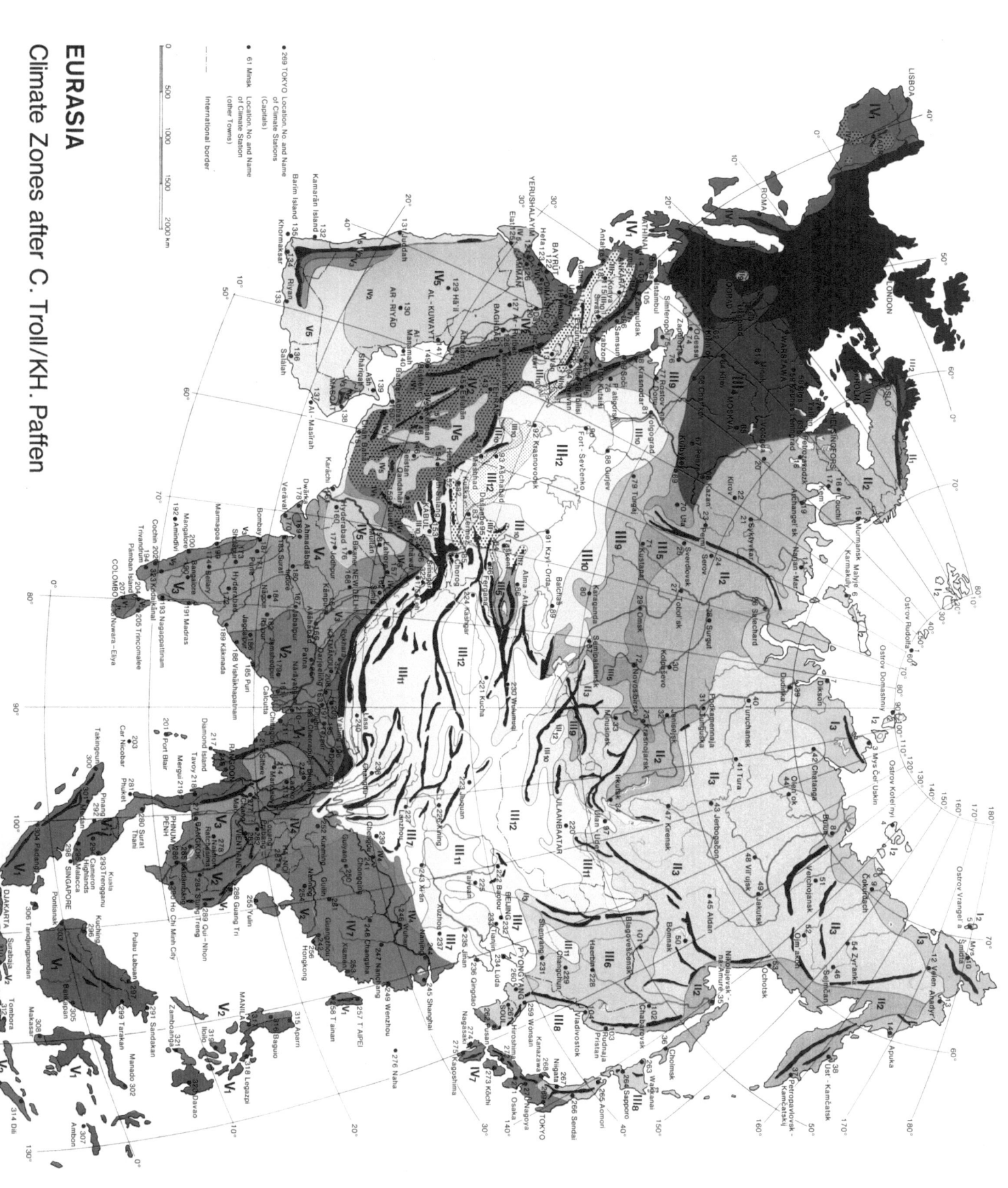

EURASIA
Climate Zones after C. Troll/KH. Paffen

185 Station/Country Puri/India

Location 19°48'N/85°49'E — Height above sealevel 6 m — Climate symbol: Köppen Aw — Troll V,2

		J	F	M	A	M	J	J	A	S	O	N	D	year	P
1 Mean daily temperature	in °C	22,2	24,4	27,2	28,6	29,7	29,4	28,6	28,9	29,2	28,3	25,0	22,2	27,0	50
2 Mean daily maximum temperature	in °C	26,7	28,3	30,0	31,1	32,2	31,7	30,6	31,1	31,7	31,7	29,4	27,2	30,0	50
3 Mean daily minimum temperature	in °C	17,8	20,6	24,4	26,1	27,2	27,2	26,7	26,7	26,7	25,0	20,6	17,2	23,9	50
4 Absolute maximum temperature	in °C	31,7	33,9	38,9	38,9	39,4	37,8	36,7	36,7	36,1	36,1	33,9	32,8	39,4	50
5 Absolute minimum temperature	in °C	13,9	12,2	15,6	18,9	16,7	19,4	22,8	21,7	17,2	16,7	13,9	10,6	10,6	50
6 Mean relative humidity	in %	74	76	80	83	84	84	84	84	81	75	78	68	79	10
7 Mean precipitation	in mm	10	23	13	15	74	191	282	277	231	183	81	5	1365	50
8 Maximum precipitation	in mm	165	120	116	108	532	571	530	632	510	606	578	135	1984	70
9 Minimum precipitation	in mm	0	0	0	0	0	23	49	47	48	0	0	0	500	70
10 Maximum precipitation in 24 h	in mm	53	109	58	81	173	201	302	188	211	318	244	89	318	50
11 Mean number of days with precipitation	> 2,5 mm	<1	1	<1	1	3	8	12	13	12	7	2	<1	82	50
12 Mean duration of sunshine	in h														
13 Mean quantity of radiation	in ly/day														
14 Mean potential evaporation	in mm	65	89	143	161	181	175	177	170	160	150	102	61	1635	
15 Mean windspeed	in m/sec	3,3	4,4	5,7	6,8	7,3	6,4	6,5	5,5	4,4	3,4	2,8	2,9	4,9	30
16 Mean predominent direction of the wind		N,S	N,S	SW,S	SW	SW	SW	SW	SW	SW,S	N,SE	N,SE	N,SE		28

186 Station/Country Jagdalpur/India

Location 19°05'N/82°02'E — Height above sealevel 553 m — Climate symbol: Köppen Aw — Troll V,2

		J	F	M	A	M	J	J	A	S	O	N	D	year	P
1 Mean daily temperature	in °C	19,7	22,5	26,7	29,7	31,4	28,6	25,3	25,3	25,6	24,4	21,4	18,9	25,0	30
2 Mean daily maximum temperature	in °C	27,8	30,6	35,0	37,2	38,3	33,3	28,3	28,3	29,4	29,4	27,8	26,7	31,1	30
3 Mean daily minimum temperature	in °C	11,7	14,4	18,3	22,2	24,4	23,9	22,2	22,2	21,7	19,4	15,0	11,1	18,9	30
4 Absolute maximum temperature	in °C	32,8	36,1	40,0	43,3	46,1	43,9	36,1	33,9	35,0	34,4	32,2	32,2	46,1	30
5 Absolute minimum temperature	in °C	4,4	6,7	10,6	16,1	17,2	17,2	19,4	16,7	18,3	11,1	6,7	4,4	4,4	30
6 Mean relative humidity	in %	58	53	42	44	45	69	84	84	83	73	66	64	64	30
7 Mean precipitation	in mm	10	28	13	46	61	246	384	414	264	112	28	5	1611	30
8 Maximum precipitation	in mm	86	202	79	111	259	498	674	905	542	374	153	40	2395	50
9 Minimum precipitation	in mm	0	0	0	2	0	34	140	186	20	12	0	0	882	50
10 Maximum precipitation in 24 h	in mm	33	119	46	53	64	132	180	203	132	137	104	23	203	30
11 Mean number of days with precipitation	> 2,5 mm	1	2	1	3	5	11	19	19	14	6	2	<1	84	30
12 Mean duration of sunshine	in h														
13 Mean quantity of radiation	in ly/day														
14 Mean potential evaporation	in mm	59	64	106	168	202	187	168	146	126	103	60	39	1428	30
15 Mean windspeed	in m/sec	0,8	1,1	1,2	1,6	1,8	2,0	2,5	2,2	1,6	1,0	0,8	0,7	1,4	30
16 Mean predominent direction of the wind		NE	SW	SW	SW	SW	SW	SW	SW	SW	NE	NE	NE		28

187 Station/Country Bombay/India

Location 18°54'N/72°49'E — Height above sealevel 11 m — Climate symbol: Köppen Aw — Troll V,3

		J	F	M	A	M	J	J	A	S	O	N	D	year	P
1 Mean daily temperature	in °C	23,9	23,9	26,1	28,1	29,7	28,9	27,2	27,0	27,0	28,1	27,2	25,6	26,9	60
2 Mean daily maximum temperature	in °C	28,3	28,3	30,0	31,7	32,8	31,7	29,4	29,4	29,4	31,7	31,7	30,6	30,6	60
3 Mean daily minimum temperature	in °C	19,4	19,4	22,2	24,4	26,7	26,1	25,0	24,4	24,4	24,4	22,8	20,6	23,3	60
4 Absolute maximum temperature	in °C	34,4	36,1	38,3	37,8	35,6	37,2	35,9	32,2	35,0	36,1	35,6	34,4	38,3	60
5 Absolute minimum temperature	in °C	11,7	11,7	16,7	20,0	22,8	21,1	22,2	22,2	21,7	21,1	17,8	12,8	11,7	60
6 Mean relative humidity	in %	66	67	69	71	71	78	83	82	76	69	66	66	73	48
7 Mean precipitation	in mm	3	3	3	2	18	485	617	340	264	64	13	3	1815	60
8 Maximum precipitation	in mm	76	43	37	64	279	1104	1500	1285	1245	507	164	29	3482	80
9 Minimum precipitation	in mm	0	0	0	0	0	92	108	83	41	0	0	0	849	80
10 Maximum precipitation in 24 h	in mm	48	41	33	28	127	409	305	287	549	150	122	25	549	60
11 Mean number of days with precipitation	> 2,5 mm	<1	<1	<1	1	1	14	21	19	13	3	1	<1	77	43
12 Mean duration of sunshine	in h	285	269	288	282	288	188	71	84	147	254	273	282	2691	
13 Mean quantity of radiation	in ly/day														
14 Mean potential evaporation	in mm	90	90	148	161	183	171	164	151	147	149	122	120	1694	
15 Mean windspeed	in m/sec	2,5	2,6	2,9	2,9	2,8	3,5	4,1	3,7	2,8	2,4	2,3	2,4	2,9	30
16 Mean predominent direction of the wind		NE,NW	NE,NW	N,NW	N,NW	W,NW	W	W	W	W	E,NW	E,NW	E,NW		28

188 Station/Country Vishākhapatnam/India

Location 17°42'N/83°18'E — Height above sealevel 38 m — Climate symbol: Köppen Aw — Troll V,2

		J	F	M	A	M	J	J	A	S	O	N	D	year	P
1 Mean daily temperature	in °C	23,3	25,0	27,2	28,9	30,3	30,3	28,9	28,9	28,3	28,1	25,6	23,6	27,4	45
2 Mean daily maximum temperature	in °C	27,2	28,3	30,6	32,2	33,3	33,3	31,7	31,7	31,1	31,1	28,9	27,2	30,6	45
3 Mean daily minimum temperature	in °C	19,4	21,7	27,9	25,6	27,2	27,2	26,1	25,6	25,6	25,0	22,2	20,0	23,9	45
4 Absolute maximum temperature	in °C	32,2	36,7	36,1	37,2	43,3	43,9	38,3	37,2	37,8	36,1	32,8	31,1	43,9	45
5 Absolute minimum temperature	in °C	14,4	16,1	18,3	18,3	20,0	21,1	21,7	21,1	22,2	20,6	15,0	15,0	14,4	45
6 Mean relative humidity	in %	72	74	75	75	75	75	76	77	78	75	65	65	74	10
7 Mean precipitation	in mm	10	23	13	18	51	104	112	132	165	198	119	15	960	40
8 Maximum precipitation	in mm	166	122	125	103	299	304	302	394	470	635	531	251	1442	60
9 Minimum precipitation	in mm	0	0	0	0	0	11	16	17	33	0	0	0		60
10 Maximum precipitation in 24 h	in mm	132	64	64	74	145	165	81	107	150	267	272	191	272	40
11 Mean number of days with precipitation	> 2,5 mm	<1	1	<1	1	3	7	9	8	9	8	4	<1	53	40
12 Mean duration of sunshine	in h														
13 Mean quantity of radiation	in ly/day	451	521	553	567	576	425	395	444	438	445	441	427	473	8
14 Mean potential evaporation	in mm	83	103	145	163	184	176	135	167	153	146	117	82	1694	
15 Mean windspeed	in m/sec	1,7	1,8	2,7	4,1	4,4	3,8	4,6	3,7	2,6	2,2	2,2	2,1	3,0	30
16 Mean predominent direction of the wind		NW,E	W,S	W,S	SW	SW	SW	SW	W,SW	W,SW	W,NE	N,E	N,E		28

189 Station/Country Käkinada/India

Location 16°57'N/82°14'E — Height above sealevel 8 m — Climate symbol: Köppen Aw — Troll V,3

		J	F	M	A	M	J	J	A	S	O	N	D	year	P	
1 Mean daily temperature	in °C	23.1	25.6	28.1	30.9	32.5	31.7	28.9	28.9	28.9	27.8	25.3	23.1	27.9	45	1
2 Mean daily maximum temperature	in °C	27.2	30.0	33.3	35.6	37.2	36.1	32.2	32.2	32.2	31.1	28.3	27.2	31.7	45	2
3 Mean daily minimum temperature	in °C	18.9	21.1	22.8	26.1	27.8	27.2	25.6	25.6	25.6	24.4	22.2	18.9	23.9	45	3
4 Absolute maximum temperature	in °C	32.8	37.8	38.9	42.8	46.7	47.8	41.7	37.8	37.2	37.2	33.3	32.2	47.8	45	4
5 Absolute minimum temperature	in °C	14.4	15.6	17.2	18.9	21.1	21.7	21.1	21.7	22.2	18.9	15.6	13.9	13.9	45	5
6 Mean relative humidity	in %	74	72	69	69	68	69	75	77	78	76	72	72	73	10	6
7 Mean precipitation	in mm	8	8	13	15	38	119	165	142	157	216	142	18	1041	60	7
8 Maximum precipitation	in mm	83	102	111	134	283	667	438	318	394	834	697	215	1826	80	8
9 Minimum precipitation	in mm	0	0	0	0	0	20	28	30	17	0	0	0	408	80	9
10 Maximum precipitation in 24 h	in mm	79	48	71	61	109	155	107	145	130	282	277	130	282	60	10
11 Mean number of days with precipitation	> 2,5 mm	<1	<1	<1	1	2	7	11	10	9	9	4	<1	57	60	11
12 Mean duration of sunshine	in h															12
13 Mean quantity of radiation	in ly/day															13
14 Mean potential evaporation	in mm	72	101	155	176	195	186	176	170	159	147	102	72	1801	45	14
15 Mean windspeed	in m/sec	2.9	2.4	2.3	2.5	3.1	3.4	3.4	3.0	2.4	2.8	3.3	3.1	2.9	30	15
16 Mean predominant direction of the wind		NE,SE	NE,SE	SW,SE	SW,SE	SW,SE	SW	SW	SW	SW	NE,SE	NE	NE,SE		28	16

190 Station/Country Bangalore/India

Location 12°57'N/77°37'E — Height above sealevel 920 m — Climate symbol: Köppen Aw — Troll V,2

		J	F	M	A	M	J	J	A	S	O	N	D	year	P	
1 Mean daily temperature	in °C	20.6	22.8	25.6	27.2	27.0	24.4	23.3	23.3	23.1	23.1	21.7	20.6	23.6	32	1
2 Mean daily maximum temperature	in °C	27.2	30.0	32.8	33.9	33.3	29.4	27.8	27.8	27.8	28.7	26.7	26.1	29.4	32	2
3 Mean daily minimum temperature	in °C	13.9	15.6	18.3	20.6	20.6	19.4	18.9	18.9	18.3	18.3	16.7	15.0	17.8	33	3
4 Absolute maximum temperature	in °C	32.2	34.4	37.2	38.3	38.9	37.8	33.3	32.2	32.8	32.2	31.1	31.1	38.9	15	4
5 Absolute minimum temperature	in °C	11.1	11.1	13.3	16.7	17.2	17.8	17.2	17.2	16.1	14.4	11.1	11.1	11.1	15	5
6 Mean relative humidity	in %	59	52	46	52	59	70	75	75	75	73	69	66	64	26	6
7 Mean precipitation	in mm	5	8	10	41	107	74	99	127	170	150	69	10	870	60	7
8 Maximum precipitation	in mm	102	90	72	166	287	219	350	344	491	522	252	86	1349	80	8
9 Minimum precipitation	in mm	0	0	0	1	1	5	6	21	8	10	0	0	544	80	9
10 Maximum precipitation in 24 h	in mm	15	53	33	91	61	58	84	94	84	117	114	69	117	30	10
11 Mean number of days with precipitation	> 1,0 mm	<1	<1	<1	3	7	6	8	9	9	9	5	1	60	60	11
12 Mean duration of sunshine	in h	267	258	295	258	229	144	93	127	156	174	216	233	2450		12
13 Mean quantity of radiation	in ly/day	472	493	596	587	556	443	354	366	405	412	388	417	466	3	13
14 Mean potential evaporation	in mm	65	79	124	147	152	112	102	100	92	76	70	65	1184	32	14
15 Mean windspeed	in m/sec	2.9	2.7	2.6	2.6	3.3	4.7	4.8	4.2	3.4	2.3	2.4	2.7	3.2	30	15
16 Mean predominant direction of the wind		E	E	SE,E	SW,E	W	W	W	W	W	W,NE	E	E		28	16

191 Station/Country Madras/India

Location 13°04'N/80°15'E — Height above sealevel 16 m — Climate symbol: Köppen Aw — Troll V,2

		J	F	M	A	M	J	J	A	S	O	N	D	year	P	
1 Mean daily temperature	in °C	24.4	25.6	27.5	30.3	33.1	32.5	30.9	30.3	29.7	28.1	25.9	24.7	28.6	60	1
2 Mean daily maximum temperature	in °C	29.4	31.1	32.8	35.0	38.3	37.8	35.6	35.0	34.4	32.2	29.4	28.9	33.3	60	2
3 Mean daily minimum temperature	in °C	19.4	20.0	22.2	25.6	27.8	27.2	26.1	25.6	25.0	23.9	22.2	20.6	23.9	60	3
4 Absolute maximum temperature	in °C	32.8	36.7	38.9	42.8	45.0	43.3	41.1	40.0	38.9	38.9	34.4	32.8	45.0	60	4
5 Absolute minimum temperature	in °C	13.9	15.0	16.7	20.0	21.1	20.6	21.7	20.6	20.6	18.7	15.0	13.9	13.9	60	5
6 Mean relative humidity	in %	77	75	74	73	65	60	64	67	73	79	81	80	72	10	6
7 Mean precipitation	in mm	36	10	8	15	25	48	91	117	119	305	358	140	1270	60	7
8 Maximum precipitation	in mm	244	183	283	191	389	195	266	336	278	892	1088	699	2135	80	8
9 Minimum precipitation	in mm	0	0	0	0	0	3	15	18	12	11	5	<1	522	80	9
10 Maximum precipitation in 24 h	in mm	213	124	64	84	132	58	117	79	99	234	236	262	262	60	10
11 Mean number of days with precipitation	> 2,5 mm	2	<1	<1	1	4	7	8	7	11	11	5		59	60	11
12 Mean duration of sunshine	in h	264	272	301	285	270	198	149	174	189	202	201	228	2734		12
13 Mean quantity of radiation	in ly/day	441	539	584	581	559	498	449	474	482	425	387	376	481	12	13
14 Mean potential evaporation	in mm	90	97	143	166	192	185	182	173	161	150	117	93	1749	60	14
15 Mean windspeed	in m/sec	2.5	2.5	2.8	2.9	3.6	4.5	4.0	3.6	3.1	2.5	3.2	3.5	3.3	30	15
16 Mean predominant direction of the wind		NW,NE	NW,E	SE	S,SE	S,SE	W,SE	W,SE	W,SE	W,SE	W,SE	W,NE	W,NE		28	16

192 Station/Country Amindivi (Amindivi Islands)/India

Location 11°07'N/72°44'E — Height above sealevel 4 m — Climate symbol: Köppen Aw — Troll V,2

		J	F	M	A	M	J	J	A	S	O	N	D	year	P	
1 Mean daily temperature	in °C	26.7	27.5	28.6	30.0	30.0	28.1	27.5	27.5	27.5	27.5	27.2	26.7	27.8	29	1
2 Mean daily maximum temperature	in °C	30.0	30.6	31.7	33.3	32.8	30.6	30.0	30.0	30.0	30.0	30.6	30.0	30.6	29	2
3 Mean daily minimum temperature	in °C	23.3	24.4	25.6	26.7	27.2	25.6	25.0	25.0	25.0	25.0	23.9	23.3	25.0	29	3
4 Absolute maximum temperature	in °C	32.8	34.4	35.0	37.2	36.7	35.6	32.8	32.8	33.3	33.9	33.9	35.0	37.2	21	4
5 Absolute minimum temperature	in °C	18.9	20.0	20.6	22.8	21.7	22.2	21.7	22.2	22.2	22.1	21.1	18.9	18.3	21	5
6 Mean relative humidity	in %	72	72	72	71	74	82	83	82	81	80	76	72	76	26	6
7 Mean precipitation	in mm	18	2	2	38	94	363	305	196	160	147	66	33	1424	30	7
8 Maximum precipitation	in mm	158	27	55	247	696	836	731	559	567	397	305	393	2550	65	8
9 Minimum precipitation	in mm	0	0	0	0	<1	145	35	12	45	22	0	0	910	65	9
10 Maximum precipitation in 24 h	in mm	74	18	30	122	140	211	180	241	218	114	66	86	241	32	10
11 Mean number of days with precipitation	> 2,5 mm	1	<1	<1	2	5	17	16	11	10	9	4	2	79	30	11
12 Mean duration of sunshine	in h															12
13 Mean quantity of radiation	in ly/day															13
14 Mean potential evaporation	in mm	139	134	159	166	177	158	157	153	147	147	139	139	1815	52	14
15 Mean windspeed	in m/sec	1.6	2.2	2.4	2.8	3.8	5.5	6.4	6.3	5.1	2.8	1.6	1.4	3.5	30	15
16 Mean predominant direction of the wind		N	N	NW	NW	NW	W	W	W	NW	NW	N	N		28	16

193 Station/Country Nagappattinam/India

Location 10°48'N/79°51'E   Height above sealevel 9 m   Climate symbol: Köppen Aw   Troll V,2

		J	F	M	A	M	J	J	A	S	O	N	D	year	P	
1	Mean daily temperature	in °C	25,0	26,1	28,1	30,0	31,4	31,4	30,9	29,7	29,4	28,1	26,4	25,0	28,5	43	1
2	Mean daily maximum temperature	in °C	28,3	29,4	31,7	33,9	36,1	36,7	35,6	34,4	33,9	31,7	29,4	27,8	32,2	43	2
3	Mean daily minimum temperature	in °C	21,7	22,8	24,4	26,1	26,7	26,1	25,0	25,0	24,4	23,3	22,2	24,4	43		3
4	Absolute maximum temperature	in °C	31,7	35,6	40,0	41,7	42,8	41,7	41,7	40,6	37,8	37,2	34,4	33,9	42,8	43	4
5	Absolute minimum temperature	in °C	16,1	15,6	16,7	20,0	20,6	20,6	21,7	20,6	20,6	20,6	17,8	16,7	15,6	43	5
6	Mean relative humidity	in %	77	74	74	73	69	62	63	70	72	77	80	79	73	10	6
7	Mean precipitation	in mm	43	15	8	15	41	33	48	91	97	267	450	290	1398	43	7
8	Maximum precipitation	in mm	571	256	158	289	573	120	234	283	388	928	1507	858	2196	80	8
9	Minimum precipitation	in mm	0	0	0	0	0	0	0	0	5	18	50	0	604	80	9
10	Maximum precipitation in 24 h	in mm	130	104	71	71	135	66	114	124	76	269	366	206	366	43	10
11	Mean number of days with precipitation	> 2,5 mm	2	1	1	1	2	2	3	5	6	10	13	9	55	43	11
12	Mean duration of sunshine	in h															12
13	Mean quantity of radiation	in ly/day															13
14	Mean potential evaporation	in mm	99	156	148	167	186	180	180	171	161	150	123	105	1826	60	14
15	Mean windspeed	in m/sec	5,2	4,4	4,0	3,6	3,5	3,5	3,1	2,9	2,7	2,4	3,7	4,9	3,7	30	15
16	Mean predominent direction of the wind		NE	NW,NE	NW,E	SW,SE	SW,S	W,SW	W,SW	W,SW	W,SE	W,NE	NW,NE	NW,NE		28	16

194 Station/Country Trivandrum/India

Location 8°29'N/76°57'E   Height above sealevel 61 m   Climate symbol: Köppen Aw   Troll V,2

		J	F	M	A	M	J	J	A	S	O	N	D	year	P	
1	Mean daily temperature	in °C	25,6	26,7	28,1	28,3	28,1	26,4	25,9	26,1	26,1	26,1	25,9	25,6	26,6	29	1
2	Mean daily maximum temperature	in °C	28,9	30,0	31,1	31,1	30,6	28,3	27,8	28,3	28,3	28,3	28,3	28,3	28,9	31	2
3	Mean daily minimum temperature	in °C	22,2	23,3	25,0	25,6	25,6	24,4	23,9	23,9	23,9	23,9	23,9	22,8	23,9	29	3
4	Absolute maximum temperature	in °C	32,2	37,8	33,9	33,9	33,2	32,2	31,1	31,7	31,7	31,1	33,9	31,1	37,8	30	4
5	Absolute minimum temperature	in °C	17,2	18,9	21,7	21,1	21,7	19,4	16,1	21,1	21,7	20,6	17,8	18,9	16,1	29	5
6	Mean relative humidity	in %	69	69	70	74	77	82	83	81	79	81	80	74	77	12	6
7	Mean precipitation	in mm	20	20	38	114	224	335	198	119	114	272	178	64	1696	100	7
8	Maximum precipitation	in mm	144	150	184	418	1055	960	557	451	444	531	568	340	3036	120	8
9	Minimum precipitation	in mm	0	0	0	2	7	74	22	16	6	57	7	0	1029	120	9
10	Maximum precipitation in 24 h	in mm	53	89	58	130	217	150	152	101	124	216	163	150	277	100	10
11	Mean number of days with precipitation	> 2,5 mm	1	1	3	7	10	17	14	9	8	12	9	4	95	69	11
12	Mean duration of sunshine	in h	257	238	257	201	183	111	127	164	177	171	177	229	2292		12
13	Mean quantity of radiation	in ly/day	503	538	563	530	470	449	416	458	499	448	425	442	478	10	13
14	Mean potential evaporation	in mm	125	126	145	155	159	140	134	133	136	136	123	124	1636		14
15	Mean windspeed	in m/sec	1,4	1,6	1,8	2,2	2,5	2,7	3,0	3,1	2,8	2,0	1,5	1,3	2,2	30	15
16	Mean predominent direction of the wind		NE,SW	NE,SW	NE,SW	N,W	N,NW	NW	NW	NW	NW	N,NW	NE,SW	NE,SW		28	16

195 Station/Country Tezpur/India

Location 26°37'N/92°47'E   Height above sealevel 79 m   Climate symbol: Köppen Aw   Troll V,2

		J	F	M	A	M	J	J	A	S	O	N	D	year	P	
1	Mean daily temperature	in °C	17,2	18,9	22,5	23,9	26,4	28,3	28,6	28,6	28,3	25,9	21,7	17,8	24,0	19	1
2	Mean daily maximum temperature	in °C	23,3	24,4	28,3	28,3	30,6	31,7	31,7	31,7	30,0	27,2	23,9	28,3	20		2
3	Mean daily minimum temperature	in °C	11,1	13,3	16,7	19,4	22,2	25,0	25,6	25,6	25,0	21,7	16,1	11,7	19,4	19	3
4	Absolute maximum temperature	in °C	27,8	30,6	36,7	36,7	36,7	36,7	36,1	36,1	36,7	36,1	33,3	27,2	36,7	20	4
5	Absolute minimum temperature	in °C	6,7	6,1	10,6	13,3	17,8	19,4	21,7	22,2	21,1	16,1	11,7	6,1	6,1	20	5
6	Mean relative humidity	in %	81	75	66	71	79	83	84	85	86	83	83	84	80	10	6
7	Mean precipitation	in mm	13	28	58	157	251	305	366	366	208	107	18	5	1882	20	7
8	Maximum precipitation	in mm	65	71	196	331	590	492	664	667	627	285	117	34	2883	60	8
9	Minimum precipitation	in mm	0	0	0	69	88	139	75	76	6	0	0	1334	60		9
10	Maximum precipitation in 24 h	in mm	30	28	61	74	147	137	130	142	104	140	61	20	147	20	10
11	Mean number of days with precipitation	> 2,5 mm	1	3	5	12	14	16	17	16	12	6	1	1	104	20	11
12	Mean duration of sunshine	in h															12
13	Mean quantity of radiation	in ly/day															13
14	Mean potential evaporation	in mm	34	43	87	108	154	174	181	173	152	124	64	37	1331	40	14
15	Mean windspeed	in m/sec	0,7	0,9	1,4	1,9	1,2	0,8	0,7	0,6	0,6	0,5	0,6	0,6	0,9	30	15
16	Mean predominent direction of the wind		NE	E	E	E	E	E	E	E	E	E	E	NE		27	16

196 Station/Country Dibrugarh/India

Location 27°28'N/94°55'E   Height above sealevel 108 m   Climate symbol: Köppen Cwa/Am   Troll V,1

		J	F	M	A	M	J	J	A	S	O	N	D	year	P	
1	Mean daily temperature	in °C	15,9	17,5	20,9	22,5	25,3	27,0	27,2	27,5	27,2	25,0	20,9	16,7	22,8	20	1
2	Mean daily maximum temperature	in °C	21,7	22,2	25,6	26,7	28,9	30,6	30,6	30,6	30,6	28,9	26,1	22,8	27,2	20	2
3	Mean daily minimum temperature	in °C	10,0	12,8	16,1	18,3	21,7	23,3	23,9	24,4	23,9	21,1	15,6	10,6	18,3	20	3
4	Absolute maximum temperature	in °C	26,1	27,8	32,8	34,4	36,1	38,9	35,6	36,1	35,6	36,1	30,6	27,2	38,9	20	4
5	Absolute minimum temperature	in °C	5,0	4,4	6,1	12,2	16,7	15,6	20,0	21,1	16,1	15,6	9,4	5,6	4,4	20	5
6	Mean relative humidity	in %	85	81	78	81	83	86	87	87	88	84	86	86	84	6	6
7	Mean precipitation	in mm	38	61	117	256	284	490	554	462	310	147	28	10	2757	20	7
8	Maximum precipitation	in mm	123	161	358	521	538	855	887	729	748	450	129	112	3300	60	8
9	Minimum precipitation	in mm	0	9	4	82	81	186	217	211	99	6	0	0	2165	60	9
10	Maximum precipitation in 24 h	in mm	36	46	56	198	114	160	165	224	107	104	43	25	224	20	10
11	Mean number of days with precipitation	> 2,5 mm	3	6	9	14	15	19	22	19	14	8	2	1	132	20	11
12	Mean duration of sunshine	in h															12
13	Mean quantity of radiation	in ly/day															13
14	Mean potential evaporation	in mm	30	38	71	90	157	158	166	159	144	116	62	32	1223	40	14
15	Mean windspeed	in m/sec	0,5	0,6	0,8	0,9	0,8	0,7	0,7	0,7	0,6	0,5	0,4	0,4	0,6	30	15
16	Mean predominent direction of the wind		E	E	E,NE	E,NE	E,NE	N	N	N	E	E	E	E		28	16

197 Station/Country Dhubri/India

Location 26°01'N/89°59'E — Height above sealevel 35 m — Climate symbol: Köppen Am — Troll V.2

		J	F	M	A	M	J	J	A	S	O	N	D	year	P	
1 Mean daily temperature	in °C	17,8	19,4	23,8	25,9	26,4	27,2	27,8	28,1	27,2	26,1	22,2	18,1	24,2	32	1
2 Mean daily maximum temperature	in °C	23,3	25,8	30,0	30,6	30,0	30,0	30,0	30,0	29,4	29,4	26,7	23,3	28,3	32	2
3 Mean daily minimum temperature	in °C	11,7	13,3	17,2	21,1	22,8	24,4	25,6	26,1	25,0	22,8	17,8	12,8	20,0	32	3
4 Absolute maximum temperature	in °C	27,2	32,2	38,8	39,4	39,4	35,6	35,0	34,4	35,0	33,3	31,1	26,7	39,4	39	4
5 Absolute minimum temperature	in °C	6,1	2,8	10,0	12,2	17,2	20,6	22,8	22,2	20,6	16,7	11,7	7,8	2,8	39	5
6 Mean relative humidity	in %	75	69	58	64	81	87	87	87	87	82	79	79	78	10	6
7 Mean precipitation	in mm	8	18	48	130	373	605	434	343	368	117	8	3	2453	40	7
8 Maximum precipitation	in mm	55	99	193	434	881	1126	1309	809	911	424	98	25	3669	80	8
9 Minimum precipitation	in mm	0	0	0	0	130	127	105	91	58	0	0	0	1704	80	9
10 Maximum precipitation in 24 h	in mm	33	61	109	86	196	368	287	185	254	155	76	18	368	39	10
11 Mean number of days with precipitation	> 2,5 mm	1	1	3	8	15	18	16	15	13	5	1	<1	97	40	11
12 Mean duration of sunshine	in h															12
13 Mean quantity of radiation	in ly/day															13
14 Mean potential evaporation	in mm	33	45	97	140	152	167	170	162	144	131	96	36	1373	60	14
15 Mean windspeed	in m/sec	1,2	1,4	2,0	2,3	2,2	1,9	1,7	1,5	1,5	1,4	1,2	1,1	1,8	30	15
16 Mean predominent direction of the wind		NE	NE	NE,SW	NE	NE	NE	NE,S	NE,S	NE,SW	NE	NE	NE		28	16

198 Station/Country Cherrapunji/India

Location 25°15'N/91°44'E — Height above sealevel 1313 m — Climate symbol: Köppen Cwb/Am — Troll V.1

		J	F	M	A	M	J	J	A	S	O	N	D	year	P	
1 Mean daily temperature	in °C	11,7	13,1	16,7	18,3	19,2	20,0	20,3	20,6	20,6	19,2	15,9	12,8	17,5	35	1
2 Mean daily maximum temperature	in °C	15,8	16,7	20,6	21,7	22,2	22,2	22,2	22,8	22,8	22,2	19,4	16,7	20,6	35	2
3 Mean daily minimum temperature	in °C	7,8	9,4	12,8	15,0	16,1	17,8	18,3	18,3	18,3	16,1	12,2	8,9	14,4	35	3
4 Absolute maximum temperature	in °C	26,7	28,9	30,6	28,3	27,8	27,8	28,3	28,3	28,9	29,4	26,7	23,3	30,6	35	4
5 Absolute minimum temperature	in °C	1,1	0,6	0,6	3,9	3,3	11,7	11,7	13,3	12,8	10,6	6,7	3,9	0,6	35	5
6 Mean relative humidity	in %	69	67	62	74	83	91	92	92	87	80	72	71	78	10	6
7 Mean precipitation	in mm	18	53	185	665	1280	2695	2446	1781	1100	493	69	13	10798	35	7
8 Maximum precipitation	in mm	104	166	776	1472	3278	5689	4563	3482	2525	1435	356	244	24000	55	8
9 Minimum precipitation	in mm	0	0	6	87	274	1276	983	685	281	22	0	0	7178	55	9
10 Maximum precipitation in 24 h	in mm	86	91	307	462	813	925	838	683	655	592	333	191	925	35	10
11 Mean number of days with precipitation	> 2,5 mm	1	3	7	16	22	25	27	26	19	8	2	<1	158	35	11
12 Mean duration of sunshine	in h															12
13 Mean quantity of radiation	in ly/day															13
14 Mean potential evaporation	in mm	30	31	56	72	85	94	96	93	89	74	52	52	804	35	14
15 Mean windspeed	in m/sec	1,9	2,8	3,3	3,3	3,2	3,3	3,2	2,4	2,0	1,7	1,5	1,5	2,5	30	15
16 Mean predominent direction of the wind		NE,SW	SW	SW	SW	SW	SW	SW	NE,SW	NE,SW	NE,SW	NE,SW	NE,SW		27	16

199 Station/Country Marmagoa(near Goa)/India

Location 15°25'N/73°47'E — Height above sealevel 62 m — Climate symbol: Köppen Am — Troll V.3

		J	F	M	A	M	J	J	A	S	O	N	D	year	P	
1 Mean daily temperature	in °C	25,3	25,6	27,2	28,9	29,7	27,8	26,7	26,4	26,1	26,7	26,4	25,3	26,8	29	1
2 Mean daily maximum temperature	in °C	29,4	29,4	30,6	31,7	32,2	30,6	28,9	28,3	28,3	28,4	30,0	29,4	30,0	30	2
3 Mean daily minimum temperature	in °C	21,1	21,7	23,9	26,1	27,2	25,0	24,4	24,4	23,9	23,9	22,8	21,1	23,9	29	3
4 Absolute maximum temperature	in °C	33,3	33,9	35,0	35,6	35,0	35,0	31,7	31,1	31,1	34,4	33,9	33,3	35,6	30	4
5 Absolute minimum temperature	in °C	15,6	16,7	16,7	20,0	22,2	21,7	21,7	21,7	21,7	20,6	17,8	16,1	15,6	29	5
6 Mean relative humidity	in %	64	69	71	73	74	85	87	88	86	80	67	61	75	5	6
7 Mean precipitation	in mm	2	2	2	18	66	752	792	404	241	97	33	5	2414	30	7
8 Maximum precipitation	in mm	36	0	5	217	581	1185	1735	1258	573	297	237	53	3500	25	8
9 Minimum precipitation	in mm	0	0	0	0	0	472	171	145	42	13	0	0	1844	25	9
10 Maximum precipitation in 24 h	in mm	15	28	8	117	307	297	269	196	142	163	74	51	307	30	10
11 Mean number of days with precipitation	> 2,5 mm	<1	<1	<1	1	3	21	25	23	15	6	2	<1	100	30	11
12 Mean duration of sunshine	in h															12
13 Mean quantity of radiation	in ly/day	484	541	579	599	588	434	329	436	467	490	467	453	490	6	13
14 Mean potential evaporation	in mm	113	111	145	161	181	161	153	144	130	142	130	121	1689	50	14
15 Mean windspeed	in m/sec	2,9	3,2	3,5	3,7	4,0	5,3	6,4	5,2	3,1	2,7	2,6	2,7	3,8	25	15
16 Mean predominent direction of the wind		E,W	E,W	NE,W	NE,W	NW,W	SW,W	W	W	W	E,W	E,W	E,W		25	16

200 Station/Country Mangalore/India

Location 12°52'N/74°51'E — Height above sealevel 22 m — Climate symbol: Köppen Am — Troll V.2

		J	F	M	A	M	J	J	A	S	O	N	D	year	P	
1 Mean daily temperature	in °C	26,7	27,0	28,3	29,4	29,4	26,7	26,1	26,1	26,1	27,0	27,2	26,7	27,2	30	1
2 Mean daily maximum temperature	in °C	31,7	31,1	32,2	32,8	32,8	29,4	28,9	28,9	28,9	30,0	31,1	31,7	30,6	30	2
3 Mean daily minimum temperature	in °C	21,7	22,8	24,4	26,1	26,1	23,9	23,3	23,3	23,3	23,9	23,3	21,7	23,3	30	3
4 Absolute maximum temperature	in °C	35,0	37,8	36,7	35,6	36,7	34,4	31,7	32,2	31,7	34,4	35,0	35,0	37,8	30	4
5 Absolute minimum temperature	in °C	16,7	16,7	18,3	21,7	18,9	20,0	20,6	20,6	21,1	20,0	18,3	17,2	16,7	30	5
6 Mean relative humidity	in %	65	70	72	71	73	87	90	89	87	82	74	66	77	5	6
7 Mean precipitation	in mm	3	3	5	38	157	942	988	597	267	206	74	13	3293	60	7
8 Maximum precipitation	in mm	61	39	90	296	752	1640	1678	1236	703	488	249	158	4703	80	8
9 Minimum precipitation	in mm	0	0	0	0	0	408	277	237	38	27	0	0	2270	80	9
10 Maximum precipitation in 24 h	in mm	28	36	38	117	361	251	269	231	185	180	99	152	361	60	10
11 Mean number of days with precipitation	> 2,5 mm	<1	<1	<1	2	7	25	27	25	15	10	5	1	120	42	11
12 Mean duration of sunshine	in h															12
13 Mean quantity of radiation	in ly/day	455	518	551	541	503	359	293	353	413	421	411	431	427	5	13
14 Mean potential evaporation	in mm	134	126	155	164	175	147	146	140	136	140	133	134	1730	50	14
15 Mean windspeed	in m/sec	2,4	2,4	2,4	2,5	2,7	2,8	2,6	2,2	1,9	2,0	2,0	2,2	2,3	30	15
16 Mean predominent direction of the wind		E,NW	E,NW	E,NW	E,NW	E,NW	E,SW	SW	SW	E,NW	E,NW	E,NW	E,NW		28	16

201 Station/Country Port Blair (Andaman Islands)/India

Location 11°40'N/92°43'E   Height above sealevel 80 m   Climate symbol: Köppen **Am**   Troll V.1

		J	F	M	A	M	J	J	A	S	O	N	D	year	P	
1 Mean daily temperature	in °C	25.8	25.6	26.1	27.8	27.2	26.4	26.4	26.1	25.9	26.1	25.9	25.9	26.3	80	1
2 Mean daily maximum temperature	in °C	28.9	29.4	30.6	31.7	30.6	28.9	28.9	28.3	28.3	28.9	28.9	28.9	29.4	80	2
3 Mean daily minimum temperature	in °C	22.2	21.7	21.7	23.9	23.9	23.9	23.9	23.9	23.3	22.8	22.8	23.3	23.3	80	3
4 Absolute maximum temperature	in °C	32.2	32.8	34.4	36.1	36.1	35.6	32.2	31.7	31.7	30.0	31.7	32.2	36.1	80	4
5 Absolute minimum temperature	in °C	16.7	17.2	18.3	19.4	18.9	19.4	20.0	20.6	20.0	17.8	19.4	18.3	16.7	80	5
6 Mean relative humidity	in %	74	74	73	75	81	85	86	86	87	84	80	76	80	9	6
7 Mean precipitation	in mm	46	28	28	61	384	551	391	414	442	318	287	201	3131	80	7
8 Maximum precipitation	in mm	584	173	206	339	1030	1054	930	925	1123	580	648	618	4027	80	8
9 Minimum precipitation	in mm	0	0	0	0	62	240	134	73	126	82	24	2	2132	80	9
10 Maximum precipitation in 24 h	in mm	208	132	66	206	264	259	150	173	178	152	173	295	295	60	10
11 Mean number of days with precipitation	>2,5 mm	2	2	2	5	16	21	21	21	22	17	14	7	150	60	11
12 Mean duration of sunshine	in h	267	269	288	261	158	117	127	124	141	195	225	245	2417		12
13 Mean quantity of radiation	in ly/day	403	486	515	492	307	306	300	247	361	346	342	374	373	5	13
14 Mean potential evaporation	in mm	120	112	137	149	154	144	148	138	127	129	122	121	1599	60	14
15 Mean windspeed	in m/sec	3,7	2,6	2,2	2,4	3,4	5,5	5,5	5,2	4,4	3,0	3,2	3,9	3,7	27	15
16 Mean predominent direction of the wind		NE	NE	NE,E	NE,E	SW	SW	SW	SW	SW	W	E	NE		25	16

202 Station/Country Cochin/India

Location 9°58'N/76°14'E   'Height above sealevel 3 m   Climate symbol: Köppen **Am**   Troll V.1

		J	F	M	A	M	J	J	A	S	O	N	D	year	P	
1 Mean daily temperature	in °C	27.0	27.8	28.9	29.7	28.9	26.7	26.1	26.1	26.7	27.2	27.5	27.2	27.5	43	1
2 Mean daily maximum temperature	in °C	31.7	32.2	33.3	33.3	32.2	29.4	28.9	28.9	30.6	31.1	31.7	31.1	31.1	43	2
3 Mean daily minimum temperature	in °C	22.2	23.3	25.0	26.1	25.6	23.9	23.3	23.9	23.9	23.9	23.9	22.8	23.9	43	3
4 Absolute maximum temperature	in °C	35.6	37.2	36.7	36.1	37.2	34.4	32.2	32.2	32.8	33.3	34.4	35.0	37.2	43	4
5 Absolute minimum temperature	in °C	16.1	18.3	21.7	21.7	19.4	18.9	20.0	20.0	21.1	20.6	20.0	17.8	16.1	43	5
6 Mean relative humidity	in %	67	71	74	75	80	87	88	86	84	81	77	71	78	5	6
7 Mean precipitation	in mm	23	20	51	124	297	724	592	328	196	383	170	41	2949	60	7
8 Maximum precipitation	in mm															8
9 Minimum precipitation	in mm															9
10 Maximum precipitation in 24 h	in mm	135	104	101	130	254	185	213	115	112	236	122	124	254	60	10
11 Mean number of days with precipitation	>2,5 mm	1	1	3	6	12	24	24	18	14	14	9	3	129	43	11
12 Mean duration of sunshine	in h															12
13 Mean quantity of radiation	in ly/day															13
14 Mean potential evaporation	in mm	122	139	160	185	167	145	141	142	140	144	141	140	1746	60	14
15 Mean windspeed	in m/sec															15
16 Mean predominent direction of the wind																16

203 Station/Country Car Nicobar (Nicobar Islands)/India

Location 9°15'N/92°48'E   Height above sealevel 8 m   Climate symbol: Köppen **Am**   Troll V.2

		J	F	M	A	M	J	J	A	S	O	N	D	year	P	
1 Mean daily temperature	in °C	27.5	27.8	28.3	28.6	27.8	27.8	27.5	27.5	27.2	27.5	27.8	27.2	27.7	13	1
2 Mean daily maximum temperature	in °C	30.0	30.6	31.7	32.2	30.6	30.0	30.0	30.0	30.0	29.4	29.4	29.4	30.6	13	2
3 Mean daily minimum temperature	in °C	25.0	25.0	25.0	25.0	25.0	25.6	25.0	25.0	24.4	23.9	24.4	25.0	25.0	13	3
4 Absolute maximum temperature	in °C	32.2	32.2	33.9	35.0	33.9	32.8	33.3	33.3	33.9	32.2	31.7	31.1	35.0	10	4
5 Absolute minimum temperature	in °C	18.9	21.1	20.0	22.8	21.7	22.2	22.8	22.8	22.2	22.8	21.7	20.6	18.9	10	5
6 Mean relative humidity	in %	78	77	78	81	89	88	88	89	90	92	88	81	85	13	6
7 Mean precipitation	in mm	99	30	53	89	318	315	236	259	328	295	290	198	2510	30	7
8 Maximum precipitation	in mm															8
9 Minimum precipitation	in mm															9
10 Maximum precipitation in 24 h	in mm	157	69	76	109	165	178	135	152	188	157	157	152	188	38	10
11 Mean number of days with precipitation	>2,5 mm	4	2	3	5	14	15	12	13	15	14	13	9	119	30	11
12 Mean duration of sunshine	in h															12
13 Mean quantity of radiation	in ly/day															13
14 Mean potential evaporation	in mm	144	134	155	158	159	156	156	154	144	147	144	140	1791	13	14
15 Mean windspeed	in m/sec															15
16 Mean predominent direction of the wind		ENE	ENE	ENE	ENE	WSW	WSW	WSW	WSW	WSW	WSW	ENE	ENE		6	16

204 Station/Country Pāmban Island/India

Location 9°16'N/79°18'E   Height above sealevel 11 m   Climate symbol: Köppen **Aw**   Troll V.2

		J	F	M	A	M	J	J	A	S	O	N	D	year	P	
1 Mean daily temperature	in °C	26.1	26.7	28.3	29.7	30.3	29.4	28.9	28.9	28.9	28.3	27.0	26.1	28.2	50	1
2 Mean daily maximum temperature	in °C	30.6	32.8	36.1	38.3	37.2	31.7	28.3	27.8	29.4	31.7	30.6	29.4	32.2	24	2
3 Mean daily minimum temperature	in °C	11.7	12.8	16.7	20.0	22.2	22.8	22.2	21.7	20.6	18.9	15.0	11.7	17.8	24	3
4 Absolute maximum temperature	in °C	33.3	33.3	34.4	37.2	36.7	35.0	37.2	34.4	35.0	35.0	33.3	33.9	37.2	50	4
5 Absolute minimum temperature	in °C	20.6	19.4	20.6	20.6	21.1	20.6	22.2	22.2	21.1	21.1	21.7	20.0	19.4	50	5
6 Mean relative humidity	in %	83	79	76	77	79	78	78	79	80	82	84	86	80	5	6
7 Mean precipitation	in mm	66	23	18	48	25	3	13	15	28	216	297	193	943	50	7
8 Maximum precipitation	in mm	283	120	148	287	217	70	174	104	93	627	692	840	1772	70	8
9 Minimum precipitation	in mm	0	0	0	0	0	0	0	0	0	20	33	6	411	70	9
10 Maximum precipitation in 24 h	in mm	127	89	64	58	101	56	130	74	109	135	137	213	213	50	10
11 Mean number of days with precipitation	>2,5 mm	3	1	1	3	2	<1	1	1	2	10	3	9	37	30	11
12 Mean duration of sunshine	in h															12
13 Mean quantity of radiation	in ly/day															13
14 Mean potential evaporation	in mm	123	127	155	164	178	167	188	165	159	153	135	123	1817	50	14
15 Mean windspeed	in m/sec	5,1	3,9	3,1	3,4	5,0	5,6	4,6	4,3	4,4	3,6	4,1	5,3	4,3	30	15
16 Mean predominent direction of the wind		N.NE	NE	NE	SE.S	S	SW.S	SW.S	SW.S	SW.S	S	N.NE	N.NE		28	16

205 Station / Country Trincomalee / Sri Lanka

Location 8°35'N/81°15'E   Height above sealevel 7 m   Climate symbol: Köppen **Aw**   Troll **V,2**

		J	F	M	A	M	J	J	A	S	O	N	D	year	P	
1	Mean daily temperature	in °C	25,3	26,1	27,0	28,6	29,7	29,7	29,4	29,2	29,2	27,8	26,4	25,6	27,8	25	1
2	Mean daily maximum temperature	in °C	26,7	27,8	29,4	31,7	33,3	33,3	33,3	33,3	33,3	31,1	28,9	27,2	30,6	25	2
3	Mean daily minimum temperature	in °C	23,9	24,4	24,4	25,6	26,1	26,1	25,6	25,0	25,0	24,4	23,9	23,9	25,0	25	3
4	Absolute maximum temperature	in °C	33,3	35,6	38,3	38,9	40,0	39,4	38,3	38,9	38,9	38,9	36,1	32,8	40,0	57	4
5	Absolute minimum temperature	in °C	18,3	18,9	19,4	19,4	19,4	21,7	21,1	20,6	21,1	20,6	19,4	18,9	18,3	57	5
6	Mean relative humidity	in %	79	71	71	69	64	60	59	61	63	71	79	80	71	6	6
7	Mean precipitation	in mm	173	66	48	58	69	28	51	107	107	221	358	363	1649	80	7
8	Maximum precipitation	in mm	740	551	282	238	434	181	219	298	289	566	954	820	2578	107	8
9	Minimum precipitation	in mm	<1	0	0	0	0	0	0	<1	3	26	55	56	886	107	9
10	Maximum precipitation in 24 h	in mm	208	122	107	74	104	112	99	107	130	127	264	256	264	59	10
11	Mean number of days with precipitation	> 1,0 mm	10	4	4	5	5	2	3	6	6	13	17	16	91	40	11
12	Mean duration of sunshine	in h	205	210	273	287	248	231	226	254	234	214	168	171	2701	10	12
13	Mean quantity of radiation	in ly/day															13
14	Mean potential evaporation	in mm	115	117	145	157	172	168	172	167	159	148	126	117	1763	30	14
15	Mean windspeed	in m/sec	5,2	4,0	2,9	2,9	4,6	6,0	5,5	5,1	4,5	3,7	3,8	5,1	4,4	30	15
16	Mean predominent direction of the wind		NE	NE	NE	SW,E	SW	SW	SW	SW	SW	SW,NE	SW,NE	N,NE		10	16

206 Station / Country Nuwara–Eliya / Sri Lanka

Location 6°58'N/80°46'E   Height above sealevel 1880 m   Climate symbol: Köppen **Am**   Troll **V,1**

		J	F	M	A	M	J	J	A	S	O	N	D	year	P	
1	Mean daily temperature	in °C	13,9	13,9	14,7	15,6	16,4	15,9	15,6	15,9	15,6	15,6	15,3	14,4	15,3	9	1
2	Mean daily maximum temperature	in °C	19,4	21,1	21,7	21,7	21,1	18,9	18,3	19,4	19,4	20,0	20,0	20,0	20,0	9	2
3	Mean daily minimum temperature	in °C	8,3	8,7	7,8	9,4	11,7	12,8	12,8	12,2	11,7	11,1	10,6	8,9	10,8	9	3
4	Absolute maximum temperature	in °C	24,4	23,9	23,9	24,4	25,6	23,9	24,4	23,3	23,9	23,9	23,3	23,3	25,6	9	4
5	Absolute minimum temperature	in °C	-2,8	-1,7	0,0	3,3	0,6	7,2	7,2	6,7	5,0	5,0	0,6	-1,1	-2,8	9	5
6	Mean relative humidity	in %	79	78	75	84	84	88	86	86	86	87	86	86	84	6	6
7	Mean precipitation	in mm	170	43	109	119	175	277	300	196	226	269	241	203	2328	9	7
8	Maximum precipitation	in mm	645	274	245	343	884	729	889	678	876	724	431	802	3448	107	8
9	Minimum precipitation	in mm	0	<1	0	11	<1	77	35	41	30	52	29	37	1321	107	9
10	Maximum precipitation in 24 h	in mm	107	119	64	56	188	76	150	64	99	142	76	203	203	20	10
11	Mean number of days with precipitation	> 2,5 mm	13	6	11	15	18	25	25	22	20	22	22	17	216	20	11
12	Mean duration of sunshine	in h	167	162	198	156	102	84	68	74	87	118	123	143	1482	10	12
13	Mean quantity of radiation	in ly/day															13
14	Mean potential evaporation	in mm	52	47	58	61	67	66	64	67	61	62	56	55	714	9	14
15	Mean windspeed	in m/sec	2,6	2,8	2,5	1,9	3,0	3,7	3,5	3,1	2,3	2,3	2,6	2,8	2,8	15	15
16	Mean predominent direction of the wind		E	E	E	E	W	W	W,NW	W	NW,W		E	E		20	16

207 Station / Country Colombo / Sri Lanka

Location 6°54'N/79°52'E   Height above sealevel 7 m   Climate symbol: Köppen **Af**   Troll **V,1**

		J	F	M	A	M	J	J	A	S	O	N	D	year	P	
1	Mean daily temperature	in °C	26,1	26,4	27,2	27,8	28,1	27,2	27,2	27,2	27,2	26,7	26,1	25,9	26,9	25	1
2	Mean daily maximum temperature	in °C	30,0	30,6	31,1	31,1	30,6	29,4	29,4	29,4	29,4	29,4	29,4	29,4	30,0	25	2
3	Mean daily minimum temperature	in °C	22,2	22,2	23,3	24,4	25,6	25,0	25,0	25,0	25,0	23,9	22,8	22,2	23,9	25	3
4	Absolute maximum temperature	in °C	34,4	35,6	36,3	33,3	32,8	31,7	31,1	31,1	31,7	31,7	32,8	35,6	36,3	25	4
5	Absolute minimum temperature	in °C	15,0	16,1	17,8	21,1	20,6	22,2	21,7	21,7	21,7	20,6	18,9	17,2	15,0	25	5
6	Mean relative humidity	in %	70	69	69	72	77	79	78	77	76	77	76	72	74	30	6
7	Mean precipitation	in mm	89	69	147	231	371	224	135	109	160	348	315	147	2345	40	7
8	Maximum precipitation	in mm	310	269	287	657	859	483	529	435	551	848	640	547	3934	68	8
9	Minimum precipitation	in mm	0	0	4	19	20	38	7	3	18	83	56	7	1360	68	9
10	Maximum precipitation in 24 h	in mm	79	84	97	183	290	94	183	127	152	256	211	114	290	40	10
11	Mean number of days with precipitation	> 1,0 mm	7	6	8	14	19	18	12	11	13	19	16	10	152	40	11
12	Mean duration of sunshine	in h	245	252	251	216	198	162	189	195	186	202	192	242	2530	30	12
13	Mean quantity of radiation	in ly/day															13
14	Mean potential evaporation	in mm	130	121	145	148	156	149	150	148	142	140	130	128	1687		14
15	Mean windspeed	in m/sec	2,5	2,1	1,9	2,1	2,8	3,0	2,8	2,9	2,8	2,2	1,9	2,4	2,5	30	15
16	Mean predominent direction of the wind		NE,N	NE,NW	E,W	E,SW	SW	SW	SW	SW	SW	SW	NE,NW	NE,N		10	16

208 Station / Country Kätmändu / Nepal

Location 27°42'N/85°12'E   Height above sealevel 1337 m   Climate symbol: Köppen **Cwa**   Troll **V,2**

		J	F	M	A	M	J	J	A	S	O	N	D	year	P	
1	Mean daily temperature	in °C	10,0	11,7	16,1	20,0	23,1	24,4	24,4	24,2	23,6	20,0	15,3	11,1	18,7	9	1
2	Mean daily maximum temperature	in °C	18,3	19,4	25,0	28,3	30,0	29,4	28,9	28,3	28,3	26,7	23,3	19,4	25,6	9	2
3	Mean daily minimum temperature	in °C	1,7	3,9	7,2	11,7	16,1	19,4	20,0	20,0	18,9	13,3	7,2	2,8	11,7	10	3
4	Absolute maximum temperature	in °C	25,0	25,0	33,3	35,0	33,9	34,4	32,8	33,3	33,3	33,3	28,3	24,4	35,0	9	4
5	Absolute minimum temperature	in °C	-2,2	-0,6	1,7	4,4	10,0	14,4	17,8	17,2	13,3	6,1	-0,6	-1,7	-2,2	10	5
6	Mean relative humidity	in %	80	79	63	61	67	76	84	86	85	84	81	78	78	10	6
7	Mean precipitation	in mm	18	17	39	48	90	248	386	285	178	78	6	1	1394	8	7
8	Maximum precipitation	in mm	31	32	80	177	110	608	494	379	321	180	20	11	1800	8	8
9	Minimum precipitation	in mm	1	1	8	24	57	75	205	155	36	34	0	0	1227	8	9
10	Maximum precipitation in 24 h	in mm	36	23	48	66	51	66	76	81	58	48	15	8	81	9	10
11	Mean number of days with precipitation	> 2,5 mm	1	5	2	6	10	15	21	20	12	4	1	<1	98	9	11
12	Mean duration of sunshine	in h	183	157	236	285	177	152	81	74	99	161	158	158	1919	30	12
13	Mean quantity of radiation	in ly/day															13
14	Mean potential evaporation	in mm	18	26	53	83	110	128	136	124	107	77	40	28	928	40	14
15	Mean windspeed	in m/sec	0,9	1,4	2,4	1,7	1,6	1,3	1,2	1,1	0,9	0,7	0,5	0,4			15
16	Mean predominent direction of the wind		SW	SW	W	W	W	SW	SW	SW	SW	SW	W	W		5	16

209 Station/Country Yatung (near Thimbu)/Bhutan

Location 27°29'N/88°55'E **Height above sealevel** 2987 m **Climate symbol: Köppen** Cwb **Troll** V,1

		J	F	M	A	M	J	J	A	S	O	N	D	year	P	
1	Mean daily temperature	in °C	0,0	1,4	4,4	8,1	11,1	13,6	15,0	14,4	13,1	8,9	4,2	1,1	7,9	19	1
2	Mean daily maximum temperature	in °C	7,8	8,9	12,2	15,0	17,2	18,3	19,4	18,3	17,8	15,6	11,7	8,9	14,4	19	2
3	Mean daily minimum temperature	in °C	-7,7	-6,1	-3,3	1,1	5,0	8,9	10,6	10,0	8,3	2,2	-3,3	-6,7	1,7	19	3
4	Absolute maximum temperature	in °C	18,3	20,0	18,3	21,7	23,3	24,4	23,9	23,9	21,7	21,1	18,9	18,9	24,4	19	4
5	Absolute minimum temperature	in °C	-20,6	-13,3	-11,7	-6,7	-1,1	2,8	6,1	5,6	-0,6	-6,7	-10,0	-11,7	-20,6	19	5
6	Mean relative humidity	in %	91	90	92	89	90	91	92	92	92	90	89	90	91	12	6
7	Mean precipitation	in mm	15	48	64	99	107	119	130	117	101	53	18	6	876	19	7
8	Maximum precipitation	in mm															8
9	Minimum precipitation	in mm															9
10	Maximum precipitation in 24 h	in mm	43	71	38	58	51	46	43	36	69	132	71	10	132	19	10
11	Mean number of days with precipitation	> 2,5 mm	2	4	6	11	11	13	15	16	11	3	1	<1	94	19	11
12	Mean duration of sunshine	in h															12
13	Mean quantity of radiation	in ly/day															13
14	Mean potential evaporation	in mm	0	0	25	46	67	83	91	85	69	47	23	6	542	45	14
15	Mean windspeed	in m/sec															15
16	Mean predominent direction of the wind																16

210 Station/Country Nārāyanganj (near Dacca)/Bangladesh

Location 23°37'N/90°30'E **Height above sealevel** 8 m **Climate symbol: Köppen** Aw **Troll** V,2

		J	F	M	A	M	J	J	A	S	O	N	D	year	P	
1	Mean daily temperature	in °C	19,2	21,5	26,1	28,3	28,6	28,6	28,6	28,6	28,9	27,5	23,9	20,0	25,8	60	1
2	Mean daily maximum temperature	in °C	25,6	27,8	32,2	33,3	32,8	31,7	31,1	31,1	31,7	31,1	28,9	26,1	30,6	60	2
3	Mean daily minimum temperature	in °C	12,8	15,0	20,0	23,3	24,4	25,6	26,1	26,1	26,1	23,9	18,9	13,9	21,7	60	3
4	Absolute maximum temperature	in °C	31,1	34,4	38,9	40,6	38,3	36,7	35,0	35,6	36,7	35,0	34,4	30,6	40,6	60	4
5	Absolute minimum temperature	in °C	7,2	6,7	10,6	16,7	14,4	19,4	21,1	21,7	21,1	18,3	12,2	8,9	6,7	60	5
6	Mean relative humidity	in %	72	63	61	67	76	83	83	83	81	76	72	73	74	20	6
7	Mean precipitation	in mm	8	30	61	137	244	315	330	338	249	135	25	5	1877	61	7
8	Maximum precipitation	in mm	95	73	149	325	544	672	727	451	605	177	24			30	8
9	Minimum precipitation	in mm	0	0	0	tr	tr	tr	tr	tr	tr	tr	0	0		30	9
10	Maximum precipitation in 24 h	in mm	46	74	135	122	130	185	236	208	178	198	180	43	236	60	10
11	Mean number of days with precipitation	> 2,5 mm	<1	2	3	7	11	15	18	18	12	6	1	<1	95	61	11
12	Mean duration of sunshine	in h	273	224	229	192	155	81	59	56	90	198	252	267	2076	30	12
13	Mean quantity of radiation	in ly/day															13
14	Mean potential evaporation	in mm	40	51	129	159	176	173	179	171	156	143	83	43	1503		14
15	Mean windspeed	in m/sec															15
16	Mean predominent direction of the wind																16

211 Station/Country Chittagong/Bangladesh

Location 22°21'N/91°50'E **Height above sealevel** 27 m **Climate symbol: Köppen** Am **Troll** V,1

		J	F	M	A	M	J	J	A	S	O	N	D	year	P	
1	Mean daily temperature	in °C	19,4	21,4	25,0	27,2	27,8	27,8	27,5	27,2	27,5	26,7	23,6	20,0	25,1	60	1
2	Mean daily maximum temperature	in °C	26,1	27,8	30,6	31,7	31,7	30,6	30,0	30,0	30,6	30,6	28,9	26,1	29,4	60	2
3	Mean daily minimum temperature	in °C	12,8	15,0	19,4	22,8	23,9	25,0	25,0	24,4	24,4	22,8	18,3	13,9	20,6	60	3
4	Absolute maximum temperature	in °C	31,7	33,9	37,2	38,9	36,7	36,7	34,4	33,9	35,0	34,4	33,9	31,1	38,9	60	4
5	Absolute minimum temperature	in °C	7,2	7,8	10,6	15,0	18,3	20,0	19,4	22,2	21,7	16,7	11,1	8,3	7,2	60	5
6	Mean relative humidity	in %	70	67	71	75	78	84	86	87	85	82	78	77	78	10	6
7	Mean precipitation	in mm	5	28	64	150	284	533	597	518	320	180	56	15	2730	60	7
8	Maximum precipitation	in mm	69	117	376	355	635	950	1527	780	759	579	321	146		30	8
9	Minimum precipitation	in mm	0	0	0	tr	tr	tr	tr	tr	tr	tr	0	0		30	9
10	Maximum precipitation in 24 h	in mm	69	236	178	180	163	384	289	272	259	422	137	117	422	60	10
11	Mean number of days with precipitation	> 2,5 mm	<1	1	2	6	11	17	19	17	13	7	2	<1	97	60	11
12	Mean duration of sunshine	in h															12
13	Mean quantity of radiation	in ly/day															13
14	Mean potential evaporation	in mm	42	58	116	148	167	165	162	158	144	137	82	50	1427	60	14
15	Mean windspeed	in m/sec	1,4			3,3			3,6			1,4					15
16	Mean predominent direction of the wind																16

212 Station/Country Bhamo/Burma

Location 24°16'N/97°17'E **Height above sealevel** 118 m **Climate symbol: Köppen** Cwa **Troll** V,2

		J	F	M	A	M	J	J	A	S	O	N	D	year	P	
1	Mean daily temperature	in °C	16,4	19,7	23,3	26,7	28,3	27,8	27,8	27,5	27,2	25,9	21,7	17,8	24,2	27	1
2	Mean daily maximum temperature	in °C	24,4	27,8	31,1	33,9	33,9	31,7	31,1	31,1	31,7	30,6	27,2	24,4	30,0	27	2
3	Mean daily minimum temperature	in °C	8,3	11,7	15,6	19,4	22,8	23,9	24,4	23,9	22,8	21,1	16,1	11,1	18,3	27	3
4	Absolute maximum temperature	in °C	30,6	32,8	37,8	39,4	40,6	38,3	36,1	36,7	36,7	36,7	32,8	28,3	40,6	27	4
5	Absolute minimum temperature	in °C	2,8	5,6	8,9	13,9	16,7	19,4	17,2	21,1	16,1	13,3	8,3	4,4	2,8	27	5
6	Mean relative humidity	in %	85	78	66	61	69	85	87	89	82	86	87	89	81	7	6
7	Mean precipitation	in mm	10	15	18	46	155	358	424	409	249	117	43	13	1857	55	7
8	Maximum precipitation	in mm	44	52	69	93	299	493	816	607	328	302	252	104	2427	30	8
9	Minimum precipitation	in mm	0	0	0	3	41	193	199	308	75	44	0	0	1276	30	9
10	Maximum precipitation in 24 h	in mm	79	28	33	58	79	119	165	221	132	101	86	53	221	27	10
11	Mean number of days with precipitation	> 2,5 mm	1	2	2	5	10	19	20	19	13	7	3	1	102	55	11
12	Mean duration of sunshine	in h															12
13	Mean quantity of radiation	in ly/day															13
14	Mean potential evaporation	in mm	32	45	94	144	171	166	170	162	149	125	64	36	1358	45	14
15	Mean windspeed	in m/sec															15
16	Mean predominent direction of the wind																16

213 Station/Country Lashio/Burma

Location 22°58'N/97°51'E — Height above sealevel 854 m — Climate symbol: Köppen Cwa — Troll V.4

		J	F	M	A	M	J	J	A	S	O	N	D	year	P	
1	Mean daily temperature in °C	15.8	17.2	21.7	24.2	25.0	25.0	24.7	24.7	24.4	22.8	19.2	16.1	21.7	21	1
2	Mean daily maximum temperature in °C	23.3	25.6	30.0	31.7	30.6	28.9	28.3	28.3	28.9	27.8	25.0	22.8	27.8	21	2
3	Mean daily minimum temperature in °C	7.8	8.4	13.3	16.7	19.4	21.1	21.1	21.1	20.0	17.8	13.3	9.4	16.1	21	3
4	Absolute maximum temperature in °C	26.1	30.0	33.9	37.2	36.7	35.0	33.3	31.1	31.7	31.1	29.4	26.1	37.2	10	4
5	Absolute minimum temperature in °C	1.7	3.9	8.3	12.2	13.9	16.7	17.8	16.7	14.4	10.6	8.3	3.3	1.7	10	5
6	Mean relative humidity in %	78	67	54	57	71	82	86	87	87	88	86	78	78	8	6
7	Mean precipitation in mm	8	8	15	56	175	249	305	323	198	145	69	23	1574	21	7
8	Maximum precipitation in mm															8
9	Minimum precipitation in mm															9
10	Maximum precipitation in 24 h in mm	15	18	10	36	64	112	99	69	107	69	43	53	122	8	10
11	Mean number of days with precipitation > 2,5 mm	1	1	1	5	12	15	27	19	14	10	5	1	111	21	11
12	Mean duration of sunshine in h															12
13	Mean quantity of radiation in ly/day															13
14	Mean potential evaporation in mm	35	44	83	117	133	132	135	127	114	93	58	37	1108	45	14
15	Mean windspeed in m/sec															15
16	Mean predominent direction of the wind															16

214 Station/Country Mandaly Burma

Location 21°58'N/96°06'E — Height above sealevel 77 m — Climate symbol: Köppen Cwa — Troll V.3

		J	F	M	A	M	J	J	A	S	O	N	D	year	P	
1	Mean daily temperature in °C	20.3	23.1	27.5	31.7	31.4	29.7	29.7	29.2	28.8	27.2	24.2	20.3	25.9	20	1
2	Mean daily maximum temperature in °C	27.8	31.1	36.1	38.3	36.7	33.9	33.9	33.3	32.8	31.7	29.4	26.7	32.8	20	2
3	Mean daily minimum temperature in °C	12.8	15.0	18.9	25.0	26.1	25.8	25.8	25.0	24.4	22.8	18.9	13.9	21.1	20	3
4	Absolute maximum temperature in °C	32.8	37.2	42.2	43.3	43.9	41.7	41.1	38.3	39.4	38.9	36.7	32.1	43.9	20	4
5	Absolute minimum temperature in °C	7.2	8.3	12.2	17.8	20.6	20.0	22.2	21.7	20.6	16.7	13.3	6.7	6.7	20	5
6	Mean relative humidity in %	68	54	42	44	60	71	72	78	79	81	79	74	67	8	6
7	Mean precipitation in mm	3	3	5	30	147	180	69	104	137	108	51	10	828	20	7
8	Maximum precipitation in mm															8
9	Minimum precipitation in mm															9
10	Maximum precipitation in 24 h in mm	18	18	20	58	135	107	135	89	79	99	99	38	135	20	10
11	Mean number of days with precipitation > 2,5 mm	<1	<1	<1	2	8	7	8	8	9	7	3	>1	64	20	11
12	Mean duration of sunshine in h															12
13	Mean quantity of radiation in ly/day															13
14	Mean potential evaporation in mm	44	68	147	179	193	180	183	173	157	143	87	45	1599	50	14
15	Mean windspeed in m/sec															15
16	Mean predominent direction of the wind	N,ESE	S	S	S	S	S	S	R	S	SSE	E			5	16

215 Station/Country Akyab/Burma

Location 20°08'N/92°55'E — Height above sealevel 9 m — Climate symbol: Köppen Am — Troll V,1

		J	F	M	A	M	J	J	A	S	O	N	D	year	P	
1	Mean daily temperature in °C	21.1	22.5	25.6	28.1	28.9	27.5	27.0	27.0	27.5	27.5	25.6	22.2	25.9	60	1
2	Mean daily maximum temperature in °C	27.2	28.9	31.1	32.2	32.2	30.0	28.9	28.9	30.0	30.6	29.4	27.2	30.0	60	2
3	Mean daily minimum temperature in °C	15.0	16.1	20.0	23.9	25.6	25.0	25.0	25.0	25.0	24.4	21.7	17.2	22.2	60	3
4	Absolute maximum temperature in °C	34.4	35.0	37.8	37.2	37.2	36.7	33.9	32.2	34.4	33.9	32.8	31.7	37.8	60	4
5	Absolute minimum temperature in °C	8.3	9.4	12.2	16.7	18.9	20.0	21.7	21.7	21.1	18.3	15.6	10.6	8.3	60	5
6	Mean relative humidity in %	73	70	74	74	78	89	91	89	87	84	82	79	81	8	6
7	Mean precipitation in mm	3	5	10	51	391	1151	1400	1133	577	287	130	18	5156	60	7
8	Maximum precipitation in mm	15	71	18	183	1138	1897	2101	2238	836	620	218	185		60	8
9	Minimum precipitation in mm	0	0	0	3	94	615	696	490	256	36	3	0			9
10	Maximum precipitation in 24 h in mm	25	58	71	292	358	381	343	404	249	330	465	135	465	60	10
11	Mean number of days with precipitation > 2,5 mm	<1	<1	<1	2	11	24	28	27	19	9	4	<1	128	60	11
12	Mean duration of sunshine in h															12
13	Mean quantity of radiation in ly/day															13
14	Mean potential evaporation in mm	57	69	127	159	174	160	158	154	148	144	111	68	1529	60	14
15	Mean windspeed in m/sec															15
16	Mean predominent direction of the wind	ENE	ENE	ENE	WSW	E	S	E	ESE	E	ENE	ENE	ENE		5	16

216 Station/Country Rangun/Burma

Location 16°46'N/96°11'E — Height above sealevel 5 m — Climate symbol: Köppen Am — Troll V,1

		J	F	M	A	M	J	J	A	S	O	N	D	year	P	
1	Mean daily temperature in °C	25.0	26.4	28.6	30.3	29.2	27.2	27.5	27.5	27.8	27.8	27.0	25.3	27.5	60	1
2	Mean daily maximum temperature in °C	31.7	33.3	35.6	36.1	33.3	30.0	29.4	29.4	30.0	31.1	31.1	31.1	31.7	60	2
3	Mean daily minimum temperature in °C	18.3	19.4	21.7	24.4	25.0	24.4	24.4	24.4	24.4	24.4	22.8	19.4	22.8	60	3
4	Absolute maximum temperature in °C	37.8	38.3	39.4	41.1	40.6	36.7	33.9	33.9	34.4	35.0	35.0	35.6	41.1	60	4
5	Absolute minimum temperature in °C	12.8	13.3	16.1	20.0	20.6	21.7	21.1	20.0	22.2	21.7	16.1	12.8	12.8	60	5
6	Mean relative humidity in %	62	62	64	68	78	86	89	89	87	80	76	68	76	8	6
7	Mean precipitation in mm	3	5	8	51	307	480	582	528	394	180	69	10	2617	60	7
8	Maximum precipitation in mm	37	20	73	102	526	953	745	870	571	403	201	118	3262	30	8
9	Minimum precipitation in mm	0	0	0	0	84	290	303	293	200	66	0	0	1940	30	9
10	Maximum precipitation in 24 h in mm	74	48	41	361	231	152	140	135	132	135	150	101	361	60	10
11	Mean number of days with precipitation > 2,5 mm	<1	<1	<1	2	14	23	28	25	20	10	3	<1	127	60	11
12	Mean duration of sunshine in h	301	263	301	306	217	105	84	93	138	186	192	260	2446	30	12
13	Mean quantity of radiation in ly/day															13
14	Mean potential evaporation in mm	108	123	158	172	176	158	157	151	145	147	132	110	1735	60	14
15	Mean windspeed in m/sec															15
16	Mean predominent direction of the wind	NNE	NNE	W	W	WSW	WSW	WSW	WSW	WSW	E	NNE	NNE		5	16

217 Station / Country **Diamond Island / Burma**

Location **15°51'N/94°19'E** — Height above sealevel **12 m** — Climate symbol: Köppen **Am** — Troll **V,1**

	J	F	M	A	M	J	J	A	S	O	N	D	year	P	
1 Mean daily temperature	in °C	25.6	25.9	27.0	28.6	28.6	27.2	26.7	26.7	27.2	27.5	27.0	25.6	27.0	60	1
2 Mean daily maximum temperature	in °C	28.9	28.9	29.4	31.1	31.1	29.4	28.9	28.9	28.4	30.0	29.4	28.3	29.4	60	2
3 Mean daily minimum temperature	in °C	22.2	22.8	24.4	26.1	26.1	25.0	24.4	25.0	25.0	24.4	22.8		24.4	60	3
4 Absolute maximum temperature	in °C	32.8	33.3	33.9	36.7	36.1	32.8	33.3	33.3	31.7	33.3	33.9	32.2	36.7	60	4
5 Absolute minimum temperature	in °C	16.1	17.8	19.4	20.6	16.7	17.8	20.6	20.6	20.0	20.6	20.0	16.7	16.1	60	5
6 Mean relative humidity	in %	69	70	73	74	78	84	86	86	86	83	77	71	78	60	6
7 Mean precipitation	in mm	3	3	5	33	284	640	701	648	439	213	127	20	3116	60	7
8 Maximum precipitation	in mm															8
9 Minimum precipitation	in mm															9
10 Maximum precipitation in 24 h	in mm	38	97	48	135	206	213	188	348	241	170	147	135	348	60	10
11 Mean number of days with precipitation	>2,5 mm	<1	<1	<1	1	11	21	23	22	17	11	5	<1	115	60	11
12 Mean duration of sunshine	in h															12
13 Mean quantity of radiation	in ly/day															13
14 Mean potential evaporation	in mm	114	114	143	158	166	156	156	150	143	146	134	116	1698	60	14
15 Mean windspeed	in m/sec															15
16 Mean predominent direction of the wind																16

218 Station / Country **Tavoy / Burma**

Location **14°07'N/98°18'E** — Height above sealevel **6 m** — Climate symbol: Köppen **Am** — Troll **V,1**

	J	F	M	A	M	J	J	A	S	O	N	D	year	P	
1 Mean daily temperature	in °C	25.0	26.4	27.8	29.2	27.8	26.7	25.9	25.9	26.1	27.2	26.1	25.0	26.6	20	1
2 Mean daily maximum temperature	in °C	31.7	32.8	33.9	34.4	31.7	29.4	28.3	28.3	28.9	31.1	31.1	31.1	31.1	20	2
3 Mean daily minimum temperature	in °C	18.3	20.0	21.7	23.9	23.9	23.9	23.3	23.3	23.3	23.3	21.1	18.9	22.2	20	3
4 Absolute maximum temperature	in °C	35.0	36.1	37.2	37.8	37.2	33.9	32.8	32.2	32.8	37.2	37.8	36.7	37.8	20	4
5 Absolute minimum temperature	in °C	11.7	12.8	14.4	18.3	21.1	21.1	20.6	20.0	21.1	18.9	12.8	8.9	8.9	20	5
6 Mean relative humidity	in %	77	78	77	78	85	91	92	92	92	87	81	76	84	8	6
7 Mean precipitation	in mm	5	10	41	66	577	1123	1250	1201	841	269	58	10	5451	32	7
8 Maximum precipitation	in mm															8
9 Minimum precipitation	in mm															9
10 Maximum precipitation in 24 h	in mm	36	48	64	97	300	307	206	244	333	198	79	41	333	32	10
11 Mean number of days with precipitation	>2,5 mm	<1	1	2	4	16	25	27	27	23	12	4	1	143	32	11
12 Mean duration of sunshine	in h															12
13 Mean quantity of radiation	in ly/day															13
14 Mean potential evaporation	in mm	115	123	151	162	163	144	139	135	134	141	126	109	1642	50	14
15 Mean windspeed	in m/sec															15
16 Mean predominent direction of the wind		ENE	ENE	ENE	ENE	S	S	S	S	S	ENE	ENE	ENE		5	16

219 Station / Country **Mergui / Burma**

Location **12°26'N/98°36'E** — Height above sealevel **20 m** — Climate symbol: Köppen **Am** — Troll **V,1**

	J	F	M	A	M	J	J	A	S	O	N	D	year	P	
1 Mean daily temperature	in °C	25.6	26.7	27.5	28.3	27.8	26.4	25.9	25.9	25.9	26.4	26.1	25.6	26.5	60	1
2 Mean daily maximum temperature	in °C	30.6	31.7	32.2	32.8	31.7	29.4	28.9	28.9	28.9	30.0	30.6	30.6	30.6	60	2
3 Mean daily minimum temperature	in °C	20.6	21.7	22.8	23.9	23.9	23.3	22.8	22.8	22.8	22.8	21.7	20.6	22.2	60	3
4 Absolute maximum temperature	in °C	35.0	36.1	37.2	36.7	36.7	34.4	32.8	32.8	33.9	34.4	34.4	37.2	37.2	60	4
5 Absolute minimum temperature	in °C	11.7	15.6	17.2	20.0	19.4	19.4	18.9	18.9	18.9	17.2	15.0	12.8	11.7	60	5
6 Mean relative humidity	in %	71	73	72	74	82	88	90	90	89	84	78	73	80	8	6
7 Mean precipitation	in mm	25	53	79	127	424	762	636	762	632	307	97	20	4124	60	7
8 Maximum precipitation	in mm	98	162	241	257	889	1234	1189	1267	981	636	130	177	4809	30	8
9 Minimum precipitation	in mm	0	0	0	38	121	289	438	468	261	159	0	0	3254	30	9
10 Maximum precipitation in 24 h	in mm	56	78	94	140	155	244	241	180	165	104	132	137	244	60	10
11 Mean number of days with precipitation	>2,5 mm	1	3	5	8	18	25	26	26	23	16	6	2	157	60	11
12 Mean duration of sunshine	in h															12
13 Mean quantity of radiation	in ly/day															13
14 Mean potential evaporation	in mm	122	124	149	155	158	142	139	134	129	138	125	116	1631	60	14
15 Mean windspeed	in m/sec															15
16 Mean predominent direction of the wind																16

220 Station / Country **Ulaanbaatar (Ulan Bator) / Mongolia**

Location **47°55'N/106°50'E** — Height above sealevel **1325 m** — Climate symbol: Köppen **BSk** — Troll **III,11**

	J	F	M	A	M	J	J	A	S	O	N	D	year	P	
1 Mean daily temperature	in °C	-25.6	-21.1	-12.8	-0.9	5.6	13.6	16.1	14.2	8.1	-0.9	-12.8	-22.2	-3.2	12	1
2 Mean daily maximum temperature	in °C	-18.9	-12.8	-3.9	6.7	12.8	20.6	21.7	20.6	14.4	6.1	-5.6	-16.1	3.9	12	2
3 Mean daily minimum temperature	in °C	-32.2	-29.4	-21.7	-6.3	-1.7	6.7	10.6	7.8	1.7	-7.7	-15.6	-28.3	-10.0	12	3
4 Absolute maximum temperature	in °C	-6.1	1.7	17.8	24.4	30.0	36.1	33.3	32.8	28.3	22.8	11.1	0.0	36.1	12	4
5 Absolute minimum temperature	in °C	-43.9	-44.4	-39.4	-23.9	-12.2	-4.4	1.1	-8.7	-10.6	-26.7	-35.6	-42.8	-44.4	12	5
6 Mean relative humidity	in %	77	72	70	53	52	56	66	63	61	63	70	82	65	4	6
7 Mean precipitation	in mm	2	2	3	5	10	28	76	51	23	5	5	3	213	12	7
8 Maximum precipitation	in mm															8
9 Minimum precipitation	in mm															9
10 Maximum precipitation in 24 h	in mm	3	3	5	8	18	28	74	61	41	8	18	18	74	10	10
11 Mean number of days with precipitation	>1,0 mm	1	1	2	2	4	5	10	8	3	2		1	41	12	11
12 Mean duration of sunshine	in h															12
13 Mean quantity of radiation	in ly/day															13
14 Mean potential evaporation	in mm	0	0	0	0	64	109	125	104	52	0	0	0	454	26	14
15 Mean windspeed	in m/sec															15
16 Mean predominent direction of the wind																16

221 Station/Country Kucha/China

Location 41°40'N/83°06'E — Height above sealevel 970 m — Climate symbol: Köppen **BWk** — Troll **III,12**

		J	F	M	A	M	J	J	A	S	O	N	D	year	P	
1 Mean daily temperature	in °C	−12,6	−3,6	7,2	13,1	18,1	21,4	24,1	22,5	18,1	12,2	1,7	−7,2	9,6	2	1
2 Mean daily maximum temperature	in °C	−5,6	3,3	15,6	20,6	26,7	30,0	32,2	30,6	27,2	20,6	9,4	−0,6	17,2	2	2
3 Mean daily minimum temperature	in °C	−19,4	−10,6	−1,1	5,6	12,6	16,1	14,4	6,9	3,9	−6,1	−13,9	1,7	2		3
4 Absolute maximum temperature	in °C	2,2	11,7	26,1	30,0	32,2	36,7	37,2	36,1	33,3	29,4	16,7	8,9	37,2	2	4
5 Absolute minimum temperature	in °C	−25,0	−20,0	−8,9	−0,6	−0,6	6,3	11,1	7,8	1,1	−6,1	−11,1	−26,7	−26,7	2	5
6 Mean relative humidity	in %															6
7 Mean precipitation	in mm	3	3	5	3	3	33	16	8	5	0	2	8	91	2	7
8 Maximum precipitation	in mm															8
9 Minimum precipitation	in mm															9
10 Maximum precipitation in 24 h	in mm															10
11 Mean number of days with precipitation	> 0,1 mm	3	1	<1	3	2	7	6	7	2	0	<1	1	33	2	11
12 Mean duration of sunshine	in h															12
13 Mean quantity of radiation	in ly/day															13
14 Mean potential evaporation	in mm	0	0	22	59	106	135	151	126	79	31	0	0	709	4	14
15 Mean windspeed	in m/sec															15
16 Mean predominent direction of the wind																16

222 Station/Country Baotou/China

Location 40°34'N/109°50'E — Height above sealevel 1044 m — Climate symbol: Köppen **BSk** — Troll **III,11**

		J	F	M	A	M	J	J	A	S	O	N	D	year	P	
1 Mean daily temperature	in °C	−11,8	−9,1	−0,1	7,2	15,5	20,9	22,7	20,7	14,5	8,0	−1,8	−10,0	6,4	7	1
2 Mean daily maximum temperature	in °C															2
3 Mean daily minimum temperature	in °C															3
4 Absolute maximum temperature	in °C	6,4	13,4	20,7	26,4	33,4	36,8	38,4	35,5	31,7	25,6	15,5	6,7	38,4		4
5 Absolute minimum temperature	in °C	−29,6	−26,9	−22,2	−2,4	3,7	9,2	4,9	−2,3	−9,9	−22,0	−32,8	−32,8			5
6 Mean relative humidity	in %	63	64	49	48	49	52	64	69	61	59	61	61	58		6
7 Mean precipitation	in mm	<1	4	4	19	31	29	82	76	31	23	4	<1	305	18	7
8 Maximum precipitation	in mm															8
9 Minimum precipitation	in mm															9
10 Maximum precipitation in 24 h	in mm	1	3	4	20	29	40	42	43	33	24	6	1	43		10
11 Mean number of days with precipitation	> 0,1 mm	1	4	3	5	6	7	10	10	6	4	3	2	61		11
12 Mean duration of sunshine	in h	227	209	258	277	285	289	281	254	271	251	184	209	2995		12
13 Mean quantity of radiation	in ly/day															13
14 Mean potential evaporation	in mm	0	0	0	38	88	122	135	123	73	37	0	0	616	3	14
15 Mean windspeed	in m/sec	2,0	2,4	2,5	3,5	2,5	2,3	2,3	2,2	2,1	2,2	2,2	1,9	2,5		15
16 Mean predominent direction of the wind		N	E	E	W	E	E	SE	ESE	ESE	ESE	ESE	NW			16

223 Station/Country Jiuquan/China

Location 39°48'N/98°34'E — Height above sealevel 1006 m — Climate symbol: Köppen **BWk** — Troll **III,12**

		J	F	M	A	M	J	J	A	S	O	N	D	year	P	
1 Mean daily temperature	in °C	−8,3	−3,3	2,2	9,1	15,9	20,6	23,6	21,7	15,9	10,3	−0,6	−6,7	8,4	6	1
2 Mean daily maximum temperature	in °C	−2,2	3,3	8,9	16,1	22,8	27,2	30,6	27,8	22,8	17,2	5,6	−0,6	15,0	6	2
3 Mean daily minimum temperature	in °C	−14,4	−10,0	−4,4	2,2	8,9	13,9	16,7	15,6	8,9	3,3	−6,7	−12,8	1,7	6	3
4 Absolute maximum temperature	in °C	12,8	17,2	21,1	27,8	32,8	37,8	37,2	36,7	34,4	26,7	17,8	13,3	37,8	7	4
5 Absolute minimum temperature	in °C	−22,8	−23,9	−17,8	−7,2	−2,8	5,6	11,7	8,3	0,0	−6,7	−19,4	−25,0	−25,0	7	5
6 Mean relative humidity	in %															6
7 Mean precipitation	in mm	2	3	3	3	3	13	13	28	8	2	3	3	84	6	7
8 Maximum precipitation	in mm															8
9 Minimum precipitation	in mm															9
10 Maximum precipitation in 24 h	in mm															10
11 Mean number of days with precipitation	> 0,1 mm	1	1	2	3	3	5	6	7	2	1	2	2	35	6	11
12 Mean duration of sunshine	in h															12
13 Mean quantity of radiation	in ly/day															13
14 Mean potential evaporation	in mm	0	0	7	44	93	126	148	124	76	40	0	0	658	2	14
15 Mean windspeed	in m/sec															15
16 Mean predominent direction of the wind																16

224 Station/Country Kashgar/China

Location 39°24'N/78°07'E — Height above sealevel 1309 m — Climate symbol: Köppen **BWk** — Troll **III,12**

		J	F	M	A	M	J	J	A	S	O	N	D	year	P	
1 Mean daily temperature	in °C	−5,3	−0,6	7,5	15,3	15,3	24,7	26,7	25,6	21,1	13,9	5,3	−2,6	12,2	27	1
2 Mean daily maximum temperature	in °C	0,6	6,1	13,3	21,7	27,2	31,7	33,3	32,2	28,3	21,7	12,2	3,3	19,4	27	2
3 Mean daily minimum temperature	in °C	−11,1	−7,2	1,8	8,9	14,4	17,8	20,0	18,9	13,9	6,1	−1,7	−8,3	6,1	27	3
4 Absolute maximum temperature	in °C	10,6	16,7	25,6	33,9	36,1	38,9	41,1	38,3	36,7	31,7	21,1	16,7	41,1	10	4
5 Absolute minimum temperature	in °C	−21,7	−20,0	−13,3	−2,8	3,9	5,6	11,7	12,2	3,9	−1,7	−17,2	−26,1	−26,1	10	5
6 Mean relative humidity	in %	76	71	57	47	46	44	49	54	55	57	67	79	58	10	6
7 Mean precipitation	in mm	15	3	13	5	8	5	10	8	3	3	5	8	86	18	7
8 Maximum precipitation	in mm															8
9 Minimum precipitation	in mm															9
10 Maximum precipitation in 24 h	in mm	8	10	15	25	23	15	13	8	5	15	5	3	25	7	10
11 Mean number of days with precipitation	> 0,25 mm	1	<1	1	1	1	1	1	2	1	<1	1	<1	13	18	11
12 Mean duration of sunshine	in h															12
13 Mean quantity of radiation	in ly/day															13
14 Mean potential evaporation	in mm	0	1	24	70	117	154	173	140	96	52	11	0	838		14
15 Mean windspeed	in m/sec															15
16 Mean predominent direction of the wind																16

116

Location   37°55'N / 112°34'E   Height above sealevel   782 m   Climate symbol: Köppen **BSk**   Troll III,11

		J	F	M	A	M	J	J	A	S	O	N	D	year	P	
1 Mean daily temperature	in °C	-7,3	-3,2	4,2	11,8	18,5	22,8	25,0	22,8	17,4	10,4	2,2	-4,8	10,0	18	1
2 Mean daily maximum temperature	in °C	-1,1	3,9	11,7	18,3	26,1	29,4	31,1	28,3	24,4	18,3	8,9	1,7	16,7	8	2
3 Mean daily minimum temperature	in °C	-13,9	-9,4	-4,4	4,4	10,6	15,6	19,4	17,2	10,6	2,8	-4,4	-10,0	3,3	8	3
4 Absolute maximum temperature	in °C	14,7	19,4	28,0	33,5	38,2	39,8	41,4	37,2	34,6	32,6	27,0	14,0	41,4		4
5 Absolute minimum temperature	in °C	-29,7	-22,3	-18,3	-15,0	0,0	4,6	8,2	6,9	-0,9	-12,0	-20,1	-24,0	-29,7		5
6 Mean relative humidity	in %	56	59	54	55	53	56	67	71	67	61	64	63	61		6
7 Mean precipitation	in mm	4	4	11	15	27	40	108	104	50	18	11	4	396	37	7
8 Maximum precipitation	in mm															8
9 Minimum precipitation	in mm															9
10 Maximum precipitation in 24 h	in mm	30	8	75	29	51	71	66	73	50	36	35	10	75		10
11 Mean number of days with precipitation	> 0,1 mm	1	2	3	3	5	7	11	9	7	3	2	2	55		11
12 Mean duration of sunshine	in h	171	163	218	206	236	234	204	201	210	220	186	156	2373		12
13 Mean quantity of radiation	in ly / day															13
14 Mean potential evaporation	in mm	0	0	11	51	104	136	157	138	82	38	5	0	722	21	14
15 Mean windspeed	in m / sec	1,2	1,5	1,9	2,2	1,9	1,7	1,4	1,0	1,3	1,2	1,2	1,1	1,5		15
16 Mean predominent direction of the wind		SSE	NW	WNW	WNW	N	N	SSE	N	NNE	SSE	WNW	WNW			16

Location   36°35'N / 101°55'E   Height above sealevel   2244 m   Climate symbol: Köppen **BSk**   Troll III,11

		J	F	M	A	M	J	J	A	S	O	N	D	year	P	
1 Mean daily temperature	in °C	-6,4	-2,8	2,9	8,6	13,3	15,7	18,3	17,8	12,9	7,9	0,1	-5,3	6,9	14	1
2 Mean daily maximum temperature	in °C															2
3 Mean daily minimum temperature	in °C															3
4 Absolute maximum temperature	in °C	12,1	19,1	25,0	27,9	30,9	32,1	32,4	32,3	27,1	25,6	20,7	13,1	32,4		4
5 Absolute minimum temperature	in °C	-23,1	-19,4	-13,4	-11,2	-2,3	2,7	6,1	5,5	0,1	-6,6	-18,9	-22,8	-23,1		5
6 Mean relative humidity	in %	50	45	40	45	48	58	61	65	68	62	56	54	54		6
7 Mean precipitation	in mm	1	2	5	18	34	46	73	92	74	27	4	2	378	13	7
8 Maximum precipitation	in mm															8
9 Minimum precipitation	in mm															9
10 Maximum precipitation in 24 h	in mm	3	5	6	25	21	27	34	47	39	25	12	3	47		10
11 Mean number of days with precipitation	> 0,1 mm	2	2	3	4	8	12	12	12	12	6	2	1	78		11
12 Mean duration of sunshine	in h	213	201	194	214	246	243	239	233	188	209	211	228	2619		12
13 Mean quantity of radiation	in ly / day															13
14 Mean potential evaporation	in mm	0	0	11	36	83	96	115	108	67	39	0	0	555	4	14
15 Mean windspeed	in m / sec	1,9	1,5	1,7	1,9	1,4	1,3	1,2	1,3	1,2	1,3	1,2	0,9	1,3		15
16 Mean predominent direction of the wind		E	E	E	E	E	E	E	E	E	E	E	SE			16

Location   36°01'N / 103°59'E   Height above sealevel   1508 m   Climate symbol: Köppen **BSk**   Troll III,11

		J	F	M	A	M	J	J	A	S	O	N	D	year	P	
1 Mean daily temperature	in °C	-6,5	-1,7	5,4	12,1	17,4	20,9	22,8	21,4	13,6	10,1	1,7	-5,3	9,5	21	1
2 Mean daily maximum temperature	in °C	0,6	7,2	11,7	18,3	24,4	27,2	28,9	27,2	21,7	16,7	8,3	3,3	15,6	8	2
3 Mean daily minimum temperature	in °C	-13,9	-8,9	-1,7	4,4	10,0	13,9	16,1	16,1	11,1	3,9	-5,0	-10,6	2,8	8	3
4 Absolute maximum temperature	in °C	13,8	17,5	26,9	33,2	35,5	36,5	38,0	37,0	31,9	28,0	21,6	10,9	38,0		4
5 Absolute minimum temperature	in °C	-23,1	-22,1	-16,3	-8,4	0,4	2,9	9,3	5,4	0,5	-6,6	-15,4	-21,6	-23,1		5
6 Mean relative humidity	in %	61	54	47	45	47	52	60	64	69	67	63	64	58		6
7 Mean precipitation	in mm	1	3	8	14	34	40	66	92	56	16	4	2	338	21	7
8 Maximum precipitation	in mm	15	10	23	28	61	104	127	218	109	51	13	10	218	18	8
9 Minimum precipitation	in mm	0	<1	0	<1	<1	5	18	33	23	<1	0	0	0	18	9
10 Maximum precipitation in 24 h	in mm	16	9	9	15	39	53	69	72	70	20	14	9	72		10
11 Mean number of days with precipitation	> 0,1 mm	1	2	4	5	7	9	10	11	12	6	2	1	70		11
12 Mean duration of sunshine	in h	160	143	147	170	202	209	224	227	159	167	166	168	2142		12
13 Mean quantity of radiation	in ly / day															13
14 Mean potential evaporation	in mm	0	0	17	52	90	119	139	121	70	39	2	0	655		14
15 Mean windspeed	in m / sec	1,0	1,2	1,8	1,9	1,9	1,9	1,8	1,7	1,5	1,3	1,0	1,0	1,5		15
16 Mean predominent direction of the wind		E	ENE	E	E	E	E	E	E	SW	E	E	E			16

Location   45°45'N / 126°38'E   Height above sealevel   143 m   Climate symbol: Köppen **Dwa**   Troll III,8

		J	F	M	A	M	J	J	A	S	O	N	D	year	P	
1 Mean daily temperature	in °C	-20,1	-15,8	-6,0	5,8	14,0	19,8	23,3	21,8	14,3	5,7	-6,6	-16,7	3,3	38	1
2 Mean daily maximum temperature	in °C	-12,2	-7,7	2,2	12,8	20,6	25,6	27,8	26,7	20,6	10,6	0,0	-10,0	9,4	9	2
3 Mean daily minimum temperature	in °C	-24,4	-13,9	-11,7	-1,1	6,1	12,8	16,7	15,6	8,3	-1,7	-11,7	-21,1	-2,8	9	3
4 Absolute maximum temperature	in °C	1,4	12,3	19,3	29,6	35,6	39,0	39,1	37,8	32,0	23,1	18,6	6,1	39,1		4
5 Absolute minimum temperature	in °C	-41,4	-39,9	-29,1	-14,0	-3,7	3,8	9,6	6,0	-3,0	-14,8	-31,6	-35,6	-41,4		5
6 Mean relative humidity	in %	79	76	66	59	58	70	77	78	75	66	78	78	71		6
7 Mean precipitation	in mm	4	6	17	23	44	92	167	119	52	36	12	5	577	41	7
8 Maximum precipitation	in mm															8
9 Minimum precipitation	in mm															9
10 Maximum precipitation in 24 h	in mm	5	11	17	31	62	79	147	113	54	27	15	8	147		10
11 Mean number of days with precipitation	> 0,1 mm	5	5	6	7	11	15	16	13	12	7	6	6	109		11
12 Mean duration of sunshine	in h	197	218	247	234	257	258	267	254	217	214	182	177	2722		12
13 Mean quantity of radiation	in ly / day															13
14 Mean potential evaporation	in mm	0	0	0	27	85	127	154	132	70	21	0	0	616	8	14
15 Mean windspeed	in m / sec	4,5	4,5	5,2	6,0	5,6	4,6	4,2	3,9	4,3	5,1	5,2	4,6	4,8		15
16 Mean predominent direction of the wind		W	S	W	W	SW	S	S	S	S	S	SSW	S			16

229 Station / Country Changchun / China

Location 43°52'N / 125°20'E   Height above sealevel 216 m   Climate symbol: Köppen Dwa   Troll III,8

		J	F	M	A	M	J	J	A	S	O	N	D	year	P	
1 Mean daily temperature	in °C	-16,8	-12,7	-3,9	6,6	14,4	20,0	23,5	21,9	14,9	6,8	-4,2	-13,8	4,7	38	1
2 Mean daily maximum temperature	in °C	-11,1	-6,1	1,7	13,3	21,1	26,7	28,9	27,8	21,7	13,3	1,1	-8,3	10,6	21	2
3 Mean daily minimum temperature	in °C	-23,3	-18,9	-10,6	0,0	7,2	13,9	18,3	16,7	8,3	0,6	-10,0	-19,4	-1,7	21	3
4 Absolute maximum temperature	in °C	3,7	13,7	19,9	30,2	34,3	39,5	38,3	37,0	33,1	29,8	24,0	7,5	39,5		4
5 Absolute minimum temperature	in °C	-35,7	-36,0	-27,7	-15,5	-3,4	4,4	9,0	6,4	-2,7	-13,4	-29,4	-34,7	-36,0		5
6 Mean relative humidity	in %	72	69	59	52	54	68	76	78	72	66	66	71	67		6
7 Mean precipitation	in mm	6	6	15	23	52	110	172	139	54	33	16	5	632	46	7
8 Maximum precipitation	in mm															8
9 Minimum precipitation	in mm															9
10 Maximum precipitation in 24 h	in mm	8	13	24	21	40	86	97	108	79	36	60	18	108		10
11 Mean number of days with precipitation	> 0,1 mm	6	4	6	7	11	15	16	13	10	8	6	5	107		11
12 Mean duration of sunshine	in h	201	208	249	243	259	264	247	249	267	221	179	186	2773		12
13 Mean quantity of radiation	in ly / day															13
14 Mean potential evaporation	in mm	0	0	0	31	86	129	156	132	73	27	0	0	634		14
15 Mean windspeed	in m / sec	3,4	3,5	4,2	4,8	4,5	3,4	2,8	2,4	2,8	3,4	3,9	3,6	3,6		15
16 Mean predominent direction of the wind		WSW	WSW	WSW	WSW	WSW	SW	SSW	SSW	S	WSW	WSW	WSW			16

230 Station / Country Wulumuqi (Urumchi) / China

Location 43°47'N / 87°37'E   Height above sealevel 913 m   Climate symbol: Köppen Dfa   Troll III,10

		J	F	M	A	M	J	J	A	S	O	N	D	year	P	
1 Mean daily temperature	in °C	-15,8	-13,6	-4,0	8,5	17,7	21,5	23,9	21,9	16,7	6,1	-6,2	-13,0	5,3	3	1
2 Mean daily maximum temperature	in °C	-10,6	-8,3	-0,6	15,6	22,2	25,6	27,8	26,7	20,6	10,0	-1,1	-8,3	10,0	6	2
3 Mean daily minimum temperature	in °C	-21,7	-19,4	-11,1	2,2	8,3	12,2	14,4	13,3	8,3	1,6	-10,6	-13,3	-1,1	6	3
4 Absolute maximum temperature	in °C	3,3	8,1	19,0	28,0	35,7	36,7	38,1	38,0	34,9	27,9	19,4	6,5	38,1		4
5 Absolute minimum temperature	in °C	-34,4	-41,5	-33,4	-12,0	-9,7	4,4	7,5	5,5	1,5	-18,3	-36,6	-36,3	-41,5		5
6 Mean relative humidity	in %	87	89	80	62	44	49	46	44	46	62	84	86	65		6
7 Mean precipitation	in mm	6	15	15	33	25	33	16	35	15	47	22	11	273	7	7
8 Maximum precipitation	in mm															8
9 Minimum precipitation	in mm															9
10 Maximum precipitation in 24 h	in mm	9	12	28	17	28	46	11	39	16	37	13	9	46		10
11 Mean number of days with precipitation	> 0,1 mm	10	11	7	8	5	6	7	5	4	9	12	11	95		11
12 Mean duration of sunshine	in h	187	148	194	203	297	271	279	268	263	213	144	146	2613		12
13 Mean quantity of radiation	in ly / day															13
14 Mean potential evaporation	in mm	0	0	0	42	98	123	143	125	74	20	0	0	625	3	14
15 Mean windspeed	in m / sec	1,1	1,5	1,6	2,6	2,9	2,3	2,0	2,3	2,3	2,2	1,6	1,6	2,0		15
16 Mean predominent direction of the wind		SW	N	N	NW	NW	NW	NW	NW	NW	NW	NW	SW			16

231 Station / Country Shenyang / China

Location 41°46'N / 123°28'E   Height above sealevel 416 m   Climate symbol: Köppen Dwa   Troll III,8

		J	F	M	A	M	J	J	A	S	O	N	D	year	P	
1 Mean daily temperature	in °C	-12,8	-9,1	-0,8	8,8	16,2	21,6	24,9	23,6	17,0	9,4	-0,9	-9,8	7,3	44	1
2 Mean daily maximum temperature	in °C	-5,6	-2,2	6,1	16,1	23,3	28,9	30,6	29,4	23,9	16,1	5,0	-3,9	13,9	10	2
3 Mean daily minimum temperature	in °C	-18,3	-14,4	-6,1	2,8	10,0	16,1	20,6	19,4	11,1	3,3	-5,6	-15,0	2,2	10	3
4 Absolute maximum temperature	in °C	8,8	14,3	20,2	30,3	33,9	39,3	38,9	38,3	33,9	30,4	25,2	11,2	39,3		4
5 Absolute minimum temperature	in °C	-33,1	-32,7	-25,0	-9,6	-2,0	6,5	16,7	9,6	-1,0	-10,0	-26,3	-32,2	-33,1		5
6 Mean relative humidity	in %	68	64	57	54	58	66	78	78	73	70	65	67	66		6
7 Mean precipitation	in mm	6	6	14	30	66	96	177	162	74	41	23	10	705	15	7
8 Maximum precipitation	in mm															8
9 Minimum precipitation	in mm															9
10 Maximum precipitation in 24 h	in mm	14	15	50	41	66	87	119	149	137	53	37	17	149		10
11 Mean number of days with precipitation	> 0,1 mm	4	3	6	6	10	12	15	13	9	7	5	4	94		11
12 Mean duration of sunshine	in h	190	200	238	245	238	257	230	232	235	224	180	173	2642		12
13 Mean quantity of radiation	in ly / day															13
14 Mean potential evaporation	in mm	0	0	0	37	92	133	164	143	81	34	0	0	684		14
15 Mean windspeed	in m / sec	2,3	2,6	3,3	3,7	3,4	2,8	2,4	2,0	2,1	2,6	2,7	2,5	2,7		15
16 Mean predominent direction of the wind		N	N	SSW	SSW	SSW	SSW	S	S	S	S	N	N			16

232 Station / Country Beijing (Peking) / China

Location 39°57'N / 116°19'E   Height above sealevel 52 m   Climate symbol: Köppen Dwa   Troll III,7

		J	F	M	A	M	J	J	A	S	O	N	D	year	P	
1 Mean daily temperature	in °C	-4,7	-1,9	4,8	13,7	20,1	24,7	26,1	24,9	19,9	12,8	3,8	-2,7	11,8	36	1
2 Mean daily maximum temperature	in °C	1,0	5,0	12,0	21,0	27,0	31,0	32,0	30,0	27,0	21,0	10,0	3,0	18,0	27	2
3 Mean daily minimum temperature	in °C	-10,0	-7,0	-1,0	7,0	13,0	18,0	22,0	21,0	15,0	7,0	-1,0	-8,0	6,0	27	3
4 Absolute maximum temperature	in °C	14,2	18,5	28,1	35,8	38,1	42,6	40,5	38,3	33,3	31,1	24,2	13,5	42,6		4
5 Absolute minimum temperature	in °C	-22,8	-17,8	-13,8	-3,3	3,4	10,1	14,9	11,3	1,6	-4,7	-13,5	-19,6	-22,8		5
6 Mean relative humidity	in %	50	50	48	46	49	56	72	74	67	59	56	51	56		6
7 Mean precipitation	in mm	4	5	8	17	35	78	243	141	58	16	11	3	619	112	7
8 Maximum precipitation	in mm	48	28	69	58	117	292	826	373	292	97	71	20	826	69	8
9 Minimum precipitation	in mm	0	0	0	0	0	<1	8	8	0	0	0	0	0	69	9
10 Maximum precipitation in 24 h	in mm	14	11	25	67	109	203	225	144	96	37	55	11	225		10
11 Mean number of days with precipitation	> 0,1 mm	3	3	4	6	8	8	13	11	7	3	3	2	66		11
12 Mean duration of sunshine	in h	206	197	237	239	265	261	220	224	229	244	193	192	2707		12
13 Mean quantity of radiation	in ly / day															13
14 Mean potential evaporation	in mm	0	0	12	57	113	159	188	145	94	44	7	0	799	95	14
15 Mean windspeed	in m / sec	2,3	2,6	2,9	3,3	2,9	2,3	1,7	1,6	1,9	2,2	2,3	2,5	2,4		15
16 Mean predominent direction of the wind		NNW	N	N	*	SSW	N	E	E	W	E	E	NW			16

* SSW/NNW

233 Station / Country Tianjin (Tientsin) / China

Location 39°06'N / 117°10'E   Height above sealevel 3 m   Climate symbol: Köppen Dwa   Troll III,7

		J	F	M	A	M	J	J	A	S	O	N	D	year	P	
1 Mean daily temperature	in °C	-4,1	-1,8	4,8	13,4	19,8	24,5	26,8	25,9	21,0	14,1	4,9	-1,9	12,8	42	1
2 Mean daily maximum temperature	in °C	0,8	3,9	11,1	20,0	26,7	31,7	32,2	30,6	26,1	20,0	9,4	2,2	17,8	24	2
3 Mean daily minimum temperature	in °C	-8,9	-6,1	-0,6	7,2	13,3	18,9	22,8	22,2	16,7	8,9	0,8	-6,1	7,2	24	3
4 Absolute maximum temperature	in °C	12,1	18,1	28,8	35,7	41,4	42,7	42,9	39,6	35,8	32,4	25,8	14,2	42,9		4
5 Absolute minimum temperature	in °C	-20,4	-18,6	-14,0	-2,5	3,6	9,4	16,2	13,7	4,7	-2,5	-12,3	-18,8	-20,4		5
6 Mean relative humidity	in %	61	58	55	52	54	62	75	76	69	63	61	59	62		6
7 Mean precipitation	in mm	3	4	9	17	28	63	177	149	45	15	13	5	528	49	7
8 Maximum precipitation	in mm	23	15	43	41	66	175	447	353	135	41	48	25		33	8
9 Minimum precipitation	in mm	0	0	0	0	<1	5	46	28	5	0	0	0		33	9
10 Maximum precipitation in 24 h	in mm	16	13	35	48	48	79	130	163	59	45	75	19	163		10
11 Mean number of days with precipitation	> 0,1 mm	2	2	3	4	6	9	13	11	6	4	3	2	65		11
12 Mean duration of sunshine	in h	184	189	233	245	284	273	253	232	233	241	191	176	2734		12
13 Mean quantity of radiation	in ly / day															13
14 Mean potential evaporation	in mm	0	0	9	51	106	146	171	151	98	49	8	0	789	32	14
15 Mean windspeed	in m / sec	2,3	2,5	3,0	3,3	3,1	2,7	2,2	1,8	2,1	2,3	2,4	2,4	2,5		15
16 Mean predominent direction of the wind		SW	E	SSE	SW	SSE	SE	SSW	SSW	SW	SSW	NNE	SSW			16

234 Station / Country Lüda / China

Location 38°54'N / 121°38'E   Height above sealevel 96 m   Climate symbol: Köppen Dwa   Troll III,7

		J	F	M	A	M	J	J	A	S	O	N	D	year	P	
1 Mean daily temperature	in °C	-5,2	-3,6	1,9	9,4	15,5	20,3	23,7	24,5	19,9	13,8	5,3	-1,9	10,3	39	1
2 Mean daily maximum temperature	in °C	-1,1	0,6	6,1	14,4	20,0	25,0	27,2	28,3	24,4	18,3	9,4	1,7	14,4	25	2
3 Mean daily minimum temperature	in °C	-8,9	-7,2	-1,7	5,0	11,1	16,1	20,6	21,7	16,1	9,4	1,1	-6,1	6,7	25	3
4 Absolute maximum temperature	in °C	10,6	14,0	18,8	28,3	32,4	35,0	36,1	35,7	31,6	27,7	23,0	13,0	36,1		4
5 Absolute minimum temperature	in °C	-19,9	-19,3	-14,8	-4,0	-1,4	10,2	14,9	14,8	5,8	-2,9	-11,6	-18,9	-19,9		5
6 Mean relative humidity	in %	63	62	80	57	61	71	83	80	70	64	62	61	66		6
7 Mean precipitation	in mm	10	7	15	25	46	49	155	133	66	33	25	14	575	39	7
8 Maximum precipitation	in mm															8
9 Minimum precipitation	in mm															9
10 Maximum precipitation in 24 h	in mm	36	30	47	48	55	83	190	162	104	70	47	51	190		10
11 Mean number of days with precipitation	> 0,1 mm	4	3	4	5	6	8	12	11	7	6	6	5	77		11
12 Mean duration of sunshine	in h	193	199	243	257	281	275	231	239	245	242	184	173	2762		12
13 Mean quantity of radiation	in ly / day															13
14 Mean potential evaporation	in mm	0	0	6	40	81	119	147	145	97	55	15	0	705	39	14
15 Mean windspeed	in m / sec	4,8	4,8	5,3	5,6	5,3	4,5	4,3	3,7	4,1	4,7	5,2	4,9	4,8		15
16 Mean predominent direction of the wind		NNW	NNW	S	SSE	S	S	SSE	SSE	S	S	NNW	NNW			16

235 Station / Country Jinan / China

Location 36°41'N / 116°58'E   Height above sealevel 55 m   Climate symbol: Köppen Cwa   Troll III,7

		J	F	M	A	M	J	J	A	S	O	N	D	year	P	
1 Mean daily temperature	in °C	-1,2	1,4	8,4	16,1	22,5	27,3	28,4	26,8	22,2	16,3	8,3	1,0	14,8	34	1
2 Mean daily maximum temperature	in °C	2,8	6,7	13,9	21,7	28,3	32,8	33,3	31,1	27,8	21,7	13,3	5,6	20,0	10	2
3 Mean daily minimum temperature	in °C	-5,6	-2,8	3,3	10,6	16,7	21,1	23,9	22,2	16,7	10,6	2,8	-3,3	9,4	10	3
4 Absolute maximum temperature	in °C	20,2	20,5	32,3	36,0	40,8	42,4	42,7	41,2	36,8	34,7	28,2	19,0	42,7		4
5 Absolute minimum temperature	in °C	-19,2	-15,2	-11,3	-2,3	2,5	10,5	16,9	13,2	4,4	-2,7	-11,3	-16,7	-19,2		5
6 Mean relative humidity	in %	65	66	58	50	54	54	74	79	69	64	66	68	64		6
7 Mean precipitation	in mm	7	9	12	21	34	72	185	174	64	21	20	12	631	27	7
8 Maximum precipitation	in mm															8
9 Minimum precipitation	in mm															9
10 Maximum precipitation in 24 h	in mm	16	16	23	67	95	67	147	125	50	40	36	24	147		10
11 Mean number of days with precipitation	> 0,1 mm	3	4	4	5	6	8	13	13	7	3	4	4	74		11
12 Mean duration of sunshine	in h	184	179	220	232	271	275	244	228	227	241	188	179	2668		12
13 Mean quantity of radiation	in ly / day															13
14 Mean potential evaporation	in mm	0	0	22	66	124	173	188	158	102	58	15	0	906	34	14
15 Mean windspeed	in m / sec	2,5	3,0	3,6	3,9	3,6	3,3	2,5	1,9	3,1	2,6	2,9	2,5	2,9		15
16 Mean predominent direction of the wind		SW	ENE	SW	SW	SW	NE	NE	NE	ENE	SW	ENE	SW			16

236 Station / Country Qingdao (Tsingtao) / China

Location 36°04'N / 120°19'E   Height above sealevel 77 m   Climate symbol: Köppen Cwa   Troll III,7

		J	F	M	A	M	J	J	A	S	O	N	D	year	P	
1 Mean daily temperature	in °C	-1,1	0,1	4,4	10,3	15,7	20,0	23,7	25,1	21,4	15,9	8,8	1,8	12,1	55	1
2 Mean daily maximum temperature	in °C	2,2	3,3	8,3	14,4	19,4	23,3	26,1	28,3	25,0	20,0	12,2	4,4	15,6	26	2
3 Mean daily minimum temperature	in °C	-4,4	-3,3	0,6	6,7	12,2	16,7	21,1	22,8	17,8	12,2	4,4	-2,2	8,9	26	3
4 Absolute maximum temperature	in °C	11,6	15,4	22,6	29,7	31,5	33,2	36,2	36,6	33,7	28,5	28,8	15,8	36,2		4
5 Absolute minimum temperature	in °C	-16,4	-12,8	-11,3	-4,3	3,2	10,9	14,5	14,6	8,7	1,1	-9,2	-14,1	-16,4		5
6 Mean relative humidity	in %	68	68	68	70	74	82	89	84	73	66	65	67	73		6
7 Mean precipitation	in mm	11	9	19	34	42	77	150	149	85	33	22	17	648	55	7
8 Maximum precipitation	in mm	61	30	109	86	132	343	300	259	147	94	53			36	8
9 Minimum precipitation	in mm	0	0	0	<1	<1	15	13	25	5	0	<1	0		36	9
10 Maximum precipitation in 24 h	in mm	33	32	56	69	67	136	132	163	225	133	74	22	225		10
11 Mean number of days with precipitation	> 0,1 mm	4	4	6	6	7	8	13	11	8	4	4	4	78		11
12 Mean duration of sunshine	in h	182	175	211	224	247	217	187	220	212	229	187	171	2462		12
13 Mean quantity of radiation	in ly / day															13
14 Mean potential evaporation	in mm	0	0	11	44	78	111	143	148	103	65	24	2	727	36	14
15 Mean windspeed	in m / sec	5,4	5,3	5,8	5,7	5,7	5,5	5,1	4,6	4,7	5,0	5,5	5,6	5,3		15
16 Mean predominent direction of the wind		N	N	S	SSE	SSE	SSE	SSE	SSE	N	N	N	N			16

119

237 Station / Country Xuzhou / China

Location 34°17'N / 117°10'E   Height above sealevel 3 m   Climate symbol: Köppen Cwa   Troll III,7

		J	F	M	A	M	J	J	A	S	O	N	D	year	P	
1 Mean daily temperature	in °C	-1,4	1,4	7,5	13,3	20,0	25,3	28,1	26,7	21,7	15,6	8,1	2,0	14,1	8	1
2 Mean daily maximum temperature	in °C	3,3	6,7	14,4	20,0	27,2	32,2	34,4	32,2	28,3	22,8	15,0	6,7	20,6	8	2
3 Mean daily minimum temperature	in °C	-6,1	-3,9	0,6	6,7	12,8	18,3	21,7	21,1	15,0	8,3	1,1	-2,8	7,8	8	3
4 Absolute maximum temperature	in °C	17,2	22,2	31,7	33,3	38,9	40,6	41,1	39,4	36,1	32,8	18,3	18,3	41,1	8	4
5 Absolute minimum temperature	in °C	-16,7	-18,3	-8,9	-2,8	1,1	11,1	13,9	10,0	3,3	-2,2	-14,4	-16,1	-18,3	8	5
6 Mean relative humidity	in %	76	75	62	71	71	67	76	80	70	63	65	77	71	8	6
7 Mean precipitation	in mm	13	18	20	61	48	112	127	135	81	25	20	33	693	8	7
8 Maximum precipitation	in mm															8
9 Minimum precipitation	in mm															9
10 Maximum precipitation in 24 h	in mm	18	18	25	48	84	99	86	104	147	46	23	25	147	8	10
11 Mean number of days with precipitation	> 1,0 mm	3	3	3	6	5	5	7	9	4	2	4	5	56	8	11
12 Mean duration of sunshine	in h															12
13 Mean quantity of radiation	in ly/day															13
14 Mean potential evaporation	in mm	0	1	19	51	106	154	182	159	103	56	17	2	850	8	14
15 Mean windspeed	in m/sec															15
16 Mean predominent direction of the wind																16

238 Station / Country Changdu / China

Location 31°11'N / 96°59'E   Height above sealevel 3200 m   Climate symbol: Köppen Cwb   Troll III, 1

		J	F	M	A	M	J	J	A	S	O	N	D	year	P	
1 Mean daily temperature	in °C	-2,5	1,4	4,7	8,2	12,3	14,5	15,7	15,0	13,2	9,1	3,1	1,4	7,8	4	1
2 Mean daily maximum temperature	in °C															2
3 Mean daily minimum temperature	in °C															3
4 Absolute maximum temperature	in °C	16,4	21,7	23,4	25,8	28,0	31,1	33,3	33,1	29,4	26,9	21,8	19,5	33,3		4
5 Absolute minimum temperature	in °C	-18,0	-15,8	-12,2	-7,6	-3,1	1,2	0,2	1,3	-6,8	-12,7	-17,8	-18,0			5
6 Mean relative humidity	in %	46	44	45	50	52	65	71	70	69	61	43	45	53		6
7 Mean precipitation	in mm	2	4	11	23	79	88	135	99	81	29	<1	4	556	4	7
8 Maximum precipitation	in mm															8
9 Minimum precipitation	in mm															9
10 Maximum precipitation in 24 h	in mm	2	5	6	12	25	26	37	24	23	23	2	2	37		10
11 Mean number of days with precipitation	> 0,1 mm	3	3	6	9	16	19	23	20	19	10	1	2	131		11
12 Mean duration of sunshine	in h	182	180	186	186	227	180	156	203	176	200	216	192	2284		12
13 Mean quantity of radiation	in ly/day															13
14 Mean potential evaporation	in mm	2	11	31	52	81	93	106	100	75	50	13	0	614		14
15 Mean windspeed	in m/sec	1,5	1,4	1,3	1,4	1,7	1,5	1,5	1,1	1,3	1,2	1,3	1,2	1,4		15
16 Mean predominent direction of the wind		var	var	var	var	var	var	var	var	var	var	var	var			16

239 Station / Country Chengdu / China

Location 30°40'N / 104°04'E   Height above sealevel 498 m   Climate symbol: Köppen Cwa   Troll III,7

		J	F	M	A	M	J	J	A	S	O	N	D	year	P	
1 Mean daily temperature	in °C	6,2	8,4	12,7	17,5	22,2	24,4	26,5	25,9	22,1	17,5	12,4	7,7	17,0	22	1
2 Mean daily maximum temperature	in °C	8,3	10,6	15,0	20,6	25,6	28,3	29,4	28,9	24,4	20,0	14,4	11,1	19,4	8	2
3 Mean daily minimum temperature	in °C	3,3	5,6	8,3	13,3	17,8	21,1	23,3	22,8	19,4	15,0	10,0	5,0	13,9	8	3
4 Absolute maximum temperature	in °C	21,3	26,1	30,0	38,4	37,9	38,3	40,1	39,0	34,5	32,6	26,2	20,8	40,1		4
5 Absolute minimum temperature	in °C	-4,0	-4,0	-0,4	0,0	9,6	13,1	17,1	15,7	12,9	7,0	0,1	-3,7	-4,0		5
6 Mean relative humidity	in %	81	80	77	76	74	80	84	84	85	86	84	83	81		6
7 Mean precipitation	in mm	7	15	25	56	98	122	304	303	139	53	18	8	1148	22	7
8 Maximum precipitation	in mm	10	25	71	114	173	180	813	609	396	94	31	13		12	8
9 Minimum precipitation	in mm	<1	<1	4	20	31	38	127	140	32	20	<1	<1		12	9
10 Maximum precipitation in 24 h	in mm	8	16	28	91	64	84	233	188	130	24	12	8	233		10
11 Mean number of days with precipitation	> 0,1 mm	6	8	11	14	15	16	17	16	16	16	9	5	149		11
12 Mean duration of sunshine	in h	62	60	82	102	134	111	101	171	77	72	60	57	1089		12
13 Mean quantity of radiation	in ly/day															13
14 Mean potential evaporation	in mm	9	13	34	65	104	130	142	140	96	61	30	15	839	19	14
15 Mean windspeed	in m/sec	1,2	1,4	1,5	1,6	1,6	1,6	1,4	1,5	1,6	1,3	1,3	1,4	1,4		15
16 Mean predominent direction of the wind		NNE	NE	NE	NE	NE	SW	NNE	NE	N	NNE	NNE	NNE			16

240 Station / Country Lasa / China

Location 29°40'N / 91°07'E   Height above sealevel 3685 m   Climate symbol: Köppen BSk   Troll III,11

		J	F	M	A	M	J	J	A	S	O	N	D	year	P	
1 Mean daily temperature	in °C	-1,7	1,1	4,7	8,1	12,2	16,7	16,4	15,6	13,7	8,9	3,9	0,0	8,8	7	1
2 Mean daily maximum temperature	in °C	6,7	8,3	11,7	15,6	19,4	23,9	23,3	22,2	21,1	16,7	12,8	8,9	16,1	7	2
3 Mean daily minimum temperature	in °C	-10,0	-6,7	-2,2	0,6	5,0	9,4	9,4	8,9	7,2	1,1	-5,0	-8,9	0,6	7	3
4 Absolute maximum temperature	in °C	16,1	22,2	20,6	24,4	26,1	31,7	28,9	27,2	25,6	23,3	20,6	16,1	31,7	7	4
5 Absolute minimum temperature	in °C	-16,1	-15,0	-10,0	-7,8	-2,8	2,2	1,7	2,8	0,0	-7,8	-12,2	-15,0	-16,1	7	5
6 Mean relative humidity	in %	71	71	72	67	59	64	71	72	71	64	71	71	69	7	6
7 Mean precipitation	in mm	2	13	8	5	25	64	122	89	66	13	3	0	410	7	7
8 Maximum precipitation	in mm															8
9 Minimum precipitation	in mm															9
10 Maximum precipitation in 24 h	in mm	2	38	13	8	20	28	28	28	23	15	23	0	38	7	10
11 Mean number of days with precipitation	> 0,25 mm	<1	<1	1	<1	3	8	13	10	7	2	<1	0	48	7	11
12 Mean duration of sunshine	in h															12
13 Mean quantity of radiation	in ly/day															13
14 Mean potential evaporation	in mm	0	4	30	50	74	98	101	86	74	44	18	3	582	3	14
15 Mean windspeed	in m/sec															15
16 Mean predominent direction of the wind																16

120

241 Station/Country Chongqing/China

Location 29°30′N/106°33′E — Height above sealevel 261 m — Climate symbol: Köppen Cwa — Troll IV,7

		J	F	M	A	M	J	J	A	S	O	N	D	year	P	
1 Mean daily temperature	in °C	8,1	9,7	14,1	18,8	22,7	25,1	28,7	28,8	24,2	18,7	14,2	9,9	18,6	29	1
2 Mean daily maximum temperature	in °C	9,4	12,8	18,3	22,8	26,7	29,4	33,9	35,0	27,8	21,7	16,1	12,8	22,2	8	2
3 Mean daily minimum temperature	in °C	5,0	7,2	11,1	15,6	19,4	22,2	24,4	25,0	21,7	16,1	11,7	7,8	15,6	8	3
4 Absolute maximum temperature	in °C	20,6	27,8	31,7	39,9	33,0	40,3	41,3	44,0	40,4	34,6	30,0	21,7	44,0		4
5 Absolute minimum temperature	in °C	-0,8	-2,5	3,1	0,9	11,6	15,0	16,6	17,8	13,3	6,8	0,7	-1,5	-2,5		5
6 Mean relative humidity	in %	83	83	81	81	82	83	80	77	82	87	87	86	83		6
7 Mean precipitation	in mm	17	21	38	97	148	182	142	120	148	109	49	21	1090	63	7
8 Maximum precipitation	in mm	51	140	89	231	315	366	351	282	465	213	152	53		45	8
9 Minimum precipitation	in mm	0	3	8	23	56	48	15	15	18	23	15	<1		45	9
10 Maximum precipitation in 24 h	in mm	29	23	41	89	208	133	154	98	198	70	93	18	208		10
11 Mean number of days with precipitation	> 0,1 mm	7	8	10	12	14	15	10	9	13	16	11	8	133		11
12 Mean duration of sunshine	in h	48	56	81	111	139	128	221	222	123	63	47	54	1293		12
13 Mean quantity of radiation	in ly/day															13
14 Mean potential evaporation	in mm	0	0	23	55	96	133	149	122	76	47	6	0	707	14	14
15 Mean windspeed	in m/sec	1,1	1,2	1,3	1,3	1,3	1,1	1,1	1,0	1,3	0,9	0,9	0,8	1,1		15
16 Mean predominent direction of the wind		N	N	N	N	N	N	N	N	N	N	N	N			16

242 Station/Country Guangzhou (Canton)/China

Location 23°00′N/113°13′E — Height above sealevel 18 m — Climate symbol: Köppen Cwa — Troll V,2

		J	F	M	A	M	J	J	A	S	O	N	D	year	P	
1 Mean daily temperature	in °C	13,6	14,2	17,2	21,6	25,6	27,3	28,8	28,2	27,2	24,0	19,7	15,7	21,9	53	1
2 Mean daily maximum temperature	in °C	17,2	17,2	20,6	25,0	30,0	31,1	32,8	32,8	31,7	28,3	25,0	20,6	26,1	14	2
3 Mean daily minimum temperature	in °C	9,4	11,1	12,8	18,9	23,3	24,4	25,6	25,6	24,4	19,4	15,6	12,2	18,3	14	3
4 Absolute maximum temperature	in °C	28,0	29,2	30,6	33,0	35,7	36,7	37,2	37,7	37,6	36,0	32,0	29,0	37,7		4
5 Absolute minimum temperature	in °C	0,0	0,0	0,0	8,9	10,6	16,7	21,0	20,4	13,7	10,0	1,1	-0,3	-0,3		5
6 Mean relative humidity	in %	74	80	82	84	83	83	80	81	77	71	76	71	78		6
7 Mean precipitation	in mm	27	65	101	185	256	292	264	249	149	49	51	34	1722	41	7
8 Maximum precipitation	in mm	168	267	259	399	457	442	536	620	356	239	239	140		34	8
9 Minimum precipitation	in mm	0	3	5	15	71	10	74	112	23	0	0	0		34	9
10 Maximum precipitation in 24 h	in mm	36	72	74	129	144	108	183	126	131	77	133	56	183		10
11 Mean number of days with precipitation	> 0,1 mm	7	12	14	18	17	20	17	17	12	6	6	7	153		11
12 Mean duration of sunshine	in h	125	79	83	81	154	154	215	214	204	209	192	158	1868		12
13 Mean quantity of radiation	in ly/day															13
14 Mean potential evaporation	in mm	20	22	44	84	152	164	178	171	148	103	59	34	1179	14	14
15 Mean windspeed	in m/sec	2,1	2,0	1,8	1,8	1,7	1,6	1,8	1,7	1,7	1,9	1,9	1,9	1,8		15
16 Mean predominent direction of the wind		N	N	N	SE	SE	SE	SE	E	N	N	N	N			16

243 Station/Country Xi'an/China

Location 34°15′N/108°35′E — Height above sealevel 412 m — Climate symbol: Köppen Cwa — Troll III,7

		J	F	M	A	M	J	J	A	S	O	N	D	year	P	
1 Mean daily temperature	in °C	-0,5	2,5	8,7	14,9	20,6	25,6	27,9	25,9	20,1	14,3	7,1	0,9	14,0	26	1
2 Mean daily maximum temperature	in °C	4,4	7,8	16,1	22,8	28,3	33,3	33,9	32,8	26,7	20,6	13,3	6,7	20,6	13	2
3 Mean daily minimum temperature	in °C	-5,0	-1,7	5,0	11,1	17,8	22,8	25,6	24,4	18,3	12,2	3,3	-2,2	11,1	13	3
4 Absolute maximum temperature	in °C	19,0	23,0	28,7	36,5	40,6	44,1	45,2	43,0	37,6	34,0	24,5	18,9	45,2		4
5 Absolute minimum temperature	in °C	-19,1	-11,5	-11,0	-2,5	1,5	9,2	14,5	12,5	4,8	-4,9	-16,8	-17,4	-19,1		5
6 Mean relative humidity	in %	68	67	63	64	62	57	65	71	77	76	75	73	68		6
7 Mean precipitation	in mm	5	11	23	41	58	49	89	97	116	57	24	8	578	21	7
8 Maximum precipitation	in mm															8
9 Minimum precipitation	in mm															9
10 Maximum precipitation in 24 h	in mm	13	14	17	32	39	56	58	62	52	52	22	13	62		10
11 Mean number of days with precipitation	> 0,1 mm	3	5	7	8	8	8	11	11	12	8	6	4	91		11
12 Mean duration of sunshine	in h	109	107	119	138	172	185	217	197	133	136	112	112	1737		12
13 Mean quantity of radiation	in ly/day															13
14 Mean potential evaporation	in mm	0	3	24	57	112	156	176	151	89	50	14	1	833	8	14
15 Mean windspeed	in m/sec	1,4	1,3	2,0	1,7	1,8	2,1	1,8	1,9	1,5	1,5	1,4	1,4	1,8		15
16 Mean predominent direction of the wind		SW	NE	NE	NE	NE	SW	NW	NE	NE	SW	NE	NE			16

244 Station/Country Nanjing (Nanking)/China

Location 32°04′N/118°47′E — Height above sealevel 62 m — Climate symbol: Köppen Cfa — Troll IV,7

		J	F	M	A	M	J	J	A	S	O	N	D	year	P	
1 Mean daily temperature	in °C	2,2	4,0	8,8	15,0	20,5	24,7	28,0	28,0	23,3	17,5	11,2	4,9	15,7	16	1
2 Mean daily maximum temperature	in °C	6,1	7,8	12,8	18,9	25,6	28,3	31,1	31,1	26,7	21,7	15,0	8,9	19,4	27	2
3 Mean daily minimum temperature	in °C	-1,7	0,0	4,4	10,0	15,0	20,6	23,9	23,3	18,9	12,8	6,1	0,0	11,1	27	3
4 Absolute maximum temperature	in °C	19,6	23,0	29,4	34,0	35,7	33,2	43,0	40,9	38,8	34,0	28,1	24,5	43,0		4
5 Absolute minimum temperature	in °C	-13,6	-7,4	-6,0	0,4	6,8	15,0	16,6	18,5	10,3	3,7	-4,5	-12,6	-13,6		5
6 Mean relative humidity	in %	74	77	70	72	74	76	80	79	78	71	71	76	75		6
7 Mean precipitation	in mm	39	56	74	82	93	118	152	164	64	47	38	52	979	25	7
8 Maximum precipitation	in mm	91	137	193	190	178	406	620	363	241	203	117	170		31	8
9 Minimum precipitation	in mm	0	8	13	5	15	33	13	0	15	0	<1	<1		31	9
10 Maximum precipitation in 24 h	in mm	29	24	55	53	79	125	199	64	81	40	31	41	199		10
11 Mean number of days with precipitation	> 0,1 mm	9	10	11	11	11	10	13	12	10	8	8	11	124		11
12 Mean duration of sunshine	in h	137	110	155	153	205	192	231	232	183	190	143	128	2059		12
13 Mean quantity of radiation	in ly/day															13
14 Mean potential evaporation	in mm	2	4	22	54	95	140	175	166	109	65	26	6	864	32	14
15 Mean windspeed	in m/sec	4,5	4,9	5,4	5,2	4,8	4,7	4,8	4,5	4,0	4,2	4,5	4,5	4,7		15
16 Mean predominent direction of the wind		NNE	NE	NE,ESE	ESE	ESE	ESE	ESE	ESE	NE	NNE,ESE	NNE	NNE			16

245 Station/Country **Shanghai/China**

Location	31°12'N/121°26'E		Height above sealevel	5 m		Climate symbol: Köppen	Cfa		Troll	IV,7					
			J	F	M	A	M	J	J	A	S	O	N	D	year	P
1 Mean daily temperature	in °C		3.4	4.3	8.2	13.7	18.9	23.1	27.1	27.2	23.0	17.7	11.6	5.8	15.3	81
2 Mean daily maximum temperature	in °C		7.8	8.3	12.8	18.9	25.0	27.8	32.2	32.2	27.8	23.3	17.2	11.7	20.4	38
3 Mean daily minimum temperature	in °C		0.6	1.1	4.4	10.0	15.0	19.4	23.3	23.3	18.9	13.9	7.2	2.2	11.6	38
4 Absolute maximum temperature	in °C		23.3	26.5	32.0	34.8	35.7	39.3	40.2	40.0	37.8	33.6	29.9	24.1	40.2	
5 Absolute minimum temperature	in °C		−12.1	−10.3	−5.8	−1.3	3.0	10.5	15.9	10.1	6.8	1.1	−5.1	−10.2	−12.1	
6 Mean relative humidity	in %		78	79	79	79	80	84	84	84	83	79	78	77	80	
7 Mean precipitation	in mm		48	58	84	94	94	180	147	142	130	71	51	36	1135	62
8 Maximum precipitation	in mm		198	180	290	239	213	493	368	343	381	310	196	152		65
9 Minimum precipitation	in mm		0	0	18	13	18	18	3	13	20	<1	3	0		65
10 Maximum precipitation in 24 h	in mm		55	39	65	57	90	161	148	155	196	108	74	45	196	
11 Mean number of days with precipitation	> 0,1 mm		10	10	12	12	14	11	11	12	9	8	8		130	
12 Mean duration of sunshine	in h		123	105	136	143	169	142	213	232	158	174	141	138	1874	
13 Mean quantity of radiation	in ly/day															
14 Mean potential evaporation	in mm		4	5	19	48	92	127	171	160	111	66	29	10	842	64
15 Mean windspeed	in m/sec		4.6	4.6	4.9	4.9	4.6	4.4	4.9	4.7	4.1	3.9	4.2	4.5	4.5	
16 Mean predominent direction of the wind			NW	NE	ESE	SE	SE	ESE	SE	ESE	ENE	ESE	NW	NW		

246 Station/Country **Wuhan/China**

Location	30°33'N/114°17'E		Height above sealevel	23 m		Climate symbol: Köppen	Cfa		Troll	IV,7					
			J	F	M	A	M	J	J	A	S	O	N	D	year	P
1 Mean daily temperature	in °C		3.8	5.4	10.4	16.3	22.1	25.8	28.9	28.7	23.9	18.4	12.0	6.1	18.8	40
2 Mean daily maximum temperature	in °C		7.8	9.4	13.9	20.6	26.1	30.6	33.9	33.9	28.9	22.8	16.7	10.6	21.1	28
3 Mean daily minimum temperature	in °C		1.1	2.2	6.1	12.8	17.8	22.8	26.1	26.1	20.6	15.6	8.9	3.3	13.3	28
4 Absolute maximum temperature	in °C		21.1	28.0	32.6	36.0	36.6	38.2	41.1	41.3	38.0	34.4	29.7	25.0	41.3	
5 Absolute minimum temperature	in °C		−13.0	−7.2	−5.0	0.3	9.5	14.3	18.4	18.1	11.6	5.2	−5.0	−7.5	−13.0	
6 Mean relative humidity	in %		76	78	76	77	75	77	77	75	74	73	75	74	76	
7 Mean precipitation	in mm		25	54	92	140	166	216	173	110	69	74	46	28	1194	50
8 Maximum precipitation	in mm															
9 Minimum precipitation	in mm															
10 Maximum precipitation in 24 h	in mm		45	43	106	166	206	220	214	113	84	85	84	67	220	
11 Mean number of days with precipitation	> 0,1 mm		8	9	12	11	13	11	10	8	8	9	8	7	114	
12 Mean duration of sunshine	in h		117	97	127	144	185	187	240	262	193	164	142	120	1978	
13 Mean quantity of radiation	in ly/day															
14 Mean potential evaporation	in mm		4	6	25	59	113	153	184	174	116	68	28	7	937	
15 Mean windspeed	in m/sec		2.0	2.0	2.0	2.0	1.8	1.7	1.9	1.9	2.1	1.8	1.9	1.9	1.9	
16 Mean predominent direction of the wind			NNE	N	NNE	NE	SE	SE	SE	NE	NE	N	N	NNE		

247 Station/Country **Nanchang/China**

Location	28°40'N/115°58'E		Height above sealevel	49 m		Climate symbol: Köppen	Cfa		Troll	IV,7					
			J	F	M	A	M	J	J	A	S	O	N	D	year	P
1 Mean daily temperature	in °C		5.5	7.0	10.5	16.4	21.4	25.1	28.9	29.5	24.6	18.9	13.6	7.2	17.4	13
2 Mean daily maximum temperature	in °C															
3 Mean daily minimum temperature	in °C															
4 Absolute maximum temperature	in °C		24.4	27.2	29.0	33.8	36.5	39.0	39.0	39.4	38.7	37.5	29.7	23.9	39.4	
5 Absolute minimum temperature	in °C		−4.5	−5.9	−0.4	6.1	10.6	13.9	20.0	20.0	14.8	7.0	0.4	−4.4	−5.9	
6 Mean relative humidity	in %		81	85	88	85	84	84	82	79	82	80	83	82	83	
7 Mean precipitation	in mm		55	108	192	250	289	295	258	111	109	57	70	70	1864	25
8 Maximum precipitation	in mm															
9 Minimum precipitation	in mm															
10 Maximum precipitation in 24 h	in mm		55	60	62	132	119	152	89	102	64	39	58	49	152	
11 Mean number of days with precipitation	> 0,1 mm		9	12	14	16	16	14	10	9	10	8	9	10	137	
12 Mean duration of sunshine	in h		121	78	92	89	109	127	195	278	279	234	185	159	1944	
13 Mean quantity of radiation	in ly/day															
14 Mean potential evaporation	in mm		6	10	24	53	98	135	180	174	117	72	34	13	916	8
15 Mean windspeed	in m/sec		3.8	4.9	3.5	2.2	1.8	1.9	2.2	2.2	2.9	2.8	3.3	3.2	2.9	
16 Mean predominent direction of the wind			N	N	N	N	N	N	SSW	N	N	N	N	N		

248 Station/Country **Changsha/China**

Location	28°15'N/112°50'E		Height above sealevel	48 m		Climate symbol: Köppen	Cfa		Troll	IV,7					
			J	F	M	A	M	J	J	A	S	O	N	D	year	P
1 Mean daily temperature	in °C		4.3	6.2	10.9	16.6	22.0	25.8	29.3	29.0	24.6	18.3	12.1	6.7	17.2	12
2 Mean daily maximum temperature	in °C		7.2	8.9	15.0	21.1	26.7	30.0	34.4	34.4	30.0	23.9	17.8	10.6	21.7	14
3 Mean daily minimum temperature	in °C		1.7	2.8	7.8	13.3	19.4	22.8	25.6	25.6	21.1	15.0	9.4	3.9	13.9	14
4 Absolute maximum temperature	in °C		26.7	26.2	28.9	35.3	36.9	38.9	40.0	43.0	39.2	38.6	31.5	28.3	43.0	
5 Absolute minimum temperature	in °C		−8.1	−3.0	−1.2	4.3	11.2	15.3	20.1	19.4	12.8	3.0	−1.5	−6.3	−8.1	
6 Mean relative humidity	in %		84	87	85	84	83	81	78	76	79	79	82	81	82	
7 Mean precipitation	in mm		64	121	122	203	212	254	118	121	80	83	85	68	1531	19
8 Maximum precipitation	in mm		142	208	213	300	361	495	345	323	185	262	193	155		27
9 Minimum precipitation	in mm		0	38	51	43	94	53	<1	23	10	13	0			27
10 Maximum precipitation in 24 h	in mm		38	55	57	92	121	195	129	141	67	77	48	46	195	
11 Mean number of days with precipitation	> 0,1 mm		12	15	16	18	17	14	10	11	9	11	12	14	159	
12 Mean duration of sunshine	in h		86	51	94	73	132	144	206	228	157	200	112	77	1560	
13 Mean quantity of radiation	in ly/day															
14 Mean potential evaporation	in mm		4	6	26	61	118	154	191	182	131	71	32	9	985	14
15 Mean windspeed	in m/sec		2.6	3.1	2.5	2.3	2.1	2.0	2.4	2.1	2.5	2.3	2.6	2.7	2.5	
16 Mean predominent direction of the wind			NNW	NNW	NNW	NW	NW	S	S	NW,S	NW	NW	*	NNW		

* NNW,NW

Station / Country Wenzhou / China

Location 28°01'N / 120°49'E   Height above sealevel 5 m   Climate symbol: Köppen Cfa   Troll IV,7

		J	F	M	A	M	J	J	A	S	O	N	D	year	P	
1	Mean daily temperature	in °C	7,7	8,4	11,6	16,7	21,4	25,2	28,9	29,0	25,7	20,9	15,9	11,1	18,5	18	1
2	Mean daily maximum temperature	in °C															2
3	Mean daily minimum temperature	in °C															3
4	Absolute maximum temperature	in °C	23,0	23,0	27,2	32,5	33,9	38,3	40,5	39,0	37,8	33,5	27,5	22,8	40,5		4
5	Absolute minimum temperature	in °C	-3,0	-2,2	-0,5	5,5	11,1	15,5	18,1	18,1	15,0	8,0	0,6	-1,7	-3,0		5
6	Mean relative humidity	in %	80	85	86	86	86	89	87	85	85	78	78	76	83		6
7	Mean precipitation	in mm	49	89	132	148	190	265	204	258	204	87	55	43	1724	61	7
8	Maximum precipitation	in mm															8
9	Minimum precipitation	in mm															9
10	Maximum precipitation in 24 h	in mm	48	71	76	73	102	107	213	243	269	287	90	92	287		10
11	Mean number of days with precipitation	> 0,1 mm	10	12	15	15	16	16	13	13	12	9	7	8	146		11
12	Mean duration of sunshine	in h	118	92	87	133	104	130	213	242	201	163	145	134	1782		12
13	Mean quantity of radiation	in ly / day															13
14	Mean potential evaporation	in mm	10	11	26	57	101	142	181	163	129	81	43	21	965	14	14
15	Mean windspeed	in m / sec	3,4	3,2	3,0	2,8	3,0	2,7	3,1	3,6	2,6	2,8	3,2	3,8	3,1		15
16	Mean predominent direction of the wind		NW	NW	NW	SE	SE	SE	SE	SE	NW	NW	NW	NW			16

Station / Country Guiyang / China

Location 26°34'N / 106°42'E   Height above sealevel 1071 m   Climate symbol: Köppen Cwa   Troll IV,7

		J	F	M	A	M	J	J	A	S	O	N	D	year	P	
1	Mean daily temperature	in °C	5,0	6,6	11,7	16,4	20,4	22,5	24,7	24,1	21,0	15,6	11,7	7,2	15,6	34	1
2	Mean daily maximum temperature	in °C															2
3	Mean daily minimum temperature	in °C															3
4	Absolute maximum temperature	in °C	26,0	30,0	31,8	35,1	35,1	35,6	39,5	37,5	38,4	32,1	29,7	26,1	39,5		4
5	Absolute minimum temperature	in °C	-9,5	-5,4	-2,7	-2,3	6,7	10,1	13,5	10,3	5,4	0,2	-5,7	-3,2	-9,5		5
6	Mean relative humidity	in %	81	80	77	77	78	79	76	79	78	80	81	81	79		6
7	Mean precipitation	in mm	21	28	40	94	200	204	187	138	120	110	47	24	1213	33	7
8	Maximum precipitation	in mm															8
9	Minimum precipitation	in mm															9
10	Maximum precipitation in 24 h	in mm	22	26	42	88	125	95	79	95	83	73	34	29	125		10
11	Mean number of days with precipitation	> 0,1 mm	14	14	13	16	19	18	17	16	13	16	14	12	182		11
12	Mean duration of sunshine	in h	58	55	92	112	141	127	186	186	138	85	78	67	1325		12
13	Mean quantity of radiation	in ly / day															13
14	Mean potential evaporation	in mm	8	11	37	64	104	117	142	129	95	56	30	14	807	34	14
15	Mean windspeed	in m / sec	1,7	1,9	2,1	2,0	1,8	1,8	1,8	1,6	1,8	1,7	1,9	1,8	1,8		15
16	Mean predominent direction of the wind		N	NE	NE	N	N	S	S	S	N	N	NE	NE			16

Station / Country Guilin (Kweilin) / China

Location 25°15'N / 110°10'E   Height above sealevel 167 m   Climate symbol: Köppen Cfa   Troll IV,7

		J	F	M	A	M	J	J	A	S	O	N	D	year	P	
1	Mean daily temperature	in °C	9,2	9,7	13,4	19,0	23,7	23,7	28,4	27,9	26,5	22,0	15,6	11,1	19,4	11	1
2	Mean daily maximum temperature	in °C															2
3	Mean daily minimum temperature	in °C															3
4	Absolute maximum temperature	in °C	31,2	31,3	32,8	34,9	38,0	39,4	39,7	39,3	39,4	39,0	32,9	27,3	39,7		4
5	Absolute minimum temperature	in °C	-5,0	-3,3	-1,9	2,9	10,2	15,0	17,5	14,5	13,5	6,0	2,0	-2,4	-5,0		5
6	Mean relative humidity	in %	77	79	82	80	84	82	82	81	70	70	74	74	78		6
7	Mean precipitation	in mm	51	79	161	223	359	370	236	200	101	87	53	47	1967	16	7
8	Maximum precipitation	in mm	94	201	193	638	807	790	592	475	287	272	163	125		23	8
9	Minimum precipitation	in mm	0	3	10	84	158	152	31	89	0	<1	<1	0		23	9
10	Maximum precipitation in 24 h	in mm	41	53	60	104	162	160	190	121	101	88	54	39	190		10
11	Mean number of days with precipitation	> 0,1 mm	12	14	19	19	20	19	19	16	8	8	11	11	176		11
12	Mean duration of sunshine	in h	79	40	57	104	117	149	186	218	206	150	127	135	1568		12
13	Mean quantity of radiation	in ly / day															13
14	Mean potential evaporation	in mm	9	13	28	70	119	155	178	167	140	94	40	21	1034	5	14
15	Mean windspeed	in m / sec	2,8	2,2	2,5	2,1	1,7	1,6	1,5	1,3	2,0	2,7	3,0	3,0	2,2		15
16	Mean predominent direction of the wind		N	N	N	N	N	N	N,S	N	N	N	N	N			16

Station / Country Kunming / China

Location 25°02'N / 102°43'E   Height above sealevel 1893 m   Climate symbol: Köppen Cwb   Troll IV,4

		J	F	M	A	M	J	J	A	S	O	N	D	year	P	
1	Mean daily temperature	in °C	9,8	11,0	14,5	17,7	19,7	19,8	20,2	19,9	18,4	15,6	12,5	9,8	15,7	18	1
2	Mean daily maximum temperature	in °C	16,1	17,8	21,1	24,4	25,6	25,0	25,0	25,0	23,9	21,1	18,3	16,7	21,7	32	2
3	Mean daily minimum temperature	in °C	2,8	3,9	6,7	10,6	13,9	16,7	16,7	16,7	15,0	11,7	7,2	3,3	10,6	32	3
4	Absolute maximum temperature	in °C	26,5	28,5	29,0	32,1	33,0	31,0	30,5	30,5	30,0	28,5	28,5	26,0	33,0		4
5	Absolute minimum temperature	in °C	-5,4	-3,1	-1,4	4,0	7,6	9,4	13,0	12,0	7,1	1,5	0,7	-2,9	-5,4		5
6	Mean relative humidity	in %	64	61	56	58	85	76	79	79	77	77	73	70	70		6
7	Mean precipitation	in mm	3	18	21	31	99	192	214	220	161	95	31	11	1096	17	7
8	Maximum precipitation	in mm	56	58	58	71	323	574	460	363	295	193	175	46		28	8
9	Minimum precipitation	in mm	0	0	0	0	15	48	91	36	28	5	0	0		29	9
10	Maximum precipitation in 24 h	in mm	13	31	40	24	70	87	100	105	102	77	44	16	105		10
11	Mean number of days with precipitation	> 0,1 mm	1	4	4	6	12	18	20	14	15	13	6	3	122		11
12	Mean duration of sunshine	in h	252	234	244	238	194	107	102	131	118	168	181	203	2172		12
13	Mean quantity of radiation	in ly / day															13
14	Mean potential evaporation	in mm	24	23	48	73	98	102	108	101	80	57	38	26	778	15	14
15	Mean windspeed	in m / sec	2,8	3,1	3,2	3,3	2,8	2,1	1,9	1,5	1,9	1,9	2,4	2,6	2,4		15
16	Mean predominent direction of the wind		SW	SW	SW	SW	SW	SSW	S	NE	E	S	SW	SW			16

253 Station/Country Xiamen (Amoy)/China

Location 24°27'N/118°04'E Height above sealevel 41 m Climate symbol: Köppen Cfa Troll V.2

		J	F	M	A	M	J	J	A	S	O	N	D	year	P	
1 Mean daily temperature	in °C	14,8	13,5	15,3	19,4	23,4	27,2	29,0	29,0	28,1	25,0	20,8	16,6	21,8	25	1
2 Mean daily maximum temperature	in °C	18,7	16,1	18,3	22,2	26,1	29,4	31,7	31,7	30,6	27,8	23,3	20,0	24,4	12	2
3 Mean daily minimum temperature	in °C	11,1	11,1	12,8	17,2	21,7	25,0	26,7	26,7	25,6	21,7	17,8	13,9	19,4	12	3
4 Absolute maximum temperature	in °C	26,7	26,1	27,8	30,5	32,5	34,5	36,1	37,9	37,5	35,0	32,5	27,8	37,9		4
5 Absolute minimum temperature	in °C	3,9	3,9	6,0	8,5	13,9	16,1	16,7	23,0	18,3	13,3	6,7	5,0	3,9		5
6 Mean relative humidity	in %	77	80	89	81	81	82	79	79	74	70	72	75	77		6
7 Mean precipitation	in mm	36	71	92	130	171	178	132	185	108	38	32	34	1187	16	7
8 Maximum precipitation	in mm	127	216	325	480	414	424	366	699	467	356	158	145		42	8
9 Minimum precipitation	in mm	0	0	0	18	20	5	10	18	<1	0	0	0		42	9
10 Maximum precipitation in 24 h	in mm	81	89	80	206	184	150	127	128	159	191	76	37	206		10
11 Mean number of days with precipitation	> 0,1 mm	7	11	12	12	13	14	10	11	8	4	4	6	112		11
12 Mean duration of sunshine	in h	125	114	85	101	155	172	254	244	189	209	193	149	1990		12
13 Mean quantity of radiation	in ly/day															13
14 Mean potential evaporation	in mm	24	21	35	64	115	159	181	174	151	110	64	37	1135	20	14
15 Mean windspeed	in m/sec	3,2	3,2	3,1	2,9	2,7	2,8	2,8	3,0	3,4	3,8	3,9	3,4	3,2		15
16 Mean predominent direction of the wind		ENE	ENE	ENE	ENE	ESE	SW	SW	SW	SE	ENE	ENE	ENE			16

254 Station/Country Nanning/China

Location 22°48'N/108°18'E Height above sealevel 75 m Climate symbol: Köppen Cfa Troll V.2

		J	F	M	A	M	J	J	A	S	O	N	D	year	P	
1 Mean daily temperature	in °C	13,6	14,4	17,9	22,5	26,7	28,0	28,5	28,3	27,6	23,5	19,7	15,6	22,2	28	1
2 Mean daily maximum temperature	in °C															2
3 Mean daily minimum temperature	in °C															3
4 Absolute maximum temperature	in °C	29,3	32,7	35,5	36,0	36,7	37,4	38,3	38,3	38,8	36,3	32,2	31,1	38,8		4
5 Absolute minimum temperature	in °C	1,7	2,3	4,3	4,3	8,0	14,8	16,0	19,5	13,5	10,4	4,4	2,5	1,7		5
6 Mean relative humidity	in %	77	77	80	78	79	81	81	84	76	71	72	75	78		6
7 Mean precipitation	in mm	32	55	48	79	167	214	217	225	109	105	37	34	1322	39	7
8 Maximum precipitation	in mm															8
9 Minimum precipitation	in mm															9
10 Maximum precipitation in 24 h	in mm	48	43	77	52	102	119	114	120	128	283	36	88	283		10
11 Mean number of days with precipitation	> 0,1 mm	8	11	11	11	14	15	16	16	9	8	6	6	131		11
12 Mean duration of sunshine	in h	71	67	76	109	164	178	214	179	181	216	153	119	1727		12
13 Mean quantity of radiation	in ly/day															13
14 Mean potential evaporation	in mm	20	23	48	91	156	183	174	167	147	98	60	30	1177	16	14
15 Mean windspeed	in m/sec	1,8	2,0	2,0	1,9	1,9	1,9	2,0	1,7	1,4	1,3	1,4	1,5	1,7		15
16 Mean predominent direction of the wind		E	E	E	E.SE	SE	SE	SE	E	E	E	E	E			16

255 Station/Country Yulin (Hainandao Island)/China

Location 18°14'N/109°32'E Height above sealevel 2 m Climate symbol: Köppen Aw Troll V.2

		J	F	M	A	M	J	J	A	S	O	N	D	year	P	
1 Mean daily temperature	in °C	21,4	22,4	24,4	26,6	28,2	28,3	28,5	28,0	27,1	25,9	23,0	21,4	25,5	5	1
2 Mean daily maximum temperature	in °C															2
3 Mean daily minimum temperature	in °C															3
4 Absolute maximum temperature	in °C	30,0	29,8	31,6	32,8	34,6	34,6	34,6	34,6	34,0	34,0	32,3	30,2	34,6		4
5 Absolute minimum temperature	in °C	9,0	12,6	16,1	19,0	20,7	20,0	21,6	22,7	21,9	15,5	10,0	6,6	6,6		5
6 Mean relative humidity	in %	75	78	79	80	82	83	83	84	85	80	77	73	80		6
7 Mean precipitation	in mm	11	7	21	28	150	197	149	189	293	190	54	43	1332	3	7
8 Maximum precipitation	in mm															8
9 Minimum precipitation	in mm															9
10 Maximum precipitation in 24 h	in mm	3	10	3	19	52	78	47	44	122	120	24	37	122		10
11 Mean number of days with precipitation	> 0,1 mm	4	3	3	4	10	14	14	16	19	12	7	7	113		11
12 Mean duration of sunshine	in h	210	201	210	211	230	254	286	235	203	243	210	196	2689		12
13 Mean quantity of radiation	in ly/day															13
14 Mean potential evaporation	in mm	66	73	105	142	168	165	173	162	144	126	82	66	1472	5	14
15 Mean windspeed	in m/sec	2,5	2,3	2,3	2,1	2,2	2,2	2,2	2,0	2,0	2,5	2,2	2,4	2,2		15
16 Mean predominent direction of the wind		NW	NW	NW	NW	NW	NW	SE	SE	NW	SE	NW	NW			16

256 Station/Country Hongkong/United Kingdom

Location 22°18'N/114°10'E Height above sealevel 33 m Climate symbol: Köppen Cwa Troll V.2

		J	F	M	A	M	J	J	A	S	O	N	D	year	P	
1 Mean daily temperature	in °C	15,6	15,0	17,5	21,7	25,6	27,5	28,1	28,1	27,2	25,0	20,9	17,5	22,5	50	1
2 Mean daily maximum temperature	in °C	17,8	17,2	19,4	23,9	27,8	29,4	30,6	30,6	29,4	27,2	23,3	20,0	25,0	50	2
3 Mean daily minimum temperature	in °C	13,3	12,8	15,6	19,4	23,3	25,6	25,6	25,6	25,0	22,8	18,3	15,0	20,0	50	3
4 Absolute maximum temperature	in °C	26,1	26,1	28,3	31,7	32,8	34,4	34,4	36,1	34,4	34,4	30,0	27,8	36,1	50	4
5 Absolute minimum temperature	in °C	0,0	3,3	7,2	11,1	15,6	19,4	22,2	22,2	18,3	13,9	6,7	5,0	0,0	50	5
6 Mean relative humidity	in %	72	78	79	82	83	82	82	82	78	69	67	69	77	20	6
7 Mean precipitation	in mm	33	46	74	292	394	381	394	361	247	114	43	30	2162	50	7
8 Maximum precipitation	in mm	213	201	292	437	1240	874	785	871	828	610	269	125		88	8
9 Minimum precipitation	in mm	0	0	3	10	15	58	114	15	15	<1	0	0		88	9
10 Maximum precipitation in 24 h	in mm	99	52	97	157	521	320	533	282	203	292	150	91	533	50	10
11 Mean number of days with precipitation	> 1,0 mm	4	5	7	8	13	18	17	15	12	6	2	3	110	20	11
12 Mean duration of sunshine	in h	145	98	95	114	158	181	210	201	198	217	186	172	1953	80	12
13 Mean quantity of radiation	in ly/day															13
14 Mean potential evaporation	in mm	32	29	46	81	131	158	166	161	143	112	76	46	1181	56	14
15 Mean windspeed	in m/sec															15
16 Mean predominent direction of the wind		E	E	E	E	E	E.SW	SW	W.E	E.NE	NE	NE	NE.E		5	16

124

257 Station / Country T'aipei / Taiwan

Location 25°02'N / 121°31'E Height above sealevel 8 m Climate symbol: Köppen Cfa Troll V,1

		J	F	M	A	M	J	J	A	S	O	N	D	year	P	
1 Mean daily temperature	in °C	15,2	14,8	16,9	20,6	24,1	26,6	28,2	27,9	26,2	23,0	19,8	16,8	21,7	44	1
2 Mean daily maximum temperature	in °C	18,9	18,3	21,1	25,0	28,3	31,7	33,3	32,8	31,1	27,2	23,9	20,6	26,1	37	2
3 Mean daily minimum temperature	in °C	12,2	11,7	13,9	17,2	20,6	22,8	24,4	23,9	22,8	19,4	16,7	13,9	18,3	37	3
4 Absolute maximum temperature	in °C	28,8	31,2	32,6	34,8	36,5	37,1	38,6	37,7	36,4	36,1	33,6	31,5	38,6		4
5 Absolute minimum temperature	in °C	2,6	-0,2	1,4	7,5	10,0	15,8	19,5	18,9	13,5	10,8	1,1	1,8	-0,2		5
6 Mean relative humidity	in %	84	84	84	83	82	81	78	78	80	81	81	83	82		6
7 Mean precipitation	in mm	91	147	164	182	205	322	269	266	189	117	71	77	2100	52	7
8 Maximum precipitation	in mm	224	343	399	594	511	688	874	940	782	658	170	262		44	8
9 Minimum precipitation	in mm	10	28	20	15	64	76	31	8	20	13	5	8		44	9
10 Maximum precipitation in 24 h	in mm	63	59	80	176	169	199	359	328	293	199	57	79	359		10
11 Mean number of days with precipitation	> 0,1 mm	17	17	18	15	16	16	14	15	14	15	15	16	188		11
12 Mean duration of sunshine	in h	86	74	90	112	138	168	225	218	191	143	107	92	1644		12
13 Mean quantity of radiation	in ly / day															13
14 Mean potential evaporation	in mm	32	30	45	81	132	162	181	171	142	101	64	41	1182		14
15 Mean windspeed	in m / sec	3,4	3,3	3,5	3,1	2,9	2,2	2,5	2,8	3,1	3,7	3,9	3,6	3,2		15
16 Mean predominent direction of the wind		E	E	E	E	E	E	E	E	E	E	E	E			16

258 Station / Country T'ainan / Taiwan

Location 23°00'N / 120°13'E Height above sealevel 13 m Climate symbol: Köppen Cfa Troll V,2

		J	F	M	A	M	J	J	A	S	O	N	D	year	P	
1 Mean daily temperature	in °C	17,0	17,0	19,8	23,3	26,2	27,4	27,8	27,3	27,1	24,8	21,6	18,5	23,1	49	1
2 Mean daily maximum temperature	in °C	23,6	23,4	25,9	28,9	30,8	31,4	32,1	31,5	31,8	30,5	27,8	24,7	28,5	30	2
3 Mean daily minimum temperature	in °C	12,8	12,5	15,3	19,0	22,1	23,8	24,3	24,2	23,3	20,6	17,2	14,1	19,1	30	3
4 Absolute maximum temperature	in °C	32,4	32,4	35,4	34,4	36,4	35,7	36,9	36,6	36,6	34,7	35,2	32,0	36,9		4
5 Absolute minimum temperature	in °C	2,6	2,4	5,1	9,9	14,7	18,9	21,1	21,2	15,4	12,6	2,9	4,3	2,4		5
6 Mean relative humidity	in %	79	79	79	79	81	84	83	85	82	76	78	79	81		6
7 Mean precipitation	in mm	20	36	48	68	171	369	431	448	156	33	16	16	1812	49	7
8 Maximum precipitation	in mm	170	254	213	376	742	795	1908	1275	831	132	119	86		44	8
9 Minimum precipitation	in mm	0	0	0	0	18	43	38	36	<1	0	0	0		44	9
10 Maximum precipitation in 24 h	in mm	101	100	87	182	252	238	398	385	382	79	43	54	398		10
11 Mean number of days with precipitation	> 0,1 mm	5	6	7	7	10	15	17	19	11	4	3	4	108		11
12 Mean duration of sunshine	in h	196	181	198	209	230	229	236	214	242	244	212	198	2589		12
13 Mean quantity of radiation	in ly / day															13
14 Mean potential evaporation	in mm	40	41	65	106	160	167	174	166	149	125	81	52	1326		14
15 Mean windspeed	in m / sec	3,9	3,9	3,6	2,9	2,5	2,6	2,7	2,7	2,5	2,7	3,2	3,6	3,1		15
16 Mean predominent direction of the wind		N	N	N	N	N	SE	SE	SE	N	N	N	N			16

259 Station / Country Wŏnsan / North Korea

Location 39°11'N / 127°26'E Height above sealevel 35 m Climate symbol: Köppen Dwa Troll III,8

		J	F	M	A	M	J	J	A	S	O	N	D	year	P	
1 Mean daily temperature	in °C	-3,8	-2,4	2,5	9,7	14,9	19,0	22,7	23,4	18,8	13,1	5,8	-0,9	10,2	46	1
2 Mean daily maximum temperature	in °C	1,1	2,2	7,2	15,0	20,6	23,9	26,7	27,2	23,3	18,3	10,6	3,3	15,0	36	2
3 Mean daily minimum temperature	in °C	-8,3	-6,7	-1,7	4,4	10,0	15,0	19,4	20,0	14,4	8,3	1,1	-5,0	6,1	36	3
4 Absolute maximum temperature	in °C	12,2	14,4	24,4	31,1	37,2	38,3	39,4	37,8	34,4	30,6	24,4	17,8	39,4	29	4
5 Absolute minimum temperature	in °C	-21,7	-19,4	-15,6	-4,4	1,1	7,2	11,7	11,1	3,9	-1,7	-13,3	-20,0	-21,7	29	5
6 Mean relative humidity	in %	53	56	59	62	68	77	83	83	67	66	59	52	66	46	6
7 Mean precipitation	in mm	29	32	46	68	86	126	273	312	178	70	61	29	1310	46	7
8 Maximum precipitation	in mm															8
9 Minimum precipitation	in mm															9
10 Maximum precipitation in 24 h	in mm	74	74	97	117	127	122	175	196	244	224	86	84	244	29	10
11 Mean number of days with precipitation	> 1,0 mm	3	4	5	5	7	9	14	13	9	5	5	3	82	46	11
12 Mean duration of sunshine	in h	205	204	232	235	236	207	173	176	198	223	191	191	2471	46	12
13 Mean quantity of radiation	in ly / day															13
14 Mean potential evaporation	in mm	0	0	7	40	80	111	140	136	90	53	16	0	673	22	14
15 Mean windspeed	in m / sec															15
16 Mean predominent direction of the wind																16

260 Station / Country P'yongyang / North Korea

Location 39°01'N / 125°49'E Height above sealevel 27 m Climate symbol: Köppen Dwa Troll III,8

		J	F	M	A	M	J	J	A	S	O	N	D	year	P	
1 Mean daily temperature	in °C	-8,1	-4,8	1,7	9,5	15,5	20,6	24,2	24,4	18,9	11,9	3,4	-4,8	9,4	43	1
2 Mean daily maximum temperature	in °C	-2,7	0,6	7,2	16,2	22,0	26,9	29,1	29,3	24,8	18,5	8,8	0,1	15,1	30	2
3 Mean daily minimum temperature	in °C	-13,3	-9,9	-3,2	3,6	9,8	15,3	20,3	20,5	13,9	6,2	-1,4	-9,6	4,4	30	3
4 Absolute maximum temperature	in °C	10,0	12,5	22,5	28,5	32,0	37,0	37,0	36,5	34,5	28,5	22,5	13,0	37,0	30	4
5 Absolute minimum temperature	in °C	-28,5	-23,0	-19,0	-4,5	1,5	7,0	12,5	12,5	2,5	-6,5	-20,0	-30,0	-30,0	30	5
6 Mean relative humidity	in %	74	70	66	63	66	71	80	80	75	73	73	74	72	43	6
7 Mean precipitation	in mm	15	11	26	46	67	76	237	228	112	45	41	21	925	43	7
8 Maximum precipitation	in mm															8
9 Minimum precipitation	in mm															9
10 Maximum precipitation in 24 h	in mm															10
11 Mean number of days with precipitation	> 1,0 mm	3	3	4	5	7	7	12	10	7	6	7	4	75	43	11
12 Mean duration of sunshine	in h	201	207	238	250	266	261	210	213	235	243	181	181	2686	43	12
13 Mean quantity of radiation	in ly / day															13
14 Mean potential evaporation	in mm	0	0	3	42	82	120	150	142	92	46	7	0	684	27	14
15 Mean windspeed	in m / sec															15
16 Mean predominent direction of the wind																16

261 Station/Country **Soul (Seoul)/South Korea**

Location 37°34'N/126°58'E — Height above sealevel **86 m** — Climate symbol: Köppen **Dwa** — Troll **III,8**

		J	F	M	A	M	J	J	A	S	O	N	D	year	P	
1 Mean daily temperature	in °C	-4,9	-1,9	3,6	10,5	16,3	20,8	24,5	25,4	20,3	13,4	6,3	-1,2	11,1	30	1
2 Mean daily maximum temperature	in °C	0,0	2,8	8,3	16,7	22,2	26,7	28,9	30,6	25,6	19,4	10,6	2,8	16,1	22	2
3 Mean daily minimum temperature	in °C	-9,4	-6,7	-1,7	5,0	10,6	16,1	21,1	21,7	15,0	7,2	0,0	-6,7	6,1	22	3
4 Absolute maximum temperature	in °C	12,2	15,6	22,2	28,3	32,2	36,7	36,7	37,2	32,8	30,0	23,3	14,4	37,2	22	4
5 Absolute minimum temperature	in °C	-22,2	-19,4	-15,0	-3,9	2,2	9,4	12,8	14,4	3,3	-3,9	-11,7	-24,4	-24,4	22	5
6 Mean relative humidity	in %	64	64	64	63	66	73	81	78	73	68	68	66	69	30	6
7 Mean precipitation	in mm	17	21	56	88	86	169	358	224	142	49	38	32	1258	30	7
8 Maximum precipitation	in mm	83	111	164	151	181	539	610	869	346	120	77	79		30	8
9 Minimum precipitation	in mm	2	3	5	6	24	33	88	37	5	4	1	8		30	9
10 Maximum precipitation in 24 h	in mm	62	66	78	66	122	202	355	210	164	100	68	38	355	30	10
11 Mean number of days with precipitation	> 1,0 mm	3	3	6	6	7	9	14	10	7	5	5	5	80	30	11
12 Mean duration of sunshine	in h	180	182	207	228	257	214	179	202	206	231	180	161	2427	30	12
13 Mean quantity of radiation	in ly/day															13
14 Mean potential evaporation	in mm	0	0	5	41	88	123	151	144	98	51	15	0	716	30	14
15 Mean windspeed	in m/sec															15
16 Mean predominant direction of the wind																16

262 Station/Country **Pusan/South Korea**

Location 35°06'N/129°02'E — Height above sealevel **69 m** — Climate symbol: Köppen **Cwa** — Troll **IV,7**

		J	F	M	A	M	J	J	A	S	O	N	D	year	P	
1 Mean daily temperature	in °C	1,8	3,5	7,3	12,5	16,7	19,8	23,7	25,4	21,6	16,6	11,1	5,0	13,8	27	1
2 Mean daily maximum temperature	in °C	6,1	7,2	11,7	16,7	20,6	23,9	27,2	29,4	25,6	21,1	15,0	8,9	17,8	36	2
3 Mean daily minimum temperature	in °C	-1,7	-0,6	2,8	8,3	12,8	16,7	21,7	22,8	18,3	12,2	6,1	0,6	10,0	36	3
4 Absolute maximum temperature	in °C	18,3	18,8	20,6	25,6	28,9	33,3	34,4	35,6	32,2	26,7	23,9	19,4	35,6	29	4
5 Absolute minimum temperature	in °C	-13,9	-11,7	-7,2	-1,7	5,6	9,4	13,9	15,6	9,4	2,2	-3,3	-12,2	-13,9	29	5
6 Mean relative humidity	in %	49	52	59	66	71	80	85	80	74	64	59	53	66	21	6
7 Mean precipitation	in mm	25	44	89	114	139	198	248	165	205	73	44	39	1383	27	7
8 Maximum precipitation	in mm	135	159	232	237	474	595	858	397	448	253	132	112		30	8
9 Minimum precipitation	in mm	-0	2	5	44	22	26	4	16	4	<1	0			30	9
10 Maximum precipitation in 24 h	in mm	209	68	74	179	144	155	253	209	202	177	51	37	253	30	10
11 Mean number of days with precipitation	> 1,0 mm	3	4	6	7	8	9	11	8	8	4	4	4	77	27	11
12 Mean duration of sunshine	in h	205	190	213	219	241	196	183	232	182	218	196	198	2473	27	12
13 Mean quantity of radiation	in ly/day															13
14 Mean potential evaporation	in mm	3	5	21	46	81	109	147	151	103	65	29	8	771		14
15 Mean windspeed	in m/sec															15
16 Mean predominant direction of the wind																16

263 Station/Country **Wakkanai/Japan**

Location 45°25'N/141°41'E — Height above sealevel **2 m** — Climate symbol: Köppen **Dfb** — Troll **III,8**

		J	F	M	A	M	J	J	A	S	O	N	D	year	P	
1 Mean daily temperature	in °C	-5,9	-5,6	-1,8	4,0	8,4	12,2	16,7	19,6	16,6	10,7	3,0	-2,9	6,2	30	1
2 Mean daily maximum temperature	in °C	-4,4	-3,9	1,1	6,7	11,1	15,6	20,0	22,8	19,4	13,9	5,6	-1,7	8,9	8	2
3 Mean daily minimum temperature	in °C	-8,9	-8,4	-3,9	0,6	5,0	9,4	13,9	17,2	13,3	7,2	0,0	-5,6	3,3	8	3
4 Absolute maximum temperature	in °C	8,3	5,4	12,3	20,2	26,9	25,1	29,5	31,3	29,0	23,5	17,0	11,5	31,3	30	4
5 Absolute minimum temperature	in °C	-19,4	-17,3	-18,1	-8,0	-2,2	2,1	6,8	8,9	3,5	-4,4	-8,8	-16,0	-19,4	30	5
6 Mean relative humidity	in %	75	77	75	77	76	84	84	80	75	67	66	71	76	2	6
7 Mean precipitation	in mm	94	62	65	63	78	70	112	105	152	129	120	112	1162	30	7
8 Maximum precipitation	in mm															8
9 Minimum precipitation	in mm															9
10 Maximum precipitation in 24 h	in mm	26	36	101	49	42	40	128	138	140	85	73	52	140	30	10
11 Mean number of days with precipitation	> 0,1 mm	22	15	14	10	9	9	9	9	10	13	17	21	158	30	11
12 Mean duration of sunshine	in h	42	79	148	184	203	177	164	180	188	157	73	38	1631	30	12
13 Mean quantity of radiation	in ly/day															13
14 Mean potential evaporation	in mm	0	0	0	21	56	82	115	126	87	55	13	0	555	30	14
15 Mean windspeed	in m/sec	5,6	5,6	5,3	5,6	5,6	4,5	4,2	4,2	4,5	4,7	5,3	5,7	5,1	30	15
16 Mean predominant direction of the wind		NNW	E	S	SSW	SSW	E	E	E	S	SSW	W	WNW		30	16

264 Station/Country **Sapporo/Japan**

Location 43°03'N/141°20'E — Height above sealevel **17 m** — Climate symbol: Köppen **Dfb** — Troll **III,8**

		J	F	M	A	M	J	J	A	S	O	N	D	year	P	
1 Mean daily temperature	in °C	-5,5	-4,7	-1,0	5,7	11,3	15,5	20,0	21,7	16,6	10,4	3,6	-2,6	7,6	30	1
2 Mean daily maximum temperature	in °C	-1,7	-0,6	2,2	10,6	16,1	20,6	23,9	26,1	21,7	15,6	7,8	0,6	11,7	30	2
3 Mean daily minimum temperature	in °C	-11,7	-10,6	-6,7	0,0	4,4	10,0	14,4	16,1	10,6	3,9	-1,7	-7,7	1,7	30	3
4 Absolute maximum temperature	in °C	11,2	10,8	16,8	25,2	31,1	31,9	35,6	34,5	32,1	25,1	21,3	14,6	35,6	30	4
5 Absolute minimum temperature	in °C	-23,9	-22,5	-17,9	-9,7	-2,2	2,5	7,2	8,5	1,7	-4,2	-12,0	-20,2	-23,9	30	5
6 Mean relative humidity	in %	78	77	75	70	72	77	81	80	80	77	75	78	78	30	6
7 Mean precipitation	in mm	111	83	67	66	59	67	100	107	145	113	112	104	1134	30	7
8 Maximum precipitation	in mm	224	177	175	169	171	186	245	357	481	300	222	197		30	8
9 Minimum precipitation	in mm	26	20	19	9	11	12	10	7	31	11	15	35		30	9
10 Maximum precipitation in 24 h	in mm	115	62	47	104	85	120	124	147	140	104	89	67	147	30	10
11 Mean number of days with precipitation	> 0,1 mm	17	14	12	10	8	9	9	10	11	12	13	15	139	30	11
12 Mean duration of sunshine	in h	99	112	158	199	213	205	190	201	173	186	112	91	1919		12
13 Mean quantity of radiation	in ly/day															13
14 Mean potential evaporation	in mm	0	0	0	30	68	98	128	130	84	48	12	0	596	30	14
15 Mean windspeed	in m/sec	2,8	3,0	3,6	4,2	4,4	3,8	3,4	3,1	3,0	2,9	2,9	2,8	3,4	30	15
16 Mean predominant direction of the wind		WNW	WNW	NW	WNW	SSE	SE	SSE	SSE	NNW	NW	NW	WNW		30	16

126

Location 40°49'N/140°47'E   Height above sealevel 4 m   Climate symbol: Köppen Cfa   Troll III,8

		in	J	F	M	A	M	J	J	A	S	O	N	D	year	P	
1	Mean daily temperature	in °C	-2,7	-2,2	0,4	6,7	12,3	16,2	20,4	22,3	18,0	11,9	6,0	0,2	9,1	30	1
2	Mean daily maximum temperature	in °C															2
3	Mean daily minimum temperature	in °C															3
4	Absolute maximum temperature	in °C	13,5	11,6	20,0	26,0	29,2	31,0	34,2	36,0	35,9	30,5	23,7	21,1	36,0	30	4
5	Absolute minimum temperature	in °C	-17,4	-18,7	-18,4	-7,5	-1,4	4,7	7,2	9,5	3,0	-2,2	-8,8	-13,5	-18,7	30	5
6	Mean relative humidity	in %															6
7	Mean precipitation	in mm	135	108	73	69	67	78	112	139	148	110	121	144	1304	30	7
8	Maximum precipitation	in mm															8
9	Minimum precipitation	in mm															9
10	Maximum precipitation in 24 h	in mm	47	82	93	82	87	101	112	188	107	75	108	52	188	30	10
11	Mean number of days with precipitation	> 0,1 mm	25	20	16	11	8	9	9	9	11	12	16	22	188	30	11
12	Mean duration of sunshine	in h	57	76	142	201	215	197	168	192	172	161	91	47	1719	30	12
13	Mean quantity of radiation	in ly / day															13
14	Mean potential evaporation	in mm	0	0	0	30	66	98	137	140	96	54	21	0	642	25	14
15	Mean windspeed	in m / sec	5,5	5,3	4,9	5,3	4,5	3,6	3,3	3,2	3,5	3,6	4,2	5,1	4,3	30	15
16	Mean predominent direction of the wind		SW	SW	SW	SW	SW	SW	SW	SW	SW	SW	SW	SW		30	16

Location 38°16'N/140°54'E   Height above sealevel 38 m   Climate symbol: Köppen Cfa   Troll III,8

		in	J	F	M	A	M	J	J	A	S	O	N	D	year	P	
1	Mean daily temperature	in °C	0,1	0,6	3,5	9,0	13,9	17,8	22,0	23,8	19,8	13,8	8,2	2,9	11,3	30	1
2	Mean daily maximum temperature	in °C															2
3	Mean daily minimum temperature	in °C															3
4	Absolute maximum temperature	in °C	13,7	14,1	21,0	26,1	30,9	33,7	35,3	36,4	35,4	28,9	23,3	21,4	36,4	30	4
5	Absolute minimum temperature	in °C	-19,8	-20,2	-15,4	-6,0	-1,2	5,4	10,4	11,5	4,1	-1,0	-5,4	-12,9	-20,2	30	5
6	Mean relative humidity	in %															6
7	Mean precipitation	in mm	37	44	62	95	100	155	167	136	191	133	61	50	1231	30	7
8	Maximum precipitation	in mm															8
9	Minimum precipitation	in mm															9
10	Maximum precipitation in 24 h	in mm	43	49	58	89	75	131	144	187	152	107	95	66	187	30	10
11	Mean number of days with precipitation	> 0,1 mm	7	6	8	9	10	13	14	12	10	9	6	8	110	30	11
12	Mean duration of sunshine	in h	151	155	192	203	206	153	134	163	129	149	142	131	1908	30	12
13	Mean quantity of radiation	in ly / day															13
14	Mean potential evaporation	in mm	0	0	22	36	68	100	140	142	100	59	28	7	702	15	14
15	Mean windspeed	in m / sec	3,2	3,3	3,4	3,5	2,9	2,4	1,9	2,1	2,2	2,5	2,7	2,9	2,8	30	15
16	Mean predominent direction of the wind		WNW	WNW	NW	WNW	SSE	SE	SSE	SSE	NNW	NW	NW	WNW			16

Location 37°55'N/139°03'E   Height above sealevel 2 m   Climate symbol: Köppen Cfa   Troll III,8

		in	J	F	M	A	M	J	J	A	S	O	N	D	year	P	
1	Mean daily temperature	in °C	1,7	1,8	4,8	10,2	15,3	19,9	24,1	25,8	21,4	15,5	9,8	4,7	12,9	30	1
2	Mean daily maximum temperature	in °C	4,4	4,4	8,3	15,0	19,4	23,9	27,8	30,0	25,6	19,4	13,9	7,8	16,7	65	2
3	Mean daily minimum temperature	in °C	-1,1	-1,1	1,1	6,1	11,1	16,1	20,6	22,2	17,8	11,7	6,1	1,1	9,4	65	3
4	Absolute maximum temperature	in °C	15,2	18,8	25,1	28,0	31,3	35,0	38,5	39,1	36,2	33,3	26,1	23,6	39,1	30	4
5	Absolute minimum temperature	in °C	-11,7	-13,0	-6,4	-2,5	2,0	6,7	11,4	14,5	7,9	3,0	-1,3	-8,0	-13,0	30	5
6	Mean relative humidity	in %	80	77	79	74	78	80	76	80	78	75	76	78	78	65	6
7	Mean precipitation	in mm	194	126	121	104	95	127	193	107	177	165	171	264	1844	30	7
8	Maximum precipitation	in mm															8
9	Minimum precipitation	in mm															9
10	Maximum precipitation in 24 h	in mm	68	44	51	73	54	112	141	133	117	82	78	77	141	30	10
11	Mean number of days with precipitation	> 0,1 mm	22	19	16	12	10	11	11	13	14	18	24		181	30	11
12	Mean duration of sunshine	in h	70	82	140	201	220	202	207	254	169	145	110	59	1659	30	12
13	Mean quantity of radiation	in ly / day															13
14	Mean potential evaporation	in mm	3	3	12	40	77	112	148	151	106	60	28	10	748	30	14
15	Mean windspeed	in m / sec	6,1	5,3	4,9	4,5	3,9	3,7	3,2	3,2	3,5	3,7	4,3	5,8	4,4	40	15
16	Mean predominent direction of the wind		NW	SSW	SSW	SSW	S	NNE	SW	NNE	S	S	SSW	SSW			16

Location 36°33'N/136°39'E   Height above sealevel 27 m   Climate symbol: Köppen Cfa   Troll IV,7

		in	J	F	M	A	M	J	J	A	S	O	N	D	year	P	
1	Mean daily temperature	in °C	2,5	2,5	5,5	11,0	16,1	20,2	24,5	25,9	21,7	15,6	10,5	5,6	13,5	30	1
2	Mean daily maximum temperature	in °C															2
3	Mean daily minimum temperature	in °C															3
4	Absolute maximum temperature	in °C	21,2	23,6	25,0	29,9	32,4	34,5	36,9	37,1	38,5	33,0	27,6	23,6	38,5	30	4
5	Absolute minimum temperature	in °C	-9,7	-9,4	-8,3	-1,6	1,5	6,8	11,0	14,5	7,6	2,2	-0,7	-6,2	-9,7	30	5
6	Mean relative humidity	in %															6
7	Mean precipitation	in mm	309	191	173	164	135	167	223	154	248	217	225	353	2559	30	7
8	Maximum precipitation	in mm															8
9	Minimum precipitation	in mm															9
10	Maximum precipitation in 24 h	in mm	72	61	69	72	83	147	179	115	156	131	89	85	179	30	10
11	Mean number of days with precipitation	> 0,1 mm	24	19	18	13	12	13	13	10	13	13	16	23	188	30	11
12	Mean duration of sunshine	in h	62	84	142	195	213	186	190	241	159	151	123	70	1818	30	12
13	Mean quantity of radiation	in ly / day															13
14	Mean potential evaporation	in mm	3	3	12	43	77	111	150	151	102	58	31	10	741	30	14
15	Mean windspeed	in m / sec	4,7	4,2	4,1	3,8	3,6	3,3	2,9	3,0	3,1	3,3	3,7	4,3	3,7	30	15
16	Mean predominent direction of the wind		ESE	ESE	ESE	ESE	E	E	E	E	E	ESE	ESE	ESE			16

269 Station/Country Tokyo/Japan

Location 35°41'N/139°46'E Height above sealevel 4 m Climate symbol: Köppen Cfa Troll IV,7

		J	F	M	A	M	J	J	A	S	O	N	D	year	P	
1 Mean daily temperature	in °C	3,7	4,3	7,6	13,1	17,6	21,1	25,1	26,4	22,8	16,7	11,3	6,1	14,7	30	1
2 Mean daily maximum temperature	in °C	8,3	8,9	12,2	17,2	21,7	24,4	28,3	30,0	21,1	20,8	15,6	11,1	18,9	60	2
3 Mean daily minimum temperature	in °C	-1,7	-0,6	2,2	7,8	12,2	17,2	21,1	22,2	18,9	12,8	6,1	0,6	10,0	60	3
4 Absolute maximum temperature	in °C	21,3	24,9	25,2	27,2	31,4	34,7	37,0	38,4	36,4	32,3	27,3	22,7	38,4	30	4
5 Absolute minimum temperature	in °C	-9,2	-7,9	-5,6	-3,1	2,2	8,5	13,0	15,4	10,5	-0,5	-3,1	-6,8	-9,2	30	5
6 Mean relative humidity	in %	61	60	64	70	74	79	80	79	80	76	71	64	72	65	6
7 Mean precipitation	in mm	48	73	101	135	131	182	146	147	217	220	101	61	1562	30	7
8 Maximum precipitation	in mm	145	227	221	211	400	649	674	420	673	542	288	188		30	8
9 Minimum precipitation	in mm	0	2	26	27	49	34	9	11	5	20	5	<1		30	9
10 Maximum precipitation in 24 h	in mm	48	91	87	81	121	278	151	172	393	164	169	85	393	30	10
11 Mean number of days with precipitation	> 0,1 mm	6	7	10	11	12	12	11	10	13	12	8	5	115	30	11
12 Mean duration of sunshine	in h	186	166	176	180	193	149	181	204	136	136	144	169	2020	30	12
13 Mean quantity of radiation	in ly/day															13
14 Mean potential evaporation	in mm	4	6	19	47	83	115	161	157	110	65	31	11	809	30	14
15 Mean windspeed	in m/sec	3,5	3,8	4,3	4,3	3,9	3,5	3,6	3,7	3,5	3,7	3,2	3,0	3,7	30	15
16 Mean predominent direction of the wind		NNW	NNW	NNW	N	S	S	S	S	N	NNW	N	NNW		30	16

270 Station/Country Nagoya/Japan

Location 35°10'N/136°58'E Height above sealevel 51 m Climate symbol: Köppen Cfa Troll IV,7

		J	F	M	A	M	J	J	A	S	O	N	D	year	P	
1 Mean daily temperature	in °C	2,9	3,6	7,1	12,7	17,5	21,5	25,7	26,6	22,7	16,5	10,9	5,6	15,4	30	1
2 Mean daily maximum temperature	in °C															2
3 Mean daily minimum temperature	in °C															3
4 Absolute maximum temperature	in °C	19,0	19,8	25,8	29,6	34,8	34,8	38,9	38,9	36,2	30,2	27,2	20,9	39,9	30	4
5 Absolute minimum temperature	in °C	-10,3	-9,5	-6,8	-2,1	2,8	8,2	14,0	14,4	9,5	1,5	-2,7	-7,2	-10,3	30	5
6 Mean relative humidity	in %															6
7 Mean precipitation	in mm	49	64	100	137	145	204	178	155	212	160	86	57	1547	30	7
8 Maximum precipitation	in mm															8
9 Minimum precipitation	in mm															9
10 Maximum precipitation in 24 h	in mm	53	101	96	105	100	184	176	172	240	167	92	54	240	30	10
11 Mean number of days with precipitation	> 0,1 mm	7	8	9	10	11	13	13	9	14	9	7	5	114	30	11
12 Mean duration of sunshine	in h	178	181	204	200	215	173	202	237	166	172	175	167	2270	30	12
13 Mean quantity of radiation	in ly/day															13
14 Mean potential evaporation	in mm	3	5	18	46	86	119	166	162	113	64	28	10	820	25	14
15 Mean windspeed	in m/sec	3,7	4,0	4,5	4,2	3,7	3,2	3,1	3,4	3,2	3,1	3,3	3,5	3,6	30	15
16 Mean predominent direction of the wind		NW	NW	NW	NNW	NW	S	SSE	SSE	N	N	N	N		30	16

271 Station/Country Osaka/Japan

Location 34°39'N/135°32'E Height above sealevel 7 m Climate symbol: Köppen Cfa Troll IV,7

		J	F	M	A	M	J	J	A	S	O	N	D	year	P	
1 Mean daily temperature	in °C	4,5	4,9	8,0	13,6	18,3	22,3	26,6	27,8	23,7	17,4	11,9	7,0	15,5	30	1
2 Mean daily maximum temperature	in °C	8,3	8,9	12,2	18,3	22,8	26,7	30,6	32,2	28,3	22,2	16,7	11,1	20,0	60	2
3 Mean daily minimum temperature	in °C	0,0	0,6	2,8	8,3	12,8	17,8	22,8	23,3	19,4	12,8	6,7	2,8	10,6	60	3
4 Absolute maximum temperature	in °C	18,0	23,7	23,8	28,8	31,8	34,3	36,7	38,2	35,1	32,5	26,1	22,2	38,2	30	4
5 Absolute minimum temperature	in °C	-7,5	-6,5	-5,2	-2,6	3,5	8,9	14,8	13,6	10,4	3,0	-2,2	-4,5	-7,5	30	5
6 Mean relative humidity	in %	69	69	69	71	71	75	76	74	75	74	72	70	72	62	6
7 Mean precipitation	in mm	43	58	96	127	122	193	177	118	171	122	81	52	1360	30	7
8 Maximum precipitation	in mm	115	160	197	287	361	612	660	333	437	340	195	131		30	8
9 Minimum precipitation	in mm	1	7	20	49	29	40	9	5	30	8	24	2		30	9
10 Maximum precipitation in 24 h	in mm	58	51	73	102	132	251	134	175	118	133	103	55	251	30	10
11 Mean number of days with precipitation	> 0,1 mm	7	7	11	10	11	12	10	7	12	9	7	5	107	30	11
12 Mean duration of sunshine	in h	150	144	175	196	209	181	213	241	175	166	156	146	2152	30	12
13 Mean quantity of radiation	in ly/day															13
14 Mean potential evaporation	in mm	5	8	19	49	87	120	166	171	117	67	28	13	850		14
15 Mean windspeed	in m/sec	3,4	3,2	3,1	3,1	2,9	2,9	2,8	3,0	2,7	2,6	2,3	2,9	2,9	30	15
16 Mean predominent direction of the wind		W	NNE	N	NNE	NNE	W	W	W	NNE	NNE	NNE	W		30	16

272 Station/Country Hiroshima/Japan

Location 34°22'N/132°26'E Height above sealevel 29 m Climate symbol: Köppen Cfa Troll IV,7

		J	F	M	A	M	J	J	A	S	O	N	D	year	P	
1 Mean daily temperature	in °C	4,2	4,7	7,6	12,7	17,1	21,0	25,4	26,6	22,7	16,7	11,5	6,6	14,7	30	1
2 Mean daily maximum temperature	in °C	8,9	9,4	12,6	17,8	22,2	25,6	29,4	31,7	27,8	22,8	16,7	11,7	19,4	60	2
3 Mean daily minimum temperature	in °C	-0,6	0,0	2,8	7,2	12,2	17,2	22,2	22,8	18,9	11,7	6,1	1,7	10,0	60	3
4 Absolute maximum temperature	in °C	18,8	19,2	22,2	25,0	29,6	30,9	35,0	36,7	33,2	30,0	25,4	20,5	36,7	30	4
5 Absolute minimum temperature	in °C	-8,3	-8,3	-7,2	-1,4	1,8	6,6	14,2	13,7	8,8	1,5	-2,6	-6,6	-8,6	30	5
6 Mean relative humidity	in %	71	69	69	71	74	75	79	74	75	70	72	73	73	10	6
7 Mean precipitation	in mm	45	70	106	158	154	249	250	116	216	115	67	51	1597	30	7
8 Maximum precipitation	in mm	142	197	238	408	400	666	611	445	509	315	201	145		30	8
9 Minimum precipitation	in mm	3	5	24	51	19	56	6	8	13	8	13	<1		30	9
10 Maximum precipitation in 24 h	in mm	58	66	75	122	109	150	152	184	340	159	76	47	340	30	10
11 Mean number of days with precipitation	> 0,1 mm	6	7	10	10	9	11	11	7	11	7	6	6	101	60	11
12 Mean duration of sunshine	in h	148	153	188	205	222	191	211	248	176	190	177	155	2262	30	12
13 Mean quantity of radiation	in ly/day															13
14 Mean potential evaporation	in mm	5	7	18	45	74	115	165	167	113	66	30	13	818	22	14
15 Mean windspeed	in m/sec	3,7	3,8	3,9	3,6	3,1	2,9	2,7	3,2	3,6	4,2	4,1	3,8	3,5	30	15
16 Mean predominent direction of the wind		NNE	NNE	NNE	NNE	NNE	NNE	NNE	NNE	NNE	NNE	NNE	NNE		30	16

273 Station/Country Kōchi/Japan

Location 33°34′N/133°33′E Height above sealevel 1 m Climate symbol: Köppen Cfa Troll IV,7

		J	F	M	A	M	J	J	A	S	O	N	D	year	P	
1 Mean daily temperature	in °C	5,2	6,3	9,6	14,4	18,5	21,8	25,7	26,3	23,5	18,0	12,9	7,8	15,8	30	1
2 Mean daily maximum temperature	in °C															2
3 Mean daily minimum temperature	in °C															3
4 Absolute maximum temperature	in °C	23,4	23,7	26,3	30,0	31,2	33,9	37,1	37,3	35,9	32,2	27,8	23,5	37,3	30	4
5 Absolute minimum temperature	in °C	-7,6	-7,0	-6,5	-0,9	3,8	9,1	14,6	15,9	11,3	2,5	-1,9	-6,6	-7,6	30	5
6 Mean relative humidity	in %															6
7 Mean precipitation	in mm	55	97	177	261	279	344	369	344	350	184	108	80	2648	30	7
8 Maximum precipitation	in mm															8
9 Minimum precipitation	in mm															9
10 Maximum precipitation in 24 h	in mm	136	128	195	195	239	263	291	364	371	212	116	166	371	30	10
11 Mean number of days with precipitation	> 0,1 mm	6	7	10	12	13	14	14	12	13	8	6	4	118	30	11
12 Mean duration of sunshine	in h	194	179	200	196	201	162	194	222	178	184	183	191	2284	30	12
13 Mean quantity of radiation	in ly/day															13
14 Mean potential evaporation	in mm	8	11	25	52	88	118	159	157	118	72	37	16	861	25	14
15 Mean windspeed	in m/sec	2,5	2,5	2,6	2,7	2,4	2,1	2,0	2,3	2,3	2,2	2,2	2,5	2,3	30	15
16 Mean predominent direction of the wind		W	W	W	W	W	W	SSE	WNW	W	W	WNW	W		30	16

274 Station/Country Nagasaki/Japan

Location 32°44′N/129°52′E Height above sealevel 27 m Climate symbol: Köppen Cfa Troll IV,7

		J	F	M	A	M	J	J	A	S	O	N	D	year	P	
1 Mean daily temperature	in °C	6,4	7,6	10,2	14,7	18,5	21,9	26,3	27,4	23,9	18,6	13,8	9,0	16,5	30	1
2 Mean daily maximum temperature	in °C	9,4	10,0	13,9	18,9	22,8	25,6	29,4	31,1	27,2	22,2	17,2	11,7	20,0	59	2
3 Mean daily minimum temperature	in °C	2,2	2,2	5,0	10,0	13,9	18,3	22,8	23,3	20,0	14,4	9,4	4,4	12,2	59	3
4 Absolute maximum temperature	in °C	21,3	22,6	24,4	27,7	30,0	34,3	36,3	37,5	34,6	31,6	27,4	23,8	37,5	30	4
5 Absolute minimum temperature	in °C	-3,2	-3,2	-2,0	2,6	7,7	12,0	18,3	18,8	12,2	5,5	1,3	-2,6	-3,2	30	5
6 Mean relative humidity	in %	67	68	65	70	75	79	82	76	75	65	67	70	72	10	6
7 Mean precipitation	in mm	70	83	116	182	207	294	288	189	253	106	83	87	1958	30	7
8 Maximum precipitation	in mm	187	150	193	383	400	749	735	510	831	334	171	182		30	8
9 Minimum precipitation	in mm	28	22	36	61	50	106	34	31	51	12	12	13		30	9
10 Maximum precipitation in 24 h	in mm	83	86	133	345	172	385	179	252	345	139	88	111	385	30	10
11 Mean number of days with precipitation	> 0,1 mm	12	9	11	11	13	13	12	9	12	6	8	8	122	30	11
12 Mean duration of sunshine	in h	114	132	177	188	202	172	210	252	186	200	170	128	2131	30	12
13 Mean quantity of radiation	in ly/day															13
14 Mean potential evaporation	in mm	9	9	27	74	83	115	156	155	113	68	34	16	859		14
15 Mean windspeed	in m/sec															15
16 Mean predominent direction of the wind																16

275 Station/Country Kagoshima/Japan

Location 31°34′N/130°33′E Height above sealevel 5 m Climate symbol: Köppen Cfa Troll IV,7

		J	F	M	A	M	J	J	A	S	O	N	D	year	P	
1 Mean daily temperature	in °C	6,6	7,7	10,8	15,1	19,0	22,6	26,8	27,1	24,4	18,9	14,0	9,0	16,8	30	1
2 Mean daily maximum temperature	in °C	11,7	12,2	15,6	20,0	23,3	25,6	29,4	30,6	28,3	23,9	18,3	13,3	21,1	30	2
3 Mean daily minimum temperature	in °C	2,8	3,3	6,1	11,1	14,4	18,9	22,8	23,3	20,6	15,0	9,4	4,4	12,8	30	3
4 Absolute maximum temperature	in °C	23,9	24,1	25,8	27,7	31,1	34,1	36,6	37,0	34,2	32,2	28,6	24,4	37,0	30	4
5 Absolute minimum temperature	in °C	-5,7	-6,7	-3,9	-1,0	3,9	9,0	15,9	16,8	9,8	2,6	-1,5	-5,5	-6,7	30	5
6 Mean relative humidity	in %	73	73	72	76	78	80	81	79	80	74	75	77	77	10	6
7 Mean precipitation	in mm	75	116	149	228	249	454	343	220	213	120	90	79	2336	30	7
8 Maximum precipitation	in mm															8
9 Minimum precipitation	in mm															9
10 Maximum precipitation in 24 h	in mm	114	90	94	145	156	306	234	217	174	167	98	169	306	30	10
11 Mean number of days with precipitation	> 0,1 mm	10	10	13	13	14	15	13	11	11	8	7	8	133	30	11
12 Mean duration of sunshine	in h	153	147	178	179	190	159	222	244	201	198	180	171	2222	30	12
13 Mean quantity of radiation	in ly/day															13
14 Mean potential evaporation	in mm	11	13	28	55	89	120	167	162	121	74	40	18	898	30	14
15 Mean windspeed	in m/sec	4,1	4,1	4,2	4,1	3,8	3,6	3,6	4,0	3,9	4,0	3,9	3,9	3,9	30	15
16 Mean predominent direction of the wind		NW	NW	NW	NW	NW	NW	NW	NW	NW	NW	NW	NW		30	16

276 Station/Country Naha (Okinawa)/Japan

Location 26°14′N/127°41′E Height above sealevel 36 m Climate symbol: Köppen Cfa Troll IV,7

		J	F	M	A	M	J	J	A	S	O	N	D	year	P	
1 Mean daily temperature	in °C	16,1	16,5	17,9	20,4	23,4	25,5	27,9	27,4	26,7	24,0	21,2	18,1	22,1	30	1
2 Mean daily maximum temperature	in °C	19,4	19,4	21,1	24,4	26,7	29,4	31,7	31,1	30,8	27,2	23,9	21,1	25,6	30	2
3 Mean daily minimum temperature	in °C	13,3	12,8	15,0	17,8	20,0	23,9	25,0	25,0	23,9	20,6	17,8	14,4	18,9	30	3
4 Absolute maximum temperature	in °C	25,8	25,7	26,9	29,1	31,2	32,6	33,8	32,6	32,6	30,4	29,7	26,4	33,8	30	4
5 Absolute minimum temperature	in °C	6,7	6,6	7,2	10,6	14,0	16,9	22,2	22,3	20,7	14,8	12,2	7,2	6,6	30	5
6 Mean relative humidity	in %	75	77	77	84	84	81	84	83	80	76	74	75	75	10	6
7 Mean precipitation	in mm	121	137	168	165	246	329	180	296	167	154	146	114	2223	30	7
8 Maximum precipitation	in mm	417	335	414	516	637	720	513	679	503	842	530	305		30	8
9 Minimum precipitation	in mm	16	20	21	18	15	34	10	35	22	18	7			30	9
10 Maximum precipitation in 24 h	in mm	95	188	180	154	194	200	167	272	203	372	259	86	372	30	10
11 Mean number of days with precipitation	> 0,1 mm	13	13	12	12	14	13	9	14	11	9	10	10	141	30	11
12 Mean duration of sunshine	in h	110	106	130	153	169	195	279	247	224	183	146	123	2065	30	12
13 Mean quantity of radiation	in ly/day															13
14 Mean potential evaporation	in mm	36	37	56	77	118	141	172	161	141	104	74	49	1166	30	14
15 Mean windspeed	in m/sec	6,4	6,1	5,7	5,7	5,1	5,4	5,0	5,8	5,6	6,0	6,3	6,0	5,8	30	15
16 Mean predominent direction of the wind		N	N	NNE	S	SSW	SSW	S	E	E	NE	NE	NE		30	16

129

277 Station/Country: Chiang Mai/Thailand

Location 18°47'N/98°59'E — Height above sealevel 314 m — Climate symbol: Köppen Aw — Troll V.2

		J	F	M	A	M	J	J	A	S	O	N	D	year	P	
1	Mean daily temperature	in °C	21,1	23,1	26,0	28,6	28,7	27,9	27,4	27,0	26,9	26,2	24,5	21,8	25,8	29	1
2	Mean daily maximum temperature	in °C	29,2	32,1	35,0	36,0	34,3	32,3	31,4	30,8	30,9	30,8	30,1	28,8	31,8	29	2
3	Mean daily minimum temperature	in °C	13,3	14,0	16,9	21,1	23,2	23,5	23,3	23,2	22,9	21,6	18,9	14,9	19,7	29	3
4	Absolute maximum temperature	in °C	34,7	37,5	39,8	41,5	39,6	39,5	36,5	35,4	35,5	35,2	34,7	33,5	41,5	29	4
5	Absolute minimum temperature	in °C	6,0	6,4	10,0	13,2	19,4	19,1	18,9	20,0	19,7	14,2	10,2	6,1	6,0	29	5
6	Mean relative humidity	in %	73	65	58	61	72	78	80	83	83	82	79	77	74	29	6
7	Mean precipitation	in mm	7	7	16	45	146	137	169	223	270	143	40	14	1217	55	7
8	Maximum precipitation	in mm	95	89	84	139	441	347	405	526	592	345	160	85	2033	55	8
9	Minimum precipitation	in mm	0	0	0	2	7	32	54	75	36	7	0	0	501	55	9
10	Maximum precipitation in 24 h	in mm	18	25	51	38	74	56	74	71	104	107	117	18	117	8	10
11	Mean number of days with precipitation	> 1,0 mm	<1	1	2	5	12	15	21	20	17	8	4	2	108	13	11
12	Mean duration of sunshine	in h	270	266	285	264	236	170	147	137	166	226	247	270	2684		12
13	Mean quantity of radiation	in ly/day															13
14	Mean potential evaporation	in mm	61	77	135	152	174	165	164	155	141	129	96	88	1517	7	14
15	Mean windspeed	in m/sec	0,4	0,6	0,7	0,9	0,9	0,7	0,6	0,6	0,8	0,5	0,4	0,4	0,6	15	15
16	Mean predominent direction of the wind		S	S	S	S	S	S	S	S	S	N	N	S		15	16

278 Station/Country: Nakhon Ratchasima/Thailand

Location 14°58'N/102°07'E — Height above sealevel 181 m — Climate symbol: Köppen Aw — Troll V.2

		J	F	M	A	M	J	J	A	S	O	N	D	year	P	
1	Mean daily temperature	in °C	23,5	26,6	28,9	29,8	29,2	28,6	28,2	28,0	27,4	26,6	25,0	23,0	27,1	29	1
2	Mean daily maximum temperature	in °C	31,6	34,1	36,1	36,4	34,7	33,6	33,1	32,9	32,0	31,2	31,0	30,1	33,1	29	2
3	Mean daily minimum temperature	in °C	15,4	19,1	21,6	23,2	23,8	23,7	23,3	23,2	23,0	22,0	19,4	16,0	21,1	29	3
4	Absolute maximum temperature	in °C	38,0	40,7	41,5	43,4	41,0	40,1	39,0	38,2	38,4	37,1	38,0	37,5	43,4	29	4
5	Absolute minimum temperature	in °C	4,9	10,6	11,6	15,7	20,7	20,9	20,6	20,5	19,6	15,1	8,4	6,4	4,9	29	5
6	Mean relative humidity	in %	67	64	65	69	77	78	76	77	83	81	78	71	74	29	6
7	Mean precipitation	in mm	8	37	47	85	170	118	120	148	241	173	35	3	1183	55	7
8	Maximum precipitation	in mm	71	306	125	267	365	292	283	758	566	566	174	37	2418	55	8
9	Minimum precipitation	in mm	0	0	0	0	45	18	13	37	105	36	0	0	731	55	9
10	Maximum precipitation in 24 h	in mm	51	71	71	76	89	38	53	56	58	66	104	10	104	5	10
11	Mean number of days with precipitation	> 1,0 mm	1	3	9	8	15	14	15	17	19	14	3	2	120	12	11
12	Mean duration of sunshine	in h	284	244	244	234	246	203	204	180	181	223	255	261	2739		12
13	Mean quantity of radiation	in ly/day															13
14	Mean potential evaporation	in mm	87	126	162	187	172	164	167	181	147	137	100	76	1666	7	14
15	Mean windspeed	in m/sec	0,8	0,9	0,8	0,8	0,8	1,2	1,2	1,1	0,8	0,9	0,9	0,8	0,9	15	15
16	Mean predominent direction of the wind		NE	NE	W	SW	SW	SW	W	W	W	NE	NE	NE		15	16

279 Station/Country: Bangkok/Thailand

Location 13°45'N/100°28'E — Height above sealevel 2 m — Climate symbol: Köppen Aw — Troll V.2

		J	F	M	A	M	J	J	A	S	O	N	D	year	P	
1	Mean daily temperature	in °C	26,0	27,8	29,2	30,1	29,7	28,9	28,5	28,4	28,0	27,7	27,0	25,7	28,1	29	1
2	Mean daily maximum temperature	in °C	32,0	33,0	34,2	34,9	34,2	33,0	32,4	32,2	31,8	31,3	31,0	30,9	32,6	29	2
3	Mean daily minimum temperature	in °C	20,1	22,6	24,3	25,3	25,1	24,9	24,5	24,5	24,2	24,1	22,9	20,5	23,8	29	3
4	Absolute maximum temperature	in °C	36,9	39,4	39,8	39,9	39,8	38,8	35,8	35,7	35,8	35,9	35,5	35,3	39,9	29	4
5	Absolute minimum temperature	in °C	9,9	16,0	16,5	18,2	21,1	21,7	20,2	21,2	21,2	20,4	18,8	12,3	9,9	29	5
6	Mean relative humidity	in %	72	75	74	75	79	80	80	81	83	83	80	74	78	29	6
7	Mean precipitation	in mm	9	30	36	82	165	153	168	183	310	239	55	8	1438	55	7
8	Maximum precipitation	in mm	94	193	225	319	278	333	368	345	553	494	188	70	2072	55	8
9	Minimum precipitation	in mm	0	0	0	36	46	87	62	116	27	0	0	0	891	55	9
10	Maximum precipitation in 24 h	in mm	58	64	79	117	114	89	114	79	175	112	142	107	175	37	10
11	Mean number of days with precipitation	> 1,0 mm	1	1	3	3	9	10	13	13	15	14	5	1	88	20	11
12	Mean duration of sunshine	in h	254	224	248	300	233	183	146	161	156	189	226	242	2556	30	12
13	Mean quantity of radiation	in ly/day															13
14	Mean potential evaporation	in mm	121	134	163	170	177	167	168	164	152	146	133	109	1804	16	14
15	Mean windspeed	in m/sec	1,3	1,6	1,6	1,6	1,5	1,6	1,4	1,5	1,3	1,2	1,2	1,1	1,4	15	15
16	Mean predominent direction of the wind		NE	S	S	S	S	S	SW	SW	SW	NE	N	N		15	16

280 Station/Country: Surat Thani (Ban Don)/Thailand

Location 9°07'N/99°17'E — Height above sealevel 3 m — Climate symbol: Köppen Aw — Troll V.2

		J	F	M	A	M	J	J	A	S	O	N	D	year	P	
1	Mean daily temperature	in °C	26,1	27,0	28,3	29,2	28,9	28,3	27,8	28,1	27,5	27,2	26,4	25,9	27,5	13	1
2	Mean daily maximum temperature	in °C	31,1	33,3	35,0	35,0	33,9	32,8	32,2	32,8	32,2	31,7	30,0	30,0	32,2	13	2
3	Mean daily minimum temperature	in °C	21,1	20,6	21,7	23,3	23,9	23,9	23,3	23,3	22,8	22,8	22,8	21,7	22,8	13	3
4	Absolute maximum temperature	in °C	36,1	37,2	38,3	39,4	37,2	36,1	36,7	36,7	35,5	35,8	35,0	34,4	39,4	13	4
5	Absolute minimum temperature	in °C	15,6	16,1	15,6	19,4	20,6	21,1	20,6	20,6	20,6	20,6	19,4	16,7	15,6	13	5
6	Mean relative humidity	in %	84	82	80	81	83	93	84	83	84	87	88	88	85	7	6
7	Mean precipitation	in mm	53	18	41	114	258	140	114	137	201	238	320	216	1846	13	7
8	Maximum precipitation	in mm															8
9	Minimum precipitation	in mm															9
10	Maximum precipitation in 24 h	in mm	168	15	61	74	107	53	86	71	79	135	170	168	170	13	10
11	Mean number of days with precipitation	> 1,0 mm	8	4	5	13	18	17	18	18	19	20	20	14	168	13	11
12	Mean duration of sunshine	in h															12
13	Mean quantity of radiation	in ly/day															13
14	Mean potential evaporation	in mm	123	127	155	161	168	158	159	159	147	144	126	123	1750	13	14
15	Mean windspeed	in m/sec															15
16	Mean predominent direction of the wind		NE	N-E	N	N	SW	SW	SW	SW	SW	SW	N.NE	N.NE			16

281 Station / Country Phuket / Thailand

Location 7°58'N/98°24'E   Height above sealevel 3 m   Climate symbol: Köppen Am   Troll V.2

		J	F	M	A	M	J	J	A	S	O	N	D	year	P	
1 Mean daily temperature	in °C	27,2	27,8	28,6	28,9	28,1	28,1	28,1	27,5	27,0	27,2	27,2	27,2	27,7	33	1
2 Mean daily maximum temperature	in °C	31,1	32,2	33,3	32,8	31,1	31,1	31,1	30,6	30,0	30,6	30,6	30,6	31,1	33	2
3 Mean daily minimum temperature	in °C	23,3	23,3	23,9	25,0	25,0	25,0	25,0	24,4	23,9	23,9	23,9	23,9	24,4	33	3
4 Absolute maximum temperature	in °C	33,9	35,0	35,0	35,0	35,0	33,3	33,3	34,4	33,3	33,3	33,3	35,0	35,0	11	4
5 Absolute minimum temperature	in °C	18,9	20,0	20,0	22,8	22,8	22,2	21,7	21,7	20,6	21,7	20,6	20,0	18,9	11	5
6 Mean relative humidity	in %	78	75	77	82	84	82	82	83	85	85	83	78	81	9	6
7 Mean precipitation	in mm	36	38	74	127	297	264	216	244	328	315	193	79	2211	20	7
8 Maximum precipitation	in mm															8
9 Minimum precipitation	in mm															9
10 Maximum precipitation in 24 h	in mm	23	56	43	79	97	229	79	89	89	117	76	53	227	27	10
11 Mean number of days with precipitation	> 1,0 mm	4	3	6	15	19	19	17	17	19	19	14	8	160	11	11
12 Mean duration of sunshine	in h															12
13 Mean quantity of radiation	in ly/day															13
14 Mean potential evaporation	in mm	144	136	154	154	155	152	153	153	143	142	136	140	1782	7	14
15 Mean windspeed	in m/sec	2,8	2,5	2,3	1,9	1,7	2,4	2,3	2,8	2,2	1,9	2,1	2,9		25	15
16 Mean predominent direction of the wind		NE	NE	E	E	W	W	W	W	W	W	NE	NE		25	16

282 Station / Country Louang Prabang / Laos

Location 19°53'N/102°08'E   Height above sealevel 287 m   Climate symbol: Köppen Aw   Troll V.2

		J	F	M	A	M	J	J	A	S	O	N	D	year	P	
1 Mean daily temperature	in °C	20,6	23,1	25,6	28,1	28,9	28,6	27,8	27,8	27,8	26,1	23,6	21,1	25,8	28	1
2 Mean daily maximum temperature	in °C	27,8	31,7	33,9	35,6	35,0	33,9	32,2	32,2	31,7	29,4	27,2	27,2	31,7	28	2
3 Mean daily minimum temperature	in °C	13,3	14,4	17,2	20,6	22,8	23,3	23,3	23,3	22,8	20,6	17,8	15,0	19,4	28	3
4 Absolute maximum temperature	in °C	39,4	38,9	41,1	45,0	43,9	40,0	38,9	40,0	37,8	38,3	36,1	32,8	45,0	28	4
5 Absolute minimum temperature	in °C	0,6	7,0	10,0	13,9	17,2	13,9	19,4	13,9	10,8	12,8	6,1	4,4	0,6	28	5
6 Mean relative humidity	in %	70	64	58	58	62	71	71	78	72	71	70	71	68	19	6
7 Mean precipitation	in mm	15	18	30	100	163	155	231	300	165	79	30	13	1308	31	7
8 Maximum precipitation	in mm	140	130	91	314	383	386	485	540	475	248	116	65	1879	58	8
9 Minimum precipitation	in mm	0	0	17	11	21	8	51	72	24	tr	0	0	483	58	9
10 Maximum precipitation in 24 h	in mm	109	74	41	104	79	101	127	89	81	117	58	51	127	31	10
11 Mean number of days with precipitation	> 1,0 mm	2	2	4	8	13	12	17	19	12	7	3	1	106	19	11
12 Mean duration of sunshine	in h															12
13 Mean quantity of radiation	in ly/day															13
14 Mean potential evaporation	in mm	44	65	105	141	158	157	153	144	133	110	71	43	1324	32	14
15 Mean windspeed	in m/sec	0,8	0,8	1,1	0,8	0,8	0,6	0,5	0,5	0,4	0,5	0,5	0,6		12	15
16 Mean predominent direction of the wind		N,NE	SW,N	SW,W	SW,W	SW,W	SW,W	SW,W	SW,W	NE,N	N,NE	N,NE	N,NE		12	16

283 Station / Country Vientiane / Laos

Location 17°59'N/102°36'E   Height above sealevel 162 m   Climate symbol: Köppen Aw   Troll V.2

		J	F	M	A	M	J	J	A	S	O	N	D	year	P	
1 Mean daily temperature	in °C	21,1	23,9	26,1	28,3	27,5	27,8	27,2	27,5	27,2	25,6	23,9	22,0	25,7	9	1
2 Mean daily maximum temperature	in °C	28,3	30,0	32,8	33,9	32,2	31,7	30,6	31,1	30,6	30,6	29,4	28,3	30,6	9	2
3 Mean daily minimum temperature	in °C	13,9	17,2	19,4	22,8	22,8	23,9	23,9	23,9	23,9	21,1	18,3	15,6	20,6	9	3
4 Absolute maximum temperature	in °C	35,0	36,7	40,0	39,4	38,9	35,6	34,4	36,7	35,0	34,4	34,4	33,3	40,0	9	4
5 Absolute minimum temperature	in °C	3,9	7,8	12,2	17,2	20,6	21,1	21,1	21,1	21,1	12,8	10,6	5,0	3,9	9	5
6 Mean relative humidity	in %	77	75	71	74	82	85	87	86	86	82	79	78	80	9	6
7 Mean precipitation	in mm	5	15	38	99	267	302	267	292	302	109	15	3	1714	27	7
8 Maximum precipitation	in mm	67	66	134	329	499	499	544	619	777	329	109	24	2138	59	8
9 Minimum precipitation	in mm	0	0	0	7	79	102	81	50	99	0	0	0	1200	59	9
10 Maximum precipitation in 24 h	in mm	36	64	53	155	145	163	193	130	117	130	69	23	193	10	10
11 Mean number of days with precipitation	> 1,0 mm	1	2	4	7	15	17	18	18	16	7	1	1	107	27	11
12 Mean duration of sunshine	in h	233	193	242	192	109	75	50	62	78	174	180	202	1790	30	12
13 Mean quantity of radiation	in ly/day															13
14 Mean potential evaporation	in mm	55	79	128	147	162	158	158	153	141	121	87	57	1444	16	14
15 Mean windspeed	in m/sec	1,9	2,0	2,0	2,3	2,2	2,0	1,8	1,8	1,7	1,8	1,8	1,9	1,9	14	15
16 Mean predominent direction of the wind		N,NE	N,SE	SE,S	S,SE	S,SE	S,SW	S,SW	S,SW	S,SW	N,S	N,NE	N,NE		14	16

284 Station / Country Stung Treng / Cambodia

Location 13°31'N/105°58'E   Height above sealevel 51 m   Climate symbol: Köppen Aw   Troll V.2

		J	F	M	A	M	J	J	A	S	O	N	D	year	P	
1 Mean daily temperature	in °C	24,2	26,5	28,8	29,6	28,1	27,3	26,6	26,5	26,3	26,1	25,3	23,8	22,6	21	1
2 Mean daily maximum temperature	in °C	31,0	33,1	34,7	35,0	32,9	31,6	30,6	30,6	30,2	30,2	30,1	29,5	31,6	28	2
3 Mean daily minimum temperature	in °C	18,8	20,8	23,5	25,0	24,8	24,2	23,9	23,6	23,7	23,6	23,0	21,5	19,0	28	3
4 Absolute maximum temperature	in °C	36,0	37,1	38,3	41,4	38,6	37,2	34,5	34,5	34,0	34,0	34,2	34,0	41,4	28	4
5 Absolute minimum temperature	in °C	9,5	14,0	17,3	20,0	18,9	19,6	20,1	19,3	19,4	17,2	14,6	10,5	9,5	28	5
6 Mean relative humidity	in %	72	66	64	68	81	85	84	88	89	86	83	79	79	28	6
7 Mean precipitation	in mm	2	11	28	76	198	276	340	317	323	184	56	12	1823	48	7
8 Maximum precipitation	in mm	15	118	125	252	460	470	860	576	576	641	192	65	2295	48	8
9 Minimum precipitation	in mm	0	0	0	0	47	134	137	100	169	35	0	0	1481	48	9
10 Maximum precipitation in 24 h	in mm															10
11 Mean number of days with precipitation	> 0,1 mm	0	1	2	5	14	15	17	20	19	12	6	1	112		11
12 Mean duration of sunshine	in h															12
13 Mean quantity of radiation	in ly/day															13
14 Mean potential evaporation	in mm	94	117	161	165	165	152	150	139	129	124	107	87	1590	21	14
15 Mean windspeed	in m/sec	1,6	1,6	1,6	1,6	1,6	1,6	1,6	1,6	1,6	1,6	1,6	1,6		17	15
16 Mean predominent direction of the wind		NE	NE	S	S	S	SW	SW	SW	SW	NE	NE	NE		15	16

285 Station/Country Båtdâmbâng / Cambodia

Location 13°06'N/103°12'E — Height above sealevel 22 m — Climate symbol: Köppen Aw — Troll V,2

		J	F	M	A	M	J	J	A	S	O	N	D	year	P	
1	Mean daily temperature	in °C	24,8	26,8	28,7	29,4	28,4	28,2	27,3	27,2	26,8	26,6	25,8	24,5	27,0	18	1
2	Mean daily maximum temperature	in °C	30,6	33,3	35,0	35,6	33,9	32,8	31,7	31,7	31,1	30,0	30,0	30,0	32,1	30	2
3	Mean daily minimum temperature	in °C	19,4	21,1	22,8	23,3	24,4	24,4	23,9	23,9	23,3	22,2	20,0	20,0	22,8	30	3
4	Absolute maximum temperature	in °C	37,7	38,2	40,8	41,0	40,3	38,4	35,5	35,3	34,5	34,1	34,0	34,8	41,0	20	4
5	Absolute minimum temperature	in °C	10,4	14,4	16,1	19,8	21,1	21,7	21,2	22,0	21,7	19,8	13,1	10,7	10,4	20	5
6	Mean relative humidity	in %	74	72	71	74	79	80	83	83	81	83	83	88	79	20	6
7	Mean precipitation	in mm	5	15	54	85	148	147	166	188	245	240	85	19	1377	39	7
8	Maximum precipitation	in mm	54	113	204	266	340	276	284	315	609	524	282	132	1970	39	8
9	Minimum precipitation	in mm	0	0	tr	5	41	23	71	70	79	70	1	0	919	39	9
10	Maximum precipitation in 24 h	in mm	53	46	104	84	89	84	79	74	150	132	109	46	150	30	10
11	Mean number of days with precipitation	> 0,1 mm	1	2	4	7	13	12	16	16	16	15	7	3	112	30	11
12	Mean duration of sunshine	in h															12
13	Mean quantity of radiation	in ly/day															13
14	Mean potential evaporation	in mm	100	125	157	163	163	157	156	151	140	136	117	98	1683		14
15	Mean windspeed	in m/sec	1,6	1,6	1,6	1,6	2,1	2,6	2,1	2,1	1,8	1,6	1,6	1,0	1,8	13	15
16	Mean predomindent direction of the wind		E	E	E	E	SW	SW	SW	SW	SW	E	NE	NE		11	16

286 Station/Country Phnum Pênh / Cambodia

Location 11°33'N/104°55'E — Height above sealevel 10 m — Climate symbol: Köppen Aw — Troll V,2

		J	F	M	A	M	J	J	A	S	O	N	D	year	P	
1	Mean daily temperature	in °C	26,0	27,5	28,9	29,6	28,6	28,1	27,5	27,6	27,2	27,1	26,7	25,6	27,5	28	1
2	Mean daily maximum temperature	in °C	31,7	32,2	33,9	34,4	33,9	32,8	31,7	31,7	31,1	30,6	30,0	30,0	32,0	30	2
3	Mean daily minimum temperature	in °C	21,1	22,2	23,3	23,9	24,4	24,4	24,4	24,4	24,4	24,4	23,3	21,7	23,5	30	3
4	Absolute maximum temperature	in °C	36,3	36,7	39,0	40,5	38,5	38,4	36,6	36,0	33,9	34,4	34,4	34,8	40,5	42	4
5	Absolute minimum temperature	in °C	13,3	15,2	19,0	17,9	20,6	21,2	20,1	22,0	21,9	20,8	16,8	14,4	13,3	42	5
6	Mean relative humidity	in %	71	69	69	72	80	80	82	82	84	86	79	75	77	42	6
7	Mean precipitation	in mm	8	9	36	74	146	142	146	154	228	256	134	39	1372	56	7
8	Maximum precipitation	in mm	57	127	193	359	395	393	359	380	443	650	298	176	2310	56	8
9	Minimum precipitation	in mm	0	0	0	0	30	27	37	44	93	63	2	0	936	56	9
10	Maximum precipitation in 24 h	in mm	28	71	142	105	98	94	98	140	109	157	108	63	157	30	10
11	Mean number of days with precipitation	> 0,1 mm	1	1	3	6	14	15	16	16	19	17	9	4	121	30	11
12	Mean duration of sunshine	in h	276	252	270	249	226	192	174	171	147	205	234	267	2663	30	12
13	Mean quantity of radiation	in ly/day															13
14	Mean potential evaporation	in mm	113	126	157	163	163	156	155	154	144	143	135	117	1726	37	14
15	Mean windspeed	in m/sec	2,6	2,6	3,1	3,1	2,6	3,1	3,1	3,1	2,6	2,6	3,1	3,1	2,9	17	15
16	Mean predomindent direction of the wind		N	S	S	S	SW	W	W	SW	W	N	N	N		16	16

287 Station/Country Ha-noi / Vietnam

Location 21°02'N/105°52'E — Height above sealevel 16 m — Climate symbol: Köppen Cwa — Troll V,2

		J	F	M	A	M	J	J	A	S	O	N	D	year	P	
1	Mean daily temperature	in °C	16,7	17,5	20,3	24,2	27,8	29,4	29,2	28,9	27,8	25,3	21,7	18,6	24,0	33	1
2	Mean daily maximum temperature	in °C	20,0	20,6	23,3	27,8	32,2	33,3	32,8	32,2	31,1	28,9	25,6	22,2	27,8	33	2
3	Mean daily minimum temperature	in °C	13,3	14,4	17,2	20,6	23,3	25,6	25,6	25,6	24,4	21,7	17,8	15,0	20,6	33	3
4	Absolute maximum temperature	in °C	33,3	34,4	36,7	39,4	42,8	40,0	40,0	38,3	37,2	35,6	36,1	36,7	42,8	33	4
5	Absolute minimum temperature	in °C	5,6	6,1	11,7	10,0	15,6	20,6	21,7	21,1	17,2	13,9	6,7	6,7	5,6	33	5
6	Mean relative humidity	in %	73	76	80	79	73	75	76	79	76	72	71	71	75	21	6
7	Mean precipitation	in mm	18	28	38	81	196	239	323	343	254	99	43	20	1682	31	7
8	Maximum precipitation	in mm	108	92	119	202	493	597	564	905	711	321	150	105	2741	30	8
9	Minimum precipitation	in mm	0	10	4	29	64	26	104	80	53	5	2	0	1275	30	9
10	Maximum precipitation in 24 h	in mm	46	48	64	104	155	244	208	262	251	157	69	30	282	31	10
11	Mean number of days with precipitation	> 1,0 mm	7	13	15	14	15	14	15	16	14	9	7	7	146	21	11
12	Mean duration of sunshine	in h	43	39	40	66	130	150	149	130	129	130	96	65	1167	11	12
13	Mean quantity of radiation	in ly/day															13
14	Mean potential evaporation	in mm	32	35	60	103	158	171	174	165	144	108	71	41	1262	40	14
15	Mean windspeed	in m/sec															15
16	Mean predomindent direction of the wind		NE	N,NE	SE	SE	SE	SE	SE	var	var,N	N	N	N			16

288 Station/Country Guang Tri / Vietnam

Location 16°44'N/107°11'E — Height above sealevel 7 m — Climate symbol: Köppen Am — Troll V,1

		J	F	M	A	M	J	J	A	S	O	N	D	year	P	
1	Mean daily temperature	in °C	20,0	20,9	22,8	26,4	28,6	29,4	29,4	29,4	27,5	25,3	23,1	20,9	25,3	32	1
2	Mean daily maximum temperature	in °C	22,8	23,9	26,1	30,6	33,3	33,9	33,3	33,9	31,7	28,3	25,6	23,3	28,9	32	2
3	Mean daily minimum temperature	in °C	17,2	17,8	19,4	22,2	23,9	25,0	25,6	25,0	23,3	22,2	20,6	18,3	21,7	32	3
4	Absolute maximum temperature	in °C	33,9	36,1	39,4	39,4	39,4	40,0	40,0	38,9	39,4	36,1	33,9	32,2	40,0	33	4
5	Absolute minimum temperature	in °C	9,4	10,6	12,2	12,8	17,8	20,0	20,0	20,0	17,8	15,6	12,2	11,1	9,4	33	5
6	Mean relative humidity	in %	85	85	84	77	70	65	63	63	75	82	84	86	77	21	6
7	Mean precipitation	in mm	170	56	69	56	99	76	89	97	396	561	566	305	2540	32	7
8	Maximum precipitation	in mm															8
9	Minimum precipitation	in mm															9
10	Maximum precipitation in 24 h	in mm	107	43	84	84	163	142	231	193	216	305	353	188	353	27	10
11	Mean number of days with precipitation	> 1,0 mm	14	11	11	9	10	6	6	7	14	19	20	18	135	21	11
12	Mean duration of sunshine	in h															12
13	Mean quantity of radiation	in ly/day															13
14	Mean potential evaporation	in mm	51	56	84	138	169	173	177	172	146	119	81	59	1425		14
15	Mean windspeed	in m/sec															15
16	Mean predomindent direction of the wind																16

289 Station / Country Qui-nhon / Vietnam

Location 13°45'N / 109°13'E Height above sealevel 6 m Climate symbol: Köppen Aw Troll V,1

		J	F	M	A	M	J	J	A	S	O	N	D	year	P	
1 Mean daily temperature	in °C	22.8	23.6	25.3	27.0	28.6	29.4	30.0	30.6	28.6	26.4	25.0	22.2	26.6	21	1
2 Mean daily maximum temperature	in °C	25.0	26.1	28.3	30.0	32.2	33.9	33.9	35.0	32.2	32.2	28.9	27.2	24.4	21	2
3 Mean daily minimum temperature	in °C	20.6	21.1	22.2	23.9	25.0	25.0	26.1	26.1	25.0	23.9	22.8	20.0	23.3	21	3
4 Absolute maximum temperature	in °C	32.8	35.0	38.3	36.1	39.4	41.1	42.2	41.1	38.9	37.2	32.8	30.6	42.2	21	4
5 Absolute minimum temperature	in °C	15.0	15.6	18.3	19.4	18.9	21.7	20.6	20.6	21.1	17.8	15.0	16.1	15.0	21	5
6 Mean relative humidity	in %	83	81	79	79	75	69	66	62	75	82	85	84	77	21	6
7 Mean precipitation	in mm	53	41	38	25	53	56	69	56	249	434	434	142	1650	21	7
8 Maximum precipitation	in mm															8
9 Minimum precipitation	in mm															9
10 Maximum precipitation in 24 h	in mm	114	89	66	97	71	140	124	91	269	305	310	130	310	22	10
11 Mean number of days with precipitation	> 1,0 mm	9	5	3	3	5	5	6	5	15	18	13	16	83	21	11
12 Mean duration of sunshine	in h															12
13 Mean quantity of radiation	in ly / day															13
14 Mean potential evaporation	in mm	74	82	114	146	169	173	179	172	151	133	108	82	1583	35	14
15 Mean windspeed	in m / sec															15
16 Mean predominent direction of the wind																16

290 Station / Country Ho Chi Minh City (Sai-gon) / Vietnam

Location 10°47'N / 106°42'E Height above sealevel 9 m Climate symbol: Köppen Aw Troll V,2

		J	F	M	A	M	J	J	A	S	O	N	D	year	P	
1 Mean daily temperature	in °C	26.4	27.2	28.6	30.0	28.9	27.8	27.5	27.5	27.2	27.2	26.7	26.1	27.6	31	1
2 Mean daily maximum temperature	in °C	31.7	32.8	33.9	35.0	33.3	31.7	31.1	31.1	31.1	31.1	30.6	30.6	32.2	31	2
3 Mean daily minimum temperature	in °C	21.1	21.7	23.3	24.4	24.4	23.9	23.9	23.3	23.3	23.3	22.8	21.7	23.3	31	3
4 Absolute maximum temperature	in °C	36.7	38.9	39.4	40.0	38.9	37.8	34.4	35.0	35.6	34.4	35.0	36.1	40.0	31	4
5 Absolute minimum temperature	in °C	13.9	16.1	17.8	20.0	21.1	20.6	19.4	20.0	20.6	20.0	17.8	13.9	13.9	31	5
6 Mean relative humidity	in %	65	61	61	62	71	78	80	78	79	79	75	70	71	20	6
7 Mean precipitation	in mm	15	3	13	43	221	330	315	269	335	269	114	58	1985	33	7
8 Maximum precipitation	in mm	111	10	129	178	561	522	595	499	507	603	286	173	2718	32	8
9 Minimum precipitation	in mm	0	0	0	0	49	180	98	118	204	82	3	13	1552	32	9
10 Maximum precipitation in 24 h	in mm	69	10	104	89	104	137	150	178	132	114	132	71	178	33	10
11 Mean number of days with precipitation	> 1,0 mm	2	1	2	4	16	21	23	21	21	20	11	7	139	20	11
12 Mean duration of sunshine	in h	143	162	158	165	124	105	121	140	156	115	111	121	1621	30	12
13 Mean quantity of radiation	in ly / day															13
14 Mean potential evaporation	in mm	120	125	150	159	161	150	149	150	139	138	130	116	1687	35	14
15 Mean windspeed	in m / sec															15
16 Mean predominent direction of the wind		E	SE	SE	SE	SW	SW-W	SW	SW	SW	SW	var	N-E			16

291 Station / Country Sandakan (Borneo) / Malaysia

Location 5°50'N / 118°07'E Height above sealevel 46 m Climate symbol: Köppen Af Troll V,1

		J	F	M	A	M	J	J	A	S	O	N	D	year	P	
1 Mean daily temperature	in °C	26.4	26.7	27.2	28.1	28.1	27.8	27.8	27.8	27.8	27.5	27.2	26.7	27.4	45	1
2 Mean daily maximum temperature	in °C	29.4	30.0	30.6	31.7	31.7	31.7	31.7	31.7	31.7	31.1	30.8	30.0	31.1	45	2
3 Mean daily minimum temperature	in °C	23.3	23.3	23.9	24.4	24.4	23.9	23.9	23.9	23.9	23.9	23.9	23.3	23.9	45	3
4 Absolute maximum temperature	in °C	32.8	32.8	34.4	35.0	35.0	37.2	35.6	35.0	36.1	34.4	33.9	33.3	37.2	32	4
5 Absolute minimum temperature	in °C	21.1	21.1	21.7	21.7	22.2	21.7	21.1	21.7	21.1	21.1	21.7	21.7	21.1	32	5
6 Mean relative humidity	in %	84	83	83	81	81	81	80	80	79	81	84	84	82	9	6
7 Mean precipitation	in mm	483	277	218	114	157	188	170	201	238	259	368	470	3141	46	7
8 Maximum precipitation	in mm	1094	1079	556	374	366	366	436	381	447	430	680	828	3794	31	8
9 Minimum precipitation	in mm	106	27	44	<1	15	59	56	94	32	60	149	186	2194	31	9
10 Maximum precipitation in 24 h	in mm	368	198	168	114	140	119	89	99	122	142	130	318	368	28	10
11 Mean number of days with precipitation	> 1,0 mm	19	13	13	11	10	14	14	12	15	18	19	20	178	9	11
12 Mean duration of sunshine	in h	158	172	223	264	239	210	208	226	210	205	156	177	2448		12
13 Mean quantity of radiation	in ly / day															13
14 Mean potential evaporation	in mm	135	128	145	153	159	153	156	154	148	148	140	139	1759	46	14
15 Mean windspeed	in m / sec															15
16 Mean predominent direction of the wind																16

292 Station / Country Pinang / Malaysia

Location 5°25'N / 100°19'E Height above sealevel 5 m Climate symbol: Köppen Af Troll V,1

		J	F	M	A	M	J	J	A	S	O	N	D	year	P	
1 Mean daily temperature	in °C	27.5	27.8	28.3	28.3	27.8	27.8	27.8	27.2	27.0	27.2	27.0	27.2	27.6	49	1
2 Mean daily maximum temperature	in °C	32.2	32.8	33.3	32.8	32.2	32.2	32.2	31.7	31.1	31.7	31.1	31.7	32.2	49	2
3 Mean daily minimum temperature	in °C	22.8	22.8	23.3	23.9	23.3	23.3	23.3	22.8	22.8	22.8	22.8	22.8	23.3	49	3
4 Absolute maximum temperature	in °C	36.7	36.1	36.7	36.1	35.6	36.1	35.0	35.6	36.9	34.4	35.0	36.7	36.9	48	4
5 Absolute minimum temperature	in °C	18.9	18.9	19.4	19.4	19.4	20.0	20.6	20.6	20.0	19.4	18.3	19.4	18.3	48	5
6 Mean relative humidity	in %	72	69	70	73	72	72	72	73	75	76	75	72	73	8	6
7 Mean precipitation	in mm	94	79	142	188	272	196	191	295	401	429	302	147	2738	50	7
8 Maximum precipitation	in mm	211	365	351	424	506	309	434	453	783	880	482	326	3710	38	8
9 Minimum precipitation	in mm	<1	3	15	17	64	65	31	92	167	141	46	19	1934	38	9
10 Maximum precipitation in 24 h	in mm	69	142	132	157	135	145	157	168	221	241	183	196	241	48	10
11 Mean number of days with precipitation	> 0,25 mm	8	7	11	14	16	12	12	15	18	21	19	11	164	46	11
12 Mean duration of sunshine	in h	239	229	236	225	202	204	208	192	182	170	168	202	2437		12
13 Mean quantity of radiation	in ly / day															13
14 Mean potential evaporation	in mm	145	136	152	152	154	153	155	150	141	147	141	147	1773		14
15 Mean windspeed	in m / sec															15
16 Mean predominent direction of the wind																16

293 Station / Country Kuala Trengganu / Malaysia

Location 5°20'N / 103°08'E   Height above sealevel 32 m   Climate symbol: Köppen Af   Troll V,1

		J	F	M	A	M	J	J	A	S	O	N	D	year	P	
1 Mean daily temperature	in °C	25.3	25.6	26.7	27.2	27.8	27.2	27.2	27.2	27.0	26.7	26.1	25.3	26.6	12	1
2 Mean daily maximum temperature	in °C	28.3	29.4	30.6	31.7	32.2	31.7	31.7	31.7	31.1	30.6	29.4	28.3	30.6	12	2
3 Mean daily minimum temperature	in °C	22.2	21.7	22.8	22.8	23.3	22.8	22.8	22.8	22.8	22.8	22.8	22.2	22.8	12	3
4 Absolute maximum temperature	in °C	31.1	31.7	33.3	33.9	35.0	35.0	34.4	33.3	33.3	33.3	32.8	31.7	35.0	12	4
5 Absolute minimum temperature	in °C	17.2	18.3	18.9	20.0	20.0	20.6	20.6	20.0	20.0	20.6	21.1	19.4	17.2	12	5
6 Mean relative humidity	in %	85	84	84	83	82	82	84	83	83	85	88	87	84	12	6
7 Mean precipitation	in mm	295	163	160	155	135	109	117	150	191	279	610	554	2918	15	7
8 Maximum precipitation	in mm	1414	640	544	617	355	247	281	328	338	819	1427	1463	4674	45	8
9 Minimum precipitation	in mm	12	8	<1	0	18	11	23	31	41	85	169	57	1746	45	9
10 Maximum precipitation in 24 h	in mm	465	124	140	213	79	64	91	124	122	401	307	290	465	15	10
11 Mean number of days with precipitation	>0,25 mm	19	13	12	10	12	10	12	12	17	20	22	23	182	15	11
12 Mean duration of sunshine	in h	177	217	264	257	229	208	216	207	193	181	143	150	2440	5	12
13 Mean quantity of radiation	in ly / day															13
14 Mean potential evaporation	in mm	114	110	139	147	153	148	149	148	140	141	119	115	1623	11	14
15 Mean windspeed	in m / sec															15
16 Mean predominent direction of the wind																16

294 Station / Country Cameron Highlands / Malaysia

Location 4°28'N / 101°23'E   Height above sealevel 1448 m   Climate symbol: Köppen Cfb   Troll V,1

		J	F	M	A	M	J	J	A	S	O	N	D	year	P	
1 Mean daily temperature	in °C	17.5	17.5	17.8	18.6	18.9	18.3	17.8	17.8	18.1	18.1	17.8	17.5	18.0	28	1
2 Mean daily maximum temperature	in °C	21.7	22.2	22.8	23.3	23.3	23.3	22.8	22.2	22.2	22.2	21.7	21.7	22.2	28	2
3 Mean daily minimum temperature	in °C	13.3	12.8	12.8	13.9	14.4	13.3	12.8	13.3	13.9	13.9	13.3	13.3	13.3	28	3
4 Absolute maximum temperature	in °C	25.0	26.1	26.1	26.7	26.1	26.7	26.1	25.6	25.6	25.0	25.6	25.0	26.7	28	4
5 Absolute minimum temperature	in °C	2.2	4.4	6.1	8.3	7.2	5.0	6.7	7.2	7.8	6.7	5.6	2.2	2.2	28	5
6 Mean relative humidity	in %	86	84	86	88	89	85	86	87	88	89	89	88	87	11	6
7 Mean precipitation	in mm	168	132	216	297	246	140	122	163	262	340	330	229	2645	28	7
8 Maximum precipitation	in mm	410	255	514	539	447	262	424	362	520	608	643	672	3681	28	8
9 Minimum precipitation	in mm	23	23	39	52	116	56	17	37	98	122	113	42	2074	28	9
10 Maximum precipitation in 24 h	in mm	84	66	79	84	69	86	74	122	97	84	107	160	160	28	10
11 Mean number of days with precipitation	>0,25 mm	17	14	19	23	22	16	15	19	22	26	24	21	238	28	11
12 Mean duration of sunshine	in h	128	132	166	150	134	159	160	148	115	120	96	119	1628	5	12
13 Mean quantity of radiation	in ly / day															13
14 Mean potential evaporation	in mm	63	58	69	72	75	70	68	69	66	69	66	64	809	10	14
15 Mean windspeed	in m / sec															15
16 Mean predominent direction of the wind																16

295 Station / Country Malacca / Malaysia

Location 2°12'N / 102°16'E   Height above sealevel 45 m   Climate symbol: Köppen Af   Troll V,1

		J	F	M	A	M	J	J	A	S	O	N	D	year	P	
1 Mean daily temperature	in °C	26.7	27.0	27.0	27.2	27.2	27.0	26.7	26.7	26.7	26.7	26.7	26.7	26.9	48	1
2 Mean daily maximum temperature	in °C	31.1	31.7	31.7	31.7	31.7	31.1	31.1	31.1	31.1	31.1	31.1	31.1	31.1	48	2
3 Mean daily minimum temperature	in °C	22.2	22.2	22.2	22.8	22.8	22.8	22.2	22.2	22.2	22.2	22.2	22.2	22.2	48	3
4 Absolute maximum temperature	in °C	33.9	35.6	35.6	35.6	36.7	35.6	37.2	34.4	37.2	33.3	33.9	34.4	37.2	35	4
5 Absolute minimum temperature	in °C	18.3	17.2	18.3	19.4	19.4	19.4	17.8	16.1	18.3	17.8	16.7	17.8	16.1	35	5
6 Mean relative humidity	in %	79	79	82	85	85	83	84	84	84	84	84	82	83	20	6
7 Mean precipitation	in mm	99	94	124	188	173	201	198	262	224	256	221	165	2205	48	7
8 Maximum precipitation	in mm	216	298	411	391	368	386	399	355	437	477	493	436	2723	44	8
9 Minimum precipitation	in mm	5	1	22	33	25	59	50	64	69	58	80	39	1408	44	9
10 Maximum precipitation in 24 h	in mm	142	147	191	264	236	155	236	267	254	180	191	254	267	48	10
11 Mean number of days with precipitation	>0,25 mm	9	8	10	13	12	12	12	15	14	16	17	14	152	27	11
12 Mean duration of sunshine	in h	188	196	239	217	213	210	205	199	190	194	179	186	2414	5	12
13 Mean quantity of radiation	in ly / day															13
14 Mean potential evaporation	in mm	138	130	148	141	146	139	140	138	132	140	134	140	1666	11	14
15 Mean windspeed	in m / sec															15
16 Mean predominent direction of the wind																16

296 Station / Country Kuching (Borneo) / Malaysia

Location 1°29'N / 110°20'E   Height above sealevel 26 m   Climate symbol: Köppen Af   Troll V,1

		J	F	M	A	M	J	J	A	S	O	N	D	year	P	
1 Mean daily temperature	in °C	25.9	26.1	27.0	27.5	27.5	27.8	27.2	27.5	27.0	27.2	26.7	26.4	27.0	5	1
2 Mean daily maximum temperature	in °C	29.4	30.0	31.1	32.2	32.2	32.8	32.2	32.8	31.7	31.7	31.1	30.6	31.7	5	2
3 Mean daily minimum temperature	in °C	22.2	22.2	22.8	22.8	22.8	22.8	22.2	22.2	22.2	22.8	22.2	22.2	22.2	6	3
4 Absolute maximum temperature	in °C	33.9	33.9	33.9	35.0	35.0	35.6	34.4	34.4	34.4	33.9	33.3	33.3	36.1	5	4
5 Absolute minimum temperature	in °C	18.9	17.8	20.6	20.6	21.1	19.4	19.4	18.3	18.9	20.6	20.6	20.6	17.8	6	5
6 Mean relative humidity	in %	75	74	73	71	70	66	66	68	70	71	74	75	71	7	6
7 Mean precipitation	in mm	610	511	328	279	262	180	196	234	218	267	358	462	3905	19	7
8 Maximum precipitation	in mm	1207	1559	598	458	432	324	446	387	409	562	634	881	5293	31	8
9 Minimum precipitation	in mm	146	100	167	85	151	115	27	66	96	143	215	264	3098	31	9
10 Maximum precipitation in 24 h	in mm	318	163	130	109	104	61	99	66	94	86	269	170	318	9	10
11 Mean number of days with precipitation	>0,1 mm	25	21	21	19	16	16	18	19	23	24	25		247	31	11
12 Mean duration of sunshine	in h	121	124	155	165	183	183	177	167	144	146	135	136	1836		12
13 Mean quantity of radiation	in ly / day															13
14 Mean potential evaporation	in mm	128	116	144	145	150	148	147	150	139	147	139	134	1707	5	14
15 Mean windspeed	in m / sec															15
16 Mean predominent direction of the wind																16

297 Station/Country Pulau Labuan/Malaysia

Location 5°17'N/115°16'E — Height above sealevel 18 m — Climate symbol: Köppen Af — Troll V,1

#		unit	J	F	M	A	M	J	J	A	S	O	N	D	year	P	
1	Mean daily temperature	in °C	27,2	27,2	27,5	28,1	28,1	27,8	28,1	27,8	27,5	27,5	27,5	27,2	27,7	21	1
2	Mean daily maximum temperature	in °C	30,0	30,0	30,6	31,7	31,7	31,1	31,1	31,1	30,6	30,6	30,6	30,0	30,8	21	2
3	Mean daily minimum temperature	in °C	24,4	24,4	24,4	24,4	24,4	24,4	25,0	24,4	24,4	24,4	24,4	24,4	24,4	21	3
4	Absolute maximum temperature	in °C	33,3	33,3	33,9	35,0	35,6	34,4	33,9	34,4	33,3	34,4	33,9	33,9	35,6	20	4
5	Absolute minimum temperature	in °C	20,0	20,0	17,2	15,6	15,0	15,6	20,0	20,0	17,8	17,2	20,6	20,6	15,0	20	5
6	Mean relative humidity	in %	83	84	83	82	82	82	82	79	80	81	81	82	82	10	6
7	Mean precipitation	in mm	112	117	150	297	345	351	318	297	417	465	419	284	3572	14	7
8	Maximum precipitation	in mm	516	869	254	333	548	392	636	500	823	749	721	499	4522	11	8
9	Minimum precipitation	in mm	31	3	10	36	77	174	81	125	23	79	224	68	2429	11	9
10	Maximum precipitation in 24 h	in mm	130	104	127	155	137	279	147	185	236	157	130	124	279	14	10
11	Mean number of days with precipitation	> 0,25 mm	9	11	10	15	19	16	15	17	18	21	21	19	191	14	11
12	Mean duration of sunshine	in h	229	213	245	258	229	213	223	205	192	226	207	220	2660	30	12
13	Mean quantity of radiation	in ly/day															13
14	Mean potential evaporation	in mm	144	131	148	153	159	151	159	154	145	148	143	144	1779	21	14
15	Mean windspeed	in m/sec															15
16	Mean predominent direction of the wind																16

298 Station/Country Singapore/Singapore

Location 1°18'N/103°50'E — Height above sealevel 10 m — Climate symbol: Köppen Af — Troll V,1

#		unit	J	F	M	A	M	J	J	A	S	O	N	D	year	P	
1	Mean daily temperature	in °C	26,4	27,0	27,5	27,5	27,8	27,5	27,5	27,2	27,2	27,0	27,0	27,0	27,2	39	1
2	Mean daily maximum temperature	in °C	30,0	31,1	31,1	31,1	31,7	31,1	31,1	30,6	30,6	30,6	30,6	30,6	30,6	39	2
3	Mean daily minimum temperature	in °C	22,8	22,8	23,9	23,9	23,9	23,9	23,9	23,9	23,9	23,9	23,3	23,3	23,3	39	3
4	Absolute maximum temperature	in °C	33,9	34,4	34,4	35,0	36,1	35,0	33,9	33,9	33,9	33,9	33,3	33,9	36,1	39	4
5	Absolute minimum temperature	in °C	20,0	18,9	19,4	21,1	21,1	21,1	21,1	20,6	20,6	20,6	20,6	20,6	18,9	39	5
6	Mean relative humidity	in %	80	74	73	76	76	78	78	75	76	75	77	80	76	10	6
7	Mean precipitation	in mm	251	173	193	188	173	173	170	196	178	208	254	256	2413	64	7
8	Maximum precipitation	in mm	819	567	528	455	387	379	527	527	427	497	522	671	3452	103	8
9	Minimum precipitation	in mm	22	8	19	38	58	57	25	18	40	32	91	63	563	103	9
10	Maximum precipitation in 24 h	in mm	218	140	168	135	216	99	185	107	99	152	183	236	236	50	10
11	Mean number of days with precipitation	> 0,25 mm	17	11	14	15	15	13	13	14	14	16	18	19	179	49	11
12	Mean duration of sunshine	in h	158	179	191	178	182	180	183	187	161	138	138	171		35	12
13	Mean quantity of radiation	in ly/day	458	506	519	495	458	449	452	465	477	454	427	387	462	11	13
14	Mean potential evaporation	in mm	135	130	147	145	151	145	150	150	143	147	140	140	1723	28	14
15	Mean windspeed	in m/sec															15
16	Mean predominent direction of the wind		N	NE	NE	NE	S	S	S	S	S	W	NE	NE		16	16

299 Station/Country Tarakan (Borneo)/Indonesia

Location 3°19'N/117°36'E — Height above sealevel 12 m — Climate symbol: Köppen Af — Troll V,1

#		unit	J	F	M	A	M	J	J	A	S	O	N	D	year	P	
1	Mean daily temperature	in °C	26,1	26,4	26,7	27,0	27,0	26,7	27,0	27,0	27,0	27,0	26,7	26,7	26,8	19	1
2	Mean daily maximum temperature	in °C	29,4	30,0	30,0	30,0	30,6	30,0	30,6	30,6	30,6	30,6	30,0	30,0	30,0	19	2
3	Mean daily minimum temperature	in °C	22,8	22,8	23,3	23,9	23,3	23,3	23,3	23,3	23,3	23,3	23,3	23,3	23,3	19	3
4	Absolute maximum temperature	in °C	32,8	34,4	32,8	32,8	33,3	32,8	33,9	33,3	33,3	33,3	32,8	34,4	34,4	19	4
5	Absolute minimum temperature	in °C	20,0	19,4	20,0	20,6	21,7	19,4	20,6	20,0	20,0	20,0	21,1	19,4	19,4	19	5
6	Mean relative humidity	in %	79	78	78	78	79	79	78	78	77	77	79	80	78	19	6
7	Mean precipitation	in mm	277	259	356	353	343	320	262	315	295	363	386	340	3869	31	7
8	Maximum precipitation	in mm															8
9	Minimum precipitation	in mm															9
10	Maximum precipitation in 24 h	in mm	64	64	69	76	76	66	58	76	69	74	79	69	79	31	10
11	Mean number of days with precipitation	> 0,1 mm	17	15	19	19	19	18	15	16	16	19	20	20	213	31	11
12	Mean duration of sunshine	in h	124	120	130	132	140	144	171	171	165	155	138	140	1730	5	12
13	Mean quantity of radiation	in ly/day															13
14	Mean potential evaporation	in mm	128	119	135	134	143	134	140	138	134	138	131	131	1605	31	14
15	Mean windspeed	in m/sec															15
16	Mean predominent direction of the wind																16

300 Station/Country Takingeun (Sumatra)/Indonesia

Location 4°40'N/95°50'E — Height above sealevel 1205 m — Climate symbol: Köppen Am — Troll V,1

#		unit	J	F	M	A	M	J	J	A	S	O	N	D	year	P	
1	Mean daily temperature	in °C	20,9	21,1	21,4	21,7	21,7	21,1	20,6	20,6	20,6	20,9	20,6	20,9	21,0	10	1
2	Mean daily maximum temperature	in °C	25,6	26,7	26,7	26,7	27,2	26,7	26,1	26,1	25,6	25,6	25,0	25,6	26,1	10	2
3	Mean daily minimum temperature	in °C	16,1	15,6	16,1	16,7	16,1	15,6	15,0	15,0	15,6	16,1	16,1	16,1	15,6	10	3
4	Absolute maximum temperature	in °C	30,6	30,6	30,6	31,1	30,6	31,1	31,1	31,1	30,6	30,0	29,4	32,2	32,2	10	4
5	Absolute minimum temperature	in °C	9,4	11,1	11,1	13,3	11,1	9,4	10,0	11,1	8,9	11,7	12,2	10,0	8,9	10	5
6	Mean relative humidity	in %	74	72	73	75	73	70	69	70	72	75	78	75	73	10	6
7	Mean precipitation	in mm	168	119	183	165	119	58	69	84	142	196	221	224	1748	37	7
8	Maximum precipitation	in mm															8
9	Minimum precipitation	in mm															9
10	Maximum precipitation in 24 h	in mm	43	33	43	41	30	20	18	25	38	41	46	51	51	37	10
11	Mean number of days with precipitation	> 0,1 mm	12	9	13	4	12	8	7	10	13	17	17	16	148	37	11
12	Mean duration of sunshine	in h	152	157	167	156	158	153	164	155	135	130	123	133	1783	6	12
13	Mean quantity of radiation	in ly/day															13
14	Mean potential evaporation	in mm	75	72	80	81	84	78	77	76	72	74	71	74	914	33	14
15	Mean windspeed	in m/sec															15
16	Mean predominent direction of the wind																16

301 Station/Country Medan (Sumatra)/Indonesia

Location 3°35'N/98°41'E Height above sealevel 23 m Climate symbol: Köppen Af Troll V,1

		J	F	M	A	M	J	J	A	S	O	N	D	year	P	
1 Mean daily temperature	in °C	25,6	26,1	26,7	27,2	27,2	27,0	27,0	27,0	26,7	26,1	26,1	25,9	26,6	17	1
2 Mean daily maximum temperature	in °C	29,4	30,6	31,1	31,7	31,7	31,7	31,7	31,7	31,1	30,0	30,0	29,4	30,6	17	2
3 Mean daily minimum temperature	in °C	21,7	21,7	22,2	22,8	22,8	22,2	22,2	22,2	22,2	22,2	22,2	22,2	22,2	17	3
4 Absolute maximum temperature	in °C	33,9	34,4	35,0	35,0	35,6	35,0	35,6	35,0	35,6	33,9	33,9	34,4	35,6	16	4
5 Absolute minimum temperature	in °C	18,3	18,3	18,3	19,4	18,3	17,2	17,8	18,3	18,9	17,8	15,6	18,3	15,6	16	5
6 Mean relative humidity	in %	80	77	77	78	78	78	76	77	80	82	79	78	78	13	6
7 Mean precipitation	in mm	137	91	104	132	175	132	135	178	211	259	246	229	2029	50	7
8 Maximum precipitation	in mm															8
9 Minimum precipitation	in mm															9
10 Maximum precipitation in 24 h	in mm	48	38	38	43	53	43	43	43	53	64	56	61	64	63	10
11 Mean number of days with precipitation	> 0,5 mm	11	7	8	10	12	9	9	13	14	17	17	15	142	50	11
12 Mean duration of sunshine	in h	112	123	136	132	143	150	158	143	126	118	108	102	1551	28	12
13 Mean quantity of radiation	in ly/day															13
14 Mean potential evaporation	in mm	115	113	132	137	145	138	140	133	123	117	114	115	1522	63	14
15 Mean windspeed	in m/sec															15
16 Mean predominent direction of the wind																16

302 Station/Country Manado (Celebes)/Indonesia

Location 1°30'N/124°50'E Height above sealevel 2 m Climate symbol: Köppen Af Troll V,1

		J	F	M	A	M	J	J	A	S	O	N	D	year	P	
1 Mean daily temperature	in °C	26,1	26,1	26,1	26,7	27,0	27,0	26,7	27,5	27,2	27,2	27,2	26,7	26,8	12	1
2 Mean daily maximum temperature	in °C	29,4	29,4	29,4	30,6	30,6	30,6	30,6	31,7	31,7	31,7	31,7	30,0	30,6	12	2
3 Mean daily minimum temperature	in °C	22,8	22,8	22,8	22,8	23,3	23,3	22,8	23,3	22,8	22,8	22,8	23,3	22,8	12	3
4 Absolute maximum temperature	in °C	32,8	33,3	32,8	33,3	33,3	33,9	33,9	35,0	35,6	34,4	33,3	32,8	35,6	12	4
5 Absolute minimum temperature	in °C	20,0	20,6	20,0	18,3	20,6	20,6	19,4	17,4	18,9	18,3	18,9	21,1	17,4	12	5
6 Mean relative humidity	in %	84	83	83	83	81	80	75	72	75	77	82	83	80	13	6
7 Mean precipitation	in mm	465	358	305	198	160	183	119	97	86	122	218	371	2608	50	7
8 Maximum precipitation	in mm															8
9 Minimum precipitation	in mm															9
10 Maximum precipitation in 24 h	in mm	185	175	300	66	84	198	122	91	66	74	124	183	300	10	10
11 Mean number of days with precipitation	> 0,5 mm	21	18	16	13	13	13	9	9	8	10	14	19	163	50	11
12 Mean duration of sunshine	in h	130	120	140	150	161	147	167	183	174	174	150	146	1842	18	12
13 Mean quantity of radiation	in ly/day															13
14 Mean potential evaporation	in mm	128	117	129	127	137	129	138	144	138	141	132	133	1593	63	14
15 Mean windspeed	in m/sec															15
16 Mean predominent direction of the wind																16

303 Station/Country Pontianak (Borneo)/Indonesia

Location 0°01'S/109°20'E Height above sealevel 3 m Climate symbol: Köppen Af Troll V,1

		J	F	M	A	M	J	J	A	S	O	N	D	year	P	
1 Mean daily temperature	in °C	27,0	28,1	27,8	27,8	28,1	28,1	27,5	27,8	28,1	27,8	27,5	27,2	27,7	20	1
2 Mean daily maximum temperature	in °C	30,6	31,7	31,7	31,7	32,2	32,2	31,7	32,2	32,2	31,7	31,1	30,6	31,7	20	2
3 Mean daily minimum temperature	in °C	23,3	24,4	23,9	23,9	23,9	23,9	23,3	23,3	23,9	23,9	23,9	23,9	23,9	20	3
4 Absolute maximum temperature	in °C	34,4	34,4	35,6	35,0	35,6	35,0	35,0	35,0	35,0	35,0	35,0	33,9	35,6	20	4
5 Absolute minimum temperature	in °C	20,6	20,6	20,0	21,7	21,1	21,1	20,0	20,6	20,6	21,7	21,7	21,1	20,0	20	5
6 Mean relative humidity	in %	80	79	79	80	80	80	78	78	78	79	81	81	79	20	6
7 Mean precipitation	in mm	274	208	241	277	282	221	165	203	229	339	389	323	3151	63	7
8 Maximum precipitation	in mm															8
9 Minimum precipitation	in mm															9
10 Maximum precipitation in 24 h	in mm	66	58	66	61	66	66	56	56	58	74	74	69	74	63	10
11 Mean number of days with precipitation	> 1,0 mm	17	13	15	16	16	13	10	12	13	19	21	19	183	63	11
12 Mean duration of sunshine	in h	146	185	186	192	192	213	220	202	165	167	174	180	2222	30	12
13 Mean quantity of radiation	in ly/day															13
14 Mean potential evaporation	in mm	134	128	144	140	146	142	145	146	139	140	130	134	1668	63	14
15 Mean windspeed	in m/sec															15
16 Mean predominent direction of the wind																16

304 Station/Country Padang (Sumatra)/Indonesia

Location 0°56'S/100°22'E Height above sealevel 7 m Climate symbol: Köppen Af Troll V,1

		J	F	M	A	M	J	J	A	S	O	N	D	year	P	
1 Mean daily temperature	in °C	27,0	27,0	27,0	27,2	27,5	27,0	25,0	25,0	26,7	26,7	26,7	26,7	26,6	17	1
2 Mean daily maximum temperature	in °C	30,6	31,7	31,7	31,7	32,2	31,7	31,7	32,2	32,2	31,7	31,1	30,6	31,7	20	2
3 Mean daily minimum temperature	in °C	23,3	24,4	23,9	23,9	23,9	23,9	23,3	23,3	23,9	23,9	23,9	23,9	23,9	20	3
4 Absolute maximum temperature	in °C	33,9	34,4	33,9	33,3	33,9	33,9	33,3	33,3	32,8	33,3	32,8	32,8	34,4	21	4
5 Absolute minimum temperature	in °C	21,1	20,6	21,1	21,7	21,7	20,0	21,1	20,6	21,1	21,1	21,1	21,1	20,0	21	5
6 Mean relative humidity	in %	78	77	77	78	76	75	74	75	76	78	79	79	77	13	6
7 Mean precipitation	in mm	351	259	307	363	315	307	277	348	152	495	518	480	3172	50	7
8 Maximum precipitation	in mm															8
9 Minimum precipitation	in mm															9
10 Maximum precipitation in 24 h	in mm	91	79	81	97	94	104	81	109	109	112	112	104	112	63	10
11 Mean number of days with precipitation	> 0,5 mm	16	13	15	17	14	12	12	14	16	20	21	20	190	50	11
12 Mean duration of sunshine	in h	208	221	223	225	248	231	229	228	207	202	186	192	1598	30	12
13 Mean quantity of radiation	in ly/day															13
14 Mean potential evaporation	in mm	136	127	141	138	144	138	138	138	130	132	128	132	1622	63	14
15 Mean windspeed	in m/sec															15
16 Mean predominent direction of the wind																16

305 Station / Country Balikpapan (Borneo) / Indonesia

Location 1°17'S/116°51'E Height above sealevel 7 m Climate symbol: Köppen Af Troll V.1

		J	F	M	A	M	J	J	A	S	O	N	D	year	P	
1	Mean daily temperature	in °C	26,1	26,4	26,4	26,1	26,4	26,1	25,6	26,1	26,1	26,4	26,1	26,1	26,1	6	1
2	Mean daily maximum temperature	in °C	29,4	30,0	30,0	29,4	29,4	28,9	28,3	28,9	28,9	29,4	29,4	29,4	29,4	6	2
3	Mean daily minimum temperature	in °C	22,8	22,8	22,8	22,8	23,3	23,3	22,8	23,3	23,3	23,3	22,8	22,8	22,8	6	3
4	Absolute maximum temperature	in °C	33,3	33,3	33,3	32,2	32,8	32,2	29,4	30,0	31,1	32,8	33,3	33,3	33,3	6	4
5	Absolute minimum temperature	in °C	21,1	21,7	21,1	21,1	21,7	15,6	20,0	20,6	19,4	20,6	21,1	21,1	15,6	6	5
6	Mean relative humidity	in %	82	81	81	82	83	83	83	80	77	78	80	79	81	7	6
7	Mean precipitation	in mm	201	175	231	208	231	193	180	163	140	132	168	206	2228	43	7
8	Maximum precipitation	in mm															8
9	Minimum precipitation	in mm															9
10	Maximum precipitation in 24 h	in mm	51	43	58	56	64	53	51	51	51	48	48	53	64	43	10
11	Mean number of days with precipitation	> 0,5 mm	14	13	15	13	13	12	11	11	9	9	12	15	147	43	11
12	Mean duration of sunshine	in h															12
13	Mean quantity of radiation	in ly / day															13
14	Mean potential evaporation	in mm	121	110	121	119	128	118	122	132	128	130	118	121	1468	42	14
15	Mean windspeed	in m / sec															15
16	Mean predominent direction of the wind																16

306 Station / Country Tandjungpandan (Belitung) / Indonesia

Location 2°45'S/107°39'E Height above sealevel 3 m Climate symbol: Köppen Af Troll V.1

		J	F	M	A	M	J	J	A	S	O	N	D	year	P	
1	Mean daily temperature	in °C	26,1	26,4	26,4	26,7	27,0	26,4	26,7	26,7	26,7	26,7	26,1	26,1	26,5	6	1
2	Mean daily maximum temperature	in °C	28,3	28,9	29,4	30,0	30,6	30,0	30,6	31,1	31,1	30,6	29,6	28,9	30,0	6	2
3	Mean daily minimum temperature	in °C	23,9	23,9	23,3	23,3	23,3	22,8	22,8	22,2	22,2	22,8	22,8	23,3	22,8	6	3
4	Absolute maximum temperature	in °C	31,7	30,6	31,7	32,8	32,8	32,8	33,3	33,3	34,4	33,9	33,3	31,7	34,4	6	4
5	Absolute minimum temperature	in °C	22,2	21,7	21,7	21,7	21,7	20,6	19,4	18,9	18,9	18,9	20,0	21,1	18,9	6	5
6	Mean relative humidity	in %	84	84	86	89	88	86	81	79	81	85	90	86	85	6	6
7	Mean precipitation	in mm	277	165	193	267	256	191	170	140	163	274	371	404	2871	63	7
8	Maximum precipitation	in mm															8
9	Minimum precipitation	in mm															9
10	Maximum precipitation in 24 h	in mm	69	53	51	61	53	48	53	43	48	61	66	76	76	63	10
11	Mean number of days with precipitation	> 0,1 mm	17	11	14	17	17	14	11	10	10	17	21	22	181	63	11
12	Mean duration of sunshine	in h															12
13	Mean quantity of radiation	in ly / day															13
14	Mean potential evaporation	in mm	131	122	131	122	132	122	126	130	126	130	119	128	1519	63	14
15	Mean windspeed	in m / sec															15
16	Mean predominent direction of the wind																16

307 Station / Country Ambon (Moluccas) / Indonesia

Location 3°42'S/128°10'E Height above sealevel 4 m Climate symbol: Köppen Af Troll V.1

		J	F	M	A	M	J	J	A	S	O	N	D	year	P	
1	Mean daily temperature	in °C	27,8	27,8	27,8	27,2	26,4	25,6	25,3	25,3	25,9	26,4	27,5	27,8	26,7	18	1
2	Mean daily maximum temperature	in °C	31,1	31,1	31,1	30,0	28,9	27,8	27,2	27,2	28,3	29,4	31,1	31,1	29,4	18	2
3	Mean daily minimum temperature	in °C	24,4	24,4	24,4	24,4	23,9	23,3	23,3	23,3	23,3	23,3	23,9	24,4	23,9	18	3
4	Absolute maximum temperature	in °C	35,6	35,6	35,0	33,9	32,2	30,6	30,0	30,6	31,1	32,8	34,4	35,6	35,6	18	4
5	Absolute minimum temperature	in °C	22,2	22,8	22,2	21,7	21,1	20,6	20,0	19,4	18,9	18,9	21,1	20,0	18,9	18	5
6	Mean relative humidity	in %	78	77	79	82	83	84	83	82	81	80	79	78	81	15	6
7	Mean precipitation	in mm	127	119	135	279	516	638	602	401	241	155	114	132	3459	50	7
8	Maximum precipitation	in mm															8
9	Minimum precipitation	in mm															9
10	Maximum precipitation in 24 h	in mm	114	89	99	241	249	249	300	249	226	185	99	117	300	41	10
11	Mean number of days with precipitation	> 0,1 mm	13	13	15	19	22	24	23	20	15	13	11	13	201	44	11
12	Mean duration of sunshine	in h	192	185	211	177	158	120	115	112	150	192	219	202	2044	30	12
13	Mean quantity of radiation	in ly / day															13
14	Mean potential evaporation	in mm	149	134	146	133	129	111	109	111	117	133	140	148	1560	63	14
15	Mean windspeed	in m / sec															15
16	Mean predominent direction of the wind																16

308 Station / Country Makasar (Celebes) / Indonesia

Location 5°08'S/119°28'E Height above sealevel 2 m Climate symbol: Köppen Am Troll V.2

		J	F	M	A	M	J	J	A	S	O	N	D	year	P	
1	Mean daily temperature	in °C	26,1	26,4	26,4	26,7	27,0	26,1	25,6	25,6	25,9	26,4	26,7	26,1	26,3	10	1
2	Mean daily maximum temperature	in °C	28,9	28,9	29,4	30,0	30,6	30,0	30,0	30,6	30,6	30,6	30,0	28,9	30,0	10	2
3	Mean daily minimum temperature	in °C	23,3	23,9	23,3	23,3	23,3	22,2	21,1	20,6	21,1	22,2	23,3	23,3	22,8	10	3
4	Absolute maximum temperature	in °C	31,1	31,7	31,7	32,8	32,8	32,8	33,3	34,4	35,0	33,3	32,8	31,7	35,0	10	4
5	Absolute minimum temperature	in °C	21,1	21,1	21,1	20,6	17,2	17,2	17,2	16,7	14,4	18,3	18,9	20,6	14,4	10	5
6	Mean relative humidity	in %	85	83	84	82	81	79	77	76	75	78	81	85	81	11	6
7	Mean precipitation	in mm	686	536	424	150	89	74	36	10	15	43	178	610	2851	50	7
8	Maximum precipitation	in mm															8
9	Minimum precipitation	in mm															9
10	Maximum precipitation in 24 h	in mm	196	193	160	163	79	122	79	10	43	36	160	168	196	10	10
11	Mean number of days with precipitation	> 0,5 mm	25	20	18	10	8	6	4	2	2	5	11	22	133	50	11
12	Mean duration of sunshine	in h	161	168	189	240	245	252	279	304	308	298	273	167	2882	30	12
13	Mean quantity of radiation	in ly / day															13
14	Mean potential evaporation	in mm	133	123	134	136	139	126	121	128	127	144	140	133	1584	63	14
15	Mean windspeed	in m / sec	1,8	1,5	1,3	1,1	1,0	1,1	1,2	1,5	1,5	1,7	1,4	1,5	1,4	10	15
16	Mean predominent direction of the wind																16

309 Station / Country Djakarta (Java) / Indonesia

Location 6°11'S / 106°50'E   Height above sealevel 8 m   Climate symbol: Köppen Am   Troll V,1

		J	F	M	A	M	J	J	A	S	O	N	D	year	P	
1 Mean daily temperature	in °C	26,1	26,1	26,7	27,2	27,2	27,0	26,7	26,7	27,2	27,0	26,7	26,4	26,8	80	1
2 Mean daily maximum temperature	in °C	28,9	28,9	30,0	30,6	30,8	30,6	30,6	30,6	31,1	30,6	30,0	29,4	30,0	80	2
3 Mean daily minimum temperature	in °C	23,3	23,3	23,3	23,9	23,9	23,3	22,8	22,8	23,3	23,3	23,3	23,3	23,3	80	3
4 Absolute maximum temperature	in °C	33,9	33,3	33,3	34,4	33,9	33,9	33,3	34,4	35,6	36,7	35,8	33,9	36,7	80	4
5 Absolute minimum temperature	in °C	20,6	20,6	20,6	20,6	21,1	19,4	19,4	19,4	18,9	20,6	20,0	19,4	18,9	80	5
6 Mean relative humidity	in %	85	85	84	83	82	80	78	76	71	72	80	82	80	80	6
7 Mean precipitation	in mm	300	300	211	147	114	97	64	43	66	112	142	203	1799	78	7
8 Maximum precipitation	in mm	779	635	369	201	185	282	135	171	162	207	303	281			8
9 Minimum precipitation	in mm	83	91	53	34	14	0	0	0	1	13	58			30	9
10 Maximum precipitation in 24 h	in mm	66	71	53	48	43	38	30	20	28	41	43	53	71	78	10
11 Mean number of days with precipitation	> 1,0 mm	18	17	15	11	9	7	5	4	5	8	12	14	125	36	11
12 Mean duration of sunshine	in h	189	182	239	255	260	255	282	295	288	279	231	220	2975	30	12
13 Mean quantity of radiation	in ly / day															13
14 Mean potential evaporation	in mm	128	115	127	135	141	128	130	135	136	144	137	134	1590	63	14
15 Mean windspeed	in m / sec	1,6	1,7	1,5	1,5	1,8	1,8	1,8	1,7	1,8	1,8	1,5	1,5	1,6	10	15
16 Mean predominent direction of the wind		NW	NW	NW	E	E	E	E	E	N	N	N	NW		10	16

310 Station / Country Surabaja (Java) / Indonesia

Location 7°38'S / 112°55'E   Height above sealevel 5 m   Climate symbol: Köppen Aw   Troll V,2

		J	F	M	A	M	J	J	A	S	O	N	D	year	P	
1 Mean daily temperature	in °C	27,2	27,2	27,2	27,2	26,9	26,1	25,9	25,9	26,7	27,5	28,1	27,2	26,9	17	1
2 Mean daily maximum temperature	in °C	31,1	31,1	31,1	31,1	31,1	30,6	30,6	30,6	31,7	32,2	32,2	31,1	31,1	17	2
3 Mean daily minimum temperature	in °C	23,3	23,3	23,3	23,3	22,8	21,7	21,1	21,1	21,7	22,8	23,3	23,3	22,8	17	3
4 Absolute maximum temperature	in °C	33,3	34,4	33,9	33,3	33,9	33,9	33,9	34,4	33,9	35,0	35,6	35,0	35,6	17	4
5 Absolute minimum temperature	in °C	21,1	21,1	20,6	18,3	16,7	15,6	14,4	16,1	16,7	17,8	19,4	20,0	14,4	17	5
6 Mean relative humidity	in %	78	78	79	78	75	72	60	66	65	64	68	74	71	10	6
7 Mean precipitation	in mm	226	279	213	137	94	56	25	5	5	18	81	165	1284	50	7
8 Maximum precipitation	in mm															8
9 Minimum precipitation	in mm															9
10 Maximum precipitation in 24 h	in mm	160	130	137	117	183	99	114	43	61	81	114	132	183	50	10
11 Mean number of days with precipitation	> 0,5 mm	17	17	15	9	6	5	2	<1	<1	1	5	13	92	50	11
12 Mean duration of sunshine	in h	143	92	155	201	223	186	245	242	231	248	219	146	2331		12
13 Mean quantity of radiation	in ly / day															13
14 Mean potential evaporation	in mm	144	128	140	139	142	132	126	132	139	152	151	151	1676	57	14
15 Mean windspeed	in m / sec	3,7	3,0	2,9	2,2	3,0	3,1	3,6	4,3	4,3	3,8	2,4	2,6	3,2	6	15
16 Mean predominent direction of the wind																16

311 Station / Country Surakarta (Java) / Indonesia

Location 7°45'S / 110°36'E   Height above sealevel 150 m   Climate symbol: Köppen Am   Troll V,1

		J	F	M	A	M	J	J	A	S	O	N	D	year	P	
1 Mean daily temperature	in °C	25,6	25,6	25,9	26,4	26,1	26,5	25,0	25,3	26,1	26,7	26,1	25,9	25,9	14	1
2 Mean daily maximum temperature	in °C															2
3 Mean daily minimum temperature	in °C															3
4 Absolute maximum temperature	in °C	31,7	32,2	32,8	33,3	32,8	32,8	32,2	32,8	34,4	35,0	35,0	32,8	35,0	14	4
5 Absolute minimum temperature	in °C	20,0	20,0	18,9	20,0	18,3	16,7	17,2	16,7	17,8	18,9	20,0	20,0	16,7	14	5
6 Mean relative humidity	in %	81	81	78	75	74	72	71	68	68	69	75	79	74	14	6
7 Mean precipitation	in mm	307	287	239	185	119	79	33	36	33	91	203	236	1849	37	7
8 Maximum precipitation	in mm															8
9 Minimum precipitation	in mm															9
10 Maximum precipitation in 24 h	in mm	64	58	56	48	41	30	15	18	15	33	51	53	64	37	10
11 Mean number of days with precipitation	> 0,1 mm	19	18	17	13	9	7	3	3	4	7	14	17	131	37	11
12 Mean duration of sunshine	in h															12
13 Mean quantity of radiation	in ly / day															13
14 Mean potential evaporation	in mm	125	112	132	131	127	126	112	116	126	146	131	136	1522	37	14
15 Mean windspeed	in m / sec															15
16 Mean predominent direction of the wind																16

312 Station / Country Tombora (Sumbawa) / Indonesia

Location 8°12'S / 117°50'E   Height above sealevel 500 m   Climate symbol: Köppen Af   Troll V,1

		J	F	M	A	M	J	J	A	S	O	N	D	year	P	
1 Mean daily temperature	in °C	22,2	22,2	22,5	22,5	22,2	21,7	21,4	21,1	21,7	22,8	22,8	22,5	22,1	4	1
2 Mean daily maximum temperature	in °C	24,4	24,4	25,0	25,6	25,0	25,0	25,0	25,0	25,6	26,1	25,6	25,0	25,0	4	2
3 Mean daily minimum temperature	in °C	20,0	20,0	20,0	19,4	19,4	18,3	17,8	17,2	17,8	19,4	20,0	20,0	18,9	4	3
4 Absolute maximum temperature	in °C	28,3	28,3	28,9	28,3	27,8	27,2	28,3	28,9	28,9	28,9	28,9	27,2	28,9	4	4
5 Absolute minimum temperature	in °C	17,2	17,2	18,3	17,8	15,0	15,6	14,4	14,4	15,6	17,8	18,7	18,7	14,4	4	5
6 Mean relative humidity	in %															6
7 Mean precipitation	in mm	937	574	564	320	142	157	71	79	71	89	292	462	3752	10	7
8 Maximum precipitation	in mm															8
9 Minimum precipitation	in mm															9
10 Maximum precipitation in 24 h	in mm	201	97	109	84	43	48	30	48	33	33	51	84	201	10	10
11 Mean number of days with precipitation	> 0,1 mm	24	21	20	16	10	7	5	3	3	6	17	21	153	10	11
12 Mean duration of sunshine	in h	78	87	118	105	121	114	118	118	120	133	114	105	1331	9	12
13 Mean quantity of radiation	in ly / day															13
14 Mean potential evaporation	in mm	89	82	91	87	86	79	75	76	81	95	93	93	1027		14
15 Mean windspeed	in m / sec															15
16 Mean predominent direction of the wind																16

313 Station/Country Kupang (Timor)/Indonesia

Location 10°10'S/123°34'E — Height above sealevel 45 m — Climate symbol: Köppen Aw — Troll V.2

		J	F	M	A	M	J	J	A	S	O	N	D	year	P
1 Mean daily temperature	in °C	27,2	27,2	27,0	27,2	27,0	26,4	26,1	26,4	27,2	28,1	28,3	27,8	27,2	21
2 Mean daily maximum temperature	in °C	30,8	30,8	30,8	31,7	31,7	31,1	31,1	31,7	32,8	33,3	33,3	31,7	31,7	21
3 Mean daily minimum temperature	in °C	23,9	23,9	23,3	22,8	22,2	21,7	21,1	21,7	21,7	22,8	23,3	23,9	22,8	21
4 Absolute maximum temperature	in °C	35,0	34,4	35,6	36,1	35,6	34,4	35,0	36,7	37,2	38,3	38,3	36,7	38,7	21
5 Absolute minimum temperature	in °C	21,1	20,0	20,6	17,2	17,8	15,6	15,6	15,6	16,7	18,3	20,0	21,1	15,6	21
6 Mean relative humidity	in %	78	79	78	68	61	59	56	53	53	58	61	74	65	21
7 Mean precipitation	in mm	289	366	221	84	28	10	5	3	3	18	89	248	1440	50
8 Maximum precipitation	in mm														
9 Minimum precipitation	in mm														
10 Maximum precipitation in 24 h	in mm	307	249	150	165	76	46	74	25	18	101	104	157	307	50
11 Mean number of days with precipitation	> 0,5 mm	18	16	12	5	3	1	<1	<1	<1	1	7	15	79	50
12 Mean duration of sunshine	in h	189	193	223	267	276	276	288	304	306	288	264	205	3066	30
13 Mean quantity of radiation	in ly/day														
14 Mean potential evaporation	in mm	149	132	143	135	134	119	115	121	136	152	154	155	1645	63
15 Mean windspeed	in m/sec														
16 Mean predominent direction of the wind															

314 Station/Country Dili (Timor)/Portugal

Location 8°35'S/125°35'E — Height above sealevel 0 m — Climate symbol: Köppen Aw — Troll V.2

		J	F	M	A	M	J	J	A	S	O	N	D	year	P
1 Mean daily temperature	in °C	28,3	28,3	28,3	28,3	28,1	27,5	26,7	26,4	28,4	27,2	28,6	28,9	27,8	10
2 Mean daily maximum temperature	in °C	31,1	31,1	31,7	31,7	31,7	31,1	30,6	30,6	31,1	32,2	32,2	31,1	31,1	10
3 Mean daily minimum temperature	in °C	25,6	25,6	25,0	25,0	24,4	23,9	22,2	22,2	22,2	23,3	25,0	25,6	24,4	10
4 Absolute maximum temperature	in °C	36,1	35,0	36,7	36,1	35,0	36,7	33,3	35,0	33,9	33,9	36,1	35,0	36,7	10
5 Absolute minimum temperature	in °C	21,1	16,2	20,0	21,7	20,6	18,9	16,1	17,2	16,1	18,3	21,1	22,8	16,1	10
6 Mean relative humidity	in %	75	76	77	75	70	67	67	67	66	68	68	72	71	20
7 Mean precipitation	in mm	127	119	137	109	86	25	13	5	8	23	51	140	843	23
8 Maximum precipitation	in mm														
9 Minimum precipitation	in mm														
10 Maximum precipitation in 24 h	in mm	91	130	112	36	274	58	20	8	18	43	66	76	274	10
11 Mean number of days with precipitation	> 1,0 mm	13	13	14	10	6	4	2	1	1	2	7	13	86	23
12 Mean duration of sunshine	in h	202	175	227	234	260	253	275	284	288	302	280	209	2989	
13 Mean quantity of radiation	in ly/day														
14 Mean potential evaporation	in mm	162	148	158	149	152	138	138	133	132	149	161	172	1790	10
15 Mean windspeed	in m/sec														
16 Mean predominent direction of the wind															

315 Station/Country Aparri/Philippines

Location 18°22'N/121°38'E — Height above sealevel 4 m — Climate symbol: Köppen Am — Troll V.1

		J	F	M	A	M	J	J	A	S	O	N	D	year	P
1 Mean daily temperature	in °C	23,2	23,8	25,3	27,1	28,1	28,5	28,1	27,9	27,5	26,5	25,2	23,9	26,3	
2 Mean daily maximum temperature	in °C	27,2	28,3	30,0	32,2	33,9	33,9	32,8	32,8	32,2	30,0	28,3	27,2	30,6	15
3 Mean daily minimum temperature	in °C	20,0	20,6	21,7	22,8	23,9	24,4	24,4	23,9	23,3	22,8	21,1		22,8	15
4 Absolute maximum temperature	in °C	33,4	33,9	36,1	38,4	37,9	38,9	37,1	37,0	36,0	35,0	34,0	33,2	38,9	
5 Absolute minimum temperature	in °C	15,6	14,8	15,8	16,0	20,0	21,1	21,0	20,0	21,0	19,0	18,3	15,0	14,8	
6 Mean relative humidity	in %	83	82	81	78	79	79	79	81	83	84	84	85	81	
7 Mean precipitation	in mm	144	90	55	49	111	144	173	234	295	367	338	218	2118	
8 Maximum precipitation	in mm														
9 Minimum precipitation	in mm														
10 Maximum precipitation in 24 h	in mm	119	72	92	137	131	168	236	806	253	237	310	181	806	
11 Mean number of days with precipitation	> 0,1 mm	16	11	8	6	11	11	13	15	15	19	19	19	163	
12 Mean duration of sunshine	in h														
13 Mean quantity of radiation	in ly/day														
14 Mean potential evaporation	in mm	77	77	112	145	165	165	166	160	145	130	102	86	1530	43
15 Mean windspeed	in m/sec	4	3	4	3	3	3	3	3	3	4	4	4	3	
16 Mean predominent direction of the wind		NE	NE	NE	NE	NE	S	S	S	S	NE	NE	NE		

316 Station/Country Baguio/Philippines

Location 16°25'N/120°36'E — Height above sealevel 1482 m — Climate symbol: Köppen Cwb — Troll V.2

		J	F	M	A	M	J	J	A	S	O	N	D	year	P
1 Mean daily temperature	in °C	16,9	17,4	18,3	18,9	19,2	18,9	18,4	18,0	18,2	18,3	18,1	17,5	18,2	
2 Mean daily maximum temperature	in °C														
3 Mean daily minimum temperature	in °C														
4 Absolute maximum temperature	in °C	29,7	28,3	29,7	29,3	28,4	28,6	26,8	27,0	26,9	27,3	27,8	28,2	29,7	
5 Absolute minimum temperature	in °C	7,3	7,3	10,3	10,0	12,8	11,8	12,5	12,8	12,9	11,3	9,2	7,6	7,3	
6 Mean relative humidity	in %	80	80	80	83	87	89	91	93	92	88	83	80	85	
7 Mean precipitation	in mm	20	19	48	122	352	422	922	1043	650	362	163	56	4178	
8 Maximum precipitation	in mm														
9 Minimum precipitation	in mm														
10 Maximum precipitation in 24 h	in mm	92	48	70	129	416	497	880	643	800	690	407	184	880	
11 Mean number of days with precipitation	> 0,1 mm	4	4	6	10	20	23	27	27	25	19	10	6	181	
12 Mean duration of sunshine	in h														
13 Mean quantity of radiation	in ly/day														
14 Mean potential evaporation	in mm	56	55	69	73	80	77	75	72	68	68	63	60	816	43
15 Mean windspeed	in m/sec	4	4	3	3	4	3	5	5	4	3	4	3	4	
16 Mean predominent direction of the wind		SE	SE	SE	SE	SE	SE	W	W	W	SE	SE	SE		

317 Station / Country Manila / Philippines

Location 14°35'N / 120°59'E   Height above sealevel 16 m   Climate symbol: Köppen Aw   Troll V,2

		J	F	M	A	M	J	J	A	S	O	N	D	year	P	
1 Mean daily temperature	in °C	25,0	25,5	26,8	28,3	28,6	27,9	27,1	27,0	26,9	26,7	25,9	25,2	26,7		1
2 Mean daily maximum temperature	in °C	30,0	31,1	32,8	33,9	33,9	32,8	31,1	30,6	31,1	31,1	30,6	30,0	31,7	61	2
3 Mean daily minimum temperature	in °C	20,6	20,6	21,7	22,8	23,9	23,9	23,9	23,9	23,9	23,3	22,2	21,1	22,8	61	3
4 Absolute maximum temperature	in °C	35,2	35,6	36,7	38,0	38,6	37,6	36,3	35,2	35,3	35,1	34,4	34,6	38,6		4
5 Absolute minimum temperature	in °C	14,5	15,6	16,2	17,2	20,0	21,6	20,8	20,6	20,8	19,5	16,8	15,7	14,5		5
6 Mean relative humidity	in %	77	73	70	69	75	80	84	84	85	83	82	80	78		6
7 Mean precipitation	in mm	23	11	17	32	128	253	414	437	353	195	138	68	2069		7
8 Maximum precipitation	in mm	195	44	100	174	488	661	810	1465	1469	457	596	347		30	8
9 Minimum precipitation	in mm	0	0	0	0	0	10	119	71	51	10	6	tr		30	9
10 Maximum precipitation in 24 h	in mm	136	44	60	143	218	253	294	324	336	194	278	128	336		10
11 Mean number of days with precipitation	> 0,1 mm	6	3	4	4	12	17	24	23	22	19	14	11	159		11
12 Mean duration of sunshine	in h	177	196	226	258	223	162	133	133	132	158	153	152	2103	30	12
13 Mean quantity of radiation	in ly / day															13
14 Mean potential evaporation	in mm	102	103	141	156	169	159	157	151	141	138	114	106	1637	81	14
15 Mean windspeed	in m / sec	2	2	2	3	3	3	4	3	2	2	2	2	3		15
16 Mean predominent direction of the wind		NE	E	SE	SE	SE	SW	SW	SW	SW	NE	NE	NE			16

318 Station / Country Legazpi / Philippines

Location 13°08'N / 123°44'E   Height above sealevel 19 m   Climate symbol: Köppen Af   Troll V,1

		J	F	M	A	M	J	J	A	S	O	N	D	year	P	
1 Mean daily temperature	in °C	25,7	25,9	26,7	27,7	28,2	28,1	27,4	27,4	27,3	27,1	26,7	26,2	27,0		1
2 Mean daily maximum temperature	in °C	28,9	30,0	31,1	32,2	33,3	32,8	31,7	31,7	31,7	31,1	30,0	29,4	31,1	15	2
3 Mean daily minimum temperature	in °C	23,3	23,3	23,9	25,0	25,0	25,0	24,4	24,4	24,4	23,9	24,4	23,9	24,4	15	3
4 Absolute maximum temperature	in °C	32,7	33,7	33,9	38,1	37,0	37,0	36,6	36,9	36,0	35,3	33,1	37,2			4
5 Absolute minimum temperature	in °C	17,2	16,7	17,0	18,9	19,9	21,4	20,8	21,1	20,4	18,4	18,6	17,5	16,7		5
6 Mean relative humidity	in %	83	82	81	80	81	82	83	83	85	84	84	84	83		6
7 Mean precipitation	in mm	366	265	218	158	178	194	235	209	252	313	479	503	3370		7
8 Maximum precipitation	in mm	579	600	651	498	414	396	337	398	534	769	856	1082		30	8
9 Minimum precipitation	in mm	159	39	51	73	46	73	117	48	93	163	257	97		30	9
10 Maximum precipitation in 24 h	in mm	239	185	154	193	203	378	224	143	218	264	297	248	278		10
11 Mean number of days with precipitation	> 0,1 mm	23	16	17	15	14	15	19	18	19	21	22	24	223		11
12 Mean duration of sunshine	in h															12
13 Mean quantity of radiation	in ly / day															13
14 Mean potential evaporation	in mm	115	109	142	152	165	159	157	154	145	141	132	124	1695	43	14
15 Mean windspeed	in m / sec	4	3	3	3	2	2	3	3	3	3	3	4	3		15
16 Mean predominent direction of the wind		NE	NE	NE	NE	NE	SW	SW	SW	SW	NE	NE	NE			16

319 Station / Country Iloilo / Philippines

Location 10°42'N / 122°34'E   Height above sealevel 14 m   Climate symbol: Köppen Am   Troll V,2

		J	F	M	A	M	J	J	A	S	O	N	D	year	P	
1 Mean daily temperature	in °C	25,8	26,1	26,9	28,2	28,3	27,7	27,2	26,9	27,1	26,9	26,6	26,1	27,0		1
2 Mean daily maximum temperature	in °C	29,4	30,6	31,1	33,3	32,8	31,7	30,6	30,6	31,1	31,1	30,6	30,0	31,1	15	2
3 Mean daily minimum temperature	in °C	22,8	23,3	23,3	24,4	25,0	24,4	24,4	24,4	24,4	23,9	23,9	23,3	23,9	15	3
4 Absolute maximum temperature	in °C	33,2	35,4	35,3	37,0	36,6	35,7	35,2	34,8	36,0	34,7	34,8	33,8	37,0		4
5 Absolute minimum temperature	in °C	18,5	18,0	19,3	21,2	21,7	21,4	19,8	20,1	20,1	20,8	20,3	18,3	18,0		5
6 Mean relative humidity	in %	81	78	76	73	78	81	83	83	84	84	84	83	81		6
7 Mean precipitation	in mm	59	38	36	52	153	265	390	370	294	283	207	121	2248		7
8 Maximum precipitation	in mm															8
9 Minimum precipitation	in mm															9
10 Maximum precipitation in 24 h	in mm	119	80	79	85	224	155	233	222	152	183	238	172	238		10
11 Mean number of days with precipitation	> 0,1 mm	11	7	7	6	14	18	21	20	19	18	15	14	170		11
12 Mean duration of sunshine	in h															12
13 Mean quantity of radiation	in ly / day															13
14 Mean potential evaporation	in mm	119	112	143	153	164	154	154	148	143	142	132	124	1688	43	14
15 Mean windspeed	in m / sec	6	5	6	5	3	3	4	4	3	3	4	5	4		15
16 Mean predominent direction of the wind		NE	NE	NE	NE	NE	SW	SW	SW	SW	NE	NE	NE			16

320 Station / Country Davao / Philippines

Location 7°04'N / 125°36'E   Height above sealevel 20 m   Climate symbol: Köppen Af   Troll V,1

		J	F	M	A	M	J	J	A	S	O	N	D	year	P	
1 Mean daily temperature	in °C	26,3	26,6	27,2	27,7	27,6	27,0	26,8	26,9	27,0	27,1	27,0	26,6	27,0		1
2 Mean daily maximum temperature	in °C															2
3 Mean daily minimum temperature	in °C															3
4 Absolute maximum temperature	in °C	34,7	36,7	36,7	36,6	37,3	35,2	35,2	36,0	34,5	35,4	36,2	34,7	37,3		4
5 Absolute minimum temperature	in °C	17,0	18,0	17,4	19,1	20,2	20,7	20,0	18,5	20,0	19,2	19,1	16,2	16,2		5
6 Mean relative humidity	in %	82	79	78	79	82	84	83	83	83	82	83	83	82		6
7 Mean precipitation	in mm	118	103	119	142	236	217	176	162	176	193	143	145	1930		7
8 Maximum precipitation	in mm															8
9 Minimum precipitation	in mm															9
10 Maximum precipitation in 24 h	in mm	75	87	118	101	132	99	95	94	124	133	89	154	154		10
11 Mean number of days with precipitation	> 0,1 mm	10	9	9	10	15	15	13	12	12	13	11	11	140		11
12 Mean duration of sunshine	in h															12
13 Mean quantity of radiation	in ly / day															13
14 Mean potential evaporation	in mm	130	124	145	147	154	144	148	148	139	145	137	136	1695		14
15 Mean windspeed	in m / sec	3	3	2	2	2	2	2	2	2	2	2	2	2		15
16 Mean predominent direction of the wind		N	N	N	N	N	S	S	S	S	N	N	N			16

321 Station/Country Zamboanga/Philippines

Location 6°54'N/122°04'E Height above sealevel 6 m Climate symbol: Köppen Aw Troll V.2

		J	F	M	A	M	J	J	A	S	O	N	D	year	P	
1 Mean daily temperature	in °C	26.3	26.4	26.8	27.1	27.1	26.8	26.6	26.8	26.7	26.6	26.6	26.6	26.7		1
2 Mean daily maximum temperature	in °C	31.1	31.1	31.7	31.1	31.1	31.1	30.6	31.1	31.1	31.1	31.1	31.7	31.1	15	2
3 Mean daily minimum temperature	in °C	22.8	22.8	23.3	23.3	23.9	23.9	23.3	23.9	23.3	23.3	23.3	22.8	23.3	15	3
4 Absolute maximum temperature	in °C	34.9	34.7	35.7	35.7	35.1	34.8	35.0	35.1	34.8	35.8	35.2	34.5	35.8		4
5 Absolute minimum temperature	in °C	16.6	15.6	17.5	19.7	20.7	20.4	20.0	19.0	19.9	19.4	18.5	16.7	15.6		5
6 Mean relative humidity	in %	83	82	81	83	84	85	85	85	85	85	86	84	84		6
7 Mean precipitation	in mm	52	53	38	53	90	113	126	113	123	155	121	88	1125		7
8 Maximum precipitation	in mm	150	226	108	250	155	257	252	272	271	608	356	259		30	8
9 Minimum precipitation	in mm	0	2	6	9	39	39	13	28	33	18	39	12		30	9
10 Maximum precipitation in 24 h	in mm	128	157	75	81	55	56	93	137	117	121	163	161	163		10
11 Mean number of days with precipitation	> 0,1 mm	8	6	7	8	11	13	14	13	12	13	13	10	128		11
12 Mean duration of sunshine	in h															12
13 Mean quantity of radiation	in ly/day															13
14 Mean potential evaporation	in mm	131	120	142	142	149	142	147	146	138	140	135	132	1664	45	14
15 Mean windspeed	in m/sec	2	2	2	2	2	2	2	2	2	2	2	2	2		15
16 Mean predominent direction of the wind		NE	N	W	N	N	N	N	W	W	W	W	W			16

322 Station/Country Srinagar(Kashmir)/India

Location 34°05'N/74°50'E Height above sealevel 1586 m Climate symbol: Köppen Cfa Troll IV.1

		J	F	M	A	M	J	J	A	S	O	N	D	year	P	
1 Mean daily temperature	in °C	1.1	3.6	8.5	13.9	17.9	22.0	24.6	23.9	20.5	14.2	7.7	3.5	13.9	30	1
2 Mean daily maximum temperature	in °C	4.4	7.9	13.4	19.3	24.6	29.5	30.8	29.9	28.3	22.6	15.5	8.8	19.5	30	2
3 Mean daily minimum temperature	in °C	-2.3	-0.8	3.5	7.4	11.2	14.4	18.4	17.9	12.7	5.7	-0.1	-1.8	7.2	30	3
4 Absolute maximum temperature	in °C	17.2	20.6	29.6	31.1	35.6	37.8	38.3	36.7	35.0	33.9	23.9	18.3	38.3	70	4
5 Absolute minimum temperature	in °C	-14.4	-20.0	-5.6	0.0	2.8	7.2	10.6	10.0	4.4	-1.7	-7.8	-11.7	-20.0	70	5
6 Mean relative humidity	in %	83	78	72	65	60	60	61	66	62	67	70	78	69	30	6
7 Mean precipitation	in mm	73	72	104	78	63	36	61	63	32	29	18	36	564	30	7
8 Maximum precipitation	in mm	270	171	229	270	162	112	194	179	181	171	102	121	1292	70	8
9 Minimum precipitation	in mm	0	5	18	9	<1	1	10	3	<1	0	0	0	399	70	9
10 Maximum precipitation in 24 h	in mm	148	66	70	65	53	66	80	67	102	60	64	65	148	70	10
11 Mean number of days with precipitation	>0,25 mm	7	6	8	7	6	4	5	5	3	2	2	3	57	30	11
12 Mean duration of sunshine	in h	78	112	140	183	239	246	254	239	240	245	201	121	2298		12
13 Mean quantity of radiation	in ly/day	175	247	386	493	593	628	583	506	427	322	278	172	401	2	13
14 Mean potential evaporation	in mm	0	3	21	50	89	118	142	131	87	45	17	4	707		14
15 Mean windspeed	in m/sec	1.0	1.2	1.6	1.5	1.2	1.1	1.1	1.0	1.0	0.9	0.8	0.9	1.1	30	15
16 Mean predominent direction of the wind		SE	SE,NW	SE,NW	SE,NW	SE,NW	SE,NW	SE,NW	SE,NW	SE,NW	SE,NW	SE,NW	SE,NW		28	16

323 Station/Country Kodaikanal/India

Location 10°14'N/77°28'E Height above sealevel 2343 m Climate symbol: Köppen Cfb Troll V.1

		J	F	M	A	M	J	J	A	S	O	N	D	year	P	
1 Mean daily temperature	in °C	12.5	13.3	14.6	15.7	16.5	15.4	14.3	14.5	14.5	14.0	13.0	12.5	14.3	30	1
2 Mean daily maximum temperature	in °C	17.0	17.9	19.2	19.8	20.4	18.5	17.1	17.5	17.7	17.0	16.1	16.4	17.9	30	2
3 Mean daily minimum temperature	in °C	8.0	8.7	10.0	11.6	12.7	12.2	11.5	11.5	11.3	10.9	9.8	8.6	10.6	30	3
4 Absolute maximum temperature	in °C	24.4	24.4	26.7	26.1	27.8	25.0	22.2	21.1	21.7	21.1	21.7	22.8	27.8	60	4
5 Absolute minimum temperature	in °C	2.8	4.4	4.4	8.2	7.8	5.0	8.9	8.3	8.9	6.1	3.9	2.8	2.8	60	5
6 Mean relative humidity	in %	69	64	59	71	75	77	85	85	85	89	88	76	77	30	6
7 Mean precipitation	in mm	63	32	64	166	157	104	117	163	161	250	377	121	1672	30	7
8 Maximum precipitation	in mm	355	166	297	463	353	224	319	407	377	66	623	399	2355	60	8
9 Minimum precipitation	in mm	0	0	0	12	31	14	25	53	16	75	2	0	1184	60	9
10 Maximum precipitation in 24 h	in mm	195	159	107	124	131	61	79	119	104	147	346	133	346	60	10
11 Mean number of days with precipitation	>0,25mm	3	2	4	9	11	10	10	12	10	16	13	7	107	30	11
12 Mean duration of sunshine	in h	223	221	242	207	186	117	87	109	126	121	138	195	1972		12
13 Mean quantity of radiation	in ly/day	508	562	555	535	510	445	374	397	421	383	405	453	462	7	13
14 Mean potential evaporation	in mm	47	46	61	69	75	65	60	60	57	55	49	45	689		14
15 Mean windspeed	in m/sec	3.6	3.4	3.9	3.4	3.3	4.4	4.6	3.8	3.8	2.8	3.4	3.5	3.6	30	15
16 Mean predominent direction of the wind		SE	NE,SE	NE,SE	NE,SE	NE,NW	NW	NW	NW	NW	NW	NE,NW	NE		28	16

324 Station/Country Pokhara Airport/Nepal

Location 28°13'N/84°00'E Height above sealevel 827 m Climate symbol: Köppen Cwa Troll V.2

		J	F	M	A	M	J	J	A	S	O	N	D	year	P	
1 Mean daily temperature	in °C	12.6	13.6	19.2	22.6	23.8	24.8	25.0	25.0	24.0	21.2	16.8	13.3	20.2		1
2 Mean daily maximum temperature	in °C	18.8	21.2	26.4	29.8	29.6	29.6	29.1	29.0	27.8	25.6	22.8	19.5	25.8		2
3 Mean daily minimum temperature	in °C	6.4	6.0	12.0	15.4	18.0	20.0	20.9	21.0	20.2	16.8	10.8	7.1	14.6		3
4 Absolute maximum temperature	in °C	22.0	28.2	33.1	37.4	35.0	33.4	32.4	32.4	31.0	29.8	27.0	23.3	37.4		4
5 Absolute minimum temperature	in °C	1.8	3.0	5.0	6.0	8.0	12.0	13.0	13.8	15.9	10.4	4.0	3.9	1.8		5
6 Mean relative humidity	in %															6
7 Mean precipitation	in mm	26	25	50	87	292	569	809	705	581	224	19	1	3388		7
8 Maximum precipitation	in mm															8
9 Minimum precipitation	in mm															9
10 Maximum precipitation in 24 h	in mm	22	40	38	55	135	158	173	205	148	136	35	8	205		10
11 Mean number of days with precipitation	> 0,1 mm															11
12 Mean duration of sunshine	in h															12
13 Mean quantity of radiation	in ly/day															13
14 Mean potential evaporation	in mm															14
15 Mean windspeed	in m/sec	0.4	0.7	0.9	1.0	1.0	0.8	0.9	0.9	0.6	0.5	0.2	0.2			15
16 Mean predominent direction of the wind		NW-SE	NNE-SE	SSE-SE	S-E	ESE-E	SE-E	E	E	SE-E	S-SE	S-SE	S-SE			16

NORTH AMERICA AND MEXICO

UNITED STATES (ALASKA)

1 Barrow
2 Barter Island
3 Nome
4 Kotzebue
5 Fairbanks
6 Mc Grath
7 Anchorage
8 Bethel
9 Yakutat
10 Kodiak
11 Juneau

CANADA

12 Isachsen
13 Mould Bay
14 Resolute
15 Sachs Harbour
16 Baker Lake
17 Coral Harbour
18 Frobisher Bay
19 Port Harrison
20 Aklavik
21 Norman Wells
22 Yellowknife
23 Whitehorse
24 Fort Smith
25 Fort Nelson
26 Churchill
27 Prince Ruppert
28 Prince George
29 Trout Lake
30 Edmonton
31 The Pas
32 Goose Bay
33 Calgary
34 Saskatoon
35 Regina
36 Fort Chimo
37 Vancouver
38 Lethbridge
39 Winnipeg
40 Kapuskasing
41 St. John's
42 Québec
43 Ottawa
44 Montreal
45 St. John
46 Halifax
47 Toronto

UNITED STATES

48 Tatoosh Island
49 Seattle
50 Spokane
51 Havre
52 Miles City
53 Williston
54 Bismarck
55 Duluth
56 Sault Ste. Marie
57 Mt. Washington
58 Burlington
59 Portland
60 Walla Walla
61 Billings
62 Minneapolis
63 Lansing
64 Binghamton
65 Boston
66 Eureka
67 Medford
68 Boise
69 Elko
70 Salt Lake City
71 Lander
72 Cheyenne
73 Rapid City
74 Huron
75 Omaha
76 Des Moines
77 Chicago
78 Detroit
79 Cleveland
80 Buffalo
81 New York City
82 Harrisburg
83 San Francisco
84 Reno
85 Fresno
86 Modena
87 Denver
88 Dodge City
89 Kansas City
90 St. Louis
91 Cairo
92 Indianapolis
93 Cincinnati
94 Evansville
95 Nashville
96 Parkersburg
97 Atlantic City

98 Washington D. C.
99 Richmond
100 Greensboro
101 Los Angeles
102 San Diego
103 Greenland Ranch
104 Phoenix
105 Winslow
106 Albuquerque
107 El Paso
108 Amarillo
109 Abilene
110 Oklohoma City
111 Dallas
112 Tulsa
113 Little Rock
114 Memphis
115 Vicksburg
116 Birmingham
117 Montgomery
118 Atlanta
119 Ashville
120 Charlotte
121 Charleston
122 Jacksonville
123 San Antonio
124 Houston
125 New Orleans
126 Appalachicola
127 Tampa
128 Miami
129 Brownsville
130 Honolulu

MEXICO

131 Guaymas
132 Monterrey
133 Ciudad Lerdo
134 La Paz
135 Mazatlán
136 Zacatecas
137 Tampico
138 Guadalajara
139 Guanajuato
140 Ciudad de México
(Mexico City)
141 Mérida
142 Isla de Cozumel
143 Acapulco
144 Salina Gruz

NORTH AMERICA AND MEXICO

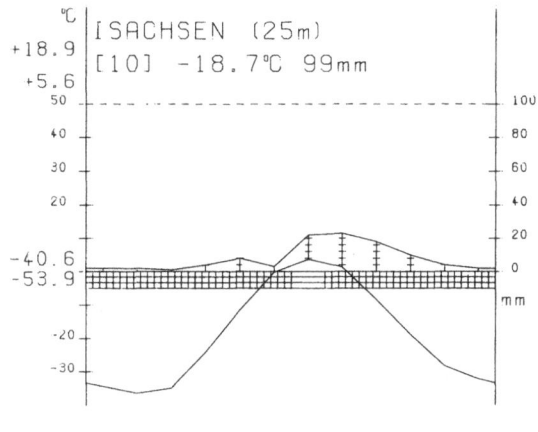

ISACHSEN (25m)
[10] -18.7°C 99mm

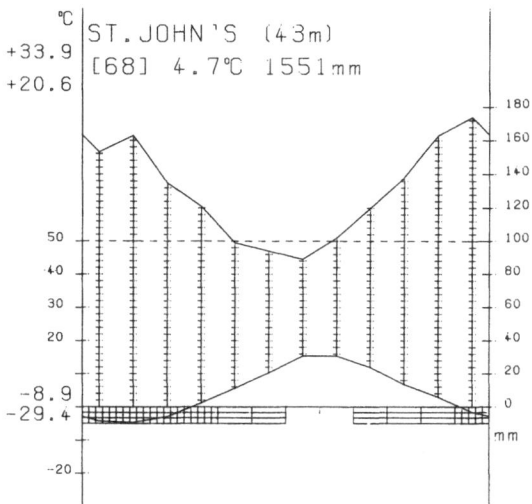

ST.JOHN'S (43m)
[68] 4.7°C 1551mm

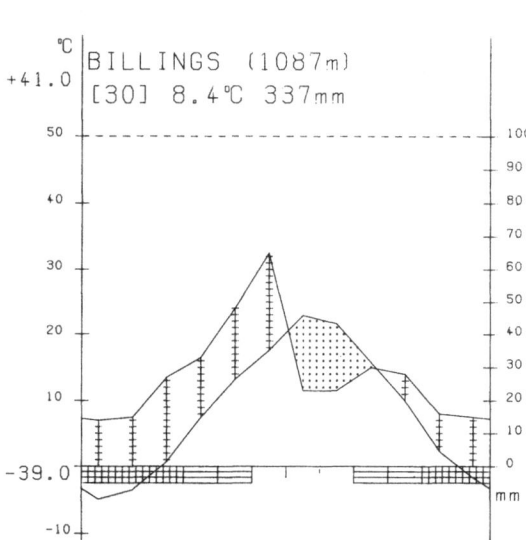

BILLINGS (1087m)
[30] 8.4°C 337mm

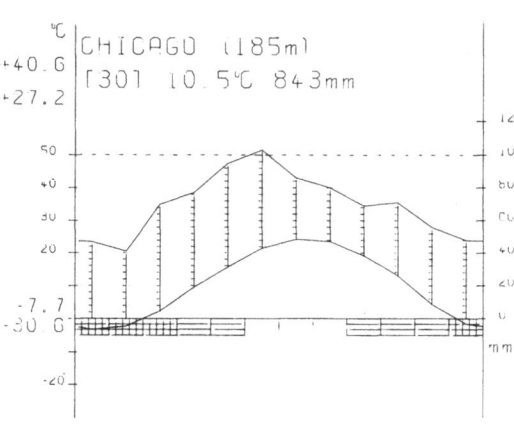

CHICAGO (185m)
[30] 10.5°C 843mm

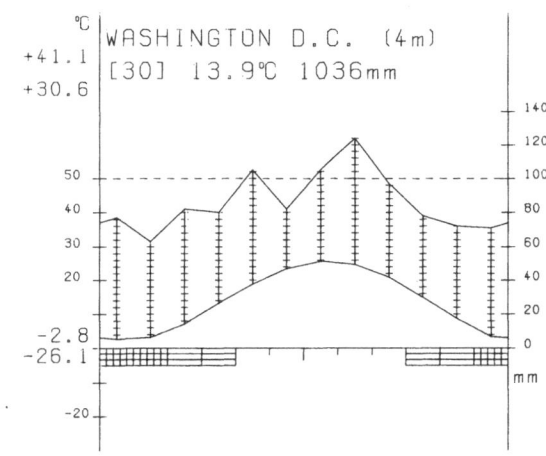

WASHINGTON D.C. (4m)
[30] 13.9°C 1036mm

NORTH AMERICA AND MEXICO

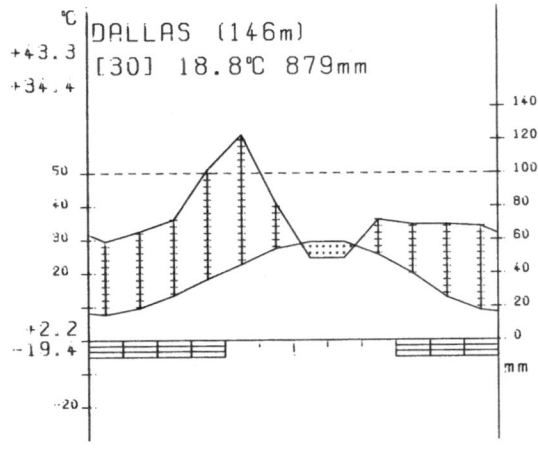

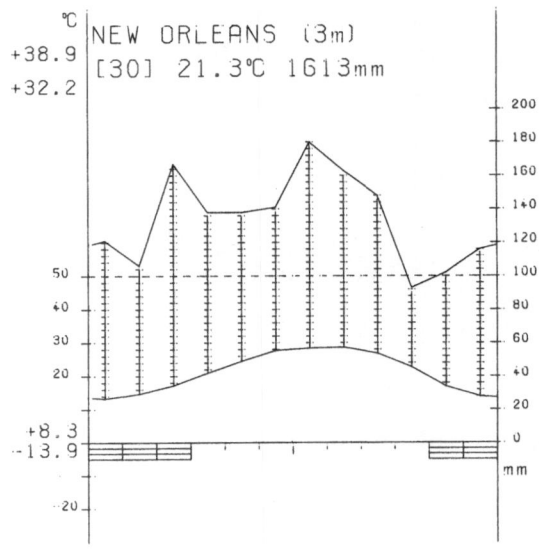

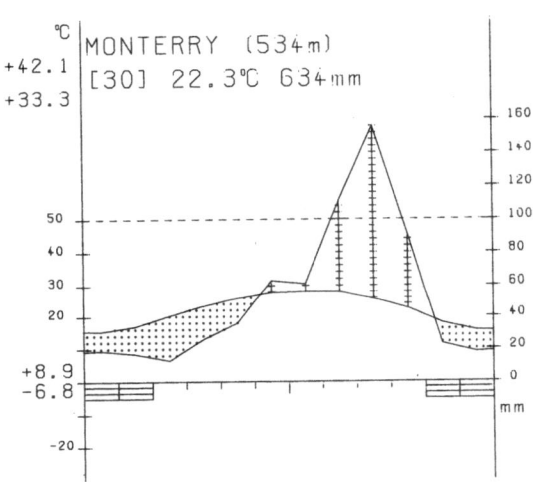

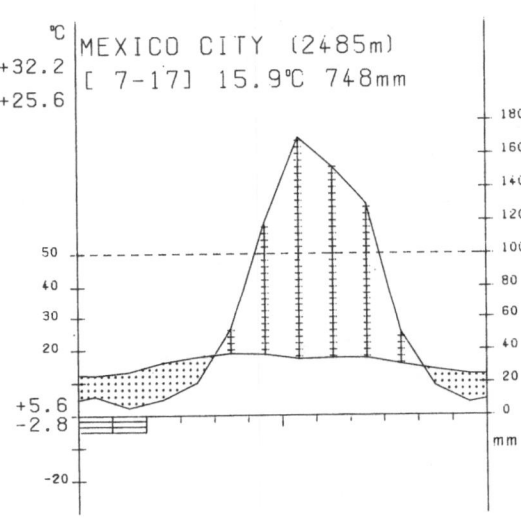

1 Station/Country Barrow (Alaska)/U.S.

Location 71°18'N/156°47'W   Height above sealevel 7 m   Climate symbol: Köppen ET   Troll I,3

		J	F	M	A	M	J	J	A	S	O	N	D	year	P	
1 Mean daily temperature	in °C	-26,8	-27,9	-25,9	-17,7	-7,6	0,6	3,9	3,3	0,8	-8,6	-18,2	-24,0	-12,4	30	1
2 Mean daily maximum temperature	in °C	-22,8	-244	-22,2	-13,9	4,4	3,9	7,8	6,7	1,1	-5,6	-13,9	-20,0	-8,9	32	2
3 Mean daily minimum temperature	in °C	-30,0	-31,7	-30,0	-22,2	-10,6	-1,7	0,6	0,6	-2,8	-11,1	-20,6	-27,2	-15,6	32	3
4 Absolute maximum temperature	in °C	0,6	-0,6	-1,1	5,6	7,2	21,1	25,6	22,8	16,7	6,1	3,9	1,1	25,6	37	4
5 Absolute minimum temperature	in °C	-47,2	-48,9	-46,7	-41,1	-27,8	-13,3	-5,8	-6,7	-17,2	-28,3	-40,0	-48,3	-48,9	37	5
6 Mean relative humidity	in %	65	61	64	74	87	93	92	93	92	85	76	66	79	13	6
7 Mean precipitation	in mm	5	4	3	3	3	9	20	23	16	13	6	4	110	30	7
8 Maximum precipitation	in mm															8
9 Minimum precipitation	in mm															9
10 Maximum precipitation in 24 h	in mm	18	8	7	5	8	21	22	16	13	25	10	7	25	30	10
11 Mean number of days with precipitation	>0,25 mm	4	4	3	3	3	4	8	10	9	9	6	4	67	37	11
12 Mean duration of sunshine	in h															12
13 Mean quantity of radiation	in ly/day															13
14 Mean potential evaporation	in mm	0	0	0	0	0	48	84	28	0	0	0	0	160	40	14
15 Mean windspeed	in m/sec	4,9	5,1	4,9	5,1	53	5,1	5,3	5,7	6,1	6,3	5,6	4,9	5,4	30	15
16 Mean predominent direction of the wind		ESE	ENE	NE	E	NE	E	SW	E	ENE	NE	NE	ENE		7	16

2 Station/Country Barter Island (Alaska) U.S.

Location 70°07'N/143°40'W   Height above sealevel 15 m   Climate symbol: Köppen ET   Troll I,3

		J	F	M	A	M	J	J	A	S	O	N	D	year	P	
1 Mean daily temperature	in °C	-26,4	-27,8	-25,7	-17,2	-6,6	1,8	4,9	4,0	-0,4	-8,1	-17,5	-23,9	-11,9		1
2 Mean daily maximum temperature	in °C															2
3 Mean daily minimum temperature	in °C															3
4 Absolute maximum temperature	in °C	2,8	1,1	-0,6	6,1	8,9	19,4	21,7	22,2	17,8	6,1	2,8	-1,1	22,2		4
5 Absolute minimum temperature	in °C	-45,6	-50,6	-45,6	-38,3	-25,6	-7,8	-3,3	-4,4	-13,9	-26,7	-46,1	-46,1	-50,6		5
6 Mean relative humidity	in %															6
7 Mean precipitation	in mm	10	9	7	5	6	11	30	30	23	26	15	12	184		7
8 Maximum precipitation	in mm															8
9 Minimum precipitation	in mm															9
10 Maximum precipitation in 24 h	in mm	31	31	11	7	19	29	30	28	57	50	11	14	57		10
11 Mean number of days with precipitation	>0,25 mm	6	5	6	7	7	7	10	11	10	14	9	6	98		11
12 Mean duration of sunshine	in h															12
13 Mean quantity of radiation	in ly/day															13
14 Mean potential evaporation	in mm	0	0	0	0	0	24	58	41	0	0	0	0	123		14
15 Mean windspeed	in m/sec	6,1	6,6	5,8	5,5	5,4	4,9	4,6	5,4	5,6	6,7	6,5	6,3	5,8		15
16 Mean predominent direction of the wind		E	W	W	W	E	ENE	ENE	E	E	E	E	W			16

3 Station/Country Nome (Alaska)/U.S.

Location 64°30'N/165°20'W   Height above sealevel 4 m   Climate symbol: Köppen ET   Troll I,3

		J	F	M	A	M	J	J	A	S	O	N	D	year	P	
1 Mean daily temperature	in °C	-15,3	-14,7	-13,4	-6,0	1,7	7,7	9,7	9,4	5,5	-1,3	-8,6	-14,3	-3,3	30	1
2 Mean daily maximum temperature	in °C	-11,7	-10,0	-8,3	-2,2	4,4	11,7	12,8	12,8	8,9	1,7	-5,6	-10,0	0,6	36	2
3 Mean daily minimum temperature	in °C	-19,4	-18,9	-17,8	-11,1	-2,8	3,3	6,7	6,1	2,2	-4,4	-12,2	-17,2	-17,2	36	3
4 Absolute maximum temperature	in °C	3,9	8,3	5,6	10,6	20,0	27,2	23,9	22,8	17,2	15,0	6,7	1,7	27,2	11	4
5 Absolute minimum temperature	in °C	-39,4	-41,1	-38,9	-32,8	-23,9	-3,9	0,6	-1,1	-8,3	-19,4	-39,4	-40,6	-41,1	11	5
6 Mean relative humidity	in %	80	76	79	81	79	81	87	87	82	80	83	78	81	11	6
7 Mean precipitation	in mm	26	24	22	20	18	24	58	97	68	43	29	25	454	30	7
8 Maximum precipitation	in mm															8
9 Minimum precipitation	in mm															9
10 Maximum precipitation in 24 h	in mm	20	11	17	14	18	52	45	61	32	58	12	28	61	11	10
11 Mean number of days with precipitation	>0,25 mm	10	9	11	9	8	9	14	18	14	9	12	10	133	11	11
12 Mean duration of sunshine	in h	72	109	193	226	285	297	204	146	142	101	67	42	1884	27	12
13 Mean quantity of radiation	in ly/day															13
14 Mean potential evaporation	in mm	0	0	0	0	30	83	94	86	48	0	0	0	341	54	14
15 Mean windspeed	in m/sec	5,4	5,0	5,0	4,9	4,5	4,2	4,4	4,8	5,1	4,9	5,2	4,4	4,8	11	15
16 Mean predominent direction of the wind		E	ENE	E	ENE	NE	SW	WSW	SW	N	NE	N	E			16

4 Station/Country Kotzebue (Alaska)/U.S.

Location 66°52'N/160°38'W   Height above sealevel 3 m   Climate symbol: Köppen Dfc   Troll II,2

		J	F	M	A	M	J	J	A	S	O	N	D	year	P	
1 Mean daily temperature	in °C	-20,9	-20,0	-18,9	-10,4	-0,6	6,6	11,5	10,3	4,9	-4,1	-13,7	-19,8	-6,3	30	1
2 Mean daily maximum temperature	in °C	-17,2	-15,6	-13,9	-6,1	2,2	9,4	15,0	12,2	7,8	-0,6	-9,4	-15,6	-2,8	10	2
3 Mean daily minimum temperature	in °C	-25,0	-23,9	-23,3	-16,1	-5,6	2,8	8,9	7,2	1,7	-6,1	-15,6	-22,2	-10,0	10	3
4 Absolute maximum temperature	in °C	2,2	1,7	1,1	5,6	23,3	27,2	27,8	23,9	17,2	10,6	3,3	1,1	27,8	15	4
5 Absolute minimum temperature	in °C	-41,7	-44,4	-44,4	-42,2	-27,8	-6,7	1,1	-0,6	-8,3	-22,2	-37,8	-43,9	-44,4	15	5
6 Mean relative humidity	in %	71	72	73	79	84	85	83	86	85	84	78	72	79	14	6
7 Mean precipitation	in mm	10	8	7	8	8	12	37	55	31	15	9	7	208	30	7
8 Maximum precipitation	in mm															8
9 Minimum precipitation	in mm															9
10 Maximum precipitation in 24 h	in mm	6	17	12	5	12	20	45	38	22	12	7	10	45	15	10
11 Mean number of days with precipitation	>0,25 mm	7	7	9	6	6	8	11	16	13	11	10	9	113		11
12 Mean duration of sunshine	in h	62	105	177	270	285	261	192	115	108	90	54	25	1743		12
13 Mean quantity of radiation	in ly/day															13
14 Mean potential evaporation	in mm	0	0	0	0	0	76	109	94	50	0	0	0	329	16	14
15 Mean windspeed	in m/sec	6,8	6,4	5,9	6,0	4,6	5,5	5,8	6,2	5,8	6,1	6,2	5,8	5,9		15
16 Mean predominent direction of the wind		E	E	E	ESE	W	W	W	W	ESE	NE	ESE	NE			16

NORTH AMERICA
Climate Zones after W. Köppen/R. Geiger

NORTH AMERICA
Climate Zones after C. Troll/KH. Paffen

● 140 MEXICO - CITY — Location, No. and Name of Climate Station (Capitals)

● 128 Miami — Location, No. and Name of Climate Station (other Towns)

International border

0 1000 km

← 130 Honolulu
V₂

5 Station/Country **Fairbanks (Alaska)/U.S.**

Location 64°49'N/147°52'W — Height above sealevel 133 m — Climate symbol: Köppen **Dfc** — Troll II,2

		J	F	M	A	M	J	J	A	S	O	N	D	year	P	
1 Mean daily temperature	in °C	-23,9	-19,4	-12,8	-1,4	8,4	14,7	15,4	12,4	6,4	-3,2	-15,6	-22,1	-3,4	30	1
2 Mean daily maximum temperature	in °C	-18,9	-11,7	-5,0	5,6	15,0	21,7	22,2	18,9	12,2	1,7	-11,1	-17,2	2,8	44	2
3 Mean daily minimum temperature	in °C	-28,9	-23,3	-20,0	-8,3	1,7	7,8	8,9	6,7	0,6	-7,7	-20,6	-26,7	-8,9	44	3
4 Absolute maximum temperature	in °C	5,6	10,0	12,8	20,6	32,2	32,8	33,9	30,6	28,9	17,8	12,2	14,4	33,9	28	4
5 Absolute minimum temperature	in °C	-54,4	-50,0	-45,0	-35,6	-17,8	-1,1	1,1	-5,0	-11,1	-33,3	-40,6	-50,6	-54,4	28	5
6 Mean relative humidity	in %	69	68	66	61	58	62	70	76	77	79	73	71	69	17	6
7 Mean precipitation	in mm	23	13	10	6	18	35	47	56	28	22	15	14	287	30	7
8 Maximum precipitation	in mm	49	44	53	21	42	89	110	157	77	47	84	58		30	8
9 Minimum precipitation	in mm	<1	2	0	0	2	5	10	10	4	2	0	0		30	9
10 Maximum precipitation in 24 h	in mm	47	20	17	17	22	39	55	59	31	30	24	19	59	28	10
11 Mean number of days with precipitation	>0,25 mm	9	7	7	4	8	11	13	15	10	11	9	8	112	28	11
12 Mean duration of sunshine	in h	54	120	224	302	319	334	274	164	122	85	71	36	2105	28	12
13 Mean quantity of radiation	in ly/day															13
14 Mean potential evaporation	in mm	0	0	0	0	71	114	118	86	48	0	0	0	437	52	14
15 Mean windspeed	in m/sec	1,4	1,7	2,1	2,6	3,0	2,9	2,6	2,5	2,5	2,2	1,7	1,3	2,2	27	15
16 Mean predominent direction of the wind		N	N	N	N	N	SW	SW	N	N	N	N	N		8	16

6 Station/Country **Mc Grath (Alaska)/U.S.**

Location 62°58'N/155°37'W — Height above sealevel 102 m — Climate symbol: Köppen **Dfc** — Troll II,2

		J	F	M	A	M	J	J	A	S	O	N	D	year	P	
1 Mean daily temperature	in °C	-22,8	-17,7	-13,2	-2,4	6,7	13,6	14,8	12,3	6,8	-2,9	-14,8	-21,7	-3,4	30	1
2 Mean daily maximum temperature	in °C	-17,8	-13,9	-6,1	2,8	12,2	18,3	20,0	16,1	10,6	1,7	-8,3	-17,2	1,7	6	2
3 Mean daily minimum temperature	in °C	-28,9	-27,8	-21,1	-10,6	1,7	7,2	9,4	7,2	1,7	-6,1	-18,7	-26,1	-9,4	6	3
4 Absolute maximum temperature	in °C	12,2	12,8	10,6	17,2	26,7	31,1	30,6	28,3	24,4	15,0	8,3	6,7	31,1	15	4
5 Absolute minimum temperature	in °C	-53,3	-53,3	-46,1	-33,3	-18,9	0,0	1,1	-2,2	-14,4	-30,0	-45,0	-51,7	-53,3	15	5
6 Mean relative humidity	in %	71	71	71	67	65	65	71	81	81	81	77	73	73	15	6
7 Mean precipitation	in mm	32	29	24	12	22	42	82	96	66	34	27	28	472	30	7
8 Maximum precipitation	in mm															8
9 Minimum precipitation	in mm															9
10 Maximum precipitation in 24 h	in mm	26	29	27	17	23	61	32	53	42	39	33	25	61	15	10
11 Mean number of days with precipitation	>0,25 mm	10	10	11	6	9	12	14	20	15	13	12	12	144	15	11
12 Mean duration of sunshine	in h															12
13 Mean quantity of radiation	in ly/day															13
14 Mean potential evaporation	in mm															14
15 Mean windspeed	in m/sec	1,2	1,6	1,8	2,2	2,5	2,5	2,4	2,3	2,2	1,7	1,1	1,1	1,9	8	15
16 Mean predominent direction of the wind		NW	NW	NW	N	E	S	S	S	N	N	ESE	NW		9	16

7 Station/Country **Anchorage (Alaska)/U.S.**

Location 61°10'N/149°59'W — Height above sealevel 27 m — Climate symbol: Köppen **Dfc** — Troll II,1

		J	F	M	A	M	J	J	A	S	O	N	D	year	P	
1 Mean daily temperature	in °C	-10,9	-7,8	-4,8	2,1	7,7	12,5	13,9	13,1	8,8	1,7	-5,4	-9,8	1,8	30	1
2 Mean daily maximum temperature	in °C	-7,2	-2,8	0,6	6,7	12,2	16,7	18,3	17,8	13,9	6,1	-1,1	-6,7	6,1	22	2
3 Mean daily minimum temperature	in °C	-15,0	-12,8	-10,6	-2,8	2,2	6,7	9,4	8,3	3,9	-1,7	-9,4	-14,4	-2,8	23	3
4 Absolute maximum temperature	in °C	13,3	13,3	13,3	17,2	27,8	30,0	28,3	27,8	22,8	17,2	15,6	11,7	30,0	35	4
5 Absolute minimum temperature	in °C	-37,2	-38,9	-30,0	-29,4	-17,2	-0,6	1,7	-0,6	-7,2	-21,1	-29,4	-36,1	-38,9	35	5
6 Mean relative humidity	in %	74	70	69	66	64	66	73	77	78	78	76	75	72	18	6
7 Mean precipitation	in mm	20	18	13	11	13	25	47	65	64	47	26	24	374	30	7
8 Maximum precipitation	in mm	32	63	25	23	41	44	73	75	117	59	69	61		30	8
9 Minimum precipitation	in mm	6	14	3	0	11	5	18	8	20	13	17	11		30	9
10 Maximum precipitation in 24 h	in mm	24	30	20	23	20	28	52	43	46	40	20	41	52	35	10
11 Mean number of days with precipitation	>0,25 mm	7	6	7	4	5	7	11	15	15	11	8	7	103	35	11
12 Mean duration of sunshine	in h	78	114	210	254	268	288	255	184	128	96	68	48	1992	18	12
13 Mean quantity of radiation	in ly/day															13
14 Mean potential evaporation	in mm	0	0	0	18	71	104	115	105	65	21	0	0	499	45	14
15 Mean windspeed	in m/sec	2,3	2,6	2,6	2,5	2,9	2,8	2,5	2,3	2,3	2,4	2,3	2,2	2,5	8	15
16 Mean predominent direction of the wind		NE	N	N	N	S	S	S	NW	NNE	N	N	NE		8	16

8 Station/Country **Bethel (Alaska)/U.S.**

Location 60°47'N/161°43'W — Height above sealevel 3 m — Climate symbol: Köppen **Dfc** — Troll II,2

		J	F	M	A	M	J	J	A	S	O	N	D	year	P	
1 Mean daily temperature	in °C	-15,6	-13,2	-11,3	-3,4	3,9	10,9	12,8	11,3	7,0	-0,3	-8,2	-15,1	-1,8	30	1
2 Mean daily maximum temperature	in °C	-8,9	-8,3	-5,6	2,2	10,0	17,2	17,2	16,1	11,7	3,3	-4,4	-9,4	3,3	16	2
3 Mean daily minimum temperature	in °C	-18,3	-17,8	-16,1	-7,7	-0,6	6,1	7,8	7,2	3,3	-4,4	-11,7	-17,8	-6,1	16	3
4 Absolute maximum temperature	in °C	6,1	8,3	8,9	13,9	23,3	28,9	30,0	27,2	21,1	18,3	8,9	5,6	30,0	15	4
5 Absolute minimum temperature	in °C	-46,7	-42,8	-41,1	-35,0	-20,6	-1,1	1,7	-1,1	-7,8	-18,9	-32,8	-42,2	-46,7	15	5
6 Mean relative humidity	in %	80	79	80	82	78	76	82	89	88	88	86	81	83	9	6
7 Mean precipitation	in mm	28	28	26	15	24	30	52	107	66	39	27	26	468	30	7
8 Maximum precipitation	in mm															8
9 Minimum precipitation	in mm															9
10 Maximum precipitation in 24 h	in mm	48	24	25	10	28	21	49	61	39	48	22	21	61	15	10
11 Mean number of days with precipitation	>0,25 mm	11	12	13	10	11	11	15	21	16	14	13	13	160	15	11
12 Mean duration of sunshine	in h															12
13 Mean quantity of radiation	in ly/day															13
14 Mean potential evaporation	in mm	0	0	0	0	50	100	108	94	57	0	0	0	409	37	14
15 Mean windspeed	in m/sec	5,0	5,5	5,1	4,6	4,3	4,2	4,1	4,4	4,3	4,6	4,9	4,8	4,6	14	15
16 Mean predominent direction of the wind		NNE	NNE	NW	NW	ESE	NW	SSW	SSW	N	NE	NNE	N		8	16

9 Station/Country   Yakutat (Alaska)/U.S.

Location   59°31'N/139°40'W   Height above sealevel   9 m   Climate symbol: Köppen Dfc   Troll II,1

	J	F	M	A	M	J	J	A	S	O	N	D	year	P	
1 Mean daily temperature	in °C	-2,6	-1,9	-0,3	2,8	7,0	10,3	12,3	12,1	9,6	5,5	1,0	-2,2	4,4	30	1
2 Mean daily maximum temperature	in °C															2
3 Mean daily minimum temperature	in °C															3
4 Absolute maximum temperature	in °C	6,1	8,9	12,8	18,3	24,4	25,6	28,9	30,0	25,0	15,6	15,0	7,8	30,0	11	4
5 Absolute minimum temperature	in °C	-30,0	-28,3	-25,0	-16,1	-5,6	-1,1	2,2	-1,1	-3,9	-11,1	-23,3	-26,1	-30,0	11	5
6 Mean relative humidity	in %	81	83	83	83	85	85	87	89	89	85	87	86	85		6
7 Mean precipitation	in mm	276	208	221	184	203	129	214	277	420	498	407	312	3348	30	7
8 Maximum precipitation	in mm															8
9 Minimum precipitation	in mm															9
10 Maximum precipitation in 24 h	in mm	130	52	85	68	86	64	63	117	118	108	181	96	181	11	10
11 Mean number of days with precipitation	>0,25 mm	18	18	17	17	18	15	15	17	21	22	22	23	223	11	11
12 Mean duration of sunshine	in h	59	51	84	81	71	84	90	71	54	50	36	28	759		12
13 Mean quantity of radiation	in ly/day															13
14 Mean potential evaporation	in mm	0	0	0	26	61	93	104	95	65	14	29	0	487		14
15 Mean windspeed	in m/sec	4,1	4,4	3,6	3,7	3,8	3,6	3,1	3,3	3,6	4,1	4,3	4,4	3,8	9	15
16 Mean predominent direction of the wind		E	E	E	E	ESE	ESE	E	ESE	ESE	E	E	E		9	16

10 Station/Country   Kodiak (Alaska)/U.S.

Location   57°30'N/152°45'W   Height above sealevel   50 m   Climate symbol: Köppen Cfc   Troll II,1

	J	F	M	A	M	J	J	A	S	O	N	D	year	P	
1 Mean daily temperature	in °C	-1,2	-0,2	-0,1	2,6	6,0	9,7	12,1	12,6	9,8	5,2	1,7	-1,3	4,8	30	1
2 Mean daily maximum temperature	in °C	2,2	2,2	2,2	6,7	9,4	13,3	16,1	16,7	13,9	8,9	5,0	2,8	8,3	10	2
3 Mean daily minimum temperature	in °C	-2,8	-2,8	-2,8	0,0	2,8	6,1	8,9	9,4	6,7	3,3	-0,6	-2,2	2,2	10	3
4 Absolute maximum temperature	in °C	10,6	13,3	11,7	15,0	24,4	30,0	26,7	27,2	24,4	15,6	12,2	10,0	30,0	30	4
5 Absolute minimum temperature	in °C	-20,6	-18,3	-19,4	-12,2	-6,7	1,1	4,4	2,2	-0,6	-7,2	-12,8	-16,1	-20,6	18	5
6 Mean relative humidity	in %	80	79	78	77	79	80	82	82	82	78	80	78	79	18	6
7 Mean precipitation	in mm	130	128	96	100	141	100	97	95	151	169	153	118	1477	18	7
8 Maximum precipitation	in mm															8
9 Minimum precipitation	in mm															9
10 Maximum precipitation in 24 h	in mm	81	115	45	70	48	55	63	51	87	92	59	47	115	18	10
11 Mean number of days with precipitation	>0,25 mm	17	15	16	15	16	15	15	15	17	17	16	16	190	18	11
12 Mean duration of sunshine	in h															12
13 Mean quantity of radiation	in ly/day															13
14 Mean potential evaporation	in mm	0	0	9	26	59	86	103	95	72	38	13	0	501	39	14
15 Mean windspeed	in m/sec	5,4	5,2	5,2	4,8	4,4	3,7	2,9	3,2	3,9	4,9	5,4	5,4	4,5	18	15
16 Mean predominent direction of the wind		NW	NW	NW	NW	E	E	E	NW	NW	NW	NW	NW		18	16

11 Station/Country   Juneau (Alaska)/U.S.

Location   58°22'N/134°35'W   Height above sealevel   5 m   Climate symbol: Köppen Dfc   Troll II,1

	J	F	M	A	M	J	J	A	S	O	N	D	year	P	
1 Mean daily temperature	in °C	-3,8	-2,9	-0,9	3,3	7,6	11,1	12,9	12,3	9,3	5,3	1,3	-2,0	4,5	30	1
2 Mean daily maximum temperature	in °C	0,0	1,7	3,9	8,3	12,8	16,7	17,8	16,7	13,3	8,9	4,4	1,7	8,9	44	2
3 Mean daily minimum temperature	in °C	-4,4	-3,3	-1,7	1,1	4,4	7,8	10,0	9,4	7,2	3,9	-0,6	-2,8	2,8	43	3
4 Absolute maximum temperature	in °C	8,9	8,9	10,6	18,9	27,8	28,3	28,9	28,3	22,2	16,1	13,3	12,2	28,9	14	4
5 Absolute minimum temperature	in °C	-28,3	-24,4	-23,9	-13,9	-3,9	-0,6	2,2	-2,2	-3,3	-8,9	-19,4	-29,4	-29,4	14	5
6 Mean relative humidity	in %	78	78	78	75	76	73	78	81	85	85	84	81	80	14	6
7 Mean precipitation	in mm	102	78	83	73	82	86	114	128	169	212	154	107	1387	30	7
8 Maximum precipitation	in mm															8
9 Minimum precipitation	in mm															9
10 Maximum precipitation in 24 h	in mm	70	60	46	40	35	49	42	61	81	118	85	90	118	14	10
11 Mean number of days with precipitation	>0,25 mm	18	17	18	17	18	15	17	18	20	23	20	21	222	14	11
12 Mean duration of sunshine	in h	71	102	171	200	230	251	193	161	123	67	60	51	1680	29	12
13 Mean quantity of radiation	in ly/day															13
14 Mean potential evaporation	in mm	0	0	0	26	71	93	104	86	65	35	6	0	486	36	14
15 Mean windspeed	in m/sec	3,8	3,8	3,9	3,9	3,4	3,4	3,2	3,2	3,4	4,0	3,7	3,9	3,7	14	15
16 Mean predominent direction of the wind		N	ESE	ESE	ESE	ESE	N	N	N	N	ESE	ESE	ESE		19	16

12 Station/Country   Isachsen (Northwest Territories)/Canada

Location   78°47'N/103°32'W   Height above sealevel   25 m   Climate symbol: Köppen ET   Troll I,2

	J	F	M	A	M	J	J	A	S	O	N	D	year	P	
1 Mean daily temperature	in °C	-34,6	-36,6	-35,0	-24,2	-11,5	-0,2	3,7	1,4	-8,4	-18,8	-28,2	-32,3	-18,7	10	1
2 Mean daily maximum temperature	in °C	-33,3	-33,9	-30,0	-20,0	-7,2	1,7	5,6	2,8	-6,1	-16,1	-23,9	-29,4	-16,1	6	2
3 Mean daily minimum temperature	in °C	-40,0	-40,6	-36,7	-28,3	-13,3	-2,8	1,1	-1,7	-12,2	-23,3	-31,7	-36,1	-22,2	6	3
4 Absolute maximum temperature	in °C	-3,9	-20,6	-8,3	-1,1	2,2	16,7	18,9	14,4	2,8	-1,7	-3,9	-9,4	18,9	10	4
5 Absolute minimum temperature	in °C	-52,8	-51,1	-53,9	-42,2	-29,4	-14,4	-3,3	-13,3	-27,2	-37,2	-45,6	-51,1	-53,9	10	5
6 Mean relative humidity	in %															6
7 Mean precipitation	in mm	2	2	1	4	8	3	22	23	18	10	4	2	99	10	7
8 Maximum precipitation	in mm															8
9 Minimum precipitation	in mm															9
10 Maximum precipitation in 24 h	in mm	1	2	1	5	5	3	15	20	20	10	6	2	20	10	10
11 Mean number of days with precipitation	>0,25 mm	4	3	3	5	9	5	9	9	12	8	4	3	74	10	11
12 Mean duration of sunshine	in h															12
13 Mean quantity of radiation	in ly/day															13
14 Mean potential evaporation	in mm	0	0	0	0	0	0	90	46	0	0	0	0	136	6	14
15 Mean windspeed	in m/sec	4,5	3,4	3,0	3,2	4,4	4,3	4,7	4,3	4,3	4,7	3,9	4,2	4,2	10	15
16 Mean predominent direction of the wind		N	N	N	N	N	N	NW	N,SW	N	N	N	N,NW		10	16

13 Station / Country **Mould Bay (Northwest Territories) / Canada**

Location 76°14'N / 119°20'W   Height above sealevel 15 m   Climate symbol: Köppen ET   Troll I,2

		J	F	M	A	M	J	J	A	S	O	N	D	year	P	
1 Mean daily temperature	in °C	−33.1	−35.4	−32.7	−22.7	−10.8	0.2	4.0	1.8	−6.0	−17.1	−26.7	−31.2	−17.4	10	1
2 Mean daily maximum temperature	in °C	−30.0	−32.2	−26.7	−16.1	−6.1	1.7	5.6	3.3	−4.4	−13.3	−22.2	−29.4	−14.4	8	2
3 Mean daily minimum temperature	in °C	−38.3	−38.9	−35.0	−26.1	−12.8	−3.3	0.6	−1.7	−9.4	−21.1	−29.4	−35.0	−21.1	8	3
4 Absolute maximum temperature	in °C	−9.4	−10.6	−10.6	−1.7	1.7	13.3	15.6	13.9	7.8	0.0	−7.2	−9.4	15.6	10	4
5 Absolute minimum temperature	in °C	−48.3	−50.0	−48.9	−41.7	−28.9	−13.3	−3.9	−10.0	−25.0	−36.1	−43.3	−52.8	−52.8	10	5
6 Mean relative humidity	in %															6
7 Mean precipitation	in mm	4	2	3	3	7	4	17	21	11	8	3	3	86	10	7
8 Maximum precipitation	in mm															8
9 Minimum precipitation	in mm															9
10 Maximum precipitation in 24 h	in mm	6	2	6	4	7	2	15	48	11	5	2	2	48	10	10
11 Mean number of days with precipitation	>0.25 mm	5	4	4	4	10	4	9	10	11	9	4	6	80	10	11
12 Mean duration of sunshine	in h															12
13 Mean quantity of radiation	in ly/day															13
14 Mean potential evaporation	in mm	0	0	0	0	0	0	88	55	0	0	0	0	143	6	14
15 Mean windspeed	in m/sec	4.2	3.6	3.4	3.6	4.8	6.6	5.2	4.8	4.9	4.8	4.3	3.6		10	15
16 Mean predominant direction of the wind		NW	NW	N	N.NW	NW	NW	NW	S.NE	NW	NW	NW	NW		10	16

14 Station / Country **Resolute (Northwest Territories) / Canada**

Location 74°41'N / 94°54'W   Height above sealevel 18 m   Climate symbol: Köppen ET   Troll I,2

		J	F	M	A	M	J	J	A	S	O	N	D	year	P	
1 Mean daily temperature	in °C	−32.6	−33.5	−31.3	−23.1	−10.7	−0.3	4.3	2.7	−4.9	−14.7	−24.2	−28.6	−16.4	24	1
2 Mean daily maximum temperature	in °C	−30.0	−31.1	−26.1	−17.2	−5.6	3.3	7.2	5.0	−2.8	−11.7	−18.9	−25.6	−12.8	6	2
3 Mean daily minimum temperature	in °C	−37.8	−38.9	−34.4	−26.7	−12.8	−1.7	1.7	0.0	−7.2	−18.3	−26.7	−32.8	−19.4	6	3
4 Absolute maximum temperature	in °C	−5.6	−14.4	−6.7	−1.1		13.9	15.6	15.0	5.6	−1.1	−2.8	−8.3	15.6	5	4
5 Absolute minimum temperature	in °C	−47.2	−47.8	−48.3	−38.9	−28.9	−13.3	−1.7	−8.3	−17.8	−31.1	−41.7	−46.1	−48.3	5	5
6 Mean relative humidity	in %	85	90	85	88	93	89	84	88	91	91	86	85	88	30	6
7 Mean precipitation	in mm	3	3	3	6	9	13	26	31	18	15	6	5	138	24	7
8 Maximum precipitation	in mm															8
9 Minimum precipitation	in mm															9
10 Maximum precipitation in 24 h	in mm	3	4	3	7	9	18	21	25	15	11	5	4	25	24	10
11 Mean number of days with precipitation	>0.25 mm	5	5	5	6	9	7	9	10	11	14	7	6	94	24	11
12 Mean duration of sunshine	in h	16	17	145	267	274	244	276	159	56	21	0	0	1475	23	12
13 Mean quantity of radiation	in ly/day															13
14 Mean potential evaporation	in mm	0	0	0	0	0	52	90	69	0	0	0	0	211	6	14
15 Mean windspeed	in m/sec	5.8	5.3	4.9	5.4	5.1	5.6	5.3	6.2	6.0	5.9	5.1	5.5	5.5	6	15
16 Mean predominant direction of the wind		NW	NW	NW	NW	N	NW	NW	NW	NW	NW	NW	NW		8	16

15 Station / Country **Sachs Harbour (Northwest Territories) / Canada**

Location 71°57'N / 124°44'W   Height above sealevel 84 m   Climate symbol: Köppen ET   Troll I,2

		J	F	M	A	M	J	J	A	S	O	N	D	year	P	
1 Mean daily temperature	in °C	−29.9	−31.5	−28.2	−19.9	−8.7	1.8	5.6	4.4	−1.6	−12.5	−24.3	−27.9	−14.4		1
2 Mean daily maximum temperature	in °C															2
3 Mean daily minimum temperature	in °C															3
4 Absolute maximum temperature	in °C	−5.0	−6.1	−10.0	2.2	7.2	17.8	18.7	15.6	15.6	0.6	−7.8	−10.6	17.8		4
5 Absolute minimum temperature	in °C	−45.6	−47.8	−43.9	−38.9	−26.7	−13.9	−3.3	−6.1	−13.9	−29.4	−36.7	−45.6	−47.8		5
6 Mean relative humidity	in %															6
7 Mean precipitation	in mm	2	2	4	3	6	5	25	17	17	11	5	4	101		7
8 Maximum precipitation	in mm															8
9 Minimum precipitation	in mm															9
10 Maximum precipitation in 24 h	in mm	1	1	5	1	5	7	22	15	12	4	3	3	22		10
11 Mean number of days with precipitation	>0.25 mm	6	4	5	5	7	4	10	8	10	15	7	7	88		11
12 Mean duration of sunshine	in h	0	59	187	280	263	344	291	202	71	46	3	0	1726		12
13 Mean quantity of radiation	in ly/day															13
14 Mean potential evaporation	in mm	0	0	0	0	0	24	66	49	0	0	0	0	139		14
15 Mean windspeed	in m/sec	5.7	4.9	4.6	4.7	5.5	5.5	5.5	5.8	6.3	6.5	5.6	5.1	5.6		15
16 Mean predominant direction of the wind		N	E	SE	E.SE	E	N.E	NW	SE	E	E	E	E			16

16 Station / Country **Baker Lake (Northwest Territories) / Canada**

Location 64°18'N / 96°00'W   Height above sealevel 4 m   Climate symbol: Köppen ET   Troll I,3

		J	F	M	A	M	J	J	A	S	O	N	D	year	P	
1 Mean daily temperature	in °C	−32.9	−32.8	−26.3	−16.4	−5.8	3.9	10.7	10.0	2.8	−7.5	−20.0	−28.2	−11.9	6	1
2 Mean daily maximum temperature	in °C	−31.1	−30.0	−21.7	−12.2	−1.7	6.1	15.6	15.0	6.1	−4.4	−15.0	−23.9	−7.7	6	2
3 Mean daily minimum temperature	in °C	−38.3	−35.0	−28.9	−21.1	−8.9	−1.1	3.9	5.0	−1.1	−12.8	−24.4	−31.1	−16.1	6	3
4 Absolute maximum temperature	in °C	−12.2	−8.9	−6.1	3.9	9.4	22.2	25.0	27.8	18.9	9.4	2.2	−4.4	27.8	4	4
5 Absolute minimum temperature	in °C	−49.4	−48.3	−50.0	−34.4	−25.6	−9.4	−1.7	−2.2	−8.9	−23.9	−36.7	−45.0	−50.0	3	5
6 Mean relative humidity	in %															6
7 Mean precipitation	in mm	5	4	6	9	8	21	40	45	34	20	9	7	208		7
8 Maximum precipitation	in mm															8
9 Minimum precipitation	in mm															9
10 Maximum precipitation in 24 h	in mm	3	5	5	5	11	14	36	21	26	32	9	5	36		10
11 Mean number of days with precipitation	>0.25 mm	7	6	7	9	8	7	10	10	10	12	10	8	104		11
12 Mean duration of sunshine	in h															12
13 Mean quantity of radiation	in ly/day															13
14 Mean potential evaporation	in mm	0	0	0	0	0	45	104	99	35	0	0	0	283	6	14
15 Mean windspeed	in m/sec	6.3	5.5	5.6	6.3	5.4	4.4	4.7	5.4	5.6	6.4	6.0	6.3	5.6		15
16 Mean predominant direction of the wind		NW	NW	NW	N	N	N	N	N	NW	N	N	NW			16

17 Station / Country Coral Harbour (Northwest Territories) / Canada

Location 64°12'N/83°22'W — Height above sealevel 18 m — Climate symbol: Köppen ET — Troll I,3

			J	F	M	A	M	J	J	A	S	O	N	D	year	P	
1	Mean daily temperature	in °C	−30,4	−29,6	−23,9	−15,7	−6,1	2,3	8,3	7,6	0,9	−8,0	−16,8	−24,8	−11,3		1
2	Mean daily maximum temperature	in °C															2
3	Mean daily minimum temperature	in °C															3
4	Absolute maximum temperature	in °C	−1	−3	−1	4	9	19	24	25	17	4	2	−4	25		4
5	Absolute minimum temperature	in °C	−52	−48	−46	−39	−38	−12	−1	−3	−13	−29	−37	−47	−52		5
6	Mean relative humidity	in %															6
7	Mean precipitation	in mm	8	9	9	14	19	26	35	38	33	29	17	12	249		7
8	Maximum precipitation	in mm															8
9	Minimum precipitation	in mm															9
10	Maximum precipitation in 24 h	in mm	8	8	12	25	23	21	29	27	23	18	8	9	29		10
11	Mean number of days with precipitation	>0,25 mm	7	8	7	8	9	8	10	9	9	12	11	9	107		11
12	Mean duration of sunshine	in h															12
13	Mean quantity of radiation	in ly / day															13
14	Mean potential evaporation	in mm	0	0	0	0	0	50	99	90	20	0	0	0	259		14
15	Mean windspeed	in m / sec	5,3	5,4	5,6	5,6	5,7	5,3	5,5	5,6	5,9	6,7	6,5	5,8	5,7		15
16	Mean predominent direction of the wind		N	N	N	N	N	N	N	N	N	N	N	N			16

18 Station / Country Frobisher Bay (Northwest Territories) / Canada

Location 63°45'N/68°33'W — Height above sealevel 7 m — Climate symbol: Köppen ET — Troll I,3

			J	F	M	A	M	J	J	A	S	O	N	D	year	P	
1	Mean daily temperature	in °C	−26,5	−25,5	−21,5	−13,7	−3,1	3,6	7,9	6,9	2,2	−4,7	−12,3	−20,5	−8,9	25	1
2	Mean daily maximum temperature	in °C	−22,2	−20,8	−17,6	−9,3	0,1	6,6	11,9	10,3	5,1	−1,9	−8,7	−16,2	−5,2	30	2
3	Mean daily minimum temperature	in °C	−30,1	−29,7	−27,1	−14,0	−6,7	0,4	3,9	3,6	0,3	−7,6	−16,2	−24,4	−12,7	30	3
4	Absolute maximum temperature	in °C	3,9	4,4	3,9	5,0	13,3	21,7	24,4	23,3	17,2	7,2	5,6	3,3	24,4	25	4
5	Absolute minimum temperature	in °C	−45,0	−45,6	−43,9	−33,9	−26,1	−8,3	−2,8	−1,1	−12,8	−22,2	−36,1	−42,2	−45,6	25	5
6	Mean relative humidity	in %	77	78	78	79	80	80	73	79	79	81	77	79	78	10	6
7	Mean precipitation	in mm	24	28	21	22	23	38	53	58	43	42	37	26	415	25	7
8	Maximum precipitation	in mm															8
9	Minimum precipitation	in mm															9
10	Maximum precipitation in 24 h	in mm	41	25	24	18	27	29	53	34	27	21	28	22	53	25	10
11	Mean number of days with precipitation	>0,25 mm	10	11	9	9	10	10	11	11	13	14	12	12	135	25	11
12	Mean duration of sunshine	in h	35	88	181	234	183	165	210	143	78	59	43	16	1435	15	12
13	Mean quantity of radiation	in ly / day															13
14	Mean potential evaporation	in mm	0	0	0	0	0	63	99	82	39	0	0	0	283	10	14
15	Mean windspeed	in m / sec	4,7	4,7	4,4	5,1	6,1	5,3	4,0	4,1	5,4	6,4	5,5	5,0	5,1	12	15
16	Mean predominent direction of the wind		NW	NW	NW	NW	NW	NW	SE	SE	NW	NW	NW	NW		12	16

19 Station / Country Port Harrison (Québec) / Canada

Location 58°27'N/78°08'W — Height above sealevel 6 m — Climate symbol: Köppen ET — Troll I,3

			J	F	M	A	M	J	J	A	S	O	N	D	year	P	
1	Mean daily temperature	in °C	−25,0	−25,3	−19,8	−10,8	−2,2	4,4	8,9	8,6	5,0	−0,4	−8,1	−18,3	−8,9		1
2	Mean daily maximum temperature	in °C	−23,3	−23,3	−16,7	−6,7	2,2	7,2	12,8	12,2	7,8	−2,2	−5,6	−15,0	−3,9	18	2
3	Mean daily minimum temperature	in °C	−32,2	−32,2	−26,1	−16,7	−5,6	0,0	3,9	4,4	1,7	−3,9	−11,1	−22,2	−11,7	18	3
4	Absolute maximum temperature	in °C	0,6		3,3	6,7	23,3	25,6	26,7	23,9	22,2	16,7	8,9	1,7	26,7	29	4
5	Absolute minimum temperature	in °C	−46,1	−49,4	−45,0	−32,2	−20,6	−9,4	−6,7	−2,2	−11,1	−22,8	−33,9	−43,3	−49,4	29	5
6	Mean relative humidity	in %				92	94	90	87	92	92	93	93			6	6
7	Mean precipitation	in mm	14	9	16	17	23	30	51	54	62	49	47	23	395		7
8	Maximum precipitation	in mm															8
9	Minimum precipitation	in mm															9
10	Maximum precipitation in 24 h	in mm	14	10	23	14	25	37	25	23	30	23	23	11	36		10
11	Mean number of days with precipitation	>0,25 mm	7	6	8	9	11	8	11	13	16	16	16	12	133		11
12	Mean duration of sunshine	in h	60	110	155	162	135	186	198	156	76	50	27	27	1342		12
13	Mean quantity of radiation	in ly / day															13
14	Mean potential evaporation	in mm	0	0	0	0	0	64	97	89	56	0	0	0	306	26	14
15	Mean windspeed	in m / sec	5,7	5,9	6,1	6,2	6,6	5,9	6,1	6,3	6,8	7,0	7,4	6,4	6,4		15
16	Mean predominent direction of the wind		W	NE	N	N	N	N	S	N	N	N	NE	N			16

20 Station / Country Aklavik (Northwest Terrritories) / Canada

Location 68°14'N/134°50'W — Height above sealevel 10 m — Climate symbol: Köppen Dfc — Troll II,3

			J	F	M	A	M	J	J	A	S	O	N	D	year	P	
1	Mean daily temperature	in °C	−28,6	−27,4	−22,3	−12,7	−0,4	9,6	13,8	10,8	3,6	−7,1	−19,5	−27,3	−8,9	22	1
2	Mean daily maximum temperature	in °C	−23,3	−22,8	−17,8	−7,2	4,4	14,4	18,9	14,4	6,7	−3,9	−16,1	−22,2	−4,4	22	2
3	Mean daily minimum temperature	in °C	−32,2	−31,1	−27,2	−18,9	−5,6	4,4	8,3	5,6	0,0	−9,4	−22,8	−31,1	−13,3	22	3
4	Absolute maximum temperature	in °C	2,2	9,4	7,2	13,3	25,0	30,0	30,6	28,9	24,4	12,8	6,7	10,0	30,6	19	4
5	Absolute minimum temperature	in °C	−48,9	−52,2	−43,3	−42,2	−25,6	−6,7	−1,1	−3,9	−11,1	−28,3	−45,6	−47,8	−52,2	19	5
6	Mean relative humidity	in %			70	72	72	64	65	74	80	85	80			10	6
7	Mean precipitation	in mm	12	11	11	8	8	18	34	36	20	32	21	24	235	22	7
8	Maximum precipitation	in mm															8
9	Minimum precipitation	in mm															9
10	Maximum precipitation in 24 h	in mm	20	25	25	25	51	30	41	44	16	19	21	15	51	37	10
11	Mean number of days with precipitation	>0,25 mm	8	7	8	7	5	6	9	12	11	11	10	9	103	22	11
12	Mean duration of sunshine	in h	14	94	154	277	297	401	276	206	95	73	22	0	1909	5	12
13	Mean quantity of radiation	in ly / day															13
14	Mean potential evaporation	in mm	0	0	0	0	0	92	116	89	38	0	0	0	333	22	14
15	Mean windspeed	in m / sec	2,0	1,8	2,6	3,0	3,3	3,5	3,2	3,1	3,0	2,7	2,1	2,2	2,7	6	15
16	Mean predominent direction of the wind		E	E	E	E	E	NW	NW	E	E	E	E	NW		6	16

21 Station / Country   Norman Wells (Northwest Territories) / Canada

Location 65°17'N / 126°48'W   Height above sealevel 20 m   Climate symbol: Köppen **Dfc**   Troll II,3

		J	F	M	A	M	J	J	A	S	O	N	D	year	P	
1 Mean daily temperature	in °C	-28,2	-26,3	-18,8	-7,8	-5,2	13,4	15,9	13,1	6,1	-4,0	-17,8	-26,0	-6,2	10	1
2 Mean daily maximum temperature	in °C	-23,9	-21,7	-12,2	-0,8	11,7	20,0	22,2	18,3	10,0	0,0	-12,8	-21,1	-0,8	10	2
3 Mean daily minimum temperature	in °C	-32,2	-30,8	-24,4	-13,9	0,0	7,8	10,0	7,2	1,7	-6,7	-20,0	-29,4	-11,1	10	3
4 Absolute maximum temperature	in °C	-1,1	0,0	1,1	6,1	8,3	18,3	25,0	19,4	13,9	11,1	2,2	-0,6	25,0	7	4
5 Absolute minimum temperature	in °C	-38,3	-40,0	-40,0	-32,2	-20,6	-11,7	-3,9	-5,6	-10,6	-15,6	-28,3	-33,9	-40,0	8	5
6 Mean relative humidity	in %	91	89	91	93	93	89	87	91	93	92	94	94	91	4	6
7 Mean precipitation	in mm	19	16	12	13	15	34	49	64	38	24	24	20	328		7
8 Maximum precipitation	in mm															8
9 Minimum precipitation	in mm															9
10 Maximum precipitation in 24 h	in mm	18	12	7	24	21	37	45	49	27	12	17	17	49		10
11 Mean number of days with precipitation >0,25 mm		13	12	10	7	7	10	11	12	12	11	14	13	132		11
12 Mean duration of sunshine	in h															12
13 Mean quantity of radiation	in ly / day															13
14 Mean potential evaporation	in mm	0	0	0	0	53	111	126	96	46	0	0	0	432	10	14
15 Mean windspeed	in m / sec	2,8	2,2	2,4	3,8	3,5	3,8	3,2	3,3	3,1	3,2	2,5	2,3	3,0		15
16 Mean predominent direction of the wind		NW	SE	NW	SE	SE	SE	NW	SE	SE	NW	NW	NW			16

22 Station / Country   Yellowknife (Northwest Territories) / Canada

Location 62°28'N / 114°20'W   Height above sealevel 215 m   Climate symbol: Köppen **Dfc**   Troll II,3

		J	F	M	A	M	J	J	A	S	O	N	D	year	P	
1 Mean daily temperature	in °C	-27,8	-26,1	-18,1	-7,8	4,2	11,4	15,9	14,2	7,5	-0,2	-13,9	-25,3	-5,5	10	1
2 Mean daily maximum temperature	in °C	-23,3	-21,1	-12,2	-1,7	8,9	16,1	20,6	18,3	11,1	2,2	-10,0	-21,1	-1,1	10	2
3 Mean daily minimum temperature	in °C	-32,2	-31,1	-23,9	-13,9	-0,6	6,7	11,1	10,0	3,9	-3,3	-17,8	-29,4	-10,0	10	3
4 Absolute maximum temperature	in °C	-5,6	-6,1	3,3	14,4	26,1	27,2	30,0	30,0	26,1	18,3	6,7	0,0	30,0	8	4
5 Absolute minimum temperature	in °C	-51,7	-51,7	-42,8	-36,1	-15,6	-2,2	0,6	1,1	-7,2	-22,2	-36,7	-48,3	-51,7	8	5
6 Mean relative humidity	in %		78	76	73	63	58	62	71	77	86	87	80		10	6
7 Mean precipitation	in mm	14	12	12	10	14	17	33	36	28	31	24	19	250	29	7
8 Maximum precipitation	in mm															8
9 Minimum precipitation	in mm															9
10 Maximum precipitation in 24 h	in mm	13	8	11	14	34	20	41	45	24	36	12	11	45	29	10
11 Mean number of days with precipitation >0,25 mm		10	10	9	8	6	9	10	10	11	14	14	14		29	11
12 Mean duration of sunshine	in h															12
13 Mean quantity of radiation	in ly / day															13
14 Mean potential evaporation	in mm	0	0	0	0	47	99	123	104	56	0	0	0	429	10	14
15 Mean windspeed	in m / sec	4,1	3,8	4,3	4,9	4,9	5,0	4,8	4,6	5,0	5,1	4,4	3,9	4,6	12	15
16 Mean predominent direction of the wind		NW	NW	NE	E	NE	S	S	E	NW	E	E	E		12	16

23 Station / Country   Whitehorse (Yukon) / Canada

Location 60°43'N / 135°04'W   Height above sealevel 2128 m   Climate symbol: Köppen **Dfc**   Troll II,2

		J	F	M	A	M	J	J	A	S	O	N	D	year	P	
1 Mean daily temperature	in °C	-18,1	-14,1	-7,6	-0,2	7,5	12,6	14,2	12,4	7,9	0,7	-8,2	-15,1	-0,7	10	1
2 Mean daily maximum temperature	in °C	-10,6	-8,9	-0,8	5,0	13,9	18,9	19,4	17,8	12,8	5,6	-6,1	-11,7	4,4	10	2
3 Mean daily minimum temperature	in °C	-19,4	-18,9	-11,1	-5,6	1,1	6,1	7,2	6,1	2,8	-2,2	-13,3	-20,0	-5,6	10	3
4 Absolute maximum temperature	in °C	8,3	10,0	12,2	15,0	30,0	31,7	32,8	28,9	26,7	18,9	10,6	8,3	32,8	10	4
5 Absolute minimum temperature	in °C	-52,2	-50,6	-37,8	-25,6	-7,2	-1,1	0,0	-4,4	-9,4	-20,0	-40,6	-47,8	-52,2	10	5
6 Mean relative humidity	in %	86	83	73	66	58	61	64	66	70	72	86	89	73	6	6
7 Mean precipitation	in mm	18	14	15	11	13	27	35	37	25	19	23	20	257		7
8 Maximum precipitation	in mm															8
9 Minimum precipitation	in mm															9
10 Maximum precipitation in 24 h	in mm	9	10	20	14	12	21	21	31	22	12	12	11	31		10
11 Mean number of days with precipitation >0,25 mm		12	10	8	8	5	8	12	10	9	9	12	12	113		11
12 Mean duration of sunshine	in h	48	74	164	246	285	295	241	219	148	114	56	28	1898		12
13 Mean quantity of radiation	in ly / day															13
14 Mean potential evaporation	in mm	0	0	0	0	70	105	110	94	56	15	0	0	450	10	14
15 Mean windspeed	in m / sec	3,9	4,0	4,0	3,9	3,9	3,8	3,3	3,5	4,1	4,7	4,1	3,9	3,9		15
16 Mean predominent direction of the wind		S	S	S	S	SE	SE	SE	SE	S,SE	S	S	S			16

24 Station / Country   Fort Smith (Northwest Territories) / Canada

Location 60°01'N / 111°58'W   Height above sealevel 62 m   Climate symbol: Köppen **Dfc**   Troll II,3

		J	F	M	A	M	J	J	A	S	O	N	D	year	P	
1 Mean daily temperature	in °C	-25,4	-22,2	-14,4	-2,8	7,8	12,9	16,2	14,2	7,9	0,2	-11,7	-21,3	-3,2		1
2 Mean daily maximum temperature	in °C	-22,2	-17,2	-9,4	3,3	13,3	19,4	23,3	20,6	12,8	3,3	-7,7	-17,8	1,7	26	2
3 Mean daily minimum temperature	in °C	-31,7	-28,9	-22,8	-10,0	0,0	5,6	8,3	6,1	1,1	-5,6	-18,1	-27,2	-10,0	26	3
4 Absolute maximum temperature	in °C	12,2	12,2	15,0	26,7	31,1	32,2	39,4	34,4	31,7	26,1	10,6	11,1	39,4	25	4
5 Absolute minimum temperature	in °C	-53,3	-51,1	-46,3	-36,7	-12,2	-5,6	-3,9	-6,7	-11,1	-10,0	-41,1	-51,1	-53,3	25	5
6 Mean relative humidity	in %			67	68	65	62	66	72	76	82	91	95		10	6
7 Mean precipitation	in mm	15	17	19	17	26	31	53	35	42	29	26	27	337		7
8 Maximum precipitation	in mm															8
9 Minimum precipitation	in mm															9
10 Maximum precipitation in 24 h	in mm	15	19	15	15	25	29	31	20	23	21	24	13	31		10
11 Mean number of days with precipitation >0,25 mm		11	11	9	7	7	8	10	10	12	12	13	14	124		11
12 Mean duration of sunshine	in h	61	117	167	240	282	317	296	268	133	66	46	27	2040		12
13 Mean quantity of radiation	in ly / day															13
14 Mean potential evaporation	in mm	0	0	0	0	63	106	127	101	54	0	0	0	451		14
15 Mean windspeed	in m / sec	2,0	2,9	3,3	3,5	3,3	3,0	3,0	2,9	3,1	3,4	3,0	2,3	3,0		15
16 Mean predominent direction of the wind		NW	SE	SE	NW,SE	SE	NW	NW	NW	NW	NW	SE	SE			16

25 Station / Country Fort Nelson (British–Columbia) / Canada

Location 58°50'N / 122°35'W **Height above sealevel** 114 m **Climate symbol: Köppen** Dfb **Troll** II,2

		J	F	M	A	M	J	J	A	S	O	N	D	year	P	
1 Mean daily temperature	in °C	-22,4	-17,6	-8,7	1,5	10,0	14,3	16,8	14,7	9,3	1,2	-12,1	-20,4	-1,1	10	1
2 Mean daily maximum temperature	in °C	-16,7	-11,7	-0,6	10,0	16,7	20,6	23,3	22,2	16,1	7,2	-10,0	-16,7	5,0	10	2
3 Mean daily minimum temperature	in °C	-25,0	-22,2	-15,0	-3,9	3,3	7,8	10,0	8,3	3,3	-2,8	-17,2	-23,9	-6,7	10	3
4 Absolute maximum temperature	in °C	7,2	13,9	18,7	24,4	31,7	31,1	36,7	33,9	32,8	25,6	11,7	7,8	36,7	10	4
5 Absolute minimum temperature	in °C	-52,2	-46,3	-36,7	-28,3	-4,4	-0,6	3,3	-1,7	-7,8	-21,7	-35,6	-47,8	-52,2	10	5
6 Mean relative humidity	in %	88	85	71	67	62	72	73	70	75	80	90	91	77	5	6
7 Mean precipitation	in mm	24	26	26	19	39	68	65	51	34	26	31	28	435		7
8 Maximum precipitation	in mm															8
9 Minimum precipitation	in mm															9
10 Maximum precipitation in 24 h	in mm	19	20	21	20	35	52	48	63	29	16	29	35	63		10
11 Mean number of days with precipitation	>0,25 mm	11	11	11	7	9	12	12	11	9	8	11	13	125		11
12 Mean duration of sunshine	in h															12
13 Mean quantity of radiation	in ly / day															13
14 Mean potential evaporation	in mm	0	0	0	15	78	109	125	100	63	12	0	0	502	13	14
15 Mean windspeed	in m / sec	1,8	2,1	2,5	2,9	2,9	2,7	2,5	2,4	2,3	2,2	1,8	1,6	2,3		15
16 Mean predominent direction of the wind		S	N	N	N	N	N	S	S	S	S	S	S			16

26 Station / Country Churchill (Manitoba) / Canada

Location 58°45'N / 94°04'W **Height above sealevel** 11 m **Climate symbol: Köppen** Dfc **Troll** II,3

		J	F	M	A	M	J	J	A	S	O	N	D	year	P	
1 Mean daily temperature	in °C	-27,5	-26,4	-19,8	-10,7	-2,3	5,8	12,0	11,6	5,7	-1,1	-11,7	-21,9	-7,2	30	1
2 Mean daily maximum temperature	in °C	-23,9	-22,2	-15,6	-4,4	3,3	11,1	17,8	16,7	9,4	1,1	-10,6	-19,4	-3,3	30	2
3 Mean daily minimum temperature	in °C	-32,8	-31,7	-26,7	-15,6	-5,6	1,1	6,1	6,1	1,1	-6,7	-18,9	-28,3	-12,8	30	3
4 Absolute maximum temperature	in °C	3,9	3,9	5,0	16,7	30,6	31,1	35,6	32,2	29,9	18,3	7,2	1,1	35,6	30	4
5 Absolute minimum temperature	in °C	-49,4	-46,7	-46,7	-32,2	-25,8	-10,6	-5,6	-3,9	-9,4	-27,2	-47,2	-43,9	-49,4	30	5
6 Mean relative humidity	in %				91	89	81	80	84	89	92	93	95		3	6
7 Mean precipitation	in mm	13	14	17	26	30	41	52	61	53	38	39	23	407		7
8 Maximum precipitation	in mm															8
9 Minimum precipitation	in mm															9
10 Maximum precipitation in 24 h	in mm	9	13	17	28	30	33	53	41	42	24	22	20	53		10
11 Mean number of days with precipitation	>0,25 mm	9	9	10	12	11	9	12	13	14	15	17	13	144		11
12 Mean duration of sunshine	in h	79	127	180	186	166	212	285	232	100	67	44	54	1732		12
13 Mean quantity of radiation	in ly / day															13
14 Mean potential evaporation	in mm	0	0	0	0	71	112	96	51	0	0	0	0	330	30	14
15 Mean windspeed	in m / sec	6,3	6,4	6,2	6,5	5,9	5,4	5,5	5,6	6,6	7,2	6,6	7,2	6,3		15
16 Mean predominent direction of the wind		NW	NW	NW	NW	N	N	N	NW	N	NW	NW	NW			16

27 Station / Country Prince Rupert (British Columbia) / Canada

Location 54°17'N / 130°23'W **Height above sealevel** 16 m **Climate symbol: Köppen** Cfb **Troll** III,3

		J	F	M	A	M	J	J	A	S	O	N	D	year	P	
1 Mean daily temperature	in °C	1,8	2,4	3,8	6,3	9,5	11,7	13,4	13,9	12,1	8,7	5,2	2,8	7,6	26	1
2 Mean daily maximum temperature	in °C	3,9	5,6	7,2	10,0	12,8	15,6	16,7	17,8	15,6	11,7	7,8	4,4	11,1	26	2
3 Mean daily minimum temperature	in °C	-1,1	-0,6	0,6	2,8	5,0	7,8	9,4	10,6	8,3	5,6	2,8	0,0	4,4	26	3
4 Absolute maximum temperature	in °C	16,7	18,9	20,0	21,7	28,9	31,1	30,6	30,0	26,1	21,7	20,0	15,6	31,1	30	4
5 Absolute minimum temperature	in °C	-19,4	-16,7	-9,4	-5,6	-1,1	1,7	0,6	3,9	-1,1	-5,6	-10,0	-17,2	-19,4	30	5
6 Mean relative humidity	in %	81	81	79	79	79	83	85	87	84	84	83	85	83	5	6
7 Mean precipitation	in mm	225	177	196	173	130	108	117	149	217	336	293	278	2399		7
8 Maximum precipitation	in mm															8
9 Minimum precipitation	in mm															9
10 Maximum precipitation in 24 h	in mm	125	65	70	71	49	57	55	49	58	141	138	126	141		10
11 Mean number of days with precipitation	>0,25 mm	20	18	20	19	17	16	17	16	18	24	22	23	230		11
12 Mean duration of sunshine	in h	37	58	80	106	140	107	108	112	85	50	35	25	943		12
13 Mean quantity of radiation	in ly / day															13
14 Mean potential evaporation	in mm	0	0	0	32	73	102	116	104	61	29	0	0	517	26	14
15 Mean windspeed	in m / sec	3,8	3,4	3,3	3,3	2,2	1,8	1,6	1,7	2,3	3,5	3,7	3,9	2,9		15
16 Mean predominent direction of the wind		SE	SE	SE	SE	SE	SE	SE	SE	SE	SE	SE	SE			16

28 Station / Country Prince George (British Columbia) / Canada

Location 53°53'N / 122°41'W **Height above sealevel** 206 m **Climate symbol: Köppen** Dfb **Troll** II,2

		J	F	M	A	M	J	J	A	S	O	N	D	year	P	
1 Mean daily temperature	in °C	-11,3	-7,5	-2,3	4,3	9,7	12,9	14,9	13,7	10,1	4,8	-2,5	-6,6	3,3	27	1
2 Mean daily maximum temperature	in °C	-5,0	-0,6	5,6	12,2	17,8	21,1	23,9	23,3	18,3	11,1	3,3	-3,9	10,6	27	2
3 Mean daily minimum temperature	in °C	-16,1	-14,4	-7,7	-2,8	1,1	5,6	6,7	6,1	-1,1	-8,1	-13,3		-3,3	27	3
4 Absolute maximum temperature	in °C	12,2	14,4	20,0	30,0	35,0	33,9	38,9	35,6	33,3	28,9	16,7	12,8	38,9	27	4
5 Absolute minimum temperature	in °C	-49,4	-46,7	-37,2	-14,4	-11,1	-4,4	-2,2	-3,9	-14,4	-20,0	-33,3	-48,9	-49,4	27	5
6 Mean relative humidity	in %	88	85	71	67	62	72	73	70	75	80	90	91	77	5	6
7 Mean precipitation	in mm	56	44	36	28	43	62	64	65	56	59	57	56	626		7
8 Maximum precipitation	in mm															8
9 Minimum precipitation	in mm															9
10 Maximum precipitation in 24 h	in mm	29	23	20	16	21	39	26	50	28	21	22	29	50		10
11 Mean number of days with precipitation	>0,25 mm	17	14	13	11	11	14	14	13	13	15	15	17	167		11
12 Mean duration of sunshine	in h	54	89	130	183	248	240	267	243	166	101	56	41	1816		12
13 Mean quantity of radiation	in ly / day															13
14 Mean potential evaporation	in mm	0	0	0	32	73	102	116	104	61	29	0	0	517	27	14
15 Mean windspeed	in m / sec	3,6	3,5	3,8	3,8	3,3	3,1	2,7	2,7	2,8	3,4	3,5	3,6	3,3		15
16 Mean predominent direction of the wind		S	S	S	S	S	S	S	S	S	S	S				16

## 29

Station / Country Trout Lake (Ontario) / Canada

Location 53°50'N / 89°52'W Height above sealevel 67 m Climate symbol: Köppen Dfc Troll II,2

		J	F	M	A	M	J	J	A	S	O	N	D	year	P	
1 Mean daily temperature	in °C	-23,9	-21,4	-14,4	-4,6	3,6	11,2	15,9	14,7	8,7	1,8	-8,9	-19,3	-3,1		1
2 Mean daily maximum temperature	in °C	-16,3	-20,0	-7,7	1,1	8,9	16,1	21,1	19,4	13,3	5,6	-5,6	-15,0	1,7	12	2
3 Mean daily minimum temperature	in °C	-29,4	-27,8	-22,2	-11,7	-2,2	5,6	10,6	10,0	5,0	-1,1	-12,2	-23,9	-8,3	12	3
4 Absolute maximum temperature	in °C	2	5	12	21	30	32	36	30	29	20	10	4	36		4
5 Absolute minimum temperature	in °C	-48	-47	-42	-32	-21	-7	0	-1	-8	-19	-34	-44	-48		5
6 Mean relative humidity	in %	87	89	86	80	75	74	77	81	82	85	91	90	83	6	6
7 Mean precipitation	in mm	25	21	17	26	48	75	100	92	75	53	47	29	608		7
8 Maximum precipitation	in mm															8
9 Minimum precipitation	in mm															9
10 Maximum precipitation in 24 h	in mm	29	15	10	33	25	49	75	84	64	29	31	15	84		10
11 Mean number of days with precipitation	>0,25 mm	12	12	9	9	12	13	15	15	15	13	17	15	157		11
12 Mean duration of sunshine	in h															12
13 Mean quantity of radiation	in ly / day															13
14 Mean potential evaporation	in mm	0	0	0	0	39	91	125	108	62	11	0	0	436		14
15 Mean windspeed	in m / sec	4,0	4,0	4,0	4,7	4,5	4,5	4,3	4,6	5,2	5,1	5,1	4,2	4,5		15
16 Mean predominent direction of the wind		NW	NW	NW	N,NW	NW	NW	S,SW	NW	NW	NW	NW	W			16

## 30

Station / Country Edmonton (Alberta) Canada

Location 53°34'N / 113°31'W Height above sealevel 206 m Climate symbol: Köppen Dfb Troll II,2

		J	F	M	A	M	J	J	A	S	O	N	D	year	P	
1 Mean daily temperature	in °C	-14,1	-11,6	-5,5	4,2	11,2	14,3	17,3	15,6	10,8	5,1	-4,2	-10,4	2,7	56	1
2 Mean daily maximum temperature	in °C	-9,4	-5,6	1,1	11,1	17,8	21,1	23,3	22,2	16,7	11,1	1,1	-6,1	8,9	56	2
3 Mean daily minimum temperature	in °C	-20,0	-17,2	-11,1	-2,2	3,3	7,2	9,4	8,3	3,3	-1,1	-8,9	-15,0	-3,9	56	3
4 Absolute maximum temperature	in °C	13,9	16,7	22,2	31,1	34,4	37,2	36,7	35,6	32,2	28,3	23,3	16,1	37,2	56	4
5 Absolute minimum temperature	in °C	-49,4	-49,4	-40,0	-26,1	-12,2	-3,9	-1,7	-3,3	-11,1	-26,1	-42,2	-43,3	-49,4	56	5
6 Mean relative humidity	in %	82	81	77	65	64	77	74	75	76	73	81	83	76	7	6
7 Mean precipitation	in mm	24	20	21	28	46	80	85	65	34	23	22	25	473		7
8 Maximum precipitation	in mm															8
9 Minimum precipitation	in mm															9
10 Maximum precipitation in 24 h	in mm	15	13	18	38	41	64	114	52	49	18	40	21	114		10
11 Mean number of days with precipitation	>0,25 mm	12	10	10	7	9	13	13	12	9	7	8	11	121		11
12 Mean duration of sunshine	in h	86	119	163	221	258	251	315	269	186	157	100	78	2203		12
13 Mean quantity of radiation	in ly / day															13
14 Mean potential evaporation	in mm	0	0	0	31	80	106	123	103	60	29	0	0	532	56	14
15 Mean windspeed	in m / sec	3,5	3,6	4,0	4,8	4,7	4,4	4,0	3,7	4,0	4,0	3,6	3,3	40		15
16 Mean predominent direction of the wind		S	S	S	S	NW	NW	NW	NW	NW	S	S	S			16

## 31

Station / Country The Pas (Manitoba) / Canada

Location 53°58'N / 101°06'W Height above sealevel 83 m Climate symbol: Köppen Dfb Troll II,2

		J	F	M	A	M	J	J	A	S	O	N	D	year	P	
1 Mean daily temperature	in °C	-21,7	-18,2	-11,4	-0,3	8,7	13,9	18,2	16,7	10,3	3,3	-7,9	-16,8	-0,4	27	1
2 Mean daily maximum temperature	in °C	-17,2	-12,2	-4,4	7,2	15,6	21,1	23,9	22,2	15,6	7,2	-3,9	-12,8	5,0	27	2
3 Mean daily minimum temperature	in °C	-27,8	-25,0	-18,9	-6,1	2,2	8,9	12,2	10,0	3,3	-3,3	-13,3	-22,2	-6,7	27	3
4 Absolute maximum temperature	in °C	6,7	13,3	15,0	29,4	33,9	37,8	35,6	32,7	26,7	14,4	8,9	37,8	27		4
5 Absolute minimum temperature	in °C	-47,8	-47,8	-41,7	-29,4	-10,6	-3,9	0,0	-5,6	-7,2	-23,3	-36,1	-46,1	-47,8	27	5
6 Mean relative humidity	in %	92	84	83	73	69	70	74	77	75	81	86	90	79	3	6
7 Mean precipitation	in mm	20	17	21	26	45	60	68	59	55	28	29	23	451		7
8 Maximum precipitation	in mm															8
9 Minimum precipitation	in mm															9
10 Maximum precipitation in 24 h	in mm	12	10	26	22	41	48	57	68	53	29	14	14	68		10
11 Mean number of days with precipitation	>0,25 mm	12	10	9	8	8	10	12	12	11	9	13	13	127		11
12 Mean duration of sunshine	in h	94	123	165	211	258	244	300	259	158	125	64	66	2067		12
13 Mean quantity of radiation	in ly / day															13
14 Mean potential evaporation	in mm	0	0	0	14	69	113	134	110	57	12	0	0	509	27	14
15 Mean windspeed	in m / sec	4,4	4,3	4,5	4,8	4,9	4,5	4,8	4,6	5,0	5,4	5,2	4,7	4,8		15
16 Mean predominent direction of the wind		W	E	W	W	N	N	W	E,W	W	W	W	W			16

Station / Country Goose Bay (Newfoundland) / Canada

Location 53°19'N / 60°25'W Height above sealevel 13 m Climate symbol: Köppen Dfb Troll II,2

		J	F	M	A	M	J	J	A	S	O	N	D	year	P	
1 Mean daily temperature	in °C	-16,6	-14,9	-8,4	-1,6	5,1	11,9	16,3	14,7	10,1	3,2	-4,4	-12,9	0,2	10	1
2 Mean daily maximum temperature	in °C	-13,3	-10,0	-3,9	2,8	9,4	16,1	21,7	19,4	15,0	7,2	-0,6	-8,9	4,4	10	2
3 Mean daily minimum temperature	in °C	-22,2	-20,6	-14,4	-6,0	0,0	5,6	11,1	9,4	5,6	-0,6	-7,7	-16,7	-5,0	10	3
4 Absolute maximum temperature	in °C	5,6	7,8	10,6	16,7	31,7	31,7	37,8	32,8	28,9	22,8	14,4	11,7	37,8	9	4
5 Absolute minimum temperature	in °C	-35,6	-37,2	-35,6	-25,0	-12,2	-1,1	3,3	0,0	-6,7	-11,7	-22,8	-31,7	-37,2	9	5
6 Mean relative humidity	in %	82	82	73	67	63	62	62	65	68	71	79	87	72	9	6
7 Mean precipitation	in mm	72	63	68	62	56	72	84	91	76	63	67	63	837		7
8 Maximum precipitation	in mm															8
9 Minimum precipitation	in mm															9
10 Maximum precipitation in 24 h	in mm	30	40	37	43	34	29	35	66	43	30	41	32	66		10
11 Mean number of days with precipitation	>0,25 mm	16	14	14	14	13	15	15	15	14	14	14	15	173		11
12 Mean duration of sunshine	in h	90	111	143	136	176	198	194	187	124	91	69	66	1585		12
13 Mean quantity of radiation	in ly / day															13
14 Mean potential evaporation	in mm	0	0	0	0	44	88	126	105	68	23	0	0	454	10	14
15 Mean windspeed	in m / sec	4,8	4,4	4,5	4,4	4,2	3,9	3,8	3,8	4,2	4,5	4,2	4,4	4,3		15
16 Mean predominent direction of the wind		W	W	W	NW	NE	NE	SW	SW,W	W	W	W	W			16

33 Station/Country: Calgary (Alberta)/Canada

Location 51°06'N/114°01'W — Height above sealevel 329 m — Climate symbol: Köppen Dfb — Troll II,2

		J	F	M	A	M	J	J	A	S	O	N	D	year	P	
1	Mean daily temperature in °C	-9.9	-8.8	-4.4	3.6	9.8	13.0	16.7	15.1	10.9	5.4	-2.2	-6.6	3.6	55	1
2	Mean daily maximum temperature in °C	-4.4	-2.2	2.8	11.7	17.2	20.6	24.4	23.3	17.8	12.2	3.3	-1.7	10.6	55	2
3	Mean daily minimum temperature in °C	-16.7	-14.4	-10.0	-2.8	2.2	6.1	8.3	7.2	2.8	-1.7	-8.3	-12.8	-3.3	55	3
4	Absolute maximum temperature in °C	16.1	18.9	23.9	29.4	32.2	35.0	36.1	35.6	32.2	29.4	21.7	19.4	36.1	67	4
5	Absolute minimum temperature in °C	-44.4	-45.0	-37.2	-25.6	-11.1	-3.3	0.0	-2.2	-13.3	-22.2	-35.0	-42.8	-45.0	67	5
6	Mean relative humidity in %	73	76	71	65	64	69	67	65	70	70	74	72	70	7	6
7	Mean precipitation in mm	17	20	26	35	52	88	58	59	35	23	16	15	444		7
8	Maximum precipitation in mm															8
9	Minimum precipitation in mm															9
10	Maximum precipitation in 24 h in mm	20	28	20	32	41	79	68	81	31	24	28	14	81		10
11	Mean number of days with precipitation >0.25 mm	9	10	11	10	11	14	11	12	8	6	7	8	117		11
12	Mean duration of sunshine in h	101	117	146	188	240	234	318	275	186	159	111	91	2166		12
13	Mean quantity of radiation in ly/day															13
14	Mean potential evaporation in mm	0	0	0	31	76	99	123	105	63	31	0	0	528	55	14
15	Mean windspeed in m/sec	4.5	4.3	4.3	5.0	5.0	4.6	4.1	4.0	4.4	4.4	4.3	4.4	4.4		15
16	Mean predominant direction of the wind	W	W	N,S	SE	NW	N,NW	NW	N	NW	S,W	W	W			16

34 Station/Country: Saskatoon (Saskatchewan)/Canada

Location 52°08'N/106°38'W — Height above sealevel 157 m — Climate symbol: Köppen Dfb — Troll III,5

		J	F	M	A	M	J	J	A	S	O	N	D	year	P	
1	Mean daily temperature in °C	-17.6	-14.9	-7.9	3.6	11.2	15.4	19.3	17.6	11.6	4.9	-5.8	-13.2	2.0	38	1
2	Mean daily maximum temperature in °C	-12.8	-10.6	-2.8	9.4	17.8	21.7	25.0	23.9	17.2	10.6	-0.6	-8.9	7.8	38	2
3	Mean daily minimum temperature in °C	-23.9	-22.2	-14.4	-3.3	3.3	8.9	11.1	9.2	3.3	-2.8	-11.1	-18.9	-5.0	38	3
4	Absolute maximum temperature in °C	10.0	12.8	22.8	32.8	37.2	40.0	40.0	37.8	33.3	32.2	20.0	14.4	40.0	38	4
5	Absolute minimum temperature in °C	-48.3	-45.0	-36.7	-22.8	-12.8	-3.3	-0.6	-2.2	-11.1	-25.6	-35.0	-40.6	-48.3	38	5
6	Mean relative humidity in %	92	92	84	68	65	68	67	68	70	74	87	90	77	5	6
7	Mean precipitation in mm	15	16	15	21	34	58	60	45	34	19	18	17	352		7
8	Maximum precipitation in mm															8
9	Minimum precipitation in mm															9
10	Maximum precipitation in 24 h in mm	25	24	20	22	38	63	55	84	44	23	19	29	84		10
11	Mean number of days with precipitation >0.25 mm	1.1	9	8	8	7	10	10	10	8	6	9	9	105		11
12	Mean duration of sunshine in h	96	129	191	226	275	270	340	293	210	166	99	86	2381		12
13	Mean quantity of radiation in ly/day															13
14	Mean potential evaporation in mm	0	0	0	20	77	113	132	110	61	21	0	0	534	38	14
15	Mean windspeed in m/sec	5.2	4.9	4.9	5.6	5.6	5.3	4.9	4.7	5.2	5.2	4.9	4.5	5.1		15
16	Mean predominant direction of the wind	S	SW	SE	NW	SE	NW	NW	SE	NW	S	SW	S			16

35 Station/Country: Regina (Saskatchewan)/Canada

Location 50°26'N/104°40'W — Height above sealevel 175 m — Climate symbol: Köppen Dfb — Troll III,10

		J	F	M	A	M	J	J	A	S	O	N	D	year	P	
1	Mean daily temperature in °C	-16.9	-14.8	-8.1	3.4	11.2	15.3	19.3	17.8	11.9	5.1	-5.4	-12.3	2.2		1
2	Mean daily maximum temperature in °C	-12.2	-10.6	-2.8	10.0	18.3	22.8	26.1	25.0	18.3	11.1	0.0	-8.9	7.8	55	2
3	Mean daily minimum temperature in °C	-23.9	-22.8	-14.4	-3.3	2.8	8.3	10.6	8.9	3.3	-2.8	-11.7	-18.3	-5.6	55	3
4	Absolute maximum temperature in °C	8.9	11.7	24.4	31.7	37.2	38.9	41.7	40.0	37.2	30.6	22.8	15.0	41.7	55	4
5	Absolute minimum temperature in °C	-47.8	-48.9	-42.2	-28.9	-13.9	-5.0	-2.2	-5.0	-12.8	-26.1	-43.9	-48.3	-48.9	55	5
6	Mean relative humidity in %	89	91	88	74	64	72	71	69	72	72	87	87	78	55	6
7	Mean precipitation in mm	19	17	21	21	40	83	55	49	34	18	20	17	394		7
8	Maximum precipitation in mm															8
9	Minimum precipitation in mm															9
10	Maximum precipitation in 24 h in mm	18	13	19	30	46	95	65	79	63	24	24	10	95		10
11	Mean number of days with precipitation >0.25 mm	12	10	11	8	8	13	10	9	8	6	10	10	115		11
12	Mean duration of sunshine in h	98	118	152	215	266	249	334	286	197	170	96	85	2266		12
13	Mean quantity of radiation in ly/day															13
14	Mean potential evaporation in mm	0	0	0	23	77	113	132	110	61	21	0	0	537	55	14
15	Mean windspeed in m/sec	5.7	5.8	5.8	6.3	6.2	5.9	5.3	5.4	5.6	5.9	5.8	5.4	5.8		15
16	Mean predominant direction of the wind	NW	SE	SE	SE	SE	SE	SE,W	SE	SE,NW	SE	SE	SE			16

36 Station/Country: Fort-Chimo (Québec)/Canada

Location 58°06'N/66°25'W — Height above sealevel 11 m — Climate symbol: Köppen Dfc — Troll II,2

		J	F	M	A	M	J	J	A	S	O	N	D	year	P	
1	Mean daily temperature in °C	-23.9	-22.7	-16.2	-9.3	0.4	7.3	11.8	10.5	5.6	-0.3	-8.4	-17.8	-5.2		1
2	Mean daily maximum temperature in °C															2
3	Mean daily minimum temperature in °C															3
4	Absolute maximum temperature in °C	6	5	8	11	31	31	32	28	26	18	10	8	32		4
5	Absolute minimum temperature in °C	-46	-43	-42	-33	-19	-6	-2	-2	-4	-15	-31	-40	-46		5
6	Mean relative humidity in %															6
7	Mean precipitation in mm	21	20	21	19	33	38	50	62	52	39	34	28	417		7
8	Maximum precipitation in mm															8
9	Minimum precipitation in mm															9
10	Maximum precipitation in 24 h in mm	18	17	18	18	28	29	22	25	23	15	20	16	29		10
11	Mean number of days with precipitation >0.25 mm	10	9	12	10	10	11	13	14	16	14	14	15	148		11
12	Mean duration of sunshine in h															12
13	Mean quantity of radiation in ly/day															13
14	Mean potential evaporation in mm	0	0	0	0	0	80	110	90	53	0	0	0	333		14
15	Mean windspeed in m/sec	4.7	4.7	4.3	4.5	4.7	4.6	4.4	4.6	5.3	5.3	5.0	4.9	4.8		15
16	Mean predominant direction of the wind	SW	SW	W	N	N	N	N	N	SW	W	SW	SW			16

155

37 Station / Country: Vancouver (British Columbia) / Canada

Location: 49°11'N / 123°10'W — Height above sealevel 2 m — Climate symbol: Köppen Csb — Troll III,2

		J	F	M	A	M	J	J	A	S	O	N	D	year	P	
1 Mean daily temperature	in °C	2,9	4,1	6,2	9,1	12,8	15,8	17,7	17,6	14,3	10,2	6,2	4,2	10,2	43	1
2 Mean daily maximum temperature	in °C	5,0	6,7	10,0	14,4	17,8	20,6	23,3	22,8	18,3	13,9	8,9	6,1	13,9	43	2
3 Mean daily minimum temperature	in °C	0,0	1,1	2,8	4,4	7,8	11,1	12,2	12,2	9,4	6,7	3,9	1,7	6,1	43	3
4 Absolute maximum temperature	in °C	15,0	16,1	20,0	26,1	28,3	33,3	32,8	33,3	29,4	25,0	23,3	15,6	33,3	43	4
5 Absolute minimum temperature	in °C	-16,7	-13,3	-9,4	-2,8	0,6	1,7	4,4	3,9	-1,1	-6,1	-12,2	-13,3	-16,7	43	5
6 Mean relative humidity	in %	89	85	81	78	76	76	76	76	82	86	88	90	82	7	6
7 Mean precipitation	in mm	140	120	96	58	49	47	26	35	54	117	138	164	1044		7
8 Maximum precipitation	in mm															8
9 Minimum precipitation	in mm															9
10 Maximum precipitation in 24 h	in mm	94	61	57	48	36	50	37	46	52	67	80	87	94		10
11 Mean number of days with precipitation	>0,25 mm	19	16	16	13	10	11	6	8	9	15	17	19	159		11
12 Mean duration of sunshine	in h	58	89	124	195	250	229	311	250	190	114	71	44	1925		12
13 Mean quantity of radiation	in ly / day															13
14 Mean potential evaporation	in mm	6	13	27	48	77	103	119	104	70	44	21	11	643	41	14
15 Mean windspeed	in m / sec	3,7	3,8	4,0	3,9	3,7	3,6	3,4	3,4	3,3	3,5	3,6	3,8	3,6		15
16 Mean predominent direction of the wind		E	E	E	E	E	E	E	E	E	E	E	E			16

38 Station / Country: Lethbridge (Alberta) / Canada

Location: 49°38'N / 112°48'W — Height above sealevel 280 m — Climate symbol: Köppen BSk — Troll III,10

		J	F	M	A	M	J	J	A	S	O	N	D	year	P	
1 Mean daily temperature	in °C	-8,2	-7,1	-2,4	5,4	11,2	14,7	18,9	17,4	12,7	7,4	-0,4	-4,6	5,4	30	1
2 Mean daily maximum temperature	in °C	-2,8	-0,6	4,4	12,8	17,8	22,2	25,6	25,0	18,9	13,3	5,6	-0,6	11,7	30	2
3 Mean daily minimum temperature	in °C	-15,0	-13,3	-8,3	-1,7	3,8	7,8	10,0	8,9	3,9	-0,6	-6,7	-11,7	-2,2	30	3
4 Absolute maximum temperature	in °C	18,3	18,9	23,3	31,1	35,6	36,7	38,9	36,1	32,8	31,7	23,3	19,4	38,9	30	4
5 Absolute minimum temperature	in °C	-42,8	-42,2	-37,8	-25,6	-11,1	-1,7	0,6	-1,7	-15,6	-26,1	-31,7	-42,8	-42,8	30	5
6 Mean relative humidity	in %	68	75	67	57	60	66	58	57	63	64	72	74	65	7	6
7 Mean precipitation	in mm	22	27	27	35	53	81	43	42	35	27	27	20	439		7
8 Maximum precipitation	in mm															8
9 Minimum precipitation	in mm															9
10 Maximum precipitation in 24 h	in mm	19	24	32	53	48	62	41	55	36	39	38	23	62		10
11 Mean number of days with precipitation	>0,25 mm	9	9	10	8	10	12	8	8	7	6	8	8	103		11
12 Mean duration of sunshine	in h	103	126	162	210	265	271	345	300	216	177	113	96	2384		12
13 Mean quantity of radiation	in ly / day															13
14 Mean potential evaporation	in mm	0	0	0	36	76	107	127	109	67	34	0	0	556	30	14
15 Mean windspeed	in m / sec	7,1	6,6	6,3	6,8	6,2	5,8	5,2	5,0	5,8	6,6	6,8	7,0	6,3		15
16 Mean predominent direction of the wind		W	W	W	W	W	W	W	W	W	W	WSW	W			16

39 Station / Country: Winnipeg (Manitoba) / Canada

Location: 49°54'N / 97°15'W — Height above sealevel 254 m — Climate symbol: Köppen Dfb — Troll II,2

		J	F	M	A	M	J	J	A	S	O	N	D	year	P	
1 Mean daily temperature	in °C	-17,7	-15,5	-7,9	3,3	11,3	16,5	20,2	18,9	12,8	6,2	-4,8	-12,9	2,5		1
2 Mean daily maximum temperature	in °C	-13,9	-11,1	-2,8	8,9	18,3	23,3	26,1	24,4	18,3	10,6	-1,1	-9,4	8,3	66	2
3 Mean daily minimum temperature	in °C	-25,0	-22,8	-15,0	-2,8	3,9	10,0	12,8	10,6	6,1	-0,6	-10,6	-19,4	-4,4	66	3
4 Absolute maximum temperature	in °C	7,5	11,7	23,4	34,0	37,5	38,1	42,3	40,4	37,2	30,0	21,8	11,4	42,3		4
5 Absolute minimum temperature	in °C	-44,2	-45,0	-38,9	-29,9	-11,7	-6,1	1,8	-0,9	-8,3	-20,7	-38,7	-47,5	-47,5		5
6 Mean relative humidity	in %	78	79	80	68	56	58	64	63	66	69	78	82	70	10	6
7 Mean precipitation	in mm	26	21	27	30	50	81	69	70	55	37	29	22	517		7
8 Maximum precipitation	in mm															8
9 Minimum precipitation	in mm															9
10 Maximum precipitation in 24 h	in mm	19	24	38	33	49	49	69	65	65	74	28	25	74		10
11 Mean number of days with precipitation	>0,25 mm	12	10	10	8	10	13	11	11	10	8	11	11	125		11
12 Mean duration of sunshine	in h	101	133	187	207	244	248	310	270	181	153	82	81	2177		12
13 Mean quantity of radiation	in ly / day															13
14 Mean potential evaporation	in mm	0	0	0	21	78	120	137	115	68	26	0	0	565	66	14
15 Mean windspeed	in m / sec	5,5	5,4	5,8	6,3	6,2	5,4	4,6	4,8	5,3	5,7	6,0	5,5	5,5		15
16 Mean predominent direction of the wind		NW	S	NW,S	NNE,S	S	S	S	S	S	S	S	S			16

40 Station / Country: Kapuskasing (Ontario) / Canada

Location: 49°25'N / 82°28'W — Height above sealevel 70 m — Climate symbol: Köppen Dfb — Troll II,2

		J	F	M	A	M	J	J	A	S	O	N	D	year	P	
1 Mean daily temperature	in °C	-17,8	-15,4	-9,3	0,2	7,9	14,3	17,3	16,0	15,5	4,8	-4,9	-14,0	0,8		1
2 Mean daily maximum temperature	in °C	-12,2	-9,4	-2,8	6,1	14,4	21,1	23,9	22,2	16,7	8,3	-1,1	-8,9	6,7	19	2
3 Mean daily minimum temperature	in °C	-25,6	-23,3	-17,2	-7,2	0,6	7,2	10,0	9,4	5,0	-0,6	-10,0	-19,4	-6,1	19	3
4 Absolute maximum temperature	in °C	8	12	19	29	33	36	35	35	33	28	19	10	36		4
5 Absolute minimum temperature	in °C	-42	-41	-42	-26	-11	-4	1	-1	-6	-16	-34	-42	-42		5
6 Mean relative humidity	in %	91	90	84	71	69	70	71	76	81	83	89	89	80	7	6
7 Mean precipitation	in mm	55	45	54	53	80	94	85	87	90	72	83	60	858		7
8 Maximum precipitation	in mm															8
9 Minimum precipitation	in mm															9
10 Maximum precipitation in 24 h	in mm	38	28	36	30	56	47	48	65	58	47	45	36	65		10
11 Mean number of days with precipitation	>0,25 mm	19	16	14	11	14	14	15	12	15	15	20	19	184		11
12 Mean duration of sunshine	in h	72	104	135	167	195	208	238	202	126	90	44	54	1635		12
13 Mean quantity of radiation	in ly / day															13
14 Mean potential evaporation	in mm	0	0	0	0	61	104	123	107	64	23	0	0	482		14
15 Mean windspeed	in m / sec	4,4	4,5	4,6	4,6	4,4	4,3	4,0	3,8	4,4	4,6	4,5	4,4	4,4		15
16 Mean predominent direction of the wind		S	NW	NW	NW	N	SW	SW	S,SW	S	S	S	NW			16

41 Station/Country St. John's (Newfoundland)/Canada

Location 47°37'N/52°45'W   Height above sealevel 43 m   Climate symbol: Köppen Dfb   Troll II,2

		J	F	M	A	M	J	J	A	S	O	N	D	year	P	
1 Mean daily temperature	in °C	-4,3	-4,7	-2,9	1,2	5,6	10,3	15,4	15,4	12,0	6,7	2,9	-1,6	4,7	68	1
2 Mean daily maximum temperature	in °C	-1,7	-2,2	0,6	5,0	10,0	16,1	20,0	20,6	16,7	11,7	5,6	1,1	8,3	68	2
3 Mean daily minimum temperature	in °C	-7,7	-8,9	-5,6	-1,1	1,7	6,7	10,6	11,7	8,3	4,4	0,0	-4,4	1,1	68	3
4 Absolute maximum temperature	in °C	15,0	13,3	19,4	22,2	27,2	30,6	32,2	33,9	28,9	30,6	20,0	15,6	33,9	68	4
5 Absolute minimum temperature	in °C	-28,3	-29,4	-25,6	-18,3	-6,7	-2,8	0,6	0,0	-1,7	-5,6	-14,4	-20,0	-29,4	68	5
6 Mean relative humidity	in %	76	77	79	81	79	77	79	80	78	80	79	79	79	16	6
7 Mean precipitation	in mm	153	163	135	121	99	94	89	102	120	138	163	174	1551		7
8 Maximum precipitation	in mm															8
9 Minimum precipitation	in mm															9
10 Maximum precipitation in 24 h	in mm	85	55	46	92	51	75	121	59	66	101	64	85	121		10
11 Mean number of days with precipitation	>0,25 mm	22	20	20	18	16	13	13	14	14	17	19	22	208		11
12 Mean duration of sunshine	in h	57	75	93	116	152	176	212	176	149	117	59	53	1432		12
13 Mean quantity of radiation	in ly/day															13
14 Mean potential evaporation	in mm	0	0	0	15	50	85	114	108	72	43	17	0	504	19	14
15 Mean windspeed	in m/sec	8,2	7,8	7,8	6,8	6,5	6,4	6,2	6,1	6,6	7,5	7,2	7,7	7,0		15
16 Mean predominent direction of the wind		W	W	W	W	SW	SW,W	SW	SW	W	W	W	W			16

42 Station/Country Québec (Québec)/Canada

Location 46°48'N/71°23'W   Height above sealevel 23 m   Climate symbol: Köppen Dfb   Troll III,4

		J	F	M	A	M	J	J	A	S	O	N	D	year	P	
1 Mean daily temperature	in °C	-11,5	-10,7	-4,9	3,2	10,8	16,4	19,3	18,2	13,3	6,9	-0,1	-8,8	4,4		1
2 Mean daily maximum temperature	in °C	-7,7	-6,7	-0,6	7,2	16,1	22,2	24,4	22,8	17,8	10,6	2,2	-5,6	8,3	68	2
3 Mean daily minimum temperature	in °C	-16,7	-15,6	-9,4	-1,7	5,0	11,1	13,9	12,2	8,3	2,8	-4,4	-12,8	-0,6	68	3
4 Absolute maximum temperature	in °C	11,1	9,4	17,8	26,7	32,8	34,4	35,6	36,1	31,1	25,0	21,7	12,2	36,1	68	4
5 Absolute minimum temperature	in °C	-36,7	-35,6	-30,0	-18,3	-6,7	-0,6	3,9	2,2	-2,8	-10,0	-28,9	-35,6	-36,7	68	5
6 Mean relative humidity	in %	79	72	77	71	68	74	75	78	81	78	80	81	76	10	6
7 Mean precipitation	in mm	79	77	70	78	75	106	107	91	101	81	94	99	1058		7
8 Maximum precipitation	in mm															8
9 Minimum precipitation	in mm															9
10 Maximum precipitation in 24 h	in mm	93	43	33	46	35	104	75	131	73	45	53	45	131		10
11 Mean number of days with precipitation	>0,25 mm	16	14	13	12	12	13	13	11	12	12	15	16	159		11
12 Mean duration of sunshine	in h	82	102	128	157	199	201	228	162	157	123	65	65	1669		12
13 Mean quantity of radiation	in ly/day															13
14 Mean potential evaporation	in mm	0	0	0	16	73	112	132	112	72	33	0	0	549	72	14
15 Mean windspeed	in m/sec	4,5	4,6	4,8	4,5	4,6	3,7	3,2	3,2	3,4	3,7	4,4	4,3	4,1		15
16 Mean predominent direction of the wind		W	W	NE	NE	NE	SW	SW	SW	W	W	W	W			16

43 Station/Country Ottawa (Ontario)/Canada

Location 45°19'N/75°40'W   Height above sealevel 38 m   Climate symbol: Köppen Dfb   Troll III,4

		J	F	M	A	M	J	J	A	S	O	N	D	year	P	
1 Mean daily temperature	in °C	-10,8	-9,8	-3,6	5,3	12,8	18,2	20,7	19,5	14,7	8,2	0,8	-8,0	5,7	65	1
2 Mean daily maximum temperature	in °C	-6,1	-5,6	0,6	10,6	18,9	24,4	27,2	25,0	20,0	12,2	3,9	-4,4	10,6	65	2
3 Mean daily minimum temperature	in °C	-16,1	-16,1	-8,9	-0,6	6,7	12,2	14,4	12,8	8,9	2,8	-3,3	-12,8	0,0	65	3
4 Absolute maximum temperature	in °C	12,2	12,2	25,6	30,0	34,4	36,1	36,7	38,9	38,9	30,6	21,7	12,8	38,9	65	4
5 Absolute minimum temperature	in °C	-35,6	-37,2	-36,7	-18,9	-6,1	0,6	3,3	1,7	-4,4	-10,0	-23,3	-36,7	-37,2	65	5
6 Mean relative humidity	in %	80	81	75	67	66	68	67	69	75	75	76	79	73	7	6
7 Mean precipitation	in mm	60	58	66	67	75	76	78	76	77	67	67	83	850		7
8 Maximum precipitation	in mm															8
9 Minimum precipitation	in mm															9
10 Maximum precipitation in 24 h	in mm	30	45	48	48	45	78	72	90	93	51	37	35	93		10
11 Mean number of days with precipitation	>0,25 mm	16	14	13	12	12	11	11	10	11	11	14	17	152		11
12 Mean duration of sunshine	in h	93	110	140	168	230	250	281	254	173	135	78	77	1989		12
13 Mean quantity of radiation	in ly/day															13
14 Mean potential evaporation	in mm	0	0	0	28	81	116	138	115	75	34	0	0	585	49	14
15 Mean windspeed	in m/sec	4,9	5,1	5,2	5,1	4,7	4,3	4,0	3,9	4,3	4,5	4,8	4,8	4,6		15
16 Mean predominent direction of the wind		NW	NW	NW	NW	NW	SW	SW	SW	SW	SW	NW	E			16

44 Station/Country Montréal (Québec)/Canada

Location 45°30'N/73°35'W   Height above sealevel 17 m   Climate symbol: Köppen Dfb   Troll III,4

		J	F	M	A	M	J	J	A	S	O	N	D	year	P	
1 Mean daily temperature	in °C	-8,7	-7,8	-2,1	6,2	13,6	18,9	21,6	20,5	15,6	9,4	2,3	-5,9	6,9		1
2 Mean daily maximum temperature	in °C	-6,1	-5,0	0,6	10,0	17,8	23,3	25,6	23,9	19,4	12,2	3,9	-3,3	10,0	67	2
3 Mean daily minimum temperature	in °C	-14,4	-13,3	-7,2	0,6	8,3	13,9	16,1	15,0	10,6	6,4	-2,8	-10,6	1,7	67	3
4 Absolute maximum temperature	in °C	12,2	10,6	20,0	28,3	34,4	34,4	36,1	35,6	32,2	26,7	21,1	15,0	36,1	67	4
5 Absolute minimum temperature	in °C	-32,8	-33,3	-28,9	-16,7	-5,0	3,3	7,8	6,1	0,0	-6,7	-27,8	-33,9	-33,9	67	5
6 Mean relative humidity	in %	76	73	69	64	63	65	64	67	70	70	75	77	69	55	6
7 Mean precipitation	in mm	87	76	86	83	81	91	102	87	95	83	88	89	1048		7
8 Maximum precipitation	in mm															8
9 Minimum precipitation	in mm															9
10 Maximum precipitation in 24 h	in mm	45	31	42	41	52	68	90	76	49	86	53	63	90		10
11 Mean number of days with precipitation	>0,25 mm	18	15	12	13	14	13	13	10	13	13	15	17	166		11
12 Mean duration of sunshine	in h	86	103	140	154	209	230	253	240	187	132	69	69	1852		12
13 Mean quantity of radiation	in ly/day															13
14 Mean potential evaporation	in mm	0	0	0	27	81	119	139	121	77	37	2	0	603	55	14
15 Mean windspeed	in m/sec	5,6	5,7	5,6	5,5	5,0	4,4	4,2	4,1	4,4	4,7	5,2	5,3	5,0		15
16 Mean predominent direction of the wind		SW	W	SW	SW	SW	SW	SW	SW	SW	SW	SW	W			16

## 45

Station / Country Saint John (New Brunswick) / Canada

Location 45°19'N/65°33'W   Height above sealevel 326 m   Climate symbol: Köppen Dfb   Troll II,2

	in	J	F	M	A	M	J	J	A	S	O	N	D	year	P	
1 Mean daily temperature	in °C	-6,9	-6,4	2,0	3,8	9,4	13,9	17,2	17,1	13,4	8,3	2,7	-4,3	5,4	61	1
2 Mean daily maximum temperature	in °C	-2,2	-2,2	2,2	6,1	13,9	17,8	20,6	20,6	17,2	12,2	6,1	0,0	9,4	61	2
3 Mean daily minimum temperature	in °C	-11,7	-11,1	-5,6	0,0	5,0	9,4	12,2	12,2	9,4	5,0	-1,1	-8,3	1,1	61	3
4 Absolute maximum temperature	in °C	12,8	10,0	16,7	23,9	30,6	31,1	31,7	32,2	33,9	28,9	17,2	13,3	33,9	63	4
5 Absolute minimum temperature	in °C	-29,4	-28,9	-23,3	-14,4	-3,3	1,7	5,0	5,6	-0,6	-6,7	-22,8	-29,4	-29,4	63	5
6 Mean relative humidity	in %	72	71	70	67	66	72	77	75	72	70	72	76	72	56	6
7 Mean precipitation	in mm	144	122	106	105	98	94	87	108	104	108	153	133	1362	61	7
8 Maximum precipitation	in mm															8
9 Minimum precipitation	in mm															9
10 Maximum precipitation in 24 h	in mm	64	72	61	100	62	72	73	123	86	72	100	68	123		10
11 Mean number of days with precipitation	>0,25 mm	16	14	14	14	14	13	13	13	12	12	15	15	165		11
12 Mean duration of sunshine	in h	93	117	150	157	203	194	231	213	168	149	91	92	1858		12
13 Mean quantity of radiation	in ly / day															13
14 Mean potential evaporation	in mm	0	0	0	25	64	93	110	104	75	43	10	0	524	56	14
15 Mean windspeed	in m / sec	4,8	5,1	4,9	4,7	4,3	3,8	3,6	3,5	4,0	4,4	4,2	4,9	4,4		15
16 Mean predominent direction of the wind		N	N	N	NW	S,SW	S	S	S	S	SW	N,NW	NW			16

## 46

Station / Country Halifax (Nova Scotia) / Canada

Location 44°39'N/63°34'W   Height above sealevel 8 m   Climate symbol: Köppen Dfb   Troll II,2

	in	J	F	M	A	M	J	J	A	S	O	N	D	year	P	
1 Mean daily temperature	in °C	-3,3	-3,6	-0,1	4,8	9,9	14,4	18,5	18,8	15,5	10,3	5,3	-0,9	7,4	75	1
2 Mean daily maximum temperature	in °C	0,0	-0,6	3,3	8,3	15,0	20,0	23,3	23,3	19,4	13,9	7,8	1,7	11,1	75	2
3 Mean daily minimum temperature	in °C	-9,4	-9,4	-5,0	-0,6	4,4	8,9	12,8	13,3	10,0	5,0	0,0	-6,1	2,2	75	3
4 Absolute maximum temperature	in °C	14,4	11,1	21,1	28,3	32,2	34,4	37,2	34,4	34,4	31,1	20,6	16,7	37,2	72	4
5 Absolute minimum temperature	in °C	-27,2	-29,4	-23,3	-13,9	-4,4	0,0	3,9	3,9	-1,7	-6,1	-15,6	-25,6	-29,4	72	5
6 Mean relative humidity	in %	76	72	69	68	69	70	73	74	74	73	76	74	73	10	6
7 Mean precipitation	in mm	141	119	113	112	109	94	94	96	117	120	143	126	1384		7
8 Maximum precipitation	in mm															8
9 Minimum precipitation	in mm															9
10 Maximum precipitation in 24 h	in mm	78	88	57	65	90	82	132	92	239	74	69	75	239		10
11 Mean number of days with precipitation	>0,25 mm	16	14	13	13	13	11	11	10	10	10	14	16	151		11
12 Mean duration of sunshine	in h	91	117	144	162	201	202	219	222	179	157	92	87	1873		12
13 Mean quantity of radiation	in ly / day															13
14 Mean potential evaporation	in mm	0	0	0	29	70	102	126	116	83	51	18	0	595	75	14
15 Mean windspeed	in m / sec	5,3	5,6	5,1	5,2	4,4	4,0	3,5	3,5	3,9	4,5	4,9	5,6	4,6		15
16 Mean predominent direction of the wind		NW	NW	NW	NW	NW,SW	SW	SW	SW	NW	NW	NW	NW			16

## 47

Station / Country Toronto (Ontario) / Canada

Location 43°40'N/79°24'W   Height above sealevel 35 m   Climate symbol: Köppen Dfb   Troll III,4

	in	J	F	M	A	M	J	J	A	S	O	N	D	year	P	
1 Mean daily temperature	in °C	-3,9	-3,8	0,2	7,0	13,2	19,0	21,9	21,1	16,6	10,6	4,3	-1,8	8,7		1
2 Mean daily maximum temperature	in °C	-1,1	-1,1	2,8	10,0	17,2	22,8	26,1	25,0	20,6	13,3	6,1	0,6	11,7	105	2
3 Mean daily minimum temperature	in °C	-8,9	-9,4	-5,0	1,1	6,7	12,2	15,0	14,4	10,6	4,4	-0,6	-6,1	2,8	105	3
4 Absolute maximum temperature	in °C	14,4	12,8	26,7	32,2	33,9	36,1	40,6	38,8	35,6	29,4	21,1	18,1	40,6		4
5 Absolute minimum temperature	in °C	-32,2	-31,7	-26,7	-15,0	-3,9	-2,2	3,9	4,4	-2,2	-8,9	-20,6	-30,0	-32,2		5
6 Mean relative humidity	in %	74	73	69	65	64	68	68	66	72	73	75	76	71	30	6
7 Mean precipitation	in mm	67	59	67	66	70	63	74	61	65	60	63	61	776		7
8 Maximum precipitation	in mm															8
9 Minimum precipitation	in mm															9
10 Maximum precipitation in 24 h	in mm	52	42	43	47	52	64	65	82	68	97	51	52	97		10
11 Mean number of days with precipitation	>0,25 mm	16	13	13	12	12	9	10	9	9	9	13	13	138		11
12 Mean duration of sunshine	in h	78	105	139	170	220	257	287	259	198	154	84	65	2026		12
13 Mean quantity of radiation	in ly / day															13
14 Mean potential evaporation	in mm	0	0	0	30	74	112	138	119	81	41	11	0	606	78	14
15 Mean windspeed	in m / sec	6,6	6,5	6,2	5,8	4,7	4,1	3,7	3,8	4,3	4,5	5,8	6,5	5,2		15
16 Mean predominent direction of the wind		W	W	NW	E,NW	E	E,SW	SW	SW	E,SW	SW	SW	SW			16

Station / Country Tatoosh Island (Washington) / U.S.

Location 48°23'N/124°44'W   Height above sealevel 31 m   Climate symbol: Köppen Csb   Troll III,1

	in	J	F	M	A	M	J	J	A	S	O	N	D	year	P	
1 Mean daily temperature	in °C	5,6	6,2	6,8	8,6	10,6	12,2	13,1	13,3	12,7	11,1	8,4	6,9	9,6	30	1
2 Mean daily maximum temperature	in °C	7,2	7,8	8,9	10,6	12,2	13,9	15,0	15,6	14,4	12,8	10,0	8,3	11,7	59	2
3 Mean daily minimum temperature	in °C	3,9	3,9	5,0	6,1	8,3	10,0	10,6	10,6	10,0	8,9	6,7	5,0	7,2	59	3
4 Absolute maximum temperature	in °C	18	18	21	24	27	29	31	26	27	25	20	16	31	58	4
5 Absolute minimum temperature	in °C	-10	-9	-4	1	2	6	7	7	4	1	-7	-7	-10	58	5
6 Mean relative humidity	in %	84	84	84	84	86	89	91	93	90	88	86	85	87	53	6
7 Mean precipitation	in mm	275	221	212	133	76	72	59	50	90	209	287	309	1973	30	7
8 Maximum precipitation	in mm	573	420	416	274	204	198	196	129	313	441	563	533		56	8
9 Minimum precipitation	in mm	47	36	61	17	14	3	<1	4	2	64	72	92		56	9
10 Maximum precipitation in 24 h	in mm	93	116	121	94	56	70	94	58	96	150	111	102	150	58	10
11 Mean number of days with precipitation	>0,25 mm	22	18	20	17	14	12	10	10	11	17	21	23	197	58	11
12 Mean duration of sunshine	in h	70	100	135	182	229	217	235	190	175	129	71	60	1793	30	12
13 Mean quantity of radiation	in ly / day															13
14 Mean potential evaporation	in mm	21	24	34	48	67	80	88	81	66	50	32	24	615		14
15 Mean windspeed	in m / sec	9	8	7	6	5	4	4	4	5	7	8	9	6	29	15
16 Mean predominent direction of the wind		E	E	E	W	W	W	S	S	S	E	E	E		21	16

## 49

Station / Country Seattle (Washington) / U.S.

Location 47°36'N / 122°20'W   Height above sealevel 4 m   Climate symbol: Köppen Csb   Troll III,2

		J	F	M	A	M	J	J	A	S	O	N	D	year	P	
1 Mean daily temperature	in °C	5,1	6,4	8,0	11,0	14,1	16,3	18,7	18,3	18,2	12,4	8,3	6,6	11,8	30	1
2 Mean daily maximum temperature	in °C	7,2	8,9	11,1	14,4	17,8	20,6	22,2	22,8	19,4	15,0	10,6	8,3	15,0	57	2
3 Mean daily minimum temperature	in °C	2,2	2,8	3,9	6,1	8,3	11,1	12,2	12,8	11,1	8,3	5,0	3,3	7,2	57	3
4 Absolute maximum temperature	in °C	19,4	21,1	27,2	30,6	33,3	36,7	37,8	35,6	33,3	27,8	20,0	18,3	37,8	57	4
5 Absolute minimum temperature	in °C	-16,1	-15,6	-6,7	-1,1	2,2	4,4	7,8	7,8	2,2	-1,7	-9,4	-11,1	-16,1	57	5
6 Mean relative humidity	in %	83	79	75	72	71	69	68	66	75	82	84	84	76	47	6
7 Mean precipitation	in mm	132	99	84	50	40	36	16	19	42	83	127	138	866	30	7
8 Maximum precipitation	in mm	278	176	184	116	119	80	46	52	88	189	239	264		25	8
9 Minimum precipitation	in mm	36	33	31	4	9	6	tr	tr	3	20	26	25		25	9
10 Maximum precipitation in 24 h	in mm	62	68	59	39	34	27	31	20	49	50	81	84	84	27	10
11 Mean number of days with precipitation	>0,25 mm	19	15	16	13	11	9	5	6	8	14	17	19	152	27	11
12 Mean duration of sunshine	in h	74	99	154	201	247	234	304	248	197	122	77	62	2019	30	12
13 Mean quantity of radiation	in ly/day	71	128	244	357	437	475	504	444	315	174	94	60	275	11	13
14 Mean potential evaporation	in mm	14	19	31	51	79	100	117	107	76	47	25	17	683	56	14
15 Mean windspeed	in m/sec	3	3	4	4	3	3	3	3	3	3	3	3	3	10	15
16 Mean predominent direction of the wind		S	S	S	S	S	SW	NNW	NNW	SSE	SSE	SSE	SSE		11	16

## 50

Station / Country Spokane (Washington) / U.S.

Location 47°37'N / 117°31'W   Height above sealevel 718 m   Climate symbol: Köppen Cfb   Troll II,2

		J	F	M	A	M	J	J	A	S	O	N	D	year	P	
1 Mean daily temperature	in °C	-2,5	-0,3	4,4	9,5	13,6	17,3	21,1	20,3	15,3	9,5	3,1	-0,5	9,5	66	1
2 Mean daily maximum temperature	in °C	0,6	3,9	9,4	15,6	20,0	23,9	28,9	28,3	22,2	15,6	6,7	2,2	15,0	66	2
3 Mean daily minimum temperature	in °C	-5,6	-4,4	-0,6	3,3	7,2	10,6	13,3	12,2	8,3	3,3	-0,6	-3,3	3,9	66	3
4 Absolute maximum temperature	in °C	16,7	15,6	23,3	32,2	36,1	37,8	42,2	40,0	36,7	30,6	21,1	15,6	42,2	66	4
5 Absolute minimum temperature	in °C	-34,4	-30,6	-23,3	-10,6	-1,7	1,1	5,0	2,8	-5,6	-12,8	-25,0	-27,8	-34,4	66	5
6 Mean relative humidity	in %	82	77	65	57	56	52	43	44	54	65	80	84	63	41	6
7 Mean precipitation	in mm	53	43	30	28	33	33	15	15	23	28	53	53	409	50	7
8 Maximum precipitation	in mm	124	83	95	78	145	72	33	36	39	103	100	110		11	8
9 Minimum precipitation	in mm	13	27	11	2	15	6	tr	tr	2	3	9	31		11	9
10 Maximum precipitation in 24 h	in mm	38	28	23	25	43	56	30	43	43	36	38	28	56	69	10
11 Mean number of days with precipitation	>0,25 mm	14	12	11	9	9	8	4	4	7	8	13	14	113	50	11
12 Mean duration of sunshine	in h	38	120	197	262	308	309	397	350	264	177	86	57	2605	30	12
13 Mean quantity of radiation	in ly/day	119	204	321	474	563	596	665	556	404	225	131	75	361		13
14 Mean potential evaporation	in mm	0	0	0	44	83	108	141	125	76	39	9	0	625		14
15 Mean windspeed	in m/sec	3,6	4,0	4,0	4,0	3,6	3,6	3,6	3,6	3,6	3,6	3,6	3,6	3,6		15
16 Mean predominent direction of the wind		NE	SSW	SSW	SSW	SSW	SSW	SW	SW	NE	SW	NE	NE			16

## 51

Station / Country Havre (Montana) / U.S.

Location 48°34'N / 109°40'W   Height above sealevel 758 m   Climate symbol: Köppen BSk   Troll III,10

		J	F	M	A	M	J	J	A	S	O	N	D	year	P	
1 Mean daily temperature	in °C	-10,1	-8,7	-2,8	5,9	12,3	16,1	21,1	19,4	13,4	7,9	-1,1	-6,2	5,6	30	1
2 Mean daily maximum temperature	in °C	-5,0	-3,3	3,9	13,3	18,9	23,3	28,3	27,2	20,6	13,9	5,0	-1,1	12,2	50	2
3 Mean daily minimum temperature	in °C	-16,1	-15,0	-8,3	0,0	5,0	10,0	12,2	11,1	5,6	0,0	-7,2	-12,2	-1,1	50	3
4 Absolute maximum temperature	in °C	18	22	25	35	37	42	42	41	37	33	23	22	42	37	4
5 Absolute minimum temperature	in °C	-49	-44	-36	-22	-10	-2	3	-3	-8	-22	-33	-38	-49	37	5
6 Mean relative humidity	in %	81	80	75	66	61	55	60	60	66	67	75	79	69		6
7 Mean precipitation	in mm	12	11	15	25	39	69	32	28	28	20	13	12	302	30	7
8 Maximum precipitation	in mm	44	27	48	94	180	150	150	99	152	72	61	67		55	8
9 Minimum precipitation	in mm	<1	<1	3	1	3	17	3	3	3	<1	tr	<1		55	9
10 Maximum precipitation in 24 h	in mm	17	13	23	48	42	75	48	57	69	35	19	25	75	37	10
11 Mean number of days with precipitation	>0,25 mm	8	7	7	7	10	12	8	7	7	5	6	6	90	57	11
12 Mean duration of sunshine	in h	136	174	234	268	311	312	384	339	260	202	132	122	2874	30	12
13 Mean quantity of radiation	in ly/day															13
14 Mean potential evaporation	in mm	0	0	0	34	79	113	142	123	69	33	0	0	593		14
15 Mean windspeed	in m/sec	4	4	4	4	4	4	3	3	3	4	4	4	4	57	15
16 Mean predominent direction of the wind		SW	SW	SW	E	E	E	E	E	SW	SW	SW	SW		37	16

Station / Country Miles City (Montana) / U.S.

Location 46°24'N / 105°50'W   Height above sealevel 729 m   Climate symbol: Köppen BSk   Troll III,10

		J	F	M	A	M	J	J	A	S	O	N	D	year	P	
1 Mean daily temperature	in °C	-8,4	-7,0	-0,3	7,6	13,4	18,9	17,6	17,2	15,6	8,9	0,9	-5,6	7,5	39	1
2 Mean daily maximum temperature	in °C	-2,8	-1,1	5,6	14,4	20,0	25,6	30,6	30,0	22,8	15,6	6,7	0,0	13,9	39	2
3 Mean daily minimum temperature	in °C	-13,9	-12,8	-6,1	0,7	6,7	12,2	15,6	14,4	8,3	2,2	-5,0	-11,1	1,1	39	3
4 Absolute maximum temperature	in °C	17,8	20,0	31,1	32,8	37,8	42,2	43,9	42,2	41,1	32,8	24,4	17,8	43,9	39	4
5 Absolute minimum temperature	in °C	-43,3	-45,0	-34,4	-21,7	-6,7	0,6	5,6	1,1	-8,3	-22,2	-32,2	-41,7	-45,0	39	5
6 Mean relative humidity	in %	77	76	69	62	59	61	55	55	60	65	71	75	66		6
7 Mean precipitation	in mm	15	13	20	25	56	69	41	25	25	23	15	13	340	53	7
8 Maximum precipitation	in mm	19	23	46	65	135	248	116	102	119	81	32	29		21	8
9 Minimum precipitation	in mm	2	2	2	3	6	23	10	3	4	tr	<1	<1		21	9
10 Maximum precipitation in 24 h	in mm	33	33	71	43	94	91	71	79	58	41	28	16	94	58	10
11 Mean number of days with precipitation	>0,25 mm	8	8	8	8	9	11	8	6	6	6	6	7	89	39	11
12 Mean duration of sunshine	in h															12
13 Mean quantity of radiation	in ly/day															13
14 Mean potential evaporation	in mm	0	0	0	34	81	113	154	135	78	39	0	0	634		14
15 Mean windspeed	in m/sec															15
16 Mean predominent direction of the wind																16

53 Station/Country Williston (North Dakota)/U.S.

Location 48°09'N/103°37'W   Height above sealevel 579 m   Climate symbol: Köppen Dfb   Troll III,10

		J	F	M	A	M	J	J	A	S	O	N	D	year	P	
1 Mean daily temperature	in °C	-12,3	-10,3	-3,7	6,0	12,9	17,3	21,8	20,4	14,3	7,8	-1,9	-7,9	5,3	30	1
2 Mean daily maximum temperature	in °C	-8,3	-6,7	1,1	12,2	18,9	23,9	27,8	27,2	20,8	13,3	2,8	-5,0	10,8	51	2
3 Mean daily minimum temperature	in °C	-20,0	-18,3	-10,8	-0,6	5,0	10,8	13,3	11,7	5,8	0,0	-8,3	-15,8	-2,2	51	3
4 Absolute maximum temperature	in °C	11,1	14,4	28,9	33,3	38,3	41,7	41,1	41,7	38,3	35,0	21,7	15,0	41,7	51	4
5 Absolute minimum temperature	in °C	-45,0	-45,0	-37,2	-20,0	-11,7	-1,1	2,2	0,0	-10,6	-19,4	-33,9	-43,3	-45,0	51	5
6 Mean relative humidity	in %	80	79	72	63	60	64	61	61	64	66	74	78	68		6
7 Mean precipitation	in mm	14	12	18	24	36	84	48	38	28	19	15	13	349	30	7
8 Maximum precipitation	in mm	38	31	55	62	119	200	154	129	92	54	38	54		42	8
9 Minimum precipitation	in mm	1	tr	2	tr	2	14	9	5	<1	tr	tr	<1		42	9
10 Maximum precipitation in 24 h	in mm	14	18	31	39	75	78	53	46	50	37	24	19	78	44	10
11 Mean number of days with precipitation	>0,25 mm	8	6	7	7	9	12	9	8	7	5	6	7	91	44	11
12 Mean duration of sunshine	in h	141	168	215	260	305	312	377	328	247	206	131	129	2819	29	12
13 Mean quantity of radiation	in ly/day															13
14 Mean potential evaporation	in mm	0	0	0	35	80	117	129	104	64	25	0	0	554		14
15 Mean windspeed	in m/sec	3	3	4	4	4	4	3	3	3	3	3	3	3	33	15
16 Mean predominent direction of the wind		W	W	SE	SE	SE	SE	SE	SE	SE	SE	W	W		33	16

54 Station/Country Bismarck (North Dakota)/U.S.

Location 46°46'N/100°45'W   Height above sealevel 511 m   Climate symbol: Köppen Dfa   Troll III,10

		J	F	M	A	M	J	J	A	S	O	N	D	year	P	
1 Mean daily temperature	in °C	-12,8	-10,8	-3,8	6,1	13,0	18,1	22,3	21,0	14,8	7,9	-2,0	-8,4	5,4	30	1
2 Mean daily maximum temperature	in °C	-7,7	-5,6	1,7	12,8	19,4	24,4	28,3	27,2	21,7	13,9	3,9	-2,8	11,7	72	2
3 Mean daily minimum temperature	in °C	-18,9	-17,2	-10,0	0,0	5,6	11,1	13,9	12,8	7,2	0,8	-7,7	-15,0	-1,7	72	3
4 Absolute maximum temperature	in °C	15,6	18,3	27,2	32,2	38,9	41,7	45,6	42,8	40,6	32,8	23,3	18,9	45,6	72	4
5 Absolute minimum temperature	in °C	-42,8	-42,8	-37,8	-19,4	-10,6	0,0	0,0	0,0	-12,2	-23,3	-33,3	-41,1	-42,8	72	5
6 Mean relative humidity	in %	77	77	71	63	61	66	63	63	66	66	73	78	69	13	6
7 Mean precipitation	in mm	11	11	20	31	50	86	56	44	30	22	15	9	385	30	7
8 Maximum precipitation	in mm	31	28	72	77	112	191	106	128	99	59	65	23		19	8
9 Minimum precipitation	in mm	<1	3	5	tr	8	31	22	8	<1	2	2	tr		19	9
10 Maximum precipitation in 24 h	in mm	17	19	33	39	48	83	45	52	51	32	25	15	83	21	10
11 Mean number of days with precipitation	>0,25 mm	8	7	8	7	9	12	9	9	7	6	7	7	96	21	11
12 Mean duration of sunshine	in h	141	170	205	236	279	294	358	307	243	198	130	125	2686	30	12
13 Mean quantity of radiation	in ly/day	159	244	350	445	532	593	607	524	388	274	161	125	367	9	13
14 Mean potential evaporation	in mm	0	0	0	31	78	115	144	124	72	31	0	0	595	56	14
15 Mean windspeed	in m/sec	5	5	5	6	6	6	4	5	5	5	5	4	5	21	15
16 Mean predominent direction of the wind		WNW	WNW	WNW	WNW	E	WNW	SSE	E	WNW	WNW	WNW	WNW		11	16

55 Station/Country Duluth (Minnesota)/U.S.

Location 46°50'N/92°11'W   Height above sealevel 435 m   Climate symbol: Köppen Dfb   Troll III,5

		J	F	M	A	M	J	J	A	S	O	N	D	year	P	
1 Mean daily temperature	in °C	-12,9	-11,5	-5,7	3,1	9,8	15,2	18,9	17,9	12,6	6,7	-2,9	-10,0	3,4	30	1
2 Mean daily maximum temperature	in °C	-8,9	-6,7	0,0	7,2	13,3	19,4	22,8	21,7	17,2	10,6	2,8	-5,0	7,8	72	2
3 Mean daily minimum temperature	in °C	-18,3	-16,1	-8,9	-1,7	3,3	8,9	12,2	12,2	8,3	2,8	-5,0	-13,3	-1,1	72	3
4 Absolute maximum temperature	in °C	12,8	14,4	27,2	29,4	35,0	36,1	41,1	36,1	34,4	29,4	22,8	13,3	41,1	72	4
5 Absolute minimum temperature	in °C	-40,6	-37,8	-32,2	-17,2	-8,9	-0,6	5,6	3,3	-3,9	-13,3	-33,9	-37,2	-40,6	72	5
6 Mean relative humidity	in %	84	83	78	75	68	73	73	74	77	78	80	85	77	13	6
7 Mean precipitation	in mm	29	24	41	60	84	108	90	97	73	55	45	29	735	30	7
8 Maximum precipitation	in mm	79	45	98	148	164	191	215	169	159	191	79	94		17	8
9 Minimum precipitation	in mm	5	6	21	20	22	30	25	34	5	3	20	4		17	9
10 Maximum precipitation in 24 h	in mm	15	18	45	58	57	103	69	74	69	71	57	54	103	11	10
11 Mean number of days with precipitation	>0,25 mm	11	10	11	10	13	13	11	11	11	9	11	11	132	19	11
12 Mean duration of sunshine	in h	125	163	221	235	268	282	328	277	203	166	100	107	2475	30	12
13 Mean quantity of radiation	in ly/day															13
14 Mean potential evaporation	in mm	0	0	0	24	66	98	127	113	75	37	0	0	540	56	14
15 Mean windspeed	in m/sec	6	6	6	7	6	5	5	5	5	6	6	6	6	11	15
16 Mean predominent direction of the wind		NW	NW	WNW	ENE	E	E	WNW	E	WNW	WNW	WNW	NW		11	16

56 Station/Country Sault Ste. Marie (Michigan)/U.S.

Location 46°28'N/84°22'W   Height above sealevel 220 m   Climate symbol: Köppen Dfb   Troll III,4

		J	F	M	A	M	J	J	A	S	O	N	D	year	P	
1 Mean daily temperature	in °C	-9,3	-9,3	-4,8	3,1	9,5	14,7	17,8	17,5	12,9	7,4	0,2	-6,4	4,4	30	1
2 Mean daily maximum temperature	in °C	-5,7	-5,2	-0,2	7,6	15,4	20,9	24,1	23,0	18,0	11,3	3,2	-3,0	9,1	30	2
3 Mean daily minimum temperature	in °C	-14,6	-15,4	-9,9	-2,3	3,4	8,0	11,3	11,2	7,7	2,4	-3,8	-10,8	-1,1	30	3
4 Absolute maximum temperature	in °C	6	7	24	28	29	33	33	37	34	27	19	11	37	19	4
5 Absolute minimum temperature	in °C	-31	-32	-31	-17	-6	-1	2	0	-4	-7	-21	-29	-32	19	5
6 Mean relative humidity	in %	82	82	79	72	71	77	78	80	83	81	83	84	79	17	6
7 Mean precipitation	in mm	53	38	46	55	70	84	63	73	97	72	85	58	794	30	7
8 Maximum precipitation	in mm	111	68	82	131	134	169	153	120	143	112	143	90		17	8
9 Minimum precipitation	in mm	22	15	17	15	31	32	24	13	22	6	39	19		17	9
10 Maximum precipitation in 24 h	in mm	35	24	31	68	62	67	60	56	46	52	38	22	68	19	10
11 Mean number of days with precipitation	>0,25 mm	18	15	13	12	11	12	10	10	13	11	18	19	162	19	11
12 Mean duration of sunshine	in h	83	123	187	217	252	269	309	256	165	133	61	62	2117	30	12
13 Mean quantity of radiation	in ly/day	130	225	356	416	523	557	573	472	322	216	105	96	333		13
14 Mean potential evaporation	in mm	0	0	0	20	66	101	123	110	75	37	0	0	532		14
15 Mean windspeed	in m/sec	5	5	5	5	5	4	4	4	4	4	5	5	4	19	15
16 Mean predominent direction of the wind		E	E	WNW	WNW	WNW	WNW	WNW	WNW	WNW	E	E	E		11	16

57 Station/Country Mount Washington (New Hampshire)/U.S.

Location 44°16'N/71°18'W   Height above sealevel 1909 m   Climate symbol: Köppen Dfc   Troll III.4

		J	F	M	A	M	J	J	A	S	O	N	D	year	P	
1	Mean daily temperature	in °C	-14,3	-14,7	-11,3	-5,0	1,7	7,2	9,5	8,7	5,0	-0,8	-6,5	-12,9	-2,8	30	1
2	Mean daily maximum temperature	in °C	-7,2	-7,7	-7,2	-1,7	4,4	10,8	13,3	12,2	7,8	3,3	-2,8	-7,7	1,7	3	2
3	Mean daily minimum temperature	in °C	-17,2	-16,1	-15,0	-8,3	-2,2	3,9	6,7	6,1	1,7	-3,3	-9,4	-15,0	-5,8	3	3
4	Absolute maximum temperature	in °C	6,7	3,3	8,9	13,9	15,8	20,8	21,7	21,7	17,8	14,4	10,0	7,2	21,7	10	4
5	Absolute minimum temperature	in °C	-42,2	-43,3	-38,3	-27,8	-18,3	-13,3	-2,2	-2,2	-11,1	-16,7	-25,0	-38,3	-43,3	10	5
6	Mean relative humidity	in %	81	80	87	87	85	87	89	92	86	82	85	86	86	2	6
7	Mean precipitation	in mm	138	132	148	150	148	165	170	169	178	157	168	160	1881	30	7
8	Maximum precipitation	in mm															8
9	Minimum precipitation	in mm															9
10	Maximum precipitation in 24 h	in mm	86	108	97	96	88	97	80	132	122	179	87	86	179	28	10
11	Mean number of days with precipitation	>0.25 mm	18	17	19	18	18	16	17	15	15	14	19	20	206	28	11
12	Mean duration of sunshine	in h	94	98	133	141	162	145	150	143	139	159	89	87	1540	18	12
13	Mean quantity of radiation	in ly/day															13
14	Mean potential evaporation	in mm	0	0	0	0	30	70	90	83	31	0	0	0	304	28	14
15	Mean windspeed	in m/sec	21	21	20	17	14	13	11	11	13	15	18	21	16	28	15
16	Mean predominent direction of the wind		W	W	W	W	W	W	W	W	W	W	W	W		22	16

58 Station/Country Burlington (Vermont)/U.S.

Location 44°28'N/73°09'W   Height above sealevel 101 m   Climate symbol: Köppen Dfb   Troll III.4

		J	F	M	A	M	J	J	A	S	O	N	D	year	P	
1	Mean daily temperature	in °C	-7,7	-7,0	-1,6	6,2	13,2	18,7	21,4	20,1	15,5	9,2	2,7	-5,0	7,2	30	1
2	Mean daily maximum temperature	in °C	-2,5	-2,1	3,9	11,7	19,5	25,3	28,0	26,8	21,8	14,7	6,8	-0,8	12,8	30	2
3	Mean daily minimum temperature	in °C	-13,2	-13,3	-6,9	-0,3	6,3	11,9	14,6	13,4	9,1	3,1	-2,0	-9,7	1,1	30	3
4	Absolute maximum temperature	in °C	17	16	29	29	33	38	37	38	34	29	24	17	38	17	4
5	Absolute minimum temperature	in °C	-34	-32	-29	-13	-4	2	7	4	-3	-6	-19	-30	-34	17	5
6	Mean relative humidity	in %	76	74	72	68	67	70	70	73	76	75	76	78	73	17	6
7	Mean precipitation	in mm	50	45	54	67	76	89	98	86	84	75	67	54	845	30	7
8	Maximum precipitation	in mm	95	106	115	148	150	252	205	293	208	171	257	134		52	8
9	Minimum precipitation	in mm	12	10	6	18	9	28	25	18	17	4	16	8		52	9
10	Maximum precipitation in 24 h	in mm	37	42	33	36	57	65	62	91	63	43	46	66	91	17	10
11	Mean number of days with precipitation	>0,25 mm	14	13	13	13	13	11	12	11	12	11	14	14	151	17	11
12	Mean duration of sunshine	in h	103	127	184	185	244	270	291	266	199	152	77	80	2178	30	12
13	Mean quantity of radiation	in ly/day															13
14	Mean potential evaporation	in mm	0	0	0	30	80	116	148	119	78	43	7	0	621		14
15	Mean windspeed	in m/sec	5	4	4	4	4	4	4	3	4	4	4	5	4	17	15
16	Mean predominent direction of the wind		S	S	S	S	S	S	S	S	S	S	S	S		17	16

59 Station/Country Portland (Oregon)/U.S.

Location 45°32'N/122°40'W   Height above sealevel 47 m   Climate symbol: Köppen Csb   Troll III.2

		J	F	M	A	M	J	J	A	S	O	N	D	year	P	
1	Mean daily temperature	in °C	4,8	6,6	8,7	11,9	15,1	17,4	20,3	20,1	18,1	13,6	8,4	6,2	12,6	30	1
2	Mean daily maximum temperature	in °C	6,7	8,9	12,2	16,1	18,9	22,2	25,0	25,0	21,7	16,7	11,7	7,8	16,1	72	2
3	Mean daily minimum temperature	in °C	1,1	2,2	3,9	6,1	8,3	11,7	13,3	13,3	11,1	8,3	5,0	2,8	7,2	72	3
4	Absolute maximum temperature	in °C	18,3	20,0	28,3	33,9	37,2	38,9	41,7	38,9	38,9	31,1	22,8	18,3	41,7	72	4
5	Absolute minimum temperature	in °C	-18,9	-13,9	-6,7	-2,2	0,0	3,9	6,1	6,1	1,7	-1,7	-11,7	-16,1	-18,9	72	5
6	Mean relative humidity	in %	82	78	73	69	68	67	64	65	70	78	83	83	73	59	6
7	Mean precipitation	in mm	161	124	121	62	52	43	10	18	44	99	153	188	1075	30	7
8	Maximum precipitation	in mm	373	290	268	158	128	108	65	86	140	246	366	443		56	8
9	Minimum precipitation	in mm	23	4	16	13	11	<1	0	tr	2	<1	9	55		56	9
10	Maximum precipitation in 24 h	in mm	117	68	64	50	46	55	34	33	73	62	113	127	127	58	10
11	Mean number of days with precipitation	>0,25 mm	20	16	17	14	12	9	3	4	8	12	17	19	151	58	11
12	Mean duration of sunshine	in h	77	97	142	203	246	249	329	275	218	134	87	65	2122	30	12
13	Mean quantity of radiation	in ly/day	97	159	289	375	491	488	527	454	360	206	111	78	301		13
14	Mean potential evaporation	in mm	9	17	34	54	81	101	123	113	81	54	28	13	706	56	14
15	Mean windspeed	in m/sec	5	4	4	3	3	3	3	3	3	4	4	3		12	15
16	Mean predominent direction of the wind		ESE	ESE	ESE	NW	NW	NW	NW	NW	NW	ESE	ESE	ESE		12	16

60 Station/Country Walla Walla (Washington)/U.S.

Location 46°02'N/118°20'W   Height above sealevel 289 m   Climate symbol: Köppen Csa   Troll III.10

		J	F	M	A	M	J	J	A	S	O	N	D	year	P	
1	Mean daily temperature	in °C	0,7	3,8	7,8	12,1	16,1	19,6	24,4	23,2	18,9	12,8	5,7	3,2	12,3	30	1
2	Mean daily maximum temperature	in °C	3,3	6,7	12,2	17,2	21,7	26,1	31,7	30,6	24,4	17,8	10,0	5,0	17,2	45	2
3	Mean daily minimum temperature	in °C	-2,8	-0,6	2,8	6,1	9,4	12,2	16,1	16,1	11,1	8,1	2,2	-1,1	6,7	45	3
4	Absolute maximum temperature	in °C	21,1	22,2	26,1	33,9	37,8	41,1	44,4	45,0	37,8	30,6	25,6	22,8	45,0	59	4
5	Absolute minimum temperature	in °C	-33,9	-28,9	-16,7	-6,7	0,0	3,9	7,2	5,0	-3,3	-5,8	-22,8	-27,2	-33,9	59	5
6	Mean relative humidity	in %	79	74	64	57	55	51	42	42	53	61	74	79	61	15	6
7	Mean precipitation	in mm	48	39	40	36	38	31	5	8	20	39	44	47	395	30	7
8	Maximum precipitation	in mm	115	101	94	93	106	77	36	47	114	107	100	104		44	8
9	Minimum precipitation	in mm	9	3	12	3	tr	1	0	tr	tr	<1	<1	15		44	9
10	Maximum precipitation in 24 h	in mm	36	34	30	40	47	51	29	26	35	35	36	34	51	46	10
11	Mean number of days with precipitation	>0,25 mm	14	11	12	9	9	7	3	3	5	9	11	14	107	46	11
12	Mean duration of sunshine	in h	72	108	194	262	317	335	411	367	280	198	92	51	2685	30	12
13	Mean quantity of radiation	in ly/day															13
14	Mean potential evaporation	in mm	0	7	28	54	89	117	158	139	87	48	17	4	748	56	14
15	Mean windspeed	in m/sec	2	2	3	3	3	2	2	2	2	2	2	2	2	45	15
16	Mean predominent direction of the wind		S	S	S	S	S	S	S	S	S	S	S	S		45	16

161

61 Station/Country Billings (Montana)/U.S.

Location 45°48'N/108°32'W Height above sealevel 1087 m Climate symbol: Köppen **Dfa** Troll III,10

		J	F	M	A	M	J	J	A	S	O	N	D	year	P	
1	Mean daily temperature	in °C	-4,9	-3,5	0,7	7,5	13,2	17,6	22,9	21,6	15,8	10,0	2,3	-1,7	8,4	30	1
2	Mean daily maximum temperature	in °C	0,2	2,3	6,3	14,4	20,2	24,7	31,1	29,8	22,8	15,9	7,3	2,2	14,8	30	2
3	Mean daily minimum temperature	in °C	-10,4	-8,9	-4,4	1,3	6,4	10,8	14,7	13,5	8,3	3,1	-3,1	-7,5	2,0	30	3
4	Absolute maximum temperature	in °C	20	21	25	33	36	39	41	40	38	30	22	21	41	26	4
5	Absolute minimum temperature	in °C	-34	-39	-28	-21	-10	0	6	4	-3	-16	-30	-27	-39	28	5
6	Mean relative humidity	in %	63	66	65	57	56	61	48	46	52	55	62	63	58	20	6
7	Mean precipitation	in mm	14	15	27	33	48	65	23	23	30	28	16	15	337	30	7
8	Maximum precipitation	in mm	39	42	69	112	125	194	79	50	127	69	33	48		24	8
9	Minimum precipitation	in mm	1	3	3	3	13	17	3	1	6	<1	tr	1		24	9
10	Maximum precipitation in 24 h	in mm	15	15	25	63	72	71	48	20	46	50	35	18	72	26	10
11	Mean number of days with precipitation	>0,25 mm	7	8	9	9	10	10	7	6	7	6	6	6	92	26	11
12	Mean duration of sunshine	in h	140	154	208	236	283	301	372	332	258	213	136	129	2762	21	12
13	Mean quantity of radiation	in ly/day															13
14	Mean potential evaporation	in mm	0	0	3	37	81	109	150	132	81	42	5	0	640	56	14
15	Mean windspeed	in m/sec	6	5	5	5	5	5	4	4	5	5	6	6	5	21	15
16	Mean predominent direction of the wind		SW	SW	SW	SW	SW	SW	SW	SW	SW	SW	SW	WSW			16

62 Station/Country Minneapolis (Minnesota)/U.S.

Location 44°53'N/93°13'W Height above sealevel 254 m Climate symbol: Köppen **Dfa** Troll III,7

		J	F	M	A	M	J	J	A	S	O	N	D	year	P	
1	Mean daily temperature	in °C	-10,9	-8,9	-2,4	7,1	14,2	19,6	22,8	21,4	15,8	9,1	-0,8	-7,9	6,6	30	1
2	Mean daily maximum temperature	in °C	-5,6	-3,9	3,3	13,3	20,0	25,0	28,3	26,7	22,2	15,0	4,4	-2,8	12,2	57	2
3	Mean daily minimum temperature	in °C	-14,4	-13,3	-5,8	2,2	8,9	14,4	17,2	16,1	11,1	5,0	-3,3	-11,1	2,2	57	3
4	Absolute maximum temperature	in °C	14,4	17,8	28,3	32,8	41,1	40,0	42,2	39,4	40,0	32,2	25,0	17,2	42,2	57	4
5	Absolute minimum temperature	in °C	-36,7	-36,1	-32,8	-14,4	-5,6	1,1	6,7	5,6	-3,3	-12,2	-25,0	-32,8	-36,7	57	5
6	Mean relative humidity	in %															6
7	Mean precipitation	in mm	18	20	39	47	81	102	83	81	62	40	36	22	631	30	7
8	Maximum precipitation	in mm															8
9	Minimum precipitation	in mm															9
10	Maximum precipitation in 24 h	in mm	20	23	36	40	67	84	105	82	90	47	74	28	105	22	10
11	Mean number of days with precipitation	>0,25 mm	8	7	11	9	11	13	10	10	9	8	8	8	112	22	11
12	Mean duration of sunshine	in h	140	166	200	231	272	302	343	296	237	193	115	112	2607	30	12
13	Mean quantity of radiation	in ly/day	159	253	359	426	487	543	555	491	361	246	146	124	346	10	13
14	Mean potential evaporation	in mm	0	0	0	37	88	124	149	127	81	39	0	0	645	56	14
15	Mean windspeed	in m/sec	5	5	5	6	5	5	4	4	5	5	5	5	5	22	15
16	Mean predominent direction of the wind		NW	NW	NW	NW	SE	SE	SE	SE	S	SE	NW	NW		11	16

63 Station/Country Lansing (Michigan)/U.S.

Location 42°44'N/84°29'W Height above sealevel 261 m Climate symbol: Köppen **Dfa** Troll III,7

		J	F	M	A	M	J	J	A	S	O	N	D	year	P	
1	Mean daily temperature	in °C	-4,3	-4,3	0,2	7,6	13,9	19,7	21,2	21,2	16,7	10,7	3,3	-2,5	8,7	30	1
2	Mean daily maximum temperature	in °C	-0,6	-0,1	5,3	13,1	19,7	25,1	28,0	26,8	22,3	15,7	7,2	0,8	13,6	30	2
3	Mean daily minimum temperature	in °C	-8,6	-8,5	-4,1	1,7	7,4	14,2	15,4	14,3	10,8	4,8	-0,8	-6,3		30	3
4	Absolute maximum temperature	in °C	17	19	28	31	34	37	39	39	36	31	24	18	39	53	4
5	Absolute minimum temperature	in °C	-27	-32	-23	-13	-5	1	6	3	-2	-7	-16	-28	-32	53	5
6	Mean relative humidity	in %	76	75	71	65	66	67	65	68	69	71	75	77	71	11	6
7	Mean precipitation	in mm	50	50	61	73	95	85	66	77	66	64	56	51	794	30	7
8	Maximum precipitation	in mm	101	190	132	162	203	190	192	234	197	186	130	108		53	8
9	Minimum precipitation	in mm	9	6	11	21	18	25	2	5	13	7	1	11		53	9
10	Maximum precipitation in 24 h	in mm	44	110	69	67	73	139	81	70	80	73	42	43	139	53	10
11	Mean number of days with precipitation	>0,25 mm	14	13	13	12	13	11	9	9	11	11	13	14	143	43	11
12	Mean duration of sunshine	in h	84	119	175	215	272	305	344	294	228	182	87	73	2378	30	12
13	Mean quantity of radiation	in ly/day	121	210	309	359	483	547	540	466	373	255	136	108	311	30	13
14	Mean potential evaporation	in mm	0	0	0	34	76	119	135	122	81	46	10	0	623		14
15	Mean windspeed	in m/sec	4	4	4	4	4	3	3	3	3	3	4	4	4	22	15
16	Mean predominent direction of the wind		SW	SW	NW	SW	SW	S	SW	SW	S	SW	SW	SW		22	16

64 Station/Country Binghamton (New York)/U.S.

Location 42°06'N/75°55'W Height above sealevel 261 m Climate symbol: Köppen **Dfa** Troll III,8

		J	F	M	A	M	J	J	A	S	O	N	D	year	P	
1	Mean daily temperature	in °C	-3,0	-3,1	1,3	8,3	14,6	19,6	22,1	21,0	16,9	11,1	4,8	-1,5	9,3	30	1
2	Mean daily maximum temperature	in °C	1,1	1,4	6,7	13,4	20,6	25,6	28,2	26,9	23,0	16,6	9,0	2,4	14,5	30	2
3	Mean daily minimum temperature	in °C	-8,1	-8,4	-3,3	2,1	7,9	13,1	15,6	14,4	10,8	4,7	0,4	-5,7	3,6	30	3
4	Absolute maximum temperature	in °C	21	23	28	33	34	38	39	38	38	33	27	20	39	70	4
5	Absolute minimum temperature	in °C	-33	-32	-24	-14	-5	-1	4	1	-4	-8	-12	-30	-33	70	5
6	Mean relative humidity	in %	77	76	73	70	71	74	75	77	78	75	77	77	75	15	6
7	Mean precipitation	in mm	82	80	78	78	92	78	97	91	82	77	71	65	931	30	7
8	Maximum precipitation	in mm	130	113	158	185	190	242	207	196	185	244	161	137		68	8
9	Minimum precipitation	in mm	19	14	17	16	11	18	22	17	10	11	12	4		68	9
10	Maximum precipitation in 24 h	in mm	49	53	50	50	77	67	108	84	116	104	71	60	116	70	10
11	Mean number of days with precipitation	>0,25 mm	15	13	14	14	13	12	12	11	10	11	13	14	152	64	11
12	Mean duration of sunshine	in h	94	119	154	170	226	256	266	230	184	158	92	79	2025	30	12
13	Mean quantity of radiation	in ly/day															13
14	Mean potential evaporation	in mm	0	0	31	37	83	114	134	118	81	43	15	0	656		14
15	Mean windspeed	in m/sec	5	5	5	5	5	4	4	4	4	4	5	5	5	12	15
16	Mean predominent direction of the wind		WSW	SSE	NW	WNW	NNW	NNW	WSW	SSW	SSW	WSW	NNW	WSW		12	16

## 65

Station / Country Boston (Massachusetts)/U-S.

Location 42°13'N/71°07'W Height above sealevel 192 m Climate symbol: Köppen Dfb Troll III,8

	J	F	M	A	M	J	J	A	S	O	N	D	year	P	
1 Mean daily temperature	in °C	-2,8	-2,8	1,6	7,6	13,7	18,4	21,6	20,8	16,9	11,5	5,6	-1,1	9,3	30	1
2 Mean daily maximum temperature	in °C	2,2	2,8	6,1	12,2	18,9	23,9	26,7	25,6	21,7	16,7	9,4	4,4	14,4	59	2
3 Mean daily minimum temperature	in °C	-6,7	-6,1	-2,2	3,3	9,4	14,4	17,2	16,7	12,8	7,8	1,7	-3,9	5,6	59	3
4 Absolute maximum temperature	in °C	21,1	20,0	30,0	31,7	36,1	37,8	40,0	38,3	38,9	32,2	25,6	20,8	40,0	77	4
5 Absolute minimum temperature	in °C	-25,0	-27,8	-22,2	-11,7	-0,6	5,0	10,0	7,8	1,1	-3,9	-18,9	-27,2	-27,8	77	5
6 Mean relative humidity	in %	68	66	64	63	66	66	69	69	70	67	69	68	67	14	6
7 Mean precipitation	in mm	114	95	115	102	88	95	83	103	100	95	115	101	1206	30	7
8 Maximum precipitation	in mm	242	180	279	199	340	219	206	434	211	220	208	247		30	8
9 Minimum precipitation	in mm	23	29	38	31	13	12	13	32	9	24	44	26		30	9
10 Maximum precipitation in 24 h	in mm	79	123	73	68	95	100	113	252	133	153	129	77	252	75	10
11 Mean number of days with precipitation	>0,25 mm	13	11	13	12	13	12	11	10	10	10	11	11	137	75	11
12 Mean duration of sunshine	in h	148	168	212	222	263	283	300	280	232	207	152	148	2615	30	12
13 Mean quantity of radiation	in ly/day	156	229	320	391	472	512	500	449	366	275	167	141	332	29	13
14 Mean potential evaporation	in mm	0	0	9	37	79	118	142	125	87	51	20	0	668	46	14
15 Mean windspeed	in m/sec	8	8	8	7	7	8	6	6	6	7	7	8		52	15
16 Mean predominant direction of the wind		W	W	W	NW	S	S	SW	SW	SW	NW	W	W		52	16

## 66

Station / Country Eureka (California)/U.S.

Location 40°48'N/124°10'W Height above sealevel 18 m Climate symbol: Köppen Csb Troll IV,1

	J	F	M	A	M	J	J	A	S	O	N	D	year	P	
1 Mean daily temperature	in °C	8,6	9,0	9,3	10,2	11,7	13,1	13,5	13,7	13,6	12,4	10,7	9,4	11,3	30	1
2 Mean daily maximum temperature	in °C	11,7	11,7	12,2	13,3	13,9	15,0	15,6	15,6	16,1	15,6	13,9	12,2	13,9	60	2
3 Mean daily minimum temperature	in °C	5,0	5,0	6,1	6,7	8,3	10,0	10,8	11,1	10,6	8,9	7,2	5,6	7,8	60	3
4 Absolute maximum temperature	in °C	25,0	29,4	25,6	26,1	28,9	29,4	24,4	26,1	29,4	28,9	27,2	21,1	29,4	60	4
5 Absolute minimum temperature	in °C	-6,7	-4,4	-1,7	-0,6	1,7	4,4	6,1	6,7	2,2	1,1	-2,8	-5,6	-6,7	60	5
6 Mean relative humidity	in %	82	83	82	82	83	84	85	87	86	85	84	83	84	57	6
7 Mean precipitation	in mm	170	140	133	68	55	19	3	3	16	81	117	170	975	30	7
8 Maximum precipitation	in mm	350	354	355	190	148	65	34	31	90	331	347	327		48	8
9 Minimum precipitation	in mm	41	13	2	8	<1	0	0	0	0	0	tr	30		48	9
10 Maximum precipitation in 24 h	in mm	112	124	78	60	57	44	30	23	30	148	116	106	148	50	10
11 Mean number of days with precipitation	>0,25 mm	17	14	15	12	9	5	2	2	5	9	12	15	118	50	11
12 Mean duration of sunshine	in h	120	138	180	209	247	261	244	205	195	164	127	108	2198	30	12
13 Mean quantity of radiation	in ly/day															13
14 Mean potential evaporation	in mm	32	32	43	50	64	76	80	75	66	55	39	31	643	56	14
15 Mean windspeed	in m/sec	3	3	3	4	4	3	3	3	2	3	3	3	3	30	15
16 Mean predominant direction of the wind		SE	SE	N	N	N	N	N	NW	N	N	SE	SE		30	16

## 67

Station / Country Medford (Oregon)/U.S.

Location 42°22'N/122°52'W Height above sealevel 396 m Climate symbol: Köppen Csa Troll III.2

	J	F	M	A	M	J	J	A	S	O	N	D	year	P	
1 Mean daily temperature	in °C	2,7	5,3	7,7	10,9	14,4	17,9	22,2	21,5	18,2	12,2	6,3	3,8	11,9	30	1
2 Mean daily maximum temperature	in °C	7,0	11,1	15,1	18,7	22,9	26,6	31,5	31,5	27,5	20,2	12,3	7,3	19,3	30	2
3 Mean daily minimum temperature	in °C	-1,3	0,7	1,8	3,8	6,8	9,9	12,6	11,9	8,4	4,7	1,2	-0,2	5,1	30	3
4 Absolute maximum temperature	in °C	20	23	30	33	38	41	46	42	42	36	24	18	46	31	4
5 Absolute minimum temperature	in °C	-19	-14	-9	-6	-2	-1	4	5	-2	-7	-9	-16	-19	31	5
6 Mean relative humidity	in %	86	79	70	64	61	58	50	50	56	72	83	88	68		6
7 Mean precipitation	in mm	80	61	45	27	37	26	5	5	15	49	66	86	502	30	7
8 Maximum precipitation	in mm	169	136	141	65	116	89	34	29	59	233	219	223		29	8
9 Minimum precipitation	in mm	13	7	11	4	2	0	0	0	tr	<1	19			29	9
10 Maximum precipitation in 24 h	in mm	81	75	40	23	42	50	24	29	34	74	76	77	81	31	10
11 Mean number of days with precipitation	>0,25 mm	14	12	12	9	9	6	1	1	4	8	11	14	101	31	11
12 Mean duration of sunshine	in h															12
13 Mean quantity of radiation	in ly/day	116	215	338	482	592	652	698	605	447	279	149	93	389	13	13
14 Mean potential evaporation	in mm	10	17	31	50	83	107	138	125	87	48	20	9	725		14
15 Mean windspeed	in m/sec	2	2	2	3	3	3	3	2	2	2	1	1	2	11	15
16 Mean predominant direction of the wind		SSE	S	NNW	NNW	NW	NW	WNW	WNW	WNW	S	SE	SSE		11	16

Station / Country Boise (Idaho)/U.S.

Location 43°34'N/116°13'W Height above sealevel 868 m Climate symbol: Köppen BSk Troll III,10

	J	F	M	A	M	J	J	A	S	O	N	D	year	P	
1 Mean daily temperature	in °C	-1,9	1,1	5,1	9,9	14,3	18,2	23,7	22,3	17,3	11,4	3,9	0,1	10,4	30	1
2 Mean daily maximum temperature	in °C	3,3	6,1	11,7	16,7	21,7	26,7	32,2	31,1	24,4	17,8	10,0	4,4	17,2	65	2
3 Mean daily minimum temperature	in °C	-5,6	-2,8	0,6	3,3	7,2	10,6	14,4	13,3	8,3	3,9	-0,6	-4,4	3,9	62	3
4 Absolute maximum temperature	in °C	16,7	20,6	28,3	33,3	37,8	42,8	45,0	44,4	39,4	35,0	29,4	21,1	45,0	83	4
5 Absolute minimum temperature	in °C	-33,3	-25,0	-20,6	-11,7	-3,9	-1,1	4,4	0,0	-5,0	-10,0	-23,3	-27,8	-33,3	83	5
6 Mean relative humidity	in %	78	69	60	54	52	49	38	37	46	56	66	79	57	32	6
7 Mean precipitation	in mm	34	34	34	29	33	23	5	4	10	21	30	34	290	30	7
8 Maximum precipitation	in mm	98	55	58	67	102	87	24	60	65	57	61	81		30	8
9 Minimum precipitation	in mm	3	5	5	2	2	<1	tr	0	0	0	9	6		30	9
10 Maximum precipitation in 24 h	in mm	38	25	28	27	38	57	24	20	22	19	18	29	57	21	10
11 Mean number of days with precipitation	>0,25 mm	11	11	10	8	9	7	2	2	3	7	10	11	90	21	11
12 Mean duration of sunshine	in h	116	144	218	274	322	352	412	378	311	232	143	104	3006	30	12
13 Mean quantity of radiation	in ly/day	148	229	341	467	586	640	679	575	467	309	179	124	397	12	13
14 Mean potential evaporation	in mm	0	5	21	47	80	112	152	133	81	46	12	0	689	51	14
15 Mean windspeed	in m/sec	4	4	5	5	4	4	4	4	4	4	4	4	4	22	15
16 Mean predominant direction of the wind		SE	SE	SE	SE	NW	NW	NW	NW	SE	SE	SE	SE		21	16

69 Station / Country Elko (Nevada) / U.S.

Location 40°50'N / 115°47'W   Height above sealevel 1547 m   Climate symbol: Köppen BSk   Troll III,7a

		J	F	M	A	M	J	J	A	S	O	N	D	year	P	
1	Mean daily temperature	in °C	-5,2	-2,2	1,9	6,8	11,1	15,6	20,9	19,4	14,4	8,3	1,2	-3,1	7,4	30	1
2	Mean daily maximum temperature	in °C	1,4	4,5	9,6	15,8	21,3	26,1	32,9	31,7	26,3	18,8	9,5	3,8	16,8	30	2
3	Mean daily minimum temperature	in °C	-12,7	-8,8	-5,1	-1,9	1,9	5,3	9,5	7,8	4,3	8,5	1,4	-2,8	7,6	30	3
4	Absolute maximum temperature	in °C	14	19	23	28	33	38	40	39	37	30	23	18	40	30	4
5	Absolute minimum temperature	in °C	-42	-38	-23	-19	-9	-5	-1	-4	-13	-13	-24	-39	-42	30	5
6	Mean relative humidity	in %	75	71	66	54	54	45	35	33	37	48	64	75	55	11	6
7	Mean precipitation	in mm	29	23	21	21	24	18	10	8	9	19	23	26	231	30	7
8	Maximum precipitation	in mm	85	74	46	52	61	64	60	70	28	70	70	71		28	8
9	Minimum precipitation	in mm	4	2	9	3	tr	<1	tr	tr	tr	tr	1	3		28	9
10	Maximum precipitation in 24 h	in mm	32	23	21	28	23	24	26	26	30	33	33	41	41	30	10
11	Mean number of days with precipitation	>0,25 mm	9	9	8	7	8	5	3	3	3	5	8	8	75	30	11
12	Mean duration of sunshine	in h															12
13	Mean quantity of radiation	in ly / day	242	330	462	558	621	704	659	614	520	394	286	224	468	10	13
14	Mean potential evaporation	in mm	0	0	12	37	71	102	133	118	72	37	7	0	589		14
15	Mean windspeed	in m / sec	2	3	3	3	3	3	3	2	2	2	2	2	3	12	15
16	Mean predominent direction of the wind		SW	SW	SW	WSW	SW	SW	SW	SW	SW	SW	SW	SW		19	16

70 Station / Country Salt Lake City (Utah) / U.S.

Location 40°46'N / 111°58'W   Height above sealevel 1286 m   Climate symbol: Köppen Csa   Troll III,10

		J	F	M	A	M	J	J	A	S	O	N	D	year	P	
1	Mean daily temperature	in °C	-2,1	0,6	4,7	9,9	14,7	19,4	24,7	23,6	18,3	11,5	3,4	-0,2	10,7	30	1
2	Mean daily maximum temperature	in °C	1,7	5,0	10,6	16,7	22,8	27,8	33,3	32,2	26,1	18,9	9,4	4,4	17,2	30	2
3	Mean daily minimum temperature	in °C	-8,3	-4,4	-0,6	3,3	7,2	11,1	16,1	15,6	9,4	4,4	-2,2	-5,6	3,9	19	3
4	Absolute maximum temperature	in °C	16,7	20,0	25,6	29,4	33,9	39,4	40,6	38,9	36,1	31,1	23,3	20,0	40,6	69	4
5	Absolute minimum temperature	in °C	-28,9	-25,0	-17,8	-7,8	-3,9	0,0	6,1	5,6	-1,7	-5,6	-18,9	-23,3	-28,9	69	5
6	Mean relative humidity	in %	70	66	57	52	47	39	38	39	42	51	59	68	52	13	6
7	Mean precipitation	in mm	34	30	40	45	36	25	15	22	13	29	33	32	354	30	7
8	Maximum precipitation	in mm	80	82	93	124	86	74	55	83	80	92	85	74		30	8
9	Minimum precipitation	in mm	4	3	3	11	tr	<1	<1	tr	tr	0	<1	10		30	9
10	Maximum precipitation in 24 h	in mm	35	27	48	61	52	48	42	50	23	37	29	26	61	32	10
11	Mean number of days with precipitation	>0,25 mm	10	9	10	9	8	5	4	6	5	6	7	9	88	32	11
12	Mean duration of sunshine	in h	137	155	227	267	329	358	377	348	306	249	171	135	3095	30	12
13	Mean quantity of radiation	in ly / day	174	256	381	490	588	654	647	569	462	328	208	151	409	9	13
14	Mean potential evaporation	in mm	0	0	15	40	79	121	160	143	90	46	12	0	706	61	14
15	Mean windspeed	in m / sec	3	4	4	4	4	4	4	4	4	4	3	3	4	31	15
16	Mean predominent direction of the wind		SSE	SE	SSE	SE	SE	SSE	SSE	SSE	SE	SE	SSE	SSE		29	16

71 Station / Country Lander (Wyoming) / U.S.

Location 42°49'N / 108°44'W   Height above sealevel 1696 m   Climate symbol: Köppen Dfb   Troll III,10

		J	F	M	A	M	J	J	A	S	O	N	D	year	P	
1	Mean daily temperature	in °C	-7,1	-4,4	0,1	6,2	11,8	16,7	21,4	20,4	15,0	8,4	-0,6	-4,9	6,9	30	1
2	Mean daily maximum temperature	in °C	0,0	2,2	6,7	12,8	17,8	24,4	28,9	27,8	22,2	14,4	6,7	0,0	13,3	39	2
3	Mean daily minimum temperature	in °C	-15,0	-12,8	-7,2	-1,7	3,3	7,2	10,6	9,4	4,4	-1,1	-7,7	-13,9	-2,2	39	3
4	Absolute maximum temperature	in °C	17,8	17,8	21,7	27,8	31,1	35,6	37,8	35,6	32,2	28,3	22,2	16,7	37,8	39	4
5	Absolute minimum temperature	in °C	-39,4	-37,2	-31,1	-19,4	-10,6	-3,3	0,0	-5,0	-13,9	-25,6	-35,0	-40,0	-40,0	39	5
6	Mean relative humidity	in %	69	68	66	61	58	53	51	51	55	60	66	70	60	13	6
7	Mean precipitation	in mm	12	18	29	62	67	35	20	12	26	31	23	11	346	30	7
8	Maximum precipitation	in mm	42	55	48	139	153	175	28	26	117	79	53	38		12	8
9	Minimum precipitation	in mm	tr	1	15	20	15	tr	3	2	<1	3	<1	<1		12	9
10	Maximum precipitation in 24 h	in mm	21	22	32	50	62	90	18	15	50	43	29	25	90	17	10
11	Mean number of days with precipitation	>0,25 mm	5	6	8	8	8	6	4	4	5	4	5	4	70	17	11
12	Mean duration of sunshine	in h	200	208	280	264	301	340	361	328	280	233	186	185	3144	30	12
13	Mean quantity of radiation	in ly / day	230	320	450	540	581	670	641	573	472	353	237	196	439	11	13
14	Mean potential evaporation	in mm	0	0	0	34	68	104	135	119	72	34	0	0	566	56	14
15	Mean windspeed	in m / sec	3	3	3	4	3	3	3	3	3	3	3	2	3	17	15
16	Mean predominent direction of the wind		SW	SW	SW	SW	SW	SW	SW	SW	SW	SW	SW	SW		17	16

72 Station / Country Cheyenne (Wyoming) / U.S.

Location 41°09'N / 104°49'W   Height above sealevel 1871 m   Climate symbol: Köppen BSk   Troll III,4

		J	F	M	A	M	J	J	A	S	O	N	D	year	P	
1	Mean daily temperature	in °C	-3,6	-2,8	0,6	5,0	10,0	15,6	19,2	18,6	13,9	7,2	1,4	-1,9	7,0	74	1
2	Mean daily maximum temperature	in °C	2,2	3,3	6,7	11,7	16,7	23,3	26,7	26,1	21,7	14,4	7,8	3,9	13,9	74	2
3	Mean daily minimum temperature	in °C	-9,4	-8,9	-5,6	-1,7	3,3	7,8	11,7	11,1	6,1	0,0	-5,0	-7,7	0,0	74	3
4	Absolute maximum temperature	in °C	17,8	18,9	25,0	27,8	31,1	36,1	37,8	35,6	32,8	29,4	23,9	20,6	37,8	74	4
5	Absolute minimum temperature	in °C	-38,9	-36,7	-29,4	-21,1	-13,3	-2,2	0,6	-3,9	-8,9	-20,6	-29,4	-33,3	-38,9	74	5
6	Mean relative humidity	in %	55	57	63	60	60	56	55	55	53	55	54	56	56		6
7	Mean precipitation	in mm	10	15	25	48	61	41	53	41	30	25	13	13	375	60	7
8	Maximum precipitation	in mm	71	55	65	128	136	135	102	77	95	91	49	43		23	8
9	Minimum precipitation	in mm	tr	1	6	9	17	9	18	<1	3	2	<1	2		23	9
10	Maximum precipitation in 24 h	in mm	36	25	48	81	69	81	119	43	74	43	33	33	119	78	10
11	Mean number of days with precipitation	>0,25 mm	6	6	8	10	12	9	11	10	6	6	6	6	94	60	11
12	Mean duration of sunshine	in h	191	197	243	237	259	304	318	286	265	242	188	170	2900	30	12
13	Mean quantity of radiation	in ly / day															13
14	Mean potential evaporation	in mm	0	0	0	23	68	102	126	114	75	37	7	0	552		14
15	Mean windspeed	in m / sec															15
16	Mean predominent direction of the wind		NW	NW	W	NW	NW	NW	W	NW	W	W	W	W		50	16

73 Station/Country Rapid City(South Dakota)/U.S.

Location 44°02'N/103°03'W

Height above sealevel 993 m Climate symbol: Köppen BSk Troll III,10

		J	F	M	A	M	J	J	A	S	O	N	D	year	P	
1 Mean daily temperature	in °C	-5.6	-4.4	-0.5	6.9	13.2	18.3	23.2	22.2	16.4	10.0	1.7	-2.7	8.2	30	1
2 Mean daily maximum temperature	in °C	0.6	2.0	5.8	13.7	19.2	24.2	29.8	29.1	23.4	16.5	8.2	2.9	14.6	30	2
3 Mean daily minimum temperature	in °C	-12.7	-11.0	-6.6	0.2	5.9	10.9	14.8	14.0	8.1	2.4	-4.6	-9.8	1.0	30	3
4 Absolute maximum temperature	in °C	23	23	28	32	34	38	43	41	40	34	25	21	43	18	4
5 Absolute minimum temperature	in °C	-33	-28	-26	-13	-8	-1	4	4	-4	-10	-28	-28	-33	18	5
6 Mean relative humidity	in %	67	67	67	59	60	63	56	55	48	53	61	65	60	8	6
7 Mean precipitation	in mm	9	12	26	42	68	78	45	31	24	20	10	8	373	30	7
8 Maximum precipitation	in mm	45	62	77	77	187	144	115	112	100	57	53	17		16	8
9 Minimum precipitation	in mm	<1	2	4	7	26	23	21	3	<1	<1	<1	1		16	9
10 Maximum precipitation in 24 h	in mm	32	25	56	76	62	81	64	36	50	36	28	13	81	18	10
11 Mean number of days with precipitation	>0,25 mm	6	7	9	9	12	13	9	8	5	5	6	5	94	18	11
12 Mean duration of sunshine	in h	164	182	222	245	278	300	348	317	286	228	184	144	2858	30	12
13 Mean quantity of radiation	in ly/day	186	274	397	481	522	587	588	538	423	310	202	170	389	12	13
14 Mean potential evaporation	in mm	0	0	3	34	78	112	144	126	78	40	7	0	620	56	14
15 Mean windspeed	in m/sec	4	5	6	6	5	5	5	5	5	5	5	5	5	10	15
16 Mean predominent direction of the wind		NNW	NNW	NNW	NNW	NNW	NNW	NNW	NNW	NNW	NNW	NNW	NNW		10	16

74 Station/Country Huron(South Dakota)/U.S.

Location 44°23'N/98°13'W

Height above sealevel 391 m Climate symbol: Köppen Dfa Troll III,9

		J	F	M	A	M	J	J	A	S	O	N	D	year	P	
1 Mean daily temperature	in °C	-10.3	-8.3	-1.3	7.8	14.4	20.1	24.2	22.9	16.9	9.9	0.1	-6.5	7.5	30	1
2 Mean daily maximum temperature	in °C															2
3 Mean daily minimum temperature	in °C															3
4 Absolute maximum temperature	in °C	17	22	32	33	37	41	43	43	39	36	24	22	43	21	4
5 Absolute minimum temperature	in °C	-37	-34	-31	-12	-7	0	4	2	-7	-12	-28	-31	-37	21	5
6 Mean relative humidity	in %															6
7 Mean precipitation	in mm	12	15	28	47	60	80	48	53	39	29	17	14	440	30	7
8 Maximum precipitation	in mm															8
9 Minimum precipitation	in mm															9
10 Maximum precipitation in 24 h	in mm	40	32	48	45	89	79	56	105	66	103	31	32	105	21	10
11 Mean number of days with precipitation	>0,25 mm	6	7	9	9	10	11	8	9	7	6	6	6	94	21	11
12 Mean duration of sunshine	in h	153	177	213	250	295	321	367	320	260	212	142	134	2844	30	12
13 Mean quantity of radiation	in ly/day															13
14 Mean potential evaporation	in mm	0	0	0	37	84	120	148	126	81	37	0	0	633		14
15 Mean windspeed	in m/sec	5	5	6	6	6	5	5	5	5	5	5	5	5	21	15
16 Mean predominent direction of the wind		SSE	NW	NW	SSE	SSE	SSE	SSE	SSE	SSE	SSE	NW	SSE		11	16

75 Station/Country Omaha(Nebraska)/U.S.

Location 41°18'N/95°54'W

Height above sealevel 337 m Climate symbol: Köppen Dfa Troll III,7

		J	F	M	A	M	J	J	A	S	O	N	D	year	P	
1 Mean daily temperature	in °C	-5.4	-3.1	2.7	10.9	17.2	22.8	25.8	24.6	19.4	13.2	3.8	-2.1	10.8	30	1
2 Mean daily maximum temperature	in °C	-1.1	1.7	8.3	16.1	22.2	27.2	30.0	28.9	24.4	17.8	8.9	1.7	15.6	58	2
3 Mean daily minimum temperature	in °C	-10.6	-8.3	-2.2	5.6	11.7	18.7	19.4	18.3	13.9	7.2	-1.1	-7.2	5.0	58	3
4 Absolute maximum temperature	in °C	19.4	25.6	32.8	34.4	37.2	40.6	42.8	43.9	38.9	33.3	26.7	21.7	43.9	58	4
5 Absolute minimum temperature	in °C	-35.6	-32.2	-22.2	-14.4	-3.9	5.6	10.0	6.7	-1.1	-13.3	-25.6	-28.9	-35.6	58	5
6 Mean relative humidity	in %	74	72	65	64	64	67	64	66	67	64	70	75	68	14	6
7 Mean precipitation	in mm	21	24	37	65	88	115	86	101	67	44	32	20	700	30	7
8 Maximum precipitation	in mm	94	66	81	164	180	275	244	184	134	102	103	84		23	8
9 Minimum precipitation	in mm	1	3	3	6	14	33	13	19	10	tr	1	tr		23	9
10 Maximum precipitation in 24 h	in mm	34	57	37	65	76	88	86	86	92	60	64	45	92	25	10
11 Mean number of days with precipitation	>0,25 mm	7	7	8	9	11	11	9	10	7	6	6	6	95	25	11
12 Mean duration of sunshine	in h	172	188	222	259	305	332	379	311	270	248	166	145	2997	30	12
13 Mean quantity of radiation	in ly/day	204	275	353	462	509	570	577	519	398	304	201	170	379	6	13
14 Mean potential evaporation	in mm	0	0	9	47	94	136	160	139	90	49	10	0	734	58	14
15 Mean windspeed	in m/sec	5	5	6	6	5	5	4	4	5	5	5	5	5	25	15
16 Mean predominent direction of the wind		NNW	N	N	NNW	SSE	SSE	SSE	SSE	SSE	SSE	SSE	SSE		12	16

76 Station/Country Des Moines (Iowa)/U.S.

Location 41°35'N/93°37'W

Height above sealevel 244 m Climate symbol: Köppen Dfa Troll III,7

		J	F	M	A	M	J	J	A	S	O	N	D	year	P	
1 Mean daily temperature	in °C	-6.1	-4.4	2.5	10.3	16.2	21.4	24.2	23.1	18.6	12.0	3.9	-3.3	9.7	52	1
2 Mean daily maximum temperature	in °C	-1.1	0.6	7.8	16.1	21.7	26.7	30.0	28.9	24.4	17.8	8.9	1.1	15.0	52	2
3 Mean daily minimum temperature	in °C	-11.1	-9.4	-2.8	4.4	10.6	16.1	18.3	17.2	12.8	6.1	-1.1	-7.7	4.4	52	3
4 Absolute maximum temperature	in °C	18.3	25.6	31.1	33.3	36.7	38.9	42.8	43.3	37.2	32.8	26.1	20.6	43.3	52	4
5 Absolute minimum temperature	in °C	-34.4	-32.2	-23.3	-11.7	-3.3	2.8	8.9	4.4	-3.3	-13.9	-23.3	-29.4	-34.4	52	5
6 Mean relative humidity	in %	76	74	68	65	63	67	65	68	69	66	72	76	69		6
7 Mean precipitation	in mm	28	28	46	74	112	122	86	91	64	38	30	30	810	54	7
8 Maximum precipitation	in mm	79	76	130	155	172	360	267	266	219	185	117	74		19	8
9 Minimum precipitation	in mm	2	3	13	27	31	33	11	18	10	<1	2	6		19	9
10 Maximum precipitation in 24 h	in mm	36	46	74	74	112	137	109	109	124	94	66	38	137	71	10
11 Mean number of days with precipitation	>0,25 mm	8	8	9	10	12	11	9	9	9	8	7	8	108	52	11
12 Mean duration of sunshine	in h	155	170	203	236	276	303	348	299	263	227	156	138	2770	30	12
13 Mean quantity of radiation	in ly/day															13
14 Mean potential evaporation	in mm	0	0	6	44	91	130	157	136	90	46	10	0	710		14
15 Mean windspeed	in m/sec	5.4	5.8	6.3	6.7	5.4	4.9	4.5	4.0	4.5	4.9	5.8	5.8	5.4		15
16 Mean predominent direction of the wind		NW	NW	NW	NW	SE	S	S	S	S	S	NW	NW			16

77 Station/Country Chicago (Illinois)/U.S.

Location 41°47'N/87°47'W **Height above sealevel** 185 m **Climate symbol: Köppen** Dfa **Troll** III.7

		J	F	M	A	M	J	J	A	S	O	N	D	year	P	
1 Mean daily temperature	in °C	-3,3	-2,3	2,4	9,5	15,6	21,5	24,3	23,6	19,1	13,0	4,4	-1,8	10,5	30	1
2 Mean daily maximum temperature	in °C	0,0	1,1	6,1	12,8	18,3	23,9	27,2	26,1	22,8	16,1	8,3	2,2	13,9	75	2
3 Mean daily minimum temperature	in °C	-7,7	-6,7	-1,7	4,4	10,0	15,8	18,9	18,3	14,4	8,3	1,1	-5,0	5,8	75	3
4 Absolute maximum temperature	in °C	18,3	20,0	27,8	32,8	36,7	38,9	40,6	38,9	37,8	31,1	25,6	20,0	40,6	77	4
5 Absolute minimum temperature	in °C	-28,9	-29,4	-24,4	-8,3	-2,8	1,7	9,4	8,3	-1,7	-10,0	-18,9	-30,8	-30,6	77	5
6 Mean relative humidity	in %	75	74	71	68	66	69	67	70	70	69	72	76	71	11	6
7 Mean precipitation	in mm	47	41	70	77	95	103	86	80	69	71	56	48	843	30	7
8 Maximum precipitation	in mm	98	85	127	212	193	163	228	150	153	306	95	169		16	8
9 Minimum precipitation	in mm	10	8	8	33	20	20	34	25	12	8	23	9		16	9
10 Maximum precipitation in 24 h	in mm	73	39	64	104	74	116	159	79	65	143	47	60	159	18	10
11 Mean number of days with precipitation	>0,25 mm	10	10	12	13	12	11	9	8	8	7	10	10	120	18	11
12 Mean duration of sunshine	in h	128	142	199	221	274	300	333	299	247	216	136	118	2611	30	12
13 Mean quantity of radiation	in ly/day	96	147	227	331	424	473	473	403	313	207	120	76	273		13
14 Mean potential evaporation	in mm	0	0	6	37	79	122	146	132	90	51	12	0	675	51	14
15 Mean windspeed	in m/sec	5	5	5	5	4	3	3	4	4	5	5	5	4	18	15
16 Mean predominent direction of the wind		W	W	W	W	SSW	SSW	SW	SW	S	S	W	W		16	16

78 Station/Country Detroit (Michigan)/U.S.

Location 42°24'N/83°00'W **Height above sealevel** 189 m **Climate symbol: Köppen** Dfa **Troll** III.7

		J	F	M	A	M	J	J	A	S	O	N	D	year	P	
1 Mean daily temperature	in °C	-2,8	-2,7	1,6	8,4	14,7	20,7	23,3	22,4	18,1	12,1	4,7	-1,2	9,9	30	1
2 Mean daily maximum temperature	in °C	-0,6	0,0	5,6	12,8	19,4	25,0	27,8	26,7	22,8	15,6	7,8	1,7	13,9	73	2
3 Mean daily minimum temperature	in °C	-7,2	-7,7	-2,8	2,8	8,9	14,4	17,2	16,7	12,8	6,7	0,6	-4,4	5,0	73	3
4 Absolute maximum temperature	in °C	18,9	18,3	27,2	31,1	35,0	40,0	40,6	40,0	37,8	31,7	23,9	18,3	40,6	77	4
5 Absolute minimum temperature	in °C	-26,7	-28,9	-21,7	-13,3	-2,2	3,3	8,9	6,1	-1,1	-5,6	-17,8	-31,1	-31,1	77	5
6 Mean relative humidity	in %	81	79	74	69	66	66	63	65	70	72	77	81	72	13	6
7 Mean precipitation	in mm	52	53	61	76	90	72	72	73	62	67	56	53	787	30	7
8 Maximum precipitation	in mm	111	126	112	175	204	119	179	191	150	198	105	117		25	8
9 Minimum precipitation	in mm	11	10	12	19	15	27	21	27	15	15	14	11		25	9
10 Maximum precipitation in 24 h	in mm	41	62	47	75	84	67	71	83	65	94	55	39	94	27	10
11 Mean number of days with precipitation	>0,25 mm	13	12	13	12	12	11	9	8	9	9	12	13	133	27	11
12 Mean duration of sunshine	in h	90	128	180	212	263	295	321	284	226	189	98	89	2375	30	12
13 Mean quantity of radiation	in ly/day															13
14 Mean potential evaporation	in mm	0	0	3	37	83	122	142	125	87	46	12	0	657	56	14
15 Mean windspeed	in m/sec	5	5	5	5	4	4	4	4	4	4	5	5	5	27	15
16 Mean predominent direction of the wind		NW	NW	NW	W	S	S	S	N	S	S	SW	SW		11	16

79 Station/Country Cleveland (Ohio)/U.S.

Location 41°24'N/81°51'W **Height above sealevel** 237 m **Climate symbol: Köppen** Cfb **Troll** III.8

		J	F	M	A	M	J	J	A	S	O	N	D	year	P	
1 Mean daily temperature	in °C	-2,4	-2,2	1,9	8,1	14,2	19,6	21,7	20,8	16,9	11,0	4,1	-1,4	9,3	20	1
2 Mean daily maximum temperature	in °C	2,2	2,4	7,3	13,9	21,1	26,9	29,2	28,1	24,3	17,9	9,5	3,0	15,5	30	2
3 Mean daily minimum temperature	in °C	-6,2	-6,2	-2,1	3,0	9,0	14,6	17,0	16,2	12,8	6,9	1,2	-4,3	5,2	30	3
4 Absolute maximum temperature	in °C	23	21	28	31	33	38	39	39	38	32	28	21	39	20	4
5 Absolute minimum temperature	in °C	-23	-22	-21	-7	-2	3	8	7	0	-4	-14	-23	-23	20	5
6 Mean relative humidity	in %	78	76	73	68	68	69	68	71	71	71	73	76	72	17	6
7 Mean precipitation	in mm	68	59	80	87	89	87	84	83	74	61	66	59	897	30	7
8 Maximum precipitation	in mm	178	118	154	150	153	154	136	132	162	241	164	142		17	8
9 Minimum precipitation	in mm	20	19	20	30	26	35	31	41	40	15	29	18		17	9
10 Maximum precipitation in 24 h	in mm	59	59	70	49	95	71	69	78	47	87	57	32	95	19	10
11 Mean number of days with precipitation	>0,25 mm	16	14	16	15	14	11	10	9	9	10	15	15	154	19	11
12 Mean duration of sunshine	in h	79	111	167	209	274	301	325	288	235	187	99	77	2352	30	12
13 Mean quantity of radiation	in ly/day	125	183	303	286	502	562	562	494	278	289	141	115	335		13
14 Mean potential evaporation	in mm	0	0	6	37	83	121	141	125	90	49	15	0	667		14
15 Mean windspeed	in m/sec	6	6	6	6	5	4	4	4	5	6	6	6	5	19	15
16 Mean predominent direction of the wind		S	S	W	S	S	S	S	S	S	S	S	S		11	16

80 Station/Country Buffalo (New York)/U.S.

Location 42°56'N/78°44'W **Height above sealevel** 211 m **Climate symbol: Köppen** Dfb **Troll** III.8

		J	F	M	A	M	J	J	A	S	O	N	D	year	P	
1 Mean daily temperature	in °C	-4,7	-4,9	-0,8	6,1	12,4	18,2	21,0	20,2	16,3	10,4	3,7	-2,7	7,9	30	1
2 Mean daily maximum temperature	in °C	-0,7	-0,6	3,6	11,6	18,6	24,0	26,8	25,9	22,0	15,7	8,1	1,3	13,0	30	2
3 Mean daily minimum temperature	in °C	-7,7	-8,2	-4,2	1,1	6,8	12,5	15,2	14,5	10,7	5,2	-0,2	-6,0	3,3	30	3
4 Absolute maximum temperature	in °C	22	18	27	31	32	35	34	37	37	31	24	19	37	17	4
5 Absolute minimum temperature	in °C	-24	-21	-19	-11	-3	2	6	6	0	-4	-14	-22	-24	17	5
6 Mean relative humidity	in %	77	76	74	70	70	70	69	71	74	73	75	76	73	19	6
7 Mean precipitation	in mm	72	69	82	76	75	65	65	77	80	76	91	76	904	30	7
8 Maximum precipitation	in mm	163	146	179	128	187	248	211	270	189	232	170	217		88	8
9 Minimum precipitation	in mm	23	12	17	15	13	3	4	1	8	2	13	18		88	9
10 Maximum precipitation in 24 h	in mm	61	59	54	37	52	44	86	99	68	79	47	52	99	10	10
11 Mean number of days with precipitation	>0,25 mm	20	17	16	15	12	9	10	12	10	10	16	20	167	10	11
12 Mean duration of sunshine	in h	110	125	180	212	274	319	338	297	239	183	97	84	2458		12
13 Mean quantity of radiation	in ly/day															13
14 Mean potential evaporation	in mm	0	0	0	30	72	111	135	122	84	48	15	0	617		14
15 Mean windspeed	in m/sec	7	7	7	6	6	5	5	5	5	6	6	7	6	21	15
16 Mean predominent direction of the wind		WSW	SW	SW	SW	SW	SW	SW	SW	S	S	WSW	WSW		11	16

166

81 Station/Country New York (New York)/U.S.

Location 40°47'N/73°58'W Height above sealevel 96 m Climate symbol: Köppen Cfa Troll III,8

		J	F	M	A	M	J	J	A	S	O	N	D	year	P	
1 Mean daily temperature	in °C	0,7	0,8	4,7	10,8	16,9	21,9	24,9	23,9	20,3	14,6	8,3	2,2	12,5	30	1
2 Mean daily maximum temperature	in °C	2,8	3,3	7,2	13,9	20,0	25,0	27,8	26,7	26,1	20,6	10,6	5,0	15,6	46	2
3 Mean daily minimum temperature	in °C	-4,4	-4,4	-1,1	5,6	11,7	15,6	18,9	18,9	15,6	9,4	2,8	-1,7	1,7	46	3
4 Absolute maximum temperature	in °C	20,0	22,8	28,9	32,8	35,0	36,1	38,9	38,9	37,8	32,2	23,9	20,6	38,9	77	4
5 Absolute minimum temperature	in °C	-21,1	-25,6	-16,1	-11,1	1,1	6,7	12,2	10,6	3,9	-2,8	-13,9	-25,0	-25,6	77	5
6 Mean relative humidity	in %	66	64	63	61	62	66	68	70	70	67	68	67	65	28	6
7 Mean precipitation	in mm	84	72	102	87	93	84	94	113	98	80	86	83	1076	30	7
8 Maximum precipitation	in mm	202	174	223	233	216	248	302	276	428	338	253	191		90	8
9 Minimum precipitation	in mm	20	12	23	24	8	<1	12	14	5	7	15	6		90	9
10 Maximum precipitation in 24 h	in mm	85	75	108	68	97	120	90	122	211	284	93	82	284	49	10
11 Mean number of days with precipitation	>0,25 mm	11	10	11	11	11	10	11	10	8	8	9	10	121	92	11
12 Mean duration of sunshine	in h	154	171	213	237	268	289	302	271	235	213	169	155	2677	30	12
13 Mean quantity of radiation	in ly/day	132	199	289	369	430	471	459	389	331	242	148	116	298	34	13
14 Mean potential evaporation	in mm	0	0	12	40	86	125	149	132	94	55	22	2	717	61	14
15 Mean windspeed	in m/sec	5	5	5	5	4	4	4	4	4	4	4	5	5	4	15
16 Mean predominent direction of the wind		NW	NW	NW	NW	SW	SW	SW	SW	SW	SW	NW	NW		41	16

82 Station/Country Harrisburg (Pennsylvania)/U.S.

Location 40°13'N/76°51'W Height above sealevel 102 m Climate symbol: Köppen Cfa Troll III,8

		J	F	M	A	M	J	J	A	S	O	N	D	year	P	
1 Mean daily temperature	in °C	-0,1	0,3	4,6	11,0	17,1	21,8	24,3	23,1	19,1	13,2	6,6	0,8	11,8	30	1
2 Mean daily maximum temperature	in °C	3,6	4,5	9,8	16,2	22,8	27,5	29,7	28,3	24,3	18,3	11,2	4,7	16,8	30	2
3 Mean daily minimum temperature	in °C	-4,7	-4,4	-0,2	4,8	10,9	15,9	18,4	17,2	13,5	7,4	2,2	-2,7	6,5	30	3
4 Absolute maximum temperature	in °C	23	24	30	33	36	38	38	38	39	36	29	22	39	22	4
5 Absolute minimum temperature	in °C	-20	-17	-13	-6	0	6	9	8	0	-4	-11	-22	-22	22	5
6 Mean relative humidity	in %	67	64	62	61	64	67	68	71	72	70	68	67	67	20	6
7 Mean precipitation	in mm	70	59	87	77	99	87	89	93	72	75	75	74	957	30	7
8 Maximum precipitation	in mm	117	102	139	152	205	142	216	230	137	142	142	133		20	8
9 Minimum precipitation	in mm	18	21	30	14	14	15	20	24	15	18	13	6		20	9
10 Maximum precipitation in 24 h	in mm	49	41	54	55	79	71	99	85	111	66	74	51	111	22	10
11 Mean number of days with precipitation	>0,25 mm	12	9	12	12	13	11	10	10	9	9	9	10	126	22	11
12 Mean duration of sunshine	in h	132	160	203	230	277	297	319	282	233	200	140	131	2604	30	12
13 Mean quantity of radiation	in ly/day															13
14 Mean potential evaporation	in mm	0	0	15	43	93	128	149	131	94	52	17	0	722		14
15 Mean windspeed	in m/sec	4	4	4	4	3	3	3	3	3	3	3	3	3	22	15
16 Mean predominent direction of the wind		WNW	WNW	WNW	W	W	W	W	W	WSW	W	WNW	WNW		11	16

83 Station/Country San Francisco (California)/U.S.

Location 37°47'N/122°25'W Height above sealevel 16 m Climate symbol: Köppen Csb Troll IV,1

		J	F	M	A	M	J	J	A	S	O	N	D	year	P	
1 Mean daily temperature	in °C	10,4	11,7	12,6	13,2	14,1	15,1	14,9	15,2	16,7	16,3	14,1	11,4	13,8	30	1
2 Mean daily maximum temperature	in °C	12,8	15,0	16,1	16,7	17,2	18,9	18,3	18,3	20,6	20,0	17,2	13,9	17,2	73	2
3 Mean daily minimum temperature	in °C	7,2	8,3	8,9	9,4	10,6	11,1	11,7	11,7	12,8	12,2	10,6	8,3	10,0	73	3
4 Absolute maximum temperature	in °C	25,6	26,7	30,0	31,7	36,1	37,8	37,2	33,3	38,3	35,6	28,3	23,3	38,3	77	4
5 Absolute minimum temperature	in °C	-1,7	0,6	0,6	4,4	5,6	7,8	8,3	7,8	8,3	6,1	3,3	-2,8	-2,8	77	5
6 Mean relative humidity	in %	77	75	72	72	74	76	80	81	76	72	72	76	75	52	6
7 Mean precipitation	in mm	116	93	74	37	16	4	tr	tr	6	23	51	108	529	30	7
8 Maximum precipitation	in mm	272	216	209	139	91	16	2	11	37	88	158	291		22	8
9 Minimum precipitation	in mm	25	1	3	tr	tr	0	tr	0	0	<1	1	9		22	9
10 Maximum precipitation in 24 h	in mm	89	59	93	60	33	15	2	6	52	39	62	80	93	24	10
11 Mean number of days with precipitation	>0,25 mm	11	11	10	6	4	2	1	1	1	4	7	10	67	24	11
12 Mean duration of sunshine	in h	165	182	251	281	314	330	300	272	267	243	189	156	2959	30	12
13 Mean quantity of radiation	in ly/day															13
14 Mean potential evaporation	in mm	31	35	49	59	70	78	79	77	75	66	48	35	702	56	14
15 Mean windspeed	in m/sec	3	3	4	4	5	5	5	5	4	3	3	3	4	24	15
16 Mean predominent direction of the wind		N	W	W	W	W	W	W	W	W	W	W	N		24	16

84 Station/Country Reno (Nevada)/U.S.

Location 39°30'N/119°47'W Height above sealevel 1342 m Climate symbol: Köppen BSk Troll III,10

		J	F	M	A	M	J	J	A	S	O	N	D	year	P	
1 Mean daily temperature	in °C	-0,1	2,3	5,0	8,6	11,9	15,6	20,1	19,2	15,7	10,1	4,1	0,8	9,4	30	1
2 Mean daily maximum temperature	in °C	6,1	8,9	12,2	16,7	20,6	25,6	31,1	30,6	25,6	19,4	12,8	7,2	18,3	60	2
3 Mean daily minimum temperature	in °C	-6,1	-3,9	-1,7	1,1	5,0	8,3	11,7	11,1	6,7	2,2	-2,2	-5,6	2,2	60	3
4 Absolute maximum temperature	in °C	20,0	24,4	26,1	31,1	36,7	37,8	41,1	39,4	36,1	31,7	26,7	20,6	41,1	59	4
5 Absolute minimum temperature	in °C	-28,3	-24,4	-19,4	-10,6	-8,9	-2,2	1,7	0,0	-4,4	-8,9	-15,0	-21,7	-28,3	60	5
6 Mean relative humidity	in %	69	65	57	51	47	43	38	38	44	51	58	70	52	13	6
7 Mean precipitation	in mm	30	26	17	14	13	9	7	4	6	13	14	27	180	30	7
8 Maximum precipitation	in mm	103	48	51	52	42	25	21	14	26	54	52	133		17	8
9 Minimum precipitation	in mm	<1	6	1	<1	tr	0	0	0	0	tr	tr	<1		17	9
10 Maximum precipitation in 24 h	in mm	60	37	31	42	30	17	20	5	20	32	31	55	60	19	10
11 Mean number of days with precipitation	>0,25 mm	6	6	6	4	5	3	3	2	2	3	4	6	47	18	11
12 Mean duration of sunshine	in h	185	199	287	306	354	376	414	391	336	273	212	170	3483	30	12
13 Mean quantity of radiation	in ly/day															13
14 Mean potential evaporation	in mm	0	7	22	40	71	101	130	117	78	43	17	2	628	52	14
15 Mean windspeed	in m/sec	3	3	3	4	3	3	3	3	2	2	2	2	3	18	15
16 Mean predominent direction of the wind		S	S	WNW	WNW	WNW	WNW	WNW	WNW	WNW	WNW	S	WNW		12	16

168

85 Station/Country Fresno (California)/U.S.

Location 36°48'N/119°42'W Height above sealevel 100 m Climate symbol: Köppen BSk Troll IV,2

		J	F	M	A	M	J	J	A	S	O	N	D	year	P	
1	Mean daily temperature	in °C	7,5	9,9	12,4	16,0	19,7	23,3	27,0	25,8	23,3	17,9	11,8	8,0	16,9	30	1
2	Mean daily maximum temperature	in °C	12,2	16,1	18,9	23,3	27,2	32,8	37,2	36,1	31,7	25,8	18,9	12,8	24,4	43	2
3	Mean daily minimum temperature	in °C	3,3	5,8	6,7	8,9	11,7	15,0	18,3	17,2	14,4	10,8	6,1	3,3	10,0	43	3
4	Absolute maximum temperature	in °C	22,8	28,9	30,8	38,3	43,3	44,4	48,1	45,0	43,9	37,8	30,0	24,4	48,1	43	4
5	Absolute minimum temperature	in °C	-8,3	-4,4	-2,2	1,1	3,3	5,8	10,0	10,8	5,8	2,2	-2,8	-5,0	-8,3	43	5
6	Mean relative humidity	in %	79	73	68	60	52	43	37	40	46	55	65	79	57	43	6
7	Mean precipitation	in mm	52	56	50	29	8	2	tr	<1	3	11	24	50	285	30	7
8	Maximum precipitation	in mm	150	128	147	69	40	8	<1	2	12	30	49	171		19	8
9	Minimum precipitation	in mm	tr	2	2	<1	0	0	0	0	0	0	0	8		19	9
10	Maximum precipitation in 24 h	in mm	62	49	41	27	24	8	1	2	23	21	34	45	62	21	10
11	Mean number of days with precipitation	>0,25 mm	8	7	7	5	2	1	<1	<1	1	2	4	8	46	21	11
12	Mean duration of sunshine	in h	153	192	283	330	389	418	435	406	355	306	221	144	3632	29	12
13	Mean quantity of radiation	in ly/day	182	285	425	549	642	701	684	621	506	374	248	158	448	34	13
14	Mean potential evaporation	in mm	13	20	37	63	99	139	180	165	114	70	31	12	43	56	14
15	Mean windspeed	in m/sec	2	2	3	3	4	4	3	3	3	2	2	2	3	21	15
16	Mean predominant direction of the wind		SE	NW	NW	NW	NW	NW	NW	NW	NW	NW	NW	SE		12	16

86 Station/Country Modena (Utah)/U.S.

Location 37°48'N/113°54'W Height above sealevel 1665 m Climate symbol: Köppen BSk Troll III,10

		J	F	M	A	M	J	J	A	S	O	N	D	year	P	
1	Mean daily temperature	in °C	-3,1	0,0	3,6	7,8	12,2	17,0	21,7	20,8	15,9	9,5	2,8	-2,0	8,9	47	1
2	Mean daily maximum temperature	in °C	3,9	6,7	11,1	16,1	21,1	27,2	31,1	29,4	25,0	18,3	11,1	5,0	17,2	47	2
3	Mean daily minimum temperature	in °C	-10,0	-6,7	-3,9	-0,6	3,3	7,8	12,2	12,2	6,7	0,6	-5,6	-8,9	0,6	47	3
4	Absolute maximum temperature	in °C	17,8	19,4	23,9	27,8	33,3	37,8	38,3	38,3	35,0	29,4	23,3	20,6	38,3	47	4
5	Absolute minimum temperature	in °C	-35,6	-32,8	-21,7	-13,9	-6,7	-5,6	-0,6	1,7	-5,6	-13,9	-23,9	-31,1	-35,6	47	5
6	Mean relative humidity	in %	66	64	57	49	43	32	37	45	42	47	58	67	51		6
7	Mean precipitation	in mm	23	25	30	23	20	10	36	41	23	23	15	15	284	30	7
8	Maximum precipitation	in mm															8
9	Minimum precipitation	in mm															9
10	Maximum precipitation in 24 h	in mm	30	43	38	28	43	30	56	53	58	51	32	38	58	49	10
11	Mean number of days with precipitation	>0,25 mm	6	6	7	6	6	3	7	8	4	4	4	5	56	30	11
12	Mean duration of sunshine	in h															12
13	Mean quantity of radiation	in ly/day															13
14	Mean potential evaporation	in mm	0	0	14	37	70	104	135	121	77	39	10	0	607		14
15	Mean windspeed	in m/sec															15
16	Mean predominant direction of the wind																16

87 Station/Country Denver ((Colorado)/U.S.

Location 39°46'N/104°53'W Height above sealevel 1610 m Climate symbol: Köppen BSk Troll III,10

		J	F	M	A	M	J	J	A	S	O	N	D	year	P	
1	Mean daily temperature	in °C	-1,1	0,3	3,3	8,6	13,7	19,4	23,0	22,2	17,5	11,3	4,0	0,6	10,2	30	1
2	Mean daily maximum temperature	in °C	5,7	7,0	9,9	15,9	21,4	27,8	31,3	30,5	26,1	19,2	10,9	7,3	17,7	30	2
3	Mean daily minimum temperature	in °C	-9,5	-7,7	-5,1	0,2	5,5	10,6	14,1	13,4	8,3	2,3	-4,7	-7,8	1,6	30	3
4	Absolute maximum temperature	in °C	22	24	27	29	36	40	40	38	36	31	26	23	40	28	4
5	Absolute minimum temperature	in °C	-31	-34	-24	-14	-6	-1	6	5	-4	-13	-22	-24	-34	28	5
6	Mean relative humidity	in %	53	57	57	53	55	57	54	54	56	51	57	56	55	30	6
7	Mean precipitation	in mm	14	18	31	54	69	37	39	33	29	26	18	12	380	30	7
8	Maximum precipitation	in mm	37	35	73	106	186	108	106	114	103	87	75	26		24	8
9	Minimum precipitation	in mm	<1	<1	3	10	15	3	4	4	tr	2	<1	1		24	9
10	Maximum precipitation in 24 h	in mm	20	26	38	61	84	52	42	87	62	43	25	18	87	28	10
11	Mean number of days with precipitation	>0,25 mm	6	6	8	9	11	9	9	8	6	6	5	4	87	28	11
12	Mean duration of sunshine	in h	207	205	247	252	281	311	321	297	274	246	200	192	3033	30	12
13	Mean quantity of radiation	in ly/day															13
14	Mean potential evaporation	in mm	0	0	12	37	74	112	141	127	81	43	12	0	641		14
15	Mean windspeed	in m/sec	4	5	5	5	5	4	4	4	4	4	4	4	4	12	15
16	Mean predominant direction of the wind		S	S	S	S	S	S	S	S	S	S	S			12	16

88 Station/Country Dodge City (Kansas)/U.S.

Location 37°46'N/99°58'W Height above sealevel 791 m Climate symbol: Köppen Cfa Troll III,10

		J	F	M	A	M	J	J	A	S	O	N	D	year	P	
1	Mean daily temperature	in °C	-1,7	0,9	5,8	12,2	17,8	22,8	25,6	24,8	20,6	13,4	6,1	-0,3	12,5	56	1
2	Mean daily maximum temperature	in °C	5,0	7,8	13,3	19,4	23,9	29,4	32,2	31,7	27,8	20,6	13,3	5,6	19,4	56	2
3	Mean daily minimum temperature	in °C	-8,3	-6,1	-1,7	5,0	10,8	16,1	18,9	17,8	13,3	6,1	-1,1	-6,1	5,6	56	3
4	Absolute maximum temperature	in °C	28,1	28,9	36,7	35,0	38,3	41,7	42,2	40,6	38,9	34,4	30,0	28,1	42,2	56	4
5	Absolute minimum temperature	in °C	-28,9	-32,2	-23,3	-10,6	-7,2	2,2	7,8	6,1	-1,1	-12,2	-25,0	-28,1	-32,2	56	5
6	Mean relative humidity	in %	67	65	61	63	66	64	66	62	63	63	64	68	64		6
7	Mean precipitation	in mm	10	18	23	48	74	81	79	68	48	36	20	15	518	56	7
8	Maximum precipitation	in mm	50	44	120	132	221	202	143	130	96	116	53	41		16	8
9	Minimum precipitation	in mm	<1	15	3	8	17	3	4	20	<1	tr	tr	tr		16	9
10	Maximum precipitation in 24 h	in mm	33	53	48	74	135	152	107	84	150	104	53	48	152	74	10
11	Mean number of days with precipitation	>0,25 mm	4	5	6	7	11	9	8	7	7	5	4	4	75	56	11
12	Mean duration of sunshine	in h	205	191	249	288	305	335	359	335	290	266	218	198	3219	30	12
13	Mean quantity of radiation	in ly/day	255	316	418	528	568	650	642	592	493	380	285	234	447		13
14	Mean potential evaporation	in mm	0	0	15	49	92	138	165	147	100	52	13	0	771		14
15	Mean windspeed	in m/sec	6,7	6,7	7,6	7,6	7,2	6,7	6,3	6,3	6,7	6,3	6,7	6,3	6,6		15
16	Mean predominant direction of the wind		S	N	N	SSE	S	S	S	S	S	S	S	N			16

89 Station/Country Kansas City (Missouri)/U.S.

Location 39°07'N/94°35'W Height above sealevel 226 m Climate symbol: Köppen Cfa Troll III,7

		J	F	M	A	M	J	J	A	S	O	N	D	year	P	
1 Mean daily temperature	in °C	-0,7	1,8	6,0	12,9	18,4	24,1	27,2	26,3	21,6	15,4	6,7	1,6	13,4	30	1
2 Mean daily maximum temperature	in °C	3,3	5,0	11,7	18,3	23,3	28,3	31,7	30,6	26,7	20,0	11,7	5,0	18,3	58	2
3 Mean daily minimum temperature	in °C	-5,6	-4,4	1,1	7,8	13,3	18,3	21,1	20,0	15,6	9,4	2,2	-3,3	7,8	58	3
4 Absolute maximum temperature	in °C	21,1	27,2	32,8	35,0	39,4	42,2	43,3	45,0	41,7	36,7	28,3	23,3	45,0	58	4
5 Absolute minimum temperature	in °C	-28,9	-30,0	-19,4	-8,9	-2,8	6,7	11,7	7,8	1,1	-8,3	-15,6	-25,0	-30,0	58	5
6 Mean relative humidity	in %	71	69	64	63	65	67	64	65	67	65	66	71	67	13	6
7 Mean precipitation	in mm	36	32	63	90	112	116	81	96	83	73	46	39	867	30	7
8 Maximum precipitation	in mm	133	83	129	268	254	279	272	202	203	303	137	95		25	8
9 Minimum precipitation	in mm	1	9	2	20	44	13	9	6	5	4	1	6		25	9
10 Maximum precipitation in 24 h	in mm	49	42	71	151	67	86	106	112	95	76	53	70	151	27	10
11 Mean number of days with precipitation	>0,25 mm	7	7	9	11	12	11	8	8	8	7	6	6	100	27	11
12 Mean duration of sunshine	in h															12
13 Mean quantity of radiation	in ly/day	182	251	335	435	482	578	562	529	443	324	219	174	375	12	13
14 Mean potential evaporation	in mm	0	0	15	53	96	141	170	152	100	55	15	0	797	56	14
15 Mean windspeed	in m/sec	5	5	5	5	5	5	4	4	4	4	5	5	5	21	15
16 Mean predominent direction of the wind		SSW	SSW	ENE	S	S	S	S	S	S	S	SSW	SSW		12	16

90 Station/Country St. Louis (Missouri)/U.S.

Location 38°38'N/90°12'W Height above sealevel 142 m Climate symbol: Köppen Cfa Troll III,7

		J	F	M	A	M	J	J	A	S	O	N	D	year	P	
1 Mean daily temperature	in °C	-0,1	1,8	6,2	13,0	18,7	24,2	26,4	25,4	21,1	14,9	6,7	1,6	13,3	30	1
2 Mean daily maximum temperature	in °C	4,4	6,1	12,2	18,3	23,9	28,9	31,1	30,6	26,7	20,0	12,2	6,1	18,3	75	2
3 Mean daily minimum temperature	in °C	-4,4	-3,3	2,2	8,3	13,9	18,9	21,7	20,6	16,7	10,0	3,3	-2,2	8,9	75	3
4 Absolute maximum temperature	in °C	23,3	28,9	33,3	33,9	35,6	40,0	43,3	42,2	39,4	33,9	28,3	23,9	43,3	76	4
5 Absolute minimum temperature	in °C	-30,0	-27,8	-16,1	-17,8	0,0	6,7	12,8	11,1	2,2	-6,1	-16,1	-26,1	-30,0	77	5
6 Mean relative humidity	in %	72	70	66	64	65	65	62	65	68	66	68	72	67	13	6
7 Mean precipitation	in mm	50	52	78	94	95	109	84	77	70	73	65	50	897	30	7
8 Maximum precipitation	in mm	206	150	242	224	259	358	269	519	254	189	194	117		23	8
9 Minimum precipitation	in mm	8	10	15	21	22	34	15	21	<1	10	6	2		23	9
10 Maximum precipitation in 24 h	in mm	99	73	70	160	86	222	176	223	106	101	54	75	223	23	10
11 Mean number of days with precipitation	>0,25 mm	9	9	11	12	12	11	8	8	7	7	8	8	110	23	11
12 Mean duration of sunshine	in h															12
13 Mean quantity of radiation	in ly/day															13
14 Mean potential evaporation	in mm	0	0	19	53	100	145	170	152	100	55	18	2	814	56	14
15 Mean windspeed	in m/sec	6	6	6	6	5	5	4	4	5	5	6	6	5	44	15
16 Mean predominent direction of the wind		NW	NW	S	S	S	S	SW	S	S	S	S	S		31	16

91 Station/Country Cairo (Illinois)/U.S.

Location 37°00'N/89°10'W Height above sealevel 96 m Climate symbol: Köppen Cfa Troll IV,7

		J	F	M	A	M	J	J	A	S	O	N	D	year	P	
1 Mean daily temperature	in °C	3,0	4,8	9,0	15,2	20,8	25,5	27,3	26,6	22,5	16,8	9,8	4,2	15,3	30	1
2 Mean daily maximum temperature	in °C	7,2	9,2	14,4	20,6	25,7	30,7	32,6	31,5	28,0	22,0	13,8	8,3	20,3	30	2
3 Mean daily minimum temperature	in °C	-1,2	0,3	4,5	10,5	15,7	20,2	22,0	21,2	17,5	11,2	4,6	0,2	10,6	30	3
4 Absolute maximum temperature	in °C	24	23	28	31	37	40	40	39	39	33	28	23	40	18	4
5 Absolute minimum temperature	in °C	-16	-21	-14	-1	4	11	12	12	6	-2	-15	-15	-21	18	5
6 Mean relative humidity	in %	75	73	69	67	70	70	71	72	72	70	69	73	71	16	6
7 Mean precipitation	in mm	113	93	122	103	112	105	81	79	76	73	98	93	1148	30	7
8 Maximum precipitation	in mm	380	226	234	210	292	260	208	207	160	166	331	203		16	8
9 Minimum precipitation	in mm	8	14	44	24	50	19	21	21	12	16	18	17		18	9
10 Maximum precipitation in 24 h	in mm	155	122	94	96	140	109	71	192	86	75	110	63	192	18	10
11 Mean number of days with precipitation	>0,25 mm	11	10	12	12	11	10	9	7	8	7	9	9	115	18	11
12 Mean duration of sunshine	in h	124	160	218	254	298	324	345	336	279	254	181	145	2918	15	12
13 Mean quantity of radiation	in ly/day															13
14 Mean potential evaporation	in mm	5	8	28	66	110	148	169	151	108	64	23	7	887		14
15 Mean windspeed	in m/sec	4	4	5	5	4	3	3	3	3	3	4	4	4	18	15
16 Mean predominent direction of the wind		SW	NE	SW	SW	SW	SW	SW	NE	NE	S	S	S		18	16

92 Station/Country Indianapolis (Indiana)/U.S.

Location 39°44'N/86°16'W Height above sealevel 241 m Climate symbol: Köppen Cfa Troll III,8

		J	F	M	A	M	J	J	A	S	O	N	D	year	P	
1 Mean daily temperature	in °C	-1,6	-0,5	3,8	10,4	16,3	21,7	24,0	23,2	19,2	13,0	4,9	-0,5	11,2	30	1
2 Mean daily maximum temperature	in °C	2,2	3,9	9,4	16,1	22,2	27,8	30,0	28,9	25,0	18,3	10,0	3,9	16,7	76	2
3 Mean daily minimum temperature	in °C	-5,6	-5,0	0,0	6,1	11,2	17,2	19,4	18,3	14,4	8,3	1,7	-3,3	6,7	76	3
4 Absolute maximum temperature	in °C	21,1	22,8	28,9	32,2	35,6	38,3	41,1	39,4	37,8	31,7	25,6	20,6	41,1	76	4
5 Absolute minimum temperature	in °C	-31,7	-27,8	-20,6	-7,2	-0,6	3,9	8,9	6,7	-1,1	-5,6	-20,6	-26,1	-31,7	76	5
6 Mean relative humidity	in %	76	74	69	64	62	64	61	65	67	67	72	77	68	14	6
7 Mean precipitation	in mm	77	58	87	95	101	117	89	77	82	67	78	68	996	30	7
8 Maximum precipitation	in mm	322	135	197	201	257	247	206	169	153	212	140	170		19	8
9 Minimum precipitation	in mm	5	9	26	27	49	29	25	26	24	15	23	12		19	9
10 Maximum precipitation in 24 h	in mm	88	59	63	59	67	94	95	69	62	99	77	46	99	18	10
11 Mean number of days with precipitation	>0,25 mm	12	10	12	12	13	11	9	8	7	8	10	10	122	21	11
12 Mean duration of sunshine	in h	118	140	193	227	278	313	342	313	265	222	139	118	2668	30	12
13 Mean quantity of radiation	in ly/day	153	218	308	403	543	552	540	489	408	297	178	140	352	12	13
14 Mean potential evaporation	in mm	0	0	12	47	93	135	152	135	94	49	12	0	729	51	14
15 Mean windspeed	in m/sec	5	6	6	6	5	4	4	3	4	4	5	5	5	12	15
16 Mean predominent direction of the wind		NW	WNW	WNW	SW	SW	SW	SW	SW	SW	SW	SW	SW		11	16

93 Station/Country Cincinnati (Ohio)/U.S.

Location 39°09'N/84°31'W — Height above sealevel 232 m — Climate symbol: Köppen Cfa — Troll III,8

	J	F	M	A	M	J	J	A	S	O	N	D	year	P	
1 Mean daily temperature	in °C	0,9	1,7	5,9	12,3	17,9	23,0	24,9	24,3	20,6	14,4	7,0	1,8	12,9	30	1
2 Mean daily maximum temperature	in °C	4,9	6,1	11,5	17,7	23,4	28,4	30,8	29,6	26,5	20,1	11,7	5,8	18,0	30	2
3 Mean daily minimum temperature	in °C	-3,8	-3,0	1,0	6,3	11,8	16,9	18,8	17,9	14,4	8,0	2,3	-2,5	7,3	30	3
4 Absolute maximum temperature	in °C	25	24	31	32	35	39	43	39	38	33	28	22	43	44	4
5 Absolute minimum temperature	in °C	-27	-23	-15	-8	0	4	10	6	0	-7	-17	-25	-27	44	5
6 Mean relative humidity	in %	76	72	68	66	68	71	71	71	71	70	71	74	71	19	6
7 Mean precipitation	in mm	93	71	99	92	97	106	91	83	69	57	75	70	1003	30	7
8 Maximum precipitation	in mm	347	158	278	219	224	230	255	166	149	242	164	176		43	8
9 Minimum precipitation	in mm	25	12	27	21	19	15	8	15	18	4	8	14		43	9
10 Maximum precipitation in 24 h	in mm	115	61	96	87	121	88	103	88	69	65	76	85	121	44	10
11 Mean number of days with precipitation	>0,25 mm	13	11	13	12	12	13	10	9	9	9	10	11	132	45	11
12 Mean duration of sunshine	in h	115	137	186	222	273	309	323	295	253	205	138	118	2574	30	12
13 Mean quantity of radiation	in ly/day															13
14 Mean potential evaporation	in mm	0	2	17	50	102	134	155	138	97	51	17	3	766	60	14
15 Mean windspeed	in m/sec	4	4	4	4	3	3	2	2	2	3	4	4	3	39	15
16 Mean predominent direction of the wind		SW	SW	SW	SW	SW	SW	SW	SW	SW	SW	SW	SW		37	16

94 Station/Country Evansville (Indiana)/U.S.

Location 38°03'N/87°32'W — Height above sealevel 116 m — Climate symbol: Köppen Cfa — Troll IV,7

	J	F	M	A	M	J	J	A	S	O	N	D	year	P	
1 Mean daily temperature	in °C	1,2	2,6	6,8	13,2	18,5	23,8	25,7	24,9	20,9	14,8	7,1	2,3	13,5	30	1
2 Mean daily maximum temperature	in °C	6,3	8,0	13,8	19,9	24,7	29,6	31,7	30,6	27,7	21,7	11,5	7,5	16,8	30	2
3 Mean daily minimum temperature	in °C	-3,3	-2,0	2,4	7,9	12,4	17,7	19,6	18,6	15,1	8,6	2,5	-2,1	8,1	30	3
4 Absolute maximum temperature	in °C	24	23	28	32	34	40	41	39	39	34	28	24	41	20	4
5 Absolute minimum temperature	in °C	-26	-31	-23	-4	1	5	8	8	-1	-6	-19	-22	-31	20	5
6 Mean relative humidity	in %	77	73	69	67	70	71	71	72	73	72	72	76	72	18	6
7 Mean precipitation	in mm	101	84	109	101	106	95	84	78	73	85	80	78	1051	30	7
8 Maximum precipitation	in mm	343	174	290	203	209	236	246	141	251	212	216	202		18	8
9 Minimum precipitation	in mm	15	7	23	55	44	29	25	3	14	7	25	18		18	9
10 Maximum precipitation in 24 h	in mm	81	80	131	100	109	80	79	66	88	59	68	52	131	20	10
11 Mean number of days with precipitation	>0,25 mm	11	9	12	12	12	10	10	7	7	7	9	10	116	20	11
12 Mean duration of sunshine	in h	123	145	199	237	294	322	342	318	274	236	156	120	2766	30	12
13 Mean quantity of radiation	in ly/day															13
14 Mean potential evaporation	in mm	0	3	15	48	92	141	169	144	100	49	15	2	778		14
15 Mean windspeed	in m/sec	4	5	5	5	4	3	3	3	3	3	4	4	4	20	15
16 Mean predominent direction of the wind		NW	NW	WNW	SSW	SSW	SW	SW	SW	SSW	NW	SSW	SSW		11	16

95 Station/Country Nashville (Tennessee)/U.S.

Location 36°07'N/86°41'W — Height above sealevel 176 m — Climate symbol: Köppen Cfa — Troll III,8

	J	F	M	A	M	J	J	A	S	O	N	D	year	P	
1 Mean daily temperature	in °C	4,4	5,6	9,5	15,3	20,3	25,2	26,8	26,2	22,7	16,4	9,2	5,2	15,6	30	1
2 Mean daily maximum temperature	in °C	8,3	10,0	15,0	20,6	25,6	30,0	31,7	31,1	27,8	22,2	14,4	9,4	20,6	75	2
3 Mean daily minimum temperature	in °C	-0,6	0,8	4,4	9,4	14,4	19,4	21,1	20,0	16,7	10,0	4,4	0,8	10,0	75	3
4 Absolute maximum temperature	in °C	25,6	26,1	31,7	32,2	35,6	38,3	41,1	40,6	40,0	33,3	29,4	23,9	41,1	75	4
5 Absolute minimum temperature	in °C	-23,3	-25,0	-16,1	-3,9	2,2	5,6	12,2	10,6	3,3	-3,3	-13,3	-18,9	-25,0	75	5
6 Mean relative humidity	in %	73	71	66	62	66	66	65	69	69	68	70	73	68	13	6
7 Mean precipitation	in mm	139	115	132	95	94	83	94	73	73	59	83	106	1146	30	7
8 Maximum precipitation	in mm	375	314	262	301	253	296	240	244	278	212	230	344		88	8
9 Minimum precipitation	in mm	29	16	22	18	21	5	12	13	3	<1	14	23		88	9
10 Maximum precipitation in 24 h	in mm	112	103	118	85	91	125	90	70	78	58	92	99	125	21	10
11 Mean number of days with precipitation	>0,25 mm	12	11	12	11	11	9	10	9	7	7	9	11	119	19	11
12 Mean duration of sunshine	in h	123	142	196	241	285	308	292	279	250	224	168	126	2634	30	12
13 Mean quantity of radiation	in ly/day	155	229	320	433	384	591	528	473	402	311	203	148	356	20	13
14 Mean potential evaporation	in mm	5	8	25	59	105	146	167	150	105	58	21	8	857	46	14
15 Mean windspeed	in m/sec	4	4	4	4	3	3	3	3	3	3	4	4	3	19	15
16 Mean predominent direction of the wind		S	S	S	S	S	S	S	S	S	S	S	S		19	16

96 Station/Country Parkersburg (West Virginia)/U.S.

Location 39°16'N/81°34'W — Height above sealevel 187 m — Climate symbol: Köppen Cfa — Troll III,8

	J	F	M	A	M	J	J	A	S	O	N	D	year	P	
1 Mean daily temperature	in °C	1,4	1,9	5,9	12,3	17,8	22,6	24,3	23,6	20,1	13,9	7,1	2,1	12,8	30	1
2 Mean daily maximum temperature	in °C	5,9	6,8	12,0	18,3	23,7	28,3	30,1	29,0	26,2	19,9	12,2	6,7	18,3	30	2
3 Mean daily minimum temperature	in °C	-3,3	-2,9	0,9	5,9	11,2	16,5	18,4	17,5	14,1	7,6	2,2	-2,2	7,2	30	3
4 Absolute maximum temperature	in °C	26	25	32	34	36	37	40	41	39	33	28	24	41	72	4
5 Absolute minimum temperature	in °C	-27	-33	-19	-9	-2	3	8	7	0	-7	-16	-23	-33	72	5
6 Mean relative humidity	in %	74	72	68	62	63	67	66	70	68	68	69	73	68		6
7 Mean precipitation	in mm	85	72	90	83	94	108	104	96	69	52	80	72	985	30	7
8 Maximum precipitation	in mm	228	179	177	171	191	219	306	265	214	165	142	140		70	8
9 Minimum precipitation	in mm	23	25	3	20	13	30	18	26	12	2	3	14		70	9
10 Maximum precipitation in 24 h	in mm	75	73	61	86	76	91	122	91	76	86	82	68	122	72	10
11 Mean number of days with precipitation	>0,25 mm	16	13	14	13	12	13	11	10	9	9	11	13	144	72	11
12 Mean duration of sunshine	in h	91	111	155	200	252	277	286	264	230	189	117	93	2265	30	12
13 Mean quantity of radiation	in ly/day															13
14 Mean potential evaporation	in mm	0	3	19	50	96	130	151	135	94	52	18	2	750		14
15 Mean windspeed	in m/sec	3	3	3	3	3	2	2	2	2	2	3	3	3	72	15
16 Mean predominent direction of the wind		NW	NW	SE	SW	NW	NW	NW	N	NW	NW	NW	W		73	16

170

97 Station/Country Atlantic City (New Jersey)/U.S.

Location 39°22'N/74°25'W | Height above sealevel 3 m | Climate symbol: Köppen Cfa | Troll III.8

		J	F	M	A	M	J	J	A	S	O	N	D	year	P	
1	Mean daily temperature	in °C	1,6	1,5	5,1	10,6	16,3	21,1	23,9	23,2	19,6	14,0	8,2	2,6	12,3	30	1
2	Mean daily maximum temperature	in °C	5,8	5,6	9,0	13,4	13,5	23,9	26,3	26,0	23,4	17,7	12,7	7,2	15,8	30	2
3	Mean daily minimum temperature	in °C	-1,7	-1,9	1,5	5,9	11,5	16,9	19,9	19,6	16,9	10,7	5,2	-0,4	8,7	30	3
4	Absolute maximum temperature	in °C	20	25	29	32	35	37	39	40	34	33	27	20	40	85	4
5	Absolute minimum temperature	in °C	-20	-23	-13	-9	1	7	11	9	3	-2	-12	-22	-23	85	5
6	Mean relative humidity	in %	74	73	72	71	74	76	77	77	76	72	72	73	74		6
7	Mean precipitation	in mm	90	80	99	87	89	72	94	124	84	81	93	82	1075	30	7
8	Maximum precipitation	in mm	207	189	192	176	224	215	260	377	374	308	230	209		85	8
9	Minimum precipitation	in mm	10	22	24	27	13	7	4	4	<1	1	10	6		85	9
10	Maximum precipitation in 24 h	in mm	76	68	80	80	82	107	137	228	232	234	131	89	234	85	10
11	Mean number of days with precipitation	>0,25 mm	12	10	12	11	11	10	10	10	8	9	9	10	122	85	11
12	Mean duration of sunshine	in h	151	173	210	233	273	287	298	271	239	218	177	153	2683	30	12
13	Mean quantity of radiation	in ly/day															13
14	Mean potential evaporation	in mm	0	3	6	50	89	126	144	138	94	55	23	0	728		14
15	Mean windspeed	in m/sec	7	7	8	7	7	6	6	6	6	7	7	7	7	37	15
16	Mean predominant direction of the wind		NW	NW	NW	NW	SW	S	SW	SW	SW	NW	NW	NW		75	16

98 Station/Country Washington D.C./U.S.

Location 38°51'N/77°03'W | Height above sealevel 4 m | Climate symbol: Köppen Cfa | Troll III.8

		J	F	M	A	M	J	J	A	S	O	N	D	year	P	
1	Mean daily temperature	in °C	2,7	3,2	7,1	13,2	18,8	23,4	25,7	24,7	20,9	15,0	8,7	3,4	13,9	30	1
2	Mean daily maximum temperature	in °C	5,6	6,7	11,7	17,8	23,9	28,3	30,6	28,9	25,6	19,4	12,8	7,2	18,3	78	2
3	Mean daily minimum temperature	in °C	-2,8	-2,2	1,7	6,7	12,2	17,2	20,0	18,9	15,0	8,9	3,3	-1,7	8,3	78	3
4	Absolute maximum temperature	in °C	25,0	28,9	33,9	35,0	36,1	38,9	41,1	41,1	40,0	35,6	28,3	23,3	41,1	79	4
5	Absolute minimum temperature	in °C	-25,6	-26,1	-15,6	-9,4	0,6	6,1	11,1	9,4	2,2	-3,3	-11,7	-25,0	-26,1	79	5
6	Mean relative humidity	in %	65	62	60	57	60	64	66	67	67	66	64	65	63	35	6
7	Mean precipitation	in mm	77	63	82	80	105	82	105	124	97	78	72	71	1036	30	7
8	Maximum precipitation	in mm	129	105	189	152	271	174	281	364	171	208	161	143		17	8
9	Minimum precipitation	in mm	8	22	16	7	28	32	27	15	16	17	24	6		17	9
10	Maximum precipitation in 24 h	in mm	44	39	87	45	110	93	75	162	92	126	64	47	162	17	10
11	Mean number of days with precipitation	>0,25 mm	11	8	12	10	12	9	10	10	8	8	8	9	115	19	11
12	Mean duration of sunshine	in h	76	97	135	182	221	214	226	186	170	123	87	66	1733	22	12
13	Mean quantity of radiation	in ly/day	163	233	350	408	514	565	502	467	370	300	194	125	349	48	13
14	Mean potential evaporation	in mm	3	5	19	53	96	141	163	142	100	55	20	5	800	30	14
15	Mean windspeed	in m/sec	5	5	5	4	4	4	4	4	4	4	4	4	4	12	15
16	Mean predominant direction of the wind		NW	NW	NW	S	S	S	SSW	S	S	SSW	SSW	NW		10	16

99 Station/Country Richmond (Virginia)/U.S.

Location 37°30'N/77°20'W | Height above sealevel 49 m | Climate symbol: Köppen Cfa | Troll III.8

		J	F	M	A	M	J	J	A	S	O	N	D	year	P	
1	Mean daily temperature	in °C	3,7	4,4	8,2	13,9	19,2	23,7	25,6	24,7	21,2	15,1	9,2	4,3	14,4	30	1
2	Mean daily maximum temperature	in °C	8,9	9,4	15,0	20,0	24,0	28,9	31,1	30,0	27,2	21,7	15,0	9,4	20,0	50	2
3	Mean daily minimum temperature	in °C	-1,1	-1,1	3,3	7,8	12,8	17,8	20,0	19,4	16,1	9,4	3,9	-0,6	8,9	50	3
4	Absolute maximum temperature	in °C	26,7	27,8	34,4	35,6	36,7	40,0	40,6	41,7	38,3	37,2	28,3	25,0	41,7	49	4
5	Absolute minimum temperature	in °C	-18,3	-19,4	-11,1	-7,2	1,7	6,1	11,1	9,4	4,4	-2,2	-10,0	-18,9	-19,4	49	5
6	Mean relative humidity	in %	73	69	66	63	63	67	68	70	70	67	68	72	68	13	6
7	Mean precipitation	in mm	88	74	87	80	94	95	142	141	93	76	77	75	1122	30	7
8	Maximum precipitation	in mm															8
9	Minimum precipitation	in mm															9
10	Maximum precipitation in 24 h	in mm	82	40	52	53	58	72	124	223	97	101	103	80	223	23	10
11	Mean number of days with precipitation	>0,25 mm	10	9	11	10	11	9	12	10	8	9	8	9	116	23	11
12	Mean duration of sunshine	in h	144	166	211	248	280	296	286	263	230	211	176	152	2663	30	12
13	Mean quantity of radiation	in ly/day															13
14	Mean potential evaporation	in mm	5	8	22	56	96	141	158	140	100	55	20	5	806	31	14
15	Mean windspeed	in m/sec	4	4	4	4	3	3	3	3	3	3	3	3	3	12	15
16	Mean predominant direction of the wind		S	WSW	W	S	SSW	S	SSW	S	S	NNE	S	SW		12	16

100 Station/Country Greensboro (North Carolina)/U.S.

Location 36°05'N/79°57'W | Height above sealevel 273 m | Climate symbol: Köppen Cfa | Troll III.8

		J	F	M	A	M	J	J	A	S	O	N	D	year	P	
1	Mean daily temperature	in °C	4,3	5,0	8,8	14,1	19,4	23,8	25,2	24,6	21,2	15,2	8,8	4,4	14,6	30	1
2	Mean daily maximum temperature	in °C	9,7	11,0	15,0	20,7	26,0	30,1	30,9	30,2	27,2	21,9	15,5	10,1	20,7	30	2
3	Mean daily minimum temperature	in °C	-1,3	-1,0	2,1	7,4	12,8	17,4	19,4	18,9	15,2	8,3	2,2	-3,0	8,3	30	3
4	Absolute maximum temperature	in °C	26	27	32	34	37	39	39	38	38	35	29	25	39	32	4
5	Absolute minimum temperature	in °C	-22	-20	-15	-6	1	6	9	8	2	-5	-12	-17	-22	32	5
6	Mean relative humidity	in %	69	64	61	66	71	74	79	79	79	75	67	71	71	6	6
7	Mean precipitation	in mm	86	84	94	87	84	88	122	117	93	69	68	80	1072	30	7
8	Maximum precipitation	in mm	209	179	183	157	153	203	249	318	337	244	196	161		41	8
9	Minimum precipitation	in mm	26	21	31	14	9	8	25	30	4	7	9	8		41	9
10	Maximum precipitation in 24 h	in mm	78	76	78	69	77	105	113	114	190	139	77	91	190	32	10
11	Mean number of days with precipitation	>0,25 mm	10	10	11	10	10	11	13	11	8	7	8	9	118	32	11
12	Mean duration of sunshine	in h	157	171	217	231	298	302	287	272	243	236	190	163	2767	30	12
13	Mean quantity of radiation	in ly/day	200	276	354	489	531	564	544	485	406	322	243	197	383		13
14	Mean potential evaporation	in mm	5	8	25	56	98	139	156	139	99	55	21	5	806		14
15	Mean windspeed	in m/sec	4	4	4	4	4	3	3	3	3	3	3	3	4	32	15
16	Mean predominant direction of the wind		SW	SW	SW	SW	SW	SW	SW	SW	NE	NE	SW	SW		12	16

101 Station / Country: Los Angeles (California) / U.S.

Location 34°03'N / 118°14'W — Height above sealevel 103 m — Climate symbol: Köppen Csa — Troll IV,2

		J	F	M	A	M	J	J	A	S	O	N	D	year	P	
1	Mean daily temperature	in °C	13.2	13.9	15.2	16.6	18.2	20.0	22.8	22.8	22.2	19.7	17.1	14.6	18.0	30	1
2	Mean daily maximum temperature	in °C	18.3	18.9	19.4	21.1	22.2	24.4	27.2	27.8	27.2	24.4	22.8	19.4	22.8	70	2
3	Mean daily minimum temperature	in °C	7.8	8.3	8.9	10.0	11.7	13.3	15.6	15.6	14.4	12.2	10.0	8.3	11.1	70	3
4	Absolute maximum temperature	in °C	32.2	33.3	37.2	37.8	39.4	40.6	42.8	41.1	42.2	38.9	35.6	33.3	42.8	69	4
5	Absolute minimum temperature	in °C	-2.2	-2.2	-0.6	2.2	4.4	7.8	9.4	9.4	6.7	4.4	1.1	-1.1	-2.2	70	5
6	Mean relative humidity	in %	57	64	64	69	73	73	72	71	67	62	50	52	64	43	6
7	Mean precipitation	in mm	78	85	57	30	4	2	tr	1	6	10	27	73	373	30	7
8	Maximum precipitation	in mm	379	315	207	153	36	8	<1	10	46	39	246	167		30	8
9	Minimum precipitation	in mm	tr	tr	0	0	0	0	0	0	0	0	0	0		30	9
10	Maximum precipitation in 24 h	in mm	155	102	87	52	27	4	<1	10	9	16	59	76	155	20	10
11	Mean number of days with precipitation	>0,25 mm	6	5	6	4	2	1	<1	1	1	2	4	5	37	20	11
12	Mean duration of sunshine	in h	224	217	273	264	292	299	352	336	295	263	249	220	3284	30	12
13	Mean quantity of radiation	in ly/day	248	325	438	503	555	578	658	592	539	391	303	257	449	11	13
14	Mean potential evaporation	in mm	34	36	49	59	78	94	117	115	96	73	52	39	840	56	14
15	Mean windspeed	in m/sec	3	3	3	3	3	3	2	2	2	3	3	3	3	20	15
16	Mean predominent direction of the wind		NE	W	W	W	W	W	W	W	W	W	W	NE		20	16

102 Station / Country: San Diego (California) / U.S.

Location 32°44'N / 117°10'W — Height above sealevel 4 m — Climate symbol: Köppen BSk — Troll IV,2

		J	F	M	A	M	J	J	A	S	O	N	D	year	P	
1	Mean daily temperature	in °C	13.1	13.7	14.7	16.1	17.5	18.2	20.9	21.5	20.8	18.7	16.3	14.2	17.2	30	1
2	Mean daily maximum temperature	in °C	17.2	17.2	17.8	18.9	19.4	20.6	22.8	23.3	22.8	21.7	20.6	18.3	20.0	72	2
3	Mean daily minimum temperature	in °C	8.3	8.9	10.0	11.7	13.3	15.0	17.2	17.8	16.7	13.9	11.1	8.9	12.8	72	3
4	Absolute maximum temperature	in °C	29.4	31.7	37.2	35.6	36.7	35.6	37.8	34.4	43.3	35.6	33.9	28.9	43.3	75	4
5	Absolute minimum temperature	in °C	-3.9	1.1	2.2	3.9	7.2	10.0	12.2	12.2	10.0	6.7	2.2	0.0	-3.9	75	5
6	Mean relative humidity	in %	70	74	73	75	77	79	80	80	79	76	69	69	75	59	6
7	Mean precipitation	in mm	51	55	40	20	4	1	tr	2	4	12	23	52	264	30	7
8	Maximum precipitation	in mm	159	135	150	91	24	7	3	22	48	74	148	193		30	8
9	Minimum precipitation	in mm	0	tr	0	0	tr	tr	tr	tr	tr	tr	tr	<1		30	9
10	Maximum precipitation in 24 h	in mm	67	43	61	31	11	7	2	21	18	30	62	78	78	20	10
11	Mean number of days with precipitation	>0,25 mm	7	7	7	5	2	1	<1	<1	1	3	4	6	43	20	11
12	Mean duration of sunshine	in h	216	212	262	242	261	253	293	277	255	234	236	217	2958	30	12
13	Mean quantity of radiation	in ly/day															13
14	Mean potential evaporation	in mm	34	36	49	59	71	82	105	104	86	67	50	39	782	56	14
15	Mean windspeed	in m/sec	3	3	3	3	3	3	3	3	3	2	2	3		20	15
16	Mean predominent direction of the wind		NE	NE	WNW	WNW	WNW	SSW	WNW	WNW	NW	WSW	NE	NE		12	16

103 Station / Country: Greenland Ranch–Death Valley (California) / U.S.

Location 36°28'N / 116°51'W — Height above sealevel –56 m — Climate symbol: Köppen BWh — Troll IV,5

		J	F	M	A	M	J	J	A	S	O	N	D	year	P	
1	Mean daily temperature	in °C	11.1	14.4	18.9	23.9	28.9	34.2	38.6	37.2	32.0	23.9	16.1	11.4	24.4	37	1
2	Mean daily maximum temperature	in °C	18.9	22.2	27.2	32.2	37.2	42.8	46.7	45.6	41.1	32.8	24.4	18.9	32.8	37	2
3	Mean daily minimum temperature	in °C	3.3	6.7	10.6	15.6	20.6	25.6	30.6	28.9	22.8	15.0	7.8	3.9	16.1	37	3
4	Absolute maximum temperature	in °C	29.4	33.3	37.8	42.8	48.9	51.1	56.7	52.8	49.4	43.3	33.9	30.0	56.7	37	4
5	Absolute minimum temperature	in °C	-9.4	-6.1	-1.1	1.7	5.6	9.4	18.7	18.3	5.0	0.0	-4.4	-7.2	-9.4	37	5
6	Mean relative humidity	in %															6
7	Mean precipitation	in mm	3	<1	3	3	5	5	8	8	5	<1	3	<1	41	20	7
8	Maximum precipitation	in mm															8
9	Minimum precipitation	in mm															9
10	Maximum precipitation in 24 h	in mm	25	25	18	15	13	15	15	15	31	13	36	18	36	38	10
11	Mean number of days with precipitation	>0,25 mm	1	1	2	1	<1	<1	<1	<1	<1	1	1	2	9	10	11
12	Mean duration of sunshine	in h															12
13	Mean quantity of radiation	in ly/day															13
14	Mean potential evaporation	in mm	5	13	46	122	205	225	229	214	179	113	23	7	1381	18	14
15	Mean windspeed	in m/sec															15
16	Mean predominent direction of the wind																16

104 Station / Country: Phoenix (Arizona) / U.S.

Location 33°26'N / 112°01'W — Height above sealevel 340 m — Climate symbol: Köppen BWh — Troll IV,5

		J	F	M	A	M	J	J	A	S	O	N	D	year	P	
1	Mean daily temperature	in °C	10.4	12.5	15.6	20.4	25.0	29.8	32.9	31.7	29.1	22.3	15.1	11.4	21.4	30	1
2	Mean daily maximum temperature	in °C	18.3	20.6	23.9	27.8	32.8	38.3	40.0	38.3	36.1	30.0	23.9	18.9	28.9	52	2
3	Mean daily minimum temperature	in °C	3.9	6.1	8.3	11.7	15.6	20.6	25.0	24.4	20.6	13.3	7.2	4.4	13.3	52	3
4	Absolute maximum temperature	in °C	28.9	33.3	35.0	39.4	45.6	47.8	47.8	48.1	45.0	40.6	35.6	28.9	47.8	52	4
5	Absolute minimum temperature	in °C	-8.9	-4.4	-1.1	1.7	3.9	9.4	17.2	14.4	9.4	2.2	-2.8	-5.6	-8.9	52	5
6	Mean relative humidity	in %	54	51	45	36	29	26	39	44	42	43	51	54	43	35	6
7	Mean precipitation	in mm	19	22	17	8	3	2	20	28	19	12	12	22	184	30	7
8	Maximum precipitation	in mm	61	57	106	53	24	24	106	141	107	68	77	101		30	8
9	Minimum precipitation	in mm	0	0	0	0	0	0	tr	<1	0	0	0	0		30	9
10	Maximum precipitation in 24 h	in mm	33	27	34	35	24	24	50	78	54	37	27	40	78	23	10
11	Mean number of days with precipitation	>0,25 mm	4	4	3	2	1	1	4	5	3	2	4		36	21	11
12	Mean duration of sunshine	in h	248	244	314	346	404	404	377	351	334	307	267	238	3832	30	12
13	Mean quantity of radiation	in ly/day	301	406	531	640	718	730	659	606	555	447	339	281	518	14	13
14	Mean potential evaporation	in mm	13	21	40	75	129	189	211	193	158	84	31	13	1157	56	14
15	Mean windspeed	in m/sec	2	2	3	3	3	3	3	3	2	2	2	2	2	15	15
16	Mean predominent direction of the wind		E	E	E	E	E	E	W	E	E	E	E	E		15	16

Station/Country Winslow (Arizona)/U.S.

Location 35°01'N/110°44'W **Height above sealevel** 1487 m **Climate symbol:** Köppen BSk **Troll** III,10

		J	F	M	A	M	J	J	A	S	O	N	D	year	P	
1 Mean daily temperature	in °C	-0,3	3,3	7,8	13,4	18,5	23,8	27,1	25,7	21,8	14,4	5,4	0,8	13,4	30	1
2 Mean daily maximum temperature	in °C	7,9	12,2	16,1	21,0	25,8	31,8	33,8	32,0	28,9	22,4	14,4	8,5	21,2	30	2
3 Mean daily minimum temperature	in °C	-7,1	-3,6	-1,1	3,3	7,4	12,0	16,5	15,7	11,7	4,4	-2,5	-5,8	4,2	30	3
4 Absolute maximum temperature	in °C	23	26	27	33	38	39	41	39	37	32	26	23	41	29	4
5 Absolute minimum temperature	in °C	-28	-22	-14	-8	-3	2	8	8	2	-7	-18	-24	-28	29	5
6 Mean relative humidity	in %	58	49	38	32	27	23	41	43	36	37	45	51	40	10	6
7 Mean precipitation	in mm	11	12	10	11	8	7	28	36	23	17	9	13	183	30	7
8 Maximum precipitation	in mm	41	30	29	40	34	54	69	68	61	62	42	50		27	8
9 Minimum precipitation	in mm	tr	tr	tr	tr	tr	0	4	5	0	0	0	tr		27	9
10 Maximum precipitation in 24 h	in mm	14	18	12	12	16	54	37	35	30	28	15	18	54	16	10
11 Mean number of days with precipitation	>0,25 mm	4	4	4	3	3	2	7	9	5	4	3	4	53	29	11
12 Mean duration of sunshine	in h															12
13 Mean quantity of radiation	in ly/day															13
14 Mean potential evaporation	in mm	0	8	22	49	87	127	159	139	99	52	13	3	758	14	14
15 Mean windspeed	in m/sec	3	4	5	5	5	5	4	4	4	3	3	3	4	17	15
16 Mean predominent direction of the wind		SE	SE	WSW	SW	WSW	SW	SW	SW	SW	SE	SE	SE		22	16

106 **Station/Country** Albuquerque (New Mexico)/U.S.

Location 35°03'N/106°37'W **Height above sealevel** 1620 m **Climate symbol:** Köppen BSk **Troll** III,7a

		J	F	M	A	M	J	J	A	S	O	N	D	year	P	
1 Mean daily temperature	in °C	1,7	4,4	7,9	13,2	18,4	23,8	25,8	24,8	21,4	14,7	6,7	2,8	13,8	30	1
2 Mean daily maximum temperature	in °C	8,0	11,2	14,9	20,6	25,7	31,4	32,9	31,1	27,9	21,5	13,4		20,7	30	2
3 Mean daily minimum temperature	in °C	-4,7	-2,5	0,4	5,7	11,1	16,2	18,8	17,9	14,2	7,4	-0,5	-3,6	6,7	30	3
4 Absolute maximum temperature	in °C	19	22	27	31	37	38	40	38	37	31	23	22	40	21	4
5 Absolute minimum temperature	in °C	-17	-21	-13	-7	1	7	13	12	4	-3	-12	-16	-21	21	5
6 Mean relative humidity	in %	51	47	39	32	29	31	45	49	44	43	49	57	43	30	6
7 Mean precipitation	in mm	10	10	12	12	19	14	30	34	24	19	10	12	206	30	7
8 Maximum precipitation	in mm	30	36	43	50	78	43	85	84	51	73	37	47		30	8
9 Minimum precipitation	in mm	tr	tr	tr	tr	tr	tr	4	tr	tr	0	0	0		30	9
10 Maximum precipitation in 24 h	in mm	14	12	18	28	27	42	29	29	49	33	19	34	49	21	10
11 Mean number of days with precipitation	>0,25 mm	4	4	4	4	4	4	9	10	5	5	2	4	58	21	11
12 Mean duration of sunshine	in h	221	218	273	299	343	365	340	317	299	279	245	219	3418	30	12
13 Mean quantity of radiation	in ly/day	305	388	509	623	694	732	682	626	548	443	332	280	514	15	13
14 Mean potential evaporation	in mm	3	8	25	52	91	134	151	136	96	52	15	3	766	56	14
15 Mean windspeed	in m/sec	4	4	5	5	5	4	4	4	4	4	3	3	4	21	15
16 Mean predominent direction of the wind		N	N	SE	SE	S	SE	SE	SE	SE	SE	N	N		12	16

107 **Station/Country** El Paso (Texas)/U.S.

Location 31°48'N/106°24'W **Height above sealevel** 1194 m **Climate symbol:** Köppen BWk **Troll** IV,5

		J	F	M	A	M	J	J	A	S	O	N	D	year	P	
1 Mean daily temperature	in °C	6,6	9,8	12,7	17,4	22,2	26,9	27,4	26,6	23,9	18,6	11,2	7,3	17,6	30	1
2 Mean daily maximum temperature	in °C	13,9	16,7	20,6	25,0	30,0	34,4	33,9	32,8	30,0	25,0	18,9	13,9	24,4	60	2
3 Mean daily minimum temperature	in °C	0,0	2,8	5,8	10,0	14,4	19,4	21,1	20,0	17,2	11,1	4,4	0,6	10,6	67	3
4 Absolute maximum temperature	in °C	25,0	30,0	33,9	35,0	38,9	41,1	40,6	39,4	37,8	34,4	29,4	25,0	41,1	60	4
5 Absolute minimum temperature	in °C	-21,1	-15,0	-10,0	-3,3	2,2	7,8	13,3	11,0	5,0	-3,3	-11,7	-20,6	-21,1	67	5
6 Mean relative humidity	in %	49	43	36	31	29	33	47	51	49	48	52	49		13	6
7 Mean precipitation	in mm	12	10	9	7	10	18	33	30	29	23	8	12	201	30	7
8 Maximum precipitation	in mm	47	36	57	31	49	68	140	104	160	109	41	44		30	8
9 Minimum precipitation	in mm	0	0	tr	0	0	tr	4	tr	tr	0	0	0		30	9
10 Maximum precipitation in 24 h	in mm	15	22	44	23	31	29	48	51	73	45	30	27	73	21	10
11 Mean number of days with precipitation	>0,25 mm	3	3	2	2	2	4	7	7	4	4	3	4	44	21	11
12 Mean duration of sunshine	in h	234	236	299	329	373	369	336	327	300	287	257	236	3583	30	12
13 Mean quantity of radiation	in ly/day	344	431	550	658	576	662	668	639	570	465	365	313	520	13	13
14 Mean potential evaporation	in mm	11	18	37	68	114	166	174	159	114	65	24	10	960	56	14
15 Mean windspeed	in m/sec	5	5	6	6	6	5	5	4	4	4	4	4	5	18	15
16 Mean predominent direction of the wind		N	N	WSW	WSW	W	W	S	S	S	N	N	N		12	16

108 **Station/Country** Amarillo (Texas)/U.S.

Location 35°14'N/101°42'W **Height above sealevel** 1099 m **Climate symbol:** Köppen BSk **Troll** III,10

		J	F	M	A	M	J	J	A	S	O	N	D	year	P	
1 Mean daily temperature	in °C	2,3	4,3	7,8	13,3	18,3	24,2	26,2	25,6	21,3	15,3	7,5	3,8	14,2	30	1
2 Mean daily maximum temperature	in °C	9,4	10,6	16,1	20,6	25,0	29,4	31,1	31,1	27,8	21,1	15,0	8,9	20,6	39	2
3 Mean daily minimum temperature	in °C	-4,4	-3,9	0,6	5,6	11,1	15,6	18,3	17,8	14,4	7,8	1,1	-3,9	6,7	39	3
4 Absolute maximum temperature	in °C	28,3	28,9	35,6	34,4	37,8	41,1	40,6	40,0	38,3	34,4	29,4	26,1	41,1	39	4
5 Absolute minimum temperature	in °C	-23,9	-26,7	-18,9	-10,6	-3,3	6,1	10,6	8,9	0,0	-9,4	-15,6	-21,1	-26,7	39	5
6 Mean relative humidity	in %	64	63	63	58	61	61	60	62	64	63	63	65	62	14	6
7 Mean precipitation	in mm	17	16	21	34	66	73	59	66	48	45	17	20	502	30	7
8 Maximum precipitation	in mm	52	46	72	95	249	129	186	131	128	194	54	96		18	8
9 Minimum precipitation	in mm	tr	tr	tr	6	11	<1	3	10	6	0	tr	tr		18	9
10 Maximum precipitation in 24 h	in mm	21	27	25	40	171	156	104	108	87	88	30	79	171	20	10
11 Mean number of days with precipitation	>0,25 mm	4	4	4	6	9	8	9	8	5	5	3	4	69	20	11
12 Mean duration of sunshine	in h	207	199	258	276	305	338	350	328	288	260	229	205	3243	30	12
13 Mean quantity of radiation	in ly/day															13
14 Mean potential evaporation	in mm	3	5	25	52	91	138	159	143	99	55	18	5	793	56	14
15 Mean windspeed	in m/sec	6	6	7	7	6	6	5	5	6	6	6	6	6	20	15
16 Mean predominent direction of the wind		SW	SW	SW	SW	S	S	S	S	S	SW	SW	SW		12	16

109 Station/Country Abilene (Texas)/U.S.

Location 32°26'N/99°41'W — Height above sealevel 506 m — Climate symbol: Köppen Cfa — Troll IV.4

		J	F	M	A	M	J	J	A	S	O	N	D	year	P	
1 Mean daily temperature	in °C	7,0	9,1	12,8	17,9	22,1	26,8	28,4	28,3	24,4	19,0	11,7	7,8	17,9	30	1
2 Mean daily maximum temperature	in °C	13,3	15,6	20,6	25,0	28,3	32,8	34,4	34,4	30,6	25,0	18,3	13,2	24,4	61	2
3 Mean daily minimum temperature	in °C	1,1	2,2	6,7	11,7	16,1	20,6	22,2	22,2	18,3	12,2	6,1	2,2	11,7	61	3
4 Absolute maximum temperature	in °C	32,2	34,4	36,7	38,9	40,6	43,3	43,3	43,9	40,0	37,2	32,8	28,3	43,9	61	4
5 Absolute minimum temperature	in °C	-22,8	-21,1	-12,8	-3,9	0,6	6,7	12,2	8,9	3,9	-5,0	-10,6	-17,2	-22,8	61	5
6 Mean relative humidity	in %	62	58	55	56	62	61	57	57	62	64	61	60	60	13	6
7 Mean precipitation	in mm	22	28	26	58	110	68	58	37	53	72	28	32	592	30	7
8 Maximum precipitation	in mm	45	72	61	166	335	158	177	147	107	228	85	65		19	8
9 Minimum precipitation	in mm	<1	1	1	11	4	<1	tr	tr	tr	0	0	<1		19	9
10 Maximum precipitation in 24 h	in mm	31	54	50	95	69	93	95	96	78	130	39	58	130	21	10
11 Mean number of days with precipitation	>0,25 mm	5	6	4	7	8	7	4	4	5	6	4	4	64	21	11
12 Mean duration of sunshine	in h	190	199	250	259	290	347	335	322	276	245	223	201	3137	13	12
13 Mean quantity of radiation	in ly/day	287	362	473	553	606	616	674	578	516	400	317	295	473	9	13
14 Mean potential evaporation	in mm	8	13	37	71	114	163	185	172	117	71	26	10	987	56	14
15 Mean windspeed	in m/sec	6	6	7	7	6	6	5	5	5	5	6	6	6	16	15
16 Mean predominent direction of the wind		S	S	S	SSE	SSE	SSE	SSE	SSE	SSE	S	S	SSW		12	16

110 Station/Country Oklahoma City (Oklahoma)/U.S.

Location 35°29'N/97°32'W — Height above sealevel 382 m — Climate symbol: Köppen Cfa — Troll III,9a

		J	F	M	A	M	J	J	A	S	O	N	D	year	P	
1 Mean daily temperature	in °C	3,1	4,8	10,3	15,6	20,0	25,0	27,5	27,2	23,3	17,0	9,8	4,2	15,9	56	1
2 Mean daily maximum temperature	in °C	8,3	10,6	16,7	21,7	25,6	30,6	33,3	33,3	29,4	22,8	15,6	9,4	21,7	56	2
3 Mean daily minimum temperature	in °C	-2,2	-1,1	3,9	9,4	14,4	19,4	21,7	21,1	17,2	11,1	3,9	-1,1	10,0	56	3
4 Absolute maximum temperature	in °C	28,3	32,2	36,1	35,6	37,2	41,7	42,8	45,0	40,6	36,1	30,0	26,1	45,0	56	4
5 Absolute minimum temperature	in °C	-23,9	-27,2	-17,2	-6,7	0,6	7,8	12,8	9,4	1,7	-8,9	-12,8	-18,9	-27,2	56	5
6 Mean relative humidity	in %	72	67	62	64	70	69	64	63	67	67	69	70	67		6
7 Mean precipitation	in mm	33	25	56	84	130	89	74	69	76	76	51	41	804	19	7
8 Maximum precipitation	in mm	144	75	182	274	242	218	192	166	243	234	110	72		19	8
9 Minimum precipitation	in mm	<1	tr	tr	19	8	16	2	9	tr	tr	tr	<1		19	9
10 Maximum precipitation in 24 h	in mm	81	58	94	94	140	178	130	114	114	201	119	79	201	59	10
11 Mean number of days with precipitation	>0,25 mm	6	5	7	8	10	8	7	7	7	6	5	6	82	40	11
12 Mean duration of sunshine	in h	175	182	235	253	290	329	352	331	282	243	201	175	3048	29	12
13 Mean quantity of radiation	in ly/day	251	319	409	494	536	615	610	593	487	377	291	240	436		13
14 Mean potential evaporation	in mm	3	5	25	59	102	149	177	167	114	61	21	5	888		14
15 Mean windspeed	in m/sec	6,7	6,7	7,2	7,2	6,3	6,3	5,4	5,4	5,8	5,8	6,3	6,3	6,3		15
16 Mean predominent direction of the wind		N	N	SSE	SSE	SE	SSE	S	S	SSE	SSE	S	S			16

111 Station/Country Dallas (Texas)/U.S.

Location 32°51'N/96°51'W — Height above sealevel 146 m — Climate symbol: Köppen Cfa — Troll IV.4

		J	F	M	A	M	J	J	A	S	O	N	D	year	P	
1 Mean daily temperature	in °C	7,7	9,7	13,4	18,3	22,7	27,4	29,4	29,4	25,5	19,9	12,7	8,9	18,8	30	1
2 Mean daily maximum temperature	in °C	12,8	15,6	19,4	23,9	27,8	32,2	34,4	34,4	31,1	25,6	18,9	13,9	23,9	34	2
3 Mean daily minimum temperature	in °C	2,2	4,4	7,8	12,8	17,3	21,7	23,9	23,3	20,0	13,9	8,3	3,3	13,3	34	3
4 Absolute maximum temperature	in °C	31,1	33,9	35,6	35,6	39,4	40,6	40,6	43,3	41,1	37,8	30,6	27,2	43,3	34	4
5 Absolute minimum temperature	in °C	-19,4	-16,7	-11,7	-1,1	6,7	11,7	13,3	13,9	2,2	-3,3	-7,2	-12,2	-19,4	34	5
6 Mean relative humidity	in %	70	67	62	64	69	67	62	62	65	66	65	67	65	17	6
7 Mean precipitation	in mm	59	65	72	102	123	82	49	49	72	69	69	68	879	30	7
8 Maximum precipitation	in mm	215	195	242	352	349	309	145	275	132	204	252	155		18	8
9 Minimum precipitation	in mm	8	14	4	28	39	7	tr	<1	tr	1	5	3		18	9
10 Maximum precipitation in 24 h	in mm	131	69	135	130	158	104	137	233	83	166	127	85	233	20	10
11 Mean number of days with precipitation	>0,25 mm	7	8	8	9	9	6	5	5	6	6	8	8	81	20	11
12 Mean duration of sunshine	in h	155	159	220	238	279	326	341	325	274	240	191	163	2911	30	12
13 Mean quantity of radiation	in ly/day	254	321	428	487	565	638	616	592	493	399	294	241	444	12	13
14 Mean potential evaporation	in mm	8	15	34	72	118	171	192	183	130	73	29	10	1035	48	14
15 Mean windspeed	in m/sec	5	5	6	6	5	6	4	4	4	4	5	5	5	20	15
16 Mean predominent direction of the wind		S	S	S	SSE	S	S	S	SSE	SE	SE	S	SSE			16

112 Station/Country Tulsa (Oklahoma)/U.S.

Location 36°11'N/95°54'W — Height above sealevel 198 m — Climate symbol: Köppen Cfa — Troll III.9

		J	F	M	A	M	J	J	A	S	O	N	D	year	P	
1 Mean daily temperature	in °C	2,9	5,1	9,2	15,2	19,9	25,2	27,9	27,8	23,5	17,7	9,5	4,8	15,7	30	1
2 Mean daily maximum temperature	in °C	8,3	11,3	16,1	21,9	25,7	30,9	33,9	33,9	29,4	23,8	15,7	10,2	21,8	30	2
3 Mean daily minimum temperature	in °C	-2,2	-0,3	3,7	10,0	14,4	19,6	21,7	21,4	17,0	11,0	3,8	-0,5	10,0	30	3
4 Absolute maximum temperature	in °C	26	28	33	34	35	39	44	43	43	37	31	27	44	22	4
5 Absolute minimum temperature	in °C	-22	-18	-19	-6	2	9	14	12	2	-3	-12	-16	-22	22	5
6 Mean relative humidity	in %	70	67	64	64	71	71	66	63	64	64	63	68	67	16	6
7 Mean precipitation	in mm	43	45	62	102	134	119	75	77	102	84	58	41	942	30	7
8 Maximum precipitation	in mm	169	100	156	234	457	284	229	190	258	419	192	100		20	8
9 Minimum precipitation	in mm	tr	10	6	13	34	40	<1	5	tr	tr	<1	10		20	9
10 Maximum precipitation in 24 h	in mm	57	38	68	82	185	127	83	106	162	139	70	44	185	22	10
11 Mean number of days with precipitation	>0,25 mm	6	8	8	9	11	9	7	7	6	7	6	7	91	22	11
12 Mean duration of sunshine	in h	152	164	200	213	244	287	314	308	281	241	207	172	2783	18	12
13 Mean quantity of radiation	in ly/day															13
14 Mean potential evaporation	in mm	3	8	28	59	102	150	186	171	117	61	21	5	911		14
15 Mean windspeed	in m/sec	5	5	6	6	5	5	4	4	4	4	5	5	5	12	15
16 Mean predominent direction of the wind		N	S	S	S	S	S	S	S	S	S	S	S		12	16

113 Station/Country Little Rock (Arkansas)/U.S.

Location 34°44'N/92°14'W — Height above sealevel 78 m — Climate symbol: Köppen Cfa — Troll IV,7

		J	F	M	A	M	J	J	A	S	O	N	D	year	P	
1 Mean daily temperature	in °C	4,8	6,9	11,0	16,9	21,4	26,1	27,7	27,4	23,5	17,3	9,7	5,5	16,5	30	1
2 Mean daily maximum temperature	in °C	10,0	12,2	17,2	22,2	26,1	30,6	32,2	32,2	28,9	23,3	16,1	11,1	21,7	67	2
3 Mean daily minimum temperature	in °C	1,1	2,2	6,7	11,7	16,1	20,6	22,2	21,7	18,3	12,2	6,1	2,2	11,7	67	3
4 Absolute maximum temperature	in °C	27,2	30,6	32,2	34,4	36,1	40,6	42,2	43,3	40,0	33,9	28,9	25,6	43,3	67	4
5 Absolute minimum temperature	in °C	-22,7	-24,4	-11,7	-2,2	3,9	10,6	14,4	11,1	2,8	-2,8	-12,2	-15,0	-24,4	67	5
6 Mean relative humidity	in %	72	70	66	66	69	69	68	69	69	69	70	72	69	43	6
7 Mean precipitation	in mm	133	110	122	125	134	92	85	72	82	73	105	104	1237	30	7
8 Maximum precipitation	in mm	318	280	241	288	294	198	193	193	229	246	242	212		17	8
9 Minimum precipitation	in mm	30	13	39	53	41	tr	22	8	7	<1	7	32		17	9
10 Maximum precipitation in 24 h	in mm	92	131	86	114	196	117	87	75	94	103	129	81	196	19	10
11 Mean number of days with precipitation	>0,25 mm	10	10	10	11	10	8	8	7	7	6	8	9	104	19	11
12 Mean duration of sunshine	in h	143	158	213	243	291	316	321	316	265	251	181	142	2840	30	12
13 Mean quantity of radiation	in ly/day	199	264	366	449	531	556	547	509	427	341	238	186	384	10	13
14 Mean potential evaporation	in mm	8	10	31	65	112	151	176	160	114	64	23	10	924	19	14
15 Mean windspeed	in m/sec	4	4	5	4	4	4	3	3	3	3	4	4	4	19	15
16 Mean predominent direction of the wind		S	SW	WNW	S	S	SSW	SW	SW	NNE	SW	SW	SW		12	16

114 Station/Country Memphis (Tennessee)/U.S.

Location 35°03'N/89°59'W — Height above sealevel 80 m — Climate symbol: Köppen Cfa — Troll IV,7

		J	F	M	A	M	J	J	A	S	O	N	D	year	P	
1 Mean daily temperature	in °C	5,6	7,0	10,9	16,6	21,6	26,1	27,7	27,2	23,6	17,6	10,3	6,4	16,7	30	1
2 Mean daily maximum temperature	in °C	10,2	11,8	16,7	22,2	26,9	31,3	32,8	32,6	29,6	24,2	16,3	11,4	22,2	30	2
3 Mean daily minimum temperature	in °C	0,5	2,1	5,6	10,8	15,4	20,1	21,8	21,0	17,3	10,8	4,3	1,0	10,9	30	3
4 Absolute maximum temperature	in °C	26	26	29	33	36	40	41	41	39	35	29	26	41	19	4
5 Absolute minimum temperature	in °C	-17	-24	-11	-2	3	10	11	9	2	-4	-13	-12	-24	19	5
6 Mean relative humidity	in %	75	72	68	66	69	70	71	71	70	70	69	73	70	19	6
7 Mean precipitation	in mm	154	119	129	118	107	93	90	75	72	69	111	125	1262	30	7
8 Maximum precipitation	in mm	446	256	331	353	339	461	192	269	275	257	369	304		87	8
9 Minimum precipitation	in mm	25	18	18	23	5	1	<1	10	0	1	12	12		87	9
10 Maximum precipitation in 24 h	in mm	98	74	91	89	125	56	84	81	118	55	83	56	125	10	10
11 Mean number of days with precipitation	>0,25 mm	10	10	11	10	8	8	9	7	6	5	8	10	102	10	11
12 Mean duration of sunshine	in h	135	152	204	244	296	321	319	314	261	243	180	139	2808	30	12
13 Mean quantity of radiation	in ly/day															13
14 Mean potential evaporation	in mm	5	10	31	65	113	152	173	180	114	64	28	10	923		14
15 Mean windspeed	in m/sec	5	5	5	5	4	4	3	3	3	4	4	5	4	10	15
16 Mean predominent direction of the wind		S	S	S	S	S	S	S	S	NE	S	S	S		12	16

115 Station/Country Vicksburg (Mississippi)/U.S.

Location 32°21'N/90°53'W — Height above sealevel 71 m — Climate symbol: Köppen Cfa — Troll IV,7

		J	F	M	A	M	J	J	A	S	O	N	D	year	P	
1 Mean daily temperature	in °C	9,4	11,0	14,2	18,7	23,0	26,5	27,7	27,6	24,8	19,7	13,4	10,2	18,8	30	1
2 Mean daily maximum temperature	in °C	13,9	15,6	20,0	23,9	27,8	31,1	32,2	32,2	30,0	25,0	18,9	14,4	23,9	73	2
3 Mean daily minimum temperature	in °C	4,4	6,1	10,0	13,3	17,8	21,1	22,8	22,2	19,4	13,9	8,3	5,6	13,9	73	3
4 Absolute maximum temperature	in °C	27,8	28,9	33,3	33,3	36,1	38,3	38,9	38,3	40,0	34,4	30,0	27,8	40,0	73	4
5 Absolute minimum temperature	in °C	-16,1	-18,3	-8,3	-0,6	6,1	11,1	15,0	12,2	5,0	-0,6	-5,6	-12,2	-18,3	73	5
6 Mean relative humidity	in %	75	72	69	69	72	73	73	74	72	70	71	74	72	13	6
7 Mean precipitation	in mm	130	135	146	125	105	88	99	76	64	52	113	125	1258	30	7
8 Maximum precipitation	in mm	351	291	369	565	336	288	278	280	267	438	414	350		87	8
9 Minimum precipitation	in mm	9	11	13	6	<1	6	2	1	tr	tr	25			87	9
10 Maximum precipitation in 24 h	in mm	87	123	253	222	98	126	115	170	77	119	135	96	253	23	10
11 Mean number of days with precipitation	>0,25 mm	10	10	10	9	8	9	10	7	7	6	8	10	104	23	11
12 Mean duration of sunshine	in h	136	141	199	232	284	304	291	297	254	244	183	140	2705	30	12
13 Mean quantity of radiation	in ly/day															13
14 Mean potential evaporation	in mm	13	21	40	75	118	159	174	166	121	71	32	16	1006	55	14
15 Mean windspeed	in m/sec	4	4	4	4	4	3	3	3	3	3	4	4	4		15
16 Mean predominent direction of the wind		N	N	S	S	S	S	S	S	N	N	N	N		22	16

116 Station/Country Birmingham (Alabama)U.S.

Location 33°34'N/86°45'W — Height above sealevel 186 m — Climate symbol: Köppen Cfa — Troll IV,7

		J	F	M	A	M	J	J	A	S	O	N	D	year	P	
1 Mean daily temperature	in °C	7,5	8,7	11,9	16,7	21,3	25,5	26,7	26,4	23,5	17,8	11,0	7,7	17,1	30	1
2 Mean daily maximum temperature	in °C	12,8	13,9	18,9	22,8	27,2	31,1	32,2	32,2	30,0	24,4	17,8	12,8	22,8	35	2
3 Mean daily minimum temperature	in °C	2,8	3,3	7,8	11,7	16,1	20,0	21,1	21,1	18,9	12,8	6,7	3,3	12,2	35	3
4 Absolute maximum temperature	in °C	25,0	27,8	32,2	32,2	37,2	38,3	41,7	39,4	41,1	34,4	28,9	25,6	41,7	35	4
5 Absolute minimum temperature	in °C	-17,2	-23,3	-11,1	-2,2	3,3	8,3	13,9	12,8	5,6	-2,8	-10,0	-15,0	-23,3	35	5
6 Mean relative humidity	in %	69	67	63	62	66	66	64	70	67	64	66	70	67	27	6
7 Mean precipitation	in mm	128	134	152	114	87	102	131	123	85	75	90	128	1347	30	7
8 Maximum precipitation	in mm	279	448	290	251	282	214	348	276	247	179	387	355		30	8
9 Minimum precipitation	in mm	38	30	45	34	29	17	28	21	0	1	11	33		30	9
10 Maximum precipitation in 24 h	in mm	148	117	150	95	74	98	80	130	100	83	124	84	150	17	10
11 Mean number of days with precipitation	>0,25 mm	12	11	11	9	9	10	12	10	7	6	9	11	117	17	11
12 Mean duration of sunshine	in h	138	152	207	248	293	294	269	265	244	234	182	136	2662	30	12
13 Mean quantity of radiation	in ly/day															13
14 Mean potential evaporation	in mm	13	15	37	69	115	155	168	157	121	67	26	13	956	51	14
15 Mean windspeed	in m/sec	4	4	5	4	3	3	3	3	3	3	4	4	4		15
16 Mean predominent direction of the wind		S	N	S	S	S	SSW	SSE	NE	ENE	ENE	N	N			16

117 Station / Country Montgomery (Alabama) / U.S.

Location 32°23'N/86°18'W — Height above sealevel 61 m — Climate symbol: Köppen Cfa — Troll IV.7

		J	F	M	A	M	J	J	A	S	O	N	D	year	P	
1	Mean daily temperature	in °C	9,4	10,9	14,5	18,6	22,8	26,7	27,5	27,2	24,7	19,2	13,4	9,7	18,6	76	1
2	Mean daily maximum temperature	in °C	14,4	16,1	20,0	24,4	28,3	32,2	32,8	32,2	30,0	25,0	18,9	14,4	23,9	76	2
3	Mean daily minimum temperature	in °C	4,4	5,6	8,9	12,8	17,2	21,1	22,2	22,2	19,4	13,3	7,8	5,0	13,3	76	3
4	Absolute maximum temperature	in °C	27,2	28,9	32,2	33,3	37,2	41,1	41,7	39,4	41,1	35,6	30,0	26,7	41,7	75	4
5	Absolute minimum temperature	in °C	-15,0	-20,6	-6,7	-1,1	6,1	8,9	16,1	14,4	7,2	-0,6	-7,7	-13,3	-20,6	76	5
6	Mean relative humidity	in %	72	71	69	66	65	67	70	71	69	66	70	72	69		6
7	Mean precipitation	in mm	130	140	160	119	99	104	119	101	79	61	89	119	1321	58	7
8	Maximum precipitation	in mm	452	299	419	405	260	396	341	396	305	281	511	262		86	8
9	Minimum precipitation	in mm	12	34	18	7	13	11	11	11	3	tr	5	12		86	9
10	Maximum precipitation in 24 h	in mm	254	132	155	152	107	198	145	183	155	119	173	130	254	77	10
11	Mean number of days with precipitation >0,25 mm		11	10	10	8	9	11	12	11	8	8	7	10	113	58	11
12	Mean duration of sunshine	in h	160	168	227	267	317	311	288	290	260	250	200	156	2894	30	12
13	Mean quantity of radiation	in ly/day															13
14	Mean potential evaporation	in mm	16	21	40	71	121	163	174	162	124	71	29	16	1008		14
15	Mean windspeed	in m/sec	3,6	3,6	4,0	3,1	2,7	2,7	2,7	2,2	2,7	2,7	3,1	3,1	3,1		15
16	Mean predominent direction of the wind		NW	S	NW	S	S	S	SW	NE	NE	NE	SW	NW			16

118 Station / Country Atlanta (Georgia) / U.S.

Location 33°39'N/84°25'W — Height above sealevel 308 m — Climate symbol: Köppen Cfa — Troll IV.7

		J	F	M	A	M	J	J	A	S	O	N	D	year	P	
1	Mean daily temperature	in °C	7,1	8,1	11,1	16,1	20,8	24,9	26,0	25,8	22,8	17,1	10,6	8,7	16,4	30	1
2	Mean daily maximum temperature	in °C	10,6	12,2	16,7	21,7	26,1	30,0	30,6	30,0	27,8	22,2	16,1	11,1	21,1	68	2
3	Mean daily minimum temperature	in °C	1,7	2,8	6,1	9,4	14,4	19,4	21,1	20,6	17,8	12,2	6,1	2,8	11,7	68	3
4	Absolute maximum temperature	in °C	24,4	25,6	30,6	33,9	36,1	38,9	39,4	38,3	38,9	34,4	27,8	23,9	39,4	69	4
5	Absolute minimum temperature	in °C	-18,9	-22,2	-13,3	-3,9	3,3	3,9	14,4	12,8	6,1	-2,2	-10,0	-17,2	-22,2	69	5
6	Mean relative humidity	in %	74	72	67	64	65	66	70	72	69	67	69	73	69	13	6
7	Mean precipitation	in mm	113	115	136	114	80	97	120	91	83	62	75	111	1197	30	7
8	Maximum precipitation	in mm	275	324	292	250	199	191	286	221	186	191	399	252		30	8
9	Minimum precipitation	in mm	36	25	69	37	8	19	30	22	7	tr	10	27		30	9
10	Maximum precipitation in 24 h	in mm	83	82	122	108	130	87	138	128	139	83	104	104	139	26	10
11	Mean number of days with precipitation >0,25 mm		11	10	12	10	9	10	12	10	7	7	8	11	117	26	11
12	Mean duration of sunshine	in h	154	165	218	266	309	304	284	285	247	241	188	160	2821	25	12
13	Mean quantity of radiation	in ly/day	218	290	380	488	533	562	532	508	418	344	268	211	396		13
14	Mean potential evaporation	in mm	11	13	31	62	100	144	157	146	111	61	26	10	880	51	14
15	Mean windspeed	in m/sec	5	5	5	5	4	4	3	3	4	4	4	5	4	22	15
16	Mean predominent direction of the wind		NW	NW	NW	NW	SW	NW	SW	NW	ENE	NW	NW	NW		11	16

119 Station / Country Ashville (North Carolina) / U.S.

Location 35°36'N/82°32'W — Height above sealevel 671 m — Climate symbol: Köppen Cfa — Troll IV.7

		J	F	M	A	M	J	J	A	S	O	N	D	year	P	
1	Mean daily temperature	in °C	4,3	4,8	7,9	13,3	18,0	22,1	23,6	23,1	19,9	14,2	8,1	4,4	13,8	30	1
2	Mean daily maximum temperature	in °C	8,9	10,0	13,9	18,9	23,3	26,7	28,3	28,9	25,6	20,0	13,3	9,4	18,9	39	2
3	Mean daily minimum temperature	in °C	-1,7	-1,1	2,2	6,7	11,1	15,0	17,2	17,8	13,9	7,2	2,2	-1,1	7,2	39	3
4	Absolute maximum temperature	in °C	25,0	26,7	30,6	31,7	33,9	36,7	37,2	35,6	35,0	32,2	26,1	25,6	37,2	44	4
5	Absolute minimum temperature	in °C	-20,6	-21,1	-13,9	-6,7	-0,6	4,4	7,8	7,2	1,7	-6,7	-15,6	-20,0	-21,1	44	5
6	Mean relative humidity	in %	72	70	66	63	66	71	72	74	74	70	70	72	70	14	6
7	Mean precipitation	in mm	81	77	95	81	73	89	109	92	71	63	56	74	961	30	7
8	Maximum precipitation	in mm	86	167	169	139	172	129	191	287	119	179	84	156		30	8
9	Minimum precipitation	in mm	44	16	66	28	44	62	82	62	30	67	33	4		30	9
10	Maximum precipitation in 24 h	in mm	96	57	64	103	44	81	61	172	90	103	64	110	172	30	10
11	Mean number of days with precipitation >0,25 mm		11	11	13	10	11	12	15	12	9	8	8	10	130	30	11
12	Mean duration of sunshine	in h	146	161	211	247	289	292	268	250	235	222	179	146	2646	30	12
13	Mean quantity of radiation	in ly/day															13
14	Mean potential evaporation	in mm	8	10	28	56	91	124	138	125	93	52	23	10	758	51	14
15	Mean windspeed	in m/sec	4	4	4	4	3	3	3	3	3	3	4	4	3	30	15
16	Mean predominent direction of the wind		NW	NW	NW	NW	NW	NW	NW	NW	NW	NW	NW	NW		30	16

120 Station / Country Charlotte (North Carolina) / U.S.

Location 35°13'N/80°56'W — Height above sealevel 224 m — Climate symbol: Köppen Cfa — Troll IV.7

		J	F	M	A	M	J	J	A	S	O	N	D	year	P	
1	Mean daily temperature	in °C	5,9	6,8	10,0	15,7	20,6	25,0	26,2	25,9	22,8	16,9	10,2	5,9	16,0	30	1
2	Mean daily maximum temperature	in °C	10,8	12,0	15,6	21,7	26,3	30,9	31,5	31,0	27,8	22,6	15,9	10,7	21,3	30	2
3	Mean daily minimum temperature	in °C	1,1	1,4	4,4	9,7	14,7	19,2	20,8	20,8	17,6	11,2	4,4	1,1	10,6	30	3
4	Absolute maximum temperature	in °C	23,3	25,6	30,0	32,8	35,0	37,2	37,2	37,8	34,4	29,4	23,3		37,8	9	4
5	Absolute minimum temperature	in °C	-15,6	-13,9	-7,7	-2,2	0,0	7,8	11,7	11,7	3,9	-4,4	-6,7	-16,7	-16,7	9	5
6	Mean relative humidity	in %	67	63	61	63	68	72	75	75	74	70	68	68	68	9	6
7	Mean precipitation	in mm	90	90	112	89	79	92	124	107	89	75	64	92	1102	30	7
8	Maximum precipitation	in mm	189	174	221	194	134	210	232	253	277	195	208	188		30	8
9	Minimum precipitation	in mm	31	22	54	25	3	17	32	22	<1	15	11			30	9
10	Maximum precipitation in 24 h	in mm	91	74	92	81	68	96	76	112	120	123	71	66	123	30	10
11	Mean number of days with precipitation >0,25 mm		10	10	11	9	9	9	12	10	7	7	7	10	110	30	11
12	Mean duration of sunshine	in h															12
13	Mean quantity of radiation	in ly/day															13
14	Mean potential evaporation	in mm	8	10	31	62	105	142	162	143	105	58	26	10	862		14
15	Mean windspeed	in m/sec	3,5	3,8	3,9	4,0	3,4	3,1	2,9	2,9	3,1	3,2	3,3	3,2	3,4	20	15
16	Mean predominent direction of the wind		SW	NE	SW	S	SW	SW	SW	S	NE	NNE	SSW	SW		14	16

121 Station/Country Charleston (South Carolina)/U.S.

Location 32°54'N/80°02'W — Height above sealevel 12 m — Climate symbol: Köppen Cfa — Troll IV.7

		J	F	M	A	M	J	J	A	S	O	N	D	year	P	
1 Mean daily temperature	in °C	10,2	10,8	13,7	17,9	22,2	25,7	26,7	26,5	24,2	19,0	13,3	10,0	18,3	30	1
2 Mean daily maximum temperature	in °C	14,4	15,0	18,9	22,8	26,7	30,0	31,1	30,6	28,3	23,9	18,9	15,0	22,8	75	2
3 Mean daily minimum temperature	in °C	6,1	6,7	10,0	13,9	18,0	22,8	23,9	23,9	21,7	16,1	10,6	6,7	15,0	75	3
4 Absolute maximum temperature	in °C	27,8	27,8	34,4	33,9	37,2	40,0	40,0	38,9	37,8	35,0	28,3	27,2	40,0	78	4
5 Absolute minimum temperature	in °C	-12,2	-13,9	-4,4	0,0	7,2	9,4	16,1	16,7	9,4	2,8	-5,0	-11,1	-13,9	78	5
6 Mean relative humidity	in %	73	72	71	68	69	71	73	75	76	71	70	73	72	13	6
7 Mean precipitation	in mm	65	84	100	73	92	127	196	168	148	72	53	72	1250	30	7
8 Maximum precipitation	in mm	170	161	282	241	236	408	489	415	440	232	139	180		30	8
9 Minimum precipitation	in mm	16	8	25	21	17	24	73	58	13	2	12	21		30	9
10 Maximum precipitation in 24 h	in mm	57ʳ	83	168	104	77	107	148	104	225	147	54	60	225	18	10
11 Mean number of days with precipitation	>0,25 mm	9	9	11	8	14	10	15	12	10	6	7	8	119	18	11
12 Mean duration of sunshine	in h	188	189	243	284	323	308	297	281	244	239	210	187	2993	30	12
13 Mean quantity of radiation	in ly/day	251	310	390	447	549	555	526	498	410	350	285	230	400	13	13
14 Mean potential evaporation	in mm	18	21	40	72	118	157	174	159	121	73	34	21	1008	50	14
15 Mean windspeed	in m/sec	4	5	5	5	4	4	4	3	4	4	4	4	4	11	15
16 Mean predominant direction of the wind		SW	NNE	SSW	SSW	SW	SW	SW	SW	NNE	NNE	N	SW		11	16

122 Station/Country Jacksonville (Florida)/U.S.

Location 30°25'N/81°39'W — Height above sealevel 7 m — Climate symbol: Köppen Cfa — Troll IV.7

		J	F	M	A	M	J	J	A	S	O	N	D	year	P	
1 Mean daily temperature	in °C	13,3	14,2	16,8	20,4	24,3	27,1	28,1	27,9	26,3	21,3	16,5	13,4	20,8	30	1
2 Mean daily maximum temperature	in °C	18,3	19,4	22,2	25,6	28,3	31,1	32,2	31,7	30,0	25,6	21,7	18,3	25,6	76	2
3 Mean daily minimum temperature	in °C	8,3	9,4	12,2	15,6	19,4	22,2	23,3	23,3	21,7	17,2	12,2	8,9	16,1	76	3
4 Absolute maximum temperature	in °C	28,9	30,0	32,8	33,3	37,2	38,3	40,0	38,3	37,2	35,0	30,0	28,3	40,0	78	4
5 Absolute minimum temperature	in °C	-9,4	-12,2	-3,9	1,1	7,8	12,2	18,3	17,8	9,4	2,8	-3,9	-10,0	-12,2	78	5
6 Mean relative humidity	in %	73	69	68	67	65	70	72	76	76	74	73	75	72	13	6
7 Mean precipitation	in mm	62	74	89	90	88	161	195	174	192	131	43	56	1355	30	7
8 Maximum precipitation	in mm	232	233	318	209	376	592	380	420	553	413	181	197		87	8
9 Minimum precipitation	in mm	<1	3	3	3	2	32	4	19	2	2	<1	tr		87	9
10 Maximum precipitation in 24 h	in mm	86	63	82	124	129	101	94	99	258	169	107	64	258	19	10
11 Mean number of days with precipitation	>0,25 mm	7	8	9	7	8	11	15	14	15	9	6	8	117	19	11
12 Mean duration of sunshine	in h	192	189	241	267	269	260	255	248	199	205	195	170	2713	30	12
13 Mean quantity of radiation	in ly/day															13
14 Mean potential evaporation	in mm	24	29	53	78	120	151	166	154	124	78	40	24	1039	51	14
15 Mean windspeed	in m/sec	4	4	4	4	4	4	3	3	4	4	4	4	4	11	15
16 Mean predominant direction of the wind		NW	WNW	NW	SE	WSW	SW	SW	SW	NE	NE	NW	NW		11	16

123 Station/Country San Antonio (Texas)/U.S.

Location 29°32'N/98°28'W — Height above sealevel 241 m — Climate symbol: Köppen Cfa — Troll IV.7

		J	F	M	A	M	J	J	A	S	O	N	D	year	P	
1 Mean daily temperature	in °C	11,1	13,0	16,1	20,1	24,1	27,7	28,9	28,8	25,9	21,4	15,3	12,1	20,4	30	1
2 Mean daily maximum temperature	in °C	16,1	18,9	22,9	26,1	30,3	33,3	35,0	35,4	32,2	28,1	21,8	17,9	26,6	30	2
3 Mean daily minimum temperature	in °C	4,8	6,8	9,7	14,1	18,5	22,2	22,9	23,8	20,4	15,5	9,2	5,5	14,3	30	3
4 Absolute maximum temperature	in °C	31	33	36	37	38	39	41	41	39	35	33	32	41	18	4
5 Absolute minimum temperature	in °C	-18	-14	-6	1	7	14	18	17	5	1	-5	-10	-18	18	5
6 Mean relative humidity	in %	69	69	64	67	71	69	64	63	67	67	66	68	67	18	6
7 Mean precipitation	in mm	44	42	42	72	88	75	53	60	89	64	35	44	708	30	7
8 Maximum precipitation	in mm	116	99	106	237	209	210	208	156	401	243	114	106		18	8
9 Minimum precipitation	in mm	0	<1	<1	4	22	7	tr	0	2	tr	3	<1		18	9
10 Maximum precipitation in 24 h	in mm	72	58	60	70	109	157	177	141	175	134	49	73	177	18	10
11 Mean number of days with precipitation	>0,25 mm	8	8	7	7	7	6	4	5	7	6	6	7	79	18	11
12 Mean duration of sunshine	in h	148	153	213	224	258	292	325	307	261	241	183	160	2765	30	12
13 Mean quantity of radiation	in ly/day	279	347	417	445	541	612	639	585	493	398	295	256	442		13
14 Mean potential evaporation	in mm	16	23	48	84	127	168	187	178	136	85	37	21	1108		14
15 Mean windspeed	in m/sec	4	4	5	5	5	5	4	4	4	4	4	4	4	18	15
16 Mean predominant direction of the wind		NE	NE	NE	SE	SE	SE	SE	SE	SE	N	N	N		12	16

124 Station/Country Houston (Texas)/U.S.

Location 24°48'N/95°22'W — Height above sealevel 12 m — Climate symbol: Köppen Cfa — Troll IV.7

		J	F	M	A	M	J	J	A	S	O	N	D	year	P	
1 Mean daily temperature	in °C	12,6	13,9	16,9	20,7	24,6	27,9	28,8	28,9	26,6	22,4	16,4	13,6	21,1	30	1
2 Mean daily maximum temperature	in °C	16,7	18,3	22,2	25,6	28,9	32,2	33,3	33,9	31,1	27,2	21,7	17,2	25,6	34	2
3 Mean daily minimum temperature	in °C	6,7	7,8	12,2	15,8	18,9	22,2	23,3	23,3	21,1	16,1	10,6	7,2	15,6	34	3
4 Absolute maximum temperature	in °C	28,9	30,6	34,4	33,3	36,7	39,4	40,0	42,2	38,3	37,2	31,7	28,3	42,2	36	4
5 Absolute minimum temperature	in °C	-15,0	-14,4	-5,0	1,1	7,2	12,8	12,8	12,2	8,3	0,6	-5,0	-9,4	-15,0	39	5
6 Mean relative humidity	in %	76	73	72	73	74	74	73	71	73	70	71	74	73	15	6
7 Mean precipitation	in mm	94	82	61	87	113	97	131	90	97	91	103	104	1150	30	7
8 Maximum precipitation	in mm	204	288	290	205	366	372	256	257	391	293	259	228		30	8
9 Minimum precipitation	in mm	9	2	2	23	16	4	2	10	4	tr	13	16		30	9
10 Maximum precipitation in 24 h	in mm	81	99	93	98	138	210	176	230	132	194	275	86	275	22	10
11 Mean number of days with precipitation	>0,25 mm	10	10	9	8	8	8	10	9	9	7	8	10	106	22	11
12 Mean duration of sunshine	in h	144	141	193	212	266	298	294	281	238	239	181	148	2633	30	12
13 Mean quantity of radiation	in ly/day															13
14 Mean potential evaporation	in mm	19	23	49	81	127	168	184	174	136	88	37	21	1107	56	14
15 Mean windspeed	in m/sec	5	5	5	5	5	4	4	4	4	4	5	5	5	12	15
16 Mean predominant direction of the wind		N	SE	SE	SE	SE	S	S	S	SE	SE	SE	SE		22	16

125 Station/Country New Orleans (Louisiana)/U.S.

Location 29°57'N/90°04'W — Height above sealevel 3 m — Climate symbol: Köppen Cfa — Troll IV,7

		J	F	M	A	M	J	J	A	S	O	N	D	year	P
1 Mean daily temperature	in °C	13,3	14,7	17,2	21,0	24,5	27,7	28,4	28,8	26,8	22,7	16,9	13,9	21,3	30
2 Mean daily maximum temperature	in °C	16,7	18,3	21,7	25,0	28,3	31,1	32,2	32,2	30,0	26,1	21,1	17,8	25,0	73
3 Mean daily minimum temperature	in °C	8,3	10,0	12,8	16,1	20,0	23,3	24,4	24,4	22,8	17,8	12,8	8,9	16,7	73
4 Absolute maximum temperature	in °C	28,3	28,9	32,2	32,2	35,6	38,9	38,9	37,8	37,2	34,4	31,7	28,9	38,9	76
5 Absolute minimum temperature	in °C	-9,4	-13,9	-2,2	3,3	11,1	14,4	18,9	17,2	12,2	4,4	-1,7	-7,2	-13,9	76
6 Mean relative humidity	in %	77	76	75	73	67	72	73	74	73	71	73	76	74	13
7 Mean precipitation	in mm	121	106	167	138	138	141	180	163	148	93	102	116	1613	30
8 Maximum precipitation	in mm	321	268	485	223	394	225	291	298	425	164	370	274		30
9 Minimum precipitation	in mm	7	23	14	27	31	23	4	13	4	0	18	17		30
10 Maximum precipitation in 24 h	in mm	119	155	278	356	231	189	135	135	273	347	197	123	356	48
11 Mean number of days with precipitation >0,25 mm		10	9	9	7	9	12	15	14	10	7	7	10	119	45
12 Mean duration of sunshine	in h	160	158	213	247	292	287	260	269	241	260	200	157	2744	30
13 Mean quantity of radiation	in ly/day	214	259	335	412	449	443	417	416	383	357	278	198	347	
14 Mean potential evaporation	in mm	22	26	49	84	127	168	180	171	139	88	40	24	1118	51
15 Mean windspeed	in m/sec	3	3	3	3	3	3	3	3	3	3	3	3	3	45
16 Mean predominent direction of the wind		NE	E	NW	SE	SE	NE	NW	SE	SE	SE	NW	SW		47

126 Station/Country Apalachicola (Florida)/U.S.

Location 29°44'N/84°59'W — Height above sealevel 4 m — Climate symbol: Köppen Cfa — Troll IV,7

		J	F	M	A	M	J	J	A	S	O	N	D	year	P
1 Mean daily temperature	in °C	12,8	13,8	16,1	19,7	23,8	26,8	27,5	27,5	26,1	21,8	16,3	13,2	20,4	30
2 Mean daily maximum temperature	in °C	16,8	17,8	19,9	23,4	27,3	30,8	30,7		29,2	25,7	20,8	17,6	24,1	30
3 Mean daily minimum temperature	in °C	8,9	9,7	12,2	16,0	19,7	23,0	23,9	24,0	22,8	17,7	12,0	9,3	16,7	30
4 Absolute maximum temperature	in °C	26	27	28	32	36	38	39	37	36	34	31	28	39	31
5 Absolute minimum temperature	in °C	-8	-6	-3	3	10	17	19	19	11	4	-	-5	-8	31
6 Mean relative humidity	in %	80	79	78	78	77	79	80	80	81	77	77	81	79	
7 Mean precipitation	in mm	80	99	115	109	73	135	201	197	217	62	66	75	1429	30
8 Maximum precipitation	in mm	162	211	355	473	201	295	375	481	704	194	229	199		36
9 Minimum precipitation	in mm	1	10	18	2		47	75	47	37	<1	1	8		36
10 Maximum precipitation in 24 h	in mm	96	95	208	192	180	136	149	144	297	143	148	105	297	31
11 Mean number of days with precipitation >0,25 mm		8	8	8	6	6	10	16	14	12	6	6	8	108	31
12 Mean duration of sunshine	in h	193	195	233	274	328	296	273	259	236	263	216	175	2941	26
13 Mean quantity of radiation	in ly/day	298	367	441	535	603	578	529	511	456	413	332	262	444	
14 Mean potential evaporation	in mm	24	29	46	75	124	161	173	164	136	82	40	26	1080	
15 Mean windspeed	in m/sec	4	4	4	4	4	3	3	3	4	4	4	4	4	27
16 Mean predominent direction of the wind		N	N	SE	SE	SE	SW	W	SW	NE	NE	N	N		22

127 Station/Country Tampa (Florida)/U.S.

Location 27°57'N/82°27'W — Height above sealevel 11 m — Climate symbol: Köppen Cfa — Troll IV,7

		J	F	M	A	M	J	J	A	S	O	N	D	year	P
1 Mean daily temperature	in °C	16,1	17,0	19,4	22,0	25,0	27,0	27,5	27,8	26,7	23,6	19,4	16,7	22,5	58
2 Mean daily maximum temperature	in °C	21,1	21,7	24,4	27,2	30,0	31,7	31,7	32,2	31,1	28,3	24,4	21,7	27,2	58
3 Mean daily minimum temperature	in °C	11,1	12,2	14,4	16,7	20,0	22,2	23,3	23,3	22,2	18,9	14,4	11,7	17,8	58
4 Absolute maximum temperature	in °C	29,4	30,0	33,3	32,8	35,6	36,7	36,7	36,1	35,6	35,0	31,1	30,0	36,7	58
5 Absolute minimum temperature	in °C	-5,0	-5,6	-0,6	3,3	11,1	15,0	17,8	18,9	12,2	6,1	-0,6	-7,2	-7,2	58
6 Mean relative humidity	in %	74	72	69	65	68	70	72	73	73	72	70	73	71	
7 Mean precipitation	in mm	66	69	69	51	76	183	221	218	160	71	43	58	1285	72
8 Maximum precipitation	in mm	204	276	251	204	239	470	394	472	481	262	137	187		69
9 Minimum precipitation	in mm	tr	1	tr	4	4	32	52	20	20	<1	1	2		69
10 Maximum precipitation in 24 h	in mm	91	130	142	109	91	264	140	137	185	165	107	99	264	60
11 Mean number of days with precipitation >0,25 mm		7	7	6	5	7	15	17	17	15	8	5	7	116	41
12 Mean duration of sunshine	in h	223	220	260	283	320	275	257	252	232	243	227	209	3001	30
13 Mean quantity of radiation	in ly/day	327	391	474	539	596	574	534	494	452	400	356	300	453	
14 Mean potential evaporation	in mm	33	37	59	90	137	160	170	166	142	103	54	35	1186	
15 Mean windspeed	in m/sec	4,0	4,0	4,5	4,5	4,0	3,6	3,1	3,6	3,6	4,0	4,0	4,0	4,0	
16 Mean predominent direction of the wind		N	E	S	ENE	E	E	E	ENE	ENE	NNE	NNE	N		

128 Station/Country Miami (Florida)/U.S.

Location 25°48'N/80°16'W — Height above sealevel 2 m — Climate symbol: Köppen Aw — Troll V,2

		J	F	M	A	M	J	J	A	S	O	N	D	year	P
1 Mean daily temperature	in °C	19,4	19,9	21,4	23,4	25,3	27,1	27,7	27,8	27,4	25,4	22,4	20,1	23,9	30
2 Mean daily maximum temperature	in °C	23,3	23,9	25,6	26,7	28,9	30,0	31,1	31,1	30,6	28,3	25,6	24,4	27,2	51
3 Mean daily minimum temperature	in °C	16,1	16,1	17,8	19,4	21,7	23,3	24,4	24,4	23,9	22,2	18,9	16,7	20,6	51
4 Absolute maximum temperature	in °C	29,4	31,1	33,3	33,9	34,4	34,4	35,6	35,6	35,0	33,9	31,1	32,8	35,6	51
5 Absolute minimum temperature	in °C	-1,7	-2,8	1,1	7,2	10,0	16,1	18,9	15,6	16,7	11,1	2,2	-1,1	-2,8	51
6 Mean relative humidity	in %	74	73	70	69	71	72	72	72	75	75	71	74	72	13
7 Mean precipitation	in mm	52	48	58	99	164	187	171	177	241	209	72	42	1520	30
8 Maximum precipitation	in mm	169	167	183	259	471	568	343	429	620	535	334	162		30
9 Minimum precipitation	in mm	1	<1	<1	2	11	46	45	42	67	38	2	3		30
10 Maximum precipitation in 24 h	in mm	64	52	180	132	214	189	116	80	193	253	201	45	253	18
11 Mean number of days with precipitation >0,25 mm		6	5	6	7	10	13	17	16	18	15	8	7	128	18
12 Mean duration of sunshine	in h	222	227	266	275	280	251	267	263	216	215	212	209	2903	30
13 Mean quantity of radiation	in ly/day	344	412	494	548	555	533	539	505	478	427	397	344	465	12
14 Mean potential evaporation	in mm	47	50	74	103	136	162	172	168	147	116	76	57	1309	21
15 Mean windspeed	in m/sec	4	4	4	5	4	3	3	4	4	4	4	4	4	11
16 Mean predominent direction of the wind		SE	SE	SE	ESE	ESE	SE	SE	SE	ESE	ENE	N	N		12

129 Station / Country **Brownsville (Texas) / U.S.**

Location 25°54'N/97°26'W — Height above sealevel 5 m — Climate symbol: Köppen **Cfa** — Troll **IV,3**

		J	F	M	A	M	J	J	A	S	O	N	D	year	P
1 Mean daily temperature	in °C	16,3	17,8	19,9	23,3	26,1	28,2	28,9	28,9	27,3	24,4	19,8	17,2	23,2	30
2 Mean daily maximum temperature	in °C	20,6	22,2	25,6	28,3	30,6	32,8	33,3	33,9	32,2	29,4	25,0	21,7	27,8	50
3 Mean daily minimum temperature	in °C	10,0	12,2	15,0	18,9	21,7	23,3	23,9	23,9	22,2	18,9	15,0	11,1	17,8	50
4 Absolute maximum temperature	in °C	32,2	34,4	38,9	37,8	38,9	39,4	38,9	40,0	38,9	37,2	36,7	33,9	40,0	50
5 Absolute minimum temperature	in °C	-7,8	-11,1	-1,1	3,9	5,0	13,3	13,9	17,2	10,6	3,3	-2,8	-7,8	-11,1	50
6 Mean relative humidity	in %	76	77	75	75	75	76	73	71	76	75	75	78	75	8
7 Mean precipitation	in mm	34	38	26	39	60	75	43	70	127	90	34	44	680	30
8 Maximum precipitation	in mm	130	260	108	149	139	332	142	180	226	435	159	240		19
9 Minimum precipitation	in mm	tr	tr	<1	<1	<1	<1	<1	6	13	14	<1	<1		19
10 Maximum precipitation in 24 h	in mm	75	126	47	94	102	208	92	112	138	169	92	145	208	21
11 Mean number of days with precipitation	>0,25 mm	7	7	5	4	4	5	4	7	10	7	7	6	73	18
12 Mean duration of sunshine	in h	147	152	187	210	272	297	328	311	246	252	165	151	2716	30
13 Mean quantity of radiation	in ly/day	286	344	400	467	561	608	624	566	467	409	297	254	440	11
14 Mean potential evaporation	in mm	30	37	65	103	150	172	179	171	147	101	55	33	1243	56
15 Mean windspeed	in m/sec	5	6	6	6	6	6	5	5	4	4	5	5	5	18
16 Mean predominent direction of the wind		SSE	SSE	SE	SE	SE	SE	SE	SE	SE	SE	SSE	NNW		11

130 Station / Country **Honolulu (Hawaii) / U.S.**

Location 21°19'N/157°52'W — Height above sealevel 10 m — Climate symbol: Köppen **Af** — Troll **V,2**

		J	F	M	A	M	J	J	A	S	O	N	D	year	P
1 Mean daily temperature	in °C	22,5	21,9	22,2	22,8	23,9	24,7	25,3	25,9	25,9	25,0	23,9	23,0	23,9	40
2 Mean daily maximum temperature	in °C	24,4	24,4	25,0	25,6	26,7	27,2	27,8	28,3	28,3	27,8	26,7	25,6	26,7	40
3 Mean daily minimum temperature	in °C	20,6	19,4	19,4	20,0	21,1	22,2	22,8	23,3	23,3	22,2	21,1	20,6	21,1	40
4 Absolute maximum temperature	in °C	28,9	28,9	28,9	30,0	30,6	31,1	31,1	31,1	31,1	32,2	30,0	29,4	32,2	56
5 Absolute minimum temperature	in °C	12,2	11,1	11,7	15,0	15,6	17,2	17,2	17,2	17,2	17,2	15,0	12,8	11,1	56
6 Mean relative humidity	in %	71	71	69	67	67	66	67	68	68	70	71	72	69	9
7 Mean precipitation	in mm	104	66	78	48	25	18	23	28	36	48	64	104	643	35
8 Maximum precipitation	in mm	374	347	528	227	184	62	51	78	70	148	374	307		30
9 Minimum precipitation	in mm	12	12	<1	<1	1	tr	1	1	2	3	1	10		30
10 Maximum precipitation in 24 h	in mm	114	163	343	254	119	76	30	53	152	117	140	155	343	46
11 Mean number of days with precipitation	>0,25 mm	14	11	13	12	11	12	14	13	13	13	15	14	154	34
12 Mean duration of sunshine	in h	227	202	250	255	278	280	293	290	279	257	221	211	3041	30
13 Mean quantity of radiation	in ly/day	363	422	516	559	617	615	615	612	573	507	424	371	516	
14 Mean potential evaporation	in mm	68	65	77	87	105	118	129	130	118	108	86	76	1187	
15 Mean windspeed	in m/sec	4,4	4,6	5,1	5,5	6,7	5,9	6,2	6,3	5,2	4,9	4,8	5,1	5,3	9
16 Mean predominent direction of the wind		SW	W	SE	NE	E	E	E	SE	NE	NE	NE	NE		27

131 Station / Country **Guaymas / Mexico**

Location 27°55'N/110°53'W — Height above sealevel 4 m — Climate symbol: Köppen **BWh** — Troll **IV,5**

		J	F	M	A	M	J	J	A	S	O	N	D	year	P
1 Mean daily temperature	in °C	17,9	19,0	21,0	23,5	26,4	29,8	31,2	31,1	30,7	27,7	22,7	19,4	25,0	30
2 Mean daily maximum temperature	in °C	22,8	23,9	26,1	28,9	31,1	33,9	34,4	35,0	35,0	30,0	27,8	23,3	29,4	6
3 Mean daily minimum temperature	in °C	12,8	13,9	15,6	17,8	20,6	24,4	26,7	26,7	25,6	22,2	17,8	13,3	20,0	6
4 Absolute maximum temperature	in °C	33,3	34,4	35,2	39,6	41,3	42,0	42,0	42,2	42,0	40,7	37,5	32,5	42,2	30
5 Absolute minimum temperature	in °C	3,2	5,2	9,5	12,0	13,5	18,0	20,6	20,5	17,7	4,7	3,8	6,0	3,2	30
6 Mean relative humidity	in %	52	53	50	48	50	57	63	64	66	67	70	72	59	14
7 Mean precipitation	in mm	12	7	3	1	0	3	39	60	51	22	7	17	222	
8 Maximum precipitation	in mm														
9 Minimum precipitation	in mm														
10 Maximum precipitation in 24 h	in mm	41	48	15	15	<1	15	57	74	152	309	43	75	309	30
11 Mean number of days with precipitation	>0,25 mm	3	1	1	<1	<1	1	6	7	4	1	1	2	27	30
12 Mean duration of sunshine	in h	202	210	238	271	306	302	260	257	242	259	225	198	2970	30
13 Mean quantity of radiation	in ly/day														
14 Mean potential evaporation	in mm	35	41	65	89	147	180	198	188	168	139	72	41	1363	22
15 Mean windspeed	in m/sec	2,1	2,5	2,6	2,4	2,1	2,0	1,9	2,1	1,9	2,0	2,1	2,0	2,1	30
16 Mean predominent direction of the wind		WNW	NW	W	W	W	SW	SW	SW	SW	W	NW	NW		30

132 Station / Country **Monterrey / Mexico**

Location 25°40'N/100°18'W — Height above sealevel 534 m — Climate symbol: Köppen **BSh** — Troll **IV,3**

		J	F	M	A	M	J	J	A	S	O	N	D	year	P
1 Mean daily temperature	in °C	15,4	17,0	20,3	23,4	25,9	27,8	28,1	28,0	25,6	22,5	17,9	15,6	22,3	30
2 Mean daily maximum temperature	in °C	20,0	22,2	24,4	28,9	30,6	32,8	32,2	33,3	30,0	26,7	21,7	18,3	26,7	11
3 Mean daily minimum temperature	in °C	8,9	11,1	13,9	16,7	20,0	21,7	21,7	22,2	21,1	17,8	12,8	10,0	16,7	11
4 Absolute maximum temperature	in °C	36,6	37,0	40,8	40,7	41,7	42,1	39,7	39,7	38,7	35,5	36,5	36,0	42,1	30
5 Absolute minimum temperature	in °C	-6,8	-3,5	0,4	7,0	11,7	13,5	16,0	18,0	12,1	8,4	-0,5	-2,5	-6,8	30
6 Mean relative humidity	in %	69	69	67	65	64	70	62	68	71	70	63	61	66	2
7 Mean precipitation	in mm	19	17	13	26	38	63	61	111	158	91	23	18	634	30
8 Maximum precipitation	in mm														
9 Minimum precipitation	in mm														
10 Maximum precipitation in 24 h	in mm	42	22	25	50	47	118	81	129	160	232	45	32	232	30
11 Mean number of days with precipitation	>0,25 mm	4	4	3	6	6	8	5	7	9	7	5	5	88	30
12 Mean duration of sunshine	in h	124	120	156	136	166	193	214	199	152	133	121	111	1825	30
13 Mean quantity of radiation	in ly/day														
14 Mean potential evaporation	in mm	28	36	68	96	148	164	168	165	125	83	44	25	1150	43
15 Mean windspeed	in m/sec	1,3	1,6	2,0	2,3	2,1	2,4	2,4	2,2	1,8	1,5	1,2	1,1	1,8	
16 Mean predominent direction of the wind		NE	E	NE	E	E	E	E	E	E	E	NE	NE		30

133 Station/Country Ciudad Lerdo (Durango)/Mexico

Location 25°32'N/103°32'W · Height above sealevel 1140 m · Climate symbol: Köppen BWk · Troll IV.5

		J	F	M	A	M	J	J	A	S	O	N	D	year	P	
1	Mean daily temperature	in °C	13,6	16,1	19,2	22,8	25,8	27,7	26,8	26,3	24,2	21,2	16,5	13,6	21,2		1
2	Mean daily maximum temperature	in °C															2
3	Mean daily minimum temperature	in °C															3
4	Absolute maximum temperature	in °C	31,0	38,4	36,5	38,5	40,4	40,8	38,0	38,2	38,0	36,2	34,2	32,0	40,8		4
5	Absolute minimum temperature	in °C	-8,7	-8,0	-3,5	-2,0	7,4	11,0	14,5	14,0	7,0	3,0	-5,0	-5,5	-8,7		5
6	Mean relative humidity	in %															6
7	Mean precipitation	in mm	6	6	2	4	16	27	40	42	57	33	7	9	248		7
8	Maximum precipitation	in mm															8
9	Minimum precipitation	in mm															9
10	Maximum precipitation in 24 h	in mm	18	33	14	21	55	49	46	51	91	70	24	28	91		10
11	Mean number of days with precipitation	>0,25 mm	2	2	1	1	3	5	6	6	6	4	2	3	41		11
12	Mean duration of sunshine	in h															12
13	Mean quantity of radiation	in ly/day															13
14	Mean potential evaporation	in mm	27	37	64	100	144	164	158	148	111	79	43	28	1101		14
15	Mean windspeed	in m/sec	1,0	1,2	1,2	1,2	1,3	1,3	1,2	1,0	0,9	0,7	0,8	0,9	1,0		15
16	Mean predominent direction of the wind																16

134 Station/Country La Paz/Mexico

Location 24°10'N/110°18'W · Height above sealevel 18 m · Climate symbol: Köppen BWh · Troll V.5

		J	F	M	A	M	J	J	A	S	O	N	D	year	P	
1	Mean daily temperature	in °C	18,3	19,0	20,7	22,8	24,9	26,8	29,4	29,8	29,0	26,4	23,1	20,0	24,2	30	1
2	Mean daily maximum temperature	in °C	22,2	23,3	26,7	28,3	31,1	33,3	35,0	33,9	33,3	31,7	27,2	23,3	28,9	6	2
3	Mean daily minimum temperature	in °C	13,9	13,3	13,3	15,6	17,8	20,6	23,9	24,4	24,4	21,7	18,9	15,0	18,3	6	3
4	Absolute maximum temperature	in °C	30,5	34,5	35,0	38,0	39,2	40,2	40,6	40,0	40,0	39,0	36,2	33,2	40,6	30	4
5	Absolute minimum temperature	in °C	2,2	1,9	6,0	8,6	10,0	12,8	16,0	16,7	17,5	14,3	12,0	7,8	1,9	30	5
6	Mean relative humidity	in %	83	82	77	75	79	75	74	73	77	76	83	85	78	2	6
7	Mean precipitation	in mm	14	14	16	17	19	21	25	26	26	23	19	17	227	30	7
8	Maximum precipitation	in mm															8
9	Minimum·precipitation	in mm															9
10	Maximum precipitation in 24 h	in mm	67	30	18	22	4	35	66	90	179	75	60	44	179	30	10
11	Mean number of days with precipitation	>0,25 mm	<1	<1	<1	<1	0	<1	1	3	4	1	1	2	14	30	11
12	Mean duration of sunshine	in h															12
13	Mean quantity of radiation	in ly/day															13
14	Mean potential evaporation	in mm	40	42	62	84	121	147	180	177	155	133	81	50	1272	22	14
15	Mean windspeed	in m/sec	2,2	1,9	1,9	2,1	2,2	2,6	2,0	1,9	1,7	1,8	2,0	1,9	2,0	30	15
16	Mean predominent direction of the wind		NE	NE	NE	SW	SW	SW	SW	SW	SW	SW	NE	NE		30	16

135 Station/Country Mazatlán/Mexico

Location 23°13'N/106°25'W · Height above sealevel 78 m · Climate symbol: Köppen Aw · Troll V.3

		J	F	M	A	M	J	J	A	S	O	N	D	year	P	
1	Mean daily temperature	in °C	19,8	19,8	20,5	22,1	24,6	27,1	28,0	28,1	27,9	27,2	24,3	21,5	24,2	30	1
2	Mean daily maximum temperature	in °C	21,7	21,7	22,8	24,4	26,7	28,9	30,0	30,0	29,4	28,7	26,7	23,9	26,1	10	2
3	Mean daily minimum temperature	in °C	16,1	16,7	17,2	18,3	21,1	24,4	25,0	25,0	25,0	24,4	21,7	18,3	21,1	10	3
4	Absolute maximum temperature	in °C	27,4	26,3	27,8	30,8	31,8	32,7	32,7	34,0	32,8	32,9	30,8	28,8	32,4	30	4
5	Absolute minimum temperature	in °C	11,0	11,2	11,8	13,9	14,8	21,0	20,5	20,0	20,0	20,5	19,0	12,4	11,0	30	5
6	Mean relative humidity	in %	70	66	65	61	63	73	72	79	76	75	71	70	70		6
7	Mean precipitation	in mm	13	6	3	1	1	33	188	227	245	57	16	15	803	30	7
8	Maximum precipitation	in mm															8
9	Minimum precipitation	in mm															9
10	Maximum precipitation in 24 h	in mm	65	27	30	5	62	80	215	214	152	111	69	44	215	30	10
11	Mean number of days with precipitation	>0,25 mm	2	1	<1	<1	<1	4	15	15	13	4	1	2	57	30	11
12	Mean duration of sunshine	in h	189	192	243	249	269	242	211	213	190	242	214	181	2635	30	12
13	Mean quantity of radiation	in ly/day															13
14	Mean potential evaporation	in mm	53	51	68	83	116	159	171	168	150	140	94	66	1318	30	14
15	Mean windspeed	in m/sec	3,4	3,4	3,1	2,9	2,6	2,6	2,7	2,7	2,6	2,4	2,6	3,1	2,8	30	15
16	Mean predominent direction of the wind		NW	NW	NW	WNW	WNW	W	W	WNW	WNW	WNW	WNW	NW		30	16

136 Station/Country Zacatecas/Mexico

Location 22°47'N/102°35'W · Height above sealevel 2612 m · Climate symbol: Köppen BSk · Troll V.4

		J	F	M	A	M	J	J	A	S	O	N	D	year	P	
1	Mean daily temperature	in °C	9,6	10,7	12,9	15,1	16,6	16,2	14,6	14,8	13,7	13,2	11,5	10,1	13,3		1
2	Mean daily maximum temperature	in °C															2
3	Mean daily minimum temperature	in °C															3
4	Absolute maximum temperature	in °C	22,0	23,3	26,0	26,5	28,7	27,4	24,8	24,8	24,8	23,9	22,8	20,5	28,7		4
5	Absolute minimum temperature	in °C	-8,1	-9,2	-3,8	-1,6	6,0	5,9	7,2	7,5	3,1	1,7	-3,2	-6,8	-9,2		5
6	Mean relative humidity	in %															6
7	Mean precipitation	in mm	7	3	2	4	11	18	64	66	54	23	10	4	265		7
8	Maximum precipitation	in mm															8
9	Minimum precipitation	in mm															9
10	Maximum precipitation in 24 h	in mm	26	10	34	23	24	56	76	34	33	31	43	15	76		10
11	Mean number of days with precipitation	>0,25 mm	2	1	<1	<1	4	9	12	12	11	6	2	2	63		11
12	Mean duration of sunshine	in h	215	216	245	248	273	227	215	230	183	211	227	206	2696		12
13	Mean quantity of radiation	in ly/day															13
14	Mean potential evaporation	in mm	35	38	55	68	89	89	80	80	69	63	45	39	750		14
15	Mean windspeed	in m/sec	7,4	7,5	7,5	7,0	5,6	4,9	5,0	4,7	5,3	5,1	5,5	6,2	6,0		15
16	Mean predominent direction of the wind		SSW	SSW	SSW	SW	SW	E	E	E	E	E	E	SSW			16

137 Station/Country Tampico/Mexico

Location 22°12'N/97°51'W   Height above sealevel 73 m   Climate symbol: Köppen Aw   Troll V,3

		J	F	M	A	M	J	J	A	S	O	N	D	year	P	
1 Mean daily temperature	in °C	19,2	20,4	22,0	24,6	26,8	28,0	28,0	28,2	27,2	25,6	22,0	19,9	24,3	30	1
2 Mean daily maximum temperature	in °C															2
3 Mean daily minimum temperature	in °C															3
4 Absolute maximum temperature	in °C	33,5	36,2	42,3	42,7	39,3	39,0	36,5	38,5	38,5	34,5	35,0	30,0	42,7	30	4
5 Absolute minimum temperature	in °C	0,0	0,7	6,0	10,8	16,0	19,2	16,9	15,6	10,3	5,0	3,5	6,5	0,0	30	5
6 Mean relative humidity	in %															6
7 Mean precipitation	in mm	43	16	12	23	47	121	153	135	281	132	46	28	1035	30	7
8 Maximum precipitation	in mm															8
9 Minimum precipitation	in mm															9
10 Maximum precipitation in 24 h	in mm	119	37	41	79	94	94	193	226	185	136	196	32	226	30	10
11 Mean number of days with precipitation	>0,25 mm	6	4	4	4	5	9	11	10	5	9	7	5	89	30	11
12 Mean duration of sunshine	in h															12
13 Mean quantity of radiation	in ly/day															13
14 Mean potential evaporation	in mm	48	59	80	111	157	165	169	167	144	119	72	56	1346	30	14
15 Mean windspeed	in m/sec	1,7	2,0	2,1	2,3	2,0	1,8	1,3	1,6	1,4	1,5	1,6	1,6	1,7	30	15
16 Mean predominent direction of the wind		N	N	ESE	E	E	ENE	E	E	NE	NE	N	N		30	16

138 Station/Country Guadalajara/Mexico

Location 20°41'N 103°20'W   Height above sealevel 1589 m   Climate symbol: Köppen Cwa   Troll V,3

		J	F	M	A	M	J	J	A	S	O	N	D	year	P	
1 Mean daily temperature	in °C	14,7	16,6	18,4	21,1	22,9	22,3	20,5	20,5	19,9	19,1	16,9	15,2	19,0	30	1
2 Mean daily maximum temperature	in °C															2
3 Mean daily minimum temperature	in °C															3
4 Absolute maximum temperature	in °C	29,2	30,0	35,0	35,0	35,8	38,0	31,1	31,0	31,0	30,8	31,0	28,8	38,0	30	4
5 Absolute minimum temperature	in °C	-5,5	-2,0	-1,0	1,6	5,4	9,9	10,3	11,0	6,4	2,4	-2,8	-3,6	-5,5	30	5
6 Mean relative humidity	in %															6
7 Mean precipitation	in mm	15	4	4	5	24	172	251	194	158	48	10	8	893	30	7
8 Maximum precipitation	in mm															8
9 Minimum precipitation	in mm															9
10 Maximum precipitation in 24 h	in mm	33	26	25	29	47	70	97	66	56	56	23	47	97	30	10
11 Mean number of days with precipitation	>0,25 mm	2	1	1	1	4	16	23	20	17	7	2	2	96	30	11
12 Mean duration of sunshine	in h	199	216	261	251	262	202	192	207	179	212	217	195	2593	30	12
13 Mean quantity of radiation	in ly/day															13
14 Mean potential evaporation	in mm	37	46	68	91	115	107	93	90	80	69	50	42	887		14
15 Mean windspeed	in m/sec	1,5	1,8	2,3	2,3	2,3	1,9	1,6	1,3	1,4	1,4	1,3	1,4	1,7	30	15
16 Mean predominent direction of the wind																16

139 Station/Country Guanajuato/Mexico

Location 21°01'N/101°15'W   Height above sealevel 2037 m   Climate symbol: Köppen Cwb   Troll V,3

		J	F	M	A	M	J	J	A	S	O	N	D	year	P	
1 Mean daily temperature	in °C	14,2	15,8	18,2	20,2	21,6	20,3	19,0	19,1	18,4	17,7	16,1	14,7	17,9	30	1
2 Mean daily maximum temperature	in °C															2
3 Mean daily minimum temperature	in °C															3
4 Absolute maximum temperature	in °C	27,4	29,0	31,8	31,6	33,6	33,3	29,8	29,6	29,5	30,0	28,0	26,9	33,6	30	4
5 Absolute minimum temperature	in °C	-1,3	-0,8	1,6	5,5	9,1	10,6	10,5	10,1	8,8	5,5	2,7	0,3	-1,3	30	5
6 Mean relative humidity	in %															6
7 Mean precipitation	in mm	14	5	5	16	33	127	138	136	124	43	15	11	667	30	7
8 Maximum precipitation	in mm															8
9 Minimum precipitation	in mm															9
10 Maximum precipitation in 24 h	in mm	28	24	25	28	26	66	72	107	70	47	54	35	107	30	10
11 Mean number of days with precipitation	>0,25 mm	3	2	2	3	7	13	14	14	12	6	3	2	81	30	11
12 Mean duration of sunshine	in h	232	245	284	265	251	216	211	235	206	246	248	233	2872	30	12
13 Mean quantity of radiation	in ly/day															13
14 Mean potential evaporation	in mm	41	46	68	86	102	91	82	81	70	64	49	43	823	22	14
15 Mean windspeed	in m/sec	1,4	1,8	1,9	1,9	2,2	2,5	2,4	2,3	2,8	2,5	1,9	1,4	2,1	30	15
16 Mean predominent direction of the wind		var	var	SW	var	NE	ENE	ENE	ENE	ENE	ENE	NE	var		30	16

140 Station/Country Ciudad de México (Mexico City)/Mexico

Location 19°24'N/99°12'W   Height above sealevel 2485 m   Climate symbol: Köppen Cwb   Troll V,3

		J	F	M	A	M	J	J	A	S	O	N	D	year	P	
1 Mean daily temperature	in °C	12,2	13,3	16,1	17,8	18,9	18,6	17,2	17,5	17,5	15,6	13,9	12,5	15,9	7	1
2 Mean daily maximum temperature	in °C	18,9	20,6	23,9	25,0	25,6	24,4	22,8	22,8	23,3	21,1	20,0	18,9	22,2	7	2
3 Mean daily minimum temperature	in °C	5,6	6,1	8,3	10,6	12,2	12,8	11,7	12,2	11,7	10,0	7,8	6,1	9,4	7	3
4 Absolute maximum temperature	in °C	23,3	27,2	28,9	32,2	31,7	30,6	28,3	27,2	25,6	25,6	25,0	22,8	32,2	7	4
5 Absolute minimum temperature	in °C	-2,8	-1,7	1,1	0,6	6,1	9,4	8,3	9,4	1,1	1,7	2,2	0,0	-2,8	7	5
6 Mean relative humidity	in %	57	50	47	48	49	65	67	68	70	65	62	59	59	6	6
7 Mean precipitation	in mm	12	5	10	20	53	119	170	152	130	51	18	8	748	17	7
8 Maximum precipitation	in mm															8
9 Minimum precipitation	in mm															9
10 Maximum precipitation in 24 h	in mm	33	13	18	36	38	53	48	58	51	64	28	28	64	12	10
11 Mean number of days with precipitation	>0,25 mm	4	5	9	14	17	21	27	27	23	13	6	4	170	12	11
12 Mean duration of sunshine	in h	208	218	220	201	217	171	189	211	165	189	192	186	2367	7	12
13 Mean quantity of radiation	in ly/day															13
14 Mean potential evaporation	in mm	41	46	65	77	87	79	75	73	65	57	46	41	752	43	14
15 Mean windspeed	in m/sec															15
16 Mean predominent direction of the wind																16

141 Station/Country **Mérida/Mexico**

Location 20°58'N/89°38'W — Height above sealevel 22 m — Climate symbol: Köppen Aw — Troll V.4

		J	F	M	A	M	J	J	A	S	O	N	D	year	P	
1 Mean daily temperature	in °C	23.0	23.8	25.8	27.1	27.8	27.7	27.3	27.4	27.1	25.9	24.2	23.0	25.8	30	1
2 Mean daily maximum temperature	in °C	28.3	29.4	31.7	33.3	34.4	33.3	33.3	32.8	32.2	30.6	29.4	27.8	31.1	22	2
3 Mean daily minimum temperature	in °C	16.7	17.2	18.9	20.6	22.2	22.8	22.8	22.8	22.8	21.7	19.4	17.8	20.6	22	3
4 Absolute maximum temperature	in °C	33.0	35.0	37.2	38.8	40.2	37.8	35.4	35.0	35.0	34.0	33.8	35.2	40.2	30	4
5 Absolute minimum temperature	in °C	11.2	9.2	11.0	14.2	17.0	18.0	18.0	17.2	18.2	15.1	13.2	10.2	9.2	30	5
6 Mean relative humidity	in %	70	66	65	61	63	73	72	74	76	75	71	70	70	10	6
7 Mean precipitation	in mm	31	24	17	21	83	134	130	148	180	91	35	34	928	30	7
8 Maximum precipitation	in mm															8
9 Minimum precipitation	in mm															9
10 Maximum precipitation in 24 h	in mm	68	102	41	49	139	118	66	189	77	70	66	68	189	30	10
11 Mean number of days with precipitation	>0.25 mm	4	4	3	3	7	12	15	14	16	10	5	5	98	30	11
12 Mean duration of sunshine	in h	162	153	189	185	216	199	205	202	174	172	157	159	2173	30	12
13 Mean quantity of radiation	in ly/day															13
14 Mean potential evaporation	in mm	73	79	118	144	168	160	163	158	143	126	94	76	1502	35	14
15 Mean windspeed	in m/sec	2.3	2.5	3.0	3.0	2.7	2.2	1.5	1.4	1.3	1.5	1.8	2.1		30	15
16 Mean predominent direction of the wind		ESE	ESE	ESE	ESE	ESE	ESE	E	ESE	E	NE	NNE	NNE		30	16

142 Station/Country **Isla de Cozumel/Mexico**

Location 20°31'N/86°57'W — Height above sealevel 3 m — Climate symbol: Köppen Aw — Troll V.3

		J	F	M	A	M	J	J	A	S	O	N	D	year	P	
1 Mean daily temperature	in °C	22.9	23.3	24.5	26.0	26.9	27.2	27.2	27.2	26.8	26.0	24.6	23.4	25.5	30	1
2 Mean daily maximum temperature	in °C															2
3 Mean daily minimum temperature	in °C															3
4 Absolute maximum temperature	in °C	31.8	32.1	35.0	35.0	35.0	35.2	35.6	35.6	35.2	35.8	33.4	32.4	35.8	30	4
5 Absolute minimum temperature	in °C	6.5	6.2	7.8	9.0	11.4	18.2	17.4	15.6	16.4	14.0	10.4	9.0	6.2	30	5
6 Mean relative humidity	in %															6
7 Mean precipitation	in mm	87	63	48	53	141	200	112	152	243	234	111	109	1553	30	7
8 Maximum precipitation	in mm															8
9 Minimum precipitation	in mm															9
10 Maximum precipitation in 24 h	in mm	180	124	92	137	156	182	89	276	176	175	112	91	276	30	10
11 Mean number of days with precipitation	>0.25 mm	8	6	4	4	8	13	12	12	16	14	10	10	117	30	11
12 Mean duration of sunshine	in h															12
13 Mean quantity of radiation	in ly/day															13
14 Mean potential evaporation	in mm	82	81	102	129	156	157	162	157	141	123	91	87	1467	30	14
15 Mean windspeed	in m/sec															15
16 Mean predominent direction of the wind		E	SE	SE	SE	SE	SE	SE	var	SE	NE	NNE	NE		30	16

143 Station/Country **Acapulco/Mexico**

Location 16°50'N/99°56'W — Height above sealevel 3 m — Climate symbol: Köppen Aw — Troll V.3

		J	F	M	A	M	J	J	A	S	O	N	D	year	P	
1 Mean daily temperature	in °C	26.7	26.5	26.7	27.5	28.5	28.6	28.7	28.8	28.1	28.1	27.7	26.7	27.7	30	1
2 Mean daily maximum temperature	in °C															2
3 Mean daily minimum temperature	in °C															3
4 Absolute maximum temperature	in °C	36.0	35.8	37.8	37.0	40.5	37.5	37.8	37.0	36.8	37.0	37.0	32.4	40.5	30	4
5 Absolute minimum temperature	in °C	10.9	18.0	18.1	18.2	20.0	21.0	21.4	21.0	20.4	20.5	19.4	18.2	10.9	30	5
6 Mean relative humidity	in %															6
7 Mean precipitation	in mm	6	1	<1	1	36	281	256	252	349	159	28	8	1377	30	7
8 Maximum precipitation	in mm															8
9 Minimum precipitation	in mm															9
10 Maximum precipitation in 24 h	in mm	44	23	8	11	148	228	208	193	235	246	224	34	246	30	10
11 Mean number of days with precipitation	>0.25 mm	1	0	0	0	3	13	14	13	16	9	2	1	72	30	11
12 Mean duration of sunshine	in h	255	245	260	212	206	194	213	221	178	216	252	256	2708	30	12
13 Mean quantity of radiation	in ly/day															13
14 Mean potential evaporation	in mm	114	109	132	147	168	165	172	166	148	146	133	121	1721	22	14
15 Mean windspeed	in m/sec	2.2	2.9	3.3	3.2	3.1	3.1	3.0	3.4	3.9	2.8	2.3	1.8	2.9	30	15
16 Mean predominent direction of the wind																16

144 Station/Country **Salina Cruz/Mexico**

Location 18°12'N/95°12'W — Height above sealevel 56 m — Climate symbol: Köppen Aw — Troll V.3

		J	F	M	A	M	J	J	A	S	O	N	D	year	P	
1 Mean daily temperature	in °C	25.8	25.9	27.0	28.4	29.5	28.3	28.4	28.5	27.6	27.4	26.7	26.0	27.4	30	1
2 Mean daily maximum temperature	in °C	29.4	29.4	30.0	31.1	32.8	31.1	31.7	31.7	30.6	30.6	29.4	30.6	30.6	10	2
3 Mean daily minimum temperature	in °C	22.2	22.2	23.3	24.4	25.6	25.0	24.4	25.0	23.9	23.9	23.3	22.2	23.9	10	3
4 Absolute maximum temperature	in °C	32.8	33.3	35.0	36.1	36.7	36.1	35.0	36.1	35.0	33.9	34.4	33.9	36.7	10	4
5 Absolute minimum temperature	in °C	16.7	17.2	17.2	18.3	21.1	18.9	20.0	20.0	20.0	18.9	16.7	17.2	16.7	10	5
6 Mean relative humidity	in %	81	83	83	81	80	81	81	79	80	77	78	80	80	10	6
7 Mean precipitation	in mm	4	4	2	<1	48	264	207	176	240	88	9	4	1046	30	7
8 Maximum precipitation	in mm															8
9 Minimum precipitation	in mm															9
10 Maximum precipitation in 24 h	in mm	23	35	13	7	134	226	242	169	287	265	66	19	287	30	10
11 Mean number of days with precipitation	>0.25 mm	<1	<1	<1	<1	3	10	8	8	9	4	<1	<1	47	30	11
12 Mean duration of sunshine	in h	259	255	281	235	222	164	200	215	167	230	256	253	2737	30	12
13 Mean quantity of radiation	in ly/day															13
14 Mean potential evaporation	in mm	110	109	137	163	203	167	190	186	148	146	125	118	1802	22	14
15 Mean windspeed	in m/sec	6.1	5.5	4.8	4.1	3.5	2.3	2.8	2.8	2.2	4.5	6.0	6.0	4.2	30	15
16 Mean predominent direction of the wind		NNE	NNE	NNE	NNE	NNE	NNE	NNE	NNE	NNE	NNE	NNE	NNE		30	16

MIDDLE AND SOUTH AMERICA

NEW PROVIDENCE

1 Nassau (Bahamas)

CUBA

2 La Habana (Havanna)
3 Cienfuegos

DOMINICAN REPUBLIC

4 Santo Domingo

U.S.

San Juan (Puerto Rico)

JAMAICA

6 Kingston

BELIZE

7 Belize City

HONDURAS

8   Swan Islands
11 Tela
13 Tegucigalpa

FRANCE

9   Pointe-à-Pitre (Guadeloupe)
12 Fort-de-France (Martinique)

GUATEMALA

10 Guatemala City

EL SALVADOR

14 San Salvador

BARBADOS

15 Bridgetown

NETHERLANDS ANTILLES

16 Curaçao

NICARAGUA

17 Managua

TRINIDAD

18 Piarco

COSTA RICA

19 San José

PANAMA

20 Cristobal
21 Balboa
22 Jaque

COLUMBIA

23 Barranquilla
24 Turbo
25 Barrancabermeja
26 Arauca
27 Bogotá
28 Andagoya
29 Villaviciencio
30 Popayán
31 Tumaco

VENEZUELA

32 Maracaibo
33 Caracas
34 Colonia Tovar
35 Barcelona
36 Mérida
37 San Fernando de Apure
38 Puerto Ayacucho

39 Santa Elena de Uairén
40 San Carlos de Rio Negro

GUYANA

41 Georgetown
42 Saint Ignatius

SURINAM

43 Paramaribo
44 Tafelberg

FRENCH GUIANA

45 Maripasoula
46 Cayenne

ECUADOR
(Galapagos Islands)

47 Puerto Baquerizo Moreno

ECUADOR

48 Quito
49 Pichilingue
50 Puyo
51 Guayaquil

PERU

52 Iquitos
53 Chiclayo
54 Cajamarca
55 Tingo Maria
56 Lima
57 Huancayo
58 Puerto Maldonado
59 Cuzco
61 San Juan
62 Arequipa

BOLIVIA

60 Apolo
63 La Paz

64 Santa Cruz
65 Oruro
110 Yacuiba

BRAZIL

66 Uaupés
67 Manaus
68 Santarém
69 Belém
70 São Luis
71 Barra do Corda
72 Quixeramobim
73 Alto Tapajós
74 Sena Madureira
75 Conceicao do Araguaia
76 Recife (Olinda)
77 Remanso
78 Monte Santo
79 Pôrto Nacional
80 Ibipetuba
81 Salvador
82 Caetité
83 Brasilia
84 Formosa
85 Cuiaba
86 Corumbá
87 Catalão
88 Belo Horizonte
89 Vitória
90 Três Lagoas
91 Bela Vista
92 Campinas

93 Rio de Janeiro
94 Curitiba
95 Guarapuava
96 Alegrete
97 Porto Alegre

CHILE

98 Arica
99 Iquique
100 Antofagasta
101 La Serena
102 Valparaiso
103 Santiago
104 Concépcion
105 Valdivia
106 Puerto Montt
107 San Pedro
108 Los Evangelistas
109 Punta Arenas

PARAGUAY

111 Marischal Estigarriba
112 Puerto Casado
114 Asunción
115 Puerto Presidente Franco

URUGUAY

116 Paso De Los Toros
117 Montevideo

ARGENTINA

113 Las Lomitas
118 La Quiaca
119 Salta
120 San Miguel De Tucumán
121 Corrientes
122 Posadas
123 Santiago del Estero
124 Catamarca
125 San Juan
126 Córdoba
127 Concordia
128 Santa Fe
129 Cristo Redentor
130 Mendoza
131 San Luis
132 Rosario
133 Buenos Aires
134 Macachín
135 Chos Malal
136 Mar del Plata
137 Cipoletti
138 San Carlos de Bariloche
139 Trewlew
140 Sarmiento
141 Comodoro Rivadavia
142 Santa Cruz
143 Ushaia

UNITED KINGDOM

144 Stanley (Falkland Islands)

SOUTH AMERICA

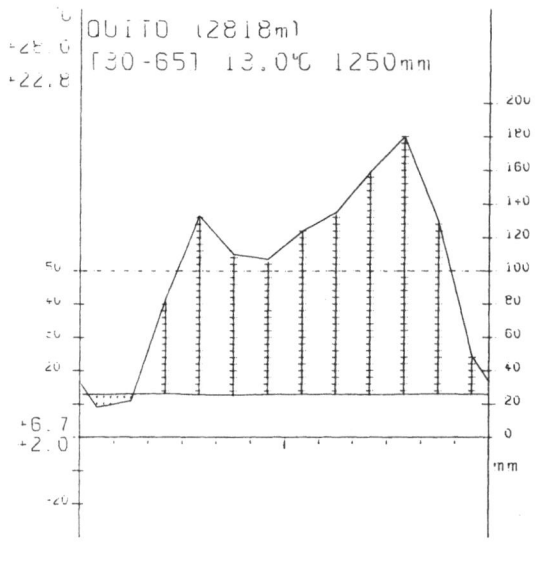

QUITO (2818m)
[30-65] 13.0°C 1250mm

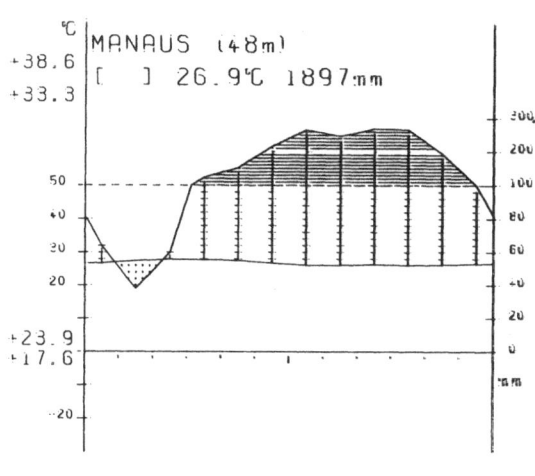

MANAUS (48m)
[ ] 26.9°C 1897mm

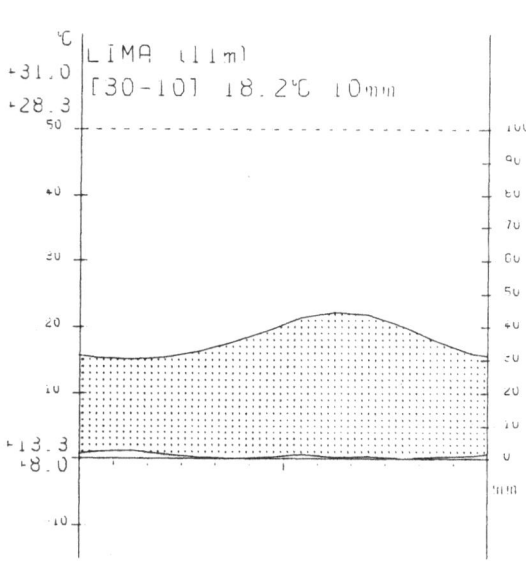

LIMA (11m)
[30-10] 18.2°C 10mm

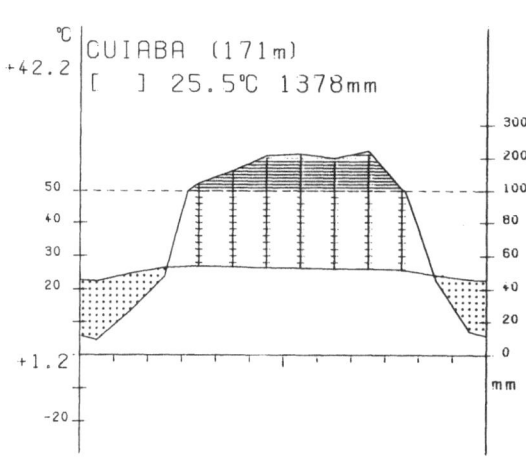

CUIABA (171m)
[ ] 25.5°C 1378mm

SOUTH AMERICA

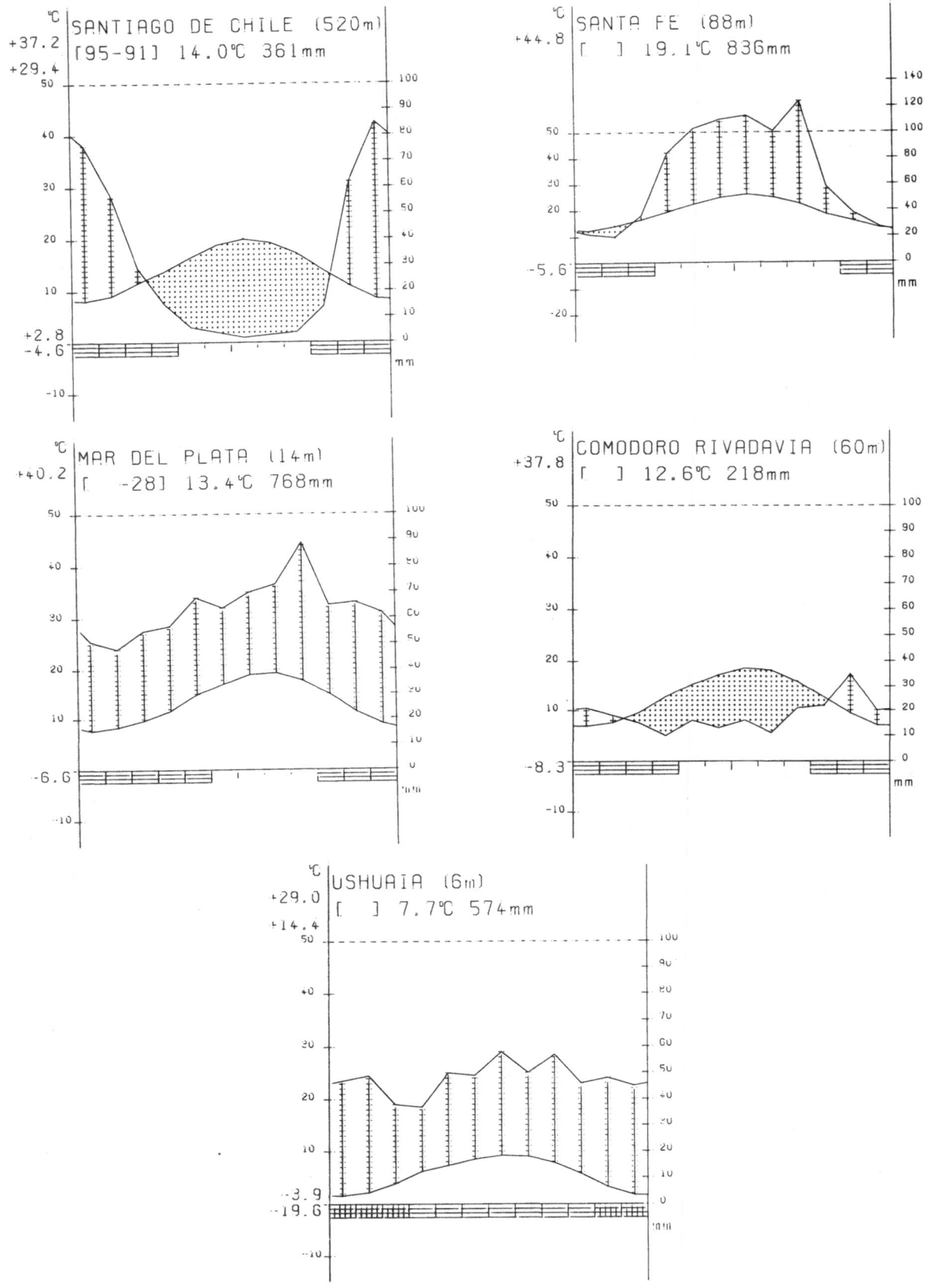

1 Station/Country Nassau (Bahamas)/New Providence

Location 25°03'N/77°28'W   Height above sealevel 3 m   Climate symbol: Köppen Aw   Troll V.2

		J	F	M	A	M	J	J	A	S	O	N	D	year	P	
1 Mean daily temperature	in °C	20.3	20.9	22.2	23.6	25.2	26.7	27.4	27.7	27.1	25.5	23.5	21.6	24.3	10	1
2 Mean daily maximum temperature	in °C	25.0	25.0	26.1	27.2	28.9	30.6	31.1	31.7	31.1	29.4	27.2	26.1	28.3	35	2
3 Mean daily minimum temperature	in °C	18.3	17.8	18.9	20.6	21.7	23.3	23.9	24.4	23.9	22.8	21.1	19.4	21.1	35	3
4 Absolute maximum temperature	in °C	27.9	28.7	29.2	30.2	31.2	32.4	32.8	33.3	32.5	31.7	30.0	28.4	33.3	10	4
5 Absolute minimum temperature	in °C	10.7	11.6	12.1	15.2	17.2	20.2	21.4	21.7	21.4	17.9	14.9	12.7	10.7	10	5
6 Mean relative humidity	in %	71	80	74	77	78	79	82	79	82	80	80	79	78	10	6
7 Mean precipitation	in mm	36	43	45	78	117	159	150	135	165	164	85	39	1216	30	7
8 Maximum precipitation	in mm															8
9 Minimum precipitation	in mm															9
10 Maximum precipitation in 24 h	in mm	19	48	30	131	78	124	61	58	58	64	40	106	131	10	10
11 Mean number of days with precipitation	>1.0 mm	6	6	5	6	11	14	16	16	17	15	9	9	130	10	11
12 Mean duration of sunshine	in h	234	229	263	259	264	234	271	256	215	210	226	218	2879	10	12
13 Mean quantity of radiation	in ly/day															13
14 Mean potential evaporation	in mm	69	66	81	99	135	162	174	168	148	134	92	70	1397	28	14
15 Mean windspeed	in m/sec															15
16 Mean predominant direction of the wind		NE	NE	NE	NE	SE	SE	SE	SE	NE	NE	NE	NE		10	16

2 Station/Country La Habana (Havana)/Cuba

Location 23°08'N/82°21'W   Height above sealevel 24 m   Climate symbol: Köppen Aw   Troll V.2

		J	F	M	A	M	J	J	A	S	O	N	D	year	P	
1 Mean daily temperature	in °C	22.2	22.2	23.3	24.8	26.1	27.2	27.8	27.8	27.5	26.1	23.9	22.8	25.1	25	1
2 Mean daily maximum temperature	in °C	26.1	26.1	27.2	28.9	30.0	31.1	31.7	31.7	31.1	29.4	27.2	26.1	28.9	25	2
3 Mean daily minimum temperature	in °C	18.3	18.3	19.4	20.6	22.2	23.3	23.9	23.9	22.8	20.6	19.4		21.1	25	3
4 Absolute maximum temperature	in °C	31.7	32.8	32.8	34.4	34.4	35.6	33.9	35.0	34.4	34.4	32.8	31.7	35.6	25	4
5 Absolute minimum temperature	in °C	10.0	10.0	11.7	12.8	15.0	18.9	18.9	20.0	19.4	17.2	12.8	10.6	10.0	25	5
6 Mean relative humidity	in %	75	73	71	71	74	76	75	76	78	78	75	74	75	43	6
7 Mean precipitation	in mm	71	46	46	58	119	165	124	135	150	173	79	58	1224	72	7
8 Maximum precipitation	in mm															8
9 Minimum precipitation	in mm															9
10 Maximum precipitation in 24 h	in mm	163	79	84	211	160	122	104	112	97	168	193	89	211	70	10
11 Mean number of days with precipitation	>0.1 mm	6	4	4	4	7	10	9	10	11	11	7	6	89	28	11
12 Mean duration of sunshine	in h															12
13 Mean quantity of radiation	in ly/day															13
14 Mean potential evaporation	in mm	73	69	90	111	144	159	177	171	147	125	88	77	1431	25	14
15 Mean windspeed	in m/sec															15
16 Mean predominant direction of the wind																16

3 Station/Country Cienfuegos/Cuba

Location 22°09'N/80°27'W   Height above sealevel 30 m   Climate symbol: Köppen Aw   Troll V.2

		J	F	M	A	M	J	J	A	S	O	N	D	year	P	
1 Mean daily temperature	in °C	22.2	22.3	23.4	24.4	25.6	26.7	27.2	27.2	26.7	26.7	23.9	22.8	24.9	12	1
2 Mean daily maximum temperature	in °C	27.2	27.8	28.9	29.4	30.6	31.7	32.2	32.2	31.7	31.1	28.3	27.8	30.0	12	2
3 Mean daily minimum temperature	in °C	17.2	16.7	17.8	19.4	20.6	21.7	22.2	22.2	21.7	21.7	19.4	17.8	20.0	12	3
4 Absolute maximum temperature	in °C	31.1	32.8	33.3	32.8	34.4	35.0	35.0	34.4	34.4	33.9	32.8	31.7	35.0	12	4
5 Absolute minimum temperature	in °C	7.8	7.2	7.2	10.0	13.3	18.9	20.0	19.4	20.0	15.6	11.1	8.9	7.2	12	5
6 Mean relative humidity	in %	73	72	71	72	74	77	76	76	78	79	76	75	75	12	6
7 Mean precipitation	in mm	18	25	33	46	119	152	122	160	173	160	41	23	1072	23	7
8 Maximum precipitation	in mm															8
9 Minimum precipitation	in mm															9
10 Maximum precipitation in 24 h	in mm	46	56	66	46	99	254	76	140	124	254	109	43	254	30	10
11 Mean number of days with precipitation	>0.1 mm	2	3	3	4	9	12	11	13	13	10	4	2	86	28	11
12 Mean duration of sunshine	in h															12
13 Mean quantity of radiation	in ly/day															13
14 Mean potential evaporation	in mm	73	69	96	108	133	155	162	157	141	137	88	78	1397	12	14
15 Mean windspeed	in m/sec															15
16 Mean predominant direction of the wind																16

4 Station/Country Santo Domingo/Dominican Republic

Location 18°29'N/69°54 W   Height above sealevel 19 m   Climate symbol: Köppen Aw   Troll V.1

		J	F	M	A	M	J	J	A	S	O	N	D	year	P	
1 Mean daily temperature	in °C	23.9	24.0	24.5	25.3	25.9	26.5	26.7	27.0	26.8	26.4	26.9	24.6	25.6	10	1
2 Mean daily maximum temperature	in °C	29.0	29.5	29.0	29.5	30.0	30.5	31.0	31.0	31.0	30.5	30.0	29.5	30.0		2
3 Mean daily minimum temperature	in °C	19.0	19.0	19.5	20.5	21.5	22.0	22.0	23.0	23.0	22.0	21.0	19.5	21.0		3
4 Absolute maximum temperature	in °C	32.7	31.1	32.1	32.1	31.6	32.1	32.1	32.1	32.7	33.2	33.2	32.7	33.2	10	4
5 Absolute minimum temperature	in °C	15.0	15.5	15.5	16.3	18.3	19.4	20.0	17.8	20.0	18.9	16.1	16.3	15.0	10	5
6 Mean relative humidity	in %	78	73	75	76	77	78	78	78	79	79	79	79	78	10	6
7 Mean precipitation	in mm	47	45	45	65	190	175	158	147	168	165	113	67	1386	10	7
8 Maximum precipitation	in mm	193	145	199	281	745	511	428	358	316	343	288	134			8
9 Minimum precipitation	in mm	9	5	0	4	16	31	73	45	71	53	0	6			9
10 Maximum precipitation in 24 h	in mm	58	36	71	226	125	137	145	162	508	117	155	96			10
11 Mean number of days with precipitation	>1.0 mm	7	6	5	7	11	12	11	11	11	11	10	8	110	10	11
12 Mean duration of sunshine	in h	178	178	229	203	185	185	188	213	198	200	185	176	2318		12
13 Mean quantity of radiation	in ly/day															13
14 Mean potential evaporation	in mm	97	90	108	121	146	148	156	155	144	138	124	109	1534	28	14
15 Mean windspeed	in m/sec	3.9	3.7	4.0	3.9	3.7	3.8	3.9	3.6	3.5	3.7	4.2	4.3	3.9	10	15
16 Mean predominant direction of the wind		N	N	N	N	N	N	N	N	N	N	N	N		10	16

## 5

Station/Country San Juan (Puerto Rico)/U.S.

Location 18°28'N/66°06'W   Height above sealevel **15 m**   Climate symbol: Köppen **Am**   Troll **V,1**

		J	F	M	A	M	J	J	A	S	O	N	D	year	P	
1	Mean daily temperature	in °C	23,9	23,9	24,4	25,0	26,0	26,6	26,7	27,1	27,1	26,8	25,9	24,8	25,7	30	1
2	Mean daily maximum temperature	in °C	26,7	26,7	27,2	27,8	28,9	29,4	29,4	29,4	30,0	29,4	28,9	27,2	28,3	48	2
3	Mean daily minimum temperature	in °C	21,1	21,1	21,1	22,2	23,3	23,9	23,9	24,4	23,9	23,9	22,8	22,2	22,8	48	3
4	Absolute maximum temperature	in °C	32,2	32,8	35,6	33,9	34,4	33,9	33,3	34,4	34,4	35,6	33,9	32,2	35,6	30	4
5	Absolute minimum temperature	in °C	17,2	16,7	17,2	18,3	18,9	18,9	21,1	20,0	20,1	20,0	18,3	16,7	16,7	30	5
6	Mean relative humidity	in %	78	77	75	75	76	77	78	77	78	78	79	77	77		6
7	Mean precipitation	in mm	105	69	53	99	182	148	153	161	153	133	154	124	1534	30	7
8	Maximum precipitation	in mm	393	207	199	260	429	299	277	358	294	242	397	259			8
9	Minimum precipitation	in mm	19	1	7	10	38	38	38	101	38	67	47	47			9
10	Maximum precipitation in 24 h	in mm	115	123	134	171	136	164	103	251	216	98	205	288	288	30	10
11	Mean number of days with precipitation	> 1,0 mm	16	10	10	10	12	13	15	16	14	15	16	16	163	30	11
12	Mean duration of sunshine	in h	217	222	255	241	239	240	259	266	227	233	210	218	2827	30	12
13	Mean quantity of radiation	in ly/day															13
14	Mean potential evaporation	in mm	90	85	97	118	141	151	155	151	140	137	113	96	1472	45	14
15	Mean windspeed	in m/sec	3,9	4,0	4,1	3,8	3,8	3,8	4,2	3,9	3,3	3,0	3,2	3,5	3,7	30	15
16	Mean predominant direction of the wind		E	E	E	E	E	E	E	E	E	SE	SE	ESE		30	16

## 6

Station/Country Kingston/Jamaica

Location 17°58'N/76°48'W   Height above sealevel **35 m**   Climate symbol: Köppen **Aw**   Troll **V,2**

		J	F	M	A	M	J	J	A	S	O	N	D	year	P	
1	Mean daily temperature	in °C	25,4	25,4	25,8	26,7	27,8	28,0	28,3	28,5	28,2	27,6	27,1	26,1	27,1	10	1
2	Mean daily maximum temperature	in °C	30,0	30,0	30,0	30,6	30,8	31,7	32,2	32,2	31,7	31,1	30,8	30,6	31,1	33	2
3	Mean daily minimum temperature	in °C	19,4	19,4	20,0	21,1	22,2	22,8	22,8	22,8	22,8	21,7	20,6	21,7	21,7	33	3
4	Absolute maximum temperature	in °C	33,9	33,2	33,8	33,9	34,4	35,0	35,5	36,1	35,5	35,5	35,5	35,5	36,1	10	4
5	Absolute minimum temperature	in °C	13,9	15,0	14,4	17,2	18,9	20,0	18,9	20,0	20,0	18,3	18,3	13,9	13,9	10	5
6	Mean relative humidity	in %	73	73	72	73	73	73	71	76	78	81	78	74	75		6
7	Mean precipitation	in mm	20	18	10	37	138	114	51	92	86	168	52	25	811	10	7
8	Maximum precipitation	in mm															8
9	Minimum precipitation	in mm															9
10	Maximum precipitation in 24 h	in mm	46	46	89	46	235	238	111	124	162	280	226	122	280	10	10
11	Mean number of days with precipitation	> 1,0 mm	3	3	3	4	6	4	3	7	7	9	6	2	57	10	11
12	Mean duration of sunshine	in h	257	249	270	281	257	234	264	263	228	226	255	239	3003		12
13	Mean quantity of radiation	in ly/day															13
14	Mean potential evaporation	in mm	94	88	111	123	153	155	159	153	144	127	110	103	1520	53	14
15	Mean windspeed	in m/sec	3,0	3,3	3,5	3,6	3,8	4,4	4,2	3,7	3,0	2,9	2,7	2,7	3,4	10	15
16	Mean predominant direction of the wind		N,SE	N,SE	SE	SE	SE	SE	SE	SE	SE	SE	N	N		10	16

## 7

Station/Country Belize/Belize

Location 17°31'N/88°11'W   Height above sealevel **5 m**   Climate symbol: Köppen **Am**   Troll **V,1**

		J	F	M	A	M	J	J	A	S	O	N	D	year	P	
1	Mean daily temperature	in °C	23,3	23,8	25,0	26,6	27,2	27,2	27,2	27,2	27,2	25,6	23,8	23,3	25,8	10	1
2	Mean daily maximum temperature	in °C	27,2	27,8	28,9	30,0	30,6	30,6	30,6	31,1	30,6	30,0	28,3	27,2	29,4	27	2
3	Mean daily minimum temperature	in °C	19,4	20,6	21,7	23,3	23,9	23,9	23,9	23,9	23,3	22,2	20,0	20,0	22,2	32	3
4	Absolute maximum temperature	in °C	32,1	33,8	34,9	36,0	35,5	36,0	34,9	35,5	36,0	35,5	34,9	33,2	36,0	10	4
5	Absolute minimum temperature	in °C	9,4	9,4	12,2	15,0	15,5	17,7	16,7	15,5	15,5	14,4	11,1	9,4	9,4	10	5
6	Mean relative humidity	in %	91	89	89	89	89	90	90	90	91	91	83	92	90		6
7	Mean precipitation	in mm	137	61	38	56	109	196	163	170	244	305	226	186	1893	10	7
8	Maximum precipitation	in mm															8
9	Minimum precipitation	in mm															9
10	Maximum precipitation in 24 h	in mm	107	74	66	79	150	158	132	157	299	130	249	122	249	10	10
11	Mean number of days with precipitation	> 0,1 mm	12	8	4	5	7	13	15	14	15	16	12	14	133	10	11
12	Mean duration of sunshine	in h															12
13	Mean quantity of radiation	in ly/day															13
14	Mean potential evaporation	in mm	81	86	114	138	157	154	158	156	142	118	89	85	1478	35	14
15	Mean windspeed	in m/sec															15
16	Mean predominant direction of the wind		NW	ESE	ESE	ESE	E	E	E	E	ESE	NW	NW	NW		10	16

Station/Country Swan Islands/Honduras

Location 17°24'N/83°56'W   Height above sealevel **9 m**   Climate symbol: Köppen **Aw**   Troll **V,2**

		J	F	M	A	M	J	J	A	S	O	N	D	year	P	
1	Mean daily temperature	in °C	25,7	25,9	26,6	27,4	27,9	28,0	27,9	28,3	28,3	27,7	26,6	25,8	27,2		1
2	Mean daily maximum temperature	in °C	28,3	28,3	29,4	30,0	31,1	30,0	30,6	31,1	31,1	30,6	29,4	28,9	30,0	8	2
3	Mean daily minimum temperature	in °C	22,8	22,8	23,3	23,9	25,0	24,4	25,0	25,0	25,0	24,4	23,9	23,3	23,9	8	3
4	Absolute maximum temperature	in °C															4
5	Absolute minimum temperature	in °C															5
6	Mean relative humidity	in %	75	75	74	75	77	82	80	79	80	79	77	75	77		6
7	Mean precipitation	in mm	87	29	16	21	84	164	107	93	132	246	188	143	1311		7
8	Maximum precipitation	in mm															8
9	Minimum precipitation	in mm															9
10	Maximum precipitation in 24 h	in mm	75	79	51	30	114	128	210	192	110	272	229	104	272	10	10
11	Mean number of days with precipitation	> 0,1 mm	15	8	6	5	9	14	13	13	14	18	16	16	147		11
12	Mean duration of sunshine	in h															12
13	Mean quantity of radiation	in ly/day	442	498	615	646	625	544	588	591	535	457	394	382	526		13
14	Mean potential evaporation	in mm															14
15	Mean windspeed	in m/sec	4,7	4,7	4,8	4,7	4,5	4,7	4,2	3,8	3,6	3,9	4,4	4,4	4,4		15
16	Mean predominant direction of the wind		E	E	SE	SE	E	E	E	E	SE	NE	NE	E			16

9 Station / Country Pointe-à-Pitre (Guadeloupe) / France

Location 16°15'N/61°31'W Height above sealevel 7 m Climate symbol: Köppen Af Troll V.1

	in	J	F	M	A	M	J	J	A	S	O	N	D	year	P	
1 Mean daily temperature	°C	23,4	23,5	24,0	26,0	26,0	26,8	26,7	26,8	26,7	26,1	26,3	24,3	25,5		1
2 Mean daily maximum temperature	°C															2
3 Mean daily minimum temperature	°C															3
4 Absolute maximum temperature	°C	29,9	30,0	31,3	31,8	37,0	32,6	32,2	32,4	32,5	32,0	31,3	30,4			4
5 Absolute minimum temperature	°C	13,5	13,0	13,9	15,7	17,8	18,9	19,6	19,8	15,4	19,0	17,2	15,6			5
6 Mean relative humidity	%															6
7 Mean precipitation	mm	91	66	66	86	135	148	179	244	196	229	231	143	1814	30	7
8 Maximum precipitation	mm															8
9 Minimum precipitation	mm															9
10 Maximum precipitation in 24 h	mm	51	58	60	46	121	115	86	109	113	81	123	103	123		10
11 Mean number of days with precipitation >1,0 mm		19	16	16	16	21	21	25	23	22	22	23	25	249		11
12 Mean duration of sunshine	h	223	219	257	239	231	227	231	232	210	220	210	220	2719		12
13 Mean quantity of radiation	ly/day															13
14 Mean potential evaporation	mm	104	117	141	138	126	129	124	107	89	88	82	90	1335		14
15 Mean windspeed	m/sec	1,9	2,4	2,5	2,5	2,4	2,7	2,5	2,0	1,9	1,7	1,7	2,0	2,2		15
16 Mean predominent direction of the wind		E	E	E	E	E	E	E	E	E	E	E	E			16

10 Station / Country Guatemala-City / Guatemala

Location 15°29'N/90°16'W Height above sealevel 1300 m Climate symbol: Köppen Cwb Troll V.2

	in	J	F	M	A	M	J	J	A	S	O	N	D	year	P	
1 Mean daily temperature	°C	16,3	17,0	18,4	19,5	19,6	18,7	18,5	18,7	18,3	17,7	16,7	16,3	18,0	30	1
2 Mean daily maximum temperature	°C	22,8	25,0	27,2	27,8	28,9	27,2	25,6	26,1	24,4	23,3	22,7		25,6	6	2
3 Mean daily minimum temperature	°C	11,7	12,2	13,9	14,4	15,6	16,1	15,6	15,6	15,6	15,6	13,9	12,8	14,4	6	3
4 Absolute maximum temperature	°C	30,9	29,5	30,0	32,2	31,6	30,0	28,8	28,3	27,8	27,8	28,3	28,3	32,2	30	4
5 Absolute minimum temperature	°C	5,0	6,1	5,0	8,3	11,1	11,1	10,6	11,1	12,2	10,0	6,7	5,0	5,0	30	5
6 Mean relative humidity	%	70	68	69	70	76	83	81	79	83	82	76	73	76	30	6
7 Mean precipitation	mm	3	2	7	19	141	265	211	187	257	159	23	7	1281	30	7
8 Maximum precipitation	mm	12	21	53	90	361	466	479	383	480	367	94	56			8
9 Minimum precipitation	mm	0	0	0	0	11	137	104	74	128	22	0	0			9
10 Maximum precipitation in 24 h	mm	11	13	31	56	48	64	76	43	86	64	28	8	86		10
11 Mean number of days with precipitation >0,1 mm		4	2	3	5	15	23	21	21	22	18	7	4	145	30	11
12 Mean duration of sunshine	h															12
13 Mean quantity of radiation	ly/day															13
14 Mean potential evaporation	mm	53	55	73	75	94	83	83	80	75	68	58	51	848	46	14
15 Mean windspeed	m/sec															15
16 Mean predominent direction of the wind																16

11 Station / Country Tela / Honduras

Location 15°43'N/87°29'W Height above sealevel 3 m Climate symbol: Köppen Af Troll V.1

	in	J	F	M	A	M	J	J	A	S	O	N	D	year	P	
1 Mean daily temperature	°C	23,2	23,7	25,1	26,2	26,8	27,1	26,7	26,8	27,2	25,9	24,5	23,5	25,6	10	1
2 Mean daily maximum temperature	°C															2
3 Mean daily minimum temperature	°C															3
4 Absolute maximum temperature	°C	27,4	28,2	29,7	30,7	31,2	31,6	30,9	31,1	31,6	30,1	28,6	27,4	31,6	10	4
5 Absolute minimum temperature	°C	19,1	19,3	20,4	21,7	22,6	22,9	22,5	22,5	22,7	21,7	20,4	19,4	19,1	10	5
6 Mean relative humidity	%															6
7 Mean precipitation	mm	255	146	72	90	103	122	181	240	207	405	410	356	2587	10	7
8 Maximum precipitation	mm															8
9 Minimum precipitation	mm															9
10 Maximum precipitation in 24 h	mm	129	128	75	254	83	101	90	116	153	187	317	279	317	10	10
11 Mean number of days with precipitation >0,1 mm		15	11	6	5	6	12	17	17	15	19	15	18	156	10	11
12 Mean duration of sunshine	h															12
13 Mean quantity of radiation	ly/day															13
14 Mean potential evaporation	mm	81	84	112	142	157	154	158	154	149	142	111	88	1532	19	14
15 Mean windspeed	m/sec	3,1	3,1	3,1	3,1	3,1	3,1	3,1	3,1	2,6	2,6	2,6	2,6	2,9	10	15
16 Mean predominent direction of the wind		NE	NE	NE	NE	NNE	NNE	NE	NNE	NW	SE	NW	NW		10	16

12 Station / Country Fort-de-France (Martinique) / France

Location 14°35'N/61°12'W Height above sealevel 144 m Climate symbol: Köppen Af Troll V.1

	in	J	F	M	A	M	J	J	A	S	O	N	D	year	P	
1 Mean daily temperature	°C	23,5	23,5	24,0	24,7	25,4	25,7	25,6	26,0	25,9	25,6	25,2	24,2	24,9	26	1
2 Mean daily maximum temperature	°C	28,3	28,9	29,4	30,0	30,6	30,0	30,0	30,6	31,1	30,6	30,0	28,9	30,0	22	2
3 Mean daily minimum temperature	°C	20,6	20,6	20,6	21,7	22,8	23,3	23,3	23,3	22,8	22,2	21,8	22,2	22,2	22	3
4 Absolute maximum temperature	°C	30,2	30,0	32,6	32,8	32,5	31,5	31,3	33,0	32,9	32,0	31,8	31,0	33,0	26	4
5 Absolute minimum temperature	°C	17,8	17,3	18,6	18,9	19,9	20,0	19,5	20,3	20,6	20,2	19,8	17,4	17,3	26	5
6 Mean relative humidity	%	84	81	80	80	81	83	84	85	85	86	87	85	84	23	6
7 Mean precipitation	mm	96	68	58	82	126	160	214	227	232	221	230	126	1840	26	7
8 Maximum precipitation	mm															8
9 Minimum precipitation	mm															9
10 Maximum precipitation in 24 h	mm	45	63	42	78	123	76	71	103	238	152	133	86	238	26	10
11 Mean number of days with precipitation >0,1 mm		21	17	18	19	21	24	27	25	23	23	22	23	263	30	11
12 Mean duration of sunshine	h	237	227	257	245	242	211	224	256	219	223	220	226	2787	26	12
13 Mean quantity of radiation	ly/day															13
14 Mean potential evaporation	mm	98	92	113	130	151	148	153	150	143	139	124	108	1549	31	14
15 Mean windspeed	m/sec	6,3	6,7	6,6	6,5	6,4	7,0	6,6	5,6	4,9	5,0	5,1	6,1	6,1	26	15
16 Mean predominent direction of the wind		ENE	ENE	ENE	ENE	ENE	ENE	ENE	ENE	ENE	ENE	NE	ENE		26	16

SOUTH AMERICA
Climate Zones after
W. Köppen/R. Geiger

● 56 LIMA   Location, No. and Name
            of Climate Station
            (Capitals)

● 11 Tela   Location, No. and Name
            of Climate Station
            (other Towns)

---------   International border

0   1000 km

NASSAU   1
LA HABANA 2
Cienfuegos 3
Aw
              4
     6   KINGSTON
SANTO
DOMINGO   San Juan 5
                 Pointe-a-Pitre 9
BELIZE   8   12
      Swan Islands   Fort-de-France
GUATEMALA CITY 11 Tela   Bridgetown 15
Cw 10   13 Af   BSh
      14   Cw
SAN SALVADOR TEGUCIGALPA
SAN SALVADOR
MANAGUA Aw   16 Curaçao
              20 Cristobal   Piarco
          91 Turbo Aw CARACAS 34 35 18
SAN JOSE   Jaqué Barrancabermeja   Colonia Barcelona
Aw   24 25   Merida Tovar San Fernando
         28 27 Arauca 37 de Apure   GEORGETOWN
Af   Andagoya BOGOTA 38 Puerto Aw PARAMARIBO
         29 Villavicencio Ayacucho Santa 43
Tumaco 30 31 Popayan   39 Elena 46 CAYENNE
         E   San Carlos de Uairen Tafelberg 45
49 48 QUITO   de Rio   42 Saint Maripasoula
50 Pichilingue Negro 66 Ignatius
BWh Puerto Puyo Uaupés Af
    Baquerizo 51 Guayaquil   Af
    Moreno 52
         Aw Iquitos 67 Manaus Santarem Belém São Luis
BS   83 Cw   68   69 70
Chiclayo 54   71
        56 Tingo 74 Sena Madureira   Barra 72
        Maria   do Corda Quixeramobim
Af   57 58 Puerto   Alto Tapajós
LIMA 55 Huancayo Maldonado 73   75 Conceicao Remanso BSh
BWh 59 Cuzco   Aw   do Araguaia 77 78 Recife
    E 61 60 Apolo   79 Pôrto 80   76
San Juan   63 LA PAZ 64   Nacional Ibipetuba Monte
    62 Arequipa 65 Santa Cruz 85 Cuiaba 83 84 Santo Af 82
Arica 98 Oruro 86 Corumbá BRASILIA Formosa Caetité Salvador 81
         BW Iquique 99 Yacuiba 111 Três Lagoas 87 88 Belo Horizonte
         BS 110 Mariscal 112 91 90 Catalao 89 Vitória
         E Estigarribia Puerto Bela
Antofagasta 100 La Quiaca 118 Casado Vista Campinas 92 93
BWkl Salta 119 Lomita 114 115 Cw Rio de Janeiro
    San Miguel 120 ASUNCION Guarapuava 95 94 Curitiba Af
    de Tucuman 123 Corrientes Presidente Cf
    124 Santiago Posadas Franco 22
    101 del Estero Alegrete 96
La Serena Catamarca 127 97 Porto Alegre
BSk 125 San Juan 126 Concordia 116 Paso
Cristo Redentor 129 Cordoba Santa Fe 132 de los 117
102 Mendoza 130 San Luis Rosario Toros
Valparaiso 131 BS 133 MONTEVIDEO
SANTIAGO 103   BUENOS
Csb 135 134 AIRES
Concepcion 104 Chos Malal Macachin 136
                              Mar del Plata
Valdivia 105 Cipoletti
106 San Carlos
Puerto Montt de Bariloche BW 139
                Trewlew
Sarmiento BS
140 141 Comodoro
        Rivadavia
San Pedro 107
BS
Cf 142
108 E Santa Cruz 144 Stanley
Los Evangelistas Cf E
Punta Arenas 109 143
    Ushaia
E

SOUTH AMERICA
Climate Zones after
C. Troll/KH.Paffen

● 56 LIMA Location No and Name of Climate Station (Capitals)

● 11 Tela Location No and Name of Climate Station (other Towns)

--- International border

0   1000 km

13 Station/Country Tegucigalpa/Honduras

Location 14°04'N/87°13'W   Height above sealevel 1007 m   Climate symbol: Köppen Aw   Troll V,2

		J	F	M	A	M	J	J	A	S	O	N	D	year	P	
1 Mean daily temperature	in °C	19,3	20,5	21,9	23,3	23,6	23,1	22,5	22,8	22,9	22,1	20,7	19,6	21,9	10	1
2 Mean daily maximum temperature	in °C	25	27	29	30	30	28	28	28	29	27	26	25	27,5	10	2
3 Mean daily minimum temperature	in °C	14	14	15	16	18	19	17	17	17	17	16	15	16,5	10	3
4 Absolute maximum temperature	in °C															4
5 Absolute minimum temperature	in °C															5
6 Mean relative humidity	in %	74	68	63	63	70	79	78	76	80	81	79	77	74		6
7 Mean precipitation	in mm	12	2	1	26	180	177	70	74	151	87	38	14	832	10	7
8 Maximum precipitation	in mm															8
9 Minimum precipitation	in mm															9
10 Maximum precipitation in 24 h	in mm	21	8	7	93	109	50	59	62	172	79	54	15	172	10	10
11 Mean number of days with precipitation	> 0,1 mm	5	3	2	2	9	13	11	11	16	13	8	7	100	10	11
12 Mean duration of sunshine	in h															12
13 Mean quantity of radiation	in ly/day															13
14 Mean potential evaporation	in mm	62	68	90	103	112	107	103	104	101	88	75	61	1073	10	14
15 Mean windspeed	in m/sec	5,7	5,1	4,6	4,6	4,1	4,1	4,1	4,1	4,1	4,1	5,1	5,1	4,6	10	15
16 Mean predominent direction of the wind		N	N	N	N	N	N	NE	N	E	N	N	N		10	16

14 Station/Country San Salvador/El Salvador

Location 13°43'N/89°12'W   Height above sealevel 700 m   Climate symbol: Köppen Aw   Troll V,2

		J	F	M	A	M	J	J	A	S	O	N	D	year	P	
1 Mean daily temperature	in °C	22,1	22,4	23,5	24,2	23,7	23,1	22,9	23,0	22,5	22,4	22,0	22,0	22,8	30	1
2 Mean daily maximum temperature	in °C	32,2	33,3	34,4	33,9	32,8	30,6	31,7	31,7	30,6	30,6	30,6	31,7	32,2	39	2
3 Mean daily minimum temperature	in °C	15,6	15,6	16,7	18,3	19,4	18,9	18,3	18,9	18,9	18,3	17,2	16,1	17,8	39	3
4 Absolute maximum temperature	in °C	38,5	39,5	40,5	40	39,5	36,5	36,5	36,5	37	38,5	39	38,5	40,5		4
5 Absolute minimum temperature	in °C	7	9,5	7	12	14,5	13,5	14,5	15,5	11,5	12	9,5	8,5	·7		5
6 Mean relative humidity	in %	63	62	62	65	73	78	75	76	80	77	68	66	70		6
7 Mean precipitation	in mm	5	3	8	60	190	322	304	297	325	220	35	7	1775	30	7
8 Maximum precipitation	in mm	41	25	67	511	324	617	493	489	611	364	124	27			8
9 Minimum precipitation	in mm	0	0	0	0	69	139	117	82	211	25	0	0			9
10 Maximum precipitation in 24 h	in mm	18	24	32	88	120	205	95	165	175	170	54	46	205	30	10
11 Mean number of days with precipitation	> 0,1 mm	1	1	1	5	13	20	20	20	20	16	4	2	123	30	11
12 Mean duration of sunshine	in h	301	277	294	243	220	174	239	257	180	211	267	294	2957	30	12
13 Mean quantity of radiation	in ly/day															13
14 Mean potential evaporation	in mm	81	78	99	109	113	98	105	104	91	88	80	81	1127	3	14
15 Mean windspeed	in m/sec	3,4	2,8	2,8	2,4	2,0	1,7	1,8	1,8	1,7	1,9	2,7	3,2	2,4	30	15
16 Mean predominent direction of the wind		N	N	SW	SW	W	SW	N	W	SW	N	N	N		30	16

15 Station/Country Bridgetown/Barbados

Location 13°08'N/59°36'W   Height above sealevel 55 m   Climate symbol: Köppen Aw   Troll V,2

		J	F	M	A	M	J	J	A	S	O	N	D	year	P	
1 Mean daily temperature	in °C	25,2	25,1	25,9	26,3	26,8	27,1	26,8	27,1	27,1	26,7	26,3	25,9	26,3	10	1
2 Mean daily maximum temperature	in °C	28,3	28,3	29,4	30,0	30,6	30,6	30,0	30,6	30,6	30,0	29,4	28,3	29,4	35	2
3 Mean daily minimum temperature	in °C	21,1	20,6	21,1	22,2	22,8	23,3	23,3	23,3	23,3	23,3	22,8	21,7	22,2	35	3
4 Absolute maximum temperature	in °C	30,5	30,5	31,6	31,6	32,2	32,2	32,2	35,0	32,8	33,2	31,6	31,1	35,0	10	4
5 Absolute minimum temperature	in °C	16,1	16,1	16,6	17,7	18,9	19,4	20,0	20,6	19,4	19,4	18,9	17,7	16,1	10	5
6 Mean relative humidity	in %	73	69	67	66	68	71	73	74	75	77	79	75	72		6
7 Mean precipitation	in mm	66	28	33	36	58	112	147	147	170	178	205	96	1276	10	7
8 Maximum precipitation	in mm	230	169	162	273	332	428	303	610	518	516	554	312			8
9 Minimum precipitation	in mm	19	19	12	16	28	37	87	61	94	81	58	34			9
10 Maximum precipitation in 24 h	in mm	63	23	23	71	63	94	66	89	94	107	152	74	152	10	10
11 Mean number of days with precipitation	> 0,1 mm	13	8	8	7	9	14	18	18	15	15	16	14	153	10	11
12 Mean duration of sunshine	in h															12
13 Mean quantity of radiation	in ly/day															13
14 Mean potential evaporation	in mm	99	94	111	124	151	148	151	149	139	132	114	108	1520	29	14
15 Mean windspeed	in m/sec	4,8	5,4	5,5	5,3	5,4	5,7	5,2	4,4	3,6	3,3	3,8	4,4	4,7	10	15
16 Mean predominent direction of the wind		ENE	ENE	E	E	E	E	E	E	E	E	E	ENE		10	16

16 Station/Country Curaçao/Netherlands Antilles

Location 12°12'N/68°58'W   Height above sealevel 8 m   Climate symbol: Köppen BSh   Troll V,4

		J	F	M	A	M	J	J	A	S	O	N	D	year	P	
1 Mean daily temperature	in °C	26,2	26,1	26,5	27,1	27,7	28,0	28,0	·28,4	28,8	28,3	27,8	26,9	27,5	10	1
2 Mean daily maximum temperature	in °C	28,3	28,9	28,9	30,0	30,0	30,6	30,6	31,1	31,7	31,1	30,0	28,9	30,0	32	2
3 Mean daily minimum temperature	in °C	23,9	23,3	23,3	24,4	25,0	25,6	25,0	25,6	25,6	25,6	24,4	23,9	24,4	32	3
4 Absolute maximum temperature	in °C	30,9	31,1	32,2	33,4	35,6	32,9	33,8	34,9	35,8	35,2	33,5	32,7	35,8	10	4
5 Absolute minimum temperature	in °C	19,0	19,9	20,0	21,4	22,2	21,0	22,6	20,4	20,4	20,0	20,0	20,1	19,0	10	5
6 Mean relative humidity	in %															6
7 Mean precipitation	in mm	68	31	14	12	18	26	34	48	31	67	98	85	532	10	7
8 Maximum precipitation	in mm															8
9 Minimum precipitation	in mm															9
10 Maximum precipitation in 24 h	in mm	75	24	33	33	43	98	42	64	91	119	157	104	157	10	10
11 Mean number of days with precipitation	> 0,1 mm	9	5	2	2	3	4	5	6	4	7	10	10	67	10	11
12 Mean duration of sunshine	in h	266	272	282	263	217	266	297	296	280	273	245	242	3199	10	12
13 Mean quantity of radiation	in ly/day															13
14 Mean potential evaporation	in mm	119	109	133	152	157	154	158	157	154	150	139	133	1715	40	14
15 Mean windspeed	in m/sec	7,3	8,0	8,0	8,2	8,2	8,5	8,1	7,6	7,3	8,3	6,1	6,7	7,5	10	15
16 Mean predominent direction of the wind		E	E	E	E	E	E	E	E	E	E	E	E		10	16

17 Station/Country Managua/Nicaragua

Location 12°08'N/86°11'W — Height above sealevel 56 m — Climate symbol: Köppen Aw — Troll V,2

		J	F	M	A	M	J	J	A	S	O	N	D	year	P	
1 Mean daily temperature	in °C	26,3	27,2	28,6	29,3	29,4	27,2	26,9	27,2	26,9	26,5	26,3	26,1	27,3	10	1
2 Mean daily maximum temperature	in °C	30	30	30	32	32	31	31	31	31	31	30	30	31		2
3 Mean daily minimum temperature	in °C	23	24	26	28	27	26	26	25	26	24	24	24	25		3
4 Absolute maximum temperature	in °C	31,0	32,1	33,8	34,3	34,0	31,4	30,9	31,4	31,3	30,8	30,8	30,8	34,3	10	4
5 Absolute minimum temperature	in °C	20,4	20,8	21,7	22,6	23,4	23,0	22,6	22,4	22,2	22,1	20,9	20,0	20,0	10	5
6 Mean relative humidity	in %	74	74	70	70	70	83	85	82	81	80	77	76	77		6
7 Mean precipitation	in mm	4	1	5	6	78	296	134	130	182	243	59	6	1142	10	7
8 Maximum precipitation	in mm															8
9 Minimum precipitation	in mm															9
10 Maximum precipitation in 24 h	in mm	9	2	29	23	92	119	89	60	119	108	45	9	119	10	10
11 Mean number of days with precipitation	> 0,1 mm	3	1	1	1	6	22	20	17	20	19	10	2	122	10	11
12 Mean duration of sunshine	in h	217	214	250	223	193	128	142	174	178	181	195	215	2310	10	12
13 Mean quantity of radiation	in ly/day															13
14 Mean potential evaporation	in mm	140	133	154	188	177	156	161	160	149	147	142	134	1821	19	14
15 Mean windspeed	in m/sec	4,3	4,8	4,8	4,7	4,1	3,1	3,5	3,4	2,7	2,3	2,8	3,5	3,7	10	15
16 Mean predominent direction of the wind		E	E	E	E	E	E	E	E	E	E	E	E		10	16

18 Station/Country Piarco/Trinidad

Location 10°37'N/61°21'W — Height above sealevel 14 m — Climate symbol: Köppen Aw — Troll V,1

		J	F	M	A	M	J	J	A	S	O	N	D	year	P	
1 Mean daily temperature	in °C	24,5	24,7	25,4	26,3	26,6	26,1	25,9	26,1	26,2	25,9	25,4	24,8	25,7	10	1
2 Mean daily maximum temperature	in °C	29,4	30,0	30,6	31,1	31,1	30,6	30,6	30,6	31,1	31,1	30,6	30,0	30,6	40	2
3 Mean daily minimum temperature	in °C	19,4	19,4	19,4	20,6	21,1	21,7	21,1	21,7	21,7	21,7	21,1	20,6	20,6	40	3
4 Absolute maximum temperature	in °C	31,1	32,1	32,7	33,9	34,4	35,5	32,7	33,2	33,9	32,7	32,7	31,6	35,5	10	4
5 Absolute minimum temperature	in °C	16,1	16,7	16,7	17,2	19,4	18,9	18,3	18,9	19,4	20,5	18,3	16,7	16,1		5
6 Mean relative humidity	in %															6
7 Mean precipitation	in mm	77	61	27	71	129	269	243	213	144	151	212	153	1750	10	7
8 Maximum precipitation	in mm															8
9 Minimum precipitation	in mm															9
10 Maximum precipitation in 24 h	in mm	58	48	25	71	46	73	89	119	61	58	79	109	119	10	10
11 Mean number of days with precipitation	> 0,1 mm	13	8	8	8	15	22	21	20	14	15	15	15	172	10	11
12 Mean duration of sunshine	in h	230	219	251	241	233	190	217	220	202	207	196	202	2608		12
13 Mean quantity of radiation	in ly/day															13
14 Mean potential evaporation	in mm	93	87	106	116	138	124	124	122	120	119	109	111	1371	65	14
15 Mean windspeed	in m/sec															15
16 Mean predominent direction of the wind		E	E	E	E	E	E	E	E	E	E	E	E		10	16

19 Station/Country San José/Costa Rica

Location 9°56'N/84°08'W — Height above sealevel 1120 m — Climate symbol: Köppen Aw — Troll V,1

		J	F	M	A	M	J	J	A	S	O	N	D	year	P	
1 Mean daily temperature	in °C	19,0	19,3	20,3	21,0	21,4	21,2	20,6	20,8	20,9	20,8	19,9	19,3	20,4	10	1
2 Mean daily maximum temperature	in °C	23,9	24,4	26,1	26,1	26,7	26,1	25,0	25,6	26,1	25,0	25,0	23,9	25,0	12	2
3 Mean daily minimum temperature	in °C	14,4	14,4	15,0	16,7	18,7	18,7	16,7	16,7	16,1	15,8	15,6	14,4	15,6	8	3
4 Absolute maximum temperature	in °C	30,5	31,1	32,7	31,6	31,0	33,2	28,9	29,4	30,0	29,4	28,7	30,5	33,2	10	4
5 Absolute minimum temperature	in °C	9,4	10,5	10,0	11,7	12,2	13,9	12,2	13,3	13,3	12,7	11,1	9,4	9,4	10	5
6 Mean relative humidity	in %	80	80	80	79	84	86	86	85	86	88	84	82	83	10	6
7 Mean precipitation	in mm	8	5	10	37	244	284	230	233	342	333	172	46	1944	30	7
8 Maximum precipitation	in mm															8
9 Minimum precipitation	in mm															9
10 Maximum precipitation in 24 h	in mm	12	8	12	25	83	88	123	60	60	75	75	22	123	30	10
11 Mean number of days with precipitation	> 0,1 mm	3	1	2	7	19	22	23	24	24	25	14	6	170	30	11
12 Mean duration of sunshine	in h	217	220	248	210	161	120	124	136	150	136	135	183	2040		12
13 Mean quantity of radiation	in ly/day															13
14 Mean potential evaporation	in mm	66	59	75	82	89	88	83	82	82	78	70	64	918	51	14
15 Mean windspeed	in m/sec															15
16 Mean predominent direction of the wind																16

20 Station/Country Cristobal/Panama

Location 9°21'N/79°55'W — Height above sealevel 12 m — Climate symbol: Köppen Am — Troll V,1

		J	F	M	A	M	J	J	A	S	O	N	D	year	P	
1 Mean daily temperature	in °C	26,8	26,8	27,0	27,2	26,8	26,4	26,6	26,6	25,6	26,5	26,1	26,4	26,6	10	1
2 Mean daily maximum temperature	in °C	28,9	29,4	29,4	30,0	30,0	30,0	29,4	29,4	30,0	30,0	28,9	29,4	29,4	36	2
3 Mean daily minimum temperature	in °C	24,4	24,4	25,0	25,0	24,4	24,4	24,4	24,4	23,9	23,9	23,9	24,4	24,4	36	3
4 Absolute maximum temperature	in °C	31,7	32,8	33,9	33,9	33,9	35,0	33,9	37,2	35,0	33,9	33,9	33,9	37,2	10	4
5 Absolute minimum temperature	in °C	17,8	17,8	17,8	18,9	21,1	21,1	21,1	20,0	20,0	20,0	20,0	20,0	17,8	10	5
6 Mean relative humidity	in %	82	81	80	82	87	89	89	89	89	89	89	86	86	20	6
7 Mean precipitation	in mm	85	40	42	105	324	307	406	387	321	397	569	302	3285	103	7
8 Maximum precipitation	in mm															8
9 Minimum precipitation	in mm															9
10 Maximum precipitation in 24 h	in mm	77	113	106	157	154	137	166	135	87	173	343	130	343	10	10
11 Mean number of days with precipitation	> 0,1 mm	12	11	12	12	21	22	24	26	22	22	25	22	231	10	11
12 Mean duration of sunshine	in h	248	235	268	243	188	156	154	156	171	164	148	206	2337	10	12
13 Mean quantity of radiation	in ly/day															13
14 Mean potential evaporation	in mm	136	123	142	144	150	144	147	150	142	134	125	135	1872	32	14
15 Mean windspeed	in m/sec	5,5	5,9	5,7	5,4	3,4	2,5	3,2	3,0	2,4	2,3	2,9	4,5	3,9	10	15
16 Mean predominent direction of the wind		NE	N	N	N	N	NW	NNE	N	NW	S	N	NE		10	16

191

21 Station / Country Balboa / Panama

Location 8°58'N/79°33'W Height above sealevel 31 m Climate symbol: Köppen Aw Troll V,2

		J	F	M	A	M	J	J	A	S	O	N	D	year	P	
1 Mean daily temperature	in °C	26,6	26,9	27,8	27,9	27,2	26,8	26,9	26,8	26,8	26,2	26,2	26,8	26,9	10	1
2 Mean daily maximum temperature	in °C	31,1	31,7	32,2	32,2	30,6	30,0	30,6	30,8	30,0	29,4	29,4	30,6	30,8	34	2
3 Mean daily minimum temperature	in °C	21,7	21,7	22,2	23,3	23,3	23,3	23,3	23,3	23,3	22,8	22,8	22,8	22,8	34	3
4 Absolute maximum temperature	in °C	33,3	34,4	35,6	35,0	35,0	33,9	33,9	33,9	33,3	33,3	33,3	33,3	35,6	10	4
5 Absolute minimum temperature	in °C	18,9	17,8	18,9	20,0	21,1	21,7	21,1	20,0	21,1	20,6	20,6	20,0	17,8	10	5
6 Mean relative humidity	in % ++	88	85	81	81	87	90	90	90	91	90	91	90	88	34	6
7 Mean precipitation	in mm	38	16	17	76	198	203	183	190	193	254	249	130	1747	10	7
8 Maximum precipitation	in mm															8
9 Minimum precipitation	in mm															9
10 Maximum precipitation in 24 h	in mm	37	48	30	74	184	107	138	112	130	125	116	66	184	10	10
11 Mean number of days with precipitation	> 0,1 mm	6	3	<1	5	17	17	20	20	15	22	24	12	162	10	11
12 Mean duration of sunshine	in h	218	249	278	271	195	172	139	133	170	158	158	214	2355	10	12
13 Mean quantity of radiation	in ly / day															13
14 Mean potential evaporation	in mm	122	115	138	141	143	136	138	135	129	122	116	121	1556	25	14
15 Mean windspeed	in m/sec	3,7	4,3	4,2	3,8	2,7	2,1	2,3	2,3	2,1	2,2	2,3	2,9	2,9	10	15
16 Mean predominent direction of the wind		NW	NW	NNW	NNW	NW	NW	NW	NW	NW	NW	NW	NW		10	16

++ 7.30 a.m.

22 Station / Country Jaque / Panama

Location 7°32'N/78°10'W Height above sealevel 5 m Climate symbol: Köppen Aw Troll V,2

		J	F	M	A	M	J	J	A	S	O	N	D	year	P	
1 Mean daily temperature	in °C	26,1	26,1	26,7	26,7	26,1	26,1	25,6	25,6	25,6	25,0	25,6	25,6	25,9		1
2 Mean daily maximum temperature	in °C															2
3 Mean daily minimum temperature	in °C															3
4 Absolute maximum temperature	in °C	33,9	35,0	33,9	35,0	31,7	31,7	30,6	31,7	30,6	30,6	30,6	32,8	35,0		4
5 Absolute minimum temperature	in °C	18,9	18,9	20,0	21,1	20,0	21,1	20,0	20,0	20,0	20,0	20,0	20,0	18,9		5
6 Mean relative humidity	in %															6
7 Mean precipitation	in mm	8	3	28	91	622	411	432	409	353	955	696	305	4313		7
8 Maximum precipitation	in mm															8
9 Minimum precipitation	in mm															9
10 Maximum precipitation in 24 h	in mm															10
11 Mean number of days with precipitation	> 1,0 mm	4	<1	3	10	22	23	.25	24	21	26	23	18	199		11
12 Mean duration of sunshine	in h															12
13 Mean quantity of radiation	in ly / day															13
14 Mean potential evaporation	in mm															14
15 Mean windspeed	in m / sec															15
16 Mean predominent direction of the wind			NNE			NE			NNE			NNE				16

23 Station / Country Barranquilla / Colombia

Location 10°41'N/74°47'W Height above sealevel 13 m Climate symbol: Köppen Aw Troll V,3

		J	F	M	A	M	J	J	A	S	O	N	D	year	P	
1 Mean daily temperature	in °C	26,8	26,9	27,4	28,4	28,9	28,7	28,7	28,9	28,6	28,2	28,1	28,0	28,1	15	1
2 Mean daily maximum temperature	in °C															2
3 Mean daily minimum temperature	in °C															3
4 Absolute maximum temperature	in °C	36,1	36,7	36,1	37,8	37,8	37,2	38,9	37,8	37,8	37,2	35,6	36,1	38,9	7	4
5 Absolute minimum temperature	in °C	18,9	17,8	19,4	20,0	18,3	18,3	21,7	20,6	21,1	20,0	20,6	19,4	17,8	7	5
6 Mean relative humidity	in %	80	77	80	79	80	82	80	81	83	85	83	81	81	15	6
7 Mean precipitation	in mm	1	0	1	11	87	103	54	102	138	202	82	65	846	23	7
8 Maximum precipitation	in mm															8
9 Minimum precipitation	in mm															9
10 Maximum precipitation in 24 h	in mm															10
11 Mean number of days with precipitation	> 0,1 mm	<1	0	0	1	7	9	6	7	12	14	7	1	64	20	11
12 Mean duration of sunshine	in h															12
13 Mean quantity of radiation	in ly / day															13
14 Mean potential evaporation	in mm	134	122	142	149	164	160	162	163	153	150	142	140	1781	27	14
15 Mean windspeed	in m/sec	6,2	6,8	6,6	4,9	3,4	2,9	3,8	3,4	2,9	2,7	3,9	5,3	4,4	34	15
16 Mean predominent direction of the wind		ENE	ENE	ENE	ENE	NE	NE	NE	NE	E	ENE	NE	NE		34	16

24 Station / Country Turbo / Colombia

Location 8°06'N/76°44'W Height above sealevel 2 m Climate symbol: Köppen Af Troll V,1

		J	F	M	A	M	J	J	A	S	O	N	D	year	P	
1 Mean daily temperature	in °C	26,4	26,4	26,4	26,3	25,4	26,1	26,3	26,4	26,6	26,5	26,6	26,2	26,3		1
2 Mean daily maximum temperature	in °C															2
3 Mean daily minimum temperature	in °C															3
4 Absolute maximum temperature	in °C	32,0	31,0	32,4	32,0	32,6	33,0	33,2	33,0	33,5	33,0	39,4	31,0	39,4	3	4
5 Absolute minimum temperature	in °C	20,0	20,0	20,2	20,0	20,0	20,0	20,2	20,2	21,0	21,0	20,0	20,0	20,0	3	5
6 Mean relative humidity	in %	85	85	85	84	85	86	85	87	87	88	87	87	86		6
7 Mean precipitation	in mm	91	70	48	200	276	236	278	257	244	245	298	242	2485		7
8 Maximum precipitation	in mm															8
9 Minimum precipitation	in mm															9
10 Maximum precipitation in 24 h	in mm															10
11 Mean number of days with precipitation	> 0,5 mm	7	7	6	12	18	16	17	17	16	13	15	14	158		11
12 Mean duration of sunshine	in h															12
13 Mean quantity of radiation	in ly / day															13
14 Mean potential evaporation	in mm	140	128	143	146	149	144	148	151	142	144	139	140	1712		14
15 Mean windspeed	in m / sec	3,8	4,7	4,0	3,6	2,9	3,8	2,9	3,1	3,1	3,8	2,7	4,7	3,6		15
16 Mean predominent direction of the wind		NNW	NNW	N	NNW	SW	SSW	SSW	SSW	S	SW	S	NNW			16

25 Station/Country Barrancabermeja/Colombia

Location 7°04'N / 73°52'W   **Height above sealevel** 107 m   **Climate symbol:** Köppen Af   **Troll** V.1

		J	F	M	A	M	J	J	A	S	O	N	D	year	P	
1 Mean daily temperature	in °C	29,3	29,8	29,7	29,3	28,9	28,9	29,1	28,8	28,6	28,2	28,6	29,1	29,0	23	1
2 Mean daily maximum temperature	in °C						·									2
3 Mean daily minimum temperature	in °C															3
4 Absolute maximum temperature	in °C	37,2	37,8	38,3	38,0	37,8	37,8	37,0	39,0	37,2	36,0	36,0	36,7	39,0	13	4
5 Absolute minimum temperature	in °C	21,7	20,6	20,0	21,7	21,1	21,7	21,1	21,1	20,6	20,0	20,6	21,1	20,0	13	5
6 Mean relative humidity	in %	79	78	79	77	77	75	75	74	74	80	77	75	77	23	6
7 Mean precipitation	in mm	71	81	135	243	312	269	186	280	347	446	295	121	2786		7
8 Maximum precipitation	in mm															8
9 Minimum precipitation	in mm															9
10 Maximum precipitation in 24 h	in mm															10
11 Mean number of days with precipitation	>1,0 mm	3	5	8	13	14	13	12	13	14	16	13	6	130		11
12 Mean duration of sunshine	in h															12
13 Mean quantity of radiation	in ly/day															13
14 Mean potential evaporation	in mm	163	149	161	158	163	163	163	163	150	151	148	155	1877		14
15 Mean windspeed	in m/sec															15
16 Mean predominent direction of the wind																16

26 Station/Country Arauca/Colombia

Location 7°04'N / 70°40'W   **Height above sealevel** 122 m   **Climate symbol:** Köppen Aw   **Troll** V.2

		J	F	M	A	M	J	J	A	S	O	N	D	year	P	
1 Mean daily temperature	in °C	27,1	27,4	28,0	27,6	26,5	26,1	25,7	26,5	26,7	27,1	27,2	27,4	26,9		1
2 Mean daily maximum temperature	in °C															2
3 Mean daily minimum temperature	in °C															3
4 Absolute maximum temperature	in °C	38,0	38,5	39,8	40,5	38,5	34,2	36,0	36,8	37,2	35,0	35,5	37,0	40,5	6	4
5 Absolute minimum temperature	in °C	18,0	17,1	18,8	18,0	19,8	19,0	18,0	19,0	18,5	18,0	18,5	17,2	17,1	6	5
6 Mean relative humidity	in %	65	61	60	72	81	82	82	80	79	78	80	73	74		6
7 Mean precipitation	in mm	12	8	24	191	249	289	273	219	214	200	87	31	1797		7
8 Maximum precipitation	in mm															8
9 Minimum precipitation	in mm															9
10 Maximum precipitation in 24 h	in mm															10
11 Mean number of days with precipitation	>1,0 mm	<1	1	3	9	14	15	15	11	10	9	6	3	96		11
12 Mean duration of sunshine	in h															12
13 Mean quantity of radiation	in ly/day															13
14 Mean potential evaporation	in mm	143	140	156	149	140	138	137	143	142	147	143	149	1727		14
15 Mean windspeed	in m/sec															15
16 Mean predominent direction of the wind																16

27 Station/Country Bogotá/Colombia

Location 4°38'N / 74°05'W   **Height above sealevel** 2556 m   **Climate symbol:** Köppen Cwb   **Troll** V.2

		J	F	M	A	M	J	J	A	S	O	N	D	year	P	
1 Mean daily temperature	in °C	12,8	13,2	13,7	13,7	13,7	13,2	12,9	12,9	12,8	12,9	13,1	13,1	13,2	5	1
2 Mean daily maximum temperature	in °C	19,4	20,0	19,4	18,9	18,9	18,3	17,8	18,3	18,9	18,9	18,9	18,9	18,9	10	2
3 Mean daily minimum temperature	in °C	8,9	9,4	10,0	10,6	10,6	10,6	10,0	10,0	9,4	10,0	10,0	9,4	10,0	10	3
4 Absolute maximum temperature	in °C	25,0	25,0	25,0	24,2	24,6	23,4	23,2	23,2	24,0	24,2	23,4	23,8	25,0	30	4
5 Absolute minimum temperature	in °C	-1,5	-5,2	2,4	1,0	3,8	4,2	0,4	1,4	1,8	2,0	0,5	1,0	-5,2	30	5
6 Mean relative humidity	in %	68	68	69	71	72	71	70	69	68	74	76	66	70		6
7 Mean precipitation	in mm	51	50	69	100	105	57	47	41	52	144	138	85	939	30	7
8 Maximum precipitation	in mm															8
9 Minimum precipitation	in mm															9
10 Maximum precipitation in 24 h	in mm	33	40	61	46	69	28	37	30	40	59	72	49	72	30	10
11 Mean number of days with precipitation	>0,1 mm	9	10	13	19	21	18	18	17	15	20	19	13	192	30	11
12 Mean duration of sunshine	in h	178	151	138	87	99	107	130	135	132	105	123	147	1532	20	12
13 Mean quantity of radiation	in ly/day	453	445	418	375	377	393	407	413	415	355	386	401	403	5	13
14 Mean potential evaporation	in mm	57	54	62	61	64	58	57	58	58	60	56	59	704	75	14
15 Mean windspeed	in m/sec															15
16 Mean predominent direction of the wind		WNW	W	WSW	SSE	SSE	SSE	SSE	SSE	SSE	SW	WNW	NW		30	16

28 Station/Country Andagoya/Colombia

Location 5°06'N / 76°40'W   **Height above sealevel** 65 m   **Climate symbol:** Köppen Af   **Troll** V.1

		J	F	M	A	M	J	J	A	S	O	N	D	year	P	
1 Mean daily temperature	in °C	26,8	27,2	27,6	27,5	27,4	27,1	27,2	27,3	27,2	27,0	26,7	27,0	27,2	23	1
2 Mean daily maximum temperature	in °C	32,2	31,7	32,2	32,2	31,7	31,7	31,7	31,7	32,2	32,2	31,1	31,1	31,7	8	2
3 Mean daily minimum temperature	in °C	23,9	23,9	23,9	23,9	23,9	23,3	23,3	23,3	23,3	23,3	23,3	23,3	23,3	8	3
4 Absolute maximum temperature	in °C	38,7	37,2	37,8	37,8	37,2	37,8	37,8	37,2	37,8	38,5	37,7	37,7	38,5		4
5 Absolute minimum temperature	in °C	20,0	18,5	20,0	20,6	20,0	20,6	20,0	20,0	18,9	20,0	16,7	18,9	16,7		5
6 Mean relative humidity	in %	74	77	75	74	82	82	85	85	86	82	82	82	81	24	6
7 Mean precipitation	in mm	554	519	557	620	655	655	572	574	561	563	563	512	6905	29	7
8 Maximum precipitation	in mm															8
9 Minimum precipitation	in mm															9
10 Maximum precipitation in 24 h	in mm	155	135	150	173	163	175	130	147	147	135	130	165	175	17	10
11 Mean number of days with precipitation	>0,25 mm	26	21	23	25	26	25	27	27	27	25	27	27	306	15	11
12 Mean duration of sunshine	in h															12
13 Mean quantity of radiation	in ly/day															13
14 Mean potential evaporation	in mm	137	126	142	142	147	142	147	145	137	139	133	137	1674		14
15 Mean windspeed	in m/sec															15
16 Mean predominent direction of the wind																16

29 Station/Country Villavicencio/Colombia

Location 4°09'N / 73°36'W   Height above sealevel 423 m   Climate symbol: Köppen Af   Troll V.1

		J	F	M	A	M	J	J	A	S	O	N	D	year	P	
1 Mean daily temperature	in °C	26,8	26,9	27,1	25,8	25,8	25,4	25,2	25,5	26,0	26,3	26,2	26,0	26,0		1
2 Mean daily maximum temperature	in °C															2
3 Mean daily minimum temperature	in °C															3
4 Absolute maximum temperature	in °C	37,0	38,5	38,0	37,0	37,0	39,0	35,5	38,0	37,0	39,5	36,5	39,5	39,5		4
5 Absolute minimum temperature	in °C	16,0	15,0	15,0	14,0	15,0	14,0	15,0	14,0	13,5	15,0	15,5	15,0	13,5		5
6 Mean relative humidity	in %	75	71	72	81	82	83	80	79	80	81	80	78	79		6
7 Mean precipitation	in mm	61	113	159	461	605	498	524	360	335	443	381	158	4096		7
8 Maximum precipitation	in mm															8
9 Minimum precipitation	in mm															9
10 Maximum precipitation in 24 h	in mm															10
11 Mean number of days with precipitation	> 1,0 mm	5	7	10	20	23	21	22	19	16	19	17	9	187		11
12 Mean duration of sunshine	in h															12
13 Mean quantity of radiation	in ly / day															13
14 Mean potential evaporation	in mm	137	131	142	140	128	115	119	116	118	131	126	132	1535		14
15 Mean windspeed	in m/sec															15
16 Mean predominent direction of the wind																16

30 Station/Country Popayán/Colombia

Location 2°27'N / 76°35'W   Height above sealevel 1789 m   Climate symbol: Köppen Csb   Troll V.1

		J	F	M	A	M	J	J	A	S	O	N	D	year	P	
1 Mean daily temperature	in °C	17,3	17,5	17,5	17,4	17,5	17,3	17,5	17,9	17,9	17,2	16,9	17,2	17,4	18	1
2 Mean daily maximum temperature	in °C															2
3 Mean daily minimum temperature	in °C															3
4 Absolute maximum temperature	in °C	28,8	30,0	28,3	28,9	29,5	28,0	29,5	29,9	31,0	29,6	27,7	28,2	31,0	18	4
5 Absolute minimum temperature	in °C	9,0	9,0	9,5	10,0	10,0	8,5	8,5	8,5	9,5	9,2	9,5	9,5	8,5	18	5
6 Mean relative humidity	in %															6
7 Mean precipitation	in mm	152	143	188	187	140	98	40	32	88	267	311	284	1911	18	7
8 Maximum precipitation	in mm															8
9 Minimum precipitation	in mm															9
10 Maximum precipitation in 24 h	in mm	60	65	74	75	63	50	48	27	51	93	80	86	93	18	10
11 Mean number of days with precipitation	> 0,1 mm	18	18	18	21	20	17	13	10	14	24	25	24	220	18	11
12 Mean duration of sunshine	in h	180	174	149	125	131	155	174	177	187	134	131	155	1852	18	12
13 Mean quantity of radiation	in ly / day															13
14 Mean potential evaporation	in mm	74	69	74	74	75	70	75	75	74	72	70	73	875	10	14
15 Mean windspeed	in m/sec															15
16 Mean predominent direction of the wind																16

31 Station/Country Tumaco/Colombia

Location 1°48'N / 78°47'W   Height above sealevel 4 m   Climate symbol: Köppen Af   Troll V.1

		J	F	M	A	M	J	J	A	S	O	N	D	year	P	
1 Mean daily temperature	in °C	25,6	25,9	26,2	26,4	26,2	26,1	26,2	26,2	25,8	25,8	25,3	25,6	25,9	17	1
2 Mean daily maximum temperature	in °C															2
3 Mean daily minimum temperature	in °C															3
4 Absolute maximum temperature	in °C	30,0	30,5	30,8	32,0	31,0	30,0	30,0	29,5	30,0	29,6	29,5	30,0	32,0	11	4
5 Absolute minimum temperature	in °C	21,4	22,8	22,8	22,4	22,0	22,0	22,8	20,0	19,8	20,0	21,8	22,0	19,8	11	5
6 Mean relative humidity	in %	86	86	87	85	86	86	85	86	86	86	85	86	86	17	6
7 Mean precipitation	in mm	430	288	242	371	442	303	203	187	168	184	136	172	3128	10	7
8 Maximum precipitation	in mm															8
9 Minimum precipitation	in mm															9
10 Maximum precipitation in 24 h	in mm															10
11 Mean number of days with precipitation	> 0,1 mm	24	19	15	21	26	25	21	21	19	17	13	18	239	10	11
12 Mean duration of sunshine	in h															12
13 Mean quantity of radiation	in ly / day															13
14 Mean potential evaporation	in mm	130	118	136	134	137	132	134	134	127	131	118	130	1561	14	14
15 Mean windspeed	in m/sec															15
16 Mean predominent direction of the wind																16

32 Station/Country Maracaibo/Venezuela

Location 10°41'N / 71°39'W   Height above sealevel 40 m   Climate symbol: Köppen BSh   Troll V.3

		J	F	M	A	M	J	J	A	S	O	N	D	year	P	
1 Mean daily temperature	in °C	26,8	26,9	27,2	27,8	28,4	28,6	28,6	28,8	28,7	28,1	27,6	27,4	27,9	14	1
2 Mean daily maximum temperature	in °C	32,2	32,2	32,8	33,3	33,3	33,9	34,4	34,4	34,4	33,3	32,8	32,8	33,3	12	2
3 Mean daily minimum temperature	in °C	22,8	22,8	23,3	24,4	25,0	25,0	24,4	25,0	25,0	24,4	24,4	23,9	24,4	13	3
4 Absolute maximum temperature	in °C	35,0	35,4	35,6	36,3	37,8	36,4	37,0	38,0	38,7	36,5	35,9	35,4	38,7	14	4
5 Absolute minimum temperature	in °C	18,9	20,9	20,0	20,2	20,5	20,8	21,0	19,5	19,0	18,9	20,4	20,7	18,9	14	5
6 Mean relative humidity	in %	75	75	74	76	77	75	74	73	74	77	78	78	75	14	6
7 Mean precipitation	in mm	2	1	9	24	78	47	44	56	49	133	77	13	533	30	7
8 Maximum precipitation	in mm															8
9 Minimum precipitation	in mm															9
10 Maximum precipitation in 24 h	in mm	64	5	66	95	93	89	94	82	53	109	166	39	166	27	10
11 Mean number of days with precipitation	> 0,1 mm	1	1	1	4	8	8	7	8	9	11	8	2	68	14	11
12 Mean duration of sunshine	in h	264	240	251	180	192	207	254	254	216	198	219	242	2717		12
13 Mean quantity of radiation	in ly / day															13
14 Mean potential evaporation	in mm	142	130	148	156	165	183	170	189	158	152	144	144	1841	36	14
15 Mean windspeed	in m/sec	3,1	3,7	3,7	3,8	2,4	2,4	2,5	2,4	2,1	2,0	2,3	2,6	2,8	14	15
16 Mean predominent direction of the wind		NE	NE	NE	NE	NE	NE	NE	NE	NE	NE	NE	NE		14	16

194

33 Station / Country Caracas/Venezuela

Location 10°30'N/66°56'W — Height above sealevel 1035 m — Climate symbol: Köppen Aw — Troll V.3

		J	F	M	A	M	J	J	A	S	O	N	D	year	P	
1 Mean daily temperature	in °C	19,2	19,7	20,7	21,7	22,0	21,5	21,1	21,8	21,8	21,5	20,8	19,9	21,0	10	1
2 Mean daily maximum temperature	in °C	23,9	25,0	26,1	27,2	26,7	25,6	25,8	26,1	26,7	26,1	25,0	25,6	25,6	30	2
3 Mean daily minimum temperature	in °C	13,3	13,3	14,4	15,6	16,7	16,7	16,1	16,1	16,1	16,1	15,6	14,4	15,6	30	3
4 Absolute maximum temperature	in °C	30,8	31,6	32,8	33,1	32,4	30,8	29,8	31,0	31,2	31,5	30,5	30,0	33,1	45	4
5 Absolute minimum temperature	in °C	8,3	7,8	7,1	10,6	11,1	11,7	11,1	11,7	11,7	12,2	10,6	8,3	7,1	45	5
6 Mean relative humidity	in %	80	78	76	77	80	83	83	82	81	82	84	82	81	25	6
7 Mean precipitation	in mm	22	15	10	32	95	106	97	112	94	122	86	44	835	30	7
8 Maximum precipitation	in mm															8
9 Minimum precipitation	in mm															9
10 Maximum precipitation in 24 h	in mm	41	73	36	74	79	71	71	92	84	81	70	48	92	58	10
11 Mean number of days with precipitation	> 0,1 mm	6	4	3	8	12	16	17	16	14	15	15	11	135	30	11
12 Mean duration of sunshine	in h	238	246	257	219	198	201	239	236	213	214	210	217	2688		12
13 Mean quantity of radiation	in ly / day															13
14 Mean potential evaporation	in mm	63	58	73	83	92	85	85	86	84	74	73	76	933	56	14
15 Mean windspeed	in m / sec	2,7	3,1	3,5	3,1	3,2	3,4	3,1	2,8	2,8	2,8	2,8	2,7	2,9	30	15
16 Mean predominent direction of the wind		SE	SE	SE	SE	SE	SE	SE	SE	SE	SE	NW	SE		30	16

34 Station / Country Colonia Tover/Venezuela

Location 10°25'N/67°17'W — Height above sealevel 1790 m — Climate symbol: Köppen Cw — Troll V.1

		J	F	M	A	M	J	J	A	S	O	N	D	year	P	
1 Mean daily temperature	in °C	15,7	16,1	17,0	17,5	17,5	16,7	16,5	16,9	17,1	17,1	16,9	16,2	16,8	14	1
2 Mean daily maximum temperature	in °C															2
3 Mean daily minimum temperature	in °C															3
4 Absolute maximum temperature	in °C	25,3	26,8	26,2	26,0	26,1	24,0	23,0	23,3	24,1	24,1	24,5	26,6	28,2		4
5 Absolute minimum temperature	in °C	3,7	2,9	4,1	6,7	8,5	8,3	7,1	7,1	8,1	8,1	7,4	2,8	2,8		5
6 Mean relative humidity	in %	80	79	77	80	85	88	88	88	88	87	86	83	84	14	6
7 Mean precipitation	in mm	48	21	25	61	142	135	165	166	143	185	140	66	1297	30	7
8 Maximum precipitation	in mm															8
9 Minimum precipitation	in mm															9
10 Maximum precipitation in 24 h	in mm	53	83	58	38	80	66	56	56	53	88	51	63	88		10
11 Mean number of days with precipitation	> 0,1 mm	9	5	5	10	16	20	23	22	22	21	20	13	186		11
12 Mean duration of sunshine	in h															12
13 Mean quantity of radiation	in ly / day															13
14 Mean potential evaporation	in mm	52	48	58	64	69	65	65	65	63	62	59	55	725		14
15 Mean windspeed	in m / sec	1,0	0,9	0,9	0,8	0,7	0,8	0,8	0,7	0,7	0,7	0,8	0,8	0,8	14	15
16 Mean predominent direction of the wind		ENE	ENE	ENE	ENE	ESE	ESE	ESE	ESE	ESE	ESE	ESE	ESE		14	16

35 Station / Country Barcelona/Venezuela

Location 10°07'N/64°41'W — Height above sealevel 7 m — Climate symbol: Köppen Aw — Troll V.3

		J	F	M	A	M	J	J	A	S	O	N	D	year	P	
1 Mean daily temperature	in °C	25,2	25,5	26,4	27,2	27,8	26,7	26,0	26,2	26,7	27,0	26,5	25,9	26,4	10	1
2 Mean daily maximum temperature	in °C															2
3 Mean daily minimum temperature	in °C															3
4 Absolute maximum temperature	in °C	34,8	35,0	35,4	36,0	37,0	37,0	35,2	37,0	36,2	36,8	35,4	34,9	37,0	13	4
5 Absolute minimum temperature	in °C	15,7	16,6	16,9	19,4	19,0	19,0	20,0	19,8	19,7	19,0	19,2	17,4	15,7	13	5
6 Mean relative humidity	in %	74	72	70	71	73	80	82	82	80	78	78	76	76	13	6
7 Mean precipitation	in mm	10	4	6	7	43	99	137	111	73	66	49	24	629	30	7
8 Maximum precipitation	in mm															8
9 Minimum precipitation	in mm															9
10 Maximum precipitation in 24 h	in mm	17	17	10	23	44	71	86	73	56	73	42	49	86	20	10
11 Mean number of days with precipitation	> 0,1 mm	2	1	1	2	6	15	17	17	13	10	9	5	98	11	11
12 Mean duration of sunshine	in h	291	280	301	261	260	210	245	239	252	257	261	273	3130		12
13 Mean quantity of radiation	in ly / day															13
14 Mean potential evaporation	in mm	137	121	140	145	154	148	140	149	148	144	140	137	1699	26	14
15 Mean windspeed	in m / sec	2,1	2,3	2,2	2,3	2,2	1,7	1,8	1,8	1,7	1,7	1,7	1,8	1,9	6	15
16 Mean predominent direction of the wind		N	N	N	N	N	N	N	N	ESE	ESE	N	N		6	16

36 Station / Country Mérida/Venezuela

Location 8°35'N/71°10'W — Height above sealevel 1495 m — Climate symbol: Köppen Am — Troll V.1

		J	F	M	A	M	J	J	A	S	O	N	D	year	P	
1 Mean daily temperature	in °C	17,7	18,2	19,0	19,2	19,4	19,1	18,8	19,3	19,4	19,0	18,4	18,0	18,8	14	1
2 Mean daily maximum temperature	in °C	22,8	23,3	23,8	23,9	24,4	23,9	24,4	24,4	24,4	23,9	22,8	22,8	23,9	14	2
3 Mean daily minimum temperature	in °C	13,3	14,4	15,0	15,6	16,1	15,6	15,0	15,0	15,0	15,6	15,0	13,9	15,0	14	3
4 Absolute maximum temperature	in °C	28,2	29,7	31,6	31,2	31,6	30,8	30,1	30,5	31,4	29,9	29,3	27,9	31,6	26	4
5 Absolute minimum temperature	in °C	9,2	10,0	10,0	10,5	11,5	11,1	10,7	10,3	12,2	11,4	10,8	9,7	9,2	26	5
6 Mean relative humidity	in %	79	77	78	82	82	82	81	78	78	82	84	82	80	17	6
7 Mean precipitation	in mm	53	50	62	164	272	163	125	138	172	265	221	95	1770	30	7
8 Maximum precipitation	in mm	102	120	80	79	210	362	237	299	334	405	145	151			8
9 Minimum precipitation	in mm	0	0	0	0	13	28	60	65	55	9	1	0			9
10 Maximum precipitation in 24 h	in mm	56	68	61	82	117	68	59	60	87	117	104	60	117	27	10
11 Mean number of days with precipitation	> 0,1 mm	9	9	10	17	23	22	21	21	20	23	20	14	209	14	11
12 Mean duration of sunshine	in h	245	220	229	186	186	174	205	208	198	189	204	229	2473		12
13 Mean quantity of radiation	in ly / day															13
14 Mean potential evaporation	in mm	63	79	73	75	79	75	77	80	76	72	67	66	882	26	14
15 Mean windspeed	in m / sec	1,8	1,9	2,4	2,2	1,7	1,6	1,6	1,9	1,9	1,7	1,5	1,8	1,8	14	15
16 Mean predominent direction of the wind																16

37 Station/Country San Fernando de Apure/Venezuela

Location 7°53'N/67°26'W   Height above sealevel 73 m   Climate symbol: Köppen **Aw**   Troll **V.2**

		J	F	M	A	M	J	J	A	S	O	N	D	year	P	
1	Mean daily temperature	in °C	26,7	27,6	28,8	29,0	27,4	25,9	25,5	26,2	27,0	27,2	27,2	26,8	27,1	14	1
2	Mean daily maximum temperature	in °C															2
3	Mean daily minimum temperature	in °C															3
4	Absolute maximum temperature	in °C	37,6	40,2	39,9	39,4	38,4	36,5	35,2	34,3	36,5	38,3	38,2	35,9	40,2	23	4
5	Absolute minimum temperature	in °C	17,5	17,2	17,9	20,0	19,9	19,1	19,5	19,8	20,0	19,4	20,0	19,2	17,2	23	5
6	Mean relative humidity	in %	71	65	62	63	76	84	86	86	83	81	80	76	76	17	6
7	Mean precipitation	in mm	1	4	14	71	186	277	303	282	171	129	42	11	1491	30	7
8	Maximum precipitation	in mm															8
9	Minimum precipitation	in mm															9
10	Maximum precipitation in 24 h	in mm	11	69	40	87	165	112	122	140	76	99	122	99	165	24	10
11	Mean number of days with precipitation	> 0,1 mm	1	1	1	5	14	21	25	22	17	11	6	3	127	14	11
12	Mean duration of sunshine	in h	276	257	279	213	195	159	183	180	207	236	258	273	2716		12
13	Mean quantity of radiation	in ly/day															13
14	Mean potential evaporation	in mm	141	133	157	158	151	136	139	144	145	149	142	142	1737	26	14
15	Mean windspeed	in m/sec	2,5	2,8	2,6	2,4	1,7	1,5	1,3	1,3	1,5	1,7	2,0	2,5	2,0	12	15
16	Mean predominant direction of the wind		ENE	NE	ENE	ENE	ENE	E	E	E	E	E	E	ENE		12	16

38 Station/Country Puerto Ayacucho/Venezuela

Location 5°41'N/67°38'W   Height above sealevel 99 m   Climate symbol: Köppen **Aw**   Troll **V.2**

		J	F	M	A	M	J	J	A	S	O	N	D	year	P	
1	Mean daily temperature	in °C	28,5	29,3	29,5	28,1	26,6	25,4	25,2	25,7	26,3	26,8	27,3	27,7	27,2	7	1
2	Mean daily maximum temperature	in °C															2
3	Mean daily minimum temperature	in °C															3
4	Absolute maximum temperature	in °C	38,0	39,6	39,4	38,6	37,2	33,8	33,8	35,4	36,1	36,3	36,5	36,8	39,6	9	4
5	Absolute minimum temperature	in °C	20,0	21,0	20,2	21,8	20,6	19,9	19,3	20,0	20,3	20,1	21,0	19,5	19,3	9	5
6	Mean relative humidity	in %	89	66	64	75	83	86	86	85	83	81	80	78	78	10	6
7	Mean precipitation	in mm	14	17	66	156	337	437	436	292	175	169	110	40	2249	16	7
8	Maximum precipitation	in mm															8
9	Minimum precipitation	in mm															9
10	Maximum precipitation in 24 h	in mm	28	33	190	149	137	115	133	98	66	104	116	56	190	10	10
11	Mean number of days with precipitation	> 0,1 mm	2	4	8	13	22	27	27	23	21	17	12	6	182	8	11
12	Mean duration of sunshine	in h															12
13	Mean quantity of radiation	in ly/day															13
14	Mean potential evaporation	in mm	154	148	167	153	152	141	141	142	143	154	148	152	1795	2	14
15	Mean windspeed	in m/sec	2,2	1,9	1,4	1,2	0,9	0,8	0,8	1,1	0,9	1,1	1,3	1,4	1,2	5	15
16	Mean predominant direction of the wind																16

39 Station/Country Santa Elena de Uairén/Venezuela

Location 4°36'N/61°07'W   Height above sealevel 907 m   Climate symbol: Köppen **Af**   Troll **V.2**

		J	F	M	A	M	J	J	A	S	O	N	D	year	P	
1	Mean daily temperature	in °C	21,6	22,0	22,4	22,3	22,0	21,4	20,0	21,1	21,4	21,8	21,5	21,7	21,7	14	1
2	Mean daily maximum temperature	in °C	30,0	30,8	31,1	30,0	28,9	27,8	27,8	28,3	28,9	29,4	30,0	28,9	29,4	10	2
3	Mean daily minimum temperature	in °C	16,1	16,7	17,8	17,8	18,3	17,8	16,7	17,2	17,2	17,2	16,7	17,2	17,2	10	3
4	Absolute maximum temperature	in °C	33,3	33,0	35,0	35,1	32,5	31,0	30,2	31,2	33,5	34,2	34,2	32,0	35,1	21	4
5	Absolute minimum temperature	in °C	9,0	10,4	11,4	10,9	11,6	12,2	12,0	12,3	9,5	11,0	11,4	10,8	9,0	21	5
6	Mean relative humidity	in %	79	77	75	80	85	89	89	88	84	81	83	82	83	17	6
7	Mean precipitation	in mm	66	69	78	145	221	248	229	182	109	106	130	115	1700	21	7
8	Maximum precipitation	in mm															8
9	Minimum precipitation	in mm															9
10	Maximum precipitation in 24 h	in mm	50	132	120	84	63	82	68	113	83	86	99	60	132	21	10
11	Mean number of days with precipitation	> 0,1 mm	13	11	12	15	24	27	26	24	16	14	16	16	214	21	11
12	Mean duration of sunshine	in h	189	184	202	183	164	147	174	189	237	229	210	195	2303		12
13	Mean quantity of radiation	in ly/day	448	479	509	512	427	416	442	485	502	479	456	415	462		13
14	Mean potential evaporation	in mm	84	78	93	91	91	82	83	85	86	90	85	85	1033	7	14
15	Mean windspeed	in m/sec	1,3	1,4	1,5	1,4	1,1	0,0	0,9	0,9	1,0	1,1	1,0	1,1	1,1	14	15
16	Mean predominant direction of the wind																16

40 Station/Country San Carlos de Rio Negro/Venezuela

Location 1°54'N/67°03'W   Height above sealevel 95 m   Climate symbol: Köppen **Af**   Troll **V.1**

		J	F	M	A	M	J	J	A	S	O	N	D	year	P	
1	Mean daily temperature	in °C	26,3	26,3	26,5	25,9	25,8	25,7	25,4	25,9	26,8	26,7	26,7	26,2	26,2	8	1
2	Mean daily maximum temperature	in °C															2
3	Mean daily minimum temperature	in °C															3
4	Absolute maximum temperature	in °C	33,3	33,6	33,7	32,9	32,3	31,8	31,7	32,7	33,6	33,6	33,6	33,2	33,8	8	4
5	Absolute minimum temperature	in °C	19,1	19,0	19,3	19,0	18,8	19,6	19,1	19,1	19,1	19,6	19,5	19,3	18,8	8	5
6	Mean relative humidity	in %															6
7	Mean precipitation	in mm	222	229	206	395	381	390	330	328	249	257	314	220	3521	8	7
8	Maximum precipitation	in mm															8
9	Minimum precipitation	in mm															9
10	Maximum precipitation in 24 h	in mm	70	72	83	115	76	70	75	107	88	97	87	69	115	8	10
11	Mean number of days with precipitation	> 0,1 mm	21	19	21	23	29	27	28	28	22	29	19	20	286	8	11
12	Mean duration of sunshine	in h															12
13	Mean quantity of radiation	in ly/day															13
14	Mean potential evaporation	in mm	133	128	148	153	135	119	119	123	119	146	138	140	1601		14
15	Mean windspeed	in m/sec															15
16	Mean predominant direction of the wind																16

41 Station/Country Georgetown/Guyana

Location 6°48'N/58°08'W Height above sealevel 2 m Climate symbol: Köppen Af Troll V,1

	J	F	M	A	M	J	J	A	S	O	N	D	year	P	
1 Mean daily temperature	in °C	26.3	26.4	26.8	27.1	27.0	26.7	26.7	27.2	27.7	27.7	27.4	26.7	27.0	30	1
2 Mean daily maximum temperature	in °C	28.9	28.9	28.9	29.4	29.4	29.4	29.4	30.0	30.8	30.8	30.0	28.9	29.4	54	2
3 Mean daily minimum temperature	in °C	23.3	23.3	23.9	24.4	23.9	23.9	23.9	23.9	24.4	24.4	24.4	23.9	23.9	54	3
4 Absolute maximum temperature	in °C	31.1	31.7	31.7	32.2	32.2	31.7	32.2	32.2	33.9	33.9	32.8	32.2	33.9	57	4
5 Absolute minimum temperature	in °C	20.0	20.6	20.6	20.6	21.1	20.6	21.1	21.1	20.6	21.1	20.6	21.1	20.0	57	5
6 Mean relative humidity	in %	81	79	77	78	82	85	84	83	81	79	79	82	81		6
7 Mean precipitation	in mm	251	122	113	178	296	348	281	185	88	98	147	313	2418	30	7
8 Maximum precipitation	in mm	1058	1114	1385	1252	888	906	747	539	125	125	423	629			8
9 Minimum precipitation	in mm	13	29	27	97	305	234	78	19	1	2	11	63			9
10 Maximum precipitation in 24 h	in mm	183	138	152	211	160	130	109	117	129	114	133	196	211	73	10
11 Mean number of days with precipitation	>0,1 mm	20	16	16	18	23	24	23	17	9	8	13	21	206	73	11
12 Mean duration of sunshine	in h	188	186	209	196	182	165	215	245	239	226	212	181	2444		12
13 Mean quantity of radiation	in ly/day															13
14 Mean potential evaporation	in mm	133	121	140	142	149	144	149	152	149	150	143	137	1709	37	14
15 Mean windspeed	in m/sec	3.1	3.3	3.5	3.4	3.0	2.5	2.2	2.2	2.5	2.6	2.6	2.8	2.8	40	15
16 Mean predominent direction of the wind		NE	NE	NE	NE	NE	NE	NE	NE	NE	NE	NE	NE		40	16

42 Station/Country Saint Ignatius/Guyana

Location 03°21'N/59°48'W Height above sealevel 99 m Climate symbol: Köppen Aw Troll V,3

	J	F	M	A	M	J	J	A	S	O	N	D	year	P	
1 Mean daily temperature	in °C	27.7	27.9	28.2	28.1	27.4	26.7	26.5	27.2	28.5	29.1	28.8	28.8	27.8		1
2 Mean daily maximum temperature	in °C															2
3 Mean daily minimum temperature	in °C															3
4 Absolute maximum temperature	in °C	35.0	35.5	36.1	37.8	38.9	34.4	33.9	35.0	36.7	37.2	37.2	36.1	38.9		4
5 Absolute minimum temperature	in °C	18.3	17.2	20.5	20.0	17.2	18.9	18.3	18.9	18.3	17.8	16.1	16.1			5
6 Mean relative humidity	in %	64	63	60	64	72	80	79	74	64	60	64	64	67	9	6
7 Mean precipitation	in mm	31	28	32	110	300	376	343	224	80	52	35	26	1636		7
8 Maximum precipitation	in mm															8
9 Minimum precipitation	in mm															9
10 Maximum precipitation in 24 h	in mm															10
11 Mean number of days with precipitation	>0,1 mm															11
12 Mean duration of sunshine	in h															12
13 Mean quantity of radiation	in ly/day															13
14 Mean potential evaporation	in mm															14
15 Mean windspeed	in m/sec															15
16 Mean predominent direction of the wind																16

43 Station/Country Paramaribo/Surinam

Location 5°51'N/55°10'W Height above sealevel 3 m Climate symbol: Köppen Af Troll V,1

	J	F	M	A	M	J	J	A	S	O	N	D	year	P	
1 Mean daily temperature	in °C	26.4	26.6	27.0	27.2	26.8	26.8	27.1	27.9	28.5	28.5	28.0	26.9	27.3	30	1
2 Mean daily maximum temperature	in °C	29.4	29.4	29.4	30.0	30.0	30.0	30.6	31.7	32.8	32.8	31.7	30.0	30.6	35	2
3 Mean daily minimum temperature	in °C	22.2	21.7	22.2	22.8	22.8	22.8	22.8	22.8	22.8	22.8	22.8	22.8	22.8	35	3
4 Absolute maximum temperature	in °C	35.9	35.2	35.1	35.3	34.8	34.7	34.6	35.2	35.9	36.7	36.0	37.3	37.3	67	4
5 Absolute minimum temperature	in °C	15.8	16.6	17.1	17.4	17.9	17.8	18.2	18.3	18.5	19.4	18.0	17.5	15.6	67	5
6 Mean relative humidity	in %	85	82	82	82	85	86	83	79	77	77	80	84	82	35	6
7 Mean precipitation	in mm	193	150	162	232	321	303	228	167	86	87	109	174	2210	30	7
8 Maximum precipitation	in mm															8
9 Minimum precipitation	in mm															9
10 Maximum precipitation in 24 h	in mm	178	202	108	148	144	127	96	112	76	82	88	81	202	71	10
11 Mean number of days with precipitation	>0,1 mm	22	18	19	20	26	27	25	19	12	12	15	22	237	53	11
12 Mean duration of sunshine	in h	174	170	195	177	167	189	239	270	279	282	228	180	2550		12
13 Mean quantity of radiation	in ly/day															13
14 Mean potential evaporation	in mm	115	106	119	120	125	121	123	143	142	144	137	118	1513	87	14
15 Mean windspeed	in m/sec	1.4	1.6	1.6	1.6	1.3	1.1	1.2	1.4	1.6	1.6	1.5	1.4	1.5	30	15
16 Mean predominent direction of the wind		ENE	ENE	ENE	ENE	ENE	ENE	E	E	E	E	E	ENE		30	16

44 Station/Country Tafelberg/Surinam

Location 3°47'N/56°03'W Height above sealevel 344 m Climate symbol: Köppen Af Troll V,1

	J	F	M	A	M	J	J	A	S	O	N	D	year	P	
1 Mean daily temperature	in °C	24.9	24.9	25.3	25.5	25.4	25.3	25.7	26.4	27.0	27.3	26.8	25.7	25.9	10	1
2 Mean daily maximum temperature	in °C															2
3 Mean daily minimum temperature	in °C															3
4 Absolute maximum temperature	in °C	32.4	31.8	32.9	32.5	32.2	31.0	31.7	32.2	33.2	33.4	33.7	33.0	33.7		4
5 Absolute minimum temperature	in °C	14.9	17.3	17.6	18.9	18.0	18.8	18.3	18.3	17.7	17.4	18.2	17.5	14.9		5
6 Mean relative humidity	in %	83	81	80	80	82	82	79	76	72	70	74	78	78	10	6
7 Mean precipitation	in mm	306	249	253	284	402	401	321	188	65	48	109	188	2814		7
8 Maximum precipitation	in mm															8
9 Minimum precipitation	in mm															9
10 Maximum precipitation in 24 h	in mm	80	106	101	193	157	65	87	76	97	54	63	70	193		10
11 Mean number of days with precipitation	>0,25 mm	17	13	13	13	20	19	16	11	4	3	7	13	148		11
12 Mean duration of sunshine	in h															12
13 Mean quantity of radiation	in ly/day															13
14 Mean potential evaporation	in mm															14
15 Mean windspeed	in m/sec	1.3	1.3	1.1	0.9	1.0	1.1	1.1	1.2	1.3	1.3	0.9	1.1	1.1	10	15
16 Mean predominent direction of the wind																16

45 Station/Country Maripasoula/French Guiana

Location 3°38'N/54°02'W   Height above sealevel 104 m   Climate symbol: Köppen Af   Troll V,1

		J	F	M	A	M	J	J	A	S	O	N	D	year	P	
1	Mean daily temperature	in °C	24,5	24,5	24,8	25,1	25,0	24,7	24,8	25,3	25,6	26,0	26,0	25,2	25,1	10	1
2	Mean daily maximum temperature	in °C															2
3	Mean daily minimum temperature	in °C															3
4	Absolute maximum temperature	in °C	32,0	31,8	32,4	32,9	33,0	33,5	33,3	33,5	31,0	35,4	34,9	33,1	35,4	10	4
5	Absolute minimum temperature	in °C	16,9	18,0	18,4	18,9	19,8	19,5	19,4	19,4	18,9	17,9	17,9	19,0	16,9	10	5
6	Mean relative humidity	in %															6
7	Mean precipitation	in mm	218	214	212	249	399	278	196	136	77	83	92	234	2388	10	7
8	Maximum precipitation	in mm															8
9	Minimum precipitation	in mm															9
10	Maximum precipitation in 24 h	in mm	50	75	70	59	115	60	45	69	75	70	35	137	137	10	10
11	Mean number of days with precipitation	> 0,1 mm	24	22	20	21	27	26	23	17	11	9	13	22	235	10	11
12	Mean duration of sunshine	in h	151	142	161	190	151	170	215	248	257	266	230	181	2362	5	12
13	Mean quantity of radiation	in ly/day															13
14	Mean potential evaporation	in mm															14
15	Mean windspeed	in m/sec	1,7	1,8	2,2	1,7	1,5	1,3	1,3	1,3	1,3	1,6	1,6	1,5	1,6	10	15
16	Mean predominent direction of the wind		NE	NE	NE	NE	NE	E	E	E	E	E	E	E		10	16

46 Station/Country Cayenne/French Guiana

Location 4°50'N/52°22'W   Height above sealevel 8 m   Climate symbol: Köppen Af   Troll V,1

		J	F	M	A	M	J	J	A	S	O	N	D	year	P	
1	Mean daily temperature	in °C	25,2	25,3	25,6	25,7	25,5	25,1	25,2	25,6	26,2	26,2	25,9	25,4	25,6	10	1
2	Mean daily maximum temperature	in °C	31	31	32	32	32	32	33	33	34	34	33	32	34		2
3	Mean daily minimum temperature	in °C	25	25	26	26	26	25	25	26	26	26	26	25	26		3
4	Absolute maximum temperature	in °C	32,8	33,9	33,3	33,3	33,3	33,9	33,9	35,6	36,1	36,1	35,0	33,9	36,1	37	4
5	Absolute minimum temperature	in °C	17,4	19,2	18,9	18,3	19,9	18,9	19,2	19,3	18,8	19,1	17,2	19,5	17,2	37	5
6	Mean relative humidity	in %															6
7	Mean precipitation	in mm	431	423	432	480	590	457	274	144	32	42	122	317	3744	30	7
8	Maximum precipitation	in mm															8
9	Minimum precipitation	in mm															9
10	Maximum precipitation in 24 h	in mm	418	375	259	596	223	148	168	88	61	95	138	152	596	37	10
11	Mean number of days with precipitation	> 0,1 mm	27	24	24	23	29	28	24	18	8	9	15	24	253	15	11
12	Mean duration of sunshine	in h	143	113	137	130	137	170	208	233	261	266	234	192	2224		12
13	Mean quantity of radiation	in ly/day															13
14	Mean potential evaporation	in mm	133	125	138	139	144	140	146	151	151	153	143	141	1704	51	14
15	Mean windspeed	in m/sec	3,4	3,8	4,0	3,8	3,2	2,8	2,8	3,2	3,7	3,9	3,8	3,4	3,5	15	15
16	Mean predominent direction of the wind		ENE	ENE	ENE	E	E	E	E	E	E	E	E	E		15	16

47 Station/Country Puerto Baquerizo Moreno (Galapagos Islands)/Ecuador

Location 0°54'S/89°37'W   Height above sealevel 6 m   Climate symbol: Köppen BWh   Troll V,4

		J	F	M	A	M	J	J	A	S	O	N	D	year	P	
1	Mean daily temperature	in °C	26,0	26,5	26,5	26,5	25,5	24,0	22,0	21,5	21,0	22,0	23,0	24,0	24,0	4	1
2	Mean daily maximum temperature	in °C															2
3	Mean daily minimum temperature	in °C															3
4	Absolute maximum temperature	in °C	31	33	32	32	31	30	29	28	28	29	29	30	33	10	4
5	Absolute minimum temperature	in °C	17	20	19	18	16	16	17	17	16	16	17	18	16	10	5
6	Mean relative humidity	in %	80	80	81	80	78	77	79	79	79	77	75	78	79	10	6
7	Mean precipitation	in mm	48	67	85	35	16	2	4	5	7	7	5	7	288	10	7
8	Maximum precipitation	in mm	135	205	488	217	84	3	12	12	14	14	10	12	785	10	8
9	Minimum precipitation	in mm	10	5	0	0	0	1	1	0	1	0	2	1	93	10	9
10	Maximum precipitation in 24 h	in mm	34	70	258	67	35	2	2	3	10	5	5	4	258	10	10
11	Mean number of days with precipitation	> 0,1 mm	12	8	7	5	4	4	12	11	13	11	9	11	107	10	11
12	Mean duration of sunshine	in h	178	218	234	233	238	227	193	174	144	162	165	179	2343	6	12
13	Mean quantity of radiation	in ly/day															13
14	Mean potential evaporation	in mm	177	151	146	128	92	66	58	63	88	93	116	140	1291	4	14
15	Mean windspeed	in m/sec	3,6	2,8	2,2	3,7	4,4	5,1	4,4	4,3	4,3	4,1	4,6	4,6	4,0	8	15
16	Mean predominent direction of the wind		SE	SE	SE	SE	SE	SE	SE	SE	SE	SE	S	SE		8	16

48 Station/Country Quito/Ecuador

Location 0°13'S/78°30'W   Height above sealevel 2818 m   Climate symbol: Köppen Cwb   Troll V,2

		J	F	M	A	M	J	J	A	S	O	N	D	year	P	
1	Mean daily temperature	in °C	13,0	13,0	12,9	13,0	13,1	13,0	12,9	13,1	13,2	12,9	12,9	13,0	13,0	30	1
2	Mean daily maximum temperature	in °C	22,2	21,7	21,7	21,1	21,1	21,7	22,2	22,8	22,8	22,2	22,2	22,2	22,2	13	2
3	Mean daily minimum temperature	in °C	7,8	8,3	8,3	8,3	8,3	7,2	6,7	7,2	7,2	7,8	7,2	7,8	7,8	13	3
4	Absolute maximum temperature	in °C	26	25	25	25	26	26	26	26	28	27	26	27	28	10	4
5	Absolute minimum temperature	in °C	2	3	4	4	4	2	3	3	2	3	3	3	2	10	5
6	Mean relative humidity	in %	81	81	82	84	80	76	70	69	72	81	82	81	78	10	6
7	Mean precipitation	in mm	124	135	159	180	130	49	18	22	83	133	110	107	1250	65	7
8	Maximum precipitation	in mm	149	219	220	285	171	93	50	75	131	223	254	122	1366	10	8
9	Minimum precipitation	in mm	19	35	44	59	47	8	1	2	37	85	22	29	890	10	9
10	Maximum precipitation in 24 h	in mm	44	47	40	54	32	24	18	27	28	54	35	34	54	10	10
11	Mean number of days with precipitation	> 0,1 mm	15	16	18	22	19	13	8	7	11	20	18	16	183	10	11
12	Mean duration of sunshine	in h	180	144	141	131	175	170	212	199	189	166	156	186	2049	8	12
13	Mean quantity of radiation	in ly/day															13
14	Mean potential evaporation	in mm	60	55	61	59	59	58	57	60	60	60	60	61	710	23	14
15	Mean windspeed	in m/sec	0,5	0,6	0,6	0,6	0,7	0,9	1,1	1,1	0,9	0,6	0,6	0,6	0,7	8	15
16	Mean predominent direction of the wind		E	E	E	W	W	W	S	E	W	W	E			8	16

49 Station/Country Pichilingue/Ecuador

Location 1°06'S/79°29'W Height above sealevel 73 m Climate symbol: Köppen Aw Troll V.3

		J	F	M	A	M	J	J	A	S	O	N	D	year	P	
1 Mean daily temperature	in °C	22,9	23,1	23,4	23,5	23,0	22,8	22,8	23,2	23,7	23,8	23,6	23,3	23,2	14	1
2 Mean daily maximum temperature	in °C															2
3 Mean daily minimum temperature	in °C															3
4 Absolute maximum temperature	in °C	34	33	34	37	34	33	31	35	34	35	35	38	37	10	4
5 Absolute minimum temperature	in °C	18	17	19	17	18	16	15	16	17	17	15	16	15	10	5
6 Mean relative humidity	in %	86	86	87	87	87	88	87	84	82	81	80	80	85	8	6
7 Mean precipitation	in mm	468	438	389	358	116	41	5	9	7	19	14	120	1984	10	7
8 Maximum precipitation	in mm	629	836	622	712	378	149	20	40	38	107	47	300	3005	10	8
9 Minimum precipitation	in mm	247	195	150	101	12	1	0	0	0	0	1	27	1236	10	9
10 Maximum precipitation in 24 h	in mm	139	144	117	135	74	39	4	40	21	102	19	88	144	9	10
11 Mean number of days with precipitation	> 0,1 mm	28	25	27	22	15	11	3	5	4	7	5	15	167	9	11
12 Mean duration of sunshine	in h	78	82	102	112	88	55	47	65	62	52	50	64	857	7	12
13 Mean quantity of radiation	in ly/day															13
14 Mean potential evaporation	in mm	94	85	100	97	94	88	90	97	110	103	97	97	1140		14
15 Mean windspeed	in m/sec	1,1	1,2	1,2	1,1	1,3	1,2	1,2	1,2	1,2	1,2	1,3	1,2	1,2	7	15
16 Mean predominent direction of the wind		SW	SW	SW	SW	SW	SW	SW	SW	SW	SE	SW	SW		7	16

50 Station/Country Puyo/Ecuador

Location 1°35'S/77°54'W Height above sealevel 950 m Climate symbol: Köppen Af Troll V.1

		J	F	M	A	M	J	J	A	S	O	N	D	year	P	
1 Mean daily temperature	in °C	21,5	21,5	21,5	21,5	21,5	20,5	20,5	21,0	21,5	21,5	22,0	22,0	21,5	5	1
2 Mean daily maximum temperature	in °C	26	26	26	26	26	25	25	26	27	27	27	27	26	5	2
3 Mean daily minimum temperature	in °C	17	17	17	17	17	16	16	16	16	16	17	17	17	5	3
4 Absolute maximum temperature	in °C	30	31	31	30	30	29	29	30	30	30	30	31	31	5	4
5 Absolute minimum temperature	in °C	11	11	13	11	12	9	12	12	11	13	13	13	9	5	5
6 Mean relative humidity	in %	89	89	89	89	89	89	89	87	87	88	88	88	88	5	6
7 Mean precipitation	in mm	299	294	391	453	324	391	339	345	354	360	367	377	4294	5	7
8 Maximum precipitation	in mm	411	389	513	556	422	453	583	470	495	429	528	564	4888	5	8
9 Minimum precipitation	in mm	203	206	287	380	204	276	236	227	231	304	295	212	4000	5	9
10 Maximum precipitation in 24 h	in mm	67	112	75	95	74	102	83	84	74	84	79	130	130	5	10
11 Mean number of days with precipitation	> 0,1 mm	26	23	27	27	27	26	27	26	26	28	26	26	315	5	11
12 Mean duration of sunshine	in h	74	75	50	61	74	79	80	101	98	98	105	90	985	5	12
13 Mean quantity of radiation	in ly/day															13
14 Mean potential evaporation	in mm	88	79	87	85	87	75	78	84	85	87	92	95	1022	5	14
15 Mean windspeed	in m/sec	0,8	0,8	0,8	0,8	0,8	0,7	0,7	0,8	0,8	0,9	0,8	0,8	0,8	5	15
16 Mean predominent direction of the wind		NE	NE	NE	E	NE	NE	SE	NE	E	E	E	NE			16

51 Station/Country Guayaquil/Ecuador

Location 2°12'S/79°53'W Height above sealevel 6 m Climate symbol: Köppen Aw Troll V.4

		J	F	M	A	M	J	J	A	S	O	N	D	year	P	
1 Mean daily temperature	in °C	25,5	26,0	26,4	26,3	25,8	24,4	23,5	23,2	23,8	24,0	24,6	25,4	24,9	30	1
2 Mean daily maximum temperature	in °C	31,1	30,8	31,1	31,7	31,1	30,6	28,9	30,0	30,6	30,0	31,1	31,1	30,8	3	2
3 Mean daily minimum temperature	in °C	21,1	21,7	22,2	21,7	20,0	20,0	19,4	18,3	18,9	20,0	20,0	21,1	20,6	2	3
4 Absolute maximum temperature	in °C	35	35	35	35	34	34	32	30	34	35	34	35	35	10	4
5 Absolute minimum temperature	in °C	19	20	20	19	19	18	17	17	17	18	17	19	17	10	5
6 Mean relative humidity	in %	78	81	81	78	78	79	78	76	75	74	74	72	77	10	6
7 Mean precipitation	in mm	217	189	231	133	38	15	0	0	0	4	1	15	843	10	7
8 Maximum precipitation	in mm	519	365	434	385	207	127	2	0	2	18	6	80	1332	10	8
9 Minimum precipitation	in mm	52	95	111	11	0	0	0	0	0	0	0	0	397	10	9
10 Maximum precipitation in 24 h	in mm	155	93	98	53	41	125	2	0	2	17	5	28	155	10	10
11 Mean number of days with precipitation	> 0,1 mm	16	16	19	13	5	2	0	0	0	1	1	4	77	10	11
12 Mean duration of sunshine	in h	106	112	141	149	165	123	130	146	161	126	115	140	1614	6	12
13 Mean quantity of radiation	in ly/day															13
14 Mean potential evaporation	in mm	136	127	144	139	136	114	107	104	108	114	121	136	1486	10	14
15 Mean windspeed	in m/sec	2,1	1,8	1,9	2,2	2,3	2,5	3,1	3,6	3,3	3,4	2,9	3,0	2,7	8	15
16 Mean predominent direction of the wind		NE	NE	NE	NE	SW	SW	SW	SW	SW	SW	SW	SW		8	16

52 Station/Country Iquitos/Peru

Location 3°46'S/73°20'W Height above sealevel 104 m Climate symbol: Köppen Af Troll V.1

		J	F	M	A	M	J	J	A	S	O	N	D	year	P	
1 Mean daily temperature	in °C	27,4	26,6	26,5	26,4	26,0	25,6	25,6	26,3	26,6	26,7	26,9	27,5	26,5	10	1
2 Mean daily maximum temperature	in °C															2
3 Mean daily minimum temperature	in °C															3
4 Absolute maximum temperature	in °C	36	36	36	38	36	36	35	37	37	37	37	38	37	22	4
5 Absolute minimum temperature	in °C	17	15	17	18	16	15	14	11	15	18	18	16	11	22	5
6 Mean relative humidity	in %	80	81	82	84	83	82	81	79	78	79	80	80	81	22	6
7 Mean precipitation	in mm	256	276	349	306	271	199	165	157	191	214	244	217	2845	22	7
8 Maximum precipitation	in mm	361	719	1069	553	620	838	345	656	336	410	620	439	5821	22	8
9 Minimum precipitation	in mm	109	63	90	176	153	61	25	34	51	99	75	86	2259	22	9
10 Maximum precipitation in 24 h	in mm	160	144	201	135	192	198	90	247	114	105	210	150	247	22	10
11 Mean number of days with precipitation	> 0,1 mm															11
12 Mean duration of sunshine	in h															12
13 Mean quantity of radiation	in ly/day															13
14 Mean potential evaporation	in mm	158	143	152	138	129	121	111	141	140	156	152	158	1699	2	14
15 Mean windspeed	in m/sec	1,0	1,0	1,0	1,0	1,0	1,0	1,0	1,0	1,0	1,0	1,5	1,0	1,0	22	15
16 Mean predominent direction of the wind		NE	NE	NE	NE	S	S	S	S	NE	NE	NE	E		22	16

53 Station/Country **Chiclayo/Peru**

Location 6°47'S/79°50'W | Height above sealevel 31 m | Climate symbol: Köppen **BWh** | Troll **V,5**

		J	F	M	A	M	J	J	A	S	O	N	D	year	P	
1	Mean daily temperature	in °C	24,8	25,8	25,9	24,4	22,3	20,1	18,8	18,6	19,3	19,9	20,7	22,4	21,9	18	1
2	Mean daily maximum temperature	in °C															2
3	Mean daily minimum temperature	in °C															3
4	Absolute maximum temperature	in °C	34	35	34	33	32	31	29	29	29	33	29	33	35	16	4
5	Absolute minimum temperature	in °C	13	15	15	14	13	13	12	10	12	10	13	14	10	16	5
6	Mean relative humidity	in %	73	73	73	75	75	76	80	80	79	79	77	76	76	16	6
7	Mean precipitation	in mm	10	2	8	3	<1	0	<1	0	<1	1	4	2	31	16	7
8	Maximum precipitation	in mm	105	17	70	12	2	0	1	0	4	6	26	18	118	16	8
9	Minimum precipitation	in mm	0	0	0	0	0	0	0	0	0	0	0	0	<1	16	9
10	Maximum precipitation in 24 h	in mm	105	5	70	4	2	0	1	0	1	3	23	18	105	10	10
11	Mean number of days with precipitation	>0,1 mm															11
12	Mean duration of sunshine	in h															12
13	Mean quantity of radiation	in ly/day															13
14	Mean potential evaporation	in mm	107	110	120	110	88	72	61	61	61	67	74	95	1026	3	14
15	Mean windspeed	in m/sec	5,7	5,1	5,1	5,7	5,7	5,1	5,1	5,7	5,7	6,2	5,7	6,2	5,7	16	15
16	Mean predominent direction of the wind		S	S	S	S	S	S	S	S	S	S	S	S		16	16

54 Station/Country **Cajamarca/Peru**

Location 7°08'S/78°28'W | Height above sealevel 2621 m | Climate symbol: Köppen **Cwa** | Troll **V,3**

		J	F	M	A	M	J	J	A	S	O	N	D	year	P	
1	Mean daily temperature	in °C	21,6	22,7	22,5	20,8	18,2	16,4	15,6	15,1	15,4	16,4	17,8	20,0	18,5	16	1
2	Mean daily maximum temperature	in °C	21,7	21,1	21,1	21,1	21,7	21,1	21,1	21,7	21,7	21,7	22,2	21,7	21,7	9	2
3	Mean daily minimum temperature	in °C	8,9	8,9	8,9	8,3	6,7	5,6	5,0	5,6	7,2	8,3	7,8	8,3	7,2	9	3
4	Absolute maximum temperature	in °C	26	27	25	25	25	26	25	26	26	29	25	25	29	12	4
5	Absolute minimum temperature	in °C	0	0	0	0	0	0	0	0	0	1	0	0	0	12	5
6	Mean relative humidity	in %	67	67	72	69	64	58	55	55	57	64	64	64	63	12	6
7	Mean precipitation	in mm	89	89	114	96	32	8	5	11	25	85	79	83	716	12	7
8	Maximum precipitation	in mm	185	183	235	212	72	25	28	44	85	126	143	176	924	12	8
9	Minimum precipitation	in mm	33	4	18	15	1	0	0	0	3	23	20	22	441	12	9
10	Maximum precipitation in 24 h	in mm	31	51	41	38	26	12	13	10	18	29	40	44	51	12	10
11	Mean number of days with precipitation	>0,1 mm	13	17	17	14	9	4	2	2	9	9	9	11	115	5	11
12	Mean duration of sunshine	in h															12
13	Mean quantity of radiation	in ly/day															13
14	Mean potential evaporation	in mm	64	58	62	59	55	51	52	54	58	63	61	65	702	12	14
15	Mean windspeed	in m/sec	1,5	1,5	1,5	1,5	1,5	2,6	3,1	3,1	2,6	2,1	1,5	1,5	2,1	12	15
16	Mean predominent direction of the wind		S	S	S	S	S	N	NE	E	E	SE	S	S		12	16

55 Station/Country **Tingo María/Peru**

Location 9°08'S/75°57'W | Height above sealevel 665 m | Climate symbol: Köppen **Af** | Troll **V,1**

		J	F	M	A	M	J	J	A	S	O	N	D	year	P	
1	Mean daily temperature	in °C	24,5	24,5	24,5	25,0	25,0	24,5	24,5	24,5	24,5	24,5	25,0	25,0	25,0		1
2	Mean daily maximum temperature	in °C	30	30	30	31	31	31	31	31	31	31	31	31	31	14	2
3	Mean daily minimum temperature	in °C	19	19	19	19	19	18	18	18	18	18	19	19	19	16	3
4	Absolute maximum temperature	in °C	38	35	36	36	35	34	34	36	36	37	38	35	37	14	4
5	Absolute minimum temperature	in °C	15	13	13	13	14	12	10	11	12	10	13	12	10	16	5
6	Mean relative humidity	in %	81	81	81	80	78	79	78	76	77	79	79	80	79	16	6
7	Mean precipitation	in mm	394	359	353	319	238	119	132	125	158	280	278	317	3072	16	7
8	Maximum precipitation	in mm	796	585	618	579	419	252	240	278	308	455	550		3860	16	8
9	Minimum precipitation	in mm	229	126	47	83	26	5		47	71	130	169	137	1994	18	9
10	Maximum precipitation in 24 h	in mm															10
11	Mean number of days with precipitation	>0,25 mm	20	19	17	15	15	10	12	9	12	16	15	18	178		11
12	Mean duration of sunshine	in h	112	102	115	129	161	159	180	208	168	164	138	121	1757		12
13	Mean quantity of radiation	in ly/day															13
14	Mean potential evaporation	in mm	114	101	111	106	102	91	88	102	105	113	111	114	1258		14
15	Mean windspeed	in m/sec	0,5	0,5	1,0	0,5	1,0	1,0	1,0	1,0	1,0	1,0	1,0	1,0	1,0	18	15
16	Mean predominent direction of the wind		N	N	N	N	N	N	N	N	N	N	N	N		18	16

56 Station/Country **Lima/Peru**

Location 12°00'S/77°07'W | Height above sealevel 11 m | Climate symbol: Köppen **BWh** | Troll **V,5**

		J	F	M	A	M	J	J	A	S	O	N	D	year	P	
1	Mean daily temperature	in °C	21,5	22,3	21,9	20,1	17,8	16,0	15,3	15,1	15,4	16,3	17,7	19,4	18,2	30	1
2	Mean daily maximum temperature	in °C	27,8	28,3	28,3	26,7	23,3	20,0	19,4	18,9	20,0	21,7	23,3	25,8	23,9	11	2
3	Mean daily minimum temperature	in °C	18,9	19,4	18,9	17,2	15,8	14,4	13,9	13,3	13,9	14,4	15,8	16,7	16,1	11	3
4	Absolute maximum temperature	in °C	31	30	29	28	27	25	23	23	22	23	27	28	31	10	4
5	Absolute minimum temperature	in °C	15	15	15	11	10	8	9	10	12	13	10	13	8	10	5
6	Mean relative humidity	in %	83	83	84	85	86	85	85	87	87	85	83	83	85	10	6
7	Mean precipitation	in mm	1	<1	<1	<1	<1	<1	2	2	1	<1	<1	<1	10	10	7
8	Maximum precipitation	in mm	11	2	3	<1	2	3	8	10	4	1	<1	1	20	10	8
9	Minimum precipitation	in mm	0	0	0	0	0	0	0	1	0	0	0	0	3	10	9
10	Maximum precipitation in 24 h	in mm	8	1	2	<1	1		5	3	1	1	<1	2	8	10	10
11	Mean number of days with precipitation	>0,1 mm	<1	<1	<1	<1	<1	1	1	2	1	<1	<1	<1	7		11
12	Mean duration of sunshine	in h	195	192	214	201	124	42	34	31	33	78	123	155	1422		12
13	Mean quantity of radiation	in ly/day															13
14	Mean potential evaporation	in mm	107	103	109	86	66	49	48	46	50	60	70	90	884	20	14
15	Mean windspeed	in m/sec	3,8	3,1	3,1	3,1	2,8	2,1	2,8	3,1	3,1	3,1	3,8	3,8	3,1	10	15
16	Mean predominent direction of the wind		S	S	S	S	S	S	S	S	S	S	S	S		10	16

57 Station/Country Huancayo/Peru

Location 12°07'S/75°20'W Height above sealevel 3380 m Climate symbol: Köppen Cwb Troll V.3

		J	F	M	A	M	J	J	A	S	O	N	D	year	P	
1 Mean daily temperature	in °C	12,5	12,5	12,0	12,0	11,0	9,5	9,5	11,0	12,5	13,0	13,0	12,5	11,8	20	1
2 Mean daily maximum temperature	in °C	18	18	18	19	19	19	19	20	20	20	20	19	19	20	2
3 Mean daily minimum temperature	in °C	7	7	6	5	3	0	0	2	5	6	6	6	4	20	3
4 Absolute maximum temperature	in °C	24	24	24	23	23	23	22	24	24	25	26	25	26	38	4
5 Absolute minimum temperature	in °C	0	0	-1	-4	-6	-9	-10	-6	-2	-3	-3	-3	-10	38	5
6 Mean relative humidity	in %	73	78	77	72	65	57	56	55	61	64	62	68	66	20	6
7 Mean precipitation	in mm	119	123	107	55	25	8	8	14	40	69	67	89	724	20	7
8 Maximum precipitation	in mm	200	203	185	94	63	35	23	68	95	125	115	144	902	20	8
9 Minimum precipitation	in mm	52	43	39	23	1	0	0	0	19	26	33	19	575	20	9
10 Maximum precipitation in 24 h	in mm	43	46	54	29	34	19	19	22	37	38	37	39	54	36	10
11 Mean number of days with precipitation	>0,25 mm	22	24	20	15	9	2	2	4	12	14	14	20	158		11
12 Mean duration of sunshine	in h	177	152	173	195	228	254	258	238	197	205	207	204	2488	16	12
13 Mean quantity of radiation	in ly/day															13
14 Mean potential evaporation	in mm	61	54	57	51	47	42	41	47	54	61	61	64	640	23	14
15 Mean windspeed	in m/sec															15
16 Mean predominent direction of the wind																16

58 Station/Country Puerto Maldonado/Peru

Location 12°38'S/69°12'W Height above sealevel 256 m Climate symbol: Köppen Aw Troll V.1

		J	F	M	A	M	J	J	A	S	O	N	D	year	P	
1 Mean daily temperature	in °C	26,0	26,0	25,5	25,5	25,0	23,0	23,0	25,0	26,0	26,0	26,5	26,0	25,3	9	1
2 Mean daily maximum temperature	in °C	31	31	31	31	31	29	30	32	33	32	32	31	31	9	2
3 Mean daily minimum temperature	in °C	21	21	20	20	19	17	16	18	19	20	21	21	19	9	3
4 Absolute maximum temperature	in °C	36	36	36	36	37	39	38	38	38	38	38	35	39	12	4
5 Absolute minimum temperature	in °C	15	14	15	12	7	10	10	10	9	14	14	14	7	12	5
6 Mean relative humidity	in %	80	81	80	77	76	76	72	67	63	72	74	78	75	9	6
7 Mean precipitation	in mm	262	271	289	118	119	54	55	53	97	140	173	296	1927	9	7
8 Maximum precipitation	in mm	372	488	494	395	420	187	99	131	177	201	267	491	2541	9	8
9 Minimum precipitation	in mm	160	184	177	33	20	4	6	4	25	74	88	105	1433	9	9
10 Maximum precipitation in 24 h	in mm															10
11 Mean number of days with precipitation	> 0,1 mm															11
12 Mean duration of sunshine	in h															12
13 Mean quantity of radiation	in ly/day															13
14 Mean potential evaporation	in mm	139	123	123	115	110	83	85	111	126	135	140	140	1429	9	14
15 Mean windspeed	in m/sec	2,6	2,6	2,1	2,6	2,1	2,1	2,1	2,6	2,6	2,6	2,6	2,6		9	15
16 Mean predominent direction of the wind		N	N	N	E	E	E	E	E	N	N	N	N		9	16

59 Station/Country Cuzco/Peru

Location 13°33'S/71°59'W Height above sealevel 3312 m Climate symbol: Köppen Cwb Troll V.2

		J	F	M	A	M	J	J	A	S	O	N	D	year	P	
1 Mean daily temperature	in °C	15,9	15,7	15,6	15,5	15,8	15,2	14,5	15,1	15,2	15,9	16,5	16,2	15,6	16	1
2 Mean daily maximum temperature	in °C	20,0	20,6	21,1	21,7	21,1	20,6	21,1	21,1	21,7	22,2	22,8	21,7	21,1	13	2
3 Mean daily minimum temperature	in °C	7,2	7,2	6,7	4,4	1,7	0,6	-0,6	1,1	4,4	6,1	6,1	6,7	4,4	13	3
4 Absolute maximum temperature	in °C	25	24	24	24	23	25	23	25	26	28	27	27	27	17	4
5 Absolute minimum temperature	in °C	1	2	1	1	-3	-4	-4	-4	-3	0	-1	3	-4	17	5
6 Mean relative humidity	in %	64	66	65	61	55	48	47	46	51	51	52	59	55	17	6
7 Mean precipitation	in mm	151	139	106	39	12	2	5	6	22	52	77	134	750	17	7
8 Maximum precipitation	in mm	253	196	165	82	27	15	42	24	48	84	183	254	982	17	8
9 Minimum precipitation	in mm	57	69	37	10	0	0	0	0	10	22	32	40	390	17	9
10 Maximum precipitation in 24 h	in mm	38	43	47	20	14	8	38	11	22	26	37	47	47	17	10
11 Mean number of days with precipitation	> 0,1 mm	18	13	11	8	3	2	2	2	7	8	12	16	102		11
12 Mean duration of sunshine	in h	143	121	170	210	239	228	257	236	195	198	185	158	2350		12
13 Mean quantity of radiation	in ly/day															13
14 Mean potential evaporation	in mm	65	58	62	53	47	39	39	40	45	54	64	67	660	14	14
15 Mean windspeed	in m/sec	2,6	2,6	2,6	2,1	2,1	2,6	3,1	3,6	3,6	3,6	3,1	3,1		17	15
16 Mean predominent direction of the wind		NW	E	NE	NW	W	NW	NW	W	NE	NE	NE	W		17	16

60 Station/Country Apolo/Bolivia

Location 14°43'S/68°30'W Height above sealevel 1382 m Climate symbol: Köppen Aw Troll V.2

		J	F	M	A	M	J	J	A	S	O	N	D	year	P	
1 Mean daily temperature	in °C	21	21	21	20	19	18	18	19,5	21	21,5	21,5	21,5	20,5	10	1
2 Mean daily maximum temperature	in °C	25	25	25	24	23	22	22	24	26	26	26	26	25	10	2
3 Mean daily minimum temperature	in °C	17	17	17	16	15	14	14	15	16	17	17	17	16	10	3
4 Absolute maximum temperature	in °C	31	30	29	29	27	26	26	29	31	31	32	32	32	10	4
5 Absolute minimum temperature	in °C	13	14	11	8	5	6	7	6	11	10	11	13	5	10	5
6 Mean relative humidity	in %															6
7 Mean precipitation	in mm	194	176	177	113	55	32	25	32	69	111	120	220	1324	10	7
8 Maximum precipitation	in mm	367	267	255	162	109	65	55	65	137	274	267	292	1650	10	8
9 Minimum precipitation	in mm	75	86	75	54	16	7	-	1	5	46	26	189	968	10	9
10 Maximum precipitation in 24 h	in mm															10
11 Mean number of days with precipitation	>0,25 mm															11
12 Mean duration of sunshine	in h															12
13 Mean quantity of radiation	in ly/day															13
14 Mean potential evaporation	in mm	99	79	90	75	73	64	70	73	85	89	93	97	987		14
15 Mean windspeed	in m/sec	3,5	3	2,5	2,5	2,5	3	3	3	3	3,5	3	3	3	10	15
16 Mean predominent direction of the wind		N	N	N	N	N	N	N	N	N	N	N	N		10	16

61 Station/Country San Juan/Peru

Location 15°22'S/75°12'W — Height above sealevel 30 m — Climate symbol: Köppen BWh — Troll IV.5

		J	F	M	A	M	J	J	A	S	O	N	D	year	P	
1 Mean daily temperature	in °C	22,0	21,8	20,8	19,2	17,7	16,0	15,1	15,0	15,8	16,2	17,1	20,0	18,0	1	1
2 Mean daily maximum temperature	in °C	26	27	26	25	22	20	18	18	19	20	22	24	22	14	2
3 Mean daily minimum temperature	in °C	18	19	18	17	16	14	13	13	13	14	15	17	16	14	3
4 Absolute maximum temperature	in °C	29	29	29	28	26	23	22	21	22	23	25	27	29	14	4
5 Absolute minimum temperature	in °C	13	15	15	13	12	11	10	8	11	11	10	12	8	12	5
6 Mean relative humidity	in %	77	76	75	75	76	78	78	79	79	79	78	78	77	14	6
7 Mean precipitation	in mm	<1	<1	<1	<1	<1	1	<1	<1	2	2	<1	<1	8	14	7
8 Maximum precipitation	in mm	2	4	<1	4	2	11	2	7	7	10	5	1	32	14	8
9 Minimum precipitation	in mm	0	0	0	0	0	0	0	0	0	0	0	0	<1	14	9
10 Maximum precipitation in 24 h	in mm															10
11 Mean number of days with precipitation	> 0,1 mm															11
12 Mean duration of sunshine	in h															12
13 Mean quantity of radiation	in ly/day															13
14 Mean potential evaporation	in mm	111	92	91	50	65	50	47	48	51	61	64	95	825	14	14
15 Mean windspeed	in m/sec	5,1	5,1	5,1	5,1	5,7	5,7	5,7	6,2	6,2	6,2	5,7	5,1	5,7	14	15
16 Mean predominent direction of the wind		S	S	S	S	S	S	S	S	S	S	S	S		14	16

62 Station/Country Arequipa/Peru

Location 16°19'S/71°33'W — Height above sealevel 2525 m — Climate symbol: Köppen BWk — Troll V.5

		J	F	M	A	M	J	J	A	S	O	N	D	year	P	
1 Mean daily temperature	in °C	13,9	13,9	13,5	14,1	13,8	13,2	13,1	13,8	14,4	13,8	13,9	14,1	13,8	33	1
2 Mean daily maximum temperature	in °C															2
3 Mean daily minimum temperature	in °C															3
4 Absolute maximum temperature	in °C	27	27	28	27	26	27	27	28	27	27	28	29	29	21	4
5 Absolute minimum temperature	in °C	2	1	1	1	0	-3	-4	-3	-2	-1	0	0	-4	21	5
6 Mean relative humidity	in %	57	63	59	48	37	29	27	26	29	30	34	44	40	21	6
7 Mean precipitation	in mm	31	48	19	0	0	0	0	0	1	0	1	4	104	21	7
8 Maximum precipitation	in mm	150	173	184	8	3	1	0	0	8	3	8	18	486	21	8
9 Minimum precipitation	in mm	0	1	0	0	0	0	0	0	0	0	0	0	13	21	9
10 Maximum precipitation in 24 h	in mm	27	59	44	5	3	1	0	0	8	2	5	13	59	21	10
11 Mean number of days with precipitation	>0,25 mm	5	25	4	1	0	0	0	0	0	0	0	2	37		11
12 Mean duration of sunshine	in h															12
13 Mean quantity of radiation	in ly/day															13
14 Mean potential evaporation	in mm	71	62	63	58	52	43	45	51	57	64	61	68	695	8	14
15 Mean windspeed	in m/sec	4,1	3,6	3,6	3,6	4,1	4,6	4,6	4,6	4,1	4,1	3,6	4,1	4,1	21	15
16 Mean predominent direction of the wind		W	W	W	W	W	W	W	W	W	W	W	W		21	16

63 Station/Country La Paz/Bolivia

Location 16°30'S/68°08'W — Height above sealevel 3632 m — Climate symbol: Köppen Cwb — Troll V.2

		J	F	M	A	M	J	J	A	S	O	N	D	year	P	
1 Mean daily temperature	in °C	17,5	16,2	15,5	14,1	11,7	10,1	9,8	10,9	14,4	15,5	17,5	17,9	14,3		1
2 Mean daily maximum temperature	in °C	17,2	17,2	17,8	18,3	17,8	16,7	16,7	17,2	17,8	18,6	18,4	18,3	17,8	31	2
3 Mean daily minimum temperature	in °C	6,1	6,1	5,6	4,4	2,8	1,1	0,8	1,7	3,3	4,4	5,6	5,8	3,9	31	3
4 Absolute maximum temperature	in °C	25	24	24	23	23	22	21	22	24	25	24	25	25	10	4
5 Absolute minimum temperature	in °C	3	3	1	-1	-1	-3	-2	-1	1	2	1	2	-3	10	5
6 Mean relative humidity	in %	68	71	65	56	49	42	40	47	56	53	55	65	56	5	6
7 Mean precipitation	in mm	92	89	62	28	11	2	4	7	34	48	85		488	10	7
8 Maximum precipitation	in mm	144	130	124	50	38	5	15	19	78	48	113	113	664	10	8
9 Minimum precipitation	in mm	38	38	24	7	2	0	0	0	3	12	10	46	388	10	9
10 Maximum precipitation in 24 h	in mm	48	41	28	33	18	10	20	25	33	25	43	38			10
11 Mean number of days with precipitation	> 0,1 mm	21	18	16	9	5	2	2	4	9	9	11	18	124		11
12 Mean duration of sunshine	in h	183	153	149	165	223	240	236	217	189	180	171	186	2292	35	12
13 Mean quantity of radiation	in ly/day															13
14 Mean potential evaporation	in mm	58	51	55	49	44	35	34	40	47	57	62	63	595	35	14
15 Mean windspeed	in m/sec	2,1	1,5	2,1	2,1	1,5	2,1	2,8	2,1	2,1	2,1	2,1	1,5	2,1	10	15
16 Mean predominent direction of the wind		E	SE	E	SE	SE	W	W	W	W	W	E	SE		10	16

64 Station/Country Santa Cruz/Bolivia

Location 17°47'S/63°11'W — Height above sealevel 437 m — Climate symbol: Köppen Aw — Troll V.2

		J	F	M	A	M	J	J	A	S	O	N	D	year	P	
1 Mean daily temperature	in °C	25,8	25,4	24,9	23,4	21,6	20,0	19,6	22,0	25,2	26,3	26,6	26,2	23,9	20	1
2 Mean daily maximum temperature	in °C															2
3 Mean daily minimum temperature	in °C															3
4 Absolute maximum temperature	in °C	38	37	37	36	35	31	32	35	37	38	40	40	40	10	4
5 Absolute minimum temperature	in °C	13	13	10	9	4	3	4	5	6	11	13	13	3	10	5
6 Mean relative humidity	in %															6
7 Mean precipitation	in mm	181	136	94	68	63	53	34	40	52	88	116	141	1046	10	7
8 Maximum precipitation	in mm	377	279	219	156	104	107	123	134	111	235	248	265	1235	10	8
9 Minimum precipitation	in mm	88	46	8	12	17	16	3	2	2	11	28	50	820	10	9
10 Maximum precipitation in 24 h	in mm															10
11 Mean number of days with precipitation	> 0,1 mm	14	10	12	9	11	8	5	4	5	7	8	11	104		11
12 Mean duration of sunshine	in h															12
13 Mean quantity of radiation	in ly/day															13
14 Mean potential evaporation	in mm	146	125	123	93	76	59	661	89	111	130	147	152	1312	10	14
15 Mean windspeed	in m/sec	5,1	4,6	4,1	4,1	4,6	6,2	6,7	6,7	6,7	6,2	5,1	5,1	5,7	10	15
16 Mean predominent direction of the wind		NW	NW	NW	NW	S	NW	NW	NW	NW	NW	NW	NW		10	16

65 Station/Country Oruro/Bolivia

Location 17°58'S/67°07'W Height above sealevel 3 708 m Climate symbol: Köppen BWh Troll V,4

		J	F	M	A	M	J	J	A	S	O	N	D	year	P	
1	Mean daily temperature	in °C	20,9	20,6	19,9	18,8	18,5	14,7	14,7	17,6	19,7	19,9	20,5	20,0	18,6	20	1
2	Mean daily maximum temperature	in °C															2
3	Mean daily minimum temperature	in °C															3
4	Absolute maximum temperature	in °C	25	23	22	20	18	19	19	24	25	25	25	25	25	10	4
5	Absolute minimum temperature	in °C	-2	-3	-8	-10	-16	-22	-20	-20	-15	-13	-8	-4	-22	10	5
6	Mean relative humidity	in %															6
7	Mean precipitation	in mm	62	60	32	11	5	2	1	9	13	11	24	52	282	10	7
8	Maximum precipitation	in mm	143	115	90	28	19	14	6	33	33	35	56	90	415	10	8
9	Minimum precipitation	in mm	13	20	0	0	0	0	0	0	1	0	4	26	152	10	9
10	Maximum precipitation in 24 h	in mm															10
11	Mean number of days with precipitation	> 0,1 mm															11
12	Mean duration of sunshine	in h															12
13	Mean quantity of radiation	in ly/day															13
14	Mean potential evaporation	in mm	69	58	63	55	37	23	27	35	48	63	73	73	624	10	14
15	Mean windspeed	in m/sec	3,1	2,6	2,6	2,1	2,1	2,1	2,1	2,6	2,6	2,6	2,6	3,1	2,6	10	15
16	Mean predominant direction of the wind		E	E	N	N	S	S	S	N	N	N	S	E		10	16

66 Station/Country Uaupés/Brazil

Location 0°08'S/67°05'W Height above sealevel 85 m Climate symbol: Köppen Af Troll V,1

		J	F	M	A	M	J	J	A	S	O	N	D	year	P	
1	Mean daily temperature	in °C	25,5	25,8	25,6	25,4	25,1	24,7	24,3	25,0	25,5	25,8	26,1	25,7	25,4		1
2	Mean daily maximum temperature	in °C	31,1	31,7	31,1	31,1	30,6	30,0	29,4	30,6	31,7	31,7	32,2	31,1	31,1	15	2
3	Mean daily minimum temperature	in °C	22,2	22,2	22,2	22,2	22,2	21,7	21,1	21,7	21,7	21,7	22,2	22,2	21,7	15	3
4	Absolute maximum temperature	in °C	36,1	36,1	36,8	36,4	36,9	36,8	35,6	35,8	38,0	39,0	37,5	36,9	39,0		4
5	Absolute minimum temperature	in °C	20,0	19,8	19,1	19,6	17,7	17,6	16,0	18,5	18,8	18,7	19,4	19,3	16,0		5
6	Mean relative humidity	in %	85	88	88	88	87	86	86	84	84	84	90	93	85	2	6
7	Mean precipitation	in mm	284	261	284	263	329	244	234	186	160	164	190	270	2869		7
8	Maximum precipitation	in mm															8
9	Minimum precipitation	in mm															9
10	Maximum precipitation in 24 h	in mm	113	81	156	97	105	80	157	60	65	113	107	99	157		10
11	Mean number of days with precipitation	> 0,1 mm	21	17	18	19	23	21	21	18	16	15	15	19	223		11
12	Mean duration of sunshine	in h	161	154	157	143	144	144	163	196	196	188	187	165	1998		12
13	Mean quantity of radiation	in ly/day															13
14	Mean potential evaporation	in mm	126	119	123	116	112	102	99	113	117	130	131	134	2323		14
15	Mean windspeed	in m/sec	0,9	1,0	1,1	1,1	1,0	0,9	1,0	1,1	1,1	1,1	1,2	1,1	1,0		15
16	Mean predominant direction of the wind		N,NW	SE	SE	SE	N	N	N	N	N	N,W	SE	N,W			16

67 Station/Country Manaus/Brazil

Location 3°08'S/60°01'W Height above sealevel 48 m Climate symbol: Köppen Am Troll V,1

		J	F	M	A	M	J	J	A	S	O	N	D	year	P	
1	Mean daily temperature	in °C	26,2	26,2	26,4	26,2	26,3	26,6	26,8	27,5	27,9	27,8	27,6	26,8	26,9		1
2	Mean daily maximum temperature	in °C	31,1	31,1	31,1	30,6	31,1	31,1	31,7	32,8	33,3	33,3	32,8	32,2	31,7	11	2
3	Mean daily minimum temperature	in °C	23,9	23,9	23,9	23,9	23,9	23,9	23,9	23,9	23,9	24,4	24,4	23,9	23,9	11	3
4	Absolute maximum temperature	in °C	37,4	36,7	36,1	36,6	35,0	35,0	35,2	36,7	37,2	37,8	37,2	38,6	38,6		4
5	Absolute minimum temperature	in °C	20,4	20,0	19,4	20,2	20,0	19,0	17,6	19,2	20,0	20,2	20,2	19,6	17,6		5
6	Mean relative humidity	in %	85	80	81	82	81	78	76	72	71	69	75	78	77		6
7	Mean precipitation	in mm	266	247	269	267	194	100	64	38	60	124	152	216	1897		7
8	Maximum precipitation	in mm	410	422	578	455	501	249	156	100	178	230	275	376			8
9	Minimum precipitation	in mm	150	157	109	139	64	22	6	2	14	10	48	108			9
10	Maximum precipitation in 24 h	in mm	96	102	129	121	103	74	69	51	52	92	100	85	129		10
11	Mean number of days with precipitation	> 0,1 mm	20	19	21	20	18	12	8	6	8	11	12	16	171		11
12	Mean duration of sunshine	in h	126	108	123	125	161	203	238	259	225	208	188	163	2127		12
13	Mean quantity of radiation	in ly/day															13
14	Mean potential evaporation	in mm	132	120	134	126	133	132	136	150	144	150	147	138	1642	31	14
15	Mean windspeed	in m/sec	1,5	1,7	1,7	1,4	1,6	1,6	1,6	1,8	1,8	1,7	1,7	1,7	1,6		15
16	Mean predominant direction of the wind		E	E	E	E	E	E	E	E	E,S	E,S	E	E			16

68 Station/Country Santarém/Brazil

Location 2°25'S/54°42'W Height above sealevel 20 m Climate symbol: Köppen Am Troll V,1

		J	F	M	A	M	J	J	A	S	O	N	D	year	P	
1	Mean daily temperature	in °C	25,7	25,3	25,4	25,4	25,3	25,1	25,1	26,0	26,6	26,9	26,9	26,4	25,8		1
2	Mean daily maximum temperature	in °C	30,0	29,4	29,4	29,4	29,4	30,0	30,6	31,7	32,8	32,8	32,8	31,7	31,1	22	2
3	Mean daily minimum temperature	in °C	22,8	22,8	22,8	22,8	22,8	22,2	21,7	22,2	22,8	22,8	22,8	22,8	22,8	22	3
4	Absolute maximum temperature	in °C	35,4	34,2	33,2	32,8	34,3	34,2	34,8	35,7	37,1	36,6	36,0	35,6	37,1		4
5	Absolute minimum temperature	in °C	19,8	20,1	20,3	20,6	20,4	19,5	18,5	19,0	20,2	20,4	20,6	19,9	18,5		5
6	Mean relative humidity	in %	85	88	89	89	90	88	84	80	77	75	76	80	84	11	6
7	Mean precipitation	in mm	168	270	331	336	286	188	101	43	36	47	60	109	1975		7
8	Maximum precipitation	in mm															8
9	Minimum precipitation	in mm															9
10	Maximum precipitation in 24 h	in mm	84	113	175	129	88	103	71	30	40	97	88	98	175		10
11	Mean number of days with precipitation	> 0,1 mm	21	23	25	26	26	22	15	10	8	6	7	12	201		11
12	Mean duration of sunshine	in h	132	96	97	110	136	163	200	233	219	216	191	174	1967		12
13	Mean quantity of radiation	in ly/day															13
14	Mean potential evaporation	in mm	126	106	118	114	116	110	114	129	136	144	140	135	1488	27	14
15	Mean windspeed	in m/sec	1,4	1,2	1,2	1,2	1,1	1,1	1,4	1,6	1,7	1,7	1,5	1,5	1,4		15
16	Mean predominant direction of the wind		NE	NE	NE	NE	NE	NE	NE	NE	NE	NE	NE	NE			16

203

69 Station/Country Belém/Brazil

Location 1°28'S/48°27'W Height above sealevel 24 m Climate symbol: Köppen Af Troll V.1

		J	F	M	A	M	J	J	A	S	O	N	D	year	P	
1 Mean daily temperature	in °C	25,2	25,0	25,1	25,5	25,8	25,8	25,8	25,9	25,8	26,1	26,3	25,9	25,7		1
2 Mean daily maximum temperature	in °C	30,6	30,0	30,6	30,6	31,1	31,1	31,1	31,1	31,7	31,7	32,2	31,7	31,1	16	2
3 Mean daily minimum temperature	in °C	22,2	22,2	22,8	22,8	22,8	22,2	21,7	21,7	21,7	21,7	21,7	22,2	22,2	16	3
4 Absolute maximum temperature	in °C	34,6	33,9	34,5	34,1	34,2	34,2	33,7	35,1	34,6	34,6	35,1	35,4	35,4		4
5 Absolute minimum temperature	in °C	20,3	20,2	19,8	21,1	20,4	19,9	18,5	19,1	19,4	18,9	19,4	19,3	18,5		5
6 Mean relative humidity	in %	92	95	94	94	93	92	92	91	90	89	89	91	92		6
7 Mean precipitation	in mm	339	408	436	344	288	175	145	127	118	92	86	175	2733		7
8 Maximum precipitation	in mm	464	604	696	581	545	289	295	220	222	197	240	349			8
9 Minimum precipitation	in mm	191	221	248	190	142	85	67	16	46	31	23	62			9
10 Maximum precipitation in 24 h	in mm	98	121	102	87	126	71	72	51	49	48	40	78	126		10
11 Mean number of days with precipitation	> 0,1 mm	27	26	28	27	26	22	17	15	16	15	13	19	253		11
12 Mean duration of sunshine	in h	156	105	118	137	201	242	275	279	255	266	248	231	2513		12
13 Mean quantity of radiation	in ly/day															13
14 Mean potential evaporation	in mm	119	104	115	118	128	124	128	128	124	131	130	128	1477	19	14
15 Mean windspeed	in m/sec	0,9	0,8	0,9	0,9	1,0	1,1	1,2	1,3	1,2	1,3	1,4	1,1	1,1		15
16 Mean predominent direction of the wind		NE	NE	NE	NE	NE	NE	NE	NE	NE	NE	NE	NE			16

70 Station/Country São Luíz/Brazil

Location 2°32'S/44°17'W Height above sealevel 20 m Climate symbol: Köppen Aw Troll V.2

		J	F	M	A	M	J	J	A	S	O	N	D	year	P	
1 Mean daily temperature	in °C	26,7	26,7	26,1	26,1	26,7	26,1	26,7	27,2	27,2	27,2	27,2	27,2	26,7		1
2 Mean daily maximum temperature	in °C	30,0	29,4	30,0	30,0	30,6	30,6	30,6	30,6	30,6	30,6	30,6	30,6	30,6	19	2
3 Mean daily minimum temperature	in °C	23,9	23,3	23,3	23,3	23,3	23,3	22,8	23,3	23,9	23,9	23,9	23,9	23,3	19	3
4 Absolute maximum temperature	in °C	32,8	32,8	34,4	32,8	33,9	33,3	33,9	33,3	35,0	33,9	33,9	35,0	35,0	19	4
5 Absolute minimum temperature	in °C	20,0	20,6	20,6	20,6	20,0	19,4	20,6	21,1	18,9	21,1	21,7	20,0	18,9	19	5
6 Mean relative humidity	in %	81	83	83	85	83	81	82	80	78	77	76	78	80	11	6
7 Mean precipitation	in mm	155	269	416	416	318	155	112	36	8	5	20	46	1956		7
8 Maximum precipitation	in mm															8
9 Minimum precipitation	in mm															9
10 Maximum precipitation in 24 h	in mm	107	127	130	252	97	79	160	38	43	76	84	89			10
11 Mean number of days with precipitation	>0,25 mm	15	19	23	24	22	17	12	5	3	2	3	5	150	19	11
12 Mean duration of sunshine	in h															12
13 Mean quantity of radiation	in ly/day															13
14 Mean potential evaporation	in mm	145	126	138	136	144	140	134	146	142	147	144	148	1690		14
15 Mean windspeed	in m/sec															15
16 Mean predominent direction of the wind																16

71 Station/Country Barra do Corda/Brazil

Location 5°30'S/45°16'W Height above sealevel 81 m Climate symbol: Köppen Aw Troll V.2

		J	F	M	A	M	J	J	A	S	O	N	D	year	P	
1 Mean daily temperature	in °C	25,5	25,4	25,4	25,6	25,2	24,8	24,2	25,4	27,2	27,4	27,0	26,2	25,8		1
2 Mean daily maximum temperature	in °C	31,7	31,1	31,1	31,7	31,7	32,2	33,3	34,4	35,0	34,4	33,9	32,8	32,8	9	2
3 Mean daily minimum temperature	in °C	21,7	21,7	21,7	21,7	20,6	18,9	17,8	18,3	20,6	22,2	21,7	21,7	20,6	9	3
4 Absolute maximum temperature	in °C	37,3	37,0	35,2	38,6	38,4	36,8	37,2	37,8	39,0	39,4	38,0	38,6	39,4	11	4
5 Absolute minimum temperature	in °C	19,0	17,6	18,0	18,8	13,6	14,2	12,2	12,0	14,8	13,6	12,5	18,6	12,0	11	5
6 Mean relative humidity	in %	81	82	82	82	80	75	73	69	68	70	73	78	76		6
7 Mean precipitation	in mm	190	208	214	144	60	16	7	7	23	41	70	117	1097		7
8 Maximum precipitation	in mm															8
9 Minimum precipitation	in mm															9
10 Maximum precipitation in 24 h	in mm	168	63	151	88	55	32	28	41	53	63	68	75	168		10
11 Mean number of days with precipitation	>0,25 mm	17	17	19	16	9	3	2	1	4	6	8	12	114		11
12 Mean duration of sunshine	in h	132	121	138	149	181	215	243	227	179	169	149	150	2053		12
13 Mean quantity of radiation	in ly/day															13
14 Mean potential evaporation	in mm	124	109	119	117	116	103	101	117	138	148	139	135	1466		14
15 Mean windspeed	in m/sec	0,7	0,7	0,7	0,7	0,8	0,9	0,9	0,9	0,9	0,9	0,8	0,7	0,8		15
16 Mean predominent direction of the wind																16

72 Station/Country Quixeramobim/Brazil

Location 5°12'S/39°18'W Height above sealevel 198 m Climate symbol: Köppen BSh Troll V.3

		J	F	M	A	M	J	J	A	S	O	N	D	year	P	
1 Mean daily temperature	in °C	28,6	27,8	27,1	26,8	26,5	26,2	26,4	27,3	28,0	28,4	28,5	28,8	27,5		1
2 Mean daily maximum temperature	in °C															2
3 Mean daily minimum temperature	in °C															3
4 Absolute maximum temperature	in °C	36,7	36,3	36,4	35,0	34,8	34,4	33,5	34,5	35,8	35,8	36,3	36,4	36,7		4
5 Absolute minimum temperature	in °C	21,2	19,8	19,7	20,4	18,4	18,4	19,0	20,2	21,4	21,8	20,7	20,7	18,4		5
6 Mean relative humidity	in %															6
7 Mean precipitation	in mm	67	108	188	169	111	54	26	9	3	2	6	21	764		7
8 Maximum precipitation	in mm															8
9 Minimum precipitation	in mm															9
10 Maximum precipitation in 24 h	in mm	119	113	180	103	138	53	64	32	22	8	27	82	180		10
11 Mean number of days with precipitation	> 0,1 mm	9	12	16	16	13	10	5	3	1	2	3	5	95		11
12 Mean duration of sunshine	in h	229	189	201	200	225	231	262	286	277	280	262	252	2894		12
13 Mean quantity of radiation	in ly/day															13
14 Mean potential evaporation	in mm	162	137	144	132	135	125	132	142	144	161	157	162	1733	31	14
15 Mean windspeed	in m/sec	3,6	2,7	2,1	2,1	2,1	2,3	2,5	3,5	4,1	4,7	4,5	4,3	3,2		15
16 Mean predominent direction of the wind		E	E	E	E	E	E,SE	E	E	E	E	E	E			16

73 Station / Country: Alto Tapajós / Brazil

Location 7°20'S / 57°30'W — Height above sealevel 140 m — Climate symbol: Köppen Am — Troll V.2

		J	F	M	A	M	J	J	A	S	O	N	D	year	P
1 Mean daily temperature	in °C	24,7	24,7	24,9	25,1	25,3	24,6	24,0	25,2	25,3	25,3	25,1	24,9	24,9	
2 Mean daily maximum temperature	in °C														
3 Mean daily minimum temperature	in °C														
4 Absolute maximum temperature	in °C	36,2	35,2	35,7	35,4	36,5	35,6	36,8	37,8	37,7	37,4	36,6	36,0	37,8	
5 Absolute minimum temperature	in °C	19,0	19,8	19,0	17,0	15,0	8,8	9,4	12,6	16,0	18,8	19,8	19,8	8,8	
6 Mean relative humidity	in %														
7 Mean precipitation	in mm	398	416	379	302	112	20	22	51	144	247	315	335	2741	
8 Maximum precipitation	in mm														
9 Minimum precipitation	in mm														
10 Maximum precipitation in 24 h	in mm	96	107	85	81	77	29	33	65	72	83	88	137	137	
11 Mean number of days with precipitation	> 0,1 mm	26	24	25	23	14	4	2	6	15	19	22	24	204	
12 Mean duration of sunshine	in h	108	91	110	130	199	248	270	261	173	161	127	117	1995	
13 Mean quantity of radiation	in ly / day														
14 Mean potential evaporation	in mm	116	104	119	114	119	106	100	119	117	123	117	123	1377	12
15 Mean windspeed	in m / sec	0,4	0,4	0,4	0,4	0,6	0,8	0,9	0,8	0,6	0,6	0,4	0,3	0,6	
16 Mean predominent direction of the wind		E	E	E	E	E	E	E	E	E	E	E	E		

74 Station / Country: Sena Madureira / Brazil

Location 9°08'S / 68°40'W — Height above sealevel 135 m — Climate symbol: Köppen Am — Troll V.1

		J	F	M	A	M	J	J	A	S	O	N	D	year	P
1 Mean daily temperature	in °C	25,4	25,3	25,2	24,9	24,1	23,3	22,9	23,9	24,9	25,3	25,5	25,5	24,7	
2 Mean daily maximum temperature	in °C	33,3	33,3	32,8	32,8	32,2	32,2	32,8	33,9	33,9	33,9	33,9	33,9	33,3	12
3 Mean daily minimum temperature	in °C	20,6	20,6	20,6	20,0	19,4	18,3	17,2	18,3	20,0	20,6	20,6	20,6	20,0	12
4 Absolute maximum temperature	in °C	37,6	37,0	37,4	37,0	37,0	37,3	37,4	37,8	38,4	38,8	37,7	38,0	38,8	
5 Absolute minimum temperature	in °C	16,4	15,2	15,8	14,2	8,8	7,3	8,0	7,9	8,0	10,3	14,0	13,6	7,3	
6 Mean relative humidity	in %	98	98	98	99	99	98	98	98	98	98	98	98	98	2
7 Mean precipitation	in mm	317	285	266	231	125	66	36	46	126	173	193	274	2138	
8 Maximum precipitation	in mm														
9 Minimum precipitation	in mm														
10 Maximum precipitation in 24 h	in mm	97	90	71	112	87	61	53	64	91	63	69	70	112	
11 Mean number of days with precipitation	> 0,1 mm	18	16	17	14	9	8	5	5	9	11	11	15	138	
12 Mean duration of sunshine	in h														
13 Mean quantity of radiation	in ly / day														
14 Mean potential evaporation	in mm	130	113	123	113	100	90	90	98	126	124	109	131	1347	27
15 Mean windspeed	in m / sec	0,5	0,4	0,4	0,5	0,5	0,5	0,6	0,6	0,5	0,5	0,4	0,4	0,5	
16 Mean predominent direction of the wind		N.NW	N	N.NW	SE	S	S	S	S	N	N	N.NW	N		

75 Station / Country: Conceicao do Araguaia / Brazil

Location 8°15'S / 49°12'W — Height above sealevel 16 m — Climate symbol: Köppen Aw — Troll V.2

		J	F	M	A	M	J	J	A	S	O	N	D	year	P
1 Mean daily temperature	in °C	25,0	25,0	25,0	25,6	25,6	25,0	25,0	26,1	26,7	25,6	25,6	25,0	25,6	
2 Mean daily maximum temperature	in °C	31,1	30,6	31,7	32,8	32,8	33,9	35,0	36,7	36,1	33,9	33,3	32,2	33,3	5
3 Mean daily minimum temperature	in °C	21,1	20,6	20,6	20,0	20,0	17,8	17,2	17,8	19,4	20,0	20,6	21,7	19,4	5
4 Absolute maximum temperature	in °C	35,0	35,6	37,2	37,2	36,1	37,2	37,2	38,9	38,9	37,2	36,1		38,9	5
5 Absolute minimum temperature	in °C	17,2	16,1	15,6	16,1	15,0	15,0	12,8	12,8	15,6	18,3	18,9	18,3	12,8	5
6 Mean relative humidity	in %	85	88	85	85	79	74	70	66	73	78	81	81	79	3
7 Mean precipitation	in mm	254	251	264	163	61	8	8	16	64	163	196	226	1674	27
8 Maximum precipitation	in mm														
9 Minimum precipitation	in mm														
10 Maximum precipitation in 24 h	in mm	104	71	99	64	64	38	36	33	66	71	66	94	104	11
11 Mean number of days with precipitation	> 0,25mm	17	18	18	13	5	1	1	2	6	10	14	13	118	11
12 Mean duration of sunshine	in h														
13 Mean quantity of radiation	in ly / day														
14 Mean potential evaporation	in mm	119	107	120	116	118	108	112	129	132	130	122	123	1436	
15 Mean windspeed	in m / sec														
16 Mean predominent direction of the wind															

76 Station / Country: Recife (Olinda) / Brazil

Location 8°01'S / 34°51'W — Height above sealevel 57 m — Climate symbol: Köppen As — Troll V.2a

		J	F	M	A	M	J	J	A	S	O	N	D	year	P
1 Mean daily temperature	in °C	27,1	27,1	27,0	26,6	25,9	25,0	24,3	24,4	25,3	26,2	26,6	26,9	26,0	
2 Mean daily maximum temperature	in °C	30,0	30,0	30,0	29,4	28,3	27,8	26,7	27,2	27,8	28,9	29,4	29,4	28,9	27
3 Mean daily minimum temperature	in °C	25,0	25,0	24,4	23,9	23,3	22,8	21,7	21,7	22,8	23,9	24,4	25,0	23,9	27
4 Absolute maximum temperature	in °C	33,2	33,4	33,8	33,9	32,2	31,8	30,5	31,0	32,3	32,8	33,0	33,2	33,9	
5 Absolute minimum temperature	in °C	21,3	20,2	20,8	20,1	19,4	18,8	17,8	18,0	18,7	18,9	20,3	18,7	17,8	
6 Mean relative humidity	in %	73	76	76	78	79	80	79	78	74	71	71	72	76	
7 Mean precipitation	in mm	47	109	157	226	260	257	186	116	52	30	28	33	1501	
8 Maximum precipitation	in mm	163	525	666	598	763	609	457	288	148	298	93	123		
9 Minimum precipitation	in mm	4	5	17	56	140	145	107	46	5	2	2	5		
10 Maximum precipitation in 24 h	in mm	110	150	105	134	154	228	84	71	52	71	68	154	228	
11 Mean number of days with precipitation	> 0,1 mm	12	14	16	21	23	24	25	22	15	11	9	10	202	
12 Mean duration of sunshine	in h	256	200	212	195	185	172	159	204	224	239	235	263	2544	
13 Mean quantity of radiation	in ly / day														
14 Mean potential evaporation	in mm	158	142	154	143	137	116	110	114	130	147	150	161	1662	48
15 Mean windspeed	in m / sec	3,0	2,8	2,6	2,9	2,8	3,6	4,0	3,3	3,4	3,2	2,1	2,0	2,9	
16 Mean predominent direction of the wind		E.SE	SE	SE	SE	SE	SE	SE	SE	SE	E.SE	NE	E		

77 Station / Country Remanso / Brazil

Location 9°41'S / 42°04'W — Height above sealevel 411 m — Climate symbol: Köppen BSh — Troll V,4

			J	F	M	A	M	J	J	A	S	O	N	D	year	P	
1	Mean daily temperature	in °C	27,2	25,9	27,1	27,3	26,7	26,2	25,7	26,2	27,2	28,0	27,9	27,2	26,8		1
2	Mean daily maximum temperature	in °C															2
3	Mean daily minimum temperature	in °C															3
4	Absolute maximum temperature	in °C	38,7	38,4	38,7	39,1	37,7	37,7	37,7	38,1	38,5	39,5	39,1	39,3	39,5		4
5	Absolute minimum temperature	in °C	10,8	16,4	15,8	16,0	15,6	14,6	14,0	14,8	15,0	16,4	14,4	9,8	9,8		5
6	Mean relative humidity	in %															6
7	Mean precipitation	in mm	78	83	88	35	22	10	11	0	8	14	56	92	497		7
8	Maximum precipitation	in mm															8
9	Minimum precipitation	in mm															9
10	Maximum precipitation in 24 h	in mm	67	64	98	64	63	2	4	0	32	56	53	64	98		10
11	Mean number of days with precipitation	> 0,1 mm	5	6	6	2	1	0	0	0	2	1	4	6	33		11
12	Mean duration of sunshine	in h															12
13	Mean quantity of radiation	in ly/day															13
14	Mean potential evaporation	in mm	152	134	148	140	139	125	123	132	141	156	153	154	1697	14	14
15	Mean windspeed	in m/sec	1,2	1,4	1,1	1,2	1,4	1,3	1,6	1,5	1,5	1,5	1,3	1,4	1,4		15
16	Mean predominent direction of the wind		E	E	E	E	E	E	E	E	E	E	E	E			16

78 Station / Country Monte Santo / Brazil

Location 10°27'S / 39°18'W — Height above sealevel 545 m — Climate symbol: Köppen Aw — Troll V,3

			J	F	M	A	M	J	J	A	S	O	N	D	year	P	
1	Mean daily temperature	in °C	25,6	25,6	25,0	23,9	22,2	21,1	20,0	20,6	22,2	23,9	25,0	25,6	23,3		1
2	Mean daily maximum temperature	in °C	32,2	32,2	32,2	30,6	28,3	27,2	26,1	27,2	29,4	32,2	32,8	32,8	30,6	15	2
3	Mean daily minimum temperature	in °C	20,0	20,6	20,6	20,0	18,9	17,8	16,7	16,7	17,8	18,9	20,0	20,0	18,9	15	3
4	Absolute maximum temperature	in °C	39,4	37,2	38,1	35,6	33,9	33,9	31,7	33,9	38,1	37,2	37,8	38,3	39,4		4
5	Absolute minimum temperature	in °C	16,1	16,1	16,7	15,6	15,0	13,9	12,2	12,8	13,9	15,0	15,0	15,6	12,2		5
6	Mean relative humidity	in %	67	69	69	72	74	77	79	74	67	59	62	64	69		6
7	Mean precipitation	in mm	56	43	71	71	66	48	53	36	16	16	71	71	618	30	7
8	Maximum precipitation	in mm															8
9	Minimum precipitation	in mm															9
10	Maximum precipitation in 24 h	in mm	66	64	61	51	64	38	48	41	33	30	71	66	71	22	10
11	Mean number of days with precipitation	> 0,25 mm	7	8	8	1	2	17	20	13	7	4	7	7	101	22	11
12	Mean duration of sunshine	in h															12
13	Mean quantity of radiation	in ly/day															13
14	Mean potential evaporation	in mm	130	110	123	101	88	69	63	70	87	92	120	129	1182		14
15	Mean windspeed	in m/sec															15
16	Mean predominent direction of the wind																16

79 Station / Country Pôrto Nacional / Brazil

Location 10°31'S / 48°43'W — Height above sealevel 237 m — Climate symbol: Köppen Aw — Troll V,2

			J	F	M	A	M	J	J	A	S	O	N	D	year	P	
1	Mean daily temperature	in °C	25,2	24,9	25,2	25,7	25,6	24,6	24,3	26,1	27,6	26,7	25,8	25,3	25,6		1
2	Mean daily maximum temperature	in °C															2
3	Mean daily minimum temperature	in °C															3
4	Absolute maximum temperature	in °C	38,0	37,7	37,7	39,7	38,2	39,9	38,0	42,8	40,1	41,0	38,4	38,4	42,8		4
5	Absolute minimum temperature	in °C	14,1	18,1	18,6	16,5	10,9	11,2	9,5	10,5	13,1	14,5	15,4	18,1	9,5		5
6	Mean relative humidity	in %															6
7	Mean precipitation	in mm	298	90	292	152	44	<1	3	8	42	150	242	292	1814		7
8	Maximum precipitation	in mm															8
9	Minimum precipitation	in mm															9
10	Maximum precipitation in 24 h	in mm	102	70	93	68	58	3	24	65	65	69	81	77	102		10
11	Mean number of days with precipitation	> 0,1 mm	20	20	20	13	4	0	0	1	5	12	17	19	131		11
12	Mean duration of sunshine	in h	177	130	157	205	272	293	307	314	239	195	158	164	2612		12
13	Mean quantity of radiation	in ly/day															13
14	Mean potential evaporation	in mm	123	108	120	122	18	101	99	127	141	153	128	125	1466	27	14
15	Mean windspeed	in m/sec	0,6	0,7	0,6	0,6	0,6	0,7	0,8	0,8	0,7	0,7	0,7	0,7	0,7		15
16	Mean predominent direction of the wind		SW	var	var	S	E	E	E	E	E	var	var	var			16

80 Station / Country Ibipetuba / Brazil

Location 11°01'S / 44°31'W — Height above sealevel 436 m — Climate symbol: Köppen Aw — Troll V,3

			J	F	M	A	M	J	J	A	S	O	N	D	year	P	
1	Mean daily temperature	in °C	24,4	24,4	24,4	24,4	23,3	21,7	21,7	22,2	25,0	26,1	25,6	25,0	23,9		1
2	Mean daily maximum temperature	in °C	32,8	32,2	31,7	32,8	33,3	33,3	32,8	34,4	36,1	36,7	33,9	32,8	33,3	10	2
3	Mean daily minimum temperature	in °C	18,9	18,9	18,3	18,3	16,1	12,8	11,7	12,2	16,1	19,4	20,0	20,0	16,7	10	3
4	Absolute maximum temperature	in °C	39,4	40,0	38,9	38,9	38,3	40,0	38,3	40,0	41,7	41,7	41,1	41,7			4
5	Absolute minimum temperature	in °C	12,8	14,4	15,0	15,0	11,1	7,2	6,1	8,9	10,0	13,3	13,9	15,6	6,1		5
6	Mean relative humidity	in %	78	81	77	78	69	64	61	62	63	67	73	74	70	3	6
7	Mean precipitation	in mm	124	145	136	74	13	1	1	1	8	53	157	198	911	30	7
8	Maximum precipitation	in mm															8
9	Minimum precipitation	in mm															9
10	Maximum precipitation in 24 h	in mm	64	94	76	51	15	0	3	0	41	30	71	81	94	10	10
11	Mean number of days with precipitation	> 0,25 mm	10	12	11	6	2	0	<1	0	1	4	9	11	66	10	11
12	Mean duration of sunshine	in h															12
13	Mean quantity of radiation	in ly/day															13
14	Mean potential evaporation	in mm	118	105	113	107	99	81	80	94	120	140	126	125	1308		14
15	Mean windspeed	in m/sec															15
16	Mean predominent direction of the wind																16

81 Station/Country Salvador/Brazil

Location 12°55'S/38°41'W   Height above sealevel 45 m   Climate symbol: Köppen Af   Troll V,1

		J	F	M	A	M	J	J	A	S	O	N	D	year	P	
1 Mean daily temperature	in °C	26,1	26,3	26,2	25,8	24,8	23,8	23,0	23,1	23,8	24,6	25,1	25,6	24,8		1
2 Mean daily maximum temperature	in °C	30,0	30,0	30,0	28,9	27,8	26,7	26,1	26,1	27,2	28,3	28,9	28,9	28,3	25	2
3 Mean daily minimum temperature	in °C	23,3	23,3	23,3	22,2	21,7	20,6	20,6	21,1	21,7	22,2	22,8	22,2	25		3
4 Absolute maximum temperature	in °C	34,4	34,8	35,2	34,2	33,2	30,7	30,0	29,9	31,0	34,8	33,5	33,1	35,2		4
5 Absolute minimum temperature	in °C	20,0	19,0	19,9	20,6	19,8	18,0	18,4	16,8	17,0	19,0	19,6	19,5	16,8		5
6 Mean relative humidity	in %	80	81	81	83	84	83	82	81	80	81	81	80	81	13	6
7 Mean precipitation	in mm	74	116	185	278	296	225	204	116	98	102	116	124	1914		7
8 Maximum precipitation	in mm															8
9 Minimum precipitation	in mm															9
10 Maximum precipitation in 24 h	in mm	80	90	158	124	156	110	103	66	95	81	120	129	158		10
11 Mean number of days with precipitation	> 0,1 mm	14	16	19	22	25	24	24	20	16	14	15	14	223		11
12 Mean duration of sunshine	in h	283	225	237	211	190	190	190	223	218	239	226	249	2661		12
13 Mean quantity of radiation	in ly/day															13
14 Mean potential evaporation	in mm	142	129	139	123	107	91	88	90	96	112	121	133	1372		14
15 Mean windspeed	in m/sec	2,8	2,7	2,8	2,8	3,1	2,8	3,2	3,1	3,0	3,3	3,3	3,0	3,0		15
16 Mean predominent direction of the wind		E	E	SE	SE	SE	SE	SE	SE	SE	E	E	E			16

82 Station/Country Caetité/Brazil

Location 14°03'S/42°37'W   Height above sealevel 878 m   Climate symbol: Köppen Aw   Troll V,3

		J	F	M	A	M	J	J	A	S	O	N	D	year	P	
1 Mean daily temperature	in °C	22,6	22,5	22,6	22,0	21,0	19,5	18,9	20,0	21,5	22,7	22,4	22,3	21,4		1
2 Mean daily maximum temperature	in °C	28,3	28,3	28,3	27,8	26,1	25,0	25,0	26,1	28,3	28,9	28,3	27,2	27,2	12	2
3 Mean daily minimum temperature	in °C	17,8	17,8	17,8	17,8	16,1	14,4	13,3	13,9	15,6	16,7	17,8	17,8	16,1	13	3
4 Absolute maximum temperature	in °C	36,9	34,3	34,8	33,9	33,8	32,2	31,7	34,6	35,4	36,7	35,8	35,6	36,9		4
5 Absolute minimum temperature	in °C	11,4	13,2	10,2	11,4	9,2	8,0	7,6	9,8	10,8	10,2	11,8	11,8	7,6		5
6 Mean relative humidity	in %	76	76	77	78	77	74	72	69	66	69	74	74	74	10	6
7 Mean precipitation	in mm	120	117	97	53	16	9	9	10	17	62	148	152	810	10	7
8 Maximum precipitation	in mm															8
9 Minimum precipitation	in mm															9
10 Maximum precipitation in 24 h	in mm	87	80	70	82	17	19	13	43	54	52	77	83	87		10
11 Mean number of days with precipitation	> 0,25 mm	9	9	8	8	6	5	5	3	2	6	12	11	84		11
12 Mean duration of sunshine	in h	198	185	212	195	206	207	225	252	234	209	159	171	2453		12
13 Mean quantity of radiation	in ly/day															13
14 Mean potential evaporation	in mm	103	91	98	88	80	62	59	69	84	103	100	101	1038		14
15 Mean windspeed	in m/sec	2,8	3,1	3,2	3,9	0,8	0,6	1,3	1,2	1,1	4,4	3,4	2,7	2,4		15
16 Mean predominent direction of the wind		SE	SE	SE	SE	C-E	C-E	SE	C-E	C-E	SE	SE	SE			16

83 Station/Country Brasília/Brazil

Location 15°32'S/47°18'W   Height above sealevel 910 m   Climate symbol: Köppen Aw   Troll V,2

		J	F	M	A	M	J	J	A	S	O	N	D	year	P	
1 Mean daily temperature	in °C	23	23	23	22,5	21	20	19,5	21	23	23,5	23	22,5	21,8		1
2 Mean daily maximum temperature	in °C	27,5	28	27,5	27,5	27	26,5	26,5	28,5	30	29	27,5	26,5	27,5		2
3 Mean daily minimum temperature	in °C	18	18	18	17	15	13	12,5	13,5	16	18	18	18	16		3
4 Absolute maximum temperature	in °C	33	33,5	32,5	33	32,5	31	31,5	34	37	35,5	35	33	37		4
5 Absolute minimum temperature	in °C	14	13	13,5	10	8	5	6	6	8,5	12,5	12	15	5		5
6 Mean relative humidity	in %	80	76	80	75	70	67	60	54	53	67	76	79	70		6
7 Mean precipitation	in mm	252	204	227	93	17	3	6	3	30	127	255	343	1560		7
8 Maximum precipitation	in mm	568	364	437	276	78	21	89	66	162	240	437	726			8
9 Minimum precipitation	in mm	78	55	67	8	0	0	0	0	0	3	75	145			9
10 Maximum precipitation in 24 h	in mm	101	85	93	78	42	18	25	46	64	103	108	125	125		10
11 Mean number of days with precipitation	> 0,1 mm	21	16	18	13	5	2	1	3	7	12	19	23	140		11
12 Mean duration of sunshine	in h															12
13 Mean quantity of radiation	in ly/day															13
14 Mean potential evaporation	in mm															14
15 Mean windspeed	in m/sec															15
16 Mean predominent direction of the wind																16

84 Station/Country Formosa/Brazil

Location 15°32'S/47°18'W   Height above sealevel 912 m   Climate symbol: Köppen Aw   Troll V,2

		J	F	M	A	M	J	J	A	S	O	N	D	year	P	
1 Mean daily temperature	in °C	21,7	21,8	21,8	21,5	20,1	19,0	18,7	20,7	22,7	22,7	21,8	21,5	21,2		1
2 Mean daily maximum temperature	in °C															2
3 Mean daily minimum temperature	in °C															3
4 Absolute maximum temperature	in °C	38,8	33,0	32,7	32,8	31,4	30,2	30,8	34,0	35,5	35,8	35,5	34,9	35,8		4
5 Absolute minimum temperature	in °C	12,9	13,4	13,6	9,9	7,8	5,1	5,5	6,1	8,8	11,0	10,7	13,0	5,1		5
6 Mean relative humidity	in %															6
7 Mean precipitation	in mm	273	227	194	99	19	4	4	9	45	138	233	350	1595		7
8 Maximum precipitation	in mm															8
9 Minimum precipitation	in mm															9
10 Maximum precipitation in 24 h	in mm	107	103	79	64	32	42	25	44	64	85	108	102	108		10
11 Mean number of days with precipitation	> 0,1 mm	21	18	18	11	4	1	1	1	5	12	19	22	133		11
12 Mean duration of sunshine	in h	189	164	205	221	276	285	295	309	228	213	144	145	2674		12
13 Mean quantity of radiation	in ly/day															13
14 Mean potential evaporation	in mm	97	89	95	82	68	56	55	78	96	103	93	92	1004	29	14
15 Mean windspeed	in m/sec	2,1	2,0	2,1	2,1	2,1	2,2	2,3	2,4	2,4	2,3	2,2	2,1	2,2		15
16 Mean predominent direction of the wind		N	N	N	S	S	S	E	E	E	N	N	N			16

85 Station/Country Cuiabá/Brazil

Location 15°35'S/56°06'W Height above sealevel 171 m Climate symbol: Köppen Aw Troll V.2

		J	F	M	A	M	J	J	A	S	O	N	D	year	P	
1 Mean daily temperature	in °C	26,4	26,2	26,2	25,9	24,3	23,0	22,5	24,8	26,6	27,0	26,8	26,5	25,5		1
2 Mean daily maximum temperature	in °C															2
3 Mean daily minimum temperature	in °C															3
4 Absolute maximum temperature	in °C	37,9	38,4	39,2	38,2	38,8	36,0	37,8	40,0	40,6	42,2	38,7	39,4	42,2		4
5 Absolute minimum temperature	in °C	17,5	15,0	15,8	13,3	6,4	1,2	5,0	5,8	7,4	12,3	12,8	16,7	1,2		5
6 Mean relative humidity	in %															6
7 Mean precipitation	in mm	213	200	222	106	48	14	9	27	48	124	162	208	1378		7
8 Maximum precipitation	in mm															8
9 Minimum precipitation	in mm															9
10 Maximum precipitation in 24 h	in mm	99	118	95	80	75	41	54	134	106	126	107	124	134		10
11 Mean number of days with precipitation	> 0,1 mm	19	18	19	12	6	3	2	2	6	11	15	19	132		11
12 Mean duration of sunshine	in h	137	123	143	181	206	200	225	210	158	189	158	129	2037		12
13 Mean quantity of radiation	in ly/day															13
14 Mean potential evaporation	in mm	144	125	132	121	97	81	79	105	132	144	141	142	1443	30	14
15 Mean windspeed	in m/sec	1,7	1,5	1,4	1,2	1,2	1,2	1,3	1,5	1,7	1,7	1,6	1,7	1,5		15
16 Mean predominent direction of the wind		N	N	N	N	S	S	S	S	N	N	N	N			16

86 Station/Country Corumbá/Brazil

Location 19°00'S/57°39'W Height above sealevel 138 m Climate symbol: Köppen Aw Troll V.2

		J	F	M	A	M	J	J	A	S	O	N	D	year	P	
1 Mean daily temperature	in °C	26,7	26,5	26,1	24,3	22,6	21,2	21,0	23,1	24,8	25,7	26,6	26,8	24,8		1
2 Mean daily maximum temperature	in °C	34,4	33,9	33,9	33,3	31,7	27,8	28,9	30,0	32,8	33,9	34,4	33,9	32,2	8	2
3 Mean daily minimum temperature	in °C	22,8	22,2	22,8	22,8	20,0	17,8	17,8	18,9	20,6	21,1	23,3	22,2	21,1	8	3
4 Absolute maximum temperature	in °C	39,2	40,0	39,0	38,0	36,3	35,8	36,3	38,6	40,6	41,8	40,2	41,0	41,8		4
5 Absolute minimum temperature	in °C	15,1	14,0	11,6	10,8	5,4	0,8	3,8	5,8	6,2	10,0	11,0	12,0	0,8		5
6 Mean relative humidity	in %															6
7 Mean precipitation	in mm	176	147	119	83	67	32	18	24	66	103	122	163	1120		7
8 Maximum precipitation	in mm															8
9 Minimum precipitation	in mm															9
10 Maximum precipitation in 24 h	in mm	180	97	118	90	104	70	74	99	78	114	104	131	180		10
11 Mean number of days with precipitation	> 0,1 mm	13	12	11	7	5	3	2	2	5	8	9	11	88		11
12 Mean duration of sunshine	in h	218	199	217	218	222	210	244	256	219	229	237	228	2697		12
13 Mean quantity of radiation	in ly/day															13
14 Mean potential evaporation	in mm	154	132	135	99	81	61	63	89	108	136	147	157	1362	30	14
15 Mean windspeed	in m/sec	1,2	1,3	1,3	1,4	1,5	1,8	1,8	1,9	1,8	1,6	1,5	1,3	1,5		15
16 Mean predominent direction of the wind		E	SE	SE	S	E	E	E	E	E	E,S	E,S	E			16

87 Station/Country Catalão/Brazil

Location 18°10'S/47°58'W Height above sealevel 830 m Climate symbol: Köppen Aw Troll V.2

		J	F	M	A	M	J	J	A	S	O	N	D	year	P	
1 Mean daily temperature	in °C	22,2	22,8	22,2	21,7	20,0	18,9	18,9	21,1	22,8	22,8	22,8	22,2	21,7		1
2 Mean daily maximum temperature	in °C	27,2	27,8	27,8	27,8	26,7	25,6	26,1	27,8	28,9	28,9	27,8	26,7	27,2		2
3 Mean daily minimum temperature	in °C	18,3	18,3	18,3	17,2	15,0	13,3	13,9	14,4	16,7	17,8	18,3	18,3	16,7		3
4 Absolute maximum temperature	in °C	32,8	31,7	32,8	30,6	31,1	30,0	30,6	32,2	35,0	34,4	32,8	32,8	35,0	8	4
5 Absolute minimum temperature	in °C	15,6	14,4	15,0	12,8	4,4	1,7	6,1	3,3	5,0	12,8	11,7	13,9	1,7	8	5
6 Mean relative humidity	in %	72	79	72	69	73	70	64	61	57	68	75	79	70	10	6
7 Mean precipitation	in mm	315	234	229	81	28	9	5	5	38	142	240	340	1666		7
8 Maximum precipitation	in mm															8
9 Minimum precipitation	in mm															9
10 Maximum precipitation in 24 h	in mm	79	69	91	64	69	10	38	28	61	64	91	84	91		10
11 Mean number of days with precipitation	> 0,25 mm	22	15	16	9	4	2	1	2	6	13	15	21	126		11
12 Mean duration of sunshine	in h															12
13 Mean quantity of radiation	in ly/day															13
14 Mean potential evaporation	in mm	102	89	95	81	64	55	55	74	93	104	100	103	1015		14
15 Mean windspeed	in m/sec															15
16 Mean predominent direction of the wind																16

88 Station/Country Belo Horizonte/Brazil

Location 19°56'S/43°56'W Height above sealevel 915 m Climate symbol: Köppen Cwa Troll V.2

		J	F	M	A	M	J	J	A	S	O	N	D	year	P	
1 Mean daily temperature	in °C	22,5	22,8	22,4	21,3	19,2	17,7	17,2	18,9	20,5	21,4	21,7	21,9	20,6		1
2 Mean daily maximum temperature	in °C	28,0	28,5	27,5	26,5	25,0	24,5	24,0	26,0	27,0	27,0	27,0	26,6	26,5		2
3 Mean daily minimum temperature	in °C	18,5	19,0	18,5	17,0	14,5	13,0	13,0	13,5	15,5	17,0	17,5	18,0	16,5		3
4 Absolute maximum temperature	in °C	35,3	34,2	32,5	32,8	32,9	30,8	30,0	34,0	34,2	35,2	35,2	35,5	35,5		4
5 Absolute minimum temperature	in °C	12,8	12,2	11,8	10,2	2,5	3,4	3,4	5,8	7,5	9,8	12,0	12,6	2,4		5
6 Mean relative humidity	in %	77	76	76	77	73	72	68	63	64	68	74	80	72		6
7 Mean precipitation	in mm	319	202	157	79	20	9	8	18	42	136	225	348	1561		7
8 Maximum precipitation	in mm	573	384	298	181	103	59	71	78	109	200	378	611			8
9 Minimum precipitation	in mm	35	47	23	0	0	0	0	0	0	36	96	180			9
10 Maximum precipitation in 24 h	in mm	155	105	110	73	35	36	21	33	58	75	78	88	155		10
11 Mean number of days with precipitation	> 0,1 mm	17	13	13	7	3	1	2	2	5	11	15	20	109		11
12 Mean duration of sunshine	in h	176	171	193	212	245	245	260	259	204	183	163	127	2428		12
13 Mean quantity of radiation	in ly/day															13
14 Mean potential evaporation	in mm	108	96	98	81	63	49	48	62	75	91	95	104	968	20	14
15 Mean windspeed	in m/sec	1,1	1,1	1,1	1,0	1,0	1,0	1,2	1,4	1,5	1,5	1,3	1,2	1,2		15
16 Mean predominent direction of the wind		NE	NE	NE	NE	NE	NE	NE	NE	NE	NE	NE	NE			16

89 Station/Country Vitória/Brazil

Location 20°19'S/40°20'W — Height above sealevel 31 m — Climate symbol: Köppen Aw — Troll V.1

	in	J	F	M	A	M	J	J	A	S	O	N	D	year	P	
1 Mean daily temperature	°C	25.3	25.6	25.3	24.2	22.6	21.5	20.5	21.0	21.9	22.7	23.6	24.6	23.2		1
2 Mean daily maximum temperature	°C															2
3 Mean daily minimum temperature	°C															3
4 Absolute maximum temperature	°C	37.3	35.2	35.4	34.4	34.0	31.8	33.1	32.3	33.7	33.9	34.4	36.5	37.3		4
5 Absolute minimum temperature	°C	17.9	16.8	19.0	16.5	14.7	12.7	10.3	11.8	14.5	9.3	16.4	16.8	9.3		5
6 Mean relative humidity	%															6
7 Mean precipitation	mm	151	118	150	130	90	66	72	48	85	135	169	191	1405		7
8 Maximum precipitation	mm															8
9 Minimum precipitation	mm															9
10 Maximum precipitation in 24 h	mm	136	101	148	93	86	71	61	74	72	111	78	126	148		10
11 Mean number of days with precipitation	>0,1 mm	14	12	15	14	11	8	12	9	10	16	17	16	154		11
12 Mean duration of sunshine	h	225	224	218	199	216	208	199	224	176	161	160	183	2393		12
13 Mean quantity of radiation	ly/day															13
14 Mean potential evaporation	mm	116	120	123	99	81	66	63	71	78	94	105	124	1142	18	14
15 Mean windspeed	m/sec	3,4	3,3	3,1	3,0	2,9	2,7	3,0	3,3	3,7	4,0	3,9	3,8	3,3		15
16 Mean predominent direction of the wind		N	N	N	E	SW	N,SW	SW	N	N	N	N	N			16

90 Station/Country Três Lagoas/Brazil

Location 20°47'S/51°42'W — Height above sealevel 312 m — Climate symbol: Köppen Aw — Troll V.2

	in	J	F	M	A	M	J	J	A	S	O	N	D	year	P	
1 Mean daily temperature	°C	25.4	25.5	25.1	23.6	20.8	19.5	19.2	20.6	22.6	24.1	25.0	25.3	23.1		1
2 Mean daily maximum temperature	°C	32.2	32.2	32.2	30.6	28.3	27.2	27.2	28.9	30.6	31.7	32.2	31.7	30.6	8	2
3 Mean daily minimum temperature	°C	21.7	21.7	20.6	18.9	15.0	13.9	12.8	13.9	16.7	18.9	20.0	21.1	17.8	8	3
4 Absolute maximum temperature	°C	40.0	39.0	39.0	38.8	35.6	37.0	35.5	38.7	41.0	39.3	39.5	39.8	41.0		4
5 Absolute minimum temperature	°C	12.5	14.5	9.8	8.0	3.8	-0.8	1.5	3.5	4.1	8.2	9.8	14.0	-0.8		5
6 Mean relative humidity	%	86	83	86	81	84	86	85	82	78	73	79	81	82	2	6
7 Mean precipitation	mm	202	168	129	90	62	50	21	35	73	109	147	200	1286		7
8 Maximum precipitation	mm															8
9 Minimum precipitation	mm															9
10 Maximum precipitation in 24 h	mm	88	75	72	69	73	55	43	57	57	57	74	69	88		10
11 Mean number of days with precipitation	>0,1 mm	15	13	10	6	4	4	3	3	6	8	10	14	96		11
12 Mean duration of sunshine	h	208	189	228	227	242	229	245	254	195	214	223	196	2649		12
13 Mean quantity of radiation	ly/day															13
14 Mean potential evaporation	mm	135	120	120	96	66	56	51	65	84	110	124	135	1162	29	14
15 Mean windspeed	m/sec	2,1	2,0	1,9	1,9	1,8	1,8	1,8	2,0	2,1	2,1	2,0	2,2	2,0		15
16 Mean predominent direction of the wind		NE	NE	NE	NE	SE	NE	NE	NE	SE	SE	SE	NE			16

91 Station/Country Bela Vista/Brazil

Location 22°32'S/55°38'W — Height above sealevel 160 m — Climate symbol: Köppen Cwa — Troll V.1

	in	J	F	M	A	M	J	J	A	S	O	N	D	year	P	
1 Mean daily temperature	°C	26.1	25.6	25.0	22.8	20.0	18.3	17.8	20.0	22.2	23.3	24.4	25.6	22.8		1
2 Mean daily maximum temperature	°C	32.6	32.6	31.7	29.4	26.1	25.6	25.0	27.8	29.4	30.6	31.7	32.6	29.4	13	2
3 Mean daily minimum temperature	°C	19.4	18.9	18.3	16.1	12.8	11.1	9.4	11.1	14.4	16.1	17.2	18.9	15.0	13	3
4 Absolute maximum temperature	°C	40.6	40.0	38.3	38.3	36.1	36.1	34.4	38.3	40.6	39.4	42.2	41.1	42.2	19	4
5 Absolute minimum temperature	°C	10.0	10.6	2.2	2.8	-2.2	-6.1	-6.7	-3.3	0.0	1.1	2.2	8.3	-6.7	19	5
6 Mean relative humidity	%	90	92	92	93	93	91	88	84	84	87	87	87	89	5	6
7 Mean precipitation	mm	173	135	119	119	122	79	33	43	79	132	147	170	1351	26	7
8 Maximum precipitation	mm															8
9 Minimum precipitation	mm															9
10 Maximum precipitation in 24 h	mm	97	69	71	66	66	99	71	64	71	66	66	66	99	20	10
11 Mean number of days with precipitation	>0,25 mm	13	11	10	8	8	7	5	5	6	9	9	11	102	20	11
12 Mean duration of sunshine	h															12
13 Mean quantity of radiation	ly/day															13
14 Mean potential evaporation	mm	150	124	123	89	63	45	42	62	84	108	122	144	1156		14
15 Mean windspeed	m/sec															15
16 Mean predominent direction of the wind																16

92 Station/Country Campinas/Brazil

Location 22°53'S/47°05'W — Height above sealevel 663 m — Climate symbol: Köppen Cwa — Troll V.1

	in	J	F	M	A	M	J	J	A	S	O	N	D	year	P	
1 Mean daily temperature	°C	22.4	22.3	21.9	20.0	17.5	16.1	16.2	17.0	18.8	20.1	21.9	22.1	19.7		1
2 Mean daily maximum temperature	°C															2
3 Mean daily minimum temperature	°C															3
4 Absolute maximum temperature	°C	34.8	35.6	33.3	32.8	31.5	35.9	30.9	33.0	35.6	36.7	35.5	36.7	36.7		4
5 Absolute minimum temperature	°C	10.5	10.4	11.5	4.3	1.5	-1.5	0.2	0.2	1.8	5.2	8.0	9.5	-1.5		5
6 Mean relative humidity	%															6
7 Mean precipitation	mm	241	199	148	61	56	53	29	36	75	121	160	215	1394		7
8 Maximum precipitation	mm															8
9 Minimum precipitation	mm															9
10 Maximum precipitation in 24 h	mm	111	89	83	85	74	70	81	40	100	80	97	98	111		10
11 Mean number of days with precipitation	>0,1 mm	17	15	12	6	6	6	3	5	8	10	12	16	116		11
12 Mean duration of sunshine	h	217	178	221	230	251	219	249	251	199	220	223	205	2663		12
13 Mean quantity of radiation	ly/day															13
14 Mean potential evaporation	mm	110	96	97	74	55	43	44	53	65	82	99	108	928	55	14
15 Mean windspeed	m/sec	1,0	1,3	1,4	1,4	1,1	1,1	1,2	1,4	1,7	1,9	1,8	1,7	1,4		15
16 Mean predominent direction of the wind		SE	SE	SE	SE	SE	SE	SE	SE	SE	SE	SE	SE			16

93 Station/Country Rio de Janeiro/Brazil

Location 22°54'S/43°10'W   Height above sealevel 31 m   Climate symbol: Köppen Aw   Troll V.1

		J	F	M	A	M	J	J	A	S	O	N	D	year	P	
1	Mean daily temperature	in °C	25,1	25,6	24,3	23,6	22,1	21,1	20,2	20,8	21,0	21,8	22,9	22,4	22,7		1
2	Mean daily maximum temperature	in °C	28,9	29,4	29,3	26,7	25,0	24,4	23,9	24,4	23,9	25,0	26,1	27,8	26,1	38	2
3	Mean daily minimum temperature	in °C	22,8	22,8	22,2	20,6	18,9	17,8	17,2	17,8	18,3	18,9	20,0	21,7	20,0	38	3
4	Absolute maximum temperature	in °C	39,1	37,8	38,4	35,0	35,2	32,6	34,1	38,4	37,6	39,0	37,5	39,0	39,1		4
5	Absolute minimum temperature	in °C	15,5	17,0	17,6	15,3	13,8	10,9	11,3	11,5	10,2	13,4	15,0	13,4	10,2		5
6	Mean relative humidity	in %	76	78	81	80	79	78	77	75	78	78	77	77	78		6
7	Mean precipitation	in mm	157	125	134	102	83	56	51	40	63	80	92	130	1093		7
8	Maximum precipitation	in mm	318	314	286	241	165	128	91	140	173	172	167	248			8
9	Minimum precipitation	in mm	28	26	9	20	25	3	4	2	3	17	40	41			9
10	Maximum precipitation in 24 h	in mm	172	136	144	223	217	206	59	51	72	52	99	151	223		10
11	Mean number of days with precipitation	> 0,1 mm	12	12	13	11	10	8	7	8	11	12	13	14	131		11
12	Mean duration of sunshine	in h	222	206	215	207	210	194	209	205	153	151	181	197	2350		12
13	Mean quantity of radiation	in ly/day															13
14	Mean potential evaporation	in mm	137	122	122	91	76	64	61	69	73	85	99	131	1130	49	14
15	Mean windspeed	in m/sec	3,1	3,0	3,2	2,7	2,8	2,7	2,7	3,1	3,5	3,9	4,0	3,8	3,2		15
16	Mean predominent direction of the wind		SSE	SSE	SSE	SSE	SSE	SSE	SSE	SSE	SSE	SSE	SSE	SSE			16

94 Station/Country Curitiba/Brazil

Location 25°26'S/49°16'W   Height above sealevel 949 m   Climate symbol: Köppen Cfb   Troll IV.7

		J	F	M	A	M	J	J	A	S	O	N	D	year	P	
1	Mean daily temperature	in °C	20,1	20,1	19,2	17,1	14,3	12,9	12,1	13,5	14,5	15,9	17,7	19,3	16,4		1
2	Mean daily maximum temperature	in °C															2
3	Mean daily minimum temperature	in °C															3
4	Absolute maximum temperature	in °C	34,3	33,4	33,2	30,3	28,4	26,4	28,1	30,7	31,0	33,4	34,4	34,3	34,4		4
5	Absolute minimum temperature	in °C	7,5	7,0	5,5	0,8	-3,6	-6,3	-6,2	-4,5	-1,9	1,5	5,3	6,2	-6,3		5
6	Mean relative humidity	in %															6
7	Mean precipitation	in mm	183	149	106	76	88	104	69	85	124	122	120	138	1364		7
8	Maximum precipitation	in mm															8
9	Minimum precipitation	in mm															9
10	Maximum precipitation in 24 h	in mm	79	68	68	64	98	164	86	58	93	112	71	80	164		10
11	Mean number of days with precipitation	> 0,1 mm	20	17	17	14	12	11	10	10	14	14	14	16	169		11
12	Mean duration of sunshine	in h	175	167	167	161	168	154	185	184	44	164	178	187	2034		12
13	Mean quantity of radiation	in ly/day															13
14	Mean potential evaporation	in mm	98	85	82	63	48	37	33	144	51	63	77	92	773	31	14
15	Mean windspeed	in m/sec	2,6	2,4	2,5	2,2	2,1	2,0	2,1	2,4	2,8	2,9	3,0	2,9	2,5		15
16	Mean predominent direction of the wind		E	E	E	E	NE,NW	NW	NE	NE,E	E	E	E	E			16

95 Station/Country Guarapuava / Brazil

Location 25°24'S/51°28'W   Height above sealevel 1095 m   Climate symbol: Köppen Cfb   Troll IV.7

		J	F	M	A	M	J	J	A	S	O	N	D	year	P	
1	Mean daily temperature	in °C	20,6	20,0	18,9	16,1	14,4	12,8	12,8	14,4	15,6	16,7	18,3	19,4	16,7		1
2	Mean daily maximum temperature	in °C	26,1	25,6	24,4	22,8	19,4	17,8	18,9	21,1	20,6	23,3	25,0	26,1	22,8	11	2
3	Mean daily minimum temperature	in °C	16,1	16,7	14,4	12,8	9,4	7,8	8,3	8,9	10,0	11,7	12,8	15,0	12,2	10	3
4	Absolute maximum temperature	in °C	33,9	32,2	32,2	30,0	28,3	28,7	28,7	28,9	29,4	29,4	31,1	34,4	34,4	8	4
5	Absolute minimum temperature	in °C	7,8	10,0	10,0	3,9	-4,4	-5,0	-2,8	0,0	1,7	2,2	6,1	1,7	-5,0	8	5
6	Mean relative humidity	in %	77	74	78	76	78	80	74	73	75	75	71	69	75	4	6
7	Mean precipitation	in mm	188	142	135	113	119	135	101	112	157	163	157	152	1674	30	7
8	Maximum precipitation	in mm															8
9	Minimum precipitation	in mm															9
10	Maximum precipitation in 24 h	in mm	64	64	69	74	61	71	43	53	71	69	46	64	74	4	10
11	Mean number of days with precipitation	> 0,25 mm	13	12	11	8	8	7	6	7	10	11	9	14	116	10	11
12	Mean duration of sunshine	in h															12
13	Mean quantity of radiation	in ly/day															13
14	Mean potential evaporation	in mm	98	82	82	60	45	37	39	50	51	66	80	96	786		14
15	Mean windspeed	in m/sec															15
16	Mean predominent direction of the wind																16

96 Station/Country Alegrete/Brazil

Location 29°46'S/55°47'W   Height above sealevel 104 m   Climate symbol: Köppen Cfa   Troll IV.6

		J	F	M	A	M	J	J	A	S	O	N	D	year	P	
1	Mean daily temperature	in °C	24,7	24,0	22,3	18,0	15,0	13,7	12,9	14,2	15,6	17,6	20,9	23,5	18,6		1
2	Mean daily maximum temperature	in °C															2
3	Mean daily minimum temperature	in °C															3
4	Absolute maximum temperature	in °C	39,4	38,3	38,6	36,2	30,8	29,5	30,4	31,4	35,7	34,3	38,8	38,6	39,4		4
5	Absolute minimum temperature	in °C	8,2	10,0	4,3	0,8	-0,6	-5,0	-5,0	-2,4	-0,6	0,7	3,0	7,2	-5,0		5
6	Mean relative humidity	in %															6
7	Mean precipitation	in mm	156	125	148	191	171	109	110	113	140	166	123	131	1683		7
8	Maximum precipitation	in mm															8
9	Minimum precipitation	in mm															9
10	Maximum precipitation in 24 h	in mm	100	82	105	145	145	124	74	87	94	110	96	154	154		10
11	Mean number of days with precipitation	> 0,1 mm	9	8	9	9	10	11	11	10	9	10	8	8	112		11
12	Mean duration of sunshine	in h	277	253	244	204	184	152	169	190	196	227	258	289	2643		12
13	Mean quantity of radiation	in ly/day															13
14	Mean potential evaporation	in mm	151	124	111	71	44	28	27	35	48	71	103	138	951	19	14
15	Mean windspeed	in m/sec	4,5	4,1	4,0	3,9	3,9	3,9	4,1	4,5	4,9	4,7	4,5	4,3	4,3		15
16	Mean predominent direction of the wind		E	E	E	E	E	E	E	E	E	E	ESE	E			16

97 Station/Country Porto Alegre/Brazil

Location 30°02'S/51°13'W — Height above sealevel 10 m — Climate symbol: Köppen Cfa — Troll IV,6

		J	F	M	A	M	J	J	A	S	O	N	D	year	P	
1 Mean daily temperature	in °C	24,7	24,3	23,0	20,2	17,1	14,8	14,3	15,1	16,5	18,6	21,0	23,6	19,5		1
2 Mean daily maximum temperature	in °C	30,6	30,6	28,3	25,6	21,7	18,9	18,9	20,0	21,1	23,3	26,7	29,4	24,4	22	2
3 Mean daily minimum temperature	in °C	19,4	20,0	18,3	15,6	12,2	.9,4	9,4	10,0	12,2	13,9	15,6	17,8	14,4	22	3
4 Absolute maximum temperature	in °C	39,2	40,4	38,9	35,9	32,1	31,4	32,9	33,3	36,0	35,3	38,0	39,8	40,4		4
5 Absolute minimum temperature	in °C	10,4	11,3	9,0	5,8	0,4	-2,0	-1,5	-0,9	2,4	4,7	6,4	7,8	-2,0		5
6 Mean relative humidity	in %	77	70	72	76	78	79	79	79	76	72	67	66	74		6
7 Mean precipitation	in mm	93	90	91	109	125	128	127	123	134	107	85	85	1298		7
8 Maximum precipitation	in mm	275	198	187	387	406	404	280	253	253	317	216	185			8
9 Minimum precipitation	in mm	15	37	21	19	30	42	11	19	34	25	5	0			9
10 Maximum precipitation in 24 h	in mm	145	68	64	115	70	91	87	92	104	101	92	86	145	10	10
11 Mean number of days with precipitation	> 0,1 mm	10	8	9	10	11	11	10	11	11	10	9	9	119		11
12 Mean duration of sunshine	in h	248	223	209	184	172	150	161	165	146	195	235	258	2346		12
13 Mean quantity of radiation	in ly/day															13
14 Mean potential evaporation	in mm	140	117	108	74	52	33	32	40	51	74	96	131	948	10	14
15 Mean windspeed	in m/sec	1,3	1,1	1,0	0,8	0,9	0,6	0,8	1,1	1,3	1,5	1,4	1,4	1,1		15
16 Mean predominent direction of the wind		SE	SE	SE	SE	SE	SE	SE	SE	SE	SE	SE	SE			16

98 Station/Country Arica/Chile

Location 18°28'S/70°22'W — Height above sealevel 29 m — Climate symbol: Köppen BWh — Troll IV,5

		J	F	M	A	M	J	J	A	S	O	N	D	year	P	
1 Mean daily temperature	in °C	22,1	22,3	21,3	19,6	18,0	16,7	15,9	15,8	16,6	17,6	19,1	20,7	18,8	40	1
2 Mean daily maximum temperature	in °C	25,6	26,1	25,0	23,3	21,1	19,4	18,9	18,9	20,0	22,2	23,9	22,2		15	2
3 Mean daily minimum temperature	in °C	17,8	18,3	17,2	15,6	14,4	13,9	12,2	12,8	13,3	14,4	15,6	16,7	15,0	17	3
4 Absolute maximum temperature	in °C	31,5	30,8	29,5	30,0	27,5	26,5	22,6	23,0	27,4	25,9	28,2	28,5	31,5	40	4
5 Absolute minimum temperature	in °C	11,0	10,1	11,5	10,2	6,6	6,8	5,2	5,8	4,2	9,0	10,0	11,0	4,2	40	5
6 Mean relative humidity	in %	67	68	70	74	74	77	77	79	78	75	70	66	73	9	6
7 Mean precipitation	in mm	0,3	tr	tr	0	tr	0,1	0,1	0,1	tr	tr	0	0,1	0,7	39	7
8 Maximum precipitation	in mm															8
9 Minimum precipitation	in mm															9
10 Maximum precipitation in 24 h	in mm	10,0	0,4	tr	0,0	tr	1,0	0,6	2,0	0,4	0,1	0,0	1,7	10,0	11	10
11 Mean number of days with precipitation	> 0,1 mm	<0,1	0	0	0	0	0	0	<0,1	0	0	0	<0,1	<0,1		11
12 Mean duration of sunshine	in h	195	167	198	219	143	105	93	81	45	149	183	189	1767		12
13 Mean quantity of radiation	in ly/day															13
14 Mean potential evaporation	in mm	106	96	94	73	62	52	48	50	54	67	72	96	870	34	14
15 Mean windspeed	in m/sec	6,0	6,0	4,0	4,0	4,0	4,0	4,0	4,0	4,0	4,0	4,0	6,0	4,0	22	15
16 Mean predominent direction of the wind		SW	SW	SW	SW	SW	SW	SW	SW	SW	SW	SW	SW		22	16

99 Station/Country Iquique/Chile

Location 20°22'S/70°11'W — Height above sealevel 9 m — Climate symbol: Köppen BWk — Troll IV,5

		J	F	M	A	M	J	J	A	S	O	N	D	year	P	
1 Mean daily temperature	in °C	20,9	20,8	19,7	18,3	17,2	16,0	15,4	15,4	16,1	17,1	18,5	19,8	17,9	29	1
2 Mean daily maximum temperature	in °C															2
3 Mean daily minimum temperature	in °C															3
4 Absolute maximum temperature	in °C	30,5	30,8	31,3	28,6	28,0	22,6	22,4	23,5	26,6	28,0	30,6	30,0	31,3	29	4
5 Absolute minimum temperature	in °C	13,0	12,0	9,0	8,4	9,0	9,0	8,6	8,0	9,0	10,0	11,2	11,5	8,0	29	5
6 Mean relative humidity	in %															6
7 Mean precipitation	in mm	0,1	tr	0	tr	tr	0,2	0,7	0,7	0,3	0,1	tr	0	2,1	39	7
8 Maximum precipitation	in mm															8
9 Minimum precipitation	in mm															9
10 Maximum precipitation in 24 h	in mm	2,3	0,8	0	0,3	1,6	4,6	12,8	3,0	4,6	2,8	tr	0	12,8	34	10
11 Mean number of days with precipitation	> 0,1 mm	0,1	<0,1	0	0,1	0,4	0,3	1,1	0,8	1,1	0,7	0,1	0	4,7		11
12 Mean duration of sunshine	in h															12
13 Mean quantity of radiation	in ly/day															13
14 Mean potential evaporation	in mm	99	86	84	67	59	49	48	50	55	65	76	92	830	44	14
15 Mean windspeed	in m/sec	2,0	2,0	2,0	2,0	2,0	2,0	2,0	2,0	2,0	2,0	2,0	2,0	2,0	17	15
16 Mean predominent direction of the wind		S	S	S	S	S	S	S	S	S	S	S	S		17	16

100 Station/Country Antofagasta/Chile

Location 23°26'S/70°28'W — Height above sealevel 119 m — Climate symbol: Köppen BWk — Troll IV,5

		J	F	M	A	M	J	J	A	S	O	N	D	year	P	
1 Mean daily temperature	in °C	19,9	20,1	18,5	16,2	15,1	13,5	13,1	13,5	14,5	15,2	16,8	18,4	16,2	10	1
2 Mean daily maximum temperature	in °C	24,4	24,4	23,3	21,1	19,4	18,3	17,2	16,7	17,8	18,9	20,6	22,2	20,6	22	2
3 Mean daily minimum temperature	in °C	17,2	17,2	16,1	14,4	12,8	11,1	10,6	11,1	11,7	12,8	14,4	15,6	13,9	22	3
4 Absolute maximum temperature	in °C	29,1	30,1	25,6	24,0	24,2	23,0	22,0	21,0	24,0	22,0	24,0	24,8	30,1	10	4
5 Absolute minimum temperature	in °C	11,0	8,4	7,0	6,0	5,3	5,0	3,2	3,0	6,0	6,5	8,2	9,6	3,0	10	5
6 Mean relative humidity	in %	74	74	75	75	76	75	77	76	74	73	73	73	74		6
7 Mean precipitation	in mm	0	0	tr	0,3	0,1	1,7	2,4	1,2	1,0	0,8	0,2	tr	7,7	12	7
8 Maximum precipitation	in mm															8
9 Minimum precipitation	in mm															9
10 Maximum precipitation in 24 h	in mm	0	0	tr	8,0	2,8	38,0	17,0	28,0	6,8	9,7	4,0	0,5	38,0	12	10
11 Mean number of days with precipitation	> 0,1 mm	0	0	0	0,1	0,1	0,2	0,5	0,4	0,5	0,2	0,3	0	2,3		11
12 Mean duration of sunshine	in h	326	291	248	129	195	183	186	170	180	170	195	276	2549		12
13 Mean quantity of radiation	in ly/day															13
14 Mean potential evaporation	in mm	104	91	86	65	53	42	39	42	49	60	72	90	793	35	14
15 Mean windspeed	in m/sec	4,0	4,0	4,0	4,0	2,0	4,0	2,0	2,0	2,0	4,0	4,0	4,0	4,0	12	15
16 Mean predominent direction of the wind		SW	SW	SW	SW	W	W	W	W	W,SW	SW	SW	SW		12	16

101 Station / Country La Serena / Chile

Location 29°54'S/71°15'W Height above sealevel 35 m Climate symbol: Köppen BWk Troll IV,2

		J	F	M	A	M	J	J	A	S	O	N	D	year	P	
1 Mean daily temperature	in °C	18.2	18.4	16.9	14.9	13.4	12.1	11.7	12.0	12.7	14.0	15.5	17.0	14.7	40	1
2 Mean daily maximum temperature	in °C															2
3 Mean daily minimum temperature	in °C															3
4 Absolute maximum temperature	in °C	27.2	27.8	29.3	25.8	23.7	24.5	23.3	24.0	26.0	25.5	25.0	25.8	29.3	40	4
5 Absolute minimum temperature	in °C	9.4	10.0	6.5	2.9	3.7	2.5	2.6	1.8	2.8	4.0	6.7	7.9	1.8	40	5
6 Mean relative humidity	in %															6
7 Mean precipitation	in mm	0.1	0.8	0.6	2.6	21.9	43.7	29.7	23.2	6.0	3.7	0.7	0.3	133.3	39	7
8 Maximum precipitation	in mm															8
9 Minimum precipitation	in mm															9
10 Maximum precipitation in 24 h	in mm	1.1	54.3	3.5	28.0	89.2	80.9	57.0	39.9	19.2	33.3	9.0	17.2	89.2	40	10
11 Mean number of days with precipitation	> 0,1 mm	5	4	5	5	4	6	5	7	7	6	4		63		11
12 Mean duration of sunshine	in h															12
13 Mean quantity of radiation	in ly/day															13
14 Mean potential evaporation	in mm	96	87	81	56	46	38	38	42	47	61	71	88	751	74	14
15 Mean windspeed	in m/sec	2.0	2.0	2.0	1.0	1.0	1.0	2.0	2.0	2.0	2.0	2.0	2.0	2.0	21	15
16 Mean predominent direction of the wind		NW	NW	NW	NW	NW	NW	NW	NW	NW	NW	NW	NW		21	16

102 Station / Country Valparaíso / Chile

Location 33°01'S/71°38'W Height above sealevel 41 m Climate symbol: Köppen Csb Troll IV,1

		J	F	M	A	M	J	J	A	S	O	N	D	year	P	
1 Mean daily temperature	in °C	18.0	17.9	16.7	14.9	13.5	12.2	11.8	12.0	12.9	14.1	15.7	17.2	14.7	52	1
2 Mean daily maximum temperature	in °C	22.2	22.2	21.1	19.4	17.2	15.8	15.6	16.1	16.7	18.3	20.6	21.7	18.9	30	2
3 Mean daily minimum temperature	in °C	13.3	13.3	12.2	11.1	10.0	8.9	8.3	8.3	8.9	10.0	11.1	12.2	10.6	30	3
4 Absolute maximum temperature	in °C	36.0	34.5	31.5	32.4	31.0	25.2	27.2	27.5	27.8	34.4	34.0	36.0	36.0	52	4
5 Absolute minimum temperature	in °C	7.8	9.0	7.0	5.0	3.8	2.2	2.0	3.0	4.0	4.5	6.5	6.7	2.0	52	5
6 Mean relative humidity	in %	75	76	76	78	80	79	79	79	78	77	73	71	77		6
7 Mean precipitation	in mm	2	2	4	18	97	128	88	67	30	16	7	3	462	40	7
8 Maximum precipitation	in mm															8
9 Minimum precipitation	in mm															9
10 Maximum precipitation in 24 h	in mm	25	24	38	72	172	186	82	104	76	41	36	25	186	40	10
11 Mean number of days with precipitation	> 0,1 mm	<1	<1	<1	1	5	7	7	5	2	2	1	<1	31		11
12 Mean duration of sunshine	in h	279	246	217	174	115	81	93	118	147	170	216	264	2120		12
13 Mean quantity of radiation	in ly/day															13
14 Mean potential evaporation	in mm	93	79	73	55	46	36	36	41	46	60	73	86	724	52	14
15 Mean windspeed	in m/sec	4.0	4.0	2.0	2.0	2.0	2.0	4.0	4.0	2.0	4.0	4.0	4.0	4.0	40	15
16 Mean predominent direction of the wind		SW	SW	SW	SW	SW	N	N	SW	SW	SW	SW	SW		40	16

103 Station / Country Santiago / Chile

Location 33°27'S/70°42'W Height above sealevel 520 m Climate symbol: Köppen Csb Troll IV,1

		J	F	M	A	M	J	J	A	S	O	N	D	year	P	
1 Mean daily temperature	in °C	20.0	19.3	17.2	13.9	10.9	8.4	8.1	9.1	11.6	13.8	16.5	18.9	14.0	95	1
2 Mean daily maximum temperature	in °C	29.4	28.8	26.7	23.3	18.3	14.4	15.0	16.7	18.9	22.2	25.6	28.3	22.2	14	2
3 Mean daily minimum temperature	in °C	11.7	11.1	9.4	7.2	5.0	2.8	2.8	3.9	5.8	7.2	8.9	10.8	7.2	14	3
4 Absolute maximum temperature	in °C	35.6	36.5	34.3	31.5	31.5	28.5	27.2	29.2	31.4	33.3	36.0	37.2	37.2	95	4
5 Absolute minimum temperature	in °C	5.9	5.2	2.0	-2.5	-3.0	-4.2	-4.6	-3.3	-2.3	-1.0	0.9	2.2	-4.6	95	5
6 Mean relative humidity	in %	54	59	64	68	75	79	76	75	72	67	57	54	67		6
7 Mean precipitation	in mm	2	3	4	14	62	85	76	57	29	15	6	4	361	91	7
8 Maximum precipitation	in mm															8
9 Minimum precipitation	in mm															9
10 Maximum precipitation in 24 h	in mm	22	53	24	62	73	103	63	76	44	26	30	37	103	19	10
11 Mean number of days with precipitation	> 0,1 mm	<0.1	<0.1	1	5	6	6	5	3	3	1	<1		31		11
12 Mean duration of sunshine	in h	332	277	271	194	116	96	114	136	161	213	264	327	2501	21	12
13 Mean quantity of radiation	in ly/day															13
14 Mean potential evaporation	in mm	115	93	81	55	36	22	23	31	41	62	82	105	748	91	14
15 Mean windspeed	in m/sec	2.0	2.0	2.0	1.0	1.0	1.0	1.0	1.0	2.0	2.0	2.0	2.0	2.0	22	15
16 Mean predominent direction of the wind		SW	SW	SW	SW	SW	SW	SW	SW	SW	SW	SW	SW		22	16

104 Station / Country Concepción / Chile

Location 38°40'S/73°03'W Height above sealevel 15 m Climate symbol: Köppen Csb Troll IV,1

		J	F	M	A	M	J	J	A	S	O	N	D	year	P	
1 Mean daily temperature	in °C	18.0	17.2	15.1	12.8	11.1	9.7	9.1	9.1	10.6	12.6	14.8	16.9	13.0	40	1
2 Mean daily maximum temperature	in °C															2
3 Mean daily minimum temperature	in °C															3
4 Absolute maximum temperature	in °C	37.5	34.0	36.8	28.6	27.5	24.1	23.8	25.2	28.1	34.0	33.5	36.5	37.5	40	4
5 Absolute minimum temperature	in °C	3.5	1.5	0.8	-0.7	-0.9	-2.5	-4.0	-5.0	-2.5	-0.5	1.1	3.8	-5.0	40	5
6 Mean relative humidity	in %															6
7 Mean precipitation	in mm	17	21	52	85	211	250	238	183	103	59	46	29	1294	91	7
8 Maximum precipitation	in mm															8
9 Minimum precipitation	in mm															9
10 Maximum precipitation in 24 h	in mm	89	54	95	92	137	126	117	115	69	83	56	57	137	40	10
11 Mean number of days with precipitation	> 0,1 mm	2	4	5	9	15	17	16	14	12	8	7	4	113		11
12 Mean duration of sunshine	in h	257	223	208	153	98	73	99	130	157	218	238	277	2131	9	12
13 Mean quantity of radiation	in ly/day															13
14 Mean potential evaporation	in mm	101	81	71	50	37	30	30	36	41	59	76	96	708	66	14
15 Mean windspeed	in m/sec	4.0	4.0	4.0	4.0	4.0	4.0	4.0	4.0	4.0	4.0	4.0	4.0	4.0	22	15
16 Mean predominent direction of the wind		SW	SW	SW	SW	N	N	N	N	SW	S	SW	SW		22	16

105 Station / Country Valdivia / Chile

Location 39°48'S / 73°14'W Height above sealevel 9 m Climate symbol: Köppen Cfb Troll III,2

	in	J	F	M	A	M	J	J	A	S	O	N	D	year	P	
1 Mean daily temperature	in °C	17,0	16,4	14,5	11,8	9,7	8,2	7,7	8,0	9,3	11,5	13,3	15,3	11,9	40	1
2 Mean daily maximum temperature	in °C	22,8	22,8	20,6	16,7	13,3	11,1	11,1	12,2	14,4	17,2	18,3	20,6	16,7	29	2
3 Mean daily minimum temperature	in °C	11,1	10,8	9,4	7,8	5,8	5,0	4,4	5,0	6,7	7,8	10,0		7,2	29	3
4 Absolute maximum temperature	in °C	34,8	36,6	32,5	28,0	21,5	17,0	19,0	22,5	26,0	29,0	32,0	34,0	36,6	40	4
5 Absolute minimum temperature	in °C	2,7	1,7	1,0	-2,0	-3,5	-3,9	-4,2	-3,8	-2,8	-0,5	0,0	-2,2	-4,2	40	5
6 Mean relative humidity	in %	76	79	82	86	91	92	92	89	84	82	79	78	84		6
7 Mean precipitation	in mm	65	69	115	212	377	414	374	301	214	119	122	107	2489	39	7
8 Maximum precipitation	in mm															8
9 Minimum precipitation	in mm															9
10 Maximum precipitation in 24 h	in mm	135	77	121	156	112	162	159	174	129	78	131	98	174	40	10
11 Mean number of days with precipitation	> 0,1 mm	7	7	11	12	21	21	20	18	13	13	10	10	163		11
12 Mean duration of sunshine	in h	257	229	205	123	68	48	65	90	111	127	189	208	1720		12
13 Mean quantity of radiation	in ly / day															13
14 Mean potential evaporation	in mm	99	81	70	47	32	25	25	29	38	56	71	89	664	75	14
15 Mean windspeed	in m / sec	2,0	2,0	2,0	2,0	2,0	4,0	2,0	4,0	2,0	1,0	2,0	2,0		40	15
16 Mean predominent direction of the wind		S	S	S	N	N	N	N	N	N	W	W	W			16

106 Station / Country Puerto Montt / Chile

Location 41°28'S / 72°57'W Height above sealevel 13 m Climate symbol: Köppen Cfb Troll III,1

	in	J	F	M	A	M	J	J	A	S	O	N	D	year	P	
1 Mean daily temperature	in °C	15,2	14,8	13,2	11,2	9,3	8,0	7,6	7,8	8,8	10,6	12,2	13,9	11,1	40	1
2 Mean daily maximum temperature	in °C															2
3 Mean daily minimum temperature	in °C															3
4 Absolute maximum temperature	in °C	28,5	29,0	28,6	24,0	19,0	17,5	20,0	19,5	25,5	23,4	26,5	28,9	29,0	40	4
5 Absolute minimum temperature	in °C	4,5	4,0	0,5	-2,6	-2,5	-2,5	-4,0	-3,0	-2,1	-1,0	-0,5	2,5	-4,0	40	5
6 Mean relative humidity	in %															6
7 Mean precipitation	in mm	90	139	139	181	236	257	209	198	158	119	131	125	1982	39	7
8 Maximum precipitation	in mm															8
9 Minimum precipitation	in mm															9
10 Maximum precipitation in 24 h	in mm	57	133	88	94	84	91	90	116	109	70	89	77	133	40	10
11 Mean number of days with precipitation	> 0,1 mm	13	12	15	20	24	24	23	21	19	17	18	16	222		11
12 Mean duration of sunshine	in h															12
13 Mean quantity of radiation	in ly / day															13
14 Mean potential evaporation	in mm	93	76	67	48	35	26	26	30	39	55	68	84	647	63	14
15 Mean windspeed	in m / sec	4,0	2,0	2,0	4,0	4,0	4,0	6,0	4,0	4,0	4,0	4,0	4,0	4,0		15
16 Mean predominent direction of the wind		S	S	N	N	N	N	N	N	N	S	S	S			16

107 Station / Country San Pedro / Chile

Location 47°43'S / 74°55'W Height above sealevel 22 m Climate symbol: Köppen Cfc Troll III,1

	in	J	F	M	A	M	J	J	A	S	O	N	D	year	P	
1 Mean daily temperature	in °C	11,2	11,2	10,9	8,5	6,8	6,1	5,7	5,5	6,4	7,8	8,8	10,2	8,2	19	1
2 Mean daily maximum temperature	in °C															2
3 Mean daily minimum temperature	in °C															3
4 Absolute maximum temperature	in °C	24,0	26,5	19,7	16,0	14,0	15,8	13,0	13,5	18,0	19,5	20,8	28,0	28,5	19	4
5 Absolute minimum temperature	in °C	2,0	0,2	-1,0	-0,5	-2,0	-3,0	-3,0	-2,5	-4,2	-0,5	-0,1	-0,2	-4,2	19	5
6 Mean relative humidity	in %															6
7 Mean precipitation	in mm	378	385	398	409	394	372	398	310	328	374	363	379	4486	37	7
8 Maximum precipitation	in mm															8
9 Minimum precipitation	in mm															9
10 Maximum precipitation in 24 h	in mm	92	113	75	114	72	98	91	116	80	98	138	137	138	19	10
11 Mean number of days with precipitation	> 0,1 mm	28	24	27	25	25	25	28	24	25	26	25	25	307		11
12 Mean duration of sunshine	in h															12
13 Mean quantity of radiation	in ly / day															13
14 Mean potential evaporation	in mm	82	68	60	43	31	24	25	29	37	52	64	77	592	11	14
15 Mean windspeed	in m / sec	7,0	4,0	4,0	9,0	9,0	9,0	9,0	7,0	9,0	7,0	7,0	9,0	7,0	19	15
16 Mean predominent direction of the wind		NW	W	W	NW	NW	N	NW	NW	NW	NW	W	NW		19	16

108 Station / Country Los Evangelistas / Chile

Location 52°24'S / 75°06'W Height above sealevel 55 m Climate symbol: Köppen ET Troll III,1

	in	J	F	M	A	M	J	J	A	S	O	N	D	year	P	
1 Mean daily temperature	in °C	8,7	8,8	8,3	7,2	6,0	4,8	4,4	4,4	4,9	5,5	6,4	7,6	6,4		1
2 Mean daily maximum temperature	in °C	10,0	10,0	10,0	8,9	7,8	6,7	6,1	6,1	6,7	7,2	7,8	8,9	7,8	16	2
3 Mean daily minimum temperature	in °C	6,7	6,7	6,1	5,0	3,3	2,2	2,2	2,2	3,3	3,9	4,4	5,6	4,4	16	3
4 Absolute maximum temperature	in °C	16,8	16,2	16,5	14,3	14,6	14,2	11,5	11,4	11,9	12,2	13,5	14,2	16,8		4
5 Absolute minimum temperature	in °C	1,0	0,2	-1,0	-3,3	-2,0	-7,2	-4,5	-3,5	-3,5	-1,0	-1,0	-1,0	-7,2		5
6 Mean relative humidity	in %	84	84	83	83	83	83	83	89	83	83	83	83	84	11	6
7 Mean precipitation	in mm	233	225	276	248	215	212	216	214	225	165	170	171	2570		7
8 Maximum precipitation	in mm															8
9 Minimum precipitation	in mm															9
10 Maximum precipitation in 24 h	in mm	73	70	74	106	73	98	83	137	123	73	56	51	137		10
11 Mean number of days with precipitation	> 1,0 mm	26	24	26	25	24	23	24	24	23	22	24	27	292		11
12 Mean duration of sunshine	in h															12
13 Mean quantity of radiation	in ly / day															13
14 Mean potential evaporation	in mm	77	82	58	44	32	23	24	28	35	48	55	72	559		14
15 Mean windspeed	in m / sec	9,0	9,0	12,0	15,0	12,0	9,0	9,0	15,0	12,0	15,0	12,0	12,0	12,0		15
16 Mean predominent direction of the wind		NW	NW	NW	NW	NW	SW	SW	NW	NW	NW	NW	NW			16

109 Station / Country   **Punta Arenas / Chile**

Location   **53°10'S / 70°54'W**   Height above sealevel **8 m**   Climate symbol: Köppen **Cfc**   Troll **III,2**

		J	F	M	A	M	J	J	A	S	O	N	D	year	P	
1 Mean daily temperature	in °C	11,7	10,6	8,9	6,7	4,2	2,6	2,5	2,8	4,6	7,1	8,5	10,2	6,7	40	1
2 Mean daily maximum temperature	in °C	14,4	14,4	12,2	10,0	7,2	5,0	4,4	5,8	7,8	10,6	12,2	13,9	10,0	15	2
3 Mean daily minimum temperature	in °C	7,2	6,7	5,0	3,9	1,7	0,6	-0,8	0,6	1,7	3,3	4,4	6,1	3,3	15	3
4 Absolute maximum temperature	in °C	29,0	24,5	22,8	19,2	16,0	15,0	13,0	13,8	19,2	20,6	25,0	25,2	29,0	40	4
5 Absolute minimum temperature	in °C	0,0	-0,1	-1,0	-3,6	-7,5	-7,3	-9,3	-7,5	-6,0	-2,5	-2,5	-0,2	-9,3	40	5
6 Mean relative humidity	in %	71	69	74	78	80	82	81	79	76	70	69	71	75	10	6
7 Mean precipitation	in mm	33	29	45	46	50	40	41	38	33	26	32	34	447	49	7
8 Maximum precipitation	in mm															8
9 Minimum precipitation	in mm															9
10 Maximum precipitation in 24 h	in mm	26	30	98	65	47	45	45	45	42	48	30	32	98	40	10
11 Mean number of days with precipitation	> 0,1 mm	6	5	7	9	8	8	6	5	5	5	5	8	75		11
12 Mean duration of sunshine	in h															12
13 Mean quantity of radiation	in ly / day															13
14 Mean potential evaporation	in mm	90	71	59	38	23	13	12	18	32	54	68	88	564	65	14
15 Mean windspeed	in m / sec	4,0	4,0	4,0	4,0	2,0	2,0	2,0	4,0	4,0	4,0	4,0	4,0	4,0	24	15
16 Mean predominent direction of the wind		W	NW	W	NW	W	W	W	W	W	W	W	W		24	16

110 Station / Country   **Yacuiba / Bolivia**

Location   **22°01'S / 63°43'W**   Height above sealevel **580 m**   Climate symbol: Köppen **Cwa**   Troll **V,4**

		J	F	M	A	M	J	J	A	S	O	N	D	year	P	
1 Mean daily temperature	in °C	24,5	24,5	22,5	20,5	18,5	15,5	15,5	17,5	20,0	22,5	23,5	24,5	20,5	7	1
2 Mean daily maximum temperature	in °C	31	30	28	26	24	21	23	26	28	29	30	31	27	7	2
3 Mean daily minimum temperature	in °C	18	19	17	15	13	10	8	9	12	16	17	18	14	7	3
4 Absolute maximum temperature	in °C	39	39	39	35	34	31	36	39	40	41	41	38	41	7	4
5 Absolute minimum temperature	in °C	10	11	7	1	-2	-3	-7	-6	-4	5	5	8	-7	7	5
6 Mean relative humidity	in %															6
7 Mean precipitation	in mm	196	175	153	105	40	11	8	8	8	62	118	129	1009	10	7
8 Maximum precipitation	in mm	424	300	388	311	115	40	24	56	25	152	208	289	1661	10	8
9 Minimum precipitation	in mm	94	96	18	20	0	0	0	0	0	3	53	77	653	10	9
10 Maximum precipitation in 24 h	in mm															10
11 Mean number of days with precipitation	> 0,25mm															11
12 Mean duration of sunshine	in h															12
13 Mean quantity of radiation	in ly / day															13
14 Mean potential evaporation	in mm	130	102	97	74	48	32	30	41	66	93	110	131	954		14
15 Mean windspeed	in m / sec	2,0	2,0	2,0	2,0	2,0	2,0	2,5	3,0	3,5	3,0	3,0	2,5	2,5	10	15
16 Mean predominent direction of the wind		S	S	S	S	N	N	N	S	S	S	S	N		10	16

111 Station / Country   **Mariscal Estigarribia / Paraguay**

Location   **22°01'S / 60°36'W**   Height above sealevel **181 m**   Climate symbol: Köppen **Aw**   Troll **V,3**

		J	F	M	A	M	J	J	A	S	O	N	D	year	P	
1 Mean daily temperature	in °C	29,7	28,9	27,4	24,3	21,8	20,0	19,8	22,8	25,4	27,0	28,4	29,5	25,4	20	1
2 Mean daily maximum temperature	in °C															2
3 Mean daily minimum temperature	in °C															3
4 Absolute maximum temperature	in °C	43,1	41,8	41,5	38,1	37,2	35,1	36,7	40,8	43,6	41,9	42,0	43,8	43,8	20	4
5 Absolute minimum temperature	in °C	10,6	9,1	7,8	3,8	-3,0	-4,5	-5,1	-2,0	-2,0	5,9	8,8	9,6	-5,1	20	5
6 Mean relative humidity	in %	54	58	60	62	66	65	57	47	47	49	51	51	56	20	6
7 Mean precipitation	in mm	112	109	80	60	42	31	17	4	26	94	84	99	758	20	7
8 Maximum precipitation	in mm															8
9 Minimum precipitation	in mm															9
10 Maximum precipitation in 24 h	in mm	125	111	64	150	99	98	42	12	80	125	97	102	150	20	10
11 Mean number of days with precipitation	> 0,1 mm	8	7	8	5	5	4	2	1	4	8	7	8	63	20	11
12 Mean duration of sunshine	in h															12
13 Mean quantity of radiation	in ly / day															13
14 Mean potential evaporation	in mm	177	149	145	87	69	45	65	75	108	134	154	173	1381	10	14
15 Mean windspeed	in m / sec	4,0	2,5	2,5	2,5	2,5	2,5	2,5	4,0	4,0	2,5	2,5	2,5	2,8	20	15
16 Mean predominent direction of the wind		N	N	N	NE	NE	NE	NE	N	N	N	N,NE	N		20	16

112 Station / Country   **Puerto Casado / Paraguay**

Location   **22°17'S / 57°52'W**   Height above sealevel **87 m**   Climate symbol: Köppen **Aw**   Troll **V,2**

		J	F	M	A	M	J	J	A	S	O	N	D	year	P	
1 Mean daily temperature	in °C	28,3	28,4	27,2	24,5	22,3	20,4	20,3	22,4	24,3	26,0	27,3	28,8	25,1	18	1
2 Mean daily maximum temperature	in °C															2
3 Mean daily minimum temperature	in °C															3
4 Absolute maximum temperature	in °C	41,1	39,7	38,3	37,8	35,8	34,7	34,6	37,8	40,1	38,8	40,0	41,7	41,7	18	4
5 Absolute minimum temperature	in °C	13,5	11,4	9,0	6,0	2,8	-0,9	0,9	1,1	0,7	8,8	9,7	10,2	-0,9	18	5
6 Mean relative humidity	in %	64	69	70	72	73	72	67	59	59	63	64	63	66	18	6
7 Mean precipitation	in mm	143	128	145	106	79	57	47	28	71	125	148	126	1203		7
8 Maximum precipitation	in mm															8
9 Minimum precipitation	in mm															9
10 Maximum precipitation in 24 h	in mm	132	116	126	134	65	61	51	31	74	87	210	100	210	18	10
11 Mean number of days with precipitation	> 0,1 mm	9	8	8	6	6	6	6	4	5	8	7	7	80	18	11
12 Mean duration of sunshine	in h															12
13 Mean quantity of radiation	in ly / day															13
14 Mean potential evaporation	in mm	175	150	145	92	76	50	57	69	105	126	150	172	1367	10	14
15 Mean windspeed	in m / sec	4,0	2,5	4,0	2,5	2,5	4,0	4,0	4,0	4,0	4,0	4,0	4,0	3,6	18	15
16 Mean predominent direction of the wind		N	N	S	S	S,N	S	S	S	S	S	S	N		18	16

113 Station / Country   Las Lomitas / Argentina

Location   24°42'S / 60°35'W   Height above sealevel **130 m**   Climate symbol: Köppen **Cwa**   Troll **V,3**

		J	F	M	A	M	J	J	A	S	O	N	D	year	P	
1 Mean daily temperature	in °C	27,1	26,5	24,6	20,7	17,9	17,0	15,7	18,4	21,2	23,3	25,3	26,7	22,0		1
2 Mean daily maximum temperature	in °C															2
3 Mean daily minimum temperature	in °C															3
4 Absolute maximum temperature	in °C	43,4	41,0	37,9	37,9	34,6	32,9	33,7	39,7	41,2	42,6	43,2	42,1	43,4		4
5 Absolute minimum temperature	in °C	9,4	10,3	9,8	0,8	-2,9	-2,4	-7,0	-5,3	-2,1	5,5	5,8	8,9	-7,0		5
6 Mean relative humidity	in %	66	70	72	77	79	79	71	59	59	63	61	65	70		6
7 Mean precipitation	in mm	113	96	112	74	40	33	18	9	36	85	88	109	812		7
8 Maximum precipitation	in mm															8
9 Minimum precipitation	in mm															9
10 Maximum precipitation in 24 h	in mm	105	145	115	105	95	75	55	35	75	115	85	125	145	30	10
11 Mean number of days with precipitation	> 1,0 mm	8	6	7	8	6	6	5	4	7	8	7	7	79		11
12 Mean duration of sunshine	in h	250	248	248	195	178	135	169	228	210	223	264	264	2630		12
13 Mean quantity of radiation	in ly / day															13
14 Mean potential evaporation	in mm	168	141	126	81	51	36	39	52	64	92	119	156	1125		14
15 Mean windspeed	in m / sec	3,1	3,1	2,8	2,8	2,8	3,1	3,3	3,9	3,9	3,9	3,6	3,1	3,2		15
16 Mean predominent direction of the wind		N	N	S	NE	NE	NE	NE	NE	E	E	E	E			16

114 Station / Country   Asunción / Paraguay

Location   25°16'S / 57°38'W   Height above sealevel **64 m**   Climate symbol: Köppen **Aw**   Troll **V,1**

		J	F	M	A	M	J	J	A	S	O	N	D	year	P	
1 Mean daily temperature	in °C	28,3	28,8	26,9	23,6	20,9	18,8	18,3	20,6	22,3	24,7	27,0	28,9	24,2	20	1
2 Mean daily maximum temperature	in °C	35,0	34,4	33,3	28,9	25,0	22,2	23,3	25,6	28,3	30,0	32,2	34,4	29,4	15	2
3 Mean daily minimum temperature	in °C	21,7	21,7	20,6	18,3	14,4	11,7	11,7	13,9	15,6	16,7	18,3	21,1	17,2	15	3
4 Absolute maximum temperature	in °C	41,8	40,8	38,9	35,8	33,0	32,0	32,6	38,0	39,1	39,1	39,5	41,2	41,8	20	4
5 Absolute minimum temperature	in °C	14,8	14,0	11,5	8,4	3,9	2,5	2,5	2,3	3,9	9,0	12,0	13,1	2,3	20	5
6 Mean relative humidity	in %	60	63	65	68	72	75	70	61	62	62	59	54	65	20	6
7 Mean precipitation	in mm	167	142	160	138	131	87	54	30	87	146	128	122	1392	20	7
8 Maximum precipitation	in mm															8
9 Minimum precipitation	in mm															9
10 Maximum precipitation in 24 h	in mm	107	202	95	128	190	94	40	40	96	97	96	142	202	20	10
11 Mean number of days with precipitation	> 0,1 mm	8	7	7	6	6	6	6	4	6	6	7	6	77	20	11
12 Mean duration of sunshine	in h	278	246	228	205	165	195	223	204	242	270	295		2803	20	12
13 Mean quantity of radiation	in ly / day															13
14 Mean potential evaporation	in mm	171	144	133	84	63	41	47	41	56	101	128	166	1175	41	14
15 Mean windspeed	in m / sec	2,5	2,5	2,5	2,5	2,5	2,5	2,5	4,0	4,0	2,5	2,5	4,0	2,8	20	15
16 Mean predominent direction of the wind		NE	NE	E	E	E	E	E	E	E	E,S	E	NE		20	16

115 Station / Country   Puerto Presidente Franco / Paraguay

Location   25°36'S / 54°34'W   Height above sealevel **125 m**   Climate symbol: Köppen **Cwa**   Troll **V,1**

		J	F	M	A	M	J	J	A	S	O	N	D	year	P	
1 Mean daily temperature	in °C	26,8	26,5	25,1	21,8	19,0	17,4	16,7	18,8	20,8	22,8	24,2	25,8	22,1		1
2 Mean daily maximum temperature	in °C															2
3 Mean daily minimum temperature	in °C															3
4 Absolute maximum temperature	in °C	40,0	39,0	38,8	35,5	32,3	30,7	30,9	35,2	36,7	37,2	39,7	39,5	40,0		4
5 Absolute minimum temperature	in °C	9,0	9,4	7,8	3,2	-3,0	-2,4	-2,0	-2,2	-1,9	4,0	5,2	9,0	-3,0		5
6 Mean relative humidity	in %	73	69	75	79	82	85	82	75	75	74	73	70	76		6
7 Mean precipitation	in mm	147	120	162	138	148	124	100	72	156	161	143	138	1609		7
8 Maximum precipitation	in mm															8
9 Minimum precipitation	in mm															9
10 Maximum precipitation in 24 h	in mm	112	85	141	102	102	92	88	67	151	115	96	82	151		10
11 Mean number of days with precipitation	> 1,0 mm	9	9	8	6	7	7	7	7	8	9	7	7	91		11
12 Mean duration of sunshine	in h															12
13 Mean quantity of radiation	in ly / day															13
14 Mean potential evaporation	in mm	159	141	128	78	56	42	39	52	7	99	126	156	1083		14
15 Mean windspeed	in m / sec	2,5	1,0	1,0	2,5	2,5	2,5	2,5	4,0	2,5	4,0	4,0	4,0	2,7		15
16 Mean predominent direction of the wind		SE,SW	SE	SE	S	S	S	S	S	SE	SE	SE	SE			16

116 Station / Country   Paso de los Toros / Uruguay

Location   32°49'S / 56°31'W   Height above sealevel **79 m**   Climate symbol: Köppen **Cfa**   Troll **IV,6**

		J	F	M	A	M	J	J	A	S	O	N	D	year	P	
1 Mean daily temperature	in °C	24,9	23,9	21,8	17,5	14,1	12,4	11,5	12,8	14,6	17,4	20,3	23,2	17,9	24	1
2 Mean daily maximum temperature	in °C															2
3 Mean daily minimum temperature	in °C															3
4 Absolute maximum temperature	in °C	42,8	40,4	38,8	36,0	32,3	28,8	27,8	32,0	35,5	36,8	39,5	40,9	42,8	24	4
5 Absolute minimum temperature	in °C	8,3	7,5	5,8	2,0	-1,7	-5,0	-5,0	-3,3	-1,6	0,4	3,9	6,3	-5,0	24	5
6 Mean relative humidity	in %	61	66	72	78	87	88	86	81	80	76	69	62	76	24	6
7 Mean precipitation	in mm	112	93	128	111	88	102	66	96	107	96	81	81	1163	24	7
8 Maximum precipitation	in mm															8
9 Minimum precipitation	in mm															9
10 Maximum precipitation in 24 h	in mm															10
11 Mean number of days with precipitation	> 0,1 mm	6	5	6	6	6	8	6	6	7	6	6	5	73	24	11
12 Mean duration of sunshine	in h															12
13 Mean quantity of radiation	in ly / day															13
14 Mean potential evaporation	in mm	146	119	102	59	38	27	24	34	45	71	97	133	896	24	14
15 Mean windspeed	in m / sec	3,6	3,3	3,3	3,1	3,1	3,3	3,3	3,9	3,9	3,6	3,6	3,6	3,6	24	15
16 Mean predominent direction of the wind		E	E	E	NE	NE	NE	NE	NE	NE	E	E	E		24	16

215

117 Station / Country Montevideo / Uruguay

Location 34°52'S/56°12'W — Height above sealevel 22 m — Climate symbol: Köppen Cfa — Troll IV,6

		J	F	M	A	M	J	J	A	S	O	N	D	year	P	
1 Mean daily temperature	in °C	22,5	22,2	20,3	17,0	13,7	10,9	10,5	11,1	12,8	15,1	18,3	21,0	16,3	50	1
2 Mean daily maximum temperature	in °C	28,3	27,8	25,8	21,7	17,8	15,0	14,4	15,0	17,2	20,0	23,3	26,1	21,1	56	2
3 Mean daily minimum temperature	in °C	18,7	18,1	15,0	11,7	8,9	6,1	6,1	6,1	7,8	9,4	12,2	15,0	11,1	56	3
4 Absolute maximum temperature	in °C	42,8	39,5	38,0	36,7	32,0	27,4	28,5	30,8	32,0	35,8	37,4	38,7	42,8	50	4
5 Absolute minimum temperature	in °C	7,8	6,8	4,8	2,0	-2,0	-4,2	-5,0	-3,8	-1,8	-1,4	2,5	5,0	-5,0	50	5
6 Mean relative humidity	in %	66	68	74	78	82	87	84	80	78	77	70	65	76	50	6
7 Mean precipitation	in mm	77	73	99	103	95	95	67	85	89	70	78	80	1011	50	7
8 Maximum precipitation	in mm															8
9 Minimum precipitation	in mm															9
10 Maximum precipitation in 24 h	in mm	104	91	135	181	151	157	108	142	79	88	90	95	181	50	10
11 Mean number of days with precipitation	> 0,1 mm	7	7	8	8	9	9	8	8	8	8	9	8	98	50	11
12 Mean duration of sunshine	in h	322	272	260	219	180	147	158	189	213	251	294	319	2824	50	12
13 Mean quantity of radiation	in ly/day															13
14 Mean potential evaporation	in mm	118	103	87	62	46	32	29	32	41	58	78	102	788	16	14
15 Mean windspeed	in m/sec	4,7	4,4	4,2	3,9	3,9	4,2	4,7	4,7	4,7	4,7	4,7	4,7	4,4	50	15
16 Mean predominent direction of the wind		ESE	NNE	NE	NE	NE	NNE	NE	NNE	NE	NE	NE	NE		50	16

118 Station / Country La Quiaca / Argentina

Location 22°06'S/65°36'W — Height above sealevel 3459 m — Climate symbol: Köppen BSk — Troll V,3

		J	F	M	A	M	J	J	A	S	O	N	D	year	P	
1 Mean daily temperature	in °C	12,4	12,4	12,2	10,3	6,8	3,9	4,0	6,4	9,2	11,1	12,3	12,8	9,5		1
2 Mean daily maximum temperature	in °C	21,1	20,6	21,1	20,8	17,2	15,8	15,8	17,8	20,0	21,7	22,8	22,2	19,4	23	2
3 Mean daily minimum temperature	in °C	5,0	5,0	3,9	0,0	-5,8	-8,9	-8,9	-6,7	-3,3	0,0	2,8	4,4	-1,1	23	3
4 Absolute maximum temperature	in °C	27,1	27,0	27,8	25,8	25,0	22,0	21,1	22,8	25,8	27,4	28,4	28,3	28,4		4
5 Absolute minimum temperature	in °C	-1,2	-1,2	-3,1	-8,7	-12,7	-15,8	-15,2	-14,6	-12,2	-10,7	-4,7	-1,2	-15,8		5
6 Mean relative humidity	in %	62	64	58	45	38	35	36	38	38	44	51	57	50		6
7 Mean precipitation	in mm	89	77	43	5	1	2	1	0	2	8	31	63	322		7
8 Maximum precipitation	in mm															8
9 Minimum precipitation	in mm															9
10 Maximum precipitation in 24 h	in mm	45	35	35	35	5	25	5	15	5	25	25	25	45		10
11 Mean number of days with precipitation	> 0,1 mm	15	12	8	2	<1	<1	<1	<1	1	2	8	12	59		11
12 Mean duration of sunshine	in h	267	238	288	291	304	282	298	304	288	310	303	291	3484		12
13 Mean quantity of radiation	in ly/day															13
14 Mean potential evaporation	in mm	72	63	64	50	31	16	17	29	45	61	69	74	591	25	14
15 Mean windspeed	in m/sec	3,6	3,3	3,1	2,8	2,8	3,3	3,1	3,6	4,2	4,4	4,4	4,2	3,6		15
16 Mean predominent direction of the wind		NE	NE	NE	NE	S	S	S	S	S	NE	NE	NE			16

119 Station / Country Salta / Argentina

Location 24°51'S/65°29'W — Height above sealevel 1226 m — Climate symbol: Köppen Cwb — Troll V,3

		J	F	M	A	M	J	J	A	S	O	N	D	year	P	
1 Mean daily temperature	in °C	21,4	20,5	19,2	16,5	13,5	11,1	10,8	12,4	15,8	18,4	20,7	21,5	16,8		1
2 Mean daily maximum temperature	in °C	28,3	27,8	26,1	23,3	21,1	20,8	21,1	21,1	25,0	26,1	26,7	27,2	24,4	10	2
3 Mean daily minimum temperature	in °C	15,0	15,8	15,0	11,7	7,2	5,0	3,9	5,8	9,4	11,7	14,4	15,8	10,6	10	3
4 Absolute maximum temperature	in °C	38,4	39,3	34,7	33,8	33,9	33,1	35,0	36,3	38,0	38,8	39,0	39,5	39,5		4
5 Absolute minimum temperature	in °C	6,1	7,7	2,8	-1,2	-4,8	-9,5	-9,9	-8,8	-3,8	-2,2	1,8	3,9	-9,9		5
6 Mean relative humidity	in %	78	82	80	75	74	74	66	58	53	58	61	67	69		6
7 Mean precipitation	in mm	176	149	94	25	8	3	2	4	5	25	61	121	671		7
8 Maximum precipitation	in mm	256	272	197	80	33	19	7	35	21	78	182	251			8
9 Minimum precipitation	in mm	82	62	16	2	0	0	0	0	0	1	6	44			9
10 Maximum precipitation in 24 h	in mm	95	115	75	55	35	15	5	5	15	45	45	95	115		10
11 Mean number of days with precipitation	> 0,1 mm	14	13	12	6	3	1	1	1	3	8	8	12	80		11
12 Mean duration of sunshine	in h	195	148	136	144	158	138	198	217	177	164	177	188	2026		12
13 Mean quantity of radiation	in ly/day															13
14 Mean potential evaporation	in mm	115	95	84	62	42	26	30	38	63	88	100	112	855	30	14
15 Mean windspeed	in m/sec	1,4	1,1	0,8	1,1	1,1	1,1	1,4	1,4	1,4	1,7	1,7	1,7	1,3		15
16 Mean predominent direction of the wind		NE	NE	NE	NE	N	N	N	NE	NE	NE	NE	NE			16

120 Station / Country San Miguel de Tucumán / Argentina

Location 26°48'S/65°12'W — Height above sealevel 481 m — Climate symbol: Köppen Cwa — Troll IV,3

		J	F	M	A	M	J	J	A	S	O	N	D	year	P	
1 Mean daily temperature	in °C	24,5	23,7	22,0	18,3	15,4	12,3	12,8	14,0	17,1	20,2	22,5	24,8	19,0		1
2 Mean daily maximum temperature	in °C															2
3 Mean daily minimum temperature	in °C															3
4 Absolute maximum temperature	in °C	44,1	40,6	41,0	34,3	37,5	29,3	37,3	39,0	41,7	43,1	41,2	44,2	44,2		4
5 Absolute minimum temperature	in °C	9,9	10,4	8,9	-0,2	-3,0	-6,0	-4,8	-5,2	-1,2	2,4	4,1	6,0	-6,0		5
6 Mean relative humidity	in %	73	80	81	81	80	81	71	58	55	65	70	71	72		6
7 Mean precipitation	in mm	183	159	162	59	29	19	10	8	12	77	108	150	976		7
8 Maximum precipitation	in mm															8
9 Minimum precipitation	in mm															9
10 Maximum precipitation in 24 h	in mm	125	105	206	65	45	25	25	15	25	75	125	95	206		10
11 Mean number of days with precipitation	> 0,1 mm	12	10	12	8	7	5	3	3	4	7	9	11	91		11
12 Mean duration of sunshine	in h	232	185	183	180	164	153	198	208	198	205	213	233	2352		12
13 Mean quantity of radiation	in ly/day															13
14 Mean potential evaporation	in mm	135	110	96	63	45	26	28	35	54	87	108	138	923		14
15 Mean windspeed	in m/sec	2,2	1,7	1,4	1,4	1,4	1,4	1,7	1,7	1,7	1,9	1,9	2,2	1,8		15
16 Mean predominent direction of the wind		SW	SW	SW	S	N	N	N	N	S	SW	SW	SW			16

121 Station / Country Corrientes / Argentina

Location 27°28'S/58°49'W   Height above sealevel 60 m   Climate symbol: Köppen Cfa   Troll IV,6

		J	F	M	A	M	J	J	A	S	O	N	D	year	P	
1 Mean daily temperature	in °C	27,4	26,7	24,9	20,9	18,8	16,0	15,7	17,1	19,2	21,5	24,0	26,5	21,5		1
2 Mean daily maximum temperature	in °C	33,9	33,3	31,7	27,2	23,9	21,1	21,7	22,8	25,8	27,8	30,6	32,8	27,8	39	2
3 Mean daily minimum temperature	in °C	21,7	21,7	20,8	17,2	13,9	11,7	11,7	11,7	13,8	15,6	18,3	20,8	16,1	39	3
4 Absolute maximum temperature	in °C	41,8	41,4	40,8	36,5	34,3	32,2	32,4	38,8	40,0	41,8	40,2	42,4	42,4		4
5 Absolute minimum temperature	in °C	12,0	11,8	10,2	5,2	2,8	-0,6	-1,1	0,2	0,9	4,8	8,0	10,9	-1,1		5
6 Mean relative humidity	in %	65	69	72	77	81	86	78	69	70	66	66	63	72		6
7 Mean precipitation	in mm	149	127	151	135	86	60	47	42	75	139	139	119	1269		7
8 Maximum precipitation	in mm															8
9 Minimum precipitation	in mm															9
10 Maximum precipitation in 24 h	in mm	135	125	115	115	125	65	105	65	75	85	85	95	135		10
11 Mean number of days with precipitation	> 0,1 mm	7	7	8	8	7	6	5	5	7	8	8	7	83		11
12 Mean duration of sunshine	in h	282	241	239	210	198	150	186	217	198	238	273	285	2615		12
13 Mean quantity of radiation	in ly/day															13
14 Mean potential evaporation	in mm	163	137	124	81	51	33	35	41	62	89	120	155	1091	20	14
15 Mean windspeed	in m/sec	2,2	2,2	1,9	2,2	2,2	2,5	2,5	2,8	3,1	2,8	2,8	2,5	2,5		15
16 Mean predominent direction of the wind		E	SE	SE	S	NE	E	NE	E	E	E	E	E			16

122 Station / Country Posadas / Argentina

Location 27°25'S/55°58'W   Height above sealevel 136 m   Climate symbol: Köppen Cfa   Troll IV,6

		J	F	M	A	M	J	J	A	S	O	N	D	year	P	
1 Mean daily temperature	in °C	26,2	25,8	24,3	20,7	18,1	16,5	15,8	17,3	18,8	20,9	23,3	25,7	21,1		1
2 Mean daily maximum temperature	in °C															2
3 Mean daily minimum temperature	in °C															3
4 Absolute maximum temperature	in °C	40,7	40,6	39,6	35,8	32,7	32,5	32,6	36,4	37,5	39,9	39,5	41,9	41,9		4
5 Absolute minimum temperature	in °C	9,8	9,1	7,9	4,3	-4,3	-2,2	-4,4	-2,4	0,5	2,6	5,5	7,2	-4,4		5
6 Mean relative humidity	in %	71	74	76	79	84	85	80	74	76	75	69	66	76		6
7 Mean precipitation	in mm	127	158	149	178	167	141	104	83	133	188	134	133	1695		7
8 Maximum precipitation	in mm															8
9 Minimum precipitation	in mm															9
10 Maximum precipitation in 24 h	in mm	115	135	135	135	115	145	75	65	115	105	85	135	145		10
11 Mean number of days with precipitation	> 0,1 mm	8	8	8	8	8	9	6	7	9	9	7	7	94		11
12 Mean duration of sunshine	in h	233	207	186	180	161	123	167	171	141	189	237	251	2246		12
13 Mean quantity of radiation	in ly/day															13
14 Mean potential evaporation	in mm	163	137	123	77	47	32	36	40	64	89	114	150	1072	39	14
15 Mean windspeed	in m/sec	2,2	2,5	2,2	2,2	2,5	2,5	2,8	2,8	2,8	2,8	2,5	2,5	2,5		15
16 Mean predominent direction of the wind		NE	NE	NE	NE	NE	NE	NE	NE	SE	SE	S	SE			16

123 Station / Country Santiago del Estero / Argentina

Location 27°46'S/64°18'W   Height above sealevel 199 m   Climate symbol: Köppen BSh   Troll IV,3

		J	F	M	A	M	J	J	A	S	O	N	D	year	P	
1 Mean daily temperature	in °C	27,3	25,8	23,4	19,5	16,4	13,5	12,9	15,2	18,8	21,9	24,4	26,8	20,5		1
2 Mean daily maximum temperature	in °C	36,1	34,4	31,7	27,8	23,9	20,8	21,1	23,9	27,8	30,6	33,3	34,4	28,9	28	2
3 Mean daily minimum temperature	in °C	20,8	20,0	18,3	15,0	10,6	6,7	6,7	7,8	11,7	15,0	17,8	19,4	14,4	28	3
4 Absolute maximum temperature	in °C	45,2	43,2	43,0	38,3	35,2	33,3	34,4	39,0	41,8	44,4	45,5	44,8	45,5		4
5 Absolute minimum temperature	in °C	8,0	9,7	6,3	0,2	-4,3	-6,7	-10,0	-7,0	-4,1	0,2	1,3	8,2	-10,0		5
6 Mean relative humidity	in %	58	63	66	67	73	75	64	54	50	52	55	54	61		6
7 Mean precipitation	in mm	90	93	92	27	14	10	3	3	8	39	80	79	518		7
8 Maximum precipitation	in mm	209	380	252	90	58	43	17	21	42	219	172	178			8
9 Minimum precipitation	in mm	24	16	19	1	0	0	0	0	0	2	8	18			9
10 Maximum precipitation in 24 h	in mm	115	135	135	105	25	35	45	15	45	105	145	135	145		10
11 Mean number of days with precipitation	> 0,1 mm	9	8	8	5	4	2	2	3	6	7	8	68			11
12 Mean duration of sunshine	in h	251	207	189	168	161	120	180	214	204	223	231	254	2408		12
13 Mean quantity of radiation	in ly/day															13
14 Mean potential evaporation	in mm	171	139	114	70	42	23	25	37	64	99	132	167	1083	25	14
15 Mean windspeed	in m/sec	3,3	2,8	2,8	2,8	2,5	2,5	2,8	3,3	3,9	4,2	3,9	3,6	3,2		15
16 Mean predominent direction of the wind		NE	NE	NE	NE	NE	S	NE	NE	NE	NE	NE	NE			16

124 Station / Country Catamarca / Argentina

Location 28°28'S/65°46'W   Height above sealevel 547 m   Climate symbol: Köppen BSh   Troll IV,3

		J	F	M	A	M	J	J	A	S	O	N	D	year	P	
1 Mean daily temperature	in °C	27,7	26,0	24,0	19,6	15,4	11,8	11,4	14,5	18,8	22,3	24,9	27,2	20,3		1
2 Mean daily maximum temperature	in °C															2
3 Mean daily minimum temperature	in °C															3
4 Absolute maximum temperature	in °C	44,5	43,9	41,3	39,6	35,2	32,0	37,4	40,1	42,2	42,3	45,2	47,2	47,2		4
5 Absolute minimum temperature	in °C	9,8	10,8	6,7	-1,3	-3,7	-5,8	-9,0	-5,5	-1,0	2,0	0,4	8,9	-9,0		5
6 Mean relative humidity	in %	48	55	58	59	65	69	60	47	41	42	45	48	53		6
7 Mean precipitation	in mm	66	82	49	20	11	6	4	4	7	29	41	51	370		7
8 Maximum precipitation	in mm															8
9 Minimum precipitation	in mm															9
10 Maximum precipitation in 24 h	in mm	65	75	45	55	15	15	15	15	25	25	65	25	75		10
11 Mean number of days with precipitation	> 0,1 mm	8	6	6	4	3	2	2	1	2	4	5	6	49		11
12 Mean duration of sunshine	in h															12
13 Mean quantity of radiation	in ly/day															13
14 Mean potential evaporation	in mm	168	131	112	74	41	18	21	37	65	103	133	162	1066	39	14
15 Mean windspeed	in m/sec	3,9	3,6	3,6	3,3	2,2	1,7	1,7	2,8	4,2	4,7	5,0	4,7	3,4		15
16 Mean predominent direction of the wind		NE	NE	NE	NE	NE	NE	NE	NE	NE	NE	NE	NE			16

125 Station / Country San Juan / Argentina

Location 31°36'S/68°33'W — Height above sealevel 630 m — Climate symbol: Köppen BWk — Troll IV.5

	in	J	F	M	A	M	J	J	A	S	O	N	D	year	P	
1 Mean daily temperature	°C	26.0	24.4	21.4	16.2	11.8	8.4	8.0	10.7	14.4	18.3	22.1	25.1	17.2		1
2 Mean daily maximum temperature	°C															2
3 Mean daily minimum temperature	°C															3
4 Absolute maximum temperature	°C	44.3	43.4	40.0	38.0	34.0	33.5	34.0	35.3	41.1	39.4	43.0	43.8	44.3		4
5 Absolute minimum temperature	°C	4.8	7.2	-1.0	-3.8	-5.0	-7.5	-7.3	-6.0	-3.0	-0.8	2.3	6.7	-7.5		5
6 Mean relative humidity	%	47	52	58	60	65	64	60	50	47	48	44	53			6
7 Mean precipitation	mm	18	12	9	5	1	2	2	2	5	9	12	10	87		7
8 Maximum precipitation	mm															8
9 Minimum precipitation	mm															9
10 Maximum precipitation in 24 h	mm	45	35	35	25	15	15	15	15	25	25	45	25	45		10
11 Mean number of days with precipitation >1,0 mm		4	3	2	1	1	1	1	1	2	2	2	21			11
12 Mean duration of sunshine	h	329	286	276	243	217	192	220	245	258	298	324	338	3226		12
13 Mean quantity of radiation	ly/day															13
14 Mean potential evaporation	mm	148	115	95	53	29	15	15	27	48	81	105	137	868		14
15 Mean windspeed	m/sec	2.5	2.2	1.9	1.7	1.7	1.4	1.7	2.2	2.2	2.5	2.5	2.8	2.1		15
16 Mean predominent direction of the wind		S	S	S	S	S	S	S	S	S	S	S	S			16

126 Station / Country Córdoba / Argentina

Location 31°24'S/64°11'W — Height above sealevel 425 m — Climate symbol: Köppen Cwa — Troll IV.3

	in	J	F	M	A	M	J	J	A	S	O	N	D	year	P	
1 Mean daily temperature	°C	24.2	23.2	20.7	16.8	13.8	11.0	10.8	12.3	15.1	17.9	20.8	23.1	17.4		1
2 Mean daily maximum temperature	°C	31.1	30.0	27.8	23.9	20.6	17.8	18.3	20.6	22.8	25.0	27.8	30.0	25.0	41	2
3 Mean daily minimum temperature	°C	16.1	15.8	14.4	10.6	6.7	3.3	3.3	4.4	7.2	10.6	13.3	15.6	10.0	41	3
4 Absolute maximum temperature	°C	45.0	42.2	41.3	38.1	35.3	34.8	33.2	37.3	38.5	42.0	42.4	44.3	45.0		4
5 Absolute minimum temperature	°C	5.7	4.8	1.6	-1.4	-5.9	-7.7	-9.2	-7.2	-3.3	0.1	2.5	4.4	-9.2		5
6 Mean relative humidity	%	57	60	67	68	72	72	62	55	52	56	56	56	61		6
7 Mean precipitation	mm	101	88	93	39	24	10	8	16	29	77	88	108	680		7
8 Maximum precipitation	mm															8
9 Minimum precipitation	mm															9
10 Maximum precipitation in 24 h	mm	95	85	75	105	65	35	45	65	45	135	75	105	135		10
11 Mean number of days with precipitation >0,1 mm		10	8	9	5	4	3	2	3	4	7	8	10	73		11
12 Mean duration of sunshine	h	285	238	220	207	177	150	186	217	218	242	264	279	2681		12
13 Mean quantity of radiation	ly/day															13
14 Mean potential evaporation	mm	131	105	91	58	36	20	22	33	49	72	97	123	837	68	14
15 Mean windspeed	m/sec	1.9	1.7	1.7	1.7	1.7	1.4	1.7	1.9	2.2	2.2	2.2	1.9	1.8		15
16 Mean predominent direction of the wind		NE	NE	NE	NE	NE	NE	NE	NE	NE	NE	NE	NE			16

127 Station / Country Concordia / Argentina

Location 31°23'S/58°02'W — Height above sealevel 38 m — Climate symbol: Köppen Cfa — Troll IV.8

	in	J	F	M	A	M	J	J	A	S	O	N	D	year	P	
1 Mean daily temperature	°C	25.7	24.7	22.8	18.3	15.5	13.2	12.5	13.8	15.1	18.3	21.4	24.2	18.8		1
2 Mean daily maximum temperature	°C															2
3 Mean daily minimum temperature	°C															3
4 Absolute maximum temperature	°C	43.5	42.4	41.1	37.0	33.5	30.5	30.8	34.8	37.2	39.4	41.4	41.7	43.5		4
5 Absolute minimum temperature	°C	10.0	8.4	6.5	0.0	-0.7	-4.8	-5.3	-3.2	-1.3	2.3	4.5	8.1	-5.3		5
6 Mean relative humidity	%	59	65	69	74	78	83	82	75	74	70	63	58	71		6
7 Mean precipitation	mm	128	109	154	136	82	87	58	65	88	117	96	96	1214		7
8 Maximum precipitation	mm															8
9 Minimum precipitation	mm															9
10 Maximum precipitation in 24 h	mm	115	105	135	155	330	75	75	85	85	85	85	135	330		10
11 Mean number of days with precipitation >0,1 mm																11
12 Mean duration of sunshine	h	282	255	242	210	180	120	158	189	192	242	285	285	2640		12
13 Mean quantity of radiation	ly/day															13
14 Mean potential evaporation	mm	160	126	108	65	40	25	28	33	48	72	104	138	947	25	14
15 Mean windspeed	m/sec	1.9	1.9	1.9	1.9	1.9	1.9	2.2	2.6	2.5	2.5	1.9	2.2	2.1		15
16 Mean predominent direction of the wind		E	E	E	NE	NE	NE	NE	NE	E	E	E	E			16

128 Station / Country Santa Fe / Argentina

Location 31°31'S/61°57'W — Height above sealevel 88 m — Climate symbol: Köppen Cfa — Troll IV.4

	in	J	F	M	A	M	J	J	A	S	O	N	D	year	P	
1 Mean daily temperature	°C	26.1	25.0	22.5	18.3	15.8	13.1	12.3	14.0	16.4	19.4	22.4	24.9	19.1		1
2 Mean daily maximum temperature	°C		.													2
3 Mean daily minimum temperature	°C															3
4 Absolute maximum temperature	°C	44.8	43.8	43.8	39.0	34.7	32.0	33.0	39.0	39.7	42.8	42.9	42.7	44.8		4
5 Absolute minimum temperature	°C	7.9	7.3	6.0	1.5	-2.3	-5.2	-5.8	-4.5	-2.8	2.0	3.2	8.0	-5.8		5
6 Mean relative humidity	%	57	64	66	74	76	78	72	63	61	63	60	58	66		6
7 Mean precipitation	mm	113	101	124	58	38	27	22	20	36	84	103	110	836		7
8 Maximum precipitation	mm															8
9 Minimum precipitation	mm															9
10 Maximum precipitation in 24 h	mm	225	85	155	125	75	45	85	55	85	120	95	115	225		10
11 Mean number of days with precipitation >0,1 mm		7	7	8	5	4	3	3	3	5	7	7	8	68		11
12 Mean duration of sunshine	h	287	248	233	198	180	138	183	205	204	245	273	276	2648	10	12
13 Mean quantity of radiation	ly/day															13
14 Mean potential evaporation	mm	146	119	106	66	42	24	24	30	46	70	100	135	908	25	14
15 Mean windspeed	m/sec	2.5	2.2	2.2	2.2	2.2	2.2	2.6	3.1	3.3	3.1	2.8	2.8	2.6		15
16 Mean predominent direction of the wind		E	E	E	NE	NE	S	NE	S	S	E	NE	E			16

129 Station/Country Cristo Redentor / Argentina

Location 32°50'S/70°05'W   Height above sealevel 3829 m   Climate symbol: Köppen ET   Troll I,2

	in	J	F	M	A	M	J	J	A	S	O	N	D	year	P	
1 Mean daily temperature	in °C	4,0	3,6	1,8	-0,6	-4,5	-5,9	-6,7	-6,5	-5,5	-3,5	-1,0	2,4	-1,7		1
2 Mean daily maximum temperature	in °C															2
3 Mean daily minimum temperature	in °C															3
4 Absolute maximum temperature	in °C	20,2	17,6	18,0	14,0	11,2	8,6	9,5	9,6	10,3	12,8	13,4	16,1	20,2		4
5 Absolute minimum temperature	in °C	-8,6	-13,0	-16,4	-14,2	-20,6	-23,5	-30,3	-22,5	-21,7	-21,7	-19,9	-13,7	-30,3		5
6 Mean relative humidity	in %															6
7 Mean precipitation	in mm	8	9	8	22	96	40	56	64	23	19	7	7	357		7
8 Maximum precipitation	in mm															8
9 Minimum precipitation	in mm															9
10 Maximum precipitation in 24 h	in mm															10
11 Mean number of days with precipitation	> 0,1 mm	5	4	6	6	9	10	10	11	8	9	7	6	91		11
12 Mean duration of sunshine	in h															12
13 Mean quantity of radiation	in ly/day															13
14 Mean potential evaporation	in mm															14
15 Mean windspeed	in m/sec	8,3	8,3	8,0	7,8	7,5	8,0	7,2	7,5	7,5	7,2	8,0	9,2	7,8	20	15
16 Mean predominent direction of the wind		SW	SW	SW	SW	SW	SW	SW	SW	SW	SV	SW	SW			16

130 Station/Country Mendoza/Argentina

Location 32°53'S/68°50'W   Height above sealevel 769 m   Climate symbol: Köppen BWk   Troll IV,5

	in	J	F	M	A	M	J	J	A	S	O	N	D	year	P	
1 Mean daily temperature	in °C	23,6	22,5	20,2	15,6	11,5	8,1	7,6	10,2	13,9	16,7	20,4	22,7	16,1	20	1
2 Mean daily maximum temperature	in °C	32,2	30,6	27,8	22,8	18,3	15,0	15,0	17,2	20,6	24,4	28,3	31,1	23,9	23	2
3 Mean daily minimum temperature	in °C	15,6	15,0	12,8	8,3	5,0	2,2	1,7	3,3	6,7	10,0	12,2	14,4	8,9	23	3
4 Absolute maximum temperature	in °C	37,5	38,9	34,9	32,1	27,9	28,1	27,9	29,8	32,7	38,3	38,2	38,5	38,2	20	4
5 Absolute minimum temperature	in °C	9,4	7,4	2,9	1,9	-2,5	-4,4	-6,2	-2,9	-1,8	0,9	3,8	6,1	-6,2	20	5
6 Mean relative humidity	in %	53	57	54	55	59	62	60	47	42	46	45	48	52	20	6
7 Mean precipitation	in mm	28	21	22	10	11	8	7	10	14	23	20	23	197	30	7
8 Maximum precipitation	in mm															8
9 Minimum precipitation	in mm															9
10 Maximum precipitation in 24 h	in mm	55	45	45	25	45	25	15	35	35	45	45	95	95	40	10
11 Mean number of days with precipitation	> 0,1 mm	5	5	4	3	2	2	2	2	2	4	4	5	40	23	11
12 Mean duration of sunshine	in h	295	255	242	210	198	171	195	217	222	251	294	301	2851	10	12
13 Mean quantity of radiation	in ly/day															13
14 Mean potential evaporation	in mm	137	103	88	52	30	15	17	27	44	71	99	122	806	75	14
15 Mean windspeed	in m/sec	1,7	1,4	1,4	1,7	1,7	1,7	1,7	1,9	1,7	1,7	1,4	1,7	1,6		15
16 Mean predominent direction of the wind		S	S	S	W	W	W	W	W	W	S	S	S		20	16

131 Station/Country San Luis/Argentina

Location 33°16'S/66°21'W   Height above sealevel 716 m   Climate symbol: Köppen BSh   Troll IV,5

	in	J	F	M	A	M	J	J	A	S	O	N	D	year	P	
1 Mean daily temperature	in °C	24,0	23,0	20,2	15,6	12,2	9,2	8,8	10,7	13,9	17,1	20,5	23,2	16,5		1
2 Mean daily maximum temperature	in °C															2
3 Mean daily minimum temperature	in °C															3
4 Absolute maximum temperature	in °C	43,4	40,0	38,3	35,3	32,8	30,6	28,6	31,9	34,0	40,8	41,6	43,0	43,4		4
5 Absolute minimum temperature	in °C	4,0	5,2	2,1	-2,1	-6,6	-7,0	-9,4	-9,3	-3,6	-4,5	0,0	2,7	-9,4		5
6 Mean relative humidity	in %	51	52	59	61	66	68	62	53	49	52	50	49	57		6
7 Mean precipitation	in mm	90	74	60	33	14	13	10	6	17	53	67	105	542		7
8 Maximum precipitation	in mm															8
9 Minimum precipitation	in mm															9
10 Maximum precipitation in 24 h	in mm	125	95	65	55	85	35	45	25	55	55	65	65	125		10
11 Mean number of days with precipitation	> 0,1 mm	8	7	5	3	3	2	2	2	3	5	6	8	54		11
12 Mean duration of sunshine	in h															12
13 Mean quantity of radiation	in ly/day															13
14 Mean potential evaporation	in mm	142	107	91	56	33	17	19	27	42	76	104	134	848	20	14
15 Mean windspeed	in m/sec	4,4	4,4	3,9	3,6	3,3	2,8	3,6	5,3	5,6	5,6	5,0	4,7	4,4		15
16 Mean predominent direction of the wind		N	N	N	N	N	E	N	N	N	N	N	E			16

132 Station/Country Rosario/Argentina

Location 32°55'S/60°47'W   Height above sealevel 27 m   Climate symbol: Köppen Cfa   Troll IV,8

	in	J	F	M	A	M	J	J	A	S	O	N	D	year	P	
1 Mean daily temperature	in °C	23,6	22,8	20,4	16,1	13,4	10,9	10,2	11,2	13,6	16,5	19,6	22,3	16,7		1
2 Mean daily maximum temperature	in °C	31,7	30,6	27,2	23,3	18,9	15,6	16,1	17,2	20,6	23,3	27,2	30,0	23,3	16	2
3 Mean daily minimum temperature	in °C	17,8	18,3	16,1	12,8	8,9	6,1	6,1	6,7	7,8	10,6	13,9	17,2	11,7	16	3
4 Absolute maximum temperature	in °C	42,4	40,3	38,2	34,4	31,5	27,5	28,0	32,6	40,0	35,4	38,4	42,1	42,4		4
5 Absolute minimum temperature	in °C	4,8	4,2	3,6	-1,5	-4,8	-10,6	-9,5	-7,3	-6,0	-0,6	1,2	3,1	-10,6		5
6 Mean relative humidity	in %	67	69	76	80	85	87	83	77	74	73	70	66	75		6
7 Mean precipitation	in mm	115	88	134	85	55	39	33	45	78	92	105	99	968		7
8 Maximum precipitation	in mm															8
9 Minimum precipitation	in mm															9
10 Maximum precipitation in 24 h	in mm	125	85	95	75	65	55	75	65	75	85	105	105	125		10
11 Mean number of days with precipitation	> 0,1 mm	7	6	8	6	5	5	4	4	6	7	8	7	73		11
12 Mean duration of sunshine	in h	304	255	248	210	195	144	174	195	204	257	297	291	2774	30	12
13 Mean quantity of radiation	in ly/day															13
14 Mean potential evaporation	in mm	143	115	96	59	39	22	24	30	44	68	95	128	863	25	14
15 Mean windspeed	in m/sec	2,5	2,5	2,5	2,2	2,8	2,8	2,8	3,3	3,3	3,3	3,1	2,8	2,8		15
16 Mean predominent direction of the wind		NE	NE	NE	NE	NE	NE	NE	NE	NE	NE	NE	NE			16

133 Station/Country Buenos Aires/Argentina

Location 34°35'S/58°29'W   Height above sealevel 25 m   Climate symbol: Köppen Cfa   Troll IV,6

		J	F	M	A	M	J	J	A	S	O	N	D	year	P	
1	Mean daily temperature	in °C	23,7	23,0	20,7	16,8	13,7	11,1	10,5	11,5	13,8	16,5	19,5	22,1	16,9		1
2	Mean daily maximum temperature	in °C	29,4	28,3	26,1	22,2	17,8	13,9	13,9	15,8	17,8	20,6	24,4	27,8	21,7	23	2
3	Mean daily minimum temperature	in °C	17,2	17,2	15,8	11,7	8,3	5,0	5,8	6,1	7,8	10,0	13,3	18,1	11,1	23	3
4	Absolute maximum temperature	in °C	43,3	38,7	37,9	33,1	31,6	28,5	27,6	30,9	34,0	33,5	36,8	39,3	43,3		4
5	Absolute minimum temperature	in °C	7,8	7,2	5,5	0,9	-2,2	-4,7	-5,3	-4,0	-1,1	0,8	3,2	5,0	-5,3		5
6	Mean relative humidity	in %	84	67	74	78	82	84	83	75	74	73	68	84	74		6
7	Mean precipitation	in mm	104	82	122	90	79	68	61	68	80	100	90	83	1027		7
8	Maximum precipitation	in mm															8
9	Minimum precipitation	in mm															9
10	Maximum precipitation in 24 h	in mm	125	195	95	125	85	75	75	95	85	105	75	95	125		10
11	Mean number of days with precipitation	> 0,1 mm	7	7	8	7	6	7	6	7	7	8	8	8	86		11
12	Mean duration of sunshine	in h	285	249	220	198	167	128	149	177	186	226	264	273	2520		12
13	Mean quantity of radiation	in ly/day															13
14	Mean potential evaporation	in mm	134	108	83	57	37	22	21	27	40	63	90	121	813	80	14
15	Mean windspeed	in m/sec	3,2	2,8	2,5	2,5	2,2	2,2	2,5	3,1	3,1	3,1	3,1	3,1	2,8		15
16	Mean predominent direction of the wind		NE	NE	NE	NE	NE	NE	NE	NE	NE	NE	NE	NE			16

134 Station/Country Macachín - Argentina

Location 37°08'S/63°41'W   Height above sealevel 142 m   Climate symbol: Köppen Cfa   Troll IV,4

		J	F	M	A	M	J	J	A	S	O	N	D	year	P	
1	Mean daily temperature	in °C	23,7	22,4	18,9	14,5	10,7	7,7	7,4	8,6	11,6	15,2	19,3	22,3	15,2		1
2	Mean daily maximum temperature	in °C															2
3	Mean daily minimum temperature	in °C															3
4	Absolute maximum temperature	in °C	43,7	42,3	39,1	35,4	30,9	25,2	26,0	32,0	34,9	37,6	39,3	43,0	43,7		4
5	Absolute minimum temperature	in °C	4,5	2,4	-0,9	-8,0	-6,7	-10,8	-11,1	-9,9	-8,2	-3,4	-2,2	2,0	-11,1		5
6	Mean relative humidity	in %	50	51	65	68	74	81	78	68	64	65	54	48	64		6
7	Mean precipitation	in mm	59	68	82	49	39	34	24	16	40	75	57	65	808		7
8	Maximum precipitation	in mm															8
9	Minimum precipitation	in mm															9
10	Maximum precipitation in 24 h	in mm	55	95	105	105	115	65	65	85	75	75	95	65	115		10
11	Mean number of days with precipitation	> 1,0 mm	6	6	7	4	4	3	4	3	5	7	7	6	62		11
12	Mean duration of sunshine	in h	326	266	266	195	167	120	149	180	198	217	294	329	2697		12
13	Mean quantity of radiation	in ly/day															13
14	Mean potential evaporation	in mm	140	108	87	52	29	16	18	24	38	65	95	128	798		14
15	Mean windspeed	in m/sec	3,3	3,3	2,8	2,8	2,8	2,5	2,5	3,6	3,6	3,3	3,3	3,9	3,1		15
16	Mean predominent direction of the wind		N	N	N	NE	N	N	N	N	N	N	N	N			16

135 Station/Country Chos Malal/Argentina

Location 37°23'S/70°17'W   Height above sealevel 848 m   Climate symbol: Köppen BSk   Troll IV,2

		J	F	M	A	M	J	J	A	S	O	N	D	year	P	
1	Mean daily temperature	in °C	21,2	20,2	17,1	12,8	8,6	6,7	6,2	7,8	9,8	14,0	17,6	20,2	13,6		1
2	Mean daily maximum temperature	in °C															2
3	Mean daily minimum temperature	in °C															3
4	Absolute maximum temperature	in °C	39,5	39,5	37,8	33,4	29,6	25,3	24,0	29,0	30,0	33,2	36,9	39,6	39,6		4
5	Absolute minimum temperature	in °C	1,3	2,4	-1,0	-9,1	-8,3	-11,4	-11,1	-9,5	-6,4	-3,1	-0,1	0,5	-11,4		5
6	Mean relative humidity	in %	33	37	41	49	53	58	58	44	37	35	33		44		6
7	Mean precipitation	in mm	9	10	12	13	41	54	32	29	13	11	7	8	237		7
8	Maximum precipitation	in mm															8
9	Minimum precipitation	in mm															9
10	Maximum precipitation in 24 h	in mm	35	65	35	35	55	65	25	45	45	35	35	15	65		10
11	Mean number of days with precipitation	> 1,0 mm	2	2	2	5	6	4	5	3	2	2	2		37		11
12	Mean duration of sunshine	in h	341	294	295	192	149	141	149	180	216	285	303	341	2886		12
13	Mean quantity of radiation	in ly/day															13
14	Mean potential evaporation	in mm	128	101	78	47	28	15	18	20	34	58	86	105	718		14
15	Mean windspeed	in m/sec	3,1	2,2	2,2	1,9	1,9	1,7	2,5	2,5	2,8	3,3	2,8	2,8	2,5		15
16	Mean predominent direction of the wind		NW	NW	NW	NW	NW	NW	NW	NW	NW	NW	NW	NW			16

136 Station/Country Mar del Plata/Argentina

Location 38°08'S/57°33'W   Height above sealevel 14 m   Climate symbol: Köppen Cfb   Troll IV,6

		J	F	M	A	M	J	J	A	S	O	N	D	year	P	
1	Mean daily temperature	in °C	19,0	19,3	17,8	14,9	11,3	8,9	7,7	8,4	9,7	11,6	14,9	17,2	13,4		1
2	Mean daily maximum temperature	in °C															2
3	Mean daily minimum temperature	in °C															3
4	Absolute maximum temperature	in °C	38,8	40,2	38,8	31,8	31,5	23,3	24,0	23,7	32,1	31,7	34,0	37,9	40,2		4
5	Absolute minimum temperature	in °C	3,2	4,0	2,0	-2,0	-3,0	-5,5	-8,6	-4,0	-4,0	-2,7	-0,6	0,2	-8,6		5
6	Mean relative humidity	in %	71	74	75	78	84	85	87	83	83	86	79	75	82		6
7	Mean precipitation	in mm	70	73	89	85	86	82	51	48	55	57	68	84	788	28	7
8	Maximum precipitation	in mm															8
9	Minimum precipitation	in mm															9
10	Maximum precipitation in 24 h	in mm	75	75	105	75	55	95	75	55	95	65	95	85	105		10
11	Mean number of days with precipitation	> 0,1 mm	8	8	9	7	9	8	8	7	8	9	8	8	97		11
12	Mean duration of sunshine	in h	279	246	223	188	155	120	140	171	177	226	264	267	2454		12
13	Mean quantity of radiation	in ly/day															13
14	Mean potential evaporation	in mm	107	90	83	60	41	25	23	26	35	50	71	94	705	53	14
15	Mean windspeed	in m/sec	5,8	6,1	5,0	4,7	4,7	4,4	5,0	5,3	5,3	5,8	6,4	6,7	5,4		15
16	Mean predominent direction of the wind		N	NE	N	N	NW	W	NW	N	S	N	NE	NE			16

220

137 Station/Country Cipoletti/Argentina

Location 38°57'S/67°59'W — Height above sealevel 265 m — Climate symbol: Köppen BWh — Troll III,12a

	J	F	M	A	M	J	J	A	S	O	N	D	year	P	
1 Mean daily temperature — in °C	21,8	20,6	17,2	12,6	8,8	5,7	5,8	7,6	10,9	14,4	18,7	20,8	13,7		1
2 Mean daily maximum temperature — in °C															2
3 Mean daily minimum temperature — in °C															3
4 Absolute maximum temperature — in °C	41,4	41,4	37,7	34,1	29,6	27,0	31,9	30,2	33,7	36,4	38,7	39,6	41,4		4
5 Absolute minimum temperature — in °C	-0,1	0,6	-3,1	-5,4	-11,2	-10,0	-10,6	-10,9	-6,6	-3,3	-0,3	-0,7	-11,2		5
6 Mean relative humidity — in %	47	52	58	65	71	74	67	59	53	50	45	46	57		6
7 Mean precipitation — in mm	17	9	14	13	22	14	13	15	11	22	13	13	176		7
8 Maximum precipitation — in mm															8
9 Minimum precipitation — in mm															9
10 Maximum precipitation in 24 h — in mm	45	55	35	25	45	25	25	35	55	35	35	35	55		10
11 Mean number of days with precipitation — > 0,1 mm	2	3	3	2	5	4	4	4	4	4	2	2	39		11
12 Mean duration of sunshine — in h	341	294	267	195	149	129	143	188	195	267	300	326	2802		12
13 Mean quantity of radiation — in ly/day															13
14 Mean potential evaporation — in mm	132	102	79	45	23	10	12	19	37	66	94	123	742	25	14
15 Mean windspeed — in m/sec	3,3	2,5	1,9	1,9	1,7	1,7	1,9	2,2	2,2	2,5	3,1	3,3	2,4		15
16 Mean predominent direction of the wind	W	W	W	W	W	W	W	W	W	W	W	W			16

138 Station/Country San Carlos de Bariloche/Argentina

Location 41°06'S/71°10'W — Height above sealevel 836 m — Climate symbol: Köppen Cfb — Troll III,2

	J	F	M	A	M	J	J	A	S	O	N	D	year	P	
1 Mean daily temperature — in °C	14,5	14,4	12,0	8,0	5,6	2,9	2,3	2,9	4,7	7,8	11,2	13,6	8,3	10	1
2 Mean daily maximum temperature — in °C															2
3 Mean daily minimum temperature — in °C															3
4 Absolute maximum temperature — in °C	32,6	34,0	31,7	25,6	18,9	15,2	16,7	19,6	19,5	24,0	27,6	32,7	34,0	10	4
5 Absolute minimum temperature — in °C	-5,7	-4,0	-8,3	-8,6	-11,1	-15,4	-14,0	-16,7	-10,7	-4,6	-8,5	-16,7		10	5
6 Mean relative humidity — in %	60	62	65	72	80	83	82	79	74	67	64	59	71	10	6
7 Mean precipitation — in mm	37	12	28	51	141	89	143	104	51	23	16	22	717		7
8 Maximum precipitation — in mm															8
9 Minimum precipitation — in mm															9
10 Maximum precipitation in 24 h — in mm															10
11 Mean number of days with precipitation — > 0,1 mm	6	4	7	9	16	16	17	15	11	6	5	5	117	10	11
12 Mean duration of sunshine — in h	335	266	248	180	124	102	124	155	186	242	297	329	2590	10	12
13 Mean quantity of radiation — in ly/day															13
14 Mean potential evaporation — in mm	97	80	67	43	28	15	14	18	29	51	64	86	592	25	14
15 Mean windspeed — in m/sec	6,9	5,8	5,3	4,7	4,7	4,4	5,0	5,3	5,0	5,8	6,7	6,4	5,6	10	15
16 Mean predominent direction of the wind	W	W	W	W	W	W	W	W	W	W	W	W		10	16

139 Station/Country Trelew/Argentina

Location 43°14'S/65°18'W — Height above sealevel 39 m — Climate symbol: Köppen BWk — Troll III,10

	J	F	M	A	M	J	J	A	S	O	N	D	year	P	
1 Mean daily temperature — in °C	20,6	20,0	17,3	13,2	9,6	6,1	6,0	7,6	10,2	14,0	17,3	19,3	13,5		1
2 Mean daily maximum temperature — in °C															2
3 Mean daily minimum temperature — in °C															3
4 Absolute maximum temperature — in °C	40,0	40,4	39,5	35,3	26,7	23,4	24,8	27,4	31,3	34,2	37,6	41,2	41,2		4
5 Absolute minimum temperature — in °C	3,2	1,7	-1,4	-2,8	-10,7	-8,0	-10,8	-8,6	-8,0	-2,2	-1,0	3,3	-10,8		5
6 Mean relative humidity — in %	37	41	48	54	62	68	65	55	50	39	38	37	48		6
7 Mean precipitation — in mm	8	14	17	11	19	11	15	13	15	17	13	14	165		7
8 Maximum precipitation — in mm															8
9 Minimum precipitation — in mm															9
10 Maximum precipitation in 24 h — in mm	25	45	25	35	35	25	25	25	35	45	25	55	55		10
11 Mean number of days with precipitation — > 1,0 mm	3	4	4	3	5	5	5	4	4	4	3	4	49		11
12 Mean duration of sunshine — in h	322	269	248	198	152	132	149	177	198	264	294	310	2713		12
13 Mean quantity of radiation — in ly/day															13
14 Mean potential evaporation — in mm	123	98	81	49	26	13	14	24	39	64	87	112	728		14
15 Mean windspeed — in m/sec	6,4	5,8	5,0	5,0	5,0	4,7	5,8	5,6	5,8	6,1	6,9	7,2	5,8		15
16 Mean predominent direction of the wind	W	W	W	W	W	W	W	W	W	W	W	W			16

140 Station/Country Sarmiento/Argentina

Location 45°35'S/69°08'W — Height above sealevel 266 m — Climate symbol: Köppen BWk — Troll III,12a

	J	F	M	A	M	J	J	A	S	O	N	D	year	P	
1 Mean daily temperature — in °C	17,3	16,9	14,3	10,8	7,0	3,9	4,0	5,5	8,0	11,6	14,3	16,4	10,8		1
2 Mean daily maximum temperature — in °C	25,6	25,0	21,1	16,7	12,2	7,8	7,2	10,6	13,9	18,9	21,1	23,3	16,7	11	2
3 Mean daily minimum temperature — in °C	11,1	10,6	8,3	5,6	2,2	-0,6	-1,7	0,6	2,2	5,0	7,6	9,4	5,0	8	3
4 Absolute maximum temperature — in °C	37,6	38,3	33,4	28,4	22,0	20,0	19,0	20,6	25,5	30,7	35,4	35,6	38,3		4
5 Absolute minimum temperature — in °C	0,6	1,5	-2,0	-5,8	-12,4	-18,9	-18,6	-11,0	-8,5	-6,7	-2,5	0,4	-18,9		5
6 Mean relative humidity — in %	42	43	47	56	65	71	69	61	52	44	41	38	52		6
7 Mean precipitation — in mm	10	8	11	15	24	16	17	15	10	8	12	9	153		7
8 Maximum precipitation — in mm															8
9 Minimum precipitation — in mm															9
10 Maximum precipitation in 24 h — in mm	15	45	25	35	25	25	25	45	15	15	15	15	45		10
11 Mean number of days with precipitation — > 0,1 mm	8	10	12	12	21	20	17	17	10	8	11	9	155		11
12 Mean duration of sunshine — in h	270	232	214	171	133	99	121	171	183	226	234	260	2314		12
13 Mean quantity of radiation — in ly/day															13
14 Mean potential evaporation — in mm	112	87	72	44	22	9	9	19	34	60	81	103	652	38	14
15 Mean windspeed — in m/sec	6,1	5,3	5,3	4,2	3,9	2,8	3,9	5,0	4,7	6,1	6,1	6,1	5,0		15
16 Mean predominent direction of the wind	W	W	W	W	W	W	W	W	W	W	W	W			16

141 Station / Country **Comodoro Rivadavia / Argentina**

Location **45°47'S/67°30'W**   Height above sealevel **60 m**   Climate symbol: Köppen **BSk**   Troll **III,12a**

		J	F	M	A	M	J	J	A	S	O	N	D	year	P	
1	Mean daily temperature	in °C	18,6	18,2	16,0	12,7	9,4	7,0	6,9	7,6	9,6	12,8	15,4	17,3	12,6		1
2	Mean daily maximum temperature	in °C															2
3	Mean daily minimum temperature	in °C															3
4	Absolute maximum temperature	in °C	37,5	37,2	35,0	28,8	26,2	22,2	21,0	23,3	29,2	31,0	37,0	37,8	37,8		4
5	Absolute minimum temperature	in °C	5,3	4,1	1,1	-4,1	-4,9	-6,8	-7,6	-8,3	-5,0	-1,6	0,5	2,3	-8,3		5
6	Mean relative humidity	in %	39	40	43	48	53	56	53	51	47	41	39	48	46		6
7	Mean precipitation	in mm	16	11	21	22	35	20	21	18	15	10	16	13	218		7
8	Maximum precipitation	in mm															8
9	Minimum precipitation	in mm															9
10	Maximum precipitation in 24 h	in mm	35	45	45	35	45	45	25	25	45	45	55	25	55		10
11	Mean number of days with precipitation	> 0,1 mm	8	10	12	12	21	20	17	17	10	8	11	9	155		11
12	Mean duration of sunshine	in h	280	218	186	150	112	102	115	136	153	208	240	248	2128		12
13	Mean quantity of radiation	in ly/day															13
14	Mean potential evaporation	in mm	174	141	123	76	52	33	34	49	78	114	148	174	1196	10	14
15	Mean windspeed	in m/sec	10,0	8,6	8,3	7,8	8,1	7,5	9,2	9,4	8,6	9,4	10,8	9,7	9,0		15
16	Mean predominent direction of the wind		W	W	W	W	W	W	W	W	W	W	W	W			16

142 Station / Country **Santa Cruz / Argentina**

Location **50°01'S/68°32'W**   Height above sealevel **12 m**   Climate symbol: Köppen **BSk**   Troll **III,12a**

		J	F	M	A	M	J	J	A	S	O	N	D	year	P	
1	Mean daily temperature	in °C	14,3	14,0	12,0	8,7	4,8	1,8	2,2	3,8	6,2	9,7	11,9	13,5	8,5		1
2	Mean daily maximum temperature	in °C															2
3	Mean daily minimum temperature	in °C															3
4	Absolute maximum temperature	in °C	34,1	34,8	32,3	28,0	24,0	17,3	15,3	20,0	23,8	31,8	32,0	34,8	34,8		4
5	Absolute minimum temperature	in °C	1,4	-3,8	-3,2	-8,0	-12,1	-16,2	-13,3	-11,6	-7,4	-4,9	-2,8	-0,7	-16,2		5
6	Mean relative humidity	in %	55	54	58	65	73	78	78	73	65	56	52	53	63		6
7	Mean precipitation	in mm	21	16	20	17	25	18	16	15	12	7	15	18	200		7
8	Maximum precipitation	in mm															8
9	Minimum precipitation	in mm															9
10	Maximum precipitation in 24 h	in mm	45	25	35	25	25	25	35	15	15	15	15	45	45		10
11	Mean number of days with precipitation	> 0,1 mm	6	5	6	5	6	5	5	5	4	3	6	6	62		11
12	Mean duration of sunshine	in h	236	204	180	144	118	96	112	143	159	205	210	223	2030		12
13	Mean quantity of radiation	in ly/day															13
14	Mean potential evaporation	in mm	104	82	65	40	20	7	7	15	34	57	77	99	607	45	14
15	Mean windspeed	in m/sec	5,0	4,7	4,2	3,3	3,3	3,1	3,6	3,9	4,4	5,0	5,6	4,7	4,2		15
16	Mean predominent direction of the wind		W	W	W	W	W	W	W	W	W	W	W	W			16

143 Station / Country **Ushuaia / Argentina**

Location **54°48'S/68°19'W**   Height above sealevel **6 m**   Climate symbol: Köppen **Cfc**   Troll **III,3**

		J	F	M	A	M	J	J	A	S	O	N	D	year	P	
1	Mean daily temperature	in °C	9,2	9,0	7,8	5,7	3,2	1,7	1,6	2,2	3,9	6,2	7,3	8,5	7,7		1
2	Mean daily maximum temperature	in °C	13,9	14,4	12,8	8,9	6,1	3,9	3,9	5,6	7,6	11,1	12,2	13,3	9,4	16	2
3	Mean daily minimum temperature	in °C	5,0	5,0	3,3	0,6	-1,7	-3,3	-3,9	-2,8	-0,6	1,7	2,2	3,9	0,6	16	3
4	Absolute maximum temperature	in °C	29,0	24,2	25,6	22,2	19,0	19,0	17,5	18,0	18,6	20,0	22,1	25,2	29,0		4
5	Absolute minimum temperature	in °C	-0,3	-4,0	-2,4	-6,3	-12,1	-12,6	-11,1	-19,6	-7,2	-5,7	-3,5	-3,3	-19,6		5
6	Mean relative humidity	in %	71	70	73	75	78	80	79	77	73	66	66	70	73		6
7	Mean precipitation	in mm	58	50	57	46	48	45	47	49	38	37	50	49	574		7
8	Maximum precipitation	in mm															8
9	Minimum precipitation	in mm															9
10	Maximum precipitation in 24 h	in mm	25	35	55	25	15	65	65	25	15	35	15	35	65		10
11	Mean number of days with precipitation	> 0,1 mm	13	12	11	12	13	20	9	8	7	11	11	13	140	21	11
12	Mean duration of sunshine	in h	171	160	130	84	62	30	43	99	144	192	186	186	1487		12
13	Mean quantity of radiation	in ly/day															13
14	Mean potential evaporation	in mm	86	69	59	38	18	7	9	16	33	52	72	82	540	20	14
15	Mean windspeed	in m/sec	4,7	4,4	3,3	3,3	3,1	2,2	2,8	3,3	4,4	5,0	5,6	5,3	3,9		15
16	Mean predominent direction of the wind		SW	SW	SW	SW	SW	SW	SW	SW	SW	SW	SW	SW			16

144 Station / Country **Stanley (Falkland Islands) / United Kingdom**

Location **51°42'S/57°51'W**   Height above sealevel **2 m**   Climate symbol: Köppen **ET**   Troll **I,4**

		J	F	M	A	M	J	J	A	S	O	N	D	year	P	
1	Mean daily temperature	in °C	9,5	8,9	8,1	6,1	3,9	2,2	1,9	2,2	3,9	5,3	7,0	8,1	5,6	25	1
2	Mean daily maximum temperature	in °C	13,3	12,8	11,7	9,4	6,7	5,0	4,4	5,0	7,2	8,9	11,1	12,2	8,9	25	2
3	Mean daily minimum temperature	in °C	5,8	5,0	4,4	2,8	1,1	0,0	-0,6	-0,6	0,6	1,7	2,8	3,9	2,2	25	3
4	Absolute maximum temperature	in °C	24,4	23,3	21,1	17,2	14,1	10,6	10,0	11,1	15,0	17,8	21,7	21,7	24,4	25	4
5	Absolute minimum temperature	in °C	-1,1	-1,1	-2,8	-6,1	-6,7	-11,1	-8,9	-11,1	-10,6	-5,6	-3,3	-1,7	-11,1	25	5
6	Mean relative humidity	in %	78	79	82	86	88	89	89	87	84	80	75	77	83	26	6
7	Mean precipitation	in mm	71	58	64	66	66	53	51	38	41	51	71	61	661	41	7
8	Maximum precipitation	in mm															8
9	Minimum precipitation	in mm															9
10	Maximum precipitation in 24 h	in mm	30	43	28	25	38	28	30	25	13	20	36	41	43	21	10
11	Mean number of days with precipitation	> 0,1 mm	17	12	15	14	15	13	13	13	12	11	12	15	162	41	11
12	Mean duration of sunshine	in h	198	161	169	115	77	57	69	90	128	189	200	198	1651		12
13	Mean quantity of radiation	in ly/day															13
14	Mean potential evaporation	in mm	6	7	6	3	0	0	0	0	0	0	4	6	34	25	14
15	Mean windspeed	in m/sec															15
16	Mean predominent direction of the wind																16

222

AFRICA

MOROCCO

1 Casablanca (Dhr-el-Beida)
2 Rabat
3 Tanger
4 Oujda
5 Marrakech

ALGERIA

6 Oran
7 Alger
8 Zaouia el Kahla (Fort Flatters)
9 Tindouf
10 Adrar
11 Tamanrasset

TUNISIA

12 Tunis
13 Gafsa

LIBYA

14 Tarābulus (Tripoli)
15 Surt
16 Banghāzi (Bengasi)
17 Darnah
18 Ghudamis
19 Sabhah
20 Al-Kufrah (Cufra)

EGYPT

21 As-Sallum
22 Al-Iskandarīyah (Alexandria)
23 Al-Qāhirah (Cairo)
24 Al-Qiysayr
25 Al-Uqsur (Luxor)
26 Mūt

MAURITANIA

27 Nouadhibou
28 Atar
29 Nouakchott
30 Néma

MALI

31 Tessalit
32 Gao
33 Mopti
34 Kayes
35 Bougouni

NIGER

36 Bilma
37 Zinder
38 Niamey

CHAD

39 Largeau
40 Abéché
41 Ndjamena
42 Moundou

SUDAN

43 Būr Sūdān (Port Sudan)
44 Dunqulah
45 Al-Khurtūm (Khartoum)
46 Al-Fāshir
47 Al-Ubayyid
48 Ar-Rusayris
49 Malakāl
50 Wāw
51 Yubo (Source Yubo)
52 Juba

ETHIOPIA

53 Asmera
54 Harar
55 Addis Abeba
56 Jima
57 Negelli

AFARS AND ISSAS

58 Djibouti

SOMALIA

59 Berbera

60 Hargeysa
61 Mogadisho

SENEGAL

62 Dakar
63 Ziguinchor

GUINEA

64 Mamou
65 Conakry

SIERRA LEONE

66 Freetown

LIBERIA

67 Monrovia

IVORY COAST

68 Bouaké
69 Abidjan

UPPER VOLTA

70 Ouagadougou
71 Bobo Dioulasso

GHANA

72 Tamale
73 Accra

BENIN

74 Kandi
75 Cotonou

NIGERIA

76 Sokoto
77 Kano

78 Maiduguri
79 Jos
80 Makurdi
81 Enugu
82 Lagos

EQUATORIAL GUINEA

83 Santa Isabel

CAMEROON

84 Garoua
85 Batouri
86 Douala

CENTRAL AFRICAN REPUBLIC

87 Birao
88 Bria
89 Bouar
90 Bangui

GABON

91 Mitzig
92 Libreville

ZAIRE

93 Bongabo
94 Yangambi
95 Eala
96 Tshibinda
97 Kinshasa (Léopoldville)
98 Kananga (Luluabourg)
99 Kalemi (Albertville)
100 Kamina
101 Lubumbashi (Elisabethville)

UGANDA

102 Gulu
103 Entebbe

BURUNDI

104 Kisozi
105 Bujumbura

KENYA

106 Marsabit
107 Garissa
108 Dagoretti (near Nairobi)
109 Mombasa

TANZANIA

110 Tabora
111 Marogoro
112 Dar-es-Salaam
113 Mbeya
114 Mtwara

ANGOLA

115 Cabinda
116 Luanda
117 Teixeira de Sousa
118 Nova Lisboa
119 Cangamba
120 Mupa

ZAMBIA

121 Kasama
122 Ndola
123 Mongu
124 Lusaka

ZIMBABWE

125 Salisbury
126 Wankie
127 Umtali
128 Bulawayo

MALAWI

129 Karonga
130 Lilongwe
131 Blantyre

MOZAMBIQUE

132 Nova Freixa
133 Mossuril
134 Beira
135 Maputo (Laurenço Marques)

MADAGASCAR

136 Diégo-Suarez
137 Majunga
138 Tamatave
139 Tananarive
140 Tuléar
141 Fort-Dauphin

NAMIBIA

142 Tsumeb
143 Windhoek
144 Swakopmund
145 Lüderitz
146 Keetmanshoop

BOTSWANA

147 Maun
148 Ghanzi
149 Mahalatswe
150 Gaborone

SOUTH AFRICA

151 Messina
152 Pietersburg
153 Upington
154 Kimberley
155 Durban
156 East London (Oos-Londen)
157 Cape Town (Kaapstad)
158 Port Elizabeth
159 Oudtshoorn

LESOTHO

160 Mokhotlong

UNITED KINGDOM

161 Jamestown (St. Helena)
162 Tristan da Cunha

MASCARENE ISLANDS

163 Rodriguez (Rodriguez Island)

FRANCE

164 Kerguelen

AFRICA

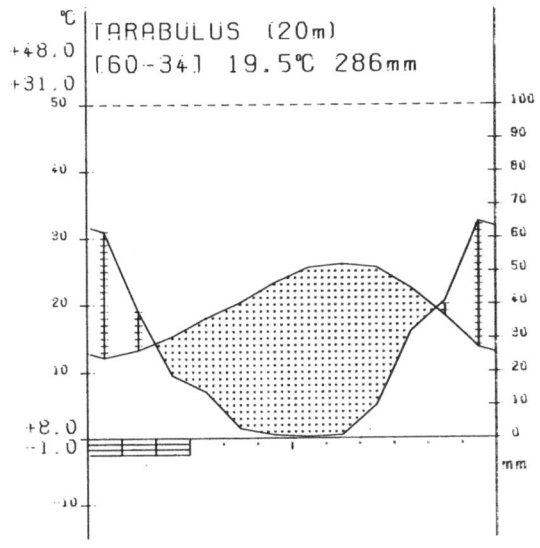

TARABULUS (20m)
[60-34] 19.5℃ 286mm

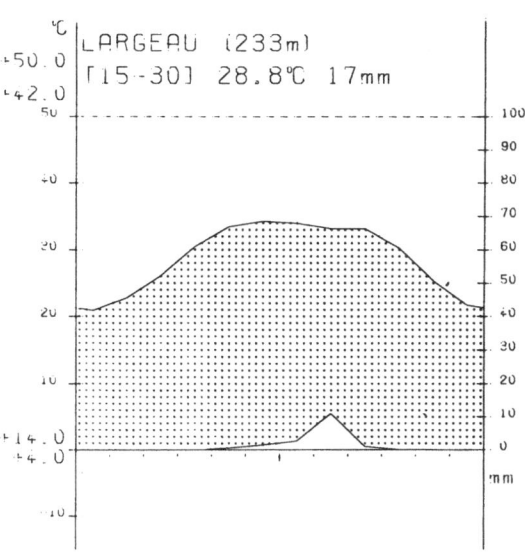

LARGEAU (233m)
[15-30] 28.8℃ 17mm

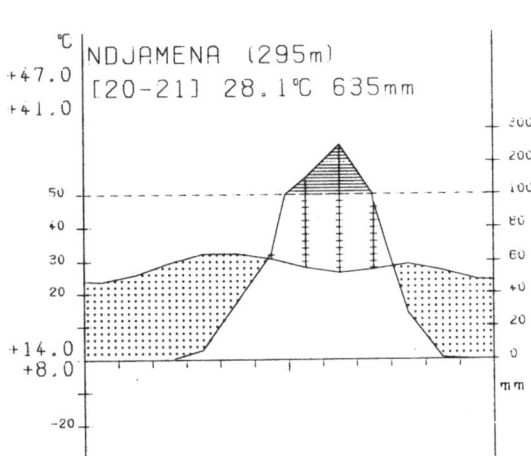

NDJAMENA (295m)
[20-21] 28.1℃ 635mm

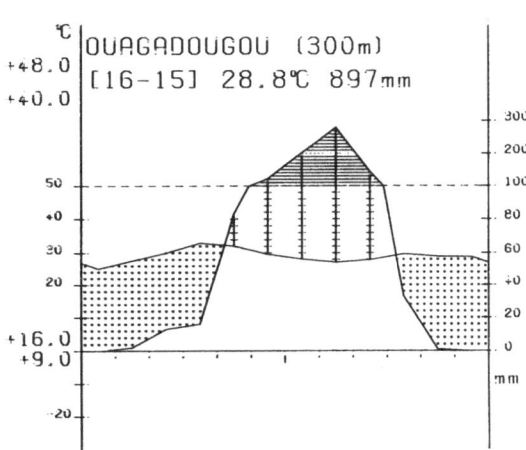

OUAGADOUGOU (300m)
[16-15] 28.8℃ 897mm

AFRICA

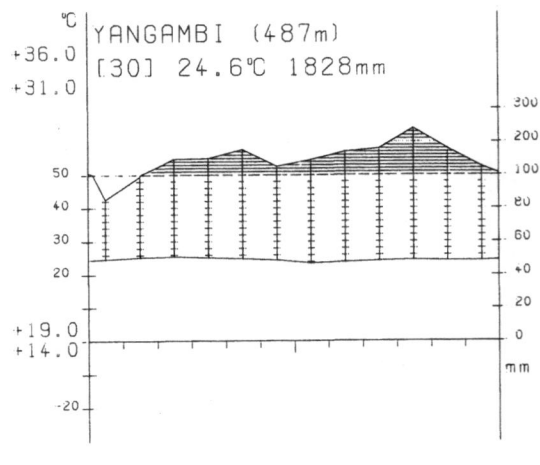

YANGAMBI (487m)
[30] 24.6°C 1828mm

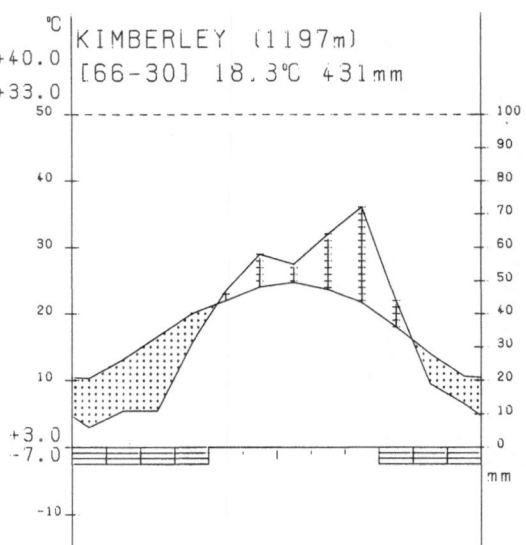

KIMBERLEY (1197m)
[66-30] 18.3°C 431mm

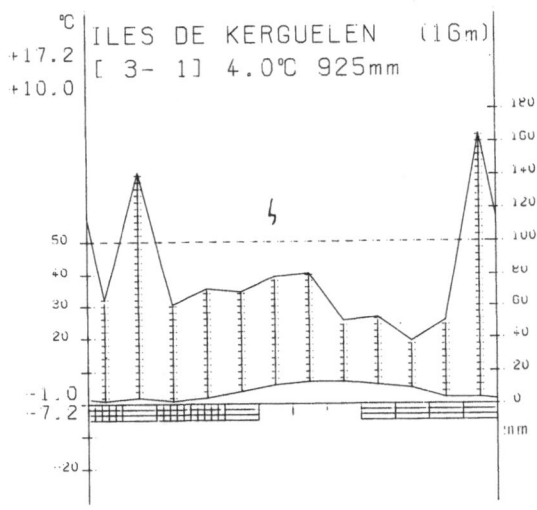

ILES DE KERGUELEN (16m)
[3-1] 4.0°C 925mm

1 Station / Country Casablanca (Dar-el-Beida) / Morocco

Location 33°34'N/7°40'W   Height above sealevel 49 m   Climate symbol: Köppen Csa   Troll IV.1

		J	F	M	A	M	J	J	A	S	O	N	D	year	P	
1 Mean daily temperature	in °C	12,4	12,9	14,6	15,9	17,6	20,4	22,6	23,0	21,8	18,5	16,2	13,5	17,5	36	1
2 Mean daily maximum temperature	in °C	17	18	19	20	22	24	26	26	26	24	20	18	21	10	2
3 Mean daily minimum temperature	in °C	9	9	11	12	15	17	19	19	15	12	10	14	10		3
4 Absolute maximum temperature	in °C	31	35	34	33	39	36	33	41	37	32	28	27	41	10	4
5 Absolute minimum temperature	in °C	2	0	5	6	9	12	14	15	14	7	6	2	0	10	5
6 Mean relative humidity	in %	78	77	77	77	76	75	78	80	80	77	76	77	77	5	6
7 Mean precipitation	in mm	80	68	68	37	22	5	0	1	3	25	77	125	511	10	7
8 Maximum precipitation	in mm	179	198	125	115	76	11	1	11	31	139	188	234	666	30	8
9 Minimum precipitation	in mm	7	0	6	0	0	0	0	0	0	1	0	7	227	30	9
10 Maximum precipitation in 24 h	in mm	50	42	50	53	72	10	1	11	24	59	45	58	72	30	10
11 Mean number of days with precipitation	> 0,1 mm	12	10	10	6	5	2	<1	<1	2	7	11	12	77	10	11
12 Mean duration of sunshine	in h															12
13 Mean quantity of radiation	in ly / day															13
14 Mean potential evaporation	in mm	30	31	43	63	79	102	122	119	99	75	45	34	832		14
15 Mean windspeed	in m/sec	0,8	1,0	1,1	1,1	1,0	1,0	0,8	0,8	0,8	0,8	1,0	1,0	0,8		15
16 Mean predominent direction of the wind		S.NE	SW	W	N	N	N	N	N	N	NE	SW,NE	SW			16

2 Station / Country Rabat / Morocco

Location 34°03'N/6°40'W   Height above sealevel 75 m   Climate symbol: Köppen Csa   Troll IV.1

		J	F	M	A	M	J	J	A	S	O	N	D	year	P	
1 Mean daily temperature	in °C	12,5	13,3	14,7	16,4	18,0	20,6	22,5	23,0	22,0	19,7	16,4	13,6	17,7	35	1
2 Mean daily maximum temperature	in °C	17	18	19	20	23	26	28	27	28	24	20	17	22		2
3 Mean daily minimum temperature	in °C	9	9	10	11	14	16	17	18	17	14	11	9	13		3
4 Absolute maximum temperature	in °C	29	36	34	33	39	41	40	44	40	36	29	28	44		4
5 Absolute minimum temperature	in °C	2	1	4	6	8	10	12	13	11	6	4	2	1		5
6 Mean relative humidity	in %	81	79	77	75	75	74	74	76	77	77	78	78	77	4	6
7 Mean precipitation	in mm	66	64	66	43	28	8	<1	1	1	48	84	86	498	20	7
8 Maximum precipitation	in mm	194	183	227	153	34	31	4	21	80	144	212	263	808	20	8
9 Minimum precipitation	in mm	21	2	7	9	0	0	0	0	0	0	0	12	259	20	9
10 Maximum precipitation in 24 h	in mm	56	40	42	151	23	23	10	21	38	55	69	54	151	20	10
11 Mean number of days with precipitation	> 0,1 mm	9	8	10	7	6	2	<1	<1	2	6	9	10	70		11
12 Mean duration of sunshine	in h	166	185	221	263	298	300	336	318	268	231	184	170	2840	20	12
13 Mean quantity of radiation	in ly / day															13
14 Mean potential evaporation	in mm	29	32	44	56	79	100	122	123	97	77	48	35	840		14
15 Mean windspeed	in m/sec	2,6	3,2	3,7	4,0	3,6	3,6	3,5	3,3	3,4	3,3	3,1	3,2	3,4		15
16 Mean predominent direction of the wind		S.W	W	W	W.N	NW	NW	NW	NW	NW	N.W	W.S	W.S			16

3 Station / Country Tanger / Morocco

Location 35°43'N/5°54'W   Height above sealevel 15 m   Climate symbol: Köppen Csa   Troll IV.1

		J	F	M	A	M	J	J	A	S	O	N	D	year	P	
1 Mean daily temperature	in °C	12,0	12,5	13,6	14,4	17,5	20,2	22,2	23,0	21,4	18,6	14,7	12,5	17,2	29	1
2 Mean daily maximum temperature	in °C	16	17	18	20	23	26	29	29	28	24	20	17	22	10	2
3 Mean daily minimum temperature	in °C	9	9	11	12	14	17	19	19	18	16	12	10	14	10	3
4 Absolute maximum temperature	in °C	23	25	28	28	31	36	39	39	31	32	28	24	39	10	4
5 Absolute minimum temperature	in °C	-1	0	3	4	8	10	11	14	11	7	3	2	-1	10	5
6 Mean relative humidity	in %	76	77	76	73	69	69	67	68	71	75	76	77	73	12	6
7 Mean precipitation	in mm	114	106	120	90	42	15	1	1	23	99	147	137	895	35	7
8 Maximum precipitation	in mm	304	208	229	187	72	39	2	11	66	193	269	374	1249	18	8
9 Minimum precipitation	in mm	22	11	7	1	0	0	0	0	0	0	24	33	439	18	9
10 Maximum precipitation in 24 h	in mm	104	109	84	111	49	94	7	5	81	112	98	122	122	24	10
11 Mean number of days with precipitation	> 0,1 mm	10	10	10	8	5	3	<1	<1	3	8	10	10	78	7	11
12 Mean duration of sunshine	in h															12
13 Mean quantity of radiation	in ly / day															13
14 Mean potential evaporation	in mm	30	31	42	53	79	102	123	123	97	72	42	32	828	30	14
15 Mean windspeed	in m/sec	6,0	6,0	6,5	6,5	6,5	6,5	6,0	6,0	5,5	6,0	6,0	6,5	6,0	10	15
16 Mean predominent direction of the wind		E	E	E	E.W	E.W	E.W	E	E	E	E	E	E		10	16

4 Station / Country Oujda / Morocco

Location 34°47'N/1°56'W   Height above sealevel 470 m   Climate symbol: Köppen Csa   Troll IV.1

		J	F	M	A	M	J	J	A	S	O	N	D	year	P	
1 Mean daily temperature	in °C	9,4	10,3	12,8	15,1	17,6	22,1	25,7	26,0	23,0	18,4	13,5	10,4	17,0	32	1
2 Mean daily maximum temperature	in °C	16	17	19	25	25	28	33	33	31	25	19	16	24		2
3 Mean daily minimum temperature	in °C	5	5	7	8	14	17	18	18	16	11	8	6	10		3
4 Absolute maximum temperature	in °C	25	33	32	31	37	42	43	41	41	34	30	28	43		4
5 Absolute minimum temperature	in °C	-5	-4	-2	1	3	8	10	10	8	0	0	-3	-5		5
6 Mean relative humidity	in %															6
7 Mean precipitation	in mm	43	31	52	45	31	15	4	1	13	15	29	58	337	30	7
8 Maximum precipitation	in mm	110	109	178	133	122	48	13	23	80	80	120	117	570	30	8
9 Minimum precipitation	in mm	1	0	0	<1	<1	0	0	0	<1	0	0	6	110	30	9
10 Maximum precipitation in 24 h	in mm	57	70	59	66	53	29	13	13	56	52	51	63	70	30	10
11 Mean number of days with precipitation	> 0,1 mm	7	7	9	8	6	5	2	1	4	5	8	10	72	30	11
12 Mean duration of sunshine	in h															12
13 Mean quantity of radiation	in ly / day															13
14 Mean potential evaporation	in mm	17	24	36	54	76	115	164	152	106	68	32	21	865		14
15 Mean windspeed	in m/sec	3,4	3,5	4,0	3,3	3,3	3,5	3,6	3,5	3,2	2,9	3,0	4,3	3,5		15
16 Mean predominent direction of the wind		W	W	W	N	N	N	N	N	N	N.W	W	W			16

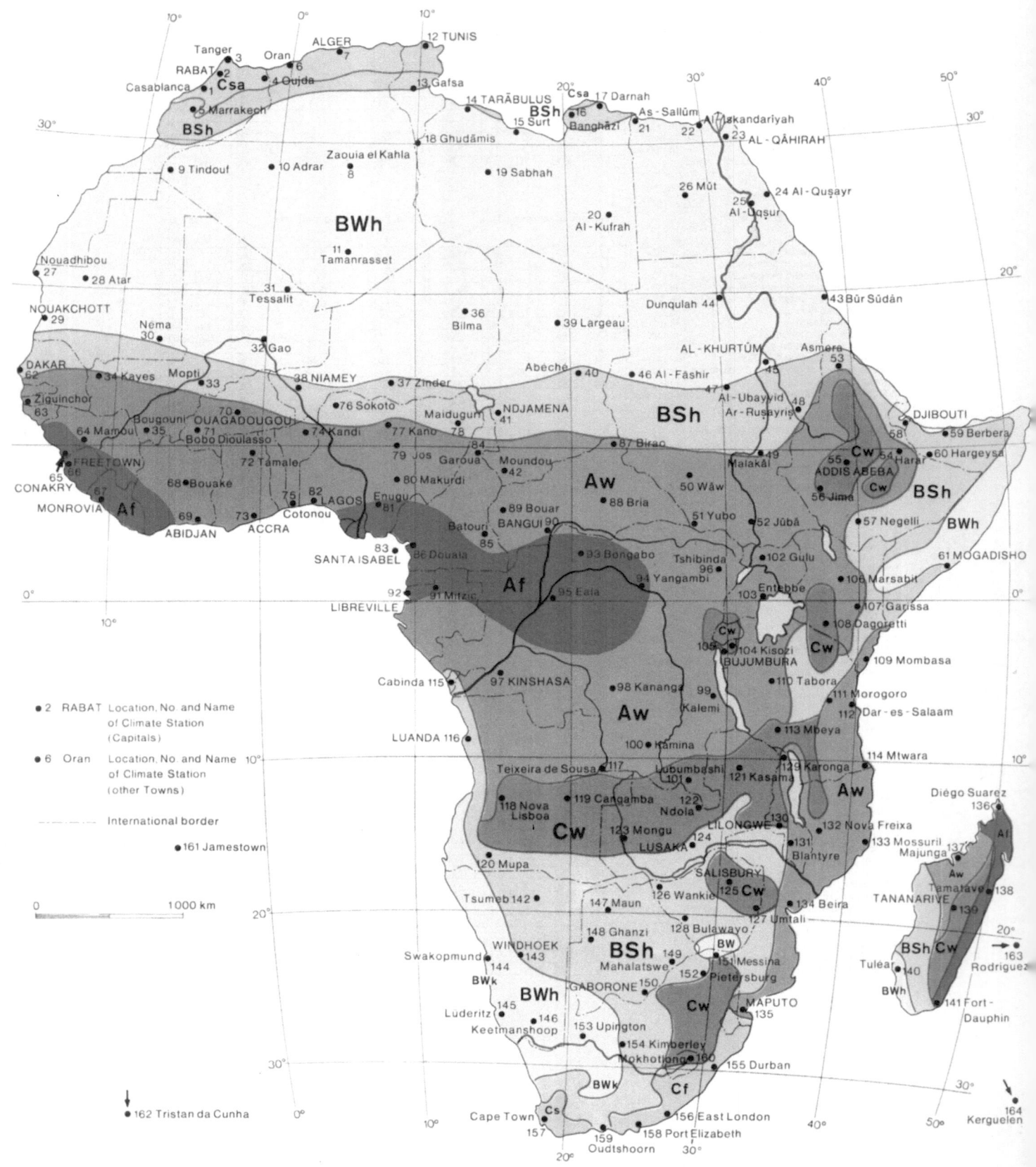

AFRICA
Climate Zones after W. Köppen/R. Geiger

Csa
BSh
BWh
BSh
Aw
Cw
BSh
BWh
Af
Aw
Cw
Aw
Af
BSh Cw
BWk
BWh
Cw
BWk
Cf
Cs

- 2 RABAT   Location, No. and Name
            of Climate Station
            (Capitals)
- 6 Oran   Location, No. and Name
            of Climate Station
            (other Towns)
- - - - -   International border
- 161 Jamestown

0   1000 km

162 Tristan da Cunha

164 Kerguelen

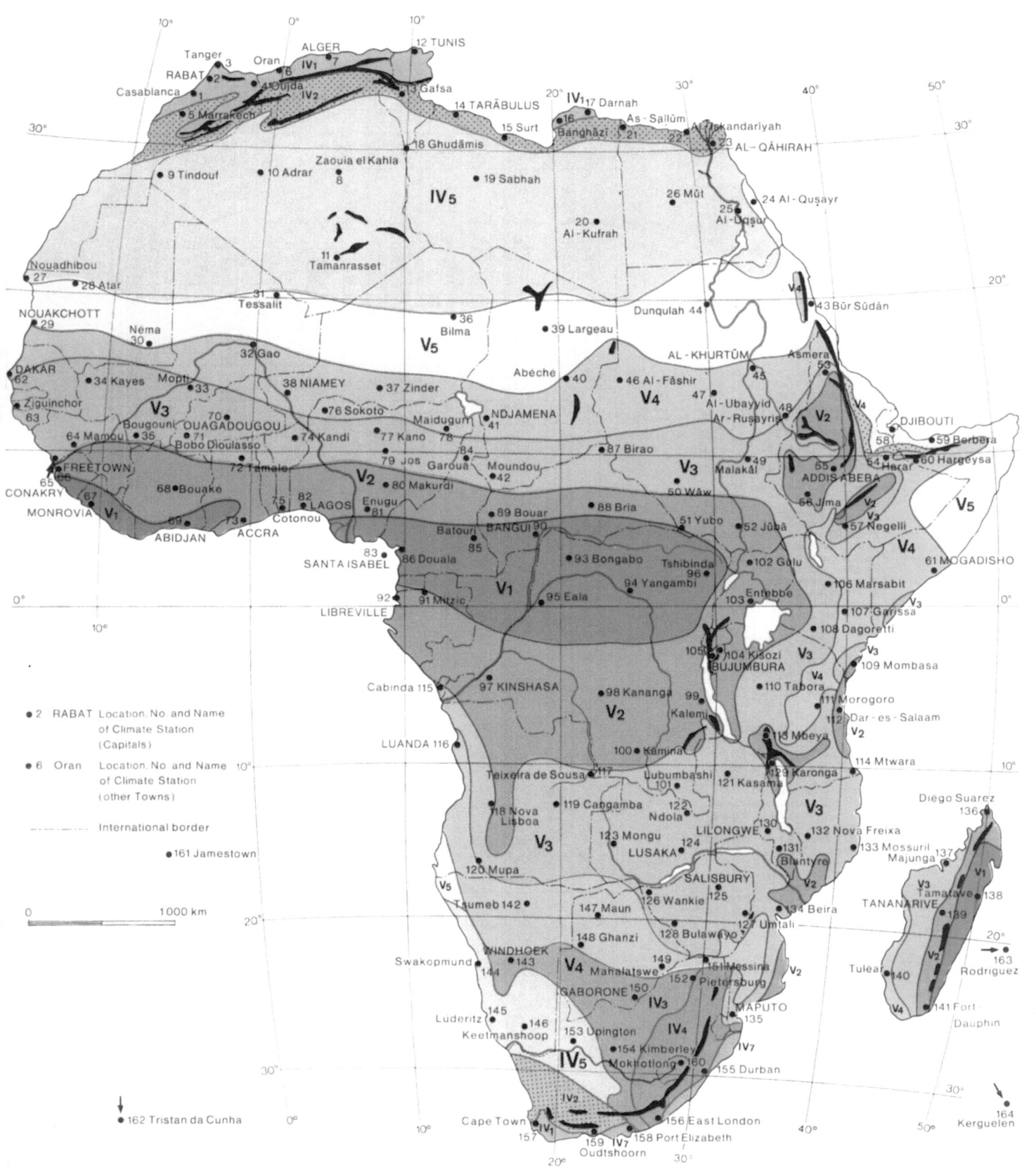

AFRICA
Climate Zones after C. Troll/KH. Paffen

10° 0° 10° 20° 30° 40° 50°

Tanger 3 Oran 6 ALGER IV₁ 7 12 TUNIS
RABAT 2 4 Oujda
Casablanca 1 IV₂ 13 Gafsa
5 Marrakech 14 TARĀBULUS IV₁ 17 Darnah
15 Surt 16 As-Sallūm Al-Iskandariyah
18 Ghudāmis Banghāzi 21 22 23 AL-QĀHIRAH
Zaouia el Kahla 19 Sabhah 26 Mût 24 Al-Quşayr
9 Tindouf 10 Adrar 8 IV₅ 25 Al-Daşur
20 Al-Kufrah 24 Al-Quşayr
11 Tamanrasset
Nouadhibou 27 28 Atar 31 43 Bûr Sûdân
NOUAKCHOTT 29 Tessalit 36 39 Largeau Dunqulah 44 43
Nema 30 Bilma V₅ Asmera
DAKAR 62 34 Kayes Mopti 32 Gao 38 NIAMEY 37 Zinder 40 46 Al-Fâshir 45 AL-KHURTŪM V₄ 53
Ziguinchor 63 V₃ 70 76 Sokoto Abéche Al-Ubayyid 48 V₂ V₄ DJIBOUTI
64 Mamou 35 Bougouni OUAGADOUGOU 71 74 Kandi NDJAMENA Maidugur 41 Ar-Rusayris 58 59 Berbera
FREETOWN 65 66 Bobo Dioulasso 72 Tamale 77 Kano 78 79 Jos Garoua 84 87 Birao V₃ Malakāl 49 55 54 Hacar 60 Hargeysa
CONAKRY 67 68 Bouake 75 82 LAGOS Enugu 80 Makurdi 42 Moundou 50 Wâw ADDIS ABEBA 56 Jima V₂
MONROVIA V₁ 69 73 Cotonou 81 V₂ BANGUI 90 89 Bouar 88 Bria 51 Yubo 52 Jûbâ 57 Negelli V₅
ABIDJAN ACCRA Batouri 85 93 Bongabo 96 Tshibinda 102 Golu V₄ 61 MOGADISHO
SANTA ISABEL 83 86 Douala 94 Yangambi 103 106 Marsabit
LIBREVILLE 92 91 Mitzic V₁ 95 Eala Entebbe 107 Garissa
108 Dagoretti V₃
105 104 Kisozi 109 Mombasa
Cabinda 115 97 KINSHASA 98 Kanange 99 BUJUMBURA 110 Tabora 111 Morogoro
Kalem V₂ 112 Dar-es-Salaam
LUANDA 116 100 Kamina 113 Mbeya V₂ 114 Mtwara
Teixeira de Sousa 117 Lubumbashi 121 Kasama 129 Karonga
118 Nova Lisboa 119 Cangamba 101 Ndola 122 V₃ Diego Suarez 136
123 Mongu 124 LILONGWE 130 132 Nova Freixa 133 Mossuril Majunga 137
120 Mupa LUSAKA 131 Blantyre TANANARIVE V₃ 138
V₅ SALISBURY 134 Beira Tamatave 139
Tsumeb 142 147 Maun 126 Wankie 125 V₂ 135 MAPUTO Tuléar 140
148 Ghanzi 128 Bulawayo 127 Umtali 151 Messina V₂ 141 Fort Dauphin
WINDHOEK 143 149 Mahalatswe 152 Pietersburg V₄
Swakopmund 144 V₄ 150 GABORONE IV₃ 163 Rodriguez
145 Luderitz 146 153 Upington IV₄ 154 Kimberley 160 IV₇
Keetmanshoop Mokhotlong 155 Durban
IV₅
IV₂ 156 East London
Cape Town 157 IV₁ 159 IV₇ 158 Port Élizabeth
Oudtshoorn

• 2 RABAT Location, No. and Name
of Climate Station
(Capitals)

• 6 Oran Location, No. and Name
of Climate Station
(other Towns)

– – – International border

• 161 Jamestown

0 _____ 1000 km

↓
• 162 Tristan da Cunha

↓
164 Kerguelen

5 Station/Country Marrakech/Morocco

Location 31°36'N/8°01'W **Height above sealevel** 460 m **Climate symbol: Köppen** BSh **Troll** IV,2

		J	F	M	A	M	J	J	A	S	O	N	D	year	P	
1 Mean daily temperature	in °C	11,8	13,2	16,0	18,3	21,3	24,8	28,7	28,8	25,5	21,4	16,3	12,3	19,9	36	1
2 Mean daily maximum temperature	in °C	18	20	22	24	28	31	36	36	32	27	21	18	26	10	2
3 Mean daily minimum temperature	in °C	7	8	10	12	15	17	20	20	18	15	10	7	13	10	3
4 Absolute maximum temperature	in °C	29	34	34	34	38	41	44	45	42	35	30	28	45	10	4
5 Absolute minimum temperature	in °C	0	0	2	4	8	12	14	14	12	7	4	-1	-1	10	5
6 Mean relative humidity	in %	77	73	70	65	60	58	53	53	57	61	65	66	63	4	6
7 Mean precipitation	in mm	28	29	32	31	17	7	2	3	10	21	28	33	241	30	7
8 Maximum precipitation	in mm	89	99	99	139	81	47	26	23	49	60	89	131	462	30	8
9 Minimum precipitation	in mm	0	0	0	0	0	0	0	0	0	0	0	<1	128	30	9
10 Maximum precipitation in 24 h	in mm	33	33	36	63	42	40	26	23	29	43	37	52	63	30	10
11 Mean number of days with precipitation	> 0,1 mm	7	6	6	6	4	2	<1	1	1	4	6	8	51	30	11
12 Mean duration of sunshine	in h	215	212	252	269	290	322	358	329	281	242	216	206	3192		12
13 Mean quantity of radiation	in ly/day															13
14 Mean potential evaporation	in mm	19	23	43	60	95	125	185	179	129	84	37	23	1002		14
15 Mean windspeed	in m/sec	2,5	3,0	3,8	3,5	3,1	2,7	3,0	3,1	2,7	2,5	2,3	2,3	2,8	10	15
16 Mean predominent direction of the wind		W,E	E,W	W	W	W	W	N,W	N,W	W	W	W	W,E			16

6 Station/Country Oran(Ouahran)/Algeria

Location 35°44'N/0°39 W **Height above sealevel** 11 m **Climate symbol: Köppen** Csa **Troll** IV,1

		J	F	M	A	M	J	J	A	S	O	N	D	year	P	
1 Mean daily temperature	in °C	11,7	12,4	13,9	15,7	18,5	21,6	24,2	24,8	22,9	19,1	15,5	12,4	17,7	22	1
2 Mean daily maximum temperature	in °C	16	17	18	21	23	25	28	29	27	23	19	17	22	25	2
3 Mean daily minimum temperature	in °C	9	10	11	13	16	18	21	22	20	16	13	9	15	25	3
4 Absolute maximum temperature	in °C	26	28	30	34	31	38	38	41	36	37	29	26	41	25	4
5 Absolute minimum temperature	in °C	1	3	6	7	9	14	17	18	13	11	7	2	1	25	5
6 Mean relative humidity	in %	74	72	70	67	71	71	74	74	71	73	71	72	72		6
7 Mean precipitation	in mm	70	54	35	33	19	7	1	3	16	43	48	67	394	30	7
8 Maximum precipitation	in mm	183	175	124	110	61	32	8	23	67	188	125	199	554	30	8
9 Minimum precipitation	in mm	0	0	2	1	0	0	0	0	0	0	0	2	288	30	9
10 Maximum precipitation in 24 h	in mm	63	81	25	30	33	20	5	23	27	43	70	52	81	15	10
11 Mean number of days with precipitation	> 0,1 mm	8	8	4	3	4	1	<1	<1	2	3	6	7	44	14	11
12 Mean duration of sunshine	in h	173	183	251	258	312	300	343	340	278	238	189	160	3005	10	12
13 Mean quantity of radiation	in ly/day															13
14 Mean potential evaporation	in mm	27	31	40	54	85	110	128	131	101	72	45	31	855	9	14
15 Mean windspeed	in m/sec	2,0	4,0	3,0	4,0	4,0	5,0	5,0	4,0	3,0	3,0	3,0	2,0	3,0	5	15
16 Mean predominent direction of the wind		SW	SW	SW	SW,N	SW,N	SW,N	SW,N	SW,N	SW,N	SW,N	SW	SW			16

7 Station/Country Alger (Algiers)/Algeria

Location 36°48'N/3°03'E **Height above sealevel** 60 m **Climate symbol: Köppen** Csa **Troll** IV,1

		J	F	M	A	M	J	J	A	S	O	N	D	year	P	
1 Mean daily temperature	in °C	12,2	12,6	13,8	16,0	18,5	22,1	24,3	25,2	23,2	20,0	16,7	13,9	18,2	6	1
2 Mean daily maximum temperature	in °C	15	16	17	20	23	27	28	29	27	23	19	16	22	25	2
3 Mean daily minimum temperature	in °C	9	9	11	13	15	18	21	22	21	17	13	11	15	25	3
4 Absolute maximum temperature	in °C	24	20	29	37	38	38	41	42	39	38	31	24	42	25	4
5 Absolute minimum temperature	in °C	1	1	3	6	7	13	17	18	12	7	4	0	0	25	5
6 Mean relative humidity	in %	71	68	65	62	66	66	67	65	68	68	68	68	67		6
7 Mean precipitation	in mm	116	78	57	65	36	14	2	4	27	84	93	117	641	30	7
8 Maximum precipitation	in mm	245	192	139	195	104	78	21	76	73	350	284	281	1087		8
9 Minimum precipitation	in mm	8	3	11	6	0	0	0	0	7	13	30		500		9
10 Maximum precipitation in 24 h	in mm	56	48	62	48	66	22	9	26	73	59	147	119	147	17	10
11 Mean number of days with precipitation	> 0,1 mm	11	9	9	5	5	2	<1	<1	4	7	11	12	78	17	11
12 Mean duration of sunshine	in h	141	159	207	227	300	301	353	324	268	197	153	146	2776		12
13 Mean quantity of radiation	in ly/day	185	237	392	426	470	553	545	486	405	283	185	145	360	3	13
14 Mean potential evaporation	in mm	26	28	42	58	85	116	149	148	113	78	44	29	916	28	14
15 Mean windspeed	in m/sec	3,0	3,0	3,0	3,0	3,0	3,0	3,0	3,0	2,0	3,0	3,0	3,0	3,0	10	15
16 Mean predominent direction of the wind		W	W	W	W	NE	NE	NE	NE	NE	W	W	W		10	16

8 Station/Country Zaouia el Kahla (Fort Flatters)/Algeria

Location 28°06'N/6°42'E **Height above sealevel** 373 m **Climate symbol: Köppen** BWh **Troll** IV,5

		J	F	M	A	M	J	J	A	S	O	N	D	year	P	
1 Mean daily temperature	in °C	10,8	14,4	19,2	23,4	28,4	33,8	34,8	33,8	31,4	25,8	18,8	12,5	23,9		1
2 Mean daily maximum temperature	in °C	19,4	22,8	28,7	32,2	36,1	41,1	43,3	42,2	38,8	33,3	26,7	21,1	32,2	15	2
3 Mean daily minimum temperature	in °C	3,3	6,1	10,0	15,0	20,0	25,0	25,8	25,0	22,8	17,2	10,0	5,0	15,6	15	3
4 Absolute maximum temperature	in °C	32,8	35,6	41,1	42,2	48,7	49,6	51,1	50,0	46,7	42,8	37,2	29,4	51,1	23	4
5 Absolute minimum temperature	in °C	-7,2	-5,0	-2,2	6,1	9,4	15,6	20,0	17,2	15,6	5,6	0,6	-5,6	-7,2	23	5
6 Mean relative humidity	in %															6
7 Mean precipitation	in mm	7	3	2	4	1	1	0	0	1	1	5	4	29	15	7
8 Maximum precipitation	in mm															8
9 Minimum precipitation	in mm															9
10 Maximum precipitation in 24 h	in mm															10
11 Mean number of days with precipitation	> 0,1 mm	<1	<1	<1	<1	<1	<1	0	<1	<1	<1	<1	1	6		11
12 Mean duration of sunshine	in h	270	271	310	321	353	342	394	372	321	304	270	239	3767		12
13 Mean quantity of radiation	in ly/day															13
14 Mean potential evaporation	in mm															14
15 Mean windspeed	in m/sec															15
16 Mean predominent direction of the wind																16

9 Station/Country: Tindouf/Algeria

Location 27°43'N/8°08'E Height above sealevel 600 m Climate symbol: Köppen BWh Troll IV,5

		J	F	M	A	M	J	J	A	S	O	N	D	year	P	
1 Mean daily temperature	in °C	15,2	17,0	20,6	24,1	27,8	29,7	35,8	35,7	31,0	25,8	19,4	15,6	24,8		1
2 Mean daily maximum temperature	in °C	21,7	25,6	28,3	32,2	34,4	38,3	45,0	43,9	38,9	31,1	26,1	21,7	32,2	15	2
3 Mean daily minimum temperature	in °C	5,0	6,7	11,7	12,8	15,6	17,2	25,0	25,6	22,2	14,4	11,1	5,6	14,4	15	3
4 Absolute maximum temperature	in °C	29,4	31,7	38,9	40,6	42,8	47,8	50,0	49,4	46,7	42,2	36,1	28,3	50,0	5	4
5 Absolute minimum temperature	in °C	1,1	1,7	4,4	8,9	9,4	12,2	16,1	12,8	13,3	10,0	3,3	1,7	1,1	5	5
6 Mean relative humidity	in %															6
7 Mean precipitation	in mm	0	0	6	0	0	0	1	11	7	4	1	3	33	32	7
8 Maximum precipitation	in mm															8
9 Minimum precipitation	in mm															9
10 Maximum precipitation in 24 h	in mm															10
11 Mean number of days with precipitation	> 0,1 mm	<1	1	<1	1	1	<1	<1	1	5	2	3	<1	18		11
12 Mean duration of sunshine	in h															12
13 Mean quantity of radiation	in ly/day															13
14 Mean potential evaporation	in mm															14
15 Mean windspeed	in m/sec															15
16 Mean predominent direction of the wind																16

10 Station/Country: Adrar/Algeria

Location 27°52'N/0°20'E Height above sealevel 258 m Climate symbol: Köppen BWh Troll IV,5

		J	F	M	A	M	J	J	A	S	O	N	D	year	P	
1 Mean daily temperature	in °C	11,8	14,7	19,8	23,8	28,4	33,6	36,7	35,4	32,0	25,2	18,6	13,0	24,5		1
2 Mean daily maximum temperature	in °C	20,6	23,3	27,6	33,3	37,2	43,3	46,1	44,4	40,6	33,3	25,6	20,6	32,8	15	2
3 Mean daily minimum temperature	in °C	3,9	6,7	10,6	15,6	19,4	25,6	27,8	26,7	23,9	17,2	10,6	5,6	16,1	15	3
4 Absolute maximum temperature	in °C	30,6	35,0	38,9	43,9	50,0	48,9	51,1	50,0	46,7	43,3	35,0	30,6	51,1	15	4
5 Absolute minimum temperature	in °C	-3,9	-1,7	0,6	7,2	10,6	18,3	20,0	18,3	16,1	16,1	1,7	-2,8	-3,9	15	5
6 Mean relative humidity	in %															6
7 Mean precipitation	in mm	0	1	2	0	1	0	1	1	1	5	5	1	18		7
8 Maximum precipitation	in mm															8
9 Minimum precipitation	in mm															9
10 Maximum precipitation in 24 h	in mm															10
11 Mean number of days with precipitation	> 0,1 mm	<1	<1	<1	<1	<1	<1	<1	<1	<1	<1	<1	1	6		11
12 Mean duration of sunshine	in h															12
13 Mean quantity of radiation	in ly/day															13
14 Mean potential evaporation	in mm															14
15 Mean windspeed	in m/sec															15
16 Mean predominent direction of the wind																16

11 Station/Country: Tamanrasset/Algeria

Location 22°42'N/5°31'E Height above sealevel 1405 m Climate symbol: Köppen BWh Troll IV,5

		J	F	M	A	M	J	J	A	S	O	N	D	year	P	
1 Mean daily temperature	in °C	12,0	13,9	17,5	21,6	25,3	28,3	28,3	27,8	26,1	22,5	18,0	13,6	21,2	26	1
2 Mean daily maximum temperature	in °C	19	22	26	30	33	35	35	34	33	29	26	21	28	35	2
3 Mean daily minimum temperature	in °C	4	6	9	13	17	21	22	21	19	15	11	6	13	35	3
4 Absolute maximum temperature	in °C	26	28	32	36	38	38	39	38	37	34	30	27	39	35	4
5 Absolute minimum temperature	in °C	-7	-4	0	3	4	15	17	17	14	8	-1	-3	-7	35	5
6 Mean relative humidity	in %	28	31	26	27	29	24	21	25	26	29	32	31	27	15	6
7 Mean precipitation	in mm	4	1	1	2	6	4	3	10	7	2	2	2	44		7
8 Maximum precipitation	in mm	55	16	18	33	95	35	18	73	96	19	34	26	159		8
9 Minimum precipitation	in mm	0	0	0	0	0	0	0	0	0	0	0	0	6		9
10 Maximum precipitation in 24 h	in mm	7	3	5	20	48	19	5	35	48	17	29	21	48	10	10
11 Mean number of days with precipitation	> 0,1 mm	1	1	1	1	2	3	2	3	3	2	1	1	21	10	11
12 Mean duration of sunshine	in h	288	252	322	312	332	300	322	319	255	279	270	245	3496		12
13 Mean quantity of radiation	in ly/day															13
14 Mean potential evaporation	in mm	19	24	52	85	146	180	179	167	136	94	51	26	1159		14
15 Mean windspeed	in m/sec															15
16 Mean predominent direction of the wind																16

12 Station/Country: Tunis/Tunisia

Location 38°50'N/10°14'E Height above sealevel 3 m Climate symbol: Köppen Csa Troll IV,1

		J	F	M	A	M	J	J	A	S	O	N	D	year	P	
1 Mean daily temperature	in °C	10,2	10,9	12,6	15,1	18,4	22,8	25,6	26,2	23,9	19,8	15,2	11,6	17,7		1
2 Mean daily maximum temperature	in °C	15	16	18	21	23	29	32	32	29	25	20	16	23	30	2
3 Mean daily minimum temperature	in °C	7	8	9	11	14	18	20	21	20	16	12	8	14	30	3
4 Absolute maximum temperature	in °C	24	30	35	32	40	41	46	47	43	37	31	25	47	30	4
5 Absolute minimum temperature	in °C	0	0	0	3	7	12	15	11	13	7	3	2	0	30	5
6 Mean relative humidity	in %	81	78	77	77	72	69	67	69	74	77	81	80	75	10	6
7 Mean precipitation	in mm	70	47	43	42	23	10	1	11	37	52	57	68	461	30	7
8 Maximum precipitation	in mm	177	130	104	131	96	66	7	47	155	215	234	278	913	30	8
9 Minimum precipitation	in mm	4	4	3	3	0	0	0	0	2	15	2		248	30	9
10 Maximum precipitation in 24 h	in mm	63	101	37	60	41	52	6	34	81	83	62	134	134	30	10
11 Mean number of days with precipitation	> 0,1 mm	13	9	9	7	5	3	1	2	5	8	10	13	85	30	11
12 Mean duration of sunshine	in h	161	166	202	239	299	319	379	351	259	205	174	153	2907	30	12
13 Mean quantity of radiation	in ly/day	194	253	361	476	579	630	630	566	452	313	226	176	405	3	13
14 Mean potential evaporation	in mm	19	22	36	54	88	129	168	161	118	76	41	22	934		14
15 Mean windspeed	in m/sec	2,6	2,6	2,1	2,6	2,6	2,6	2,6	2,6	2,1	2,1	2,1	2,1	2,6	10	15
16 Mean predominent direction of the wind		W	W	var	var	var	W.NE	W.NE	W.NE	W	W	W	W		10	16

229

13 Station/Country **Gafsa/Tunisia**

Location 34°25'N/8°49'E   Height above sealevel 313 m   Climate symbol: Köppen **BSh**   Troll **IV,2**

		J	F	M	A	M	J	J	A	S	O	N	D	year	P	
1 Mean daily temperature	in °C	9,1	10,8	13,9	17,8	22,2	26,7	29,7	29,4	25,8	20,8	14,7	9,7	19,2	50	1
2 Mean daily maximum temperature	in °C	14	17	20	24	28	34	37	36	32	26	20	15	25	30	2
3 Mean daily minimum temperature	in °C	4	5	8	11	15	19	21	22	19	15	9	5	13	30	3
4 Absolute maximum temperature	in °C	25	29	37	35	40	45	44	46	42	39	30	25	46	30	4
5 Absolute minimum temperature	in °C	-8	-4	1	3	8	11	15	16	11	5	-1	-4	-8	30	5
6 Mean relative humidity	in %	68	62	56	54	51	44	41	48	54	65	68	72	57	10	6
7 Mean precipitation	in mm	15	14	20	18	11	6	2	6	12	21	20	15	160	30	7
8 Maximum precipitation	in mm	76	55	94	82	47	37	27	50	36	126	78	54	289	30	8
9 Minimum precipitation	in mm	0	0	0	0	0	0	0	0	0	0	0	0	71	30	9
10 Maximum precipitation in 24 h	in mm	50	21	38	48	28	33	20	38	36	67	34	30	67	30	10
11 Mean number of days with precipitation	> 0,1 mm	3	3	4	3	2	1	1	1	2	3	3	3	29	30	11
12 Mean duration of sunshine	in h	211	224	249	258	315	307	366	359	286	245	219	213	3252	10	12
13 Mean quantity of radiation	in ly/day															13
14 Mean potential evaporation	in mm	10	16	35	66	113	165	201	185	134	77	33	16	1051		14
15 Mean windspeed	in m/sec	2,6	2,6	2,6	3,1	3,1	2,6	2,6	2,6	2,6	2,1	2,1	2,1	2,6	10	15
16 Mean predominent direction of the wind																16

14 Station/Country **Taräbulus(Tripoli)/Libya**

Location 32°54'N/13°11'E   Height above sealevel 20 m   Climate symbol: Köppen **BSh**   Troll **IV,2**

		J	F	M	A	M	J	J	A	S	O	N	D	year	P	
1 Mean daily temperature	in °C	12,2	13,3	15,3	18,0	20,3	23,3	25,6	26,1	25,6	22,5	18,3	13,6	19,5	60	1
2 Mean daily maximum temperature	in °C	17	18	20	23	25	29	30	31	30	28	23	18	24	34	2
3 Mean daily minimum temperature	in °C	8	9	10	13	16	19	21	22	21	18	13	9	15	34	3
4 Absolute maximum temperature	in °C	28	35	40	44	46	47	48	48	47	41	37	31	48	34	4
5 Absolute minimum temperature	in °C	-1	-1	1	3	5	7	13	14	12	7	3	0	-1	34	5
6 Mean relative humidity	in %	64	66	61	60	60	64	63	66	67	62	60	60	63	5	6
7 Mean precipitation	in mm	62	38	19	14	3	1	<1	1	10	32	65	86	288	34	7
8 Maximum precipitation	in mm	205	182	98	81	34	17	9	28	109	196	293	378	758	85	8
9 Minimum precipitation	in mm	1	0	0	0	0	0	0	0	0	0	0	0	114	85	9
10 Maximum precipitation in 24 h	in mm	53	125	52	50	20	11	4	28	78	81	130	71	130	34	10
11 Mean number of days with precipitation	> 0,1 mm	11	7	4	1	1	<1	0	<1	1	4	8	10	45	34	11
12 Mean duration of sunshine	in h	171	190	228	255	307	297	357	338	258	226	186	164	2956	12	12
13 Mean quantity of radiation	in ly/day															13
14 Mean potential evaporation	in mm	24	28	33	66	93	125	151	152	132	93	55	29	981	47	14
15 Mean windspeed	in m/sec	6,0	7,0	6,0	7,0	7,0	6,0	7,0	6,0	7,0	6,0	6,0	6,0	6,0	8	15
16 Mean predominent direction of the wind																16

15 Station/Country **Surt/Libya**

Location 31°12'N/16°35'E   Height above sealevel 20 m   Climate symbol: Köppen **BSh**   Troll **IV,2**

		J	F	M	A	M	J	J	A	S	O	N	D	year	P	
1 Mean daily temperature	in °C	12,6	13,8	16,1	18,6	20,8	23,8	25,6	26,0	25,2	23,1	17,8	14,2	19,8	12	1
2 Mean daily maximum temperature	in °C	18	19	22	23	26	28	29	31	31	28	24	19	25	7	2
3 Mean daily minimum temperature	in °C	8	9	11	14	17	19	22	23	21	18	14	10	15	7	3
4 Absolute maximum temperature	in °C	26	37	38	41	42	46	46	45	43	40	36	31	46	7	4
5 Absolute minimum temperature	in °C	3	4	4	7	11	13	16	18	16	12	6	4	3	7	5
6 Mean relative humidity	in %	68	70	65	68	71	76	78	77	75	69	67	65	71	7	6
7 Mean precipitation	in mm	40	23	16	2	3	<1	0	0	15	16	33	38	187	12	7
8 Maximum precipitation	in mm	95	65	79	24	20	3	0	0	41	67	122	101	349	17	8
9 Minimum precipitation	in mm	2	0	0	0	0	0	0	0	0	0	0	0	66	17	9
10 Maximum precipitation in 24 h	in mm	11	19	24	4	5	2	0	0	9	67	43	37	67	12	10
11 Mean number of days with precipitation	> 0,1 mm	4	4	4	2	1	<1	0	0	1	2	4	6	28	12	11
12 Mean duration of sunshine	in h	198	227	236	225	277	327	357	347	248	202	204	167	2993	2	12
13 Mean quantity of radiation	in ly/day															13
14 Mean potential evaporation	in mm	24	28	47	71	99	129	150	157	131	103	60	32	1031	14	14
15 Mean windspeed	in m/sec	4,4	4,1	3,9	3,9	3,6	3,3	3,3	3,1	3,1	3,3	3,6	3,9	3,6	6	15
16 Mean predominent direction of the wind																16

16 Station/Country **Banghazi/Libya**

Location 32°06'N/20°04'E   Height above sealevel 25 m   Climate symbol: Köppen **BSh**   Troll **IV,1**

		J	F	M	A	M	J	J	A	S	O	N	D	year	P	
1 Mean daily temperature	in °C	12,0	12,8	14,5	18,5	22,7	25,0	26,0	26,4	25,0	22,4	18,0	13,6	19,7		1
2 Mean daily maximum temperature	in °C	17	18	21	24	28	29	29	30	27	26	24	20	24	20	2
3 Mean daily minimum temperature	in °C	8	10	11	13	16	19	22	22	18	17	14	11	15	20	3
4 Absolute maximum temperature	in °C	26	32	38	38	45	45	42	34	41	38	37	30	45	20	4
5 Absolute minimum temperature	in °C	2	0	1	4	8	10	15	14	10	11	6	4	0	20	5
6 Mean relative humidity	in %															6
7 Mean precipitation	in mm	85	35	22	5	7	<1	0	<1	4	21	32	67	258	60	7
8 Maximum precipitation	in mm	257	141	90	29	46	5	4	4	45	91	200	250	426	60	8
9 Minimum precipitation	in mm	7	0	0	0	0	0	0	0	0	0	1	1	106	60	9
10 Maximum precipitation in 24 h	in mm	34	27	38	23	5	3	4	1	30	40	43	30	43	15	10
11 Mean number of days with precipitation	> 0,1 mm	14	8	7	2	2	<1	0	<1	1	4	7	12	57	30	11
12 Mean duration of sunshine	in h	201	219	245	264	325	336	391	366	291	248	222	171	3279	4	12
13 Mean quantity of radiation	in ly/day															13
14 Mean potential evaporation	in mm	27	30	45	68	101	130	143	146	118	94	60	36	998	29	14
15 Mean windspeed	in m/sec	1,4	1,7	1,7	1,7	1,7	1,4	1,4	1,1	1,1	1,4	1,4	1,4	1,4	8	15
16 Mean predominent direction of the wind																16

17 Station/Country Darnah/Libya

Location 32°49'N/22°38'E   Height above sealevel 7 m   Climate symbol: Köppen BSh   Troll IV,1

		J	F	M	A	M	J	J	A	S	O	N	D	year	P	
1 Mean daily temperature	in °C	13,3	13,3	15,3	17,8	20,0	22,8	24,7	25,3	24,1	22,2	18,6	15,0	19,4	37	1
2 Mean daily maximum temperature	in °C	18	18	18	22	24	27	28	29	28	26	23	19	23	9	2
3 Mean daily minimum temperature	in °C	11	10	13	14	16	19	22	24	22	19	16	12	17	9	3
4 Absolute maximum temperature	in °C	31	33	41	38	44	44	42	41	41	39	38	33	44	20	4
5 Absolute minimum temperature	in °C	4	4	5	7	9	8	10	18	15	10	8	7	4	20	5
6 Mean relative humidity	in %	68	67	69	70	72	71	74	72	72	69	70	71	71	5	6
7 Mean precipitation	in mm	48	34	28	5	10	0	<1	1	9	11	57	55	256	10	7
8 Maximum precipitation	in mm	143	143	107	56	40	3	1	11	61	189	173	184	455	20	8
9 Minimum precipitation	in mm	5	1	1	0	0	0	0	0	0	0	0	0	125	9	9
10 Maximum precipitation in 24 h	in mm	40	23	23	14	25	1	1	11	50	60	62	50	62	10	10
11 Mean number of days with precipitation	> 0,1 mm	8	5	5	1	1	0	0	0	1	4	5	8	38	10	11
12 Mean duration of sunshine	in h	152	188	205	231	282	297	316	298	237	223	189	146	2764	4	12
13 Mean quantity of radiation	in ly/day															13
14 Mean potential evaporation	in mm	28	28	43	63	87	114	140	140	114	90	58	35	938	14	14
15 Mean windspeed	in m/sec	5,9	6,2	5,9	5,1	4,1	4,4	4,9	5,7	4,4	3,9	4,9	5,9	5,1	6	15
16 Mean predominant direction of the wind																16

18 Station/Country Ghudāmis/Libya

Location 30°08'N/9°40'E   Height above sealevel 360 m   Climate symbol: Köppen BWh   Troll IV,5

		J	F	M	A	M	J	J	A	S	O	N	D	year	P	
1 Mean daily temperature	in °C	10,7	12,7	17,0	21,9	26,9	30,9	32,6	31,2	28,3	22,9	17,1	11,2	21,9	5	1
2 Mean daily maximum temperature	in °C	18	21	26	32	37	42	43	42	38	32	24	19	31		2
3 Mean daily minimum temperature	in °C	3	4	8	13	18	22	22	22	19	15	9	4	13		3
4 Absolute maximum temperature	in °C	32	34	41	48	52	55	53	52	50	48	39	31	55		4
5 Absolute minimum temperature	in °C	-7	-3	4	7	14	15	13	10	3	2	-3	-7	-7		5
6 Mean relative humidity	in %	58	50	47	39	31	28	28	33	35	45	48	51	41	5	6
7 Mean precipitation	in mm	5	3	5	2	2	1	0	<1	1	2	5	1	27	18	7
8 Maximum precipitation	in mm	41	35	21	22	10	8	0	4	20	23	35	26	79	8	8
9 Minimum precipitation	in mm	0	0	0	0	0	0	0	0	0	0	0	0	6		9
10 Maximum precipitation in 24 h	in mm	8	4	17	11	10	0	<1	4	5	17	5	4	17		10
11 Mean number of days with precipitation	> 0,1 mm	1	1	1	<1	<1	<1	0	0	<1	1	2	1	7	18	11
12 Mean duration of sunshine	in h															12
13 Mean quantity of radiation	in ly/day															13
14 Mean potential evaporation	in mm	8	16	42	91	167	202	210	197	159	96	36	12	1236	18	14
15 Mean windspeed	in m/sec	4,1	3,8	3,3	4,6	4,5	4,4	4,1	3,7	4,1	3,5	3,2	4,1	4,0	10	15
16 Mean predominant direction of the wind		W	W	W,N	E	E	NE	NE	NE	NE	NE	SW	W		10	16

19 Station/Country Sabahah/Libya

Location 27°01'N/14°28'E   Height above sealevel 444 m   Climate symbol: Köppen BWh   Troll IV,5

		J	F	M	A	M	J	J	A	S	O	N	D	year	P	
1 Mean daily temperature	in °C	11,7	14,4	17,8	23,3	27,8	31,1	30,8	30,6	28,3	25,0	17,8	12,8	22,8		1
2 Mean daily maximum temperature	in °C	17,8	22,2	26,1	31,7	36,1	40,0	38,9	37,8	38,1	32,8	26,7	16,7	30,0	4	2
3 Mean daily minimum temperature	in °C	5,0	7,8	10,6	15,6	19,4	23,3	23,3	23,3	21,1	17,8	12,8	7,2	15,6	3	3
4 Absolute maximum temperature	in °C	28,3	36,1	37,2	42,8	44,4	45,6	45,6	43,9	48,9	40,6	37,8	31,7	48,9	12	4
5 Absolute minimum temperature	in °C	-2,2	-4,4	2,8	6,7	10,6	14,4	17,8	16,1	15,6	11,1	2,2	0,0	-4,4	12	5
6 Mean relative humidity	in %															6
7 Mean precipitation	in mm	1	1	0	1	2	0	0	0	0	0	2	1	8		7
8 Maximum precipitation	in mm															8
9 Minimum precipitation	in mm															9
10 Maximum precipitation in 24 h	in mm															10
11 Mean number of days with precipitation	> 0,1 mm	<1	<1	1	<1	1	<1	0	0	<1	<1	<1	<1	4		11
12 Mean duration of sunshine	in h															12
13 Mean quantity of radiation	in ly/day															13
14 Mean potential evaporation	in mm	11	21	50	100	169	196	197	191	162	112	48	20	1277		14
15 Mean windspeed	in m/sec															15
16 Mean predominant direction of the wind																16

20 Station/Country Al-Kufrah (Cufra)/Libya

Location 24°13'N/23°20'E   Height above sealevel 381 m   Climate symbol: Köppen BWh   Troll IV,5

		J	F	M	A	M	J	J	A	S	O	N	D	year	P	
1 Mean daily temperature	in °C	12,4	14,6	18,8	23,4	27,9	29,7	30,3	30,7	28,1	24,5	19,0	14,2	22,8	10	1
2 Mean daily maximum temperature	in °C	21	23	27	33	37	39	38	38	36	32	27	22	31	22	2
3 Mean daily minimum temperature	in °C	5	7	10	15	20	22	23	23	21	18	11	6	15	22	3
4 Absolute maximum temperature	in °C	32	34	39	42	48	50	43	46	42	42	38	33	50	11	4
5 Absolute minimum temperature	in °C	-3	-2	1	7	10	16	17	18	15	8	4	-1	-3	11	5
6 Mean relative humidity	in %	47	41	33	28	27	24	26	26	31	35	43	48	34	22	6
7 Mean precipitation	in mm	0	<1	0	0	<1	0	0	<1	<1	<1	0	<1	2	22	7
8 Maximum precipitation	in mm	<1	7	1	<1	3	<1	0	11	7	2	1	2	13	22	8
9 Minimum precipitation	in mm	0	0	0	0	0	0	0	0	0	0	0	0	0	22	9
10 Maximum precipitation in 24 h	in mm	0	4	1	0	3	<1	0	11	5	1	1	2	11	22	10
11 Mean number of days with precipitation	> 0,1 mm	0	<1	0	0	0	0	0	0	<1	<1	0	0	1	22	11
12 Mean duration of sunshine	in h	276	255	295	285	347	353	378	375	309	307	297	260	3637	4	12
13 Mean quantity of radiation	in ly/day															13
14 Mean potential evaporation	in mm	16	28	56	111	183	191	195	189	153	114	58	25	1319	7	14
15 Mean windspeed	in m/sec	2,1	2,6	2,6	3,1	3,1	3,1	3,6	3,1	3,1	2,6	2,1	2,1	2,6	5	15
16 Mean predominant direction of the wind		N,E	N	N	N	N	N	N	N	N	NE	E	NE		5	16

21 Station / Country As–Sallūm / Egypt

Location 31°53′N / 25°11′E — Height above sealevel 170 m — Climate symbol: Köppen BWh — Troll IV.2

		J	F	M	A	M	J	J	A	S	O	N	D	year	P	
1 Mean daily temperature	in °C	12,2	13,3	15,3	17,8	20,3	23,0	25,0	24,7	23,6	22,0	18,8	14,4	19,2	10	1
2 Mean daily maximum temperature	in °C	19	20	21	24	26	30	31	30	29	27	25	20	25	20	2
3 Mean daily minimum temperature	in °C	9	10	11	13	17	20	21	22	20	18	15	11	16	20	3
4 Absolute maximum temperature	in °C	27	33	41	42	43	48	44	43	42	40	36	32	48	20	4
5 Absolute minimum temperature	in °C	3	3	4	6	9	14	16	16	14	12	7	5	3	20	5
6 Mean relative humidity	in %	64	63	61	58	62	59	65	70	68	65	66	64	64	9	6
7 Mean precipitation	in mm	12	12	12	1	3	0	0	0	1	5	28	21	95	20	7
8 Maximum precipitation	in mm	67	37	59	10	23	1	tr	0	9	73	227	70	324	35	8
9 Minimum precipitation	in mm	tr	tr	0	0	0	0	0	0	0	0	tr	tr	4		9
10 Maximum precipitation in 24 h	in mm	38	17	30	8	18	1	tr	0	6	58	121	33	121	20	10
11 Mean number of days with precipitation	> 0,1 mm	7	5	4	1	2	0	0	0	<1	3	2	6	30	20	11
12 Mean duration of sunshine	in h	217	210	279	297	307	369	394	378	327	300	219	211	3508	9	12
13 Mean quantity of radiation	in ly / day															13
14 Mean potential evaporation	in mm	25	38	44	65	94	123	145	138	110	92	58	33	955	9	14
15 Mean windspeed	in m / sec	5,0	5,0	5,0	4,0	4,0	4,0	5,0	4,0	4,0	3,0	3,0	5,0	4,0	9	15
16 Mean predominent direction of the wind																16

22 Station / Country Al–Iskandarīyah (Alexandria) / Egypt

Location 31°12′N / 29°51′E — Height above sealevel 7 m — Climate symbol: Köppen BWh — Troll IV.2

		J	F	M	A	M	J	J	A	S	O	N	D	year	P	
1 Mean daily temperature	in °C	14,1	14,7	16,2	18,6	21,3	24,0	25,9	26,5	25,5	23,6	20,2	16,2	20,6	60	1
2 Mean daily maximum temperature	in °C	18	19	21	24	27	28	30	31	29	28	24	20	25	24	2
3 Mean daily minimum temperature	in °C	9	9	11	13	16	20	23	23	21	18	15	11	16	24	3
4 Absolute maximum temperature	in °C	26	36	40	42	45	44	38	40	40	39	36	29	45	24	4
5 Absolute minimum temperature	in °C	2	2	6	7	10	12	17	17	15	12	7	4	2	24	5
6 Mean relative humidity	in %	65	63	62	63	67	70	73	69	65	64	64	65	66	26	6
7 Mean precipitation	in mm	49	31	12	3	2	tr	0	<1	<1	9	29	56	191	24	7
8 Maximum precipitation	in mm	101	91	48	18	10	<1	tr	tr	4	34	100	154	316	34	8
9 Minimum precipitation	in mm	1	6	tr	0	0	0	0	0	0	0	0	5	33	34	9
10 Maximum precipitation in 24 h	in mm	48	28	13	13	9	<1	tr	tr	4	24	32	37	48	24	10
11 Mean number of days with precipitation	> 0,1 mm	11	7	6	2	1	0	0	0	0	3	7	10	47	24	11
12 Mean duration of sunshine	in h	217	218	279	318	338	357	372	369	333	307	246	208	3562	24	12
13 Mean quantity of radiation	in ly / day															13
14 Mean potential evaporation	in mm	30	30	48	88	111	136	160	158	136	109	70	39	1095	45	14
15 Mean windspeed	in m / sec	4,0	4,0	5,0	4,0	4,0	4,0	4,0	4,0	4,0	3,0	3,0	4,0	4,0	9	15
16 Mean predominent direction of the wind		NW	NW	N	N	N	N	N	W	N	N	N	NW		14	16

23 Station / Country Al–Qāhira (Cairo) / Egypt

Location 30°08′N / 31°34′E — Height above sealevel 95 m — Climate symbol: Köppen BWh — Troll IV.2

		J	F	M	A	M	J	J	A	S	O	N	D	year	P	
1 Mean daily temperature	in °C	13,3	14,7	17,5	21,1	25,0	27,5	28,3	28,3	26,1	24,1	20,0	15,0	21,7	41	1
2 Mean daily maximum temperature	in °C	19	21	24	28	33	35	35	35	32	30	26	21	28	25	2
3 Mean daily minimum temperature	in °C	9	9	11	14	18	20	22	22	20	18	14	10	16	25	3
4 Absolute maximum temperature	in °C	30	35	40	42	47	46	42	41	43	40	32		47	25	4
5 Absolute minimum temperature	in °C	2	1	4	7	10	14	18	16	15	11	5	4	1	25	5
6 Mean relative humidity	in %	55	48	45	38	34	38	45	49	50	49	53	56	47	9	6
7 Mean precipitation	in mm	4	5	3	1	1	0	0	0	0	1	1	8	24	25	7
8 Maximum precipitation	in mm	22	21	15	5	17	tr	tr	<1	tr	8	13	54	63	35	8
9 Minimum precipitation	in mm	0	0	0	0	0	0	0	0	0	0	0	0	3	35	9
10 Maximum precipitation in 24 h	in mm	9	11	11	3	10	tr	tr	<1	tr	4	12	44	44	25	10
11 Mean number of days with precipitation	> 0,1 mm	3	2	2	<1	<1	0	0	0	0	<1	1	3	10	25	11
12 Mean duration of sunshine	in h	236	238	281	318	353	384	391	375	333	304	258	236	3717	24	12
13 Mean quantity of radiation	in ly / day															13
14 Mean potential evaporation	in mm	22	26	48	82	142	168	184	173	136	105	61	23	1170	37	14
15 Mean windspeed	in m / sec	4,0	4,0	4,0	4,0	4,0	3,0	3,0	3,0	3,0	3,0	3,0	3,0	3,0	25	15
16 Mean predominent direction of the wind		NE SW	NE SW	N	N	N	N	N	N	NE	NE	NE	NE		14	16

24 Station / Country Al–Quṣayr / Egypt

Location 26°08′N / 34°18′E — Height above sealevel 10 m — Climate symbol: Köppen BWh — Troll V.5

		J	F	M	A	M	J	J	A	S	O	N	D	year	P	
1 Mean daily temperature	in °C	17,8	18,4	20,7	23,4	26,8	28,9	29,8	30,3	28,7	26,7	23,4	19,6	24,5	19	1
2 Mean daily maximum temperature	in °C	22	23	25	27	30	32	33	34	32	30	27	24	28	25	2
3 Mean daily minimum temperature	in °C	14	14	18	19	23	25	26	27	25	23	19	16	21	25	3
4 Absolute maximum temperature	in °C	33	35	38	43	41	48	42	41	38	39	34	31	48	25	4
5 Absolute minimum temperature	in °C	4	6	7	13	16	21	21	21	19	17	11	9	4	25	5
6 Mean relative humidity	in %	53	51	51	52	51	52	54	52	54	55	56	54	53	13	6
7 Mean precipitation	in mm	0	0	<1	<1	0	0	0	0	0	1	2	<1	4	25	7
8 Maximum precipitation	in mm	1	tr	9	2	1	tr	0	0	tr	11	20	4	34	35	8
9 Minimum precipitation	in mm	0	0	0	0	0	0	0	0	0	0	0	0	tr	35	9
10 Maximum precipitation in 24 h	in mm	1	tr	9	2	1	tr	0	0	tr	11	20	4	20	25	10
11 Mean number of days with precipitation	> 0,1 mm	0	0	<1	<1	0	0	0	0	0	<1	<1	0	1	25	11
12 Mean duration of sunshine	in h															12
13 Mean quantity of radiation	in ly / day															13
14 Mean potential evaporation	in mm	37	39	64	100	162	179	190	185	158	138	86	51	1389	19	14
15 Mean windspeed	in m / sec	5,0	5,0	5,0	5,0	5,0	5,0	4,0	4,0	5,0	4,0	5,0	4,0	5,0	5	15
16 Mean predominent direction of the wind		W	W	N	N	N	N	N	W	N	N	NW	W		5	16

25 Station/Country Al–Uqsur (Luxor)/Egypt

Location 25°40'N/32°42'E   Height above sealevel 95 m   Climate symbol: Köppen BWh   Troll IV,5

		J	F	M	A	M	J	J	A	S	O	N	D	year	P	
1 Mean daily temperature	in °C	14,4	18,4	20,0	25,0	30,3	31,1	32,2	32,0	30,6	27,5	21,4	16,4	24,8	20	1
2 Mean daily maximum temperature	in °C	23	25	29	35	39	41	41	41	38	35	30	25	34	18	2
3 Mean daily minimum temperature	in °C	6	7	11	16	20	23	24	24	21	18	12	8	16	18	3
4 Absolute maximum temperature	in °C	32	35	42	46	48	49	48	47	45	43	38	35	49	18	4
5 Absolute minimum temperature	in °C	0	-2	2	6	11	17	20	17	18	10	3	0	-2	18	5
6 Mean relative humidity	in %	59	48	39	30	27	30	32	35	42	42	50	57	41	5	6
7 Mean precipitation	in mm	<1	<1	tr	tr	<1	0	0	0	0	<1	<1	<1	1	18	7
8 Maximum precipitation	in mm	2	3	<1	tr	6	0	0	0	1	1	1	1	6	35	8
9 Minimum precipitation	in mm	0	0	0	0	0	0	0	0	0	0	0	0	tr	35	9
10 Maximum precipitation in 24 h	in mm	2	3	<1	tr	6	0	0	0	1	1	1	1	6	18	10
11 Mean number of days with precipitation	> 0,1 mm	0	0	0	0	<1	0	0	0	0	<1	<1	<1	1	25	11
12 Mean duration of sunshine	in h															12
13 Mean quantity of radiation	in ly/day															13
14 Mean potential evaporation	in mm	17	25	57	123	187	194	202	194	166	144	60	25	1394	11	14
15 Mean windspeed	in m/sec	2,0	2,0	2,0	3,0	2,0	2,0	2,0	2,0	2,0	1,0	2,0	2,0	2,0	19	15
16 Mean predominent direction of the wind		N	N	NW	NW	NW	NW	NW	NW	N	N	N	N		14	16

26 Station/Country Mût/Egypt

Location 25°29'N/29°00'E   Height above sealevel 110 m   Climate symbol: Köppen BWh   Troll IV,5

		J	F	M	A	M	J	J	A	S	O	N	D	year	P	
1 Mean daily temperature	in °C	11,9	13,9	18,1	23,2	28,4	30,4	30,8	30,5	27,7	24,6	18,9	13,6	22,7	22	1
2 Mean daily maximum temperature	in °C	22	24	28	33	38	39	39	39	36	33	28	23	32	25	2
3 Mean daily minimum temperature	in °C	4	5	9	14	19	23	23	23	20	17	11	6	14	25	3
4 Absolute maximum temperature	in °C	36	39	43	47	48	50	49	48	44	44	42	35	50	25	4
5 Absolute minimum temperature	in °C	-3	-4	0	2	7	13	16	16	11	8	2	-3	-4	25	5
6 Mean relative humidity	in %	47	44	37	33	30	28	28	30	34	39	42	48	35	15	6
7 Mean precipitation	in mm	0	<1	0	<1	0	0	0	0	0	0	<1	<1	<1	25	7
8 Maximum precipitation	in mm	<1	10	<1	tr	3	tr	0	0	0	1	tr	1	11	35	8
9 Minimum precipitation	in mm	0	0	0	0	0	0	0	0	0	0	0	0	tr	35	9
10 Maximum precipitation in 24 h	in mm	<1	8	<1	tr	3	tr	0	0	0	1	tr	1	8	25	10
11 Mean number of days with precipitation	> 0,1 mm	0	1	0	0	1	0	0	0	0	0	0	1	1	25	11
12 Mean duration of sunshine	in h															12
13 Mean quantity of radiation	in ly/day															13
14 Mean potential evaporation	in mm	16	23	53	106	174	190	194	188	155	124	56	22	1301	26	14
15 Mean windspeed	in m/sec	2,0	2,0	2,0	2,0	2,0	2,0	2,0	2,0	2,0	2,0	2,0	1,0	2,0	5	15
16 Mean predominent direction of the wind																16

27 Station/Country Nouadhibou/Mauritania

Location 20°56'N/17°03'W   Height above sealevel 4 m   Climate symbol: Köppen BWh   Troll IV,5

		J	F	M	A	M	J	J	A	S	O	N	D	year	P	
1 Mean daily temperature	in °C	19,3	19,9	20,3	20,6	21,2	22,8	23,0	24,4	25,5	24,5	22,6	20,3	22,0	20	1
2 Mean daily maximum temperature	in °C	26	26	26	27	28	29	27	29	31	30	28	25	28	13	2
3 Mean daily minimum temperature	in °C	13	13	14	14	16	17	18	19	20	18	16	14	16	13	3
4 Absolute maximum temperature	in °C	31	37	36	43	39	46	39	40	42	40	38	36	46	13	4
5 Absolute minimum temperature	in °C	7	8	8	8	9	10	13	12	14	11	11	8	7	13	5
6 Mean relative humidity	in %	67	67	66	70	71	71	78	78	72	73	70	65	66	10	6
7 Mean precipitation	in mm	2	1	2	1	<1	<1	<1	3	8	5	5	2	27	33	7
8 Maximum precipitation	in mm	33	13	28	18	3	18	1	15	58	83	61	66		33	8
9 Minimum precipitation	in mm	0	0	0	0	0	0	0	0	0	0	0	0		33	9
10 Maximum precipitation in 24 h	in mm	15	10	28	18	2	18	1	10	53	83	10	66	83	33	10
11 Mean number of days with precipitation	> 0,1 mm	<1	<1	<1	0	0	0	<1	1	1	1	<1	1	5	33	11
12 Mean duration of sunshine	in h	254	255	291	306	322	303	279	288	261	257	249	248	3313		12
13 Mean quantity of radiation	in ly/day															13
14 Mean potential evaporation	in mm	57	63	73	78	85	99	118	138	141	123	98	70	1143	6	14
15 Mean windspeed	in m/sec	1,9	2,5	2,5	2,5	2,8	2,8	2,5	2,2	1,9	1,9	1,7	1,7	2,2	5	15
16 Mean predominent direction of the wind		NE	N	N	N	N	N	N	N	N	N	N	NE		5	16

28 Station/Country Atar/Mauritania

Location 20°31'N/13°04'W   Height above sealevel 225 m   Climate symbol: Köppen BWh   Troll V,5

		J	F	M	A	M	J	J	A	S	O	N	D	year	P	
1 Mean daily temperature	in °C	19,6	21,7	24,6	27,6	30,6	34,2	34,6	34,1	33,2	30,2	25,6	20,7	28,1	10	1
2 Mean daily maximum temperature	in °C	27	29	34	36	39	42	42	41	40	37	33	28	37	29	2
3 Mean daily minimum temperature	in °C	12	14	17	20	23	27	27	27	27	23	18	14	21	29	3
4 Absolute maximum temperature	in °C	38	39	45	45	47	48	48	49	49	45	40	40	49	29	4
5 Absolute minimum temperature	in °C	4	5	9	11	11	17	17	14	12	12	7	4	4	29	5
6 Mean relative humidity	in %	39	34	33	29	27	28	39	45	40	36	38	41	36	10	6
7 Mean precipitation	in mm	1	1	<1	<1	1	8	8	33	35	8	10	1	106	33	7
8 Maximum precipitation	in mm	15	18	21	5	21	98	48	86	121	119	56	8			8
9 Minimum precipitation	in mm	0	0	0	0	0	0	<1	0	0	0	0	0		33	9
10 Maximum precipitation in 24 h	in mm	3	3	13	5	5	13	35	28	68	46	15	8	68	33	10
11 Mean number of days with precipitation	> 0,1 mm	<1	<1	<1	0	<1	<1	1	4	3	1	<1	1	10	33	11
12 Mean duration of sunshine	in h	251	260	310	327	322	306	298	285	252	257	255	254	3387	10	12
13 Mean quantity of radiation	in ly/day															13
14 Mean potential evaporation	in mm	39	51	113	151	195	203	212	201	182	161	78	39	1623	6	14
15 Mean windspeed	in m/sec	3,0	2,8	3,1	2,8	3,3	3,9	4,0	4,4	4,5	3,5	3,1	3,1	3,5		15
16 Mean predominent direction of the wind		NE	N	N	N	N	N	W,N	N	N	N	N				16

29 Station / Country Nouakchott / Mauritania

Location 18°07'N/15°36'W Height above sealevel 21 m Climate symbol: Köppen BWh Troll V,5

		J	F	M	A	M	J	J	A	S	O	N	D	year	P	
1 Mean daily temperature	in °C	21,0	22,7	24,8	25,8	27,0	27,8	28,0	28,7	30,0	29,3	26,2	22,0	26,1	20	1
2 Mean daily maximum temperature	in °C	27	30	35	38	42	43	42	40	41	38	33	27	36	20	2
3 Mean daily minimum temperature	in °C	7	9	13	17	21	23	23	23	21	16	11	8	16	20	3
4 Absolute maximum temperature	in °C	37	40	45	46	47	49	47	48	45	43	41	37	49	20	4
5 Absolute minimum temperature	in °C	-3	0	1	8	13	14	15	16	12	7	3	-1	-3	20	5
6 Mean relative humidity	in %	41	41	42	48	55	62	73	74	69	56	45	42	54	10	6
7 Mean precipitation	in mm	<1	0	0	<1	1	1	3	10	5	2	0	0	22	27	7
8 Maximum precipitation	in mm	4	2	0	2	12	15	23	62	33	49	<1	0		27	8
9 Minimum precipitation	in mm	0	0	0	0	0	0	0	0	0	0	0	0		27	9
10 Maximum precipitation in 24 h	in mm	4	0	0	1	8	10	30	33	49	0	0		49	37	10
11 Mean number of days with precipitation	> 0,1 mm	<1	<1	0	0	0	<1	2	4	4	1	<1	<1	12	25	11
12 Mean duration of sunshine	in h	242	246	298	318	310	285	273	264	249	257	258	248	3248	10	12
13 Mean quantity of radiation	in ly/day															13
14 Mean potential evaporation	in mm	60	73	105	126	156	163	168	170	165	156	117	68	1527	20	14
15 Mean windspeed	in m/sec	2,8	2,9	3,6	3,3	3,3	4,5	4,4	3,6	3,4	3,6	3,3	3,6	3,5	10	15
16 Mean predominent direction of the wind		NE	NE	N	NW	NW	NW	NW	WNW	WNW	NW	N	NE		10	16

30 Station / Country Néma / Mauritania

Location 18°36'N / 7°16'W Height above sealevel 265 m Climate symbol: Köppen BWh Troll V,5

		J	F	M	A	M	J	J	A	S	O	N	D	year	P	
1 Mean daily temperature	in °C	23,4	26,4	30,2	33,4	35,5	34,3	32,0	29,9	31,4	32,6	29,4	24,4	30,2		1
2 Mean daily maximum temperature	in °C	30	33	37	40	42	42	38	35	37	39	36	31	37	23	2
3 Mean daily minimum temperature	in °C	17	19	23	26	29	28	26	24	25	26	23	18	24	23	3
4 Absolute maximum temperature	in °C	39	42	44	46	49	47	46	43	43	45	44	40	49	23	4
5 Absolute minimum temperature	in °C	8	11	15	15	20	17	12	18	18	16	15	9	8	23	5
6 Mean relative humidity	in %	18	16	14	14	22	39	58	68	59	30	20	22	32	10	6
7 Mean precipitation	in mm	<1	<1	1	3	10	30	63	111	55	15	<1	<1	288	30	7
8 Maximum precipitation	in mm	18	5	5	23	92	117	185	312	117	58	16	25	507	30	8
9 Minimum precipitation	in mm	0	0	0	0	0	0	15	38	10	0	0	0	184	30	9
10 Maximum precipitation in 24 h	in mm	10	5	3	10	38	81	46	125	33	25	8	18	125	30	10
11 Mean number of days with precipitation	> 0,1 mm	<1	0	<1	<1	1	3	7	7	5	2	<1	<1	25	30	11
12 Mean duration of sunshine	in h	276	266	304	297	288	267	270	264	261	273	273	257	3296	10	12
13 Mean quantity of radiation	in ly/day															13
14 Mean potential evaporation	in mm	81	117	167	189	205	196	197	177	174	177	149	89	1918	30	14
15 Mean windspeed	in m/sec	3,0	2,9	2,5	2,0	2,6	1,9	1,5	1,9	1,2	1,6	2,8	2,8	2,2		15
16 Mean predominent direction of the wind		SE	NE	NE	ENE	ENE	E	SW	SW	SW	SE	NE	NE			16

31 Station / Country Tessalit / Mali

Location 20°12'N/0°59'E Height above sealevel 520 m Climate symbol: Köppen BWh Troll V,5

		J	F	M	A	M	J	J	A	S	O	N	D	year	P	
1 Mean daily temperature	in °C	18,8	21,6	25,2	29,7	33,0	35,2	36,0	33,7	32,7	30,3	24,8	19,8	28,4	6	1
2 Mean daily maximum temperature	in °C	27	30	32	37	40	43	42	40	40	37	33	28	37	10	2
3 Mean daily minimum temperature	in °C	12	14	18	22	25	28	27	26	26	24	19	14	21	10	3
4 Absolute maximum temperature	in °C	34	38	41	44	45	48	46	45	43	42	38	35	48	10	4
5 Absolute minimum temperature	in °C	4	4	8	12	17	21	18	16	18	11	7	3	3	10	5
6 Mean relative humidity	in %	25	20	21	19	18	22	33	45	34	23	23	24	26	6	6
7 Mean precipitation	in mm	<1	<1	<1	<1	1	2	15	51	23	2	1	<1	96	20	7
8 Maximum precipitation	in mm	1	2	3	1	8	27	53	185	61	3	4	1		10	8
9 Minimum precipitation	in mm	0	0	0	0	0	0	0	12	0	0	0	0		10	9
10 Maximum precipitation in 24 h	in mm	1	2	3	1	8	15	25	63	38	3	2	1	63		10
11 Mean number of days with precipitation	> 0,1 mm												0		17	11
12 Mean duration of sunshine	in h	298	266	319	315	316	276	298	291	279	298	276	279	3511	10	12
13 Mean quantity of radiation	in ly/day															13
14 Mean potential evaporation	in mm	29	49	111	167	200	203	209	200	181	165	89	39	1642	6	14
15 Mean windspeed	in m/sec	3,0	3,0	3,0	3,0	3,0	3,0	3,0	2,0	2,0	3,0	3,0	3,0	3,0	5	15
16 Mean predominent direction of the wind		NE	NE	NE	NE	NE	SW	SW	SW	SW	NE	NE	NE		5	16

32 Station / Country Gao / Mali

Location 16°16'N/0°03'W Height above sealevel 270 m Climate symbol: Köppen BWh Troll V,4

		J	F	M	A	M	J	J	A	S	O	N	D	year	P	
1 Mean daily temperature	in °C	22,0	25,0	28,8	32,4	34,6	34,5	32,3	29,8	31,8	31,9	28,4	23,3	29,8	20	1
2 Mean daily maximum temperature	in °C	28	33	38	41	41	39	36	34	37	38	34	31	36	15	2
3 Mean daily minimum temperature	in °C	14	17	22	25	27	28	27	25	26	26	21	17	23	15	3
4 Absolute maximum temperature	in °C	44	43	45	47	47	48	46	42	44	44	42	39	48	15	4
5 Absolute minimum temperature	in °C	7	7	9	15	18	18	19	17	18	15	11	8	7	15	5
6 Mean relative humidity	in %	27	23	21	21	28	42	56	68	62	40	28	25	37	8	6
7 Mean precipitation	in mm	<1	0	<1	<1	8	23	71	127	38	3	<1	<1	270	33	7
8 Maximum precipitation	in mm	15	0	8	25	61	73	159	259	154	23	7	6	431	30	8
9 Minimum precipitation	in mm	0	0	0	0	0	0	10	5	0	0	0	0	134	30	9
10 Maximum precipitation in 24 h	in mm	8	1	8	20	38	33	76	120	35	10	2	2	120	33	10
11 Mean number of days with precipitation	> 0,1 mm	0	0	<1	<1	1	4	8	8	4	1	<1	0	26	33	11
12 Mean duration of sunshine	in h	291	269	298	297	301	261	285	282	282	301	294	288	3449	10	12
13 Mean quantity of radiation	in ly/day															13
14 Mean potential evaporation	in mm	48	116	168	183	202	197	188	177	173	174	141	73	1840	5	14
15 Mean windspeed	in m/sec	2,0	1,7	2,2	1,8	2,1	2,8	2,4	1,6	1,8	1,3	1,7	1,7	1,9	8	15
16 Mean predominent direction of the wind		ENE	N	N	N	N	SW	SW	SW	SW	NE	N	N		8	16

33 Station/Country Mopti/Mali

Location 14°30'N/4°12'W Height above sealevel 280 m Climate symbol: Köppen BSh Troll V,3

		J	F	M	A	M	J	J	A	S	O	N	D	year	P	
1 Mean daily temperature	in °C	22,6	25,2	29,0	31,8	32,8	31,2	28,6	27,3	28,3	28,8	26,8	23,1	27,9	20	1
2 Mean daily maximum temperature	in °C	30	33	37	40	40	37	34	31	32	34	33	30	34	20	2
3 Mean daily minimum temperature	in °C	14	16	20	23	25	25	23	23	24	24	20	16	21	20	3
4 Absolute maximum temperature	in °C	38	42	44	45	46	45	43	37	38	40	41	39	46	20	4
5 Absolute minimum temperature	in °C	7	7	11	14	16	19	17	19	19	18	12	8	7	20	5
6 Mean relative humidity	in %	36	30	26	29	44	60	74	81	79	68	44	39	51	6	6
7 Mean precipitation	in mm	<1	<1	1	5	23	56	147	198	94	18	1	<1	543	34	7
8 Maximum precipitation	in mm	10	0	10	41	109	157	292	441	206	78	5	5	964	30	8
9 Minimum precipitation	in mm	0	0	0	0	0	10	46	74	31	0	0	0	360	30	9
10 Maximum precipitation in 24 h	in mm	5	0	2	10	103	43	90	127	74	23	5	0	127	30	10
11 Mean number of days with precipitation	> 0,1 mm	<1	<1	<1	<1	2	7	11	12	8	2	<1	<1	43	30	11
12 Mean duration of sunshine	in h	251	238	282	258	248	246	233	217	240	273	267	226	2979	14	12
13 Mean quantity of radiation	in ly/day															13
14 Mean potential evaporation	in mm	48	87	145	174	193	176	173	158	152	156	114	65	1641	6	14
15 Mean windspeed	in m/sec	1,9	2,2	2,8	2,5	2,8	2,5	2,8	2,2	1,9	1,9	2,2	2,5	2,2	5	15
16 Mean predominent direction of the wind		NNE	NNE	NE	NE	NE,SW	SW	SW	SW	SW	SW	NE	NE		5	16

34 Station/Country Kayes/Mali

Location 14°26'N/11°26'W Height above sealevel 30 m Climate symbol: Köppen Aw Troll V,3

		J	F	M	A	M	J	J	A	S	O	N	D	year	P	
1 Mean daily temperature	in °C	25,2	27,8	31,5	34,2	35,4	31,9	28,9	27,8	28,5	29,7	28,9	25,6	29,6	36	1
2 Mean daily maximum temperature	in °C	34	37	40	43	43	38	34	32	33	35	36	34	37	21	2
3 Mean daily minimum temperature	in °C	17	19	22	26	28	26	25	22	23	23	21	17	22	21	3
4 Absolute maximum temperature	in °C	41	44	47	48	47	46	42	39	39	41	41	41	48	21	4
5 Absolute minimum temperature	in °C	11	13	16	19	19	17	18	18	18	17	15	10	10	21	5
6 Mean relative humidity	in %	25	19	21	22	38	62	78	83	83	73	46	32	49	6	6
7 Mean precipitation	in mm	<1	<1	1	1	23	96	170	244	160	46	5	<1	748	34	7
8 Maximum precipitation	in mm	18	18	3	31	122	223	343	516	370	122	84	5	1127	34	8
9 Minimum precipitation	in mm	0	0	0	0	0	23	69	56	54	2	0	0	494	34	9
10 Maximum precipitation in 24 h	in mm	18	10	3	31	46	63	65	122	68	46	48	2	122	34	10
11 Mean number of days with precipitation	> 0,1 mm	0	0	0	<1	2	8	10	14	11	4	<1	<1	50	34	11
12 Mean duration of sunshine	in h	257	246	304	294	280	234	208	183	210	245	258	229	2928	10	12
13 Mean quantity of radiation	in ly/day															13
14 Mean potential evaporation	in mm	96	130	174	189	202	188	171	158	156	164	148	93	1869	16	14
15 Mean windspeed	in m/sec	1,9	2,2	2,5	2,2	2,5	2,5	2,2	1,9	1,4	1,4	1,4	1,9	1,9	10	15
16 Mean predominent direction of the wind		NE	E	NE	N	W	W	W	SW	W	W	ENE	ENE			16

35 Station/Country Bougouni/Mali

Location 11°25'N/7°30'W Height above sealevel 370 m Climate symbol: Köppen Aw Troll V,3

		J	F	M	A	M	J	J	A	S	O	N	D	year	P	
1 Mean daily temperature	in °C	25,3	27,0	30,0	30,6	29,7	27,2	25,3	25,3	25,8	26,7	26,7	24,7	27,0	13	1
2 Mean daily maximum temperature	in °C	36	36	38	38	36	33	31	29	31	33	35	34	34	22	2
3 Mean daily minimum temperature	in °C	17	20	23	25	24	22	22	21	21	21	20	17	21	22	3
4 Absolute maximum temperature	in °C	40	43	43	45	42	39	39	41	36	39	39	39	45	22	4
5 Absolute minimum temperature	in °C	11	10	16	20	18	18	16	18	18	17	14	11	10	22	5
6 Mean relative humidity	in %															6
7 Mean precipitation	in mm	<1	<1	5	17	68	140	231	335	210	61	10	<1	1078	34	7
8 Maximum precipitation	in mm	18	23	33	152	229	371	640	660	691	330	187	21	1874	35	8
9 Minimum precipitation	in mm	0	0	0	1	3	71	132	150	93	3	0	0	805	35	9
10 Maximum precipitation in 24 h	in mm	2	21	19	81	93	76	140	199	170	89	28	15	199	18	10
11 Mean number of days with precipitation	> 0,1 mm	0	0	1	3	6	11	14	16	14	8	1	<1	74	34	11
12 Mean duration of sunshine	in h															12
13 Mean quantity of radiation	in ly/day															13
14 Mean potential evaporation	in mm	107	126	167	170	173	149	118	116	122	138	131	95	1612	13	14
15 Mean windspeed	in m/sec	0,8	0,8	1,1	1,4	1,4	0,8	0,8	0,8	0,3	0,8	0,3	0,8	0,8	5	15
16 Mean predominent direction of the wind		S	S	S	S	S	S	S	S	S	S	N,S	N		5	16

36 Station/Country Bilma/Niger

Location 18°39'N/13°23'E Height above sealevel 355 m Climate symbol: Köppen BWh Troll V,5

		J	F	M	A	M	J	J	A	S	O	N	D	year	P	
1 Mean daily temperature	in °C	17,0	18,9	23,9	28,8	32,1	33,2	33,4	32,8	31,4	27,3	23,2	17,6	26,6	10	1
2 Mean daily maximum temperature	in °C	27	30	35	38	42	43	52	40	41	38	43	27	38	20	2
3 Mean daily minimum temperature	in °C	7	9	13	17	21	23	23	23	21	16	11	8	16	20	3
4 Absolute maximum temperature	in °C	37	40	45	46	47	49	47	48	45	43	41	37	49	20	4
5 Absolute minimum temperature	in °C	-3	0	1	8	13	14	15	16	12	7	3	-1	-3	20	5
6 Mean relative humidity	in %	29	23	20	16	22	22	29	44	29	25	29	31	27	5	6
7 Mean precipitation	in mm	<1	0	0	<1	1	1	3	10	5	2	0	0	22	27	7
8 Maximum precipitation	in mm	4	2	0	2	12	15	23	62	33	49	<1	0		27	8
9 Minimum precipitation	in mm	0	0	0	0	0	0	0	0	0	0	0	0	0	27	9
10 Maximum precipitation in 24 h	in mm	4	0	0	1	8	10	10	30	33	49	0	0	49	37	10
11 Mean number of days with precipitation	> 0,1 mm	0	0	0	<1	<1	<1	1	<1	0	0		2	27		11
12 Mean duration of sunshine	in h	290	279	300	298	322	317	335	320	305	320	305	292	3681	11	12
13 Mean quantity of radiation	in ly/day															13
14 Mean potential evaporation	in mm	14	30	90	161	197	195	205	195	174	141	75	23	1500	10	14
15 Mean windspeed	in m/sec	2,2	2,1	1,7	1,4	1,3	1,7	1,3	1,0	1,0	0,6	0,8	0,9	1,3	10	15
16 Mean predominent direction of the wind		NE	NE	NE	E	E	E	E	W	E	E	E	NE		10	16

37 Station/Country Zinder/Niger

Location 13°48'N/8°59'E Height above sealevel 510 m Climate symbol: Köppen BSh Troll V,4

		J	F	M	A	M	J	J	A	S	O	N	D	year	P	
1	Mean daily temperature	in °C	22,2	24,9	29,4	32,8	33,1	31,7	28,6	26,4	28,5	29,8	27,2	23,0	28,2	20	1
2	Mean daily maximum temperature	in °C	31	34	38	41	41	39	35	32	35	38	36	32	36	29	2
3	Mean daily minimum temperature	in °C	14	17	21	25	25	24	22	22	22	22	19	15	21	29	3
4	Absolute maximum temperature	in °C	41	42	47	47	48	45	42	40	42	43	41	40	47	29	4
5	Absolute minimum temperature	in °C	4	5	12	17	19	17	15	17	18	17	11	8	4	29	5
6	Mean relative humidity	in %	24	21	18	20	42	52	68	77	69	41	28	27	40	6	6
7	Mean precipitation	in mm	0	<1	<1	1	23	48	160	218	69	10	<1	0	529	19	7
8	Maximum precipitation	in mm	1	<1	<1	21	158	114	259	418	182	89	10	0	662	19	8
9	Minimum precipitation	in mm	0	0	0	0	0	8	54	81	2	0	0	0	330	19	9
10	Maximum precipitation in 24 h	in mm	1	1	1	2	51	48	89	119	63	31	2	0	119	19	10
11	Mean number of days with precipitation	> 0,1 mm	0	0	0	<1	4	9	12	7	<1	0	0		38	19	11
12	Mean duration of sunshine	in h	270	253	285	260	274	258	241	213	254	291	279	274	3132	8	12
13	Mean quantity of radiation	in ly/day															13
14	Mean potential evaporation	in mm	67	127	174	184	198	184	170	153	181	184	143	78	1803	11	14
15	Mean windspeed	in m/sec	2,5	1,4	1,7	1,4	1,7	1,7	1,7	1,1	1,4	1,4	1,7	1,7	1,7	6	15
16	Mean predominent direction of the wind		E	E,SE	E	E	WNW	WNW	W	W	WNW	E	E	E		6	16

38 Station/Country Niamey/Niger

Location 13°30'N/2°07'E Height above sealevel 220 m Climate symbol: Köppen BSh Troll V,4

		J	F	M	A	M	J	J	A	S	O	N	D	year	P	
1	Mean daily temperature	in °C	23,8	26,8	30,3	34,0	34,0	31,6	28,8	27,0	29,0	30,6	28,2	24,7	29,1	13	1
2	Mean daily maximum temperature	in °C	34	37	41	42	41	38	34	32	34	38	38	34	37	10	2
3	Mean daily minimum temperature	in °C	14	17	21	25	27	25	23	23	23	23	18	15	21	10	3
4	Absolute maximum temperature	in °C	39	43	44	46	46	46	40	38	41	43	43	40	46	10	4
5	Absolute minimum temperature	in °C	8	10	11	17	19	19	18	17	19	18	12	9	8	10	5
6	Mean relative humidity	in %	22	20	19	28	48	59	70	80	75	59	35	27	45	7	6
7	Mean precipitation	in mm	0	<1	<1	7	36	87	138	206	88	21	1	0	584	37	7
8	Maximum precipitation	in mm	1	5	53	84	140	170	328	490	231	96	16	0	980	35	8
9	Minimum precipitation	in mm	0	0	0	0	0	14	28	76	18	0	0	0	452	35	9
10	Maximum precipitation in 24 h	in mm	1	2	53	35	46	61	81	173	54	59	10	0	173	35	10
11	Mean number of days with precipitation	> 0,1 mm	0	0	<1	1	4	6	9	13	7	2	0	0	42	25	11
12	Mean duration of sunshine	in h	280	264	264	251	257	251	238	203	228	285	285	276	3087	8	12
13	Mean quantity of radiation	in ly/day															13
14	Mean potential evaporation	in mm	82	128	168	187	198	186	174	180	181	188	145	94	1849	11	14
15	Mean windspeed	in m/sec	1,4	1,4	1,7	1,7	1,9	1,9	1,9	1,4	1,1	1,1	1,1	1,4	1,4	6	15
16	Mean predominent direction of the wind		E	E	E	var	WSW	WSW	SW	SW	SSW	W	E	E		6	16

39 Station/Country Largeau/Chad

Location 18°00'N/19°10'E Height above sealevel 233 m Climate symbol: Köppen BWh Troll V,5

		J	F	M	A	M	J	J	A	S	O	N	D	year	P	
1	Mean daily temperature	in °C	20,9	22,8	26,1	30,5	33,5	34,3	34,0	33,2	33,2	30,3	25,4	21,7	28,8	15	1
2	Mean daily maximum temperature	in °C	25	30	34	39	41	42	41	40	39	37	33	28	36	28	2
3	Mean daily minimum temperature	in °C	14	15	18	21	24	26	25	26	25	23	19	15	21	28	3
4	Absolute maximum temperature	in °C	39	42	44	50	50	50	47	46	46	41	38	50	50	28	4
5	Absolute minimum temperature	in °C	4	5	8	11	18	18	15	15	17	12	8	8	4	28	5
6	Mean relative humidity	in %	37	34	27	26	30	31	40	51	38	29	34	39	35	5	6
7	Mean precipitation	in mm	0	tr	0,1	tr	0,6	1,5	2,7	11	1	0,1	0	0	17	30	7
8	Maximum precipitation	in mm	0	0	0	0	2	48	13	35	23	3	1	0	48	30	8
9	Minimum precipitation	in mm	0	0	0	0	0	0	0	0	0	0	0	0	tr	30	9
10	Maximum precipitation in 24 h	in mm	0	0	0,1	0	1,5	48	13	45	23	2	0	0	48	30	10
11	Mean number of days with precipitation	> 0,1 mm	0	0	<1	0	<1	<1	1	2	<1	<1	0	0	4	30	11
12	Mean duration of sunshine	in h															12
13	Mean quantity of radiation	in ly/day															13
14	Mean potential evaporation	in mm	44	54	122	172	204	200	203	196	182	165	95	45	1682	7	14
15	Mean windspeed	in m/sec	2,0	2,0	1,9	1,5	1,5	1,3	1,1	1,0	1,4	1,6	1,6	1,7	1,8	10	15
16	Mean predominent direction of the wind		NE	NE	NE	NE	NE	NE	NE	NE	NE	ENE	NE	NNE		10	16

40 Station/Country Abéché/Chad

Location 13°51'N/20°51'E Height above sealevel 550 m Climate symbol: Köppen BSh Troll V,4

		J	F	M	A	M	J	J	A	S	O	N	D	year	P	
1	Mean daily temperature	in °C	26,8	28,1	31,2	33,3	33,2	32,2	28,7	27,0	28,4	30,0	28,5	25,9	29,4	14	1
2	Mean daily maximum temperature	in °C	35	36	39	41	40	38	34	31	34	37	36	35	36	14	2
3	Mean daily minimum temperature	in °C	15	18	22	25	25	24	23	21	21	21	17	21	21	14	3
4	Absolute maximum temperature	in °C	45	49	50	49	50	48	45	39	45	46	46	43	50	14	4
5	Absolute minimum temperature	in °C	8	9	11	11	11	15	14	17	15	15	11	10	8	14	5
6	Mean relative humidity	in % +	25	23	21	24	39	47	66	79	66	37	28	29	40	5	6
7	Mean precipitation	in mm	0	0	<1	1	24	26	141	232	67	14	0	0	505	30	7
8	Maximum precipitation	in mm	0	0	2	8	74	69	273	454	171	85	0	0	898	30	8
9	Minimum precipitation	in mm	0	0	0	<1	3	6	12	17	7	1	0	0	48	30	9
10	Maximum precipitation in 24 h	in mm	0	0	2	8	40	31	68	138	62	36	0	0	138	30	10
11	Mean number of days with precipitation	> 0,1 mm	0	0	<1	1	6	12	17	7	1	0	0		48	30	11
12	Mean duration of sunshine	in h	313	292	295	297	307	258	236	201	261	310	308	318	3392	10	12
13	Mean quantity of radiation	in ly/day	141	152	189	206	203	175	152	119	138	164	156	132	1928	9	13
14	Mean potential evaporation	in mm	137	131	174	187	198	191	175	150	157	166	148	129	1943	10	14
15	Mean windspeed	in m/sec	3,3	3,6	3,6	3,1	3,1	2,8	2,5	2,2	1,9	2,8	3,1	3,1	2,8	10	15
16	Mean predominent direction of the wind		NE	NE	NE	E	E	S	S	S	E	ENE	NE	ENE		10	16

+ 9,30 a.m.

236

41 Station/Country Ndjamena/Chad

Location 12°08'N/15°02'E   Height above sealevel 295 m   Climate symbol: Köppen BSh   Troll V.4

		J	F	M	A	M	J	J	A	S	O	N	D	year	P	
1 Mean daily temperature	in °C	23,7	25,9	28,8	32,4	32,4	30,8	28,1	26,5	27,6	29,2	27,1	24,3	28,1	20	1
2 Mean daily maximum temperature	in °C	34	38	39	41	40	38	34	31	33	37	37	34	36	20	2
3 Mean daily minimum temperature	in °C	14	16	20	23	25	24	23	22	22	22	18	15	20	20	3
4 Absolute maximum temperature	in °C	42	43	47	48	48	43	41	38	38	43	42	40	47	20	4
5 Absolute minimum temperature	in °C	8	10	13	14	18	18	17	17	19	14	11	8	8	20	5
6 Mean relative humidity	in %	32	25	21	25	48	57	72	83	77	62	32	33	47	5	6
7 Mean precipitation	in mm	0	0	0	6	35	65	151	254	95	28	1	0	635	21	7
8 Maximum precipitation	in mm	0	<1	<1	53	141	137	281	582	206	95	17	<1		30	8
9 Minimum precipitation	in mm	0	0	0	0	2	3	49	115	17	0	0	0		30	9
10 Maximum precipitation in 24 h	in mm	0	0	0	5	25	41	84	155	48	46	0	0	155	5	10
11 Mean number of days with precipitation	> 0,1 mm	<1	<1	<1	<1	<1	<1	0	<1	<1	<1	<1	<1	6	10	11
12 Mean duration of sunshine	in h	239	224	220	231	233	216	183	180	189	248	249	257	2669	15	12
13 Mean quantity of radiation	in ly/day															13
14 Mean potential evaporation	in mm	81	103	163	182	191	180	166	144	148	160	136	92	1746	11	14
15 Mean windspeed	in m/sec															15
16 Mean predominent direction of the wind																16

42 Station/Country Moundou/Chad

Location 8°37'N/16°04'E   Height above sealevel 420 m   Climate symbol: Köppen Aw   Troll V.3

		J	F	M	A	M	J	J	A	S	O	N	D	year	P	
1 Mean daily temperature	in °C	25,2	27,7	30,7	31,5	29,5	27,6	26,0	25,7	25,7	26,9	26,9	25,0	27,6	8	1
2 Mean daily maximum temperature	in °C	34	37	39	37	35	33	30	30	30	32	35	34	34	13	2
3 Mean daily minimum temperature	in °C	15	17	22	24	23	22	21	21	21	21	19	15	20	13	3
4 Absolute maximum temperature	in °C	43	45	46	46	44	45	43	41	37	38	39	40	46	13	4
5 Absolute minimum temperature	in °C	10	11	15	17	19	19	18	18	17	13	13	9	9	13	5
6 Mean relative humidity	in %	47	44	45	54	65	74	80	82	80	74	60	50	63	9	6
7 Mean precipitation	in mm	0	4	2	40	118	171	244	303	250	96	4	0	1228	30	7
8 Maximum precipitation	in mm	0	9	52	119	295	293	612	490	441	328	35	2	2188	30	8
9 Minimum precipitation	in mm	0	0	0	19	85	101	134	93	18	6	0		868	30	9
10 Maximum precipitation in 24 h	in mm	0	6	26	61	81	107	140	121	79	78	31	2	140	30	10
11 Mean number of days with precipitation	> 0,1 mm	0	4	1	4	9	11	15	17	15	8	1	0	80	30	11
12 Mean duration of sunshine	in h	288	258	250	213	239	198	183	158	177	228	285	298	2773	10	12
13 Mean quantity of radiation	in ly/day															13
14 Mean potential evaporation	in mm	105	130	170	176	172	151	123	127	122	141	135	102	1654	8	14
15 Mean windspeed	in m/sec	1.9	2.5	2.2	2.2	2.2	0.3	1.7	1.4	1.1	1.1	1.1	1.4	1.7	10	15
16 Mean predominent direction of the wind																16

43 Station/Country Bûr Sûdân (Port Sudan)/Sudan

Location 19°35'N/37°13'E   Height above sealevel 5 m   Climate symbol: Köppen BWh   Troll V.5

		J	F	M	A	M	J	J	A	S	O	N	D	year	P	
1 Mean daily temperature	in °C	23,6	23,2	24,3	26,8	29,7	32,4	34,3	34,3	32,7	29,4	27,4	25,3	28,7	18	1
2 Mean daily maximum temperature	in °C	27	27	29	32	35	38	41	41	38	34	31	29	34	30	2
3 Mean daily minimum temperature	in °C	20	19	20	21	24	26	28	29	27	25	24	22	24	30	3
4 Absolute maximum temperature	in °C	32	32	35	40	44	47	47	48	46	39	36	32	48	30	4
5 Absolute minimum temperature	in °C	10	12	13	14	17	20	22	21	22	20	18	14	10	30	5
6 Mean relative humidity	in %	68	67	65	58	51	42	42	44	50	66	68	69	58	25	6
7 Mean precipitation	in mm	4	1	1	1	2	tr	9	3	tr	12	52	25	110	30	7
8 Maximum precipitation	in mm	41	6	24	10	20	6	55	28	2	156	182	145	422	30	8
9 Minimum precipitation	in mm	0	0	0	0	0	0	0	0	0	0	0	0	19	30	9
10 Maximum precipitation in 24 h	in mm	17	5	22	8	19	6	48	21	1	78	112	106	112	30	10
11 Mean number of days with precipitation	> 0,1 mm	1	1	1	1	1	<1	1	<1	1	1	4	3	11	30	11
12 Mean duration of sunshine	in h	214	230	282	312	338	309	307	298	300	307	249	236	3382	20	12
13 Mean quantity of radiation	in ly/day	354	450	539	613	625	577	564	552	547	486	383	340	505	10	13
14 Mean potential evaporation	in mm	74	61	88	141	182	194	207	203	178	160	134	102	1724	30	14
15 Mean windspeed	in m/sec	5.1	5.1	4.6	4.6	4.1	3.6	4.1	3.6	3.6	3.6	4.1	4.6	4.1	9	15
16 Mean predominent direction of the wind																16

44 Station/Country Dunqulah/Sudan

Location 19°10'N/30°29'E   Height above sealevel 225 m   Climate symbol: Köppen BWh   Troll V.5

		J	F	M	A	M	J	J	A	S	O	N	D	year	P	
1 Mean daily temperature	in °C	19,1	21,1	24,1	29,4	31,7	33,9	33,9	33,9	33,3	29,7	25,8	20,0	28,0	42	1
2 Mean daily maximum temperature	in °C	28	29	34	38	42	43	42	42	41	39	33	29	27	18	2
3 Mean daily minimum temperature	in °C	9	10	14	18	22	24	25	26	25	21	15	11	19	18	3
4 Absolute maximum temperature	in °C	37	40	45	47	48	48	47	47	47	45	40	38	48	18	4
5 Absolute minimum temperature	in °C	3	3	4	10	14	17	19	19	15	10	6	4	3	18	5
6 Mean relative humidity	in %															6
7 Mean precipitation	in mm	0	tr	0	0	1	tr	9	13	tr	tr	0	0	23	30	7
8 Maximum precipitation	in mm	0	tr	0	tr	20	tr	50	58	3	3	2	tr	60	23	8
9 Minimum precipitation	in mm	0	0	0	0	0	0	0	0	0	0	0	0	0	23	9
10 Maximum precipitation in 24 h	in mm	6	tr	0	tr	18	tr	32	36	3	3	2	tr	36	30	10
11 Mean number of days with precipitation	> 0,1 mm	0	<1	0	0	<1	0	1	2	<1	<1	<1	0	3	30	11
12 Mean duration of sunshine	in h															12
13 Mean quantity of radiation	in ly/day															13
14 Mean potential evaporation	in mm	29	31	74	156	200	199	205	209	176	163	93	39	1577	6	14
15 Mean windspeed	in m/sec															15
16 Mean predominent direction of the wind		NE	NE	NE	NE	NE	NE	NE	NE	NE	NE	NE	NE	NE	6	16

## 45

Station / Country Al–Khurtum (Khartoum) / Sudan

Location 15°36′N / 32°32′E **Height above sealevel** 380 m **Climate symbol: Köppen** BWh **Troll** V,4

	J	F	M	A	M	J	J	A	S	O	N	D	year	P	
1 Mean daily temperature in °C	22,8	23,9	27,7	31,2	33,6	33,9	31,4	30,4	31,5	31,7	27,8	24,0	29,2	60	1
2 Mean daily maximum temperature in °C	32	34	37	40	42	42	38	36	38	40	36	33	37	30	2
3 Mean daily minimum temperature in °C	16	17	19	23	26	27	26	25	25	25	21	17	22	30	3
4 Absolute maximum temperature in °C	40	43	45	47	47	48	46	43	45	45	42	40	48	30	4
5 Absolute minimum temperature in °C	7	8	12	14	19	20	19	18	17	19	13	6	6	30	5
6 Mean relative humidity in %	28	20	16	14	14	29	45	55	42	29	26	30	30	41	6
7 Mean precipitation in mm	0	tr	tr	1	5	7	48	72	27	4	tr	tr	164	30	7
8 Maximum precipitation in mm	0	<1	<1	21	15	60	159	175	96	23	3	<1	382	30	8
9 Minimum precipitation in mm	0	0	0	0	0	0	2	7	1	0	0	0	76	30	9
10 Maximum precipitation in 24 h in mm	0	tr	tr	21	10	26	70	80	66	19	3	0	80	30	10
11 Mean number of days with precipitation > 0,1 mm	0	0	0	<1	1	1	5	7	3	1	<1	0	18	30	11
12 Mean duration of sunshine in h	329	300	307	312	319	291	267	270	300	319	324	329	3667	20	12
13 Mean quantity of radiation in ly/day															13
14 Mean potential evaporation in mm	70	82	151	177	199	194	192	176	180	176	141	90	1828	30	14
15 Mean windspeed in m/sec	4,4	4,4	4,2	3,9	4,4	3,9	4,4	4,2	3,1	3,6	3,9	4,4	4,2	6	15
16 Mean predominent direction of the wind	N	N	N	N	N	S	S	SSW	S	.	N	N		6	16

## 46

Station / Country Al–Fāshir / Sudan

Location 13°38′N / 25°21′E **Height above sealevel** 730 m **Climate symbol: Köppen** BWh **Troll** V,4

	J	F	M	A	M	J	J	A	S	O	N	D	year	P	
1 Mean daily temperature in °C	20,6	22,2	25,0	28,3	30,0	30,6	28,9	27,2	27,8	27,8	23,9	21,1	26,1	30	1
2 Mean daily maximum temperature in °C	31,1	32,8	36,1	38,9	39,4	38,9	35,6	33,3	36,1	37,2	31,7		35,6	17	2
3 Mean daily minimum temperature in °C	10,0	11,1	14,4	17,8	20,6	21,7	21,1	20,6	20,0	17,8	13,3	10,0	16,7	17	3
4 Absolute maximum temperature in °C	38,3	41,1	42,8	45,0	44,0	43,0	43,0	44,4	41,1	41,1	39,5	38,0	45,0	17	4
5 Absolute minimum temperature in °C	2,7	0,6	4,4	10,0	11,5	14,4	16,1	15,0	12,8	10,5	5,0	2,0	0,6	17	5
6 Mean relative humidity in %															6
7 Mean precipitation in mm	0	0	1	1	10	15	101	135	36	5	0	0	304	30	7
8 Maximum precipitation in mm															8
9 Minimum precipitation in mm															9
10 Maximum precipitation in 24 h in mm	0	0	6	4	25	36	115	68	35	33	0	0	115		10
11 Mean number of days with precipitation > 0,1 mm	0	0	<1	<1	1	3	10	13	5	<1	0	0	34		11
12 Mean duration of sunshine in h	319	305	310	306	318	273	232	236	264	316	327	304	3508		12
13 Mean quantity of radiation in ly/day	480	550	590	600	610	560	540	520	570	530	490	470	543		13
14 Mean potential evaporation in mm	51	63	113	153	178	176	169	151	150	146	84	57	1491		14
15 Mean windspeed in m/sec															15
16 Mean predominent direction of the wind															16

## 47

Station / Country Al–Ubayyid / Sudan

Location 13°11′N / 30°14′E **Height above sealevel** 585 m **Climate symbol: Köppen** BSh **Troll** V,4

	J	F	M	A	M	J	J	A	S	O	N	D	year	P	
1 Mean daily temperature in °C	21,2	22,3	26,5	29,7	30,8	30,3	27,7	26,2	27,1	27,9	25,5	22,3	26,5		1
2 Mean daily maximum temperature in °C	30,6	32,2	35,6	38,9	38,9	37,2	33,9	32,8	34,4	36,7	34,4	31,7	35,0	40	2
3 Mean daily minimum temperature in °C	11,7	12,8	15,6	20,0	22,2	22,8	22,2	21,1	21,1	20,6	16,7	12,8	18,3	40	3
4 Absolute maximum temperature in °C	39,7	41,1	43,0	44,4	44,4	45,6	42,8	40,0	46,1	40,0	41,1	39,4	46,1	40	4
5 Absolute minimum temperature in °C	-0,6	2,2	6,1	7,8	12,2	13,3	13,7	13,3	13,3	12,2	6,7	4,4	-0,6	40	5
6 Mean relative humidity in %															6
7 Mean precipitation in mm	0	0	0	2	23	41	108	127	66	15	0	0	383	30	7
8 Maximum precipitation in mm															8
9 Minimum precipitation in mm															9
10 Maximum precipitation in 24 h in mm	0	0	20	15	39	56	73	87	73	39	2	0	87		10
11 Mean number of days with precipitation > 0,1 mm	0	0	<1	<1	2	4	10	10	7	2	<1	0	36		11
12 Mean duration of sunshine in h	322	297	304	303	313	246	211	211	243	298	321	329	3398		12
13 Mean quantity of radiation in ly/day															13
14 Mean potential evaporation in mm	59	68	140	140	186	178	163	150	151	159	120	72	1586		14
15 Mean windspeed in m/sec															15
16 Mean predominent direction of the wind															16

Station / Country Ar–Rusayris / Sudan

Location 11°51′N / 34°23′E **Height above sealevel** 465 m **Climate symbol: Köppen** Aw **Troll** V,3

	J	F	M	A	M	J	J	A	S	O	N	D	year	P	
1 Mean daily temperature in °C	28,4	27,5	30,0	31,4	31,1	28,6	26,7	26,1	27,0	27,0	27,8	26,4	28,0	46	1
2 Mean daily maximum temperature in °C	37	38	40	41	38	35	32	31	32	36	38	37	36	30	2
3 Mean daily minimum temperature in °C	17	18	20	24	24	22	22	21	21	20	18	16	20	30	3
4 Absolute maximum temperature in °C	42	44	45	46	44	43	38	37	40	40	41	42	46	30	4
5 Absolute minimum temperature in °C	10	9	12	13	17	18	18	18	18	16	11	10	9	30	5
6 Mean relative humidity in % +	41	34	27	31	48	66	79	83	80	70	49	42	54	26	6
7 Mean precipitation in mm	0	tr	1	11	58	126	166	221	152	36	5	0	770	30	7
8 Maximum precipitation in mm															8
9 Minimum precipitation in mm															9
10 Maximum precipitation in 24 h in mm	0	6	4	35	70	69	116	87	112	43	25	1	116	30	10
11 Mean number of days with precipitation > 0,1 mm	0	0	<1	2	7	13	15	17	11	4	<1	0	75	30	11
12 Mean duration of sunshine in h															12
13 Mean quantity of radiation in ly/day															13
14 Mean potential evaporation in mm	134	130	165	179	151	161	146	137	136	151	144	135	1769	26	14
15 Mean windspeed in m/sec															15
16 Mean predominent direction of the wind															16

+ 8.30 a.m.

49 Station/Country Malakāl/Sudan

Location 9°33'N/31°39'E Height above sealevel 385 m Climate symbol: Köppen Aw Troll V,3

		J	F	M	A	M	J	J	A	S	O	N	D	year	P	
1 Mean daily temperature	in °C	25,3	27,2	29,4	29,8	27,9	26,0	24,8	24,6	25,5	26,1	26,3	25,5	26,5	40	1
2 Mean daily maximum temperature	in °C	36	37	39	39	38	33	31	31	32	34	36	35	35	30	2
3 Mean daily minimum temperature	in °C	19	20	22	24	23	22	22	21	22	22	20	18	21	30	3
4 Absolute maximum temperature	in °C	41	43	43	43	42	41	36	36	38	39	40	41	43	30	4
5 Absolute minimum temperature	in °C	12	11	15	18	19	18	18	16	18	17	13	12	11	30	5
6 Mean relative humidity	in %	23	20	22	35	55	66	73	76	73	66	41	27	48		6
7 Mean precipitation	in mm	tr	tr	3	24	95	115	153	167	144	77	6	1	783	30	7
8 Maximum precipitation	in mm	4	3	39	63	270	211	254	312	345	245	43	21	1175	30	8
9 Minimum precipitation	in mm	0	0	0	0	27	31	73	74	15	18	0	0	508	30	9
10 Maximum precipitation in 24 h	in mm	tr	3	19	42	120	102	91	107	176	117	37	21	176	30	10
11 Mean number of days with precipitation	> 0,1 mm	0	4	1	3	8	10	14	16	11	7	4	4	70	30	11
12 Mean duration of sunshine	in h	302	260	279	249	239	165	165	183	180	226	258	313	2824	18	12
13 Mean quantity of radiation	in ly/day	421	521	527	536	510	448	448	486	484	487	487	477	490	7	13
14 Mean potential evaporation	in mm	141	138	170	174	172	155	142	138	143	150	142	139	1804	30	14
15 Mean windspeed	in m/sec	3,9	3,9	2,8	2,8	2,8	2,2	2,2	1,4	1,4	1,4	2,2	3,3	2,5	8	15
16 Mean predoment direction of the wind		NNE	NNE	N	S	S	S	S	S	var	var	N	NNE		10	16

50 Station/Country Wāw/Sudan

Location 7°42'N/28°03'E Height above sealevel 440 m Climate symbol: Köppen Aw Troll V,2

		J	F	M	A	M	J	J	A	S	O	N	D	year	P	
1 Mean daily temperature	in °C	26,7	28,3	30,0	30,0	28,9	27,2	26,1	26,1	26,7	27,8	27,8	27,2	27,8	30	1
2 Mean daily maximum temperature	in °C	36	37	38	37	35	33	32	32	33	34	36	36	35	38	2
3 Mean daily minimum temperature	in °C	18	19	21	22	22	21	21	21	21	21	19	18	20	38	3
4 Absolute maximum temperature	in °C	41	44	43	46	42	41	40	40	39	40	40	40	46	38	4
5 Absolute minimum temperature	in °C	10	11	12	14	14	14	14	15	15	16	12	10	10	38	5
6 Mean relative humidity	in %	35	29	36	45	64	73	78	80	76	71	55	46	57	30	6
7 Mean precipitation	in mm	0	4	20	69	132	170	199	234	197	130	8	0	1145	30	7
8 Maximum precipitation	in mm	6	37	75	156	218	347	331	397	297	256	33	3	1487	30	8
9 Minimum precipitation	in mm	0	0	<1	<1	79	89	70	83	62	30	0	0	894	30	9
10 Maximum precipitation in 24 h	in mm	17	25	63	101	102	86	92	102	86	103	71	12	103	33	10
11 Mean number of days with precipitation	> 0,1 mm	<1	<1	3	5	10	11	13	14	12	9	2	<1	80	30	11
12 Mean duration of sunshine	in h	300	260	268	228	254	237	177	192	210	229	267	304	2927		12
13 Mean quantity of radiation	in ly/day															13
14 Mean potential evaporation	in mm	140	137	166	166	165	147	135	129	137	150	145	141	1758		14
15 Mean windspeed	in m/sec	1,4	0,8	1,4	1,4	1,4	0,8	0,8	0,8	0,8	0,8	0,8	0,8	1,1	10	15
16 Mean predoment direction of the wind		NNE	NNE	var	S	S	SW	SW	SW	var	var	NNE	NNE		10	16

51 Station/Country Yubo (Source Yubo)/Sudan

Location 5°24'N/27°15'E Height above sealevel 509 m Climate symbol: Köppen Aw Troll V,2

		J	F	M	A	M	J	J	A	S	O	N	D	year	P	
1 Mean daily temperature	in °C	25,6	26,7	26,7	25,0	25,0	23,3	23,3	23,3	23,9	24,4	25,0	25,6	25,0	13	1
2 Mean daily maximum temperature	in °C	32,8	33,3	32,8	30,6	30,0	28,3	27,8	27,8	28,9	30,0	31,1	32,2	30,6	13	2
3 Mean daily minimum temperature	in °C	18,3	19,4	20,0	19,4	19,4	18,3	18,3	18,3	18,3	18,9	18,9	18,3	18,9	13	3
4 Absolute maximum temperature	in °C	37,2	38,3	38,3	35,6	35,6	32,8	31,1	32,2	32,8	33,9	35,6	41,1	41,1	16	4
5 Absolute minimum temperature	in °C	12,2	15,0	16,1	13,9	13,9	15,6	12,8	15,6	15,6	16,1	15,0	13,9	12,2	16	5
6 Mean relative humidity	in %	57	65	71	82	86	87	88	89	88	85	78	66	78	13	6
7 Mean precipitation	in mm	5	23	64	101	188	221	170	213	234	170	51	15	1455	13	7
8 Maximum precipitation	in mm															8
9 Minimum precipitation	in mm															9
10 Maximum precipitation in 24 h	in mm	56	51	58	89	61	84	51	122	79	71	51	61	122	13	10
11 Mean number of days with precipitation	> 1,0 mm	1	2	6	9	13	12	11	14	14	13	5	1	101	13	11
12 Mean duration of sunshine	in h															12
13 Mean quantity of radiation	in ly/day															13
14 Mean potential evaporation	in mm															14
15 Mean windspeed	in m/sec															15
16 Mean predoment direction of the wind																16

+ 8.00 a.m.

52 Station/Country Jūbā/Sudan

Location 4°51'N/31°37'E Height above sealevel 460 m Climate symbol: Köppen Aw Troll V,2

		J	F	M	A	M	J	J	A	S	O	N	D	year	P	
1 Mean daily temperature	in °C	28,5	29,5	30,0	29,0	27,5	26,5	25,5	25,5	26,0	27,0	27,5	28,0	26,0		1
2 Mean daily maximum temperature	in °C	37	38	37	35	33	32	31	31	32	34	35	36	31	30	2
3 Mean daily minimum temperature	in °C	20	21	23	23	22	21	20	20	20	20	20	21	21	30	3
4 Absolute maximum temperature	in °C	42	43	43	42	44	38	37	36	38	40	40	41	44	30	4
5 Absolute minimum temperature	in °C	11	12	16	18	17	16	17	16	16	14	13	14	11	30	5
6 Mean relative humidity	in %	43	41	50	63	85	77	82	83	77	71	63	52	66	30	6
7 Mean precipitation	in mm	5	10	43	107	157	116	136	154	105	101	35	13	982	30	7
8 Maximum precipitation	in mm	42	72	211	211	318	244	297	286	216	194	111	62	1317	30	8
9 Minimum precipitation	in mm	0	<1	3	30	38	27	29	46	27	22	<1	0	679	30	9
10 Maximum precipitation in 24 h	in mm	33	50	55	111	110	89	79	92	73	80	55	29	111	30	10
11 Mean number of days with precipitation	> 0,1 mm	1	2	6	11	12	11	11	12	10	7	4	2	93	30	11
12 Mean duration of sunshine	in h	288	230	202	195	215	225	183	214	225	238	237	251	2788		12
13 Mean quantity of radiation	in ly/day	456	464	463	454	479	459	415	466	509	480	457	448	462		13
14 Mean potential evaporation	in mm	151	144	163	163	152	138	117	118	126	143	143	143	1701		14
15 Mean windspeed	in m/sec	1,4	1,4	1,4	1,4	1,4	0,8	0,8	0,8	0,8	0,8	0,8	0,8	1,1	7	15
16 Mean predoment direction of the wind		NNE	NNE	S	S	S	S	S	S	S	S	S	NNE		7	16

53 Station/Country Asmera/Ethiopia

Location 15°17'N/38°55'E Height above sealevel 2300 m Climate symbol: Köppen BSk Troll V,3

		J	F	M	A	M	J	J	A	S	O	N	D	year	P	
1 Mean daily temperature	in °C	15,0	16,1	17,0	18,0	18,6	18,6	16,7	16,7	18,0	17,0	15,6	16,9		8	1
2 Mean daily maximum temperature	in °C	23	24	25	25	26	26	22	22	24	22	22	22	23	25	2
3 Mean daily minimum temperature	in °C	7	9	10	11	12	12	12	12	10	9	9	8	10	25	3
4 Absolute maximum temperature	in °C	31	31	31	31	31	29	29	27	30	26	26	31		25	4
5 Absolute minimum temperature	in °C	0	0	1	4	5	6	6	6	5	3	2	3	0	25	5
6 Mean relative humidity	in %	40	39	43	45	44	45	75	75	50	49	52	50	50	10	6
7 Mean precipitation	in mm	1	1	10	37	38	32	170	127	33	7	10	2	468	29	7
8 Maximum precipitation	in mm	22	31	36	107	208	116	421	363	119	23	76	20	910	34	8
9 Minimum precipitation	in mm	0	0	0	0	0	0	27	61	1	0	0	0	314	34	9
10 Maximum precipitation in 24 h	in mm	12	17	25	57	92	60	70	107	64	22	32	13	107	27	10
11 Mean number of days with precipitation	> 0,1 mm	0	<1	3	5	5	5	17	14	5	2	2	1	60	17	11
12 Mean duration of sunshine	in h															12
13 Mean quantity of radiation	in ly/day															13
14 Mean potential evaporation	in mm	60	62	77	82	82	73	76	70	65	60	56	49	812	3	14
15 Mean windspeed	in m/sec															15
16 Mean predominent direction of the wind		E	E	E	E	E	NE	NW	NW	NE	E	E	E		5	16

54 Station/Country Harar/Ethiopia

Location 9°39'N/36°51'E Height above sealevel 1750 m Climate symbol: Köppen Aw Troll V,3

		J	F	M	A	M	J	J	A	S	O	N	D	year	P	
1 Mean daily temperature	in °C	18,9	19,7	20,6	20,8	20,8	20,0	18,9	18,6	19,4	20,0	19,4	19,4	19,7	11	1
2 Mean daily maximum temperature	in °C	25	26	27	27	27	26	24	23	24	26	26	26	26	10	2
3 Mean daily minimum temperature	in °C	13	14	14	15	15	14	14	14	14	13	13	14	14	10	3
4 Absolute maximum temperature	in °C	28	29	32·	30	31	31	29	28	29	29	28	29	32	10	4
5 Absolute minimum temperature	in °C	7	11	11	13	13	12	11	11	12	10	8	7	7	10	5
6 Mean relative humidity	in %	52	50	56	59	78	73	76	76	74	55	49	52	62		6
7 Mean precipitation	in mm	11	32	60	109	121	101	142	137	98	48	23	10	888	22	7
8 Maximum precipitation	in mm	78	117	191	258	217	223	314	268	186	127	198	94	1557	22	8
9 Minimum precipitation	in mm	0	0	0	3	18	30	54	35	40	0	0	0	520	22	9
10 Maximum precipitation in 24 h	in mm	77	61	54	91	93	69	51	48	71	37	37	25	93	15	10
11 Mean number of days with precipitation	> 0,1 mm	2	4	7	12	12	12	14	15	14	5	2	1	100	22	11
12 Mean duration of sunshine	in h	254	231	202	236	222	198	135	131	192	221	285	317	2619	2	12
13 Mean quantity of radiation	in ly/day															13
14 Mean potential evaporation	in mm	66	66	81	83	87	81	71	68	72	76	70	71	892		14
15 Mean windspeed	in m/sec	2.6	2.1	2.1	2.6	1.5	2.1	2.1	2.1	2.1	1.5	2.1	3.1	2.1	8	15
16 Mean predominent direction of the wind		SE.NW	SE	SE	SE	SE	SE	SE	SE	SE	SE	NW	NW		8	16

55 Station/Country Addis Abeba/Ethiopia

Location 9°02'N/38°45'E Height above sealevel 2450 m Climate symbol: Köppen Cwb Troll V,2

		J	F	M	A	M	J	J	A	S	O	N	D	year	P	
1 Mean daily temperature	in °C	15,9	16,4	17,9	17,6	17,8	16,6	15,0	15,0	15,6	15,8	15,2	15,6	16,2	14	1
2 Mean daily maximum temperature	in °C	23	24	25	25	25	23	20	20	21	22	22	22	23	14	2
3 Mean daily minimum temperature	in °C	6	7	9	10	9	10	11	11	10	7	4	5	8	14	3
4 Absolute maximum temperature	in °C	28	30	29	31	33	34	31	29	27	28	27	27	34	14	4
5 Absolute minimum temperature	in °C	-2	0	1	3	2	6	7	7	6	2	0	0	-2	14	5
6 Mean relative humidity	in %	63	66	66	71	67	74	83	85	79	69	66	69	72		6
7 Mean precipitation	in mm	16	44	70	86	95	138	282	294	192	21	15	6	1256	42	7
8 Maximum precipitation	in mm	104	174	250	318	302	376	475	476	570	143	96	69	1937	62	8
9 Minimum precipitation	in mm	0	0	0	0	3	47	123	187	51	0	0	0	933	62	9
10 Maximum precipitation in 24 h	in mm	36	73	81	61	78	53	72	76	74	73	26	51	81	30	10
11 Mean number of days with precipitation	> 0,1 mm	4	4	5	7	7	11	14	16	13	3	1	2	87	14	11
12 Mean duration of sunshine	in h	270	238	248	213	217	156	88	84	148	267	267	267	2343	10	12
13 Mean quantity of radiation	in ly/day															13
14 Mean potential evaporation	in mm	64	52	74	64	75	59	52	58	53	60	62	59	732		14
15 Mean windspeed	in m/sec	3.1	2.6	3.1	3.6	3.1	2.6	2.1	2.1	2.6	3.6	4.1	3.6	3.1	10	15
16 Mean predominent direction of the wind		SSE	E	E	E	E	S	S	S	S	E	SE	E		10	16

56 Station/Country Jima/Ethiopia

Location 7°39'N/36°51'E Height above sealevel 1750 m Climate symbol: Köppen Cwb Troll V,2

		J	F	M	A	M	J	J	A	S	O	N	D	year	P	
1 Mean daily temperature	in °C	18,7	18,6	20,0	19,4	19,1	19,1	17,5	18,0	19,4	19,1	17,0	16,4	18,4	4	1
2 Mean daily maximum temperature	in °C	29	27	27	26	25	24	22	22	23	25	26	28	25	6	2
3 Mean daily minimum temperature	in °C	7	13	13	14	14	14	14	14	14	12	10	10	13	6	3
4 Absolute maximum temperature	in °C	35	32	31	30	30	28	26	30	27	27	32	30	32	8	4
5 Absolute minimum temperature	in °C	0	1	6	8	8	11	11	10	9	5	2	2	1	6	5
6 Mean relative humidity	in %	74	76	78	79	82	73	84	83	82	79	78	77	79		6
7 Mean precipitation	in mm	95	48	82	180	150	220	231	214	192	87	39	37	1529	15	7
8 Maximum precipitation	in mm		140	203	264	225	306	399	263	287	213	123	109	1927	15	8
9 Minimum precipitation	in mm		3	3	63	88	171	135	138	58	27	1	3	1338	15	9
10 Maximum precipitation in 24 h	in mm		56	51	52	78	52	73	75	75	99	35	37	99	15	10
11 Mean number of days with precipitation	> 0,1 mm		5	7	9	10	12	11	14	12	6	4	3	96	15	11
12 Mean duration of sunshine	in h		172	208	189	198	153	105	124	171	205	186	226	2190	7	12
13 Mean quantity of radiation	in ly/day															13
14 Mean potential evaporation	in mm	66	75	85	83	81	71	67	71	71	69	51	64	854		14
15 Mean windspeed	in m/sec		2.1	2.6	2.1	1.5	2.1	1.5	2.1	1.5	1.0	1.5	1.0	1.5	10	15
16 Mean predominent direction of the wind			E	SE	E	E	SE	SSE	SE	E	E	SSE	E		10	16

240

57 Station/Country Negelli/Ethiopia

Location 5°07'N/39°26'E Height above sealevel 1500 m Climate symbol: Köppen BSh Troll V,3

		J	F	M	A	M	J	J	A	S	O	N	D	year	P	
1 Mean daily temperature	in °C	20,8	21,1	20,6	20,0	19,4	18,3	17,5	17,8	19,4	19,1	19,7	19,7	19,5	4	1
2 Mean daily maximum temperature	in °C	28	28	28	25	24	24	24	24	26	25	25	26	28	9	2
3 Mean daily minimum temperature	in °C	12	13	13	14	12	12	12	12	12	13	12	11	13	10	3
4 Absolute maximum temperature	in °C	31	34	32	31	30	28	28	29	30	29	29	30	34	9	4
5 Absolute minimum temperature	in °C	5	8	7	6	6	8	4	3	8	8	6	7	3	10	5
6 Mean relative humidity	in %	49	48	52	67	73	72	72	66	58	69	58	48	61	3	6
7 Mean precipitation	in mm	8	4	33	172	102	8	6	7	16	119	52	23	550	11	7
8 Maximum precipitation	in mm	69	41	155	241	204	46	33	15	53	237	174	89	740	11	8
9 Minimum precipitation	in mm	0	0	0	100	35	0	0	0	0	32	0	0	291	11	9
10 Maximum precipitation in 24 h	in mm	49	41	46	54	53	13	9	15	44	57	42	21	57	7	10
11 Mean number of days with precipitation	> 0,1 mm	3	2	9	15	20	2	7	4	4	17	7	2	92	3	11
12 Mean duration of sunshine	in h															12
13 Mean quantity of radiation	in ly/day															13
14 Mean potential evaporation	in mm	83	78	83	77	76	65	60	60	73	71	71	73	870	4	14
15 Mean windspeed	in m/sec															15
16 Mean predominent direction of the wind		NE	N	NE	NE	NE	S	S	S	S	E	E	NW		4	16

58 Station/Country Djibouti/Afars and Issas

Location 11°36'N/43°09'E Height above sealevel 7 m Climate symbol: Köppen Aw Troll V,5

		J	F	M	A	M	J	J	A	S	O	N	D	year	P	
1 Mean daily temperature	in °C	25,7	26,3	27,4	28,9	31,1	33,7	35,5	35,0	33,1	29,7	27,8	26,4	30,1	22	1
2 Mean daily maximum temperature	in °C	29	29	31	32	34	38	41	39	36	33	31	28	32	16	2
3 Mean daily minimum temperature	in °C	23	24	25	26	28	30	31	29	29	27	25	23	27	16	3
4 Absolute maximum temperature	in °C	34	34	36	38	44	47	47	47	44	39	38	34	47	11	4
5 Absolute minimum temperature	in °C	19	18	21	21	21	23	22	22	23	21	18	17	17	11	5
6 Mean relative humidity	in %	74	72	74	76	70	53	39	44	55	67	71	73	69		6
7 Mean precipitation	in mm	10	13	25	12	5	1	2	8	10	22	13		129	64	7
8 Maximum precipitation	in mm	89	155	211	181	43	21	79	60	65	79	224	83	300	64	8
9 Minimum precipitation	in mm	0	0	0	0	0	0	0	0	0	0	0	0	0	64	9
10 Maximum precipitation in 24 h	in mm	71	155	211	181	43	21	25	60	64	79	156	83	211	64	10
11 Mean number of days with precipitation	> 0,1 mm	3	2	2	1	1	<1	1	1	1	1	2	2	17	64	11
12 Mean duration of sunshine	in h	254	249	279	279	310	237	236	276	282	310	291	276	3279	10	12
13 Mean quantity of radiation	in ly/day															13
14 Mean potential evaporation	in mm	99	109	149	164	190	198	203	198	186	163	138	113	1910		14
15 Mean windspeed	in m/sec															15
16 Mean predominent direction of the wind																16

59 Station/Country Berbera/Somalia

Location 10°26'N/45°02'E Height above sealevel 8 m Climate symbol: Köppen Aw Troll V,5

		J	F	M	A	M	J	J	A	S	O	N	D	year	P	
1 Mean daily temperature	in °C	24,2	25,1	26,4	28,5	31,6	35,9	36,8	36,1	34,3	28,6	26,2	25,0	30,1	9	1
2 Mean daily maximum temperature	in °C	28	29	31	32	36	42	42	42	39	33	30	29	34	30	2
3 Mean daily minimum temperature	in °C	21	22	23	25	28	31	32	32	29	24	22	22	26	30	3
4 Absolute maximum temperature	in °C	31	32	33	36	43	46	46	46	44	38	33	31	46	40	4
5 Absolute minimum temperature	in °C	16	16	20	21	23	25	27	24	23	20	19	17	16	40	5
6 Mean relative humidity	in %	78	79	79	81	73	49	44	45	51	62	74	76	67	4	6
7 Mean precipitation	in mm	8	2	5	12	8	1	<1	2	<1	2	5	5	49	30	7
8 Maximum precipitation	in mm	66	57	145	89	66	21	19	19	18	31	47	69	178	40	8
9 Minimum precipitation	in mm	0	0	0	0	0	0	0	0	0	0	0	0	2	40	9
10 Maximum precipitation in 24 h	in mm	46	38	132	59	66	21	19	16	18	28	47	33	132	40	10
11 Mean number of days with precipitation	> 0,1 mm	<1	<1	<1	1	1	<1	<1	<1	<1	<1	<1	<1	5	40	11
12 Mean duration of sunshine	in h															12
13 Mean quantity of radiation	in ly/day															13
14 Mean potential evaporation	in mm	84	89	154	164	186	194	199	197	183	165	137	91	1843		14
15 Mean windspeed	in m/sec															15
16 Mean predominent direction of the wind																16

60 Station/Country Hargeysa/Somalia

Location 9°31'N/44°06'E Height above sealevel 1370 m Climate symbol: Köppen Aw Troll V,4

		J	F	M	A	M	J	J	A	S	O	N	D	year	P	
1 Mean daily temperature	in °C	18,0	19,7	22,0	23,0	24,1	24,4	23,3	23,3	23,9	21,7	19,7	18,0	21,8	11	1
2 Mean daily maximum temperature	in °C	24	27	29	29	31	31	30	29	31	28	28	25	28	30	2
3 Mean daily minimum temperature	in °C	12	13	16	16	18	18	17	18	17	15	13	12	15	30	3
4 Absolute maximum temperature	in °C	29	32	33	33	34	34	34	33	33	31	29	28	34	30	4
5 Absolute minimum temperature	in °C	9	9	10	13	14	14	16	14	14	7	4	6	4	30	5
6 Mean relative humidity	in %	63	58	58	56	57	54	53	53	53	52	58	66	57	3	6
7 Mean precipitation	in mm	3	8	25	61	61	58	42	81	58	10	8	1	416	30	7
8 Maximum precipitation	in mm	69	120	229	188	121	189	121	179	135	64	72	17	812	40	8
9 Minimum precipitation	in mm	0	0	0	0	4	18	12	13	17	0	0	0	259	40	9
10 Maximum precipitation in 24 h	in mm	46	43	61	58	49	51	51	61	49	33	38	15	61	30	10
11 Mean number of days with precipitation	> 0,1 mm	<1	<1	2	4	6	8	5	9	8	3	<1	<1	47	30	11
12 Mean duration of sunshine	in h															12
13 Mean quantity of radiation	in ly/day															13
14 Mean potential evaporation	in mm	54	60	90	102	113	113	118	107	106	86	65	53	1061	11	14
15 Mean windspeed	in m/sec															15
16 Mean predominent direction of the wind		N	N	NE	S.NE	SW.NE	SW	SW	SW	SW	NE	N	N		5	16

61 Station/Country Mogadisho/Somalia

Location 2°02'N/45°21'E   Height above sealevel 17 m   Climate symbol: Köppen Aw   Troll V,4

		J	F	M	A	M	J	J	A	S	O	N	D	year	P	
1	Mean daily temperature	in °C	26,4	26,7	27,8	28,9	28,3	26,4	25,8	25,8	26,1	27,2	27,2	27,0	26,9	17	1
2	Mean daily maximum temperature	in °C	30	30	31	32	31	30	29	29	29	30	31	31	30	48	2
3	Mean daily minimum temperature	in °C	23	23	25	26	25	24	23	23	23	24	24	23	24	48	3
4	Absolute maximum temperature	in °C	39	40	37	40	35	34	34	36	36	37	39	37	40	48	4
5	Absolute minimum temperature	in °C	19	19	20	18	18	20	17	18	18	17	18	17	16	48	5
6	Mean relative humidity	in %	85	85	84	83	83	85	85	83	83	84	85	85	84	7	6
7	Mean precipitation	in mm	1	0	9	58	56	82	58	40	23	27	38	9	399	46	7
8	Maximum precipitation	in mm	9	3	258	245	324	349	240	182	239	192	179	78	997	46	8
9	Minimum precipitation	in mm	0	0	0	0	1	0	1	0	0	0	0	0	57	46	9
10	Maximum precipitation in 24 h	in mm	8	3	85	95	150	133	134	82	100	110	74	38	150	46	10
11	Mean number of days with precipitation	> 0,1 mm	0	0	1	5	7	13	13	10	5	4	4	2	64	46	11
12	Mean duration of sunshine	in h	268	251	282	260	273	218	226	253	265	267	261	259	3082	13	12
13	Mean quantity of radiation	in ly/day	606	615	614	581	558	492	503	567	613	615	588	570	580	5	13
14	Mean potential evaporation	in mm	149	136	156	160	153	137	126	126	133	147	144	146	1713		14
15	Mean windspeed	in m/sec	5,0	4,0	4,0	3,0	4,0	4,0	4,0	4,0	3,0	3,0	3,0	4,0	4,0	8	15
16	Mean predominent direction of the wind																16

62 Station/Country Dakar/Senegal

Location 14°44'N/17°30'W   Height above sealevel 23 m   Climate symbol: Köppen BSh   Troll V,4

		J	F	M	A	M	J	J	A	S	O	N	D	year	P	
1	Mean daily temperature	in °C	21,3	21,3	21,5	22,1	23,5	26,7	27,7	27,5	27,8	28,0	26,3	23,2	24,7	16	1
2	Mean daily maximum temperature	in °C	26	27	27	27	29	31	31	31	32	32	30	27	29	20	2
3	Mean daily minimum temperature	in °C	18	17	18	18	26	23	24	24	24	24	23	19	21	20	3
4	Absolute maximum temperature	in °C	39	38	43	38	38	38	37	37	38	38	37	35	43	20	4
5	Absolute minimum temperature	in °C	13	14	15	16	16	16	21	21	21	21	18	12	12	20	5
6	Mean relative humidity	in %	70	75	77	78	79	77	78	80	82	81	75	64	76	10	6
7	Mean precipitation	in mm	<1	<1	<1	<1	1	17	88	254	132	38	2	8	540	32	7
8	Maximum precipitation	in mm	10	21	6	8	16	94	290	476	330	249	30	98	901	54	8
9	Minimum precipitation	in mm	0	0	0	0	0	0	3	54	56	0	0	0	273	54	9
10	Maximum precipitation in 24 h	in mm	2	5	3	2	2	56	150	213	103	96	25	8	213	24	10
11	Mean number of days with precipitation	> 0,1 mm	0	<1	<1	0	0	2	7	13	11	3	1	<1	38	32	11
12	Mean duration of sunshine	in h	219	261	282	295	247	195	216	181	195	209	216	213	2719	9	12
13	Mean quantity of radiation	in ly/day	427	518	590	616	600	554	497	443	447	464	426	424	500	5	13
14	Mean potential evaporation	in mm	58	67	78	88	113	151	164	157	152	151	108	83	1370	18	14
15	Mean windspeed	in m/sec	3,9	4,4	4,7	6,1	4,2	3,3	3,1	2,8	3,1	3,1	3,6	3,9	3,9	10	15
16	Mean predominent direction of the wind		N	N	N	N	N	W	W	WNW	NW	N	N	N		10	16

63 Station/Country Ziguinchor/Senegal

Location 12°35'N/16°16'W   Height above sealevel 10 m   Climate symbol: Köppen Aw   Troll V,3

		J	F	M	A	M	J	J	A	S	O	N	D	year	P	
1	Mean daily temperature	in °C	24,0	25,4	27,2	28,0	28,4	28,2	26,8	26,4	27,0	27,7	27,0	24,4	26,9	26	1
2	Mean daily maximum temperature	in °C	33	35	37	37	36	34	31	30	31	33	33	32	33	30	2
3	Mean daily minimum temperature	in °C	15	17	18	19	21	23	23	23	22	23	21	17	20	30	3
4	Absolute maximum temperature	in °C	39	42	44	45	44	41	37	37	39	38	37	37	45	30	4
5	Absolute minimum temperature	in °C	8	8	12	15	15	18	17	18	18	17	13	10	8	30	5
6	Mean relative humidity	in %	58	56	60	63	70	77	86	88	86	83	76	63	72	10	6
7	Mean precipitation	in mm	<1	1	0	<1	11	143	407	558	338	159	8	<1	1626	30	7
8	Maximum precipitation	in mm	35	13	0	2	42	242	544	887	668	346	42	21	2031	30	8
9	Minimum precipitation	in mm	0	0	0	0	0	23	145	196	188	43	0	0	968	30	9
10	Maximum precipitation in 24 h	in mm	21	12	0	1	31	71	151	216	131	119	35	1	216	14	10
11	Mean number of days with precipitation	> 0,1 mm	0	0	0	0	1	9	20	22	18	9	1	0	80	30	11
12	Mean duration of sunshine	in h	213	228	293	299	273	170	125	93	154	206	234	179	2467	12	12
13	Mean quantity of radiation	in ly/day															13
14	Mean potential evaporation	in mm	83	86	112	130	148	152	148	130	134	142	114	87	1466		14
15	Mean windspeed	in m/sec	2,2	1,9	2,8	2,8	2,5	2,2	2,2	2,5	1,7	1,4	1,7	2,2	2,2	9	15
16	Mean predominent direction of the wind		ENE	var	var	W	WNW	W	W	W	W	N	NE	NE		9	16

64 Station/Country Mamou/Guinea

Location 10°22'N/12°04'W   Height above sealevel 730 m   Climate symbol: Köppen Aw   Troll V,2

		J	F	M	A	M	J	J	A	S	O	N	D	year	P	
1	Mean daily temperature	in °C	23,5	25,4	26,5	26,2	25,1	23,6	22,6	22,1	22,9	23,5	23,6	22,7	24,0	20	1
2	Mean daily maximum temperature	in °C	31	33	33	32	29	27	26	25	26	27	29	30	29	10	2
3	Mean daily minimum temperature	in °C	16	18	19	21	21	19	19	19	19	19	19	16	19	10	3
4	Absolute maximum temperature	in °C	37	38	38	38	37	32	31	33	33	31	34	34	38	14	4
5	Absolute minimum temperature	in °C	7	8	11	12	15	15	16	15	16	15	10	8	7	14	5
6	Mean relative humidity	in %	53	50	63	72	79	86	90	93	88	85	76	59	74	7	6
7	Mean precipitation	in mm	5	8	23	101	180	236	301	439	368	234	58	10	1963	34	7
8	Maximum precipitation	in mm	35	54	99	225	408	567	495	668	647	488	185	74	2579	34	8
9	Minimum precipitation	in mm	0	0	0	10	27	82	185	249	101	102	0	0	1798	34	9
10	Maximum precipitation in 24 h	in mm	35	32	42	93	170	71	125	108	125	59	52	37	170	15	10
11	Mean number of days with precipitation	> 0,1 mm	<1	1	4	9	12	19	22	25	21	18	6	1	138	15	11
12	Mean duration of sunshine	in h															12
13	Mean quantity of radiation	in ly/day															13
14	Mean potential evaporation	in mm	84	94	128	129	119	96	88	82	85	86	82	75	1148		14
15	Mean windspeed	in m/sec	0,6	0,8	1,1	1,1	0,8	0,8	0,8	0,8	0,6	0,6	0,8	0,8	0,8	8	15
16	Mean predominent direction of the wind		E	E,W	W	W	W	W	SW	SW	SW	var	E	ESE			16

65 Station/Country Conakry/Guinea

Location 9°31'N/13°43'W   Height above sealevel 17 m   Climate symbol: Köppen Am   Troll V,3

			J	F	M	A	M	J	J	A	S	O	N	D	year	P	
1	Mean daily temperature	in °C	26,4	27,1	27,5	27,8	27,8	26,6	25,5	25,0	25,8	26,2	26,9	26,7	26,6	13	1
2	Mean daily maximum temperature	in °C	31	31	31	32	31	29	28	27	29	30	31	31	30	13	2
3	Mean daily minimum temperature	in °C	22	23	23	23	24	23	22	22	23	23	24	23	23	13	3
4	Absolute maximum temperature	in °C	34	34	36	35	35	33	32	31	32	33	33	34	36	12	4
5	Absolute minimum temperature	in °C	18	17	20	20	19	18	19	20	18	18	19	19	17	18	5
6	Mean relative humidity	in %	74	73	72	83	78	86	91	92	89	85	84	76	81	10	6
7	Mean precipitation	in mm	1	2	5	17	154	564	1321	1057	713	330	122	10	4296	23	7
8	Maximum precipitation	in mm	15	7	64	108	360	1010	1940	1626	1130	667	301	76	5741	28	8
9	Minimum precipitation	in mm	0	0	0	0	3	272	793	725	257	129	18	0	3320	28	9
10	Maximum precipitation in 24 h	in mm	13	6	56	43	78	162	313	259	300	154	96	56	313	18	10
11	Mean number of days with precipitation	> 0,1 mm	<1	<1	1	3	10	22	29	27	24	18	8	1	143	15	11
12	Mean duration of sunshine	in h	155	190	257	210	162	105	62	43	114	164	171	121	1754	30	12
13	Mean quantity of radiation	in ly/day															13
14	Mean potential evaporation	in mm	126	122	141	147	152	138	122	111	121	132	132	124	1568	2	14
15	Mean windspeed	in m/sec	0,8	1,7	2,2	1,9	1,9	1,9	2,2	2,2	1,9	1,4	1,1	1,1	1,7	5	15
16	Mean predominent direction of the wind		NW	NW	WNW	NW	WNW	SW	SW	SW	SW	SW	SW	NW		5	16

66 Station/Country Freetown/Sierra Leona

Location 8°37'N/13°12'W   Height above sealevel 20 m   Climate symbol: Köppen Am   Troll V,2

			J	F	M	A	M	J	J	A	S	O	N	D	year	P	
1	Mean daily temperature	in °C	27,0	27,5	27,7	27,7	27,3	26,5	25,6	25,2	25,9	26,4	27,0	27,0	26,7	80	1
2	Mean daily maximum temperature	in °C	30	31	31	31	31	29	28	27	28	29	30	30	30	14	2
3	Mean daily minimum temperature	in °C	23	23	24	25	24	23	23	23	23	22	23	23	23	14	3
4	Absolute maximum temperature	in °C	33	36	33	37	33	32	31	30	31	32	34	33	37	14	4
5	Absolute minimum temperature	in °C	15	18	19	21	21	20	19	19	20	19	20	16	15	14	5
6	Mean relative humidity	in %	73	72	72	79	81	85	88	86	89	84	82	78	81	15	6
7	Mean precipitation	in mm	8	6	28	68	214	522	1190	1078	800	333	148	38	4373	30	7
8	Maximum precipitation	in mm	105	106	154	336	723	1219	1299	1553		837	333	187	5245	87	8
9	Minimum precipitation	in mm	0	0	0	0	67	128	429	293	310	135	1	0	2603	87	9
10	Maximum precipitation in 24 h	in mm	69	15	78	104	177	101	187	256	204	318	78	134	318	12	10
11	Mean number of days with precipitation	> 0,1 mm	1	1	2	5	13	20	25	24	23	19	11	5	146	23	11
12	Mean duration of sunshine	in h	251	230	239	210	195	159	87	88	120	192	198	217	2166		12
13	Mean quantity of radiation	in ly/day															13
14	Mean potential evaporation	in mm	138	130	148	151	154	145	128	121	124	136	136	138	1649	55	14
15	Mean windspeed	in m/sec	0,8	1,1	1,4	1,1	0,8	0,8	0,8	1,1	0,8	0,8	0,6	0,6	0,8	4	15
16	Mean predominent direction of the wind																16

67 Station/Country Monrovia/Liberia

Location 6°18'N/10°45'W   Height above sealevel 25 m   Climate symbol: Köppen Am   Troll V,1

			J	F	M	A	M	J	J	A	S	O	N	D	year	P	
1	Mean daily temperature	in °C	26,4	26,1	27,0	26,7	26,1	25,0	24,4	24,7	24,7	25,3	26,1	26,4	25,7	4	1
2	Mean daily maximum temperature	in °C	31	31	32	32	31	30	29	29	29	30	31	31	31	10	2
3	Mean daily minimum temperature	in °C	21	22	22	21	20	21	20	21	21	21	21	21	21	10	3
4	Absolute maximum temperature	in °C	33	33	36	34	33	35	34	31	32	33	33	34	36	10	4
5	Absolute minimum temperature	in °C	19	18	20	18	18	19	18	19	18	18	20	19	18	9	5
6	Mean relative humidity	in %	86	88	87	87	88	89	87	87	90	88	87	87	88	7	6
7	Mean precipitation	in mm	51	71	122	154	442	958	797	354	720	598	237	122	4624	10	7
8	Maximum precipitation	in mm	101	257	304	364	732		1460	712	948	866	397	304		10	8
9	Minimum precipitation	in mm	15	1	27	10	232	480	304	101	546	264	91	28		10	9
10	Maximum precipitation in 24 h	in mm	99	108	182	101	164	308	363	223	271	284	78	76	363	13	10
11	Mean number of days with precipitation	> 0,1 mm	4	3	8	12	22	24	21	17	24	22	16	9	182	8	11
12	Mean duration of sunshine	in h	158	165	198	195	155	105	84	81	96	121	147	155	1760	4	12
13	Mean quantity of radiation	in ly/day															13
14	Mean potential evaporation	in mm	132	132	147	148	151	129	121	122	121	138	129	141	1611	4	14
15	Mean windspeed	in m/sec	1,7	1,7	1,7	1,7	1,9	3,1	3,6	3,6	3,1	2,2	2,2	1,7	2,2	6	15
16	Mean predominent direction of the wind		W	W	W	W	S	S	S	SW	S	S	S	W		6	16

68 Station/Country Bouaké/Ivory Coast

Location 7°42'N/5°00'W   Height above sealevel 365 m   Climate symbol: Köppen Aw   Troll V,2

			J	F	M	A	M	J	J	A	S	O	N	D	year	P	
1	Mean daily temperature	in °C	27,1	28,0	28,4	27,9	27,2	26,1	24,8	24,5	25,5	26,1	26,7	26,7	26,6	33	1
2	Mean daily maximum temperature	in °C	33	34	34	33	33	31	29	29	30	32	33	33	32	12	2
3	Mean daily minimum temperature	in °C	20	21	21	21	22	21	20	20	20	20	21	20	21	12	3
4	Absolute maximum temperature	in °C	37	39	40	38	38	37	37	36	37	37	37	38	40	15	4
5	Absolute minimum temperature	in °C	15	15	15	14	16	15	16	16	16	16	18	14	14	15	5
6	Mean relative humidity	in %	68	68	76	79	82	86	88	88	88	85	83	76	81	9	6
7	Mean precipitation	in mm	13	46	92	140	154	135	99	108	225	140	35	23	1210	32	7
8	Maximum precipitation	in mm	65	142	186	319	306	400	298	249	398	495	135	133	1673	42	8
9	Minimum precipitation	in mm	0	0	5	17	51	48	2	5	53	42	0	0	784	42	9
10	Maximum precipitation in 24 h	in mm	56	66	102	86	125	118	156	74	137	101	50	35	156	15	10
11	Mean number of days with precipitation	> 0,1 mm	1	3	6	9	10	8	8	12	10	3	2		78	17	11
12	Mean duration of sunshine	in h	194	196	202	185	187	116	88	73	113	165	169	167	1855	8	12
13	Mean quantity of radiation	in ly/day															13
14	Mean potential evaporation	in mm	139	135	154	150	152	126	107	98	106	122	127	131	1547	9	14
15	Mean windspeed	in m/sec	1,7	1,7	2,2	2,2	1,9	1,9	2,5	2,2	1,9	1,4	1,9	2,2	1,9	7	15
16	Mean predominent direction of the wind		E	SW	SW	SW	S	S	SW	SW	SSW	SW	SW	SW		7	16

69 Station / Country: Abidjan / Ivory Coast

Location 5°15'N / 3°56'W Height above sealevel 7 m Climate symbol: Köppen Am Troll V.1

		J	F	M	A	M	J	J	A	S	O	N	D	year	P
1 Mean daily temperature	in °C	27,1	28,0	28,2	28,3	27,8	26,4	25,7	25,1	25,7	26,7	27,5	27,0		15
2 Mean daily maximum temperature	in °C	31	32	32	32	31	29	28	28	28	29	31	31	30	13
3 Mean daily minimum temperature	in °C	23	24	24	24	24	23	23	22	23	23	23	23	23	13
4 Absolute maximum temperature	in °C	34	35	36	35	34	34	33	31	32	33	34	35	36	13
5 Absolute minimum temperature	in °C	15	18	19	20	20	20	18	17	18	19	19	17	15	13
6 Mean relative humidity	in %	85	83	82	82	85	88	86	87	88	86	83	84	85	10
7 Mean precipitation	in mm	28	42	120	169	366	608	200	34	55	225	188	111	2144	24
8 Maximum precipitation	in mm	110	177	230	279	685	1038	663	254	185	459	345	175	3131	24
9 Minimum precipitation	in mm	0	0	20	45	76	200	5	5	7	8	81	5	1850	24
10 Maximum precipitation in 24 h	in mm	48	152	71	73	232	222	274	89	53	180	104	61	274	15
11 Mean number of days with precipitation	> 0,1 mm	3	4	7	9	16	19	10	6	8	13	13	7	116	16
12 Mean duration of sunshine	in h	187	196	228	284	184	110	130	116	130	189	214	197	2086	
13 Mean quantity of radiation	in ly / day														
14 Mean potential evaporation	in mm	146	139	158	156	154	133	127	124	126	144	144	146	1697	
15 Mean windspeed	in m / sec	2,2	2,5	2,8	2,8	2,5	2,8	3,3	3,6	3,9	3,6	2,8	2,2	2,8	
16 Mean predominant direction of the wind		SW	SW	SW	SW	SW	SW	SW	SW	SW	SW	SW	SSW		

70 Station / Country: Ouagadougou / Upper Volta

Location 12°22'N / 1°31'W Height above sealevel 300 m Climate symbol: Köppen Aw Troll V.3

		J	F	M	A	M	J	J	A	S	O	N	D	year	P
1 Mean daily temperature	in °C	25,1	27,8	30,0	33,0	32,1	29,6	28,1	27,1	27,9	29,7	28,8	26,2	28,8	16
2 Mean daily maximum temperature	in °C	33	37	40	39	38	36	33	31	32	35	36	35	35	10
3 Mean daily minimum temperature	in °C	16	20	23	28	26	24	23	22	23	23	22	17	22	10
4 Absolute maximum temperature	in °C	45	45	45	47	48	44	41	38	39	41	42	45	48	10
5 Absolute minimum temperature	in °C	9	12	15	15	19	17	18	14	19	18	16	11	9	10
6 Mean relative humidity	in %	29	25	27	42	57	69	76	81	79	64	44	31	52	10
7 Mean precipitation	in mm	0	2	13	16	83	122	203	280	144	33	1	0	897	15
8 Maximum precipitation	in mm	5	25	45	83	241	203	308	414	270	141	10	0	1134	15
9 Minimum precipitation	in mm	0	0	0	0	8	48	125	137	61	0	0	0	408	15
10 Maximum precipitation in 24 h	in mm	5	25	38	31	120	56	99	91	69	42	5	0	120	15
11 Mean number of days with precipitation	> 0,1 mm	<1	1	2	3	6	9	11	14	10	4	1	<1	61	15
12 Mean duration of sunshine	in h	277	252	281	250	264	241	227	177	213	287	282	280	3031	11
13 Mean quantity of radiation	in ly / day														
14 Mean potential evaporation	in mm	75	135	177	184	188	165	161	128	140	156	148	113	1770	6
15 Mean windspeed	in m / sec	1,7	1,7	1,9	2,5	2,5	2,2	2,5	1,7	1,4	1,7	0,8	0,8	1,7	6
16 Mean predominant direction of the wind		ENE	ENE	ENE	SW	SW	SW	SW	SW	SSW	SW	NE	ENE		6

71 Station / Country: Bobo Dioulasso / Upper Volta

Location 11°10'N / 4°15'W Height above sealevel 435 m Climate symbol: Köppen Aw Troll V.3

		J	F	M	A	M	J	J	A	S	O	N	D	year	P
1 Mean daily temperature	in °C	24,9	27,3	29,8	30,6	29,3	27,5	26,2	25,2	25,9	27,4	27,6	25,3	27,3	22
2 Mean daily maximum temperature	in °C	33	36	38	37	36	33	31	29	31	32	34	33	34	11
3 Mean daily minimum temperature	in °C	14	17	20	22	22	21	21	21	21	21	19	16	20	11
4 Absolute maximum temperature	in °C	39	40	46	42	41	37	35	40	37	38	38	39	46	11
5 Absolute minimum temperature	in °C	8	11	12	13	16	13	14	15	11	15	12	9	8	11
6 Mean relative humidity	in %	29	32	39	57	69	77	82	85	82	73	58	37	62	10
7 Mean precipitation	in mm	1	2	17	48	108	130	208	308	208	74	10	1	1113	50
8 Maximum precipitation	in mm	21	33	137	142	264	287	422	610	386	254	111	33	1552	46
9 Minimum precipitation	in mm	0	0	0	0	5	33	23	56	31	5	0	0	672	46
10 Maximum precipitation in 24 h	in mm	21	18	121	69	71	53	162	157	81	49	100	17	162	35
11 Mean number of days with precipitation	> 0,1 mm	0	4	2	4	7	9	11	16	13	7	1	0	71	46
12 Mean duration of sunshine	in h	272	261	267	226	240	229	197	147	188	259	265	272	2823	10
13 Mean quantity of radiation	in ly / day														
14 Mean potential evaporation	in mm	95	131	164	166	187	149	126	114	122	148	134	108	1622	8
15 Mean windspeed	in m / sec	1,1	1,1	2,5	2,2	2,8	2,5	2,2	1,9	1,4	1,1	0,6	0,3	1,7	8
16 Mean predominant direction of the wind		SSE	E	S.E	S	S	S	SSW	S	S	S	E	NNE		8

72 Station / Country: Tamale / Ghana

Location 9°25'N / 0°53'W Height above sealevel 201 m Climate symbol: Köppen Aw Troll V.2

		J	F	M	A	M	J	J	A	S	O	N	D	year	P
1 Mean daily temperature	in °C	28,3	30,0	30,6	30,0	28,3	26,7	26,1	25,6	27,2	27,2	28,3	27,2	27,8	55
2 Mean daily maximum temperature	in °C	35,6	37,2	37,2	36,1	33,3	31,1	29,4	28,9	30,0	32,2	34,4	35,0	33,3	13
3 Mean daily minimum temperature	in °C	20,6	22,8	24,4	24,4	23,9	22,2	22,2	21,7	21,7	21,7	21,7	20,0	22,2	13
4 Absolute maximum temperature	in °C	39,4	40,0	40,6	41,1	38,9	36,1	34,4	33,3	33,3	35,6	37,2	37,8	41,1	13
5 Absolute minimum temperature	in °C	15,0	17,2	18,9	20,0	18,9	18,9	18,3	19,4	18,9	18,9	16,1	15,0	15,0	13
6 Mean relative humidity	in %	28	40	50	66	75	81	83	85	85	80	60	41	65	30
7 Mean precipitation	in mm	3	8	53	84	122	142	142	197	226	94	15	3	1089	55
8 Maximum precipitation	in mm														
9 Minimum precipitation	in mm														
10 Maximum precipitation in 24 h	in mm	23	13	71	79	81	66	79	94	81	69	18	20	94	
11 Mean number of days with precipitation	> 0,1 mm	<1	<1		6	10	12	14	16	19	13	1	<1	93	
12 Mean duration of sunshine	in h	287	235	251	231	248	204	161	140	156	260	282	273	2780	30
13 Mean quantity of radiation	in ly / day														
14 Mean potential evaporation	in mm	145	150	173	169	165	144	132	123	122	143	144	144	1754	
15 Mean windspeed	in m / sec														
16 Mean predominant direction of the wind															

244

## 73

Station / Country Accra / Ghana

Location 5°36'N / 0°12'W   Height above sealevel 65 m   Climate symbol: Köppen Aw   Troll V.2

		J	F	M	A	M	J	J	A	S	O	N	D	year	P	
1 Mean daily temperature	in °C	27.5	28.1	28.2	28.3	27.7	26.3	25.2	25.0	25.8	26.7	27.5	27.9	27.0	20	1
2 Mean daily maximum temperature	in °C	32	32	32	32	31	29	27	27	28	30	31	31	30	30	2
3 Mean daily minimum temperature	in °C	23	24	24	24	23	23	22	21	22	23	21	23	23	30	3
4 Absolute maximum temperature	in °C	34	38	38	34	35	33	32	32	32	32	33	34	38	17	4
5 Absolute minimum temperature	in °C	15	17	20	19	21	20	19	18	20	19	21	17	15	17	5
6 Mean relative humidity	in %	78	79	79	81	82	86	87	87	84	84	82	81	82	9	6
7 Mean precipitation	in mm	16	37	73	82	145	193	49	16	40	80	38	18	787		7
8 Maximum precipitation	in mm	95	158	223	223	346	608	197	108	224	201	139	133	1197		8
9 Minimum precipitation	in mm	0	0	0	0	7	2	0	0	0	0	0	0	275		9
10 Maximum precipitation in 24 h	in mm	89	108	108	137	150	303	103	94	114	139	94	76	303		10
11 Mean number of days with precipitation	> 0.1 mm	1	2	4	6	9	10	4	3	4	6	3	2	54	45	11
12 Mean duration of sunshine	in h	211	204	214	219	211	138	146	155	171	223	240	242	2374	30	12
13 Mean quantity of radiation	in ly / day															13
14 Mean potential evaporation	in mm	139	134	152	149	148	129	109	104	111	131	143	144	1593	5	14
15 Mean windspeed	in m / sec															15
16 Mean predominent direction of the wind																16

## 74

Station / Country Kandi / Benin

Location 11°08'N / 2°56'E   Height above sealevel 290 m   Climate symbol: Köppen Aw   Troll V.3

		J	F	M	A	M	J	J	A	S	O	N	D	year	P	
1 Mean daily temperature	in °C	25.0	27.8	30.6	32.2	30.0	27.8	25.6	25.6	26.1	27.2	26.7	24.4	27.2		1
2 Mean daily maximum temperature	in °C	34.5	36.7	38.8	38.6	36.0	33.0	30.7	29.5	31.1	33.8	35.6	34.6	34.4		2
3 Mean daily minimum temperature	in °C	15.7	19.0	22.9	25.2	24.3	22.8	21.9	21.8	21.6	21.7	18.4	15.8	20.9		3
4 Absolute maximum temperature	in °C	40.9	42.9	44.3	46.0	42.4	39.0	37.3	35.8	35.8	38.2	38.8	39.0	46.0		4
5 Absolute minimum temperature	in °C	9.1	10.2	15.0	17.5	17.5	17.0	18.3	17.5	18.0	16.6	12.1	9.8	9.1		5
6 Mean relative humidity	in %															6
7 Mean precipitation	in mm	1	1	8	30	97	142	191	300	216	53	3	0	1042	34	7
8 Maximum precipitation	in mm															8
9 Minimum precipitation	in mm															9
10 Maximum precipitation in 24 h	in mm	1	12	37	45	82	59	72	90	81	68	17	0	90		10
11 Mean number of days with precipitation	> 0.1 mm	<1	<1	1	3	9	10	13	17	15	5	<1	0	74		11
12 Mean duration of sunshine	in h	282	260	267	246	257	234	195	158	195	276	282	282	2934		12
13 Mean quantity of radiation	in ly / day															13
14 Mean potential evaporation	in mm															14
15 Mean windspeed	in m / sec															15
16 Mean predominent direction of the wind																16

## 75

Station / Country Cotonou / Benin

Location 6°21'N / 2°26'E   Height above sealevel 10 m   Climate symbol: Köppen Aw   Troll V.2

		J	F	M	A	M	J	J	A	S	O	N	D	year	P	
1 Mean daily temperature	in °C	27.5	28.3	28.9	28.9	27.9	26.3	26.1	26.0	26.4	26.7	27.9	27.7	27.4	9	1
2 Mean daily maximum temperature	in °C	29	30	30	30	29	27	27	27	27	28	29	29	29	11	2
3 Mean daily minimum temperature	in °C	24	25	26	25	24	23	23	23	23	24	24	24	24	11	3
4 Absolute maximum temperature	in °C	33	35	35	35	35	32	32	32	32	33	34	35	35	11	4
5 Absolute minimum temperature	in °C	19	21	20	21	20	18	19	20	20	20	20	18	18	11	5
6 Mean relative humidity	in %	82	81	80	81	84	88	86	85	85	87	85	83	84	10	6
7 Mean precipitation	in mm	36	51	104	134	201	338	120	22	82	164	68	19	1339	30	7
8 Maximum precipitation	in mm	59	130	278	304	448	691	540	144	223	317	230	55	1585	33	8
9 Minimum precipitation	in mm	0	0	18	30	60	103	1	0	1	12	0	0	1114	33	9
10 Maximum precipitation in 24 h	in mm	84	86	76	118	152	170	85	56	69	117	112	25	170	15	10
11 Mean number of days with precipitation	> 0.1 mm	2	2	5	8	11	13	6	4	7	10	6	2	78	17	11
12 Mean duration of sunshine	in h	217	210	220	210	211	138	121	155	159	195	237	229	2302	30	12
13 Mean quantity of radiation	in ly / day															13
14 Mean potential evaporation	in mm	147	140	161	159	156	131	130	129	130	142	146	145	1716	9	14
15 Mean windspeed	in m / sec	2.5	3.3	4.4	3.9	3.6	4.2	5.3	4.7	4.4	3.3	2.8	2.5	3.9		15
16 Mean predominent direction of the wind		SW	SW	SW	SW	SW	SW	SW	SW	SW	SW	SW	SW			16

Station / Country Sokoto / Nigeria

Location 13°01'N / 5°15'E   Height above sealevel 345 m   Climate symbol: Köppen BSh   Troll V.4

		J	F	M	A	M	J	J	A	S	O	N	D	year	P	
1 Mean daily temperature	in °C	23.7	26.1	30.2	32.8	32.8	30.1	27.6	26.5	27.5	28.5	27.1	24.4	28.1	10	1
2 Mean daily maximum temperature	in °C	33	36	39	41	39	36	33	31	33	37	36	33	36	41	2
3 Mean daily minimum temperature	in °C	16	17	22	26	26	24	23	22	22	22	19	16	21	41	3
4 Absolute maximum temperature	in °C	42	43	46	47	46	43	39	40	40	43	43	41	47	31	4
5 Absolute minimum temperature	in °C	7	8	12	16	16	16	17	16	16	15	13	9	7	31	5
6 Mean relative humidity	in %	22	17	21	32	49	58	73	79	76	60	33	25	46		6
7 Mean precipitation	in mm	<1	0	2	10	42	93	152	244	132	13	1	0	689	41	7
8 Maximum precipitation	in mm	2	0	25	107	153	180	315	469	322	92	2	0	1030	37	8
9 Minimum precipitation	in mm	0	0	0	0	0	10	61	127	37	0	0	0	400	37	9
10 Maximum precipitation in 24 h	in mm	2	0	21	93	114	86	88	147	81	38	2	0	147	41	10
11 Mean number of days with precipitation	> 0.1 mm	0	0	<1	1	5	9	11	16	10	2	0	0	55	30	11
12 Mean duration of sunshine	in h	279	269	282	255	279	282	229	198	243	307	297	288	3218	5	12
13 Mean quantity of radiation	in ly / day															13
14 Mean potential evaporation	in mm	86	116	165	182	194	178	158	137	142	154	135	94	1741	5	14
15 Mean windspeed	in m / sec															15
16 Mean predominent direction of the wind																16

77 Station/Country Kano/Nigeria

Location 12°03′N/8°32′E — Height above sealevel 470 m — Climate symbol: Köppen Aw — Troll V,3

#	Parameter	unit	J	F	M	A	M	J	J	A	S	O	N	D	year	P
1	Mean daily temperature	in °C	21,2	23,7	27,8	30,9	30,7	28,4	26,2	25,3	26,4	27,0	24,8	21,9	26,2	56
2	Mean daily maximum temperature	in °C	30	33	37	38	37	34	31	29	31	34	33	31	33	23
3	Mean daily minimum temperature	in °C	13	15	19	24	24	23	22	21	21	19	16	13	19	23
4	Absolute maximum temperature	in °C	41	43	44	46	44	41	37	36	38	41	42	43	46	34
5	Absolute minimum temperature	in °C	6	9	10	13	17	17	17	16	17	13	11	7	6	34
6	Mean relative humidity	in %	31	27	26	35	53	69	76	81	77	64	38	34	50	
7	Mean precipitation	in mm	0	1	2	8	71	119	209	311	137	14	1	0	873	30
8	Maximum precipitation	in mm	1	7	35	61	224	267	369	499	276	115	4	0	1234	49
9	Minimum precipitation	in mm	0	0	0	0	7	41	39	135	31	0	0	0	488	49
10	Maximum precipitation in 24 h	in mm	1	7	35	55	162	76	91	112	84	45	4	0	162	44
11	Mean number of days with precipitation	> 0,1 mm	0	0	1	2	8	11	17	21	14	2	<1	0	76	26
12	Mean duration of sunshine	in h	276	255	267	252	273	261	233	186	237	295	294	285	3114	10
13	Mean quantity of radiation	in ly/day														
14	Mean potential evaporation	in mm	63	81	144	171	180	166	142	120	128	141	106	69	1511	48
15	Mean windspeed	in m/sec														
16	Mean predominent direction of the wind		NE	NE	NE	NE,SW	SW	SW	SW	SW	SW	SW	NE	NE		5

78 Station/Country Maiduguri/Nigeria

Location 11°51′N/13°05′E — Height above sealevel 350 m — Climate symbol: Köppen BSh — Troll V,4

#	Parameter	unit	J	F	M	A	M	J	J	A	S	O	N	D	year	P
1	Mean daily temperature	in °C	22,6	24,9	28,8	32,4	32,6	30,6	27,9	26,5	27,5	28,5	26,1	23,2	27,6	48
2	Mean daily maximum temperature	in °C	32	34	38	40	38	36	32	30	33	36	35	32	35	15
3	Mean daily minimum temperature	in °C	12	14	18	22	25	24	23	22	22	20	15	12	19	15
4	Absolute maximum temperature	in °C	39	42	43	44	44	41	39	37	39	41	40	38	44	15
5	Absolute minimum temperature	in °C	6	9	9	13	17	16	18	16	18	13	9	6	6	15
6	Mean relative humidity	in %	36	31	27	31	49	70	74	82	78	61	43	41	51	
7	Mean precipitation	in mm	1	<1	1	4	34	78	180	227	112	23	<1	0	659	30
8	Maximum precipitation	in mm	13	8	10	42	86	198	336	373	233	78	8	0	886	37
9	Minimum precipitation	in mm	0	0	0	0	1	7	53	16	11	0	0	0	394	37
10	Maximum precipitation in 24 h	in mm	13	8	10	25	54	94	104	117	95	68	8	0	117	43
11	Mean number of days with precipitation	> 0,1 mm	0	0	<1	2	7	9	16	19	10	3	0	0	67	30
12	Mean duration of sunshine	in h	291	280	285	264	276	264	214	189	222	291	300	298	3174	5
13	Mean quantity of radiation	in ly/day														
14	Mean potential evaporation	in mm	65	85	144	175	186	174	155	132	141	146	107	72	1582	36
15	Mean windspeed	in m/sec														
16	Mean predominent direction of the wind															

79 Station/Country Jos/Nigeria

Location 9°52′N/8°54′E — Height above sealevel 1260 m — Climate symbol: Köppen Aw — Troll V,3

#	Parameter	unit	J	F	M	A	M	J	J	A	S	O	N	D	year	P
1	Mean daily temperature	in °C	21,1	22,8	25,9	25,3	23,6	22,5	21,1	20,8	22,0	23,0	22,8	21,4	27,7	24
2	Mean daily maximum temperature	in °C	28	29	31	30	28	26	24	24	26	27	28	27	27	10
3	Mean daily minimum temperature	in °C	14	15	18	18	18	17	17	17	17	16	15	14	16	10
4	Absolute maximum temperature	in °C	32	33	34	34	33	31	28	29	29	31	32	31	34	10
5	Absolute minimum temperature	in °C	4	7	12	13	14	14	14	14	12	10	9	7	4	10
6	Mean relative humidity	in %	24	23	37	59	75	80	87	88	82	65	36	28	57	
7	Mean precipitation	in mm	2	4	24	93	205	229	318	274	219	39	5	2	1414	30
8	Maximum precipitation	in mm	34	68	112	196	351	389	569	500	334	139	46	34	1760	40
9	Minimum precipitation	in mm	0	0	0	21	35	105	185	122	80	0	0	0	1080	40
10	Maximum precipitation in 24 h	in mm	25	68	70	63	87	98	140	79	83	69	46	32	140	40
11	Mean number of days with precipitation	> 0,1 mm	<1	1	3	11	16	17	23	22	21	11	1	<1	126	
12	Mean duration of sunshine	in h	307	274	260	220	208	201	152	127	171	242	294	313	2769	10
13	Mean quantity of radiation	in ly/day														
14	Mean potential evaporation	in mm	75	79	109	122	112	90	79	72	78	87	82	73	1058	31
15	Mean windspeed	in m/sec														
16	Mean predominent direction of the wind															

80 Station/Country Makurdi/Nigeria

Location 7°42′N/8°35′E — Height above sealevel 111 m — Climate symbol: Köppen Aw — Troll V,2

#	Parameter	unit	J	F	M	A	M	J	J	A	S	O	N	D	year	P
1	Mean daily temperature	in °C	27,2	28,9	30,0	30,0	27,8	26,7	26,1	26,1	26,1	26,7	27,2	26,1	27,2	11
2	Mean daily maximum temperature	in °C	34,4	35,8	36,1	35,0	32,2	30,6	29,4	29,4	30,0	31,1	33,3	33,9	32,8	11
3	Mean daily minimum temperature	in °C	19,4	21,7	23,9	24,4	22,8	22,2	22,2	22,2	21,7	21,7	21,1	17,8	21,7	11
4	Absolute maximum temperature	in °C	37,8	39,4	40,0	41,1	37,2	33,9	32,2	32,2	33,3	34,4	35,6	37,2	41,1	7
5	Absolute minimum temperature	in °C	12,2	12,2	14,4	19,4	17,8	18,3	19,4	19,4	18,3	19,4	13,3	8,3	8,3	7
6	Mean relative humidity	in %	57	49	62	69	79	83	84	85	86	84	71	68	73	5
7	Mean precipitation	in mm	5	10	20	124	226	193	185	193	279	150	18	<3	1405	7
8	Maximum precipitation	in mm														
9	Minimum precipitation	in mm														
10	Maximum precipitation in 24 h	in mm	38	58	53	69	74	74	119	86	117	79	43	<3	119	7
11	Mean number of days with precipitation	> 0,25 mm	<1	1	3	7	12	13	15	18	17	13	1	0	100	7
12	Mean duration of sunshine	in h	236	226	223	216	223	177	143	118	144	198	228	248	2380	
13	Mean quantity of radiation	in ly/day														
14	Mean potential evaporation	in mm														
15	Mean windspeed	in m/sec														
16	Mean predominent direction of the wind															

246

81 Station / Country Enugu / Nigeria

Location 6°38'N / 7°33'E   **Height above sealevel** 140 m   **Climate symbol: Köppen** Aw   **Troll** V.2

		J	F	M	A	M	J	J	A	S	O	N	D	year	P	
1	Mean daily temperature	in °C	27,6	28,3	28,9	28,8	27,6	26,7	26,4	26,4	26,4	26,8	27,2	26,8	27,3	11	1
2	Mean daily maximum temperature	in °C	32	33	33	33	31	29	28	27	29	31	32	32	31	11	2
3	Mean daily minimum temperature	in °C	22	23	24	23	23	22	22	22	22	22	22	21	22	11	3
4	Absolute maximum temperature	in °C	36	37	37	37	35	33	32	32	32	34	35	34	37	12	4
5	Absolute minimum temperature	in °C	13	18	20	19	19	19	19	19	18	19	16	13	13	12	5
6	Mean relative humidity	in %	63	61	69	77	79	82	81	80	83	80	77	71	76		6
7	Mean precipitation	in mm	18	25	68	154	260	267	192	175	305	252	53	15	1784	43	7
8	Maximum precipitation	in mm	124	108	201	336	494	601	384	584	714	398	155	109	2300	50	8
9	Minimum precipitation	in mm	0	0	0	34	94	138	43	6	104	98	2	0	1266	50	9
10	Maximum precipitation in 24 h	in mm	125	80	104	158	125	137	108	115	313	122	83	68	313	43	10
11	Mean number of days with precipitation	> 0,1 mm	1	2	5	9	13	13	13	14	19	16	4	1	110	43	11
12	Mean duration of sunshine	in h	208	204	192	192	202	159	124	105	117	174	213	229	2119		12
13	Mean quantity of radiation	in ly / day															13
14	Mean potential evaporation	in mm	144	139	153	153	147	128	116	112	115	129	140	136	1612	35	14
15	Mean windspeed	in m / sec															15
16	Mean predominent direction of the wind																16

82 Station / Country Lagos / Nigeria

Location 6°27'N / 3°24'E   **Height above sealevel** 3 m   **Climate symbol: Köppen** Aw   **Troll** V.2

		J	F	M	A	M	J	J	A	S	O	N	D	year	P	
1	Mean daily temperature	in °C	27.0	27.9	28.3	28.0	27.4	26.1	25.3	25.1	25.6	26.2	27.2	27.3	26.8	20	1
2	Mean daily maximum temperature	in °C	31	33	33	32	31	29	27	27	28	29	31	32	30	10	2
3	Mean daily minimum temperature	in °C	22	23	23	23	22	22	22	21	22	22	23	22	22	10	3
4	Absolute maximum temperature	in °C	35	36	36	36	35	32	31	31	31	33	33	34	36	10	4
5	Absolute minimum temperature	in °C	14	18	19	20	20	18	17	16	19	19	20	17	14	10	5
6	Mean relative humidity	in %	82	81	82	84	87	90	89	86	90	89	87	83	86		6
7	Mean precipitation	in mm	40	57	100	115	215	338	150	59	214	222	77	41	1625	60	7
8	Maximum precipitation	in mm	155	180	286	325	549	763	786	580	424	450	183	150	2934	60	8
9	Minimum precipitation	in mm	0	0	5	34	90	138	2	2	10	75	4	0	1039	60	9
10	Maximum precipitation in 24 h	in mm	123	95	105	133	158	254	177	108	158	163	107	109	254	60	10
11	Mean number of days with precipitation	> 0,1 mm	4	4	8	10	18	23	15	10	17	15	8	3	135	60	11
12	Mean duration of sunshine	in h	183	190	198	189	174	120	90	93	93	152	195	205	1882	10	12
13	Mean quantity of radiation	in ly / day															13
14	Mean potential evaporation	in mm	148	142	158	149	147	129	119	118	122	141	147	154	1674	49	14
15	Mean windspeed	in m / sec															15
16	Mean predominent direction of the wind																16

83 Station / Country Santa Isabel (Fernando Póo) / Equatorial Guinea

Location 3°46'N / 8°46'E   **Height above sealevel** 12 m   **Climate symbol: Köppen** Am   **Troll** V.1

		J	F	M	A	M	J	J	A	S	O	N	D	year	P	
1	Mean daily temperature	in °C	25.2	25.8	25.7	25.7	25.4	24.8	24.0	24.0	24.4	24.4	25.0	25.0	25.0	10	1
2	Mean daily maximum temperature	in °C	31	32	31	32	31	29	29	29	30	30	30	31	30	8	2
3	Mean daily minimum temperature	in °C	19	21	21	21	22	21	21	21	21	21	22	21	21	6	3
4	Absolute maximum temperature	in °C	35	35	36	36	39	35	34	36	36	36	33	34	39	6	4
5	Absolute minimum temperature	in °C	16	18	19	19	19	19	17	17	18	18	19	18	16	6	5
6	Mean relative humidity	in %	87	85	85	86	88	90	91	92	92	91	90	89	89		6
7	Mean precipitation	in mm	32	64	107	182	238	281	189	167	243	264	89	42	1898	16	7
8	Maximum precipitation	in mm															8
9	Minimum precipitation	in mm															9
10	Maximum precipitation in 24 h	in mm	35	79	53	76	80	88	85	62	83	103	74	72	103	16	10
11	Mean number of days with precipitation	> 0,1 mm	4	5	10	12	18	19	18	15	21	20	11	4	157	16	11
12	Mean duration of sunshine	in h	149	151	109	120	118	69	47	59	48	88	99	140	1277	8	12
13	Mean quantity of radiation	in ly / day															13
14	Mean potential evaporation	in mm	117	114	127	125	124	108	101	101	103	105	117	119	1361	10	14
15	Mean windspeed	in m / sec	1,1	1,4	1,4	1,4	1,4	1,1	1,4	1,4	1,4	1,4	1,4	1,4	1,4	6	15
16	Mean predominent direction of the wind		SW	SW	SW	SW.S	SW.S	SW	SW.W	SW.W	SW.W	SW	SW	var		6	16

84 Station / Country Garoua / Cameroon

Location 9°20'N / 13°23'E   **Height above sealevel** 249 m   **Climate symbol: Köppen** Aw   **Troll** V.3

		J	F	M	A	M	J	J	A	S	O	N	D	year	P	
1	Mean daily temperature	in °C	26.7	28.3	31.7	32.8	28.3	25.0	25.0	25.0	24.4	26.1	27.2	26.1	27.2	49	1
2	Mean daily maximum temperature	in °C	35.4	37.2	39.6	39.6	36.1	32.8	30.7	30.1	30.1	33.1	36.3	35.2	34.8		2
3	Mean daily minimum temperature	in °C	17.2	19.8	23.7	25.4	24.1	22.6	22.2	22.0	21.8	21.9	19.1	16.8	21.4		3
4	Absolute maximum temperature	in °C	42.0	42.0	43.5	44.0	42.9	39.9	35.8	31.2	35.0	38.4	39.7	39.6	44.0		4
5	Absolute minimum temperature	in °C	13.2	14.2	14.7	19.2	19.0	15.8	18.0	18.2	14.5	16.2	14.0	12.2	12.2		5
6	Mean relative humidity	in %															6
7	Mean precipitation	in mm	0	0	5	43	119	155	183	224	213	71	1	1	1015	49	7
8	Maximum precipitation	in mm															8
9	Minimum precipitation	in mm															9
10	Maximum precipitation in 24 h	in mm	1	6	34	73	91	95	86	150	96	58	27	9	150		10
11	Mean number of days with precipitation	> 0,1 mm	0	0	0	4	11	11	12	14	19	7	1	0	74		11
12	Mean duration of sunshine	in h	281	269	260	236	256	219	194	170	195	266	287	299	2932		12
13	Mean quantity of radiation	in ly / day															13
14	Mean potential evaporation	in mm															14
15	Mean windspeed	in m / sec															15
16	Mean predominent direction of the wind																16

85 Station / Country Batouri / Cameroon

Location 4°35'N/14°24'E · Height above sealevel 650 m · Climate symbol: Köppen Aw · Troll V,1

		J	F	M	A	M	J	J	A	S	O	N	D	year	P	
1 Mean daily temperature	in °C	23,3	24,1	25,0	25,3	24,4	23,3	22,5	22,8	23,3	23,8	23,9	23,0	23,5	15	1
2 Mean daily maximum temperature	in °C	30	31	31	31	30	29	27	27	28	29	29	30	29	27	2
3 Mean daily minimum temperature	in °C	17	18	20	20	19	19	19	19	19	19	18	17	19	27	3
4 Absolute maximum temperature	in °C	34	36	37	35	35	34	32	32	32	32	34	34	37	27	4
5 Absolute minimum temperature	in °C	11	12	12	15	14	15	14	14	15	13	14	12	11	27	5
6 Mean relative humidity	in %	77	73	72	76	80	81	84	83	81	81	79	78	79	10	6
7 Mean precipitation	in mm	31	50	119	153	184	173	112	161	217	274	117	34	1625	25	7
8 Maximum precipitation	in mm	94	154	260	272	348	373	325	358	375	409	200	193	1913	25	8
9 Minimum precipitation	in mm	0	1	28	54	72	56	26	56	79	166	20	0	1104	25	9
10 Maximum precipitation in 24 h	in mm	93	59	65	120	81	104	94	169	101	102	94	62	169	25	10
11 Mean number of days with precipitation	>0,1 mm	3	5	9	12	14	14	10	12	18	22	11	3	133	25	11
12 Mean duration of sunshine	in h	161	165	156	174	192	139	105	94	112	138	178	183	1797	22	12
13 Mean quantity of radiation	in ly/day															13
14 Mean potential evaporation	in mm	95	92	117	119	111	98	89	91	94	99	98	92	1193	15	14
15 Mean windspeed	in m/sec	0,8	0,8	0,8	1,1	0,8	0,6	0,6	0,8	1,1	0,6	1,1	0,8	0,8	10	15
16 Mean predominant direction of the wind		W	W	W	W	var	W	W	W	W	W	var	W		10	16

86 Station / Country Douala / Cameroon

Location 4°01'N/9°43'E · Height above sealevel 11 m · Climate symbol: Köppen Am · Troll V,1

		J	F	M	A	M	J	J	A	S	O	N	D	year	P	
1 Mean daily temperature	in °C	26,7	27,0	26,8	26,6	26,3	25,4	24,3	24,1	24,7	25,0	26,0	26,4	25,8	15	1
2 Mean daily maximum temperature	in °C	31	32	32	32	31	29	27	27	28	29	30	31	30	27	2
3 Mean daily minimum temperature	in °C	23	23	23	23	23	23	22	22	23	22	23	23	23	27	3
4 Absolute maximum temperature	in °C	34	35	34	36	35	33	31	32	32	33	34	34	36	27	4
5 Absolute minimum temperature	in °C	19	20	20	20	19	20	20	20	20	20	19	19	19	27	5
6 Mean relative humidity	in %	81	80	80	80	82	84	88	87	86	83	83	82	83	10	6
7 Mean precipitation	in mm	57	82	216	243	337	486	725	776	638	388	150	52	4150	30	7
8 Maximum precipitation	in mm	183	185	426	349	599	862	1154	1240	980	602	298	184	5328	30	8
9 Minimum precipitation	in mm	1	5	58	130	141	226	277	248	315	259	36	4	3238	30	9
10 Maximum precipitation in 24 h	in mm	93	72	193	123	160	217	223	238	193	167	120	76	238	30	10
11 Mean number of days with precipitation	>0,1 mm	7	10	17	18	22	24	29	29	28	26	16	8	234	30	11
12 Mean duration of sunshine	in h	124	140	134	149	132	88	41	39	68	110	122	127	1274	25	12
13 Mean quantity of radiation	in ly/day															13
14 Mean potential evaporation	in mm	141	127	141	132	137	113	100	100	104	106	120	132	1453	13	14
15 Mean windspeed	in m/sec	1,1	1,1	0,8	0,8	0,6	1,1	0,8	0,8	0,8	0,6	0,6	0,6	0,8	10	15
16 Mean predominant direction of the wind		SW	SW	SW	var	var	var	var	var	SW	SW	SW	SW			16

87 Station / Country Birao / Central African Republic

Location 10°17'N/22°47'E · Height above sealevel 465 m · Climate symbol: Köppen Aw · Troll V,3

		J	F	M	A	M	J	J	A	S	O	N	D	year	P	
1 Mean daily temperature	in °C	24,4	26,7	29,4	29,7	30,0	27,8	26,1	25,8	26,7	25,6	24,7	22,2	26,6	6	1
2 Mean daily maximum temperature	in °C	35	37	39	39	37	34	31	30	32	34	35	35	35	11	2
3 Mean daily minimum temperature	in °C	12	15	19	21	23	22	21	21	21	20	14	12	18	11	3
4 Absolute maximum temperature	in °C	41	42	44	43	42	40	36	34	36	38	39	39	44	11	4
5 Absolute minimum temperature	in °C	3	7	8	11	15	18	18	18	18	13	7	5	3	11	5
6 Mean relative humidity	in %	45	37	40	44	60	69	76	79	77	72	59	50	59	8	6
7 Mean precipitation	in mm	0	0	2	19	97	112	217	204	171	37	1	0	860	30	7
8 Maximum precipitation	in mm	0	0	15	66	234	159	289	362	245	80	12	0	1018	10	8
9 Minimum precipitation	in mm	0	0	0		35	47	124	137	112	6	0	0	702	10	9
10 Maximum precipitation in 24 h	in mm	0	0	11	34	59	58	97	63	65	27	12	0	97	10	10
11 Mean number of days with precipitation	>0,1 mm	0	0	1	2	8	11	15	18	13	4	<1	0	72	10	11
12 Mean duration of sunshine	in h															12
13 Mean quantity of radiation	in ly/day															13
14 Mean potential evaporation	in mm	107	137	167	157	164	141	123	121	138	121	107	83	1566	6	14
15 Mean windspeed	in m/sec															15
16 Mean predominant direction of the wind																16

88 Station / Country Bria / Central African Republic

Location 6°32'N/21°59'E · Height above sealevel 584 m · Climate symbol: Köppen Aw · Troll V,2

		J	F	M	A	M	J	J	A	S	O	N	D	year	P	
1 Mean daily temperature	in °C	25,0	26,7	27,8	26,7	26,1	25,0	24,4	24,4	25,0	25,0	25,0	24,4	25,6	4	1
2 Mean daily maximum temperature	in °C	35,0	36,1	35,0	33,3	31,7	30,0	29,4	29,4	30,0	30,6	32,2	33,3	32,2	4	2
3 Mean daily minimum temperature	in °C	15,0	17,2	20,0	20,0	20,6	19,4	19,4	19,4	19,4	18,9	17,2	13,9	18,3	4	3
4 Absolute maximum temperature	in °C	39,4	40,0	40,6	38,9	38,1	33,9	33,3	32,2	33,3	33,9	33,9	36,7	40,6	4	4
5 Absolute minimum temperature	in °C	7,8	11,1	11,7	14,4	17,2	17,2	17,2	17,8	16,7	16,7	11,1	8,3	7,8	4	5
6 Mean relative humidity	in %	57	58	66	78	78	81	82	84	81	79	73	64	73	4	6
7 Mean precipitation	in mm	10	10	104	117	206	173	251	277	208	211	66	<3	1636	4	7
8 Maximum precipitation	in mm															8
9 Minimum precipitation	in mm															9
10 Maximum precipitation in 24 h	in mm	33	18	104	48	58	84	91	66	56	66	66	5	104	4	10
11 Mean number of days with precipitation	>0,1 mm	<1	2	7	9	15	13	14	17	16	15	5	<1	114	4	11
12 Mean duration of sunshine	in h															12
13 Mean quantity of radiation	in ly/day															13
14 Mean potential evaporation	in mm															14
15 Mean windspeed	in m/sec															15
16 Mean predominant direction of the wind																16

## 89

Station / Country Bouar / Central African Republic

Location 5°56′N / 15°35′E   **Height above sealevel** 929 m   **Climate symbol: Köppen** Aw   **Troll** V,2

		J	F	M	A	M	J	J	A	S	O	N	D	year	P	
1 Mean daily temperature	in °C	25,6	26,1	25,6	25,6	24,4	23,3	23,3	22,8	23,3	23,3	25,0	22,8	24,4	4	1
2 Mean daily maximum temperature	in °C	32,8	32,8	31,7	31,1	30,0	28,3	28,3	27,2	28,3	28,9	31,1	32,2	30,0	4	2
3 Mean daily minimum temperature	in °C	18,3	19,4	19,4	19,4	18,9	18,3	18,3	18,3	17,8	17,8	18,3	18,3	18,3	4	3
4 Absolute maximum temperature	in °C	36,7	37,2	36,1	35,6	37,8	40,6	36,1	32,8	32,8	32,2	33,3	35,6	40,6	4	4
5 Absolute minimum temperature	in °C	14,4	15,6	15,6	15,6	15,6	15,0	16,1	15,0	15,0	15,6	14,4	15,0	14,4	4	5
6 Mean relative humidity	in %	42	57	68	73	78	81	84	84	80	57	39		69	4	6
7 Mean precipitation	in mm	5	43	76	107	140	188	201	315	239	226	30	3	1572	4	7
8 Maximum precipitation	in mm															8
9 Minimum precipitation	in mm															9
10 Maximum precipitation in 24 h	in mm	15	43	41	43	64	56	86	112	71	81	33	5	112	4	10
11 Mean number of days with precipitation	> 0,1 mm	1	5	11	11	15	17	17	19	22	19	3	<1	140	4	11
12 Mean duration of sunshine	in h															12
13 Mean quantity of radiation	in ly / day															13
14 Mean potential evaporation	in mm															14
15 Mean windspeed	in m / sec															15
16 Mean predominent direction of the wind																16

## 90

Station / Country Bangui / Central African Republic

Location 4°22′N / 18°34′E   **Height above sealevel** 385 m   **Climate symbol: Köppen** Aw   **Troll** V,1

		J	F	M	A	M	J	J	A	S	O	N	D	year	P	
1 Mean daily temperature	in °C	26,0	27,2	27,5	27,2	26,4	26,0	25,2	25,4	25,6	25,7	25,7	25,8	26,1	20	1
2 Mean daily maximum temperature	in °C	33	34	33	33	32	31	29	30	31	31	32	32	32	31	2
3 Mean daily minimum temperature	in °C	20	20	21	21	21	21	20	20	20	20	20	19	20	31	3
4 Absolute maximum temperature	in °C	37	39	40	38	39	36	34	34	35	35	37	38	40	31	4
5 Absolute minimum temperature	in °C	13	13	16	18	16	17	15	17	17	17	17	14	13	31	5
6 Mean relative humidity	in %	70	68	74	77	79	82	83	82	82	81	80	75	77	11	6
7 Mean precipitation	in mm	21	47	124	128	173	135	185	225	185	202	101	34	1560	30	7
8 Maximum precipitation	in mm	77	106	280	235	315	296	324	371	314	328	239	85	1911	30	8
9 Minimum precipitation	in mm	0	0	22	44	65	32	94	64	62	97	30	0	1252	30	9
10 Maximum precipitation in 24 h	in mm	47	59	88	89	98	86	129	122	107	105	83	88	129	30	10
11 Mean number of days with precipitation	> 0,1 mm	2	5	9	10	14	12	14	17	16	17	10	3	130	30	11
12 Mean duration of sunshine	in h	203	201	191	184	193	158	138	138	143	158	171	220	2098	11	12
13 Mean quantity of radiation	in ly / day															13
14 Mean potential evaporation	in mm	129	130	147	144	139	123	114	121	116	116	121	118	1519	22	14
15 Mean windspeed	in m / sec															15
16 Mean predominent direction of the wind																16

## 91

Station / Country Mitzic / Gabon

Location 0°47′N / 11°34′E   **Height above sealevel** 580 m   **Climate symbol: Köppen** Aw   **Troll** V,2

		J	F	M	A	M	J	J	A	S	O	N	D	year	P	
1 Mean daily temperature	in °C	24,8	24,6	24,9	24,9	24,4	23,2	22,2	22,4	23,9	24,0	24,0	24,3	24,0	10	1
2 Mean daily maximum temperature	in °C	29	29	30	30	29	27	26	26	28	29	28	28	28	15	2
3 Mean daily minimum temperature	in °C	20	20	20	20	20	19	18	18	19	19	19	20	19	15	3
4 Absolute maximum temperature	in °C	33	34	34	34	33	31	31	31	32	33	32	32	34	15	4
5 Absolute minimum temperature	in °C	15	16	16	17	16	15	13	13	16	16	17	16	13	15	5
6 Mean relative humidity	in %	84	85	85	84	86	88	87	86	85	85	85	86	86	5	6
7 Mean precipitation	in mm	118	110	226	207	222	46	10	14	150	346	247	148	1842	30	7
8 Maximum precipitation	in mm	182	174	358	423	397	141	38	41	215	505	318	328	2380	15	8
9 Minimum precipitation	in mm	31	69	108	80	134	3	0	0	51	154	118	31	1355	15	9
10 Maximum precipitation in 24 h	in mm	94	53	82	75	123	68	34	37	86	92	80	102	123	15	10
11 Mean number of days with precipitation	> 0,1 mm	8	10	16	16	18	6	3	5	14	24	19	11	149	15	11
12 Mean duration of sunshine	in h															12
13 Mean quantity of radiation	in ly / day															13
14 Mean potential evaporation	in mm	109	103	112	113	111	98	82	82	98	107	99	108	1222	6	14
15 Mean windspeed	in m / sec															15
16 Mean predominent direction of the wind		SW	SW	W	W	SW	SW	SW	SW	SW	SW	SW	SW		15	16

Station / Country Libreville / Gabon

Location 0°27′N / 9°25′E   **Height above sealevel** 12 m   **Climate symbol: Köppen** Am   **Troll** V,2

		J	F	M	A	M	J	J	A	S	O	N	D	year	P	
1 Mean daily temperature	in °C	26,7	26,6	26,8	27,2	26,7	25,3	24,3	24,9	25,6	26,0	25,6	26,3	26,0	5	1
2 Mean daily maximum temperature	in °C	30	31	31	31	30	29	28	28	29	29	30	30	30	17	2
3 Mean daily minimum temperature	in °C	24	23	23	23	24	23	22	22	23	23	23	24	23	17	3
4 Absolute maximum temperature	in °C	34	35	34	34	34	33	31	33	31	32	32	32	35	17	4
5 Absolute minimum temperature	in °C	20	20	20	20	20	19	16	19	19	20	20	18	16	17	5
6 Mean relative humidity	in %	86	85	85	86	84	81	79	80	83	87	88	86	84	15	6
7 Mean precipitation	in mm	206	291	264	395	244	40	1	11	106	359	416	260	2592	18	7
8 Maximum precipitation	in mm	523	470	670	630	550	210	30	41	370	670	722	771	3614		8
9 Minimum precipitation	in mm	40	120	170	170	10	0	0	0	10	92	150	25	1850		9
10 Maximum precipitation in 24 h	in mm	138	158	213	176	248	54	2	12	181	108	132	158	248	16	10
11 Mean number of days with precipitation	> 0,1 mm	14	15	17	19	15	3	<1	5	13	22	21	16	161	18	11
12 Mean duration of sunshine	in h	174	170	170	171	150	122	126	114	92	106	131	170	1695	10	12
13 Mean quantity of radiation	in ly / day															13
14 Mean potential evaporation	in mm	146	133	149	144	146	121	106	114	123	138	134	145	1599	17	14
15 Mean windspeed	in m / sec	0,8	0,8	0,8	0,8	0,8	1,1	1,1	1,1	1,4	1,1	1,1	1,1	1,1	8	15
16 Mean predominent direction of the wind		E.SW	E.SW	E.SW	E.SW	E.SW	E.SW	E.SW	E.SW	E.SW	E.SW	E.SW	E.SW			16

93 Station/Country Bongabo/Zaire

Location 3°06'N / 20°32'E Height above sealevel 450 m Climate symbol: Köppen Af Troll V,1

		J	F	M	A	M	J	J	A	S	O	N	D	year	P	
1	Mean daily temperature	in °C	24,5	25,5	26	26	25,5	25	24	24,5	24,5	24,5	24,5	24,5	25	9	1
2	Mean daily maximum temperature	in °C	31	32	32	32	31	30	29	29	30	30	30	31	31	9	2
3	Mean daily minimum temperature	in °C	18	19	20	20	20	20	19	20	19	19	18	19	19	9	3
4	Absolute maximum temperature	in °C	35	37	38	35	35	34	33	33	34	34	34	34	38	9	4
5	Absolute minimum temperature	in °C	12	10	15	17	17	17	16	17	16	16	11	10	9	5	5
6	Mean relative humidity	in %	83	78	82	83	85	86	88	87	85	85	86	85	84	3	6
7	Mean precipitation	in mm	38	63	135	167	189	180	186	250	207	209	130	56	1810	20	7
8	Maximum precipitation	in mm	91	146	233	252	365	343	319	376	392	320	317	128	2210	20	8
9	Minimum precipitation	in mm	0	12	58	69	104	72	68	81	104	140	27	0	1412	20	9
10	Maximum precipitation in 24 h	in mm	85	75	76	86	90	106	73	90	92	104	68	55	106	20	10
11	Mean number of days with precipitation	> 0,1 mm	3	5	10	11	13	11	13	16	14	15	9	5	124	20	11
12	Mean duration of sunshine	in h	220	202	199	193	191	155	153	142	162	175	165	213	2170	8	12
13	Mean quantity of radiation	in ly/day	401	440	466	499	485	413	356	389	439	443	395	419	427	3	13
14	Mean potential evaporation	in mm															14
15	Mean windspeed	in m/sec															15
16	Mean predominent direction of the wind		WSW	WSW	SW	SW	SW	WSW	WSW	WSW	SW	SW	SW	WSW			16

94 Station/Country Yangambi/Zaire

Location 0°49'N / 24°29'E Height above sealevel 487 m Climate symbol: Köppen Af Troll V,1

		J	F	M	A	M	J	J	A	S	O	N	D	year	P	
1	Mean daily temperature	in °C	24,7	25,3	25,5	25,2	24,9	24,5	23,6	23,9	24,3	24,5	24,3	24,3	24,6	30	1
2	Mean daily maximum temperature	in °C	30	31	31	30	30	30	29	28	29	29	29	29	30	10	2
3	Mean daily minimum temperature	in °C	20	19	20	20	20	20	19	20	19	20	20	20	20	10	3
4	Absolute maximum temperature	in °C	36	35	35	35	35	33	32	33	33	34	33	34	36	10	4
5	Absolute minimum temperature	in °C	15	14	16	18	17	17	17	17	17	17	18	14	14	10	5
6	Mean relative humidity	in %	84	81	84	86	86	87	88	88	87	87	86	88	86	3	6
7	Mean precipitation	in mm	85	99	148	150	177	126	146	170	180	241	180	126	1828	30	7
8	Maximum precipitation	in mm	229	188	362	264	342	342	266	291	319	381	317	217	2929	30	8
9	Minimum precipitation	in mm	18	34	58	37	85	25	43	55	79	113	54	0	1220	30	9
10	Maximum precipitation in 24 h	in mm	80	112	89	74	146	81	170	104	80	134	76	119	170	30	10
11	Mean number of days with precipitation	> 0,1 mm	8	8	13	14	14	11	13	13	15	18	16	12	155	30	11
12	Mean duration of sunshine	in h	207	189	186	182	184	164	156	135	155	157	165	176	2056	12	12
13	Mean quantity of radiation	in ly/day	411	447	454	446	441	391	345	363	409	412	422	373	410	3	13
14	Mean potential evaporation	in mm	114	110	125	117	116	106	97	102	104	110	104	107	1312	10	14
15	Mean windspeed	in m/sec	0,8	0,8	1,1	1,1	0,8	0,8	0,8	1,1	1,1	1,1	0,8	0,8	0,8	4	15
16	Mean predominent direction of the wind		NE	SW,SE	SW,SE	SW,SE	SW,SE	SW,SE	SW,SE	SW,SE	SW,SE	SW,SE	SW,SE	NE		4	16

95 Station/Country Eala/Zaire

Location 0°03'N / 18°18'E Height above sealevel 340 m Climate symbol: Köppen Af Troll V,1

		J	F	M	A	M	J	J	A	S	O	N	D	year	P	
1	Mean daily temperature	in °C	25,8	26,7	26,7	26,1	26,1	25,8	24,5	24,5	25,9	25,9	25,6	25,6	25,6	14	1
2	Mean daily maximum temperature	in °C	31,1	32,2	32,2	31,7	31,7	30,6	30,0	29,4	30,6	30,6	30,6	30,8	31,1	14	2
3	Mean daily minimum temperature	in °C	19,4	20,0	20,0	20,0	20,0	19,4	17,8	17,8	19,4	19,4	19,4	19,4	19,4	14	3
4	Absolute maximum temperature	in °C	35,0	37,2	37,2	37,2	35,0	34,4	35,0	33,9	34,4	35,0	36,7	37,2	14		4
5	Absolute minimum temperature	in °C	13,9	16,1	13,9	15,0	13,9	16,7	15,0	16,1	16,1	16,7	15,0	15,8	13,9	14	5
6	Mean relative humidity	in %															6
7	Mean precipitation	in mm	84	107	127	178	157	145	71	178	178	216	193	160	1794	12	7
8	Maximum precipitation	in mm															8
9	Minimum precipitation	in mm															9
10	Maximum precipitation in 24 h	in mm	112	84	89	66	71	84	79	89	86	99	79	79	112	10	10
11	Mean number of days with precipitation	> 0,1 mm	9	9	14	15	16	17	8	13	16	17	17	13	164	12	11
12	Mean duration of sunshine	in h															12
13	Mean quantity of radiation	in ly/day															13
14	Mean potential evaporation	in mm	130	123	139	135	135	120	112	113	116	125	120	126	1494		14
15	Mean windspeed	in m/sec															15
16	Mean predominent direction of the wind																16

96 Station/Country Tshibinda/Zaire

Location 2°19'S / 28°45'E Height above sealevel 2055 m Climate symbol: Köppen Cw Troll V,1

		J	F	M	A	M	J	J	A	S	O	N	D	year	P	
1	Mean daily temperature	in °C	16,5	16,6	16,4	16,2	16,0	15,5	15,3	16,0	16,4	16,2	16,1	16,3	16,1		1
2	Mean daily maximum temperature	in °C	22	22	22	21	21	21	21	22	22	22	22	22	22	8	2
3	Mean daily minimum temperature	in °C	11	12	11	12	12	11	9	10	11	11	11	11	11	8	3
4	Absolute maximum temperature	in °C	26	25	25	24	23	24	23	25	27	25	24	25	27	8	4
5	Absolute minimum temperature	in °C	8	8	9	9	8	8	5	7	8	8	8	9	5	8	5
6	Mean relative humidity	in %	84	86	84	88	89	84	76	75	74	83	85	87	83	3	6
7	Mean precipitation	in mm	165	175	195	215	164	56	34	57	145	214	201	212	1833	30	7
8	Maximum precipitation	in mm	301	354	311	364	295	112	170	165	252	424	398	358	2306	30	8
9	Minimum precipitation	in mm	52	50	113	122	61	0	0	0	33	99	92	33		30	9
10	Maximum precipitation in 24 h	in mm	72	74	94	71	68	45	63	73	84	104	91	78	104	30	10
11	Mean number of days with precipitation	> 0,1 mm	17	17	18	21	17	7	4	7	15	19	21	20	183	30	11
12	Mean duration of sunshine	in h	164	136	162	146	131	182	209	196	180	152	147	144	1949	8	12
13	Mean quantity of radiation	in ly/day	442	452	492	463	399	397	404	421	481	454	484	465	446	4	13
14	Mean potential evaporation	in mm	64	58	64	63	62	55	57	62	62	64	62	64	737		14
15	Mean windspeed	in m/sec	1,1	1,1	1,1	1,1	1,1	1,1	1,1	1,4	1,4	1,1	1,1	1,1	1,1	2	15
16	Mean predominent direction of the wind		W	W	SE	SE	SE	SE	SE	SE	SE	SE	ENE	W			16

250

97 Station / Country Kinshasa (Léopoldville) / Zaire

Location 4°20'S/15°16'E   Height above sealevel 358 m   Climate symbol: Köppen Aw   Troll V,2

		J	F	M	A	M	J	J	A	S	O	N	D	year	P	
1	Mean daily temperature	in °C	26,0	26,2	26,7	26,8	26,0	23,4	22,0	23,3	25,8	26,2	26,1	25,9	25,3	9	1
2	Mean daily maximum temperature	in °C	30	31	31	32	30	28	27	28	30	30	30	30	30	10	2
3	Mean daily minimum temperature	in °C	22	22	22	22	22	19	17	18	20	22	22	22	21	10	3
4	Absolute maximum temperature	in °C	34	35	35	34	34	33	32	34	36	35	34	33	36	10	4
5	Absolute minimum temperature	in °C	18	19	19	20	19	15	13	11	16	17	19	20	11	10	5
6	Mean relative humidity	in %	82	80	80	82	83	81	76	69	69	76	81	83	79	3	6
7	Mean precipitation	in mm	128	139	181	209	134	5	1	4	33	137	236	171	1378	30	7
8	Maximum precipitation	in mm	321	330	429	379	280	38	34	24	100	282	348	327	1824	30	8
9	Minimum precipitation	in mm	2	49	58	59	22	0	0	0	2	20	84	47	1124	30	9
10	Maximum precipitation in 24 h	in mm	128	105	108	131	111	29	34	24	73	155	147	73	155	30	10
11	Mean number of days with precipitation	> 0,1 mm	10	10	13	15	11	1	0	1	4	11	16	14	106	30	11
12	Mean duration of sunshine	in h	132	134	150	186	140	139	125	146	129	132	142	135	1670	9	12
13	Mean quantity of radiation	in ly/day	385	427	448	445	374	332	319	364	375	381	418	380	387	7	13
14	Mean potential evaporation	in mm	130	122	141	136	131	94	76	86	110	131	122	130	1409	10	14
15	Mean windspeed	in m/sec	1,1	1,4	1,1	1,1	1,1	1,1	1,4	1,7	1,4	1,4	1,4	1,1	1,4	1	15
16	Mean predominant direction of the wind		WSW	WSW	WSW	WSW	WSW	WSW	WSW	WSW	WSW	WSW	WSW	WSW			16

98 Station / Country Kananga (Luluabourg) / Zaire

Location 5°53'S/22°25'E   Height above sealevel 660 m   Climate symbol: Köppen Aw   Troll V,2

		J	F	M	A	M	J	J	A	S	O	N	D	year	P	
1	Mean daily temperature	in °C	24,5	24,6	25,1	25,0	25,1	24,5	23,6	24,1	24,3	24,3	24,6	24,5	24,5	20	1
2	Mean daily maximum temperature	in °C	29	29	30	30	31	31	30	30	30	30	29	29	30	9	2
3	Mean daily minimum temperature	in °C	20	20	20	20	20	18	18	19	19	19	20	20	19	9	3
4	Absolute maximum temperature	in °C	35	35	35	35	35	34	34	35	34	33	34	33	35	9	4
5	Absolute minimum temperature	in °C	17	16	17	18	16	14	13	14	15	17	17	17	13	9	5
6	Mean relative humidity	in %	84	82	84	84	78	76	77	79	82	81	84	85	80	3	6
7	Mean precipitation	in mm	128	123	204	177	89	16	17	50	118	165	238	247	1572	20	7
8	Maximum precipitation	in mm	245	259	394	305	195	69	73	121	200	411	413	378	1962	20	8
9	Minimum precipitation	in mm	66	39	71	104	13	0	0	5	20	98	126	165	1274	20	9
10	Maximum precipitation in 24 h	in mm	81	72	98	67	58	65	60	63	61	179	92	118	179	20	10
11	Mean number of days with precipitation	> 0,1 mm	12	12	15	15	8	2	2	5	10	13	16	16	126	20	11
12	Mean duration of sunshine	in h	148	132	154	162	218	257	226	191	183	191	160	139	2161	9	12
13	Mean quantity of radiation	in ly/day	386	419	430	419	433	399	279	371	445	446	447	421	416	3	13
14	Mean potential evaporation	in mm	111	101	119	114	116	103	95	102	102	110	108	112	1411	20	14
15	Mean windspeed	in m/sec															15
16	Mean predominant direction of the wind		ESE	ESE	ESE	ENE	ENE	ENE	ENE	ENE	ENE	NNW	ESE	ESE		3	16

99 Station / Country Kalemi (Albertville) / Zaire

Location 5°53'S/29°11'E   Height above sealevel 790 m   Climate symbol: Köppen Aw   Troll V,2

		J	F	M	A	M	J	J	A	S	O	N	D	year	P	
1	Mean daily temperature	in °C	23,4	23,8	23,7	23,6	23,2	21,5	21,0	22,1	23,9	24,8	23,5	23,2	23,1	30	1
2	Mean daily maximum temperature	in °C	28	28	28	28	28	27	27	28	29	30	28	27	28	9	2
3	Mean daily minimum temperature	in °C	20	20	20	20	19	16	15	17	19	20	20	20	19	9	3
4	Absolute maximum temperature	in °C	33	32	33	31	31	30	30	32	34	34	33	32	34	9	4
5	Absolute minimum temperature	in °C	17	18	17	17	15	12	12	11	14	15	17	17	11	9	5
6	Mean relative humidity	in %	82	80	81	82	78	72	68	65	64	68	77	82	75	7	6
7	Mean precipitation	in mm	110	77	137	207	96	9	<1	7	43	51	146	181	1064	9	7
8	Maximum precipitation	in mm															8
9	Minimum precipitation	in mm															9
10	Maximum precipitation in 24 h	in mm	80	41	59	90	72	31	4	22	65	61	63	59	90	9	10
11	Mean number of days with precipitation	> 0,1 mm	13	12	14	17	8	1	<1	1	4	6	16	17	108	9	11
12	Mean duration of sunshine	in h	158	158	183	195	256	284	308	288	239	219	164	149	2599	5	12
13	Mean quantity of radiation	in ly/day	212	214	222	230	248	243	243	236	233	226	208	216	228	9	13
14	Mean potential evaporation	in mm	116	101	113	103	87	70	68	99	109	123	113	111	1213	5	14
15	Mean windspeed	in m/sec	0,4	0,4	0,4	0,4	0,6	0,6	0,6	0,6	0,6	0,6	0,6	0,4	0,6	4	15
16	Mean predominant direction of the wind																16

100 Station / Country Kamina / Zaire

Location 8°44'S/25°00'E   Height above sealevel 1105 m   Climate symbol: Köppen Aw   Troll V,2

		J	F	M	A	M	J	J	A	S	O	N	D	year	P	
1	Mean daily temperature	in °C	22,3	22,6	23,0	22,6	22,1	21,4	21,2	22,6	23,7	23,2	22,6	22,4	22,5	9	1
2	Mean daily maximum temperature	in °C	27	27	28	28	29	29	29	30	30	29	28	27	29	9	2
3	Mean daily minimum temperature	in °C	18	18	18	18	16	14	14	16	14	18	18	18	17	9	3
4	Absolute maximum temperature	in °C	31	32	32	32	32	32	32	34	35	34	32	31	35	9	4
5	Absolute minimum temperature	in °C	15	15	10	16	12	10	10	10	13	12	15	15	10	9	5
6	Mean relative humidity	in %	85	85	84	81	68	50	43	47	63	73	81	85	70	3	6
7	Mean precipitation	in mm	201	193	202	119	18	1	1	5	38	121	191	253	1343	20	7
8	Maximum precipitation	in mm	341	313	338	276	66	23	15	26	89	265	402	373	1681	20	8
9	Minimum precipitation	in mm	91	116	98	40	0	0	0	0	6	16	88	159	1072	20	9
10	Maximum precipitation in 24 h	in mm	87	67	85	76	66	18	10	22	79	70	82	165	165	20	10
11	Mean number of days with precipitation	> 0,1 mm	12	12	13	12	13	0	0	1	3	8	12	15	86	20	11
12	Mean duration of sunshine	in h	122	108	150	207	292	309	320	287	222	195	157	126	2495	7	12
13	Mean quantity of radiation	in ly/day	381	353	428	457	496	461	475	453	459	440	441	411	438	3	13
14	Mean potential evaporation	in mm	94	87	101	89	85	76	76	92	105	105	94	98	1102	9	14
15	Mean windspeed	in m/sec															15
16	Mean predominant direction of the wind		ESE	ESE	ESE	ESE	ESE	SE	SE	SE	SE	ESE	ESE	ESE			16

101 Station/Country Lubumbashi (Elisabethville)/Zaire

Location 11°39'S/27°28'E — Height above sealevel 1290 m — Climate symbol: Köppen Cwa — Troll V.3

		J	F	M	A	M	J	J	A	S	O	N	D	year	P	
1 Mean daily temperature	in °C	21,8	21,9	22,0	21,0	19,0	16,8	16,4	18,6	22,0	23,7	23,0	22,0	20,5	40	1
2 Mean daily maximum temperature	in °C	27	27	27	27	27	25	25	28	31	32	29	27	28	10	2
3 Mean daily minimum temperature	in °C	17	17	17	15	12	9	8	11	14	16	17	17	14	10	3
4 Absolute maximum temperature	in °C	31	31	31	30	31	30	29	34	36	37	36	32	37	10	4
5 Absolute minimum temperature	in °C	13	14	12	10	3	4	4	3	9	9	13	13	3	10	5
6 Mean relative humidity	in %	83	85	81	75	68	62	54	45	42	44	74	83	66	3	6
7 Mean precipitation	in mm	256	264	210	53	3	0	0	0	3	27	166	262	1244	30	7
8 Maximum precipitation	in mm	427	422	378	155	41	4	0	15	31	72	396	479	1554	30	8
9 Minimum precipitation	in mm	149	130	89	9	0	0	0	0	0	0	43	113	868	30	9
10 Maximum precipitation in 24 h	in mm	116	100	112	94	23	3	0	12	27	56	88	81	116	30	10
11 Mean number of days with precipitation	> 0,1 mm	24	23	21	9	2	0	0	0	1	5	17	24	126	30	11
12 Mean duration of sunshine	in h	129	120	174	234	291	295	315	318	294	282	194	132	2778	10	12
13 Mean quantity of radiation	in ly/day	451	413	449	485	480	488	477	523	555	554	493	445	481	5	13
14 Mean potential evaporation	in mm	98	87	93	78	63	42	44	56	85	107	103	100	956	10	14
15 Mean windspeed	in m/sec	1,4	1,4	1,7	1,9	1,9	1,9	1,7	1,7	2,5	2,2	1,4	1,1	1,7	3	15
16 Mean predominent direction of the wind		ESE	ESE	ESE	SE	SE	SE	ESE	ESE	ESE	var	ESE	ESE		3	16

102 Station/Country Gulu/Uganda

Location 2°45'N/32°20'E — Height above sealevel 1109 m — Climate symbol: Köppen Aw — Troll V.2

		J	F	M	A	M	J	J	A	S	O	N	D	year	P	
1 Mean daily temperature	in °C	24,4	25,0	24,4	23,6	23,0	22,5	21,4	21,7	22,2	22,8	23,0	23,3	23,1	11	1
2 Mean daily maximum temperature	in °C	32	32	31	29	28	27	27	27	28	29	30	30	29	32	2
3 Mean daily minimum temperature	in °C	16	17	18	18	18	17	17	17	17	16	16	17		32	3
4 Absolute maximum temperature	in °C	36	37	37·	37	33	32	29	30	31	33	34	33	37	32	4
5 Absolute minimum temperature	in °C	11	12	12	14	15	15	13	13	14	13	12	9	9	32	5
6 Mean relative humidity	in %	48	53	59	69	74	73	76	76	71	65	61	54	65	11	6
7 Mean precipitation	in mm	12	43	89	173	172	148	186	231	127	165	97	47	1470	52	7
8 Maximum precipitation	in mm	55	229	182	319	393	370	387	431	323	395	320	163	2144	52	8
9 Minimum precipitation	in mm	0	0	10	69	88	53	28	119	36	15	13	1	869	52	9
10 Maximum precipitation in 24 h	in mm	36	77	60	78	77	83	83	109	68	69	106	52	109	52	10
11 Mean number of days with precipitation	> 0,1 mm	4	6	11	17	17	14	17	19	17	16	11	7	156	52	11
12 Mean duration of sunshine	in h	276	244	248	228	242	231	192	198	243	251	246	267	2866	6	12
13 Mean quantity of radiation	in ly/day															13
14 Mean potential evaporation	in mm	113	107	113	101	92	86	81	81	85	92	92	99	1142	14	14
15 Mean windspeed	in m/sec	4,3	4,3	4,3	4,1	3,9	3,3	3,3	3,3	3,6	3,9	3,9	4,1	3,9	25	15
16 Mean predominent direction of the wind		N.E	N	E	S	S	S	N	N	N	E	E	E		25	16

103 Station/Country Entebbe/Uganda

Location 0°03'N/32°27'E — Height above sealevel 1146 m — Climate symbol: Köppen Af — Troll V.2

		J	F	M	A	M	J	J	A	S	O	N	D	year	P	
1 Mean daily temperature	in °C	21,9	22,0	22,0	21,6	21,3	21,0	20,5	20,6	21,0	21,5	21,5	21,5	21,4	60	1
2 Mean daily maximum temperature	in °C	27	27	27	26	26	25	25	25	26	26	26	26			2
3 Mean daily minimum temperature	in °C	17	17	18	18	18	17	16	16	16	17	17	17			3
4 Absolute maximum temperature	in °C	31	32	31	30	29	28	28	29	30	30	32	30	32		4
5 Absolute minimum temperature	in °C	13	11	14	12	14	14	10	12	13	14	14	10			5
6 Mean relative humidity	in %	74	75	78	79	81	79	78	79	77	74	76	76	77		6
7 Mean precipitation	in mm	100	86	141	280	257	98	65	91	87	168	146	126	1585		7
8 Maximum precipitation	in mm	358	199	309	425	463	264	121	184	274	265	385	339	2303		8
9 Minimum precipitation	in mm	17	18	62	182	99	47	1	13	8	20	42	5	1128		9
10 Maximum precipitation in 24 h	in mm	103	84	85	107	283	61	63	112	113	75	91	68	283		10
11 Mean number of days with precipitation	> 0,1 mm	9	9	15	22	17	11	10	9	9	13	14	14	152		11
12 Mean duration of sunshine	in h	233	204	205	180	192	186	198	195	195	201	198	211	2398	31	12
13 Mean quantity of radiation	in ly/day															13
14 Mean potential evaporation	in mm	92	83	92	86	79	76	78	82	87	87	86		1012	15	14
15 Mean windspeed	in m/sec	3,6	3,9	4,1	4,1	4,1	4,3	4,1	3,9	4,1	3,9	3,6	3,3	3,9	7	15
16 Mean predominent direction of the wind		S	S	S	S	S	S	S	S	S	S	S	S		7	16

104 Station/Country Kisozi/Burundi

Location 3°33'S/29°41'E — Height above sealevel 2155 m — Climate symbol: Köppen Cwb — Troll V.2

		J	F	M	A	M	J	J	A	S	O	N	D	year	P	
1 Mean daily temperature	in °C	16,2	16,3	16,3	16,2	15,7	14,6	14,8	15,6	16,4	16,5	16,1	16,0	16,0	10	1
2 Mean daily maximum temperature	in °C	22	22	22	22	21	21	22	23	23	23	22	22	22	9	2
3 Mean daily minimum temperature	in °C	12	12	12	12	11	9	9	10	11	11	12	12	11	9	3
4 Absolute maximum temperature	in °C	23	23	24	24	23	22	23	24	25	24	23	23	25	9	4
5 Absolute minimum temperature	in °C	11	11	11	11	10	8	7	8	9	11	11	11	7	9	5
6 Mean relative humidity	in %	83	84	84	88	87	78	70	67	62	72	80	85	78	3	6
7 Mean precipitation	in mm	167	190	196	228	120	12	6	16	64	115	174	189	1447	30	7
8 Maximum precipitation	in mm	295	248	386	324	313	56	36	75	177	194	348	349	1709	30	8
9 Minimum precipitation	in mm	71	45	93	58	40	0	0	0	6	39	72	71	1154	30	9
10 Maximum precipitation in 24 h	in mm	61	69	72	103	51	28	31	36	45	67	70	60	103	30	10
11 Mean number of days with precipitation	> 0,1 mm	21	19	22	23	17	3	1	3	8	14	22	22	175	30	11
12 Mean duration of sunshine	in h	149	126	166	152	150	217	237	231	203	183	152	138	2104	6	12
13 Mean quantity of radiation	in ly/day	438	434	445	487	465	471	469	454	481	461	429	424	453	2	13
14 Mean potential evaporation	in mm	67	60	62	62	58	55	60	66	68	69	64	65	760	9	14
15 Mean windspeed	in m/sec	1,9	1,9	1,9	1,9	1,9	2,2	2,2	2,5	2,2	2,2	1,9	1,9	1,9	2	15
16 Mean predominent direction of the wind		NE	NE	NE	SE	SE	SE.SW	SE.SW	NE.SW	SE	SE	NE	NE		2	16

252

105 Station / Country — Bujumbura / Burundi

Location 3°23'S / 29°21'E — Height above sealevel 805 m — Climate symbol: Köppen Aw — Troll V.2

		J	F	M	A	M	J	J	A	S	O	N	D	year	P	
1 Mean daily temperature	in °C	23,4	23,1	23,3	23,4	23,3	23,0	22,9	23,9	24,8	24,7	23,3	23,0	23,5		1
2 Mean daily maximum temperature	in °C	28	28	28	28	28	29	29	30	31	30	28	28	29	10	2
3 Mean daily minimum temperature	in °C	19	19	19	19	19	18	17	18	19	20	19	19	19	10	3
4 Absolute maximum temperature	in °C	30	30	30	30	29	29	30	31	32	32	30	29	32	10	4
5 Absolute minimum temperature	in °C	19	18	19	18	19	17	16	17	18	19	19	19	16	10	5
6 Mean relative humidity	in %	79	79	81	82	78	67	62	55	59	65	75	78	72	3	6
7 Mean precipitation	in mm	94	109	121	125	57	11	5	11	37	64	100	114	848	30	7
8 Maximum precipitation	in mm	161	223	225	226	123	104	42	89	106	121	169	194	1106	30	8
9 Minimum precipitation	in mm	23	48	40	53	12	0	0	0	3	18	49	40	632	30	9
10 Maximum precipitation in 24 h	in mm	65	128	60	54	49	83	33	39	30	34	64	62	128	30	10
11 Mean number of days with precipitation	> 0,1 mm	15	14	17	18	10	3	1	2	8	12	19	19	138	30	11
12 Mean duration of sunshine	in h	159	140	171	153	196	242	269	247	202	174	141	148	2242	9	12
13 Mean quantity of radiation	in ly / day	421	410	415	416	441	455	432	440	467	407	420	410	428	2	13
14 Mean potential evaporation	in mm	101	86	97	96	96	90	93	103	126	131	95	95	1209		14
15 Mean windspeed	in m / sec	1,7	1,7	1,7	1,7	2,2	1,9	2,2	2,2	1,4	2,2	1,9	1,7	1,9	2	15
16 Mean predominent direction of the wind		SW	SW	SSW	SSW	SSW	SSW	SSW	SSW	WSW	W	W	SW		2	16

106 Station / Country — Marsabit / Kenya

Location 2°05'N / 37°59'E — Height above sealevel 1433 m — Climate symbol: Köppen Aw — Troll V.3

		J	F	M	A	M	J	J	A	S	O	N	D	year	P	
1 Mean daily temperature	in °C	20,6	21,1	21,7	21,1	20,6	19,4	18,9	18,3	19,4	20,0	20,0	19,4	20,0	45	1
2 Mean daily maximum temperature	in °C	25,6	26,1	25,6	25,0	24,4	23,9	23,9	23,3	24,4	24,4	23,9	23,3	24,4	9	2
3 Mean daily minimum temperature	in °C	16,1	16,1	16,7	16,7	16,1	14,4	13,3	13,3	13,3	15,0	16,1	15,6	15,0	9	3
4 Absolute maximum temperature	in °C	31,1	29,4	29,4	29,4	28,9	27,2	27,2	27,8	28,9	28,9	26,7	28,3	31,1	9	4
5 Absolute minimum temperature	in °C	13,3	13,9	13,9	13,3	12,8	10,6	10,6	8,9	9,4	12,2	12,8	11,7	8,9	9	5
6 Mean relative humidity	in %															6
7 Mean precipitation	in mm	36	16	79	208	104	8	18	20	13	119	142	71	834	45	7
8 Maximum precipitation	in mm															8
9 Minimum precipitation	in mm															9
10 Maximum precipitation in 24 h	in mm	147	25	91	127	165	20	15	48	43	152	61	46	165		10
11 Mean number of days with precipitation	> 0,1 mm	2	3	9	16	9	3	5	7	2	9	16	11	92		11
12 Mean duration of sunshine	in h															12
13 Mean quantity of radiation	in ly / day															13
14 Mean potential evaporation	in mm	83	80	88	81	83	72	69	66	71	79	73	72	917		14
15 Mean windspeed	in m / sec															15
16 Mean predominent direction of the wind																16

107 Station / Country — Garissa / Kenya

Location 0°29'S / 39°38'E — Height above sealevel 128 m — Climate symbol: Köppen Aw — Troll V.4

		J	F	M	A	M	J	J	A	S	O	N	D	year	P	
1 Mean daily temperature	in °C	28,9	29,4	30,6	30,3	29,1	27,2	26,7	27,0	27,8	28,9	29,4	28,9	28,7	7	1
2 Mean daily maximum temperature	in °C	35	36	37	36	35	33	32	32	34	35	34	34	34	23	2
3 Mean daily minimum temperature	in °C	22	23	24	24	23	22	21	21	21	23	22	23	23	23	3
4 Absolute maximum temperature	in °C	38	39	41	44	46	42	42	38	37	38	38	37	46	23	4
5 Absolute minimum temperature	in °C	16	19	20	21	19	17	16	16	14	19	19	19	14	23	5
6 Mean relative humidity	in %	58	60	60	61	59	59	58	59	58	58	62	69	60	7	6
7 Mean precipitation	in mm	10	6	27	59	16	5	1	6	6	21	77	64	298	32	7
8 Maximum precipitation	in mm	55	39	251	251	91	28	6	39	58	169	419	181	757	32	8
9 Minimum precipitation	in mm	0	0	0	3	0	0	0	0	0	0	0	1	69	32	9
10 Maximum precipitation in 24 h	in mm	32	36	116	82	58	23	6	35	14	50	127	69	127	32	10
11 Mean number of days with precipitation	> 0,1 mm	2	1	3	5	2	1	1	1	2	3	7	7	35	32	11
12 Mean duration of sunshine	in h															12
13 Mean quantity of radiation	in ly / day															13
14 Mean potential evaporation	in mm	160	150	172	165	162	144	141	144	147	161	159	159	1864	6	14
15 Mean windspeed	in m / sec	1,4	1,7	1,4	1,9	2,5	2,8	2,8	2,8	2,5	2,2	1,4	1,4	1,9	21	15
16 Mean predominent direction of the wind		SE	S	S	S	S	S	S	S	S	SE	SE	E		21	16

108 Station / Country — Dagoretti (near Nairobi) / Kenya

Location 1°18'S / 36°45'E — Height above sealevel 1798 m — Climate symbol: Köppen Cwb — Troll V.2

		J	F	M	A	M	J	J	A	S	O	N	D	year	P	
1 Mean daily temperature	in °C	17,8	18,5	18,7	17,9	16,9	15,7	14,9	15,3	16,7	17,8	17,3	17,1	17,1	32	1
2 Mean daily maximum temperature	in °C	25	26	26	24	23	22	21	22	24	25	23	23	24	8	2
3 Mean daily minimum temperature	in °C	11	11	12	14	13	11	9	10	10	12	13	12	12	8	3
4 Absolute maximum temperature	in °C	30	30	29	29	26	26	26	28	28	28	28	27	30	8	4
5 Absolute minimum temperature	in °C	3	5	7	8	7	4	2	3	4	5	6	6	2	8	5
6 Mean relative humidity	in %	69	59	65	72	74	71	72	70	66	64	71	70	68		6
7 Mean precipitation	in mm	88	70	96	155	189	29	17	20	34	64	189	115	1086	9	7
8 Maximum precipitation	in mm	253	201	207	224	380	82	85	42	62	164	623	379	1632	9	8
9 Minimum precipitation	in mm	7	12	23	102	85	2	2	1	9	12	41	18	818	9	9
10 Maximum precipitation in 24 h	in mm	76	59	60	63	85	25	47	23	54	45	64	112	112	9	10
11 Mean number of days with precipitation	> 0,1 mm	9	7	13	17	18	5	5	5	7	8	16	11	121	9	11
12 Mean duration of sunshine	in h	273	263	270	219	183	177	133	130	174	220	210	251	2503	8	12
13 Mean quantity of radiation	in ly / day															13
14 Mean potential evaporation	in mm	80	80	88	82	75	63	57	60	69	80	77	77	888	13	14
15 Mean windspeed	in m / sec	5,1	5,1	5,1	3,9	2,8	2,6	2,3	2,6	3,1	4,3	5,1	5,4	3,9	8	15
16 Mean predominent direction of the wind		NE	NE	NE	NE	E	E,S	SE	SE	E,SE	E,NE	E,NE	NE		8	16

109 Station / Country   Mombasa / Kenya

Location 4°02'S / 39°37'E   Height above sealevel 55 m   Climate symbol: Köppen Aw   Troll V.2

		J	F	M	A	M	J	J	A	S	O	N	D	year	P	
1 Mean daily temperature	in °C	27.2	27.5	28.0	27.2	25.8	25.3	24.4	24.4	25.0	26.1	26.7	27.0	26.2	45	1
2 Mean daily maximum temperature	in °C	32	32	33	31	29	29	28	28	29	30	31	32	30	17	2
3 Mean daily minimum temperature	in °C	23	24	24	24	23	21	20	20	21	22	23	23	23	17	3
4 Absolute maximum temperature	in °C	37	36	36	35	33	32	31	31	32	33	36	36	37	17	4
5 Absolute minimum temperature	in °C	18	21	21	22	19	18	18	14	18	18	20	19	14	17	5
6 Mean relative humidity	in %	71	70	71	76	79	77	78	75	73	73	73	73	74	9	6
7 Mean precipitation	in mm	26	15	61	200	319	112	89	65	68	83	93	60	1191	73	7
8 Maximum precipitation	in mm	189	83	174	608	1043	389	299	235	323	310	703	262	1887	73	8
9 Minimum precipitation	in mm	0	0	0	15	46	4	6	10	7	4	3	0	561	73	9
10 Maximum precipitation in 24 h	in mm	38	37	106	101	139	43	37	35	150	103	86	69	150	17	10
11 Mean number of days with precipitation	> 0,1 mm	5	3	7	15	19	14	14	15	12	10	9	9	133	73	11
12 Mean duration of sunshine	in h	254	255	282	231	201	231	211	248	255	273	276	270	2987	14	12
13 Mean quantity of radiation	in ly / day															13
14 Mean potential evaporation	in mm	155	144	160	147	135	114	107	107	112	140	146	154	1621	14	14
15 Mean windspeed	in m / sec	4.6	4.6	4.1	4.3	5.1	5.4	4.9	4.9	4.9	4.3	3.9	4.1	4.6	17	15
16 Mean predominent direction of the wind		E	E	E	S	S	S	S	S	S	E	E			17	16

110 Station / Country   Tabora / Tanzania

Location 5°02'S / 32°49'E   Height above sealevel 1265 m   Climate symbol: Köppen Aw   Troll V.3

		J	F	M	A	M	J	J	A	S	O	N	D	year	P	
1 Mean daily temperature	in °C	22.0	22.1	22.0	21.9	21.9	21.5	21.5	23.0	24.4	25.3	24.2	22.3	22.7	40	1
2 Mean daily maximum temperature	in °C	28	28	28	28	28	28	28	29	31	32	31	28	29	25	2
3 Mean daily minimum temperature	in °C	17	17	17	17	16	15	15	16	18	19	19	18	17	25	3
4 Absolute maximum temperature	in °C	34	34	33	33	32	31	32	32	34	35	35	34	35	25	4
5 Absolute minimum temperature	in °C	14	15	15	15	11	11	10	12	14	16	15	15	10	25	5
6 Mean relative humidity	in %	70	71	72	74	61	54	49	46	43	42	51	67	58	15	6
7 Mean precipitation	in mm	132	129	166	134	27	2	0	1	7	17	103	174	892	69	7
8 Maximum precipitation	in mm	228	323	378	328	145	39	3	11	87	68	432	371	1390	69	8
9 Minimum precipitation	in mm	57	41	25	28	1	2	0	0	1	0	22	34	354	69	9
10 Maximum precipitation in 24 h	in mm	74	79	84	74	71	27	0	6	13	39	71	87	87	69	10
11 Mean number of days with precipitation	> 0,1 mm	16	14	15	13	4	0	0	0	1	3	13	20	99	69	11
12 Mean duration of sunshine	in h	211	196	229	240	279	300	329	313	288	288	249	220	3142	24	12
13 Mean quantity of radiation	in ly / day															13
14 Mean potential evaporation	in mm	91	84	90	86	85	74	76	88	106	128	114	94	1116	15	14
15 Mean windspeed	in m / sec	2.1	2.1	2.3	3.6	3.9	4.1	4.1	4.6	4.9	4.1	3.1	2.1	3.3	18	15
16 Mean predominent direction of the wind		E	E	E	E	E	S	S	S	E.S	E.S	E	E		18	16

111 Station / Country   Morogoro / Tanzania

Location 6°51'S / 37°40'E   Height above sealevel 579 m   Climate symbol: Köppen Aw   Troll V.4

		J	F	M	A	M	J	J	A	S	O	N	D	year	P	
1 Mean daily temperature	in °C	26.4	26.4	26.1	25.3	23.6	21.4	21.1	22.2	23.0	24.4	25.6	26.7	24.4	8	1
2 Mean daily maximum temperature	in °C	32	32	31	30	28	27	27	28	30	31	32	32	30	15	2
3 Mean daily minimum temperature	in °C	21	21	21	20	19	16	15	16	17	18	19	21	19	15	3
4 Absolute maximum temperature	in °C	36	37	36	35	32	31	31	33	33	35	36	36	37	15	4
5 Absolute minimum temperature	in °C	17	17	17	17	14	11	10	9	13	14	16	16	9	15	5
6 Mean relative humidity	in %	70	71	73	80	79	74	71	67	63	61	61	70	70	15	6
7 Mean precipitation	in mm	94	104	167	208	96	27	15	10	17	27	54	73	892	57	7
8 Maximum precipitation	in mm	301	261	500	386	402	143	119	66	102	168	238	229	1536	57	8
9 Minimum precipitation	in mm	4	2	34	98	26	1	0	0	0	1	0	5	564	57	9
10 Maximum precipitation in 24 h	in mm	63	100	93	64	40	23	38	34	62	34	75	77	100	57	10
11 Mean number of days with precipitation	> 0,1 mm	10	9	13	21	15	6	4	5	4	8	10		110	57	11
12 Mean duration of sunshine	in h	177	165	183	129	121	129	127	130	141	177	186	180	1845	16	12
13 Mean quantity of radiation	in ly / day															13
14 Mean potential evaporation	in mm	142	126	136	112	96	72	70	80	86	113	128	141	1302	5	14
15 Mean windspeed	in m / sec	1.5	1.5	1.3	0.8	0.8	0.8	1.3	1.8	1.5	1.8	2.3	2.3	1.5	14	15
16 Mean predominent direction of the wind		E	NE	NE	var	W	var	S	S	S	S	NE	E		14	16

112 Station / Country   Dar-es-Salaam / Tanzania

Location 6°50'S / 39°18'E   Height above sealevel 14 m   Climate symbol: Köppen Aw   Troll V.2

		J	F	M	A	M	J	J	A	S	O	N	D	year	P	
1 Mean daily temperature	in °C	27.8	28.0	27.5	26.4	25.6	24.4	23.6	23.6	23.9	25.0	26.1	27.2	25.8	10	1
2 Mean daily maximum temperature	in °C	30	31	31	30	30	29	29	29	29	29	30	30	30	14	2
3 Mean daily minimum temperature	in °C	25	25	24	23	22	20	19	19	19	20	22	24	22	14	3
4 Absolute maximum temperature	in °C	32	34	35	34	33	31	31	31	32	32	34	33	35	14	4
5 Absolute minimum temperature	in °C	20	20	21	21	18	16	16	15	16	17	19	21	15	14	5
6 Mean relative humidity	in %	74	75	77	82	78	73	72	70	68	67	71	73	73	10	6
7 Mean precipitation	in mm	71	64	120	280	303	35	33	25	29	49	79	91	1179	70	7
8 Maximum precipitation	in mm	260	201	346	525	600	161	221	108	71	235	331	285	1531	70	8
9 Minimum precipitation	in mm	1	1	12	44	1	1	1	1	1	2	5	1	438	70	9
10 Maximum precipitation in 24 h	in mm	60	105	115	136	152	45	37	25	37	118	70	76	152	16	10
11 Mean number of days with precipitation	> 0,1 mm	6	5	9	16	12	5	3	5	3	5	7	8	84	70	11
12 Mean duration of sunshine	in h	251	241	217	156	192	231	251	242	243	254	270	267	2815	15	12
13 Mean quantity of radiation	in ly / day															13
14 Mean potential evaporation	in mm	154	139	148	137	125	100	91	92	92	110	129	151	1468	6	14
15 Mean windspeed	in m / sec	4.1	3.3	2.3	1.8	2.3	2.8	3.1	3.1	3.3	3.1	3.1	3.1	3.1	14	15
16 Mean predominent direction of the wind		NE	NE	NE	S	S	S	S	SE	E	E	NE			14	16

254

113 Station / Country Mbeya / Tanzania

Location 8°56'S / 33°28'E — Height above sealevel 1736 m — Climate symbol: Köppen Cwb — Troll V.3

		J	F	M	A	M	J	J	A	S	O	N	D	year	P	
1 Mean daily temperature	in °C	18,5	18,4	18,2	17,3	16,3	14,8	14,2	15,6	17,7	19,5	20,1	19,1	17,5	19	1
2 Mean daily maximum temperature	in °C	23	23	23	23	22	21	21	22	25	27	26	24	23	30	2
3 Mean daily minimum temperature	in °C	14	14	14	13	11	9	8	9	11	12	14	14	12	30	3
4 Absolute maximum temperature	in °C	28	27	27	27	27	26	26	28	31	31	31	31	31	30	4
5 Absolute minimum temperature	in °C	10	10	10	9	6	3	2	-1	6	7	9	10	-1	30	5
6 Mean relative humidity	in %	78	77	78	77	71	66	67	59	55	56	60	72	68	14	6
7 Mean precipitation	in mm	199	165	161	116	17	1	1	1	3	15	52	152	883	31	7
8 Maximum precipitation	in mm	341	269	267	303	54	9	6	14	24	69	269	397	1190	31	8
9 Minimum precipitation	in mm	113	37	38	45	0	1	0	0	0	0	6	8	564	31	9
10 Maximum precipitation in 24 h	in mm	57	69	69	88	40	7	6	14	20	30	51	59	88	31	10
11 Mean number of days with precipitation	> 0,1 mm	22	20	20	17	4	0	0	0	1	3	5	17	109	31	11
12 Mean duration of sunshine	in h	127	101	118	165	257	273	304	295	270	264	204	143	2521	4	12
13 Mean quantity of radiation	in ly / day															13
14 Mean potential evaporation	in mm	76	68	72	64	58	48	45	54	67	82	84	81	799	14	14
15 Mean windspeed	in m / sec	2,8	2,6	2,3	2,8	3,3	3,9	4,1	4,3	5,1	6,2	4,9	3,3	3,9	24	15
16 Mean predominent direction of the wind		W	W	E	E	S	S	S	S	S	S	E,S	E		24	16

114 Station / Country Mtwara / Tanzania

Location 10°16'S / 40°11'E — Height above sealevel 113 m — Climate symbol: Köppen Aw — Troll V.3

		J	F	M	A	M	J	J	A	S	O	N	D	year	P	
1 Mean daily temperature	in °C	27.0	27.0	27.0	26.9	25.7	24.7	24.1	24.3	24.7	25.7	26.9	27.3	25.9	13	1
2 Mean daily maximum temperature	in °C	31	31	31	31	30	30	29	30	30	31	32	31	31	6	2
3 Mean daily minimum temperature	in °C	23	23	23	22	21	19	19	19	19	21	22	23	21	6	3
4 Absolute maximum temperature	in °C	34	35	34	34	33	32	32	33	33	34	35	34	35	6	4
5 Absolute minimum temperature	in °C	21	20	21	21	15	16	16	15	16	18	19	20	15	6	5
6 Mean relative humidity	in %															6
7 Mean precipitation	in mm	218	151	165	197	51	11	15	11	65	24	33	218	1159	13	7
8 Maximum precipitation	in mm	541	452	233	416	123	55	51	43	468	65	118	441	1504	13	8
9 Minimum precipitation	in mm	51	17	103	15	7	0	1	2	4	7	1	4	780	13	9
10 Maximum precipitation in 24 h	in mm	200	169	83	152	109	46	26	23	91	31	38	184	200	13	10
11 Mean number of days with precipitation	> 0,1 mm	13	12	12	14	5	2	3	3	3	4	5	8	84	13	11
12 Mean duration of sunshine	in h															12
13 Mean quantity of radiation	in ly / day															13
14 Mean potential evaporation	in mm	149	134	145	137	121	98	93	97	102	127	145	155	1503	13	14
15 Mean windspeed	in m / sec	4,6	4,1	3,9	4,6	5,9	6,5	6,7	6,2	5,1	5,1	4,6	4,6	5,1	13	15
16 Mean predominent direction of the wind		N	N	N	SE	S	SE	SE	SE	NE	NE	NE	N		13	16

115 Station / Country Cabinda / Angola

Location 5°33'S / 12°11'E — Height above sealevel 20 m — Climate symbol: Köppen BSh — Troll V.3

		J	F	M	A	M	J	J	A	S	O	N	D	year	P	
1 Mean daily temperature	in °C	26.4	26.7	26.7	27.0	25.8	23.0	22.0	22.5	23.9	25.6	26.1	25.8	25.1	8	1
2 Mean daily maximum temperature	in °C	30	31	31	30	29	26	26	26	27	28	29	28	28	13	2
3 Mean daily minimum temperature	in °C	23	23	23	23	23	21	18	19	21	23	23	23	22	13	3
4 Absolute maximum temperature	in °C	32	33	33	33	33	30	28	29	30	33	33	34	34	13	4
5 Absolute minimum temperature	in °C	20	19	18	19	19	17	14	15	15	19	19	19	14	13	5
6 Mean relative humidity	in %	79	79	78	80	82	75	77	77	77	78	80	80	79	21	6
7 Mean precipitation	in mm	59	109	85	117	56	<1	<1	<1	6	34	114	89	670		7
8 Maximum precipitation	in mm	233	236	371	301	259	3	1	5	11	124	333	180	1331	15	8
9 Minimum precipitation	in mm	4	1	46	46	2	0	0	0		9	14	6	332	15	9
10 Maximum precipitation in 24 h	in mm	70	76	81	103	56	2	1	2	10	90	135	96	135		10
11 Mean number of days with precipitation	> 0,1 mm	5	6	7	8	3	0	0	4	2	6	8	6	52		11
12 Mean duration of sunshine	in h	117	115	147	144	105	95	97	69	61	62	99	105	1217		12
13 Mean quantity of radiation	in ly / day															13
14 Mean potential evaporation	in mm	143	123	141	135	118	80	69	76	87	119	123	132	1346	13	14
15 Mean windspeed	in m / sec															15
16 Mean predominent direction of the wind		SW	SW	SW	SW	SW	SW	SW	SW	SW	SW	SW	SW			16

116 Station / Country Luanda / Angola

Location 8°49'S / 13°13'E — Height above sealevel 45 m — Climate symbol: Köppen BWh — Troll V.4

		J	F	M	A	M	J	J	A	S	O	N	D	year	P	
1 Mean daily temperature	in °C	25,6	26,3	26,5	26,2	24,8	21,9	20,1	20,1	21,6	23,6	24,9	25,3	23,9	39	1
2 Mean daily maximum temperature	in °C	30	31	31	31	29	27	24	24	26	28	29	30	28	20	2
3 Mean daily minimum temperature	in °C	24	24	24	24	23	20	18	18	20	22	23	23	22	20	3
4 Absolute maximum temperature	in °C	33	35	35	35	36	32	30	29	29	32	34	33	36	20	4
5 Absolute minimum temperature	in °C	19	20	20	21	18	15	14	14	17	18	20	19	14	20	5
6 Mean relative humidity	in %	79	78	79	82	82	82	82	83	82	81	81	80	81	20	6
7 Mean precipitation	in mm	26	35	97	124	19	0	0	1	2	6	34	23	367	30	7
8 Maximum precipitation	in mm	163	152	299	404	100	6	1	15	8	25	159	135	864	30	8
9 Minimum precipitation	in mm	0	0	0	15	0	0	0	0	0	0	0	0	62	30	9
10 Maximum precipitation in 24 h	in mm	96	88	94	158	59	6	1	14	8	20	54	63	158	30	10
11 Mean number of days with precipitation	> 0,1 mm	2	3	7	9	2	0	0	<1	1	2	3	3	32	30	11
12 Mean duration of sunshine	in h	219	208	213	199	233	223	175	150	145	164	199	212	2341		12
13 Mean quantity of radiation	in ly / day															13
14 Mean potential evaporation	in mm	137	134	147	134	116	78	60	62	74	105	121	132	1298	12	14
15 Mean windspeed	in m / sec	3,6	3,9	3,6	3,3	3,3	3,3	3,1	3,1	3,3	4,2	4,2	3,6	3,6		15
16 Mean predominent direction of the wind		W	W	W	W	W	W	SW	W	W	W	W	W			16

117 Station / Country Teixeira de Sousa / Angola

Location 10°43'S / 22°13'E   Height above sealevel 1100 m   Climate symbol: Köppen Aw   Troll V.3

		J	F	M	A	M	J	J	A	S	O	N	D	year	P	
1 Mean daily temperature	in °C	23,3	23,3	23,3	22,8	21,7	20,0	20,0	22,8	23,9	24,4	23,9	23,3	22,8	9	1
2 Mean daily maximum temperature	in °C	30,6	30,0	30,0	30,6	30,6	29,4	30,0	32,2	33,3	32,8	31,7	30,0	31,1	9	2
3 Mean daily minimum temperature	in °C	16,1	16,1	16,1	15,0	12,8	10,0	10,0	12,8	14,4	15,6	16,1	16,1	14,4	9	3
4 Absolute maximum temperature	in °C	38,9	35,0	37,2	34,4	35,0	33,3	32,8	35,6	37,8	37,2	38,3	35,0	38,9	9	4
5 Absolute minimum temperature	in °C	10,0	10,0	10,0	8,9	4,4	2,8	3,9	5,0	5,0	10,0	10,0	10,0	2,8	9	5
6 Mean relative humidity	in % +	78	78	79	72	52	42	43	41	49	62	74	76	62	8	6
7 Mean precipitation	in mm	224	218	236	132	13	0	<3	8	18	79	201	213	1341	10	7
8 Maximum precipitation	in mm															8
9 Minimum precipitation	in mm															9
10 Maximum precipitation in 24 h	in mm	112	79	79	58	51	0	3	25	23	51	66	79	112	10	10
11 Mean number of days with precipitation	> 0,1 mm	15	17	16	12	1	0	<1	<1	3	7	14	16	102	9	11
12 Mean duration of sunshine	in h															12
13 Mean quantity of radiation	in ly / day															13
14 Mean potential evaporation	in mm															14
15 Mean windspeed	in m / sec															15
16 Mean predominent direction of the wind																16

+ 9.30 a.m.

118 Station / Country Nova Lisboa / Angola

Location 12°48'S / 15°45'E   Height above sealevel 1700 m   Climate symbol: Köppen Cwb   Troll V.3

		J	F	M	A	M	J	J	A	S	O	N	D	year	P	
1 Mean daily temperature	in °C	19,7	19,9	19,9	19,7	18,1	16,3	16,7	18,9	20,9	21,0	19,9	19,8	19,2	20	1
2 Mean daily maximum temperature	in °C	25	25	25	25	26	25	25	27	29	27	25	25	26		2
3 Mean daily minimum temperature	in °C	14	14	15	14	11	8	8	10	13	14	14	15	12		3
4 Absolute maximum temperature	in °C	31	31	30	29	29	28	29	31	32	32	31	30	32		4
5 Absolute minimum temperature	in °C	9	8	10	7	5	2	2	5	8	11	8	9	2		5
6 Mean relative humidity	in %	73	71	75	68	52	41	35	31	41	62	75	74	58		6
7 Mean precipitation	in mm	209	179	231	144	16	0	0	1	19	124	231	233	1386		7
8 Maximum precipitation	in mm	422	373	500	422	113	0	1	9	71	216	331	477	2350	20	8
9 Minimum precipitation	in mm	58	24	88	50	0	0	0	0	0	35	82	79	962	20	9
10 Maximum precipitation in 24 h	in mm	86	79	89	86	56	0	1	6	18	53	70	78	89		10
11 Mean number of days with precipitation	> 0,1 mm	16	14	17	10	2	0	0	3	13	18	17	110			11
12 Mean duration of sunshine	in h	141	139	142	171	243	289	268	256	201	165	134	140	2268		12
13 Mean quantity of radiation	in ly / day															13
14 Mean potential evaporation	in mm	82	70	76	73	63	51	52	72	85	84	77	81	866	13	14
15 Mean windspeed	in m / sec	1,9	1,9	1,7	1,7	1,7	1,7	1,7	1,9	2,2	2,2	1,9	1,9	1,9		15
16 Mean predominent direction of the wind		W	W	var	E	E	E	E	E	E	N	N	N			16

119 Station / Country Cangamba / Angola

Location 13°41'S / 19°52'E   Height above sealevel 1325 m   Climate symbol: Köppen Cwa   Troll V.3

		J	F	M	A	M	J	J	A	S	O	N	D	year	P	
1 Mean daily temperature	in °C	22,8	22,8	23,9	23,2	21,2	17,8	18,0	18,2	22,1	22,9	23,0	22,4	21,5	7	1
2 Mean daily maximum temperature	in °C	29	29	31	32	32	30	28	31	32	31	29	29	30	5	2
3 Mean daily minimum temperature	in °C	17	17	17	14	11	9	8	8	13	15	16	16	23	5	3
4 Absolute maximum temperature	in °C	33	38	41	43	42	41	37	39	37	37	37	34	43	5	4
5 Absolute minimum temperature	in °C	13	12	11	5	1	-3	-7	0	4	7	9	12	-7	5	5
6 Mean relative humidity	in % +	86	85	80	75	76	76	78	73	68	72	75	80	77	7	6
7 Mean precipitation	in mm	225	187	172	46	1	0	0	5	5	41	130	215	1027	7	7
8 Maximum precipitation	in mm															8
9 Minimum precipitation	in mm															9
10 Maximum precipitation in 24 h	in mm	58	51	56	30	2	0	0	10	8	42	71	63	71	7	10
11 Mean number of days with precipitation	> 0,1 mm	17	13	13	6	1	0	0	1	1	5	12	14	83	7	11
12 Mean duration of sunshine	in h															12
13 Mean quantity of radiation	in ly / day															13
14 Mean potential evaporation	in mm	107	94	110	97	80	48	53	54	8	7	106	104	868	7	14
15 Mean windspeed	in m / sec															15
16 Mean predominent direction of the wind																16

+ 9.30 a.m.

120 Station / Country Mupa / Angola

Location 16°07'S / 15°53'E   Height above sealevel 1215 m   Climate symbol: Köppen BSh   Troll V.3

		J	F	M	A	M	J	J	A	S	O	N	D	year	P	
1 Mean daily temperature	in °C	24,1	23,3	23,3	22,5	20,3	17,5	17,2	19,7	23,3	25,6	25,0	24,4	22,1		1
2 Mean daily maximum temperature	in °C	31,1	29,4	30,0	30,6	29,4	27,2	27,8	30,6	33,9	34,4	33,3	31,7	30,6	11	2
3 Mean daily minimum temperature	in °C	17,2	17,2	16,7	14,4	11,1	7,8	6,7	8,9	12,8	16,7	16,7	17,2	13,3	11	3
4 Absolute maximum temperature	in °C	41,1	37,8	37,2	35,6	35,6	32,8	33,9	36,1	38,9	40,0	41,1	41,7	41,7	11	4
5 Absolute minimum temperature	in °C	10,6	11,1	10,0	6,7	0,6	0,6	-1,7	2,8	2,8	8,3	6,1	11,1	-1,7	11	5
6 Mean relative humidity	in %															6
7 Mean precipitation	in mm	104	180	145	61	3	0	0	0	1	28	58	132	712		7
8 Maximum precipitation	in mm															8
9 Minimum precipitation	in mm															9
10 Maximum precipitation in 24 h	in mm	88	76	71	77	10	0	<1	0	6	14	45	67	88		10
11 Mean number of days with precipitation	> 0,1 mm	8	14	11	5	<1	0	<1	0	<1	2	6	10	57		11
12 Mean duration of sunshine	in h															12
13 Mean quantity of radiation	in ly / day															13
14 Mean potential evaporation	in mm	118	97	100	86	66	43	43	64	96	132	121	125	1091		14
15 Mean windspeed	in m / sec															15
16 Mean predominent direction of the wind																16

121 Station / Country Kasama / Zambia

Location 10°13'S / 31°08'E   Height above sealevel 1382 m   Climate symbol: Köppen **Cwa**   Troll V.3

		J	F	M	A	M	J	J	A	S	O	N	D	year	P	
1 Mean daily temperature	in °C	20,3	20,8	20,8	20,7	19,4	17,2	17,0	18,7	21,6	23,6	22,7	21,2	20,3	27	1
2 Mean daily maximum temperature	in °C	26	26	26	26	26	24	25	28	30	31	29	27	27	25	2
3 Mean daily minimum temperature	in °C	16	16	16	16	13	10	10	11	14	16	17	16	14	25	3
4 Absolute maximum temperature	in °C	31	30	30	29	29	28	29	33	34	35	35	33	35	25	4
5 Absolute minimum temperature	in °C	14	13	13	11	7	4	4	3	6	12	13	14	3	25	5
6 Mean relative humidity	in %	81	82	80	76	66	60	55	48	43	41	60	78	64	10	6
7 Mean precipitation	in mm	267	251	259	69	8	0	0	1	1	17	135	237	1245	30	7
8 Maximum precipitation	in mm	429	378	354	181	81	1	0	12	7	78	282	402	1574	30	8
9 Minimum precipitation	in mm	142	97	122	3	0	0	0	0	0	0	13	69	830	30	9
10 Maximum precipitation in 24 h	in mm	103	81	93	90	27	1	0	12	7	37	81	87	103	30	10
11 Mean number of days with precipitation	> 0,1 mm	21	19	18	7	1	0	0	0	0	2	13	19	100	30	11
12 Mean duration of sunshine	in h	127	118	167	234	282	291	316	310	285	264	219	167	2780	22	12
13 Mean quantity of radiation	in ly/day															13
14 Mean potential evaporation	in mm	83	80	86	80	68	50	51	64	87	111	97	92	949	10	14
15 Mean windspeed	in m/sec	1,7	1,6	1,6	2,1	2,6	2,8	3,0	3,2	3,4	2,8	2,0	1,6	2,4	21	15
16 Mean predominant direction of the wind		NNW	NNW	ENE	E	ESE	ESE	ESE	ESE	ESE	E	ENE	NNW		17	16

122 Station / Country Ndola / Zambia

Location 13°00'S / 28°37'E   Height above sealevel 1269 m   Climate symbol: Köppen **Cwa**   Troll V.3

		J	F	M	A	M	J	J	A	S	O	N	D	year	P	
1 Mean daily temperature	in °C	21,4	21,4	21,1	20,3	17,8	14,7	15,0	17,0	20,8	23,3	22,8	21,7	19,8	10	1
2 Mean daily maximum temperature	in °C	26	26	27	27	26	24	25	27	30	32	29	27	27	20	2
3 Mean daily minimum temperature	in °C	17	17	16	13	9	6	6	9	12	15	17	17	13	20	3
4 Absolute maximum temperature	in °C	32	30	31	31	30	29	29	31	34	36	35	33	36	20	4
5 Absolute minimum temperature	in °C	12	11	10	7	1	0	-2	-1	3	9	11	12	-2	20	5
6 Mean relative humidity	in %	80	82	77	71	63	58	52	45	40	40	62	79	62	10	6
7 Mean precipitation	in mm	289	252	184	39	5	0	0	1	1	19	130	249	1169	30	7
8 Maximum precipitation	in mm	527	417	319	125	41	1	0	14	7	55	222	427	1621	30	8
9 Minimum precipitation	in mm	58	124	57	3	0	0	0	0	0	0	14	108	782	30	9
10 Maximum precipitation in 24 h	in mm	86	99	77	69	40	1	0	14	6	30	61	106	106	30	10
11 Mean number of days with precipitation	> 0,1 mm	20	19	15	5	0	0	0	0	0	4	12	19	94	30	11
12 Mean duration of sunshine	in h	136	120	180	240	270	267	288	295	279	276	192	140	2683	20	12
13 Mean quantity of radiation	in ly/day															13
14 Mean potential evaporation	in mm	94	84	88	74	55	37	39	53	81	112	105	99	921	5	14
15 Mean windspeed	in m/sec	1,3	1,3	1,3	1,5	1,5	1,8	2,3	2,6	2,7	2,6	1,7	1,3	1,8	17	15
16 Mean predominant direction of the wind		NNE	NNE	E	E	ESE	ESE	ESE	ESE	E	E	ENE	NE		14	16

123 Station / Country Mongu / Zambia

Location 15°15'S / 23°10'E   Height above sealevel 1052 m   Climate symbol: Köppen **Aw**   Troll V.3

		J	F	M	A	M	J	J	A	S	O	N	D	year	P	
1 Mean daily temperature	in °C	24,2	24,4	24,2	23,9	21,5	19,2	19,1	21,8	25,8	27,2	26,1	24,2	23,5	11	1
2 Mean daily maximum temperature	in °C	28	28	28	29	28	26	26	29	33	34	31	29	29	22	2
3 Mean daily minimum temperature	in °C	19	19	18	17	13	10	9	12	16	18	18	18	16	22	3
4 Absolute maximum temperature	in °C	34	33	34	33	33	32	32	36	37	39	38	37	39	22	4
5 Absolute minimum temperature	in °C	13	14	12	9	2	0	-2	2	7	7	12	13	-2	22	5
6 Mean relative humidity	in %	77	80	75	65	56	50	42	34	31	40	64	75	57	10	6
7 Mean precipitation	in mm	217	211	145	37	1	0	0	0	2	35	102	222	972	30	7
8 Maximum precipitation	in mm	419	483	341	165	22	16	0	11	28	135	264	380	1379	30	8
9 Minimum precipitation	in mm	68	73	6	0	0	0	0	0	0	0	17	92	604	30	9
10 Maximum precipitation in 24 h	in mm	85	114	89	51	8	14	0	11	28	41	71	84	114	30	10
11 Mean number of days with precipitation	> 0,1 mm	17	16	12	4	0	0	0	0	0	4	11	17	81	30	11
12 Mean duration of sunshine	in h	167	154	205	267	301	291	304	310	282	260	206	177	2924	12	12
13 Mean quantity of radiation	in ly/day															13
14 Mean potential evaporation	in mm	109	97	102	89	67	46	50	74	115	141	119	109	1118	10	14
15 Mean windspeed	in m/sec	2,4	2,3	2,5	2,8	3,3	3,3	3,8	4,0	4,0	3,1	2,4	2,4	3,0	7	15
16 Mean predominant direction of the wind		NNE	NNW	ENE	E	E	E	E	E	E	ENE	NE	NNE		7	16

124 Station / Country Lusaka / Zambia

Location 15°25'S / 28°19'E   Height above sealevel 1274 m   Climate symbol: Köppen **Cwa**   Troll V.3

		J	F	M	A	M	J	J	A	S	O	N	D	year	P	
1 Mean daily temperature	in °C	21,4	21,7	21,1	20,6	18,6	16,4	16,1	18,3	22,0	24,4	23,3	22,0	20,5	17	1
2 Mean daily maximum temperature	in °C	26	26	26	26	25	23	23	26	29	31	29	27	26	23	2
3 Mean daily minimum temperature	in °C	17	17	16	15	12	10	10	12	15	18	18	17	15	23	3
4 Absolute maximum temperature	in °C	35	31	33	32	30	28	28	34	35	38	38	34	38	23	4
5 Absolute minimum temperature	in °C	14	12	12	10	8	4	4	7	11	12	14		4	23	5
6 Mean relative humidity	in %	82	86	79	71	63	59	34	46	41	40	59	78	62	10	6
7 Mean precipitation	in mm	218	196	106	21	4	0	0	0	0	15	91	186	837	30	7
8 Maximum precipitation	in mm	414	530	250	65	37	6	2	4	6	137	184	304	1134	30	8
9 Minimum precipitation	in mm	97	46	5	0	0	0	0	0	0	0	1	60	518	30	9
10 Maximum precipitation in 24 h	in mm	83	72	56	32	25	6	1	4	4	92	72	75	92	30	10
11 Mean number of days with precipitation	> 0,1 mm	17	16	10	2	0	0	0	0	0	2	8	16	71	30	11
12 Mean duration of sunshine	in h	155	143	202	258	282	267	285	301	285	279	204	174	2835	23	12
13 Mean quantity of radiation	in ly/day															13
14 Mean potential evaporation	in mm	94	82	86	74	59	43	44	59	88	122	110	101	962	8	14
15 Mean windspeed	in m/sec	2,3	2,3	2,9	3,7	3,8	3,9	4,2	4,3	4,5	4,2	3,2	2,5	3,5	23	15
16 Mean predominant direction of the wind		ENE	ENE	E	E	E	E	ESE	E	E	E	E	E		20	16

125 Station / Country Salisbury / Zimbabwe

Location 17°50'S / 31°01'E   Height above sealevel 1470 m   Climate symbol: Köppen Cwb   Troll V,3

		J	F	M	A	M	J	J	A	S	O	N	D	year	P	
1	Mean daily temperature	in °C	20,8	20,5	20,0	18,7	16,0	13,8	13,6	15,6	18,9	21,5	21,5	20,8	18,5	60	1
2	Mean daily maximum temperature	in °C	26	26	26	25	23	21	21	23	27	29	27	26	25	30	2
3	Mean daily minimum temperature	in °C	16	16	14	12	9	7	7	8	"12	15	15	16	12	30	3
4	Absolute maximum temperature	in °C	32	31	32	32	29	27	28	31	33	35	35	33	35	30	4
5	Absolute minimum temperature	in °C	8	9	8	6	2	0	-1	-1	3	7	8	9	-1	30	5
6	Mean relative humidity	in %	75	75	72	64	58	54	50	45	42	44	57	68	58	20	6
7	Mean precipitation	in mm	213	173	101	39	11	5	1	3	5	30	100	186	868	30	7
8	Maximum precipitation	in mm	514	414	298	93	42	22	11	45	19	94	236	429	1291	30	8
9	Minimum precipitation	in mm	74	18	9	0	0	0	0	0	0	0	33	71	550	30	9
10	Maximum precipitation in 24 h	in mm	76	156	68	50	31	14	11	28	19	38	71	96	156	30	10
11	Mean number of days with precipitation	> 0,1 mm	15	13	9	4	2	1	0	0	1	4	10	14	73	30	11
12	Mean duration of sunshine	in h	195	179	214	246	267	252	273	291	282	282	207	189	2977	30	12
13	Mean quantity of radiation	in ly/day	514	507	558	535	483	439	434	517	586	597	533	499	517	7	13
14	Mean potential evaporation	in mm	95	82	82	70	50	36	36	48	70	96	96	97	858		14
15	Mean windspeed	in m/sec	2,6	2,6	2,6	2,8	2,8	3,0	3,3	3,6	3,9	3,8	3,3	2,9	3,1	28	15
16	Mean predominent direction of the wind		ENE	ENE	E	ENE	ENE	E	E	ENE	ENE	ENE	ENE	ENE		28	16

126 Station / Country Wankie / Zimbabwe

Location 18°22'S / 26°29'E   Height above sealevel 782 m   Climate symbol: Köppen BSh   Troll V,3

		J	F	M	A	M	J	J	A	S	O	N	D	year	P	
1	Mean daily temperature	in °C	26,4	26,2	25,9	25,3	22,0	19,2	18,9	21,7	26,1	29,5	28,7	27,0	24,8	15	1
2	Mean daily maximum temperature	in °C	32,2	31,7	31,7	32,2	30,0	27,2	27,2	30,0	33,9	36,7	35,6	32,8	31,7	15	2
3	Mean daily minimum temperature	in °C	20,6	20,6	20,0	18,3	13,9	11,1	10,6	13,3	18,3	22,2	21,7	21,1	17,8	15	3
4	Absolute maximum temperature	in °C	40,0	38,9	37,2	37,2	35,6	32,8	32,8	36,7	38,9	42,8	43,3	41,7	43,3	15	4
5	Absolute minimum temperature	in °C	15,0	15,0	14,4	10,0	6,7	2,2	5,0	6,7	8,9	12,8	13,9	12,8	2,2	15	5
6	Mean relative humidity	in %	73	75	69	62	56	56	52	43	36	35	51	64	56	15	6
7	Mean precipitation	in mm	147	147	79	18	5	<3	0	<3	<3	18	58	119	592	15	7
8	Maximum precipitation	in mm															8
9	Minimum precipitation	in mm															9
10	Maximum precipitation in 24 h	in mm	135	104	79	48	36	3	0	<3	10	56	48	81	135	44	10
11	Mean number of days with precipitation	> 0,1 mm	16	13	9	3	1	<1	0	<1	<1	3	9	13	68	15	11
12	Mean duration of sunshine	in h															12
13	Mean quantity of radiation	in ly/day															13
14	Mean potential evaporation	in mm	151	131	132	110	74	44	45	70	128	171	164	159	1379		14
15	Mean windspeed	in m/sec															15
16	Mean predominent direction of the wind																16

+ 8.30 a.m.

127 Station / Country Umtali / Zimbabwe

Location 18°58'S / 32°40 E   Height above sealevel 1117 m   Climate symbol: Köppen Cwa   Troll V,3

		J	F	M	A	M	J	J	A	S	O	N	D	year	P	
1	Mean daily temperature	in °C	22,5	22,2	21,7	20,3	17,5	15,0	14,7	16,7	19,1	22,0	22,5	22,2	19,7	15	1
2	Mean daily maximum temperature	in °C	28	27	26	26	24	21	21	23	26	29	28	28	26	20	2
3	Mean daily minimum temperature	in °C	17	17	16	15	11	9	9	10	13	15	16	17	14	20	3
4	Absolute maximum temperature	in °C	36	35	33	35	31	29	29	32	36	39	39	37	39	20	4
5	Absolute minimum temperature	in °C	11	11	11	8	3	1	0	2	4	6	8	11	0	20	5
6	Mean relative humidity	in %	77	79	76	70	63	59	57	55	54	56	66	73	65		6
7	Mean precipitation	in mm	171	134	99	26	10	9	7	11	10	27	91	161	756	30	7
8	Maximum precipitation	in mm	469	365	346	86	60	26	19	55	61	101	169	274	1239	30	8
9	Minimum precipitation	in mm	42	14	11	0	0	0	0	0	0	0	17	32	401	30	9
10	Maximum precipitation in 24 h	in mm	116	85	82	66	28	17	14	38	55	46	69	90	116	30	10
11	Mean number of days with precipitation	> 0,1 mm	13	12	9	4	2	2	2	2	2	3	9	13	73	30	11
12	Mean duration of sunshine	in h	226	196	223	243	251	225	233	257	264	276	219	220	2833	20	12
13	Mean quantity of radiation	in ly/day															13
14	Mean potential evaporation	in mm	110	95	91	75	54	39	38	50	69	102	105	107	935		14
15	Mean windspeed	in m/sec	2,2	2,1	2,0	1,9	1,7	1,5	1,9	2,5	3,3	3,7	3,0	2,4	2,3	13	15
16	Mean predominent direction of the wind		ESE	ESE	ESE	E	ESE	ESE	E	E	E	E	E	E		13	16

128 Station / Country Bulawayo / Zimbabwe

Location 20°09'S / 28°37'E   Height above sealevel 1344 m   Climate symbol: Köppen Cwa   Troll V,3

		J	F	M	A	M	J	J	A	S	O	N	D	year	P	
1	Mean daily temperature	in °C	21,8	21,2	20,4	19,0	16,3	14,0	13,8	16,0	19,6	22,4	22,4	21,9	19,1	60	1
2	Mean daily maximum temperature	in °C	27	27	26	26	24	21	21	24	27	30	28	27	26	30	2
3	Mean daily minimum temperature	in °C	16	16	15	13	10	7	7	9	12	15	16	16	13	30	3
4	Absolute maximum temperature	in °C	36	34	34	33	31	28	28	32	36	36	37	35	37	30	4
5	Absolute minimum temperature	in °C	9	8	9	3	-1	0	0	3	7	9	11	-2	30	5
6	Mean relative humidity	in %	74	76	71	63	56	58	50	43	42	44	60	69	59	10	6
7	Mean precipitation	in mm	134	112	65	21	9	3	0	1	5	25	89	124	589	30	7
8	Maximum precipitation	in mm	308	368	183	87	34	33	2	10	29	116	241	273	1093	30	8
9	Minimum precipitation	in mm	0	0	1	0	0	0	0	0	0	0	8	3	199	30	9
10	Maximum precipitation in 24 h	in mm	132	90	102	47	24	22	2	10	12	49	106	90	132	30	10
11	Mean number of days with precipitation	> 0,1 mm	10	9	5	3	1	0	0	0	1	3	8	10	50	30	11
12	Mean duration of sunshine	in h	220	207	239	255	282	273	285	301	285	267	222	217	3053	30	12
13	Mean quantity of radiation	in ly/day	564	537	528	486	452	401	422	491	545	574	529	529	505	8	13
14	Mean potential evaporation	in mm	103	88	84	71	50	34	35	50	74	104	105	87	885		14
15	Mean windspeed	in m/sec	3,5	3,5	3,6	3,3	3,5	3,8	4,0	4,2	4,3	4,0	3,4	3,2	3,7	25	15
16	Mean predominent direction of the wind		ESE	ESE	ESE	ESE	ESE	ESE	ESE	ESE	ESE	ESE	ESE	ESE		30	16

129 Station/Country Karnoga/Malawi

Location 9°56'S/33°56'E   Height above sealevel 482 m   Climate symbol: Köppen Aw   Troll V,3

		J	F	M	A	M	J	J	A	S	O	N	D	year	P	
1	Mean daily temperature	in °C	25,8	25,6	25,6	25,3	23,9	21,7	21,1	22,0	23,3	25,8	27,5	26,7	24,5	8	1
2	Mean daily maximum temperature	in °C	30	30	29	29	28	27	27	28	30	32	33	31	29	20	2
3	Mean daily minimum temperature	in °C	22	22	21	21	19	16	15	16	17	19	21	22	19	20	3
4	Absolute maximum temperature	in °C	35	34	34	32	32	31	30	32	34	37	37	36	37	20	4
5	Absolute minimum temperature	in °C	18	18	18	17	14	11	11	11	12	15	17	19	11	20	5
6	Mean relative humidity	in %	81	83	85	87	78	71	69	69	65	60	64	76	74	10	6
7	Mean precipitation	in mm	183	163	316	187	35	6	1	2	1	4	39	144	1081	30	7
8	Maximum precipitation	in mm	432	345	599	475	132	82	4	14	11	57	210	421	1826	30	8
9	Minimum precipitation	in mm	38	50	119	44	0	0	0	0	0	0	0	31	649	30	9
10	Maximum precipitation in 24 h	in mm	103	132	137	111	48	17	3	13	11	45	131	116	137	30	10
11	Mean number of days with precipitation	> 0,1 mm	12	13	18	12	5	1	0	0	0	0	3	12	76	30	11
12	Mean duration of sunshine	in h	177	162	220	210	257	258	264	319	306	322	267	167	2929	22	12
13	Mean quantity of radiation	in ly/day															13
14	Mean potential evaporation	in mm	135	124	122	112	100	69	65	74	90	132	144	145	1312	11	14
15	Mean windspeed	in m/sec	1,3	1,3	1,3	2,2	2,3	2,0	2,5	2,2	2,2	2,7	2,8	1,6	2,0	6	15
16	Mean predominent direction of the wind		E	ENE	ESE	SE	SE	SE	SE	SE	ESE	ESE	ESE	ESE		6	16

130 Station/Country Lilongwe/Malawi

Location 13°58'S/33°42'E   Height above sealevel 1134 m   Climate symbol: Köppen Cwa   Troll V,3

		J	F	M	A	M	J	J	A	S	O	N	D	year	P	
1	Mean daily temperature	in °C	21,5	20,8	20,8	20,3	18,0	16,2	15,5	17,4	20,4	22,7	23,0	22,5	19,9	4	1
2	Mean daily maximum temperature	in °C	27	27	27	27	26	24	24	25	25	30	30	28	27	20	2
3	Mean daily minimum temperature	in °C	17	17	16	14	10	8	6	8	10	15	17	18	13	20	3
4	Absolute maximum temperature	in °C	32	31	32	30	31	30	29	31	33	36	34	34	36	20	4
5	Absolute minimum temperature	in °C	13	12	9	4	3	-1	-3	-2	2	9	12	11	-3	20	5
6	Mean relative humidity	in %	82	85	81	77	71	67	61	58	54	50	59	76	68	10	6
7	Mean precipitation	in mm	208	207	132	37	5	1	0	2	3	5	70	175	845	30	7
8	Maximum precipitation	in mm	428	379	356	115	17	18	2	41	79	38	188	290	1197	30	8
9	Minimum precipitation	in mm	67	17	30	4	0	0	0	0	0	0	0	73	465	30	9
10	Maximum precipitation in 24 h	in mm	83	89	77	93	17	18	2	24	79	21	75	99	99	30	10
11	Mean number of days with precipitation	> 0,25mm	14	12	9	4	1	1	0	0	0	1	4	12	58	30	11
12	Mean duration of sunshine	in h	161	148	180	237	254	231	248	284	304	234	171	171	2699	15	12
13	Mean quantity of radiation	in ly/day	425	453	457	473	449	434	395	470	575	562	495	432	468	6	13
14	Mean potential evaporation	in mm	102	89	94	75	56	38	39	48	70	98	108	109	926	11	14
15	Mean windspeed	in m/sec	2,0	1,8	1,9	2,2	2,2	2,4	2,6	2,9	3,2	3,2	2,8	2,1	2,4	10	15
16	Mean predominent direction of the wind		E	SE	SE	SE	SE	SE	SE	ESE	E	E	E	E		10	16

131 Station/Country Blantyre/Malawi

Location 15°41'S/34°58'E   Height above sealevel 766 m   Climate symbol: Köppen Aw   Troll V,3

		J	F	M	A	M	J	J	A	S	O	N	D	year	P	
1	Mean daily temperature	in °C	24,0	24,0	23,0	22,5	20,5	18,5	18,5	20,0	23,0	29,0	25,5	24,5	22,0		1
2	Mean daily maximum temperature	in °C	28	28	27	28	26	24	24	26	29	31	31	29	27	22	2
3	Mean daily minimum temperature	in °C	20	20	19	17	15	13	13	14	17	19	20	20	17	22	3
4	Absolute maximum temperature	in °C	35	35	36	32	32	31	29	34	35	38	38	36	38	22	4
5	Absolute minimum temperature	in °C	15	14	14	13	10	8	8	8	10	13	15	16	8	22	5
6	Mean relative humidity	in %	79	80	80	75	70	69	63	57	52	50	60	74	67	10	6
7	Mean precipitation	in mm	200	179	125	43	9	4	3	1	5	20	81	164	834	20	7
8	Maximum precipitation	in mm	405	315	337	97	48	26	14	11	35	71	197	336	1145	22	8
9	Minimum precipitation	in mm	30	40	3	1	0	0	0	0	0	0	11	47	408	22	9
10	Maximum precipitation in 24 h	in mm	152	95	80	37	46	25	14	6	35	46	55	76	152	22	10
11	Mean number of days with precipitation	> 0,1 mm	14	12	11	5	1	0	1	1	1	2	8	12	68	22	11
12	Mean duration of sunshine	in h	186	174	189	231	251	213	226	248	243	270	219	189	2639	17	12
13	Mean quantity of radiation	in ly/day															13
14	Mean potential evaporation	in mm	101	89	93	80	64	48	47	56	80	109	113	103	983		14
15	Mean windspeed	in m/sec	2,4	2,4	2,8	3,0	3,1	3,3	3,7	3,8	3,9	4,0	3,2	2,7	3,2	16	15
16	Mean predominent direction of the wind		SE	SSE	SSE	SSE	SSE	SSE	SSE	SE	SE	ESE	ESE	SE		16	16

132 Station/Country Nova Freixa/Mozambique

Location 14°48'S/36°52'E   Height above sealevel 587 m   Climate symbol: Köppen Aw   Troll V,3

		J	F	M	A	M	J	J	A	S	O	N	D	year	P	
1	Mean daily temperature	in °C	25,7	25,8	25,4	24,6	22,0	20,0	19,8	21,6	24,6	27,3	28,0	26,5	24,2	30	1
2	Mean daily maximum temperature	in °C	31	31	31	31	29	28	28	29	32	35	35	33	31	30	2
3	Mean daily minimum temperature	in °C	20	20	20	18	15	12	12	14	17	20	21	20	18	30	3
4	Absolute maximum temperature	in °C	39	40	37	36	39	33	34	37	37	42	44	44	44	30	4
5	Absolute minimum temperature	in °C	10	10	8	8	7	3	3	6	9	10	10	10	3	30	5
6	Mean relative humidity	in %	69	69	65	60	56	53	50	46	48	40	48	64	56	9	6
7	Mean precipitation	in mm	246	205	146	25	7	2	1	2	5	15	50	186	889	30	7
8	Maximum precipitation	in mm															8
9	Minimum precipitation	in mm															9
10	Maximum precipitation in 24 h	in mm	135	60	145	37	30	22	7	19	90	95	56	76	145	30	10
11	Mean number of days with precipitation	> 0,1 mm	14	13	10	3	1	0	0	0	0	1	5	12	59	30	11
12	Mean duration of sunshine	in h	192	195	220	228	260	243	233	254	273	295	228	195	2816	9	12
13	Mean quantity of radiation	in ly/day															13
14	Mean potential evaporation	in mm	140	122	128	107	83	61	63	83	119	159	159	151	1375	16	14
15	Mean windspeed	in m/sec															15
16	Mean predominent direction of the wind		NE	var	S	S	S	S	S	S	S,NE	NE	NE	NE		10	16

259

133 Station / Country   Mossuril / Mozambique

Location 14°57'S/40°40'E — Height above sealevel 15 m — Climate symbol: Köppen Aw — Troll V,3

		J	F	M	A	M	J	J	A	S	O	N	D	year	P
1 Mean daily temperature	in °C	28,0	27,8	27,5	26,5	24,7	23,0	22,3	22,7	24,0	25,7	27,7	28,1	24,8	12
2 Mean daily maximum temperature	in °C	32	33	32	31	30	28	28	28	30	31	33	33	31	30
3 Mean daily minimum temperature	in °C	24	24	23	22	20	18	17	18	19	21	23	26	21	30
4 Absolute maximum temperature	in °C	38	37	36	35	35	32	31	34	35	36	37	39	39	30
5 Absolute minimum temperature	in °C	19	15	20	18	13	12	13	13	14	16	19	21	12	30
6 Mean relative humidity	in %	79	80	81	80	78	79	77	77	73	70	70	74	77	30
7 Mean precipitation	in mm	214	205	146	102	24	37	18	15	9	6	28	34	939	30
8 Maximum precipitation	in mm	333	466	524	387	139	109	28	72	48	5	102	247	1216	10
9 Minimum precipitation	in mm	123	40	41	30	0	11	2	0	0	0	2	51	70	10
10 Maximum precipitation in 24 h	in mm	127	203	163	100	96	88	29	43	45	70	76	108	203	30
11 Mean number of days with precipitation	>0,1 mm	13	11	11	7	3	5	4	2	1	1	5	8	69	30
12 Mean duration of sunshine	in h	199	177	169	236	251	198	227	258	268	307	292	241	2823	9
13 Mean quantity of radiation	in ly/day														
14 Mean potential evaporation	in mm	160	138	146	128	101	68	72	77	97	134	153	164	1438	28
15 Mean windspeed	in m/sec														
16 Mean predominent direction of the wind		NE	var	SW	SW	SW	SW	SW	SW	SW	NE	NE	NE		30

134 Station / Country   Beira / Mozambique

Location 19°50'S/34°51'E — Height above sealevel 8 m — Climate symbol: Köppen Aw — Troll V,3

		J	F	M	A	M	J	J	A	S	O	N	D	year	P
1 Mean daily temperature	in °C	27,4	27,5	26,5	25,2	22,8	20,8	20,3	21,2	23,0	24,9	26,2	26,8	24,4	47
2 Mean daily maximum temperature	in °C	31	32	31	30	28	26	25	26	28	29	30	31	29	30
3 Mean daily minimum temperature	in °C	26	24	25	22	19	17	16	17	19	21	22	23	20	30
4 Absolute maximum temperature	in °C	40	38	37	37	37	33	35	35	40	42	43	41	43	30
5 Absolute minimum temperature	in °C	19	19	18	16	13	9	9	10	12	13	16	17	9	30
6 Mean relative humidity	in %	70	71	71	71	72	72	74	72	75	67	68	74	71	
7 Mean precipitation	in mm	265	225	244	105	58	42	37	30	27	29	133	234	1429	30
8 Maximum precipitation	in mm	852	773	677	330	142	133	115	133	108	145	545	780	2288	48
9 Minimum precipitation	in.mm	11	26	22	27	4	2	0	0	0	1	6	38	821	48
10 Maximum precipitation in 24 h	in mm	196	219	196	184	101	88	94	65	63	102	218	222	222	30
11 Mean number of days with precipitation	>0,1 mm	12	12	12	7	6	5	4	3	3	3	7	10	84	30
12 Mean duration of sunshine	in h	245	224	243	245	253	223	231	253	243	258	229	237	2883	30
13 Mean quantity of radiation	in ly/day														
14 Mean potential evaporation	in mm	163	143	143	112	79	57	55	66	86	138	140	158	1340	41
15 Mean windspeed	in m/sec	3,9	3,9	3,9	3,3	3,3	3,3	3,3	3,6	4,2	4,7	4,4	3,9	3,9	30
16 Mean predominent direction of the wind															

135 Station / Country   Maputo (Lourenço Marques) / Mozambique

Location 25°58'S/32°36'E — Height above sealevel 64 m — Climate symbol: Köppen Aw — Troll V,3

		J	F	M	A	M	J	J	A	S	O	N	D	year	P
1 Mean daily temperature	in °C	25,4	25,5	24,6	23,1	20,6	18,5	18,2	19,2	20,6	22,2	23,4	24,7	22,2	50
2 Mean daily maximum temperature	in °C	30	30	30	29	27	25	25	26	27	28	28	30	28	30
3 Mean daily minimum temperature	in °C	22	22	21	19	16	14	14	15	16	18	20	21	18	30
4 Absolute maximum temperature	in °C	43	41	40	40	38	34	35	38	45	45	44	44	45	30
5 Absolute minimum temperature	in °C	16	17	15	12	8	7	7	9	10	12	13	16	7	30
6 Mean relative humidity	in %	69	71	72	70	67	66	66	65	65	67	68	69	68	
7 Mean precipitation	in mm	130	124	97	64	28	27	13	13	38	48	86	103	788	30
8 Maximum precipitation	in mm	368	373	560	356	145	250	87	75	138	150	379	244	1425	51
9 Minimum precipitation	in mm	2	1	0	0	0	0	0	0	4	9	10	277	51	
10 Maximum precipitation in 24 h	in mm	123	201	127	161	68	166	52	40	77	102	200	87	201	30
11 Mean number of days with precipitation	>0,1 mm	9	8	9	5	3	2	2	2	3	5	7	9	64	42
12 Mean duration of sunshine	in h	223	210	225	229	253	248	256	252	228	210	198	220	2748	30
13 Mean quantity of radiation	in ly/day														
14 Mean potential evaporation	in mm	145	125	119	90	67	48	48	58	71	95	110	134	1110	44
15 Mean windspeed	in m/sec														
16 Mean predominent direction of the wind															

136 Station / Country   Diégo-Suarez / Madagascar

Location 12°21'S/49°18'E — Height above sealevel 105 m — Climate symbol: Köppen Aw — Troll V,2

		J	F	M	A	M	J	J	A	S	O	N	D	year	P
1 Mean daily temperature	in °C	27,5	27,5	27,9	27,9	27,3	25,8	25,0	25,0	25,3	26,3	27,5	28,1	26,8	20
2 Mean daily maximum temperature	in °C	30	30	30	31	30	29	28	28	29	30	31	31	30	9
3 Mean daily minimum temperature	in °C	23	23	23	23	21	20	20	19	20	21	23	23	22	9
4 Absolute maximum temperature	in °C	36	36	35	36	35	34	34	34	32	34	37	37	37	35
5 Absolute minimum temperature	in °C	20	20	21	19	16	16	15	15	17	19	20		15	35
6 Mean relative humidity	in %	82	84	83	77	71	72	68	67	68	68	72	78	74	6
7 Mean precipitation	in mm	276	211	187	56	8	8	6	7	5	11	28	111	915	30
8 Maximum precipitation	in mm	517	468	885	196	42	35	23	44	36	61	172	489	1812	30
9 Minimum precipitation	in mm	95	38	27	0	0	0	0	0	0	0	2	8	384	30
10 Maximum precipitation in 24 h	in mm	231	177	508	97	42	23	14	37	34	26	74	255	508	30
11 Mean number of days with precipitation	>0,1 mm	20	18	14	6	3	4	4	4	4	6	12		97	30
12 Mean duration of sunshine	in h	188	190	253	325	290	256	260	289	317	348	290	273	3279	14
13 Mean quantity of radiation	in ly/day														
14 Mean potential evaporation	in mm	157	142	158	140	144	120	109	111	115	143	153	186	1658	30
15 Mean windspeed	in m/sec	3,9	3,6	4,2	5,3	5,8	6,4	6,9	7,5	7,8	8,1	6,7	4,4	5,8	9
16 Mean predominent direction of the wind		W	W	E	ESE	ESE	ESE	ESE	ESE	ESE	ESE	ESE	E		9

137 Station / Country   Majunga / Madagascar

Location   15°40'S / 46°20'E   Height above sealevel   22 m   Climate symbol: Köppen   Aw   Troll   V.3

		J	F	M	A	M	J	J	A	S	O	N	D	year	P	
1 Mean daily temperature	in °C	27,2	27,4	27,7	28,0	26,7	25,3	24,8	25,4	26,3	27,3	27,9	27,8	26,8	20	1
2 Mean daily maximum temperature	in °C	31	31	31	32	32	31	30	31	32	32	32	31	31	9	2
3 Mean daily minimum temperature	in °C	24	24	24	23	20	19	18	18	20	22	23	24	21	9	3
4 Absolute maximum temperature	in °C	37	36	38	37	36	34	34	36	36	37	39	38	39	36	4
5 Absolute minimum temperature	in °C	18	20	17	18	15	13	14	15	16	18	18	18	13	36	5
6 Mean relative humidity	in %	82	84	82	74	68	66	63	62	63	66	72	80	72	9	6
7 Mean precipitation	in mm	466	370	282	57	8	3	1	2	3	24	110	243	1567	30	7
8 Maximum precipitation	in mm	1125	781	1034	237	58	42	7	16	16	118	274	436	2692	30	8
9 Minimum precipitation	in mm	148	30	46	1	0	0	0	0	0	0	11	61	1003	30	9
10 Maximum precipitation in 24 h	in mm	274	180	321	88	58	32	7	14	14	84	84	117	321	30	10
11 Mean number of days with precipitation	> 0,1 mm	20	18	15	5	1	1	1	1	1	3	8	15	88	30	11
12 Mean duration of sunshine	in h	183	173	221	291	317	297	303	320	318	339	290	228	3280	16	12
13 Mean quantity of radiation	in ly/day															13
14 Mean potential evaporation	in mm	158	143	154	146	136	107	101	115	136	154	161	166	1677		14
15 Mean windspeed	in m/sec	3,3	3,3	3,3	3,1	3,3	3,3	3,9	4,4	4,4	4,2	3,3	3,8		9	15
16 Mean predominent direction of the wind		NW	NW	E	E	ESE	ESE	ESE	ESE	ESE	NW	NW	NNW		9	16

138 Station / Country   Tamatave / Madagascar

Location   18°07'S / 49°24'E   Height above sealevel   5 m   Climate symbol: Köppen   Af   Troll   V.1

		J	F	M	A	M	J	J	A	S	O	N	D	year	P	
1 Mean daily temperature	in °C	26,6	26,4	26,0	24,9	23,3	21,5	20,7	20,9	22,0	23,3	24,9	25,7	24,9	20	1
2 Mean daily maximum temperature	in °C	30	30	29	28	27	25	24	25	26	27	29	29	27	12	2
3 Mean daily minimum temperature	in °C	23	23	22	21	19	18	17	17	17	18	20	22	20	12	3
4 Absolute maximum temperature	in °C	37	35	36	33	30	29	29	28	29	30	32	35	37	35	4
5 Absolute minimum temperature	in °C	20	20	19	17	15	11	13	12	13	13	16	16	11	35	5
6 Mean relative humidity	in %	84	85	87	86	86	86	86	85	84	83	83	85	85	12	6
7 Mean precipitation	in mm	420	441	528	404	302	300	257	208	134	87	184	259	3526	30	7
8 Maximum precipitation	in mm	865	1065	1069	934	991	771	474	408	365	250	578	517	4911	30	8
9 Minimum precipitation	in mm	137	122	108	80	84	88	91	69	36	15	33	53	2398	30	9
10 Maximum precipitation in 24 h	in mm	233	247	442	376	154	180	129	134	122	104	217	135	442	30	10
11 Mean number of days with precipitation	> 0,1 mm	21	20	22	21	20	22	23	23	18	15	16	19	240	30	11
12 Mean duration of sunshine	in h	226	185	184	199	197	172	170	188	216	231	242	235	2445	17	12
13 Mean quantity of radiation	in ly/day															13
14 Mean potential evaporation	in mm	154	136	139	116	92	73	66	69	79	99	120	149	1292		14
15 Mean windspeed	in m/sec	4,2	3,9	4,2	4,2	4,2	4,2	4,2	3,3	3,3	3,1	3,3	3,1	3,8	12	15
16 Mean predominent direction of the wind		E	SSE	S	SSW	SSW	SSW	S	E	E	E	E	E		12	16

139 Station / Country   Tananarive / Madagascar

Location   18°54'S / 47°32'E   Height above sealevel   1310 m   Climate symbol: Köppen   Cwb   Troll   V.2

		J	F	M	A	M	J	J	A	S	O	N	D	year	P	
1 Mean daily temperature	in °C	19,8	19,9	19,4	18,2	16,4	14,0	13,3	13,8	15,8	17,8	19,3	19,5	17,3	69	1
2 Mean daily maximum temperature	in °C	25	26	25	24	22	21	20	20	22	25	26	25	24	13	2
3 Mean daily minimum temperature	in °C	16	16	16	15	12	10	10	10	11	12	15	16	13	13	3
4 Absolute maximum temperature	in °C	30	30	29	29	28	26	25	29	29	32	31	30	32	30	4
5 Absolute minimum temperature	in °C	12	11	12	9	3	3	3	2	3	6	8	10	2	30	5
6 Mean relative humidity	in %	82	81	83	80	78	79	78	76	73	71	75	81	78	13	6
7 Mean precipitation	in mm	255	187	263	42	8	9	17	13	16	47	170	366	1393	13	7
8 Maximum precipitation	in mm	427	346	604	110	88	39	52	64	113	152	361	504	1663	13	8
9 Minimum precipitation	in mm	58	54	102	4	1	1	1	1	0	1	73	252	1111	13	9
10 Maximum precipitation in 24 h	in mm	83	89	134	67	19	32	29	33	33	68	91	105	134	30	10
11 Mean number of days with precipitation	> 0,1 mm	19	14	19	7	5	7	9	7	5	5	14	22	133	13	11
12 Mean duration of sunshine	in h	204	205	194	236	231	212	218	239	249	272	224	208	2690	19	12
13 Mean quantity of radiation	in ly/day															13
14 Mean potential evaporation	in mm	97	87	89	74	58	43	40	46	59	79	93	99	864		14
15 Mean windspeed	in m/sec	1,9	1,7	1,9	1,9	1,9	1,9	1,9	1,9	1,9	1,9	1,9	1,7	1,9	13	15
16 Mean predominent direction of the wind		ESE	SE	SE	ESE	ESE	ESE	ESE	E	E	E	E	E		13	16

140 Station / Country   Tuléar / Madagascar

Location   23°23'S / 43°44'E   Height above sealevel   9 m   Climate symbol: Köppen   BWh   Troll   V.4

		J	F	M	A	M	J	J	A	S	O	N	D	year	P	
1 Mean daily temperature	in °C	27,3	27,2	26,3	24,4	22,2	19,9	19,4	20,2	22,0	23,1	24,7	26,3	23,6	20	1
2 Mean daily maximum temperature	in °C	32	32	32	31	29	27	27	27	29	29	30	31	30	11	2
3 Mean daily minimum temperature	in °C	23	23	22	20	17	15	14	15	16	18	20	22	19	11	3
4 Absolute maximum temperature	in °C	39	40	39	37	36	33	32	35	38	39	40	38	40	36	4
5 Absolute minimum temperature	in °C	16	16	14	10	8	7	6	6	8	10	12	12	6	36	5
6 Mean relative humidity	in %	78	78	76	76	74	75	74	74	74	76	75	79	76	11	6
7 Mean precipitation	in mm	71	71	42	6	18	11	4	3	10	14	34	57	341	30	7
8 Maximum precipitation	in mm	338	191	360	40	139	45	21	51	48	103	127	193	666	30	8
9 Minimum precipitation	in mm	5	2	0	0	0	0	0	0	0	0	0	0	129	30	9
10 Maximum precipitation in 24 h	in mm	104	70	58	30	67	26	19	48	31	102	83	105	105	30	10
11 Mean number of days with precipitation	> 0,1 mm	7	6	5	1	2	1	1	1	1	1	2	5	34	30	11
12 Mean duration of sunshine	in h	305	285	296	302	306	277	291	298	307	324	313	306	3610	11	12
13 Mean quantity of radiation	in ly/day															13
14 Mean potential evaporation	in mm	168	145	142	106	76	53	51	61	77	98	128	158	1263		14
15 Mean windspeed	in m/sec	3,9	3,6	3,9	3,6	3,6	3,6	3,6	3,9	4,2	4,2	4,2	3,9	3,9	11	15
16 Mean predominent direction of the wind		SW	SW	SW	SW	SSW	SSW	SSW	SSW	SSW	SSW	SW	SW		11	16

141 Station/Country Fort–Dauphin/Madagascar

Location 25°02'S/46°58'E   **Height above sealevel** 7 m   **Climate symbol: Köppen** Af   **Troll** V.2

		J	F	M	A	M	J	J	A	S	O	N	D	year	P	
1 Mean daily temperature	in °C	25,1	24,9	24,1	23,1	21,2	19,6	18,8	19,1	20,0	21,4	22,9	24,1	22,0	10	1
2 Mean daily maximum temperature	in °C	29	30	28	27	26	24	23	24	26	27	29	27	27	13	2
3 Mean daily minimum temperature	in °C	23	23	22	20	18	18	16	16	17	19	20	22	19	13	3
4 Absolute maximum temperature	in °C	34	35	35	33	31	29	29	29	31	32	33	35	35	36	4
5 Absolute minimum temperature	in °C	18	18	16	15	11	10	9	9	10	13	15	16	9	36	5
6 Mean relative humidity	in %	82	80	82	81	80	81	81	79	78	79	81	81	80	13	6
7 Mean precipitation	in mm	202	184	236	113	117	135	108	94	61	73	91	124	1537	30	7
8 Maximum precipitation	in mm	562	437	588	262	288	379	306	325	264	208	222	314	2059	30	8
9 Minimum precipitation	in mm	27	55	20	16	2	15	21	10	4	1	8	56	988	30	9
10 Maximum precipitation in 24 h	in mm	219	112	283	111	94	148	103	161	124	86	98	78	283	30	10
11 Mean number of days with precipitation	> 0,1 mm	15	15	17	13	12	13	13	11	9	8	12	14	152	30	11
12 Mean duration of sunshine	in h	226	232	200	215	236	210	201	239	230	251	232	230	2702	13	12
13 Mean quantity of radiation	in ly/day															13
14 Mean potential evaporation	in mm	143	125	121	101	75	59	56	64	70	92	116	136	1158		14
15 Mean windspeed	in m/sec	5,8	5,8	5,8	5,8	4,7	4,4	4,7	5,8	6,4	7,2	6,4	5,8	5,8	9	15
16 Mean predominent direction of the wind		NE	ENE	ENE	NE	NE	NE	NE	NE	NE	NE	NE	ENE		9	16

142 Station/Country Tsumeb/Namibia

Location 19°14'S/17°43'E   **Height above sealevel** 1311 m   **Climate symbol: Köppen** BSh   **Troll** V.4

		J	F	M	A	M	J	J	A	S	O	N	D	year	P	
1 Mean daily temperature	in °C	24,6	23,1	22,8	21,4	18,2	15,5	15,4	18,2	22,6	25,2	25,3	24,8	21,4	15	1
2 Mean daily maximum temperature	in °C	31	30	30	29	27	25	25	28	32	34	33	32	30	30	2
3 Mean daily minimum temperature	in °C	18	18	17	15	10	7	7	10	15	18	18	18	14	30	3
4 Absolute maximum temperature	in °C	39	38	38	36	35	30	31	34	38	39	40	40	40	30	4
5 Absolute minimum temperature	in °C	10	10	7	7	1	-4	-4	1	3	7	12	8	-4	30	5
6 Mean relative humidity	in %	56	57	65	57	45	37	35	27	22	27	38	48	43	5	6
7 Mean precipitation	in mm	119	139	79	40	6	0	0	0	1	19	53	97	553	30	7
8 Maximum precipitation	in mm	335	371	254	186	45	3	2	0	20	58	132	207	989	30	8
9 Minimum precipitation	in mm	24	18	2	4	0	0	0	0	0	0	7	2	235	30	9
10 Maximum precipitation in 24 h	in mm	75	123	79	121	19	3	2	0	11	33	69	81	123	30	10
11 Mean number of days with precipitation	> 0,1 mm	12	12	9	5	1	0	0	0	0	3	6	11	59	30	11
12 Mean duration of sunshine	in h	264	157	276	264	291	309	335	332	315	285	282	295	3405	1	12
13 Mean quantity of radiation	in ly/day															13
14 Mean potential evaporation	in mm	127	99	101	79	52	33	34	53	93	130	131	130	1062	15	14
15 Mean windspeed	in m/sec															15
16 Mean predominent direction of the wind																16

143 Station/Country Windhoek/Namibia

Location 22°34'S/17°06'E   **Height above sealevel** 1728 m   **Climate symbol: Köppen** BSh   **Troll** IV.3

		J	F	M	A	M	J	J	A	S	O	N	D	year	P	
1 Mean daily temperature	in °C	23,4	22,2	21,1	18,9	15,8	13,5	13,2	15,6	18,7	21,6	22,3	23,1	19,1	44	1
2 Mean daily maximum temperature	in °C	30	28	27	26	23	20	20	23	29	29	30	30	26	30	2
3 Mean daily minimum temperature	in °C	17	16	15	13	9	7	6	9	11	15	16	16	12	30	3
4 Absolute maximum temperature	in °C	36	35	34	31	29	26	25	30	33	35	36	36	36	30	4
5 Absolute minimum temperature	in °C	8	7	4	2	-2	-3	-3	-4	-1	2	0	3	-4	30	5
6 Mean relative humidity	in %	40	46	51	43	32	31	27	22	18	19	32	31	33	10	6
7 Mean precipitation	in mm	77	73	81	38	6	1	1	0	1	12	33	47	370	30	7
8 Maximum precipitation	in mm	229	223	312	170	52	19	11	2	12	41	151	134	745	30	8
9 Minimum precipitation	in mm	4	9	5	0	0	0	0	0	0	0	0	0	91	30	9
10 Maximum precipitation in 24 h	in mm	86	46	78	76	20	19	11	1	10	25	62	44	86	30	10
11 Mean number of days with precipitation	> 0,1 mm	8	9	8	3	1	0	0	0	0	2	4	6	41	30	11
12 Mean duration of sunshine	in h	288	252	282	273	310	309	326	341	321	319	297	285	3603	5	12
13 Mean quantity of radiation	in ly/day	641	575	527	492	447	409	441	506	585	624	669	647	547	8	13
14 Mean potential evaporation	in mm	128	96	89	65	44	29	29	45	68	93	104	123	913		14
15 Mean windspeed	in m/sec	1,7	1,7	1,4	1,4	1,7	2,2	1,9	2,2	2,2	2,2	2,2	1,9	1,9	5	15
16 Mean predominent direction of the wind																16

144 Station/Country Swakopmund/Namibia

Location 22°41'S/14°31'E   **Height above sealevel** 12 m   **Climate symbol: Köppen** BWk   **Troll** IV.5

		J	F	M	A	M	J	J	A	S	O	N	D	year	P	
1 Mean daily temperature	in °C	17,2	18,1	17,5	15,7	15,1	15,2	13,0	12,1	12,6	13,7	14,9	16,4	15,1	11	1
2 Mean daily maximum temperature	in °C	20	21	20	18	18	20	18	16	16	18	19	18	18	15	2
3 Mean daily minimum temperature	in °C	15	16	15	13	11	11	9	9	10	11	13	14	12	15	3
4 Absolute maximum temperature	in °C	25	29	40	40	38	36	36	40	29	41	24	27	41	15	4
5 Absolute minimum temperature	in °C	4	9	11	7	5	5	3	4	5	5	8	8	3	15	5
6 Mean relative humidity	in %	87	88	89	90	84	77	80	88	90	87	88	89	86	15	6
7 Mean precipitation	in mm	1	2	2	2	0	0	0	0	0	0	1	0	10	15	7
8 Maximum precipitation	in mm	5	18	11	9	5	4	3	1	0	4	18	4	29	15	8
9 Minimum precipitation	in mm	0	0	0	0	0	0	0	0	0	0	0	0	0	15	9
10 Maximum precipitation in 24 h	in mm	3	17	11	9	5	4	1	1	0	3	18	2	18	15	10
11 Mean number of days with precipitation	> 0,1 mm	0	0	0	1	0	0	0	0	0	0	0	0	1	15	11
12 Mean duration of sunshine	in h	233	188	211	237	251	231	236	220	189	228	210	214	2648	5	12
13 Mean quantity of radiation	in ly/day															13
14 Mean potential evaporation	in mm	85	74	77	58	60	53	48	38	47	54	63	77	732	14	14
15 Mean windspeed	in m/sec															15
16 Mean predominent direction of the wind																16

145 Station / Country Lüderitz / Namibia

Location 26°38'N / 15°06'E Height above sealevel 23 m Climate symbol: Köppen **BWk** Troll **IV,5**

		J	F	M	A	M	J	J	A	S	O	N	D	year	P	
1	Mean daily temperature	in °C	17,5	18,0	18,0	18,5	15,5	15,5	14,0	13,5	14,0	15,0	16,5	17,0	16,0	20	1
2	Mean daily maximum temperature	in °C	21	22	22	20	19	19	18	17	17	18	20	21	20	20	2
3	Mean daily minimum temperature	in °C	14	14	14	13	12	11	10	10	11	12	13	13	12	20	3
4	Absolute maximum temperature	in °C	33	30	34	35	33	32	31	33	35	35	38	31	38	20	4
5	Absolute minimum temperature	in °C	10	5	8	8	7	0	3	5	7	8	8	4	0	20	5
6	Mean relative humidity	in %	82	81	82	80	79	72	74	78	80	80	80	80	79	20	6
7	Mean precipitation	in mm	1	3	4	1	3	2	1	1	1	0	0	1	18	20	7
8	Maximum precipitation	in mm	13	36	56	8	22	13	6	5	5	2	3	7	59	20	8
9	Minimum precipitation	in mm	0	0	0	0	0	0	0	0	0	0	0	0	1	20	9
10	Maximum precipitation in 24 h	in mm	13	24	31	7	20	12	5	3	2	2	3	7	31	20	10
11	Mean number of days with precipitation	> 0,1 mm	0	0	0	0	1	0	0	0	0	0	0	0	1	20	11
12	Mean duration of sunshine	in h	140	132	236	168	143	102	161	152	195	171	147	109	1856	3	12
13	Mean quantity of radiation	in ly / day															13
14	Mean potential evaporation	in mm	78	70	72	58	50	44	41	41	45	57	67	75	698	20	14
15	Mean windspeed	in m / sec															15
16	Mean predominent direction of the wind																16

146 Station / Country Keetmanshoop / Namibia

Location 26°34'S / 18°07'E Height above sealevel 1 066 m Climate symbol: Köppen **BWh** Troll **IV,5**

		J	F	M	A	M	J	J	A	S	O	N	D	year	P	
1	Mean daily temperature	in °C	26,9	26,6	24,7	20,9	17,2	14,2	13,4	15,0	17,8	21,6	24,2	25,8	20,7	19	1
2	Mean daily maximum temperature	in °C	35	34	32	29	25	22	21	24	27	30	33	34	29	30	2
3	Mean daily minimum temperature	in °C	18	19	17	14	10	7	6	7	10	13	15	17	13	30	3
4	Absolute maximum temperature	in °C	42	42	40	37	39	32	31	34	36	40	41	42	42	30	4
5	Absolute minimum temperature	in °C	8	9	7	2	-1	-3	-4	-3	1	5	3	-4	-4	30	5
6	Mean relative humidity	in %	31	37	40	38	36	39	35	31	26	27	28	28	33	15	6
7	Mean precipitation	in mm	22	30	35	13	5	2	1	1	2	5	14	17	147	30	7
8	Maximum precipitation	in mm	87	114	182	54	35	31	8	11	45	21	56	114	333	30	8
9	Minimum precipitation	in mm	3	0	0	0	0	0	0	0	0	0	0	0	43	30	9
10	Maximum precipitation in 24 h	in mm	33	65	56	23	30	14	4	10	23	21	33	99	99	30	10
11	Mean number of days with precipitation	> 0,1 mm	3	4	4	2	1	0	0	0	0	1	2	2	19	30	11
12	Mean duration of sunshine	in h	360	297	310	306	307	291	307	326	324	350	354	375	3807	10	12
13	Mean quantity of radiation	in ly / day	718	646	553	500	411	368	392	456	558	660	725	741	561	4	13
14	Mean potential evaporation	in mm	163	138	120	78	45	26	25	32	54	93	121	150	1045	19	14
15	Mean windspeed	in m / sec	3,9	3,3	3,3	3,3	3,9	4,2	4,2	4,2	4,7	4,2	4,4	4,4	4,2	5	15
16	Mean predominent direction of the wind																16

147 Station / Country Maun / Botswana

Location 19°59'S / 23°25'E Height above sealevel 942 m Climate symbol: Köppen **BSh** Troll **V,4**

		J	F	M	A	M	J	J	A	S	O	N	D	year	P	
1	Mean daily temperature	in °C	25,2	24,7	23,9	22,2	18,5	15,6	15,3	18,1	22,9	26,2	26,4	25,6	22,1	30	1
2	Mean daily maximum temperature	in °C	32	31	31	31	28	25	25	29	33	35	34	32	30	30	2
3	Mean daily minimum temperature	in °C	19	19	17	14	10	6	6	9	13	18	19	19	14	30	3
4	Absolute maximum temperature	in °C	40	39	39	37	36	33	32	36	39	43	43	42	43	30	4
5	Absolute minimum temperature	in °C	9	9	10	3	-3	-6	-5	-4	-1	6	9	11	-6	30	5
6	Mean relative humidity	in %	71	74	67	58	50	46	40	33	30	34	52	66	51	10	6
7	Mean precipitation	in mm	110	102	85	26	22	1	0	0	1	15	46	80	471	30	7
8	Maximum precipitation	in mm	380	339	275	93	34	10	1	0	12	61	116	233	776	30	8
9	Minimum precipitation	in mm	14	0	0	0	0	0	0	0	0	0	0	16	285	30	9
10	Maximum precipitation in 24 h	in mm	90	103	120	61	26	10	1	0	8	33	55	71	120	30	10
11	Mean number of days with precipitation	> 0,1 mm	10	10	7	3	1	0	0	0	0	3	5	8	47	30	11
12	Mean duration of sunshine	in h	245	207	257	282	310	300	313	332	315	291	264	214	3330	10	12
13	Mean quantity of radiation	in ly / day	553	508	526	483	453	417	425	500	559	579	585	551	510	5	13
14	Mean potential evaporation	in mm	141	116	116	83	51	32	30	48	86	141	149	143	1136	11	14
15	Mean windspeed	in m / sec	1,7	1,4	1,4	1,4	1,4	1,7	1,7	1,7	2,5	1,8	1,9	1,7	1,7	6	15
16	Mean predominent direction of the wind																16

148 Station / Country Ghanzi / Botswana

Location 21°42'S / 21°39'E Height above sealevel 1131 m Climate symbol: Köppen **BSh** Troll **V,4**

		J	F	M	A	M	J	J	A	S	O	N	D	year	P	
1	Mean daily temperature	in °C	25,0	24,4	22,8	20,6	16,7	13,9	13,9	16,7	20,6	24,4	25,0	25,0	21,1	20	1
2	Mean daily maximum temperature	in °C	32	32	30	29	26	23	24	27	31	33	33	33	29	30	2
3	Mean daily minimum temperature	in °C	18	18	16	13	8	4	4	6	10	15	17	18	12	30	3
4	Absolute maximum temperature	in °C	42	38	39	36	33	32	31	38	38	39	42	41	42	30	4
5	Absolute minimum temperature	in °C	7	7	8	-1	-6	-7	-6	-3	-1	1	6	7	-7	30	5
6	Mean relative humidity	in %															6
7	Mean precipitation	in mm	98	94	74	39	8	1	0	0	2	21	43	66	446	30	7
8	Maximum precipitation	in mm	389	227	274	185	45	6	1	3	21	111	115	141	787	30	8
9	Minimum precipitation	in mm	5	17	3	0	0	0	0	0	0	0	0	6	200	30	9
10	Maximum precipitation in 24 h	in mm	130	124	108	52	34	6	1	3	11	83	56	81	130	30	10
11	Mean number of days with precipitation	> 0,1 mm	9	7	6	4	1	0	0	0	0	3	6	7	43	30	11
12	Mean duration of sunshine	in h	295	280	313	282	316	306	318	319	308	279	300	267	3579	1	12
13	Mean quantity of radiation	in ly / day															13
14	Mean potential evaporation	in mm	136	118	105	74	44	28	29	42	76	122	130	137	1041		14
15	Mean windspeed	in m / sec															15
16	Mean predominent direction of the wind																16

149 Station / Country Mahalatswe / Botswana

Location 23°04'S / 26°48'E Height above sealevel 1001 m Climate symbol: Köppen BSh Troll V,4

	J	F	M	A	M	J	J	A	S	O	N	D	year	P	
1 Mean daily temperature — in °C	25,5	24,5	23	20,5	17	13,5	13,5	16,5	20	24	25	25	21	30	1
2 Mean daily maximum temperature — in °C	32	31	29	28	26	23	23	26	29	32	32	32	29	30	2
3 Mean daily minimum temperature — in °C	19	18	17	13	8	4	4	7	11	16	18	18	13	30	3
4 Absolute maximum temperature — in °C	42	39	38	37	34	31	31	35	39	40	41	41	42	30	4
5 Absolute minimum temperature — in °C	9	10	9	0	-6	-4	-7	-4	-3	2	8	8	-7	30	5
6 Mean relative humidity — in %															6
7 Mean precipitation — in mm	84	95	77	30	10	6	3	2	7	28	72	97	511	30	7
8 Maximum precipitation — in mm	226	424	239	153	104	50	40	19	61	95	171	264	928	30	8
9 Minimum precipitation — in mm	16	18	6	0	0	0	0	0	0	0	0	12	193	30	9
10 Maximum precipitation in 24 h — in mm	85	128	110	126	35	37	38	18	29	63	66	125	128	30	10
11 Mean number of days with precipitation > 0,1 mm	7	7	5	3	2	1	0	0	1	3	6	8	43	30	11
12 Mean duration of sunshine — in h															12
13 Mean quantity of radiation — in ly / day															13
14 Mean potential evaporation — in mm															14
15 Mean windspeed — in m / sec															15
16 Mean predominent direction of the wind															16

150 Station / Country Gaborone / Botswana

Location 24°45'S / 25°55'E Height above sealevel 1007 m Climate symbol: Köppen BSh Troll IV,3

	J	F	M	A	M	J	J	A	S	O	N	D	year	P	
1 Mean daily temperature — in °C	25.0	25.0	22.8	19.4	15.0	12.2	12.2	15.0	18.3	22.8	23.9	24.4	20.0		1
2 Mean daily maximum temperature — in °C	32.5	31.6	30.0	28.1	24.9	22.3	22.4	25.3	28.8	31.7	31.9	32.4	28.5		2
3 Mean daily minimum temperature — in °C	17.7	17.8	15.5	11.0	5.5	1.9	1.5	4.1	8.9	14.0	15.9	16.8	10.9		3
4 Absolute maximum temperature — in °C	41.7	38.9	37.2	35.0	33.3	29.4	29.4	32.8	37.8	38.9	40.6	43.9	43.9		4
5 Absolute minimum temperature — in °C	10.1	8.9	6.7	-2.2	-3.9	-5.6	-6.7	-1.4	-4.4	0.0	5.0	5.6	-6.7		5
6 Mean relative humidity — in %															6
7 Mean precipitation — in mm	89	89	86	36	14	5	5	5	13	41	64	94	541	40	7
8 Maximum precipitation — in mm															8
9 Minimum precipitation — in mm															9
10 Maximum precipitation in 24 h — in mm	81	67	170	46	28	36	17	18	72	72	35	110	170		10
11 Mean number of days with precipitation > 0,1 mm	10	8	7	4	2	<1	<1	<1	2	6	8	9	58		11
12 Mean duration of sunshine — in h															12
13 Mean quantity of radiation — in ly / day															13
14 Mean potential evaporation — in mm	138	113	100	65	39	22	21	34	68	110	123	133	966		14
15 Mean windspeed — in m / sec															15
16 Mean predominent direction of the wind															16

151 Station / Country Messina / South Africa

Location 22°20'S / 30°03'E Height above sealevel 549 m Climate symbol: Köppen BWh Troll V,4

	J	F	M	A	M	J	J	A	S	O	N	D	year	P	
1 Mean daily temperature — in °C	26.9	26.4	25.5	23.9	20.8	17.9	17.8	19.8	22.7	25.2	26.1	26.8	23.3	41	1
2 Mean daily maximum temperature — in °C	32	32	31	30	28	25	25	27	29	31	32	32	29	30	2
3 Mean daily minimum temperature — in °C	21	21	20	18	14	11	11	13	16	19	20	21	17	30	3
4 Absolute maximum temperature — in °C	42	41	42	39	36	34	34	36	41	43	43	41	43	30	4
5 Absolute minimum temperature — in °C	15	15	14	7	6	3	4	4	9	11	11	15	3	30	5
6 Mean relative humidity — in %	58	59	58	56	51	54	53	47	48	47	50	54	53	5	6
7 Mean precipitation — in mm	78	55	40	15	4	4	3	1	7	21	42	70	340	30	7
8 Maximum precipitation — in mm	487	132	241	104	23	41	27	11	56	83	122	141	685	30	8
9 Minimum precipitation — in mm	2	0	1	0	0	0	0	0	0	0	0	4	76	30	9
10 Maximum precipitation in 24 h — in mm	167	71	94	48	23	41	21	10	30	56	63	107	167	30	10
11 Mean number of days with precipitation > 0,1 mm	6	5	4	2	1	1	0	0	1	2	5	6	32	30	11
12 Mean duration of sunshine — in h	245	230	226	249	285	267	273	291	270	264	246	267	3113	8	12
13 Mean quantity of radiation — in ly / day															13
14 Mean potential evaporation — in mm	155	133	126	90	61	40	39	55	87	128	140	158	1212	28	14
15 Mean windspeed — in m / sec															15
16 Mean predominent direction of the wind															16

152 Station / Country Pietersburg / South Africa

Location 23°51'S / 29°27'E Height above sealevel 1230 m Climate symbol: Köppen BSk Troll IV,3

	J	F	M	A	M	J	J	A	S	O	N	D	year	P	
1 Mean daily temperature — in °C	21.4	20.8	19.4	17.5	13.6	10.8	11.1	13.3	16.4	19.1	20.3	21.4	17.1	45	1
2 Mean daily maximum temperature — in °C	27	27	26	25	22	20	20	22	25	27	27	27	25	30	2
3 Mean daily minimum temperature — in °C	16	16	15	12	8	4	4	6	10	14	15	16	11	30	3
4 Absolute maximum temperature — in °C	35	34	34	31	30	27	28	32	34	36	37	36	37	30	4
5 Absolute minimum temperature — in °C	8	8	6	2	-1	-4	-4	-2	-1	5	6	9	-4	30	5
6 Mean relative humidity — in %	71	73	73	71	65	62	59	56	57	60	66	70	65	30	6
7 Mean precipitation — in mm	84	71	54	33	12	6	3	1	13	34	76	98	485	30	7
8 Maximum precipitation — in mm	218	214	139	159	70	47	33	8	56	82	173	252	792	30	8
9 Minimum precipitation — in mm	17	6	8	0	0	0	0	0	0	10	0	18	280	30	9
10 Maximum precipitation in 24 h — in mm	65	109	57	122	57	29	14	7	33	55	77	86	122	30	10
11 Mean number of days with precipitation > 0,1 mm	8	7	6	3	2	1	1	0	2	4	7	9	50	30	11
12 Mean duration of sunshine — in h	248	210	233	243	267	287	267	276	261	270	246	251	3039	13	12
13 Mean quantity of radiation — in ly / day															13
14 Mean potential evaporation — in mm	111	93	87	61	41	27	26	37	60	87	95	109	834	4	14
15 Mean windspeed — in m / sec	3.3	3.3	2.8	2.5	2.5	2.8	3.1	3.1	3.1	3.6	3.9	3.3	3.1	7	15
16 Mean predominent direction of the wind															16

153 Station / Country Upington / South Africa

Location 28°26'S / 21°16'E | Height above sealevel 809 m | Climate symbol: Köppen BWh | Troll IV,5

		J	F	M	A	M	J	J	A	S	O	N	D	year	P	
1 Mean daily temperature	in °C	27,8	25,7	24,0	19,7	14,9	11,7	10,6	13,4	17,4	21,0	24,2	26,1	19,7	11	1
2 Mean daily maximum temperature	in °C	36	34	32	28	24	21	20	23	26	30	32	34	28	30	2
3 Mean daily minimum temperature	in °C	20	21	17	13	8	4	3	5	9	12	16	18	12	30	3
4 Absolute maximum temperature	in °C	43	42	39	37	33	30	30	34	37	41	42	42	43	30	4
5 Absolute minimum temperature	in °C	9	9	7	2	-3	-8	-7	-7	-3	-1	7	8	-8	30	5
6 Mean relative humidity	in %	32	37	43	47	50	52	46	39	31	32	33	31	39	13	6
7 Mean precipitation	in mm	25	40	37	27	11	3	5	4	3	11	16	22	204	30	7
8 Maximum precipitation	in mm	139	123	125	167	74	10	73	25	41	88	70	87	566	30	8
9 Minimum precipitation	in mm	0	0	0	0	0	0	0	0	0	0	0	0	88	30	9
10 Maximum precipitation in 24 h	in mm	68	63	60	119	43	9	28	16	27	36	31	28	119	30	10
11 Mean number of days with precipitation	> 0,1 mm	3	4	4	2	2	1	1	1	1	1	2	2	24	30	11
12 Mean duration of sunshine	in h	366	294	301	291	282	273	285	310	318	335	348	366	3763	10	12
13 Mean quantity of radiation	in ly/day															13
14 Mean potential evaporation	in mm	117	141	124	74	42	22	18	33	52	92	120	164	1059	48	14
15 Mean windspeed	in m/sec	2,5	2,5	2,2	1,7	1,7	1,7	2,2	2,2	2,5	2,2	2,8	2,8	2,2	5	15
16 Mean predominent direction of the wind																16

154 Station / Country Kimberley / South Africa

Location 28°48'S / 24°46'E | Height above sealevel 1197 m | Climate symbol: Köppen BSh | Troll IV,3

		J	F	M	A	M	J	J	A	S	O	N	D	year	P	
1 Mean daily temperature	in °C	24,8	23,7	21,7	18,0	14,0	10,6	10,3	13,2	16,6	20,0	22,0	24,1	18,3	66	1
2 Mean daily maximum temperature	in °C	33	31	29	25	21	19	19	22	25	28	30	32	26	30	2
3 Mean daily minimum temperature	in °C	18	17	15	11	6	3	3	5	8	12	14	16	11	30	3
4 Absolute maximum temperature	in °C	40	38	36	34	30	26	28	31	34	38	39	39	40	30	4
5 Absolute minimum temperature	in °C	6	6	5	-1	-5	-7	-7	-6	-4	-1	2	6	-7	30	5
6 Mean relative humidity	in %	29	34	36	38	36	34	32	27	24	25	25	27	31	30	6
7 Mean precipitation	in mm	55	64	72	44	19	13	6	11	11	31	47	58	431	30	7
8 Maximum precipitation	in mm	171	235	220	141	73	104	40	117	63	135	131	172	710	30	8
9 Minimum precipitation	in mm	9	2	18	3	0	0	0	0	0	2	0	3	179	30	9
10 Maximum precipitation in 24 h	in mm	58	54	68	52	30	67	27	58	27	59	49	65	68	30	10
11 Mean number of days with precipitation	> 0,1 mm	6	9	8	5	3	1	1	1	4	5	6	6	50	30	11
12 Mean duration of sunshine	in h	301	266	276	270	267	284	282	301	291	301	308	326	3461	14	12
13 Mean quantity of radiation	in ly/day	655	590	501	436	350	315	338	430	522	598	664	675	508	7	13
14 Mean potential evaporation	in mm	149	117	98	63	37	20	20	33	53	92	107	136	925	36	14
15 Mean windspeed	in m/sec	3,9	3,6	2,8	3,3	3,3	3,3	3,6	4,2	3,9	4,2	4,4	4,2	3,6	10	15
16 Mean predominent direction of the wind																16

155 Station / Country Durban / South Africa

Location 29°50'S / 31°02'E | Height above sealevel 5 m | Climate symbol: Köppen Cfa | Troll IV,4

		J	F	M	A	M	J	J	A	S	O	N	D	year	P	
1 Mean daily temperature	in °C	24,7	24,9	23,9	22,1	20,0	18,3	17,9	18,7	19,8	20,9	22,3	23,8	21,4	60	1
2 Mean daily maximum temperature	in °C	27	28	27	26	24	22	22	22	23	23	25	26	25	30	2
3 Mean daily minimum temperature	in °C	20	21	20	18	14	11	11	12	15	17	18	19	16	30	3
4 Absolute maximum temperature	in °C	33	38	32	37	35	32	33	36	42	40	39	35	42	30	4
5 Absolute minimum temperature	in °C	14	15	14	11	7	5	4	5	8	8	10	13	4	20	5
6 Mean relative humidity	in %	81	82	83	81	77	73	73	77	79	81	81	81	79	20	6
7 Mean precipitation	in mm	118	128	113	91	59	36	28	39	63	85	121	124	1003	30	7
8 Maximum precipitation	in mm	383	358	267	315	280	358	109	138	143	251	278	363	1397	30	8
9 Minimum precipitation	in mm	10	22	23	8	5	0	1	2	7	25	21	41	631	30	9
10 Maximum precipitation in 24 h	in mm	177	151	83	146	161	240	55	82	95	78	135	100	240	30	10
11 Mean number of days with precipitation	> 0,1 mm	11	9	9	7	4	3	3	4	6	10	11	12	88	30	11
12 Mean duration of sunshine	in h	202	188	192	204	229	207	214	214	174	161	171	186	2342	10	12
13 Mean quantity of radiation	in ly/day	520	480	434	365	266	275	285	324	402	413	459	495	393	7	13
14 Mean potential evaporation	in mm	134	116	110	86	62	45	46	54	65	84	100	123	1025	40	14
15 Mean windspeed	in m/sec	3,3	3,1	3,1	2,5	2,2	2,2	2,5	3,3	3,6	3,9	4,2	3,6	3,1	10	15
16 Mean predominent direction of the wind																16

156 Station / Country East London (Oos–Londen) / South Africa

Location 33°02'S / 27°52'E | Height above sealevel 125 m | Climate symbol: Köppen Cfb | Troll IV,4

		J	F	M	A	M	J	J	A	S	O	N	D	year	P	
1 Mean daily temperature	in °C	21,2	21,3	20,6	19,1	17,2	15,6	14,8	15,7	16,6	17,7	19,1	20,4	18,3	60	1
2 Mean daily maximum temperature	in °C	25	26	25	23	23	21	21	21	21	21	23	24	23	21	2
3 Mean daily minimum temperature	in °C	18	18	17	15	13	11	10	11	12	14	15	17	14	21	3
4 Absolute maximum temperature	in °C	36	43	35	36	36	33	33	37	41	41	40	36	43	21	4
5 Absolute minimum temperature	in °C	9	11	11	7	4	3	2	3	5	7	9	9	2	21	5
6 Mean relative humidity	in %	82	82	82	78	72	66	66	71	77	77	81	80	76	21	6
7 Mean precipitation	in mm	69	78	99	68	48	35	32	42	97	111	93	87	880	21	7
8 Maximum precipitation	in mm	177	183	243	242	152	136	199	101	292	465	227	224	1348	21	8
9 Minimum precipitation	in mm	17	29	20	9	7	0	0	4	16	33	1	16	500	21	9
10 Maximum precipitation in 24 h	in mm	107	122	116	119	89	109	122	66	113	265	122	127	265	21	10
11 Mean number of days with precipitation	> 0,1 mm	9	8	9	8	5	2	3	5	7	9	8	10	81	21	11
12 Mean duration of sunshine	in h	236	204	220	216	226	237	242	236	225	211	228	229	2710	10	12
13 Mean quantity of radiation	in ly/day															13
14 Mean potential evaporation	in mm	107	92	89	69	54	41	40	48	56	72	86	104	858	53	14
15 Mean windspeed	in m/sec	4,7	4,7	4,4	4,2	4,2	4,4	4,7	5,3	5,0	5,0	5,6	5,0	4,7	10	15
16 Mean predominent direction of the wind																16

157 Station / Country Cape Town (Kaapstad) / South Africa

Location 33°54'S / 18°32'E — Height above sealevel 17 m — Climate symbol: Köppen Csb — Troll IV.1

		J	F	M	A	M	J	J	A	S	O	N	D	year	P
1 Mean daily temperature	in °C	21,2	21,5	20,3	17,5	15,1	13,4	12,6	13,2	14,5	16,3	18,3	20,1	17,0	79
2 Mean daily maximum temperature	in °C	26	26	25	22	19	18	17	18	19	21	23	25	22	30
3 Mean daily minimum temperature	in °C	16	16	14	12	10	8	7	8	9	11	13	15	12	30
4 Absolute maximum temperature	in °C	38	38	39	39	35	30	29	32	35	33	34	38	39	30
5 Absolute minimum temperature	in °C	7	5	5	2	-1	-2	-2	-1	0	1	4	5	-2	30
6 Mean relative humidity	in %	72	74	77	81	78	80	84	82	80	77	74	72	78	25
7 Mean precipitation	in mm	12	8	17	47	84	82	85	71	43	29	17	11	506	30
8 Maximum precipitation	in mm	58	61	53	150	183	173	202	156	123	102	59	37	756	30
9 Minimum precipitation	in mm	0	1	4	4	13	18	14		3	4	2	0	347	30
10 Maximum precipitation in 24 h	in mm	25	30	30	44	62	55	49	47	38	49	20	26	62	30
11 Mean number of days with precipitation	> 0,1 mm	2	2	3	6	9	9	10	10	7	5	3	2	68	30
12 Mean duration of sunshine	in h	338	294	282	207	183	180	177	198	216	276	297	344	2992	30
13 Mean quantity of radiation	in ly/day	732	627	525	369	261	223	241	306	437	562	667	722	473	9
14 Mean potential evaporation	in mm	111	96	87	63	48	34	32	38	47	66	81	104	807	31
15 Mean windspeed	in m/sec	6,1	5,8	4,7	4,4	3,9	3,6	4,4	4,7	5,0	5,6	5,8	5,8	5,0	9
16 Mean predominant direction of the wind															

158 Station / Country Port Elizabeth / South Africa

Location 33°58'S / 25°36'E — Height above sealevel 58 m — Climate symbol: Köppen Cfb — Troll IV.7

		J	F	M	A	M	J	J	A	S	O	N	D	year	P
1 Mean daily temperature	in °C	20,8	21,1	20,1	18,4	16,3	14,8	14,2	14,6	15,4	16,6	18,2	19,8	17,6	75
2 Mean daily maximum temperature	in °C	25	25	24	23	22	20	19	20	20	21	22	24	22	30
3 Mean daily minimum temperature	in °C	16	17	16	13	10	8	7	8	10	12	14	15	12	30
4 Absolute maximum temperature	in °C	39	40	40	38	35	32	32	37	40	41	40	38	41	30
5 Absolute minimum temperature	in °C	7	8	7	5	0	0	0	0	0	4	6	7	0	30
6 Mean relative humidity	in %	80	81	82	80	78	76	76	78	80	81	80	78	79	24
7 Mean precipitation	in mm	37	33	44	44	65	56	56	59	68	61	61	42	632	
8 Maximum precipitation	in mm	134	91	154	129	238	183	110	172	295	200	200	188	814	30
9 Minimum precipitation	in mm	9	10	7	1	5	0	1	5	12	17	1	9	469	30
10 Maximum precipitation in 24 h	in mm	88	73	109	51	57	70	62	133	135	91	54	100	135	30
11 Mean number of days with precipitation	> 0,1 mm	9	10	7	1	5	0	1	5	12	17	1	6	74	30
12 Mean duration of sunshine	in h	264	230	233	228	214	207	220	238	222	239	248	276	2818	10
13 Mean quantity of radiation	in ly/day	598	586	454	342	263	237	254	322	413	505	610	634	435	4
14 Mean potential evaporation	in mm	106	95	88	68	53	40	39	46	52	67	81	104	839	70
15 Mean windspeed	in m/sec	4,7	4,4	3,9	3,9	3,3	2,8	3,3	4,2	4,4	5,0	5,6	5,0	4,1	9
16 Mean predominant direction of the wind															

159 Station / Country Oudtshoorn / South Africa

Location 33°35'S / 22°12'E — Height above sealevel 335 m — Climate symbol: Köppen BSh — Troll IV.2

		J	F	M	A	M	J	J	A	S	O	N	D	year	P
1 Mean daily temperature	in °C	23,6	23,7	22,1	18,4	14,8	12,0	11,7	12,9	15,5	17,5	20,0	22,0	17,9	11
2 Mean daily maximum temperature	in °C	32	32	30	26	22	20	19	21	23	26	28	30	26	30
3 Mean daily minimum temperature	in °C	15	15	14	10	7	4	3	5	7	10	12	14	10	30
4 Absolute maximum temperature	in °C	44	43	43	38	34	30	31	36	38	40	42	43	44	30
5 Absolute minimum temperature	in °C	6	7	6	3	-2	-3	-3	-2	0	2	4	4	-3	30
6 Mean relative humidity	in %	52	56	59	64	67	67	64	65	59	56	56	52	59	20
7 Mean precipitation	in mm	18	19	27	24	21	14	21	21	20	24	28	18	254	30
8 Maximum precipitation	in mm	119	72	106	84	72	36	73	66	49	69	97	57	406	30
9 Minimum precipitation	in mm	0	0	2	2	0	0	0	0	1	0	0	0	153	30
10 Maximum precipitation in 24 h	in mm	107	49	40	37	26	30	40	44	27	35	44	35	107	30
11 Mean number of days with precipitation	> 0,1 mm	2	3	4	3	3	3	3	3	4	4	3	2	37	30
12 Mean duration of sunshine	in h														
13 Mean quantity of radiation	in ly/day														
14 Mean potential evaporation	in mm	135	115	99	65	38	25	24	34	48	68	90	115	856	11
15 Mean windspeed	in m/sec														
16 Mean predominant direction of the wind															

160 Station / Country Mokhotlong / Lesotho

Location 29°17'S / 29°05'E — Height above sealevel 2375 m — Climate symbol: Köppen Cwb — Troll IV.4

		J	F	M	A	M	J	J	A	S	O	N	D	year	P
1 Mean daily temperature	in °C	16,6	16,1	14,4	11,5	7,6	4,6	4,6	7,2	10,6	13,4	14,5	16,2	11,5	6
2 Mean daily maximum temperature	in °C	24	24	22	20	16	14	14	17	20	22	23	24	20	25
3 Mean daily minimum temperature	in °C	9	9	7	4	-1	-4	-5	-2	2	6	7	9	3	25
4 Absolute maximum temperature	in °C	33	35	28	27	26	23	23	24	27	31	31	33	35	25
5 Absolute minimum temperature	in °C	2	1	-1	-8	-12	-12	-12	-13	-8	-7	-4	0	-13	25
6 Mean relative humidity	in %														
7 Mean precipitation	in mm	96	85	63	34	26	5	10	15	20	57	83	92	586	25
8 Maximum precipitation	in mm	151	153	141	75	131	21	48	71	121	128	179	177	806	25
9 Minimum precipitation	in mm	55	24	22	5	0	0	0	0	0	17	32	32	285	25
10 Maximum precipitation in 24 h	in mm	33	45	33	31	78	9	23	33	35	76	34	35	78	25
11 Mean number of days with precipitation	> 0,1 mm	13	12	10	6	3	1	2	2	3	8	12	12	84	25
12 Mean duration of sunshine	in h														
13 Mean quantity of radiation	in ly/day														
14 Mean potential evaporation	in mm	89	74	67	46	25	13	14	26	45	67	71	86	623	6
15 Mean windspeed	in m/sec														
16 Mean predominant direction of the wind															

161 Station / Country Jamestown / St. Helena

Location 15°55′S / 5°43′W — Height above sealevel 12 m — Climate symbol: Köppen BWh — Troll V,5

		J	F	M	A	M	J	J	A	S	O	N	D	year	P	
1	Mean daily temperature	in °C	23.6	24.2	24.8	24.2	21.9	20.8	19.7	19.7	19.7	20.3	20.8	21.8	21.8	7	1
2	Mean daily maximum temperature	in °C	26.7	27.2	27.8	27.2	24.4	23.3	22.2	22.2	22.2	22.8	23.3	24.4	24.4	7	2
3	Mean daily minimum temperature	in °C	20.6	21.1	21.7	21.1	19.4	18.3	17.2	17.2	17.2	17.8	18.3	18.9	18.9	7	3
4	Absolute maximum temperature	in °C	31.7	32.2	33.3	33.9	28.3	27.2	26.1	25.6	25.6	25.6	26.7	27.8	33.9	7	4
5	Absolute minimum temperature	in °C	17.2	18.9	18.9	17.2	16.1	16.1	14.4	15.0	14.4	15.6	16.7	15.6	14.4	7	5
6	Mean relative humidity	in %	63	65	63	64	71	71	72	75	72	71	71	65	69	3	6
7	Mean precipitation	in mm	8	10	20	10	18	18	8	10	5	4	0	3	113	7	7
8	Maximum precipitation	in mm															8
9	Minimum precipitation	in mm															9
10	Maximum precipitation in 24 h	in mm	13	8	23	8	18	20	15	8	10	5	0	5	23	7	10
11	Mean number of days with precipitation	> 0,1 mm	4	4	5	3	4	6	8	3	2	1	0	1	41	7	11
12	Mean duration of sunshine	in h															12
13	Mean quantity of radiation	in ly / day															13
14	Mean potential evaporation	in mm	110	104	117	109	84	66	65	67	67	76	81	93	1039	7	14
15	Mean windspeed	in m / sec															15
16	Mean predominent direction of the wind																16

162 Station / Country Tristan da Cunha / United Kingdom

Location 37°03′S / 12°19′W — Height above sealevel 23 m — Climate symbol: Köppen Cfb — Troll IV,7

		J	F	M	A	M	J	J	A	S	O	N	D	year	P	
1	Mean daily temperature	in °C	17.0	17.8	17.6	15.9	13.7	12.5	12.0	11.3	11.3	12.8	14.2	15.9	14.3	5	1
2	Mean daily maximum temperature	in °C	18.9	20.0	18.9	17.8	15.6	14.4	13.9	13.3	13.3	15.0	16.1	17.8	16.1	5	2
3	Mean daily minimum temperature	in °C	15.0	15.6	14.4	13.9	11.7	10.6	10.0	9.4	9.4	10.6	12.2	13.9	12.2	5	3
4	Absolute maximum temperature	in °C	23.9	23.3	22.8	22.2	19.4	18.3	17.8	16.7	17.8	18.3	20.0	22.2	23.9	5	4
5	Absolute minimum temperature	in °C	8.3	8.9	9.4	7.8	4.4	5.0	5.0	4.4	3.3	5.0	6.7	7.2	3.3	5	5
6	Mean relative humidity	in %	81	81	77	79	80	81	81	80	81	80	80	80	80	5	6
7	Mean precipitation	in mm	89	89	163	119	180	150	155	175	201	147	109	101	1678	5	7
8	Maximum precipitation	in mm															8
9	Minimum precipitation	in mm															9
10	Maximum precipitation in 24 h	in mm	33	36	91	43	69	46	46	48	145	48	48	28	145	5	10
11	Mean number of days with precipitation	> 1,0 mm	12	13	11	12	17	19	20	21	18	16	13	13	185	5	11
12	Mean duration of sunshine	in h															12
13	Mean quantity of radiation	in ly / day															13
14	Mean potential evaporation	in mm	90	79	80	62	48	36	36	36	39	55	68	84	711	5	14
15	Mean windspeed	in m / sec															15
16	Mean predominent direction of the wind																16

163 Station / Country Rodriguez / Rodriguez Island

Location 19°41′S / 63°27′E — Height above sealevel 43 m — Climate symbol: Köppen Aw — Troll V,2

		J	F	M	A	M	J	J	A	S	O	N	D	year	P	
1	Mean daily temperature	in °C	27	27	27	26	24.5	23	22	22	22.5	23.5	25.5	26	24.7	9	1
2	Mean daily maximum temperature	in °C	30	30	30	29	27	26	25	26	26	27	29	29	28	9	2
3	Mean daily minimum temperature	in °C	24	24	24	23	22	20	19	19	19	20	22	23	22	9	3
4	Absolute maximum temperature	in °C	32.2	32.8	33.9	31.7	30.6	28.9	27.8	27.2	28.3	30.6	31.1	32.8	33.9	9	4
5	Absolute minimum temperature	in °C	20.6	21.1	21.7	20.0	17.8	16.7	16.7	15.0	15.6	17.2	18.9	18.3	15.0	9	5
6	Mean relative humidity	in %	73	76	75	75	72	70	70	70	69	68	69	69	71	10	6
7	Mean precipitation	in mm	140	203	211	155	130	76	86	101	48	79	30	64	1324	9	7
8	Maximum precipitation	in mm															8
9	Minimum precipitation	in mm															9
10	Maximum precipitation in 24 h	in mm	140	180	404	208	163	69	48	94	18	236	20	101	404	9	10
11	Mean number of days with precipitation	> 0,25 mm	18	19	20	19	20	21	23	22	18	15	13	15	222	9	11
12	Mean duration of sunshine	in h															12
13	Mean quantity of radiation	in ly / day															13
14	Mean potential evaporation	in mm	157	138	145	125	101	82	74	77	84	104	131	148	1366	9	14
15	Mean windspeed	in m / sec															15
16	Mean predominent direction of the wind		E	E	E	E	E	E	E	E	E	E	E	E		5	16

164 Station / Country Kerguelen / France

Location 49°25′S / 69°53′E — Height above sealevel 16 m — Climate symbol: Köppen ET — Troll I,4

		J	F	M	A	M	J	J	A	S	O	N	D	year	P	
1	Mean daily temperature	in °C	7	7	6	5	2	2	1	2	1	2	4	6	4	3	1
2	Mean daily maximum temperature	in °C	10	10	9	7	4	3	3	3	3	5	7	9	6	3	2
3	Mean daily minimum temperature	in °C	4	4	3	3	0	1	-1	1	-1	-1	3	3	1	3	3
4	Absolute maximum temperature	in °C	17.2	16.7	16.1	11.7	10.0	9.4	7.2	10.6	11.7	12.2	13.3	14.4	17.2	3	4
5	Absolute minimum temperature	in °C	1.1	0.6	-0.6	-1.7	-3.9	-3.9	-6.7	-5.6	-7.2	-3.3	-1.7	0	-7.2	3	5
6	Mean relative humidity	in %	78	74	75	78	80	77	80	81	76	74	78	76	77	3	6
7	Mean precipitation	in mm	81	51	53	38	51	165	64	142	61	71	69	79	925	1	7
8	Maximum precipitation	in mm															8
9	Minimum precipitation	in mm															9
10	Maximum precipitation in 24 h	in mm	15	10	23	10	10	18	13	25	15	8	15	20	25	1	10
11	Mean number of days with precipitation	> 0,25 mm	19	21	21	18	27	28	26	26	22	29	25	21	283	1	11
12	Mean duration of sunshine	in h															12
13	Mean quantity of radiation	in ly / day															13
14	Mean potential evaporation	in mm	73	60	49	35	14	12	7	16	8	21	42	63	401	1	14
15	Mean windspeed	in m / sec															15
16	Mean predominent direction of the wind																16

AUSTRALIA AND OCEANIA

AUSTRALIA

1 Thursday Island
2 Coen
3 Darwin
4 Wyndham
5 Townsville
6 Willis Islets
7 Normanton
8 Broome
9 Halls Creek
10 Cloncurry
11 Charleville
12 Geraldton
13 Bourke
14 Hay
15 Adelaide
16 Eyre
17 Onslow
18 Tennant Creek
19 Alice Springs
20 Mundiwindi
21 Wiluna
22 Giles
23 Windorah
24 Yalgoo
25 Tarcoola
26 Farina
27 Kalgoorlie
28 Broken Hill
29 Port Augusta
30 Perth
31 Albany
32 Esperance

33 Rockhampton
34 Brisbane
35 Armidale
36 Lord Howe Island
37 Sydney
38 Canberra
39 Mount Gambier
40 Melbourne
41 Launceston (Tasmania)
42 Hobart (Tasmania)
43 Norfolk Island

NEW ZEALAND

44 Auckland
45 Tauranga
46 New Plymouth
47 Wellington
48 Dunedin

OCEANIA

49 Fanning Island (United Kingdom)
50 Nauru (Nauru)

TERRITORY OF NEW GUINEA

51 Rabaul (New Britain)

U.S. AND NEW ZEALAND

52 Tongareva

U.S.

53 Saipan

FIJI

54 Suva (Viti Levu)

INDONESIA

55 Manokwari

PAPUA-NEW GUINEA

56 Mandang
57 Kikori
58 Daru
59 Samarai
60 Bwagaoia

269

AUSTRALIA

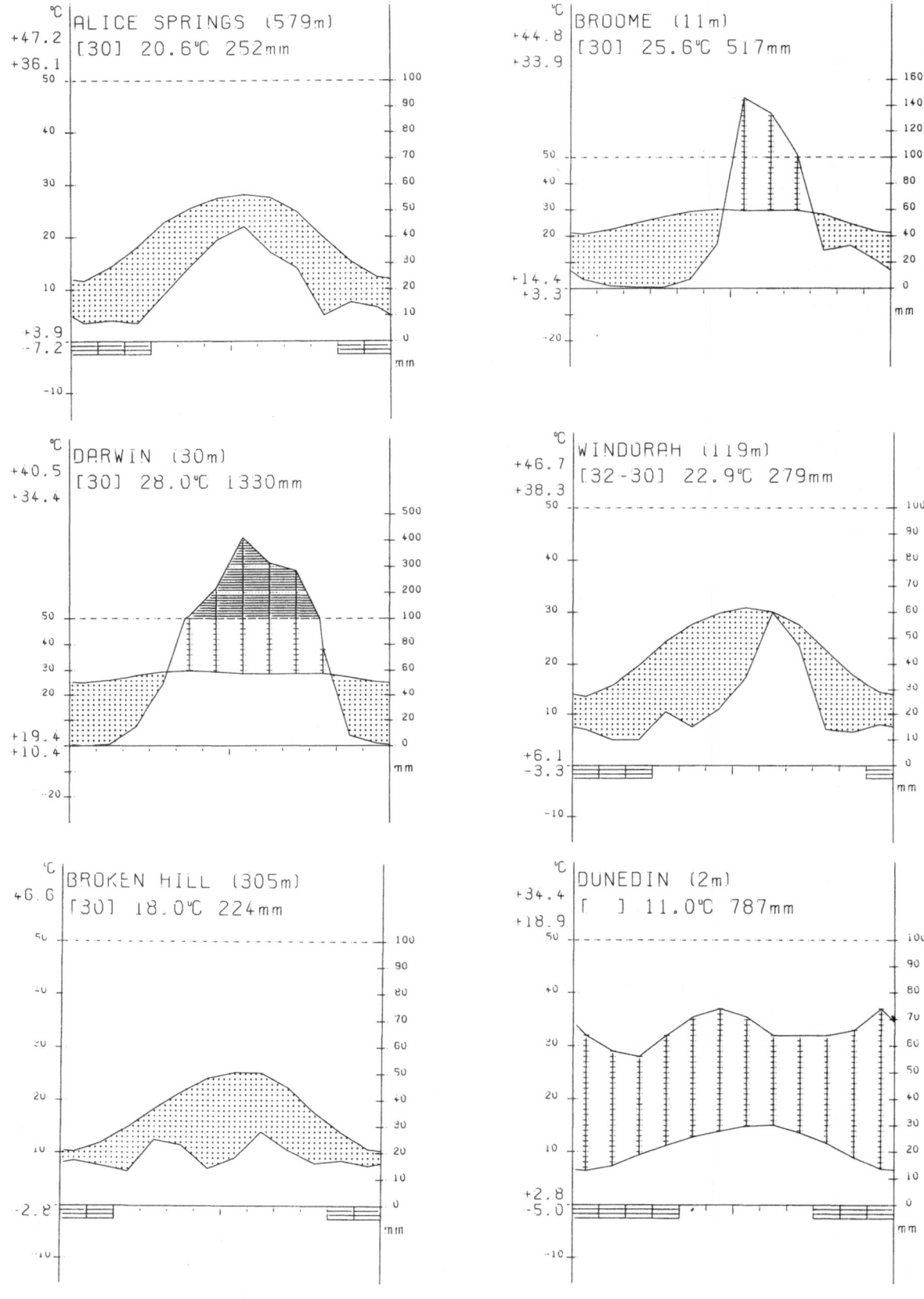

1 Station / Country Thursday Island / Australia

Location 10°34'S / 142°12'E   Height above sealevel 5 m   Climate symbol: Köppen Aw   Troll V,3

		J	F	M	A	M	J	J	A	S	O	N	D	year	P	
1 Mean daily temperature	in °C	28,0	27,7	27,7	27,6	26,9	26,1	25,3	25,3	26,0	27,2	28,3	28,8	27,1	30	1
2 Mean daily maximum temperature	in °C	30,8	30,6	30,6	30,0	29,4	28,9	27,8	27,8	28,9	30,0	31,1	31,7	30,0	31	2
3 Mean daily minimum temperature	in °C	25,0	25,0	25,0	25,0	24,4	23,3	22,8	22,8	23,3	24,4	25,0	25,6	24,4	31	3
4 Absolute maximum temperature	in °C	36,1	34,4	33,9	34,4	32,8	31,7	32,2	31,7	32,8	33,9	35,6	36,7	36,7	49	4
5 Absolute minimum temperature	in °C	21,1	21,1	21,1	21,1	18,9	17,8	17,8	20,0	20,0	21,1	21,7	21,1	17,8	49	5
6 Mean relative humidity	in %	83	84	83	81	80	80	79	78	77	75	75	77	79	30	6
7 Mean precipitation	in mm	441	378	350	203	41	16	13	5	3	5	36	198	1687	30	7
8 Maximum precipitation	in mm	895	791	649	721	209	68	51	48	18	78	190	526	2520	53	8
9 Minimum precipitation	in mm	164	130	86	13	1	1	0	0	0	0	0	1	815	53	9
10 Maximum precipitation in 24 h	in mm	178	173	148	217	75	28	35	6	6	48	92	120	217	49	10
11 Mean number of days with precipitation	>0,25 mm	21	20	20	14	10	8	7	5	4	3	4	13	129	46	11
12 Mean duration of sunshine	in h															12
13 Mean quantity of radiation	in ly / day															13
14 Mean potential evaporation	in mm	164	147	160	151	143	130	120	121	133	153	163	168	1753	24	14
15 Mean windspeed	in m / sec	2,1	2,1	2,1	3,1	4,6	4,1	4,1	4,6	4,1	4,6	3,1	2,1	3,6	5	15
16 Mean predominent direction of the wind		NW	NW	SE,NW	SE	SE	SE	SE	SE	SE	SE	SE	SE,NW		5	16

2 Station / Country Coen / Australia

Location 13°57'S / 143°12'E   Height above sealevel 600 m   Climate symbol: Köppen Aw   Troll V,3

		J	F	M	A	M	J	J	A	S	O	N	D	year	P	
1 Mean daily temperature	in °C	27,0	26,4	26,0	25,1	23,6	22,5	21,7	22,3	23,9	25,8	27,3	27,7	24,9	26	1
2 Mean daily maximum temperature	in °C															2
3 Mean daily minimum temperature	in °C															3
4 Absolute maximum temperature	in °C	40,0	38,9	36,7	33,3	32,2	33,3	30,6	30,6	31,7	34,4	40,0	40,6	40,6	31	4
5 Absolute minimum temperature	in °C	19,4	17,2	19,4	16,1	12,2	10,6	8,3	10,6	11,1	14,4	16,1	16,1	8,3	31	5
6 Mean relative humidity	in %	81	84	83	80	79	77	76	72	69	66	67	74	76		6
7 Mean precipitation	in mm	247	272	272	80	11	13	7	3	1	8	44	187	1145	30	7
8 Maximum precipitation	in mm	579	782	748	563	78	54	44	44	11	106	400	604	1818	51	8
9 Minimum precipitation	in mm	23	36	48	2	0	0	0	0	0	0	0	6	482	51	9
10 Maximum precipitation in 24 h	in mm	356	170	343	282	104	94	23	43	18	66	69	196	356	29	10
11 Mean number of days with precipitation	>0,25 mm	15	15	15	7	1	3	1	1	1	2	4	9	74		11
12 Mean duration of sunshine	in h	180	160	180	180	190	180	220	215	235	235	230	205	2410	+	12
13 Mean quantity of radiation	in ly / day															13
14 Mean potential evaporation	in mm	157	133	136	114	90	78	74	83	106	142	154	167	1434	24	14
15 Mean windspeed	in m / sec															15
16 Mean predominent direction of the wind		SE	SE	SE	SE	SE	SE	SE	SE	SE	SE	SE	SE,NE		5	16

+ estimated from maps

3 Station / Country Darwin / Australia

Location 12°28'S / 130°51'E   Height above sealevel 30 m   Climate symbol: Köppen Aw   Troll V,3

		J	F	M	A	M	J	J	A	S	O	N	D	year	P	
1 Mean daily temperature	in °C	28,7	28,6	28,7	28,8	27,4	25,8	25,1	26,2	28,1	29,4	29,8	29,4	28,0	30	1
2 Mean daily maximum temperature	in °C	32,2	32,2	32,8	33,3	32,8	31,1	30,6	31,7	32,8	33,9	34,4	33,3	32,8	58	2
3 Mean daily minimum temperature	in °C	25,0	25,0	25,0	24,4	22,8	20,6	19,4	21,1	23,3	25,0	25,6	25,6	23,3	58	3
4 Absolute maximum temperature	in °C	37,8	38,3	38,9	40,0	39,0	37,0	36,7	36,7	38,9	40,5	39,8	38,9	40,5	80	4
5 Absolute minimum temperature	in °C	20,4	17,2	19,2	16,0	15,1	12,9	10,4	13,9	17,2	20,3	19,3	20,3	10,4	80	5
6 Mean relative humidity	in %	80	80	79	68	60	55	55	61	65	69	79	73	68	57	6
7 Mean precipitation	in mm	411	314	284	78	8	2	0	1	15	49	110	218	1330	30	7
8 Maximum precipitation	in mm	708	653	556	603	356	39	65	76	69	338	400	568		92	8
9 Minimum precipitation	in mm	57	11	21	0	0	0	0	0	0	0	10	25		92	9
10 Maximum precipitation in 24 h	in mm	297	134	182	168	56	34	43	27	51	95	120	200	297	92	10
11 Mean number of days with precipitation	>0,25 mm	20	18	17	6	1	1	<1	<1	2	5	10	15	95	30	11
12 Mean duration of sunshine	in h	180	162	205	294	288	300	307	322	309	291	288	211	3151		12
13 Mean quantity of radiation	in ly / day	443	458	474	444	467	456	464	526	554	561	512	475	486	9	13
14 Mean potential evaporation	in mm	173	153	169	162	146	133	112	138	152	170	173	176	1857	49	14
15 Mean windspeed	in m / sec	3,3	4,1	3,1	3,1	3,9	4,1	3,9	3,6	3,9	3,9	3,1	3,1	3,6	3	15
16 Mean predominent direction of the wind		NW	W	SE,W	SE,E	SE,E	SE	SE	SE,NW	SE,NW	S,NW	W,NW	NW			16

4 Station / Country Wyndham / Australia

Location 15°27'S / 128°07'E   Height above sealevel 7 m   Climate symbol: Köppen BSh   Troll V,4

		J	F	M	A	M	J	J	A	S	O	N	D	year	P	
1 Mean daily temperature	in °C	31,2	30,8	30,8	29,9	27,3	24,9	24,2	26,1	29,0	31,3	32,2	31,9	29,1	30	1
2 Mean daily maximum temperature	in °C	35,6	35,6	35,0	35,0	32,2	30,0	29,4	31,7	34,4	36,1	36,7	36,1	33,9	40	2
3 Mean daily minimum temperature	in °C	26,7	26,7	26,1	25,0	22,2	20,0	18,9	21,1	23,9	26,7	27,2	27,2	24,4	40	3
4 Absolute maximum temperature	in °C	45,3	43,9	42,2	41,7	39,4	37,8	35,7	38,9	41,1	43,9	45,3	44,4	45,3	70	4
5 Absolute minimum temperature	in °C	18,7	16,7	18,3	17,2	11,1	10,0	8,9	8,3	15,6	18,3	14,4	18,3	8,3	70	5
6 Mean relative humidity	in %	66	67	63	46	41	40	38	40	44	52	55	60	51	30	6
7 Mean precipitation	in mm	202	163	122	34	10	10	5	<1	2	9	42	104	703	30	7
8 Maximum precipitation	in mm	718	364	446	238	58	120	133	21	35	82	132	291	1353	55	8
9 Minimum precipitation	in mm	13	14	3	0	0	0	0	0	0	0	1	7	365	55	9
10 Maximum precipitation in 24 h	in mm	307	150	317		69	113	86	11	35	57	85	110		80	10
11 Mean number of days with precipitation	>0,25 mm	13	11	9	3	1	0	0	0	0	2	6	10	55	30	11
12 Mean duration of sunshine	in h															12
13 Mean quantity of radiation	in ly / day															13
14 Mean potential evaporation	in mm	189	163	173	159	139	96	87	132	155	182	185	193	1853	33	14
15 Mean windspeed	in m / sec	3,6	3,3	3,1	3,1	3,3	3,6	3,3	3,1	3,3	3,9	3,6	3,1	3,3	5	15
16 Mean predominent direction of the wind		SW,N	SW,N	SW,N	SW,N	SE,N	SE	SE,N	SE,N	NW,N	SW,N	SW,N	SW,N		5	16

5 Station / Country Townsville / Australia

Location 19°14'S / 146°51'E   Height above sealevel 22 m   Climate symbol: Köppen Aw   Troll V.2

		J	F	M	A	M	J	J	A	S	O	N	D	year	P	
1	Mean daily temperature	in °C	27.5	27.2	26.8	24.9	22.5	20.1	19.4	20.3	22.3	24.7	26.5	27.5	24.2	30	1
2	Mean daily maximum temperature	in °C	30.6	30.6	30.0	28.9	27.2	25.0	23.9	25.0	26.7	28.3	29.4	30.6	27.8	31	2
3	Mean daily minimum temperature	in °C	24.4	23.9	22.8	21.1	18.3	16.1	15.0	16.1	16.9	21.7	23.3	24.6	20.6	31	3
4	Absolute maximum temperature	in °C	39.6	43.4	36.7	36.1	31.7	30.5	31.6	33.3	35.4	37.1	37.9	39.6	43.4	34	4
5	Absolute minimum temperature	in °C	18.7	17.9	16.7	11.7	6.2	4.5	3.7	1.1	6.7	8.2	14.9	18.7	1.1	34	5
6	Mean relative humidity	in %	75	76	73	70	67	66	64	66	67	68	69	73	70	23	6
7	Mean precipitation	in mm	284	346	230	72	29	31	21	11	8	21	62	102	1215	30	7
8	Maximum precipitation	in mm	831	871	601	592	206	181	137	113	178	274	335	616	2482	72	8
9	Minimum precipitation	in mm	15	2	1	0	0	0	0	0	0	0	0	0	268	72	9
10	Maximum precipitation in 24 h	in mm	347	291	366	237	122	110	90	59	66	229	152	167	366	72	10
11	Mean number of days with precipitation	>0.25 mm	15	12	10	6	5	4	3	3	2	4	5	12	81	10	11
12	Mean duration of sunshine	in h	223	199	214	255	226	240	273	285	282	295	276	260	3038	6	12
13	Mean quantity of radiation	in ly / day	514	484	481	451	386	372	401	471	559	595	610	600	494	12	13
14	Mean potential evaporation	in mm	159	133	139	120	88	61	55	66	90	124	143	154	1332	25	14
15	Mean windspeed	in m / sec	3.6	4.1	4.1	4.4	2.8	3.6	3.3	3.6	4.1	4.1	4.4	4.4	3.9	5	15
16	Mean predominant direction of the wind		E.NE	SE.NE	SE.NE	SE.NE	S.NE	S.NE	S.NE	S.NE	SE.NE	E.NE	E.NE	E.NE		5	16

6 Station / Country Willis Islets / Australia

Location 16°18'S / 149°59'E   Height above sealevel 9 m   Climate symbol: Köppen Aw   Troll V.2

		J	F	M	A	M	J	J	A	S	O	N	D	year	P	
1	Mean daily temperature	in °C	28.1	27.8	27.4	26.8	25.4	24.3	23.6	23.8	24.6	26.6	26.9	27.8	26.0	16	1
2	Mean daily maximum temperature	in °C	30.6	30.0	30.0	28.9	27.8	26.1	25.6	26.1	27.2	28.3	29.4	30.6	28.3	20	2
3	Mean daily minimum temperature	in °C	25.6	25.6	25.0	24.4	23.2	22.2	21.7	21.7	22.2	23.3	24.4	25.0	23.9	20	3
4	Absolute maximum temperature	in °C	33.9	33.3	32.8	31.7	30.6	28.9	28.3	28.3	30.0	31.1	32.2	33.3	33.9	20	4
5	Absolute minimum temperature	in °C	21.7	21.7	21.1	20.0	20.0	17.2	17.8	18.9	18.9	18.9	20.6	21.1	17.2	20	5
6	Mean relative humidity	in %	82	80	81	80	80	77	74	73	72	72	75	79	78	20	6
7	Mean precipitation	in mm	167	283	175	177	67	72	49	18	22	17	36	92	1175	20	7
8	Maximum precipitation	in mm	476	684	326	510	313	320	186	109	207	102	171	274	1596	20	8
9	Minimum precipitation	in mm	10	23	7	20	4	3	4	0	0	0	1	0	739	20	9
10	Maximum precipitation in 24 h	in mm	118	329	150	230	246	100	161	46	73	62	120	171	329	20	10
11	Mean number of days with precipitation	>0.25 mm	8	10	10	8	4	4	3	2	2	1	3	4	59	20	11
12	Mean duration of sunshine	in h															12
13	Mean quantity of radiation	in ly / day	489	519	441	464	353	358	392	465	546	584	606	581	483	1	13
14	Mean potential evaporation	in mm	171	147	151	131	109	88	86	89	99	149	150	168	1538	20	14
15	Mean windspeed	in m / sec	5.9	5.7	6.7	7.5	8.0	7.7	6.9	7.7	6.7	6.7	5.9	4.9	6.7	5	15
16	Mean predominant direction of the wind		SE	SE	SE	SE	SE	SE	SE	SE	SE	SE	SE	SE		20	16

7 Station / Country Normanton / Australia

Location 17°39'S / 141°05'E   Height above sealevel 10 m   Climate symbol: Köppen Aw   Troll V.4

		J	F	M	A	M	J	J	A	S	O	N	D	year	P	
1	Mean daily temperature	in °C	29.8	29.2	29.2	27.9	25.2	22.8	21.8	23.4	26.6	29.1	30.6	30.6	27.2	23	1
2	Mean daily maximum temperature	in °C	35.0	33.9	34.4	33.9	31.7	29.4	28.9	31.1	33.9	36.1	37.2	36.1	33.3	20	2
3	Mean daily minimum temperature	in °C	25.0	25.0	23.9	21.7	18.9	16.1	15.0	16.1	19.4	22.2	24.4	25.0	21.1	20	3
4	Absolute maximum temperature	in °C	43.3	40.6	41.7	40.0	38.3	35.6	36.1	36.7	40.6	43.9	43.3	43.3	43.9	32	4
5	Absolute minimum temperature	in °C	16.7	17.8	15.6	12.2	15.0	7.2	5.6	7.8	10.6	10.6	12.2	12.8	5.6	32	5
6	Mean relative humidity	in %	70	72	65	50	49	51	48	42	43	48	52	59	54	18	6
7	Mean precipitation	in mm	295	249	170	35	7	16	3	1	2	6	46	122	954	30	7
8	Maximum precipitation	in mm	670	770	548	305	198	106	85	34	51	99	213	619	1530	66	8
9	Minimum precipitation	in mm	2	18	0	0	0	0	0	0	0	0	0	0	354	66	9
10	Maximum precipitation in 24 h	in mm	272	280	299	174	48	78	60	25	50	72	137	173	299	50	10
11	Mean number of days with precipitation	>0.25 mm	11	12	10	7	6	6	6	4	4	5	8	10	89	26	11
12	Mean duration of sunshine	in h															12
13	Mean quantity of radiation	in ly / day															13
14	Mean potential evaporation	in mm	179	159	169	152	123	77	66	94	138	166	175	186	1684	22	14
15	Mean windspeed	in m / sec	2.6	2.8	3.1	3.6	3.6	3.3	2.8	2.8	2.6	3.1	3.1	2.3	2.8	5	15
16	Mean predominant direction of the wind		NW	NW	E	SE	SE	SE	SE	SE	SE	SE.NW	NW	NW		5	16

8 Station / Country Broome / Australia

Location 17°57'S / 122°15'E   Height above sealevel 11 m   Climate symbol: Köppen BSh   Troll V.4

		J	F	M	A	M	J	J	A	S	O	N	D	year	P	
1	Mean daily temperature	in °C	29.6	29.7	29.7	28.1	24.7	21.7	20.8	22.5	25.0	27.4	29.3	30.1	25.6	30	1
2	Mean daily maximum temperature	in °C	33.3	33.3	33.9	33.9	31.1	27.8	27.8	29.4	31.7	32.8	33.9	33.9	31.7	41	2
3	Mean daily minimum temperature	in °C	26.1	26.1	25.0	22.2	18.3	15.6	14.4	15.6	18.3	22.2	24.4	26.1	21.1	41	3
4	Absolute maximum temperature	in °C	44.2	42.7	41.7	41.7	38.3	36.2	35.0	38.1	39.7	42.8	44.0	44.8	44.8	74	4
5	Absolute minimum temperature	in °C	17.8	15.0	12.8	12.2	7.3	5.6	3.3	4.8	8.9	11.6	14.7	17.2	3.3	74	5
6	Mean relative humidity	in %	75	76	73	56	54	53	52	54	56	64	66	70	63	30	6
7	Mean precipitation	in mm	146	134	102	29	33	21	7	2	1	1	7	34	517	30	7
8	Maximum precipitation	in mm	825	599	800	232	131	246	59	95	22	10	279	368	1084	52	8
9	Minimum precipitation	in mm	3	2	0	0	0	0	0	0	0	0	0	0	142	52	9
10	Maximum precipitation in 24 h	in mm	358	302	269	181	119	143	55	37	21	10	140	143	356	74	10
11	Mean number of days with precipitation	>0.25 mm	10	8	7	2	2	1	1	<1	<1	<1	1	6	38	30	11
12	Mean duration of sunshine	in h															12
13	Mean quantity of radiation	in ly / day															13
14	Mean potential evaporation	in mm	181	157	187	146	98	58	53	71	108	152	168	185	1544	35	14
15	Mean windspeed	in m / sec	3.3	3.3	3.1	3.1	2.8	3.3	3.1	3.3	3.9	4.1	3.9	3.9	3.3	5	15
16	Mean predominant direction of the wind		SW.W	SW	SE.SW	E.SE	SE.S	SE.S	SE.S	SE.SW	SE.SW	S.SW	SW.W	SW.W		5	16

9 Station/Country Halls Creek/Australia

Location 18°13'S/127°46'E Height above sealevel 374 m Climate symbol: Köppen BSh Troll V,4

		J	F	M	A	M	J	J	A	S	O	N	D	year	P	
1 Mean daily temperature	in °C	30,3	29,8	28,6	25,3	21,4	18,6	17,7	20,6	24,3	28,8	30,7	30,8	25,6	30	1
2 Mean daily maximum temperature	in °C	36,7	36,1	35,0	33,3	29,4	27,2	27,2	30,0	33,9	36,7	37,8	37,2	33,3	33	2
3 Mean daily minimum temperature	in °C	23,9	23,3	21,7	17,8	13,3	10,6	8,9	11,1	15,0	20,6	23,3	23,9	17,8	33	3
4 Absolute maximum temperature	in °C	44,3	43,8	41,9	39,9	37,2	35,0	34,0	37,8	40,2	42,8	43,7	44,2	44,3	41	4
5 Absolute minimum temperature	in °C	15,6	12,2	11,0	7,2	2,4	0,2	-1,1	0,4	3,0	8,9	11,7	12,1	-1,1	41	5
6 Mean relative humidity	in %	54	54	50	40	41	44	42	39	35	38	38	46	45	30	6
7 Mean precipitation	in mm	123	109	51	15	12	8	9	3	3	9	22	59	423	30	7
8 Maximum precipitation	in mm	578	373	369	164	167	87	80	56	53	104	200	230	1068	41	8
9 Minimum precipitation	in mm	5	3	0	0	0	0	0	0	0	0	0	3	214	41	9
10 Maximum precipitation in 24 h	in mm	211	130	174	147	61	38	48	52	31	36	50	120	211	41	10
11 Mean number of days with precipitation	>0,25 mm	12	10	7	2	1	1	1	1	1	3	6	9	54	30	11
12 Mean duration of sunshine	in h	242	218	251	270	282	270	298	304	309	295	282	257	3278	1	12
13 Mean quantity of radiation	in ly/day	416	500	456	504	444	433	442	522	616	596	568	528	502	+	13
14 Mean potential evaporation	in mm	183	158	157	117	63	37	33	55	103	164	178	189	1437	33	14
15 Mean windspeed	in m/sec															15
16 Mean predominent direction of the wind																16

+ estimated from maps

10 Station/Country Cloncurry/Australia

Location 20°43'S/140°30'E Height above sealevel 193 m Climate symbol: Köppen BSh Troll V,4

		J	F	M	A	M	J	J	A	S	O	N	D	year	P	
1 Mean daily temperature	in °C	30,9	29,9	28,8	25,8	21,8	18,7	17,8	19,9	23,7	27,6	30,1	31,3	25,5	30	1
2 Mean daily maximum temperature	in °C	37,2	35,6	35,0	32,2	28,3	25,0	25,0	27,8	31,1	35,0	36,7	37,8	32,2	32	2
3 Mean daily minimum temperature	in °C	25,0	23,9	22,8	19,4	15,6	12,2	10,6	12,8	16,1	20,0	22,8	24,4	18,9	32	3
4 Absolute maximum temperature	in °C	52,8	46,1	43,9	39,4	37,2	37,2	35,6	39,4	41,1	44,9	48,3	51,7	52,8	32	4
5 Absolute minimum temperature	in °C	15,0	15,0	11,7	8,9	5,0	2,2	1,7	5,0	10,6	12,2	10,0		1,7	32	5
6 Mean relative humidity	in %	40	49	46	37	39	43	40	32	30	31	34	38	39	28	6
7 Mean precipitation	in mm	120	101	47	16	12	20	6	3	4	11	40	48	429	30	7
8 Maximum precipitation	in mm	498	444	279	90	122	179	83	57	112	92	130	515	1047	54	8
9 Minimum precipitation	in mm	4	0	0	0	0	0	0	0	0	0	0	0	134	54	9
10 Maximum precipitation in 24 h	in mm	63	90	122	13	-34	26	33	<1	17	28	19	38	122	7	10
11 Mean number of days with precipitation	>0,25 mm	8	6	5	1	2	1	1	<1	1	2	3	7	37	59	11
12 Mean duration of sunshine	in h															12
13 Mean quantity of radiation	in ly/day															13
14 Mean potential evaporation	in mm	187	162	165	132	73	40	35	56	101	166	179	190	1486	24	14
15 Mean windspeed	in m/sec															15
16 Mean predominent direction of the wind		SE	SE	SE	SE	SE	SE	SE	SE	SE	SE	SE.S	SE.S	.	5	16

11 Station/Country Charleville/Australia

Location 26°25'S/146°13'E Height above sealevel 294 m Climate symbol: Köppen BSh Troll IV,3

		J	F	M	A	M	J	J	A	S	O	N	D	year	P	
1 Mean daily temperature	in °C	29,0	28,4	25,8	21,2	16,6	13,2	12,3	14,2	18,2	22,7	26,1	28,1	21,3	30	1
2 Mean daily maximum temperature	in °C	36,1	35,6	32,8	28,9	24,4	20,6	20,0	22,8	26,7	31,1	33,9	35,6	28,9	33	2
3 Mean daily minimum temperature	in °C	21,7	21,1	18,3	13,3	8,3	5,6	4,4	5,6	9,4	14,4	17,8	20,0	13,3	33	3
4 Absolute maximum temperature	in °C	47,0	46,1	43,3	38,8	33,9	28,9	30,7	34,3	38,9	43,2	47,2	47,8	47,8	35	4
5 Absolute minimum temperature	in °C	11,1	9,4	5,0	1,0	-2,8	-5,0	-5,0	-4,4	-1,7	0,9	4,4	6,7	-5,0	35	5
6 Mean relative humidity	in %	44	47	50	52	59	65	61	53	47	42	42	43	49	29	6
7 Mean precipitation	in mm	75	70	67	33	28	27	31	20	18	41	40	49	498	30	7
8 Maximum precipitation	in mm	307	399	382	198	171	128	220	125	127	130	190	235	1202	60	8
9 Minimum precipitation	in mm	3	0	0	0	0	0	0	0	0	0	0	0	202	60	9
10 Maximum precipitation in 24 h	in mm	109	130	159	81	69	70	86	44	57	86	83	181	181	35	10
11 Mean number of days with precipitation	>0,25 mm	6	5	5	3	4	4	3	2	3	4	5	6	52	65	11
12 Mean duration of sunshine	in h															12
13 Mean quantity of radiation	in ly/day															13
14 Mean potential evaporation	in mm	177	150	122	77	44	23	20	29	61	103	149	162	1117	24	14
15 Mean windspeed	in m/sec															15
16 Mean predominent direction of the wind		NE.E	NE.E	N.SE	NE.SE	E.SE	SE.S	NE.S	NE.SW	NE.SW	N.NW	N.SW	N.SE		5	16

12 Station/Country Geraldton/Australia

Location 28°45'S/114°36'E Height above sealevel 4 m Climate symbol: Köppen BSh Troll IV,1

		J	F	M	A	M	J	J	A	S	O	N	D	year	P	
1 Mean daily temperature	in °C	24,1	24,3	23,5	21,5	18,6	16,6	15,4	15,8	16,8	18,1	20,8	22,5	19,8	30	1
2 Mean daily maximum temperature	in °C															2
3 Mean daily minimum temperature	in °C															3
4 Absolute maximum temperature	in °C	47,7	46,1	44,3	39,4	34,8	28,8	27,2	31,6	35,8	40,3	42,7	45,0	47,7	70	4
5 Absolute minimum temperature	in °C	8,9	10,6	8,3	5,4	2,1	0,8	0,8	1,7	1,8	3,3	5,6	7,7	0,8	70	5
6 Mean relative humidity	in %	61	61	61	60	64	67	68	68	67	65	65	63	65	42	6
7 Mean precipitation	in mm	7	10	16	30	66	113	98	64	26	18	7	5	457	30	7
8 Maximum precipitation	in mm	96	117	169	93	328	328	205	242	105	70	40	32	855	67	8
9 Minimum precipitation	in mm	0	0	0	0	6	34	18	8	1	0	0	0	262	67	9
10 Maximum precipitation in 24 h	in mm	79	82	94	69	78	109	51	39	43	70	36	51	109	80	10
11 Mean number of days with precipitation	>0,25 mm	2	2	3	5	10	13	15	12	8	7	3	2	82	29	11
12 Mean duration of sunshine	in h															12
13 Mean quantity of radiation	in ly/day															13
14 Mean potential evaporation	in mm	127	114	108	79	54	40	35	39	47	61	82	111	897	34	14
15 Mean windspeed	in m/sec	6,7	5,9	5,7	5,4	4,9	4,4	3,9	4,1	4,9	5,4	5,9	6,4	5,4	5	15
16 Mean predominent direction of the wind		S	S	SE.S	E.S	E.S	E.S	E.S	E.W	E.SW	S.SW	S	S		5	16

273

AUSTRALIA and OCEANIA
Climate Zones after W. Köppen/R. Geiger

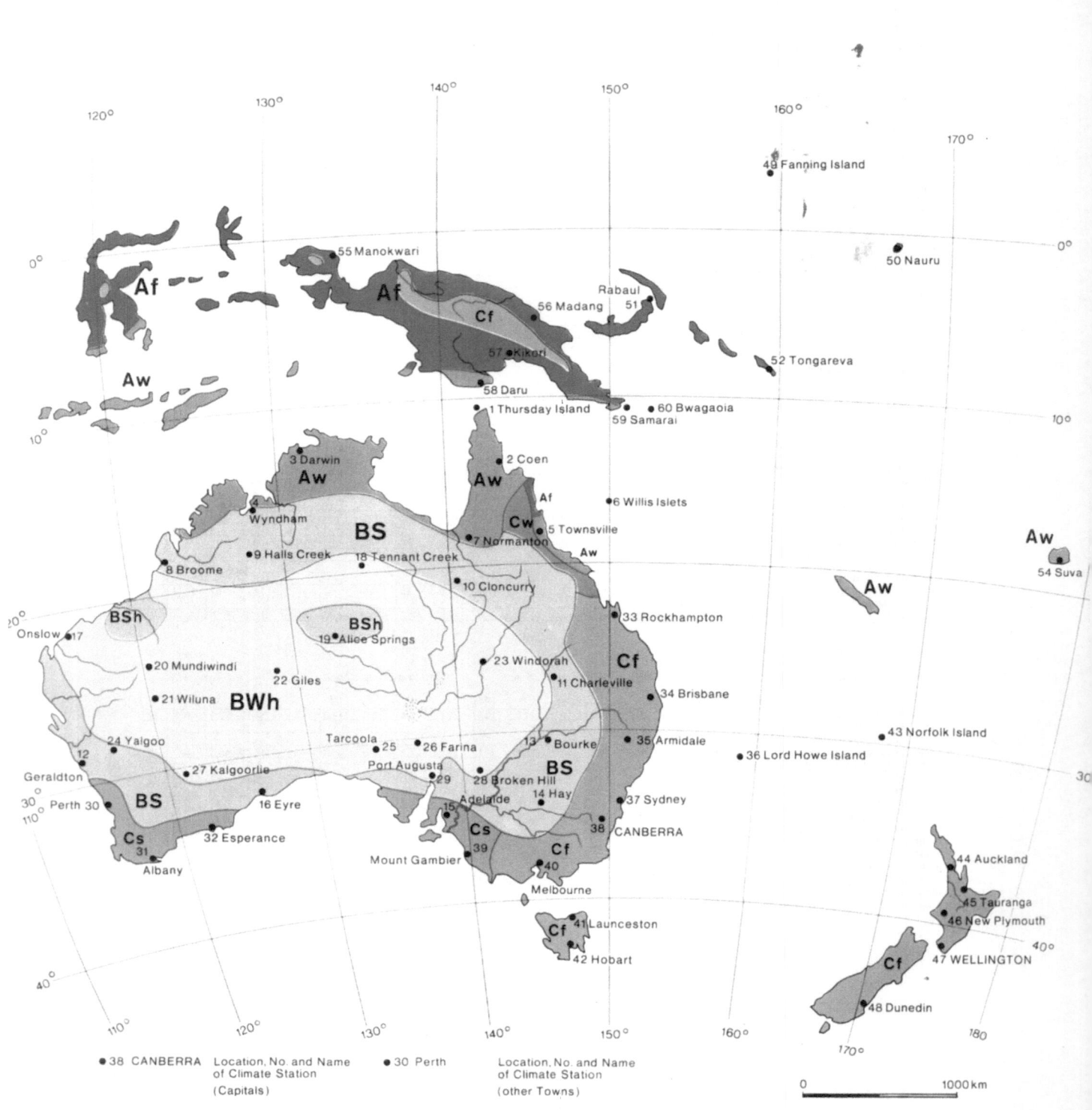

● 53 Saipan

49 Fanning Island

0°

50 Nauru

55 Manokwari

Af

Af

Cf

Rabaul
56 Madang 51

57 Kikori

52 Tongareva

58 Daru

Aw

1 Thursday Island

60 Bwagaoia
59 Samarai

3 Darwin

2 Coen

Aw

Aw

Af

6 Willis Islets

4
Wyndham

BS

Cw

5 Townsville

7 Normanton

Aw

Aw

8 Broome

9 Halls Creek

18 Tennant Creek

54 Suva

10 Cloncurry

Aw

33 Rockhampton

BSh

BSh

Onslow 17

19 Alice Springs

23 Windorah

Cf

20 Mundiwindi

22 Giles

11 Charleville

34 Brisbane

21 Wiluna

BWh

43 Norfolk Island

24 Yalgoo

Tarcoola

25

26 Farina

13

Bourke

35 Armidale

36 Lord Howe Island

12

27 Kalgoorlie

Port Augusta

29

28 Broken Hill

BS

Geraldton

Adelaide

14 Hay

37 Sydney

Perth 30

16 Eyre

15

38 CANBERRA

BS

Cs

32 Esperance

Cs

39

44 Auckland

Cs

31
Albany

Mount Gambier

Cf

40

45 Tauranga

Melbourne

46 New Plymouth

Cf
41 Launceston

47 WELLINGTON

42 Hobart

Cf

48 Dunedin

AUSTRALIA and OCEANIA
Climate Zones after C. Troll / K.H. Paffen

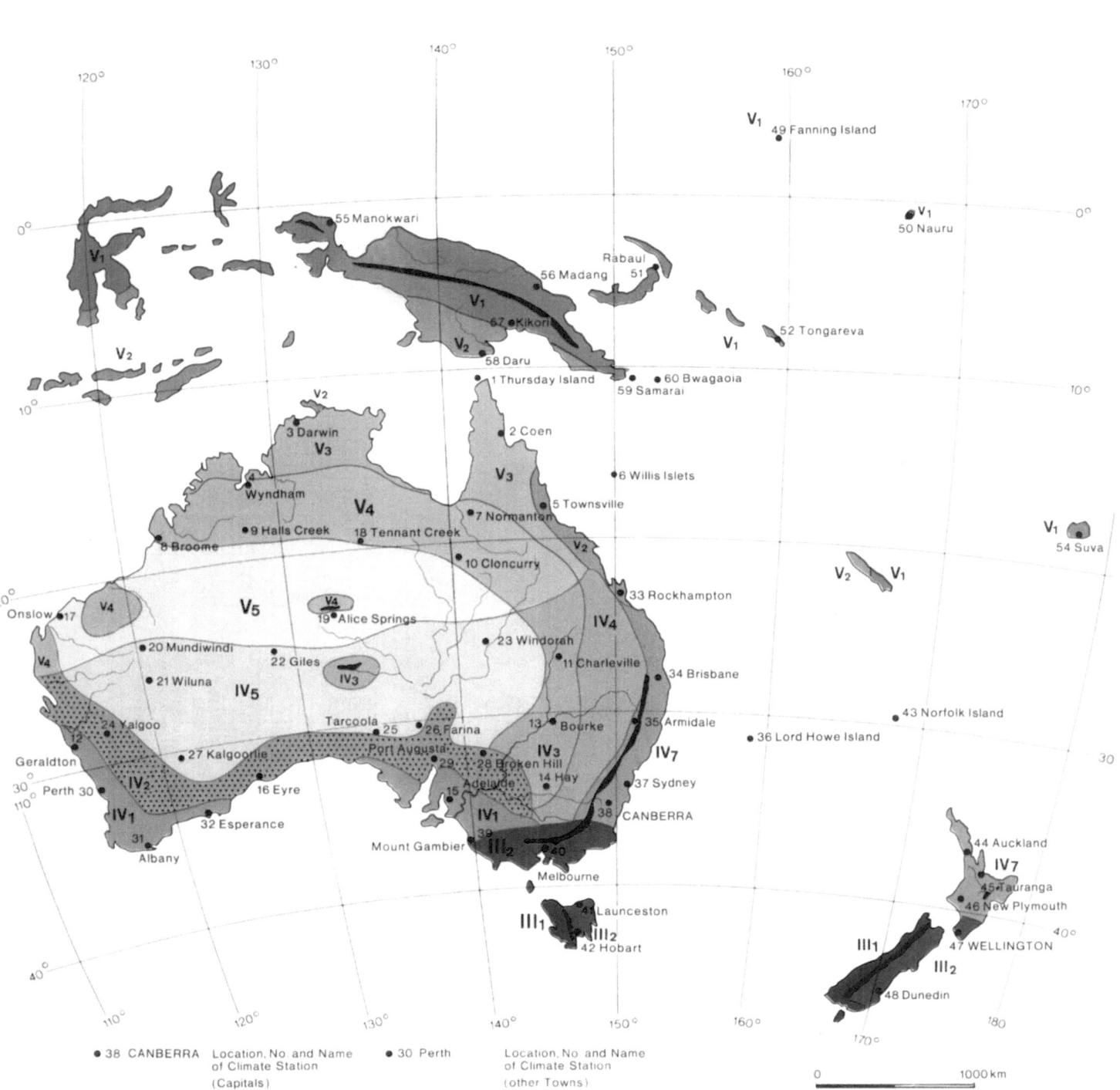

V_1
● 53 Saipan

V_1
49 ● Fanning Island

V_1
● 50 Nauru

55 ● Manokwari

Rabaul
56 ● Madang   51 ●

V_1

67 ● Kikori

V_2

52 ● Tongareva

58 ● Daru

V_1

1 ● Thursday Island

● 60 Bwagaoia
59 ● Samarai

V_2

3 ● Darwin
V_3

2 ● Coen

● 6 Willis Islets

Wyndham

V_4

7 ● Normanton

5 ● Townsville

V_3

9 ● Halls Creek   18 ● Tennant Creek

V_2

8 ● Broome

10 ● Cloncurry

33 ● Rockhampton

V_1
54 Suva

Onslow ● 17 V_4

V_5

V_4
19 ● Alice Springs

23 ● Windorah

IV_4

V_2 V_1

● 20 Mundiwindi   22 ● Giles

11 ● Charleville

34 ● Brisbane

V_4

● 21 Wiluna IV_3

IV_5

● 24 Yalgoo   Tarcoola

13 ● Bourke

35 ● Armidale

● 43 Norfolk Island

● 36 Lord Howe Island

Geraldton

● 27 Kalgoorlie   Port Augusta IV_3
25   26 ● Farina

29 ● 28 ● Broken Hill
Adelaide   14 ● Hay

IV_7

37 ● Sydney

12

Perth 30 ● IV_2 16 ● Eyre
IV_1 32 ● Esperance

15

38 ● CANBERRA

31 ●
Albany

IV_1
39
III_2 ● 40

Mount Gambier

44 ● Auckland
IV_7

45 ● Tauranga
46 ● New Plymouth

Melbourne

III_1

41 ● Launceston
III_2
42 ● Hobart

47 ● WELLINGTON

III_1

III_2

48 ● Dunedin

13 Station/Country Bourke/Australia

Location 30°13'S/145°58'E — Height above sealevel 110 m — Climate symbol: Köppen BSh — Troll IV,3

		J	F	M	A	M	J	J	A	S	O	N	D	year	P	
1 Mean daily temperature	in °C	28,7	28,2	25,2	19,9	15,4	11,9	11,3	13,4	17,1	21,3	24,9	27,2	20,4	30	1
2 Mean daily maximum temperature	in °C	37,2	36,1	32,8	27,8	22,8	18,3	18,3	21,1	25,0	29,4	33,9	36,1	28,3	63	2
3 Mean daily minimum temperature	in °C	21,1	20,8	17,8	12,8	8,3	5,8	4,4	6,1	9,4	13,3	17,2	19,4	12,8	63	3
4 Absolute maximum temperature	in °C	52,8	48,8	47,1	41,6	34,8	30,0	28,9	34,4	37,8	44,4	48,0	49,3	52,8	90	4
5 Absolute minimum temperature	in °C	8,8	9,4	1,9	1,9	-2,9	-3,9	-3,4	-2,9	-1,8	1,9	3,2	5,3	-3,9	90	5
6 Mean relative humidity	in %	37	41	44	57	58	66	64	56	47	42	40	41	40	30	6
7 Mean precipitation	in mm	33	45	38	26	27	31	23	18	17	31	28	33	348	30	7
8 Maximum precipitation	in mm	311	284	211	183	134	128	158	94	98	109	159	157	754	72	8
9 Minimum precipitation	in mm	0	0	0	0	0	0	0	0	0	0	0	0	102	72	9
10 Maximum precipitation in 24 h	in mm	115	78	93	114	45	54	38	32	48	50	72	73	115	90	10
11 Mean number of days with precipitation >0,25 mm		4	5	4	3	4	4	5	4	3	5	4	4	49	30	11
12 Mean duration of sunshine	in h															12
13 Mean quantity of radiation	in ly/day															13
14 Mean potential evaporation	in mm	173	148	127	73	39	22	18	28	50	99	144	164	1085	53	14
15 Mean windspeed	in m/sec															15
16 Mean predominent direction of the wind		NE,SE	NE,SE	NE,SE	SE	SE,SW	SE	SW	SW	NE,SW	NE,SW	NE,SW	NE,SW		5	16

14 Station/Country Hay/Australia

Location 34°30'S/144°56'E — Height above sealevel 94 m — Climate symbol: Köppen BSk — Troll IV,3

		J	F	M	A	M	J	J	A	S	O	N	D	year	P	
1 Mean daily temperature	in °C	23,9	24,4	21,2	16,6	13,0	9,8	9,2	10,8	13,5	16,8	20,2	22,8	16,8	39	1
2 Mean daily maximum temperature	in °C	32,8	32,8	29,4	23,9	19,4	15,6	15,0	17,2	20,6	25,0	28,9	31,7	24,4	64	2
3 Mean daily minimum temperature	in °C	16,1	16,1	13,3	9,4	6,1	4,4	3,3	4,4	6,1	9,4	12,2	14,4	9,4	64	3
4 Absolute maximum temperature	in °C	47,9	46,9	42,5	36,9	30,3	26,9	25,0	29,2	34,9	40,8	43,1	46,3	47,9	84	4
5 Absolute minimum temperature	in °C	5,1	5,0	1,7	0,6	-3,4	-5,1	-4,4	-4,9	-3,4	-0,8	0,8	2,2	-5,1	84	5
6 Mean relative humidity	in %	44	45	51	60	67	74	73	69	63	55	49	47	56	30	6
7 Mean precipitation	in mm	29	34	31	28	32	35	29	26	26	40	23	19	352	30	7
8 Maximum precipitation	in mm	99	123	108	141	120	116	99	107	106	93	152	115	683	64	8
9 Minimum precipitation	in mm	0	0	0	0	0	2	1	0	3	1	0	0	157	64	9
10 Maximum precipitation in 24 h	in mm	81	64	66	59	45	46	51	35	46	49	115	55	115	84	10
11 Mean number of days with precipitation >0,25 mm		3	3	4	5	6	7	8	8	6	7	5	3	65	30	11
12 Mean duration of sunshine	in h															12
13 Mean quantity of radiation	in ly/day															13
14 Mean potential evaporation	in mm	145	122	96	59	34	19	18	24	43	70	113	143	886	50	14
15 Mean windspeed	in m/sec															15
16 Mean predominent direction of the wind		SW	SW	NE,SW	NE,SW	SW	SW	SW	SW	SW	SW	SW	SW		5	16

15 Station/Country Adelaide/Australia

Location 34°58'S/138°35'E — Height above sealevel 43 m — Climate symbol: Köppen Csa — Troll IV,1

		J	F	M	A	M	J	J	A	S	O	N	D	year	P	
1 Mean daily temperature	in °C	22,6	21,0	20,9	17,2	14,6	12,1	11,2	12,0	13,4	16,0	18,5	20,7	16,7	30	1
2 Mean daily maximum temperature	in °C	30,0	30,0	27,2	22,8	18,9	16,1	15,0	16,7	18,9	22,8	26,1	28,3	22,8	86	2
3 Mean daily minimum temperature	in °C	16,1	16,7	15,0	12,8	10,0	8,3	7,2	7,8	8,9	10,6	12,8	15,0	11,7	86	3
4 Absolute maximum temperature	in °C	47,6	45,3	43,8	37,0	31,9	25,6	23,3	29,4	35,1	39,4	45,2	45,9	47,8	106	4
5 Absolute minimum temperature	in °C	7,3	7,5	6,8	4,2	2,7	0,3	0,0	0,2	0,4	2,3	4,9	6,1	0,0	106	5
6 Mean relative humidity	in %	41	43	45	56	63	71	71	67	61	52	45	43	53	30	6
7 Mean precipitation	in mm	23	23	21	50	66	61	61	59	49	47	36	27	523	30	7
8 Maximum precipitation	in mm	84	155	117	148	196	218	138	157	148	133	113	101	784	124	8
9 Minimum precipitation	in mm	0	0	0	0	3	6	10	8	7	4	2	0	289	124	9
10 Maximum precipitation in 24 h	in mm	58	141	89	80	70	54	44	57	40	57	75	61	141	124	10
11 Mean number of days with precipitation >0,25 mm		5	5	5	10	13	15	16	16	13	10	8	6	122	30	11
12 Mean duration of sunshine	in h	310	280	242	180	149	123	133	161	183	220	255	291	2507	81	12
13 Mean quantity of radiation	in ly/day															13
14 Mean potential evaporation	in mm	137	118	100	67	43	28	25	31	43	68	98	128	884	74	14
15 Mean windspeed	in m/sec	4,4	4,0	3,7	3,5	3,6	3,7	3,8	4,2	4,2	4,4	4,4	4,4	4,0		15
16 Mean predominent direction of the wind		SW	NE,SW	S,SW	NE,SW	NE,NW	NE,N	NE,NW	NE,SW	NE,SW	NE,SW	SW	SW		30	16

16 Station/Country Eyre/Australia

Location 32°14'S/126°22'E — Height above sealevel 5 m — Climate symbol: Köppen BSk — Troll IV,2

		J	F	M	A	M	J	J	A	S	O	N	D	year	P	
1 Mean daily temperature	in °C	20,7	21,2	19,9	17,9	15,2	12,9	11,9	12,6	14,6	16,3	18,1	19,5	16,7	28	1
2 Mean daily maximum temperature	in °C	26,1	26,1	25,6	23,3	21,1	18,3	17,8	18,9	20,6	22,2	23,9	25,0	22,2	29	2
3 Mean daily minimum temperature	in °C	15,6	16,1	14,4	12,2	9,4	7,2	6,1	6,7	7,8	10,0	12,2	13,9	11,1	29	3
4 Absolute maximum temperature	in °C	47,2	46,7	42,8	40,6	35,6	30,6	31,7	32,2	38,3	40,6	46,6	46,1	47,2	29	4
5 Absolute minimum temperature	in °C	4,4	3,3	2,8	1,1	0,0	-1,7	-3,9	-3,3	-3,3	-1,7	0,6	2,8	-3,9	29	5
6 Mean relative humidity	in %	62	64	65	66	71	71	71	68	64	62	62	63	66	27	6
7 Mean precipitation	in mm	13	11	22	23	39	43	32	31	21	19	18	16	289	48	7
8 Maximum precipitation	in mm	60	82	204	115	156	122	84	88	82	81	71	87	506	42	8
9 Minimum precipitation	in mm	0	0	1	1	11	6	4	4	0	1	0	0	132	42	9
10 Maximum precipitation in 24 h	in mm	41	41	71	50	37	39	22	20	41	56	27	32	71	42	10
11 Mean number of days with precipitation >0,25 mm		3	4	5	6	10	10	10	9	7	6	4	4	78	40	11
12 Mean duration of sunshine	in h															12
13 Mean quantity of radiation	in ly/day															13
14 Mean potential evaporation	in mm	114	103	95	89	49	34	31	36	48	68	81	103	829	28	14
15 Mean windspeed	in m/sec	4,9	4,9	4,6	4,1	4,4	3,6	4,1	4,1	4,4	4,4	4,6	4,4	4,4	5	15
16 Mean predominent direction of the wind		SE	SE	SE	SE	NE,SE	NW,SW	NW,SW	NW,SW	NW,W	N,S	SW,SE	SE		5	16

274

17 Station / Country   Onslow / Australia

Location   21°43'S / 114°57'E   Height above sealevel   4 m   Climate symbol: Köppen   BWh   Troll   V,5

		J	F	M	A	M	J	J	A	S	O	N	D	year	P	
1	Mean daily temperature	in °C	29,6	29,7	29,1	26,4	22,4	19,0	18,0	19,3	21,6	23,8	26,6	28,3	24,5	30	1
2	Mean daily maximum temperature	in °C	36,1	35,6	35,6	33,3	28,9	25,6	25,0	26,7	29,4	31,7	34,4	35,6	31,7	44	2
3	Mean daily minimum temperature	in °C	23,3	23,9	22,8	19,4	15,6	12,2	11,1	11,7	13,9	16,1	18,9	21,1	17,2	44	3
4	Absolute maximum temperature	in °C	47,7	48,3	46,4	43,8	38,3	32,2	32,3	35,3	38,3	44,6	46,1	47,5	48,3	50	4
5	Absolute minimum temperature	in °C	15,8	16,6	14,7	10,0	5,6	2,9	3,1	4,4	5,5	7,4	10,0	12,5	3,1	50	5
6	Mean relative humidity	in %	57	59	58	56	58	60	58	55	51	50	52	55	56	42	6
7	Mean precipitation	in mm	21	46	62	18	50	40	20	9	<1	<1	3	3	274	30	7
8	Maximum precipitation	in mm	261	323	415	279	259	157	221	107	107	15	30	61	717	57	8
9	Minimum precipitation	in mm	0	0	0	0	0	0	0	0	0	0	0	0	14	57	9
10	Maximum precipitation in 24 h	in mm	237	314	356	157	238	111	90	64	7	7	30	50	356	79	10
11	Mean number of days with precipitation	>0,25 mm	3	3	4	2	3	3	2	2	0	0	0	1	23	30	11
12	Mean duration of sunshine	in h															12
13	Mean quantity of radiation	in ly / day															13
14	Mean potential evaporation	in mm	185	161	164	132	75	42	38	48	70	107	151	175	1348	34	14
15	Mean windspeed	in m/sec	4,6	4,9	4,4	4,4	4,1	4,4	4,1	4,4	5,4	4,9	4,9	4,9	4,6	5	15
16	Mean predominent direction of the wind		S,SW	S,W	S,W	S,N	E,S,N	E,S,N	S,N	S,N,W	S,W	S,W	S,W	S,W		5	16

18 Station / Country   Tennant Creek / Australia

Location   19°34'S / 134°13'E   Height above sealevel   327 m   Climate symbol: Köppen   BWh   Troll   V,4

		J	F	M	A	M	J	J	A	S	O	N	D	year	P	
1	Mean daily temperature	in °C	30,7	29,8	28,5	25,2	21,3	18,1	17,4	19,9	23,7	27,5	29,7	30,9	25,2	30	1
2	Mean daily maximum temperature	in °C	36,7	35,6	34,4	31,7	27,2	25,0	24,4	27,2	31,7	35,0	36,7	37,2	31,7	21	2
3	Mean daily minimum temperature	in °C	24,4	23,9	22,2	19,4	15,0	11,7	10,6	12,2	16,1	20,0	22,8	23,9	18,3	21	3
4	Absolute maximum temperature	in °C	46,1	43,7	43,9	40,0	37,8	34,2	33,3	35,6	38,9	42,8	44,2	44,9	46,1	90	4
5	Absolute minimum temperature	in °C	15,6	11,1	12,1	10,0	5,1	2,8	2,4	2,8	5,6	10,6	13,3	11,7	2,4	90	5
6	Mean relative humidity	in %	41	45	40	32	36	40	36	31	29	29	32	36	34	31	6
7	Mean precipitation	in mm	103	90	53	9	5	9	6	2	3	10	27	35	351	30	7
8	Maximum precipitation	in mm	282	379	430	196	99	70	94	83	71	78	106	219	864	77	8
9	Minimum precipitation	in mm	0	0	0	0	0	0	0	0	0	0	0	0	94	77	9
10	Maximum precipitation in 24 h	in mm	187	234	117	58	66	43	34	15	28	36	75	90	234	90	10
11	Mean number of days with precipitation	>0,25 mm	7	5	3	1	1	1	1	0	1	2	3	5	30	30	11
12	Mean duration of sunshine	in h															12
13	Mean quantity of radiation	in ly / day															13
14	Mean potential evaporation	in mm	186	162	162	121	65	36	36	53	105	153	176	188	1443	21	14
15	Mean windspeed	in m/sec															15
16	Mean predominent direction of the wind		NE,E	E	E	E	E	E,SE	E,SE	E,SE	E	E	E,SE	E		5	16

19 Station / Country   Alice Springs / Australia

Location   23°42'S / 133°53'E   Height above sealevel   579 m   Climate symbol: Köppen   BWh   Troll   V,5

		J	F	M	A	M	J	J	A	S	O	N	D	year	P	
1	Mean daily temperature	in °C	28,1	27,5	24,7	19,8	15,3	12,3	11,8	14,3	18,2	22,8	25,5	27,4	20,6	30	1
2	Mean daily maximum temperature	in °C	36,1	35,0	32,2	27,2	22,8	19,4	19,4	22,8	27,2	31,1	33,9	35,6	28,3	62	2
3	Mean daily minimum temperature	in °C	21,1	20,6	17,2	12,2	7,8	5,0	3,9	6,1	9,4	14,4	17,8	20,0	12,8	62	3
4	Absolute maximum temperature	in °C	46,7	45,6	45,0	39,3	38,3	34,2	33,1	35,8	37,6	45,1	46,1	47,2	47,2	90	4
5	Absolute minimum temperature	in °C	10,0	8,5	3,9	1,9	-2,8	-5,6	-7,2	-3,9	-1,1	2,4	4,4	7,8	-7,2	90	5
6	Mean relative humidity	in %	33	36	38	41	49	54	49	40	34	30	31	32	37	60	6
7	Mean precipitation	in mm	44	34	28	10	15	13	7	8	7	18	29	39	252	30	7
8	Maximum precipitation	in mm	281	236	227	117	109	74	106	158	90	115	139	288	726	77	8
9	Minimum precipitation	in mm	0	0	0	0	0	0	0	0	0	0	0	0	60	77	9
10	Maximum precipitation in 24 h	in mm	98	84	147	72	39	51	50	63	31	58	68	118	147	90	10
11	Mean number of days with precipitation	>0,25 mm	4	3	3	2	2	2	1	2	1	3	4	4	31	30	11
12	Mean duration of sunshine	in h	319	291	288	276	248	240	276	304	300	301	303	310	3456	16	12
13	Mean quantity of radiation	in ly / day															13
14	Mean potential evaporation	in mm	175	155	123	74	37	21	18	31	80	139	150	179	1162	2	14
15	Mean windspeed	in m/sec	2,3	2,4	1,9	1,8	1,4	1,3	1,2	1,3	1,6	1,9	2,1	2,2	1,8	5	15
16	Mean predominent direction of the wind		E,SE	SE	E	SE	SE	SE	W,SE	E,SE	E,SE	E,SE	E,SE	SE		5	16

20 Station / Country   Mundiwindi / Australia

Location   23°52'S / 120°09'E   Height above sealevel   408 m   Climate symbol: Köppen   BWh   Troll   IV,5

		J	F	M	A	M	J	J	A	S	O	N	D	year	P	
1	Mean daily temperature	in °C	30,6	29,8	27,5	21,1	18,1	13,8	13,2	15,4	19,6	23,3	27,5	29,4	22,6	31	1
2	Mean daily maximum temperature	in °C	38,3	37,2	34,4	30,6	25,0	21,7	21,1	23,9	28,3	31,7	36,1	37,8	30,6	15	2
3	Mean daily minimum temperature	in °C	17,8	22,8	20,6	16,1	10,6	6,7	5,2	7,2	10,6	14,4	19,4	21,7	15,0	15	3
4	Absolute maximum temperature	in °C	44,4	44,4	42,3	40,6	36,4	29,8	30,6	37,3	37,2	40,7	43,3	44,4	44,4	31	4
5	Absolute minimum temperature	in °C	13,9	12,8	9,4	3,9	-1,7	-4,4	-5,3	-3,3	-1,7	3,3	7,8	11,7	-5,3	31	5
6	Mean relative humidity	in %	29	31	36	35	43	51	47	40	29	24	23	27	33	29	6
7	Mean precipitation	in mm	31	29	33	28	24	22	11	7	1	2	7	20	215	30	7
8	Maximum precipitation	in mm	321	325	267	138	121	113	70	53	61	93	53	160	816	31	8
9	Minimum precipitation	in mm	0	0	0	0	0	0	0	0	0	0	0	0	26	31	9
10	Maximum precipitation in 24 h	in mm	70	71	175	57	56	40	43	39	34	53	31	114	175	31	10
11	Mean number of days with precipitation	>0,25 mm	6	5	6	3	3	3	2	2	0	1	3	4	38	38	11
12	Mean duration of sunshine	in h	292	255	260	262	250	237	272	285	298	320	322	318	3371	+	12
13	Mean quantity of radiation	in ly / day															13
14	Mean potential evaporation	in mm	194	183	151	92	44	20	18	29	59	104	163	189	1226	15	14
15	Mean windspeed	in m/sec															15
16	Mean predominent direction of the wind																16

+ estimated from maps

275

21 Station/Country Wiluna/Australia

Location 26°37'S/120°15'E — Height above sealevel 518 m — Climate symbol: Köppen BWh — Troll IV.5

		J	F	M	A	M	J	J	A	S	O	N	D	year	P	
1	Mean daily temperature	in °C	30,1	29,3	26,9	21,7	16,7	13,1	12,1	14,2	17,9	21,6	25,5	28,7	21,5	30	1
2	Mean daily maximum temperature	in °C															2
3	Mean daily minimum temperature	in °C															3
4	Absolute maximum temperature	in °C	46,7	45,6	43,3	40,1	35,0	30,0	30,6	34,4	37,6	41,1	44,4	45,0	46,7	66	4
5	Absolute minimum temperature	in °C	9,4	12,2	7,1	3,9	-0,3	-3,3	-3,2	-2,3	0,6	1,1	3,6	7,8	-3,3	66	5
6	Mean relative humidity	in %	36	37	41	45	53	57	56	50	43	39	35	34	42	36	6
7	Mean precipitation	in mm	38	34	45	18	24	24	12	9	3	4	6	17	234	30	7
8	Maximum precipitation	in mm	214	134	154	138	131	156	51	77	36	26	94	104	516	50	8
9	Minimum precipitation	in mm	0	0	0	0	0	0	0	0	0	0	0	0	49	50	9
10	Maximum precipitation in 24 h	in mm	82	97	77	74	59	68	30	42	32	78	32	72	97	66	10
11	Mean number of days with precipitation	>0,25 mm	4	3	5	3	4	4	4	3	1	1	2	3	37	10	11
12	Mean duration of sunshine	in h															12
13	Mean quantity of radiation	in ly/day															13
14	Mean potential evaporation	in mm	210	178	142	84	42	21	18	28	54	93	144	202	1216	28	14
15	Mean windspeed	in m/sec															15
16	Mean predominent direction of the wind		NE,SE	NE,SE	NE,SE	NE,E	N,E	NW	NE,W	NE,NW	NE,SW	NE,SW	NE,SW	NE,E		5	16

22 Station/Country Giles/Australia

Location 25°02'S/128°17'E — Height above sealevel 580 m — Climate symbol: Köppen BWh — Troll IV.5

		J	F	M	A	M	J	J	A	S	O	N	D	year	P	
1	Mean daily temperature	in °C	30,7	39,7	27,9	23,3	17,4	14,9	13,3	15,5	20,2	24,4	27,6	28,7	22,8	9	1
2	Mean daily maximum temperature	in °C															2
3	Mean daily minimum temperature	in °C															3
4	Absolute maximum temperature	in °C	45,0	46,1	43,3	39,4	33,3	31,7	31,7	33,9	38,3	40,0	43,9	43,9	46,1	13	4
5	Absolute minimum temperature	in °C	10,0	12,2	11,1	5,0	-0,6	-2,8	-4,4	-1,7	1,1	3,9	7,2	9,4	-4,4	13	5
6	Mean relative humidity	in %	24	28	30	31	37	42	38	30	21	24	23	23	29	5	6
7	Mean precipitation	in mm	19	21	8	12	19	10	15	18	3	7	11	31	174	9	7
8	Maximum precipitation	in mm															8
9	Minimum precipitation	in mm															9
10	Maximum precipitation in 24 h	in mm	31	79	102	76	18	41	5	5	5	10	48	46	102	2	10
11	Mean number of days with precipitation	>0,25 mm	4	3	4	3	4	3	2	2	2	3	4	3	37	11	11
12	Mean duration of sunshine	in h	310	265	280	275	245	240	260	270	290	300	315	320	3370	+	12
13	Mean quantity of radiation	in ly/day															13
14	Mean potential evaporation	in mm															14
15	Mean windspeed	in m/sec	1,8	1,9	1,9	2,0	1,3	1,4	1,0	1,0	1,3	1,7	1,8	1,7	1,4	4	15
16	Mean predominent direction of the wind		E,SE	E,SE	E,SE	E,SE	E,SE	NE,SE	N,E	E,SE	S,E	S,E	S,E	S,E		4	16

23 Station/Country Windorah/Australia

Location 25°26'S/142°36'E — Height above sealevel 119 m — Climate symbol: Köppen BWh — Troll IV.5

		J	F	M	A	M	J	J	A	S	O	N	D	year	P	
1	Mean daily temperature	in °C	30,9	30,1	27,6	22,7	17,9	14,5	13,5	15,9	19,8	24,4	27,7	29,8	22,9	32	1
2	Mean daily maximum temperature	in °C	38,3	37,2	34,4	30,0	25,0	21,1	21,1	23,9	28,3	32,8	35,6	37,2	30,6	29	2
3	Mean daily minimum temperature	in °C	23,3	23,3	20,6	15,0	10,6	7,2	6,1	7,8	11,7	16,1	19,4	22,2	15,0	29	3
4	Absolute maximum temperature	in °C	46,7	45,6	43,2	39,7	35,6	35,0	32,0	35,6	40,0	44,4	45,0	46,7	35		4
5	Absolute minimum temperature	in °C	10,6	12,8	8,3	1,6	1,0	-3,3	-3,3	-0,7	-1,1	3,9	7,0	9,9	-3,3	35	5
6	Mean relative humidity	in %	34	38	39	40	47	54	51	41	34	32	34	33	38	28	6
7	Mean precipitation	in mm	35	60	47	14	13	16	14	10	10	21	15	22	279	30	7
8	Maximum precipitation	in mm	140	188	378	167	111	78	141	125	96	89	102	186	751	50	8
9	Minimum precipitation	in mm	0	0	0	0	0	0	0	0	0	0	0	0	91	50	9
10	Maximum precipitation in 24 h	in mm	130	186	103	89	86	52	42	47	64	63	50	96	186	80	10
11	Mean number of days with precipitation	>0,25 mm	4	5	4	2	2	2	2	2	2	3	3	3	34	30	11
12	Mean duration of sunshine	in h															12
13	Mean quantity of radiation	in ly/day															13
14	Mean potential evaporation	in mm	189	162	154	91	47	24	20	33	69	118	166	190	1263	24	14
15	Mean windspeed	in m/sec															15
16	Mean predominent direction of the wind		N,NE	N,SE	NE,SE	E,SE	E,SE	E,SE	E,SE	E,SE	N,SE	N,SE	N,SE	N,NE		5	16

24 Station/Country Yalgoo/Australia

Location 28°21'S/116°41'E — Height above sealevel 318 m — Climate symbol: Köppen BSh — Troll IV.2

		J	F	M	A	M	J	J	A	S	O	N	D	year	P	
1	Mean daily temperature	in °C	28,8	28,6	26,1	21,6	16,6	13,4	12,4	13,5	16,6	19,6	23,9	27,1	20,7	30	1
2	Mean daily maximum temperature	in °C	36,7	36,1	32,8	28,9	22,8	18,9	18,3	19,4	23,9	27,2	32,2	35,6	27,8	34	2
3	Mean daily minimum temperature	in °C	20,6	20,6	18,3	14,4	10,0	7,8	6,1	6,7	8,3	11,1	15,0	18,3	13,3	34	3
4	Absolute maximum temperature	in °C	46,1	46,8	43,9	40,6	34,4	28,9	27,2	33,9	35,6	40,0	42,8	45,0	46,8	58	4
5	Absolute minimum temperature	in °C	10,0	10,6	8,9	4,4	0,6	-1,7	-1,7	-0,6	1,1	-1,1	6,1	9,4	-1,7	58	5
6	Mean relative humidity	in %	38	41	45	52	61	68	70	66	55	47	41	37	49	43	6
7	Mean precipitation	in mm	14	20	24	21	28	40	38	26	9	7	9	10	246	30	7
8	Maximum precipitation	in mm	101	178	120	70	141	105	82	112	64	37	55	34	497	29	8
9	Minimum precipitation	in mm	0	0	0	0	0	5	9	4	1	2	0	0	94	29	9
10	Maximum precipitation in 24 h	in mm	61	84	99	89	64	51	30	61	38	36	33	51	99	58	10
11	Mean number of days with precipitation	>0,25 mm	3	3	4	3	5	8	8	7	3	3	2	2	51	30	11
12	Mean duration of sunshine	in h	350	300	280	255	230	210	230	250	270	305	330	360	3370	+	12
13	Mean quantity of radiation	in ly/day															13
14	Mean potential evaporation	in mm	181	152	135	82	42	24	20	27	45	71	121	169	1069	34	14
15	Mean windspeed	in m/sec															15
16	Mean predominent direction of the wind																16

+estimated from maps

25 Station/Country Tarcoola/Australia

Location 30°42'S/134°34'E Height above sealevel 120 m Climate symbol: Köppen BWh Troll IV,5

		J	F	M	A	M	J	J	A	S	O	N	D	year	P	
1 Mean daily temperature	in °C	25,9	25,5	23,3	18,3	14,6	11,2	10,8	12,7	15,8	18,8	22,2	24,7	18,7	28	1
2 Mean daily maximum temperature	in °C	34,4	34,4	31,7	26,1	22,2	18,3	18,3	20,0	23,9	27,8	31,1	33,3	26,7	18	2
3 Mean daily minimum temperature	in °C	17,2	17,2	15,0	10,6	7,2	3,9	3,3	5,0	7,2	10,6	13,3	15,6	10,6	18	3
4 Absolute maximum temperature	in °C	47,7	46,1	45,6	38,9	33,3	28,1	28,9	34,4	36,1	42,2	45,0	48,9	48,9	28	4
5 Absolute minimum temperature	in °C	5,6	6,7	4,4	1,7	-1,7	-4,7	-4,4	-3,3	-1,7	0,6	2,2	5,0	-4,7	28	5
6 Mean relative humidity	in %	40	42	44	53	60	65	64	57	50	46	40	40	48	25	6
7 Mean precipitation	in mm	10	21	12	7	12	14	11	15	11	18	11	14	156	20	7
8 Maximum precipitation	in mm	55	167	48	62	90	73	46	54	61	68	56	97	350	46	8
9 Minimum precipitation	in mm	0	0	0	0	0	0	0	0	0	0	0	0	68	46	9
10 Maximum precipitation in 24 h	in mm	44	141	38	53	72	37	23	27	45	36	30	55	141	28	10
11 Mean number of days with precipitation	>0,25 mm	2	2	2	2	3	4	4	4	3	3	3	2	34	39	11
12 Mean duration of sunshine	in h	330	280	275	250	230	195	230	250	260	300	310	330	3240	+	12
13 Mean quantity of radiation	in ly/day															13
14 Mean potential evaporation	in mm	151	142	129	73	39	22	19	28	54	100	144	178	1087	8	14
15 Mean windspeed	in m/sec															15
16 Mean predominent direction of the wind		E,SW	NE,SW	E,SW	E	NE,W	NE	NE	NE,SW	SW	SW	NE,SW	SW		3	16

+ estimated from maps

26 Station/Country Farina/Australia

Location 30°05'S/138°08'E Height above sealevel 93 m Climate symbol: Köppen BWh Troll IV,5

		J	F	M	A	M	J	J	A	S	O	N	D	year	P	
1 Mean daily temperature	in °C	27,6	27,7	24,5	19,3	14,8	11,6	10,7	12,7	16,1	20,3	24,0	26,6	19,7	50	1
2 Mean daily maximum temperature	in °C	35,6	35,6	31,7	26,7	21,7	17,8	17,2	20,0	23,9	28,3	32,2	34,4	27,2	42	2
3 Mean daily minimum temperature	in °C	20,0	20,0	17,2	12,2	7,8	5,6	3,9	5,6	8,3	12,8	16,1	18,9	12,2	42	3
4 Absolute maximum temperature	in °C	45,3	42,1	40,4	36,2	30,0	25,8	24,2	31,4	35,6	36,7	42,5	43,9	45,3	21	4
5 Absolute minimum temperature	in °C	3,9	5,9	5,6	2,2	-0,3	-2,8	-3,6	-1,8	-0,9	1,4	3,9	2,2	-3,6	21	5
6 Mean relative humidity	in %	34	35	40	49	57	65	62	55	47	39	36	34	44	50	6
7 Mean precipitation	in mm	14	15	13	10	15	15	9	7	8	14	13	13	146	30	7
8 Maximum precipitation	in mm	102	114	102	77	76	132	52	52	42	84	74	88	365	71	8
9 Minimum precipitation	in mm	0	0	0	0	0	0	0	0	0	0	0	0	47	71	9
10 Maximum precipitation in 24 h	in mm	66	107	74	41	39	49	24	48	54	23	41	42	107	21	10
11 Mean number of days with precipitation	>0,25 mm	2	2	2	2	3	3	2	2	2	3	3	2	28	30	11
12 Mean duration of sunshine	in h	330	285	280	255	230	205	230	250	270	295	310	340	3270	+	12
13 Mean quantity of radiation	in ly/day															13
14 Mean potential evaporation	in mm	180	156	128	69	35	20	16	26	46	87	121	164	1048	42	14
15 Mean windspeed	in m/sec															15
16 Mean predominent direction of the wind		SE	SE	SE	SE	SE,SW	N,NW	N	NW	NW,SW	SE,SW	SE,SW	SE,SW			16

+ estimated from maps

27 Station/Country Kalgoorlie/Australia

Location 30°45'S/121°30'E Height above sealevel 380 m Climate symbol: Köppen BWh Troll IV,5

		J	F	M	A	M	J	J	A	S	O	N	D	year	P	
1 Mean daily temperature	in °C	25,7	24,9	23,0	18,7	14,7	12,0	10,8	12,3	15,3	18,2	21,4	24,3	18,4	30	1
2 Mean daily maximum temperature	in °C	33,9	33,3	30,0	25,6	20,6	17,2	16,7	18,3	22,8	25,6	30,6	33,3	25,6	30	2
3 Mean daily minimum temperature	in °C	17,8	17,8	16,1	12,8	9,4	7,2	6,1	6,7	8,9	11,1	14,4	16,7	12,2	30	3
4 Absolute maximum temperature	in °C	45,8	46,1	43,9	39,2	33,3	27,7	27,2	30,6	35,6	40,7	43,7	45,0	46,1	69	4
5 Absolute minimum temperature	in °C	8,4	8,9	5,3	1,7	-1,8	-2,0	-3,3	-2,4	-0,6	-1,0	3,4	7,5	-3,3	69	5
6 Mean relative humidity	in %	46	40	55	59	66	74	74	66	53	48	44	43	56	44	6
7 Mean precipitation	in mm	24	27	24	18	22	25	24	23	13	14	15	13	244	30	7
8 Maximum precipitation	in mm	204	315	166	103	96	119	82	81	98	80	70	65	458	69	8
9 Minimum precipitation	in mm	0	0	0	0	0	0	2	0	0	0	0	0	121	69	9
10 Maximum precipitation in 24 h	in mm	96	178	71	69	45	57	28	31	44	62	65	37	178	69	10
11 Mean number of days with precipitation	>0,25 mm	3	2	4	4	5	6	7	6	3	4	3	2	49	57	11
12 Mean duration of sunshine	in h +	341	294	267	231	202	171	202	220	240	279	324	357	3128	30	12
13 Mean quantity of radiation	in ly/day															13
14 Mean potential evaporation	in mm	159	132	115	72	39	25	22	29	48	75	118	155	989	34	14
15 Mean windspeed	in m/sec															15
16 Mean predominent direction of the wind		SE	E,SE	SE	E	NW	NW,W	NW	SE,W	SE	SE	E			5	16

+ No 12=Merredin(31°29'S/118°17'E)

28 Station/Country Broken Hill/Australia

Location 31°57'S/141°28'E Height above sealevel 305 m Climate symbol: Köppen BWh Troll IV,2

		J	F	M	A	M	J	J	A	S	O	N	D	year	P	
1 Mean daily temperature	in °C	25,3	25,2	22,4	17,6	13,8	10,7	10,2	11,9	14,9	18,4	21,5	24,2	18,0	30	1
2 Mean daily maximum temperature	in °C	32,5	32,3	29,2	23,7	19,2	15,6	15,3	17,6	21,1	25,1	28,1	31,5	24,3	30	2
3 Mean daily minimum temperature	in °C	18,0	18,1	15,6	11,5	8,3	5,8	5,1	6,2	8,7	11,6	14,4	17,0	11,7	30	3
4 Absolute maximum temperature	in °C	46,1	46,6	45,5	37,7	31,0	26,1	26,7	28,9	34,4	39,9	43,8	45,5	46,6	85	4
5 Absolute minimum temperature	in °C	7,2	5,6	4,4	1,1	-0,8	-2,8	-1,9	-1,7	0,4	1,8	4,4	5,0	-2,8	85	5
6 Mean relative humidity	in %	36	40	44	53	63	70	67	57	48	41	38	38	47	30	6
7 Mean precipitation	in mm	18	28	21	16	17	15	17	15	13	25	23	14	224	30	7
8 Maximum precipitation	in mm	90	109	71	114	93	148	76	91	73	102	122	115	447	56	8
9 Minimum precipitation	in mm	0	0	0	0	0	0	0	0	0	0	0	0	57	56	9
10 Maximum precipitation in 24 h	in mm	60	67	60	93	62	48	33	46	48	55	103	60	103	85	10
11 Mean number of days with precipitation	>0,25 mm	1	2	0	1	1	1	1	2	1	3	1	2	16		11
12 Mean duration of sunshine	in h															12
13 Mean quantity of radiation	in ly/day															13
14 Mean potential evaporation	in mm	150	131	111	61	34	20	19	27	41	80	111	153	938	40	14
15 Mean windspeed	in m/sec															15
16 Mean predominent direction of the wind		S	S	S	S	W,SW	W	W	S,W	N,W	S	S	S		5	16

29 Station/Country: Port Augusta/Australia

Location 32°29'S/137°45'E — Height above sealevel 6 m — Climate symbol: Köppen BWh — Troll IV,2

		J	F	M	A	M	J	J	A	S	O	N	D	year	P	
1	Mean daily temperature	in °C	25,2	25,4	23,2	19,2	15,6	12,6	11,8	13,4	16,1	19,2	22,1	24,2	19,0	62	1
2	Mean daily maximum temperature	in °C	32,2	32,2	29,4	25,6	21,1	17,8	17,2	18,9	22,2	26,1	28,9	31,1	25,0	52	2
3	Mean daily minimum temperature	in °C	18,9	18,9	16,7	13,3	10,0	7,8	6,7	7,8	10,0	12,8	15,6	17,8	12,8	52	3
4	Absolute maximum temperature	in °C	48,3	47,2	43,9	37,8	33,3	27,2	26,7	32,3	35,0	41,1	43,3	46,1	48,3	55	4
5	Absolute minimum temperature	in °C	10,0	9,4	9,4	5,0	0,6	0,0	-0,6	0,0	3,3	4,4	6,1	7,8	-0,6	55	5
6	Mean relative humidity	in %	50	47	49	55	61	68	66	62	54	49	45	53	49		6
7	Mean precipitation	in mm	15	18	16	14	26	26	20	26	21	22	18	17	236	30	7
8	Maximum precipitation	in mm	81	114	143	101	97	95	59	87	104	104	141	82	469	84	8
9	Minimum precipitation	in mm	0	0	0	0	0	1	0	0	1	0	0	0	56	84	9
10	Maximum precipitation in 24 h	in mm	52	54	118	48	44	64	19	32	52	64	53	32	118	42	10
11	Mean number of days with precipitation	>0,25 mm	3	3	3	4	7	8	8	7	6	5	4	3	60	76	11
12	Mean duration of sunshine	in h															12
13	Mean quantity of radiation	in ly/day															13
14	Mean potential evaporation	in mm	153	133	126	72	44	28	24	33	51	86	114	146	1010	42	14
15	Mean windspeed	in m/sec	1,5	2,1	2,1	1,5	1,0	1,0	1,0	1,5	2,1	2,1	2,1	2,1	1,5	5	15
16	Mean predominent direction of the wind		S	S	S	S	var	N	N	N	N,S	S	S			5	16

30 Station/Country: Perth/Australia

Location 37°57'S/115°51'E — Height above sealevel 60 m — Climate symbol: Köppen Csa — Troll IV,1

		J	F	M	A	M	J	J	A	S	O	N	D	year	P	
1	Mean daily temperature	in °C	23,4	23,9	22,2	19,2	16,1	13,7	13,1	13,5	14,7	16,3	19,2	21,5	18,1	30	1
2	Mean daily maximum temperature	in °C	29,4	29,4	27,2	24,4	20,6	17,8	17,2	17,8	19,4	21,1	24,4	27,2	22,8	44	2
3	Mean daily minimum temperature	in °C	17,2	17,2	16,1	13,9	11,7	10,0	8,9	8,9	10,0	11,7	13,9	16,1	12,8	44	3
4	Absolute maximum temperature	in °C	43,7	44,6	41,3	37,6	32,4	27,6	24,7	27,8	32,7	37,2	40,3	42,2	44,6	67	4
5	Absolute minimum temperature	in °C	9,2	8,7	7,7	4,1	1,3	1,6	1,2	1,9	2,6	4,4	5,6	8,6	1,2	67	5
6	Mean relative humidity	in %	47	48	52	55	65	70	70	66	62	58	50	49	58	67	6
7	Mean precipitation	in mm	7	12	22	52	125	192	183	135	69	54	23	15	889	30	7
8	Maximum precipitation	in mm	55	166	145	149	307	477	425	318	199	200	71	81	1250	87	8
9	Minimum precipitation	in mm	0	0	0	0	20	55	61	12	9	4	0	0	508	87	9
10	Maximum precipitation in 24 h	in mm	44	87	77	67	76	99	76	74	46	44	39	47	99	87	10
11	Mean number of days with precipitation	>0,25 mm	3	3	5	8	15	17	19	19	15	12	7	5	128	30	11
12	Mean duration of sunshine	in h	322	280	273	219	180	144	164	189	213	251	291	322	2848	30	12
13	Mean quantity of radiation	in ly/day	650	603	493	349	258	219	235	321	427	538	633	672	449	12	13
14	Mean potential evaporation	in mm	130	112	105	77	49	33	33	36	47	61	88	122	893	34	14
15	Mean windspeed	in m/sec	7,5	7,5	6,9	6,2	5,7	5,7	5,7	6,2	6,7	7,2	7,5	7,5	6,7	10	15
16	Mean predominent direction of the wind		E,SW	NE,SW	E,SW	NE,SW	NE,W	N,NW	NE,W	N,W	NE,SW	NE,SW	E,SW	E,SW		30	16

31 Station/Country: Albany/Australia

Location 35°02'S/117°55'E — Height above sealevel 13 m — Climate symbol: Köppen Csb — Troll IV,1

		J	F	M	A	M	J	J	A	S	O	N	D	year	P	
1	Mean daily temperature	in °C	19,2	19,4	18,7	16,9	14,7	13,1	12,1	12,4	13,4	14,6	16,4	17,9	15,7	29	1
2	Mean daily maximum temperature	in °C	23,3	23,3	22,2	21,1	18,9	16,7	16,1	16,7	17,2	18,9	20,6	22,2	20,0	39	2
3	Mean daily minimum temperature	in °C	15,0	15,0	13,9	12,8	10,6	8,9	7,8	8,3	8,9	10,0	11,7	13,9	11,1	39	3
4	Absolute maximum temperature	in °C	41,7	44,6	40,8	37,7	35,2	24,6	23,1	27,2	30,6	36,2	41,1	41,1	44,8	50	4
5	Absolute minimum temperature	in °C	5,7	5,0	3,7	4,2	1,7	1,7	0,1	1,3	1,1	2,3	4,8	5,1	0,1	50	5
6	Mean relative humidity	in %	73	73	73	75	77	76	76	76	76	76	74	74	75	42	6
7	Mean precipitation	in mm	35	26	45	74	135	138	152	138	108	83	42	31	1008	42	7
8	Maximum precipitation	in mm	217	161	186	234	290	292	269	285	202	187	170	117	1393	42	8
9	Minimum precipitation	in mm	1	0	3	5	44	40	52	50	20	14	5	2	637	42	9
10	Maximum precipitation in 24 h	in mm	88	57	90	57	104	72	61	113	79	53	78	82	113	42	10
11	Mean number of days with precipitation	>0,25 mm	8	7	11	13	18	20	21	20	18	16	11	9	172	30	11
12	Mean duration of sunshine	in h	217	193	189	145	140	97	121	140	160	189	193	217	2001		12
13	Mean quantity of radiation	in ly/day															13
14	Mean potential evaporation	in mm	105	91	84	65	48	37	37	40	48	61	74	96	786	34	14
15	Mean windspeed	in m/sec	6,7	6,4	6,4	5,9	7,2	7,2	8,2	8,2	8,0	6,7	6,9	6,4	7,2	5	15
16	Mean predominent direction of the wind		SE	E,SE	E,SE	E,SE	NW,SE	N,W	NW	NW,SW	NW,SW	NW,SW	SE,SW	E,SE		5	16

32 Station/Country: Esperance/Australia

Location 33°50'S/121°55'E — Height above sealevel 4 m — Climate symbol: Köppen Csb — Troll IV,1

		J	F	M	A	M	J	J	A	S	O	N	D	year	P	
1	Mean daily temperature	in °C	20,1	20,5	19,6	17,4	14,9	12,8	12,1	12,6	13,9	15,2	17,3	18,9	16,3	30	1
2	Mean daily maximum temperature	in °C	25,0	25,6	24,4	22,2	20,0	17,2	16,7	17,2	18,9	20,0	22,2	23,9	21,1	44	2
3	Mean daily minimum temperature	in °C	15,6	15,6	14,4	12,2	10,0	8,3	7,2	7,8	8,9	10,0	12,2	13,9	11,1	44	3
4	Absolute maximum temperature	in °C	47,2	43,9	43,9	38,9	33,3	27,2	26,1	31,7	35,6	40,0	41,1	42,8	47,2	47	4
5	Absolute minimum temperature	in °C	5,0	5,0	5,0	3,3	1,7	0,0	-0,6	0,0	1,1	1,1	3,3	4,4	-0,6	47	5
6	Mean relative humidity	in %	70	69	72	75	77	77	77	77	72	73	71	70	73	30	6
7	Mean precipitation	in mm	20	18	32	49	92	100	107	95	66	52	27	22	679	30	7
8	Maximum precipitation	in mm	133	120	124	176	179	273	240	284	143	146	73	83	921	60	8
9	Minimum precipitation	in mm	0	0	0	5	20	30	24	19	11	14	0	0	438	60	9
10	Maximum precipitation in 24 h	in mm	69	39	44	126	52	72	55	59	116	45	51	71	126	60	10
11	Mean number of days with precipitation	>0,25 mm	5	4	7	9	15	15	16	15	13	12	7	6	124	30	11
12	Mean duration of sunshine	in h															12
13	Mean quantity of radiation	in ly/day															13
14	Mean potential evaporation	in mm	108	95	91	66	48	35	34	39	48	63	81	105	813	34	14
15	Mean windspeed	in m/sec	5,7	5,4	5,1	4,9	5,1	4,6	4,6	6,4	5,7	5,1	5,1	5,1	5,4	5	15
16	Mean predominent direction of the wind		NE,SW	NE,SE	NE,SE	SW	SW	SW	SW	SW	SW	SW	SW,SE	SW		5	16

33 Station/Country Rockhampton/Australia

Location 23°24'S / 150°30'E   Height above sealevel 11 m   Climate symbol: Köppen Cfa   Troll V,2

		J	F	M	A	M	J	J	A	S	O	N	D	year	P	
1 Mean daily temperature	in °C	27.3	26.8	25.8	23.6	20.4	17.9	16.9	18.2	21.2	23.8	25.7	26.9	22.9	30	1
2 Mean daily maximum temperature	in °C	31.7	31.7	30.8	28.9	26.1	23.3	22.8	25.0	27.8	30.0	31.7	32.2	28.3	50	2
3 Mean daily minimum temperature	in °C	22.2	22.2	21.1	18.3	14.4	11.7	10.0	11.7	15.0	17.8	19.4	21.7	17.2	50	3
4 Absolute maximum temperature	in °C	41.6	42.4	40.4	36.7	34.8	31.2	33.2	34.8	37.9	39.3	42.2	44.2	44.2	48	4
5 Absolute minimum temperature	in °C	15.6	15.6	10.1	6.3	4.3	0.4	1.4	2.4	4.4	6.1	12.2	15.1	0.4	49	5
6 Mean relative humidity	in %	68	69	69	67	67	68	65	64	63	64	66	66	66	29	6
7 Mean precipitation	in mm	154	187	118	44	44	41	50	19	20	50	68	93	887	30	7
8 Maximum precipitation	in mm	873	924	637	551	246	266	496	95	107	171	188	493	2081	71	8
9 Minimum precipitation	in mm	9	6	1	0	0	0	0	0	0	0	0	5	399	71	9
10 Maximum precipitation in 24 h	in mm	267	389	162	357	108	174	293	77	88	86	74	136	389	68	10
11 Mean number of days with precipitation	>0.25 mm	11	12	11	7	6	5	6	4	4	6	7	9	88	30	11
12 Mean duration of sunshine	in h															12
13 Mean quantity of radiation	in ly/day															13
14 Mean potential evaporation	in mm	158	136	143	102	67	45	39	51	76	120	140	159	1236	24	14
15 Mean windspeed	in m/sec	3.6	3.6	4.1	2.6	2.8	2.1	3.1	3.1	3.1	3.6	3.1	3.1	3.1	5	15
16 Mean predominent direction of the wind		E	E	SE.E	SE.E	SE.E	SE	SE.E	SE.E	SE.E	E	E	E			16

34 Station/Country Brisbane/Australia

Location 27°28'S / 153°02'E   Height above sealevel 42 m   Climate symbol: Köppen Cfa   Troll IV,7

		J	F	M	A	M	J	J	A	S	O	N	D	year	P	
1 Mean daily temperature	in °C	25.0	24.7	23.6	21.2	18.2	15.8	15.0	16.1	18.1	20.7	22.5	24.3	20.4	30	1
2 Mean daily maximum temperature	in °C	29.4	29.4	27.8	26.1	23.3	20.6	20.0	21.7	24.4	26.7	27.8	29.4	25.8	53	2
3 Mean daily minimum temperature	in °C	20.6	20.0	18.9	16.1	13.3	10.6	9.4	10.0	12.8	15.6	17.8	19.4	15.6	53	3
4 Absolute maximum temperature	in °C	43.2	40.9	37.4	35.1	32.4	31.6	29.1	32.8	38.3	40.7	41.2	41.1	43.2	74	4
5 Absolute minimum temperature	in °C	14.9	14.7	11.3	6.9	4.8	2.4	2.3	3.0	4.8	6.3	9.2	13.5	3.3	74	5
6 Mean relative humidity	in %	69	72	72	71	69	67	66	64	64	64	66	67	68	30	6
7 Mean precipitation	in mm	143	183	147	78	57	56	49	30	45	77	92	136	1092	30	7
8 Maximum precipitation	in mm	704	1026	864	388	351	356	218	372	137	290	315	441	2240	109	8
9 Minimum precipitation	in mm	8	15	0	0	0	0	0	0	3	1	0	9	411	109	9
10 Maximum precipitation in 24 h	in mm	465	269	284	139	142	163	89	124	63	135	113	168	465	109	10
11 Mean number of days with precipitation	>0.25 mm	13	14	15	12	10	8	8	7	8	9	10	12	126	80	11
12 Mean duration of sunshine	in h	236	207	217	213	205	189	211	245	246	260	246	254	2729	30	12
13 Mean quantity of radiation	in ly/day															13
14 Mean potential evaporation	in mm	140	130	115	84	55	38	33	43	62	93	115	146	1054	44	14
15 Mean windspeed	in m/sec	3.9	4.1	3.9	3.1	3.3	2.8	3.3	3.3	3.6	3.9	3.9	4.4	3.6	5	15
16 Mean predominent direction of the wind		SE.NE	SE.NE	S.E	S.E	SW.SE	SW	SW	SW.NE	SW.NE	S.NE	SE.NE	SE.NE		30	16

35 Station/Country Armidale/Australia

Location 30°39'S / 151°38'E   Height above sealevel 1015 m   Climate symbol: Köppen Cfb   Troll IV,7

		J	F	M	A	M	J	J	A	S	O	N	D	year	P	
1 Mean daily temperature	in °C	20.4	19.8	17.6	13.9	10.1	7.3	6.6	7.7	10.8	14.3	17.3	19.3	13.8	30	1
2 Mean daily maximum temperature	in °C															2
3 Mean daily minimum temperature	in °C															3
4 Absolute maximum temperature	in °C	39.7	37.8	34.4	30.1	26.7	24.4	21.1	25.7	28.3	32.5	36.4	37.7	39.7	80	4
5 Absolute minimum temperature	in °C	4.4	3.3	-0.6	-3.9	-6.7	-8.3	-10.0	-8.1	-5.6	-3.3	0.0	2.2	-10.0	80	5
6 Mean relative humidity	in %	60	63	64	69	69	68	61	64	72	59	58	60	68	30	6
7 Mean precipitation	in mm	107	105	67	42	36	55	52	56	56	72	78	88	815	30	7
8 Maximum precipitation	in mm	274	279	235	236	140	190	152	209	146	183	232	289	1507	79	8
9 Minimum precipitation	in mm	8	3	1	4	3	5	1	4	<1	1	4	<1	422	79	9
10 Maximum precipitation in 24 h	in mm	96	75	86	64	80	94	59	155	76	64	65	61	155	99	10
11 Mean number of days with precipitation	>0.25 mm	10	11	11	9	9	11	10	9	9	10	9	11	119	30	11
12 Mean duration of sunshine	in h															12
13 Mean quantity of radiation	in ly/day															13
14 Mean potential evaporation	in mm	113	86	80	52	31	19	16	22	39	64	80	110	712	52	14
15 Mean windspeed	in m/sec															15
16 Mean predominent direction of the wind		SE	SE.W	SE	SE	W	W	W	W	W	W	W	E.W		5	16

36 Station/Country Lord Howe Island/Australia

Location 31°31'S / 159°07'E   Height above sealevel 5 m   Climate symbol: Köppen Cfa   Troll IV,7

		J	F	M	A	M	J	J	A	S	O	N	D	year	P	
1 Mean daily temperature	in °C	22.5	22.5	22.0	20.3	18.3	16.7	15.6	15.9	16.7	18.0	19.4	21.1	19.2	27	1
2 Mean daily maximum temperature	in °C	25.6	25.6	25.0	23.3	21.1	19.4	18.6	18.8	19.9	21.4	22.9	24.4	22.2	27	2
3 Mean daily minimum temperature	in °C	19.3	19.6	18.9	17.2	15.3	13.8	13.4	13.4	14.8	16.3	18.0		16.1	27	3
4 Absolute maximum temperature	in °C	30.6	31.7	29.4	28.3	25.8	23.9	23.9	22.8	25.0	27.2	27.8	29.4	31.7	27	4
5 Absolute minimum temperature	in °C	11.7	12.2	12.8	10.6	8.3	7.8	6.1	6.1	6.7	7.8	9.4	10.6	6.1	27	5
6 Mean relative humidity	in %	77	75	75	77	77	77	77	75	76	78	78	78	77	28	6
7 Mean precipitation	in mm	125	106	127	171	158	195	196	135	134	131	114	123	1715	45	7
8 Maximum precipitation	in mm	291	337	375	702	376	387	496	269	322	337	292	339	2866	42	8
9 Minimum precipitation	in mm	10	24	51	38	45	82	80	13	27	36	14	17	996	42	9
10 Maximum precipitation in 24 h	in mm	140	161	111	305	117	148	131	79	247	170	127	106	305	30	10
11 Mean number of days with precipitation	>0.25 mm	11	11	14	18	20	21	22	19	14	13	13	11	187	15	11
12 Mean duration of sunshine	in h															12
13 Mean quantity of radiation	in ly/day															13
14 Mean potential evaporation	in mm	85	83	96	84	79	64	58	59	56	62	63	76	863		14
15 Mean windspeed	in m/sec	4.5	5.4	4.9	5.7	5.7	6.2	5.9	6.4	5.7	5.9	5.7	5.1	5.7	4	15
16 Mean predominent direction of the wind		SE	SE	SE	SE.SW	SE.SW	SE.SW	SE.SW	SW.NW	NW	NW.SE	SE.NW	NE		4	16

37 Station/Country Sydney/Australia

Location 33°51′S/151°31′E   Height above sealevel 42 m   Climate symbol: Köppen Cfa   Troll IV,7

		J	F	M	A	M	J	J	A	S	O	N	D	year	P	
1 Mean daily temperature	in °C	22.0	21.9	20.8	18.3	15.1	12.8	11.8	13.0	15.2	17.6	19.5	21.1	17.4	104	1
2 Mean daily maximum temperature	in °C	25.6	25.6	24.4	21.7	18.9	18.1	15.6	17.2	19.4	21.7	23.3	25.0	21.1	87	2
3 Mean daily minimum temperature	in °C	18.3	18.3	17.2	14.4	11.1	8.9	7.8	8.9	10.6	13.3	15.6	17.2	13.3	87	3
4 Absolute maximum temperature	in °C	45.3	42.1	39.2	33.0	30.0	26.9	25.7	30.4	33.5	37.4	40.3	42.2	45.3	104	4
5 Absolute minimum temperature	in °C	10.6	9.6	9.3	7.0	4.6	2.1	2.2	2.7	4.9	5.7	7.7	9.1	2.1	104	5
6 Mean relative humidity	in %	68	71	74	75	77	76	74	69	64	62	63	65	69	86	6
7 Mean precipitation	in mm	104	125	129	101	115	141	94	83	72	80	77	86	1205	30	7
8 Maximum precipitation	in mm	388	564	521	622	585	643	336	378	356	282	518	402	2102	104	8
9 Minimum precipitation	in mm	6	3	11	2	4	4	3	1	2	5	2	6	546	104	9
10 Maximum precipitation in 24 h	in mm	180	226	281	191	212	132	198	135	145	162	133	121	281	104	10
11 Mean number of days with precipitation	>0.25 mm	14	13	14	14	13	12	12	11	12	12	12	13	152	87	11
12 Mean duration of sunshine	in h	226	185	195	183	180	189	189	214	216	229	228	229	2463	42	12
13 Mean quantity of radiation	in ly/day	536	454	438	317	258	215	251	315	394	514	584	554	402		13
14 Mean potential evaporation	in mm	122	99	92	64	41	28	25	33	53	76	93	112	838	73	14
15 Mean windspeed	in m/sec	4.1	4.0	3.4	3.2	3.2	3.5	3.5	3.5	3.7	3.9	3.9	4.0	3.7	86	15
16 Mean predominant direction of the wind		S,ENE	NE	W,NE	W,NE	N,S	W	W	W	W,NE	W,NE	W,NE	S,NE		48	16

38 Station/Country Canberra/Australia

Location 35°17′S/149°08′E   Height above sealevel 559 m   Climate symbol: Köppen Cfb   Troll IV,7

		J	F	M	A	M	J	J	A	S	O	N	D	year	P	
1 Mean daily temperature	in °C	20.7	20.2	18.0	13.4	9.6	6.7	6.0	7.4	10.1	13.1	16.0	19.1	13.4	20	1
2 Mean daily maximum temperature	in °C	27.8	27.8	24.4	19.4	15.6	11.7	11.1	12.8	16.1	20.0	23.9	26.7	20.0	23	2
3 Mean daily minimum temperature	in °C	12.8	12.8	10.6	6.7	2.8	1.1	0.6	1.7	3.3	6.1	8.9	11.7	6.7	23	3
4 Absolute maximum temperature	in °C	41.9	37.7	37.3	32.1	22.6	18.3	17.5	21.7	27.5	32.2	38.6	39.7	41.9	33	4
5 Absolute minimum temperature	in °C	3.3	1.7	1.6	-1.7	-5.3	-7.7	-6.7	-6.1	-3.8	-1.7	0.1	2.2	-7.7	33	5
6 Mean relative humidity	in %	53	59	66	71	79	81	81	75	66	60	55	51	66	28	6
7 Mean precipitation	in mm	54	55	63	55	52	49	41	50	41	70	54	47	632	20	7
8 Maximum precipitation	in mm	170	153	322	132	156	155	129	120	115	177	113	223		33	8
9 Minimum precipitation	in mm	1	0	0	2	2	5	7	9	3	9	7	4		33	9
10 Maximum precipitation in 24 h	in mm	82	82	65	64	99	59	51	53	44	132	62	58	132	33	10
11 Mean number of days with precipitation	>0.25 mm	7	7	7	7	7	9	10	11	9	11	8	8	101	26	11
12 Mean duration of sunshine	in h	260	204	223	201	161	126	149	180	216	242	246	264	2472	27	12
13 Mean quantity of radiation	in ly/day															13
14 Mean potential evaporation	in mm	115	96	83	49	28	17	16	22	34	58	83	112	713	14	14
15 Mean windspeed	in m/sec	2.1	1.8	1.6	1.6	1.3	1.6	1.5	1.8	1.9	1.9	2.1	2.1	1.8	27	15
16 Mean predominant direction of the wind		NW	E,NW	E,NW	NW	NW	NW	NW	NW	NW	NW	NW	NW		27	16

39 Station/Country Mount Gambier/Australia

Location 37°50′S/140°46′E   Height above sealevel 65 m   Climate symbol: Köppen Csb   Troll IV,1

		J	F	M	A	M	J	J	A	S	O	N	D	year	P	
1 Mean daily temperature	in °C	18.1	18.5	17.6	14.4	12.1	10.3	9.6	10.3	11.6	13.3	14.9	16.6	13.9	50	1
2 Mean daily maximum temperature	in °C															2
3 Mean daily minimum temperature	in °C															3
4 Absolute maximum temperature	in °C	44.8	43.1	41.3	36.8	28.1	22.2	21.4	25.0	31.7	35.0	40.0	42.0	44.8	50	4
5 Absolute minimum temperature	in °C	0.6	1.1	0.6	-1.2	-3.2	-4.8	-4.6	-2.5	-2.0	-0.9	-0.6	1.1	-4.8	50	5
6 Mean relative humidity	in %	65	65	70	76	80	81	79	79	77	74	71	67	73	50	6
7 Mean precipitation	in mm	34	29	38	60	85	97	105	98	77	65	45	41	774	50	7
8 Maximum precipitation	in mm	158	119	153	177	206	201	226	177	164	148	130	277	1414	89	8
9 Minimum precipitation	in mm	2	0	2	<1	21	15	21	14	24	15	9	1	457	89	9
10 Maximum precipitation in 24 h	in mm	51	45	36	46	60	37	33	43	33	33	56	44	60	42	10
11 Mean number of days with precipitation	>0.25 mm	7	7	9	13	17	19	20	20	17	15	12	9	165	50	11
12 Mean duration of sunshine	in h															12
13 Mean quantity of radiation	in ly/day															13
14 Mean potential evaporation	in mm	101	86	77	54	40	29	27	34	43	59	69	92	711	65	14
15 Mean windspeed	in m/sec	6.7	6.4	6.4	6.9	6.4	5.9	6.4	7.7	7.7	7.2	6.9	6.9	6.9	5	15
16 Mean predominant direction of the wind		SE	SE	SE	N,SE	NW	N	NW	NW	NW	W	SE,W	SE		5	16

40 Station/Country Melbourne/Australia

Location 37°49′S/144°58′E   Height above sealevel 35 m   Climate symbol: Köppen Cfb   Troll III,2

		J	F	M	A	M	J	J	A	S	O	N	D	year	P	
1 Mean daily temperature	in °C	19.9	19.7	18.4	15.1	12.5	10.2	9.6	10.5	12.4	14.3	16.2	18.4	14.8	30	1
2 Mean daily maximum temperature	in °C	25.6	25.6	23.9	20.0	16.7	13.9	13.3	15.0	17.2	19.4	21.7	23.9	19.6	88	2
3 Mean daily minimum temperature	in °C	13.9	13.9	12.8	10.6	8.3	6.7	5.6	6.1	7.8	8.9	10.6	12.2	10.0	88	3
4 Absolute maximum temperature	in °C	45.6	43.1	41.7	34.9	28.7	22.4	20.7	25.0	31.4	36.9	40.9	43.7	45.6	105	4
5 Absolute minimum temperature	in °C	5.6	4.6	2.8	1.6	-1.2	-2.2	-2.8	-2.1	-0.6	0.1	2.5	4.4	-2.8	105	5
6 Mean relative humidity	in %	57	60	62	69	73	75	75	71	67	63	61	59	65	30	6
7 Mean precipitation	in mm	45	59	50	69	54	52	54	50	58	74	70	58	691	30	7
8 Maximum precipitation	in mm	169	196	191	195	142	115	178	110	201	193	206	182		105	8
9 Minimum precipitation	in mm	0	1	4	0	4	15	14	12	13	7	6	3		105	9
10 Maximum precipitation in 24 h	in mm	75	87	90	80	44	69	44	49	67	76	73	100	100	105	10
11 Mean number of days with precipitation	>0.25 mm	9	8	9	13	14	16	17	17	15	14	13	11	156	30	11
12 Mean duration of sunshine	in h	242	207	202	150	127	102	115	143	165	180	186	217	2036	35	12
13 Mean quantity of radiation	in ly/day	585	504	393	270	180	155	160	231	324	422	492	545	355	12	13
14 Mean potential evaporation	in mm	108	96	85	54	38	26	23	32	43	63	82	105	755	76	14
15 Mean windspeed	in m/sec	3.9	3.8	3.4	3.1	3.3	3.2	3.9	3.7	3.8	3.8	3.8	3.9	3.6	15	15
16 Mean predominant direction of the wind		S	N,S	N,S	N,S	N	N	N	N	N	N,S	S	S		30	16

41 Station/Country Launceston (Tasmania)/Australia

Location 41°27'S/147°10'E | Height above sealevel 81 m | Climate symbol: Köppen Cfb | Troll III,2

		J	F	M	A	M	J	J	A	S	O	N	D	year	P	
1 Mean daily temperature	in °C	17,7	18,2	16,1	12,9	10,2	8,1	7,4	8,5	10,4	12,3	14,6	16,5	12,7	51	1
2 Mean daily maximum temperature	in °C	24,4	25,0	22,2	18,3	15,0	12,8	12,2	13,3	15,6	17,8	20,6	22,8	18,3	46	2
3 Mean daily minimum temperature	in °C	11,1	11,7	10,0	7,2	5,0	3,9	2,8	3,3	5,0	6,7	8,3	10,0	7,2	46	3
4 Absolute maximum temperature	in °C	37,8	38,3	34,4	28,9	23,8	19,0	19,0	20,0	23,9	31,6	32,8	36,1	38,3	50	4
5 Absolute minimum temperature	in °C	1,1	0,9	-0,6	-2,3	-4,4	-5,6	-6,1	-3,9	-4,4	-3,9	0,3	-0,3	-6,1	50	5
6 Mean relative humidity	in %	60	63	67	72	77	75	77	77	75	72	65	63	69	33	6
7 Mean precipitation	in mm	41	50	40	62	73	71	86	80	65	68	56	50	740	30	7
8 Maximum precipitation	in mm	90	81	157	150	232	167	194	161	146	170	112	151	1018	49	8
9 Minimum precipitation	in mm	8	<1	3	0	1	13	15	9	15	7	5	0	467	49	9
10 Maximum precipitation in 24 h	in mm	43	31	71	104	38	36	56	48	36	31	43	69	104	27	10
11 Mean number of days with precipitation	>0,25 mm	8	7	9	11	13	16	18	17	15	14	11	10	149	30	11
12 Mean duration of sunshine	in h															12
13 Mean quantity of radiation	in ly/day															13
14 Mean potential evaporation	in mm	105	90	73	50	33	23	21	27	40	55	74	‹99	890	81	14
15 Mean windspeed	in m/sec	6,4	5,7	5,9	5,7	5,9	4,6	5,7	5,9	6,2	6,2	5,9	5,7	5,7	5	15
16 Mean predominent direction of the wind		NW	NW	NW	NW	NW	NW	NW	NW	NW	NW	NW	NW		5	16

42 Station/Country Hobart/Australia

Location 42°53'S/147°20'E | Height above sealevel 54 m | Climate symbol: Köppen Cfb | Troll III,2

		J	F	M	A	M	J	J	A	S	O	N	D	year	P	
1 Mean daily temperature	in °C	16,3	16,1	15,1	12,4	10,5	8,3	7,8	8,8	10,6	11,8	13,6	15,1	12,2	30	1
2 Mean daily maximum temperature	in °C	21,7	21,7	20,0	17,2	14,4	11,7	11,1	12,8	15,0	17,2	18,9	20,6	16,7	70	2
3 Mean daily minimum temperature	in °C	11,7	11,7	10,6	8,9	6,7	5,0	4,4	5,0	6,1	7,8	8,9	10,6	8,3	70	3
4 Absolute maximum temperature	in °C	40,6	40,2	37,3	30,6	25,4	20,7	18,9	22,0	27,6	33,3	36,8	40,7	40,7	91	4
5 Absolute minimum temperature	in °C	4,5	3,9	1,8	0,7	-1,6	-1,6	-2,4	-1,7	-0,6	0,0	1,7	3,3	-2,4	91	5
6 Mean relative humidity	in %	58	61	64	66	70	73	72	68	63	61	60	60	64	30	6
7 Mean precipitation	in mm	44	45	53	61	47	62	51	50	51	68	58	64	652	30	7
8 Maximum precipitation	in mm	150	232	193	216	162	207	153	258	181	169	227	229	1102	78	8
9 Minimum precipitation	in mm	4	3	7	2	4	7	4	8	10	10	8	4	392	78	9
10 Maximum precipitation in 24 h	in mm	75	56	88	133	44	147	64	58	156	66	94	85	156	78	10
11 Mean number of days with precipitation	>0,25 mm	13	10	13	14	14	16	17	18	17	18	16	14	180	30	11
12 Mean duration of sunshine	in h	239	199	198	150	136	120	136	158	177	189	216	226	2144	30	12
13 Mean quantity of radiation	in ly/day															13
14 Mean potential evaporation	in mm	90	75	72	44	36	23	26	31	43	56	70	84	650	54	14
15 Mean windspeed	in m/sec	4,1	3,6	3,3	3,3	2,6	2,6	2,6	3,2	4,1	4,0	4,1	4,1	3,3	30	15
16 Mean predominent direction of the wind		NW,SE	NW,SE	NW	NW	NW	NW	NW	NW	NW	NW,SE	NW,SE	NW,SE		30	16

43 Station/Country Norfolk Island/Australia

Location 29°04'S/167°59'E | Height above sealevel 106 m | Climate symbol: Köppen Cfa | Troll IV,7

		J	F	M	A	M	J	J	A	S	O	N	D	year	P	
1 Mean daily temperature	in °C	22,5	22,5	22,0	20,6	18,3	17,5	16,1	15,9	16,7	18,0	19,4	21,5	19,2	19	1
2 Mean daily maximum temperature	in °C	25,6	25,0	24,4	22,8	20,6	19,4	18,3	18,3	19,4	20,6	22,2	23,5	21,7	19	2
3 Mean daily minimum temperature	in °C	19,4	20,0	19,4	18,3	16,1	15,6	13,9	13,3	13,9	15,6	16,7	18,3	16,7	19	3
4 Absolute maximum temperature	in °C	31,7	31,7	29,4	28,3	27,2	23,3	22,8	26,7	27,2	25,6	30,0	30,6	31,7	22	4
5 Absolute minimum temperature	in °C	14,4	15,0	12,8	12,2	10,0	8,3	7,2	7,8	7,8	9,4	9,4	12,2	7,2	20	5
6 Mean relative humidity	in %	76	76	79	79	77	79	78	76	76	·76	75	77	77	22	6
7 Mean precipitation	in mm	84	109	94	127	145	140	155	137	94	94	66	86	1331	43	7
8 Maximum precipitation	in mm															8
9 Minimum precipitation	in mm															9
10 Maximum precipitation in 24 h	in mm	119	81	163	226	241	127	122	122	74	173	76	114	241	30	10
11 Mean number of days with precipitation	>0,25 mm	3	4	6	8	9	8	9	8	7	6	5	5	78	10	11
12 Mean duration of sunshine	in h															12
13 Mean quantity of radiation	in ly/day															13
14 Mean potential evaporation	in mm	146	121	118	91	69	54	49	51	63	83	102	126	1073		14
15 Mean windspeed	in m/sec															15
16 Mean predominent direction of the wind																16

44 Station/Country Auckland/New Zealand

Location 36°51'S/174°46'E | Height above sealevel 49 m | Climate symbol: Köppen Cfb · | Troll IV,7

		J	F	M	A	M	J	J	A	S	O	N	D	year	P	
1 Mean daily temperature	in °C	19,2	19,6	18,4	16,4	13,8	11,8	10,8	11,3	12,6	14,3	15,9	17,7	15,2		1
2 Mean daily maximum temperature	in °C	22,8	22,8	21,7	19,4	16,7	14,4	13,3	14,4	15,6	17,2	18,9	21,1	18,3	36	2
3 Mean daily minimum temperature	in °C	15,6	15,6	15,0	13,3	10,6	8,9	7,8	7,8	9,4	11,1	12,2	13,9	11,7	36	3
4 Absolute maximum temperature	in °C	32,2	32,4	29,9	27,2	23,6	21,0	19,2	19,4	21,7	23,9	27,2	31,8	32,4		4
5 Absolute minimum temperature	in °C	7,3	8,6	5,6	4,1	1,9	-0,1	0,7	1,1	1,2	2,2	5,0	6,1	-0,1		5
6 Mean relative humidity	in %	66	67	70	74	76	79	80	76	72	70	68	67	72	30	6
7 Mean precipitation	in mm	84	104	71	109	122	140	140	109	97	107	81	79	1242	30	7
8 Maximum precipitation	in mm	213	359	268	359	348	286	305	324	229	259	253	225	1917	30	8
9 Minimum precipitation	in mm	4	0	2	11	24	24	34	29	19	10	12	3	669	30	9
10 Maximum precipitation in 24 h	in mm	97	162	116	132	91	91	86	77	55	64	80	137	162		10
11 Mean number of days with precipitation	>0,25 mm	10	10	12	15	19	20	20	19	16	17	15	12	185		11
12 Mean duration of sunshine	in h	233	194	190	155	138	123	129	151	168	181	207	224	2094		12
13 Mean quantity of radiation	in ly/day															13
14 Mean potential evaporation	in mm	103	80	82	61	45	31	30	34	43	59	74	90	732	76	14
15 Mean windspeed	in m/sec	4,9	4,4	4,3	4,4	4,2	4,3	4,2	4,6	4,7	5,4	5,0	4,8	4,6		15
16 Mean predominent direction of the wind		WSW	WSW	SW	SW	WSW	WSW	WSW	WSW	WSW	WSW	WSW	WSW			16

45 Station / Country Tauranga / New Zealand

Location 37°40'S/176°12'E   Height above sealevel 4 m   Climate symbol: Köppen Cfb   Troll IV.7

		J	F	M	A	M	J	J	A	S	O	N	D	year	P	
1	Mean daily temperature	in °C	18,6	19,0	17,5	15,0	12,4	10,1	9,3	10,1	11,6	13,3	15,2	17,0	14,1		1
2	Mean daily maximum temperature	in °C															2
3	Mean daily minimum temperature	in °C															3
4	Absolute maximum temperature	in °C	33,3	31,3	29,7	27,9	23,9	21,9	22,8	19,6	24,7	25,6	28,3	30,6	33,3		4
5	Absolute minimum temperature	in °C	3,3	1,7	0,7	-0,6	-5,3	-4,6	-4,2	-3,4	-4,6	-2,3	0,6	-0,3	-5,3		5
6	Mean relative humidity	in %															6
7	Mean precipitation	in mm	89	89	97	127	124	140	127	122	97	117	84	86	1300		7
8	Maximum precipitation	in mm															8
9	Minimum precipitation	in mm															9
10	Maximum precipitation in 24 h	in mm	134	160	96	239	118	164	135	76	156	129	67	132	239		10
11	Mean number of days with precipitation	>0,25 mm	10	9	11	12	14	14	15	15	14	14	13	12	153		11
12	Mean duration of sunshine	in h	255	210	207	176	155	141	152	164	186	201	226	243	2316		12
13	Mean quantity of radiation	in ly/day															13
14	Mean potential evaporation	in mm	99	85	78	56	39	28	27	33	42	59	74	91	711	29	14
15	Mean windspeed	in m/sec	3,8	3,4	3,5	3,5	3,2	3,4	3,3	3,5	3,9	4,2	4,3	4,2	3,7		15
16	Mean predominent direction of the wind		W	W	W	W,SW	W,SW	W	W	W	W	W	W	W			16

46 Station / Country New Plymouth / New Zealand

Location 39°04'S/174°05'E   Height above sealevel 49 m   Climate symbol: Köppen Cfb   Troll IV.7

		J	F	M	A	M	J	J	A	S	O	N	D	year	P	
1	Mean daily temperature	in °C	17,1	17,8	16,6	14,7	12,4	10,3	9,6	10,1	11,3	12,7	14,1	15,6	13,5		1
2	Mean daily maximum temperature	in °C	21,1	21,1	20,6	18,9	16,1	13,9	12,8	13,3	14,4	16,1	17,8	20,0	17,2	18	2
3	Mean daily minimum temperature	in °C	12,8	13,3	12,2	11,1	8,3	6,7	6,1	6,1	7,8	9,4	10,6	12,2	10,0	18	3
4	Absolute maximum temperature	in °C	29,1	30,0	27,5	28,4	22,2	20,4	21,4	18,6	21,3	23,9	25,8	27,2	30,0	40	4
5	Absolute minimum temperature	in °C	4,6	5,1	1,6	1,8	-0,4	-1,6	-1,4	-1,6	-0,2	0,6	1,3	2,3	-1,6	40	5
6	Mean relative humidity	in %	78	80	78	78	78	82	79	80	79	78	79	79	79	12	6
7	Mean precipitation	in mm	119	104	94	127	135	163	157	150	122	147	114	122	1554		7
8	Maximum precipitation	in mm															8
9	Minimum precipitation	in mm															9
10	Maximum precipitation in 24 h	in mm	123	185	104	99	112	95	93	72	106	114	70	121	185		10
11	Mean number of days with precipitation	>0,25 mm	9	10	11	13	15	18	17	17	14	17	14	14	169		11
12	Mean duration of sunshine	in h	245	194	196	165	145	116	135	154	168	175	201	216	2110		12
13	Mean quantity of radiation	in ly/day															13
14	Mean potential evaporation	in mm	93	81	76	59	42	30	28	33	42	59	70	87	700	22	14
15	Mean windspeed	in m/sec	5,3	5,2	4,6	5,0	4,8	5,0	5,3	5,0	5,6	5,8	5,8	5,3	5,1		15
16	Mean predominent direction of the wind		W	W	SE	SE	SE	SE	SE	SE	W	W	W	W			16

47 Station / Country Wellington / New Zealand

Location 41°17'S/174°46'E   Height above sealevel 126 m   Climate symbol: Köppen Cfb   Troll III,2

		J	F	M	A	M	J	J	A	S	O	N	D	year	P	
1	Mean daily temperature	in °C	16,2	16,4	15,4	13,5	10,9	8,8	8,1	8,8	10,2	11,7	13,3	15,1	12,4		1
2	Mean daily maximum temperature	in °C	20,6	20,6	19,4	17,2	14,4	12,8	11,7	12,2	13,9	15,6	17,2	19,4	16,1	66	2
3	Mean daily minimum temperature	in °C	13,3	13,3	12,2	10,6	8,3	6,7	5,6	6,1	7,8	8,9	10,0	12,2	9,4	66	3
4	Absolute maximum temperature	in °C	29,4	31,1	27,2	27,3	21,9	20,6	18,9	20,0	20,6	24,2	26,9	29,1	31,1	99	4
5	Absolute minimum temperature	in °C	4,1	4,7	3,9	2,1	-0,7	-1,2	-1,9	-1,6	-0,6	1,1	1,7	3,4	-1,9	99	5
6	Mean relative humidity	in %	70	73	73	78	79	80	79	77	76	75	73	72	75	13	6
7	Mean precipitation	in mm	74	91	79	94	119	122	130	135	97	122	81	107	1250		7
8	Maximum precipitation	in mm	250	251	256	309	301	280	309	269	281	320	254	386	1795	30	8
9	Minimum precipitation	in mm	3	1	5	16	18	13	38	26	15	8	15	0	707	30	9
10	Maximum precipitation in 24 h	in mm	114	161	145	126	145	87	83	95	98	105	67	152	161	99	10
11	Mean number of days with precipitation	>0,25 mm	10	9	11	13	16	17	18	17	15	14	13	12	165	79	11
12	Mean duration of sunshine	in h	234	195	189	151	118	106	108	139	170	183	197	222	2012		12
13	Mean quantity of radiation	in ly/day															13
14	Mean potential evaporation	in mm	93	78	72	52	38	27	26	30	41	56	68	86	667	80	14
15	Mean windspeed	in m/sec	5,2	5,0	4,7	5,0	5,4	5,1	5,7	4,8	5,6	5,5	5,6	5,3	5,2		15
16	Mean predominent direction of the wind		NW	NW	NW	NW	NW	NW	NW	NW	NW	NW	NW	NW			16

48 Station / Country Dunedin / New Zealand

Location 45°55'S/170°31'E   Height above sealevel 2 m   Climate symbol: Köppen Cfb   Troll III,2

		J	F	M	A	M	J	J	A	S	O	N	D	year	P	
1	Mean daily temperature	in °C	14,9	15,1	13,6	11,6	8,9	6,8	6,4	7,3	9,4	11,2	12,8	13,9	11,0		1
2	Mean daily maximum temperature	in °C	18,9	18,9	17,2	15,0	11,7	9,4	8,9	10,6	12,8	15,0	16,7	18,3	14,4	77	2
3	Mean daily minimum temperature	in °C	10,0	10,0	8,9	7,2	5,0	3,9	2,8	3,3	5,0	5,8	7,2	8,9	6,7	77	3
4	Absolute maximum temperature	in °C	34,4	33,5	29,4	29,4	24,9	20,8	19,4	21,1	25,0	28,3	30,0	31,1	34,4	96	4
5	Absolute minimum temperature	in °C	2,2	2,8	1,1	-1,2	-2,2	-4,4	-5,0	-3,9	-1,7	-1,1	0,0	1,7	-5,0	96	5
6	Mean relative humidity	in %	69	70	72	74	76	77	76	73	71	68	69	72	72	12	6
7	Mean precipitation	in mm	71	64	64	64	66	74	64	58	56	64	71	74	787		7
8	Maximum precipitation	in mm	158	173	217	294	218	'158	164	132	159	139	145	219	1001	30	8
9	Minimum precipitation	in mm	22	6	11	20	6	17	9	8	6	8	17	20	510	30	9
10	Maximum precipitation in 24 h	in mm	87	137	134	229	103	95	98	117	113	58	114	138	229	109	10
11	Mean number of days with precipitation	>0,25 mm	14	12	13	13	13	14	14	12	12	14	15	16	162		11
12	Mean duration of sunshine	in h	194	166	152	122	110	94	110	132	147	166	174	167	1734		12
13	Mean quantity of radiation	in ly/day															13
14	Mean potential evaporation	in mm	95	77	68	51	34	22	21	31	43	62	74	90	668	20	14
15	Mean windspeed	in m/sec	4,0	3,8	3,4	3,2	2,6	2,8	2,2	2,8	3,4	3,8	4,1	4,2	3,4	+	15
16	Mean predominent direction of the wind		NE	NE	NE	NE	NE	W	W	W	NE	NE	NE	NE		+	16

+ 9.00 a.m.

49 Station/Country Fanning Island / United Kingdom

Location 3°54'N / 159°23'W   Height above sealevel 6 m   Climate symbol: Köppen Af   Troll V,1

		J	F	M	A	M	J	J	A	S	O	N	D	year	P	
1 Mean daily temperature	in °C	27,5	27,5	27,8	27,8	27,8	28,1	28,1	28,3	28,1	28,3	28,3	27,8	28,1	21	1
2 Mean daily maximum temperature	in °C	30,6	30,6	30,6	30,6	30,6	31,1	31,1	31,7	31,7	31,7	31,7	31,1	31,1	21	2
3 Mean daily minimum temperature	in °C	24,4	24,4	25,0	25,0	25,0	25,0	25,0	25,0	24,4	25,0	25,0	24,4	25,0	21	3
4 Absolute maximum temperature	in °C	35,0	36,1	36,1	36,7	36,1	37,2	36,7	36,7	37,8	37,2	37,8	36,1	37,8	17	4
5 Absolute minimum temperature	in °C	21,7	20,0	21,1	21,7	21,7	21,7	21,1	22,2	20,6	21,1	20,0	20,6	20,0	17	5
6 Mean relative humidity	in % +	77	78	79	81	79	77	75	72	69	69	69	74	75	16	6
7 Mean precipitation	in mm	276	267	272	358	320	254	208	112	81	91	74	203	2516	23	7
8 Maximum precipitation	in mm															8
9 Minimum precipitation	in mm															9
10 Maximum precipitation in 24 h	in mm	152	178	196	196	152	109	229	102	79	102	79	112	229	16	10
11 Mean number of days with precipitation	>0,25 mm	13	15	18	20	19	16	14	10	8	8	8	11	160	21	11
12 Mean duration of sunshine	in h															12
13 Mean quantity of radiation	in ly/day															13
14 Mean potential evaporation	in mm	147	135	150	149	157	159	159	157	154	157	150	151	1825	14	14
15 Mean windspeed	in m/sec															15
16 Mean predominent direction of the wind																16

+ 9.00 a.m.

50 Station/Country Nauru / Nauru

Location 0°32'S / 166°55'E   Height above sealevel 27 m   Climate symbol: Köppen Af   Troll V,1

		J	F	M	A	M	J	J	A	S	O	N	D	year	P	
1 Mean daily temperature	in °C	27,2	27,5	27,8	28,1	28,1	27,8	27,5	27,5	28,1	27,8	27,8	27,2	27,2	15	1
2 Mean daily maximum temperature	in °C	31,1	31,1	31,7	32,2	32,2	32,2	31,7	31,7	32,2	32,2	32,2	31,7	31,7	15	2
3 Mean daily minimum temperature	in °C	23,3	23,9	23,9	23,9	23,9	23,3	23,3	23,3	23,9	23,3	23,3	23,3	23,3	15	3
4 Absolute maximum temperature	in °C	34,4	33,9	34,4	34,4	35,0	34,4	34,4	34,4	35,0	34,4	35,0	34,4	35,0	15	4
5 Absolute minimum temperature	in °C	20,0	18,9	20,6	20,6	20,0	18,9	20,6	18,9	18,9	17,2	19,4	19,4	17,2	15	5
6 Mean relative humidity	in %	75	74	74	72	71	71	72	70	69	68	70	72	72	18	6
7 Mean precipitation	in mm	315	206	180	94	53	99	155	193	122	99	152	239	1907	20	7
8 Maximum precipitation	in mm															8
9 Minimum precipitation	in mm															9
10 Maximum precipitation in 24 h	in mm	165	99	241	66	61	114	76	264	137	94	104	254	264	16	10
11 Mean number of days with precipitation	>0,25 mm	15	11	9	6	5	8	11	10	6	5	7	13	106	10	11
12 Mean duration of sunshine	in h															12
13 Mean quantity of radiation	in ly/day															13
14 Mean potential evaporation	in mm	149	136	152	148	154	148	151	151	148	153	148	152	1790	15	14
15 Mean windspeed	in m/sec															15
16 Mean predominent direction of the wind																16

51 Station/Country Rabaul (New Britain) / Territory of New Guinea

Location 4°13'S / 152°15'E   Height above sealevel 13 m   Climate symbol: Köppen Af   Troll V,1

		J	F	M	A	M	J	J	A	S	O	N	D	year	P	
1 Mean daily temperature	in °C	27,5	27,5	27,5	27,5	27,5	27,5	27,2	27,0	27,8	28,1	27,8	27,5	27,5	20	1
2 Mean daily maximum temperature	in °C	32,2	32,2	32,2	32,2	32,2	32,2	31,7	31,7	32,8	33,3	32,8	32,2	32,2	19	2
3 Mean daily minimum temperature	in °C	22,8	22,8	22,8	22,8	22,8	22,8	22,8	22,2	22,8	22,8	22,8	22,8	22,8	20	3
4 Absolute maximum temperature	in °C	36,1	35,0	35,0	35,0	34,4	34,4	35,0	36,1	36,7	37,8	37,2	37,2	37,8	19	4
5 Absolute minimum temperature	in °C	21,1	20,6	20,0	21,1	20,6	19,4	20,0	18,3	20,6	20,6	20,0	20,6	18,3	20	5
6 Mean relative humidity	in %	77	76	77	77	75	74	74	73	69	70	73	76	74	21	6
7 Mean precipitation	in mm	376	264	259	254	132	84	137	94	89	130	180	257	2256	24	7
8 Maximum precipitation	in mm															8
9 Minimum precipitation	in mm															9
10 Maximum precipitation in 24 h	in mm	262	206	211	132	137	124	160	97	79	130	91	114	262	24	10
11 Mean number of days with precipitation	>0,25 mm	14	13	15	11	6	6	8	8	5	8	10	13	117	10	11
12 Mean duration of sunshine	in h															12
13 Mean quantity of radiation	in ly/day															13
14 Mean potential evaporation	in mm	152	137	151	145	148	142	143	144	145	154	152	154	1767		14
15 Mean windspeed	in m/sec															15
16 Mean predominent direction of the wind																16

52 Station/Country Tongareva / U.S. and New Zealand

Location 9°01'S / 158°03'W   Height above sealevel 2 m   Climate symbol: Köppen Af   Troll V,1

		J	F	M	A	M	J	J	A	S	O	N	D	year	P	
1 Mean daily temperature	in °C	28,0	28,0	28,6	28,9	28,6	28,3	27,8	27,8	28,1	28,3	28,3	28,0	28,3	5	1
2 Mean daily maximum temperature	in °C	30,6	30,6	30,6	31,1	30,6	30,0	29,4	29,4	30,0	30,6	31,1	30,6	30,6	5	2
3 Mean daily minimum temperature	in °C	25,6	25,6	26,7	26,7	26,7	26,7	26,1	26,1	26,1	26,1	25,6	25,6	26,1	5	3
4 Absolute maximum temperature	in °C	33,9	34,4	33,9	33,9	32,8	31,1	32,8	32,2	33,3	33,3	34,4	34,4	34,4	5	4
5 Absolute minimum temperature	in °C	22,8	22,8	23,3	23,9	23,3	22,2	22,8	22,8	23,3	22,2	22,8	22,2	22,2	5	5
6 Mean relative humidity	in %	81	82	82	82	83	83	80	80	79	80	78	81	81	5	6
7 Mean precipitation	in mm	325	267	211	163	122	130	175	203	84	178	178	231	2267	5	7
8 Maximum precipitation	in mm															8
9 Minimum precipitation	in mm															9
10 Maximum precipitation in 24 h	in mm	94	112	74	74	66	51	145	152	48	119	191	119	191	5	10
11 Mean number of days with precipitation	>0,25 mm	14	14	14	10	10	10	7	9	8	10	11	15	132	5	11
12 Mean duration of sunshine	in h															12
13 Mean quantity of radiation	in ly/day															13
14 Mean potential evaporation	in mm	175	157	172	168	170	160	164	167	167	173	171	175	2019		14
15 Mean windspeed	in m/sec															15
16 Mean predominent direction of the wind																16

53 Station / Country Saipan / Belau

Location 15°14'N / 145°46'E Height above sealevel 222 m Climate symbol: Köppen Af Troll V,1

		in	J	F	M	A	M	J	J	A	S	O	N	D	year	P	
1	Mean daily temperature	in °C	24,7	24,7	25,3	25,9	26,1	26,4	25,9	26,4	25,9	26,1	26,1	25,6	25,9	8	1
2	Mean daily maximum temperature	in °C	27,2	27,2	27,8	28,3	28,9	28,9	28,3	28,9	28,3	28,3	28,3	27,8	28,3	8	2
3	Mean daily minimum temperature	in °C	22,2	22,2	22,8	23,3	23,3	23,9	23,3	23,9	23,3	23,3	23,9	23,3	23,3	8	3
4	Absolute maximum temperature	in °C	29,4	29,4	30,0	30,6	30,6	30,6	31,1	30,6	31,7	30,6	29,4	29,4	31,7	8	4
5	Absolute minimum temperature	in °C	19,4	20,6	20,0	20,6	20,6	21,1	20,6	20,6	21,1	21,1	21,1	19,4	19,4	8	5
6	Mean relative humidity	in %	79	77	77	78	78	79	79	78	80	79	78	78	78	6	6
7	Mean precipitation	in mm	69	91	97	71	94	129	254	333	338	290	188	137	2091	19	7
8	Maximum precipitation	in mm															8
9	Minimum precipitation	in mm															9
10	Maximum precipitation in 24 h	in mm	36	53	155	71	135	94	208	226	137	155	333	76	333	19	10
11	Mean number of days with precipitation	>0,25 mm	12	11	13	14	13	17	23	23	22	22	19	19	208	13	11
12	Mean duration of sunshine	in h															12
13	Mean quantity of radiation	in ly / day															13
14	Mean potential evaporation	in mm	101	94	113	128	140	141	139	139	123	125	118	112	1473	3	14
15	Mean windspeed	in m / sec															15
16	Mean predominent direction of the wind																16

54 Station / Country Suva (Viti Levu) / Fiji

Location 18°08'S / 178°26'E Height above sealevel 6 m Climate symbol: Köppen Aw Troll V,1

		in	J	F	M	A	M	J	J	A	S	O	N	D	year	P	
1	Mean daily temperature	in °C	26,7	26,7	26,7	25,8	24,7	23,8	23,0	23,0	23,8	24,2	25,0	26,1	25,0	43	1
2	Mean daily maximum temperature	in °C	30,0	30,0	30,0	28,9	27,8	26,7	26,1	26,1	26,7	27,2	28,3	29,4	28,3	43	2
3	Mean daily minimum temperature	in °C	23,3	23,3	23,3	22,8	21,7	20,6	20,0	20,0	20,6	21,1	21,7	22,8	21,7	43	3
4	Absolute maximum temperature	in °C	34,4	34,4	34,4	34,4	34,4	34,4	34,4	34,4	34,4	34,4	34,4	34,4	35,0	15	4
5	Absolute minimum temperature	in °C	19,4	19,4	18,9	16,1	16,1	14,4	12,8	13,9	13,9	13,9	12,8	16,7	12,8	15	5
6	Mean relative humidity	in %	75	74	74	72	71	71	72	70	69	68	70	~ 72	72	18	6
7	Mean precipitation	in mm	315	206	180	94	53	99	155	191	312	99	152	239	2096	20	7
8	Maximum precipitation	in mm															8
9	Minimum precipitation	in mm															9
10	Maximum precipitation in 24 h	in mm	165	155	241	66	61	114	76	264	137	94	104	254	264	16	10
11	Mean number of days with precipitation	> 1,0 mm	18	18	21	19	16	13	14	15	16	15	15	18	198	25	11
12	Mean duration of sunshine	in h	195	168	171	156	124	129	152	130	144	161	162	208	1900	30	12
13	Mean quantity of radiation	in ly / day															13
14	Mean potential evaporation	in mm	144	132	138	116	101	85	81	84	89	105	117	139	1331	33	14
15	Mean windspeed	in m / sec															15
16	Mean predominent direction of the wind																16

55 Station / Country Manokwari / Indonesia

Location 0°52'S / 134°05'E Height above sealevel 19 m Climate symbol: Köppen Af Troll V,1

		in	J	F	M	A	M	J	J	A	S	O	N	D	year	P	
1	Mean daily temperature	in °C	26,1	26,1	26,4	26,4	26,7	26,4	26,4	26,7	26,7	27,0	27,0	26,7	26,5	5	1
2	Mean daily maximum temperature	in °C	29,4	29,4	29,4	29,4	30,0	29,4	29,4	30,0	30,0	30,6	30,6	30,0	29,8	5	2
3	Mean daily minimum temperature	in °C	22,8	22,8	23,3	23,3	23,3	23,3	23,3	23,3	23,3	23,3	23,3	23,3	23,2	5	3
4	Absolute maximum temperature	in °C	32,8	32,2	32,8	32,8	32,8	31,1	32,2	32,8	33,3	33,3	33,9	32,8	33,9	5	4
5	Absolute minimum temperature	in °C	20,0	21,1	20,6	21,1	21,1	21,1	20,0	21,1	21,1	21,1	21,1	21,1	20,0	5	5
6	Mean relative humidity	in %	81	79	80	84	79	82	81	79	79	82	81	85	81	5	6
7	Mean precipitation	in mm	287	249	338	277	203	193	145	140	124	109	163	269	2497	27	7
8	Maximum precipitation	in mm															8
9	Minimum precipitation	in mm															9
10	Maximum precipitation in 24 h	in mm	236	107	163	160	124	109	97	94	89	124	109	137	236	18	10
11	Mean number of days with precipitation	> 0,1 mm	16	13	16	14	11	12	10	10	9	9	10	14	144	27	11
12	Mean duration of sunshine	in h	155	123	155	177	217	219	195	226	216	217	195	167	2262	30	12
13	Mean quantity of radiation	in ly / day															13
14	Mean potential evaporation	in mm	129	118	130	128	136	128	134	132	134	140	137	136	1580	40	14
15	Mean windspeed	in m / sec															15
16	Mean predominent direction of the wind																16

56 Station / Country Mandang / Papua-New Guinea

Location 5°14'S / 145°45'E Height above sealevel 6 m Climate symbol: Köppen Af Troll V,1

		in	J	F	M	A	M	J	J	A	S	O	N	D	year	P	
1	Mean daily temperature	in °C	27,3	27,0	27,3	27,2	27,5	27,2	27,2	27,2	27,2	27,5	27,5	27,5	27,3	12	1
2	Mean daily maximum temperature	in °C	30,6	30,0	30,6	31,1	31,1	31,1	31,1	31,1	31,1	31,1	31,1	31,1	30,9	12	2
3	Mean daily minimum temperature	in °C	23,9	23,9	23,3	23,3	23,9	23,3	23,3	23,3	23,3	23,9	23,9	23,9	23,6	12	3
4	Absolute maximum temperature	in °C	34,4	33,3	33,9	33,9	35,6	35,6	36,7	33,3	33,9	33,9	33,9	33,9	36,7	20	4
5	Absolute minimum temperature	in °C	21,1	21,1	21,1	21,1	21,1	20,6	16,7	21,1	20,0	20,0	21,1	20,0	16,7	20	5
6	Mean relative humidity	in %	82	82	81	82	81	80	80	79	79	78	79	81	80	14	6
7	Mean precipitation	in mm	307	302	378	429	384	274	193	122	135	254	338	368	3484	20	7
8	Maximum precipitation	in mm															8
9	Minimum precipitation	in mm															9
10	Maximum precipitation in 24 h	in mm	150	132	178	251	262	244	124	135	226	185	330	152	330	22	10
11	Mean number of days with precipitation	>0,25 mm	15	14	17	16	15	10	10	8	9	10	12	16	152	10	11
12	Mean duration of sunshine	in h	155	123	155	177	217	219	195	226	216	217	195	167	2262	30	12
13	Mean quantity of radiation	in ly / day															13
14	Mean potential evaporation	in mm	148	133	147	142	147	142	145	146	144	152	147	151	1744	10	14
15	Mean windspeed	in m / sec															15
16	Mean predominent direction of the wind																16

284

57 Station/Country Kikori/Papua-New Guinea

Location 7°17'S/144°11'E   Height above sealevel 24 m   Climate symbol: Köppen Af   Troll V.1

			J	F	M	A	M	J	J	A	S	O	N	D	year	P	
1	Mean daily temperature	in °C	28,0	28,0	28,0	26,7	26,1	25,3	25,0	24,5	25,6	26,6	27,5	28,5	26,5	5	1
2	Mean daily maximum temperature	in °C	32,2	32,2	32,2	30,6	29,4	28,3	27,8	27,2	28,9	31,1	32,2	32,2	30,4	5	2
3	Mean daily minimum temperature	in °C	22,8	22,8	22,8	22,8	22,8	22,2	22,2	21,7	22,2	22,2	22,8	22,8	22,5	5	3
4	Absolute maximum temperature	in °C	37,2	36,7	37,8	35,0	35,0	35,0	32,2	32,8	34,4	35,0	35,6	37,8	37,8	21	4
5	Absolute minimum temperature	in °C	18,3	17,8	20,0	17,8	20,6	19,4	18,1	18,3	18,9	17,8	17,2	16,1	16,1	21	5
6	Mean relative humidity	in %	86	85	84	86	91	94	93	94	91	89	86	88	89	11	6
7	Mean precipitation	in mm	297	318	378	442	749	775	665	554	627	465	358	297	5923	25	7
8	Maximum precipitation	in mm															8
9	Minimum precipitation	in mm															9
10	Maximum precipitation in 24 h	in mm	115	115	165	142	216	236	241	175	256	297	211	119	297	16	10
11	Mean number of days with precipitation	>0,25 mm	14	14	16	17	22	22	24	22	22	17	13	13	216	10	11
12	Mean duration of sunshine	in h															12
13	Mean quantity of radiation	in ly/day															13
14	Mean potential evaporation	in mm	154	138	151	139	132	113	102	100	115	145	150	158	1597	5	14
15	Mean windspeed	in m/sec															15
16	Mean predominent direction of the wind		.														16

58 Station/Country Daru/Papua-New Guinea

Location 9°04'S/143°12'E   Height above sealevel 8 m   Climate symbol: Köppen Aw   Troll V.2

			J	F	M	A	M	J	J	A	S	O	N	D	year	P	
1	Mean daily temperature	in °C	27,8	27,5	27,5	27,2	27,0	25,9	25,3	25,6	25,8	27,2	27,8	27,8	26,8	16	1
2	Mean daily maximum temperature	in °C	31,1	30,6	30,6	30,0	28,9	27,8	27,2	27,8	28,3	30,0	31,1	31,1	29,5	16	2
3	Mean daily minimum temperature	in °C	24,4	24,4	24,4	24,4	25,0	23,9	23,3	23,3	23,3	24,4	24,4	24,4	24,1	16	3
4	Absolute maximum temperature	in °C	36,7	35,6	34,4	34,4	32,2	31,1	30,0	30,0	32,2	33,9	35,0	34,4	36,7	24	4
5	Absolute minimum temperature	in °C	20,0	18,9	20,6	18,3	21,1	18,3	17,8	17,2	18,3	20,0	17,8	17,2	17,2	24	5
6	Mean relative humidity	in %	82	83	84	82	81	82	83	81	78	76	77	79	81	20	6
7	Mean precipitation	in mm	302	264	318	320	239	97	76	56	46	58	117	206	2099	30	7
8	Maximum precipitation	in mm	155	127	140	130	145	101	135	97	56	64	163	145	163	15	8
9	Minimum precipitation	in mm															9
10	Maximum precipitation in 24 h	in mm															10
11	Mean number of days with precipitation	>0,25 mm	12	13	13	13	12	10	7	7	5	3	6	7	108	10	11
12	Mean duration of sunshine	in h															12
13	Mean quantity of radiation	in ly/day															13
14	Mean potential evaporation	in mm	153	139	149	141	139	125	112	113	121	148	148	161	1649	5	14
15	Mean windspeed	in m/sec															15
16	Mean predominent direction of the wind																16

59 Station/Country Samarai/Papua-New Guinea

Location 10°37'S/150°40'E   Height above sealevel 6 m   Climate symbol: Köppen Af   Troll V.1

			J	F	M	A	M	J	J	A	S	O	N	D	year	P	
1	Mean daily temperature	in °C	27,8	28,1	27,5	26,9	26,4	25,6	25,3	25,0	25,6	25,8	26,7	27,5	26,5	19	1
2	Mean daily maximum temperature	in °C	30,6	31,1	30,6	30,0	28,9	27,8	27,2	27,8	28,3	28,3	29,4	30,6	29,1	19	2
3	Mean daily minimum temperature	in °C	25,0	25,0	24,4	23,9	23,9	23,3	23,3	22,8	23,3	23,3	23,9	24,4	23,9	22	3
4	Absolute maximum temperature	in °C	38,9	37,2	36,7	34,4	35,6	32,8	32,2	32,2	32,8	35,0	36,7	37,2	38,9	25	4
5	Absolute minimum temperature	in °C	20,0	21,7	20,6	20,6	20,0	20,6	20,0	17,8	19,4	20,0	18,9	20,0	17,8	25	5
6	Mean relative humidity	in %	75	76	77	79	82	84	83	82	82	79	78	76	80	24	6
7	Mean precipitation	in mm	178	198	254	249	305	287	206	218	256	221	213	155	2740	29	7
8	Maximum precipitation	in mm															8
9	Minimum precipitation	in mm															9
10	Maximum precipitation in 24 h	in mm	91	86	84	132	150	267	101	117	165	244	165	170	267	17	10
11	Mean number of days with precipitation	>0,25 mm	10	10	13	10	13	14	13	13	10	12	10	7	135	10	11
12	Mean duration of sunshine	in h															12
13	Mean quantity of radiation	in ly/day															13
14	Mean potential evaporation	in mm	168	150	158	146	140	127	118	118	132	151	153	168	1729	9	14
15	Mean windspeed	in m/sec															15
16	Mean predominent direction of the wind																16

60 Station/Country Bwagaoia/Papua-New Guinea

Location 10°44'S/152°50'E   Height above sealevel 30 m   Climate symbol: Köppen Af   Troll V.1

			J	F	M	A	M	J	J	A	S	O	N	D	year	P	
1	Mean daily temperature	in °C	27,0	27,5	27,2	26,4	26,1	25,8	25,8	25,8	25,4	27,5	27,5	27,8	26,7	7	1
2	Mean daily maximum temperature	in °C	31,1	31,7	31,1	29,4	28,9	28,3	28,3	28,3	28,9	30,6	31,1	31,7	30,0	7	2
3	Mean daily minimum temperature	in °C	22,8	23,3	23,3	23,3	23,3	23,3	23,3	23,3	23,9	24,4	23,9	23,9	23,5	7	3
4	Absolute maximum temperature	in °C	33,3	33,9	33,9	33,9	32,8	32,2	32,2	32,2	31,7	35,0	33,3	35,0	35,0	12	4
5	Absolute minimum temperature	in °C	16,1	17,8	16,1	16,7	15,6	15,6	15,6	17,8	18,3	20,0	18,3	15,6	15,6	12	5
6	Mean relative humidity	in %	79	80	80	82	82	83	81	81	79	79	80	78	80	11	6
7	Mean precipitation	in mm	282	340	267	262	310	226	213	246	218	269	264	211	3108	19	7
8	Maximum precipitation	in mm															8
9	Minimum precipitation	in mm															9
10	Maximum precipitation in 24 h	in mm	109	178	160	152	112	109	135	155	112	145	203	183	203	14	10
11	Mean number of days with precipitation	>0,25 mm	14	13	13	12	12	13	11	11	10	12	12	9	142	10	11
12	Mean duration of sunshine	in h															12
13	Mean quantity of radiation	in ly/day															13
14	Mean potential evaporation	in mm	150	140	148	128	120	115	119	121	129	153	151	162	1636		14
15	Mean windspeed	in m/sec															15
16	Mean predominent direction of the wind																16

POLAR REGIONS

ARCTIC OCEAN

1 Drifting Station „Alpha"
2 North Pole

CANADA

3 Alert

GREENLAND /DENMARK

4 Nord
6 Myggbukta
8 Upernavik
9 Umanak

10 Scoresbysund (Cape Tobin)
11 Godhavn
12 Jakobshavn
13 Angmagssalik
14 Godthåb
15 Ivigtut
16 Nanortalik

SPITSBERGEN /NORWAY

5 Isfjord Radio

BEAR-ISLAND /NORWAY

7 Björnöya

ANTARCTICA

1 South Pole
2 Plateau Station
3 Wostok
4 General Bergrano
5 McMurdo
6 Halley Bay
7 Novolazarewskaja
8 Norway Station /Sanae
9 Syowa Base
10 Mawson
11 Dumont d'Urville
12 Mirnyj
13 Argentine Island
14 Wilkes
15 Hope Bay (Bahia Esperanza)
16 Orkadas /South Orkney Islands

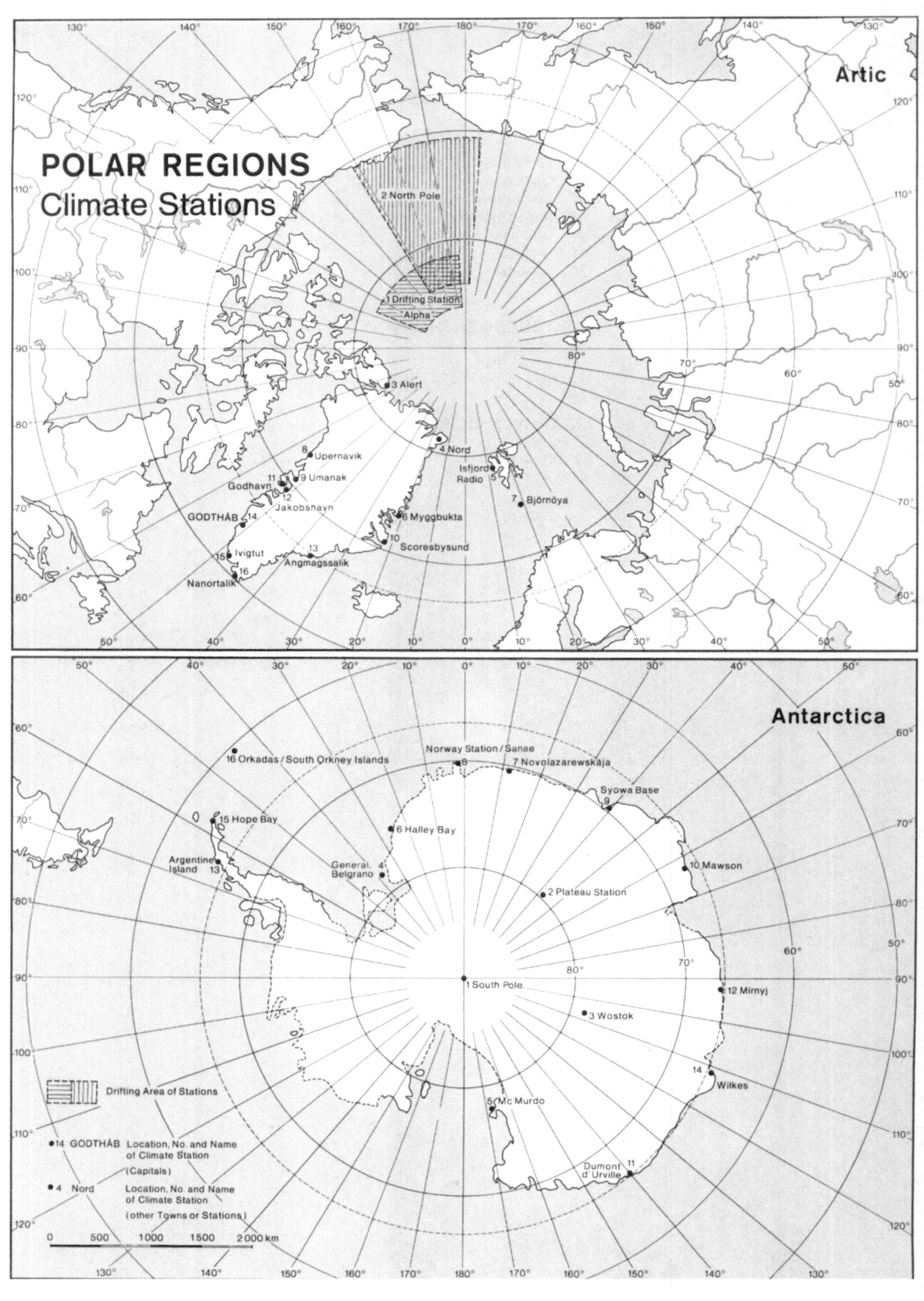

POLAR REGIONS
Climate Stations

Artic

2 North Pole

1 Drifting Station Alpha

3 Alert

4 Nord

8 Upernavik

9 Umanak

Isfjord Radio 5

11 Godhavn

12

Jakobshavn

7 Björnöya

GODTHÅB 14

6 Myggbukta

10 Scoresbysund

15 Ivigtut

13

Angmagssalik

16

Nanortalik

Antarctica

16 Orkadas / South Orkney Islands

Norway Station / Sanae

8

7 Novolazarewskaja

Syowa Base

9

15 Hope Bay

6 Halley Bay

Argentine
Island 13

General 4
Belgrano

10 Mawson

2 Plateau Station

1 South Pole

12 Mirnyj

3 Wostok

14

Wilkes

Drifting Area of Stations

5 Mc Murdo

- 14 GODTHÅB Location, No. and Name
of Climate Station
(Capitals)

Dumont
d'Urville 11

- 4 Nord Location, No. and Name
of Climate Station
(other Towns or Stations)

0   500   1000   1500   2 000 km

POLAR REGIONS

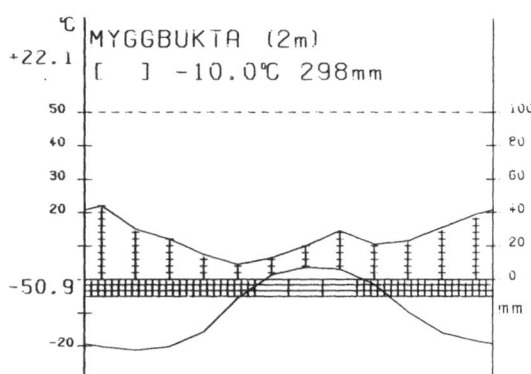

MYGGBUKTA (2m)
[ ] -10.0°C 298mm
+22.1
-50.9

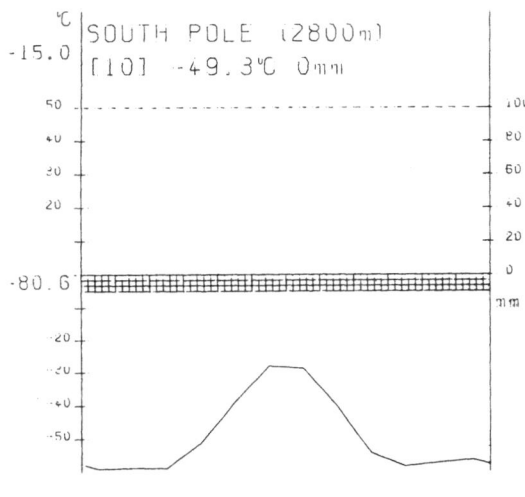

SOUTH POLE (2800m)
[10] -49.3°C 0mm
-15.0
-80.6

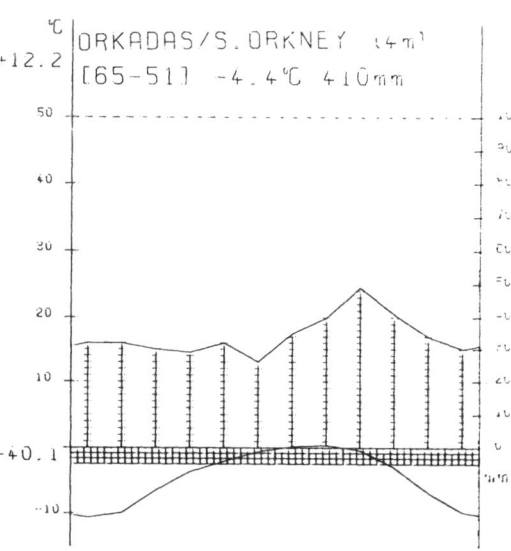

ORKADAS/S.ORKNEY (4m)
[65-51] -4.4°C 410mm
+12.2
-40.1

5 Station / Country   Isfjord Radio / Spitsbergen / Norway

Location   78°08'N / 13°38'E   Height above sealevel   9 m   Climate symbol: Köppen   ET   Troll I,2

		J	F	M	A	M	J	J	A	S	O	N	D	year	P	
1	Mean daily temperature	in °C	−10,9	−11,2	−12,1	−8,8	−3,3	1,7	4,5	4,2	1,1	−2,7	−6,2	−9,0	−4,4	23	1
2	Mean daily maximum temperature	in °C															2
3	Mean daily minimum temperature	in °C															3
4	Absolute maximum temperature	in °C	3,5	4,4	3,8	5,6	13,1	12,5	15,6	14,3	12,0	8,5	6,2	4,1	15,6	23	4
5	Absolute minimum temperature	in °C	−30,9	−32,2	−29,0	−28,2	−19,6	−8,2	−1,3	−2,0	−9,0	−15,5	−26,9	−28,1	−32,2	23	5
6	Mean relative humidity	in %	83	83	85	83	83	86	89	87	85	82	82	84		23	6
7	Mean precipitation	in mm	29	30	33	17	20	24	30	38	38	46	39	34	378	23	7
8	Maximum precipitation	in mm															8
9	Minimum precipitation	in mm															9
10	Maximum precipitation in 24 h	in mm															10
11	Mean number of days with precipitation		14	13	14	12	11	11	13	14	14	15	14	13	156	23	11
12	Mean duration of sunshine	in h															12
13	Mean quantity of radiation	in ly / day															13
14	Mean potential evaporation	in mm															14
15	Mean windspeed	in m / sec	8,8	9,1	8,3	7,5	6,3	5,1	5,3	5,6	6,5	7,5	8,3	9,6	7,3	23	15
16	Mean predominent direction of the wind		NE	NE	NE	NE	NE	NE,S	S,NE	NE,S	NE,S	NE	NE	NE		23	16

6 Station / Country   Myggbukta / Greenland / Denmark

Location   73°29'N / 21°34'E   Height above sealevel   2 m   Climate symbol: Köppen   ET   Troll I,1

		J	F	M	A	M	J	J	A	S	O	N	D	year	P	
1	Mean daily temperature	in °C	−20,2	−21,3	−20,3	−15,7	−5,6	1,4	3,7	3,1	−1,4	−9,6	−16,0	−18,5	−10,0	+	1
2	Mean daily maximum temperature	in °C															2
3	Mean daily minimum temperature	in °C															3
4	Absolute maximum temperature	in °C	1,2	5,0	7,1	7,0	12,4	22,1	21,2	20,3	14,5	10,2	7,0	3,2	22,1	+	4
5	Absolute minimum temperature	in °C	−50,9	−47,5	−47,5	−37,7	−30,2	−10,3	−6,3	−6,7	−24,7	−30,3	−42,5	−42,5	−50,9	+	5
6	Mean relative humidity	in %															6
7	Mean precipitation	in mm	44	30	24	15	9	13	20	29	21	23	31	39	298	+	7
8	Maximum precipitation	in mm															8
9	Minimum precipitation	in mm															9
10	Maximum precipitation in 24 h	in mm	61	43	46	35	9	38	26	29	30	29	28	43	61	+	10
11	Mean number of days with precipitation		10	8	9	8	7	6	7	7	8	9	9	10	96	+	11
12	Mean duration of sunshine	in h															12
13	Mean quantity of radiation	in ly / day															13
14	Mean potential evaporation	in mm															14
15	Mean windspeed	in m / sec	2,7	2,4	2,2	1,8	1,6	1,9	2,1	2,1	2,3	2,5	2,5	2,5	2,2	+	15
16	Mean predominent direction of the wind		var	var	var	var	var	var	SE	SE,E	N,SE	var	var	var		+	16

+ 19−20 years

7 Station / Country   Björnöya / Baer-Island / Norway

Location   74°31'N / 19°01'E   Height above sealevel   14 m   Climate symbol: Köppen   ET   Troll I,2

		J	F	M	A	M	J	J	A	S	O	N	D	year	P	
1	Mean daily temperature	in °C	−6,8	−6,9	−7,5	−5,2	−1,4	1,8	4,2	4,3	3,0	0,5	−2,4	−5,3	−1,8	20	1
2	Mean daily maximum temperature	in °C															2
3	Mean daily minimum temperature	in °C															3
4	Absolute maximum temperature	in °C	5,1	3,5	6,2	5,6	16,5	23,6	19,1	14,4	13,9	10,0	6,1	6,4	23,6	20	4
5	Absolute minimum temperature	in °C	−29,8	−29,1	−29,7	−22,6	−17,8	−8,4	−6,4	−1,6	−6,3	−13,1	−21,0	−25,2	−29,8	20	5
6	Mean relative humidity	in %	89	89	89	90	90	92	93	93	92	89	88	90	90	20	6
7	Mean precipitation	in mm	30	25	26	20	20	26	25	37	50	42	27	29	357	20	7
8	Maximum precipitation	in mm															8
9	Minimum precipitation	in mm															9
10	Maximum precipitation in 24 h	in mm															10
11	Mean number of days with precipitation		19	19	20	17	16	15	15	17	22	22	20	21	223	20	11
12	Mean duration of sunshine	in h															12
13	Mean quantity of radiation	in ly / day															13
14	Mean potential evaporation	in mm															14
15	Mean windspeed	in m / sec	8,8	8,8	7,7	6,7	6,1	5,8	5,6	6,1	7,0	7,5	8,3	8,5	7,2	20	15
16	Mean predominent direction of the wind		NE	NE	NE,E	E	NE,SE	E	E	E	E	E,SE	NE	NE		20	16

8 Station / Country   Upernavik / Greenland / Denmark

Location   72°47'N / 56°10'W   Height above sealevel   35 m   Climate symbol: Köppen   ET   Troll I,3

		J	F	M	A	M	J	J	A	S	O	N	D	year	P	
1	Mean daily temperature	in °C	−17,0	−19,6	−18,4	−12,3	−2,4	3,0	6,0	5,7	1,5	−3,3	−7,3	−12,6	−6,4	+	1
2	Mean daily maximum temperature	in °C															2
3	Mean daily minimum temperature	in °C															3
4	Absolute maximum temperature	in °C	8,0	9,9	11,5	8,0	18,0	16,0	19,0	16,5	14,7	9,0	8,0	7,4	19,0	+	4
5	Absolute minimum temperature	in °C	−38,0	−40,0	−39,0	−35,0	−25,0	−13,0	−7,0	−9,0	−13,0	−17,0	−25,0	−34,0	−40,0	+	5
6	Mean relative humidity	in %															6
7	Mean precipitation	in mm	9	11	9	11	11	9	21	24	30	23	17	11	186	+	7
8	Maximum precipitation	in mm															8
9	Minimum precipitation	in mm															9
10	Maximum precipitation in 24 h	in mm	6	11	14	14	18	13	34	27	35	29	37	10	37	+	10
11	Mean number of days with precipitation		14	11	12	9	11	4	5	5	9	3	15	13	120	+	11
12	Mean duration of sunshine	in h															12
13	Mean quantity of radiation	in ly / day															13
14	Mean potential evaporation	in mm															14
15	Mean windspeed	in m / sec	1,9	1,8	1,6	1,7	1,7	1,6	1,7	1,8	1,8	2,0	2,1	2,0	1,8	+	15
16	Mean predominent direction of the wind		E,NE	var	var	var	var	var	var	var	var	E	E,NE	E,NE		+	16

+ 10−25 years

9 Station/Country Umanak/Greenland/Denmark

Location 70°41'N/52°07'W   Height above sealevel 8 m   Climate symbol: Köppen ET   Troll I,3

		J	F	M	A	M	J	J	A	S	O	N	D	year	P	
1 Mean daily temperature	in °C	−12,9	−15,3	−14,0	−10,0	−1,0	4,8	7,8	7,0	2,7	−2,0	−5,4	−9,2	−4,0	+	1
2 Mean daily maximum temperature	in °C															2
3 Mean daily minimum temperature	in °C															3
4 Absolute maximum temperature	in °C	9,2	10,0	10,5	9,5	14,3	18,0	18,0	16,3	16,0	17,0	11,0	9,2	18,0	+	4
5 Absolute minimum temperature	in °C	−32,0	−35,0	−35,2	−30,0	−20,5	−7,0	−1,5	−2,0	−6,3	−12,5	−18,0	−28,8	−35,2	+	5
6 Mean relative humidity	in %															6
7 Mean precipitation	in mm	25	15	12	13	12	12	12	12	21	18	25	24	201	+	7
8 Maximum precipitation	in mm															8
9 Minimum precipitation	in mm															9
10 Maximum precipitation in 24 h	in mm	17	16	11	9	11	17	13	19	20	15	13	12	20	+	10
11 Mean number of days with precipitation																11
12 Mean duration of sunshine	in h															12
13 Mean quantity of radiation	in ly/day															13
14 Mean potential evaporation	in mm															14
15 Mean windspeed	in m/sec	2,1	1,9	1,4	1,4	1,5	1,7	1,3	1,7	2,2	2,5	2,5	2,2	1,9	+	15
16 Mean predominent direction of the wind																16

+ 19−25 years

10 Station/Country Scoresbysund (Cape Tobin)/Greenland/Denmark

Location 70°25'N/21°58'W   Height above sealevel 17 m   Climate symbol: Köppen ET   Troll I,2

		J	F	M	A	M	J	J	A	S	O	N	D	year	P	
1 Mean daily temperature	in °C	−15,3	−16,8	−16,1	−11,7	−2,9	2,4	4,7	3,7	0,8	−5,6	−10,7	−13,7	−8,7	+	1
2 Mean daily maximum temperature	in °C															2
3 Mean daily minimum temperature	in °C															3
4 Absolute maximum temperature	in °C	3,0	10,8	4,5	7,0	11,8	15,8	16,8	13,0	15,0	12,6	5,0	4,5	16,8	+	4
5 Absolute minimum temperature	in °C	−43,7	−41,6	−39,5	−32,5	−25,6	−7,5	−5,0	−4,4	−11,5	−19,2	−33,0	−39,0	−43,7	+	5
6 Mean relative humidity	in %															6
7 Mean precipitation	in mm	29	29	23	21	12	26	38	33	53	56	44	64	428	+	7
8 Maximum precipitation	in mm															8
9 Minimum precipitation	in mm															9
10 Maximum precipitation in 24 h	in mm	17	23	17	16	13	34	30	23	38	114	43	70	114	+	10
11 Mean number of days with precipitation		10	8	7	7	4	5	6	6	7	11	11	13	95	+	11
12 Mean duration of sunshine	in h															12
13 Mean quantity of radiation	in ly/day															13
14 Mean potential evaporation	in mm															14
15 Mean windspeed	in m/sec	1,7	1,8	1,8	1,4	1,1	1,2	1,2	1,5	1,8	1,7	1,7	1,6	1,5	+	15
16 Mean predominent direction of the wind		var	var	var	var	var	var	var	var	var	var	var	var		+	16

+ 17−26 years

11 Station/Country Godhavn/Greenland/Denmark

Location 69°14'N/53°31'W   Height above sealevel 11 m   Climate symbol: Köppen ET   Troll I,3

		J	F	M	A	M	J	J	A	S	O	N	D	year	P	
1 Mean daily temperature	in °C	−11,8	−13,9	−12,7	−7,3	0,0	5,1	8,0	7,2	3,2	−2,2	−5,5	−8,9	−3,2	+	1
2 Mean daily maximum temperature	in °C															2
3 Mean daily minimum temperature	in °C															3
4 Absolute maximum temperature	in °C	9,0	8,8	9,7	10,8	18,3	17,9	17,3	16,0	14,6	11,6	7,6	7,5	18,3	+	4
5 Absolute minimum temperature	in °C	−33,2	−32,0	−33,8	−25,5	−17,5	−6,7	−0,7	−0,2	−6,1	−12,1	−19,8	−26,7	−33,8	+	5
6 Mean relative humidity	in %															6
7 Mean precipitation	in mm	12	22	15	15	32	30	47	45	60	53	38	22	391	+	7
8 Maximum precipitation	in mm															8
9 Minimum precipitation	in mm															9
10 Maximum precipitation in 24 h	in mm	17	32	22	17	40	22	37	42	102	54	31	21	102	+	10
11 Mean number of days with precipitation																11
12 Mean duration of sunshine	in h															12
13 Mean quantity of radiation	in ly/day															13
14 Mean potential evaporation	in mm															14
15 Mean windspeed	in m/sec															15
16 Mean predominent direction of the wind																16

+ 14−21 years

12 Station/Country Jakobshavn/Greenland/Denmark

Location 69°13'N/51°05'W   Height above sealevel 31 m   Climate symbol: Köppen ET   Troll I,3

		J	F	M	A	M	J	J	A	S	O	N	D	year	P	
1 Mean daily temperature	in °C	−31,5	−14,4	−12,5	−7,8	0,6	5,9	8,2	6,8	2,5	−3,7	−7,6	−10,8	−3,9	+	1
2 Mean daily maximum temperature	in °C															2
3 Mean daily minimum temperature	in °C															3
4 Absolute maximum temperature	in °C	8,3	9,1	10,2	12,3	17,1	20,8	19,4	18,9	17,0	12,5	10,0	8,2	20,8	+	4
5 Absolute minimum temperature	in °C	−36,6	−34,8	−36,4	−28,2	−20,5	−5,8	0,0	−3,2	−11,1	−21,4	−26,5	−28,8	−36,4	+	5
6 Mean relative humidity	in %															6
7 Mean precipitation	in mm	10	13	14	15	20	19	35	34	41	29	21	18	269	+	7
8 Maximum precipitation	in mm															8
9 Minimum precipitation	in mm															9
10 Maximum precipitation in 24 h	in mm	10	13	22	16	16	43	39	34	24	27	29	32	43	+	10
11 Mean number of days with precipitation		10	10	9	10	9	7	8	9	12	10	11	9	115	+	11
12 Mean duration of sunshine	in h															12
13 Mean quantity of radiation	in ly/day															13
14 Mean potential evaporation	in mm															14
15 Mean windspeed	in m/sec	3,2	2,9	2,7	2,7	2,3	2,4	1,9	2,1	2,6	3,1	3,3	3,4	2,7	+	15
16 Mean predominent direction of the wind		E.SE	E.SE	E.SE	E.SE	var	W.N	var	var	var	E.SE	E.SE	E.SE		+	16

+ 19−25 years

13 Station/Country: Angmagssalik/Greenland/Denmark

Location 65°37'N/37°39'W Height above sealevel 29 m Climate symbol: Köppen ET Troll I,3

		J	F	M	A	M	J	J	A	S	O	N	D	year	P	
1 Mean daily temperature	in °C	−6,8	−7,2	−5,7	−2,7	2,1	5,8	7,4	6,6	4,2	−0,1	−3,1	−5,1	−0,4	+	1
2 Mean daily maximum temperature	in °C															2
3 Mean daily minimum temperature	in °C															3
4 Absolute maximum temperature	in °C	6,6	9,0	8,5	12,2	17,9	25,3	22,9	19,9	20,2	14,9	10,2	7,2	25,3	+	4
5 Absolute minimum temperature	in °C	−28,7	−26,0	−22,5	−20,3	−13,7	−8,8	−3,5	−3,5	−7,1	−18,3	−18,1	−21,9	−28,7	+	5
6 Mean relative humidity	in %															6
7 Mean precipitation	in mm	58	82	62	53	54	44	35	62	76	90	86	68	770	+	7
8 Maximum precipitation	in mm															8
9 Minimum precipitation	in mm															9
10 Maximum precipitation in 24 h	in mm	44	67	32	83	38	38	40	45	79	71	70	84	84	+	10
11 Mean number of days with precipitation		12	13	13	11	10	9	8	9	11	13	13	12	134	+	11
12 Mean duration of sunshine	in h															12
13 Mean quantity of radiation	in ly/day															13
14 Mean potential evaporation	in mm															14
15 Mean windspeed	in m/sec	2,0	1,7	1,6	1,5	1,3	1,4	1,3	1,1	1,4	1,6	1,7	1,8	1,5	+	15
16 Mean predominent direction of the wind		var	var	var	var	var	var	var	var	var	var	var	var		+	16

+ 16−25 years

14 Station/Country: Godthåb/Greenland/Denmark

Location 64°10'N/51°45'W Height above sealevel 20 m Climate symbol: Köppen ET Troll I,3

		J	F	M	A	M	J	J	A	S	O	N	D	year	P	
1 Mean daily temperature	in °C	−7,7	−7,3	−5,8	−3,5	2,1	5,7	7,6	6,9	4,1	−0,3	−3,8	−6,2	−0,7	+	1
2 Mean daily maximum temperature	in °C															2
3 Mean daily minimum temperature	in °C															3
4 Absolute maximum temperature	in °C	12,0	11,6	14,2	13,4	18,5	23,0	20,2	19,0	20,4	13,8	12,0	12,2	23,0	+	4
5 Absolute minimum temperature	in °C	−25,2	−26,0	−25,2	−22,0	−17,8	−4,3	−6,0	−5,2	−7,4	−13,9	−16,0	−23,0	−26,0	+	5
6 Mean relative humidity	in %															6
7 Mean precipitation	in mm	26	24	18	25	29	46	59	69	84	71	44	20	515	+	7
8 Maximum precipitation	in mm															8
9 Minimum precipitation	in mm															9
10 Maximum precipitation in 24 h	in mm	50	32	49	27	31	51	35	81	67	63	74	44	81	+	10
11 Mean number of days with precipitation		8	7	6	6	6	8	10	10	12	10	9	7	97	+	11
12 Mean duration of sunshine	in h															12
13 Mean quantity of radiation	in ly/day															13
14 Mean potential evaporation	in mm															14
15 Mean windspeed	in m/sec	3,1	3,3	3,0	2,8	2,2	2,2	2,1	2,2	2,4	2,8	3,0	3,0	2,7	+	15
16 Mean predominent direction of the wind		E.NE	NE.E	NE.E	×××	N.SW	var	var	var	var	×	××	E.NE		+	16

+ 19−25 years   × E,SW,NE   ×× E,NE,SW   ××× NE,SW,N

15 Station/Country: Ivigtut/Greenland/Denmark

Location 61°12'N/48°10'W Height above sealevel 30 m Climate symbol: Köppen ET Troll I,3

		J	F	M	A	M	J	J	A	S	O	N	D	year	P	
1 Mean daily temperature	in °C	−5,4	−4,5	−2,8	0,2	5,3	8,5	9,8	8,8	5,8	1,6	−1,8	−4,0	1,8	+	1
2 Mean daily maximum temperature	in °C															2
3 Mean daily minimum temperature	in °C															3
4 Absolute maximum temperature	in °C	12,0	14,5	14,0	16,5	19,5	23,1	22,4	19,2	21,0	17,0	17,0	15,5	23,1	+	4
5 Absolute minimum temperature	in °C	−21,8	−22,2	−23,4	−19,1	−11,0	−2,6	1,4	−0,6	−7,2	−9,8	−16,4	−18,3	−23,4	+	5
6 Mean relative humidity	in %															6
7 Mean precipitation	in mm	92	129	87	79	89	96	82	97	162	172	148	77	1038	+	7
8 Maximum precipitation	in mm															8
9 Minimum precipitation	in mm															9
10 Maximum precipitation in 24 h	in mm	72	169	59	80	60	168	73	75	89	128	99	73	169	+	10
11 Mean number of days with precipitation		11	11	11	9	9	7	10	10	12	11	11	10	121	+	11
12 Mean duration of sunshine	in h															12
13 Mean quantity of radiation	in ly/day															13
14 Mean potential evaporation	in mm															14
15 Mean windspeed	in m/sec	1,5	1,6	1,2	1,2	1,1	1,1	0,8	0,8	0,9	0,9	1,1	1,2	1,1	+	15
16 Mean predominent direction of the wind		var	var	var	var	var	var	var	var	var	var	var	var		+	16

+ 19−25 years

16 Station/Country: Nanortalik/Greenland/Denmark

Location 60°08'N/45°11'W Height above sealevel 7 m Climate symbol: Köppen ET Troll I,3

		J	F	M	A	M	J	J	A	S	O	N	D	year	P	
1 Mean daily temperature	in °C	−3,3	−2,4	−1,7	0,7	4,0	5,5	6,5	7,0	5,8	2,7	−1,0	−2,2	1,8	+	1
2 Mean daily maximum temperature	in °C															2
3 Mean daily minimum temperature	in °C															3
4 Absolute maximum temperature	in °C	11,5	11,6	10,5	15,0	16,7	20,1	18,5	17,1	16,7	14,7	13,7	17,4	20,1	+	4
5 Absolute minimum temperature	in °C	−17,2	−13,4	−14,3	−14,3	−8,5	−3,1	−2,9	−0,7	−2,8	−6,0	−10,7	−13,2	−17,2	+	5
6 Mean relative humidity	in %															6
7 Mean precipitation	in mm	64	71	41	59	45	80	53	92	119	125	94	52	895	+	7
8 Maximum precipitation	in mm															8
9 Minimum precipitation	in mm															9
10 Maximum precipitation in 24 h	in mm	42	61	25	40	28	104	30	51	143	73	75	38	143	+	10
11 Mean number of days with precipitation		9	11	11	10	10	7	11	10	14	12	10	11	126	+	11
12 Mean duration of sunshine	in h															12
13 Mean quantity of radiation	in ly/day															13
14 Mean potential evaporation	in mm															14
15 Mean windspeed	in m/sec	3,8	3,8	2,7	2,9	2,5	2,5	2,1	2,4	2,6	3,0	3,2	3,3	2,9	+	15
16 Mean predominent direction of the wind		N.W	N.NE.W	N.NE.W	N.W	W.N.NE	W.N	N.W	W.N	N.W	N.W	N.W	N		+	16

+ 9−15 years

1 Station/Country South Pole / Antarctica

Location 90°00'S/– Height above sealevel 2800 m Climate symbol: Köppen EF Troll I,1

		J	F	M	A	M	J	J	A	S	O	N	D	year	P	
1 Mean daily temperature	in °C	−28,8	−40,1	−54,4	−58,5	−57,4	−56,5	−59,2	−58,9	−59,0	−51,3	−38,9	−28,1	−49,3	10	1
2 Mean daily maximum temperature	in °C															2
3 Mean daily minimum temperature	in °C															3
4 Absolute maximum temperature	in °C	−15,0	−22,2	−28,9	−31,7	−35,0	−29,4	−35,6	−32,8	−37,8	−30,0	−19,4	−18,9	−15,0	10	4
5 Absolute minimum temperature	in °C	−40,6	−56,1	−70,0	−72,2	−73,3	−76,1	−80,8	−77,2	−77,2	−67,2	−53,9	−38,3	−80,6	10	5
6 Mean relative humidity	in %															6
7 Mean precipitation	in mm															7
8 Maximum precipitation	in mm															8
9 Minimum precipitation	in mm															9
10 Maximum precipitation in 24 h	in mm															10
11 Mean number of days with precipitation																11
12 Mean duration of sunshine	in h															12
13 Mean quantity of radiation	in ly/day	800	500	100							300	700	100			13
14 Mean potential evaporation	in mm															14
15 Mean windspeed	in m/sec	4,2	5,2	6,3	6,4	6,9	7,3	7,4	7,3	7,3	6,8	5,1	4,4	6,2	10	15
16 Mean predominent direction of the wind		031	040	043	054	035	035	037	039	039	033	039	026		10	16

2 Station/Country Plateau Station / Antarctica

Location 79°15'S/40°30'E Height above sealevel 3625 m Climate symbol: Köppen EF Troll I,1

		J	F	M	A	M	J	J	A	S	O	N	D	year	P	
1 Mean daily temperature	in °C	−33,9	−44,4	−57,2	−65,8	−66,4	−69,0	−68,0	−71,4	−65,0	−59,5	−44,4	−32,3	−56,4	3	1
2 Mean daily maximum temperature	in °C															2
3 Mean daily minimum temperature	in °C															3
4 Absolute maximum temperature	in °C	−18,5	−24,9	−35,9	−42,7	−38,9	−32,8	−43,9	−41,2	−37,8	−37,1	−26,7	−20,6	−18,5	3	4
5 Absolute minimum temperature	in °C	−48,9	−60,8	−75,3	−78,0	−80,6	−82,2	−86,2	−85,0	−84,4	−80,0	−66,1	−47,8	−86,2	3	5
6 Mean relative humidity	in %															6
7 Mean precipitation	in mm															7
8 Maximum precipitation	in mm															8
9 Minimum precipitation	in mm															9
10 Maximum precipitation in 24 h	in mm															10
11 Mean number of days with precipitation																11
12 Mean duration of sunshine	in h															12
13 Mean quantity of radiation	in ly/day															13
14 Mean potential evaporation	in mm															14
15 Mean windspeed	in m/sec	3,0	4,2	5,0	5,2	5,4	5,0	5,8	5,9	5,7	5,1	4,6	3,8	4,9	3	15
16 Mean predominent direction of the wind		N	NW	N	NNE	N	N	N	NNW	NNW	NNW	N	NNW		3	16

3 Station/Country Wostok / Antarctica

Location 78°28'S/106°48'E Height above sealevel 3488 m Climate symbol: Köppen EF Troll I,1

		J	F	M	A	M	J	J	A	S	O	N	D	year	P	
1 Mean daily temperature	in °C	−33,4	−44,2	−57,4	−65,7	−66,2	−66,0	−66,7	−68,4	−65,6	−57,4	−43,6	−32,7	−55,6	9	1
2 Mean daily maximum temperature	in °C															2
3 Mean daily minimum temperature	in °C															3
4 Absolute maximum temperature	in °C	−22,3	−24,3	−32,5	−42,4	−43,0	−39,5	−36,1	−44,9	−42,1	−39,9	−31,7	−21,0	−21,0	9	4
5 Absolute minimum temperature	in °C	−48,3	−64,0	−75,0	−81,8	−82,0	−83,0	−81,1	−88,3	−82,8	−75,7	−63,1	−48,0	−88,3	9	5
6 Mean relative humidity	in %	75	72	71	68	68	68	69	69	69	71	72	74	70	9	6
7 Mean precipitation	in mm															7
8 Maximum precipitation	in mm															8
9 Minimum precipitation	in mm															9
10 Maximum precipitation in 24 h	in mm															10
11 Mean number of days with precipitation																11
12 Mean duration of sunshine	in h															12
13 Mean quantity of radiation	in ly/day	800	500	200						100	400	800	900			13
14 Mean potential evaporation	in mm															14
15 Mean windspeed	in m/sec	4,4	4,5	5,4	5,3	5,4	5,3	5,3	5,2	5,5	5,3	5,0	4,7	5,1	9	15
16 Mean predominent direction of the wind		SW	SW	SW	SW	SW	SW	SW	W	SW	SW	SW	SW		6	16

4 Station/Country General Belgrano / Antarctica

Location 77°58'S/38°48'W Height above sealevel 50 m Climate symbol: Köppen EF Troll I,1

		J	F	M	A	M	J	J	A	S	O	N	D	year	P	
1 Mean daily temperature	in °C	−6,0	−13,2	−21,0	−27,1	−30,0	−31,7	−32,7	−32,9	−31,4	22,4	−12,9	−6,0	−22,3	12	1
2 Mean daily maximum temperature	in °C															2
3 Mean daily minimum temperature	in °C															3
4 Absolute maximum temperature	in °C	6,8	6,6	−1,1	−3,0	−2,6	−9,0	−7,2	−6,7	−6,6	−1,3	0,6	5,7	6,8	12	4
5 Absolute minimum temperature	in °C	−26,6	−36,0	−42,3	−55,0	−52,8	−52,8	−53,3	−57,2	−56,0	−48,6	−37,1	−21,5	−57,2	12	5
6 Mean relative humidity	in %															6
7 Mean precipitation	in mm															7
8 Maximum precipitation	in mm															8
9 Minimum precipitation	in mm															9
10 Maximum precipitation in 24 h	in mm															10
11 Mean number of days with precipitation																11
12 Mean duration of sunshine	in h															12
13 Mean quantity of radiation	in ly/day															13
14 Mean potential evaporation	in mm															14
15 Mean windspeed	in m/sec	3,6	4,2	5,9	5,4	6,5	6,3	6,5	6,6	6,7	5,6	4,9	3,6	5,5	12	15
16 Mean predominent direction of the wind		S	S	S	S	S	S	S	S	S	S	S	S		12	16

5 Station/Country McMurdo/Antarctica

Location 77°53'S/166°44'E   Height above sealevel 24 m   Climate symbol: Köppen EF   Troll I,1

	in	J	F	M	A	M	J	J	A	S	O	N	D	year	P	
1 Mean daily temperature	°C	−3,4	−8,3	−18,9	−21,2	−23,9	−23,4	−25,5	−27,8	−24,1	−19,9	−8,8	−3,8	−17,4	13	1
2 Mean daily maximum temperature	°C															2
3 Mean daily minimum temperature	°C															3
4 Absolute maximum temperature	°C	5,6	4,4	−3,3	−5,0	−7,2	−7,8	−4,4	−7,8	−7,8	−4,5	2,8	5,1	5,6	10	4
5 Absolute minimum temperature	°C	−15,6	−21,8	−43,3	−39,4	−44,4	−40,0	−50,6	−49,4	−41,1	−38,3	−27,8	−16,6	−50,6	10	5
6 Mean relative humidity	%															6
7 Mean precipitation	mm															7
8 Maximum precipitation	mm															8
9 Minimum precipitation	mm															9
10 Maximum precipitation in 24 h	mm															10
11 Mean number of days with precipitation																11
12 Mean duration of sunshine	h															12
13 Mean quantity of radiation	ly/day															13
14 Mean potential evaporation	mm															14
15 Mean windspeed	m/sec	5,3	7,0	7,3	6,1	6,9	7,2	6,5	6,4	6,9	6,2	5,4	6,5	6,5	10	15
16 Mean predominent direction of the wind		E	E	E	E	E	E	E	ENE,E	E	E	E	E		6	16

6 Station/Country Halley Bay/Antarctica

Location 75°30'S/26°39'W   Height above sealevel 30 m   Climate symbol: Köppen EF   Troll I,1

	in	J	F	M	A	M	J	J	A	S	O	N	D	year	P	
1 Mean daily temperature	°C	−5,0	−10,0	−16,0	−21,1	−25,0	−26,8	−27,5	−28,5	−26,7	−19,8	−12,3	−5,7	−18,7	12	1
2 Mean daily maximum temperature	°C															2
3 Mean daily minimum temperature	°C															3
4 Absolute maximum temperature	°C	2,8	4,6	−2,0	−2,1	−2,3	−6,2	−5,0	−4,4	−5,0	−1,8	−1,5	2,2	4,6	12	4
5 Absolute minimum temperature	°C	−22,2	−29,6	−37,6	−46,4	−49,0	−51,0	−50,6	−52,5	−50,3	−45,3	−32,3	−14,6	−52,5	12	5
6 Mean relative humidity	%															6
7 Mean precipitation	mm															7
8 Maximum precipitation	mm															8
9 Minimum precipitation	mm															9
10 Maximum precipitation in 24 h	mm															10
11 Mean number of days with precipitation																11
12 Mean duration of sunshine	h	256	167	114	48				15	75	201	225	258	1359	3	12
13 Mean quantity of radiation	ly/day															13
14 Mean potential evaporation	mm															14
15 Mean windspeed	m/sec	4,5	4,6	4,9	5,0	4,8	4,6	5,1	4,7	5,4	5,5	4,4	4,3	4,8	6	15
16 Mean predominent direction of the wind		E	E	E	E	E	E	E	E	E	E	E	E		6	16

7 Station/Country Novolazarewskaja/Antarctica

Location 70°46'S/11°49'E   Height above sealevel 87 m   Climate symbol: Köppen EF   Troll I,1

	in	J	F	M	A	M	J	J	A	S	O	N	D	year	P	
1 Mean daily temperature	°C	−1,2	−3,9	−8,7	−13,4	−13,5	−14,1	−17,9	−18,0	−17,6	−13,7	−6,4	−0,8	−10,8	7	1
2 Mean daily maximum temperature	°C															2
3 Mean daily minimum temperature	°C															3
4 Absolute maximum temperature	°C	6,6	4,0	2,9	−1,1	−1,3	−2,1	−1,0	−1,3	−3,5	0,8	5,2	5,3	6,6	6	4
5 Absolute minimum temperature	°C	−11,7	−15,1	−22,3	−32,2	−30,8	−36,9	−35,5	−41,0	−40,6	−30,3	−24,2	−10,9	−41,0	6	5
6 Mean relative humidity	%	58	51	52	46	49	50	51	53	54	51	53	60	52	6	6
7 Mean precipitation	mm															7
8 Maximum precipitation	mm															8
9 Minimum precipitation	mm															9
10 Maximum precipitation in 24 h	mm															10
11 Mean number of days with precipitation																11
12 Mean duration of sunshine	h															12
13 Mean quantity of radiation	ly/day	700	400	200						100	400	600	700			13
14 Mean potential evaporation	mm															14
15 Mean windspeed	m/sec	7,4	9,6	10,5	10,5	12,1	13,0	10,7	11,4	9,8	10,9	10,1	7,5	10,3	6	15
16 Mean predominent direction of the wind		SE	SE	SE	SE	SE	SE	SE	SE	SE	SE	SE	SE		5	16

8 Station/Country Norway Station/Sanae/Antarctica

Location 70°30'S/3°32'W   70°19'S/2°21'W   Height above sealevel 56 m / 52 m   Climate symbol: Köppen EF   Troll I,1

	in	J	F	M	A	M	J	J	A	S	O	N	D	year	P	
1 Mean daily temperature	°C	−4,4	−8,9	−14,6	−20,2	−22,7	−22,3	−26,1	−26,8	−26,0	−18,3	−10,7	−5,2	−17,2	11	1
2 Mean daily maximum temperature	°C															2
3 Mean daily minimum temperature	°C															3
4 Absolute maximum temperature	°C	7,3	2,8	−0,8	−2,8	−2,0	−5,2	−5,5	−3,8	−1,9	−0,8	3,1	5,2	7,3	11	4
5 Absolute minimum temperature	°C	−25,3	−31,3	−35,5	−43,9	−51,0	−47,1	−47,1	−50,0	−48,3	−42,9	−33,0	−21,4	−51,0	11	5
6 Mean relative humidity	%	78	79	81	78	79	76	73	74	74	78	84	79	78	6	6
7 Mean precipitation	mm															7
8 Maximum precipitation	mm															8
9 Minimum precipitation	mm															9
10 Maximum precipitation in 24 h	mm															10
11 Mean number of days with precipitation																11
12 Mean duration of sunshine	h	238	148	106	72	19	0	<1	43	100	188	228	247	1389	6	12
13 Mean quantity of radiation	ly/day															13
14 Mean potential evaporation	mm															14
15 Mean windspeed	m/sec	6,1	6,9	7,3	7,6	7,7	8,8	8,1	7,6	7,2	7,9	7,3	6,1	7,4	9	15
16 Mean predominent direction of the wind																16

Sanae: since April 1962

9 Station / Country Syowa Base / Antarctica

Location 69°00′S / 39°35′E   Height above sealevel 15 m   Climate symbol: Köppen EF   Troll I,1

		J	F	M	A	M	J	J	A	S	O	N	D	year	P	
1 Mean daily temperature	in °C	−1,0	−3,5	−6,5	−10,2	−14,1	−15,3	−18,2	−19,2	−19,4	−13,0	−6,8	−1,7	−10,7	8	1
2 Mean daily maximum temperature	in °C															2
3 Mean daily minimum temperature	in °C															3
4 Absolute maximum temperature	in °C	7,8	4,5	3,6	0,4	−2,4	−0,7	−3,6	−3,9	−3,9	−1,6	3,5	8,1	8,1	8	4
5 Absolute minimum temperature	in °C	−11,6	−17,0	−22,1	−29,1	−36,2	−35,0	−42,7	−39,6	−42,1	−29,4	−23,9	−12,2	−42,7	8	5
6 Mean relative humidity	in %	71	65	75	77	72	72	72	74	74	75	76	70	73	7	6
7 Mean precipitation	in mm															7
8 Maximum precipitation	in mm															8
9 Minimum precipitation	in mm															9
10 Maximum precipitation in 24 h	in mm															10
11 Mean number of days with precipitation																11
12 Mean duration of sunshine	in h	322	237	110	57	28		9	57	136	199	269	425	1849	5	12
13 Mean quantity of radiation	in ly/day															13
14 Mean potential evaporation	in mm															14
15 Mean windspeed	in m/sec	4,7	4,4	6,9	7,3	6,2	7,0	6,6	5,8	5,0	5,3	6,6	5,0	5,9	8	15
16 Mean predominent direction of the wind		NE	ENE	NE	ENE	ENE	NE	NE	ENE	NE	NE	NE	NE		8	16

10 Station / Country Mawson / Antarctica

Location 67°36′S / 62°53′E   Height above sealevel 8 m   Climate symbol: Köppen EF   Troll I,1

		J	F	M	A	M	J	J	A	S	O	N	D	year	P	
1 Mean daily temperature	in °C	−0,2	−4,2	−9,9	−14,7	−15,5	−16,4	−17,6	−18,5	−17,7	−13,2	−5,5	−0,4	−11,2	15	1
2 Mean daily maximum temperature	in °C															2
3 Mean daily minimum temperature	in °C															3
4 Absolute maximum temperature	in °C	7,7	6,3	0,6	−0,3	−2,2	−1,4	1,7	−2,7	−0,7	−0,6	4,4	8,8	8,8	14	4
5 Absolute minimum temperature	in °C	−8,4	−15,0	−25,0	−28,7	−34,5	−30,6	−34,6	−35,4	−33,4	−25,4	−20,0	−16,7	−35,4	14	5
6 Mean relative humidity	in %															6
7 Mean precipitation	in mm															7
8 Maximum precipitation	in mm															8
9 Minimum precipitation	in mm															9
10 Maximum precipitation in 24 h	in mm															10
11 Mean number of days with precipitation																11
12 Mean duration of sunshine	in h	271	215	144	122	41	0	15	87	151	228	261	260	1795	11	12
13 Mean quantity of radiation	in ly/day															13
14 Mean potential evaporation	in mm															14
15 Mean windspeed	in m/sec	8,8	11,0	12,0	10,6	11,9	11,3	11,7	11,6	11,1	10,8	10,8	9,2	10,9	14	15
16 Mean predominent direction of the wind		ESE	ESE	SE	ESE	ESE	ESE	ESE	ESE	ESE	ESE	ESE	ESE		9	16

11 Station / Country Dumont d'Urville / Antarctica

Location 66°42′S / 140°00′E   Height above sealevel 41 m   Climate symbol: Köppen EF   Troll I,1

		J	F	M	A	M	J	J	A	S	O	N	D	year	P	
1 Mean daily temperature	in °C	−1,1	−4,2	−8,5	−12,0	−15,3	−17,6	−17,1	−16,9	−16,8	−13,2	−7,1	−2,4	−11,0	12	1
2 Mean daily maximum temperature	in °C															2
3 Mean daily minimum temperature	in °C															3
4 Absolute maximum temperature	in °C	6,0	4,5	2,7	1,1	0,2	−2,3	−0,2	−0,7	−1,1	0,1	3,5	6,1	6,1	12	4
5 Absolute minimum temperature	in °C	−9,0	−14,9	−22,3	−25,1	−32,3	−33,4	−33,4	−33,2	−36,5	−28,5	−22,4	−10,6	−36,5	12	5
6 Mean relative humidity	in %															6
7 Mean precipitation	in mm															7
8 Maximum precipitation	in mm															8
9 Minimum precipitation	in mm															9
10 Maximum precipitation in 24 h	in mm															10
11 Mean number of days with precipitation																11
12 Mean duration of sunshine	in h	268	182	145	93	42	10	12	63	147	230	316	328	1836	10	12
13 Mean quantity of radiation	in ly/day															13
14 Mean potential evaporation	in mm															14
15 Mean windspeed	in m/sec	10,2	12,1	12,8	12,5	11,6	10,8	9,7	11,3	10,2	9,9	10,2	9,3	10,9	12	15
16 Mean predominent direction of the wind		SE	SE	SE	SSE	SE	SE	SE	SSE	SE	SE	SE	SE		12	16

12 Station / Country Mirnyj / Antarctica

Location 66°33′S / 93°01′E   Height above sealevel 30 m   Climate symbol: Köppen EF   Troll I,1

		J	F	M	A	M	J	J	A	S	O	N	D	year	P	
1 Mean daily temperature	in °C	−1,8	−5,1	−10,0	−13,8	−15,5	−16,4	−16,8	−17,3	−17,1	−13,8	−7,3	−2,7	−11,5	12	1
2 Mean daily maximum temperature	in °C															2
3 Mean daily minimum temperature	in °C															3
4 Absolute maximum temperature	in °C	8,0	5,2	−0,1	−1,0	0,0	0,0	5,2	−2,9	−2,0	−0,8	5,0	8,0	8,0	12	4
5 Absolute minimum temperature	in °C	−14,1	−18,6	−29,0	−31,3	−40,0	−32,8	−36,9	−40,3	−37,3	−32,0	−21,6	−16,2	−40,4	12	5
6 Mean relative humidity	in %	69	67	68	71	72	71	73	73	72	68	68	71	70	12	6
7 Mean precipitation	in mm															7
8 Maximum precipitation	in mm															8
9 Minimum precipitation	in mm															9
10 Maximum precipitation in 24 h	in mm															10
11 Mean number of days with precipitation																11
12 Mean duration of sunshine	in h															12
13 Mean quantity of radiation	in ly/day	700	500	200	100					200	400	600	700			13
14 Mean potential evaporation	in mm															14
15 Mean windspeed	in m/sec	7,9	9,5	11,2	12,8	14,1	13,5	13,1	13,5	12,8	10,4	10,1	8,8	11,5	12	15
16 Mean predominent direction of the wind		E	SE	SE	SE	SE	SE	SE	SE	SE	SE	SE	E		10	16

13 Station / Country Argentine Island / Antarctica

Location 65°15'S/64°15'W   Height above sealevel 11 m   Climate symbol: Köppen ET   Troll I,2

		J	F	M	A	M	J	J	A	S	O	N	D	year	P	
1	Mean daily temperature	in °C	0,2	-0,2	-1,3	-4,7	-6,6	-9,6	-11,2	-11,9	-9,0	-5,2	-2,7	-0,4	-5,2	21	1
2	Mean daily maximum temperature	in °C															2
3	Mean daily minimum temperature	in °C															3
4	Absolute maximum temperature	in °C	10,0	11,7	7,8	7,2	8,7	6,1	4,4	7,2	5,0	6,1	8,7	6,7	11,7	14	4
5	Absolute minimum temperature	in °C	-10,6	-12,2	-16,1	-34,4	-34,4	-34,4	-40,6	-43,3	-38,9	-28,9	-22,2	-11,7	-43,3	14	5
6	Mean relative humidity	in %	84	85	87	85	83	86	86	87	87	88	85	87	86	9	6
7	Mean precipitation	in mm															7
8	Maximum precipitation	in mm															8
9	Minimum precipitation	in mm															9
10	Maximum precipitation in 24 h	in mm															10
11	Mean number of days with precipitation																11
12	Mean duration of sunshine	in h															12
13	Mean quantity of radiation	in ly / day															13
14	Mean potential evaporation	in mm															14
15	Mean windspeed	in m / sec	3,1	3,4	3,8	4,4	4,0	4,0	4,2	3,8	4,6	5,3	3,7	2,8	3,9	9	15
16	Mean predoment direction of the wind																16

14 Station / Country Wilkes / Antarctica

Location 66°15'S/110°35'E   Height above sealevel 12 m   Climate symbol: Köppen EF   Troll I,1

		J	F	M	A	M	J	J	A	S	O	N	D	year	P	
1	Mean daily temperature	in °C	-0,3	-2,3	-6,2	-11,5	-13,9	-16,3	-15,3	-15,0	-14,9	-11,4	-5,4	-1,3	-9,4	11	1
2	Mean daily maximum temperature	in °C															2
3	Mean daily minimum temperature	in °C															3
4	Absolute maximum temperature	in °C	7,8	6,1	3,6	3,3	3,9	5,0	1,7	0,6	1,1	0,8	5,6	6,6	7,8	11	4
5	Absolute minimum temperature	in °C	-8,3	-13,9	-20,7	-32,2	-33,9	-33,3	-37,2	-36,7	-35,6	-26,7	-27,2	-15,6	-37,2	11	5
6	Mean relative humidity	in %															6
7	Mean precipitation	in mm															7
8	Maximum precipitation	in mm															8
9	Minimum precipitation	in mm															9
10	Maximum precipitation in 24 h	in mm															10
11	Mean number of days with precipitation		6	7	10	7	8	10	9	12	11	11	7	6	104	5	11
12	Mean duration of sunshine	in h															12
13	Mean quantity of radiation	in ly / day															13
14	Mean potential evaporation	in mm															14
15	Mean windspeed	in m / sec	5,1	6,1	6,5	6,5	7,3	7,2	8,0	7,8	7,5	7,5	6,7	5,3	6,9	11	15
16	Mean predoment direction of the wind																16

15 Station / Country Hope Bay (Bahia Esperanza) / Antarctica

Location 63°24'S/56°59'W   Height above sealevel 11 m   Climate symbol: Köppen ET   Troll I,2
63°23'S/57°00'W   7 m

		J	F	M	A	M	J	J	A	S	O	N	D	year	P	
1	Mean daily temperature	in °C	0,4	-1,0	-5,1	-6,8	-9,8	-9,8	-9,9	-9,1	-6,8	-4,0	-1,3	-0,3	-5,3	5	1
2	Mean daily maximum temperature	in °C															2
3	Mean daily minimum temperature	in °C															3
4	Absolute maximum temperature	in °C	14,2	7,6	5,4	14,2	8,4	10,5	8,9	8,9	9,5	14,6	12,4	13,0	14,6	5	4
5	Absolute minimum temperature	in °C	-8,5	-11,8	-17,6	-23,2	-27,6	-28,0	-28,2	-32,1	-24,4	-25,4	-11,0	-10,9	-32,1	5	5
6	Mean relative humidity	in %	84	86	85	80	78	79	80	81	83	82	80	85	82	5	6
7	Mean precipitation	in mm															7
8	Maximum precipitation	in mm															8
9	Minimum precipitation	in mm															9
10	Maximum precipitation in 24 h	in mm															10
11	Mean number of days with precipitation																11
12	Mean duration of sunshine	in h															12
13	Mean quantity of radiation	in ly / day															13
14	Mean potential evaporation	in mm															14
15	Mean windspeed	in m / sec	4,4	5,7	6,5	6,7	6,1	6,6	6,4	6,7	6,5	7,4	7,0	6,6	6,4	4	15
16	Mean predoment direction of the wind		SW	SSW	SSW	SSW	SW	SW	SSW	WSW	WSW	WSW	SSW	SSW		5	16

16 Station / Country Orkadas (South Orkney Islands) / Antarctica

Location 60°44'S/44°44'W   Height above sealevel 4 m   Climate symbol: Köppen ET   Troll I,2

		J	F	M	A	M	J	J	A	S	O	N	D	year	P	
1	Mean daily temperature	in °C	0,2	0,4	-0,4	-3,0	-6,8	-9,8	-10,6	-9,8	-6,5	-3,7	-2,1	-0,6	-4,4	65	1
2	Mean daily maximum temperature	in °C															2
3	Mean daily minimum temperature	in °C															3
4	Absolute maximum temperature	in °C	12,2	9,0	10,8	8,2	9,2	6,4	7,8	8,2	6,5	8,7	8,8	9,6	12,2	53	4
5	Absolute minimum temperature	in °C	-7,0	-9,8	-15,1	-31,5	-31,9	-38,3	-36,9	-40,1	-32,6	-31,2	-20,4	-13,2	-40,1	53	5
6	Mean relative humidity	in %															6
7	Mean precipitation	in mm	35	40	49	41	34	30	32	32	30	29	32	26	410	51	7
8	Maximum precipitation	in mm															8
9	Minimum precipitation	in mm															9
10	Maximum precipitation in 24 h	in mm															10
11	Mean number of days with precipitation		14	15	16	17	16	15	15	16	15	15	15	13	182	44	11
12	Mean duration of sunshine	in h	48	38	34	25	17	10	18	44	66	69	55	63	487	48	12
13	Mean quantity of radiation	in ly / day															13
14	Mean potential evaporation	in mm															14
15	Mean windspeed	in m / sec	4,1	4,8	5,0	5,4	5,3	4,8	5,1	5,3	5,5	5,6	4,8	4,0	5,0	47	15
16	Mean predoment direction of the wind																16

INDEX

Climate Zones after W. Köppen/R. Geiger

Af = Tropical rainforest climate

Aw = Tropical savanna climate

BS = Steppe climate

BW = Desert climate

Cw = Warm temperate rainy climate with dry winter

Cs = Warm temperate rainy climate with dry summer

Cf = Temperate rainy climate, moist in all seasons

Dw = Cold snowy forest climate with dry winter

Df = Cold snowy forest climate, moist in all seasons

ET = Tundra climate

EF = Climate of perpetual frost

E = Ice climates

f = Moist all year

m = Rainforest climate despite a dry season (monsoon cycle)

s = Dry summer, wet winter

w = Dry winter, wet summer

a = Hot summers

b = Warm summers

c = Moderate or cool summers

d = Very cold winters

h = Hot

k = Cold